# PROJECT MANAGEMENT HANDBOOK

SECOND EDITION

Edited by

David I. Cleland
*Professor*
*Engineering Management*
*School of Engineering*
*University of Pittsburgh*

William R. King
*University Professor*
*Katz Graduate School of Business*
*University of Pittsburgh*

JOHN WILEY & SONS, INC.

New York Chichester Weinheim Brisbane Singapore Toronto

This book is printed on acid-free paper. ♾



Published simultaneously in Canada.



**Library of Congress Cataloging-in-Publication Data:**

Project management handbook.
Includes index.
1. Industrial project management. I. Cleland, David I. II. King, William Richard
HD69.P75P75 1988 658.4'04 87-23151
ISBN 0-471-29384-9

Printed in the United States of America.

16 15 14

# Preface



When the first edition of this handbook was honored with the Institute of Industrial Engineer's Book-of-the-Year Award, we as editors felt that we had accomplished our objective, ". . . to provide project managers and those individuals concerned with project management in both public and private organizations a reference guide for the fundamental concepts and techniques of managing projects."

However, in the fast-moving field of project management, there is a need for continuing updating as new techniques, tools, understandings and experiences are developed. Hence, this second edition has been fashioned to build on the strengths of the first edition and to add insights that were unavailable when the first editon was published.

Like all good handbooks, this is a reference source for practical how-to-do-it information. A manager or professional who has a problem with project management can turn to this handbook and find the help needed to solve the problem.

However, there are other important uses to which this handbook can be put. The field of project management has been growing so rapidly in recent years that anyone who wishes to learn more about the discipline is faced with an abundance of published information. The handbook is organized so that it can be used either as a reference or for sustained study.

Thus, in such a rapidly developing field, even experienced project managers are faced with the challenge of keeping up with current developments and selecting those concepts and techniques that are most appropriate for their needs, will find it to be useful.

Those who are not experienced project managers, but who must play a role in the process of project management—functional managers, general managers, project team members and support staff—have an even more difficult task, for they must "keep up" in a rapidly expanding field that is not their special area of expertise. So, the *Project Management Handbook* is also addressed to their needs.

Students of project management may use the *Handbook* as a self-study aid, for it has been organized to facilitate an overall program of learning about the field as well as to provide a quick reference source on specific topics.

The *Project Management Handbook* seeks to provide guidance for all varieties of projects—from the largest and most complex systems development efforts, to the personal "research project." Its emphasis, however, is organizational in the sense that much of the material deals with the integration of projects into an overall managerial framework.

In addition to serving the needs of those who are directly concerned with project management, this book will be useful to top managers who wish to motivate and establish a philosophy of project management within their organizations. There are different types of project management, ranging from the simplistic use of expediters to sophisticated multiproject organizational approaches; the high-level manager who believes that one of these forms will be useful will find the *Handbook* to be a sound guide for planning for the evolution of project management in the organization.

The handbook provides information on both the theory and practice of project management. While primary emphasis is on the pragmatic aspects of managing projects, this pragmatism is casted in a sound, theoretical framework of managerial thought.

In the editors' opinion, the proportion of project participants who require access to a handbook in this field is greater than for any other managerial group. This is because of the relative newness of the field, the lack of adequate training programs, and the general awareness that, unlike some other disciplines, there is a practical body of knowledge which can serve to support all aspects of the project management process.

As one leafs through the pages of the *Project Management Handbook,* the number and variety of the factors and forces with which the project manager must deal become clear. Their very number appears at first to make it impossible for one individual to master the art and science of managing projects. However, in reading further one becomes aware of the creativity and ingenuity which the authors have brought to bear on project management. For most of the problem situations that a project participant will face, this handbook has information that can be of help. Certainly, no one would claim that the state-of-the-art of this field has stabilized. Further evolution will continue. Nevertheless, the reader will find that there are workable solutions to the situations that arise in project management.

This handbook is the result of the cooperative efforts of a large number of people. The qualifications of the individual contributors are clear from the biographical sketch given on the title page of each article. The topic content of the handbook is broadly designed to be relevant to the general organizational contexts in which project management is found. Accordingly, some of the parochial subjects of project management in specific industries are not included. For example, configuration management and value engineering—two key concerns of project managers in the aerospace industry—are not treated. The editors believe that the parochial interests of a particular industry's project manager's needs can be best served by studying the literature of that industry.

Whatever its value to the reader, the *Project Management Handbook* reflects the experiences and considered judgments of many qualified individuals about the pivotal factors and forces surrounding project management. Eight interdependent areas of project management are developed:

1. *An Overview of Project and Matrix Management.* The frameword of practice and theory in which contemporary project and matrix management is found.
2. *The Project Organization.* The alignment of resources to support project objectives, particularly in terms of the matrix organization.
3. *Organizational Strategy and Project Management.* The deployment of resources to support broader organization missions, objectives, and goals.
4. *Life Cycle Management.* The management of projects as they fit into broader and longer-range organizational cycles.
5. *Project Planning.* Planning to include the development of goals, strategies, and actions to allocate project resources.
6. *Project Implementation.* The actions and constraints that are relevant to getting the project that has been planned actually carried out.
7. *Project Control.* The means to determine the harmony of actual and planned cost, schedule, and performance goals, including the use of computers.
8. *Behavioral Dimensions and Teamwork in Project Management.* The development of a climate whereby the project people work together with economic, social, and psychological satisfaction.
9. *The Successful Application of Project Management.* An examination of what counts for success in project management.

We thank the contributing authors who have given so importantly and unselfishly through their practical how-to-do-it presentations of the forces and factors involved in project management.

The editors are deeply indebted to Claire Zubritzky, who managed the administration involved in the development and production of this handbook. We are also indebted to Olivia Harris, whose contributions, both to the *Handbook* and to the milieu in which we work, were substantial. Karen Bursic's efforts on the index and Ellen Hufnagel's proofreading are also appreciated.

We thank the late Dr. Albert G. Holzman, former Chairman of the Industrial Engineering Department, Dr. Harvey Wolfe, the current chairman, Dr. Charles A. Sorber, Dean of the School of Engineering, and Dr. H. J. Zoffer, Dean of the Katz Graduate School of Business, all of the University of Pittsburgh, who provided us with the environment to pursue this effort.

DAVID I. CLELAND
WILLIAM R. KING
Pittsburgh, PA

# Contents

## SECTION VI / PROJECT IMPLEMENTATION / 477



## SECTION VII / PROJECT CONTROL / 557



## SECTION VIII / BEHAVIORAL DIMENSIONS AND TEAMWORK IN PROJECT MANAGEMENT / 737

# Section I

# Overview of Project and Matrix Management

This introductory section of the handbook presents an overview of project and matrix management. Project management is viewed as a powerful tool that is particularly useful in terms of the management of the many interfaces that exist within an organization, and between an organization and its environment.

However, despite the power of the concept and its history of successful application, project management is not presented as a panacea. Rather, it is a tool which, *when properly used under appropriate circumstances,* can aid the organization in the achievement of its major goals.

In the first chapter, Robert D. Gilbreath demonstrates how rapid change has created a need for project management. He distinguishes the modern project environment from the traditional "operations" of continuous repetitive processes.

In Chapter 2, Peter W. G. Morris explains the need for project and matrix management in an insightful chapter on "interface management." He discusses project and matrix management in conceptual terms and provides, as well, numerous real-world illustrations and prescriptions for the successful management of interfaces.

In the third chapter, Linn C. Stuckenbruck discusses project integration in the matrix organization by emphasizing the proactive nature of integration; that it does not just happen, but must be made to happen. He discusses how a project management system can be implemented in the organization and what the project manager must do to properly begin the project.

# 1. Working with Pulses, not Streams: Using Projects to Capture Opportunity

Robert D. Gilbreath*

The phenomena of rapid social, economic, technological, and political change have revised our view of business activity, destroyed the viability of accepted business models, and led to new definitions of ourselves and our effect. Although our products or our services may be similar to those of the past, how we bring them about and how reliable that effect can be are no longer taken for granted. A greater understanding and more intelligent response to change forces us to reconsider the very foundations of our work, our contribution to it, and its place in the realm of business. Change is reshaping the nature and meaning of work.

Nowhere is this more pronounced than through the *"projectization"* of work, its centering in unique, temporary packets of effort. Whereas industrial production and all its social and cultural spin-offs relied upon linear, sequential arrays of highly specialized and synchronized effort (i.e., the automobile assembly line), this new, change-directed concept of work involves parallel, unsynchronized, and generalized effort not tied to or dependent upon any established tools or techniques.

If the concept of work dominant in the past has been symbolized by a *line,* the new representation of work is more apt to be a *circle*—a circle that encloses a comprehensive interaction of concurrent, temporary, and accomplishment-oriented tasks. If the old term for work was *"operations,"* the new term is *"projects."*

In times of change the project orientation dominates all operational frameworks. The logic supporting this conclusion is inescapable, and we see it manifested with great frequency by business examples all about us. Perceptive managers know, then, that in times of change, for today and tomorrow, they will more often than not be managing *projects.*

---

* Robert D. Gilbreath advises corporations on new management perspectives and evolving executive issues. He is the author of *Managing Construction Contracts* and *Winning At Project Management* (John Wiley & Sons, 1983, 1986). This chapter is excerpted with permission from his third management book *Forward Thinking: The Pragmatist's Guide to Today's Business Trends* (McGraw-Hill Book Co., Copyright 1987).

We cannot simply count the growing number of efforts our companies now designate as *"projects"* and hold this increase up as the sole result of a projectization phenomenon. Much more is happening to change the nature of our work and our perception of our role in it. Subtle, project-influenced changes are even taking place in operational settings. Most of all, operationally oriented managers are beginning to understand the unique aspects and condition of project assignments. Here are some other signposts indicating a widespread turn toward projects.

## SIGNPOSTS OF CHANGE

- ***Adaptation of the "project manager" designation.*** Once regarded as an organizational oddity, the project manager is now invading even the most nonproject-oriented companies. In a future dominated by change, every manager, at one time or another, will be a project manager.
- ***More temporary, results-directed organizations.*** The days of the concrete, tiered organization are gone. We now view all organizations as temporary, goal-directed contrivances—necessary evils rather than structures of intrinsic beauty. Projects cannot be managed by pyramids. They demand clusters of people gathered around a challenge.
- ***More common use of outsiders (consultants, subcontractors, joint ventures, temporaries, etc.) for specific efforts.*** Organizations are being built and destroyed on the bases of risk and pragmatism. The old notions of insiders and outsiders don't fit anymore. We'll use whichever players we need, regardless of what uniforms they wear.
- ***More local and perishable procedures, plans, standards.*** Standards don't work well in times of flux. Like organizations, they will be disposable, situationally responsive, and full of room for discretion.
- ***Emphasis on "people skills" among management.*** Project management is, first, people management. To coalesce disparate interests, transcend goal conflicts, and create binding mutuality, we will need those who hold people skills.
- ***Constant creation and dismantling of management scaffolding.*** Procedures, policies, reports, information systems, and progress measurements are all elements of managerial scaffolding. When everything remains static, the scaffolding can be built of steel and anchored in bedrock. In times of change, it must be as temporary and mobile as a tent.
- ***Devaluation of tradition, of "what worked last time."*** In the world of projects, there simply is no "last time." This is the world of change—the phenomenon that mocks the past.

- ***The emergence of pragmatism and resourcefulness over perfection and compliance as favorable management attributes.*** We will need scroungers, tinkerers, masters of the extemporaneous, and those who can make it happen, regardless of the rules, the odds, or the inevitable second guess. Project positions are contribution-based and need-justified. Our position and authority will be functions of what we *are doing,* not what we *have done* or *who we are.*

## ILLUSTRATING THE DISTINCTIONS*

One must clearly understand and fully appreciate the bold and fine differences between *operations* and *projects* in order to comprehend the powerful influence of change on business. By shifting our needs from operationally-achieved work to work accomplished through projects, change has exerted one of its most observable and profound effects. Its impacts reach our notion of work itself, our roles in work, the needs and limitations of management—affecting both our senses of identity and our value systems.

As we draw and give examples of some of the many distinctions between these two approaches to work, we will not only be depicting differences in work orientations but also describing the particular nature and far-reaching impact of change. To understand the differences between operations and projects is to understand the differences between a changeless and a changing world—that is, to understand change itself.

## WHY OPERATIONS AND PROJECTS DIFFER

There are countless differences between operations and projects as ways to organize and manage business effort, and they include concept, context, intent, and application. Rather than defining each and every difference, our approach will be to point out the most significant differences and those which most directly illustrate the effect of change.

Operations are based on the concept of *using* existing systems, properties, processes, and capabilities in a continuous, fairly repetitive fashion. If our business is automobile production, our operational base is comprised of physical plant (the factory), tools, equipment, information and control systems, knowledge, and production skills. Our operational objective is to use this fixed potential as efficiently and effectively as possi-

---

* In this chapter I draw from some distinctions made in my earlier book, *Winning At Project Management: What Works, What Fails, and Why* (Wiley, New York, 1986), not to identify all project aspects but to present only those that illustrate business responses to change.

ble. Operations are aimed at making the best use of what exists, over and over again.

## FEWER PRESUMPTIONS

Projects, in contrast to operations, presuppose no fixed tools, techniques, or capability. They seek to create a limited impact through temporary and expedient means. The design and construction of an automobile assembly plant is a good example. As such, this "project" is unique and apart from any other undertaken. We produce one "product," the plant, rather than a series of similar products (the cars that will result from operation of the finished plant). Uniqueness of effort and result are the hallmarks of project situations. Consistency and uniformity are typical of operations.

Operations are geared to maintain and exploit, while projects are conceived to create and make exploitation available. Projects, therefore, typically precede operations in the normal business cycle.

Projects are temporary and expedient exercises, while operations are more sustained and continuous, and therefore more amenable to optimization. Economies of scale help to optimize operations, as do trial and error, but given the temporary and unique world of each project, neither of these approaches is very useful in a project. If a successful operation can be imagined as a continuous, uninterrupted *stream* of effort yielding a predictable collection of similar results, we must view each project as a temporary *pulse* of activity yielding a unique, singular result.

Projects represent, then, one-time-only configurations of resources, people, tools, and management expectations, while operations presuppose continuity of the conversion process well into the future. We seldom consider the end of an operation, but we always consider the end of a project as soon as we conceive of its initiation. We expect projects to be completed, to be finished, and, like cruise missiles, to be self-consuming once their singular purposes have been accomplished. Operations may outlive their results, but projects expire when their result is achieved.

## FREE VARIABLES AND SOFT LINKS

In order to exploit induced change, operations rely on circumstances in which most variables are fixed and the rest are manipulated. Working conditions are fairly constant; we are usually enclosed in a factory, office, or shop; and our resources are of like nature and consistent in appearance and quality. Project work, though, is hostage to many free variables, few of which are presumed to be fixed. The fixing of free variables must take place every time a project is initiated, as when staffs are hired, organizations are created, procedures are adopted, plans are made, and lines of communication are strung. Most of these variables are fixed and already in place by the time operational work begins.

Since synchronization is essential to any effort involving more than one participant, both operations and projects need to meet the challenge of synchronization. They differ in how the challenge is met, though. With operations, the links connecting serial work steps are usually "hard" or mechanical, inherent in the equipment, line, material, or techniques used. If metal must be rolled before stamping, we simply position the rollers in advance of the stamping press in the production line. This is a *"hard link,"* designed into the operation and therefore difficult to avoid or circumvent.

With project work, activity cannot be so easily synchronized. We often must depend on forced, artificial, or human linkage, the so-called soft links that tie otherwise disparate work elements together. Soft links include contract terms and conditions ("All steel shall be rolled before stamping"), written procedures, management inspection, and supervision. Soft links require constant enforcement and monitoring, for they may be easily bypassed or ignored. They are much less dependable as guarantees of synchronicity. Again, with project work less is fixed (or fixable) and less can be taken for granted.

Most astute executives understand the tension between process and product-dominated business approaches and have seen how a product-oriented effort better fits changing conditions. This also applies to a project effort, where we begin with an expected set of results and seek to find or build a collection of processes to bring it about (*result drives process*). With operations the reverse is true. We begin with a process (factory, plant, refinery, line, etc.) and search for materials to feed it and markets in which to discharge the result (*process drives product*).

Project work is also amenable to the pressure change exerts toward generalization and away from specialty restrictions. Project work, by its nature, begins with nothing and relies on synthesis to proceed. It requires coalescence: bringing elements together and creating mutuality among them. Operations tend to divide labor and its contribution into discrete, incremental stages. In fact, division of labor is a hallmark of the industrial revolution.

## CHANGE SENSITIVITY OR DEFIANCE

Because of their fixed conditions, hard links, stationary components, and hard-wired process steps, most operational efforts are rightly classified as insensitive to change or defiant of it. They shield themselves as much as possible from incurred change and stick to their original goals, methods, and results. Project work cannot be so shielded. It is conceived, born, lives, and expires in change. It evolves continuously, because of both induced and incurred change. It takes a different approach to the challenge of change immunity; it opts for disposability or adaptation rather than durability.

Finally, we must recognize that operations take time, money, and a great deal of effort to establish, long before the first product rolls off the end of the line. All the capital investment, design, hard wiring, and variable fixing they employ must precede their use. Once erected, operational apparatus cranks out a great deal of product at lower and lower unit cost. The question is, "Do we have the time to spend and the resources to sink before the first stream of products starts to flow?" Unless we exist in a relatively static environment, the answer is, "No, we have neither."

When conditions are in a state of flux, opportunities are fleeting, and our businesses must seize the moment, the rapid deployment of a project effort is much more suitable. If the targets of opportunity remain similar and stationary for long periods of time, if they are unthreatening and fixed, we can design and erect a very efficient method of shooting them down. This is the operational solution. If instead we are running through a jungle of beasts, some pursued and some pursuing, the ability to quickly access and deploy *any* weapon is much more important to us. Selecting the *available,* albeit not often *ideal,* weapon is the project solution. Changing conditions favor the pragmatic.

## THE PROJECT ANOMALY

Projects are a special type of business anomaly. They often evolve, change shape, and resist definition (as force waves, for example, do), but then again, they are always composed of discrete items such as people, resources, and conversion processes (as are particles of mass). And we can never point to, look at, or "see" a project (as we can never see subatomic matter). We can see project activity and results, and we can see project managers and production workers doing their work. But even when considered all together, these factors do not actually make up the project model; they are simply parts of it. Projects follow the concept of *holism* or *synergy* in that each is a collection of entities or objects that can generate a larger reality not analyzable in terms of the components themselves.

The best way to model a project is not to use a visual representation at all. A project is a *mutual effort,* using a collection of resources in an orchestrated way to achieve a joint goal. As such, projects are like "waves"—forces and bundles of energy moving through time—each with its own identity, culture, methods of conversion, and contrived cohesion. Upon accomplishment of the project goal, this contrived cohesion no longer serves to bind the project together (thus making it an anachronism); instead, it dissolves the project—dissipating the project wave upon the beach of success.

It's easy to identify project work in architecture or engineering, for the physical result symbolizes the project itself. A new power plant, a factory, a shopping mall, and an aircraft carrier are examples. But the project model applies to many other fields and to efforts most of us are involved with at one time or another. Here is a partial list of other efforts that might fit the definition of "project":

Performing a heart transplant
Designing a new weapons system
Producing a stage play
Developing a strategic business plan
Researching and writing a book
Conducting a political campaign
Renovating an antique automobile
Producing a motion picture
Establishing a small business
Throwing a party
Introducing a new product
Taking a vacation trip
Designing and installing a computer network
Creating an occupational training program

There is subtle commonality in all these examples. They each involve (1) working with few existing standards, (2) the need for creativity and synthesis, (3) a temporary pulse of effort, and (4) a keen sense of, if not reliance on, the phenomenon of change.

## THE DIRECTION OF CHANGE

Each element characterizing a project, and differentiating it from operations, can be seen as a change-directed quality. Projects are the perfect response to change. They are no less than change-responsive bundles of effort. Because they more closely represent waves, and not incompressible particles, projects can expand; shrink; bend; accelerate and slow down; change shape and direction; and escape the burdens of capitalization, process addiction, and hardness. While operations try to withstand the impacts of change, projects ride along with it: business waves upon the sea of time.

## FORCES AND FACTORS FAVORING PROJECT ORIENTATIONS

Once we understand the project concept and couple it with what we know about change, it's easy to see why projects are becoming more and more the accepted context of work. Here we should briefly mention why this is true.

Rapid creation and deployment allow project efforts to respond quickly to changed conditions. Their independence from capital burdens lets our work flex with or dodge unforeseen changes in technology or process methodology. The fleeting nature of opportunity gives its rewards only to the most adaptive, mobile pursuer. Project orientations provide that mobility; we can move our effort to the opportunity rather than attempt to entice it into our operating environment, in terms of space and time.

This explains the sudden and successful emergence of *entrepreneurism* in today's business culture. Entrepreneurs succeed where large, static organizations fail—because of their flexibility, mobility, and pursuit. Entrepreneurs are lean and responsive, able to quickly detect and pursue the opportunities that change strews all about. Even with the most expensive, powerful mechanism, a large supertanker needs miles of space and huge amounts of time to execute a complete 180-degree turn, whereas a small speedboat can turn on a dime.

Changing markets, resources, prices, and needs do not allow us the benefit of time or of the fixed variables that operations presuppose. Project work is amenable to dynamic conditions, while operations demand static ones. The tendency of change to emphasize purpose over process also runs counter to the operational premise (process in search of purpose) and diminishes the value of pursuing process enhancements.

If change serves to make any given method or technique less lasting and more perishable, then it also devalues fixed versions of processes and operations. If an operation is dependent on technique and that technique is made obsolete through change, the operation becomes obsolete. Project work has few dependencies, and those it suffers are seldom exclusive.

## THE CONTRIBUTION OF BUSINESS MEGACHANGE

Megachanges in the business world itself, each caused partly by recognition of the power of change, also seem to be running in the project direction. These are (1) *variation* in enterprise among megacompanies, (2) *diversification* of risk and assets, (3) use of more *distributed* information and effort (outsourcing, subcontracting, joint ventures, prefabrication elsewhere, containerization, process packaging, etc.), and (4) *decentralization* of business authority and management effect.

Each of these megachanges should have had and has had a tremendous effect on our reevaluation of work and how we relate to it. And each favors the project orientation. Whether they are effects of change or causes of change itself, they signal a new business culture, more cognizant of and more responsive to change. It we are going to manage in the future, we are going to have to manage in this culture—this culture which is impacted by change and in which work is more often than not "projectized."

## THE NEW BUSINESS CULTURE

The emergence and proliferation of the project model of work, of work as integrated pulses rather than a continuous line, is helping to define a new and very different business culture. This culture will be dominated by a need to generalize, to spread awareness and ability over large spaces and times, rather than to focus both on ever-narrowing fields of specialty. If generalization represents the prevailing condition of the new business culture, then *synthesis* represents the activity that will bring it about. The ability to synthesize—to coalesce or bring together the forces and resources needed to equip a project effort—is quickly becoming the critical ability of the future. "Synthesis" and "adaptation" are fast replacing "optimization" as the watchwords of today's management.

Management in change will more often be judged by its results than by its inherent sophistication, complexity, or level of detail. A business culture ruled by pragmatism will tend to value accomplishment over refinement, attainment of goals over perfection of a limited ability to attain them. To be truly adaptive, we will embrace independence of any certain set of plans, tools, techniques, or conversion processes. We will view such factors as simply means to an end, and means of temporary and utilitarian value. No business will transcend time and change totally intact, for no business is immune to these forces.

If we are to adapt, our methods and scaffolding must be malleable, disposable, and expedient. We can no longer shackle our companies to any given feature, no matter how well it has served us in the past or how attached or dependent upon it our managers have become. Managers in change are independent managers, serving not technique or tool but higher principles which do transcend momentary changes. These principles are pragmatism, expediency, reason, adaptability, independence, and human value.

We will no longer treasure managers who continue to divide, to separate, and to polarize. Instead, our new leaders will be *integrators,* skilled in human synthesis and mutuality of intent and effort. Only integrators will achieve mutually beneficial results.

The new business culture is evident among entrepreneurial companies today, where flexibility and maneuverability dominate fixity and steadfastness. Other companies that learn from entrepreneurs will quickly trim their concepts and scaffolding to increase their capabilities. Entrepreneurs aren't successful because they have the best ideas or the greatest intelligence, but because they possess the agility to pursue fleeting opportunity and are lean enough to survive the chase. They can succeed because they are not tied to the legacy of the past: massive capital investment, ingrained management approaches, and obsolete tools and techniques. Their efforts are almost always new, and because they are new they have been created to embody the new culture and to treasure the new values. These are the efforts most aligned with the concept of change, for they are being forged in the white-hot furnace of business-change: today's business climate.

In this climate we will continue to take less and less for granted. Fixed variables are the relics of the past, and nostalgia is beneficial only when it sharpens our awareness of the difference between the past and the present. Because less will be taken for granted, our managers will have to more frequently contrive "mental gathering points," selected reference points at which all members of a project team are allowed to regroup, synchronize their intentions, and recognize a set of mutual accomplishments and goals. Project managers need to coalesce the raw material of their efforts not only at the beginning of a project, but periodically throughout the life of the project, and they need to align not only forces and things but also concepts, energy, and understandings. Again, project managers must be creators and ensurers of *sustained mutuality*.

The new business culture will also favor the quick, the rapid, and the immediate. This increases our dependence on the project mode of work, for projects surpass operations in this regard. They can be marshalled almost instantaneously and implemented with little cost and a minimum amount of time.

Project orientations allow us to *capture* opportunity, while subsequent operations allow us to *exploit* what we have captured. As exploitation times become shorter with change, and capture becomes dependent on alert and mobile management, the abilities to listen and to move quickly become more valuable. The business culture of tomorrow, then, will involve more frequent and more urgent pursuit. And the project, the *energy pulse,* is ideally created to meet that need.

As change becomes more apparent and active, and as company managements recognize the impetus change gives to project efforts and skills, we will see more managers shifting from operations to projects. Even the most operationally entrenched managers will have to encounter and accomplish project tasks more frequently than ever before. The line between projects and operations will begin to blur, and the distinctions will eventually evaporate, for most work will be viewed as project work. As work becomes "projectized," so shall managers.

This being the case, it is best to become acquainted with project skills and project management tenets, no matter what our present alignment may be. We will all need them. If you are going to profit in change, you will profit as some sort of project manager.

The essence of project managers is cohesion. Project managers provide this by acting as organizational "glue," and they exert a strong directing influence upon what they help to bind together. This includes the frequent use of and referral to plans, benchmarks, standards of accomplishment, and temporary achievements, as well as those "mental gathering points" that ensure periodic synchronicity. *Cohesion* and *direction* will replace the old operational needs of *drive* and *efficiency*.

Project managers must be the guardians of plans and of the integrity of objectives, while operationally oriented counterparts are needed to safeguard the productive capacity: the machines, line, or process. The new manager will most often exhibit irreverence for scaffolding, technique, and tradition, whereas the manager in the past defended and reinforced these qualities. In each project experience, the new manager will seek to quickly find and harness the affective essence of the project combination. In contrast, the essence of an operation is fairly obvious to its managers, who seek only to squeeze more and more incremental value from it (to achieve optimization).

The position of project manager is by its nature both tenuous and demanding. No project manager has value without a project, and each demonstrates no strengths except through the project effort. Moreover, project managers cannot hide behind a particular process, concept, or technique. The position of project management is perishable and must be constantly renewed. With operations, management value is embedded in the operational process once synchronization and optimization are attained. The project manager, tomorrow's manager, will have value embedded only in himself or herself. To the degree that this confidence is deserved, it will be acknowledged.

Project managers will seek to create their own policies and rules, to create their own procedures and methods, rather than to comply with those established by others. Project-oriented managers, because of their twin drives for expediency and pragmatism, will value only methods that contribute, that work. They will shun the rest as burdens or obstacles. They should be judged by their resourcefulness and adaptation, in contrast to operational managers of the past, who were punished when they exhibited these traits but rewarded for strict compliance.

In the end, project managers will reap the rewards or penalties of their own embedded capital and that which they nurture in others. They will not be able to depend on fixed assets or scaffolding, or to cherish attached capital, as is so often done in operational settings. Their challenge will be to transcend differences and limitations rather than to depend on presumed fixity. If they do this successfully, they will become the magicians of the new business world: the project-oriented world of work.

## DEVELOPING CONTRASTS

To enhance our appreciation of the fundamental differences between projects and operations, we can contrast their expectations—what they require of us as a price for our allegiance:

| **Operations Demand That We:** | **Projects Demand That We:** |
|---|---|
| *Use them repetitively* | *Create and abandon them at will* |
| *Bring opportunity to them* | *Use them to pursue opportunity* |
| *Harness similarity* | *Harvest diversity* |
| *Let them define expectations* | *Let them achieve expectations* |
| *Steer the phenomenon of change* | *Steer through or around the phenomenon of change* |
| *Wrap the work around the tools* | *Wrap the tools around the work* |
| *Chain people to the process* | *Chain the process to people* |

## THE PRICES AND RISKS OF MANAGING PROJECTS, NOT OPERATIONS

Nothing in this chapter should be construed to diminish the principles of operational management or to suggest that these will no longer be needed. Operational orientations and skills have created the business climate and bounty which most industrialized nations now enjoy, and they will be needed far into the future. What is implied, however, is that, given the nature and emerging awareness of change, the significance of operational models and behavior will diminish, and that of project-oriented approaches will increase.

People who agree with the notion of project suitability and strength in times of change should be aware of the cultural changes it brings and the management ramifications involved. Most of these have been presented in this chapter. In terms of risk and price, however, a few need amplification.

One price, at least to some managements, of a shift to project work is the constant state of organizational flux it requires. We cannot simply reorganize one time and hope this will change our company from a functional, operating organization to a project-sensitive one. Project organizations need to be constantly created, modified, and destroyed. This is often done in concert with modifications and reorganizations on the functional side of the company, where operations occur. Reshaping of project organizations is not done in a vacuum, for it commonly involves movement of people and groups across project-operations borders. Although various forms of matrix organizations have been contrived to facilitate this transition, "tension" is still the best word we can use to describe it.

Process-dependent, technique-addicted, or specialty-limited managers will not be comfortable with the new project environment, just as they are uncomfortable with other aspects of change. Most of the demands of this environment will run counter to their skills and proclivities. Managers who cannot make the attitudinal shift to the new business culture will have to be reassigned to scarcer operational settings, or they will become handicaps to our adaptability.

Executive managers will need to sharpen their own project-related skills—to enhance their ability to harness and nurture strong personalities, independent managers, and creative and innovative pragmatists. Such people are extremely difficult to orchestrate, and simple reliance on procedures or convention will never suffice. Project managers, we must remember, love to create and hate to comply. They will not follow unless they are led. To harvest the bounties they bring, our executive management must reshape its own values and abilities in view of change. Sometimes this price is too high for them.

Project work requires *agility of pursuit* and *agility of perspective* as well. This means we must constantly shift our attention, focus our vision, tune our listening, and most importantly, reshape our understanding. Frequent reestablishment and modification of view and attitude is difficult for many of us. By our natures, we prefer fixity. Unfortunately, change does not allow us this luxury.

Projects require unique views as well as unique efforts. They exist in and for change, and change affects not only our conversion processes and our particular company attitudes and cultures, but the entire fabric of business enterprise, the culture of the business world. The emergence of projects is but one aspect of this new culture. It affects not only that which surrounds us, but ourselves as well, our group and individual identities and values. To some, this reshaping is painful and frightening. To managers attuned to change, it is exciting, challenging, and very rewarding.

# 2. Managing Project Interfaces—Key Points for Project Success

Peter W. G. Morris*

One of the most important qualities of a project manager is a mature understanding of the way projects develop. This allows the nature of project activities to be better understood, problems to be seen in perspective, and needs to be assessed ahead of time.

To some extent this understanding of project development is intuitive, though it clearly also depends upon specialist knowledge of the project's technology and industry. It can, however, also be acquired in large part from formal study of the development process of projects, since all projects, regardless of size or type, follow a broadly similar pattern of development.

The organizational framework underlying a project's development is the subject of this chapter. The intent of the chapter is to illustrate the types of issues that are normally encountered as a project develops and to suggest ways in which these issues should be handled. The extent to which these issues can be related to project success is then discussed.

## THE SYSTEMS PERSPECTIVE AND PROJECT MANAGEMENT

The most pervasive intellectual tradition to project management, whether in organization, planning, control, or other aspects, is without doubt the systems approach.

---

* Dr. Peter Morris is Director of the Major Projects Association, a member of Oxford University's Faculty of Physical Sciences and an Associate Fellow of Templeton College, the Oxford Centre for Management Studies, Oxford, England. He undertook research into project management at Manchester University, England, in the late 1960s, gaining his Ph.D. in 1972. He has worked both as a manager and as a consultant on a variety of projects around the world ranging from telecommunications and petrochemical projects in the Mideast, North Africa, and Europe; steel projects in Latin America; and construction, MIS, and aerospace projects in North America and Europe. Prior to returning to England in 1984, he was responsible for the international business of the Arthur D. Little Program Systems Management Company, Cambridge, Massachusetts. He is the co-author of *The Anatomy of Major Projects*, published by John Wiley & Sons Ltd., 1987.

A system is an assemblage of people, things, information, or other attributes, grouped together according to a particular system "objective." Thus, one has an electrical system, the digestive system, a high-pressure weather system, an air conditioning system, a weapons system, a system for winning at cards.

A system may be logically broken down into a number of subsystems, that is, assemblages of people, things, information, or organizations required to achieve a defined system *sub*objective, like the switching, outside plant, building, transmission, and subscriber subsystems in a telephone system. The subsets of each subsystem may then be identified—cables, poles, microwave, and transmission and distribution equipment for the transmission subsystem—thereby creating sub-subsystems. Subsets of these subsets may then be identified, and so on.

Properly organized and managed, the overall system acts in a way that is greater than the sum of its parts. The systems approach emphasizes treating the system as a whole.

The systems approach has its origin in the late 1920s and 1930s. Biologists noticed similarities in the way that living organisms interacted with and controlled their environments. At the same time, similar patterns were observed by Gestalt psychologists in the way the human mind organized sensory data. Both the mind and living organisms have to adapt to changes in their environment. Systems of this type are known as "open" systems. Before long it was seen that all social systems operate as open systems.[1]

During the 1950s, work in economics, psychology, sociology, anthropology, and other disciplines developed these open-system ideas by elaborating such concepts as self-organization, purposive systems, the importance of goals and objectives, the hierarchical classification of systems and subsystems, and the importance of systems' boundaries and interfaces (2)*. At the same time, this "systemic" view of the world was enriched by a parallel (but initially separate) set of disciplines which had their origin in the industrial and military applications of the scientific method during and immediately after World War II. This was the essentially numeric set of disciplines, such as cybernetics, control theory, oper-

---

[1] Open systems are "open" to the effects of their environment. On the other hand, closed systems, which are the other major system type (including, for example, much of physical chemistry and many types of machines), operate independently of their environment. In open systems, events rather than things are structured; there is a constant energy and information exchange between the system and its environment; the system organizes to minimize entropic decay; equilibrium with the environment is achieved through a process known as homeostasis; and there is a tendency towards differentiation. Closed systems operate in almost exactly the opposite manner (1).*

* Numbered references are given at the end of this chapter.

ations research, systems analysis, and systems engineering, concerned with modeling real-life situations so that complex behavior could be more accurately described and forecast. Slowly both streams merged, encouraged greatly by the enormous growth in the ability of the computer to apply these systems ideas with powerful effectiveness, so that the systems approach is now an established and vigorous influence on management and research.

The systems perspective has contributed substantially to the development of project management. Firstly, the systems emphasis on viewing a system as a whole has frequently been behind the recognition of the need for an across-the-board integrating role—that is, for project management itself (3).[2]

Secondly, systems thinking has shown how projects should work as successfully regulated organizations, for example, the need for clearly defined objectives, the recognition that projects are organizations in constant change, and the need to define and manage major subsystems and their interfaces. A third important contribution is that the dynamic control needs of projects are now better understood—the importance of feedback, the progressive development of information and multilevel project control. And a fourth contribution is the widespread use of systems techniques—systems analysis, systems engineering, work breakdown structures, and simulation models.

Interface Management, as it is used in project management today,[3] is an outgrowth of the first two of these influences of systems thinking on project management. Interface Management identifies the following:

- The subsystems to be managed on a project,
- The principal subsystem interfaces requiring management attention,

---

[2] The development of project management by the U.S. military is an illustration: the systems ideas developed initially for technical purposes were adapted to generate the organizational flexibility and control missing in the existing military bureaucracy. This can be seen in each of the steps in the U.S. military's development of project management—the development of the USAF and Navy program management practices (4), particularly for the ICBM programs (5) and for Polaris (6) around 1952–1955; Peck and Scherer's study of the U.S. and Soviet weapons procurement processes in the late 1950s (7); the development of PERT on Polaris in 1958; the introduction of project organizations in the Navy, Air Force, and Army in the late 1950s/early 1960s; McNamara's extensive study and implementation of program management and project control techniques in the early 1960s; and Laird's and Packard's process-oriented focus on the needs of the total project life cycle in the late 1960s/early 1970s (8).

[3] Interface Management is generally used now in a broader sense than it was ten or twenty years ago. In the 1960s and early 1970s Interface Management generally referred simply to ensuring that system interfaces matched (i.e., had the same specifications, were not missing any equipment, data, etc.). Today it is used in the sense of defining systems—organizational, managerial, and technical—and of actively managing their interrelationships. The term "interface management" is relatively common in high-technology projects, such as information systems and aerospace; it is much rarer on construction projects.

- The ways in which these interactions should be managed successfully.

The emphasis on identifying key interfaces and on focusing on interface performance has grown as it has been increasingly realized that all projects share a common pattern of interfaces derived from a common pattern of subsystem interaction. This is true no matter what the type of project, be it a theater production, an aid program, an election or a major capital investment program.

There are three sets of subsystems on any project: those deriving from the project's life cycle, its management levels, and its operational characteristics.

## PROJECT LIFE CYCLE[4a]

Project management teaches that to achieve the desired project objective one must go through a specific process. There is no exception to this rule. The process is known as the project "life cycle."

Projects (like people) have a life cycle that involves a gradual buildup as definitions are established and working characteristics developed, a full-bodied implementation as the work is accomplished, and a phasing out as the work is completed and the project winds down. While there are various definitions of the project (or program) life cycle (Figure 2-1), the essential sequencing is invariant, although (as with people) sometimes not fully recognized or respected.

A project starts as an incipient idea which is explored for financial and technical feasibility in the *Prefeasibility/Feasibility Stage*. Capacity is decided, locations chosen, financing arranged, overall schedule and budget agreed, and preliminary organizations set up. At the end of this phase there should be a formal "go/no-go" decision. In the next phase, *Design,* the work is organizationally and managerially similar to the first phase only it is more comprehensive and detailed. The technical definition of the project is expanded (albeit generally still at a fairly strategic level); schedule, budget, and financing are reappraised; contracting strategy is defined; permits are sought; and infrastructure and logistics systems are defined.

---

[4] Note that an interface is technically defined as the space between interacting subsystems. Even though there might be a common set of subsystems on all projects, this does not necessarily mean there will be a common set of interfaces. The extent that there is depends on the commonality of subsystem interaction. This chapter will show that subsystem interaction does in fact follow a common pattern on most projects.

[4a] The implications of the life cycle are treated in greater detail in Chapter 9.

Kelley's model [9] is essentially that of a program life cycle. He differentiates product control from project control, product control being concerned with the definitions of the end product parameters and project control being concerned with the process of making that product.

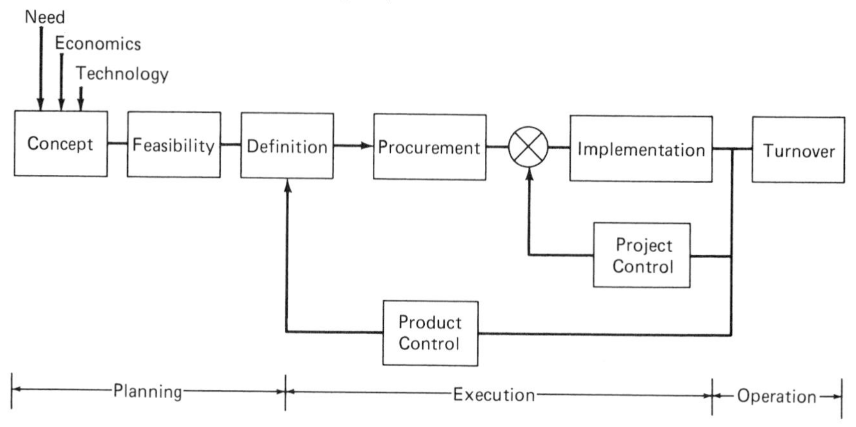


Wearne's model [10] is typical of industrial projects. "The nature and scale of activities change at each stage but the stages usually overlap." Of particular interest are the Demand and the Use, Maintenance, and Records & Experience stages, since these are often overlooked in project management. Iteration within and between the early stages is not unlikely.

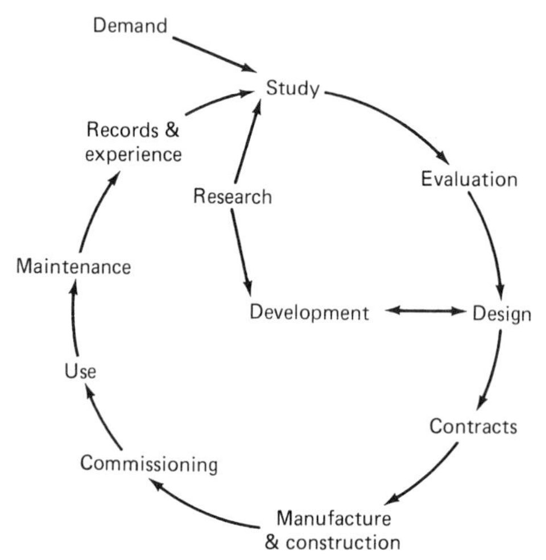


Figure 2-1. Views of the project/program life cycle.

In phase three, *Manufacture, Construction, and Installation* (or *Production*), equipment is procured, civil work is undertaken, and equipment and facilities are installed. This phase differs dramatically from the previous two. First, whereas the *Design* and *Feasibility* phases were organic and evolutionary in character, the *Production* phase is highly mechanistic

Morris' [11] concentrated on the feasibility-design-implementation phases to make several points about the nature of work within and between phases. The design phase is basically "organistic," for example, while the production phase is more "mechanistic." Crossing the design-production interface, with the letting of implementation contracts, is a major transition point in the project life cycle.

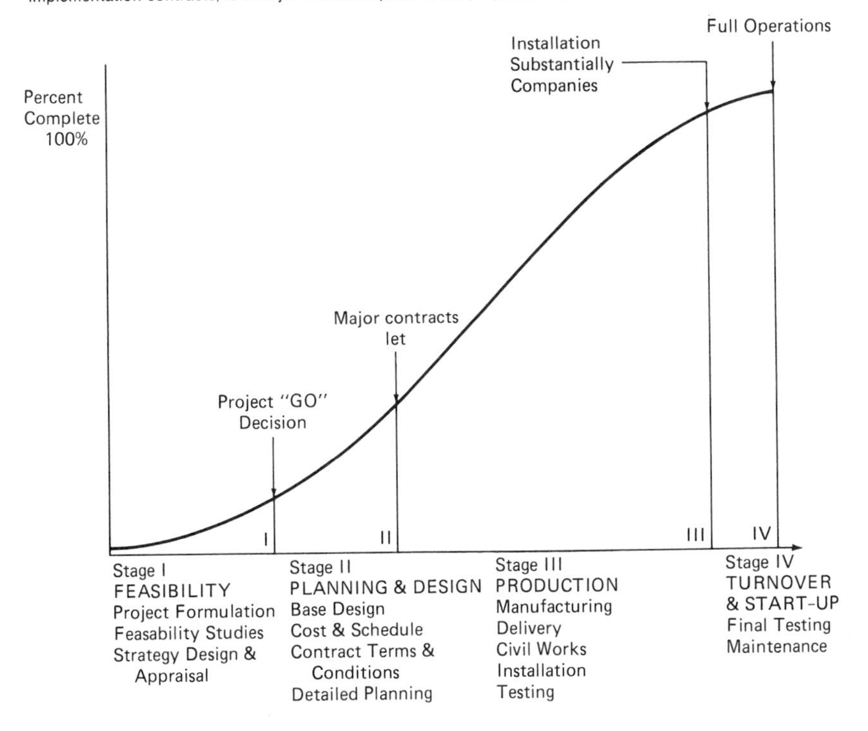


Kerzner's diagram [12] illustrates the financial life of a product development program (me). Kerzner comments on the way the life cycle varies between systems, projects and products; apart from differences in terminology, a major difference is in the overlapping of phases.

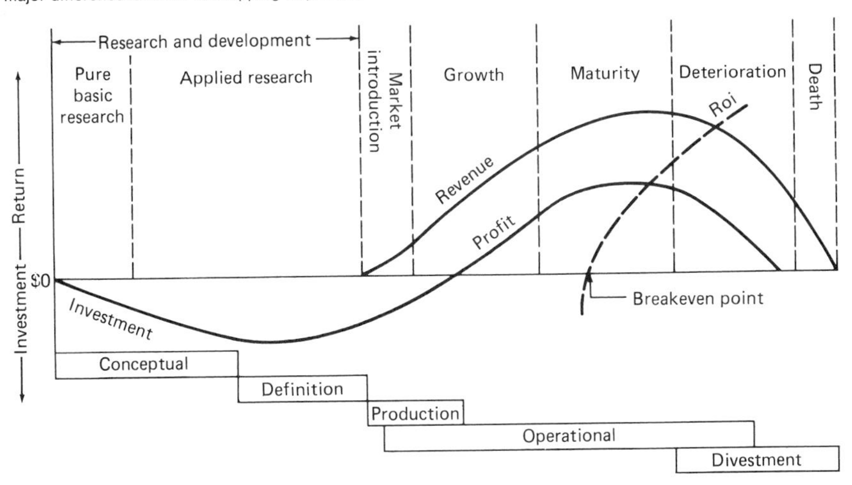


Figure 2-1. *(continued)*

(13). The aim is not to develop new technical options but to build as efficiently as possible the thing which has been defined in the *Design* phase. Second, there is a large—sometimes vast—expansion in organization (whereas there may have been only dozens or hundreds of persons active in the first two phases, there may be thousands or even tens of thousands involved in this third phase.) And third, the characteristic mode of control changes from one of "estimating" costs and durations to one of tight "monitoring" of quality, schedule, and cost to keep actual performance within the target estimates.

The fourth and final phase, *Turn-Over and Start-Up,* overlaps the third phase and involves planning all the activities necessary for acceptance and operation of the project. Successfully synchronizing phases three and four can prove a major management exercise. The cost of capital locked up in the yet uncommissioned plant, and the opportunity costs of both underutilized operating systems such as sales, operating plant, personnel, etc., and a possible diminishing strategic advantage while competitors develop rival products, can prove enormous.

Between each of these life-cycle phases there are distinct "change points" (what shall later be called "dynamic project interfaces"):

- From *Prefeasibility/Feasibility* to *Design:* the "go" decision.
- From *Design* to *Production.*
- From *Production* to *Turn-Over & Start-Up.*

The project on either side of these change points is dramatically different—in mission, size, technology, scale, and rate of change—and these differences create their own particular different characteristics of work, personal behavior, and direction and control needs. Thus, importantly, the management style of each of the four main life-cycle phases is significantly different.

## PROJECT MANAGEMENT LEVELS

The four phases have a set and important managerial relation to each other. The work of the *Prefeasibility/Feasibility* stage is highly "institutional" (top management) in kind—decisions taken in this phase will later have an overriding impact on the health of the investing enterprises. In *Design,* the work is of a "strategic" nature, laying the axes upon which the detailed, "tactical" work, of the third, *Production* phase will rest. Interestingly, the fourth phase, *Turn-Over & Start-Up,* exhibits a mixture of all three managerial levels of work: institutional, strategic, and tactical.

These three levels of management activity have been recognized as distinct levels of management since at least the time of Talcott Parsons, the eminent American sociologist. Parsons made the point that each of the three levels has an essential role to play in any successfully regulated enterprise: the technical/tactical level (III) manufactures the product; middle management (II) coordinates the manufacturing effort; at the institutional level (I) top management connects the enterprise to the wider social system (14). Each of these three management levels has a fundamental role to play in the management of every project (although it is true that the levels tend to become more blurred on the smaller projects). Yet surprisingly, most project management literature deals only with Levels II and III. There is little in the literature that treats such Level I issues as: the role of the owner and his financer; relations with the media, local and federal government, regulatory agencies, lobbyists, and community groups; the sizing and timing of the project in relation to product demand and the cost of finance—all issues that became crucially important during the 1970s and are particularly so in the 1980s.[5]

The distinction between Levels II and I is quite critical since it is essentially the distinction between the project and its outside world (Figure 2-2). Levels II and III deal almost exclusively with such familiar project activities as engineering, procurement, installation, testing, and start-up—Level III providing the technical input, Level II providing both a buffer from the outside world and guidance in how to avoid external pitfalls. But no project exists in isolation from outside events. Level I provides the coordination of the project with outside events and institutions. Level I actors typically include the project owner and his finance team, government agencies, community groups, very senior project management, and one or two special project executives specifically charged with external affairs, such as Public Relations and Legal Counsel.

The involvement of each of these Levels is different during each of the major phases of the project life cycle. During the *Prefeasibility/Feasibility* stage, the owner and his team (Levels I and II) have to make crucial decisions about the technical performance and business advantages they are to get for their investment—and indeed, whether the project should "go" at all. Once the decision to go ahead is taken the weight of the work moves to the design team (Levels II and III). During *Production* engineering reaches a detailed level. Both project management (Level II) and technical staff (Level III) are now at full stretch, while top management (Level I) takes a more reduced "monitoring" role. Finally, during *Turn-Over and Start-Up*, all three Levels are typically highly active as engi-

[5] Such "institutional" project management issues are preeminently the concern of the UK Major Projects Association (15). They are treated in this handbook in Chapters 8, 13, 18, and 22.

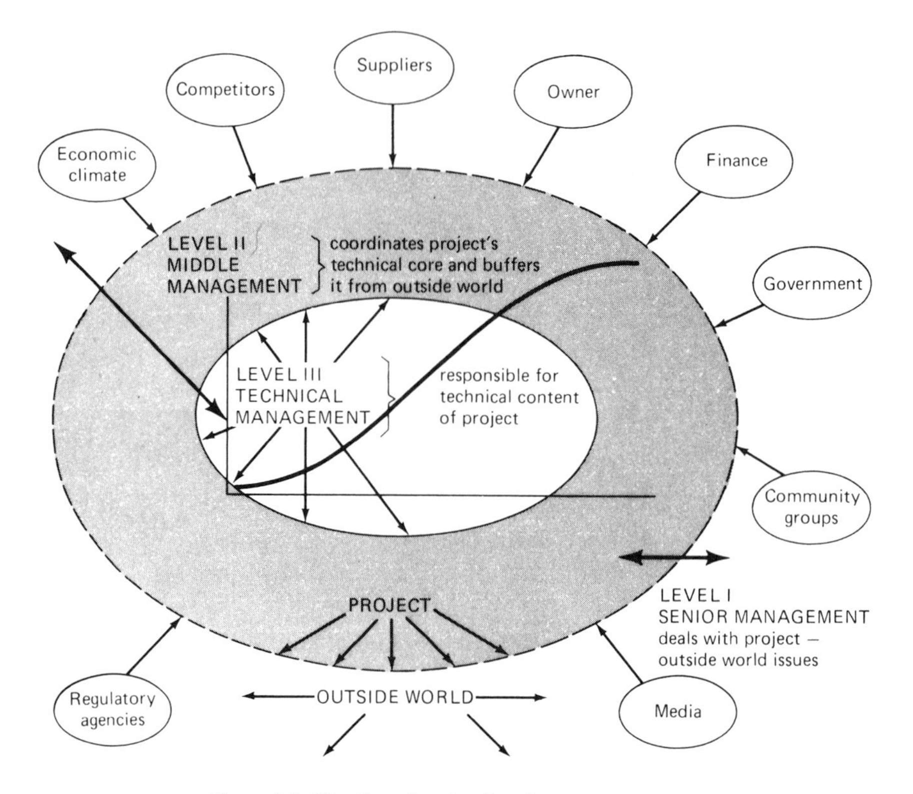


Figure 2-2. The three levels of project management.

neering work gets completed (Level III), often under intense management pressure (Level II), while high-level coordination is required at the owner level in coordinating the initiation of Start-Up activities.

The responsibilities of these Levels thus focus on two main areas of activity: Levels II and III on the technical and middle management work within the project, and Level I on the senior management work at the project/outside world interface.

## PROJECT OPERATIONAL SUBSYSTEMS

The work of these two essentially distinct levels of project management activity tends to follow a pattern which is similar on many projects.

At the project/outside world level, the concern is to ensure that the project is commercially viable and, as far as possible, is provided with the conditions and resources necessary to succeed. The principal areas of work at this level are

- Ensuring satisfactory *Project Definition* (which includes both technical content, cost and schedule requirements)
- Preparing for *Operations & Maintenance*
- Preparing for *Sales & Marketing*
- Ensuring appropriate *Organization* structures and systems, both for the project and for operations
- Facilitating relations with important *Outside Groups* such as government and community groups, financial institutions, and the media
- Ensuring appropriately skilled *Manpower* for both the project and future operations
- Ensuring that the total enterprise is *Commercially Sound* and "adequately financed."

Work within the project, on the other hand, focuses more on accomplishing the tasks within the strategic parameters developed and managed by senior management. At the intra-project level, the principal subsystems are

- Realizing the desired *Project Definition*—i.e., assuring that the project is produced to technical specification, on time, and in budget
- Creating the *Organization* needed to execute the project—this includes both the formal organization structures, contractual relationships, systems of information flow and control procedures, and also informal patterns of working relationships and communication
- Minimizing external disruptions from the *Environment*—by, for example, acquiring adequate materials inventory to provide buffer stocks against delivery disruptions, handling union negotiations, obtaining necessary regulatory approvals, or warning top management of future financing needs[6]
- Providing adequate *Infrastructure and Logistics* to accomplish the project (facilities, transportation, communication, utilities).

[6] The effects of environment on a project can be profound, and continue to provoke keen theoretical analysis and discussion. Of particular interest is the problem of how organizations behave in a constantly and rapidly changing environment. Theorists describe such environments as turbulent and call the type of systems that operate in such environments "multistable" (16). Large or complex projects in particular suffer many of the consequences predicted for multistable systems, such as large subsystem interaction, continuous objectives redefinition, rich internal feedback processes, high impact of external factors (often causing the subsystems to have to act in an apparently less-than-rational way), and substantial organizational change, often of a step-function size. These characteristics can be found on major projects such as the Concorde and North Sea Oil projects, nuclear power projects, and many defense projects (17).

## STATIC AND DYNAMIC INTERFACES

The likely existence of these subsystems in a project, no matter how it unfolds, enables us to categorize certain interfaces as on-going or "static"—they are not a function of the way the project develops but represent relationships between on-going subsystems (like engineering and procurement, or Level I and Level II). There is another group of interfaces, however, which arise only as a function of the pattern of activity interdependencies generated by the way the project develops. These we may identify as life-cycle or "dynamic" interfaces.

Dynamic interfaces between life-cycle (or activity) subsystems are of the utmost importance in project management, first because of the continuous importance of the clock in all projects, and second because early subsystems (like *Design*) have a managerially dominant role on subsequent ones (like *Manufacturing*). Dynamic interrelationships require careful handling if minor mistakes in early systems are not to pass unnoticed and snowball into larger ones later in the project.

Boundaries should be positioned where there are major discontinuities in technology, territory, time, or organization (18). Major breakpoints in the project life cycle—as, for instance, between each of the four major phases, and also between activity subsystems within each phase (for example between manufacture, inspection, delivery, warehousing, installation, and testing)—provide important dynamic interfaces. These serve as "natural" check points at which management can monitor performance.

Most major dynamic interfaces are in fact used in this way: for example, the Project Feasibility Report, the initial Project Technical Design, the formulation and negotiation of the "Production" contracts, and Testing and Hand-Over. Review points such as design-freeze points, estimates-to-complete, and monthly progress reports may also be introduced for purely control purposes without there being any "natural" discontinuity. Each in its own way represents a response by project management to control the project's momentum across its dynamic interfaces.

Whereas the important dynamic interfaces are relatively sharply defined for Level II and III management, at Level I they are less distinct. Level I management is certainly partly driven by the anatomy of the project's internal development, but it also has its own dynamic interfaces for each of its own principal subsystems. Operation, Sales, many of the Outside Groups, Manpower, and Finance and Commercial issues each have their own often distinct life cycles. (For example: the process of recruiting and training manpower; preparing annual financial plans.) Thus at Level I, dynamic interfaces do not become less important; rather they become more varied and less clearly defined. They are still crucial to the project's success.

Static interfaces too are less clearly defined at Level I than at Levels II and III, partly due to the wider scope of concern of Level I (which gives rise to much multifacted subsystem interaction, as, for example, between Operations, Sales, Manpower, and Finance) and partly due to the disruptive effect of the outside environment.

Figure 2-3 sketches the three principal sets of project subsystems which have now been identified: the three levels of management, static subsystems, and dynamic subsystems.

## PROJECT INTEGRATION

Some interfaces are clearly larger and more important than others. Organization theorists describe the size of an interface not in terms of, for instance, a small change point or a major one, but in terms of the degree of *differentiation* between subsystems. Typical measures of differentiation include differences in

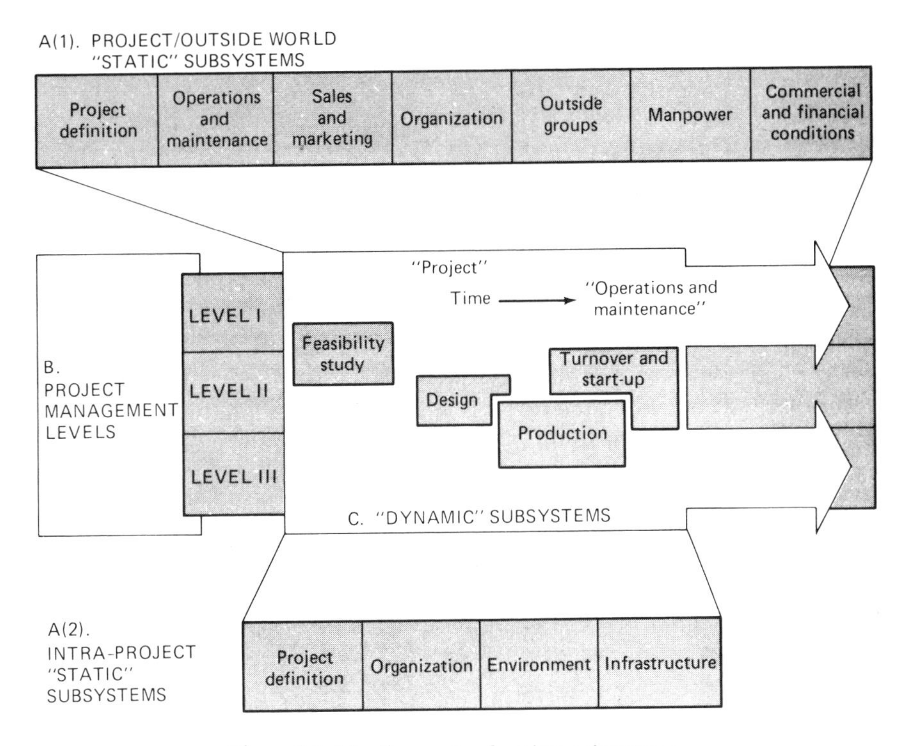


Figure 2-3. The three sets of project subsystems.

- Organization structure
- Interpersonal orientations
- Time horizons
- Goals and objectives (19).

Thus, a mechanized infantry brigade can be differentiated clearly from a local community opposition group on all dimensions. The R&D wing of a company can be similarly differentiated from the marketing wing. The architect on a building project can be differentiated from the building contractor.

A mechanized infantry brigade will go to some lengths to avoid having to integrate with a local opposition group but R&D has to integrate with marketing quite frequently, and it is inevitable that architects must integrate with contractors. Why? Because the activities of the groups create certain technical, organizational, and environmental *interdependencies*. These interdependencies may be almost accidental or may be deliberately organized. *Integration* becomes important when the degree of organizational interdependence becomes significant. Research has shown that tighter organizational integration is necessary when

- The goals and objectives of an enterprise bring a need for different groups to work closely together
- The environment is complex or changing rapidly
- The technology is uncertain or complex
- The enterprise is changing quickly
- The enterprise is organizationally complex (20).

The amount of integration actually required at an interface depends both on the size of differentiation across the interface and on how much "pulling together" the interfacing subsystems need.

Certain project subsystems can be differentiated from one another quite markedly. For example, the project/outside groups interface is marked by very strong differences in time horizons, goals and objectives, interpersonal orientations, and structure—this is why the conflict over many environmental and regulatory issues is so drastic on many large projects: the aims and mores of the environmentalists are far, far removed from those who are trying to build the project. The design group often functions quite differently from the construction group—the former's interest might be elegant engineering, time might not be money, and quality might be paramount; the construction crew might be of less elitist thinking, have strong incentives not to waste time, and might often work in a tougher organizational milieu. Similarly, there are major differences in perspective between operations and project personnel, between the project finance team and project engineering, between a construction manager and project management, and so on. It is, in short, possible to establish the degree of differentiation between each one of the project's interacting subsystems, and in so doing thereby establish which are the principal project interfaces (21).

Despite the insights of management theorists, choosing the degree of integration—the amount of "pulling together"—required across an interface still calls for considerable judgment. This is inevitable. There is no easy answer to the question, "How much management is enough?" There are some pointers, however. James D. Thompson, in a classic book (22), observed that there are three kinds of interdependency, each requiring its own type of integration. The simplest, "pooled," only requires that people obey certain rules or standards. The second form, "sequential," requires that interdependencies be *scheduled.* "Reciprocal" interdependence, the most complex kind, requires *mutual adjustment* between parties (Figure 2-4). In project terms, subsystems which are in continuous

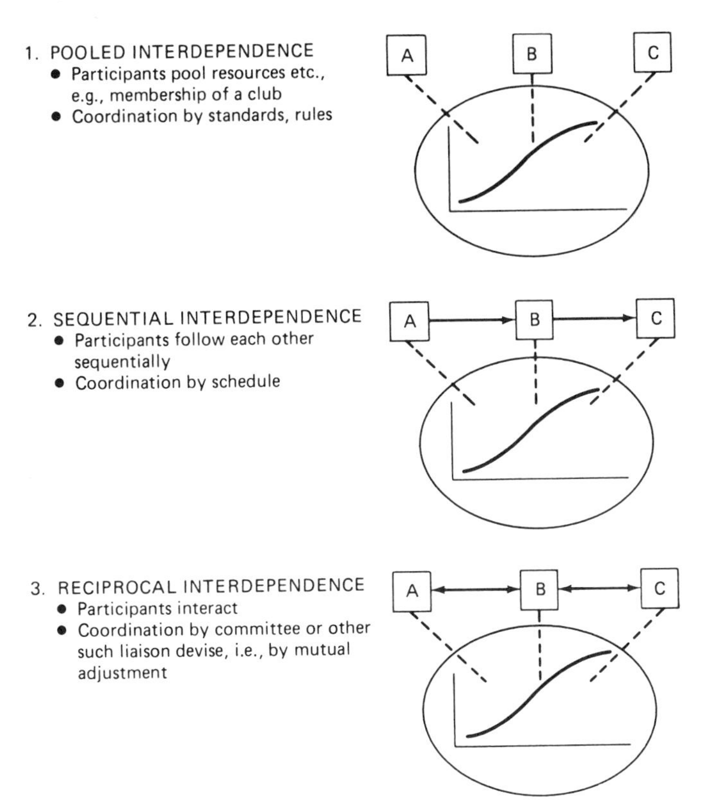


Figure 2-4. The three types of interdependence.

interaction require liaison in order to achieve the necessary integration, whereas those that just follow on from one another can follow plans and schedules.

There is a range of devices which can be used to achieve liaison (23):

- Liaison positions
- Task forces
- Special teams
- Coordinators (or permanent integrators)
- Full project management
- Matrix organizations.

Each of these options provides stronger integration than the last.

The primary function of *liaison positions* is to facilitate communication between groups. Other than this, the liaison position carries no real authority and little responsibility. *Task forces* are much stronger. Task forces provide mission-oriented integration: a group is formed specifically for a particular task and upon completion of the task the group disbands. *Special teams* are like task forces but attend to regularly recurring types of problems rather than specific issues. A *coordinator,* or permanent integrator, provides a similar service as a liaison position but has some formal authority. He exercises this authority over the decision-making processes, however, not over the actual decision makers themselves. This is a subtle point, but an important one, and it often causes difficulties in projects. The coordinator cannot command the persons he is coordinating to take specific actions. That authority rests with their functional manager. He can, however, influence their behavior and decisions, either through formal means such as managing the project's budget and schedule, approving scope changes, etc., or through informal means such as his persuasive and negotiating skills. The full *project manager* role upgrades the authority and responsibility of the integration function to allow cross-functional coordination. The integrator—the project manager—now has authority to order groups directly to take certain actions or decisions. *Matrix organizations* are, by general consent, considered about the most complex form of organization structure. Matrix structures provide for maximum information exchange, management coordination, and resource sharing (24). Matrixes achieve this by having staff account simultaneously to both the integrating (project) managers and the functional managers whose work is being integrated. Both project managers and functional managers have authority and responsibility over the work, albeit there is a division of responsibility: the functional manager is responsible for the "what" and "by whom"; the project manager decides the

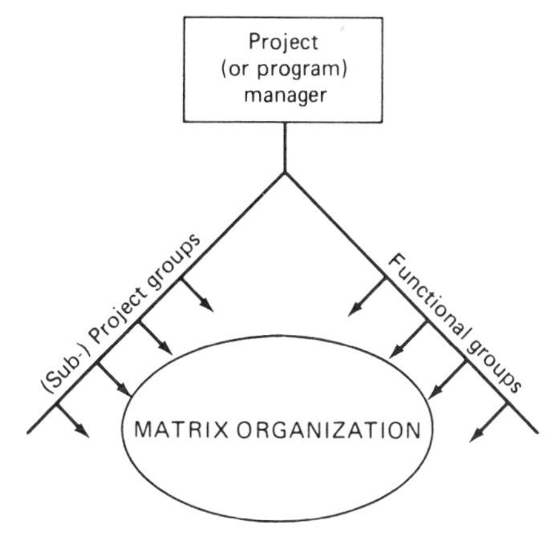


Figure 2-5. Typical use of the project management and matrix organizations simultaneously.

"when" and "for how much." Unfortunately the person who often comes off the worst in the matrix is the poor soul (at Level III) who is actually doing the work. He reports to two bosses—his project manager and his functional manager—which is not in itself necessarily a problem except when, as often happens, the project manager and functional manager are themselves in conflict (for instance, over how much should be spent on the project). Matrix structures generate considerable conflict and suffer from constantly changing boundaries and interfaces (25).

The relative merits of the matrix organization vis-à-vis the full-fledged project organization is one of those hardy perennials of project management. Various writers at various times have offered all kinds of reasons why one or other form is better. Three points seem to stand out, however. First, the full project management role—with a project manager in overall command of the project—does offer stronger leadership and better unity of command.[7] It is better for achieving the big challenge. Second, the matrix organization is more economical on resources. For this reason alone it is often almost unavoidable on large or complex projects. Third, it is quite common to find a full-fledged project manager sitting on top of a matrix structure (Figure 2-5)—the two forms are not incompatible but in fact fit rather well together: the top project manager (Level I) providing

[7] Note, then, that if one were to list the integrating devices in terms of ascending project management authority, the last three forms would change order to become "project coordinator, matrix organization, full project form" (26).

the leadership and ultimate decision-making authority, the matrix providing maximum middle management (Levels II and III) integration.

The challenge in moving through this range of liaison devices is very clear. Achieving greater integration requires increased attention to interfacing parties. Interfaces tend to become increasingly difficult to manage as one moves through the continuum. Let us look now at some experience of managing project interfaces.

## MANAGING PROJECT INTERFACES

Interface Management is not, it must be admitted, a well-developed theory of management well supported by a tight body of research and experience. It is more a way of looking at project management which is particularly useful especially on large, complex, or urgent projects. The insights which are offered below are therefore illustrative rather than comprehensive in their exposition of Interface Management.

### Keep Static Interfaces Clearly Defined

On projects, problems require solutions within short time frames, organizational conflicts abound, and compromises are inevitable. In such an environment, boundaries can blur. It is therefore a fundamental principle of Interface Management to maintain the static interfaces clearly defined.

In the Apollo program there was a constant need to reinforce organizational boundaries. When General Phillips was appointed director of the Apollo program in 1963, he found that the program was organized entirely along project grounds: one group for the Lunar Excursion Module, one for the rocket, and so on. This created a number of problems, particularly with the wide geographic dispersion of the program. The program was therefore reorganized to stress its functional and geographic needs as well as its project requirements. Five functional divisions were created—systems engineering, checkout and test, flight operations, reliability and quality assurance, and program control—with project offices in Houston, Huntsville, and Cape Canaveral. A matrix organization was thus created which reported to a strong but small program office in Washington, D.C. (Figure 2-6). This office, of only about 120 persons, managed a program which consisted at times of upwards of 300,000 persons. It did so by very clearly defining lines of responsibility and authority and program interface relationships, and by insisting that work be delegated and accounted for strictly in accordance with these lines and procedures (27).

Organizational checks and balances also help keep organization interfaces clearly demarcated. There are four groups which must always be

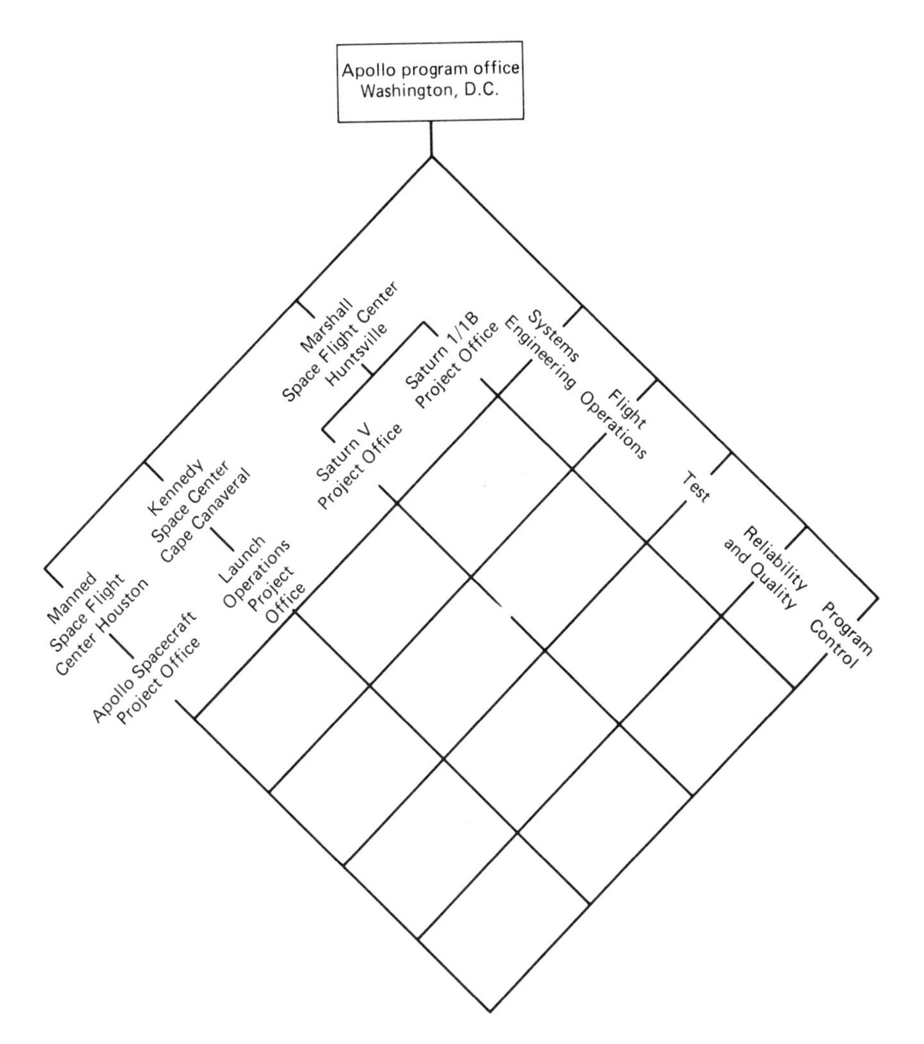


Figure 2-6. Apollo program organization.

organizationally distinct on projects: project management, project control, the functional groups, and subprojects. Project management should be separate since its role as an integrator requires it to maintain an independent viewpoint and power base. The Project Control Office should be independent since its job is to report accurate and objective data on project progress. If positioned as part of another group, say project management or construction, there will be a tendency to downplay poor performance because management will inevitably hope that things will improve. Functional groups—engineering, contracting, production, testing, reliability, contracts, etc.—represent the "engine room" of the project: the place where technical progress is accomplished. On a large project or program which is divided into subprojects there is usually a number of important schedule interlinkages and competition for scarce resources between the subprojects. Often budget and personnel are swapped between the subprojects. Subproject boundaries should be clearly defined and their interfaces closely monitored by senior management if subproject performance is to be properly controlled.

The organization structure used for the $3.5 billion Açominas steel mill project shows these four principal groups very clearly (Figure 2-7), as does the Apollo organization shown in Figure 2-6. (The Açominas case is described in more detail below.)

**Early Firm Control of Technical Definition Is ESSENTIAL to Project Success.**

Research has shown that time and again projects fail because the technical content of the program is not controlled strictly enough or early enough (28).

Software development projects (particularly large, complex ones) are extremely difficult to manage at the best of times since their work content (residing for most of the project in the project team's heads) is not as tangible as in other projects. Unless the system design is very carefully defined and communicated, the system often ends up technically inadequate, late, and very costly. The software development life cycle consists of five basic phases: concept definition, design, development, evaluation, and operation. The first phase involves problem definition and feasibility study; the second, specification of user functions and technical system design; the third, coding, integration testing, user documentation, etc.[8] Many software projects rush the first two phases and move too quickly into coding. Subsystem interfaces are then wrongly designed and code is inappropriately written. Project management techniques are now being increasingly applied to software projects. Configuration management is being used to help specify the technical content of the system as it develops and to control all changes as they arise. Software development techniques are also now emphasizing the careful, top-down evolution of the system's design and programming.

Figure 2-8 shows a model of the building process life cycle (29). Interestingly, in most building projects there is in practice no obvious checkpoint at the interface between Sketch Design and Detailed Design. There

---

[8] An "expanded" version of the information systems development life cycle is treated in Chapter 12.

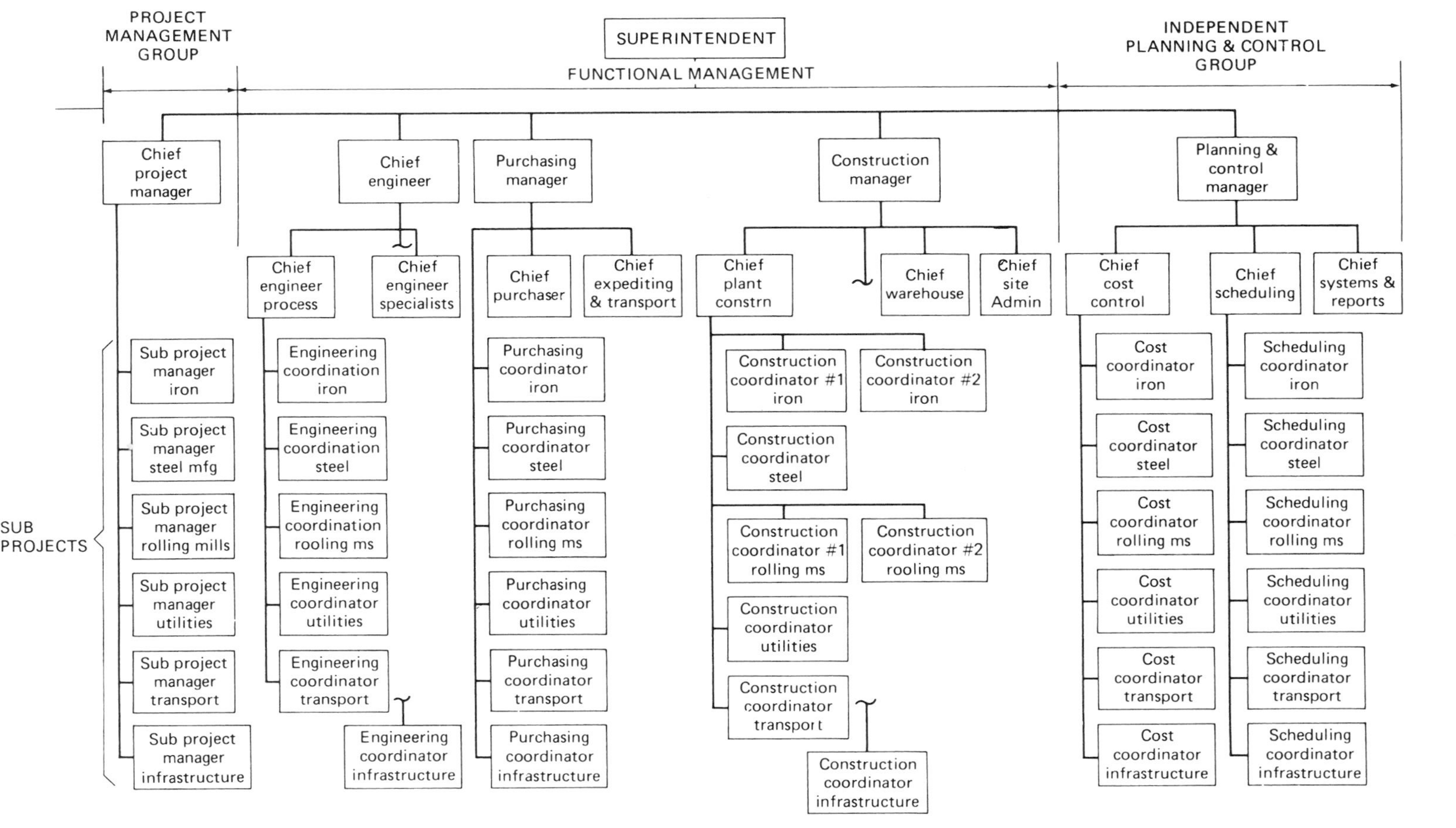


Figure 2-7. The four principal project groups on the Açominas project.

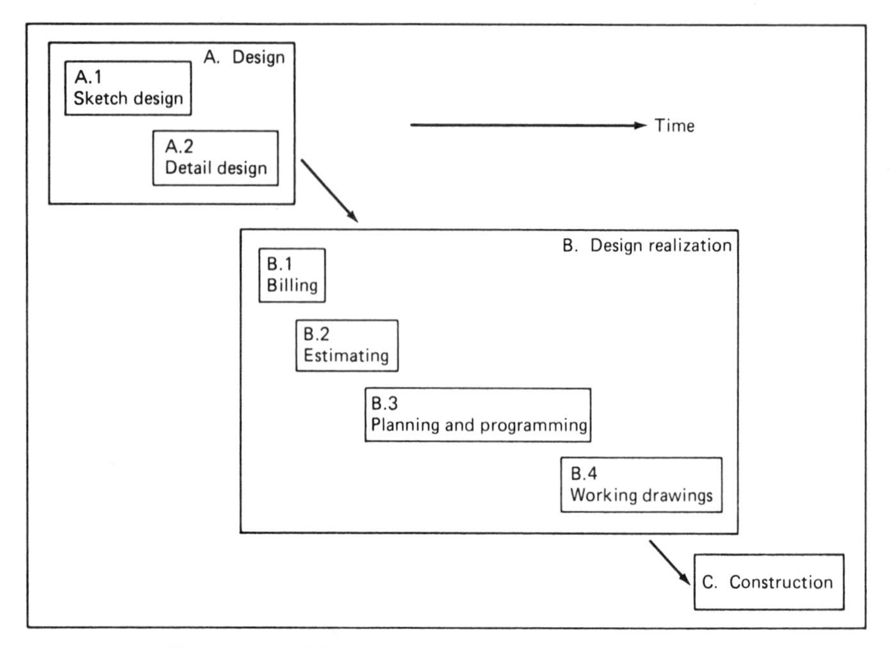


Figure 2-8. Model of the "official" British building process.

are no clearly recognized discontinuities in either technology, territory, or time at this interface,[9] nor are there any major organizational changes. Yet the outline design developed in the Sketch Design phase determines the character of the building, and thus lays the foundations for many of the technical problems which may subsequently arise during the project. If this dynamic interface is not properly controlled—and sadly, often it is not—the danger of design errors cropping up unexpectedly later in the project is high (31).

Churchill Falls is often lauded as an example of a successful project. The praise invariably centers on its tight control of design. An ambitious project in northern Canada, the project consisted of retaining the flow of the Churchill River through a series of vast reservoirs and dams over a 2500-square-mile basin in Upper Labrador. The project was begun in earnest in 1966 for a budget of $550 million and was completed eight and a half years later on budget and ahead of schedule. The project was marked by an early and very intense coordination between project management, engineering, construction management, and finance (including insur-

[9] Miller has shown that boundaries should be formed where there are discontinuities in time, technology, or territory (30).

ance)—each of which was conducted by quite separate companies. At the time of arranging project financing the state of project documentation was such that there were "virtually no questions unanswered" (32). Following this exhaustive initial design there was continuous close review by construction management of the engineering design as it developed, and intensive engineering to achieve cost savings wherever possible.

One way of achieving firm control of design is through configuration management. Configuration management documents the technical design of the project, ensures regular design reviews, rigorously checks the technical cost and schedule impact of all changes before approving them, and ensures that all parties working on the project are using up-to-date documentation. Configuration management has been used primarily in the U.S. aerospace and information systems sectors; it is only slowly being applied to other types of projects. It is not used in the building and civil engineering industries, for example, though its potential applicability on the larger and more complex of such projects is quite strong.

### The Skills Required in Managing Dynamic Interfaces Vary Depending on the Management Level and Stage of the Project

The Trans Alaskan Pipeline (TAPS) remains one of the largest and most ambitious of recent major projects. Although constructed in the three years between 1974 and 1976 (at a cost of approximately $8 billion), the project had in fact been on and off since 1968. Senior management was required to concentrate on a series of strategic issues of startling variation: firstly, on engineering (how to prevent hot oil damaging the Alaskan permafrost and how to design for seismic damage), and then moving through political support (the project manager actually moved to Washington, D.C. to advise the political effort that eventually resulted in the 1973 TAPS Act), infrastructure development (transportation, camps, equipment supply, and union negotiations), organization issues (the development of a highly decentralized matrix organization once construction began), environmental regulations, and finally, engineering and construction again. The sequence of issues is interesting: firstly, achieve agreement on the technical concept and political support for the project; secondly, assure adequate infrastructure and organization; thirdly, resolve environmental, construction, and engineering issues as the project is built. This is essentially the institutional, strategic, and tactical sequence already noted as typical for all projects. At the middle management level, however (where managers were responsible for anything up to $4 billion of work!), activity centered either on resolving engineering and construction problems or on issues of organization and leadership. Nowhere was the organizational concern more clearly evident than in the change at about 15% of the way through construction from a 9-tiered, centralized, functional organization to a 4-tiered, decentralized, matrix organization (Figure 2-9). The result was a highly flexible construction organization relying, like Apollo, on a small cadre of senior managers (33). Emphasis was on leadership, horizontal and informal communication, simple structures, and tight reporting relationships—and getting the job done.

The TAPS pipeline matrix organization tells of an experience very simi-

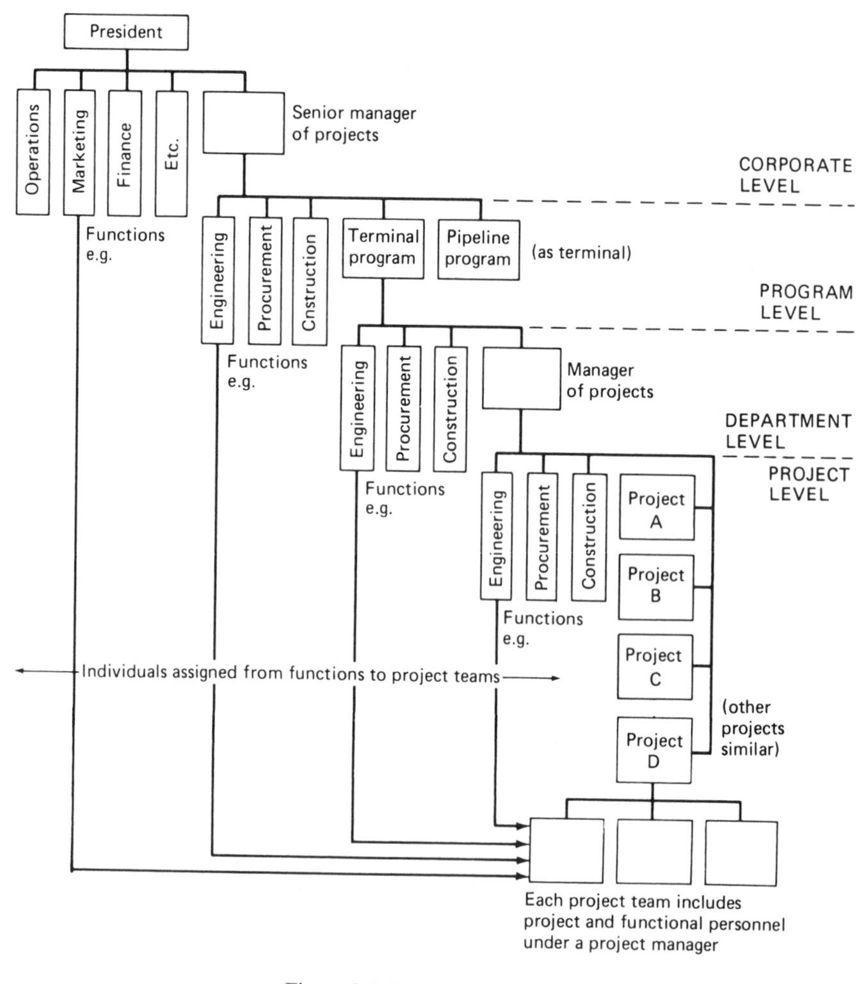


Figure 2-9. The TAPS matrix.

lar to that of Apollo and Açominas. Açominas is a steel plant recently completed at a cost of over $7.5 billion in central Brazil. Initially, the project schedule was, like TAPS and Apollo, tight. The project staff, numbering about 400, were organized on a matrix basis, operating simultaneously at three distinct levels (Figure 2-10). Like TAPS, Açominas was initially organized along primarily functional lines. Functional managers took the lead in developing the engineering design, planning the project, and negotiating the contracts. As contracts were signed and the project moved into the production phase, however, responsibility was delegated

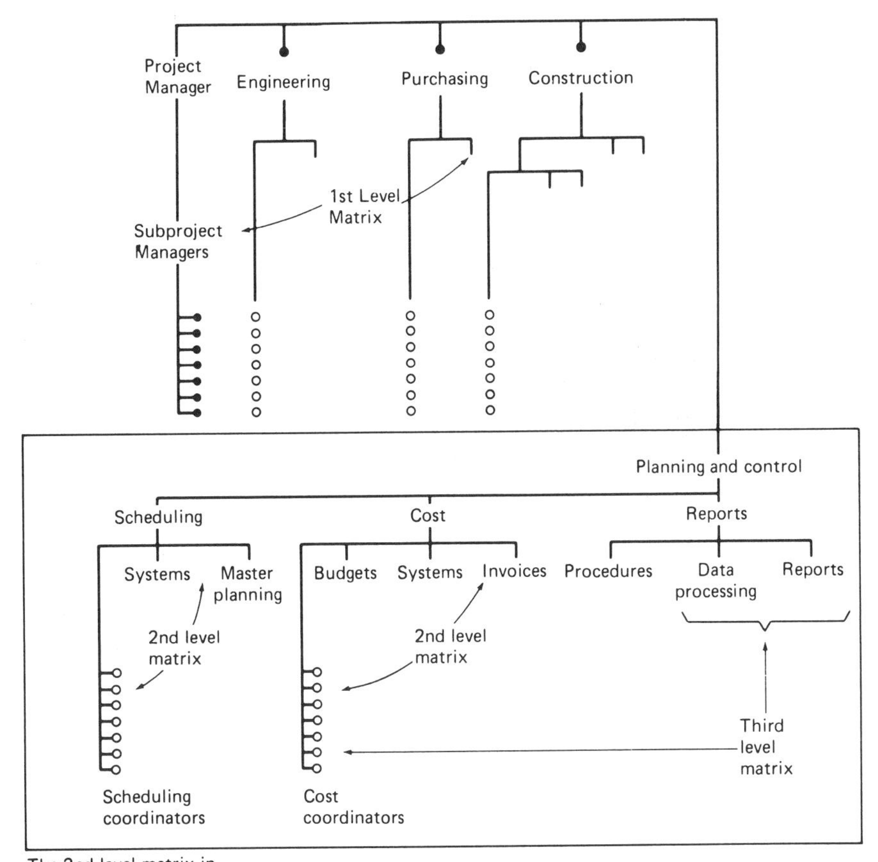


Figure 2-10. The multilevel Açominas matrix.

to the project management teams: as the project moved from its design phase to production, there was a "swing" of the matrix towards a greater "project orientation (34).

The TAPS/Acominas organization development leads to three important observations about the development of projects.

1. **Typically, Large Projects Require a Decentralized Organization During Production with Centralization Before and After.** Both projects exhibited the same pattern of centralization—decentralization—centralization. (The final centralized phase during *Turn-Over and Start-Up* has not been discussed here.) The initial, design phase requires unified strategic decision making. During production the volume of work becomes so great that responsibility must be delegated: the organization becomes decentralized under the project and functional matrix control. Finally, at *Turn-Over,* the volume of work decreases while the need for unified integration with Operations' *Start-Up* creates the need for centralization once again.

2. **The Project Organization Must Change According to the Needs of the Project's Size, Speed, and Complexity.** The Açominas matrix was planned to change at the onset of production—about one and a half years after it was set up, which fits with the time it usually takes to "grow" a matrix (35). Research suggests that while the timing of the organization change is a function of the project's schedule, the severity of the change depends on the project's size, speed, and complexity (36).

3. **Once Decentralized, Projects Require a Substantial Management Superstructure to Effect the Necessary Coordination.** Projects decentralize, essentially, to ease the pressure on decision making. Once decentralized, informal controls and communication tend to proliferate and there is a rapid growth in the number of meetings and committees. With this growth in informal decision making there is more need than ever for careful configuration management and budgetary and schedule control. Formal reporting will clearly lag actual events considerably, but in an informal organization there is a danger of assuming that things are happening when in fact they may not be. (Also, informal reporting will tend to concentrate on strategically important items only; formal reports should provide a regular update on all aspects of project progress.) It is, therefore, vital to ensure full, regular reporting during this decentralized phase.

The character of a project thus varies both at different stages of its development and at different levels of management. The skills which are required in managing the project's evolution vary depending on the level of management and stage of the project.

### Each Major Project Change Point Requires Its Own Distinctive Total Management

Changing from one major life-cycle phase to another—from *Prefeasibility/Feasibility* to *Design, Design* to *Production,* and *Production* to *Turn-Over and Start-Up*—is a major event. The *Prefeasibility/Feasibility–Design* transfer is economically the most important step in the project's life. Major federal acquisitions have long placed great importance on the need for very thorough feasibility studies. Thorough agency needs-analysis and exploration of alternative systems is now mandatory federal practice. Yet while the importance of a thorough feasibility study is now generally recognized, it is surprising how many projects do become committed to and move into *Design* on the basis of a totally inadequate feasibility study. Two of the most notorious of recent projects exhibit this clearly. Concorde was conceived almost entirely by the British aircraft establishment, largely on the wings of technological fascination with only the minimum of financial analysis (37). The proposal was championed ardently by one or two senior British ministers who effectively resisted Treasury pressure to review the financial assumptions. Once the French government joined the project the political momentum became virtually unstoppable. Final commitment to the project was made on the basis of a twenty-page report which was "little more than a sketch" (38). At this stage the research and development was estimated to cost £150 million to £170 million; the final cost is some £2 billion. Likewise, the Sydney Opera House was committed to on the basis of a totally sketchy design backed by strong political support. New South Wales' prime minister, John Cahill, saw the Opera House as an imaginative political act (39). A design competition was held and Utzon's design was selected as winner. The design was, however, little more than diagrammatic. There was little evidence of structural feasibility and no cost estimate. A quantity surveyor was therefore asked to prepare one, which he did "under duress in a few hours" (40) arriving at a figure of $A 7 million. The final cost, after drastic redesign (resulting in so reducing the scenery space that opera cannot be fully staged in the building) was $A 102 million.

The transition from *Design* to *Production* is less clear-cut than that from *Prefeasibility/Feasibility* to *Design*. It is also much broader in scope and involves much fuller management attention. At this interface, management must be fully active in all the major project subsystems: project definition, organization, environment, and infrastructure. The overriding preoccupation should be that the strategic parameters are properly set as the interface is crossed, since once *Production* begins the scale and pace of events increase dramatically. So important is this interface to project implementation success that it requires "total" management attention: planning, organizing, directing, and controlling. Contracting—the key interface activity in fact—offers a good example. The contracting process must be supported by project management through integrated planning, thorough negotiating, and through close monitoring. Often contracting is not managed but just happens, thus swamping the project with work and delaying it considerably. (Açominas had to sign 400 contracts during an 18–24-month period. Accomplishing this was a major management achievement in its own right.)

*Turn-Over and Start-Up* probably receives considerable non-project planning and control but generally from places other than within the project team. The meshing of the two important phases of project construction and operations is complex and has large cost implications. All Level I subsystems must be complete and activated as soon as *Turn-Over* occurs. Despite the obvious importance of smoothly transitioning from *Production* to *Turn-Over,* it appears rare, especially in the construction industries, that a project planning group prepares and monitors integrated plans for crossing this interface. This might be partly because of the substantial differences in *Start-Up* between aerospace programs and large new capital expansion facilities (factories, telecommunication systems, ports, etc.) which may have depressed the evolution of defense/aerospace program management ideas on this important area of work.

### Planning Must Be Phased to the Stage of the Project Life Cycle

The differing nature and requirements of the various project life-cycle phases require that different issues be addressed as the project unfolds. Project planning cannot be done comprehensively, once and for all, at the beginning of the project. The uncertainty during the early stages of a project is too great. Instead, planning must be incremental (41). Initial planning must concentrate on building viable planning bases for each principal subsystem, detail being added later in phase with the project schedule (Figure 2-11).

The Apollo Mission's "Phased Project Planning" explicitly recognized this (42)—major life-cycle phases being identified and planning review checkpoints positioned along the life cycle (Figure 2-12)—as have subse-

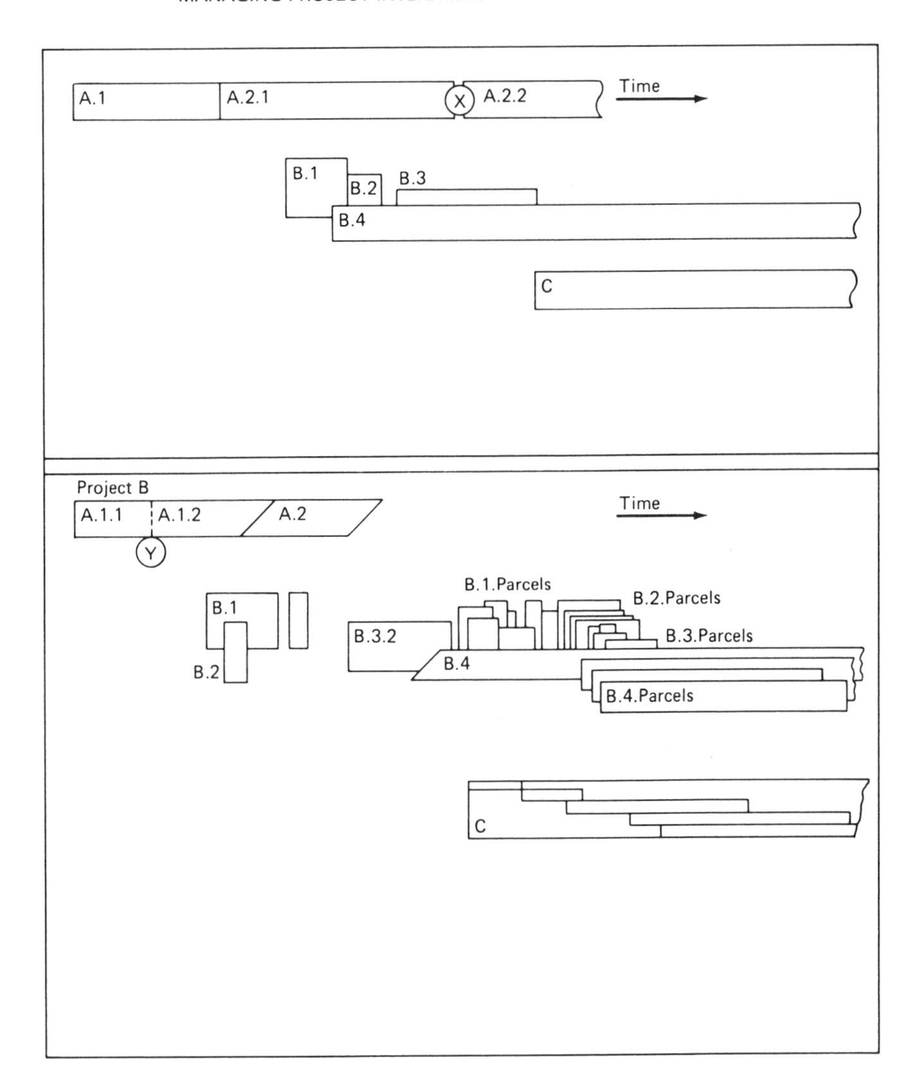


Figure 2-11. Project planning development.

quent U.S. aerospace manuals (e.g., DOD 3200.9, DOD 5000.1, AFSC 800-3, to mention just three).

Apollo was fortunate in that it had nearly a decade for NASA to develop its systems and program planning. Many projects are less fortunate. Early North Sea oil projects, for example, were implemented with great urgency, and made worse by the short summer weather window for towing and positioning platforms in the North Sea. Oil companies found

| PLANNING SYSTEM \ PLANNING STAGE | FEASIBILITY | PROJECT STRATEGY | DESIGN | PRODUCTION | TURN-OVER & START-UP |
| --- | --- | --- | --- | --- | --- |
| ECONOMIC EVALUATION | • Benefits<br>• Risk | • Continue Appraisal with View to Changing Project Specifications if Necessary | • Impact on other Business Functions Assessed<br>• Adjustments Made as Necessary | | • Assess Project Cost for Product Pricing Purposes |
| PROJECT DEFINITION | • System Specs<br>• Base Technology<br>• $ Estimate<br>• Project Schedule | • Outline Design<br>• Configuration Definition<br>• Budget by Major Areas<br>• Milestone Schedule<br>• Detailed "Planning" Schedule | • Futher development of Outline Design, Schedule and Budget | • Detailed Contract Specs and Drawings<br>• Overall Schedule Requirements<br>• Detailed Budget/Contract Bids | • Operating Manuals<br>• Training<br>• Primary Materials Preparation<br>• Hand-over Schedules<br>• Test Schedules<br>• Move |
| FINANCE | • Potential Sources | • Principal Sources<br>• Major Payments Schedule | • Detailed Sources<br>• More Detailed Cash Requirements | • Detailed Payments Schedule by Creditors & Currency | • Annual Financial Operating Plan |
| ENVIRONMENT | • Initial Impact Assessment | • Definition of Environmental Impact Statement<br>• National & Local Government Support Assessed<br>• Local Population Attitude Assessed<br>• Supplier Situation Assessed | • Schedule of Approvals Required<br>• Government or Community Support Groups Identified | • Permit Expediting System<br>• Expediting Schedules<br>• Pubic Relations | • Marketing<br>• Personnel<br>• Inventory Planning<br>• Safety Procedures<br>• Outstanding Legal Issues |

| | | | | | |
|---|---|---|---|---|---|
| ORGANIZATION & SYSTEMS | • Initial Project Outline | • Overall Concepts for:<br>—Contractor Strategies<br>—Design Fabrication, Construction<br>—Labor & Materials Sources<br>• Principal Responsibilities Determined<br>• Major Information Systems Identified<br>• Key Personnel Identified | • Contract Negotiating Plan<br>• Some Major Contracts Signed<br>• Union Discussions<br>• Possibly Some Long Lead Materials Ordered<br>• Responsibilities Matrix<br>• Manpower Plan<br>• Systems Design Schedule | • Contract Terms and Conditions<br>• Owner Organization Detailed<br>• Detailed Staffing Plans | • Operations Organization Development and Start-up<br>• Project Organization Phase-out<br>• Wind-down of project personnel |
| INFRASTRUCTURE & SUPPORT | • Assess Extent of Support Required | • Preliminary Plans for:<br>—Labor Relations<br>—Camps<br>—Logistics | • Further Definition of:<br>—Labor Relations<br>—Camps<br>—Administration<br>—Transport, Logistics & Warehousing<br>• Support Organization Outlined<br>• Permits Requested | • Detailed Definition of:<br>—Labor Relations<br>—Camps<br>—Transport, Logistics, etc.<br>• Construction Schedules/Contracts for Camps, Power, Transport, etc.<br>• Service Contracts Identified<br>• Support Organization Defined | • Wind-down and sell-off of project camps, etc.<br>• Plan for housing, transport, physical & social welfare of operating personnel |

Figure 2-12. Apollo's phased program planning.

themselves having to develop a new generation of rigs, which involved the use of new technology, working in a harsh and poorly documented environment, without adequate codes of practice or regulations—all within a very tight schedule. A slippage of just a few weeks could result in the delay of the whole program for nearly a year. To speed up the program, projects were often sanctioned on the basis of preliminary design data and manufacturing was overlapped with design as much as possible. While these early projects could probably have been managed more efficiently if there had been a longer start-up time to acquire environmental design data, develop codes, and put project management systems and procedures in place (as has been suggested (43)), the economic pressure on distributing North Sea oil as soon as possible effectively precluded this option. In these projects, the project life cycle not only determined the sequence and degree of planning appropriate at a given stage, it set an absolute limit on the time available for planning.

### Ensure Full Working Out of the Static Subsystems at Each Stage of the Project Life Cycle

"Static" project subsystems (technical definition, organization, environment, and infrastructure) must be fully worked out at each phase of the project's life cycle. Unfortunately this does not always happen, often because the habits of an industry or organization have institutionalized a culture of neglecting certain "essential" subsystem considerations.

Movie productions are notorious, for example, for overrunning budget. The most common reason for their doing so is that there is a culture of allowing the director to work out how the film will develop as he shoots it—there is neither "design" nor "schedule." (No one on the set, including the director, knew how *Casablanca* was going to end until the end of shooting; Francis Ford Coppola did not have the ending of *Apocalypse Now* worked out even as he entered the editing room.)

Sometimes, the effects of urgency are so great that it is not possible to work out fully the static subsystems before moving on to the next phase. Many defense projects cannot avoid modifying their performance requirements (because of a changed Threat Analysis) yet must still keep to schedule (44). This is the situation known as concurrency (45): there is a substantial body of knowledge available showing that concurrency is invariably associated with difficulties and overruns (46). Concurrency has proved unavoidable on several occasions in the nuclear power industry, basically when changes have been required at an advanced stage of construction or during commissioning because of regulatory changes or technical problems (47).

The tendency for architectural design to dominate the building process has already been commented upon. Figure 2-13 compares two basically well managed, large, complex building projects (48). The first (A) was managed by architects working under the RIBA's Plan of Work (49). Since the Plan of Work assumes competitive bidding by the main contractor, it does not mention the use of any form of construction management advice prior to construction bidding. Thus the architects did not schedule the project or seek any form of production advice during the early design phases. Project B, however, used a large U.S. A/E firm which was familiar with project management practice and employed both systems design techniques and a management contractor. As a result there was early project scheduling and production advice so that the technical definition subsystem was fully worked out early in the project.

### Control Needs Vary Depending on the Level of Control and Stage of the Project

As the tasks of the different levels of project management vary, so do their control needs. Level III management uses frequent, often daily, control of key performance factors such as earth moved, concrete poured, pipe laid, vessels installed, tests completed, together with basic cost data. At Level II further data are required in addition to the Level III "key drivers": for example, inventories, drawing approvals, transporta-

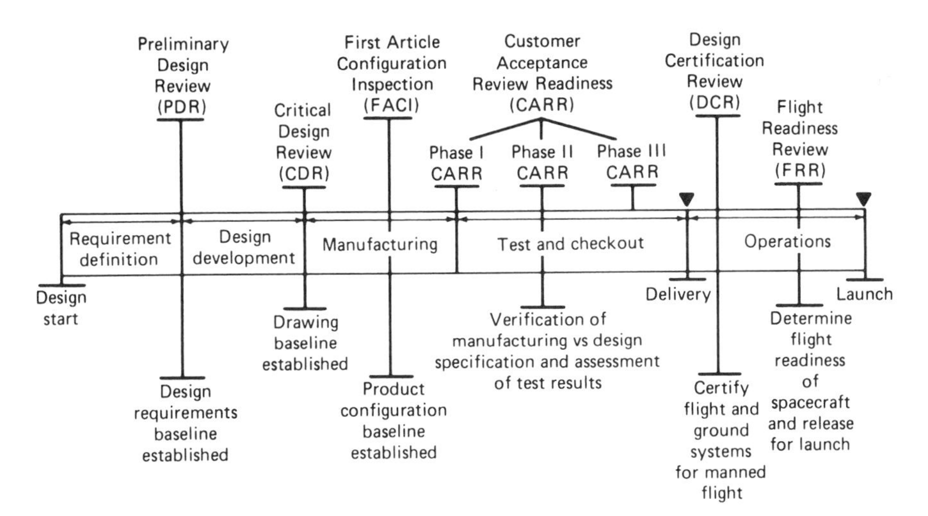


Figure 2-13. Comparison of two British building projects. (For Subsystem Coding refer to Figure 2-8.)

tion, camp capacity, security, accidents, contract management, changes pending and approved, contingency reserves. These are data not needed on such a frequent basis as the "key drivers" data. Level I interests are broader still: exception reporting on progress (i.e., report problems and poor trends only), interface relationships with other subsystems (e.g., between subprojects, between operations planning and project progress), training, cash flows, etc. (50).

"Control" has a meaning which is greater than merely monitoring. It is used in the broader context of setting standards, monitoring, and correcting for deviations between actual and planned performance. This more complete interpretation of control is the one used in cybernetics. (The word "cybernetics" itself derives from the Greek "to steer".) The nature of control during *Prefeasibility/Feasibility* and *Design* is different from that during *Production* and *Turn-Over*. As has been already noted, the need during the early stages of a project to plan, design, and estimate correctly is very large. The costs during these early phases are small compared with the total project cost, and so the need to monitor them (at least from a project as opposed to design point of view) is correspondingly small. Later, during *Production*, the crucial control need is the monitoring of performance to ensure that quality is being achieved and resources are being deployed on schedule and in budget. Hence the nature of control changes during the project life cycle from predicting to monitoring.

"Control" in this broader sense also varies depending upon the interests and objectives of the persons doing the controlling. Different groups, with their different interests, will have different control needs. A project owner, for example, will want to monitor the economic worth of the project as much as its technical, cost, and schedule performance. A contractor might be particularly concerned with cash flow control.

### Personnel Issues Will Vary, Again Depending on the Level of Control and the Stage of the Project

Conflict is inherent in every project, not least since the primary project objectives—quality, cost, and schedule—are themselves in conflict. Quality costs money and requires time; working more quickly costs money. Also, projects often engender contractual and community conflict.

Studies in the mid-1970s (51) have shown that the pattern of conflict varies during the project life cycle (Figure 2-14): schedule and priorities dominate the early phases, with technical issues coming to the fore later (and with cost as a consistently low-conflict item). One should not assume that this pattern applies for all projects, however—one would normally

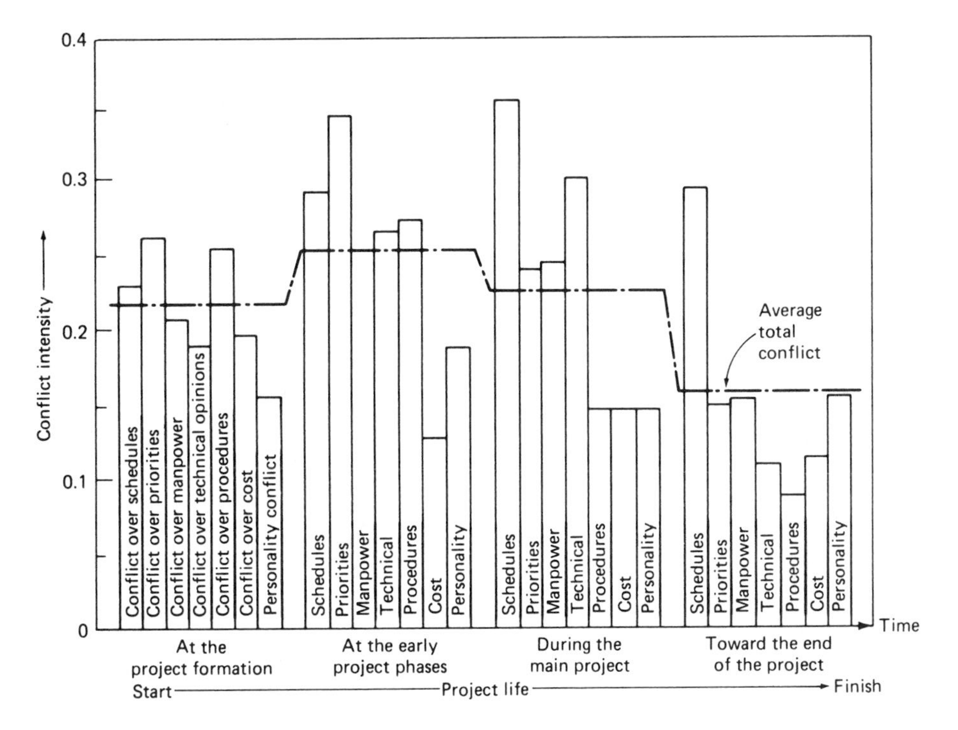


Figure 2-14. Relative intensity of conflict over the project life cycle.

expect greater conflict over technical issues earlier in the project; the conflict pattern might vary by type of project (and contract type); cost pressures may be generally more dominant than they were on the projects studied; and personal issues are probably stronger on matrix and overseas projects. Despite such necessary caveats, the findings are extremely valuable: they provide solid evidence that the type of conflict varies according to the stage of the project life cycle.

Similar research (52) has also studied how the factors which are (a) most important and (b) most inhibiting to project success vary with the life-cycle stage (Figure 2-15). It too has shown that personal issues vary with the stage of project life cycle.

The nature of conflict also differs according to the level of management. All the research on project conflict undertaken to date either concentrates on Level II/III management or has been in the aerospace industries, which have typically been more sheltered from external pressures. Conflict at Level I is usually totally different, requiring other modes of resolution. Most of the behaviorial work in projects to date assumes a normative, mechanistic view of the world: a world where people make rational

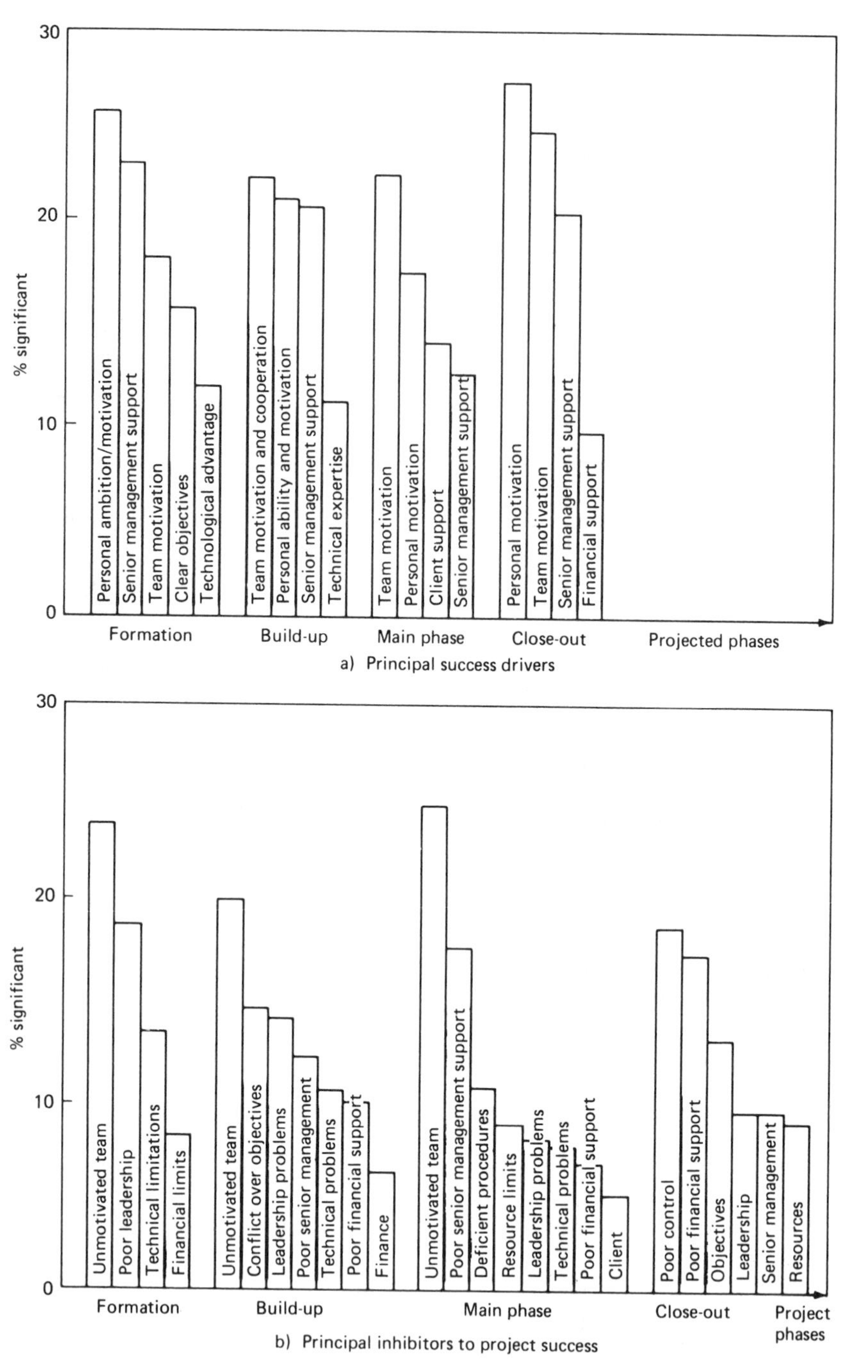


Figure 2-15. Behavioral drivers during the project life cycle.

decisions based on trade-offs of costs and benefits, and where open dialogue between men of goodwill leads to amicable solutions. While this approach is largely valid for most intra-project conflict and behavioral issues, it is often inappropriate for dealing with outside-world issues. The Level I manager will as often as not find himself having to deal with people having completely different value systems from those of the project's. Hence, while the more mechanistic approach to personal issues is appropriate to Levels II and III, Level I often requires a more political approach.

Many practitioners comment on the importance of leadership to project success. Strangely, however, the research evidence on this is at best skimpy (53). There would appear to be little doubt, however, that leaders' abilities are associated with success or failure, although it is generally difficult to isolate the effect of leadership from other factors such as top management support and authority.

Leaders have a particularly important role to play in creating the context for success. Leaders create climates: their personality can sway the ability to evaluate a project objectively; they can often negotiate or communicate things which otherwise would not be agreed. These abilities are particularly important at Level I in working with outside parties.

## RELATION TO PROJECT SUCCESS

To what extent are these findings relevant to that most important issue, the question of project success? The first point to note in addressing this question is that the notion of project success is, as has already been observed, relative. In recent research on this question, I used four distinct measures: functionality—does the project function in the way the sponsors expected?; project management—was the project completed on schedule, in budget, to technical specification?; contractors' long-term commercial success; and, where appropriate, termination efficiency (54). This said, there would seem to be considerable evidence that several of the issues raised in this chapter bear strongly on the chances of success.

Firstly, the discipline of Interface Management itself was seen to be important where there is significant interdependence on projects (55). Organizational issues have been shown to relate to project success (56). Issues of design management, concurrency, and planning have a potentially enormous impact on projects if not accomplished effectively. The evidence on the relationship between human factors and project success is, by comparison, considerably softer, as was just noted (57).

Insofar as Interface Management relates to clear planning and the orderly management of relevant project systems and dimensions, it is almost self-evident that it must contribute to project success. The insights related in this chapter are, almost without exception, proved through quoted experience or academic research to have an effect on the final outcome of a project. The twist, however, is that the definition of success varies from individual company to individual company; and that as one moves from Level III to Level I, the diversities and uncertainties of the many groups operating in the project's environment are likely to make judgments about the causes and effects of project success increasingly difficult to make.

## REFERENCES

1. Katz, D. and Kahn, R. L. *The Social Psychology Of Organizations* (Wiley. New York, 1966), pp. 14–29.
2. Kast, F. E. and Rosenzweig, J. E. *Organization and Management. A Systems Approach* (McGraw-Hill. New York, 1970).
3. Lorsch, J. W. and Lawrence, P. R. *Studies In Organization Design* (Irwin Dorsey. Homewood, Ill., 1970).
4. Putnam W. D. "The Evolution of Air Force System Acquisition Management," RAND, R-868-PR, Santa Monica, Calif., 1972.
5. Beard, E. *Developing the ICBM* (Columbia University Press. New York, 1976).
6. Sapolsky, H. *The Polaris System Development: Bureaucratic and Programmatic Success in Government* (Harvard University Press. Cambridge, Mass., 1972).
7. Peck, M. J. and Scherer, F. M. *The Weapons Acquisition Process; An Economic Analysis* (Harvard University Press. Cambridge, Mass., 1962).
8. Acker, D. D. "The Maturing of the DOD Acquisition Process." *Defense Systems Management Review,* Vol. 3(3) (Summer, 1980).
9. Kelley, A. J. "The New Project Environment" in *New Dimensions of Project Management* (Lexington Books. Lexington, Mass., 1982).
10. Wearne, S. H. *Principles of Engineering Organization* (Edward Arnold. London, 1973).
11. Morris, P. W. G. "Interface Manage—An Organization Theory Approach to Project Management." *Project Management Quarterly,* Vol. 10(2) (June, 1979).
12. Kerzner, H. *Project Management: A Systems Approach to Planning, Scheduling and Controlling* (Van Nostrand Reinhold. New York, 1982).
13. The organistic/mechanistic classification of organization types was developed by Burns, T. and Stalker, G. M. in *The Management of Innovation* (Tavistock. London, 1961).
14. Parsons, T. *Structure and Process in Modern Societies* (Free Press. Glencoe, Ill., 1960).
15. Morris, P. W. G. and Hodgson, P. "The Major Projects Association and Other Macro-Engineering Societies: Their Activities and Potential Contribution to the Development of Project Management," in *8th World Congress on Project Management* (Internet. Rotterdam, 1985).
16. See, for example: Metcalf, J. L. "Systems Models, Economic Models and the Causal Texture of Organizational Environments: An Approach to Macro-Organizational Theory." *Human Relations,* Vol. 27 (1974), pp. 639–663. Also, Emery, F. E. and Trist, E. L. "Sociotechnical Systems," in *Systems Thinking,* ed. Emery, F. E. (Penguin. Harmondsworth, 1969), pp. 241–257.
17. Morris, P. W. G. and Hough, G. H. *Preconditions of Success and Failure in Major Projects* (Major Projects Association, Templeton College. Oxford, September, 1986).

18. Miller, E. J. "Technology, Territory and Time: The Internal Differentiation of Complex Production Systems." *Human Relations,* Vol. 12(3) (1959), pp. 270–304. Also, Miller, E. J. and Rice, A. K. *Systems of Organization, The Control of Task and Sentient Boundaries* (Tavistock. London, 1967).
19. Lawrence, P. R. and Lorsch, J. W. *Organization and Environment; Managing Differentiation and Integration* (Harvard University Press. Cambridge, Mass., 1967).
20. Morris, P. W. G. "Organizational Analysis of Project Management in the Building Industry." *Build International,* Vol. 6(6) (1973), pp. 595–616.
21. Ibid.
22. Thompson, J. D. *Organizations in Action* (McGraw-Hill. New York, 1967).
23. This list is based on Galbraith, J. R. *Organization Design* (Addison-Wesley. Reading, Mass., 1973).
24. Davis, P. and Lawrence, P. R. *Matrix* (Addison-Wesley. Reading, Mass., 1977).
25. Ibid.
26. Youker, R. "Organizational Alternatives for Project Management." *Project Management Quarterly,* Vol. 8(1) (1977), pp. 18–24.
27. See for instance Baumgartner, J. S. "A Discussion with the Apollo Program Director, General Sam Phillips," in *Systems Management,* ed. Baumgartner, J. S. (The Bureau of National Affairs. Washington, D.C., 1979).
28. Alexander, A. J. and Nelson, J. R. "Measuring Technological Change: Aircraft Turbine Engines" Rand Corporation, R-1017-ARPA/PR, Santa Monica, Ca (June 1972); Cochran, E. G., Patz, A. L. and Rowe, A. J. "Concurrency and Disruption in New Product Development" *California Management Review* (Fall, 1978); Department of Energy, "North Sea Costs Escalation Study" (Her Majesty's Stationery Office, London, 1976); General Accounting Office, "Why Some Weapons Systems Encounter Production Problems While Others Do Not: Six Case Studies" GAO/NSIAD-85-34, Washington DC, 24 May 1985; Harman A. J. assisted by Henrichsen S "A Methodology for Cost Factor Comparison and Prediction" Rand Corporation, R-6269-ARPA, Santa Monica, Ca (August 1970); Marshall, A. W. and Meckling, W. H. "Predictability of the Costs, Time and Success of Development" Rand Corporation, P-1821, Santa Monica, Ca (December, 1959); Merrow, E., Chapel, S. W. and Worthing, C. A. "A Review of Cost Estimation in New Technologies: Implications for Energy Process Plants" Rand Corporation, R-2481-DOE, Santa Monica, Ca (July, 1979); National Audit Office "Ministry of Defence: Control and Management of the Development of Major Equipment", report by the Comptroller and Auditor General (Her Majesty's Stationery Office, London, 12 August 1986); Murphey, D. C., Baker, B. N. and Fisher, D. "Determinants of Project Success," National Technical Information Services, Springfield, Va.; Myers, C. W. and Devey, M. R., "How Management Can Affect Project Outcomes: An Exploration of the PPS Data Base," RAND, N-2106, Santa Monica, Calif., 1984; Perry, R. L. et al. "System Acquisition Experience" Rand Corporation, RM-6072-PR, Santa Monica, Ca (November, 1969); Pugh, P. G. "Who Can Tell What Might Happen? Risks and Contingency Allowances." Royal Aeronautic Society Management Studies Group, 1985.
29. Royal Institute of British Architects, *Plan of Work, Handbook of Architectural Practice and Management* (RIBA. London, 1963).
30. Miller, E. J., op. cit. (18).
31. Morris, P. W. G. "Systems Study of Project Management." *Building,* Vol. CCXVI(6816 and 6817) (1974), pp. 75–80 and 83–88.
32. Warnock, J. G. "A Giant Project Accomplished—Design Risk and Engineering Management," in *Successfully Accomplishing Giant Projects,* ed. Sykes, A. (Willis Faber. London, 1979), pp. 31–61.

33. Moolin, F. P. and McCoy, F. "The Organization and Management of the Trans Alaskan Pipeline: The Significance of Organizational Structure and Organization Changes." *Proceedings of the Project Management Institute Conference, Atlanta, 1980* (Project Management Institute. Drexel Hill, Pa., 1980).
34. Reis de Carvalho, E. and Morris, P. W. G. "Project Matrix Organizations, Or How To Do The Matrix Swing." *Proceedings of the Project Management Institute Conference, Los Angeles, 1979* (Project Management Institute. Drexel Hill, Pa., 1979).
35. See, for instance Davis, P. and Lawrence, P. R. *Matrix*, op. cit. (24) Also, Whitmore, K. R. *Matrix Organizations in Conventional Manufacturing-Marketing Companies*, M.S. Thesis (Sloan School of Management, MIT. Cambridge, Mass., 1975).
36. See Reis de Carvalho, E. and Morris, P. W. G., op. cit. (34).
37. Hall, P. *Great Planning Disasters* (Weidenfeld and Nicolson. London, 1980), pp. 87–108; Morris.
38. See Hall, P., op. cit. (37); and Edwards, C. E. *Concorde: Ten Years and a Billion Pounds Later* (Pluto Press. London, 1972).
39. Kouzmin, A. "Building the New Parliament House: An Opera House Revisited?" *Human Futures*, Vol. 3(1) (Spring 1980), pp. 51–74.
40. Hall, P., op. cit. (37), p. 141.
41. Horwitch, M. "Designing and Managing Large-Scale, Public-Private Technological Enterprises: A State of the Art Review." *Technology in Society*, Vol. 1 (1979), pp. 179–192.
42. Seamans, R. and Ordway, F. I. "The Apollo Tradition: An Object Lesson for the Management of Large Scale Technological Endeavors." *Interdisciplinary Science Review*, Vol. 2 (1977), pp. 270–304.
43. Department of Energy, op. cit. (28).
44. Morris, P. W. G. and Hough, G. H., op. cit. (17).
45. Acker, D. D., op. cit. (8); also Harvey, T. E. "Concurrency Today in Acquisition Management." *Defense Systems Management Review*, Vol. 3(1) (Winter, 1980), pp. 14–18.
46. Cochran, E. G., Patz, A. L. and Rowe, A. J. op. cit. (28); General Accounting Office, ibid; Patz, A. L., *Innovation Pitfalls and Management Solutions in High Technology Industries* (University of Southern California. Los Angeles, Calif., 1984).
47. Kutner, S. "The Impact of Regulatory Agencies on Superprojects," in *Planning, Engineering and Constructing the Superprojects*, American Society of Civil Engineers' Conference, Pacific Grove, 1970 (ASCE. New York, 1978); Monopolies and Mergers Commission, *Central Electricity Generating Board, A Report on the Operation of the Board of Its System for the Generation and Supply of Electricity in Bulk* (Her Majesty's Stationery Office. London, 1981).
48. Morris, P. W. G. op. cit. (20, 31).
49. Royal Institute of British Architects, op. cit. (29).
50. Morris, P. W. G. "The Use and Management of Project Control Systems in the 80's." *Project Management Quarterly*, Vol. XI(4) (December, 1980), pp. 25–28.
51. Thamhain, H. J. and Wilemon, D. L. "Conflict Management in Project Life-Cycles." *Sloan Management Review* (Summer, 1975).
52. Dugan, H. S., Thamhain, H. J. and Wilemon, D. L. "Managing Change Through Project Management." *Proceedings of the Project Management Institute Conference, Atlanta, 1980* (Project Management Institute. Drexel Hill, Pa., 1980).
53. Gemmil, G. and Thamhain, H. J. "Project Performance as a Function of the Leadership Styles of Project Managers." *Project Management Institute Conference, 1972* (Project Management Institute. Drexel Hill, Pa.); Honadle, G. and Van Sant, J. *Implementation*

*for Sustainability. Lessons from Integrated Rural Development* (Kumarian Press. West Hartford, Conn., 1985); Murphy, D. C. et al., op. cit. (28); Myers, C. W. and Devey, M. R., ibid; Rubin, I. M. and Sellig, W. "Experience as a Factor in the Selection and Performance of Project Managers." *IEEE Transactions on Engineering Management,* Vol. EM-131(35) (September, 1967), pp. 131–135; Ruskin, A. M. and Lerner, R. "Forecasting Costs and Completion Dates for Defense Research and Development Contracts" *IEEE Transactions on Engineering Management,* Vol. EM-19(4) (November, 1972), pp. 128–133.

54. Morris, P. W. G. and Hough, G. H., op. cit. (17).
55. Ibid.
56. Paulson, B. C., Fondahl, J. W. and Parker, H. W. "Development of Research in the Construction of Transportation Facilities." 'Technical Report No. 223, The Construction Institute, Department of Civil Engineering, Stanford University, Stanford, Ca., 1977.
57. Gemmil, G. and Thamhain, H. J., op. cit. (53).

# 3. Integration: The Essential Function of Project Management

Linn C. Stuckenbruck*

## INTRODUCTION

Project management has achieved almost universal recognition as the most effective way to ensure the success of large, complex, multidisciplinary tasks. The success of project management is based on the simple concept that the sole authority for the planning, the resource allocation, and the direction and control of a single time- and budget-limited enterprise is vested in a single individual. This single-point authority and responsibility constitutes the greatest strength of project management, but it also constitutes its greatest weakness. The pressures for the completion of an often almost impossible task must of necessity be focused on how effectively the project manager carries out his or her job (20).

Therefore, project management is not a panacea, and unfortunately it does not always work. Its use does not guarantee the success of a task; rather, it takes a great deal more. It takes great dedication and considerable effort on the part of an experienced and talented project manager leading an equally experienced and talented project team to ensure that a project will be a success. However, even these proven ingredients are not always enough—projects occasionally fail.

Determining the real or basic cause or causes of project failure can be a frustrating experience. It can be very difficult to pin down the basic causes because they are seldom simple or clear-cut. The problems will be numerous, extremely complex, very much interrelated, and often deeply hidden. It is all too easy to pick a scapegoat, and the project manager is usually the handiest person. Of course the project manager may not be at

---

* Dr. Linn C. Stuckenbruck is with the Institute of Safety and Systems Management at the University of Southern California where he teaches project management and other management courses. Prior to this he spent seventeen years with the Rocketdyne Division of Rockwell International where he held various management positions. He holds a Ph.D. from the State University of Iowa, and is the author of the book *The Implementation of Project Management—The Professional's Handbook,* published by Addison-Wesley Publishing Company.

fault, but there are definitely any number of things that project managers can do wrong. Among the many pitfalls that the unwary or inexperienced project manager can fall into is a failure to completely understand some of the basic aspects of the job.

Project management can, of course, be perceived as just another job requiring an experienced and conscientious manager. But, is just any experienced manager prepared for the job of project management?

It is a management axiom that the overall job of every manager is to create within the organization an environment which will facilitate the accomplishment of its objectives (11).

Certainly the job of the project manager fits this role very well. In addition, all managers, including project managers, are responsible for the universally accepted managerial functions of planning, organizing, staffing, directing, and controlling. It can then be asked whether project management is really significantly different from management in general. An old management cliche states, "A manager is a manager," or putting it another way, "A good manager can manage anything." This statement implies that there is little real difference between the job of the project manager and that of any line or disciplinary manager. However, there is one extremely important, very real, and significantly different aspect of the job of the project manager which makes it different from general management.

By definition projects are complex and multidisciplinary tasks; therefore, project managers must of necessity be very much aware of or even in some cases completely preoccupied with the problem of integrating their projects. This problem, which is of major importance to all but the simplest projects, seldom confronts line managers. This chapter will discuss this essential function of project integration and indicate the various actions that are necessary to achieve a fully integrated project.

## SYSTEMS INTEGRATION

The term systems integration is usually applicable to most projects because inevitably a project is a system. This term is used to indicate the process of integrating any system being utilized or developed, whether it is hardware, software, an organization, or some other type of system. This process of systems integration has been identified as an important management function which has been described by Lawrence and Lorch. They pointed out that with the rapid advances in technology and the increased complexity of systems to be managed, there is an increased need both for greater specialization (differentiation) and for tighter coordination (integration) (13). An effective manager has a need for both; however, since these two needs are essentially antagonistic, one can usually be achieved only at the expense of the other (14). It can be described as a trade-off between these two needs as shown in Figure 3-1.

Referring to Figure 3-1, it has been suggested that the ideal high-performance manager falls on the arrow midway between differentiation and integration, and probably is typical of high-performance top management. It is also true that line or discipline management usually falls closer to the differentiation arrow, and that the truly effective project manager falls closer to the integration arrow. This model emphasizes the importance of the project manager's role as an integrator.

Systems integration is related to what Koontz and O'Donnell call "the essence of management-coordination, or the purpose of management is the achievement of harmony of individual effort toward the accomplishment of group goals" (12). However, doesn't every manager have this function? Yes, but the project manager has to be preoccupied with it. The project manager's major responsibility is assuring that a particular system or activity is assembled so that all of the components, parts, subsystems and organizational units, and people fit together as a functioning, integrated whole according to plan. Carrying out this responsibility comprises the function of systems integration.

## INTEGRATING THE PROJECT

Every project is a system in that it consists of many interrelated and interconnected parts or elements which must function together as a "whole." Projects vary greatly in size, complexity, and urgency; however, all but the simplest projects have a common element in that they

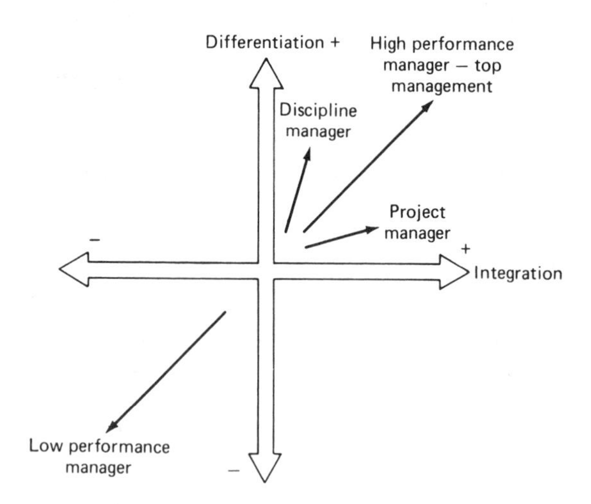


Figure 3-1. Measuring managerial performance.

must be integrated. Project integration can then be described as the process of ensuring that all elements of the project—its tasks, subsystems, components, parts, organizational units, and people—fit together as an integrated whole which functions according to plan. All levels of management ascribe to this goal, but project managers must be preoccupied with it since they have the direct responsibility to ensure that it occurs on every project. These project elements will not automatically come together; the project manager must make a concerted effort and take a number of specific actions to ensure that integration occurs.

The principal precaution that the project manager must take is to make certain that adequate attention is given to every element of the project system. It is easy to be trapped into thinking of the project as consisting entirely of the hardware or other system being designed, developed, or constructed. Many elements of the project may have little direct relationship to the system being worked on, but they may be critical to ultimate project success. Most projects involve a number of different organizational units, many only in a service or support capacity, and an infinite variety of people may be stakeholders in some aspect of the project. The total project system consists of everything and everyone that has anything to do with the project. The diversity of the project system is indicated in Figure 3-2.

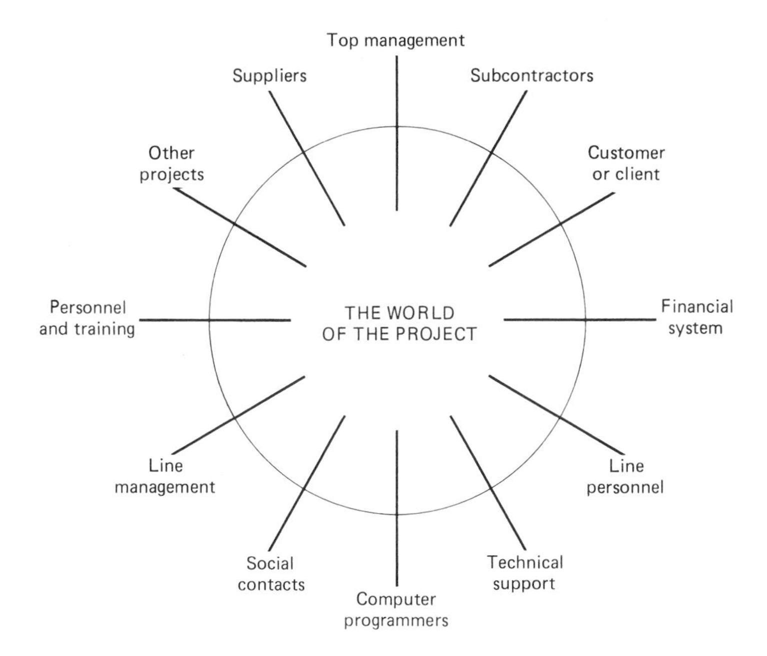


Figure 3-2. The total project system.

## INTEGRATION IN THE MATRIX

The job of project integration is most important and most difficult when the project is organized in the matrix mode. The matrix is a complex organizational form that can become extremely complicated in very large projects. The matrix is complex because it evolved to meet the needs of our increasingly complex society with its very large problems and resulting very large projects. The conventional hierarchical functional management structure usually finds itself in difficulty when dealing with large projects. The pure project organization is a solution when the project is very large, but it is not always applicable to smaller projects. Therefore, management, in an effort to obtain the advantages of both project and functional organizational forms, has evolved the matrix, which is actually a superimposition of project organizations upon a functional organization. The matrix is not for everyone (23). It should only be utilized if its advantages outweigh the resulting organizational complexity.

Why is systems integration difficult in the matrix organization? What is so different about the matrix? Since the matrix is such a complex organizational form, all decisions and actions of project managers become very difficult, primarily because they must constantly communicate and interact with many functional managers. The project manager discovers that the matrix organization is inherently a conflict situation. The matrix brings out the presence of conflicting project and functional goals and objectives. In addition, the project manager finds that many established functional managers who must contribute to the project feel threatened, and continual stresses and potential conflicts result.

The matrix organization has evolved to cope with the basic conflict inherent in any large organization—the needs of specialization versus the needs of coordination (18). These divergent needs in the hierarchical organizational structure lead to inevitable conflict between functional and top management, and often lead to nonoptimizing decisions. All major decisions must be made by top management who may have insufficient information. The matrix organization was a natural evolution growing out of the need for someone who could work problems through the experts and specialists. The project manager has assumed the role of "decision broker charged with the difficult job of solving problems through the experts" (18), all of whom know more about their particular field than he or she.

The role of the project manager in the matrix organization has been analyzed by Galbraith (8, 9), Lawrence and Lorsch (13, 14, 15, 16), and Davis and Lawrence (5). They point out that the horizontal communication in a matrix organization requires an open, problem-solving climate. However, as pointed out by Galbraith (8, 9), when the subtasks in an organization are greatly differentiated a matrix structure may be required to achieve integration. The integrator coordinates the decision processes across the interfaces of differentiation. The project manager must function as an integrator to make the matrix work.

Problem solving and decision making are critical to the integration process since most project problems occur at subsystem or organizational interfaces. The project manager is the only person in the key position to solve such interface problems. The project manager provides "1. a single point of integrative responsibility, and 2. integrative planning and control" (2). The project manager is faced with three general types of problems and with the subsequent necessity for decision making:

1. Administrative problems involving the removal of roadblocks, the setting of priorities, or the resolution of organizational conflicts involving people, resources, or facilities.
2. Technical problems involving the making of decisions, and scope changes; making key trade-offs among cost, schedule, or performance; and selecting between technical alternatives.
3. Customer or client problems which involve interpretation of and conformance to specifications and regulatory agency documents.

Matrix organizations will not automatically work, and an endless number of things can go wrong. Recognizing that the matrix is a complex organizational form is the first step. The next step is getting this complex organization to function. Its successful operation, like that of any management function, depends almost entirely on the actions and activities of the various people involved. In a matrix, however, the important actions and activities are concentrated at the interfaces between the various organizational units. The most important of these interfaces are between the project manager and top management, and between the project manager and the functional managers supporting the project. Moreover, most matrix problems occur at the interfaces between the project manager and functional managers. Project managers must effectively work across these interfaces if they are going to accomplish their integrative function.

Project managers carry out their function of project integration primarily by carefully managing all of the many diverse interfaces within their projects. Archibald indicates that "the basic concept of interface management is that the project manager plans and controls (manages) the points of interaction between various elements of the project, the product, and the organizations involved'' (1). He defines interface management as consisting of identifying, documenting, scheduling, communicating, and monitoring interfaces related to both the product and the project (1).

The complexity that results from the use of a matrix organization gives the project manager even more organizational and project interfaces to manage. These interfaces are a problem for the project manager, since whatever obstacles he or she encounters, they are usually the result of two organizational units going in different directions. An old management cliche says that all the really difficult problems occur at organizational interfaces. The problem is complicated by the fact that the organizational units are usually not under the direct management of the project manager, and some of the most important interfaces may even be completely outside of the company or enterprise.

## Types of Interfaces

There are many kinds of project interfaces. Archibald divides them into two types—product and project—and then further divides them into subgroups, of which management interfaces are a major division (2). The problem of the overall project/functional interface is thoroughly discussed by Cleland and King, who point out the complementary nature of the project and the functional or discipline-oriented organization. ''They are inseparable and one cannot survive without the other'' (3).

Another way of describing the various interfaces that the project manager must continually monitor for potential problems is (a) personal or people interfaces, (b) organizational interfaces, and (c) system interfaces (2). In other words, project management is more than just management interfaces; it involves all three of the above types.

1. Personal Interfaces—These are the ''people'' interfaces within the organization whether the people are on the project team or outside it. Whenever two people are working on the same project there is a potential for personal problems and even for conflict. If the people are both within the same line or discipline organization, the project manager may have very limited authority over them, but he or she can demand that the line supervision resolve the personal problem or conflict. If the people are not in the same line or discipline organization, the project manager must play the role of mediator, with the ultimate alternative of insisting that line management resolve the problem or remove one or both of the individuals from the project team. Personal interface problems become even more troublesome and difficult to solve when they involve two or more managers.

2. Organizational Interfaces—Organizational interfaces are the most troublesome since they involve not only people but also varied organizational goals, and conflicting managerial styles and aspirations. Each organizational unit has its own objectives, its own disciplines or specialties, and its own functions. As a result of these differences, each organizational unit has its own jargon, often difficult for other groups to understand or appreciate. It is thus apparent that misunderstandings and conflict can easily occur at the interfaces. These interfaces are more than purely management interfaces since much day-to-day contact is at the working level. Purely management interfaces exist whenever important management decisions, approvals, or other actions that will affect the project must be made. Organizational interfaces also involve units outside the immediate company or project organization such as the customer, subcontractors, or other contractors on the same or related systems.
3. System Interfaces—System interfaces are the product, hardware, facility, construction, or other types of nonpeople interfaces inherent in the system being developed or constructed by the project. These will be interfaces between the various subsystems in the project. The problem is intensified because the various subsystems will usually be developed by different organizational units. As pointed out by Archibald (1), these system interfaces can be actual physical interfaces existing between interconnecting parts of the system, or performance interfaces existing between various subsystems or components of the system. System interfaces may actually be scheduled milestones involving the transmission of information developed in one task to another task by a specific time, or the completion of a subsystem on schedule.

### Management Interfaces

Each of the three types of interfaces that have been described pose important problems. Problems become particularly troublesome when personal and organizational interfaces are combined into what may best be called management interfaces (17). Management interfaces have personal aspects because normally two individuals are concerned, such as a project manager and a particular functional manager. Management interfaces, however, also have organizational aspects because the respective managers lead organizations which probably have conflicting goals and aspirations.

There is a great difference between the conventional organization chart (whether it be hierarchical or matrix) and the actual operation of a real-world organization. The conventional hierarchical organization charts or matrix organization charts clearly show many of the management interfaces, such as superior/subordinate and project management/worker relationships. However, conventional management charts only suggest some of the other really important interfaces. These important interfaces, as shown by the double-ended arrows in Figure 3-3, consist of project manager/functional manager interfaces, project manager/top management interfaces, functional manager/functional manager interfaces, and sometimes even project manager/project manager interfaces.

Most important are the interfaces between the project managers and the various functional managers supporting the project. These relationships are almost inevitably adversary since they involve a constantly shifting balance of power between two managers on essentially the same reporting level.

The interface with top management is important because it represents the project manager's source of authority and responsibility. The project manager must not only have the real and unqualified support of top management, but must also have a clear and readily accessible communication link with them. The project manager must be able to get the "ear" of top management whenever necessary.

The interfaces between the various functional managers are important because they are the least visible to project managers who might not be immediately aware of trouble spots.

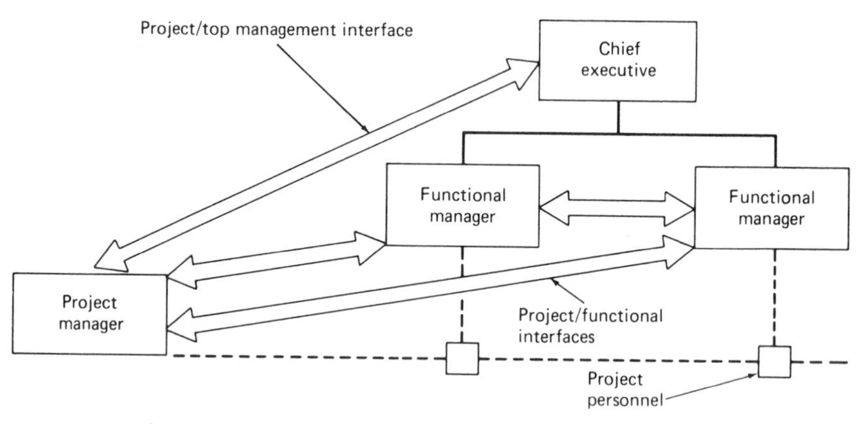


Figure 3-3. The multiple management interfaces in the matrix.

**The Balance of Power**

Having implemented project management, top management must recognize that they have placed a new player in the management game—the project manager. Problems are to be expected, particularly in a matrix organization where a new situation has been created with natural conflict or adversary roles between the project managers and the functional managers who support the projects. This managerial relationship can best be described as a balance of power between the two managers involved as illustrated by Figure 3-4. This relationship has also been described as a balance of interest and a sharing of power (6). But this does not imply that the shared power is ever truly balanced, because in reality the balance of power is a dynamic, constantly changing condition that cannot be static even if so desired.

There is no way to assure a balance of power at every managerial interface. Theoretically, it should be possible to divide the authority and responsibility more or less equally between the project and functional managers, which implies a very clear balance of power between the two managers. This is not only very difficult, but it doesn't happen very often. Various authors have attempted to clearly delineate the authority and responsibilities of both project and functional management so as to assure a balance of power (3). Certainly such a delineation can indicate where

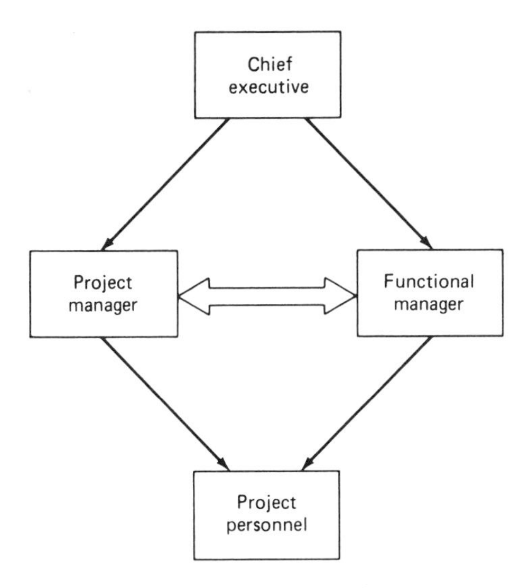


Figure 3-4. The balanced matrix.

major responsibilities lie, but cannot guarantee a balance of power. In fact, there are many reasons why it is almost impossible to have a true balance of power between functional and project management. Not the least of these reasons is the fact that a matrix consists of people, and all people—including managers—are different from each other. Managers have differing personalities and differing management styles. Some management styles depend on the persuasive abilities of the manager, while others depend on or tend to fall back on strong support from top management.

Since projects, programs, or products are usually the most important of all of a company's activities, project managers are very important persons. They are the persons who put the company in a position where it can lose money or make a profit. Therefore, in terms of the balance of power, it would seem that projects would always have the scale of power tipped in their direction, particularly with the firm support of top management. Not necessarily so! In fact, not usually so, at least in a matrix organization. In a pure project organization, there is no question as to who holds the power. But in a matrix organization the functional managers have powerful forces on their side. The functional manager is normally perceived by project personnel to be the real boss. This is inevitable since functional management is part of the management ladder in the hierarchy which goes directly up to the president of the company, and it is therefore perceived to be "permanent" by the employees. After all, the functional organization represents the "home base" to which project personnel expect to return after the completion of the project.

Very strong top management support for the project manager is necessary to get the matrix to work, and even very strong support will not guarantee project success. However, the matrix will not work without it. Project managers must get the job done by any means at their disposal even though they may not be perceived as the real boss.

### The Project/Functional Interface

The secret of the successfully functioning matrix can thus be seen to be not just a pure balance of power, but more a function of the interface or interface relationships between the project and individual functional managers. Every project decision and action must be negotiated across this interface. This interface is a natural conflict situation since many of the goals and objectives of project and functional management are so very different. Depending on the personality and dedication of the respective managers, this interface relationship can be one of smooth-working cooperation or bitter conflict. A domineering personality or power play usually is not the answer. The overpowering manager may win the local skirmish, but usually manages sooner or later to alienate everyone working on the project. Cooperation and negotiation are the keys to successful decision making across the project/functional interface. Arbitrary and one-sided decisions by either the project or functional manager can only lead to or intensify the potential for conflict. Unfortunately for project managers, they can accomplish little by themselves, they must depend on the cooperation and support of the functional manager. The old definition of successful management—"getting things done by working through others"—is essential for successful project management in the matrix organization.

The most important interface that the project manager has in a matrix organization is with the functional managers. The conventional matrix two-boss model does not adequately emphasize this most important relationship. Obviously, neither the project manager nor the functional manager can simply sit in his or her office and give orders. The two managers must be communicating with each other on at least a daily basis, and usually more often. The organizational model shown in Figure 3-4 shows the managerial relationship as a double-ended arrow indicating that the relationship is a two-way street. Consultation, cooperation, and constant support are necessary on the part of both the project and functional managers. This is a very important relationship, key to the success of any matrix organization, and one which must be carefully nurtured and actively promoted by both project and functional management.

## Strong Versus Weak Matrices

Achieving an equal balance of power between project and functional management may be a desirable goal; certainly it should be a way of minimizing potential power struggles and possible conflicts. There is no certain way to assure that there is an "equal" balance of power, and it is probably seldom really achieved. However, it can be approached by assuming that the project managers have the full support of top management and that they report at a high enough level in the management hierarchy. In fact top management can, whenever desirable, tilt the scales of power in either direction.

In many situations it may not be desirable to have an equal balance of power. For instance, a project may be so important to the company, or the budget and schedule so tight that top management feels that the project manager must be in a very strong position. Or perhaps the project managers feel that they must tilt the organizational balance of power in their favor to obtain better project performance. On the other hand, top management may feel that functional management needs more backing. In either case, the balance of power can be tilted in either direction by changing any one or any combination of the following three factors:

1. The Administrative Relationship—The levels at which the project and involved functional managers report, and the backing which they receive from top management.
2. The Physical Relationship—The physical distances between the various people involved in the project.
3. The Time Spent on the Project—The amount of time spent on the project by the respective managers.

These three factors can be used to describe whether the matrix is strong or weak. The strong matrix is one in which the balance of power is definitely on the side of project management. This can be shown by the model in Figure 3-5. A weak matrix has been described by project managers as one in which the balance of power tilts decisively in the direction of line or functional management.

The managerial alternatives have been described as a continuum ranging from pure project to functional as shown in Figure 3-6 (7). The matrix falls in the middle of the continuum, and can range from very weak to very strong depending on the relative balance of power.

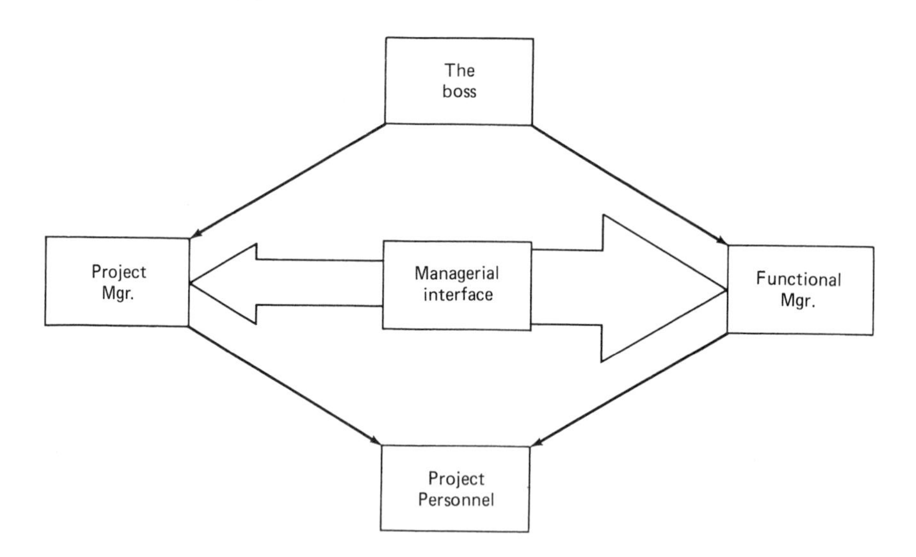


Figure 3-5. The balance of power in a strong matrix.

## THE INTEGRATION PROCESS

As previously indicated, project integration doesn't just happen, it must be made to happen. It is more than just fitting components together; the system has to function as a whole. The integration process consists of all of the specific actions that project managers must initiate to ensure that their projects are integrated. Integration cannot be an afterthought, and it does not consist only of actions that can be accomplished after the subsystems have been completed. Therefore, the critical actions leading to integration must take place very early in the life cycle of the project, particularly during the implementation phase, to ensure that integration takes place. In "pure" project organizations there is no question as to who initiates these actions, project managers run their own empires. In matrix organizations, however, project managers encounter particular difficulties and problems in carrying out their integrative functions.

## THE CRITICAL ACTIONS OF INTEGRATION

The integration process is difficult to separate from general good management practice; however, there are a number of critical actions which are uniquely important to the job of project management. These actions must be initiated and continually monitored by project managers if project integration is to occur. The project manager is the single point of integrative responsibility, and is the only person who can initiate these actions. These critical actions are of two types: (a) those which are essentially just

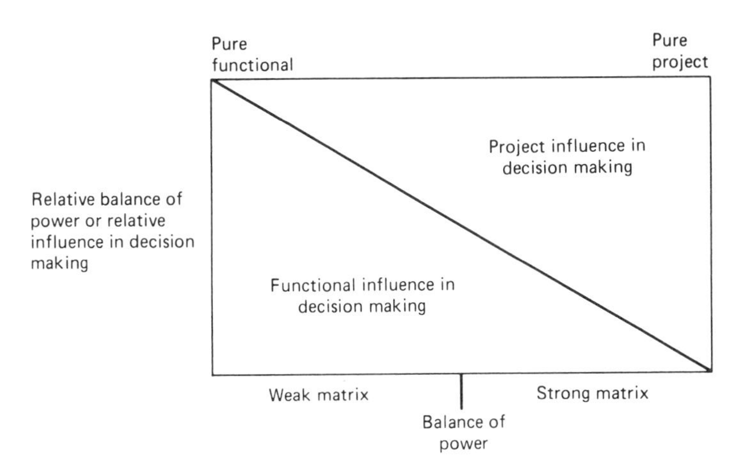


Figure 3-6. The balance of power in weak and strong matrices.

good project management practice and which must extend over the entire life of the project, and (b) specific one-time actions which must be taken by some member of management (usually the project manager or a member of top management) to ensure that the project is integrated. The most important of these actions are as follows (22):

1. Getting started on the right foot.
2. Planning for project integration.
3. Developing an integrated Work Breakdown Structure, schedule, and budget.
4. Developing integrated project control.
5. Managing conflict.
6. Removing roadblocks.
7. Setting priorities.
8. Facilitating project transfer.
9. Establishing communication links.

### Getting Started on the Right Foot

To achieve successful project integration, it is of course very important that the project get started on the right foot. There are a number of specific things that must be done, both by top management and by the project manager (22). The secret of project success is dependent on making these critical actions very early in the project life cycle. For the most part these actions are inseparable from the normal actions that must be taken to implement any successful project; however, they must be made during the project implementation phase. If the right decisions are made at this time, the project can be expected to run smoothly and the integration process will proceed as planned.

The most important decisions and resulting actions are those taken by top management, and many of these actions must be taken well before the project is actually started. Not all of these actions are directly concerned with the integration function, but they are all necessary for the successful implementation of project management. The most critical of the actions which must be taken by top management are the following:

1. Completely selling the project management concept to the entire organization.
2. Choosing the type or form of project organization to be utilized.
3. Issuing a project charter to completely delineate all project and functional authority and responsibilities.
4. Choosing the project manager or project managers.

5. Choosing the right functional managers to participate in the project and/or matrix organization.
6. Supplying adequate resources to the project organization such as finances, equipment, personnel, computer support, etc.
7. Continuing strong support for the project and for the project manager.

The above list of actions is more or less in the order that the actions must be taken, and most of them must be taken prior to the actual implementation of the project.

After top management has successfully implemented project management and has given it full support, the action passes to the newly appointed project managers. There are a number of specific actions that the project managers must now initiate to start their projects on the road to success, and to ensure project integration. The project manager is the single point of integrative responsibility, and is the only person who can initiate and monitor these actions. The most critical of these actions are as follows:

1. Issuance of the Project Implementation Plan.
2. Creation of the project Work Breakdown Structure (WBS).
3. Development of the project organization.
4. Issuance of the Project Procedures Guide.
5. Issuance of a Project Material Procurement Forecast.
6. Issuance of Work Authorizations.

These actions are more or less sequential, although they are strongly interrelated and must be worked on at the same time. The most important consideration is that documentation implementing the above actions be issued as early in the project life cycle as possible. Much of this effort should have been accomplished prior to the initiation of the project, such as during proposal preparation. Even so, a great deal of effort is required during the "front end" of a project to accomplish these actions, and to ensure that project integration takes place.

## Planning for Project Integration

Integration doesn't just happen—it must be planned. The project manager must develop a detailed planning document that can be used to get the project initiated, and to assure that all project participants understand their roles and responsibilities in the project organization.

The project manager is the only person in the key position of having an overview of the entire project system, preferably from its inception, and therefore can best foresee potential interface or other integration problems. After identifying these key interfaces, the project manager can keep a close surveillance on them to catch and correct any integration problems when they first occur. Particularly important in the project plan is a clear delineation of the project requirements for reporting, hardware delivery, completion of tests, facility construction, and other important milestones.

An important part of the project plan should be the integration plan. This plan is a subset of the project plan and may even be a separate document if a single department or even a separate contractor is responsible for system or project integration. In any case, the integration plan should define and identify all interface events, interrelationships between tasks and hardware subsystems, and potential interface problems. The integration plan should then analyze the interrelationships between tasks and the scheduled sequence of events in the project.

Project managers must continually review and update both the administration and technical portions of their project plans to provide for changes in scope and direction of their projects. They must assure that budget and resource requirements are continually reviewed and revised so that project resources are utilized in the most effective manner to produce an integrated system.

The most complete and well integrated project plan is worthless if no one uses it. Only the project manager can ensure that all task managers are aware of their roles and responsibilities in the attainment of project success. But continuous follow-up by the project manager is necessary to assure adherence to the project plan, and awareness of any necessary revision.

### Developing an Integrated Work Breakdown Structure, Schedule, and Budget

Solving the project manager's problems starts with the fact that every project must be broken down into subdivisions or tasks which are capable of accomplishment. Creating this Work Breakdown Structure (WBS) is the most difficult part of preparing a project plan because the project manager must ensure that all of the tasks fit together in a manner that will result in the development of an integrated workable system. The WBS can be considered to be the "heart" of the project integration effort. Too often a WBS is prepared by breaking up the project along easily differentiated organizational lines with very little thought as to how the final system fits together. However, the WBS is the system "organization chart" which schematically portrays the products (hardware, software, services, and other work tasks) that completely define the system (4). Therefore, it is best to prepare the WBS by breaking down the project first into subsystems and then into components and finally into tasks that can readily be accomplished. These lower-level tasks or "work packages" can be most effectively estimated and carried out if they are within single organizational units.

This process of breaking down a project into tasks or work packages, that is, creating a WBS, is just the first step. The WBS must then be carefully integrated with the schedule and budget if the project is to succeed. Each work package must have an integrated cost, scheduled start, and scheduled completion point. The WBS serves as the project framework for preparing detailed project plans, network schedules, detailed costing, and job responsibilities. A realistic WBS assures that project integration can truly be achieved.

### Developing Integrated Project Control

The most prolific project planning is useless if project control is ineffective. Whatever type of planning and control technique is used, all the important interfaces and interface events must be identified. Interface events such as hardware or facility completions will be important project milestones. The project network plan must be based on the interface events in order to facilitate analysis of the entire project on an integrated basis. Resource allocation and reporting periods can then be coordinated with interface events, and schedules and budgets can be designed on an integrated basis.

### Managing Conflict

Project managers have been described as conflict managers (10). This does not mean that they should constantly be fire fighters; however, they cannot avoid this role in resolving conflicts, particularly when the conflict involves project resources such as project personnel. Conflicts are very likely to occur in the temporary project environment where the project manager is often the new player who has not had time to develop good working relationships with project team members or with supporting functional managers. The conflict potential is also increased by the great differences between project and functional goals and objectives, and by the unavoidable competition between projects for resources.

It is inevitable that problems occur at organizational and subsystem interfaces. These problems may or may not result in actual open conflict between individuals or organizations. A common situation is personal conflict between the two managers involved at an interface. Conflict situations result primarily from the concerned groups or managers losing sight of the overall project goals or having differing interpretations of how to get the job accomplished. Project managers must continually be on the lookout for real and potential conflict situations and resolve them immediately if they expect to have an integrated project.

### Removing Roadblocks

Roadblocks are inevitable whenever there are separate organizational units which must support project efforts, particularly if the projects are matrixed. Roadblocks are inevitable in such a complex organization, and are the inevitable result of conflict situations. Resolving the conflict will eliminate many roadblocks, but there are always other roadblocks set up intentionally or unintentionally by managers and other personnel not directly involved with the project. These roadblocks may be the result of conflicting needs for resources and personnel, or conflicting priorities for the use of facilities and equipment. Administrative roadblocks often occur because managers outside the project do not understand or sympathize with the project urgency. Such roadblocks are difficult to deal with, and the project manager may be forced to go to top management to get a satisfactory resolution.

### Setting Priorities

In order to resolve or prevent conflict situations, the project manager is continually faced with the problem of setting priorities. There are two types of priorities that are of major concern to project managers:

1. The overall company or organizational priorities which relate project needs to the needs of other projects within the organization, and to overall organizational needs.
2. The priorities within projects for the utilization of personnel, equipment, and facilities.

The first type of priority may be beyond the control of project managers, but it is a problem with which they must be continually concerned. Pity the poor project manager who is so busy getting the job done that he or she forgets to cement a working and personal relationship with members of top management. The result may be a low project priority that dooms the project to failure. The second type of priority is within the project organization and therefore completely within the control of the project manager. These priority problems must be handled on a day-to-day basis, but in a manner that will promote the integration of the project system.

### Facilitating Project Transfer

Project transfer is the movement of a project through the company organizations from the conceptual phase to final delivery to the customer. Project transfer doesn't just happen, it must be carefully planned and provided for in the scheduling and budgeting of the project. The project manager has the responsibility of ensuring that project transfer takes place without wasteful effort and on schedule. The steps in a typical project transfer are shown in Figure 3-7.

If the product or system is to be delivered to the customer on schedule, it must move from block to block as indicated in Figure 3-7, which involves crossing a number of organizational interfaces. This transfer process must be expedited or even forced by the project manager if it is to be completed on schedule. The basic problem is that of making certain that the project is transferred quickly, without organizational conflict, without unnecessary redesign or rework, and without loss of relevant technology or other information. Experience has shown that the best method of ensuring effective project transfer is to utilize people who can move with the project across organizational interfaces. The project manager has two alternatives to facilitate project transfer: (a) the designation of suitable qualified personnel who can move forward with the project, that is, change their role as indicated by the left to right dashed arrows in Figure 3-7, (b) the utilization of personnel who can move backward in the organization and serve as consultants or active working members of the project team. When the project moves forward they serve as transfer agents in moving the project forward in the organization (22). Various possible personnel transfers are shown by the right to left solid arrows in Figure 3-

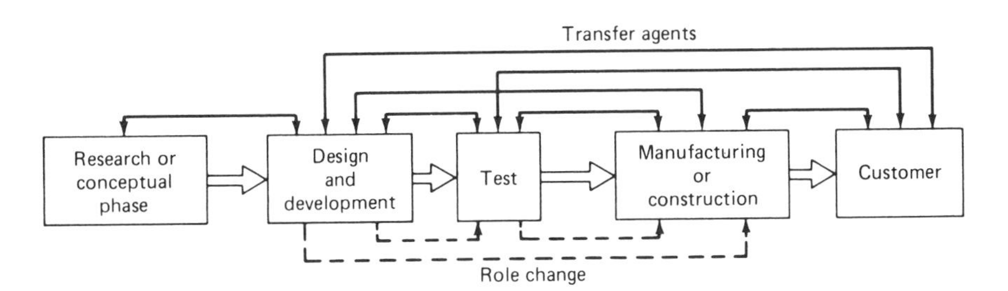


Figure 3-7. Project transfer.

7. Great importance must be placed on having customer, manufacturing, and/or construction representatives take part in the design phase of the project.

### Establishing Communication Links

The last of the integration actions, that of constantly maintaining communication links, is perhaps the most difficult and troublesome because it involves the necessity for considerable "people" skills on the part of project managers. Most project managers find that they spend at least half of their time talking to people—getting information, delegating, clarifying directives, and resolving conflicts and misunderstandings. Much of this time is involved with project managers' critical responsibilities for maintaining all communication links within and outside their projects in order to ensure project integration. Internal communication links must be maintained between each subdivision of the project, and the project managers must make sure that all project team members talk with each other. In addition, the project manager is personally responsible for maintaining communication linkages outside of the project. Many of the external communication links can be personally expedited by the project manager, and in most cases the communication consists of written documents.

Communication linkages internal to the project, however, must function continuously, with or without documentation, and whether the project manager is personally involved or not. These internal communication linkages are most important to the health of the project since they involve the technical integration of the subsystems of the product or project. However, there are usually very real barriers to effective communications across any two such subsystem interfaces. In order to assure that problems don't accumulate and build up at these interfaces, the project manager must act as a transfer agent or a communications expediter. The model shown in Figure 3-8 illustrates the interface problem.

The project manager must serve as the bridge to make sure that the communication barriers do not occur. Communication barriers can be caused by a variety of circumstances and occurrences which the project manager must watch for. A communication barrier may or may not result in actual conflict depending upon the individuals involved, but the possibility always exists.

The project manager is the one person always in a position to expedite communication linkages. He or she can be considered to be a transfer agent who expedites the completion of the communication link by personally transferring information and project requirements across the interface. Considering the number of interfaces in a complex, multidisciplinary

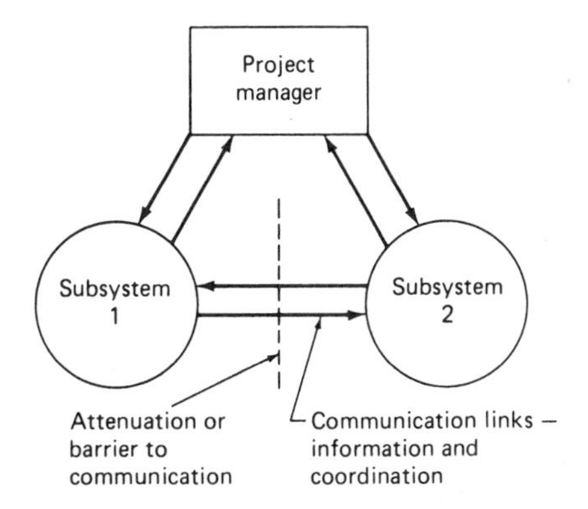


Figure 3-8. The project manager as communications expediter.

matrix-organized project, this process becomes a major effort for the project manager. The only saving grace is that many of these interfaces will be trouble free, and communication problems will not all occur at the same time.

Communication barriers may be caused by a variety of circumstances and occurrences. Some of the causes of communication barriers are as follows (19, 21):

1. Differing perceptions as to the goals and objectives of the overall company or organizational system can cause problems. In addition, a lack of understanding of project objectives is one of the most frequent and troublesome causes of misunderstanding. It can be directly attributed to insufficient action on the part of the project manager, since he or she has the major responsibility for defining project objectives. Even when these objectives are clearly stated by the project manager, they may be perceived differently by various project team members.
2. Differing perceptions of the scope and goals of the individual subsystem organizations can likewise restrict communications. Again it is the responsibility of the project managers to clarify these problems, at least as to how they impact their projects.
3. Competition for facilities, equipment, materials, manpower, and other resources can not only clog communication routes but can also lead to conflict.
4. Personal antagonisms or actual personality conflicts between managers and/or other personnel will block communication flow. There may also be antagonism toward project managers by line managers who perceive them as a threat to their authority or their empire.

5. Resistance to change or the NIH (not invented here) attitude may also detrimentally affect communication links between organizational units.

As indicated in Figure 3-9, the project manager has four important communication links: (a) upward to top management, (b)downward to the people working on the project, (c) outward to line managers and other projects at the same managerial level, and (d) outward to the customer or client. The project manager has a major responsibility for maintaining communications with the chief executives in the organization who must be provided with timely, up-to-date progress reports on the technical and financial status of the project. Similar reports must be provided to the client or customer, particularly if the customer is outside of the company, such as a governmental agency.

The other important communication link is with the people working on the project. The project manager must keep them informed by means of project directives and personal communications. In addition, there is a continual stream of reports from the discipline/line-organization managers and specialists who are working on the project.

Many of these reports concern project and administrative details and can be evaluated by administrators and assistant project managers. How-

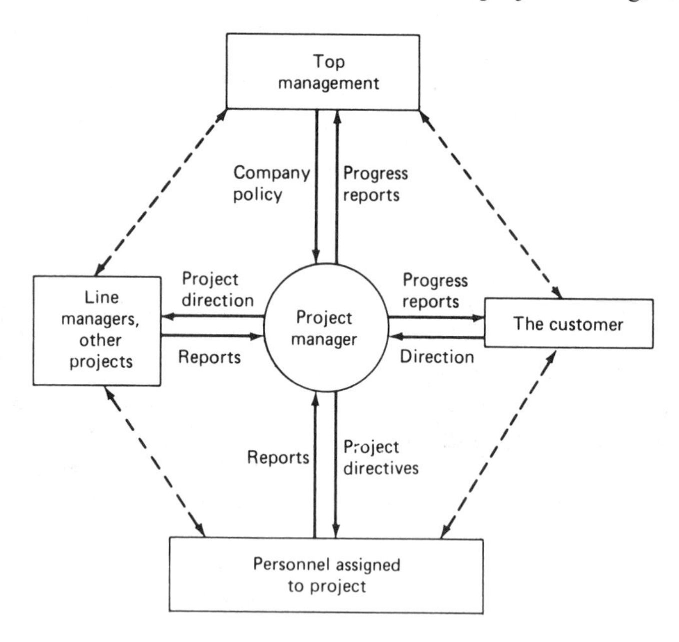


Figure 3-9. The project manager's communication links.

ever, the ultimate decision as to the worth of a report, and as to whether it should be included in progress reports to the customer and/or top management, is in the hands of the project manager. His or her communicative skills, therefore, must include the ability to accurately and rapidly evaluate, condense, and act on information from many sources.

Attenuation in these communication links at the organizational interfaces must be minimized. This means that project managers must have an open line to top management. Conversely, they cannot have too many line managers interpreting their instructions and project objectives to the people working on the project. Without open communication links, project managers will surely fail. There are also a number of important communication links outside the immediate scope of the project. The four most important such links are shown by the dashed arrows in Figure 3-9. For instance, the customer will at times talk directly to top management without going through the project manager. Project managers have to recognize the existence and the necessity for these sometimes bothersome communication links; and rather than fight them, they should endeavor to make use of them.

## CONCLUSIONS

Project integration consists of ensuring that the pieces of the project come together as a "whole" at the right time and that the project functions as an integrated unit according to plan. In other words, the project must be treated as a system. Project managers carry out their job of project integration in spite of project and system complexity, and of course their job is the most difficult in a matrix organization.

To accomplish the integration process, project managers must take a number of positive actions to ensure that integration takes place. The most important of these actions is that of maintaining communication links across the organizational interfaces and between all members of the project team. Project managers must be continually expediting communication links throughout their projects. Of almost equal importance is the need for the project manager to develop a Work Breakdown Structure which ensures project integration by providing a "framework" on which to build the total project. These integrative actions are every bit as important as the project manager's other principal function of acting as a catalyst to motivate the project team.

Project integration is just another way of saying interface management since it involves continually monitoring and controlling (i.e., managing) a large number of project interfaces. The number of interfaces can increase exponentially as the number of organizational units increases; and the life of a project manager in a matrix organization can become very complex indeed. Interfaces usually involve a balance of power between the two managers involved. This balance of power can be tilted in favor of either manager, depending on the desires of top management. Project managers must continually keep their eyes on the various managerial interfaces affecting their projects. They must take prompt action to ensure that power struggles don't degenerate into actual conflict. It takes very little foot dragging to sabotage even the best project. Integration doesn't just automatically occur. The project manager must put forth great effort to ensure that it happens.

## REFERENCES

1. Archibald, Russell D. *Managing High-Technology Programs and Projects* (Wiley. New York, 1977), p. 66.
2. Ibid., p. 5.
3. Cleland, David I. and King, William R. *Systems Analysis and Project Management,* 2nd Ed. (McGraw-Hill. New York, 1975), p. 237.
4. Ibid., p. 343.
5. Davis, Stanley M. and Lawrence, Paul R. *Matrix.* (Addison-Wesley. Reading, Mass., 1977).
6. Davis, Stanley M. "Two Models of Organization: Unity of Command versus Balance of Power." *Sloan Management Review* (Fall, 1974), pp. 29–40.
7. Galbraith, Jay R. "Matrix Organization Design." *Business Horizons* (February, 1971), pp. 29–40.
8. Galbraith, Jay R. *Designing Complex Organizations* (Addison-Wesley. Reading, Mass., 1973).
9. Galbraith, Jay R. *Organization Design* (Addison-Wesley. Reading, Mass., 1977).
10. Kerzner, Harold. *Project Management: A Systems Approach to Planning, Scheduling and Controlling* (Van Nostrand Reinhold. New York, 1979), p. 247.
11. Koontz, Harold and O'Donnell, Cyril. *Principles of Management: An Analysis of Managerial Functions* (McGraw-Hill. New York, 1972), p. 46.
12. Ibid., p. 50.
13. Lawrence, Paul R. and Lorsch, Jay W. *Organization and Environment: Managing Differentiation and Integration* (Harvard University, Division of Research, Graduate School of Business Administration. Boston, 1967).
14. Lawrence, Paul R. and Lorsch, Jay W. "New Management Job: The Integrator." *Harvard Business Review* (November–December, 1967), pp. 142–151.
15. Lawrence, Paul R. and Lorsch, Jay W. *Developing Organizations: Diagnosis and Action* (Addison-Wesley. Reading, Mass., 1969).
16. Lorsch, Jay W. and Morse, John J. *Organizations and Their Members: A Contingency Approach* (Harper & Row. New York, 1974), pp. 79–80.
17. Morris, Peter W. G. "Managing Project Interfaces—Key Points for Project Success," Chapter 1 in *Project Management Handbook,* ed. Cleland, David L. and King, William R. (Van Nostrand Reinhold. New York, 1st Ed. 1983), pp. 3–36.
18. Sayles, Leonard R. "Matrix Management: The Structure with a Future." *Organizational Dynamics* (Autumn, 1976), pp. 2–17.

19. Stickney, Frank A. and Johnson, William R. "Communication: The Key to Integration." *1980 Proceedings of the Project Management Institute Annual Seminar/Symposium,* Phoenix, Ariz. (Project Management Institute. Drexel Hill, Pa., 1980), pp. I-A.1-13.
20. Stuckenbruck, Linn C. "Project Manager—The Systems Integrator." *Project Management Quarterly* (September, 1978), pp. 31–38.
21. Stuckenbruck, Linn C. "The Integration Function in the Matrix." *1979 Proceedings of the Project Management Institute Annual Seminar/Symposium,* Atlanta, Ga. (Project Management Institute. Drexel Hill, Pa., 1979), pp. 481–492.
22. Stuckenbruck, Linn C. *The Implementation of Project Management: The Professionals Handbook* (Addison-Wesley. Reading, Mass., 1981), Chapter 6.
23. Stuckenbruck, Linn C. "Interface Management—Or Making The Matrix Work," in *Matrix Management Systems Handbook,* ed. Cleland, David I. (Van Nostrand Reinhold. New York, 1984). pp. 330–343.

# Section II

# The Project Organization

This section emphasizes the organizational dimensions of project and matrix management.

Chapter 4 presents Russell D. Archibald's depiction of the project office and project team—integral elements of the organization for projects that warrant a full-time project manager. He presents detailed specifications of the duties of the various project participants.

In Chapter 5 Richard L. Patterson expands on the role of the assistant project manager in terms of sets of needs—those of the project, the client, the contractor, and the individual. While these are presented specifically in terms of structuring a role for the assistant project manager, they are more generally relevant to understanding project organizational needs. Various projects in which this role was made operational are described.

The project team is also discussed in terms of its behavioral dimensions in Chapters 30, 31, and 32 of Section VIII.

# 4. Organizing the Project Office and Project Team: Duties of Project Participants*

Russell D. Archibald†

The approach described in this chapter is based on typical major project situations involving the design, manufacture, assembly, and testing of complex hardware and software systems. This presumes the following conditions:

- The project (or program) warrants a full-time project manager.
- The project office is held to minimum size, with maximum use of functional contributors in existing departments.

This chapter summarizes the functions of the project office and the project team under these conditions, describes the duties of key persons involved in the project, and discusses their relationships.

The situation frequently occurs wherein one project manager is responsible for several projects, or, when multiple small projects exist, the general manager will retain the project manager responsibility himself. In still other situations, a manager of projects is appointed. In such multiproject cases, centralized project planning and control is usually desirable. The positions and duties described here are still required, but they may be organized differently.

---

* Adapted from Chapter 6, "Organizing the Project Office and Project Team," *Managing High Technology Programs and Projects,* by Russell D. Archibald (John Wiley & Sons, Inc., New York, 1976).

† Russell D. Archibald is President, Archibald Associates, Los Angeles, California, consultants in project management, international business development, and strategic growth management. Mr. Archibald has directed major domestic and international programs with the Bendix Corporation and ITT; has consulted on program management to numerous large and small companies in eleven countries, as well as the U.S. Air Force; and has written and lectured extensively on this subject over the past 26 years.

## FUNCTIONS OF THE PROJECT OFFICE AND PROJECT TEAM

The project office supports the project manager in carrying out his responsibilities. Thus his basic charter, organizational relationship, and the nature of the project itself will influence the makeup of the project office. The presence or absence of other projects, and of a central project planning office, will also affect the organization of the project office.

The *project team* includes all functional contributors to the project, as well as the members of the project office. The general functions to be carried out during completion of the overall project by members of the project team are the following:

- Project and task management.
- Product design and development.
- Product manufacture.
- Purchasing and subcontracting.
- Product installation and test.

The relationships of these functions to the project manager are shown in the generalized project organization chart in Figure 4-1. Each of these functions is discussed in the following paragraphs.

### Project and Task Management

The management functions are simply those necessary to enable the project manager to fulfill his basic responsibility: overall direction and coordination of the project through all its phases to achieve the desired results within established budget and schedule, at the project level. Management of each functional task is the responsibility of the appropriate functional manager or staff member.

### Product Design and Development

The basic purpose of this general function is to produce adequate documentation (and often a prototype product or system) so that the product may be manufactured in the quantity required within the desired cost and schedule. These functions may be defined as

- Systems analysis, engineering, and integration.
- Product design.
- Product control (quality, cost, configuration).

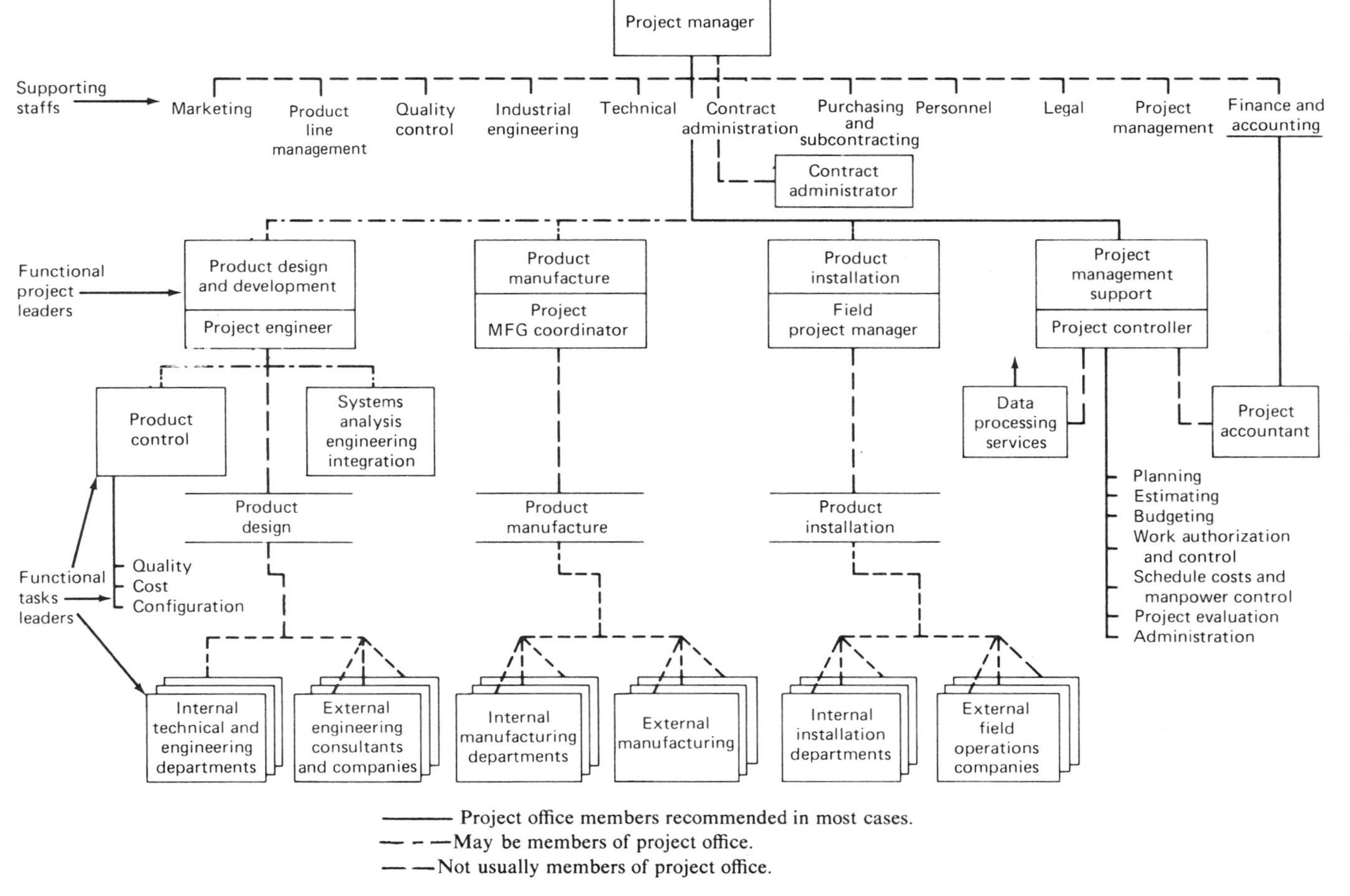


Figure 4-1. Generalized organization of the project team.

*Systems analysis, engineering, and integration* functions include system studies; functional analysis and functional design of the system or product; and coordination and integration of detailed designs, including functional and mechanical interfaces between major subsystems or components of the product.

*Product design* functions include the detailed engineering design and development functions needed to translate the functional systems design into specifications, drawings, and other documents which can be used to manufacture, assemble, install, and test the product. This may also include the manufacture and test of a prototype or first article system or product, using either model shop or factory facilities on a subcontract basis.

*Product control* functions include product quality control, using established staff specialists and procedures; product cost control, including value engineering practices; product configuration control, including design freeze practices (to establish the "base line" design), engineering change control practices, and documentation control practices.

It should be noted that the term "product" refers to *all* results of the project: hardware, software, documentation, training or other services, facilities, and so on.

The project office in a specific situation may perform none, a few, or all of these product design and development functions, depending on many factors. Generally a larger share of these functions will be assigned to the project office (together with adequate staff) when the product is new or unusual to the responsible unit, when the project is large and of long duration, or when there is little confidence that the work will be carried out efficiently and on schedule within established engineering departments. When several engineering departments, for example, from different product lines or different companies, are contributing to the product design and development, the functions of systems analysis, engineering and integration, and of product control, should be assigned to the project office.

Except in the situation described above, these functions should be performed by project team members within existing engineering departments, under the active coordination of the project manager.

### Product Manufacture

This general function is to purchase materials and components, fabricate, assemble, test, and deliver the equipment required to complete the project. These functions are carried out by the established manufacturing departments within the project's parent company or by outside companies on a subcontract or purchase order basis.

The project manager, however, must coordinate and integrate the manufacturing functions with product design and development on one hand and field operations (if any) on the other. *The lack of proper integration between these areas is the most common cause of project failure.*

In order to achieve this integration, it is necessary to appoint a project manufacturing coordinator or equivalent who will, in effect, act as a project manager for product manufacture. He is a key project team member. He may be assigned full time to one major project, or he may be able to handle two or more projects at one time, if the projects are small.

It is recommended that the project manufacturing coordinator remain within the appropriate manufacturing department. When two or more divisions or companies perform a large part of the product manufacture, each must designate a project manufacturing coordinator, with one designated as the lead division for manufacture. If it is not possible to designate a lead division, then the coordination effort must be accomplished by the project office.

### Purchasing and Subcontracting

This function is sometimes included with the product manufacture area, but it is normally important enough to warrant full functional responsibility.

A separate project purchasing and subcontracting coordinator with status equivalent to that of the manufacturing coordinator should be appointed to handle all purchasing and subcontracting matters for the project manager. This person should be a part of the purchasing department where he can maintain day-to-day contact with all persons carrying out the procurement functions.

### Product Installation, Test, and Field Support

Many projects require field installation and test of the system or equipment, and some include continuing field support for a period of time. In these cases, a field project manager (or equivalent) is required.

When field operations are a part of the project, this phase is usually clearly recognized as being of a project nature and requiring one person to be in charge. This field project manager is almost always a member of an established installation department (or equivalent) if such a department exists within the responsible company. Since engineering and manufacturing operations frequently overlap the installation phase, the overall project manager's role continues to be of critical importance to success while field operations are in progress. However, in the relationship between the project manager and the field project manager, the project manager retains the overall responsibility for the coordination of the entire project. For projects involving major construction, a Field Project Manager will be required whose duties and responsibilities are equivalent to the project manager as outlined earlier.

### Assignment of Persons to the Project Office

As a general rule, it is recommended that the number of persons assigned to a project office under the supervision of the project manager be kept as small as possible. This will emphasize the responsibility of each functional (line) department or staff for their contribution to the project and retain to the maximum degree the benefits of specialized functional departments. It will also increase flexibility of functional staffing of the project, avoid unnecessary payroll costs to the project, and minimize reassignment problems when particular tasks are completed. This will enable the project manager to devote maximum effort to the project itself, rather than to supervisory duties related to a large staff.

With adequate project planning and control procedures, a highly qualified project staff can maintain the desired control of the project. In the absence of adequate planning and control procedures to integrate the functional contributions, it is usually necessary to build up a larger staff with as many functional contributors as possible directly under the project manager in order to achieve control. Experience indicates that this is an expensive and frequently awkward approach, and it aggravates the relationships between the project manager and contributing functional managers.

The persons who should be assigned (transferred) permanently to the project office are those who

- Deal with the management aspects of the project.
- Are needed on a full-time basis for a period of at least six months.
- Must be in frequent close contact with the project manager or other members of the project office in the performance of their duties.
- Cannot otherwise be controlled effectively, because of organizational or geographic consideration.

Persons may be physically moved to the Project Office while retaining their permanent reporting relationships to a functional manager. This is a more frequent arrangement than officially transferring them out of the functional department to the project manager.

The recommended assignment location of each of the key people on the project team follows.

**Project Manager.** The project manager is always considered the manager of the project office (which could be a one-man office).

**Project Engineer.** The project engineer may be assigned to the project office in charge of product design and development where the product is new to the company or where several divisions are involved, as discussed earlier. Otherwise, he should always remain within the lead engineering department.

**Contract Administrator.** The contract administrator should remain a member of the contract administration staff, except on very large programs extending over a considerable period of time. He may be located physically in the project office while remaining with his parent organization.

**Project Controller.** The project controller should always be assigned to the project office, except where he is not needed full time or where a centralized planning and control function adequately serves the project manager.

**Project Accountant.** The project accountant should remain a member of the accounting department, except on very large programs extending over a considerable period of time. Like the contract administrator, he may physically be located in the project office while remaining with his parent organization.

**Manufacturing Coordinator.** The manufacturing coordinator should remain a member of the manufacturing organization, preferably on the staff of the manufacturing manager or within production control. When more than one division is to contribute substantially to product manufacture, it may be necessary to assign him to the project office to enable effective coordination of all contributors.

**Purchasing and Subcontracting Coordinator.** This coordinator should remain a member of the purchasing department in most cases.

**Field Project Manager.** The field project manager should remain a member of the installation or field operations department, if one exists, except under unusual circumstances that would require him to be assigned to the project office.

### Project Team Concept

Whether a person is assigned to the project office or remains in a functional department or staff, all persons holding identifiable responsibilities for direct contributions to the project are considered to be members of the project team. Creating awareness of membership in the project team is a primary task of the project manager, and development of a good project team spirit has proven to be a powerful means for accomplishing difficult objectives under tight time schedules.

### The Project Organization Chart

Figure 4-2 shows a typical representation of a project team in the format of a classic organization chart. This type of representation can be confusing if not properly understood, but it can also be useful to identify the key project team members and show their relationships to each other and to the project manager *for project purposes*. This recognizes that such a chart does not imply permanent superior-subordinate relationships portrayed in the company organization charts.

## PROJECT MANAGER DUTIES

The following description of project manager duties is presented as a guide for development of specific duties on a particular project. Some of

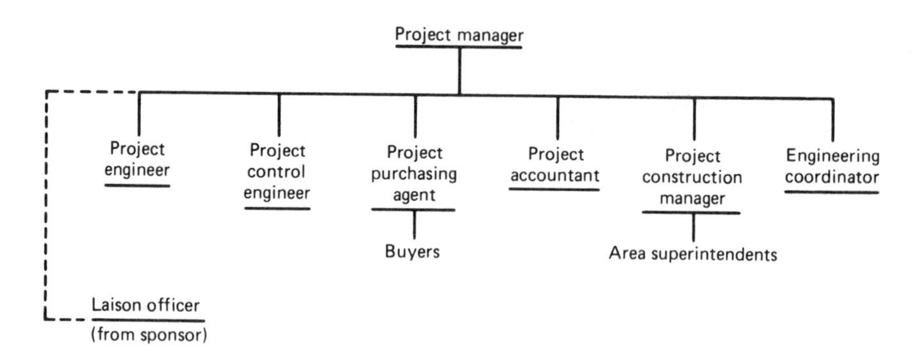


Figure 4-2. Typical construction project task force.

the duties listed may not be practical, feasible, or pertinent in certain cases, but wherever possible it is recommended that all items mentioned be included in the project manager's duties and responsibilities, with appropriate internal documentation and dissemination to all concerned managers.

### General

- Rapidly and efficiently start up the project.
- Assure that all equipment, documents, and services are properly delivered to the customer for acceptance and use within the contractual schedule and costs.
- Convey to all concerned departments a full understanding of the customer requirements of the project.
- Participate with responsible managers in developing overall project objectives, strategies, budgets, and schedules.
- Assure preparation of plans for all necessary project tasks to satisfy customer and management requirements, with emphasis on interfaces between tasks.
- Assure that all project activities are properly and realistically scheduled, budgeted, provided for, monitored, and reported.
- Identify promptly all deficiencies and deviations from plan.
- Assure that actions are initiated to correct deficiencies and deviations, and monitor execution of such actions.
- Assure that payment is received in accordance with contractual terms.
- Maintain cognizance over all contacts with the customer and assure that proper staff members participate in such contacts.
- Arbitrate and resolve conflicts and differences between functional departments on specific project tasks or activities.
- Maintain day-to-day liaison with all functional contributors to provide communication required to assure realization of their commitments.
- Make or force required decisions at successively higher organizational levels to achieve project objectives.
- Maintain communications with higher management regarding problem areas and project status.

### Customer Relations

In close cooperation with the customer relations or marketing department:

- Receive from the customer all necessary technical, cost, and scheduling information required for accomplishment of the project.
- Establish good working relationships with the customer on all levels: management, contracts, legal, accounts payable, system engineering, design engineering, field sites, and operations.
- Arrange and attend all meetings with customer (contractual, engineering, operations).
- Receive and answer all technical and operational questions from the customer, with appropriate assistance from functional departments.

### Contract Administration

- Identify any potential areas of exposure in existing or potential contracts and initiate appropriate action to alert higher management and eliminate such exposure.
- Prepare and send, or approve prior to sending by others, all correspondence on contractual matters.
- Coordinate the activities of the project contract administrator in regard to project matters.
- Prepare and participate in contract negotiations.
- Identify all open contractual commitments.
- Advise engineering, manufacturing, and field operations of contractual commitments and variations allowed.
- Prepare historical or position papers on any contractual or technical aspect of the program, for use in contract negotiations or litigation.

### Project Planning, Control, Reporting, Evaluation, and Direction

- Perform, or supervise the performance of, all project planning, controlling, reporting, evaluation, and direction functions.
- Conduct frequent, regular project evaluation and review meetings to identify current and future problems and initiate actions for their resolution.
- Prepare and submit weekly or monthly progress reports to higher management, and to the customer if required.
- Supervise the project controller and his staff.

### Marketing

Maintain close liaison with Marketing and utilize customer contacts to acquire all possible marketing intelligence for future business.

### Engineering

- Insure that Engineering fulfills its responsibilities for delivering, on schedule and within product cost estimates, drawings and specifications usable by manufacturing and field operations, meeting the customer specifications.
- In cooperation with the Engineering, Drafting, and Publications Departments define and establish schedules and budgets for all engineering and related tasks. After agreement release funding allowables and monitor progress on each task in relation to the overall project.
- Act as the interface with the customer for these departments (with their assistance as required).
- Assure the control of product quality, configuration, and cost.
- Approve technical publications prior to release to the customer.
- Coordinate engineering support related to the project for Manufacturing, Installation, Legal, and other departments.
- Participate (or delegate participation) as a voting member in the Engineering Change Control Board on matters affecting the project.

### Manufacturing

- Insure that Manufacturing fulfills its responsibility for on-schedule delivery of all required equipment, meeting the engineering specifications within estimated manufacturing costs.
- Define contractual commitments to Production Control.
- Develop schedules to meet contractual commitments in the most economical fashion.
- Establish and release manufacturing and other resource and funding allowables.
- Approve and monitor production control schedules.
- Establish project priorities.
- Approve, prior to implementation, any product changes initiated by Manufacturing.
- Approve packing and shipping instructions based on type of transportation to be used and schedule for delivery.

### Purchasing and Subcontracting

- Insure that Purchasing and Subcontracting fulfill their responsibilities to obtain delivery of materials, equipment, documents, and services on schedule and within estimated cost for the project.

- Approve make-or-buy decisions for the project.
- Define contractual commitments to Purchasing and Subcontracting.
- Establish and release procurement funding allowables.
- Approve and monitor major purchase orders and subcontracts.
- Specify planning, scheduling, and reporting requirements for major purchase orders and subcontracts.

### Installation, Test, and Other Field Operations

- Insure that Installation and Field Operations fulfill their responsibilities for on-schedule delivery to the customer of materials, equipment, and documents within the cost estimates for the project.
- Define contractual commitments to Installation and Field Operations.
- In cooperation with Installation and Field Operations, define and establish schedules and budgets for all field work. After agreement, release funding allowables and monitor progress on each task in relation to the overall project.
- Coordinate all problems of performance and schedule with Engineering, Manufacturing, and Purchasing and Subcontracting.
- Except for customer contacts related to daily operating matters, act as the customer interface for Installation and Field Operations departments.

### Financial

In addition to the financial project planning and control functions described:

- Assist in the collection of accounts receivable.
- Approve prices of all change orders and proposals to the customer.

### Project Closeout

- Insure that all required steps are taken to present adequately all project deliverable items to the customer for acceptance and that project activities are closed out in an efficient and economical manner.
- Assure that the acceptance plan and schedule comply with the customer contractual requirements.
- Assist the Legal, Contract Administration, and Marketing or Commercial Departments in preparation of a closeout plan and required closeout data.

- Obtain and approve closeout plans from each involved functional department.
- Monitor closeout activities, including disposition of surplus materials.
- Notify Finance and functional departments of the completion of activities and of the project.
- Monitor payment from the customer until all collections have been made.

## PROJECT ENGINEER DUTIES

### General

The project engineer is responsible for the technical integrity of his project and for cost and schedule performance of all engineering phases of the project. Specifically, the responsibilities of the project engineer are the following:

- Insure that the customer performance requirements are fully understood and that the company is technically capable of meeting these requirements.
- Define these requirements to the smallest subsystem to the functional areas so that they can properly schedule, cost, and perform the work to be accomplished.
- Insure that the engineering tasks so defined are accomplished within the engineering schedules and allowables (manpower, materials, funds) of the contract.
- Provide technical direction as necessary to accomplish the project objectives.
- Conduct design review meetings at regular intervals to assure that all technical objectives will be achieved.
- Act as technical advisor to the project manager and other functional departments, as requested by the project manager.

In exercising the foregoing responsibilities, the project engineer is supported by the various engineering departments.

### Proposal Preparation and Negotiation

During the proposal phase, the project engineer will do the following:

- Coordinate and plan the preparation of the technical proposal.
- Review and evaluate the statement of work and other technical data.

- Establish an engineering proposal team or teams.
- Within the bounds of the overall proposal schedule, establish the engineering proposal schedule.
- Reduce customer engineering requirements to tasks and subtasks.
- Define in writing the requirements necessary from Engineering to other functional areas, including preliminary specifications for make or buy, or subcontract items.
- Coordinate and/or prepare a schedule for all engineering functions, including handoff to and receipt from Manufacturing or Purchasing.
- Review and approve all Engineering subtask and task costs, schedules, and narrative inputs.
- Coordinate and/or prepare overall engineering cost.
- Participate in preliminary make-or-buy decisions.
- Participate in overall cost and schedule review.
- Participate, as required, in negotiation of contract.
- Bring problems between the project engineer and engineering functional managers to appropriate engineering directors for resolution.

### Project Planning and Initiation

The project engineer is responsible for the preparation of plans and schedules for all engineering tasks within the overall project plan established by the project manager. In planning the engineering tasks, he will compare the engineering proposal against the received contract. Where the received contract requirements dictate a change in cost, schedule, or technical complexity for solution, he will obtain approval from the director of engineering and the project manager to make the necessary modifications in engineering estimates of the proposal. During this phase, the project engineer will:

- Update the proposal task and subtask descriptions to conform with the contract, and within the engineering allowables prepare additional tasks and subtasks as required to provide a complete engineering implementation plan for the project.
- Prepare a master engineering schedule in accordance with the contractual requirements.
- Prepare, or have prepared, detailed task and subtask definitions and specifications. Agree on allowables, major milestones, and evaluation points in tasks with the task leaders and their functional managers.
- Through the functional engineering managers, assign responsibility for task and subtask performance, and authorize the initiation of

work against identified commitments based on cost and milestone schedules, with approval of the project manager.
- Using contract specifications as the base line, prepare, or have prepared, specifications for subcontract items.
- Participate and provide support from appropriate engineering functions in final make-or-buy decisions and source selection.
- Prepare, or have prepared, hardware and system integration and acceptance test plan. Review the test plan with Quality Assurance and advise them as to the required participation of other departments.

#### Project Performance and Control

The project engineer is responsible for the engineering progress of the project and compliance with contract requirements, cost allowables, and schedule commitments. Within these limits, the project engineer, if necessary, may make design changes and task requirement changes in accordance with his concept and assume the responsibility for the change in concert with the functional engineering managers and with the knowledge of the project manager. No changes may be made that affect other functional departments without the knowledge of that department, documentation to the project manager, and the inclusion of the appropriate chargeback of any variance caused by change. He maintains day-to-day liaison with the project manager for two-way information exchange. Specific responsibilities of the project engineer are the following:

- Prepare and maintain a file of all project specifications related to the technical integrity and performance.
- Prepare and maintain updated records of the engineering expenditures and milestones and conduct regular reviews to insure engineering performance as required.
- Initiate and prepare new engineering costs-to-complete reports as required.
- Establish work priorities within the engineering function where conflict exists: arbitrate differences and interface problems within the engineering function, and request through functional managers changes in personnel assignments if deemed necessary.
- Plan and conduct design review meetings and design audits as required, and participate in technical reviews with customer.
- Prepare project status reports as required.
- With the project manager and other functional departments, partici-

pate in evaluation and formulation of alternate plans as required by schedule delays or customer change requests.

- Assure support to Purchasing and Subcontracting, Manufacturing, Field Operations, and support activities by providing liaison and technical assistance within allowables authorized by the project manager.
- Modify and reallocate tasks and subtasks, open and close cost accounts, and change allowable allocations within the limits of the approved engineering allowables, with the concurrence of the functional managers involved. Provide details to the project manager of all such actions prior to change.
- As requested by the project manager, support Legal and Contracts Administration by providing technical information.
- Review and approve technical aspects of reports for dissemination to the customer.
- Authorize within the approved allowables the procurement of material and/or services as required for the implementation of the engineering functional responsibility.
- Adjudicate technical problems and make technical decisions within scope of contractual requirements. Cost and schedule decisions affecting contractual requirements or interface with other functions are to be approved by the appropriate engineering function manager with the cognizance of the director of engineering (or his delegate) and the project manager.
- Approve all engineering designs released for procurement and/or fabrication for customer deliverable items.
- Bring problems arising between the project engineer and engineering functional managers to the engineering director for resolution.
- Bring problems arising between the project engineer and functions outside engineering to the project manager for resolution, with the cognizance of the director of engineering and the director of the other functions.

## CONTRACT ADMINISTRATOR DUTIES

### General

Contract administration is a specialized management function indispensable to effective management of those projects carried out under contract with customers. This function has many legal implications and serves to protect the company from unforeseen risks prior to contract approval and during execution of the project. Experience dictates that well-qualified, properly organized contract administration support to a project manager is vital to the continuing success of companies responsible for major sales contracts.

Contract administration is represented both on the project manager's team and on the general manager's staff. A director of contract administration has the authority to audit project contract files and to impose status reporting requirements that will disclose operational and contractual problems relating to specific projects. The director of contract administration is also available to provide expertise in the resolution of contract problems beyond the capability of the contract administrator assigned to a given project.

The project contract administrator is responsible for day-to-day administration of (a) the contract(s) that authorize performance of the project and (b) all subcontracts with outside firms for equipment, material, and services to fulfill project requirements.

#### Proposal Preparation

- When participation of an outside subcontractor is required, assure that firm quotations are obtained based on terms and conditions compatible with those imposed by the customer.
- Review with the Legal and Financial Department all of the legal and commercial terms and conditions imposed by the customer.
- Review the proposal prior to submittal to assure that all risks and potential exposures are fully recognized.

#### Contract Negotiation

- Lead all contract negotiations for the project manager.
- Record detailed minutes of the proceedings.
- Assure that all discussions or agreements reached during negotiations are confirmed in writing with the understanding that they will be incorporated into the contract during the contract definition phase.
- Assure that the negotiating limits established by the Proposal Review Board (or equivalent) are not exceeded.

#### Contract Definition

- Expedite the preparation, management review, and execution of the contract, as follows:
  —Clarify the contract format with the customer.
  —Establish the order of precedence of contract documents incorporated in reference.

—Set the date by which the contract will be available in final form for management review prior to execution.
—Participate with the project manager in final briefing of management on the contract terms and conditions prior to signature.

**Project Planning Phase**

- Establish channels of communication with the customer and define commitment authority of project manager, contract administrator, and others.
- Integrate contract requirements and milestones into the project plan and schedule; including both company and customer obligations.
- Establish procedures for submission of contract deliverables to customer.
- Establish mechanics for monthly contract status reports for the customer and management.

**Project Execution Phase**

- Monitor and follow up all contract and project activities to assure fulfillment of contractual obligations by both the company and the customer.
- Assure that all contract deliverables are transmitted to the customer and that all contractually required notifications are made.
- Record any instance where the customer has failed to fulfill his obligations and define the cost and schedule impact on the project of such failure.
- Identify and define changes in scope and customer-caused delays and force majeure, including:
  —Early identification and notification to customer.
  —Obtaining of customer's agreement that change of scope or customer-caused delay or force majeure case has actually occurred.
  —In coordination with the project manager and the project team, preparation of a proposal that defines the scope of the change(s) and resulting price and/or schedule impact for submittal to the customer for eventual contract modification.
- Assist in negotiation and definition of contract change orders.
- Participate in project and contract status reviews and prepare required reports.
- Arrange with the customer to review the minutes of joint project review meetings to assure that they accurately reflect the proceedings.

- Assure that the customer is notified in writing of the completion of each contractual milestone and submission of each contract deliverable item, with a positive assertion that the obligation has been fulfilled.
- Where the customer insists on additional data or work before accepting completion of an item, monitor compliance with his requirement to clear such items as quickly as possible.

### Project Closeout Phase

- At the point where all contractual obligations have been fulfilled, or where all but longer-term warranties or spare parts deliveries are complete, assure that this fact is clearly and quickly communicated in writing to the customer.
- Assure that all formal documentation related to customer acceptance as required by the contract is properly executed.
- Expedite completion of all actions by the company and the customer needed to complete the contract and claim final payment.
- Initiate formal request for final payment.
- Where possible, obtain certification from the customer acknowledging completion of all contractual obligations and releasing the company from further obligations, except those under the terms of guaranty or warranty, if any.

### Project/Contract Record Retention

Prior to disbanding the project team, the project contract administrator is responsible for collecting and placing in suitable storage the following records, to satisfy legal and internal management requirements:

- The contract file, which consists of:
  Original request for proposal (RFP) and all modifications.
  All correspondence clarifying points in the RFP.
  Copy of company's proposal and all amendments thereto.
  Records of negotiations.
  Original signed copy of contract and all documents and specifications incorporated in the contract by reference.
  All contracts and modifications (supplemental agreements).
  A chronological file of all correspondence exchanged between the parties during the life of the program. This includes letters, telexes, records of telephone calls, and minutes of meetings.

Acceptance documentation.
Billings and payment vouchers.
Final releases.

- Financial records required to support postcontract audits, if required by contract or governing statutes.
- History of the project (chronology of all events—contractual and noncontractual).
- Historical cost and time records that can serve as standards for estimating future requirements.

## PROJECT CONTROLLER DUTIES

The primary responsibility of the project manager is to plan and control his project. On some smaller or less complex projects, he may be able to perform all the planning and controlling functions himself. However, on most major projects, it will be necessary to provide at least one person on his staff who is well qualified in project planning and control and who can devote his full attention to these specialized project management needs. This person is the project controller. (A number of other equivalent job titles are in use for this position.) On very large or complex programs or projects, the project controller may require one or more persons to assist him in carrying out his duties and responsibilities.

If a centralized operations planning and control function exists in the company, that office may provide the needed planning and control services to the project manager. In that case the project controller would be a member of the Operations Planning and Control Office and would have available to him the specialists in that office. In other situations, the project controller may be transferred from Operations Planning and Control to the project office for the duration of the project.

The duties of the project controller are described in the following sections.

### General

- Perform for the project manager the project planning, controlling, reporting, and evaluation functions as delegated to him, so that the project objectives are achieved within the schedule and cost limits.
- Assist the project manager to achieve clear visibility of all contract tasks so that they can be progressively measured and evaluated in sufficient time for corrective action to be taken.

### Project Planning and Scheduling

- In cooperation with responsible managers, define the project systematically so that all tasks to be performed are identified and hierarchically related to each other, including work funded under contract or by the company, using the project breakdown structure or similar technique.
- Identify all elements of work (tasks or work packages) to be controlled for time, manpower, or cost, and identify the responsible and performing organizations and project leaders for each.
- Define an adequate number of key milestones for master planning and management reporting purposes.
- Prepare and maintain a graphic project master plan and schedule, based on the project breakdown structure, identifying all tasks or work packages to be controlled in the time dimension, and incorporating all defined milestones.
- Prepare more detailed graphic plans and schedules for each major element of the project.

### Budgeting and Work Authorization

- Obtain from the responsible manager for each task or work package a task description, to include:
  Statement of work.
  Estimate of resources required (man days, computer hours, etc.).
  Estimate of labor, computer, and other costs (with assistance of the project accountant).
  Estimate of start date, and estimated total duration and duration between milestones.
- Prepare and maintain a task description file for the entire project.
- Summarize all task manpower and cost estimates, and coordinate needed revisions with responsible managers and the project manager to match the estimates with available and allocated funds for the project in total, for each major element, and for each task.
- Prepare and release, on approval of the project manager and the responsible functional manager, work authorization documents containing the statement of work, budgeted labor, and cost amounts; scheduled dates for start, completion, and intermediate milestones; and the assigned cost accounting number.
- Prepare and release, with approval of the project manager, revised work authorization documents when major changes are required or

have occurred, within the authorized funding limits and the approval authority of the project manager.

**Work Schedules**

- Assist each responsible manager or project leader in developing detailed plans and schedules for assigned tasks, reflecting the established milestone dates in the project master plan.
- Issue current schedules to all concerned showing start completion dates of tasks and occurrence dates of milestones.

**Progress Monitoring and Evaluation**

- Obtain weekly reports from all responsible managers and project leaders of:
  Activities started and completed.
  Milestones completed.
  Estimates of time required to complete activities or tasks under way.
  Changes in future plans.
  Actual or anticipated delays, additional costs, or other problems that may affect other tasks, the schedule, or project cost.
- Record reported progress on the project master plan and analyze the effect of progress in all tasks on the overall project schedule.
- Identify major deviations from schedule and determine, with the responsible managers and the project manager, appropriate action to recover delays or take advantage of early completion of tasks.
- Obtain monthly cost reports and compare to the estimates for each current task, with summaries for each level of the project breakdown structure and the total project.
- Through combined evaluation of schedule and cost progress compared to plan and budget, identify deviations that require management action and report these to the project manager.
- Participate in project review meetings, to present the overall project status and evaluate reports from managers and project leaders.
- Record the minutes of project review meetings and follow up for the project manager all resulting action assignments to assure completion of each.
- Advise the project manager of known or potential problems requiring his attention.
- Each month or quarter obtain from each responsible manager an estimate of time, manpower, and cost to complete for each incomplete task or work package; and prepare, in cooperation with the

project accountant, a revised projection of cost to complete the entire project.

### Schedule and Cost Control

- When schedule or budget revisions are necessary, due to delay or changes in the scope of work, prepare, negotiate, and issue new project master plan and schedules and revised work authorization documents, with approval of the project manager, within the authorized funding limits and the approval authority of the project manager.
- In coordination with the project accountant, notify the Finance Department to close each cost account and reject further charges when work is reported complete on the related task.

### Reporting

- Prepare for the project manager monthly progress reports to management and the customer.
- Provide cost-to-complete estimates and other pertinent information to the project accountant for use in preparing contract status reports.
- Prepare special reports as required by the project manager.

## PROJECT ACCOUNTANT DUTIES

The basic function of the project accountant is to provide to the project manager the specialized financial and accounting assistance and information needed to forecast and control manpower and costs for the project. The project accountant duties are as follows:

- Establish the basic procedure for utilizing the company financial reporting and accounting system for project control purposes to assure that all costs are properly recorded and reported.
- Assist the project controller in developing the project breakdown structure to identify the tasks or project elements that will be controlled for manpower and cost.
- Establish account numbers for the project and assign a separate number to each task or work element to be controlled.
- Prepare estimates of cost, based on manpower and other estimates provided by the controller, for all tasks in the project when required to prepare revised estimates to complete the project.
- Obtain, analyze, and interpret labor and cost accounting reports, and

provide the project manager, project controller, and other managers in the project with appropriate reports to enable each to exercise needed control.

- Assure that the information being recorded and reported by the various functional and project departments is valid, properly charged, and accurate, and that established policies and procedures are being followed for the project.
- Identify current and future deviation from budget of manpower or funds, or other financial problems, and in coordination with the project controller notify the project manager of such problems.
- Prepare, in coordination with the project manager and the project controller, sales contract performance reports as required by division or company procedures on a monthly basis for internal management purposes, and for submission to any higher headquarters.

## MANUFACTURING COORDINATOR DUTIES

### General

The general duties and responsibilities of the manufacturing coordinator (sometimes called the project leader—manufacturing) are to plan, implement, monitor, and coordinate the manufacturing aspects of his assigned project (or projects, where it is feasible for him to coordinate more than one contract).

### Specific Duties

- Review all engineering releases before acceptance by manufacturing to insure they are complete and manufacturable (clean releases), and that all changes are documented by a formal written engineering change request.
- Participate in the development of project master schedules during proposal, negotiation, and execution phases, with particular emphasis on determination of requirements for engineering releases, critical parts lists, equipment requirements, and so on, to insure meeting delivery requirements.
- Monitor all costs related to assigned projects to assure adherence to manufacturing costs and cost schedules, analyze variances and recommend corrective action, collect needed information and prepare manufacturing cost to complete.
- Develop or direct the development of detailed schedules for assigned projects, coordinating the participation of manufacturing and product

support engineering, material planning, fabrication, purchasing, material stores, assembly, test, quality control, packing and shipping, in order to insure completion of master project schedule within budget limits; provide information and schedules to different functional groups in order for action to be initiated.

- Approve all shipping authorizations for assigned projects.
- Provide liaison between the project manager and Manufacturing; diligently monitor manufacturing portions of assigned projects and answer directly for manufacturing performance against schedules; prepare status reports and provide information needed to prepare costs to complete as required.
- Take action within area of responsibility and make recommendations for corrective action in manufacturing areas to overcome schedule slippages; obtain approval from the project manager for incurring additional manufacturing costs.
- Coordinate requests for clarifications of the impact of contract change proposals on manufacturing effort.
- Participate in the preparation and approval of special operating procedures.
- Review and approve for manufacturing all engineering releases and engineering change notices affecting assigned projects, and participate in Change Control Board activity.
- Represent project manager on all Make/Buy Committee actions.

## FIELD PROJECT MANAGER DUTIES

### General

The field project manager (or equivalent) has overall responsibility for constructing required facilities and installing, testing, and maintaining for the specified time period, and handing over to the customer, all installed equipment and related documentation as specified by the contract. This includes direct supervision of all company and subcontractor field personnel, through their respective managers or supervisors.

### Specific Duties

- Participate in the development of project master schedules during proposal, negotiation, and execution phases, with particular emphasis on determination of equipment delivery schedules and manpower and special test equipment needs.

- Monitor all field operations costs for the project to assure adherence to contract allowables. Analyze variances and recommend corrective actions. Collect needed information and prepare field operations cost to complete.
- Develop or direct the development of detailed schedules for all field operations; coordinate the equipment delivery schedules from Manufacturing and subcontractors with field receiving, inspection, installation, testing, and customer acceptance procedures, with due regard for transportation and import/export requirements, to insure completion of the master project schedule within budget limits; provide information and schedules to different functional groups or departments in order for action to be initiated.
- Provide liaison between the project manager and Installation and Field Operations; diligently monitor field operations portion of the project and answer for performance against schedules; prepare status reports.
- Take action and make recommendations for corrective action in field operations and other areas to overcome schedule slippages; obtain approval of the project manager for incurring additional installation costs.
- Coordinate requests for clarifications of the impact of contract change proposals on field operations.

# 5. Developing the Role of the Assistant Project Manager by Assessing the Needs of Project Clients

Richard L. Patterson*

## INTRODUCTION

In the many discussions and papers on the subject of project management and related management concepts, techniques, and procedures there appears to be a dearth of information on the role of the Assistant Project Manager.

The reasons for this condition are not necessarily clear nor are they really germane to our discussion—but isn't it about time we took a closer look at this fellow? Shouldn't we delve into questions like who needs him and why? What function does the Assistant Project Manager perform? Where does he fit into the organization? And finally, what is his personal stake in his role?

Three basic principles evolve almost immediately during any analysis of the role of the Assistant Project Manager, namely:

- The role must provide for the accomplishment of substantial and productive functions within the scope of the total project.
- The role must fulfill certain basic needs of both the Client and the Contractor.
- The role should provide a meaningful challenge to the holder and at the same time encourage his personal growth and development.

---

* Mr. Patterson spent nine years as a project manager for Bechtel Power Corporation. During this time he managed the design, procurement, and construction of two multi-unit, coal-fired power generating stations, each costing over $500 million. These assignments gave him first-hand experience at effectively utilizing assistant project managers. He is a registered Civil Engineer and a member of the ASCE and the Project Management Institute.

## SPECIFIC NEEDS OF THE PROJECT

A Project Manager is primarily concerned with the successful completion of his project. It is from the real-life world of the project itself that the specific needs for an Assistant Project Manager arise and from which his basic role is ultimately defined.

Essentially most of our larger engineering and construction projects today, and especially those in the power generation field, challenge the Project Manager's span of control at their very outset. Some of the more significant factors which strain this span include the following:

- The extreme physical size of many of today's projects.
- The large capital costs and lengthy schedules of the projects.
- The significant financial exposure of the Client during the project cycle.
- The range and complexities of the technologies involved.
- Special services required of the Contractor in addition to his normal offerings and demonstrated capabilities.
- Geographic spread of the basic project and the major subcontractors' efforts.
- Special procurement considerations such as
  —Client/Contractor legal relationships and divisions in procurement responsibilities or activities.
  —Extensive supplier/subcontractor qualification, inspection, expediting, or test requirements.
  —Extensive commercial evaluations and negotiations to obtain satisfactory escalation provisions.
  —International as well as national procurement activities with attendant financial, legal, or administrative considerations.
- The tremendous impact of regulatory agencies and their proliferation at national, state, and local levels.
- The increasing public interest and the scrutiny of intervenors into more and more elements of a project.
- Construction and construction-related activities including such factors as
  —Labor relations and union matters.
  —Site logistics and support.
  —Housing and associated socioeconomic considerations.
  —Insurance.

These typical and basic strains on project management generally prevail throughout the average power project life cycle and obviously require the continuing attention of the Contractor's and the Client's management.

But it is also important to note that most large and long-term projects go through several major changes in emphasis and many shifts in the tenor of their operations during their life cycles.

For example, in the early stages of a power project we may be concerned with such factors as site selection, housing requirements, fuel transport, major long-lead hardware procurement, environmental reports, criteria development, and basic estimates and schedules associated with the conceptual design. As the final design is developed, the project emphasis shifts to detailed engineering demands with particular attention devoted to engineering calculations, drawings, and the preparation of detailed specifications. Subsequently, project concern will center upon the qualification of vendors and subcontractors, the preparation of bid packages, bid evaluations, and material availability, while the design, costs, and schedules undergo continuing and strengthening refinements. In the meantime, field mobilization must be planned and the site prepared. Then, with the pouring of the first major concrete, the activity in the field increases very rapidly. Now special attention must be focused on the timely delivery of hardware and materials, and getting the necessary engineering drawings to the field to ensure orderly and effective progression in a total construction effort that may involve several thousand workers. As construction advances, the project emphasis again shifts; this time to the installation and checkout of equipment, the resolution of equipment problems, and the progressive start-up of the plant.

These typical project patterns are not exempt from the unexpected or unusual problems which also serve to change management plans and expectations. Further, it should be emphasized that with each changing project pattern there is usually both a qualitative and quantitative impact upon project management. The special skills, technology, or professional requirements change with each new pattern. Similarly, the intensity of project management attention and devotion to certain project areas must shift with each change in the project pattern.

Naturally, both the Contractor and the Client desire that the Project Manager bring to the project the widest (and deepest) possible personal exposure and professional experience in all facets of the job. However, it is a rare individual who can fulfill all the demands of today's major projects.

In such cases the capabilities and experience of an Assistant Project Manager can be used to complement and supplement those of the Project Manager in one or more areas of concern and during one or more major time frames of the project. Proper matching of skills and background coupled with timely assignment can greatly strengthen the Contractor's project management and the true span of control.

## GENERAL NEEDS OF THE CLIENT

The Client desires that the needs of the project be satisfied as discussed in the previous section. He also has some very specific needs of his own.

The Client has a concern in the day-to-day availability of the Project Manager. In the Project Manager's absence, the Client invariably requires assurance that adequate management attention is being given to the project operations and that a designated and capable individual—within management—is available to respond to his needs.

Similarly, the Client has an overriding interest in the continuity of his project and generally looks with favor on the presence of a designated assistant in the Contractor's organization who can eventually (and smoothly) move into the Project Manager's shoes, if and when required. Some Clients may actively participate in the actual selection of this assistant by establishing certain qualifications or criteria for the role, or by other means.

The very fact that the Client has brought a Contractor on board is presumptive of his needs. Normally the statement of work and related contract documents will identify these needs adequately. However, it is not always possible to fully define certain key concerns or overriding needs, especially in regard to personnel qualifications, staffing, and organization.

While the Client may have no direct interest in the personal growth or development of each individual within the Contractor's organization, he does have a general and continuing concern for the overall strength and depth of the Contractor's project team.

In this respect, it is not uncommon for the Client to view the Contractor's organization as an extension of his very own. In this context then, the Client frequently demonstrates a strong advocacy for furthering the growth and development of certain of the Contractor's people in parallel with his own in-house plans. Specifically, the establishment and development of the role of Contractor's Assistant Project Manager usually fits very nicely into the Client's personnel plans and policies.

## GENERAL NEEDS OF THE CONTRACTOR

The Contractor's senior management, as well as many others within and without the Contractor's organization, will normally view the Project Manager as the primary contact (and expert) on his assigned job. And rightly so!

It naturally follows that the Project Manager is frequently sought out or solicited for reports, reviews, audits, special information or documents, and similar efforts by his own management. In addition, he is frequently called upon to head up meetings or conferences or to participate to a greater or lesser degree in related activities sponsored by others, both within and without his own organization. Obviously, he must be particularly concerned with those key management plans and operations which interface directly with his assignment.

Progressive companies will provide opportunities for deserving Project Managers to participate in executive management training courses and seminars to prepare them for positions of greater responsibility.

These demands upon the Project Manager's time make him less available to his own project organization which must deal with a multitude of questions on a day-to-day basis. Under such circumstances the assignment of an Assistant Project Manager to fill in for the Project Manager in his absence can ensure timely monitoring of operations, lend mature judgment and guidance to project personnel, provide for appropriate approvals (signature) where required, and do much to alleviate any work delays or encumbrances which might otherwise be brought about by the Project Manager's absence.

On many key projects the Contractor may have a further and naturally selfish need for the presence of an Assistant Project Manager. On long-term projects it is especially prudent for management to plan for the ultimate replacement of the Project Manager as the latter may move on to further challenges. The assignment of an Assistant Project Manager on a timely basis then becomes the Contractor's primary tool for achieving project continuity. Subsequently, an efficient transition or change in project management, with no unsettling Client or project team disturbances, can be made when required.

The successful Contractor is also alert to foster the personal growth and professional development of his own people. It is usually the more talented and motivated employees who actively seek out the promotional routes to management and who must be reassured, at timely intervals, by appropriate assignments of increasing responsibility and growth potential.

The role of an Assistant Project Manager, if properly structured, on a selected project can provide excellent training for the development of future Project Managers and at the same time maintain the dedication and motivation of the individual employee by providing him with meaningful challenges and opportunities. We stress the fact that the role must be properly structured, and for the benefit of all—the Contractor, the Client, the Project, and the Individual.

## THE NEEDS OF THE INDIVIDUAL

Senior engineering and construction personnel on their way toward a management career will be attracted to the role of Assistant Project Manager if the position will

- Provide challenging work assignments.
- Utilize their professional skills, talents, and experience.
- Call for (tangible) productive work to achieve recognizable goals.
- Permit their professional development and growth.

A stimulating work environment must be afforded the Assistant Project Manager. His care and feeding, particularly if the role represents his first assignment with "manager" in the title, is of particular concern and may markedly affect project operations.

From the individual's standpoint—and ideally—all the foregoing factors should be equally represented in the assigned role. Regrettably, not all projects necessarily present such a balanced offering. However, it is extremely fortunate that most of today's large projects do offer a wide variety of experience to the young managerial candidate. By ensuring that the Assistant Project Manager is given adequate exposure to diverse elements of the project to enhance his growth and development, the Contractor's management can frequently compensate for occasional shortcomings in other specific areas of concern to the individual.

## STRUCTURING THE ROLE

Structuring the role of an Assistant Project Manager on a specific project may turn out to be a challenging task. The interplay among the management, personal, and project variables and trade-offs must be carefully evaluated.

The individual candidates for the role will normally have clear-cut concepts of their personal requirements. However, as we pointed out earlier, it is a rare situation or project which will completely satisfy the personal needs of every candidate for the role or neatly supplement or complement the personal capabilities of the Project Manager.

While the general needs of the Contractor and the Client loom large and important, the specific needs of the project become the more compelling and should invariably establish the basic structure of the Assistant Project Manager's role.

However, within the total project needs there will usually be several marked variations in the relative priorities assigned to the various requirements. Moreover, there may be a wide disparity between disciplinary or functional needs on the project which cannot be easily reconciled or encompassed by the placement of any one individual. Finally, not all requirements can be neatly compartmentalized or effectively segmented from many other portions of the project to provide a nicely fenced-in area for the role of the Assistant Project Manager.

The timing of the assignment of the Assistant Project Manager is influenced primarily by the requirements of the project, which vary with time.

For the most part, the project scope and schedule can serve to identify the need dates for special assistance or added attention to certain project activities or events. In the case of very large and complex projects it may be desirable to staff the position coincident with the assignment of the Project Manager. On smaller or less complex projects, and especially those on a low burner in their early phases, it will usually be more prudent and cost effective to delay the assignment. In either case a project learning curve for the Assistant Project Manager should be considered so that when he is required to perform as Project Manager, he will have had time to assimilate the necessary project background and experience to enable him to perform satisfactorily.

While we have so far discussed the role of the Assistant Project Manager as that of a single individual, it should now be apparent that on certain demanding and complex projects two or more Assistant Project Managers may very well be required.

Throughout our discussion we have also consistently used the title "Assistant Project Manager." However, there is no mandatory requirement for this usage. Other titles appropriately matched to the Contractor's and Client's job classification and organizational concepts would also be suitable. Examples might include

- Deputy Project Manager.
- Associate Project Manager.

## INTERPRETING THE ROLE

When the basic structure and background needs of the role of the Assistant Project Manager have been reconciled, his position can be implemented and integrated into the project organization.

Essentially there are three elementary organizational arrangements which should be considered, as shown in Figure 5-1. In the first case, the Assistant Project Manager is placed in a staff position relative to the Project Manager. This position is indicative of a more confined or segmented role and connotes a less than full-time backup for the Project Manager. This arrangement is most beneficial when an experienced staff person is assigned to the project to concentrate his expertise on specific

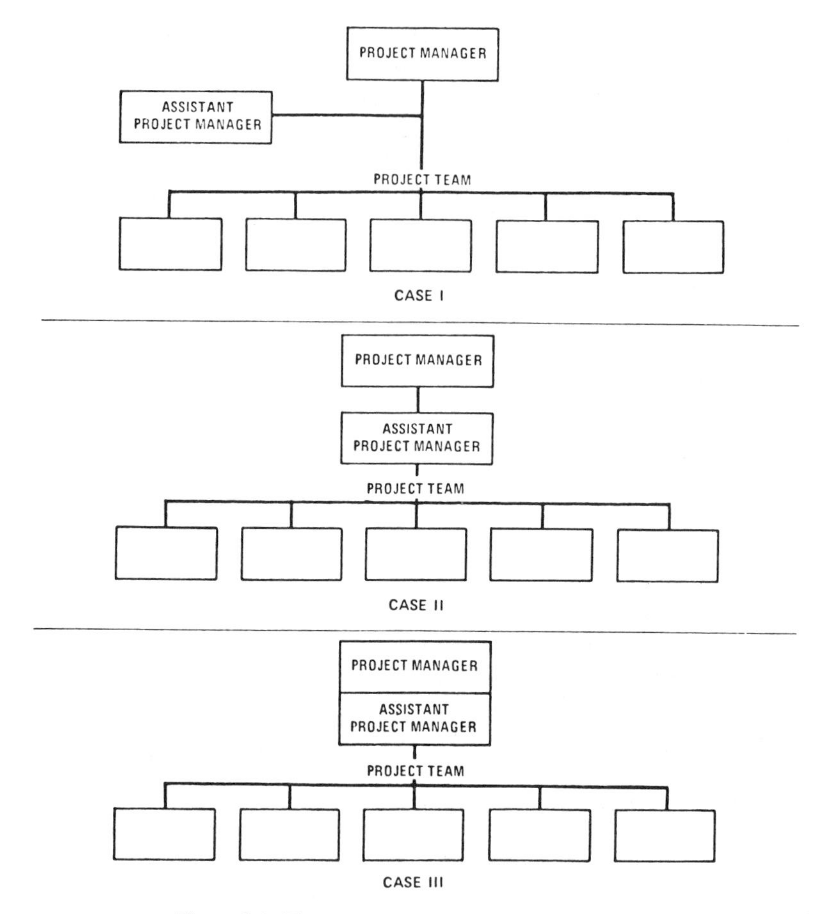


Figure 5-1. Elementary organizational arrangements.

problems. It provides only a minimum opportunity for training the Assistant Project Manager to take over the role of a Project Manager.

In the second case, the Assistant Project Manager is placed in a separate box directly below the Project Manager and with all line functions passing through his office. This organization is more indicative of a strong and rigid role for the Assistant Project Manager; it encompasses all project functions and full-time backup for the Project Manager. However, because of its rigidity this organizational approach may not necessarily increase the Project Manager's span of control and could tend to make the position of Assistant Project Manager a bottleneck.

In the third case, the Assistant Project Manager is placed in the same box with the Project Manager and directly beneath him. This arrangement reflects full-time backup for the Project Manager and a stronger measure of togetherness. The opportunity for increasing the Project Manager's span of control is greater through a sharing of the many duties. At the same time, the Assistant Project Manager may give special attention to certain functional, disciplinary, or other aspects of the project.

There are obvious merits in each approach and there can be several variations of these organizational concepts. In any case, the overriding concern of management should be to design or select the organizational arrangement which will respond best to the project needs and contribute most to project and Contractor/Client Team performance.

Some variations in the role of the Assistant Project Manager can be illustrated by a brief description and discussion of their actual employment in the engineering and construction of three current power generating projects.

## ARIZONA NUCLEAR POWER PROJECT (ANPP)

This project for Arizona Public Service (and six other participating utilities) consists of three 1270-MW nuclear units (Palo Verde 1, 2, and 3) sited in south-central Arizona about 55 miles from Phoenix.

A unique feature of the project will be the use of reclaimed sewerage water from the City of Phoenix and six adjoining cities, which will be transported to the site by a large pipeline, to meet the extensive project cooling water makeup requirements.

A simplified version of the basic organization chart is shown in Figure 5-2. The water reclamation and pipeline effort is being handled by another

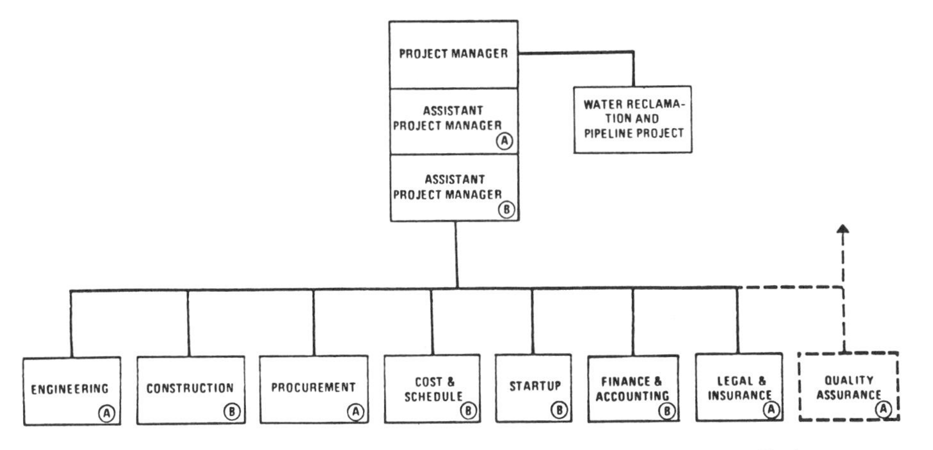


Figure 5-2. Simplified basic organization chart, Arizona Nuclear Power Project.

division of the company under the supervision of a separate Project Manager. This Project Manager reports to the ANPP Project Manager for coordination with the Prime Contract and its supporting documents and for coordination of his efforts on other contractual matters, budgets, procedures, procurement, and field support.

The scope, complexity, cost, and schedule of the total project merit the full-time attention of two Assistant Project Managers. The two were used differently in various stages of the project.

**Final Design and Construction Phase.** In this arrangement, Assistant Project Manager A, with a strong background in engineering and project field engineering, directs his primary attention to

- Engineering.
- Procurement.
- Legal and Insurance.
- Quality Assurance.

Assistant Project Manager B, with extensive experience in cost engineering, subcontract management, and planning and scheduling, including home office and field efforts, concentrates his attention on construction, start-up, finance, accounting and cost, and schedule matters.

Both Assistant Project Managers are exposed to the entire range of project operations.

**Construction Completion, Plant Start-up, and Contract Closeout.** Currently during this phase and in this arrangement, Assistant Project Manager A (with a background in engineering, business development, and office management) directs his primary attention to closeout efforts on the original Engineering-Procurement-Construction Contract which involves the supervision of Construction, Engineering, and Procurement departments.

Assistant Project Manager B, with extensive experience in start-up, concentrates his attention on completion of engineering, procurement, and construction leading to fuel load of Unit 3 and providing personnel to ANPP to support operation of Units 1 and 2.

## SAN ONOFRE UNITS 2 AND 3

This project for Southern California Edison and San Diego Gas and Electric consists of two 1100-MW nuclear units sited on the coast of California about 60 miles south of Los Angeles. The project has been exposed to

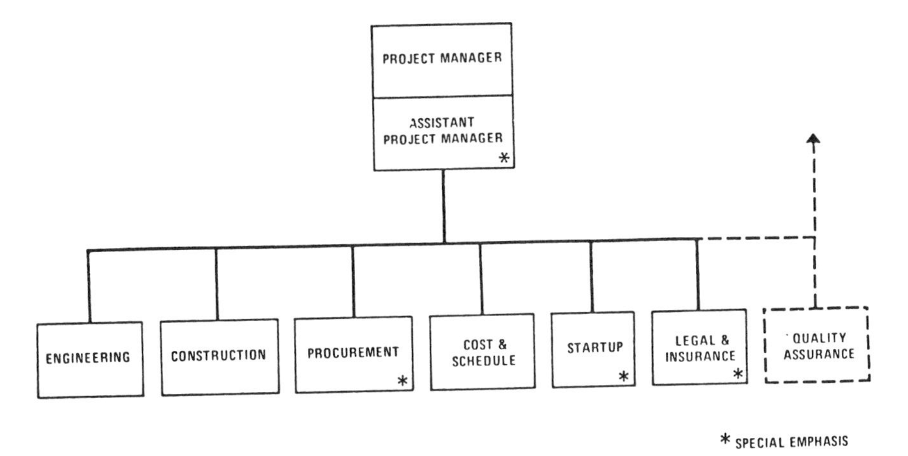


Figure 5-3. Simplified basic organization chart, San Onofre units 2 and 3.

numerous major criteria changes due to the many new regulatory guides issued by the NRC since its inception, and was exposed to extensive public scrutiny by various intervenors and the California Coastal Commission in its early design and licensing stages.

While the project site and Client and Contractor headquarters are within reasonable commuting distances of each other, there are a few significant geographical spreads in the project—for example, the turbine generators are furnished by English Electric.

A simplified version of the basic organization chart is shown in Figure 5-3. There is one very experienced Assistant Project Manager on this project who functions throughout the complete spectrum of the project, but is used extensively on procurement, start-up, and legal and insurance matters. The Assistant Project Manager has also recently been engaged in extensive restructuring/renegotiating of certain major subcontracts with companies experiencing serious financial and production difficulties which posed major problems to the project.

The Assistant Project Manager has a strong background in estimating, cost and schedule control, finance and accounting matters, and engineering administration. Nonetheless, he participates fully in all phases of the project and is a full backup for the Project Manager.

In the later stages of the project a Senior Project Manager was brought in to provide added management support during critical construction completion and start-up activities. The Senior Project Manager is in a direct line position over the Project Manager and Assistant Project Manager. With the original management team intact to direct the day-to-day operation of the project, the Senior Project Manager has the freedom to concentrate his efforts on specific problem areas and make key company, client, and supplier contacts as necessary.

This type of an arrangement shows how the project management concept and organization can be kept flexible to respond to the specific needs of the project.

## CORONADO UNITS 1 AND 2

This project for Salt River Project consists of two 350-MW coal-fired units sited in eastern Arizona. Coal sources located along an existing main line railroad are being used. There was a separate design and construction management contract for a 43-mile railroad to handle unit trains of coal from the main line to the site. The railroad effort was designed and the construction managed by another division of the company which had the contract for the power plant.

A simplified version of the basic organization chart is shown in Figure 5-4.

In this situation the primary Project Manager functions as a manager of projects. At one time he had responsibility for three projects.

The primary long-term project is the design, procurement, and construction of the Coronado Generating Station. During the early stages of Coronado, construction and start-up work were being completed on the three 700-MW coal-fired units of the Navajo Project for the same client. A separate Assistant Project Manager for the completion of Navajo worked under the general guidance of the Coronado Project Manager to provide central contact and consistency of approach with the client.

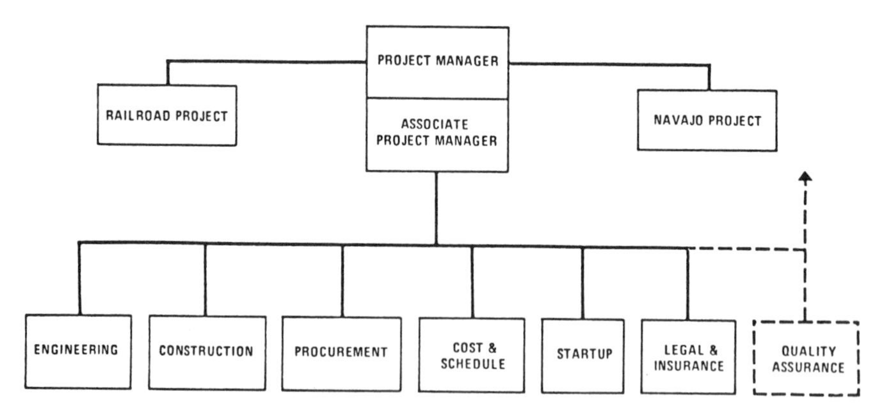


Figure 5-4. Simplified basic organization chart, Coronado units 1 and 2.

**ENGINEERING**

- KEEP CURRENT ON ALL STATUS AND PROGRESS
- PROVIDE CONTINUOUS MONITORING TO ENSURE THAT THE DESIGN IS CONSISTENT WITH COST REDUCTION GOALS
- REVIEW MAJOR STUDIES AND BID EVALUATIONS
- REVIEW AND CONTINUOUS ENGINEERING AUDIT OF PAST JOB PROBLEMS
- ATTEND DESIGN MEETINGS AND ASSIST IN CLIENT PROJECT ENGINEERING RELATIONS

**CONSTRUCTION**

- KEEP CURRENT ON ALL STATUS AND PROGRESS INCLUDING HOUSING FOR CONSTRUCTION PERSONNEL

**PROCUREMENT**

- REVIEW AND COORDINATE INSPECTION, EXPEDITING AND EQUIPMENT DELIVERIES
- PARTICIPATE IN AND CONDUCT BIDDER REVIEW OR PRE-AWARD CONFERENCES AS REQUIRED

**COST AND SCHEDULE**

- CONDUCT INITIAL REVIEW FOR PROJECT MANAGEMENT APPROVAL:
  - TRENDS
  - CHANGE NOTIFICATIONS
  - PROJECT FINANCIAL STATUS REPORT
- REVIEW, COORDINATE AND MONITOR FOR FULL IMPLEMENTATION:
  - COST AND SCHEDULE PROGRAMS (EMPHASIS ON ENG & H.O. COST CONTROL AND ENGINEERING PROGRESS REPORTING)
  - SCHEDULE PREPARATION
- CONTINUOUSLY MONITOR THE TREND PROGRAM FOR IMPLEMENTATION AND IDENTIFICATION OF COST TRENDS AND COST REDUCTION ITEMS
- COORDINATE IMPLEMENTATION OF SUGGESTED COST REDUCTION AND EFFICIENCY MEASURES
- INSURE THAT NON-REIMBURSABLE COSTS ARE BUDGETED AND CONTROLLED

**COST AND SCHEDULE (Continued)**

- PREPARE CURRENT COMMENTS FOR PROJECT FINANCIAL STATUS REPORT
- COORDINATE PREPARATION OF PFSR AND FORECASTS
- CLIENT COMMENTS ON BILLINGS

**STARTUP**

- KEEP CURRENT ON ALL STATUS

**QUALITY ASSURANCE**

- REVIEW AND COORDINATE

**LABOR RELATIONS AND SAFETY**

- REVIEW AND COORDINATE PREPARATION OF AFFIRMATIVE ACTION PROGRAM

**INSURANCE**

- REVIEW AND COORDINATE

**ADMINISTRATION AND CORRESPONDENCE**

- COMMUNICATION WITH CLIENT AS REQUIRED
- REVIEW AND COORDINATE PREPARATION AND FULL IMPLEMENTATION:
  - EXTERNAL PROCEDURES
  - INTERNAL PROCEDURES
- PREPARE MINUTES OF PROJECT REVIEW MEETINGS
- PREPARE MINUTES OF EXECUTIVE STAFF MEETINGS
- PREPARE MONTHLY REPORT
- PREPARE WEEKLY REPORT TO MANAGEMENT
- COORDINATE AND IMPLEMENT CONTRACT CHANGES INCLUDING NECESSARY COST ESTIMATES
- PROJECT TEAM SPACE ARRANGEMENTS
- USE OF COMPUTER IN FIELD OFFICE
- PUBLISH ACTION ITEMS REPORT

Figure 5-5. Associate Project Manager routine duties.

The other division of the company assigned a Project Manager to direct the design, procurement, and construction management of the railroad spur. This Project Manager also worked under the general guidance of the Coronado Project Manager in order to maintain consistency in client relations and project procedures, and also to obtain close coordination between the field activities.

The Project Manager on the Coronado Project is assisted by an Associate Project Manager who is thoroughly integrated into the basic project. The Associate Project Manager has previous management experience and a strong background in project engineering. In day-to-day operations he concentrates upon the technical engineering interface with the Client and engineering-construction interfaces. The latter effort helps to achieve proper understanding and effective operations between these two major components of the project. Typical routine duties assigned the Associate Project Manager are listed in Figure 5-5.

## SUMMARY

While we may have overlooked the role of the Assistant Project Manager in our formal discussions in the past, it is readily apparent that he can be a key man in the organization and can do much to improve the total management and performance of the project team.

The role of the Assistant Project Manager may vary considerably from project to project. There does not appear to be a universal approach which can satisfy all projects. Some of the major considerations in defining the role include the following:

- Special needs of the project.
- Client needs.
- Contractor needs.
- Capabilities and availability of the Project Manager.
- Capabilities and interests of the Assistant Project Manager.
- Geographic spread of activities.
- Various contractual arrangements and their implications.

Notwithstanding the many differences in the three examples we have cited, two very consistent factors stand out:

- All the Project Managers encourage the Assistant Project Manager to actively participate in all facets of the project.

- The experience and capabilities of the Assistant (Associate) Project Managers are being effectively concentrated on those areas of their expertise, while being exposed to the total project.

In the final analysis, the key requirement for the role of the Assistant Project Manager is to provide flexibility on the part of Contractor's management and the Client in best meeting the needs of all.

# Section III

# Organizational Strategy and Project Management

Project management is a tool for executing overall organizational strategy. Therefore it is inadequate to view project management only within the confines of the project. It must be considered within the context of the overall organization and *its* strategy.

In Chapter 6, William R. King describes the interrelationships of the various elements of business strategy. He demonstrates a method that can be used to ensure that the projects embarked on by an organization are those that are most compatible with its overall mission, strategy, and goals.

William E. Souder's Chapter 7 deals with techniques for evaluating and selecting projects. Project selection is the mechanism by which the organization ensures that it selects the "right," or the "best," projects for funding. A wide range of techniques are reviewed, each of which addresses a different objective and set of measures that may be applied. (The "strategic program evaluation" approach of Chapter 6 is another technique that may be used.)

In Chapter 8, the "owner's" role in strategically managing projects is delineated in specific terms that are as disparate as "resources" and "styles." This chapter integrates the many different dimensions that are necessarily brought into alignment if a project is to effectively contribute to the accomplishment of the organization's mission.

# 6. The Role of Projects in the Implementation of Business Strategy

William R. King*

There is a good deal of anecdotal evidence concerning business strategies that have failed because they were not implemented or because they were inappropriately implemented. Since projects and programs are the vehicles through which strategy is implemented, such failures strike at the heart of the value of project management to the organization.

In an audit of the existing and planned programs in the central research laboratory of a major diversified firm, the author found

- Programs and projects that could not be associated with any business or corporate objective or strategy.
- Programs and projects which apparently fell outside the stated mission of the corporation or the charter of the laboratory.
- Projects whose funding levels could not reasonably be justified in terms of the expected benefits to be produced.

Such observations have so frequently emanated from less formal analyses in other companies as to suggest the existence of a faulty linkage between corporate plans and strategy and the programs and projects through which they should be implemented.

---

* William R. King is University Professor in the Katz Graduate School of Business at the University of Pittsburgh. He is the author of more than a dozen books and 150 technical papers that have appeared in the leading journals in the fields of management science, information systems, and strategic planning. Among his major honors are the McKinsey Award (jointly with D. I. Cleland) for the "outstanding contribution to management literature" represented by their book *Systems Analysis and Project Management,* the IIE Book-of-the-Year Award for the first edition of this book, and designation as a fellow of the Decision Sciences Institute. Further biographical information is available in *Who's Who in the World* and *Who's Who in America.*

## THE CHOICE ELEMENTS OF CORPORATE STRATEGY

Because of the semantics jungle which exists in the area of business policy and strategy, it is necessary to rather precisely define the terms to be used. The *choice elements of corporate strategy*—those choices that must be explicitly or implicitly made in the corporate strategic planning process—are the *Organization's:*

*Mission*—the "business" that the organization is in.
*Objectives*—desired future positions on roles for the organization.
*Strategy*—the *general direction* in which the objectives are to be pursued.
*Goals*—specific targets to be sought at specified points in time.
*Programs/Projects*—resource-consuming sets of activities through which strategies are implemented and goals are pursued.
*Resource Allocations*—allocations of funds, manpower, etc., to various units, objectives, strategies, programs, and projects.

These informal definitions are meant to provide a common framework for communication rather than to define the "correct" terminology. Various firms may use different terminology, but none can escape the need to make choices of each variety. (These strategic choice elements are treated in more detail elsewhere.*)

Most organizations conduct planning processes which are aimed at explicitly choosing all or some of these strategic choice elements. However, many firms fail to deal with all of the choice elements in the detail and specificity which each deserves.

Often, for instance, missions are dealt with implicitly, as in the case of the firm that responds to the mission concept by stating their mission to be: "We make widgets." Such a product-oriented view of the organization's business ignores new market opportunities and, perhaps, the firm's generic strengths. It is these opportunities and strengths which form the most likely areas for future success. Thus, it is these opportunities and strengths, rather than the current product line, which should define the mission.

Strategies are almost always explicitly chosen by firms, but often strategies are thought of in output, rather than input, terms. In such instances, strategies may be described in terms of expected sales and profits rather

---

* See William R. King and David I. Cleland, *Strategic Planning and Policy*, New York, Van Nostrand Reinhold, 1978, Chapters 3 and 6–9, from which portions of this material are adapted with the permission of the publisher.

than in terms of strategic directions such as product redesign, new products, or new markets.

Thus, the elements of strategic choice are inescapable in the sense that the avoidance of an explicit choice about any of the elements means that it is chosen implicitly. However, many firms make poor or inappropriate choices, both explicitly and implicitly, because they do not have a clear awareness of the relationships among the strategic choice elements and their innate interdependence.

## RELATIONSHIPS AMONG THE STRATEGIC CHOICE ELEMENTS

One of the most important conditions for the effective implementation of plans has to do with the relationships among the strategic choice elements. If these relationships are well defined and carefully analyzed and conceived, the plan is likely to be implemented. If they are not, the plan is likely to be a voluminous document that requires substantial time and energy to prepare, but which is filed on the shelf until the next planning cycle commences. Indeed, many plans are so treated precisely because they do not carefully spell out the relationships among various strategic choice elements and therefore do not provide the appropriate information that is necessary to guide the many decisions which must be made to implement the plan and to develop and manage the projects and programs which are the operational essence of the plan.

Figure 6-1 shows the elements of strategic choice in the form of a triangle which illustrates that the mission and objectives are the highest-level elements. They are supported by the other elements—the strategies, goals, programs, and projects. The strategic resource allocations underlie each of these elements.

Figure 6-1. Relationship of strategic choice elements.

Figure 6-2 shows an illustration of these concepts in terms of a business firm. The mission chosen is that of "supplying system components to a worldwide nonresidential air conditioning market." Note that while this mission statement superficially appears to be product-oriented, it identifies the nature of the product (system components), and the market (worldwide nonresidential air conditioning) quite specifically. By exclusion, it guides managers in avoiding proposals for overall systems and strategies that would be directed toward residential markets. However, it does identify the world as the company's territory and (in an elaboration not shown here) defines air conditioning to include "air heating, cooling, cleaning, humidity control, and movement."

Supporting the base of the triangle are strategies, goals, and programs. The firm's strategies are stated in terms of a three-phase approach. First, the company will concentrate on achieving its objectives through existing

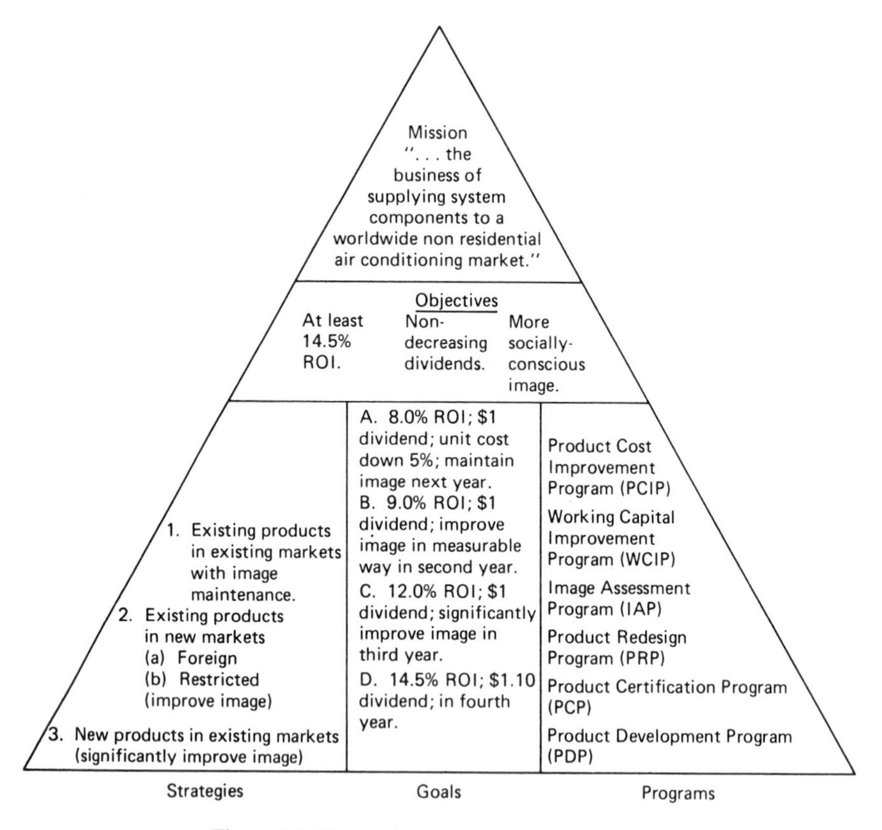


Figure 6-2. Illustrative strategic choice elements.

products and markets while maintaining its existing image. Then, it will give attention to new markets for existing products, foreign and restricted, while improving the company's image. "Restricted" markets may be thought of as those that require product-safety certification before the product can be sold in that market. Finally, it will focus on new products in existing markets while *significantly* improving its image.

Clearly, this is a staged strategy; one that focuses attention first on one thing and then on another. This staging does not imply that the first strategy element is carried through completely before the second is begun; it merely means that the first element is given primary and earliest attention, then the second and third in turn. In effect, the first element of the strategy has its implementation *begun* first. This will be made more clear in terms of goals and programs.

At the right base of the triangle, a number of the firm's programs are identified. Each of these programs is made up of a variety of projects or activities. Each program serves as a focus for various activities having a common goal. For instance, in the case of the Product Cost Improvement Program, the associated projects and activities might be as follows:

- Quality Control Project.
- Production Planning Improvement Project.
- Production Control System Development Project.
- Plant Layout Redesign Project.
- Employee Relations Project.

All of these projects and activities are focused toward the *single* goal of product cost improvement.

In the case of the Working Capital Improvement Program, the various projects and activities might include a "terms and conditions" study aimed at revising the terms and conditions under which goods are sold, an inventory reduction project, etc. Each of the other programs would have a similar collection of projects and activities focused on some single well-defined goal.

The goals are listed in the middle-lower portion of the triangle in Figure 6-2. Each goal is stated in specific and timely terms related to the staged strategy and the various programs. These goals reflect the desire to attain 8.0% ROI (a step along the way to the 14.5% objective) next year, along with a $1 dividend (the current level), a unit cost improvement of 5%, while maintaining image. For subsequent years, the goals reflect a climb to 14.5% ROI, a steady and then increasing dividend, and an increasing and measurable image consistent with the staged strategy that places image improvements later in the staged sequence. This is also consistent with the program structure, which includes an "Image Assessment Program," a program designed to develop methods and measures for quantitatively assessing the company's image.

Figure 6-3 shows the same elements as does Figure 6-2, with each being indicated by number, letter, or acronym. For instance, the block labeled 1 in Figure 6-3 represents the first stage of the strategy in Figure 6-2, the letter A represents next year's goal, etc.

The arrows in Figure 6-3 represent *some* illustrative relationships among the various objectives, programs, strategy elements, and goals. For instance, the arrows a,b, and c reflect direct relationships between specific timely goals and broad timeless objectives:

a. A, next year's goals primarily relate to the objective of nondecreasing dividends.
b. B, the second year's goals relate to the "more socially conscious image" objective.
c. D, the quantitative ROI figure is incorporated as a goal in the fourth year.

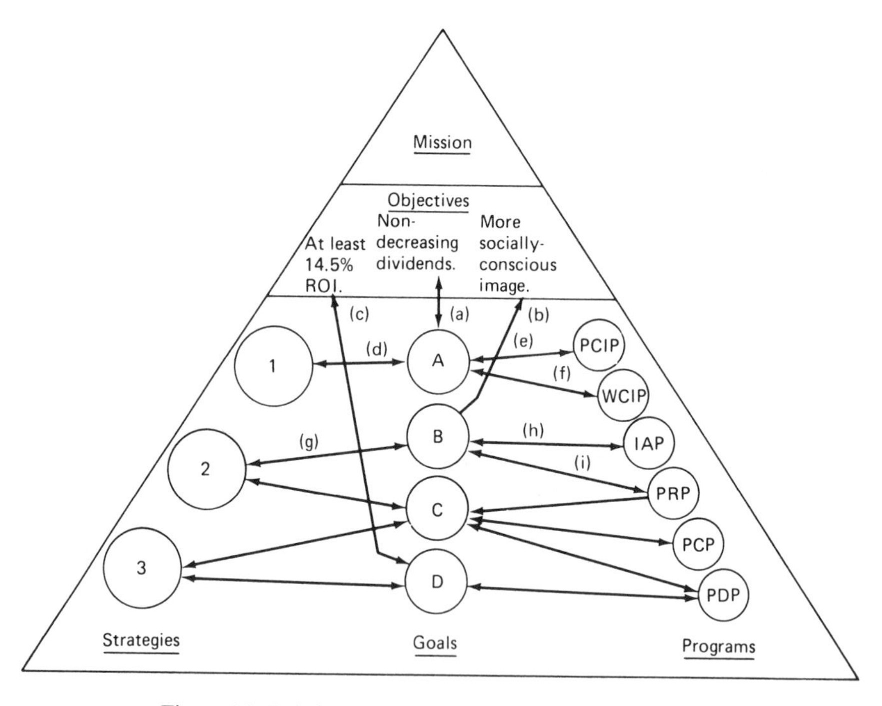


Figure 6-3. Relationships among illustrative choice elements.

Of course, each year's goals relate implicitly or explicitly to all objectives. However, these relationships are some of the most direct and obvious.

Similarly, arrow d in Figure 6-3 relates the first year's goals to the first element of the overall strategy in that these goals for next year are to be attained primarily through the strategy element involving "existing products in existing markets." However, arrows e and f also show that the Product Cost Improvement Program (PCIP) and the Working Capital Improvement Program (WCIP) are also expected to contribute to the achievement of next year's goals.

The second year's goals will begin to reflect the impact of the second strategy element (existing products in new markets) as indicated by arrow g in Figure 6-3. The effect of the Product Redesign Program (PRP) is also expected to contribute to the achievement of these goals (arrow i) as is the Image Assessment Program (IAP) expected to provide an ability to measure image by that time. The other arrows in Figure 6-3 depict other rather direct relationships whose interpretation is left to the reader.

From this figure, relationships among the various strategic decision elements can be seen:

1. Goals are specific steps along the way to the accomplishment of broad objectives.
2. Goals are established to reflect the expected outputs from strategies.
3. Goals are directly achieved through programs.
4. Strategies are implemented by programs.

Thus, the picture shown in Figure 6-3 is that of an interrelated set of strategic factors that demonstrate *what* the company wishes to accomplish in the long run, *how* it will do this in a sequenced and sensible way, and *what performance levels* it wishes to achieve at various points along the way.

## STRATEGIC PROGRAM EVALUATION

Figures 6-1 to 6-3 make it clear that the various elements of strategic choice are mutually supportive. However, there remains the question of how this high degree of interdependence can be effectively attained.

Certainly, an understanding of the logical relationships that are depicted in these figures should itself lead to better subjective choices in the planning process. However, while subjective judgment can be appropriately applied at the higher levels of strategic choice because these elements are tractable, it is an inadequate basis for choice at the lower levels of program/project selection and funding.

In other words, no formal techniques are needed in choosing among alternative missions and objectives because these choices must inherently be made on a primitive basis of the personal values and goals of management and other stakeholders. At this level, there are only a few viable options from which choices must be made.

At the level of programs, projects, and resource allocations, quite the opposite is the case. There are many contenders and combinations of contenders to be considered. Thus, some formal approach may be useful. Indeed, such an approach is not only practically useful, but it forms the integrating factor in the array of strategic choice elements.

The integrating factor is a strategic program evaluation approach *which directly utilizes the results of the higher-level strategic choices to evaluate alternative programs, projects, and funding levels.* "Project selection" approaches are well known and widely used in industry for the selection of engineering projects, R&D projects, and new product development projects. However, if program/project evaluation is to be the key link in unifying the array of organizational strategic choice elements, the evaluation framework must itself be an integral element of the strategic plan.

Thus, potential projects and programs must be "filtered" through the application of strategic criteria that are based on the higher-level choices that have previously been made—the organization's mission, objectives, and strategies. The output of this filtering process is a set of rank-ordered project and program opportunities that can serve as a basis for the allocation of resources.

Other important criteria must come into play in implementing this evaluation process. These criteria are those that are *implicit* in a good specification of the organization's mission, objectives, and strategy. However, they must be *specifically addressed* if program and projects are to truly reflect corporate strategy. These criteria are as follows:

1. Does the opportunity take advantage of a *strength* that the company possesses?
2. Correspondingly, does it avoid a dependence on something that is a *weakness* of the firm?
3. Does it offer the opportunity to attain a *comparative advantage* over competitors?
4. Does it contribute to the *internal consistency* of the existing projects and programs?
5. Does it address a *mission-related opportunity* that is presented by the evolving market environment?
6. Is the level of *risk* acceptable?
7. Is it consistent with the established *policy guidelines?*

### A Strategic Program Evaluation Illustration

A strategic program/project evaluation framework based on these criteria is shown as Table 6-1. In the leftmost column of the table is a set of evaluation criteria that relates to the example in Figures 6-2 and 6-3. The body of the table shows how a proposed new program to begin manufacturing of system components in Europe might be evaluated.

The "criteria weights" in the second column of the table reflect their relative importance and serve to permit the evaluation of complex project characteristics within a simple framework. A base weight of 20 is used here for the major criteria related to mission, objectives, strategy, and goals. Weights of 10 are applied to the other criteria.

Within each major category, the 20 "points" are judgmentally distributed to reflect the relative importance of subelements or some other characteristic of the criterion. For instance, the three stages of strategy and the four subgoals are weighted to ensure that earlier stages and goals are treated to be more important than later ones. This implicitly reflects the *time value of money* without requiring a complex discounting calculation.

The first criterion in Table 6-1 is the "fit with mission." The proposal is evaluated to be consistent with both the "product" and "market" elements of the mission and is thereby rated to be "very good," as shown by the 1.0 entries in the upper left.

In terms of "consistency with objectives," the proposal is rated to have a 20% chance of being "very good" in contributing to the ROI element of the objectives (see Figure 6-2), a 60% chance of being "good," and a 20% chance of being only "fair," as indicated by the likelihoods entered into the third row of the table. The proposed project is rated more poorly with respect to the "dividends" and "image" elements.

The proposal is also evaluated in terms of its expected contribution to each of the three stages of the strategy as outlined in Figure 6-2. In this case, the proposed project is believed to be one which would principally contribute to stage 2 of the strategy. (Note that only certain assessments may be made in this case since the stages are mutually exclusive and exhaustive.)

The proposal is similarly evaluated with respect to the other criteria.

The overall evaluation is obtained as a weighted score that represents the sum of products of the likelihoods (probabilities) as the 8, 6, 4, 2, 0 arbitrary level weights that are displayed at the top of the table. For instance, the "consistency with objectives—ROI" expected level weight is calculated as

$$0.2(8) + 0.6(6) + 0.2(4) = 6.0$$

Table 6-1. An Example of Strategic Program Evaluation.

| PROGRAM PROJECT EVALUATION CRITERIA | | CRITERIA WEIGHTS | VERY GOOD (8) | GOOD (6) | FAIR (4) | POOR (2) | VERY POOR (0) | EXPECTED LEVEL SCORE | WEIGHTED SCORE |
|---|---|---|---|---|---|---|---|---|---|
| Fit with mission | Product | 10 | 1.0 | | | | | 8.0 | 80 |
| | Market | 10 | 1.0 | | | | | 8.0 | 80 |
| Consistency with objectives | ROI | 10 | 0.2 | 0.6 | 0.2 | | | 6.0 | 60 |
| | Dividends | 5 | | 0.2 | 0.6 | 0.2 | | 4.0 | 20 |
| | Image | 5 | | | 0.8 | 0.2 | | 3.6 | 18 |
| Consistency with strategy | Stage 1 | 10 | | | | | 1.0 | 0 | 0 |
| | Stage 2 | 7 | 1.0 | | | | | 8.0 | 56 |
| | Stage 3 | 3 | | | | | 1.0 | 0 | 0 |
| Contribution to goals | Goal A | 8 | | | | | 1.0 | 0 | 0 |
| | Goal B | 6 | 0.8 | 0.2 | | | | 7.6 | 45.6 |
| | Goal C | 4 | | 0.8 | 0.2 | | | 5.6 | 22.4 |
| | Goal D | 2 | | | | | 1.0 | 0 | 0 |
| Corporate *strength* base | | 10 | | | | 0.8 | 0.2 | 1.6 | 16 |
| Corporate *weakness* avoidance | | 10 | | | | 0.2 | 0.8 | 0.4 | 4 |
| *Comparative advantage* level | | 10 | 0.7 | 0.3 | | | | 7.4 | 74 |
| Internal consistency level | | 10 | 1.0 | | | | | 8.0 | 80 |
| Mission-related opportunity | | 10 | 1.0 | | | | | 8.0 | 80 |
| Risk level acceptability | | 10 | | | | 0.7 | 0.3 | 1.4 | 14 |
| Policy guideline consistency | | 10 | | | 1.0 | | | 4.0 | 40 |
| Total score | | | | | | | | | 690 |

This is then multiplied by the criterion weight of 10 to obtain a weighted score of 60. The weighted scores are then summed to obtain an overall evaluation of 690.

Of course, this number in isolation is meaningless. However, when various programs and projects are evaluated in terms of the same criteria, their overall scores provide a reasonable basis for developing a ranking of projects that reflects their consistency with strategy. Such a ranking can be the basis for resource allocation since the top-ranked program is presumed to be the most worthy, the second-ranked is the next most worthy, etc.

## SUMMARY

The strategic program evaluation framework that is developed and demonstrated here provides the integrating factor that is necessary if strategic plans are to be effectively implemented. The critical element of the evaluation approach is its use of criteria which ensure that programs will be integrated with the mission, objectives, strategy, and goals of the organization as well as criteria that reflect critical elements of strategy such as business strengths, weaknesses, comparative advantages, internal consistency, opportunities, and policies.

# 7. Selecting Projects That Maximize Profits**

William E. Souder*

## INTRODUCTION

Project selection is one of the most important decisions managers make. No matter how well it is managed, a poorly chosen project can never be a winner. If a better project could have been selected, then a real opportunity has been forgone and real profits have been lost.

Most organizations will never see these lost profits. This is especially true if the project is well managed, achieves its target dates, and does not overrun its budget. Any thoughts about lost opportunities will be obscured by the euphoria that accompanies a job well done. However, the passing of time will dramatically distinguish those organizations that select the best projects. They will survive.

Thus, it is vitally important to the long-term future of every organization to select only the very best projects. Inferior projects must be identified and screened out as early as possible in the decision-making processes.

This chapter presents the state of the art in project selection techniques. All of these techniques are relatively inexpensive to use, compared with the benefits they provide.

---

** Selected portions of the material herein originally appeared in Wm E. Souder, "A System for using R&D Project Evaluation Methods," *Research Management* Vol 21, No 5, September 1978, pp. 29–37, by permission.

* William E. Souder is Professor of Industrial Engineering and Engineering Management, and Director of the Technology Management Studies Institute, Department of Industrial Engineering, University of Pittsburgh, Pittsburgh, Pennsylvania 15261.

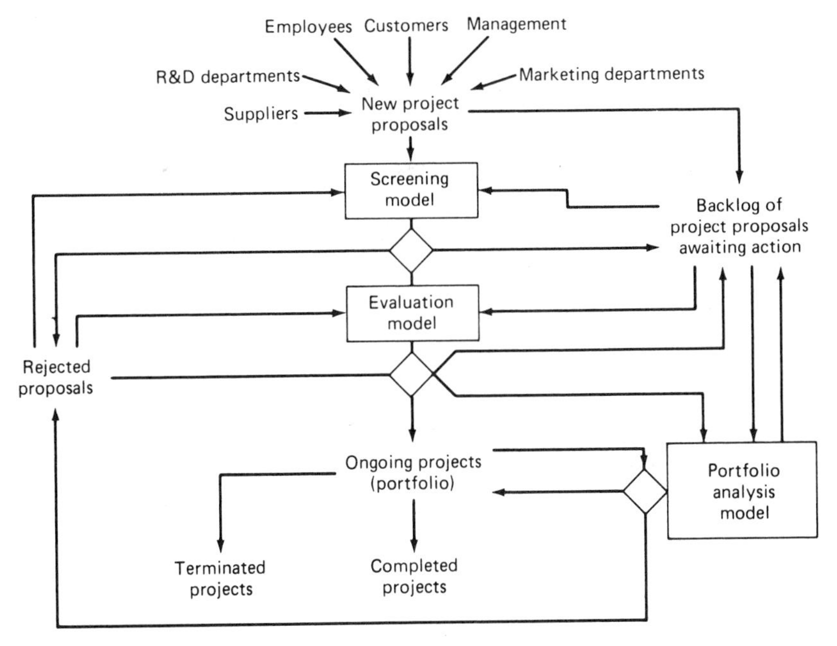


Figure 7-1. Illustration of a project selection decision process.

## PROJECT SELECTION

Figure 7-1 depicts the range of project selection activities that normally occurs continuously in most organizations. Screening models provide useful preliminary information for distinguishing candidate projects, on the basis of a few prominent criteria. Evaluation models provide a more rigorous and comprehensive analysis of candidates which survive the screening model. Portfolio models can be used to determine an optimum budget allocation among those projects which survive the evaluation model.

Projects that survive these models may be backlogged to await the release of critical manpower or other resources. Backlogged projects will normally be retrieved at some later point in time; rejected projects will not. However, new information or changed circumstances may suddenly make a previously rejected project more attractive, or may cause a previously backlogged project to be rejected.

Thus, project selection processes are very dynamic. As time passes, screening, evaluation, and portfolio decisions may be repeated many times in response to changing information states, changes in the available

Table 7-1. Example of a Profile Model

| CRITERIA OR REQUIREMENTS | EXTENT TO WHICH PROJECTS X AND Y MEET THE CRITERIA | | |
|---|---|---|---|
| | High | Medium | Low |
| Reliability | X | | Y |
| Maintainability | Y | X | |
| Safety | | X | Y |
| Cost-Effectiveness | X | Y | |
| Durability | X | | Y |

X = project X's score  Y = project Y's score

resources and funds, changes in project achievements, or the arrival of new project proposals.

## SCREENING MODELS

### Profile Models

An example of a profile model is shown in Table 7-1. Note that the ratings are qualitative in nature. No numerical assessments are made. Rather, the project proposals are compared on the basis of a subjective evaluation of their attributes. These evaluations could be done by one individual or by group consensus. Alternatively, the profiles developed by several informed individuals could be compared (1, 2).

Profile models are simple and easy to use. They display the project characteristics and ratings in such a way that they are easily communicated and readily visualized. For instance, in Table 7-1 it is apparent at a glance that project X is generally a high performer, superior to project Y on all the criteria but one.

On the other hand, a profile model does not tell us anything about the trade-offs among the criteria. For example, the profile model in Table 7-1 does not tell us if the high performances of project X on reliability, cost-effectiveness, and durability compensate for its medium performances on maintainability and safety. Thus, there is no way to get a single overall score or rating for each project.

### Checklists

Table 7-2 shows an example of a checklist. This type of model assumes that the decision maker can distinguish between several finite levels of the

Table 7-2. Example of a Checklist

| CRITERIA OR REQUIREMENTS | TOTAL SCORE | CRITERION SCORES[a] −2 | −1 | 0 | +1 | +2 |
|---|---|---|---|---|---|---|
| *Project X* | +5 | | | | | |
| Reliability | | | | | | ✓ |
| Maintainability | | | | ✓ | | |
| Safety | | | | ✓ | | |
| Cost-Effectiveness | | | | | | ✓ |
| Durability | | | | | ✓ | |
| *Project Y* | −2 | | | | | |
| Reliability | | ✓ | | | | |
| Maintainability | | | | | | ✓ |
| Safety | | | ✓ | | | |
| Cost-Effectiveness | | | | ✓ | | |
| Durability | | | ✓ | | | |
| *Project Z* | +5 | | | | | |
| Reliability | | | | | | ✓ |
| Maintainability | | | | | | ✓ |
| Safety | | | ✓ | | | |
| Cost-Effectiveness | | | | | ✓ | |
| Durability | | | | | ✓ | |

[a] *Scoring Scale:*
+2 = Best possible performance
+1 = Above average performance
0 = Average performance
−1 = Below average performance
−2 = Worst possible performance

criteria or requirements (3, 4). Each candidate proposal or project is then subjectively evaluated by the decision maker and assigned a criterion score on each requirement. The criterion score is ascertained from a predesignated scoring scale that translates subjective evaluations into numerical scores. A total score is obtained for each project by summing its criterion scores. In general, for a checklist model

$$T_j = \sum_i s_{ij} \tag{1}$$

Here, $T_j$ is the total score for the $j^{th}$ project and $s_{ij}$ is the score for project $j$ on the $i^{th}$ requirement or criterion.

Checklist models improve on profile models by providing both a graphic profile of check marks and an overall total score for each candidate project. An analysis of target achievements and a comparison of several candidate projects is facilitated by the total scores. For instance, a total score of +2 or greater may be specified as a cut-off point for acceptable proposals. Projects could be priority classified by specifying total score ranges, for example, $T_j > +3$ is a high-priority project, $+1 \leq T_j \leq +3$ is a medium-priority project, etc.

## Scoring Models

It is a short step from checklist models to scoring models. In a scoring model, each of $j = 1, \ldots, n$ candidate projects are scored on each of $i = 1, \ldots, m$ performance requirements or criteria. The criterion scores for each project are then combined with their respective criterion importance weights $w_i$ to achieve a total score $T_j$ for each project. Projects may then be ranked according to their $T_j$ values.

For example, a simple additive scoring model would be

$$T_j = \sum_i w_i s_{ij} \tag{2}$$

where $s_{ij}$ is the score for project $j$ on the $i^{th}$ criterion, and $w_i$ is the criterion weight. This model is illustrated in Table 7-3.

The influence of the weights becomes apparent if one compares the results for the Weighted Scores in Table 7-3 with the results for the Total Scores in Table 7-2. The Criterion Scores in Table 7-3 contain the same information as the Criterion Scores in Table 7-2. The difference is simply a scale transformation; each Criterion Score in Table 7-3 is +3 larger than its counterpart in Table 7-2. The Weighted Scores in Table 7-3 show that projects X and Z do indeed differ. The checklist model (Table 7-2) did not show any difference between the Total Scores for these two projects. Scoring models are more accurate because they take the trade-offs between the criteria into account, as defined by the criterion weights (2, 4, 5).

## Frontier Models

Figure 7-2 illustrates the outputs from a frontier model for seven different projects. The projects are plotted in such a way as to show their relative risks and returns. "Risk" expresses the project's chances of failure. This may be measured as $1 - p$, where $p$ is the project's probability of success. Or it may be measured in terms of the likelihood that the project will *not* achieve some desired level of output, profit, etc. "Return" expresses the project's anticipated profits, sales, or some other measure of value which the decision maker wishes to use.

Table 7-3. Example of an Additive Scoring Model

| CRITERION, $i$ | CRITERION WEIGHT, $w_i$ | × | CRITERION SCORE,* $s_{ij}$ | = | WEIGHTED SCORE |
|---|---|---|---|---|---|
| *Project X:* | | | | | |
| Reliability | 4 | | 5 | | 20 |
| Maintainability | 2 | | 3 | | 6 |
| Safety | 3 | | 3 | | 9 |
| Cost-Effectiveness | 5 | | 5 | | 25 |
| Durability | 1 | | 4 | | 4 |
| | | | | $T_1$ = | 64 |
| *Project Y:* | | | | | |
| Reliability | 4 | | 1 | | 4 |
| Maintainability | 2 | | 5 | | 10 |
| Safety | 3 | | 2 | | 6 |
| Cost-Effectiveness | 5 | | 3 | | 15 |
| Durability | 1 | | 2 | | 2 |
| | | | | $T_2$ = | 37 |
| *Project Z:* | | | | | |
| Reliability | 4 | | 5 | | 20 |
| Maintainability | 2 | | 5 | | 10 |
| Safety | 3 | | 2 | | 6 |
| Cost-Effectiveness | 5 | | 4 | | 20 |
| Durability | 1 | | 4 | | 4 |
| | | | | $T_3$ = | 60 |

*Scale: 5 = Excellent, . . . , Poor = 1

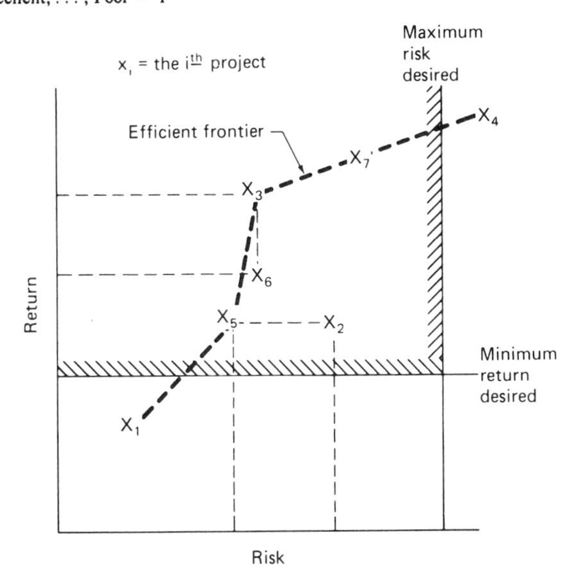


Figure 7-2. Illustration of a frontier model.

The efficient frontier in Figure 7-2 tracks the path of the most efficient return/risk ratios. For example, project 5 (denoted as $X_5$ in Figure 7-2) is more return/risk efficient than project 2 (denoted as $X_2$). Project 5 has the same return as project 2, but it has a lower risk level. Similarly, project 3 is more return/risk efficient than project 6 because of its higher return at the same risk level as project 6. The maximum desired risk and the minimum desired return levels established by the organization are also depicted in Figure 7-2. Acceptable projects must fall in the region formed by these boundaries. Thus, Figure 7-2 shows that a decision maker should accept projects 3, 5, and 7 and reject the others.

Frontier models are often very useful for examining returnrisk tradeoffs within the organizational objectives. For instance, Figure 7-2 shows that the high-risk and high-return project 4 is ruled out by its high risk level. Yet its incremental return/risk ratio is the same as the acceptable projects 3 and 7. (All of these projects lie on the same line.) Thus, the decision maker may want to make an exception and retain project 4 for further study and analyses.

Frontier models may be used to indicate the need for greater diversification in idea generation and project proposals (3, 5). For example, Figure 7-2 shows that the acceptable projects are primarily of the medium- to high-risk variety. Whether or not the portfolio ought to be more diversified must be resolved on the basis of the organization's goals and objectives. A frontier model can only point out trends and situations for further analysis (3, 6, 7).

#### Using Screening Models

Screening models are very useful for weeding out those projects which are the least desirable. Since screening models are quick and inexpensive to use, they can economize on the total evaluation efforts by reducing the number of projects to be further evaluated. Because they require a relatively small amount of input data, they can be used where the projects are not well understood or where a minimum of data are available.

However, screening models don't provide much depth of information. And they usually are not sensitive to many of the finer distinctions between the projects. Rather, screening models are like a coarse sieve that provides a partial separation but permits some undesirables to pass through. Thus, screening models can be very useful for some applications. But the decision maker should not expect them to provide a comprehensive or complete analysis.

## EVALUATION MODELS

### Economic Index Models

An index model is simply a ratio between two variables, and the index is their quotient. Changing the values of the variables changes the value of the quotient, or the index.

An example of a commonly used index model is the return on investment (ROI) index model

$$\text{ROI Index} = \frac{\sum_i R_i/(1 + r)^i}{\sum_i I_i/(1 + r)^i} \tag{3}$$

where $R_i$ is the net dollar returns expected from the project in the $i^{th}$ year, $I_i$ is the investment expected to be made in the $i^{th}$ year, and $r$ is an interest rate. The numerator of equation 3 is the present worth of all future revenues generated by the project, and the denominator is the present worth cost of all future investments.

Some other examples of index models are shown in Table 7-4. Ansoff's model uses both dollar values and index numbers as input data. The index numbers $T$ and $B$ are judgments. Olsen's index is a variation on Ansoff's index that uses all dollar input data. Viller's index is a kind of return on investment model, discounted by the compound likelihood of the project's success. Disman's index looks at the expected earnings over and above the cost to complete the project (2, 5).

Table 7-4. Examples of Index Models

*Ansoff's Index*

$$\text{Project Figure of Merit} = \frac{rdp(T + B)E}{\text{Total Investment}}$$

*Olsen's Index*

$$\text{Project Value Index} = \frac{rdp\ SP\ n}{\text{Project Cost}}$$

*Viller's Index*

$$\text{Project Index} = rdp\left(\frac{E - R}{\text{Total Investment}}\right)$$

*Disman's Index*

$$\text{Project Return} = rp(E - R)$$

*Key:* $r$ = the probability of research success, $d$ = the probability of development success, $p$ = the probability of market success, $T$ and $B$ are respective indexes of technical and business merit, $E$ = the present worth of all future earnings from the project, $S$ = annual sales volume in units, $P$ = unit profit, $n$ = number of years of product life, $R$ = present worth cost of research and development activities to complete the project

The single-number index or score that is produced by an index model can be used to rate and rank candidate projects. An example of the use of an index model is shown in Table 7-5. The index model is

$$V = \frac{P \times R}{C} \tag{4}$$

where $V$ is the index. Four projects are evaluated using this model, and their relative rankings on the index $V$ are shown in the last column of Table 7-5. Two projects, project 4 and project 5, are tied for first place in the rankings.

These hypothetical results point up some of the weaknesses of index models. One such weakness is the implicit trade-offs that often occur. For example, in computing the $V$ index, project 5's lower cost compensates for its lower probability of success. This is why project 5 is as good as project 4 on the $V$ index. However, any decision maker who wishes to avoid high risks would never rank project 5 as high as project 4. Note that project 5 has a risk of failure of $1 - P_5 = 1 - .4 = .6$. In fact, instead of ranking it first, the risk-averse decision maker might completely eliminate project 5 from any consideration at all. Thus, the index model in Table 7-5 may be completely inappropriate for some decision makers. It could lead them to make completely wrong decisions relative to their objectives.

This example shows that all index models should be carefully examined for their internal trade-offs. Unless the trade-offs are representative of those the decision maker would actually be willing to make, the model is inappropriate.

Another weakness of many index models is the sensitivity of the index to changes in some of their parameters. As an illustration, let us examine what happens to the $V$ index as one goes from project 4 to project 1 in Table 7-5. The return increases by 50% (from \$80,000 to \$120,000). The

Table 7-5. Example of the Use of an Index Model

| | RETURN ($R$) | COST ($C$) | PROBABILITY OF SUCCESS ($P$) | $V = \frac{P \times R}{C}$ | RANKING |
|---|---|---|---|---|---|
| Project 4 | \$ 80,000 | \$2,000 | .7 | 28 | 1st } tie |
| Project 5 | 70,000 | 1,000 | .4 | 28 | 1st |
| Project 1 | 120,000 | 2,000 | .2 | 12 | 2nd |
| Project 3 | 10,000 | 1,000 | .7 | 7 | 3rd |
| Project 2 | 10,000 | 1,000 | .3 | 3 | 4th |

risk goes from $1 - P_4 = .3$ to $1 - P_1 = .8$, for a 167% increase. Yet the $V$ index falls by only 57%: from $V_4 = 28$ to $V_1 = 12$. Thus, these analyses show that this index model is relatively insensitive to risks. In fact, this is a biased model; it is biased toward obscuring risks.

Still another weakness of index models lies in their inability to consider multiple objectives. Because of this, an index model may be inappropriate. For example, suppose that the decision maker also wishes to diversify the portfolio, in addition to achieving high $V$ values. Then, the decision maker might accept project 3 (Table 7-5) because it is a relatively inexpensive way (low-cost project) to get a high-probability project. Having some high-probability projects in the portfolio may be important. This may be especially true if the high-cost and high-risk project 1 is included in the portfolio. Yet the index model ranked project 3 next to last, because it could not incorporate this other objective for diversification into its analyses.

Of course, no index model can include everything. Index models are appealing because of their simplicity and ease of use. That is, they are attractive because they don't include everything. But the decision maker should be wary; index models can be deceptively appealing. Before placing great faith in the outputs from an index model, the decision maker should make sure that the model is unbiased and appropriate.

### Risk Analysis Models

A risk analysis model provides a complete picture of the distribution of outcomes for each alternative project. An illustration of a risk analysis approach to the comparison of two candidate projects is shown in Figure 7-3. Project 1 has a most likely lifetime profit of $100 million, and project 2 has a most likely lifetime profit of $150 million. However, there is only a .4 probability that project 2 will in fact achieve the $150 million level. There is a .8 probability that project 1 will achieve the $100 million level. Project 2 provides an opportunity to achieve a larger profit than project 1. But it also carries some downside risk relative to project 1. In fact, there is a .3 probability that project 2 will yield lower profits than project 1, as shown in Figure 7-3. Given these data, a risk averter would be inclined to select project 1. Project 1 has a high chance of achieving a moderate profit, with very little chance of anything less or greater. A gambler would be more inclined to select project 2, which has a small chance at a larger profit. Thus, the risk analysis approach makes the risk-averter and gambler strategies more visible, thereby permitting a decision maker to consciously select decisions consistent with one of these chosen strategies.

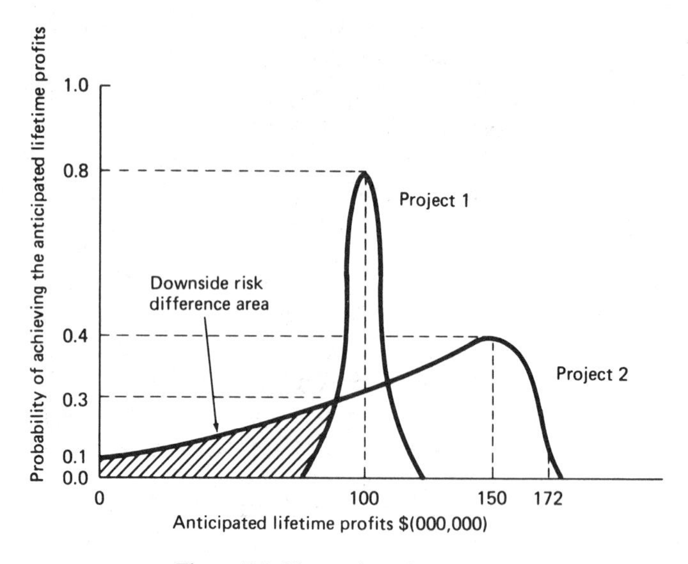


Figure 7-3. Illustration of risk analysis.

A picture like Figure 7-3 is usually not difficult to construct from a relatively small amount of data. Common methods for developing performance distributions for risk analysis include curve-fitting techniques, Monte Carlo simulation methods and modeling techniques (2, 8).

**Value-Contribution Models**

An example of a value-contribution (V-C) model is given in Table 7-6. Value-contribution models permit the decision maker to examine the degree of contribution which a project makes to the organization's hierarchy of goals.

To develop a V-C model, first list the organizational goals as a nested hierarchy. For instance, as shown in Table 7-6, there are two supergoals: short-range and long-range. Within each of these two supergoals, there are several subgoals. Within the short-range supergoal, the organization desires to achieve new product dominance and a profitability target, and to reduce their present environmental impacts. Within the long-range supergoal, the organization desires to maintain their technological state-of-art and market share.

The second step in developing a V-C model is value-weighting the goals. In the model illustrated in Table 7-6, the long-range and short-range supergoals are respectively value-weighted as $V = 60$ and $V = 40$. Note

Table 7-6. Value-Contribution Model*

| | PROJECT COSTS $ (000) | SHORT RANGE ORGANIZATIONAL GOALS (V = 60) | | | LONG RANGE ORGANIZATIONAL GOALS (V = 40) | | TOTAL VALUE-CONTRIBUTION SCORE |
|---|---|---|---|---|---|---|---|
| | | ACHIEVE NEW PRODUCT DOMINANCE (V = 30) | ACHIEVE THE PROFITABILITY TARGET (V = 20) | REDUCE ENVIRONMENTAL IMPACTS (V = 10) | MAINTAIN THE TECHNOLOGICAL STATE-OF-ART (V = 25) | MAINTAIN MARKET SHARE (V = 15) | |
| | | SCORES | | | | | |
| Project A | $100 | 30 | 20 | 5 | 15 | 5 | 75 |
| Project B | 200 | 15 | 10 | 10 | 20 | 10 | 65 |
| Project C | 150 | 25 | 10 | 5 | 15 | 10 | 65 |

*Normalized Value-Contribution:*

Project A: 75 ÷ $100,000 = $75.0 × $10^{-5}$

Project B: 65 ÷ $200,000 = $32.5 × $10^{-5}$

Project C: 65 ÷ $150,000 = $43.3 × $10^{-5}$

*Rankings:*

Project A 1st

Project C 2nd

Project B 3rd

*V = the goal value-weight

that these values must sum to 100. That is, the value-weights are determined by allocating a total of 100 points among the supergoals according to their relative importance. In Table 7-6, the value-weights indicate that the short-range supergoal is one and one-half times as important as the long-range supergoal. Within each supergoal, the total points are similarly spread among the subgoals, in such a way as to indicate their relative importance. The complete set of value-weights thus indicates the level of value contribution which a project could make. For instance, a perfect project would score 30 on "Achieve New Product Dominance." Thus, a project with a perfect contribution to all the goals would have a total value contribution score of 100 points.

The actual scaling and scoring of the candidate projects within a V-C model can be done individually or by consensus. Value-weights and scoring scales can be constructed using value assessment methods or scoring model techniques (2, 3).

In the illustration in Table 7-6, project A is short-range oriented, project B is more long-range oriented, and project C is about evenly oriented to both the long and the short range. Project A has perfect scores on the new product dominance and profitability subgoals. It has less-than-perfect scores on the other goals. But because project A is more oriented toward the short range, it contributes more towards these higher-valued subgoals. Thus, it has the highest overall total value-contribution (last column of Table 7-6). Since the total costs of the projects vary, the total value-contribution scores must be normalized by dividing them by their respective project costs. These resulting normalized value-contribution scores may then be used to rank the candidates, as shown in the lower half of Table 7-6.

V-C models permit the decision maker to think in terms of the goal-orientedness of the candidate projects, and the levels of goal achievements. V-C models may also be useful when the decision maker is trying to assemble a balanced portfolio of several projects. For instance, the results in Table 7-6 show that projects A and B together provide the maximum contributions to the short-range subgoals, and they jointly make major contributions to the long-range subgoals.

### Using Evaluation Models

Evaluation models are useful when the decision maker feels a need to have a more detailed and in-depth analysis than screening models can provide. Evaluation models permit the decision maker to make much finer discriminations between the candidate projects. On the other hand, evaluation models generally require a much greater volume and detail of data than screening models. Some evaluation models require finite numbers for life-cycle sales volumes, probabilities of success, and other parameters that may be very difficult to estimate.

In spite of the difficulties in applying them, evaluation models clearly have a place. There are times when it is difficult to make a decision without the kind of data and information that go into an evaluation model. Thus, by using the model as a guideline, the decision maker will be urged to more carefully search out and analyze the proper information. In many cases, using an evaluation model with only approximate data and rough estimates can be revealing and helpful to the decision maker.

## PORTFOLIO MODELS

### The Portfolio Problem

Table 7-7 illustrates the use of a portfolio model. The objective is to determine the best allocation of the available funds among the three alternative candidate projects. Projects A, B, and C each have four alternative funding levels: $0, $100,000, $200,000, and $300,000. The expected profits from the projects vary with these funding levels, as shown in Table 7-7. The higher funding levels result in improved products, which yield higher expected profits.

Several alternative allocations of the available $300,000 are possible. For instance, the funds can all be allocated to project C, for an expected profit return of $350 million. In this case, the other two projects would be zeroed out—no money would be spent on them. The available funds could also be spread evenly across the three projects. This would yield an expected profit return of $100 million + $120 million + $10 million = $230

Table 7-7. Illustration of a Portfolio Model

AVAILABLE FUNDS = $300,000

| ALTERNATIVE FUNDING LEVELS FOR EACH PROJECT | EXPECTED PROFITS ($M) | | |
|---|---|---|---|
| | PROJECT A | PROJECT B | PROJECT C |
| $ 0 | $ 0 | $ 0 | $ 0 |
| 100,000 | 100 | 120 | 10 |
| 200,000 | 250 | 285 | 215 |
| 300,000 | 310 | 335 | 350 |

| *Optimum Portfolio* | | *Expected Profits* |
|---|---|---|
| Project A | $100,000 | $100M |
| Project B | 200,000 | 285M |
| | $300,000 | $385M |

million. This is inferior to the above alternative of funding only project C at its upper limit. Continued searching will show that the optimum allocation is to fund project A at its $100,000 level, to fund project B at its $200,000 level, and to zero out project C. This portfolio yields the largest possible total expected profits, as shown in Table 7-7. There is no other allocation of the available funds that will achieve higher total expected profits.

It should be clear from this illustration that there are occasions when it may be more fruitful to purposely fund some projects at their lowest levels (project A) or to completely reject other projects (project C), in order to marshal funds for more productive uses (project B). The simple problem shown in Table 7-7 can be readily solved by enumerating and comparing all the alternative allocations. But when there are many candidate projects or alternative funding levels, operations research techniques and mathematical programming models are often used. These models have the advantage that various constraints may be included to insure that the portfolio is balanced for risk, or that exploratory research projects will not be disadvantaged in competing with other projects.

### Mathematical Programming Methods for Portfolio Problems

In a portfolio model, candidate projects are implicitly prioritized by the amount of funds allocated to them. The general format of all such models is

$$\max \sum_j v_j(x_j) \tag{5}$$

$$\text{subject to} \sum_j x_j \leq B \tag{6}$$

where $x_j$ is a project expenditure, $B$ is the total budget for $j = 1, \ldots, n$ candidates (projects) for funding, and the value function, $v_j(x_j)$, can be nonlinear, linear, or single-valued. In the single-valued case (one value of $v_j$ and one cost $x_j$ for each $j^{th}$ project), the portfolio model is an index model with $v_j$ as the prioritizing index.

A variety of "values" may be used in equation 5 above. Many portfolio models use expected values, so that equation 5 becomes

$$\max \sum_j v_j p_j(x_j) \tag{7}$$

where $p_j(x_j)$ is the probability of achieving $v_j$. Other portfolio models use a total score, for example, a $T_j$ "value" from a scoring model. In addition to equation 6, a typical constraint is

$$b_j^- \leq x_j \leq b_j^+ \quad (8)$$

where $b_j^-$ and $b_j^+$ are lower and upper project expenditure bounds. Also, portfolio models have been developed for multiple time periods, for example,

$$\max \sum_{ij} v_{ij}(x_{ij}) \quad (9)$$

$$\text{subject to } \sum_{ij} x_{ij} \leq B \quad (10)$$

where $i = 1, \ldots, m$ time periods.

Literally hundreds of portfolio models have been proposed in the literature. Several literature reviews are available which summarize and evaluate these models (5, 9, 10, 11, 12).

## GROUP AND ORGANIZATIONAL MODELS

### Need for Structured Group Processes

Project selection decisions that are performed in organizational and group settings are often deeply infuenced by many human emotions, desires, and departmental loyalties. Many different parties normally become involved in the project selection decision-making process, either as suppliers of decision data and information, as champions of projects, as influencers, or as decision makers. Unless a spirit of trust and openness is felt by these parties, it is not likely that essential information will be completely and openly exchanged. Each involved party must come to appreciate the interpersonal needs of the other participants, and the larger missions of the organization vis-$\alpha$a-vis their own wants. In order to achieve a total organizational consensus and commitment to a final decision, those involved must fully comprehend the nature of the proposed projects. This means that they must have a depth of factual knowledge. It also means that the parties must have a complete awareness of their own feelings, since much of the decision data are highly personal. Many decision settings fail because the participants' feelings are not crystallized and they have not fully exchanged their feelings. Thus, there is a need for a technique that bridges these behavioral gaps which are peculiar to organizational and group decision-making settings. A structured decision-making approach called the QS/NI process has been found to meet this need (2, 3).

### The QS/NI Process

Though complex psychometric phenomena underly it (1, 2, 3), the mechanics of the Q-sorting (QS) method are relatively simple, as outlined in Table 7-8. Using this procedure, each participant sequentially sorts the projects into five priority categories.

Each individual who "Q-sorts" a set of candidate projects does so according to his own perceptions and understandings of their relative value. The result is a kind of prioritizing of the candidate projects, according to their perceived value (1, 2).

The nominal-interacting (NI) decision process begins with a "nominal" period in which each individual in the group silently and anonymously Q-

Table 7-8. The Q-Sorting Method

| STEPS | RESULTS AT EACH STEP |
|---|---|
| 1. For each participant in the exercise, assemble a deck of cards, with the name and description of one project on each card. | Original deck |
| 2. Instruct each participant to divide the deck into two piles, one representing a high priority, the other a low priority level. (The piles need not be equal). | High level; Low level |
| 3. Instruct each participant to select cards from each pile to form a third pile representing the medium priority level. | High level; Medium level; Low level |
| 4. Instruct each participant to select cards from the high level pile to yield another pile representing the very high level of priority; select cards from the low level pile representing the very low level of priority. | V. high level; High level; Medium level; Low level; V. low level |
| 5. Finally, instruct each participant to survey the selections and shift any cards that seem out of place until the classifications are satisfactory. | |

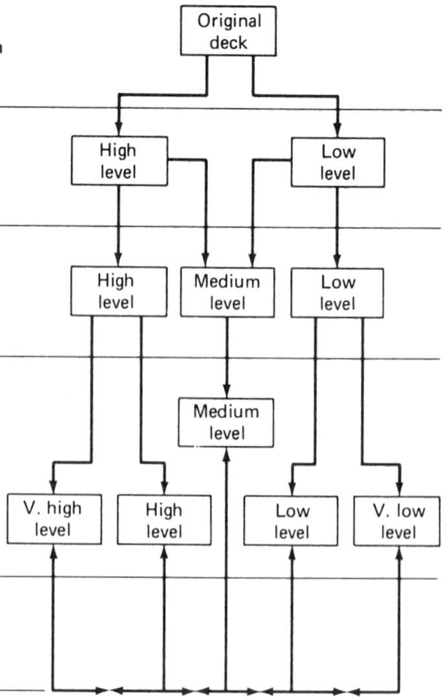

sorts the candidate projects. These results are then tabulated in a tally chart and displayed to the entire group. The tally chart focuses on the group consensus process and the agreement-disagreement statistics, without revealing who voted for what.

The group is then given an "interacting" period in which they discuss the results in the tally chart. During this period, they may share and exchange data and rationales, they may challenge each other, etc. To help guide the group in their accommodation patterns, group process measures may be taken and periodically fed back to the group. These measures generally indicate whether the group is becoming more or less cohesive and suggest what they can do to improve their team potency. It is left up to the group to decide whether or not to take these potency-improving actions (2, 13, 14).

This QS/NI sequence of an individual Q-sort in a nominal setting followed by a group discussion or interacting period can be repeated for several rounds. Experience shows that two or three rounds are needed to stimulate complete information exchange, but more than four rounds dissipates the participants. The first nominal Q-sort period permits individuals to document their own thoughts and value judgments. The subsequent first interacting period confronts the group with a diversity of opinions to be resolved. The second nominal Q-sort period permits each individual to privately restructure his or her thoughts. The second interacting period provides an opportunity to refine opinions and work toward consensus. A third nominal Q-sort period provides the environment for closure and consensus. A consensus will usually emerge as the members adopt ideas and opinions from each other, acquire more information and interpersonal understandings, or become influenced by the enthusiasm of the group. The tally chart itself is consensus-inducing for those members who identify with the group effort (1, 2, 13, 14).

Table 7-9 presents an illustration of the tally charts for two rounds of the QS/NI process, for a twenty-person group, voting on seven projects. The arrows trace the changes in the individual Q-sorts from the first to the second nominal period. Note that the degree of consensus actually declined during this part of the exercise for project G. In this case, the discussion revealed a heretofore hidden lack of information and a fundamental lack of comprehension of this project by some of the subjects. This proposal was returned to the submitter for additional work, followed by resubmittal. A consensus was reached on this resubmitted project at the end of a third round of the QS/NI process. As shown in Table 7-9, the other projects rapidly converged to a strong consensus. Note the high incidence of "block voting" or coalition voting among these data, in

Table 7-9. Illustration of Results from the QS-NI Process

| | Projects | | A | B | C | D | E | F | G |
|---|---|---|---|---|---|---|---|---|---|
| Categories | Very high priority | | (II) | I | | (IIII) | II (II) | (III) | |
| | High priority | | (卌) | 卌 | 卌 (II) | (III) (III) | (III) | I (II) | (III) II 卌 III |
| | Intermediate priority | | II IIII | (卌) (IIII) | 卌 卌 卌 | (III) | 卌 | (卌 II) (I) I | II (III) (II) |
| | Low priority | | (卌) | (II) | (III) | (III) | (IIII) | (II) | |
| | Very low priority | | (II) | (III) | | IIII | (IIII) | (II) I | |
| 1st round | K.S. test[a] | D = | .10 | .15 | .25 | .10 | .05 | .15 | .40 |
| | | p = | > .20 | > .20 | < .15 | > .20 | > .20 | > .20 | < .05 |
| | Consensus?[b] | | No | No | T | No | No | No | T |
| 2nd round | K.S. test[a] | D = | .40 | .60 | .40 | .65 | .40 | .35 | .25 |
| | | p = | < .05 | < .01 | < .05 | < .01 | < .05 | < .05 | < .15 |
| | Consensus?[b] | | Yes | Yes | Yes | Yes | Yes | Yes | No |

[a]Kolmogorov-Smirnov one-sample test of significance. The null hypothesis is that the cumulative observed distribution of votes (for that project) is not different from the cumulative rectangular distribution 4, 8, 12, 16, 20. $D$ is the largest absolute difference between the observed and rectangular distributions for any category, divided by $N$. See: Sidney Siegel, *Nonparametric Statistics* (McGraw-Hill: New York, 1965), pp. 47–52.
[b]Group consensus for a single category exists where it contains 50% more votes than any other category and $p \leqslant .10$ in the K. S. test ($p \leqslant .10$ can exist for bimodally distributed votes). $T$ indicates a tendency for consensus, in that two adjacent categories contain $\geqslant 2/3$ of the votes.

which small clusters of three to five persons are voting alike and changing their votes in a like manner. This is a common phenomenon in QS/NI exercises. The QS/NI process usually reveals a great deal about group interaction patterns and interpersonal power play strategies. Coalitions and advocate and adversary positions are usually made very visible by the QS/NI process (13, 14).

## INTEGRATED SYSTEMS

Challenged by the need to explicitly take organizational and group processes into account, project selection model builders have taken two approaches: behavioral decision aids (BDA) and decentralized hierarchical modeling (DHM). In the BDA (behavioral decision aid) approach, group process techniques are overlain on classical and/or portfolio models to make them more organizationally effective. Note that Q-sorting is a BDA. So is the NI process. Overlaying project selection models with group processes in this way changes the entire philosophy of their use. Without a BDA, project selection models are inherently viewed as *means* to decide on the *best* projects. With a BDA, the models become *aids* to intergroup *communication and interpersonal interaction*. Decisions about the best projects become obvious once all the parties have complete information and their biases and fears have been dissipated. BDAs are vehicles for achieving this.

In the DHM (decentralized hierarchical modeling) approach, the parties to the decision dialogue via computer terminals until a consensus portfolio is arrived at. As illustrated in Figure 7-4, the process is initiated by top management sending budgetary guidelines to the divisional managers.

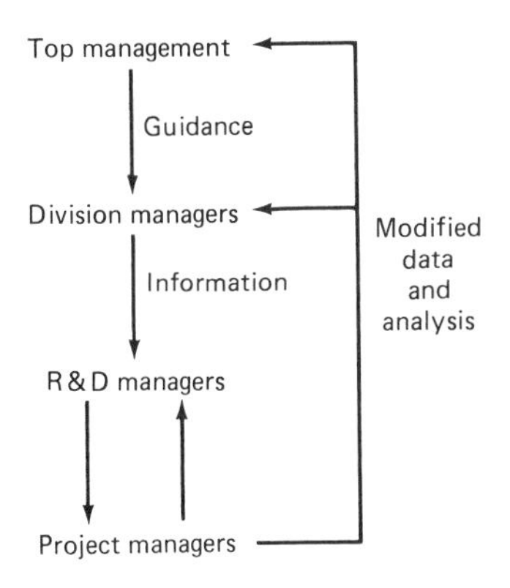


Figure 7-4. A DHM process.

Divisional managers react to these guidelines, modify them to fit their perceived circumstances, suggest prioritized program areas, and send all this information to the R&D managers. The R&D managers and the project management staff then confer to develop a proposed portfolio that fits the guidelines, and to calculate the value of their proposed portfolio vis-à-vis the given guidelines and goals. Cost-benefit calculations and project selection models are often used here. These results are sent back up by the hierarchy, with comments and analyses solicited at each stage. Top management reviews all this information and sends modified guidelines back down the hierarchy for another iteration. This process may be recycled several times, until all the parties come to consensus. The computer programs perform all the calculations, permitting all the information to be recalled in various formats (15, 17).

Because this process is carried out on a network of computer terminals, the number of face-to-face meetings is minimized and productivity increases. The process combines the features of electronic mail and interactive decision making, while still leaving the participants free to think and react in the privacy of their own offices. The process is a natural one: it follows the flow of most organizational budgeting exercises. Because it minimizes the number of face-to-face meetings, there are indications that this process avoids the social pressures and bandwagon effects that so often characterize face-to-face meetings. The DHM process appears to foster a more open and complete exchange of information, resulting in the selection of more effective projects and the enthusiastic support of these programs.

BDA and DHM systems elevate project selection models to their proper role as real decision aids. Integrated BDA/DHMs allow project selection models to be used as a *laboratory* for testing policies, sharing opinions, asking "what if" questions,and stimulating interdepartmental interactions throughout the *entire organization*.

## SUMMARY

The selection of the best projects is a very important decision problem for project managers. Today's projects entail very large commitments which have the potential to become enormous regrets if an inferior project is selected.

Techniques, models, and integrated systems are available to aid in project screening, evaluation, and selection decision making. As Table 7-10 shows, the choice of one over another will depend on the nature of the projects being assessed and the decision problem at hand (1, 2, 3, 5, 15, 16, 17).

Table 7-10. Guide to Applying Project Selection Models.
X = This model or process is appropriate here.

| | CHECK-LISTS | PRO-FILES | SCORING | FRONTIER | INDEX | RISK ANALYSIS | VALUE-CONTRI-BUTION | PORT-FOLIO | QS-NI | BCA/DHM SYSTEMS |
|---|---|---|---|---|---|---|---|---|---|---|
| Exploratory | X | X | X | | | | | | X | X |
| Applied | | | | X | X | X | | | X | X |
| Development | | | | | | X | X | X | X | X |
| Type of Decision | | | | | | | | | | |
| Problem: | | | | | | | | | | |
| Screening | X | X | X | X | X | | | | X | X |
| Prioritizing | | | | | X | X | X | X | X | X |
| Resource Allocation | | | | | | | X | X | | X |

## REFERENCES


1. Souder, W. E. "Field Studies with a Q-Sort/Nominal Group Process for Selecting R&D Projects." *Research Policy, Vol. 5*(4) (1975), pp. 172–188.
2. Souder, W. E. *Management Decision Methods for Managers of Engineering and Research* (Van Nostrand Reinhold. New York, 1980), pp. 27–36, 64–73, 137–162.
3. Souder, W. E. "A System for Using R&D Project Evaluation Methods." *Research Management, Vol. 21*(5) (1978), pp. 29–37.
4. Moore, J. R. and Baker, N. R. "An Analytical Approach to Scoring Model Design: Application to Research and Development Project Selection." *IEEE Transactions on Engineering Management,* Vol. EM-16(3) (1969), pp. 90–98.
5. Souder, W. E. "Project Selection, Planning and Control," in *Handbook of Operations Research: Models and Applications,* ed. Moder, J. J. and Elmaghraby, S. E. (Van Nostrand Reinhold. New York, 1970), pp. 301–344.
6. Markowitz, H. *Portfolio Selection* (Wiley. New York, 1960).
7. Sharpe, W. F. "A Simplified Model for Portfolio Analysis." *Management Science,* Vol. 9(1) (1963), pp. 277–293.
8. Hertz, D. B. "Risk Analysis in Capital Investment."*Harvard Business Review,* Vol. 42(1) (1964), pp. 95–106.
9. Dean, B. V. *Project Evaluation: Methods and Procedures* (American Management Association. New York, 1970).
10. Baker, N. R. and Freeland, J. "Recent Advances in R&D Benefit Measurement and Project Selection Methods." *Management Science,* Vol. 21(10) (1975), pp. 1164–1175.
11. Souder, W. E. "Analytical Effectiveness of Mathematical Programming Models for Project Selection." *Management Science.* Vol. 19(8) (1973), pp. 907–923.
12. Souder, W. E. "Utility and Perceived Acceptability of R&D Project Selection Models." *Management Science,* Vol. 19(12) (1973), pp. 1384–1894.
13. Souder, W. E. "Effectiveness of Nominal and Interacting Group Decision Processes for Integrating R&D and Marketing." *Management Science,* Vol. 23(6) (1977), pp. 595–605.
14. Souder, W. E. "Achieving Organizational Consensus with Respect to R&D Project Selection Criteria." *Management Science,* Vol. 21(6) (1975), pp. 669–691.
15. Souder, W. E. and Mandakovic, Tomislav. "R&D Project Selection Models: The Dawn of a New Era." *Research Management,* Vol. 24(4) (July–August, 1986), pp. 36–41.
16. Souder, W. E. *Managing New Product Innovations* (D.C. Heath/Lexington Books. Lexington, Mass., 1987).
17. Souder, W. E. *Project Selection and Economic Appraisal* (Van Nostrand Reinhold. New York, 1983).


## BIBLIOGRAPHY


Ansoff, H. I. "Evaluation of Applied Research in a Business Firm." in *Technical Planning on the Corporate Level,* J. R. Bright ed. Harvard University Press, Cambridge, Mass., 1962.

Augood, Derek. "A Review of R&D Evaluation Methods." *IEEE Transactions on Engineering Management, EM-20,*(4):114–120 (1973).

Baker, N. R. and Freeland, J. "Recent Advances in R&D Benefit Measurement and Project Selection Methods." *Management Science, 21*(10):1164–1175 (1975).

Cetron, M. J. and Roepcke, L. H. "The Selection of R&D Program Content." *IEEE Transactions on Engineering Management, EM-14:*4–13 (December, 1967).

Clarke, T. C. "Decision Making in Technologically Based Organizations: A Literature Survey of Present Practice." *IEEE Transactions on Engineering Management, EM-21*(1):9–23 (1974).
Dean, B. V. and Sengupta, S. S. "On a Method for Determining Corporate Research and Development Budgets." In *Management Science Models and Techniques,* C. W. Churchman and M. Verhulst, eds. Pergamon Press, New York, 1960, pp. 210–225.
Gear, A. E., Lockett, A. G., and Pearson, A. W. "Analysis of Some Portfolio Selection Models for R&D." *IEEE Transactions on Engineering Management, EM-18*(2):66–76 (1971).
Gee, R. E. "A Survey of Current Project Selection Practices." *Research Management, 14*(5):38–45 (September, 1971).
Harris, J. S. "New Product Profile Chart." *Chemical and Engineering News, 39*(16):110–118 (April 17, 1961).
Hart, A. "Evaluation of Research and Development Projects." *Chemistry and Industry,* No. 13:549–554 (March 27, 1965).
Hess, S. W. "A Dynamic Programming Approach to R&D Budgeting and Project Selection." *IRE Transactions on Engineering Management, EM-9*:170–179 (December, 1962).
Merrifield, Bruce. "Industrial Project Selection and Management." *Industrial Marketing Management, 7*(5):324–331 (1978).
Murdick, R. G. and Karger, D. W. "The Shoestring Approach to Rating New Products." *Machine Design,* January 25, 1973:86–89.
Rosen, E. M. and Souder, Wm. E. "A Method for Allocating R&D Expenditures." *IEEE Transactions on Engineering Management, EM-12*:87–93 (September, 1965).
Rubenstein, A. H. "Studies of Project Selection in Industry." In *Operations Research in Research and Development,* B. V. Dean, ed. Wiley, New York, 1963, pp. 189–205.
Souder, Wm. E. "R&D Project Selection: A Budgetary Approach." *Transactions CCDA*:25–43 (Spring, 1966).
———. "Planning R&D Expenditures with the Aid of a Computer." *Budgeting, XIV*:25–32 (March, 1966).
———. "Solving Budgeting Problems with O.R." *Budgeting, XIV*:9–11 (July/August 1967).
———. "Selecting and Staffing R&D Projects Via Op Research." *Chemical Engineering Progress, 63*:27 + (November, 1967) (reprinted in *Readings in Operations Research,* W. C. House, Auebach, 1970).
———. "Experiences with an R&D Project Control Model." *IEEE Transactions on Engineering Management, EM-15*:39–49 (March, 1968).
———. "Suitability and Validity of Project Selection Models." Ph.D. Dissertation, St. Louis University, St. Louis, Missouri (August, 1970).
———. "$R^2$: Some Results from Studies of the Research Management Process." *Proceedings AMIF*:121–130 (March, 1971).
———. "A Comparative Analysis of Risky Investment Planning Algorithms." *AIIE Transactions, 4*(1):56–62 (March, 1972).
———. "A Scoring Methodology for Assessing the Suitability of Management Science Models." *Management Science, 18*(10):526–543 (June, 1972).
———. "An R&D Planning and Control Servosystem: A Case Study." *R&D Management, 3*(1):5–12 (October, 1972).
———. "Effectiveness of Mathematical Programming Models for Project Selection: A Computational Evaluation." *Management Science, 19*(8):907–923 (April, 1973).
———. "Acceptability and Utility of Project Selection Models in Development R&D." *Management Science, 19*(12):1384–1394 (August, 1973).
———. "Autonomy, Gratification and R&D Outputs: A Small Sample Field Study." *Management Science, 20*(8):1147–1156 (April, 1974).

———. "Achieving Organizational Consensus with Respect to R&D Project Selection Criteria." *Management Science, 21*(6):669–681 (February, 1975).

———. "Experimental Test of a Q-Sort Procedure for Prioritizing R&D Projects." *IEEE Transactions on Engineering Management, EM-21*(4):159–164 (November, 1974).

———. "Field Studies with a Q-Sort/Nominal Group Process for Selecting R&D Projects." *Research Policy, 5*(4):172–188 (April, 1975).

———. "Effectiveness of Nominal and Interacting Group Decision Processes for Integrating R&D and Marketing." *Management Science, 23*(6):595–605 (February, 1977).

———. "A System for Using R&D Project Evaluation Models in Organizations." *Research Management, 21*(5):29–37 (September, 1978).

———. "An Appraisal of Eight R&D Project Evaluation Methods" to appear in *Corporate Strategy and Product Innovation,* 2nd ed. R. Rothbert, ed. Macmillan, New York, 1985.

———. *Management Decision Methods.* Van Nostrand Reinhold, New York, 1980.

———.*Project Selection and Economic Appraisal.* Van Nostrand Reinhold, New York, 1983.

———. *Managing New Product Innovations.* D.C. Heath/Lexington Books, Lexington, Mass., 1987.

———. and Mandakovic, Tomislav. "R&D Project Selection: The Dawn of a New Era." *Research Management, 24*(4):36–41 (July–August 1986).

Sullivan, C. I. "CPI Looks at R&D Project Evaluation." *Industrial and Engineering Chemistry, 53*(9):42A-46A (September, 1961).

Villers, Raymond. *Research and Development: Planning and Control,* Financial Executives Research Institute, Inc., 1964, pp. 30–38.

Watters, L. D. "Research and Development Project Selection: Interdependence and Multiperiod Probabilistic Budget Constraints." Ph.D. Dissertation, Arizona State University, Tempe, Arizona (1967).

# 8. Project Owner Strategic Management of Projects*

David I. Cleland†
William R. King‡

The purpose of this chapter is to prescribe a general approach for project owners to use in strategically managing capital projects.

## PROJECT OWNER RESPONSIBILITY

Recent attention to the quality of project management in key energy-producing industries has raised a significant question: What are the responsibilities of the project owner for the management of capital projects, the building blocks of organizational strategy? For in determining "prudent and reasonable" management in these industries, much of the focus

---

* Portions of this chapter have been adapted from D. I. Cleland, "Project Owners: Beware," *Project Management Journal*, December 1986, pp. 83–93, and D. I. Cleland, "Pyramiding Project Management Productivity," paper presented at Project Management Institute Seminar/Symposium, Houston, Texas, October 1983.

† David I. Cleland is currently Professor of Engineering Management in the Industrial Engineering Department at the University of Pittsburgh. He is the author/editor of 15 books and has published many articles appearing in leading national and internationally distributed technological, business management, and educational periodicals. Dr. Cleland has had extensive experience in management consultation, lecturing, seminars, and research. He is the recipient of the "Distinguished Contribution to Project Management" award given by the Project Management Institute in 1983, and in May 1984, received the 1983 Institute of Industrial Engineers (IIE)-Joint Publishers Book-of-the-Year Award for the *Project Management Handbook* (with W. R. King). In 1987 Dr. Cleland was elected a Fellow of the Project Management Institute.

‡ William R. King is University Professor in the Katz Graduate School of Business at the University of Pittsburgh. He is the author of more than a dozen books and 150 technical papers that have appeared in the leading journals in the fields of management science, information systems, and strategic planning. Among his major honors are the McKinsey Award (jointly with D. I. Cleland) for the "outstanding contribution to management literature" represented by their book *Systems Analysis and Project Management*, the IIE Book-of-the-Year Award for the first edition of this book, and designation as a fellow of the Decision Sciences Institute. Further biographical information is available in *Who's Who in the World* and *Who's Who in America*.

must be on the issue of the project management role of the owners to adequately plan and control the use of resources on the project.

All too often senior managers who "own" a project fail to recognize the key role that a project plays in the design and implementation of strategy. Such failure leads the project owner to neglect the proactive management of projects in their strategic management of the enterprise. These failures can be costly. Concern about the adequacy of corporate management's performance in project management is evident in the following sampling of recent situations:

- *Forbes* magazine claims that the failure of the U.S. nuclear power program ranks as the largest managerial disaster in business (1).
- $1.2 billion of LILCO's increased costs for the Shoreham project were recommended for exclusion from the rate base as having been imprudently incurred (2).
- The State of Alaska alleged before the Federal Energy Regulatory Commission that $1.6 billion in imprudent management costs were associated with the design, engineering, and construction of the $8 billion Trans-Alaska Pipeline System. A settlement on this case was reached on February 13, 1986. The agreement provides that (a) the rate base will be reduced by $450 million in recognition of the State's allegations of imprudent management; (b) the oil companies will pay $35 million for the State's legal expenses in the proceedings; (c) the owners will refund about $750 million for excessive tariffs between 1981 and 1984; (d) the tariffs will be reduced immediately from about $6.20 per barrel to about $5.00; (e) tariffs will continue to decline throughout the term of the agreement based on an established formula; and (f) the terms of the settlement will apply even if the Federal Energy Regulatory Commission or Congress at some point decides to deregulate oil pipelines (3).
- The State of Missouri Public Service Commission found that the design of the Union Electric Company's Callaway nuclear plant was not sufficiently complete when construction began and that the problem continued throughout the project causing inefficiencies and delays (4).
- In a study of quality in the design and construction of nuclear power plants, it was found that the root cause for initial quality-detached problems was a failure of the utility to implement a management system that ensured adequate control over all aspects of the project (5).

In addition to the above examples, there are many critical comments of a more general nature such as Davis's remark that capital expenditure overruns and poor performance are symptoms of a widespread problem affecting pioneer projects (6), and Bates's statement that owners have paid inadequate attention to the soaring construction costs and the reasons for them (7). Even the Department of Defense has come under sharp criticism for the excessive costs of equipment for defense projects.

As a result of such widespread criticism, many owners in the utility industry have responded by building up personnel and developing better management systems. Such involvement has enabled the owners to obtain closer control over projects and reduce the risk they have assumed (8). But owners' recognition of the strategic responsibility they have for a project should begin at an earlier time: when a capital project is selected for funding. It is at this time that the project should be recognized as a basic building block in strategic management of the enterprise.

A project owner is expected to take charge, to provide the leadership required to see that the right things are done on the project. Owner responsibility starts before the project becomes a reality—during the formative stage where a "vision" is developed which lays on the requirement for a project development to support the organizational strategy. More specifically, the senior managers of the project owner organization have a responsibility to

- Justify and establish the project as a building block of organizational strategy.
- Communicate the need for and manner by which the project will be managed.
- Position the resources around the project through an organizational form capable of managing the project as an integrated entity.
- Maintain a perspective on the project while strategically managing the organization.
- Provide a management and organizational system capable of providing effective strategic planning and management.
- Use a *project management system* as a model for project strategy and management philosophy.
- Use proven, contemporary project management theory and practice in planning, organizing, leading, and controlling the use of project resources.

Archibald posits that strategically managing a company requires the following (9):

- A *vision* of the future of the organization at the top level.
- *Consensus and commitment* within the power structure of the organization on the mission and future direction of the organization.

- *Documentation* of the key objectives and strategies to be employed in fulfilling the mission and moving toward the future direction.
- *Implementation* or execution of specific programs and projects to carry out the stated strategies and reach the desired objectives.

Once a project owner recognizes the key role that projects play in affecting the enterprise strategy, then adequate leadership and management systems will be provided to ensure the proactive and effective management of all enterprise projects, both large and small. Adequate leadership starts by the project owners recognizing their strategic management responsibility.

## OWNER STRATEGIC MANAGEMENT

*Strategic management* is concerned with the design of the organizational *mission, objectives,* and *goals* and the implementation *strategy* whereby enterprise purposes are attained.

A strategy is a series of prescriptions that provide the means, through the allocation of resources, for accomplishing organizational goals, objectives, and mission. In addition to allocating and committing resources for the future, a strategy also provides the general direction for the organization to pursue in reaching desired purposes. A strategy stipulates *what* resources are required, *why* they are required, *when* they are required, *where* they are needed, and *how* they will be used to accomplish ends. Resource allocations include anticipated expenditures for people, fixed assets, equipment, material, supplies, working capital, information, and management systems.

Strategy is the planned means for taking an organization from its present state to a desired future state. The purpose of a strategy is to provide the means to create something that does not currently exist. More specifically, the nature of strategy can be described thusly:

- It focuses on the organization as an entity.
- It emphasizes the key responsibility of the senior executive of the organization as a strategist to develop a sense of direction for the organization's future.
- It relates the organization's sense of direction in terms of identity, character, and purpose expounded in mission, objectives, and goals.
- It encourages the development of a consistent plan of action to execute the allocation of resources to gain strategic response to a changing competitive and environmental future.
- It provides for a balanced view of short-range and long-range organizational purposes.

- It integrates organizational policies, procedures, *programs, projects,* and action plans into a balanced approach to prepare for the future.

Programs and projects play a pivotal role as building blocks in attaining organizational purposes.

A program is (a) a related series of activities that continue over a broad period of time (normally years) and that are designed to accomplish broad goals or increase knowledge; (b) a related series of projects performed over time to accomplish a greater task or achieve a goal or objective; (c) a set of tasks, performed over time, which accomplish a specific purpose.

A project is a combination of human and nonhuman resources pulled together to accomplish a specified purpose in support of organizational strategies. Projects usually take the form of interrelated tasks performed by various organizations. A project has a well-defined objective, a target schedule, and a target cost (or budget). The attainment of a project objective is a tangible contribution to an organizational mission. *If a project overruns its cost and schedule, or fails to accomplish its technical performance objective, the implementation of an organizational strategy will be impaired.*

Effective strategic management means that organizational missions, objectives, goals, and strategies have been defined; an organizational design has been selected; and functional supporting plans, policies, systems, and procedures have been developed in response to changing environmental conditions and enterprise resources.

## A PROJECT MANAGEMENT MODEL

A key part of the strategic management of an enterprise is the philosophical approach to the management of projects. Figure 8-1 provides a model of an effective approach. This model derived from but different than that provided by King in Chapter 6 (10).

### Organizational Mission

At the apex of our triangular model is the organization's mission, the culminating strategic point of all organizational activity. An organization's mission is the most general strategic choice that must be made by its managers. An organization's mission tells what it is, why it exists, and the unique contribution it can make. The organization's mission answers the basic question: "What business are we in?"

The mission of an organization should provide the driving force to design suitable implementation strategies. Unfortunately many project organizations do have a concept of their mission, but fail to develop a comprehensive strategy for the consumption of resources to accomplish that mission. They fail to "work" the organization down through the successive levels of the model depicted in Figure 8-1.

### Project Objectives

While the mission is the common thread that binds together the resources and activities of an organization, a *project objective* designates the future positions or destinations that it wishes to reach in its "projects" business. Project objectives are the end result of managing the financial, schedule, and technical performance work packages of the project in consonance with the project plan. The accomplishment of project objectives contributes directly to the mission of the organization. This contribution can be measured. The proper selection and management of project objectives are essential steps in the strategic management of the organization. Such objectives are the building blocks of the project management organization's mission. Project objectives are supported by project goals.

### Project Goals

The distinctive features of project goals are their specificity and measurements on time-based points that the project team intends to meet in pursuit of its project objectives. For instance, in the management of a project, the completion of a work package in the project work breakdown structure means that progress has been made toward the objective of delivering the project on time, within budget, and in satisfaction of its operational objectives.

Mission, objectives, and goals are the triad of organizational direction. But this triad is not enough. The execution of organizational resources in support of this triad is contained in the project strategies.

### Project Strategy

A project strategy is the design of the means to accomplish results. An expressed project strategy is a project plan which provides general direction on how resources will be used to attain project goals and objectives. A project plan should cover the following: (a) project scope; (b) objectives such as technical, profit, other; (c) technical and management approach; (d) deliverables; (e) end item specifications; (f) schedules; (g) resources; (h) contributions; (i) finances; (j) risk areas such as subcontractor default, technical breakthroughs, etc. (11).

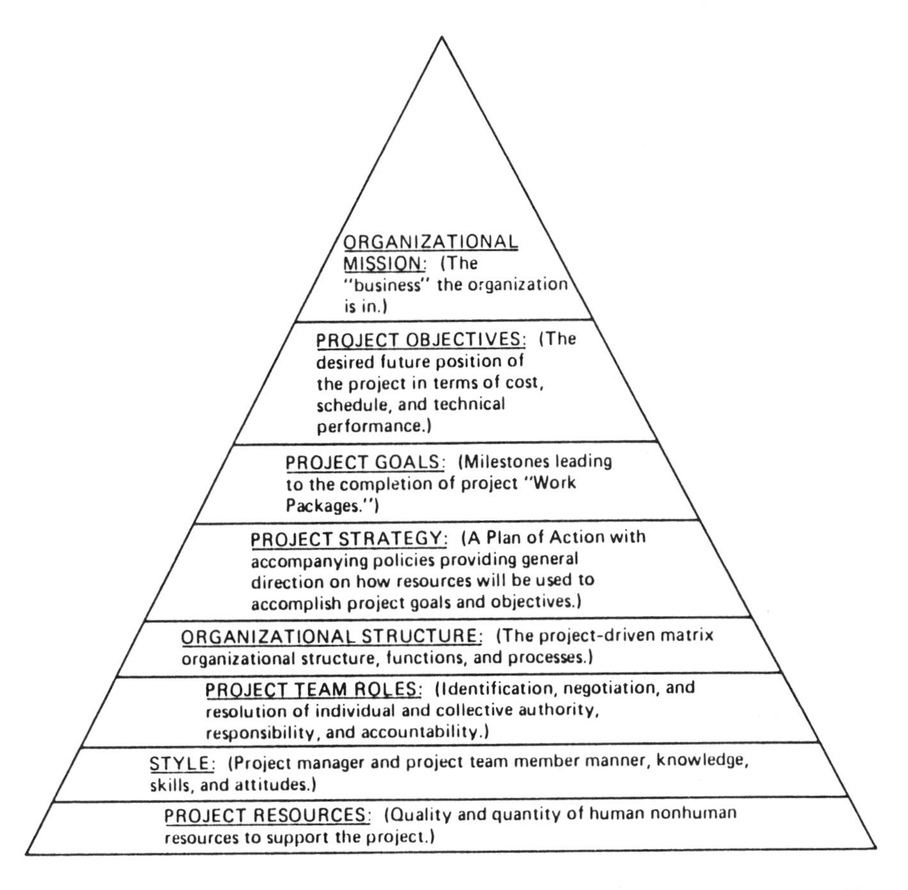


Figure 8-1. Elements in the project-driven matrix management system. (From Harold Kerzner and Daivd I. Cleland, *Project/Matrix Management Policy and Strategy,* Van Nostrand Reinhold, New York, 1985, p. 220.)

The nature of the project plan will vary depending on the project. The test of an adequate plan is its usefulness as a standard to judge project progress.

### Organizational Structure

The organizational structure is the manner by which the resources are aligned into departments based on principles of functional, product, process, geographic, or customer bases. Introduction of project management provides the opportunity to realign the organization structure on some form of the matrix organization. The organization structure becomes a project-driven matrix, a complex organization design which requires a specific delineation of project team roles.

### Project Team Roles

The structure of an organization defines the major territories that are assigned to each manager. Within each territory—production, finance, marketing, and so forth—specific roles require identification and negotiation, particularly in terms of the interaction of individuals with peers, subordinates, and supervisors. These role interrelationships come to focus through work packages and are held together by accepted authority, responsibility, and accountability of the project management team. Work packages are major work elements at the hierarchical levels of the work breakdown structure within the organization or within a project. They are used to identify and control work flows in the organization and have the following characteristics:

1. A work package represents a discrete unit of work at the appropriate level of the work breakdown structure (WBS) where work is assigned.
2. Each work package is clearly distinguished from all other work packages.
3. The primary responsibility of completing the work package on schedule and within budget can always be assigned to a member of the project team and never to more than one organizational unit.
4. A work package can be integrated with other work packages at the same level of the WBS to support the work packages at a higher level of the WBS.

Work packages are level-dependent, becoming increasingly more general at each higher level of the WBS and increasingly more specific at each lower level. A general manager would be expected to have primary responsibility to set project objectives, whereas a contract manager would be expected to take the lead in corporate maintenance agreements for a profit center manager to use in supporting equipment that has been delivered to a customer. An individual is designated as having primary responsibility for each work package. Others who have collateral responsibility involving the work package are also designated. When these collective roles have been designated, there remains no place to hide in the organization. If the work package is not completed on time or does not meet the performance standards, someone can readily be identified and held responsible and accountable for that work. Responsibility, authority, and accountability—the triad of personal performance in organized life—is not left to doubt when individual and collective roles have been adequately defined. This definition can be carried out through the process of Linear Responsibility Charting (LRC) (12).

Project team roles are affected by the management style that is followed in the organization.

### Management Style

The style of the managers and professionals associated with the project is an important influence on project success or failure. Style depends on an individual's knowledge, skills, and attitudes, manifesting itself primarily in the individual's interpersonal relationships and management philosophies. The management styles of senior managers are echoed down through the organizational structure to the project team. Style is dependent on individual and collective knowledge, skills, and attitudes expressed in the cultural ambience of the project team.

### Project Resources

Ultimately, the quality of the project end product and the schedule and cost to develop that product is dependent on the quality and quantity of the resources available and being applied to the project. The project owners have a key responsibility to provide the necessary resources to support the project as well as to maintain ongoing surveillance over the use of those resources.

### Management Strategies

A strategy is a series of prescriptions that provide the means and set the general direction for accomplishing organizational goals, objectives, and mission. These prescriptions stipulate

- What resource allocations are required?
- Why?
- When?
- Where?
- How?

These allocations include anticipated expenditures for fixed assets, equipment, working capital, people, information, and management systems. In general, strategies include the following:

- Programs/program plans.
- Projects/project plans.
- Operational plans.
- Contingency plans.
- Policies
- Procedures
- Organizational design.
- General prescribed courses of action.
- Prescriptions for resource utilization.
- Strategic performance standards.

*Projects* are an essential building block in an enterprise strategy. Projects require active management to fulfill their role in moving the organization from its present position to a desired future position at some predetermined point in time. The project owner is a key member of the project team. Only the owner can determine the strategic fit of a project in the organization's strategy and establish the priorities and performance standards for the management of the project. Project owners have the ultimate obligation to see that certain responsibilities are accomplished to support both organizational and project ends. These responsibilities are embodied in the *strategic planning and management* role of the owners, in the *project management system* used to manage the project, as well as in the quality of support provided by the functional specialists working on the project. These responsibilities include

- Development and implementation both of adequate strategic plans for the enterprise and of project plans to support the project's technical performance objective, schedules, costs, and execution strategies.
- Development of an organizational design which delegates appropriate authority, responsibility, and accountability to the managers and professionals working on the project.
- Regular ongoing surveillance by responsible managers and professionals to monitor the use of resources on the project and the appropriate reallocation of resources as required to keep the project objectives on time and within budget.
- Design and use of policies, procedures, roles, and guidelines to facilitate the management of the project.
- Provision of knowledgeable and skillful people to work on the project.
- Development of the necessary information systems to support managers and professionals working on the project.

- Facilitation of an organizational culture that fosters, recognizes, and rewards prudent and reasonable project management.
- Rewarding of people for productive and quality results on the project.
- Selection of periodic internal and external audits to determine the efficacy of project management and to verify project status.
- Surveillance and ongoing communication with project "stakeholders" to ascertain and influence their perceptions of the project.
- Ensuring of the use of a contemporaneous body of knowledge and skill in the management of the project.
- Setting of the tone for leader and follower style within the culture of the corporation in the design and execution of projects.
- Replanning, recycling, and redesigning, as necessary, of the management systems used in the management of the project.

## OWNERS' MANAGEMENT SYSTEM

Owners require a management system which enables them to play a proactive role in planning, organizing, and controlling resources used to support enterprise purposes. The starting point for an owner's management system is a clear statement of the enterprise's mission, that is, the "business" that the enterprise pursues. One utility company's mission is stated in the following way:

> Manage and direct the nuclear activities of the GPU System to provide the required high level of protection for the health and safety of the public and the employees. Consistent with the above, generate electricity from the GPU nuclear stations in a reliable and efficient manner in conformance with all applicable laws, regulations, licenses, and other requirements and the directions and interests of the owners.(13)

Once the mission of the enterprise is established through the operation of a *strategic planning system,* planning can be extended to select and develop organizational *objectives, goals,* and *strategies.* Projects are planned for and implemented through a *project management system* composed of the following subsystems:

- A *matrix-oriented organization subsystem* used in the management of the project.
- A *project management information subsystem,* which contains the intelligence essential to the effective planning, organization, and control of the project.

- The *planning subsystem,* which involves the development of project technical performance objectives, goals, schedules, costs, and strategies.
- The *project control subsystem,* which selects performance standards for the project schedule, budget, and technical performance objectives, and uses information feedback to compare actual progress with planned progress.
- The *cultural subsystem,* which concerns the perceptions, attitudes, and leader and follower style of the people working on the project.
- The *human subsystem,* which involves communications, negotiations, motivation, leadership, and the behavior patterns of the people working on the project.

Many of the above subsystems use techniques such as PERT, CPM, and related resource allocation methodologies.

Figure 8-2 depicts this project management system with all its subsystems. The utility owners who are responsible and accountable for the effective management of the project work through their board of directors and senior management with the project manager, functional managers, and functional specialists.

In the case of the Shoreham Project, the responsibility and accountability of the senior executives of the project owner's organization was made clear by the administrative law judges who concluded that

> . . . Lilco (Long Island Lighting Company) failed to develop a project plan adequate to oversee S&W management of the project. . . . To identify roles and responsibilities, to develop accurate and timely reporting systems which would enable it to monitor, measure and control costs and scheduling, to adequately staff monitoring groups or to adequately prepare for its critical owner oversight role.
>
> We conclude that, throughout Shoreham's construction, Lilco failed to staff adequately its prime area of responsibility as owner of the plant—cost and schedule control.
>
> Lilco's measurement and reporting systems continually and repeatedly failed to accurately depict cost and schedule status at Shoreham. Lilco managers were unable to use Lilco's measurement systems to gain an accurate picture of what was happening on site and complained that Lilco's reporting systems were confused and cluttered.(2)

The law judges left no doubt as to the overall responsibility of the Lilco Board of Directors for the Shoreham Project:

> We conclude that the limited information presented to the Board was inadequate for it to determine project status on the reasonableness of key management decision or to provide requisite guidance and direction to Lilco management.(2)

On the more positive side, there should be no doubt by project owners that effective planning and control contribute to the success of a project. For example, project success at Fluor Utah, Inc., is related to the quality of planning and control.

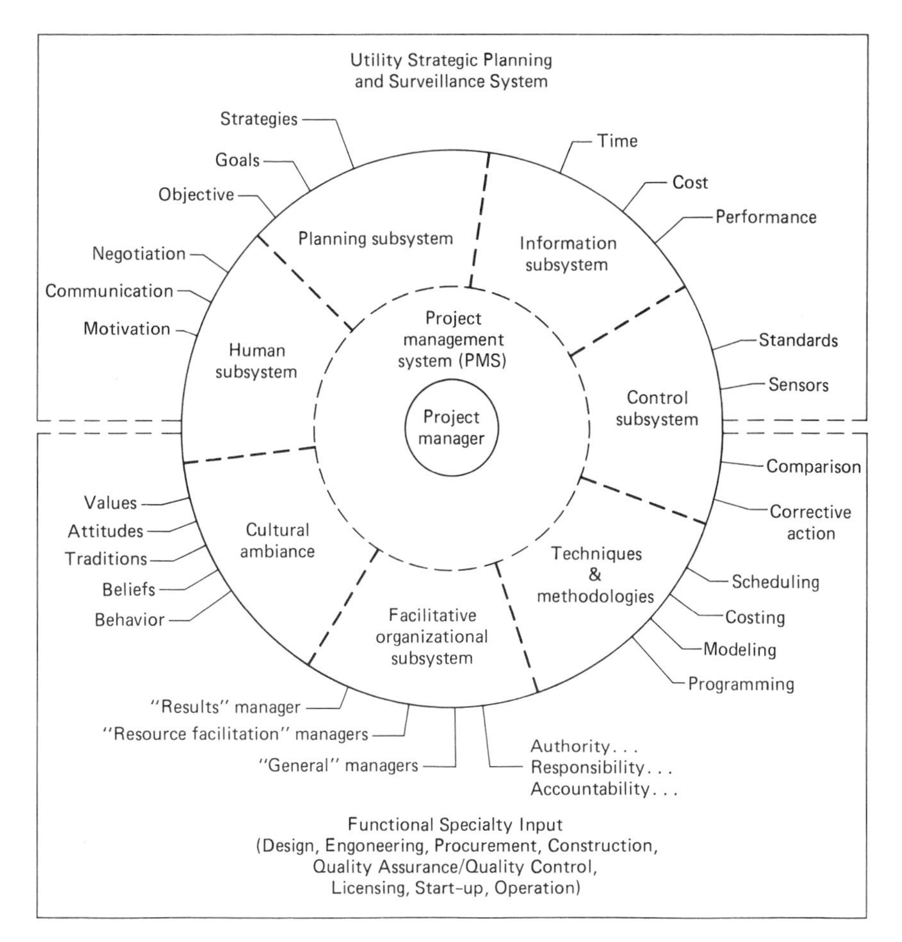


Figure 8-2. The project management system. (Adapted from D. I. Cleland, "Defining a Project Management System," *Project Management Quarterly*, December 1977).

During the early 1960s, after hundreds of projects had been completed at Fluor Utah, Inc., it became apparent that many projects successfully achieved their basic project objectives, while some failed to achieve budget, schedule, and performance objectives originally established.

The history of many of these projects was carefully reviewed to identify conditions and events common to successful projects, vis-à-vis those conditions and events that occurred frequently on less successful projects.

Common identifiable elements on most successful projects were the quality and depth of early planning by the project management group. Another major reason for the project success was the bolstering of the execution of a plan by strong management control over identifiable phases of the project (14).

In a study of the underlying causes of major quality-related problems in the construction of some nuclear power plants, the Nuclear Regulatory Commission concluded that the root cause was the failure or inability of some utility managements to implement a management system that ensured adequate control over all aspects of the project (5).

An important part of owner responsibility is the design of a suitable organizational approach to manage the project.

## ORGANIZATIONAL DESIGN

An effective organizational design should take into consideration the roles (authority, responsibility, and accountability) of executives at the following organizational levels:

*The Board of Directors* for the exercise of trusteeship in the husbanding of corporate assets used in capital projects.

*The CEO and Staff* functioning as a "plural executive" for strategic and operating responsibility for the corporate entity through the optimal use of resources to achieve corporate mission, objective, and goals and to finish capital projects on time and within budget.

*A Senior Executive* who functions in the capacity of a "Manager of Projects" for directing individual project managers' activities; proposing, planning, and facilitating the implementation of project management plans, policies, procedures, techniques, and methodologies; and evaluating and controlling project progress.

*The Project Manager* who has residual responsibility and accountability for project results on time and within budget.

*Functional Managers* who provide specialized resources to support project needs.

*Work Package Managers* who are responsible for project work package budgets, schedules, and technical performance objectives.

*Project Professionals* working on the project who, depending on the nature of the matrix organization, report to both the functional manager and the project manager.

Appropriate policy documentation should exist which portrays in specific terms the organizational structure and the fixing of authority, responsibility, and accountability of managers at each organizational level.

The impact of an inadequate organizational design for the Shoreham Project was noted by the administrative law judges who said, "A major planning and management defect was confusion over roles and responsibilities"(2). They went on to say, "Lilco's organizational arrangement left lines of authority and responsibility blurred and unclear from the start"(2, p. 77).

A McKinsey & Company survey of twelve multibillion-dollar projects built during the last two decades emphasizes the necessity for an organizational plan to define the organizational structure for the project and to ensure the allocation of specific responsibilities within the structure. This survey states that the formal, written organizational plan should be prepared at the front end of the project to obtain mutual agreement from all parties and to make certain that all required responsibilities will be addressed. The survey concludes that an owner group's inability to let go of project management, even when it recognized significant gaps in its own ability to manage, was usually "rewarded" with significant difficulties and problems in completing the project (15).

A RAND Corporation study of "new technology" process plant construction finds that the most prominently mentioned management-related reason for increased costs is "diffuse decision-making responsibility for a project." The study concludes that the "general wisdom for construction projects" dictates that "one person needs to be given broad authority for all routine project decisions and a reasonable scope for fairly important decisions on schedules, allocation of monies, and all but major modifications." The study finds that it is "standard industry practice to appoint a project manager—in the case of a pioneer plant project, a project manager of long experience—who is responsible for the undertaking from shortly after the time that the project emerges from development until an operating plant is on-line"(16).

An adequate organizational design helps to facilitate adequate owner surveillance of a project.

## OWNER ONGOING SURVEILLANCE

Ongoing surveillance of the project in order to ensure the results and control the application of organizational resources is essential to successful project management. Measurement of project results requires the development of certain standards, information feedback, and methods for comparing project plans with project results. It also requires ongoing exercise of "hindsight" in the initiation of corrective action to realign resources in order to accomplish the project objective and goals. Project owners are responsible for ensuring that regular, ongoing assessment of a project's progress is carried out to determine the status of that project. The following concepts and philosophies are essential elements for the assessment of project results:

1. The objective is to develop measurements of project trends and results through information arising out of the management of the project work breakdown structure.
2. Performance measurements are always to be tempered by the judgment of the managers and professionals doing the measurement.
3. The use of common measurement factors arises out of the status of project work packages consistent with the organizational decentralization of the project.
4. Measurements should be kept to a minimum relevant to each work package in the project work breakdown structure.
5. Measurements of work packages must be integrated into measurement of the project as a whole.
6. Measurements should be developed that are applicable to both current project results and future projections to project completion.
7. Measurement should be conducted around previously planned key result areas.

The *key result areas* of the project are those areas which are of sufficient basic importance to act as "direction indicators" of the project. To illustrate, these areas include

- Technical performance objectives.
- Cost objectives.
- Schedule objectives.
- Strategic synergistic fit with organizational product strategies.
- Potential financial return.
- Productive use of resources.
- Competitiveness.

The strategic plans of the enterprise should define these key result areas to provide performance standards for the ongoing evaluation of the project, and to seek answers to such questions as the following:

- What is going right on the project?
- What is going wrong?
- What problems are emerging?
- What opportunities are emerging?
- Where is the project with respect to schedule, cost, and technical performance objectives?
- Does the project continue to have a strategic fit with the enterprise's mission?
- Is there anything that should be done that is not being done?
- Is there any reallocation of corporate resources required to support the project needs?
- What explains the difference between planning and actual project progress?
- What replanning and reorganizing of corporate resources are required to support the project?

If owner senior managers are not actively involved and do not have sufficient knowledge to ask these questions and seek answers to them, then they are not getting the feedback they need on project progress in order to be prudent and reasonable in their management of the project. An independent audit of the project conducted on a periodic basis will also help the project owners to get the informed and intelligent answers they need.

## PROJECT MANAGEMENT AUDIT

Once the decision is made to conduct an audit, an audit team should be formed and provided with the authority and resources to carry out the audit. All key functional and general managers should be committed to help facilitate the audit by making people and information available to the audit team. An organizational policy document should exist which outlines, as a minimum, the *purpose, scope, policy, responsibilities, and procedures* for the audit. If such a policy document exists, the culture of the organization will better support an audit philosophy. The project, functional, and general managers, expecting to be audited, will be encouraged to do a better job. There is considerable support in the project management literature for an independent project audit. For example, Chilstrom states:

> Management audit of projects provides top management a means of independent appraisal in determining the effectiveness of the organization to successfully accomplish a project. This has become more impor-

> tant in recent years where most projects have matrix management requiring the functional organizations to meet the needs of many projects. Project success requires a capable project team that has responsive support from functional areas, and it is management's task to allocate needed resources and achieve the integration of all elements. In addition, the project team will directly benefit from the results of the audit, since findings and recommended actions will concentrate on both internal and external factors that are preventing the achievement of the project goals and plans.(17)

In noting the success of a large nuclear plant construction project, the project manager credited an ongoing critique as a significant contributor to the success of the project. Many times during the life of this project, independent groups were brought in to review it in order to help ensure that significant problems were not being overlooked (18).

In commenting on the role of independent reviews, Cabano notes:

> . . . We believe the evidence to be conclusive that although project responsibilities and objectives can be met using only traditional project management techniques, programmatic adoption of "Independent Project Reviews" (IPR) as described herein can add measurably to the confidence, visibility, coordination, communications and overall synchronization of a project and substantially enhances its success potential.(19)

In a study of over 50 process industry large-to mega- scale projects, Cabano found that the use of an independent project review (audit) can measurably enhance the success potential of a project. In the project histories reviewed, teams of experts were used to help resolve major problems judged sufficient to jeopardize the stability and progress of the project. The study concludes that judicious project audit process can help to or provide early warning of problems, give more time for the development of remedial strategy, and even prevent major problems. The study suggests, however, that a review of this kind should not replace effective full-time surveillance by responsible individuals but should be used to supplement the existing management system (19).

How often should an audit be conducted considering that a thorough audit takes time and money? Generally, audits should be carried out at key points in the life cycle of a project, and at times in those phases of the life cycle that represent "go/no go" trigger points such as preliminary design, final engineering design, start of construction, start of licensing, start-up, and operation.

The conduct of independent audits by utility management was an established practice in the early to mid-1970s. For example, in 1975 the legislature of the state of North Carolina enacted a law to initiate a full and complete management audit of any public utility company (PUC) once every five years. Similar actions have been taken by PUCs in Missouri, Pennsylvania, New York, Oregon, Arizona, and Connecticut. "*It* [management auditing] *is an emerging trend of great significance* to regulators, consumers, and regulated alike"(20).

The organizational policy document is one instance of how the owners can work to gain the support of the culture of the organization for an audit philosophy. On a broader scale, the owner's attitudes and approach to the project set the cultural ambience in which the entire project is managed.

## THE CULTURAL AMBIENCE

Culture is a set of refined behaviors that people strive toward in their society. It includes the whole complex of a society—knowledge, beliefs, art, ethics, morals, law, custom, and other habits and attitudes acquired by the individual as a member of society. Anthropologists have used the concept of culture in describing primitive societies. Modern-day sociologists have borrowed this anthropological usage to describe a way of life of a people.

The term "culture" is used to describe the synergistic set of shared ideas and demonstrated beliefs that are associated with a way of life in an organization using project management in the execution of its corporate strategy. An organization's culture reflects the composite management style of its executives, a style that has much to do with the organization's ability to adapt to such a change as the use of project management in corporate strategy. Arnold and Capella remind us that achieving the right kind of corporate culture is critical and that businesses are human institutions (21). Culture is that integrated pattern of human behavior that includes thought, attitudes, values, action, artifacts; it is the way things are done in organization.

An organization's culture consists of shared agreements, explicit or implicit, among organizational members as to what is important in behavior and attitudes expressed in values, beliefs, standards, and social and management practices. The culture that is developed and becomes characteristic of an organization affects strategic planning and implementation, project management, and all else.

It is possible to identify common cultural features that positively and negatively influence the practice of management and the conduct of technical affairs in an organization. Such cultural features develop out of and are influenced by

- The management leadership-and-follower style practiced by key managers and professionals.
- The example set by leaders of the organization.
- The attitudes displayed and communicated by key managers in their management of the organization.
- Manager and professional competencies.
- Assumptions held by key managers and professionals.
- Organizational plans, policies, procedures, rules, and strategies.
- The political, legal, social, technological, and economic systems with which the members of an organization interface.
- The perceived and/or actual characteristics of the organization.
- Quality and quantity of the resources (human and nonhuman) consumed in the pursuit of the organization's mission, objectives, goals, and strategies.
- The knowledge, skills, and experience of members of the organization.
- Communication patterns.
- Formal and informal roles.

Insight into the effect of a key executive's attitude was cited in a report to the Congress on Quality Assurance in the nuclear power plant industry:

> One chief executive termed his utility's first planned nuclear plant as "just another tea kettle," i.e., just an alternative way to generate steam (this was before major quality problems arose at his project).(5)

The failure of some licensees (owners) to "treat quality assurance as a management tool, rather than as a paperwork exercise (5, p.3-11) affected the outcome of the project. Policies can affect the cultural ambience of an organization. For example:

> . . . A characteristic of the projects that had not experienced quality problems was a constructive working relationship with and understanding of the NRC. For example, Florida Power and Light established a special office in Bethesda staffed by engineers to facilitate exchange of information with the NRC during the St. Lucie 2 licensing process. Also, senior management of Arizona Power Service has established the following *policies* concerning the NRC:
>
> Don't treat NRC as an adversary; NRC is not here to bother us—they see many more plants than the licensee sees; inform NRC of what we (APS) are doing and keep everything up front; and nuclear safety is more important than schedule.(5, p. 3-21)

Attitudes of key managers play an important role in successfully completing a nuclear power plant. A management commitment to quality and a management view that NRC requirements are not the ultimate goals for performance carry great weight. For example:

> Of the projects studied there tended to be a direct correlation between the project's success and the utility's view of NRC requirements: more successful utilities tended to view NRC requirements as minimum levels of performance, not maximum, and they strove to establish and meet increasingly higher, self-imposed goals. This attitude covered all aspects of the project, including quality and quality assurance.(5, p. 3-19)

Manager/professional experience affects competence and this in turn affects the culture. A common thread running through four projects studied in a report to Congress was a lack of prior nuclear experience of some members of the project team, that is, owner utility, architect-engineer, construction manager, and construction. In three of the four cases this lack of experience was a major contributor to the quality-related problems that developed on these projects. Owner's inexperience is important because in at least three of the four cases the owner underestimated the complexity and difficulty of the nuclear project and treated it much as it would have another fossil project. The effect of inexperience was significant:

> . . . Generally, the utilities' lack of experience in and understanding of nuclear construction manifested itself in some subset of the following characteristics:
>
> 1. inadequate staffing for the project, in numbers, in qualifications, and in applicable nuclear experience
> 2. selection of contractors who may have been used successfully in building fossil plants but who had very limited applicable nuclear construction experience
> 3. over-reliance on these same contractors in managing the project and evaluating its status and progress
> 4. use of contracts that emphasized cost and schedule to the detriment of quality
> 5. lack of management commitment to and understanding of how to achieve quality
> 6. lack of management support for the quality program

7. oversight of the project from corporate headquarters with only a minimal utility presence at the construction site
8. lack of appreciation of ASME codes and other nuclear-related standards
9. diffusion of project responsibility and diluted project accountability
10. failure to delegate authority commensurate with responsibility
11. misunderstanding of the NRC, its practices, its authority, and its role in nuclear safety
12. tendency to view NRC requirements as performance goals, not lower thresholds of performance
13. inability to recognize that recurring problems in the quality of construction were merely symptoms of much deeper, underlying programmatic deficiencies in the project, including project management.(5, pp. 3-8, 3-9)

In some cases a poorly functioning QA Program had its roots in management's lack of appreciation or support for the quality function. Part of this lack of appreciation was attributed to management's unawareness of vital construction quality information which was known to the quality assurance staff. The existence of many organizational levels through which information flowed was "severely attenuated" when it reached senior management (5, p. 3-12).

The critical ambience of an organization using project management is subtle yet very real. In the project-driven organization the attitudes, values, beliefs, and management systems tend to become more participative and democratic. Owners can affect that culture to support successful project management.

## SUMMARY

The owner cannot abdicate responsibility for the project to others, even experienced A&E firms, project management contractors, or constructors. Successful project management depends on a commitment by the owner to use contemporaneous project management theory and practice in designing and using appropriate management systems to proactively manage the project. Prudent owners must maintain close surveillance over project progress and remain in close touch with all project participants to glean their input into the status of the project. This means that an owner must be an active and knowledgeable participant on the project team. The owner's most critical commitments are that the project will be assertively managed, and that all project participants are provided strategic leadership for completion of the project on time and within budget so that enterprise strategy is enhanced.

Project owners who seek to recover investments through rate adjustments should find the use of the project management concepts depicted in this chapter useful to demonstrate their prudent and reasonable management of capital projects.

Project owners who seek to recover investments through rate adjustments should find the project management concepts described in this chapter useful to demonstrate their prudent and reasonable management of capital projects.

## REFERENCES

1. "Nuclear Follies." *Forbes,* February 11, 1986.
2. Recommended Decision by Administrative Law Judges Wm. C. Levey and Thomas R. Matias on Case 27563, Long Island Lighting Company—Shoreham Prudence Investigation, State of New York Public Service Commission, March 11, 1985.
3. Rogovin, Huge & Lenzner, Law Offices, 1730 Rhode Island Avenue, N.W. Washington, D.C., 20036, Ltr., February 13, 1986.
4. Case No. ER-85-160 & EO-85-17, Determination of In-Service Criteria for the Union Electric Company's Callaway Nuclear Plant and Callaway Rate Base and Authority to file Tariffs Increasing Rates for Electric Service to Customers in Missouri, before the State of Missouri Public Service Commission, March 29, 1985.
5. U. S. Nuclear Regulatory Commission (NUREG-1055). *Improving Quality and the Assurance of Quality in the Design and Construction of Nuclear Power Plants* (Washington, D.C., 20555, May 1984).
6. Davis, David. "New Projects: Beware of False Economies." *Harvard Business Review* (March-April, 1985).
7. Bates, G. Stan. "Construction Industry Cost Effectiveness Project National CICE Activities—Update" *1983 Proceedings of the Project Management Institute* (October, 1983), p. V-D-2.
8. Theodore Barry & Associates. *A Survey of Organizational and Contractual Trends in Power Plant Construction, March, 1979.*
9. Archibald, Russell D. "Implementing Business Strategies through Projects," in *Strategic Planning and Management Handbook,* 2nd Ed., ed. King, W. R. and Cleland, D. I. (Van Nostrand Reinhold. New York, 1986).
10. Paraphrased from Kerzner, Harold and Cleland, David I. *Project/Matrix Management Policy and Strategy* (Van Nostrand Reinhold. New York, 1985), pp. 219–224.
11. Paraphrased from Archibald, Russell D. *Managing High-Technology Programs and Projects* (Wiley. New York, 1976).
12. See Chapter 16 of this handbook.
13. Clark, Philip R. "Looking Beyond the Lessons: A Utility Manager's Perspective." *Nuclear News* (April 1984), p. 64.
14. Duke, Robert, Wholsen, H. Frederick and Mitchell, Douglas R. "Project Management at Fluor Utah, Inc.," in *The State-of-the-Art of Project Management 1976–1977* (The Northern California Chapter of the Project Management Institute. San Francisco, 1977), pp. 28–37.

15. Anderson, J. *Organizing for Large Project Management—The Client's Needs* (McKinsey & Company. October, 1978).
16. Rand Corporation. *A Review of Cost Estimation in New Technologies: Implications for Energy Process Plants.* (July, 1978).
17. Chilstrom, Kenneth O. "Project Management Audits," in *Project Management Handbook,* ed. Cleland, D. I. and King, W. R. (Van Nostrand Reinhold. New York, 1983), p. 465.
18. Derrickson, W. B. "St. Lucie Unit 2—A Nuclear Plant Built on Schedule." *1983 Proceedings of the Project Management Institute (October, 1983), pp. V-E-1–V-E-14.*
19. Cabano, Louis J. "Independent Project Reviews." *Project Management Institute Seminar/Symposium* (October, 1984), p. 8.
20. Alden, Raymond M. "Utility Management Audits from a Managerial Viewpoint." *Public Utilities Fortnightly* (October 7, 1976) (emphasis added).
21. Arnold, Danny R. and Capella, Louis M. "Corporate Culture and the Marketing Concept: A Diagnostic Instrument for Utilities." *Public Utilities Fortnightly* (October 17, 1985), pp. 32–38.

# Section IV

# Life-Cycle Management

One of the important reasons for the efficacy of project management is the changing mix of resources that is demanded over the life cycle of a project.

In Chapter 9, William R. King and David I. Cleland portray the project life cycle as an important rationale for project management. They present various life-cycle concepts and show how the life cycle places demands on organizations that require a "new" form of management—the project management approach.

In Chapter 10, John R. Adams and Stephen E. Barndt review a set of organizational variables in terms of their impact on projects in various stages of the life cycle. They present a series of propositions that are based on their assessments of the results of studies of more than 20 R&D projects. These propositions allow one to predict the behavior of projects throughout their life cycle.

In Chapter 11, Herbert F. Spires considers an important, and often neglected, phase of the project life cycle—divestment. The phasing out of a project may be either a "natural" part of the life cycle or it may be extraordinary. In either case, phase-down creates unique problems that are associated with no other phases and with few other endeavors in life.

William R. King and Ananth Srinivasan integrate the life-cycle notions of this section with the strategic project context of Section III by demonstrating, in Chapter 12, how the traditional systems development life cycle has evolved into a broader life cycle for the systems of the organization. Although this chapter focuses on the important information systems domain, this integration is taking place for technology life cycles, sales life cycles, and other systems cycles as well.

# 9. Life-Cycle Management*

William R. King†
David I. Cleland‡

"Life cycle management" is a term that describes project management in terms of one of the most salient project characteristics—the life cycle. The life cycle of a project is an important factor in determining the need for, and value of, a project management approach.

## BASIC LIFE-CYCLE CONCEPTS

There are a variety of life-cycle concepts that are in common use. These life cycles serve to illustrate the need for life cycle management.

### Sales Life Cycles

Perhaps the best known life cycle is the sales life cycle. A product moves through various phases of sales life cycle after it has been placed on the

* Portions of this chapter have been paraphrased from *Systems Analysis and Project Management,* 2nd Edition, (McGraw-Hill Book Company, New York, 1975), by David I. Cleland and William R. King.

† William R. King is University Professor in the Katz Graduate School of Business at the University of Pittsburgh. He is the author of more than a dozen books and 150 technical papers that have appeared in the leading journals in the fields of management science, information systems, and strategic planning. Among his major honors are the McKinsey Award (jointly with D. I. Cleland) for the "outstanding contribution to management literature" represented by their book *Systems Analysis and Project Management,* the IIE Book-of-the-Year Award for the first edition of this book, and designation as a fellow of the Decision Sciences Institute. Further biographical information is available in *Who's Who in the World* and *Who's Who in America.*

‡ David I. Cleland is currently Professor of Engineering Management in the Industrial Engineering Department at the University of Pittsburgh. He is the author/editor of 15 books and has published many articles appearing in leading national and internationally distributed technological, business management, and educational periodicals. Dr. Cleland has had extensive experience in management consultation, lecturing, seminars, and research. He is the recipient of the "Distinguished Contribution to Project Management" award given by the Project Management Institute in 1983, and in May 1984, received the 1983 Institute of Industrial Engineers (IIE)-Joint Publishers Book-of-the-Year Award for the *Project Management Handbook* (with W. R. King). In 1987 Dr. Cleland was elected a fellow of the Project Management Institute.

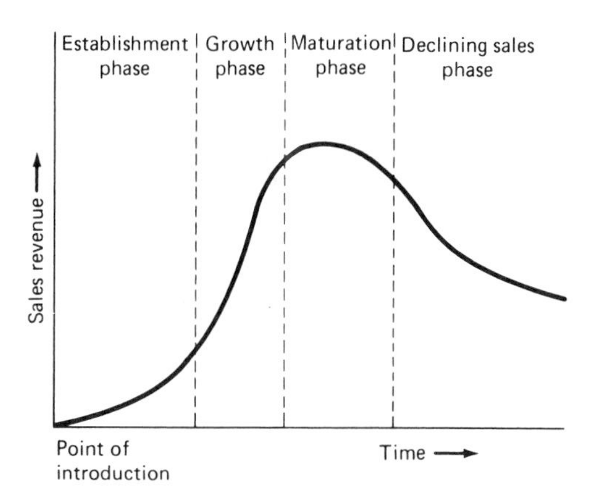


Figure 9-1. Product sales life cycle.

market. One of the authors* has referred to these life-cycle phases as *establishment, growth, maturation,* and *declining sales* phases. Figure 9-1 shows these phases in terms of the sales revenue generated by the product during its period of slow establishment in the marketplace, followed by a period of rapid sales increase, a peaking, and a long, gradual decline. Virtually every product displays these dynamic characteristics, although some may have a sales life cycle which is so long or short that the various phases are not readily distinguishable. For example, a faddish product such as "super balls" or "hula hoops" will have a very high-peaked sales curve with a rapid decline. Many such products will have a long, slow decline after an initially rapid decline from the peak. With other products, the maturation phase is very long and the declining sales phase very gradual. But the general life-cycle concept is virtually unavoidable for a successful product. Without product improvements, competition will eventually lure away customers, and consumers' attitudes, habits, and needs will change as time passes.

Of course, the sales portion of the life cycle of a product is really only one aspect of its entire life. Indeed, only products which are marketing successes ever get to experience the sales life cycle of Figure 9-1.

### Systems Development Life Cycle

All products—sales successes or otherwise—begin as a gleam in the eye of someone and undergo many different phases of development before

---

* William R. King, *Quantitative Analysis for Marketing Management* (McGraw-Hill. New York, 1967), p. 113.

being marketed and subjected to the sales life-cycle considerations of Figure 9-1. For instance, the U.S. Department of Defense (DOD) and the National Aeronautics and Space Administration (NASA) have extensively defined and detailed phases which should be encountered with hardware systems development. Their system development life-cycle concept recognizes a natural order of thought and action which is pervasive in the development of many kinds of systems—be they commercial products, space exploration systems, or management systems.

New products, services, or roles for the organization have their genesis in ideas evolving within the organization. Typically, such "systems" ideas go through a distinct life cycle, that is, a natural and pervasive order of thought and action. In each phase of this cycle, different levels and varieties of specific thought and action are required within the organization to assess the efficacy of the system. The "phases" of this cycle serve to illustrate the systems development life-cycle concept and its importance.

**The Conceptual Phase.** The germ of the idea for a system may evolve from other research, from current organizational problems, or from the observation of organizational interfaces. The conceptual phase is one in which the idea is conceived and given preliminary evaluation.

During the conceptual phase, the environment is examined, forecasts are prepared, objectives and alternatives are evaluated, and the first examination of the performance, cost, and time aspects of the system's development is performed. It is also during this phase that basic strategy, organization, and resource requirements are conceived. The fundamental purpose of the conceptual phase is to conduct a "white paper" study of the requirements in order to provide a basis for further detailed evaluation. Table 9-1 shows the details of these efforts.

There will typically be a high mortality rate of potential systems during the conceptual phase of the life cycle. Rightly so, since the study process conducted during this phase should identify projects that have high risk and are technically, environmentally, or economically infeasible or impractical.

**The Definition Phase.** The fundamental purpose of the definition phase is to determine, as soon as possible and as accurately as possible, cost, schedule, performance, and resource requirements and whether all elements, projects, and subsystems will fit together economically and technically.

The definition phase simply tells in more detail what it is we want to do, when we want to do it, how we will accomplish it, and what it will cost. The definition phase allows the organization to fully conceive and define

Table 9-1. Conceptual Phase.

1. Determine existing needs or potential deficiencies of existing systems.
2. Establish system concepts which provide initial strategic guidance to overcome existing or potential deficiencies.
3. Determine initial technical, environmental, and economic feasibility and practicability of the system.
4. Examine alternative ways of accomplishing the system objectives.
5. Provide initial answers to the questions:
   a What will the system cost?
   b When will the system be available?
   c What will the system do?
   d How will the system be integrated into existing systems?
6. Identify the human and nonhuman resources required to support the system.
7. Select initial system designs which will satisfy the system objectives.
8. Determine initial system interfaces.
9. Establish a project organization.

the system before it starts to physically put the system into its environment. Simply stated, the definition phase dictates that one stop and take time to look around to see if this is what one really wants before the resources are committed to putting the system into operation and production. If the idea has survived the end of the conceptual phase, a conditional approval for further study and development is given. The definition phase provides the opportunity to review and confirm the decision to continue development, create a prototype system, and make a production or installation decision.

Decisions that are made during and at the end of the definition phase might very well be decisions to cancel further work on the system and redirect organizational resources elsewhere. The elements of this phase are described in Table 9-2.

Table 9-2. Definition Phase.

1. Firm identification of the human and nonhuman resources required.
2. Preparation of final system performance requirements.
3. Preparation of detailed plans required to support the system.
4. Determination of realistic cost, schedule, and performance requirements.
5. Identification of those areas of the system where high risk and uncertainty exist, and delineation of plans for further exploration of these areas.
6. Definition of intersystem and intrasystem interfaces.
7. Determination of necessary support subsystems.
8. Identification and initial preparation of the documentation required to support the system, such as policies, procedures, job descriptions, budget and funding papers, letters, memoranda, etc.

**Production or Acquisition Phase.** The purpose of the production or acquisition phase is to acquire and test the system elements and the total system itself using the standards developed during the preceding phases. The acquisition process involves such things as the actual setting up of the system, the fabrication of hardware, the allocation of authority and responsibility, the construction of facilities, and the finalization of supporting documentation. Table 9-3 details this phase.

**The Operational Phase.** The fundamental role of the manager of a system during the operational phase is to provide the resource support required to accomplish system objectives. This phase indicates the system has been proven economical, feasible, and practicable and will be used to accomplish the desired ends of the system. In this phase the manager's functions change somewhat. He is less concerned with planning and organizing and more concerned with controlling the system's operation along the predetermined lines of performance. His responsibilities for planning and organization are not entirely neglected—there are always elements of these functions remaining—but he places more emphasis on motivating the human element of the system and controlling the utilization of resources of the total system. It is during this phase that the system may lose its identity per se and be assimilated into the institutional framework of the organization.

If the system in question is a product to be marketed, the operational stage begins the sales life cycle portion of the overall cycle, for it is in this phase that marketing of the product is conducted. Table 9-4 shows the important elements of this phase.

**The Divestment Phase.** The divestment phase is the one in which the organization "gets out of the business" which it began with the concep-

Table 9-3. Production Phase.

1. Updating of detailed plans conceived and defined during the preceding phases.
2. Identification and management of the resources required to facilitate the production processes such as inventory, supplies, labor, funds, etc.
3. Verification of system production specifications.
4. Beginning of production, construction, and installation.
5. Final preparation and dissemination of policy and procedural documents.
6. Performance of final testing to determine adequacy of the system to do the things it is intended to do.
7. Development of technical manuals and affiliated documentation describing how the system is intended to operate.
8. Development of plans to support the system during its operational phase.

Table 9-4. Operational Phase.

1. Use of the system results by the intended user or customer.
2. Actual integration of the project's product or service into existing organizational systems.
3. Evaluation of the technical, social and economic sufficiency of the project to meet actual operating conditions.
4. Provision of feedback to organizational planners concerned with developing new projects and systems.
5. Evaluation of the adequacy of supporting systems.

tual phase. Every system—be it a product system, a weapons system, a management system, or whatever—has a finite lifetime. Too often this goes unrecognized, with the result that outdated and unprofitable products are retained, inefficient management systems are used, or inadequate equipment and facilities are "put up with." Only by the specific and continuous consideration of the divestment possibilities can the organization realistically hope to avoid these contingencies. Table 9-5 relates to the divestment phase.

Taken together, Tables 9-1 through 9-5 provide a detailed outline of the overall systems development life cycle. Of course, the terminology used in these tables is not applicable to every system which might be under development, since the terminology generally applied to the development of consumer product systems is often different from that applied to weapons systems. Both, in turn, are different from that used in the development of a financial system for a business firm. However, whatever the terminology used, the concepts are applicable to all such systems.

Table 9-5. Divestment Phase.

1. System phasedown.
2. Development of plans transferring responsibility to supporting organizations.
3. Divestment or transfer of resources to other systems.
4. Development of "lessons learned from system" for inclusion in qualitative-quantitative data base to include:
   a Assessment of image by the customer
   b Major problems encountered and their solution
   c Technological advances
   d Advancements in knowledge relative to department strategic objectives
   e New or improved management techniques
   f Recommendations for future research and development
   g Recommendations for the management of future programs, including interfaces with associate contractors
   h Other major lessons learned during the course of the system.

For instance, Curling* has identified several phases of a major Canadian weapons acquisition program comprised of a series of decisions organized around the complete system life-cycle objective. These phases are *Conception, Definition, Acquisition, Service,* and finally, *Disposal.* Within these phases Curling identifies the following elements:

a. Policy Planning and Project Initiation.
b. Project Definition.
c. Full-Scale Project Development.
d. Project Systems Integration.
e. Project Test and Evaluation.
f. Project Production.**

Pandia describes a project as typically consisting of yet other phases:

- Identification
- Formulation
- Evaluation
- Detailed Planning
- Design and Engineering
- Procurement
- Construction/Execution
- Completion
- Post-completion activities***

## LIFE-CYCLE MANAGEMENT

Life-cycle management refers to the management of systems, products, or projects throughout their life cycle. In the context of the sales life cycle, life-cycle management is usually called "product management." In the development life cycle, it is usually called "project management." In all cases, life-cycle management is needed because the *life cycle reflects very different management requirements at its various stages.*

*The traditional hierarchical organization is not designed to cope with the constantly changing management requirements dictated by life cy-*

---

* David H. Curling, "A Personal Perspective of Acquisition (Equipment) Project Management," *1985 PMI Proceedings,* Vol. 2, Denver, pp. 1–12.
** Ibid., p. 3.
*** Rajeev M. Pandia, "Excellence in Integration of Project Phases," *1985 PMI Proceedings,* Vol. 2, Denver, pp. 1–2.

*cles*. It is established to effectively direct and control a much less dynamic milieu.

### Variability of Input and Output Measures for Various Stages of the Life Cycle

The dynamism that is inherent in the life cycle is made apparent when one considers the variability in the measures that may be used to appropriately describe the inputs to, and outputs from, a system as it goes through its life cycle.

Such measures vary widely. For instance, in developing a new product, one might characterize the various phases of the project life cycle in terms of the proportional composition of the work force assigned to the activity. In the beginning, research personnel predominate; subsequently, their role diminishes and engineers come to the forefront; finally, marketing and sales personnel become most important.

Basic life-cycle concepts hold for all projects and systems. Thus an organizational system develops and matures according to a cycle which is much like that of a product. The measures used to define various phases of an organization's life cycle might focus on its product orientation, for example, defense versus nondefense, its personnel composition, for example, scientists versus nonscientists, its per-share earnings, etc. For a management information system, the life cycle might be characterized by the expenditure level during the developmental phase together with the performance characteristics of the system after it becomes operational.

A hardware system displays no sales performance after it is in use, but it does display definite phases of operation. For example, Figure 9-2 shows a typical failure rate curve for the components making up a com-

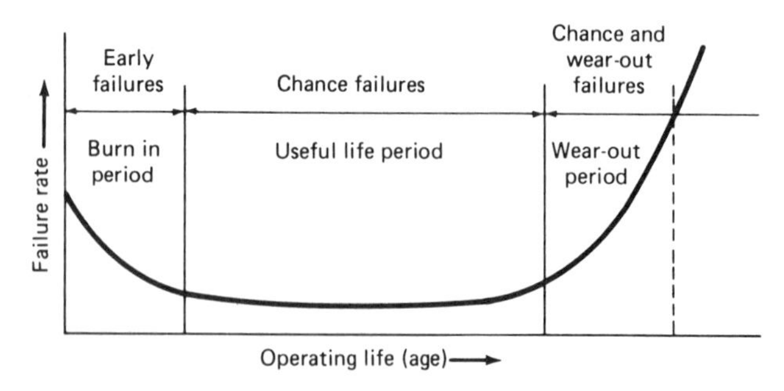


Figure 9-2. Component failure rate in a system as a function of age.

plex system. As the system is first put into operation, the failure rate is rather high because of "burn in" failures of weak components. After this period is passed, a relatively constant failure rate is experienced for a long duration; then, as wear-outs begin to occur, the component failure rate rises dramatically.

Perhaps a comparison of Figures 9-1 and 9-2 best illustrates the pervasiveness of life-cycle concepts and the importance of assessing the life cycle properly. Figure 9-1 represents a sales life cycle for a product. The most appropriate measure to be applied to this product's sales life cycle is "sales rate." Figure 9-2 shows the operating life cycle of a hardware system—for instance, a military weapons system. The concept is the same as that of Figure 9-1, but the appropriate measurement is different. In Figure 9-2 the "failure rate" is deemed to be the most important assessable aspect of the life cycle for the purpose for which the measurement will be used.

### Life-Cycle Management Dimensions

The variability of the various input and output measures and the fact that different measures may be more appropriate at one stage of the life cycle than at another suggest that project management must focus on certain *critical generic project dimensions*. These dimensions are *cost, time and performance*.

*Cost* refers to the resources being expended. One would want to assess cost sometimes in terms of an expenditure rate (e.g., dollars per month) and sometimes in terms of total cumulative expenditures (or both).

*Time* refers to the timeliness of progress in terms of a schedule which has been set up. Answers to such questions as: "Is the project on schedule?" and "How many days must be made up?" reflect this dimension of progress.

The third dimension of project progress is *performance:* that is, how is the project meeting its objectives or specifications? For example, in a product development project, performance would be assessed by the degree to which the product meets the specifications or goal set for it. Typically, products are developed by a series of improvements which successively approach a desired goal, for example, soap powder with the same cleaning properties but less sudsiness. In the case of an airplane, certain requirements as to speed, range, altitude capability, etc., are set and the degree to which a particular design in a series of successive refinements meets these requirements is an assessment of the performance dimension of the aircraft design project.

### Managing Over the Life Cycle

Since the mix of resources (inputs) and outputs associated with a project varies through the life cycle, the implication is strong that the appropriate techniques and strategies of management also vary during the various phases. Indeed, the need for management flexibility across the life cycle is one of the primary reasons that the traditional hierarchical organization is inadequate in dealing with project-intensive management situations.

The specific implications to project management are presented elsewhere in this volume. Table 9-6 shows a broad set of management strategies, developed by Fox,* that are associated with a five-stage life cycle:

1. Precommercialization.
2. Introduction.
3. Growth.
4. Maturity.
5. Decline.

The first stage of Fox's life cycle may be roughly thought of as the development life cycle that has itself previously been treated in terms of a number of stages. The remaining four stages represent the sales life cycle.

Table 9-6 clearly indicates the extreme variability in management strategy and outlook that is necessitated by the dynamics of the life cycle. The prospect of such flexibility being developed in the context of a traditional hierarchical organization, designed primarily to ensure efficiency and control, is remote. Therefore, the implications of life cycles to both the need for, and practice of, project management are straightforward.

### Overall Organization Management Implications

An organization can be characterized at any instant in a given time by a "stream of projects" that place demands on its resources. The combined effect of all the "projects" facing an organization at any given time determines the overall status of the organization at that time.

The projects facing a given organization at a given time typically are diverse in nature—some products are in various stages of their sales life cycles, other products are in various stages of development, management subsystems are undergoing development, organizational subsystems are

---

* Harold W. Fox, "A Framework for Functional Coordination," *Atlanta Economic Review*, Vol. 23, No. 6 (1973), pp. 10–11. Used with permission.

Table 9-6. Fox's Hypotheses About Appropriate Business Strategies over the Product Life Cycle

| | FUNCTIONAL FOCUS | R&D | PRODUCTION | MARKETING | PHYSICAL DISTRIBUTION |
|---|---|---|---|---|---|
| Precommercialization | Coordination of R&D and other functions | Reliability tests<br>Release blueprints | Production design<br>Process planning<br>Purchasing dept. lines up vendors & subcontractors | Test marketing<br>Detailed marketing plan | Plan shipping schedules, mixed carloads<br>Rent warehouse space, trucks |
| Introduction | Engineering: debugging in R&D production, and field | Technical corrections (Engineering changes) | Subcontracting<br>Centralize pilot plants; test various processes; develop standards. | Induce trial; fill pipelines; sales agents or commissioned salesmen; publicity | Plan a logistics system |
| Growth | Production | Start successor product | Centralize production<br>Phase out subcontractors<br>Expedite vendors ouput; long runs | Channel commitment<br>Brand emphasis<br>Salaried sales force<br>Reduce price if necessary | Expedite deliveries<br>Shift to owned facilities |
| Maturity | Marketing and logistics | Develop minor variants<br>Reduce costs thru value analysis<br>Originate major adaptations to start new cycle | Many short runs<br>Decentralize<br>Import parts, low-priced models<br>Routinization<br>Cost reduction | Short-term promotions<br>Salaried salesmen<br>Cooperative advertising<br>Forward integration | Reduce costs and raise customer service level<br>Control finished goods inventory |

Table 9-6. Fox's Hypotheses About Appropriate Business Strategies over the Product Life Cycle (continued)

| | FUNCTIONAL FOCUS | R&D | PRODUCTION | MARKETING | PHYSICAL DISTRIBUTION |
|---|---|---|---|---|---|
| | | | | Routine marketing research; panels, audits | |
| Decline | Finance | Withdraw all R&D from initial version | Revert to subcontracting; simplify production line<br>Careful inventory control; buy foreign or competitive goods; stock spare parts | Revert to commission basis; withdraw most promotional support<br>Raise price<br>Selective distribution<br>Careful phase-out, considering entire channel | Reduce inventory and services |

| | PERSONNEL | FINANCE | MANAGEMENT ACCOUNTING | OTHER | CUSTOMERS | COMPETITION |
|---|---|---|---|---|---|---|
| Precommercial-ization | Recruit for new activities<br>Negotiate operational changes with unions | LC plan for cash flows, profits, investments, planning; full costs, revenues<br>Determine optimum lengths of LC stages thru present-value method | Final legal clearances (regulatory hurdles, patents)<br>Appoint LC coordinator | | Panels & other test respondents | Neglects opportunity or is working on similar idea |

| | | | | | | |
|---|---|---|---|---|---|---|
| Introduction | Staff and train middle management<br>Stock options for executives | Accounting deficit; high net cash outflow<br>Authorize large production facilities | Help develop production & distribution standards<br>Prepare sales aids like sales management portfolio | | Innovators and some early adopters | (Monopoly)<br>Disparagement of innovation<br>Legal & extra-legal interference |
| Growth | Add suitable personnel for plant<br>Many gievances<br>Heavy overtime | Very high profits, net cash outflow still rising<br>Sell equities | Short-term analyses based on return per scarce resource | | Early adopters & early majority | (Oligopoly): A few imitate, improve, or cut prices |
| Maturity | Transfers, advancements; incentives for efficiency, safety, and so on<br>Suggestion system | Declining profit rate but increasing net cash inflow | Analyze differential costs revenue<br>Spearhead cost reduction, value analysis, and efficiency drives | Pressure for resale price maintenance<br>Price cuts bring price wars; possible price collusion | Early adopters, early & late majority, some laggards; first discontinued by late majority | (Monopoly competition<br>First shakeout, yet many rivals |
| Decline | Find new slots<br>Encourage early retirement | Administer system, retrenchment<br>Sell unneeded equipment<br>Export the machinery | Analyze escapable costs<br>Pinpoint remaining outlays | Accurate sales forecast very important | Mainly laggards | (Oligopoly)<br>After 2nd shakeout, only few rivals |

in transition, major decision problems such as merger and plant location decisions have been "projectized" for study and solution, etc.

Moreover, at any given time each of these projects will typically be in a different phase of its life cycle. For instance, one product may be in the conceptual phase undergoing feasibility study, another may be in the definition phase, some are being produced, and some are being phased out in favor of oncoming models.

The typical situation with products which are in the sales portion of their overall life cycle is shown in Figure 9-3, as projected through 1995 for the sales levels of three products, A, B, and C. Product B is expected to begin sales in 1988 and to be entering the declining sales phase of its cycle after 1991. Product A is already in the midst of a long declining sales phase. Product C is in development and will not be marketed until 1990. At any moment in time, each is in a different state. In 1991, for example, A is in a continuing decline, B is beginning a rather rapid decline, and C is just expanding rapidly.

Whatever measure is chosen to represent the activity level or state of completion of each of the projects in the stream facing an organization—be they products, product-oriented projects, management system-development projects, or decision-oriented projects—the aggregate of all of the projects facing the organization represents a stream of projects which it must pursue. Although the same measures (e.g., revenues, resources employed, percent completed, etc.) will not normally be applicable to all projects, the idea of a stream of projects—each at a different phase of its life cycle—is applicable to assessing the state of any dynamic organization.

The overall management implications of the stream of projects are clear from Figure 9-3. Top managers must plan in terms of the project stream.

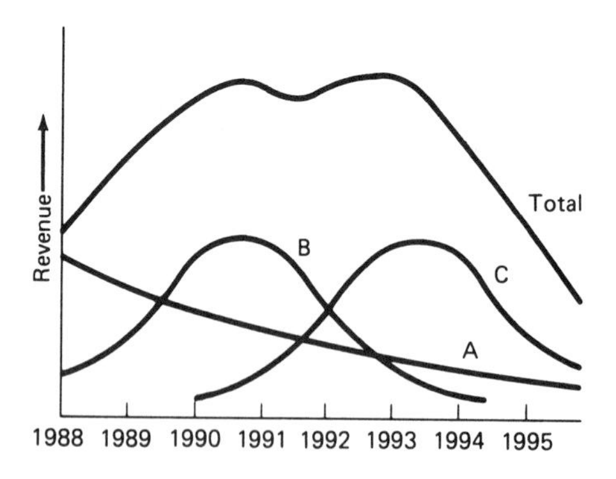


Figure 9-3. Life cycles for several products.

Since overall results are the sum of the results produced by the various projects, this planning must not only be in terms of long-run goals, but also in terms of the various steps along the way.

Most organizations do not wish to have their overall results appear to be erratic, so they must be concerned with the "sum" of the project stream at various points in time. A number of projects, each of which is pursuing a future goal quite well, might in sum not appear to be performing well at some points before they have reached their respective goals. In each case, for instance, although the progress of each may be adequate, each may still be in a phase of its life cycle where it is consuming resources more than it is producing results.

Thus, the problems associated with the overall management of an organization that is involved in a stream of projects are influenced by life cycles just as are the problems associated with managing individual projects.

# 10. Behavioral Implications of the Project Life Cycle*

John R. Adams†
Stephen E. Barndt‡

## INTRODUCTION

During the last several decades, the technology needed to design and develop major new products has become increasingly diverse and complex. The increasing diversity arises from the need to integrate an ever-widening variety of professional and technical specialties into new product design. The increasing complexity results from the rapid rate at which knowledge has grown in each major technical specialty contributing to the product. This growth has spawned a large number of professional subareas as specialists strive to remain current in ever-narrowing technical fields. The increasing technical sophistication of major new products is widely recognized. What is not well known, however, is that the increasing diversity and complexity have caused extensive innovation in man-

---

* Portions of the material presented in this chapter were published earlier in "Organizational Life Cycle Implications for Major R & D Projects," *Project Management Quarterly*, Vol. IX, No. 4 (December 1978), pp. 32–39.

† Dr. John R. Adams holds the Ph.D. in Business Administration from Syracuse University. He is currently an Associate Professor of Organization and Management at Western Carolina University, Cullowhee, North Carolina and Director of their Master of Project Management Degree Program. Dr. Adams has published a number of articles on various aspects of project management, risk and uncertainty analyses, weapon systems acquisition, and logistics management in nationally distributed journals, and is a frequent speaker at national professional and at Department of Defense symposia. His experience includes management of major weapon systems acquisition programs and supervision of a major Air Force research laboratory.

‡ Stephen E. Barndt is A Professor of Management at the School of Business Administration, Pacific Lutheran University, Washington. Dr. Barndt, who earned his Ph.D. degree from The Ohio State University, has directed research into the behavioral aspects of project management and has published articles on that subject, among others. In addition, he has co-authored texts on project management and operations management. Dr. Barndt's project management experience includes performing as an assistant program manager and as an R&D project administrator.

agement systems for developing those products, resulting in the application of project management techniques to most major, advanced-technology, nonrepetitive efforts aimed at designing and developing new products or services.

The form of management known as project management was designed to ". . . provide sustained, intensified, and integrated management of the complex ventures" (1) and to pull together a combination of human and nonhuman resources into ". . . a temporary organization to achieve a specified purpose" (2). A project organization is established for a limited period of time to accomplish a well defined and specified set of objectives—to bring a new idea for a product through its conceptual and developmental phases to the point where the new product is available for use. When these carefully defined objectives are accomplished, the project is completed and the project organization is terminated. Thus a project has a clear, finite, and well-defined life cycle, a fact which has long been used to differentiate "projects" from the more traditional, long-term "functional" organization.

The field of management, as applied to complex organizations, has been the beneficiary of a growing body of knowledge. In particular, a great deal of research has been conducted in recent years to define organizational variables and evaluate their effect on the ongoing, functional organization. Little of this general material has reached the project management literature, however, and little specific research has been conducted to identify the specific organizational factors crucial to the project management field. This is not too surprising, since the modern concept of a project cutting across corporate, industry, and governmental boundaries to develop advanced-technology products is not much more than two decades old. A number of detailed topics relevant to project organizations, such as selecting an appropriate project manager (3), developing an effective, cross-functional, network-based management information system (4), and improving productivity in project management (5), have recently been more or less intensively investigated; and research efforts on such specialized project management topics continue. Little has been done, however, to understand the broader implications of the project life cycle. In particular, no comprehensive study exists which investigates how the project life cycle may influence and change the anticipated results of the organizational variables traditionally used to analyze functional organizations.

This chapter reviews a set of accepted organization theory variables for their potential impact on projects across the project life cycle. It then integrates the results of several independently conducted but mutually supporting cross-sectional studies involving over 20 major research and development (R&D) projects. The studies were designed specifically to analyze the impact of accepted organization theory variables over the life cycle of major R&D projects. The purpose is to suggest a set of propositions which will allow the practitioner and theoretician to predict the behavior of projects through their life cycles in terms of accepted organization theory variables. The results should improve our ability to plan for and manage the unique organizational problems likely to be experienced on projects because the project life cycle exists.

## RESEARCH PROGRAMS AND PROJECTS

The concept of advanced-technology research and development projects has resulted from the need to develop ever-larger and more complex products for military, space, and commercial applications. The production and marketing strategies for such products frequently fail to fit within the constraints of a purely functional organization structure. The largest of these projects, for example the manned moon landing program, involves efforts which are simply too large for any single organization to deal with. These programs are typically sponsored and funded by a government organization such as the National Aeronautics and Space Administration (NASA), the Department of Transportation (DOT), the Department of Energy (DOE), or one of the Department of Defense agencies (the Army, Navy, or Air Force). The government agency provides the funds and overall managerial coordination. Private corporations, on the other hand, act as contractors or subcontractors, and each develops its own individual project which is responsible for achieving some portion of the overall program's goals. For example, when a new aircraft is being developed, there may be one major company responsible for designing the airframe, another for producing the engine, and a third for the avionics system, while still another develops the maintenance and support subsystems for the overall program. Similarly, in large privately sponsored projects such as developing new commercial aircraft, oceangoing ships, or offshore resource locating and extracting platforms, one firm might typically perform as the "prime," or integrating, contractor. Specific hardware and other development tasks are then performed on a contract basis by other firms or other divisions of the same firm. Each contributing organization thus supports its own major project whose output must contribute to the overall program objectives, while the sponsoring agency concentrates on coordinating the activities of the contributing organizations to meet overall schedule, budget, and performance objectives. In this situation the term "program" refers to the overall effort to achieve the end objective, a new aircraft or a "man on the moon," while the term "project" refers to an individual organization's activities leading to accomplishing its specialized goals in support of the program. The basic theory of the life cycle applies to both projects and programs, with the program milestones reflecting major accomplishments in one or more of the supporting projects.

## THE PROJECT LIFE CYCLE

Special-purpose project organizations are molded around the specific goal or task to be accomplished. The essence of project management lies in planning and controlling one-time efforts, thus encompassing the managerial aspects of both projects and programs. The project organization exists only to solve some specified problem, generally one in which the "parent" or sponsoring organization has little or no prior experience. This description summarizes most current major developmental efforts, and therefore explains the dependence of such efforts on the concepts of project management. Both projects and programs draw from the same management theory base. The term "project management" is thus used in this chapter to apply to the management of both "projects" and "programs."

In current major project management efforts, the sponsor usually needs to develop some new product or system within critical predetermined (a) performance specifications, (b) time constraints, and (c) budget limitations. These, then, define the project's goal. Once the goal is satisfied, the project loses its purpose for existing and is dissolved. This is why the project organization exhibits a predictable life cycle: it is frequently said to be "born" when the sponsoring organization accepts responsibility for the problem and decides to accomplish the goal through project management; it "grows" and expands through the planning and initial execution phases as larger increments of money, personnel, production facilities, managerial time, and other such resources are devoted to the effort; it declines as the goal nears completion and resources that are no longer required are reassigned to other work efforts; and it "dies" when responsibility for the new product or system is turned over to the ongoing functional organization—the ultimate "customer" of the entire project. The project organization itself exists primarily to focus the undivided attention of key management and technical specialists on the task of resolving the specified problem across the life span of that problem's existence (6).

As a project proceeds through its life cycle it passes through an identifiable sequence of phases, distinguished from each other by the type of tasks characteristic of each phase and frequently by formal decision points at which it is determined if the project has been sufficiently suc-

Table 10-1. Managerial actions by project phase.

| PHASE I CONCEPTUAL | PHASE II PLANNING | PHASE III EXECUTION | PHASE IV TERMINATION |
|---|---|---|---|
| Determine that a project is needed. | Define the project organization approach. | Perform the work of the project, (i.e., design, construction, production, site activation, testing, delivery, etc.) | Assist in transfer of project product. |
| Establish goals. | Define project targets. | | Transfer human and nonhuman resources to other organizations. |
| Estimate the resources the organization is willing to commit. | Prepare schedule for execution phase. | | Transfer or complete commitments |
| "Sell" the organization on the need for a project organization. | Define and allocate tasks and resources. | | Terminate project |
| Make key personnel appointments. | Build the project team. | | Reward personnel |

cessful in the earlier phases to continue on into the next (7). Different authors identify from three to six separate phases, and there is no agreement on terminology. Nevertheless, general agreement does exist to indicate that each project phase involves different management considerations and presents different tasks to be performed (8). It should be noted that this involves two distinctly different views of the project. Table 10-1 identifies four project phases and specifies the general actions that must be taken by the sponsoring organization's management, and later by the senior project management, during each phase. Figure 10-1, on the other hand, identifies the same life-cycle phases but defines them in terms of the type of tasks that must be accomplished in that phase to prepare for transition into the next. In modern, major, high-technology research and development programs, the transition points between phases may be marked by formal program reviews held by the highest level of management in the sponsoring organization. These reviews are designed to authorize the resource expenditures necessary for the project to proceed into the next phase. It thus appears reasonable to classify projects according to the phase of the life cycle they are engaged in at the time of study, and to analyze major organizational variables as they affect the projects in the various life-cycle phases.

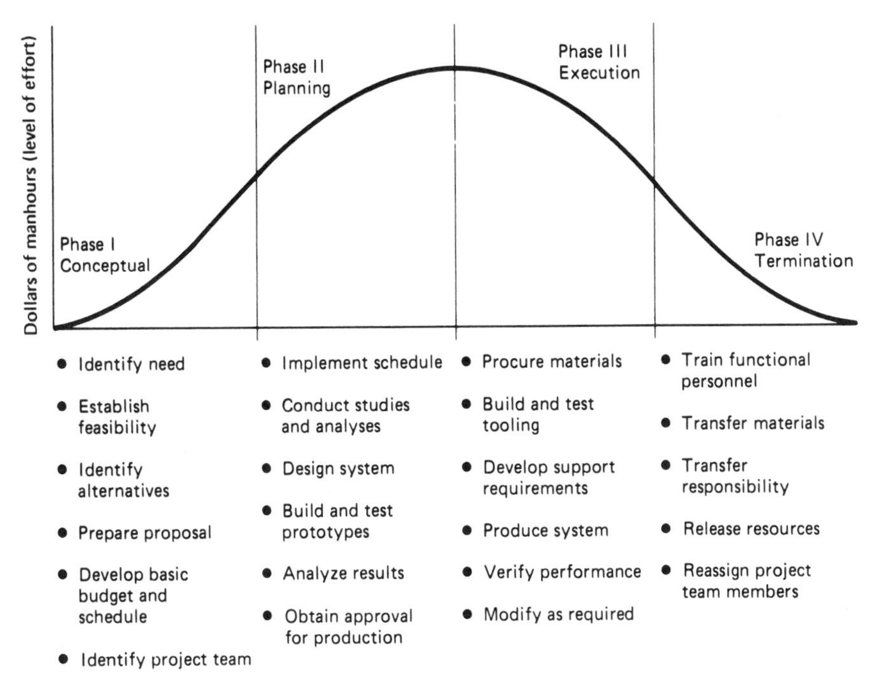


Figure 10-1. Tasks accomplished by project phase.

## RELATIVE ORGANIZATIONAL VARIABLES

A large number of variables have been investigated over the years for their effect on the ongoing, traditional type of organization. The variables discussed below were analyzed for their relevance to project management because of their wide acceptance as important variables in analyzing organizations and because a research-based body of knowledge has developed concerning each one. All of these varaibles—organizational climate, conflict, satisfaction, size, and structure (level of bureaucracy)—are free to change as a project progresses through its life cycle. In addition, the project-peculiar variable "phase of project life cycle," discussed previously, requires further elaboration as it relates to the more universally applicable variables.

### Organizational Climate

Organizational climate is a description of the organization as a whole (9). Litwin and Stringer defined organizational climate as ". . . a set of measurable properties of the work environment perceived directly or indirectly by the people who live and work in this environment and assumed to influence their motivation and behavior'' (10). These authors went on to suggest that both satisfaction and performance are affected by climate. Hellriegel and Slocum in their review of the literature cited several studies that clearly indicated a relationship between job satisfaction and organizational climate (11). However, the nature of the relationship between climate and performance is less clear. Although Likert (12) and Marrow, Bowers, and Seashore (13) found a more positive climate to be associated with higher productivity, Hellriegel and Slocum (11) cited both support and nonsupport for this finding and concluded there was no consistent relationship. The linkage, if any exists, may be indirect. Other variables may intervene between climate and performance. In addition to job satisfaction, intervening variables of particular interest may include conflict sources, conflict intensity, and conflict resolution modes. Climate may influence individuals to perceive more or less satisfaction with their work and to consequently be motivated more or less to perform to their capability. With respect to conflict, differing sources and intensities of conflict may result from or influence the climate. The combination of climate and conflict sources may in essence dictate or at least constrain the appropriateness of conflict resolution techniques. Use of these techniques may in turn influence climate and satisfaction.

## Conflict

The essence of project management is that it is interfunctional and is frequently in conflict with ''normal'' organization structure and procedures, leading to a natural conflict system (14). The ability of the project manager to foster useful conflict, or to convert disruptive to useful conflict, can often determine his degree of success in achieving the project's goals (15). Thus one of the project manager's key functions is to maintain, in the face of conflicting objectives, a reasonable degree of harmony among the many organizational elements contributing to the project. Research conducted by Evan is important in confirming that differences in conflict do exist between the traditional functional organization and the project organization (16). Since both size and formalization of project organizations may vary over the life cycle, it would appear logical to investigate the changes in conflict that could also develop. Thamhain and Wilemon have done so for a variety of small, industrial projects. They found that the mean intensity of conflict from all sources, the pattern of conflict arising from various specified sources, and the conflict resolution modes used by project managers all vary systematically over the project life cycle (17). The purpose of one study (18) reported later in this chapter was to extend the Thamhain and Wilemon findings to determine their relevance to the major R&D project environment.

### Job Satisfaction

Payne et al. described job satisfaction as an individual's affective response to his job (9). Although long a subject of research, the relationship of job satisfaction to performance is by no means settled. The preponderance of evidence indicates that the ties between job satisfaction and productivity or other measures of performance are weak or inconclusive (19). However, even the weak indicators of such a relationship should not be put aside lightly (20). In addition, there is conceptual appeal that such a link ought to exist in many situations.

### Organization Size

Size has been shown to have a strong effect on perceived organizational climate in several manufacturing organizations. Payne and Mansfield, using a modified Business Organization Climate Index, reported a relatively strong positive relationship between size and most climate scales (21). Particularly noteworthy were the reported strong relationships between size and readiness to innovate, task orientation, job challenge, and scientific and technical orientation. All of these are climate dimensions that could be expected in many project organizations. Size may also be related to climate and other behavioral variables indirectly through its influence on the nature of the organization. Research has generally shown the existence of a positive relationship between size and organizational formalization (22). Increased size may dictate more links in the scalar chain, requiring greater formalization of communication and reporting systems. Increased size may also permit economies through greater functional specialization. As a consequence, the larger project organization may tend to be more functionally structured or mechanistic in nature.

### Level of Bureaucracy

The level of bureaucracy may be defined as a continuum ranging from a mechanistic to an organic organizational structue. A mechanistic structure refers to an organization with communication directed primarily downward, high formalization of rules and procedures, adherence to the chain of command, low intergroup cooperation, and infrequent task feedback. An organic structure is characterized by high intergroup cooperation, frequent task feedback, open communication channels, low formalization of rules and procedures, and a lack of adherence to the chain of command. The latter characteristics describe the usual conception of a project organization. However, in large projects, particularly those related to major advanced-technology research and development programs, managers such as those producing program control documentation or running a project's information system may work in an environment differing little from that of the mechanistic organization where authority generally matches responsibility. A pure project manager, on the other hand, operating in an organic environment, may have responsibilities that far outreach his formal authority to marshal and direct the needed resources (15). Major project organizations may thus display a mixture of organic and mechanistic characteristics which could vary over the life cycle and have a major influence on the effectiveness of managerial actions.

### Phase of Life Cycle

As a project progresses from the conceptual phase through the termination phase, the relative degree of uncertainty associated with the determination and performance of tasks decreases and the extent to which routine is applied to task accomplishment increases as shown in Figure 10-2. Studies of other types of organizations have shown a general tendency for organizations with either task routineness or reduced environmental un-

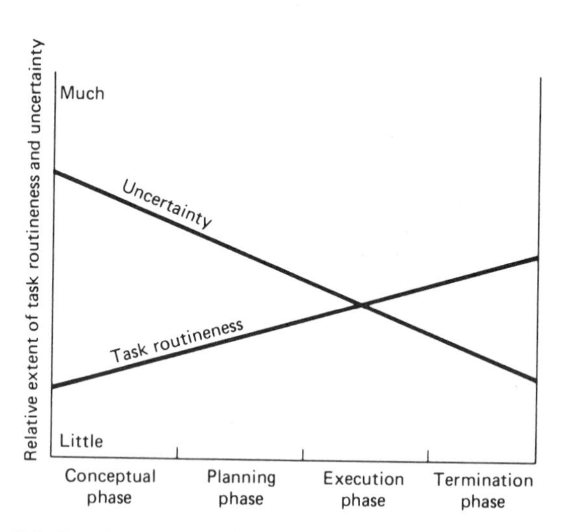


Figure 10-2. Routineness and uncertainty across the project life cycle.

certainty to become structurally more formalized and centralized (22). If the same relationship holds true in project organizations, and without the additional influence of organization size, the expectation would be that formalization (or level of bureaucracy) would increase as the project progresses. Further, since the major actions and activities change among the various phases, with differing pressures and problems arising, it should be expected that organizational climate, conflict sources, and conflict intensity would vary.

Based on the support from the literature cited above, and on several years of the authors' personal observation and research experience, Figure 10-3 was developed to demonstrate the relationships predicted to exist among the variables. As implied in the figure, size, degree of formalization (level of bureaucracy), organizational climate, and conflict source and intensity are at least partly a function of the peculiar problems and tasks that differentiate the various life-cycle phases. Organizational climate is also probably influenced by the degree of formalization of structure (level of bureaucracy), the general size of the organization itself, and the sources and intensities of conflict. Conversely, the intensity of conflict can be expected to increase as the organizational climate becomes less favorable. The degree of formalization of structure is expected to be a determinant of the sources of conflict as well as its intensity. The modes for resolving conflict should depend on the sources of those conflicts, their intensities, and the favorableness of the organizational climate. Job or work satisfaction is expected to be influenced by the overall organizational climate, the sources of conflict, the intensity of conflict, and the

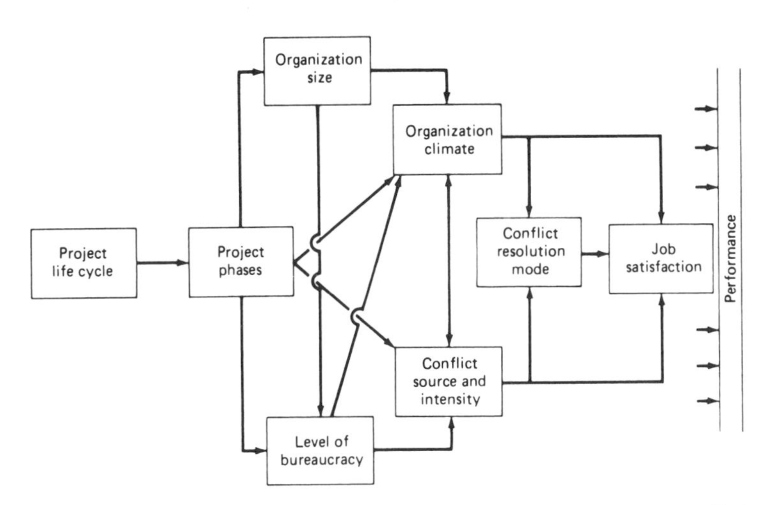


Figure 10-3. Predicted relationships among the organizational variables studied.

methods employed to resolve those conflicts. In general, the appropriateness of structure for task accomplishment; the extent and sources of conflict; the appropriateness, effectiveness, and acceptability of the methods employed in conflict resolution; the degree of favorableness or organizational climate; and the extent to which team members are satisfied with their work and work situation may all be expected to impact relative project performance.

## PROJECT-SPECIFIC STUDIES

The authors have conducted or directed a number of studies designed to analyze the impacts of these traditionally accepted organizational variables over the several life-cycle phases of major R&D projects in terms of three phases distinguishable from each other by the type of tasks being performed, as well as by clearly defined, formal project reviews resulting in authorization for the project to proceed into the next phase. In several cases it is possible to logically divide the third phase into a third and fourth phase, two separate phases based on the type of tasks being performed. The separate research efforts culminated in the four notable reports summarized in Table 10-2. All data were obtained from the same organizational environment, although at different times. The sources of data were program offices of the United States Air Force charged with managing the research and development activities associated with acquiring new aircraft weapon systems. The responsibilities of these offices included conceptual studies, concept validation, hardware demonstration, prototype development, test article development and fabrication, test and evaluation, production, modification, and initial support activities. Program offices ranged in size from very small, approximately five individuals, to an office of more than two hundred.

Most data analyzed in the study was generated through use of questionnaires. Standard instruments, modified as necessary, were used to measure satisfaction, organizational climate, source of conflict, conflict intensity, and method of conflict resolution. Measures of organizational climate were obtained through use of the short-form version of the Likert Profile of Organizational Characteristics (12). The summed score of all questionnaire items may be considered an indicator of the individual's perception of the general style or system of management prevalent in the organization. The average of the scores for all project team personnel sampled in an organization or group of organizations was considered to represent the climate relative to that of other organizations and relative to an ideal climate of openness, support, trust, and participation. Job satisfaction was measured by use of the satisfaction scales from the Job Diag-

Table 10-2. Research Data.

| SOURCE | SAMPLE | DATA COLLECTION | PERTINENT FINDINGS |
|---|---|---|---|
| Lempke and Mann (26) | 142 program managers (95% response) randomly drawn from 13 program offices representing three phases of project life cycle. | Questionnaire, personally distributed, yielded data on organizational nature of tasks, phase of life cycle, and size of organization. | Organizations are most project oriented in early phase of project life, least project oriented in middle phase of project life. Organizations are smallest in early phase, largest in middle phase. |
| Barndt, Larsen, and Ruppert (24) and Haddox and Long(25) | 185 program managers (80% response) randomly drawn from 13 program offices representing three phases of project life cycle. | Questionnaire, mailed to subject, yielded data on organizational climate, satisfaction, organizational size, and phase of life cycle. | 1. Significant differences in organizational climate exist among phases. 2. Significant differences in organizational climate exist among program offices of different sizes. Organizational climate is correlated with satisfaction. |
| Eschman and Lee(18) | 136 program managers (68% response) randomly drawn from 20 program offices representing four phases of project life cycle. | Questionnaire, personally distributed, yielded data on sources of conflict, intensity of conflict, method of conflict resolution, and phase of life cycle. | Conflict intensity changed across program life cycle, Air Force program managers perceived less intensity of conflict than civilian project managers, and Air Force and civilian project managers agreed on conflict resolution modes across life cycle phase. |

1. Findings of the Barndt, Larsen, and Ruppert study.
2. Findings of the Haddox and Long study.

nostic Survey short form (23). The seven scales indicated in Table 10-3 provide separate measures of the individual's affective reactions or feelings obtained from actually performing at his job. Sources of conflict, conflict intensity, and method of conflict resolution were measured using a questionnaire developed by Thamhain and Wilemon (17) modified to fit the Air Force program environment. The questionnaire includes the seven potential conflict sources and five conflict-handling modes identified in Table 10-3, essentially measuring the frequency of occurrence of each. Finally, the level of bureaucracy was measured with a number of specifically designed questions.

The synthesis presented in this chapter involved extracting values of the major variables from the various studies and matching them with phases of the project life cycle. Analysis was necessarily restricted to identifying and demonstrating differences across life-cycle phases. No attempt is made to support the existence of cause-and-effect relationships.

## BEHAVIORAL CHARACTERISTICS OF PROJECTS BY PHASE

The combination of study results supports the existence of marked similarities in the organizational environments of major R&D projects within the identified life-cycle phases. They also document several significant differences in the organizational environments characteristic of the separate phases. These similarities and differences are summarized in Table 10-3. Those studies which did not distinguish the execution phase from the termination phase drew most of their data for this portion of the life cycle from projects involved in tasks more descriptive of the execution phase, so it is assumed that the data presented for phases III (execution) and IV (termination) combined more nearly represent projects in phase III of the life cycle.

### Phase I

Data from projects in the conceptual phase consistently indicate that the teams are small and use a relatively organic type of structure. The overall climate is rated low in the System 4 or participative management portion of Likert's scale. Conflict intensity is highest in this phase. Manpower resources is the most noteworthy source of this conflict, with the next four sources listed grouped closely behind as generators of conflict. While confrontation (in a problem-solving mode) is the favored means of conflict resolution, smoothing and compromise are also well-used techniques. The highest levels of satisfaction are found in this phase for six of the seven measures. Thus the organizational environment of Phase I is indicative of small participative work groups, with the members working together under considerable conflict and with a relatively informal set of work rules. They resolve their differences in a generally collegial manner with apparent concern for the feelings of others, and they derive considerable satisfaction from the work.

These findings are consistent with the theory of the life cycle and with observations of projects in action. This phase is basically concerned with preplanning activities, deciding that a project is required, and establishing the overall objectives and goals. These activities take place either before or during the identification of key personnel for assignment to the project. Thus there are only a few people knowledgeable of the project, and they must work together as a small, cooperative team to identify the work that needs to be performed. This is a truly innovative portion of the project effort, and it is generally accepted that conflict and innovation are necessary partners (15).

### Phase II

Data from projects in the planning phase indicate a substantial increase in project organization size, a multiple of five to eight relative to the sizes encountered in Phase I. The type of structure is generally organic but with significant mechanistic characteristics, while the climate is rated midrange in a System 3, or consultative, type of organization. Conflict intensity is lower than in Phase I. Program priorities is the predominant conflict source, with the next four listed sources grouped closely together but in a clearly subordinate position. Confrontation and compromise are the preferred conflict resolution modes, while smoothing has decreased and forcing has increased in importance relative to Phase I. Internal work motivation rates high in this phase but is not supported by the other job satisfaction measures, indicating that overall satisfaction is not high in relative terms. The organizational environment of Phase II can be characterized as a relatively large work group organized along semiorganic lines with mechanistic tendencies—a consultative system. The members work together under considerable conflict which arises predominantly from project-oriented priorities, schedules, and technical issues. Differences are resolved in a generally collegial manner, but job satisfaction is not particularly high.

Here is where the project gets planned in detail, where budgets are defined and priorities are established. The work breakdown structure is developed to break the project effort into its individual tasks, while the planning and control networks are designed for imposing project priorities. The work group is expanding rapidly, so many relative strangers must work together. Simultaneously, the group is breaking into subunits to accomplish different aspects of the task, and these subgroups must immediately compete with others for priorities and resources. There

Table 10-3. Structural and Behavioral Characteristics of Phases.

| VARIABLE | PHASE I CONCEPTUAL | PHASE II PLANNING | PHASE III EXECUTION | PHASE IV TERMINATION |
|---|---|---|---|---|
| *SIZE* (average number of managerial and technical personnel) | 15<br>(range 11 to 18) | 114<br>(range 49 to 169) | 102<br>(range 42 to 207) | 38<br>(range 30 to 46) |
| *LEVEL OF BUREAUCRACY* (average score between pure mechanistic, 1.0, and organistic, 7.0) | 5.26 | 4.70 | 5.21 | |
| *ORGANIZATIONAL CLIMATE* (average score, scale 0–720) | 550.6-low system 4<br>(participative) | 439.9-mid system 3<br>(consultative) | 485.3-high system 3<br>(consultative) | |
| *CONFLICT INTENSITY* (on scale 0.0 to 3.0) | .704 | .672 | .621 | .443 |
| *CONFLICT SOURCES* (rank order of sources by intensity of conflict) | 1. manpower resources<br>2. program priorities<br>3. technical issues<br>4. schedules<br>5. admin matters<br>6. cost objectives<br>7. personalities | 1. program priorities<br>2. manpower resources<br>3. technical issues<br>4. schedules<br>5. admin matters<br>6. cost objectives<br>7. personalities<br>NOTE: Numbers 2 and 3 tied. | 1. program priorities<br>2. technical issues<br>3. admin matters<br>4. manpower resources<br>5. schedules<br>6. cost objectives<br>7. personalities | 1. program priorities<br>2. admin matters<br>3. schedules<br>4. technical issues<br>5. manpower resources<br>6. cost objectives<br>7. personalities |

| | | | | |
|---|---|---|---|---|
| *CONFLICT RESOLUTION MODES* (rank order, most to least used) | 1. confrontation<br>2. smoothing<br>3. compromise<br>4. withdrawal<br>5. forcing | 1. confrontation<br>2. compromise<br>3. smoothing<br>4. forcing<br>5. withdrawal | 1. confrontation<br>2. compromise<br>3. forcing<br>4. smoothing<br>5. withdrawal | 1. confrontation<br>2. compromise<br>3. smoothing<br>4. withdrawal<br>5. forcing |
| *SATISFACTION* (average score for general satisfaction, internal work motivation, pay satisfaction, security satisfaction, social satisfaction, supervisory satisfaction, growth satisfaction on a scale of 0–7) | Gen Sat - 5.83<br>IWM - 5.86<br>Pay Sat - 5.84<br>Sec Sat - 5.75<br>Soc Sat - 5.86<br>Sup Sat - 5.75<br>Growth Sat - 5.63 | Gen Sat - 5.35<br>IWM - 5.98<br>Pay Sat - 5.50<br>Sec Sat - 5.30<br>Soc Sat - 5.41<br>Sup Sat - 5.70<br>Growth Sat - 5.10 | Gen Sat - 5.29<br>IWM - 5.88<br>Pay Sat - 5.58<br>Sec Sat - 5.32<br>Soc Sat - 5.60<br>Sup Sat - 5.61<br>Growth Sat - 5.33 | |

should be little surprise that conflict is high in this phase. Further, since this is only the planning and design phase, many of the participants must recognize that they will not be available several years in the future to see the results of their work. Thus commitment to the project may be difficult to obtain, and consequently job satisfaction may also be difficult to generate and sustain.

### Phase III

Data from projects in the execution phase indicate that project sizes are generally comparable to but reflect a wider range than those in Phase II. The type of organization is organic with some mechanistic tendencies, while the climate is rated near the high area of System 3, a consultative but near-participative type of organization. Conflict intensity is lower than in Phases I or II, but is still relatively high. Program priorities, technical issues, and administrative procedures are closely grouped as principal sources of conflict, clearly dominating the remaining sources. While confrontation and compromise remain the preferred conflict resolution modes, forcing is also an important technique in this phase. In general, job satisfaction in Phase III appears to be relatively low. It should be remembered in interpreting these data that the level of bureaucracy, organizational climate, and the satisfaction values were generated from data sources somewhat contaminated with Phase IV-type work tasks. The organizational environment of Phase III can be characterized as a relatively large work group organized along semiorganic lines with some mechanistic overtones. The members work together under a conflict situation arising from priorities and technical issues combined with the administrative procedures necessary to resolve them. Use of power and authority to resolve differences (forcing) is increased in Phase III, while job satisfaction is reduced.

The use of power and authority to resolve differences has long been associated with a relatively low level of job satisfaction. In this phase the job must actually be accomplished. Project personnel are "under the gun" to meet the schedules, budget limits, and performance criteria that earlier planners built into the project as goals. Any mistakes made in earlier projections show up here and must be resolved, along with all technical problems that have developed. Pressures to achieve the goals are intense. Conflict would be expected to be high in this situation. Job satisfaction may be reduced as the current participants see themselves responsible for resolving situations created by the errors and optimism of earlier project personnel.

### Phase IV

The data from projects in the termination phase indicate a marked, significant reduction in project size from those in Phases II and III. Conflict intensity is relatively low in this phase, with program priorities, administrative procedures, and schedules being dominant contributions. Confrontation, compromise, and smoothing are the preferred conflict resolution modes. Although not complete, these data indicate some significant differences in the organizational environment of Phase IV relative to Phases II and III. The environment can be characterized as medium-sized groups working under relatively low conflict intensities. In terminating their projects, the participants find that the principal conflicts are generated from project priorities and schedules, with the needed administrative procedures taking on increased significance. Differences are resolved in the collegial mode as was done in the earliest project phases.

This phase represents the end of the project. Those few personnel who remain are involved in turning the completed product over to someone else. Further, they are likely to be preoccupied with finding themselves new jobs, since the ones they currently hold are in the process of being eliminated. At this point, individuals are likely to experience less pressure and to perceive less need to quickly resolve conflicts through forcing. The task is essentially complete, and no amount of effort at this point is likely to change the results. In this situation, low levels of conflict are to be expected.

### Reviewing the Life Cycle

Comparing the findings for each specific project life-cycle phase relative to each other reveals some interesting relationships. In the most general terms, the life-cycle theory is supported, with marked differences occurring in the organizational environments of projects from different phases. More specific analysis indicates that project size clearly is quite different across the phases. The planning and execution phases having by far the largest project teams, the conceptual phase the smallest, the termination phase has intermediate-sized project teams. The level of bureaucracy parallels this pattern, with the greater bureaucracy corresponding to the greater size, as would be expected. The level of bureaucracy measure demonstrates statistically significant differences between the planning phases and the conceptual and execution phases at above 95% level of confidence. Organizational climate also changes markedly across phases, with the early and later phases having projects more representative of

System 4, while the middle phases are more System 3-oriented. Statistical tests of the organizational climate scale indicated that all scores were significantly different from each other at the 95% level of confidence (24). Conflict intensity decreases consistently across the phases. The differences between alternate phases are statistically significant above the 95% confidence level, but those between adjacent phases are not (18). Thus there would appear to be a slowly declining trend in conflict intensity across life-cycle phases. Both the sources of conflict and the resolution modes change across phases in a manner consistent with the changes in size, level of bureaucracy, and organizational climate. Finally, job satisfaction in general seems to be highest for the smallest, most organic organizations and lower for those organizations most mechanistic in nature.

## GENERALIZATIONS

The data referred to in this chapter were drawn from a variety of research efforts using different samples collected at different times over a two-year period. As such, the findings are not directly relatable to one another, and in some cases the observed differences are not statistically significant or cannot be tested for significance. Despite these methodological shortcomings, the synthesis, by noting important differences between projects in different phases, has served to strengthen the belief that there may be extensive variability in internal organizational environments over the life cycle of major projects. The fact that these findings are supported in the available literature as well as by the logic of careful observation lends credence to these documented results. The findings clearly indicate several differences between the projects representing various phases, and suggest others. Based on these differences several very tentative conclusions concerning the internal environments of projects over their life cycle were reached and are presented in the form of the following propositions:

- Individual project organizations tend to be relatively small in the early and late phases of their life cycle, and much larger in their middle phases. This may be a function of the different types of tasks being performed in each specific phase.
- Project organizations tend to be more mechanistic in nature and exhibit less favorable organizational climates in their middle phases than in either the early or late phases of the life cycle. The most favorable organizational climate and the most organic type of organization is found in the initial phase of the project life cycle. This may

be related to the size of the work groups found in the individual phases and to the resulting differences in organization structure.

- As the project progresses in its life cycle, the overall intensity of conflict decreases. Administrative matters and program priorities become relatively more important as sources of conflict, while manpower resources become less important sources of conflict. Cost objectives and individual personalities are relatively unimportant sources of conflict across the life cycle, although the conflicts they generate may be among the most difficult to resolve.
- As conflict resolution modes, smoothing decreases while compromise and forcing increase in relative use over most of the project life cycle. This trend reverses itself in the termination phase of the life cycle. This pattern of changes in conflict resolution modes may be associated with the changes in level of bureaucracy, size, and organizational climate which occur over the life cycle.
- Project organization size is negatively related to the extent of organic (project) orientation in the work group, perceived organizational climate, and the team member's job satisfaction.
- The perceived organizational climate in project organizations is positively related to the extent of organic (project) orientation in the work group and to the perceived job satisfaction of the team members.
- The smaller the project, the more closely it reflects the characteristics classically recognized as representing project teams—participative, dynamic, and collegial team efforts. Larger efforts clearly display the characteristics of more bureaucratic organizations.

The above relationships suggest that major changes may occur in the organizational and behavioral environments of the single project as it progresses through the phases of its life cycle. Such changes could have numerous implications for managers of project managers and for the project managers themselves.

### Manager of Project Managers

One major implication of the project life cycle for the manager of project managers is that the idea of choosing a single project manager to see the project completely through its life cycle may need to be discarded, at least for the major, advanced-technology projects discussed here. Rather, it may be much more appropriate at the major project phase points to select a new project manager who is familiar with the types of tasks to be performed during the succeeding phase, and who may be best suited to the project environment anticipated to exist during that phase.

While his ideas are by no means universally accepted, Fiedler has shown that the relationship-motivated leader needs to achieve the best performance where tasks are unstructured, leader-member relations are either very good or very poor, and member behavior is influenced by the leader either by direct chain-of-command or through example, esteem, and expertise (27). The implication for the large research project is that a relationship-motivated project manager would achieve the best results in the conceptual or planning phases of the project, where the conditions closely match those specified by Fiedler's work. On the other hand, when tasks are better structured, the leader-member relations are relatively good, and the leader (because of a weak formal structure) can rely only to a limited extent on the direct and formal chain of command to effectively accomplish objectives, a task-motivated leader tends to obtain the best results (27). This set of conditions roughly parallels the situation in the execution and termination phases where organization climate is relatively favorable, tasks are relatively well structured, and the project manager has a less than mechanistic type of organization. Here, then, the best results might be expected form a task-oriented project manager. In addition, differences in the primary sources of conflict between the early and late phases, that is, manpower resources and program priorities respectively, further indicate the possibility that different managerial traits and different background experience and preparation may be called for in the project manager during different phases of the project life cycle.

### Project Managers

During the early stages of the project, the characteristics of small size, varied tasks, a high degree of uncertainty, and the less formally structured organization appear to foster the more favorable organizational climates and higher levels of job satisfaction. The project manager is thus able to take advantage of the task commitment, the challenge, and the informality of the organic type of organization. Primary managerial functions of the project manager at this time should be to act as a communicator and facilitator, and to provide the various team members with information. The intent is to encourage participation and a team commitment to confronting and resolving conflicts. The goal should be for all team members to cooperate in accomplishing the project's goals, rather than to win an individual's point at the expense of the project.

In the later stages of the project, the project team diminishes from very large during the execution phase to very small toward the end of the termination phase. Here the project manager experiences a moderately formalized structure, perhaps as a legacy from the planning phase where rapid growth, high conflict intensities, and a great deal of environmental turbulence foster formalization in the effort to "get control of the situation." The degree of formalization in the latter phases can also be at least partly attributed to the routine nature of the tasks during the execution phase, and to the increased importance of technical issues as a source of conflict. A lower level of satisfaction is also experienced during this period, probably due to the higher levels of routine in the work itself, the lack of glory involved in "finishing the job," and personal concerns over future employment. Organizational climate, however, remains relatively formal despite these negative influences. In these later phases, the project manager should carry out the same managerial functions necessary early in the project and, in addition, should devote attention to reducing structural formalization as the project diminishes. This must be done with great care, however, to avoid undue shuffling of personnel or the appearance of demoting professionals unnecessarily. Reducing structure, it should be noted, is not an easy task. The project manager must counteract and overcome the "natural" bureaucratic tendencies of organizations to establish formalized sets of rules and procedures for almost every activity that can remotely be considered repetitive, and by this time the project has had several years to establish such procedures. This may well explain the well-known difficulty, expressed in many texts, in terminating or closing down a project (28).

The planning phase is characterized by a more formalized mechanistic-type structure and by large size, yet it also demonstrates high levels of uncertainty and conflict. This presents particularly challenging behavioral problems to the project manager. First, the tendency to overstructure the organization must be avoided to prevent hampering the cooperation and participative problem solving so necessary to successful projects. Second, the project manager must respond to the strong demands for establishing effective communication links. A dynamic project situation requires that project personnel generate and transmit information quickly in the face of new developments, establishing and encouraging relatively informal channels. Third, in order to facilitate a team approach to confronting and solving conflicts, the project manager needs to develop an identification with the project effort among the participants, to visibly use confrontation techniques himself, and to reward others for using these techniques. This implies a high degree of visibility and personal leadership.

## CONCLUSION

In concluding this discussion of the project life cycle's behavioral implications for project managers and the managers of project managers, the authors offer these key suggestions:

- The project team size should be kept as small as possible, consistent with being able to accomplish the tasks. This requires a conscious and continuing effort, as there is a tendency to resolve problems by building a larger organization. The rationale may be to provide visibility in the parent organization, to make sure there are sufficient people to be "on top" of the situation, or simply to increase the project's power base. In any event, the increased number of people complicates communications and severely compounds the problems of effectively managing the project.
- Increased formalization of the project's structure (e.g., specialized groups, formal reports, chain of command, specified procedures) should be avoided whenever possible. The project manager should recognize and exercise the art of trading off the advantages of specialization and its resulting efficiency with the disadvantages of unfavorable organizational climate and poor job satisfaction. These disadvantages may be indicated by reduced identification with the project and a lack of initiative on the part of project team members, a situation which can be very costly to the project manager.
- Team members should be encouraged to work jointly to resolve conflicts in a manner that is best for the project as a whole, rather than for any one team member. This involves leadership by example, and places the greatest demands on the project manager. The project manager must establish open communication channels, take time out to listen, create challenging tasks, and praise good performance. This also means good management! The project manager should be prepared to spend a large share of the available time in leadership and communication tasks. If too little time is left for tracking technical, schedule, and budget issues, then the preferred solution would be to secure the services of a competent assistant manager to deal with such detail.

## REFERENCES

1. Butler, A. G., Jr. "Project Management: A Study in Organizational Conflict." *Academy of Management Journal,* Vol. 16 (March, 1973), pp. 84–101.
2. Cleland, David I. and King, William R. *Systems Analysis and Project Management,* 3rd ed. (McGraw-Hill. New York, 1983).
3. Adams, J. R. and Barndt, S. E. "A Contingency Model for Project Manager Selection," in *Realities of Project Management.* Proceedings of the 9th Annual Project Management Institute Symposium, Chicago, 1977, pp. 435–442.
4. Woodworth, B. M. and Willie, C. T. "A Time Constrained Approach to Resource Leveling in Multiproject Scheduling." *Project Management Quarterly,* Vol. 7 (June, 1976), pp. 26–33.

5. Cleland, David I. *Pyramiding Project Management Productivity,* Vol. 15 (June, 1984), pp. 88–95.
6. Cable, Dwayne and Adams, John R. *Organizing for Project Management (Project Management Institute. Drexel Hill, Pa., 1982).*
7. Archibald, Russell D. *Managing High-Technology Programs and Projects* (Wiley. New York, 1976).
8. Roman, Daniel D. *Research and Development Management: The Economics and Administration of Technology* (Appleton-Century-Crofts. New York, 1968).
9. Payne, R. L., Fineman, S. and Wall, T. D. "Organizational Climate and Job Satisfaction: A Conceptual Synthesis." *Organizational Behavior and Human Performance,* Vol. 16 (1976), pp. 45–62.
10. Litwin, George H. and Stringer, Robert A., Jr. *Motivation and Organizational Climate* (Harvard University. Boston, 1968).
11. Hellriegel, D. and Slocum, J. W., Jr. "Organizational Climate: Measures, Research and Contingencies." *Academy of Management Journal,* Vol. 17 (June, 1974), pp. 255–280.
12. Likert, Rensis. *The Human Organization: Its Management and Value* (McGraw-Hill. New York, 1967).
13. Marrow, A., Bowers, D. and Seashore, S. *Management by Participation* (Harper & Row. New York, 1967).
14. Kirchof, N. S. and Adams, John R. *Conflict Management for Project Managers* (Project Management Institute. Drexel Hill, Pa., 1982).
15. Kerzner, Harold. *Project Management: A Systems Approach to Planning, Scheduling, and Controlling* (Van Nostrand Reinhold. New York, 1984), pp. 343–385.
16. Evan, William M. "Conflict and Performance in R & D Organizations: Some Preliminary Findings." *Industrial Management Review,* Vol. 7 (Fall, 1965), pp. 37–46.
17. Thamhain, Hans J. and Wilemon, David L. "Conflict Management in Project-Oriented Work Environments." *Sloan Management Review,* Vol. 16 (Spring, 1975), pp. 31–50.
18. Eschmann, Karl J. and Lee, Jerry S. H. "Conflict in Civilian and Air Force Program/Project Organizations: A Comparative Study." Unpublished master's thesis, School of Systems and Logistics, Air Force Institute of Technology (AU), Wright-Patterson AFB, Ohio (1977).
19. For example, see Vroom, H. Victor. *Work and Motivation* (Wiley. New York, 1964).
20. Organ, D. W. "A Reappraisal and Reinterpretation of the Satisfaction-causes-performance Hypothesis." *The Academy of Management Review,* Vol. 2 (January, 1977), pp. 46–53.
21. Payne, R. L. and Mansfield, R. "Relationships of Perceptions of Organizational Climate to Organizational Structure, Context, and Hierarchical Position." *Administrative Science Quarterly,* Vol. 18 (December, 1973), pp. 515–516.
22. Ford, Jeffrey D. and Slocum, John W., Jr. "Size, Technology, Environment, and the Structure of Organizations." *Academy of Management Review,* Vol. 2 (October, 1977), pp. 561–575. Also see Hendrick, Hal W., "Organizational Design," in *Handbook of Human Factors,* ed. Gavriel Salvendy (Wiley. New York, 1987) pp. 470–495.
23. Hackman, R. J. and Oldham, G. R. "The Job Diagnostic Survey: An Instrument for the Diagnosis of Jobs and the Evaluation of Job Redesign Projects." Technical Report No. 4., Department of Administrative Sciences, Yale University (1974).
24. Brandt, S. E., Larsen, J. C. and Ruppert, P. J. "Organizational Climate Changes in the Project Life Cycle." *Research Management,* Vol. 20 (September, 1977), pp. 33–36.
25. Haddox, Donald L. and Long, Neal A. "A Study of Relationships Among Selected Organizational Variables in System Program Offices During the Weapon System Acqui-

sition Process.'' Unpublished master's thesis, School of Systems and Logistics, Air Force Institute of Technology, Wright-Patterson AFB, Ohio (1976).

26. Lempke, Roger P. and Mann, Greg A. "The Effects of Tenure and Task Organization on Air Force Program Managers' Role Stress.'' Unpublished master's thesis, School of Systems and Logistics, Air Force Institute of Technology, Wright-Patterson AFB, Ohio (1976).
27. Fiedler, Fred E. and Chemers, Martin M. *Leadership and Effective Management* (Scott, Foresman. Glenview, Ill., 1974).
28. Kerzner, Harold. *Project Managers for Executives* (Van Nostrand Reinhold. New York, 1982), pp. 320–370.

# 11. Phasing Out the Project

Herbert F. Spirer*
David H. Hamburger†

It is much harder to finish a project than to start it. Start-up is a time of excitement: the team is being formed and resources allocated; the client/customer‡ is enthusiastic; and planning efforts are supported by the natural high spirits and optimism that go with beginning a new enterprise. But the finish of a project is a time of decline. Both the client/customer and the project personnel look to new ventures, the plans which brought the project to this point are now outdated, and the entrepreneurial spirit which motivated and involved the project personnel has been replaced by

* Herbert F. Spirer is Professor of Information Management in the School of Business Administration of the University of Connecticut at Stamford. Holding degrees in engineering physics from Cornell University and in operations research from New York University he was an engineer, project manager, and engineering manager prior to joining the faculty of the University. His home study courses in project management for engineering, software development, construction, and engineering department management and quality control have been adopted by over twenty engineering and professional societies. He has been a frequent lecturer and seminar leader on project management and consults to many corporations and financial institutions on project management and its integration with strategic planning. He has given papers on project management at meetings of the Decision Sciences Institute and the Project Management Institute.

† David H. Hamburger, PE, MBA, MME, a principle with David Hamburger, Management Consultant, Inc., has over thirty years of diverse experience as a management consultant, operations manager, project manager, engineer, and educator. He has acquired a broad range of practical experience in both line and project management functions with AMF and Dorr-Oliver Corp.—successfully executing projects in the military, industrial, municipal, and international sectors of the economy. As a management consultant and educator he has shared this experience nationally with numerous universities, federal and municipal government agencies, and industrial firms; providing consulting services and conducting seminars on project management, technical management, financial management, and cost estimating and control. He is also an Adjunct Professor in the MBA program of the School of Business Administration of the University of Connecticut at Stamford. He has written a home study course on cost estimating and a project management text and seminar workbook. In addition, several of his papers on project management have been published in various technical journals and presented to various professional societies.

‡ The term "client/customer" refers to the internal or external customer (or "user") who contracts for the project and wants achievement of its goals.

a sense of dullness as the staff deals with a seemingly endless set of details. But successful and complete closure is important: the success of future projects can be affected by the way in which the project is phased out; many contracts allow the client/customer to withhold a disproportionate share of the payments due until items such as spare parts lists and procurement drawings are delivered. How project personnel are phased out affects their performance in future projects, and proper closure can provide input to the postperformance audit as well as the basis for rational estimates and sources of extensions into new products and projects.

## WHY IS THE PROJECT BEING TERMINATED?

The *natural* termination of a project occurs when the project's stated goals have been met. But it may not yet be time to celebrate.

- The project staff may have successfully completed acceptance testing in the manufacturing environment for a microprocessor-based pharmaceutical product test system in time to support market entry, and matched changes in scope by time and cost allocations, and had no cost overruns. But the project is unfinished. The operator's manual is a copy of an edited manuscript, the parts list does not include standard designations for all parts, engineering drawings have not been updated, several required forms have not been completed, and so forth. These are the details which the closure team must complete.

Some projects end less successfully. This *unnatural* termination occurs when some constraint has been violated, when performance is inadequate, or when the project's goals are no longer relevant or desired. The most frequent causes for unnatural termination are time and money.

- Although soil borings indicated that the foundations were solid, the partially erected structure has shifted and the cost of correcting the situation exceeds available resources.
- The project to develop a particular experiment for a space vehicle will not be completed until two weeks after the scheduled launch date. Since this spacecraft has a narrow window of time for meeting its planetary target, a project which is two minutes late is valueless.

Even when the reason for terminating a project is beyond the control of project personnel (government funds are cut, the price of fossil fuels doubles or halves, the client goes bankrupt), there must be a closure process, both to clean up the loose ends and to meet the contractual requirements of phaseout.

## THE PROJECT MUST BE TERMINATED EFFICIENTLY

The participants, both client and project personnel, may overlook the importance of efficient project phaseout as they lose interest in the project or actively seek new assignments or, fearing that they will be out of a job when the project ends, "drag their feet." But failure to terminate the project efficiently can mean the difference between financial success or failure, a happy client and the potential for additional work or an unhappy client who will look elsewhere for future services, or an organization that benefits from its technical and administrative achievements or one that must start over each time a new project is initiated. The person managing the phaseout effort must understand the significance of these points and motivate the team accordingly.

When a project is completed late or inefficiently, the organization's "bottom line" can be seriously affected. The project's cost can increase if the termination process is stretched out, the staff is not reduced in an orderly manner, or the client withholds payment awaiting contractual compliance (meeting specifications, satisfying objectives, furnishing the specified deliverables, etc.). In addition, if the client, dissatisfied with the closeout effort, becomes uncooperative and refuses to negotiate back-charges, penalties, or liquidated damages claims, or the validity, scope, and cost of changes, project costs rise. When termination is dragged out, the organization's posture in closing out its open purchase orders and subcontracts is affected. And as time passes and the right people and information are no longer available, costs will continue to accumulate.

In the phaseout effort, project personnel must always remember that a satisfied client is final proof of project success. Client satisfaction depends as much on supporting documentation, training, field assistance, spare parts, and other services as on the major deliverables, because without these support functions the major deliverables might be misused or abused. If the client is dissatisfied, goodwill diminishes and the supplier's reputation deteriorates with consequent loss of future business and a possible increase in warranty claims.

The client's profitability can be affected if the client does not receive the operating instructions or information required to comply with government regulations in a timely fashion. Added operating costs, lost revenue, fines, and the cost of correcting a noncompliance situation are among the factors which can reduce the client's profitability.

Warranty service, an element of a contract with an external client, is important to both the client and supplier, and mishandling during project closeout can create problems for both organizations. A warranty start date, which is established by the project termination manager and accepted by the client, is needed to limit the supplier's postproject financial liability. If the appropriate project data (drawings, vendor documentation, specifications, manuals, contract terms, relevant correspondence) are transferred to the organizational entity responsible for warranty management, performance will be more effective and long-term client relations will be enhanced.

The lessons learned in executing a project are also proof of project success, and effective transfer of this information from the project team to the functional departments is a major responsibility of the termination project manager. The manager must transfer the technical data before the project team is dispersed, taking with them the knowledge and techniques they developed during the course of the project. It is also important to add the actual project financial performance data to the organization's estimating data base. Proper cataloging and storage of the project files is essential in case of warranty claims, orders for replacement parts, similar future contracts, or litigation.

All of this means that, while the phaseout process may not seem as exciting as start-up, it is as essential for project success and client satisfaction.

## THE TERMINATION OF A PROJECT IS A PROJECT

Since managing a project's termination is as important as managing its start-up, many organizations assign a specific management team to attend to phaseout details. And the task of termination fits the classical definition of a project as a one-time unique goal with specific resource constraints (3, p. 4; 1, p. xi).* Research (4) has shown that conflict increases on projects when project objectives are unclear to the project personnel.

When management sights the end of a project, the project manager must determine the "profile" of project management tools and techniques to be used to bring this subproject to successful completion. As noted, closure has special problems and needs which determine plans, schedules, and personnel use.

Just as the tools should be tailored to the special characteristics of phaseout, so must the choice of project manager for this phase. The project manager and team which deal successfully with start-up and completion of major goals may not be ideally suited for the special problems, both technical and emotional, which accompany phaseout. A project manager who understands these problems and is temperamentally compatible with project closure must be chosen. Choosing the phaseout team

* Numbered references are given at the end of this chapter.

and transferring project authority to it is an important mission for management.

## PEOPLE COUNT—IN PROJECT TERMINATION

We divide our concerns about project phaseout into emotional ("affective") and intellectual ("cognitive") parts which management must deal with when a project is being terminated. Figure 11-1, a structured tree

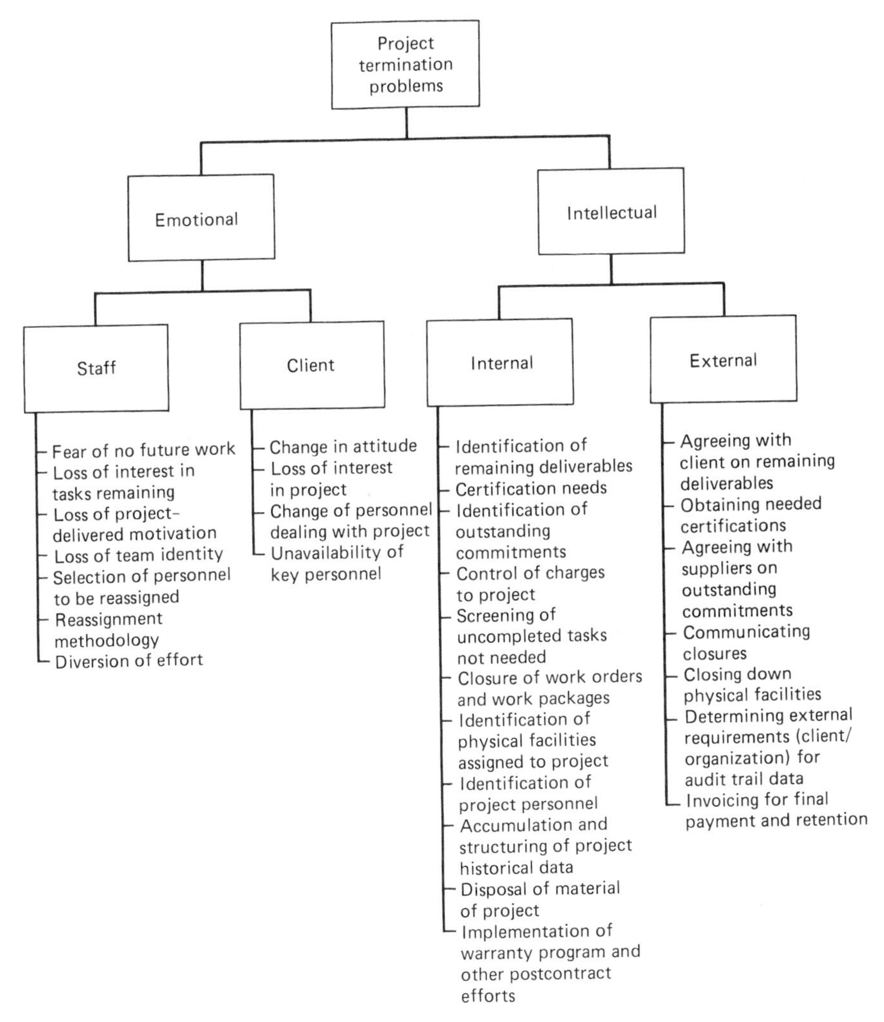


Figure 11-1. Work breakdown structure for problems of project termination.

diagram analogous to a work breakdown structure, illustrates both the general and detailed nature of phaseout issues. This figure shows that emotional issues have to do primarily with *spirit* and intellectual issues primarily with *detail.* Combining spirit with attention to detail accounts for many of the phaseout problems faced by the project management staff.

The following emotional issues concern the project staff, both those reporting directly to the project manager and those with matrix relationships:

- *Fear of no future work.* That the project is ending is no mystery to anyone. What is less clear is whether there is more work out there for the project staff. Even when there are other projects needing support, many members of the team may fear for their continued employment. This may result in a "philosophy of incompletion," and "foot-dragging" is often encountered as no one wants the project to end. Design documents take remarkably long to retrieve, instructions are repeatedly misunderstood or not acted upon, tools disappear or are not available, and tasks are stopped dead when the slightest impediment occurs. The staff seems to be working in slow motion.
- *Loss of interest in tasks remaining.* Start-up involves challenging tasks: problems to be solved, new methods to be applied, resources to be allocated. The end of a project involves familiar and often tedious tasks: completion of documents, refinement of known techniques and products, and the withdrawal of resources. The "fun" has gone out of these tasks. As a result, technical people lose interest, and many project personnel spend more time—often successfully—looking for new assignments on new or ongoing projects with more interesting tasks than they spend on completion of their current project. The result can be poor performance on the essential phaseout tasks.
- *Loss of project-derived motivation.* The concept of a project-derived mission, shared among the project team members, is recognized as one of the advantages of a project structure (2). As the project staff is reduced and activities focus on details that do not seem to relate directly to the project's goals, motivation diminishes and the sense of teamwork is lost.
- *Loss of team identity.* During phaseout new personnel with different skills are brought on board (e.g., technical writers for the operator's manuals, contract administrators) and long-term members of the team are assigned elsewhere. The perception of the project team as an ongoing group diminishes. The "team" is less interested in doing the tasks and meeting the deadlines.

- *The effect of personnel reassignment.* In almost all projects, termination is accompanied by a reduction in personnel needs. A key issue for the project manager is whom to reassign and whom to keep. The project manager rarely has the luxury of making this decision solely on the basis of project need. The needs of the organization must be balanced against those of the project, and project members may be resentful about the decisions made about their future.
- *Reassignment methodology.* Reassigning personnel must be carefully staged. Will there be a project reassignment office? Are personnel to be reassigned on an *ad hoc* basis, or will there be single or multiple mass transfers? If reassignment is not carried out consistent with the needs of termination, both reassignment and the ongoing phaseout activities can be adversely affected. Other managers may make "raids" on the staff, saying that "this project is running down anyway." Project team members may seek new projects while charging time to the project and neglecting their assigned tasks. Inept reassignments can destroy morale and any spirit of teamwork on the remaining tasks.
- *Diversion of effort.* When a project is running down, other tasks have a way of seeming much more important than those concerned with phaseout. Priorities change and it becomes harder to convince the client that project personnel are actually working towards a satisfactory finish.

Clients are also affected by the transition to termination, and the project manager must deal with this as well.

- *Changes in attitude.* The "steam" has gone out of the project and the client now worries about the problems of past performance on future success. Even where performance of the major deliverables has been satisfactory, the customer recalls when compromises were made, milestones missed—memory of the petty difficulties takes over.
- *Loss of interest.* As the staff loses interest, so does the customer. The excitement, the new challenges and, not unimportantly, the opportunity to allocate resources generously, shifts to new projects elsewhere. The client is involved in new projects, focusing on these instead of the almost finished job. Many clients regard the phaseout process as a nuisance, cluttered with detail and dull decision making.
- *Change of personnel on the project.* In both technical and managerial areas, the "first team" withdraws. The lead designers and managers move to new projects; the best inspectors, contract administrators, technicians, and project officers are moved elsewhere, and the client must now deal with a new team concerned with inspection and docu-

mentation. The customer, comfortable with the old team, must make new, and short-lived, relationships.

- *Unavailability of client's key personnel.* Personnel at the client location whose specialized skills are important to those technical tasks which remain may be unavailable because of reassignment (and perhaps, geographical changes). The customer's engineer who knows the source of specifications for exotic materials may be assigned to another project and does not return phone calls; the contract administrator who can give definitive interpretations of contractual boilerplate has gone elsewhere. Locating these people and making them available takes time and costs money, as does "making do" without them.

## RESOLUTION OF THE EMOTIONAL ISSUES

Most of the emotional problems caused by termination are similar for both client and project staff since they arise from the same causes. But the project manager must deal with clients and project staff differently.

The project manager has authority over the project staff and internal issues and can exercise influence and power to resolve problems. But the manager has no authority, minimal influence, and little power over the client/customer. The project manager does have two client-motivating factors: to "get the project off the books" with the minimum difficulty and administrative problems and to make sure that documentation, spare parts, and similar support functions do not become problems.

Within the performing organization, the project manager has all the managerial and leadership tools that were available for the start-up and middle phases; these tools can be modified for phaseout. Using these tools effectively will help ensure that future projects function smoothly.

The project manager must recognize, first and foremost, that a guaranteed paycheck is a primary concern of most employees. No matter how spirited the staff, how committed the organization to long-term employment for all employees, or how large the backlog of project work, the project manager can never assume that all employees are unconcerned about future work. Even in organizations where there has never—in decades—been a layoff or reduction in work, at the end of a project many employees worry about the future. And where there has been a history of layoffs at the end of projects, employee concern is guaranteed. Even if there is no overt concern about future employment, the project manager should discuss the prospects for future work for the project staff. Conversely, if there is no future work, the project manager should "level" with the staff and then take special measures to ensure that staff morale is reinforced so the job at hand can be successfully completed.

Honesty is the best policy, and the project manager who is honest with his staff about future prospects will gain a great deal of support for project closeout work. The project manager should tell the staff as a group and individually the nature of the backlogged work and the pattern of reassignment. Open sessions where the manager meets with the staff and fields questions about future prospects are an effective management tool. The session should be scheduled when the project termination phase is defined (see below) and repeated when the manager senses rising concern about these issues. Indications are: failures to close out individual tasks and to follow agreed-upon instructions ("I don't remember you asking for that"), absences and tardiness, disappearances at work and slow performance. The project manager, sensing employee fears, should be positive, not punitive or authoritarian. These fears, whether stemming from a rational base or not, are genuine, and the project leader must deal with them.

A second major task is to offset the loss of interest in tasks, project-derived motivation, and team identity. The following specific management tools have been found helpful:

- *Define the project termination as a project.* Make it clear that closeout has its own project identity. Some project managers give the closeout its own project name. "Start-up" meetings for the beginning of termination help establish the concept that there *is* a well-defined goal to be met—closing out the job properly.
- *Provide a team identity.* A project name provides a base for the team. If the team is large or separated geographically, a closeout newsletter can give a feeling of identity. Some organizations issue T-shirts, caps, and other identifying insignia. In some cases, where the organization didn't support these modest outlays, project managers have been known to pay for them because the cost was small compared with the long-term benefits the manager got from the increased motivation and successful completion of the project.
- *Bring the team together frequently.* As closeout requires different skills, the staff changes and there is a tendency to allow the new staff to operate loosely as individuals rather than as members of a team with a common goal. To offset this, and to improve communications, the manager should schedule regular get-togethers. These are not "meetings." A meeting implies a lengthy sit-down session. Such sessions have their place in closeout, but not as a tool to maintain team spirit and improve motivation. These are stand-up sessions, limited in length (for psychological reasons), which give the project manager the opportunity to introduce new members, announce reassignments, talk about new work following phaseout, and deal with

problems and schedules—all within a team framework. Weekly or even daily stand-up meetings, limited to ten minutes, can become so much a part of project identity that a chance omission results in a flood of inquiries.

- *Get out to the project staff.* It is not always possible to bring the entire staff together on a regular basis. Some projects are geographically dispersed, or job requirements may prevent full attendance. The project manager offsets this limitation by getting out of the office and meeting on an individual basis with the staff. This presence provides a sense of identity with the team and the project office as well as a communications link.

Reassignment presents special problems. The method for selecting which project members are to be retained and which reassigned, and how reassigned personnel are oriented to their new assignments, affects phaseout. During phaseout, you need to retain those people with the greatest flexibility, the most independence, the best sense for detail, and the highest level of skill. Flexibility and skill are needed because of the variety tasks that arise during phaseout. Independence is essential as teamwork and close supervision are lessened, and a feel for detail helps assure that no pieces are left hanging.

The method of reassignment should reassure personnel of their future with the organization. And it must also be consistent with keeping a highly motivated work force. Below are some guidelines for the project manager facing this task.

- *Make each reassignment decision a conscious, deliberate choice.* Think through each reassignment decision, weighing the factors in each case. Don't make blanket decisions (''drop all electrical engineers''); they will come back to haunt you.
- *Hold the right personnel.* It's natural to want the best people for the job remaining. But closeout personnel need different qualities from those who began the project. Careful choices are essential.
- *Carry out reassignments openly.* Make sure that the project staff knows the reassignment plans; if the staff finds out from some other source, it creates resentment and it is easier for misunderstandings to occur.
- *Play an active role in reassignment.* The project manager should play an active role and not wait for the ''mechanics'' of reassignment to happen. Is the Human Resources Department playing an active role? The manager should get to them early and offer support. And other key people in the organization should be contacted.

Communication is important. As the team reduces in size and perceived importance, the project manager must work with both individual staff members and functional managers to ensure that everyone knows the importance—both to them and the organization—of an orderly and complete closeout.

The project manager is responsible, within the parameters of his job assignment, for staff morale. But this manager has neither authority nor the capacity to reward the client. The client's interest and support can be maintained, however, through an appeal to the mutual benefits of a quick and well-managed closeout. The project manager can assure client cooperation at all levels by stressing the following benefits:

- *Personal and organizational credit for closure.* Both the individuals and their organization gain credit when there are no loose ends and a project is "wiped off the books" in an orderly manner.
- *Availability of future support for the project's deliverables.* Proper termination means that spare parts will be available in future years, manuals are complete and drawings up-to-date. If the client is assured that these phaseout items will be delivered, relations between client and supplier will remain at a high level.
- *Identification of warranty obligations and the start/completion dates.* Advising the client of warranty support will aid postcloseout planning.
- *Effective and equitable closeout negotiations.* If both parties are well prepared and willing to compromise, the project closeout will be easier to achieve and this, in the long run, will serve both client and supplier.

## INTELLECTUAL PROBLEMS OF PROJECT TERMINATION

The concern for *detail* is dominant among the intellectual issues. The intellectual branch of the work breakdown in Figure 11-1 illustrates the myriad of details involved in closeout, separated into two categories: internal and external. Internal issues are those concerned with the project itself and its staff:

- *Identification of remaining deliverables.* The contract, or other governing document, specifies deliverables such as tooling, test procedures, spare parts, spare parts lists, drawings, manuals, fixtures, shipping containers, restoration of modified facilities to original form, etc. To identify what is still undone, the project manager must match *delivered* items against contractual deliverables. If this is not done, contract administrators, auditors, or other client representatives may find undelivered items at a time when the costs of completion will be higher.

- *Certification needs.* Certificates of conformance with environmental or regulatory standards may be a part of the contract requirements (or implicit, such as UL approval). Some test procedures require multiple certifications, and the project manager must know of these requirements.
- *Identification of outstanding commitments.* It can happen in any project, especially those of large scope and lengthy time: project closeout is almost finished and a vendor delivers a surprise carload of components which are not needed. The best of commitment records can be incomplete, cancellations can be mishandled and never identified, or a genuine mistake can be made. Less dramatic, but just as difficult to resolve, is the commitment which is properly recorded, still outstanding—but no longer needed. The manager must resolve these difficulties or relations with vendors will deteriorate.
- *Control of charges to the project.* By the closeout phase, the project's charge accounts are common knowledge to an army of employees. Deliberately or inadvertently, they may charge to the project although not working on it.
- *Screening of incompleted tasks not needed.* Not all tasks being worked on may be needed. These "tag ends" can persist to the very end of the project.
- *Closure of work orders and work packages.* Once these uncompleted and unnecessary tasks have been identified, their formal authorizations must be ended. In addition, tasks which have been completed but are still carried on the records as open to charges must be identified and closed.
- *Identification of physical facilities assigned to the project.* During the course of the project, physical facilities—buildings, warehouses, typewriters, test equipment, machine tools, cars, trucks, etc.—may have been assigned to the project. The project manager is responsible for their care, and after they are no longer needed they should be redirected to projects where they can provide additional benefits to offset their cost and prevent possible continued charges to the completed project.
- *Identification of project personnel.* The manager of compensation for a major company reports that the first thing he does when entering a new facility is to ask the line manager for a list of personnel, which he then compares with a physical census of employees in the operation. The two lists rarely agree. While it is possible that some team mem-

bers may be working in remote locations and others carrying out tasks which require no supervision, it is important that the labor charges be correct—the lists should agree as closely as possible.

- *Accumulation and structuring of project historical data.* A project history puts technical and managerial achievements on record, making them available to others in the organization, is a guide to the management of future projects, is the basis for improved cost estimates for future projects and audits, can provide support in postproject disputes, and gives credit where credit is due.
- *Disposal of project material.* A project accumulates quantities of expendables, raw materials, components, partially finished assemblies, rejected units, files, catalogs, etc. The project team must dispose of these, for use in other projects if possible, or for scrap if this is impossible.
- *Implementation of the warranty program and other postcontract efforts.* Permanent functional groups, not the project team, are responsible for executing the warranty program and other postcontract activities such as in-service training, system and operating staff performance appraisal, periodic inspections, operator retraining, etc. But the project team is responsible for an orderly transfer of these functions at project termination. The responsible groups must be informed of their specific obligations and all relevant project data (contracts, drawings, purchase orders, manuals, client communications, etc.) must be transferred. Information about the client and the working relationships is also relevant. The client should be advised of these new contacts for each postcontract task. Both formal introductions and informal get-togethers can reinforce the new relationships.

External issues are those concerned with the client, vendors, subcontractors, and any other project-related entities not within the purview of the organization.

- *Agreeing with client on remaining deliverables.* Contractual statements may need interpretation. Project management knows what are the deliverable items and the specific requirements to be imposed on the remaining deliverables as well as which contractual items are no longer needed. There may be negotiations about the exact nature of the deliverables, as well as possible deletions and the trade-offs of additions against the deletions. If the specifications cannot be satisfied, financial settlement may be necessary. Every such modification or clarification is a potential change of scope and must be treated (via contract office, sales, etc.) and documented (change orders, change of scope, contract modification, letter of agreement, etc.).

- *Obtaining needed certifications.* Each certification has the potential of being a project in its own right. The deliverables—the specific documents or models required—must be identified and assembled. Finding the appropriate path for gaining certification can be a demanding task.
- *Agreeing with suppliers on outstanding commitments.* As with internal factors, some modification or cancellation of outstanding commitments is likely during phaseout. If costs are to be kept to a minimum and vendors kept satisfied, negotiations will be necessary.
- *Communicating closures.* The project manager must ensure that the closure of work orders and packages is fully understood and then carried out. Shrinking the project reduces staff contact and, to assure that work has been stopped and closeout requirements met (such as accumulation of charges, test results, delivery of fixtures, etc.), the project manager must make a conscious effort to communicate with the staff.
- *Closing down physical facilities.* Closing down the facilities often calls for a concentrated effort. Retrieving capital equipment which has been lent and operating equipment which has been installed can be difficult, even if the contract is considered watertight. For example, in certain space projects where equipment and facilities were maintained in other countries, closure was hampered by governments refusing to allow removal of the goods—a form of hostage for which a ransom had to be paid.
- *Determining external requirements for audit trail data.* Different clients/customers have different requirements for record retention for use in postproject audits. These may not be available as part of the contract. Often they must be determined through references given in the contract, or by specific agreement with the client.
- *Invoicing for final payment and retention.* The final invoice is *the* final item in project termination, signifying completion of the contract work. The resulting cash flow has a direct impact on the cost of financing the project; expediting this milestone should be the principal concern of the project manager and staff. In addition, reduction of any retention (payments withheld by a client to protect against nonperformance) should be addressed at periodic intervals throughout the termination phase as the risks perceived by the client diminish.

## RESOLUTION OF THE INTELLECTUAL PROBLEMS

To deal with these problems, the project manager needs an array of analytical tools—mostly graphical—and a special set of personal skills. First we present the analytical tools for termination:

- *Tree diagrams.* Tree diagrams are hierarchical models which are useful in organizing project entities when discussing termination with the project staff, other departments, and the client. Figure 11-1 is a tree diagram which may be useful to some project managers in organizing and communicating the closeout tasks. Figure 11-2 is another tree diagram—similar to the work breakdown structure—for organizing elements of work in a project termination. The tree must fit the particular project, and this example is actually a fragment of a particular project, used here to illustrate the tree's use. Tree diagrams may be used to track deliverables remaining. If there was a work breakdown structure for the start of the project, the delivered items are crossed off; the remaining deliverables are now easily defined. In the absence of such a work breakdown structure, the project manager can create one from the contractual documents and cross off delivered items. Similarly, tree diagrams can be used to track outstanding work orders. Using the first level of the tree to represent the performing departments, closed work orders can be crossed off.
- *Matrices.* Matrix models are useful when two or three entities must be related. For example, in determination of certification needs and

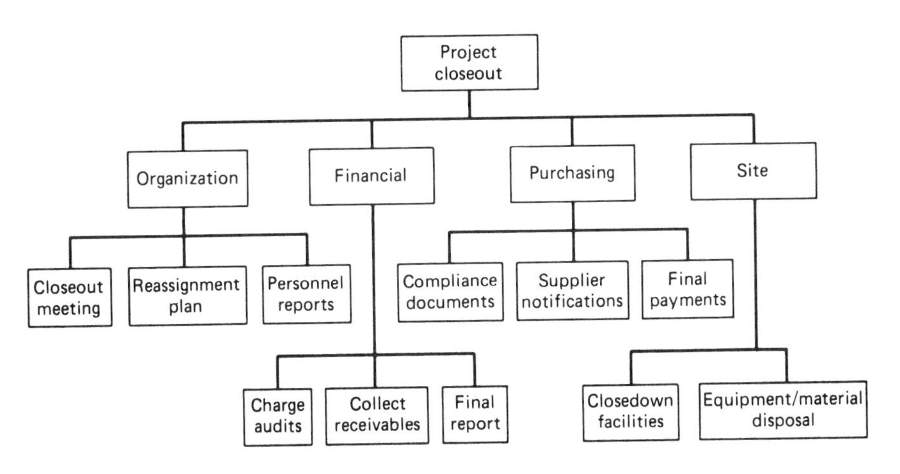


Figure 11-2. Tree diagram for a project termination.

managing the completion of certification, the two entities may be *product* and *governing specification*. Figure 11-3 shows how a matrix model may be used to collect information concerning the relationship between these two entities. Once created, in the process of collecting and organizing the information, the matrix becomes a reference for everyone working on the job and a means of communication. Checking the relationships for accuracy is quite easy with a model such as Figure 11-3. A typical three-entry case occurs in dealing with outside commitments, as shown in Figure 11-4. Here, the status of the commitment, as well as the vendor, is of interest; status is signified by the code letter in the box corresponding to the intersection of vendor and commitment. Matrix models for three entities are used to show responsibilities of differing levels. Figure 11-5 is used for assigning the levels of different individuals in meeting the records of the project. Here, the nature of responsibility and input function is shown by a code letter at the intersection points. The need for this type of matrix model is greatest when project personnel are shifting, when departmental boundaries are crossed, when it is important not to miss any details, etc., as occurs in project termination. The use of matrix models is limited only by the manager's imagination.

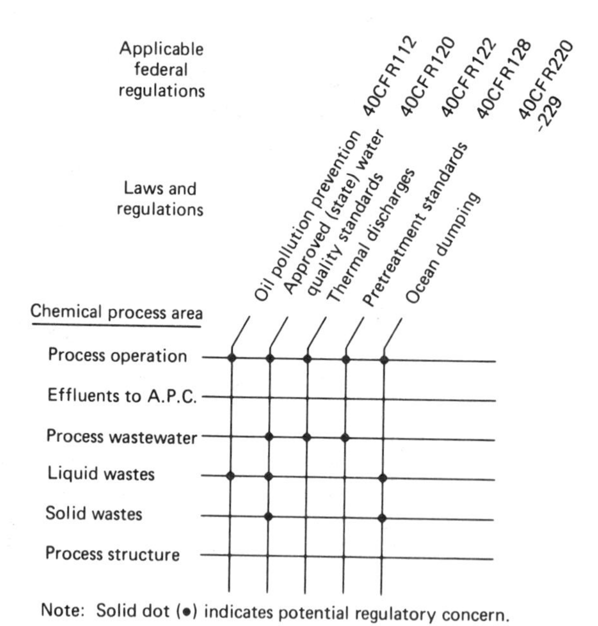


Figure 11-3. Matrix model for product versus governing specification (a portion).

| | Amtel | Matarol | Pilog | State semi | Macropolis |
|---|---|---|---|---|---|
| Z80 chips | N | D | | | C |
| 8080 chips | C | | C | | I |
| 96K RAM | I | | C | | N |
| 48K RAM | D | | | N | D |
| 4K ROM | D | | C | | I |
| EPROMs | N | | C | | D |

Note:
D, to take delivery.
N, in negotiation.
I, awaiting information.
C, closed.
Blank indicates no commitment, past or present.

Figure 11-4. Matrix model for relationship among commitment, vendor, and status.

| | Vice president operations | Quality assurance manager | Director of projects | Project manager | Task engineer | Automatic equip. manager |
|---|---|---|---|---|---|---|
| Quality assurance manual | A | P | | | | |
| QA procedure index | A | P | | | | |
| QA audit schedule | | | P/E | R | | |
| Drawings auto equipment | A | | | | P | R |
| Drawings calibration blocks | | | | A | P | |
| Spare parts specifications | | | A | R | P | |
| Calculation notebooks | | | A | R | P | P |

Note:
P – Prepare
A – Approve
R – Review
E – Execute

Figure 11-5. Matrix model for assignment of responsibility for records in project termination.

- *Lists*. Lists are the project manager's primary planning and control tool. This is a direct result of the termination process where *every* item must be accounted for. There can be no tasks left outstanding. By screening documents and talking with those concerned (internal and external to the project), the manager can produce listings of tasks to be done. At higher levels, these listings can be converted to trees (to provide structure and graphic communication) and matrices (where there is interaction). Where task interdependence is important, a network plan (CPM or PERT) may be prepared, but for the short term of project termination, as well as for weeky and daily supervision, the manager can work from lists. For this, checklists are convenient for showing what has been left undone. Figure 11-6 shows part of a checklist.

It is not unusual for the project manager to have a file or notebook of such checklists, one for each area of activity.* To provide a good audit trail, the dates when each task is completed should be marked on the checklist. Computers can be used to maintain such lists using data base management programs of modest size and cost, providing continually available listings which are free from recurring costs and the difficulties of dealing with a centralized information system. A "punchlist" is often used by the client to formally detail the contractual deficiencies which must be corrected before the scope of supply is accepted.

The personal attributes needed by the project manager responsible for phaseout include the following:

- *Knowledge of financial systems and accounting*. The concerns of the project manager are primarily managerial, especially in those financial areas of cost and accounting. The manager must understand cost accounting principles and systems, going beyond the organization's system to encompass the vendor's accounting system since this may be essential to getting agreement on outstanding commitments. Control of changes similarly calls for understanding of cost accounting.
- *Technical knowledge of the project*. The project manager must understand the design work even if he or she is not capable of actually carrying it out. To specify deliverables, screen tasks for need, close down facilities, and safely dispose of equipment and material, the project manager must know what is going on technically; managerial skills alone are not enough.

---

* An example of a checklist for project closeout appears in Russell D. Archibald, *Managing High-Technology Programs and Projects* (Wiley, New York, 1976), Appendix C.

| FINANCIAL | Responsible person/dept | Due date | Remarks |
|---|---|---|---|
| Close work orders | | | |
| Close task accounts | | | |
| Audit charges | | | |
| Close payables | | | |
| Collect receivables | | | |
| Terminate commitments | | | |
| Prepare final cost summary | | | |
| **DOCUMENTATION** | | | |
| Spare parts list | | | |
| Design drawings | | | |
| Procurement specs | | | |
| Equipment specs | | | |
| Test procedures | | | |
| Parts lists | | | |
| Maintenance manuals | | | |

Figure 11-6. Section of checklist for project termination.

- *Negotiating skills*. Both internally and externally, negotiation plays a large part in closure. Project termination is about rapidly removing obstacles at the best cost—and if the manager has no negotiating skills, minor (and major) obstacles could result in delays and added cost. Without negotiating skills, the manager will have trouble assisting personnel with reassignment, disposing of project material, getting agreement on deliverables and commitments, obtaining certifications, closing down physical facilities, and determining audit requirements. There are many successful *styles* of negotiation, and the project manager only needs to master one.
- *A sense of urgency concerning details*. There are many times during the start-up and middle phases when the best strategy is to focus on the "big picture" and bypass the details. But termination is about the trees, not the forest—details are what it's about. Some managers are good in dealing with the forest, others with the trees, and a rare few are good at both.

## THE END IS THE BEGINNING

A complete project termination is necessary for maximum goodwill and minimum cost. Because both project personnel and purpose are highly result-oriented, it is common to have the major items delivered on time and in working order but the project's end delayed for months and even years. The willingness and ability to bring the job to a satisfactory close are essential to the success of project organizations. A timely, complete closeout of a project shows good management and sets the stage for further relationships between the client/customer and the project organization.

The project manager's last task is to seek opportunities for either extensions of the project or new business. By seeking such opportunities and documenting them, the project manager ensures that the end of one project is the beginning of others.

## REFERENCES

1. Lock, Dennis. *Project Management* (Gower Press. Epping, 1979).
2. Middleton, C. J. "How to Set Up a Project Organization." *Harvard Business Review Reprints Series,* No. 67208.
3. Mulvaney, John. *Analysis Bar Charting* (Management Planning and Control Systems. Washington, 1977).
4. Thamhain, Hans J. and Wilemon, David L. "Diagnosing Conflict Determinants in Project Management." *IEEE Transactions on Management,* Vol. EM-22 (1) (February 1975).

# 12. The Evolution of the Systems Development Life Cycle: An Information Systems Perspective

William R. King*
Ananth Srinivasan†

Rapidly changing technology is becoming increasingly important to successful organizational strategy. Whether the technology be that which is developed in laboratories as the basis for new products or the computer technology that must be integrated into a modern manufacturing system, overall business results are increasingly sensitive to the effective integration of technology into strategy.

Nowhere is this change more apparent than in the information systems area, in which the information systems development life cycle (ISDLC) has been a fundamental management tool. A variety of changes have been occurring in the ISDLC that have led organizations to develop a better ability to manage information as a resource and to integrate information systems into their "organizational behavior."

Despite recent concern that the modern "end-user computing" environment may represent the death-knell of the ISDLC, it appears that there is, and will continue to be, a major role for systems development life cycle (SDLC) notions as guiding frameworks for information systems (IS)

---

* William R. King is University Professor in the Katz Graduate School of Business at the University of Pittsburgh. He is the author of more than a dozen books and 150 technical papers that have appeared in the leading journals in the fields of management science, information systems, and strategic planning. Among his major honors are the McKinsey Award (jointly with D. I. Cleland) for the "outstanding contribution to management literature" represented by their book *Systems Analysis and Project Management,* the IIE Book-of-the-Year Award for the first edition of this book, and designation as a fellow of the Decision Sciences Institute. Further biographical information is available in *Who's Who in the World* and *Who's Who in America.*

† Ananth Srinivasan is Associate Professor at Case Western Reserve University's Weatherhead School of Management. He received his Ph.D. at the University of Pittsburgh and has served as Associate Editor of the *MIS Quarterly*. His research has been published in the *MIS Quarterly, Communications of the ACM,* and the *Academy of Management* Journal, among other journals.

development. Of course, the modern SDLC has evolved to become more expansive and more robust than the "traditional" SDLC (11).

Here, we discuss this evolving role of the systems development life-cycle concept in the context of computer-based information support. The concept itself has been changing and expanding, partly in response to reported failures of information systems projects to live up to popular expectations. More importantly, the recognition of information as a corporate resource (37) and its integral role in organizational decision making has also led to this evolving role of the life-cycle concept. The implied rigidity of the traditional definitions of the SDLC concept, coupled with rapidly changing organizational environments, have caused some to express the need for flexibility to be built into the system development process (1). This would allow for the modification of the traditional life-cycle definitions to suit the contingencies faced by specific organizations in given situations.

This chapter focuses on these modifications by elaborating on classic life-cycle notions and discussing "new" phases that precede and follow those that were traditionally defined. We also discuss prototyping as a system development methodology, primarily to contrast it with the life-cycle concept. There has been considerable interest in prototyping as an "alternative" to the SDLC. Here, we show that it is complementary, rather than antithetical, to the modern version of the SDLC.

## THE SYSTEMS DEVELOPMENT LIFE CYCLE

The systems development life cycle (SDLC) is one of the fundamental concepts of the field of complex systems. The general notion of life cycles is widely used in management, engineering and elsewhere. For instance, the product life cycle is a fundamental marketing concept (17), and the organizational life cycle has been studied extensively (36).

The systems development life cycle traces its roots to those concepts and techniques that were developed in the weapons systems development context. There, the complexity of the systems under development dictated a need for management concepts and ideas that could aid in the management of great complexity (27).

The SDLC idea was quickly recognized to have a natural applicability in the development of complex computer systems. Thus, it became one of the fundamental concepts that provides a framework for much of the thinking, practice, and research in the area. It is difficult to conceive of an area of information systems practice and research that does not either deal directly with the systems development life cycle or in some fashion take account of the relevant stage or stages of the life cycle.

## THE IMPORTANCE OF THE SDLC

Most people who are first introduced to the SDLC concept see it as a natural, logical, and intuitively appealing description of the complex systems development process. Because it has such descriptive appeal, it is easy to ignore or forget its significant implications.

The notion of applying an ever-changing mix of resources—people, money, and skills—to a large-scale effort in an organized fashion is rather new. It is largely in the past few decades that this has been routinely and pervasively practiced in highly technical and complex contexts. Previously, organizations were more static and operated over time with basically the same set of resources. If the nature of the task facing the organization changed, it adapted or it did not survive. Changes in the nature or amount of organization resources were largely accomplished on a reactive basis to accommodate to uncontrollable changes in the environment or in the task. The idea that such adaptation should be expected, anticipated, and planned for is the innovation that is represented by the SDLC.

While the SDLC may be viewed as descriptive of what *does* happen in order to efficiently and effectively accomplish large-scale system development efforts, it has vast normative implications. The SDLCs that were developed in the weapon systems context by the U.S. Air Force and NASA prescribed a set of activities that were to be conducted in each phase of the life cycle as well as a set of outputs to be produced as a consequence of the activities in each phase (10) and Chapter 12.

The normative information SDLC may be thought of as providing the framework for the specification of the management actions and practices that are necessary to successfully address complexity and to administer constantly changing requirements for resources.

The pragmatic organizational implications of the life cycle reflect the fact that quite distinct and different varieties of resources, amounts of resources, and varieties of expertise are required at various stages in the life cycle (17).

In the IS context, the need to apply and manage different varieties and amounts of resources and different varieties of skills in clear (28). For example, while analytic expertise may predominate in the systems design stage, interpersonal and organizational expertise is most important at the point at which the finished model or system is being placed into routine usage (61), and while the total level of effort may be small in the early conceptual states, it is necessarily much larger and more diverse in the detailed design and programming phases.

Various authors have prescribed the specific managerial inputs, outputs and actions for the various stages of the ISDLC (8, 11). Here, we review these only as it is necessary to focus on the important evolving changes that have taken place in the conceptualization and application of the ISDLC.

## THE EVOLUTION OF THE ISDLC

The ISDLC description that best suits the objective of demonstrating the evolution that has taken place is that of Davis (11). He defines three broad phases, described as follows:

*Definition*—". . . the process which defines the requirements for a feasible cost/effective system."
*Physical Design*—the translation of requirements into ". . . a physical system of forms, procedures, programs, etc., by systems design, computer programming and procedure development."
*Implementation*—the phase in which the ". . . resulting system is tested and put into operation" (11).

This "classic" information systems development life cycle, as described by Davis (11), is represented in the center portion of Figure 12-1 in terms of the "definition," "physical design," and "implementation" phases. Although many more detailed descriptions of the classic ISDLC have been developed and used (8) this three-phase description is adequate to provide a basis for describing evolving life-cycle concepts and applications.

This "classic ISDLC" has undergone significant evolutionary change in recent years. In part, this reflects a response to the relative lack of success that has been achieved in meeting some of the early optimistic goals and forecasts for the computer support of management (49, 63).

However, more importantly, it reflects an evolving appreciation of the significant role of information as a resource and of computerized information systems as an integral part of the organization.

In its entirety, Figure 12-1 shows the expanded systems development life cycle that has evolved over the past several decades. The evolution of the ISDLC has been of three basic varieties:

1. *Extensions* of the "classic ISDLC," as shown in the middle of Figure 12-1, to include stages that precede and succeed those of the classic cycle.
2. Greater concern with information as a resource and with the information *function* in the organization rather than merely with specific systems development *projects*.

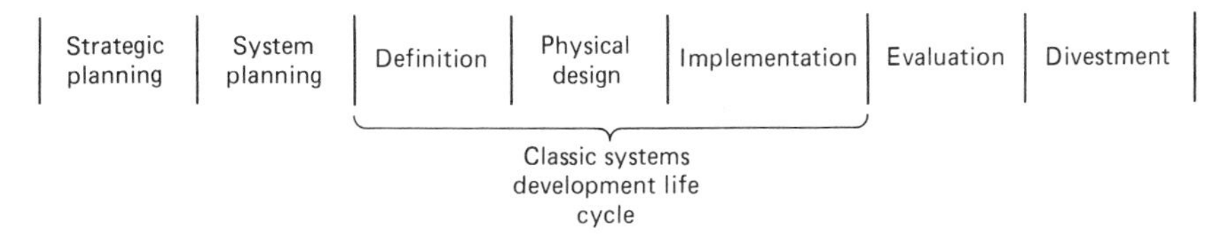


Figure 12-1. Expanded systems development life cycle.

3. A *blurring of the clear distinctions* that once existed among the phases.

### Life-Cycle Extensions

Figure 12-1 shows how the traditional life cycle has been "extended" through the addition of both prior and post phases. These additions have had the purpose of viewing information more as a resource and less as merely a service. The result has been a recognition of the need to better integrate information systems into the organization. For the first time in the computer age, this integration promises to make such systems an integral part of the management process. If this does indeed continue to occur, the old view of computers as a technical oddity operated by specialists to provide a specific service to the "real" organization will change to a view of computers as an integral element of day-to-day organizational management.

These extensions of the classic life cycle are described in Figure 12-1 in terms of two "prior" phases—labeled "Strategic planning" and "Systems planning"—and two "post" phases—labeled "Evaluation" and "Divestment." (The next major section of this chapter discusses each of these phases.)

### Functional Orientation

Many diverse computer systems and applications have been developed in most organizations in recent decades. This has led to a recognition that concern must be given to this *organizational function,* rather than merely to a series of distinct development projects.

If this were not done, redundant systems and applications would undoubtedly be developed by various managers and departments. Similarly, without such a functional orientation, the organization would be likely to proceed haphazardly with its system development efforts.

### Distinctions Among the Life-Cycle Phases

A value of, and rationale for, the ISDLC concept has been the need for different mixes of skills and resource levels during each of the phases. For instance, in the definition phase, there is generally less of a need for technical computer skills (e.g., programmers) and more need for systems analysts who can assess decision problems and information requirements (10). In the physical design phase, these relative needs shift.

The evolving view of the ISDLC has served to blur these traditional distinctions. For instance, the definitional phase of the traditional life cycle has changed in terms of the recognition of the need for inclusion of "implementation" *criteria* into the general systems design phase. Thus, considerations from a later phase in the life cycle enter into an earlier one (61). This tends to blur the conceptual distinction between the two phases as well as to alter the required mix of resources in each phase. This is so because if implementation criteria are to be developed in the definitional stage, some of the skills and resources that are usually thought of a being required primarily in the later implementation phase come to be required in the earlier definitional phase.

One such "resource," for instance, is "user involvement"—the role played, and time spent, by prospective systems users in the design process. In the early days of the computer era, user involvement was generally limited to the implementation phase, together with having users be consulted about information needs early in the definitional phase (15). Now, "user involvement" has spread to every phase of the process; even to the physical design phase, which has always been the "technical heart" of the ISDLC that was solely reserved for those with technical expertise.

This involvement of users in physical design reflects both the influence of the implementation phase and the latter evaluation phase on the earlier phases (45). This blurring of the distinctions among the phases does not mean that the ISDLC is outmoded. Rather, it is now understood to be more complex and multidimensional than it was previously thought to be.

## THE PHASES OF THE EXPANDED ISDLC

The classic ISDLC, as represented by the middle three phases in Figure 12-1, forms the core of the expanded version of the life cycle. The beginning point of the expanded cycle is a phase—the "strategic planning" phase—that is an extension of the classic cycle as well as a reflection of the need for a more functionally oriented approach.

### The Strategic Planning Phase

The purpose of this phase is to ensure the integration of management decision support systems into the organization. In this phase, the organization's purposes are used as a basis for deciding on a role to be played by information and selecting a mission or charter for the IS function.

King (39) has prescribed a process of "strategic planning for information systems" that relates various organizational "stakeholders" (5), their objectives and goals, the organization's mission, objectives, and strategies, and other salient characteristics to an "information systems strategy set." The information systems strategy set is the *product* of a planning process in that it is derived from the organization's "strategy set." This transformation is achieved by specifying a desired role for information, an IS mission and set of information *systems* objectives that are congruent with the overall organization's *business* strategy and objectives as well as with strategic organizational attributes (such as its degree of decentralization). This ensures that conflicts between the information system function and other elements of the organization are kept at a minimum.

This process is depicted in Figure 12-2 in terms of the translation of an "Organizational strategy set" into a "IS strategy set." This approach, developed by King (39) in the MIS context and adopted by IBM (25), produces the specification of a chosen role for such systems in the organization, a strategy for their development, and general objectives and constraints under which the strategy will be implemented.

Detailed discussion of the process shown in Figure 12-2 is given by King (39, 42). The process is executed by explicitly deriving each element of the IS strategy set from the collection of elements in the organizational

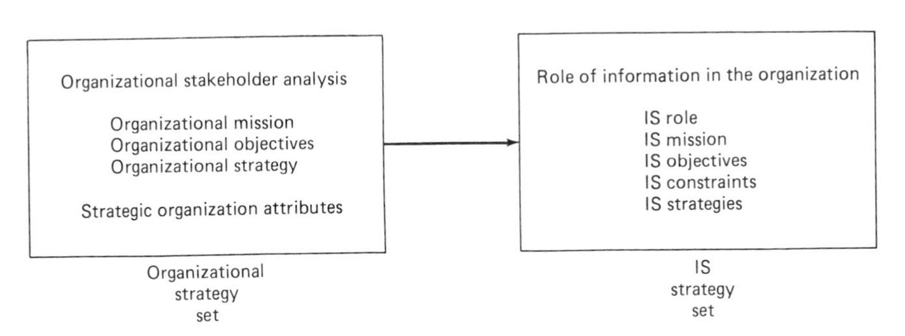


Figure 12-2. The process of strategic planning for information systems.

strategy set (as well, in some cases, as from other elements of the IS strategy set).

This is best demonstrated by an illustration adapted from King (39, 42) that relates primarily to the development of informational support for the strategic level of management. Suppose that a firm has selected, through its organizational strategic planning process, a number of *business* objectives and strategies as shown in Table 12-1. Further, suppose that a number of strategic organizational attributes have also been identified by system planners as being potentially relevant to the various kinds of strategic information systems that the organization might develop. These illustrative objectives, strategies, and attributes constitute the organizational strategy set, each element of which is identified in Table 12-1 by appropriate subscripted letter designations.

Table 12-2 shows elements of an illustrative "IS strategy set" that is derived from the organizational strategy set of Table 12-1. It specifies a role for information and IS in the organization that is chosen on the basis of providing the greatest level of support for the fulfillment of the business mission, objectives, etc., as specified in Table 12-1. More specifically, it addresses the specific mission of "strategic information systems."

The IS strategy set also entails a set of strategic information system

Table 12-1. Illustrative Organizational Strategy Set.

| ORGANIZATIONAL OBJECTIVES | ORGANIZATIONAL STRATEGIES | STRATEGIC ORGANIZATIONAL ATTRIBUTES |
|---|---|---|
| $O_1$: To increase earnings by 10% per year | $S_1$: Diversification into new businesses | $A_1$: Management is highly sophisticated |
| $O_2$: To improve cash flow | $S_2$: Improvements in credit practices | $A_2$: Poor recent performance has fostered a recognition of the need for change |
| $O_3$: To maintain a high level of customer good will | $S_3$: Product redesign | $A_3$: Most managers are experienced users of computer services |
| $O_4$: To be perceived as socially responsible | $S_4$: Improvement in productivity | $A_4$: There is a high degree of decentralization of management authority |
| $O_5$: To produce high-quality, safe products | | $A_5$: Business is highly sensitive to the business cycle |
| $O_6$: To eliminate vulnerability to the business cycle | | |

Table 12-2. IS Strategy Set.

*Role of Information in Organization*

Information is to be viewed as a resource that is used and managed in a fashion similar to other organizational resources. The organization's information systems will therefore no longer be merely a service; rather they will directly support all levels of management as well as be considered as providing a resource that is to be exploited to its maximum potential.

*Strategic Information Systems Mission*

To permit the efficient and effective assessment of alternative business strategies in terms of organization's mission and objectives using the best of available data to complement managerial judgment.

*Strategic Information Systems Objectives*

$SO_1$: To permit the prediction and assessment of the potential performance of prospective new products based on historical data and the proposed characteristics of the products.

$SO_2$: To facilitate the effective identification and assessment of potential acquisition candidates.

$SO_3$: To provide a capability for the continuous monitoring of overall performance and the degree to which organizational objectives are being achieved.

*Strategic Information Systems Constraints*

$SC_1$: The availability of funds for systems development may be significantly reduced in periods of business downturns.

$SC_2$: The overall strategic decision support system must provide the capability for managers at corporate, business unit, and divisional level to obtain and *use* performance data at the relevant organizational level.

*Strategic Information Systems Strategies*

$SS_1$: Design systems on modular basis. Design so that benefits may be directly derived from each module as it is completed.

$SS_2$: Develop "performance and objectives data base" and related subsystems initially.

objectives. Each of these objectives is directly related to, and derived from, elements of the organizational strategy set. For instance, the second systems objective—that of facilitating the identification and assessment acquisition candidates—is directly based on three of the organization's business objectives ($O_1$, $O_2$, and $O_6$ from Table 12-1), one of its strategies ($S_1$ in Table 12-1), and one strategic organizational attribute ($A_2$ in Table 12-1).

Table 12-2 also shows a strategic constraint ($SC_1$) under which IS development must occur. Because of a strategic attribute of the business ($A_5$ in Table 12-1), the availability of funds may be reduced in periods of business downturn. This, in turn, leads to a systems development strategy that directs that strategic systems be developed on a modular basis so that if funds are cut off, benefits will be derived from those modules that are already developed.

This organizational process, called "strategic planning for IS," is a precursor to the "systems planning" phase that is more commonly discussed in the literature and that is sometimes itself referred to as "strategic planning for IS." For instance, Ein-Dor and Segev (16) essentially view the two phases as a single one in that some elements of both the "strategic planning" and "systems planning" phases in Figure 12-1 are dealt with under the rubric of "strategic planning for IS." However much question there may be about terminology, though, there is no question that the consideration of the strategic elements of the organization, which have long been recognized to be essential to the fullest realization of IS potential (48), are now being made operational (25).

### The Systems Planning Phase

The "beginning" of the classic information systems development life cycle is usually described in terms of the identification of a need or the preparation of a preliminary proposal for a single new system. This is usually taken to be the first activity performed in the "definition" phase of the classic life cycle. The recognition of the limitations of this as a starting point in the life cycle reflects the need for *planned integration* of the various systems that are directed toward accomplishing the objectives of management.

Many organizations develop "successful" systems to perform a wide variety of functions without due regard for their integration. King and Cleland (10), for example, describe bank systems for checking, savings, loans, etc., that were not sufficiently well integrated to routinely provide management with a list of customers that reflected which bank services were being used and which were not used by each customer. Separate systems adequately performed their specific transaction-processing functions, but the data produced could not be readily integrated in a fashion that would facilitate more active management (such as through focused promotion of specific services to those who were already customers for other services).

The need for hardware and software compatibility is another important aspect of systems integration. Many companies are now experiencing the same incompatibilities in the new technologies of "office automation" as they experienced some years ago with computers. Two units of a firm may each purchase or lease an item of equipment and successfully put it to use only to subsequently discover that the technical incompatibility of the hardware, software, or both, prevents them from integrating the two systems. Thus, if consolidation of the same function in the two organizational units is desired by higher-level management, the systems hinder, rather than facilitate, the organization's integration of its information systems (35).

The "systems planning" phase of Figure 12-1 is directed toward the resolution of such difficulties. In this organizational planning phase, conducted prior to the commencement of specific information systems development projects, the proposed system is viewed in the context of other organizational systems. Viewed in a somewhat more proactive way, the systems planning phase is that in which the *need for* a new system or the *possibility of* new systems should be identified.

The systems planning phase is therefore the operationalization of the idea of planning for the wide variety of individual systems that may comprise the organization's overall information system. In terms of the modern concept of "information resource management," this may entail such diverse entities as electronic mail, automated offices, and telecommunications, as well as the more traditional MIS and EDP systems (7).

Sometimes the phase described here as "systems planning" is, in fact, inappropriately termed "strategic planning for MIS" (e.g., Head (24)) because it *takes into account* some of environmental factors that make up *one* of the dimensions that is an inherent element of strategic planning (43). However, as described in McLean and Soden (53), the starting point for this variety of planning is most often the IS mission or charter and environmental factors are considered to be constraints, or in McLean and Soden's terms ". . . opportunities and risks to be considered" (53, p. 24).

### The Definition Phase

The initial, or "definition," phase of the classical systems development life cycle begins with the "IS strategy set" as it applies to the particular system to be developed. Often, when the previous stages are not, in fact, carried out, this stage was prescribed to begin with a survey of user requirements (23). Now, "user involvement" is believed to be an essential and integral part of this phase. The potential system users are not merely passive respondents to a survey, but active participants in developing the general systems design (58, 59).

The other subphases of the "system definition" phase are well recognized (11). The output of the phase should be a "general design" for a specific IS to be subsequently translated into greater physical detail (40). The general design is understood to be a road map for the development of the detailed design.

However, in recent years the definitional phase has been enriched and broadened by the addition of criteria related to the later implementation phase. Much of the early literature of IS development treated "implementation" as a technical objective—for example, the testing of the system under operating conditions (21).

When systems were found to "fail," or to go unused, or not to meet usage criteria because of nontechnical considerations, the idea of implementation was broadened to include organizational and behavioral dimensions. Prescriptions to resolve this difficulty ranged from simple ideas of "selling" the system, to more sophisticated ones involving the systems designer acting as an agent of change (66), to improved training of the system's users so that they would be better able to appreciate the benefits of computerization (55). In all instances, this early view of implementation required *changes on the part of the system users.*

The evolving view of implementation sees this later stage in the life cycle as the source for a set of *goals* that are to be attended to in *each of the prior phases of the life cycle*. Schultz and Slevin (61) have prescribed "technical validity" and "organizational validity" as parallel goals to be sought in developing a model of a decision, or a system. King (41) has discussed the inherent trade-offs that will often be necessary if systems are to have a high likelihood of being accepted and used for the purpose intended by the intended users.

The traditional relegation of implementation *issues* to the implementation *phase* suggested that the process was one of developing the "optimal" (technical) system, and *then,* of having the designer consider what must be done to get the design accepted by the user. This "after-thought" view of implementation issues serves to implicitly relegate them to a low level of relative importance as well as to explicitly restrict the feasible range of implementation alternatives.

The modern view is to deal with implementation issues as early as possible in the life cycle and to continue to raise them throughout. The underlying premise of this broader concept of system optimality is simple: that it is better to develop a good system that is used for the needed purpose than to develop one which is theoretically optimal but which is not implemented.

While this criterion, or guiding principle, is more philosophical than operational, Cleland and King (10) have provided a specific IS system design process based on the criterion. Others (50, 54) have studied and prescribed other approaches to achieving greater concern for implementation in the definition phase.

Ginzberg (19) deals with this issue in terms of trying to predict systems failures at an early stage in the development cycle. By assessing users' expectations about the system in the definition stage, the management of system development projects is more easily facilitated, thereby increasing the probability of a successful system. The results obtained in his study seem to indicate that it is important for systems designers and managers to know how realistic user expectations about the system are at the definition stage of the ISDLC. In this manner, critical implementation issues are dealt with very early in the life cycle of the system. This "broader" view of implementation is also the central notion of the view of system implementation in terms of an organizational change process (47).

### The Physical Design Phase

While the techniques of physical design have evolved as havehardware and software, perhaps the most significant evolution in this phase has been in terms of user involvement and of multistage design.

In the user involvement dimension, this highly technical phase was previously viewed as the sole province of the analyst. However, the analyst is now viewed more as a catalyst who translates the user's functional and logical designs into physical reality (51). The "political" dimensions of the physical design process are also of concern since desired changes in the system may place users and designers in conflict (33) or create conflicts among user groups who have different sets of needs (18, 46).

These complicating factors, together with the inherent imprecision in the trade-offs among technical and implementation-oriented criteria which must be applied in this phase, have led to an emerging multistage view of the physical design phase.

Such trade-offs require a continuing process of adjustment and readjustment because many of the factors are so imprecise. In effect, tentative design parameters of the system must be evaluated on various bases of both a technical and organizational nature *before* they are made permanent. This means that the evolving systems design must be evaluated on a continuing basis. Thus, the evaluation phase of the life cycle, formerly thought of as being one of the latter stages, becomes a part of the physical design stage.

King and Rodriguez (45) have presented a methodology for such an ongoing evaluation process. In applying this methodology in a DSS context, Dutta and King (14) found that certain design features of an evolving DSS were viewed to be overly complex by the intended users. This led to technical changes in the systems design while it was still in the prototype stage.

Such a view of the physical design phase requires that it be multistage in some sense. This may be operationalized as modular design (40), prototyping* (31), or in some other way. However, the continuing assessment of the systems in both technical and organizational terms requires that proposed system characteristics be identified and subjected to scrutiny. Whether this is done in the form of "experiments" with users on a prototype system (45) or in some other fashion, the ongoing reconfiguration of the system in response to those assessments must be made feasible. This can only be achieved through a multistage view of the physical design phase.

Sprague (64) has described this process as it applies to decision support systems as a "collapsing" of the traditional life-cycle phases into a single phase that is iteratively applied. Keen (31) has similarly shown how "prototyping" can be used as a basis for assessing and predicting the value of decision support systems.

### The Implementation Phase

Implementation of management science and MIS was one of the "hot topics" of the 1970s. The papers presented at the first implementation conference in Pittsburgh (61) reflect the dominant view of implementation that was then held. Although some authors were proposing notions of an "adaptive approach" to systems implementation, the empirical perspective afforded to the concept of implementation centered around "putting the model or system into use." Traditional approaches to implementation in the MIS area also reflect this stress on systems use. Davis (11) represents the classical view of MIS implementation where he defines the implementation stage as that which begins after the development efforts have been accomplished and the system has commenced operation. This view of implementation was perhaps best captured by the framework of Anderson and Hoffman (4) who viewed the implementation effort as comprising three distinct phases: installation (which was traditionally the systems analyst's view of implementation); implementation, which encompasses the process of using the system (which was traditionally the user's view of implementation); and the integration of the model with its results into management behavior.

Contrary to this classical view of implementation, some researchers were proposing a new view of the implementation process based on a theory of organizational change. This "broader" perspective on implementation was deemed appropriate in light of the narrowness in the (now)

* Since prototyping is seen by some to be an "alternative" to the ISDLC, we shall deal with it in more detail in a subsequent section.

classical definitions. Specifically, it was argued that by neglecting to consider implementation issues in the earlier phases of the life cycle, a major portion of the causes of user satisfaction or dissatisfaction was being ignored.

This significant evolution in thinking is based on substantial evidence concerning the implementation failures that have occurred (51). Much of the study and research devoted to implementation has focused on "organizational change" (20, 32, 70), the role of the user (15, 44, 67), user characteristics (9, 69), and organizational characteristics (18) that may be taken into account in the system design.

Alter (3) extensively summarizes the many factors that have been identified as resulting in increased risk of implementation failure. Schultz and Slevin (61) and Doktor, Schultz, and Slevin (13) have provided compendiums of the evolving thinking in the area.

With respect to the elements of the implementation phase itself, as distinct from the integration of implementation concerns and criteria into other phases of the life cycle, various approaches have been taken. For instance, Gremillion (22) has broadly focused on the training and integration aspects of the implementation phase.

### The Evaluation Phase

While even the most traditional view of the information systems life cycle involved some evaluation component (usually as an element of the implementation phase), it was generally *post hoc* and relatively insignificant. Much of the evaluation of earlier periods can be characterized as post mortems of system failures (12, 51, 52).

A popular approach to evaluation involved using the extent of system usage as a surrogate for system success. The argument proposed in favor of this approach was that in cases where use of the system was nonmandatory, increased system usage is brought about because the user believes that such use has improved his performance. The inability of researchers to clearly establish this link between system use and decision performance led to different conceptualizations of the system evaluation phase, although some studies have indeed shown a positive correlation between "perceived system worth" and system usage (60).

Expanded life-cycle thinking integrates the evaluative phase into the prior stages of the life cycle just as it does with the implementation phase (15). Moreover, it extends the dimensions of the system to be evaluated well beyond technical considerations and the degree of usage to the domain of perceived values and beyond. For instance, King and Rodriguez (45) have assessed a Decision Support System (DSS) in terms of attitudes, values, information usage behavior, and decision impact. Ginzberg (19) also uses both behavioral (usage) and attitudinal measures. Keen and Scott-Morton (34) discuss a wider variety of evaluation criteria that may be considered for DSSs. Keen (31) discusses a "value analysis" approach that emphasizes the qualitative benefits that are derived from such systems.

The evaluation phase must also account for the changes and enhancements that will inevitably be proposed after a system is put into operation. Any such changes or enhancements demand that evaluations be made, for if the costs and benefits of proposed changes are not evaluated, resources may be consumed that could otherwise be put to better use.

### The Divestment Phase

The "last" stage of the life cycle is divestment. The formalization of this phase recognizes that "everything must end" and that the phaseout of a system should not represent failure. This variety of thinking has been applied to products and to entire businesses during the past decade. As the strategic planning and systems planning phases of the expanded systems development life cycle have been developed, the use of this sort of thinking in business has naturally led to its application in the systems context.

Despite the fact that the divestment phase for a system cannot generally be viewed as even potentially "profitable," as can the same phase of a business or product life cycle, the idea is rather new that the declining effectiveness and productivity of a system is inevitable, although not uncontrollable, and that it should therefore be planned for (10, 40).

Thus, when a system is perceived in this fashion, it can be viewed as an asset to be "milked" via judicious and limited planned investments rather than as one to either be "saved" through large incremental investment or "junked" and replaced by a costly alternative system.

The inclusion of this viewpoint into the earlier "systems planning" phase can lead to the extension of the "natural" life of some systems through cost-effective enhancements, the evolution of some systems into others that are of significantly greater power and effectiveness, and the eventual planned joint phaseout and phasein of systems that are directed toward the same objectives.

### Prototyping

One outcome from the need for a modified and evolutionary concept of system development has been the advocacy of prototyping as a development methodology. An excellent review of the literature on this subject may be found in Jenkins and Lauer (26). As Naumann and Jenkins (56) point out, a number of phrases have been used in the literature to essentially connote the prototyping approach in software development. Among the phrases are "heuristic development," "infological simulation," and "middle-out design."

Prototyping has long been used in manufacturing design as a means to work toward a final product by the production of intermediate (and hence imperfect and incomplete) versions that would be evaluated and modified over several iterations. Inherent in the prototyping approach is the idea of iterative design in system development. The application of the concept to information system development is primarily a reflection of the availability of new technology (and consequent enhanced development tools), the premise that information systems can be developed and used by senior executives, and the emergence of applications designed for relatively small user groups.

There have been some claims that prototyping is an alternative to the SDLC methodology, and studies have been performed to compare the viability of the two approaches. Alavi (2) reported that in an experimental study that compared the two approaches, the prototyping approach received a more favorable overall evaluation than did the SDLC approach to system development. Further, reduced conflict was reported between users and designers involved in the prototyping approach and a more adequate level of user participation was reported in the process that used prototyping. These results appear to indicate that prototyping is a "better" approach to system development compared to the ISDLC approach. It is our position that this conclusion is not necessarily valid.

The study mentioned above assumes that the two approaches are mutually exclusive of each other and, further, that either one may be used as the system development methodology in the context of developing an application system. This is not valid. There are clearly some applications where prototyping, when used as the *dominant* method of development, may work very well (see Keen's (30) discussion about adaptive design). Similarly, there exist large, complex applications for which the ISDLC approach has to be used in order to manage and control the intricacies that are involved. Prototyping could clearly be used here in the overall life-cycle context to deal with well-defined subproblems but certainly not as a replacement of it. This is especially true of applications development for the federal government, the aerospace industry, and other applications that tend to be of considerable size. Ironically, the results of Alavi's study also show that the ISDLC process was easier from a management control point of view compared to the prototyping process.

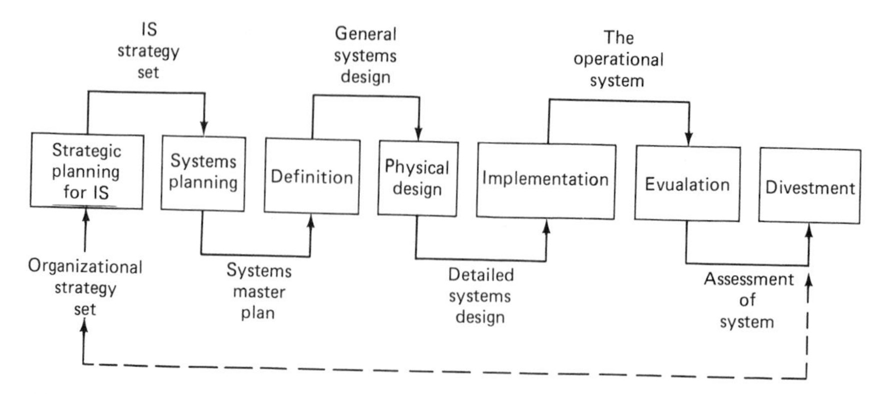


Figure 12-3. Input-output description of expanded information systems development life cycle.

In this chapter, our view of the expanded ISDLC notion incorporates the *spirit* of prototyping. Our concept of the modified life cycle takes into account the flexibility concern described earlier by providing for iterations in the process (see Figure 12-3). While we do not claim that this is the ideal process for *all* applications, it would certainly be well suited for large applications that are fairly complex and include a substantial user base.

## SUMMARY

The evolving phases of the systems development life cycle in an organization are portrayed in Figure 12-3 in terms of their inputs and outputs. The output of each phase can be seen to be the input to the subsequent phase.

The "organizational strategy set" is the output of the organization's strategic *business* planning. It is, in turn, the input that begins the initial "strategic planning for information systems" phase of the systems life cycle. The strategic planning phase converts this into an "IS strategy set" that gives details of the role to be played by information and information systems in the organization, developmental strategies for systems, etc.

The "systems planning" phase converts the systems strategy set into a systems master plan. The master plan describes the objectives, roles, and time frames for the development of the various systems that comprise the organization's overall decision support function.

In the definitional phase for a particular system, a "general design" for the system, which fulfills that system's role and objectives as reflected in the master plan, is developed. This is then converted into a detailed design in the "physical design" phase.

The "implementation" phase involves the conversion of the design into reality and into the organization's behavior. The "evaluation" phase produces a formal assessment of how well the system is meeting the goals established earlier and whether it is playing the organizational role that was established for it. The last phase is the "divestment" phase, in which the system is converted or phased out in accordance with the organization's master plan.

The dotted line linking the first and last phases of the expanded cycle reflects the integration of assessments made in the latter stages with planning in the earlier ones. In effect, the class life cycle has been expanded into an ongoing organizational process in which new systems are being developed to play new roles as well as to replace outdated ones that are being phased out.

The evolution of the classic ISDLC into this expanded version has had the effect of converting information and computer systems into a resource that is "managed"—planned for, evaluated, divested, etc.—in much the same way as are the organization's other resources. The impact of this has been an increase in the level of integration of computer information systems into organizations. While this is probably an inevitable development in the long run, the timeliness and manner in which it is being accomplished will undoubtedly have a great influence on managerial productivity and on managerial effectiveness.

## REFERENCES

1. Ahituv, Neumann N., S., and Hadass, M. "A Flexible Approach to Information System Development." *MIS Quarterly,* Vol. 8(2) (June, 1984), pp. 69–78.
2. Alavi, M. "An Assessment of the Prototyping Approach to Information Systems Development." *Communications of the ACM,* Vol. 27(6) (June, 1984), pp. 556–563.
3. Alter, S. *Decision Support Systems: Current Practice and Continuing Challenges* (Addison-Wesley. Reading, Mass., 1980).
4. Anderson, J. C. and Hoffman, T. R. "A Perspective on the Implementation of Management Science." *Academy of Management Review,* Vol. 3(3) (1978), pp. 563–571.
5. Ansoff, I. *Corporate Strategy* (McGraw-Hill. New York, 1965).
6. Anthony, R. N. *Planning and Control Systems: A Framework for Analysis.* Division of Research, Harvard Graduate School of Business Administration, 1965.
7. Bell, D. "Communications Technology: for Better or for Worse." *Harvard Business Review* (May–June, 1979).
8. Benjamin, R. *Control of the Information Systems Development Cycle* (Wiley. New York, 1971).
9. Carlson, E. D., Grace, B. F. and Sutton, J. A. "Case Studies of End User Requirements for Interactive Problem Solving Systems." *MIS Quarterly,* Vol. 1(1) (1977), pp. 51–63.
10. Cleland, D. I. and King, W. R. *Systems Analysis and Project Management,* 2nd Ed. (McGraw-Hill. New York, 1975).

11. Davis, G. B. *Management Information Systems: Conceptual Foundations, Structure, and Development* (McGraw-Hill. New York, 1974).
12. Diebold, J. "Bad Decisions on Computer Use." *Harvard Business Review* (January–February, 1969).
13. Doktor, R., Schultz, R. L. and Slevin, D. P. (eds.). *Implementation of Management Science* (TIMS Studies in Management Science, North Holland, 1979).
14. Dutta, B. K. and King, W. R. "A Competitive Scenario Modeling System." *Management Science,* Vol. 26(3) (March, 1980), pp. 261–273.
15. Edstrom, A. "User Influence and Success of MIS Projects: A Contingency Approach." *Human Relations,* Vol. 30(7) (1977), pp. 589–607.
16. Ein-Dor, P. and Segev, E. "Strategic Planning for MIS." *Management Science,* Vol. 24(15) (1978), pp. 1631–1641.
17. Fox, H. W. "A Framework for Functional Coordination." *Atlanta Economic Review,* Vol. 23(6), (1973).
18. Ghymn, K. I. and King, W. R. "Design of a Strategic Planning MIS." *OMEGA,* Vol. 4(5) (1976), pp. 595–607.
19. Ginzberg, M. J. "Early Diagnosis of MIS Implementation Failure: Promising Results and Unanswered Questions." *Management Science,* Vol. 27(4) (1981), pp. 459–478.
20. Ginzbergr, M. J. "A Process Approach to Management Science Implementation." Unpublished Ph.D. dissertation, M.I.T., 1975.
21. Gregory, R. H. and Van Horn, R. L. *Automatic Data-Processing Systems,* 2nd Ed. (Wadsworth. Belmont, Calif., 1969).
22. Gremillion, L. L. "Managing the Implementation of Standardized Computer Bases Systems." *MIS Quarterly,* Vol. 4(4)(December, 1980), pp. 51–59.
23. Hall, T. P. "Systems Life Cycle Model." *Journal of Systems Management,* Vol. 31(4) (April, 1980).
24. Head, R. V. "Strategic Planning for Information Systems." *Infosystems* (October, 1978), pp. 46–54.
25. IBM. *Business Systems Planning: Information Systems Planning Guide,* 1981.
26. Jenkins, A. M. and Lauer, T. W. "An Annotated Bibliography on Prototyping." Unpublished Discussion Paper #228, Indiana University, Division of Research, School of Business, 1983.
27. Johnson, R. A., Kast, F. and Rosenzweig, J. E. *The Theory and Management of Systems* (McGraw-Hill. New York, 1967).
28. Kaiser, K. and King, W. R. "The Manager-Analyst Interface in Systems Development." *MIS Quarterly,* Vol. 6(1) (1982).
29. Keen, P. G. W. "Decision Support Systems: Translating Analytical Techniques into Useful Tools." *Sloan Management Review* (Spring 1980a), pp. 33–44.
30. Keen, P. G. W. "Adaptive Design for Decision Support Systems." *Database,* Vol. 12(1), (Fall, 1980b), pp. 15–25.
31. Keen, P. G. W. "Value Analysis: Justifying Decision Support Systems." *MIS Quarterly* (1981a), pp. 1–66.
32. Keen, P. G. W. "Information Systems and Organizational Change." *Communications of the ACM,* Vol. 24(1) (1981b), pp. 24–33.
33. Keen, P. G. W. and Gerson, E. M. "The Politics of Software Engineering." *Datamation* (November, 1977), pp. 80–86.
34. Keen, P. G. W. and Soctt-Morton, M. S. *Decision Support Systems: An Organizational Perspective* (Addison-Wesley. Reading, Mass., 1978).
35. Ketron, R. W. "Four Roads to Office Automation." *Datamation* (November, 1930).
36. Kimberly, J. H. and Miles, R. H. *The Organizational Life Cycle* (Jossey-Bass. San Francisco, 1980).

37. King, W. R. "Exploiting Information as a Strategic Business Resource." *International Journal on Policy and Information,* Vol. 8(1) (June, 1984), pp. 1–8.
38. King, W. R. and Zmud, R. W. "Managing Information Systems: Policy Planning, Strategic Planning, and Operational Planning," in *Proceedings of the Second International Conference on Information Systems,* ed. Ross, C. A. (Cambridge, Mass., 1981), pp. 299–308.
39. King, W. R. "Strategic Planning for MIS." *MIS Quarterly,* Vol. 2(1) (1978), pp. 27–37.
40. King, W. R. *Marketing Management Information Systems* (Petrocelli/Charter. New York, 1977).
41. King W. R. "Methodological Optimality in Operations Research." *OMEGA,* Vol. 4(1) (1976), pp. 9–12.
42. King, W. R. "Integrating Computerized Planning Systems into the Organization." *Managerial Planning* (1982).
43. King, W. R. and Cleland, D. I. *Strategic Planning and Policy* (Van Nostrand Reinhold. New York, 1978).
44. King, W. R. and Cleland, D. I. "Manager-Analyst Teamwork in Management Information Systems." *Business Horizons* (April, 1971), pp. 59–68.
45. King, W. R. and Rodriguez, J. I. "Evaluating Management Information Systems." *MIS Quarterly* (September, 1978), pp. 43–51.
46. Kling, R. "The Organizational Context of User-Centered Software Designs." *MIS Quarterly* (December, 1977), pp. 41–52.
47. Kolb, D. A. and Frohman, A. L. "An Organization Development Approach to Consulting." *Sloan Management Review,* Vol. 12(1) (1970), pp. 51–65.
48. Kreibel, C. H. "The Strategic Dimension of Computer Systems Planning." *Long Range Planning,* Vol. 1(1) (1968), pp. 7–12.
49. Leavitt, H. J. and Whisler, T. L. "Management in the 1980s." *Harvard Business Review* (November–December, 1958).
50. Lucas, H. "Empirical Evidence for a Descriptive Model of Implementation." *MIS Quarterly* (June, 1978), pp. 27–41.
51. Lucas, H. *Why Information Systems Fail* (Columbia University Press. New York, 1975).
52. McKinsey and Co. "Unlocking the Computer's Profit Potential." *Computers and Automation* (April, 1969).
53. McLean, E. R. and Soden, J. V. (eds.). *Strategic Planning for MIS* (Wiley. New York, 1977).
54. Mintzberg, H. "Impediments to the Use of Management Information." (National Association of Accountants. New York, 1975).
55. Murdick, R. G. and Ross, J. E. "Management Information Systems: Training for Businessmen." *Journal of Systems Management* (October, 1969).
56. Naumann, J. D. and Jenkins, A. M. "Prototyping: The New Paradigm for Systems Development." *MIS Quarterly,* Vol. 6(3) (September, 1982), pp. 28–44.
57. Norton, D. P. "Information System Centralization: The Issues," in *Information Systems Administration,* ed. McFarlan, F. W., Nolan, R. and Norton, D. (Holt, Rinehart and Winston. New York, 1973), Chapter 12.
58. Powers, R. F. and Dickson, G. W. "MIS Project Management: Myths, Opinions, and Reality." *California Management Review,* Vol. 15(3) (Spring, 1973), pp. 147–156.
59. Reisman, A. and de Kluyver, C. A. "Strategies for Implementing Systems Studies," in *Implementing OR/MS,* ed. Schultz, R. L. and Slevin, D. P. (North Holland. 1975), pp. 291–309.
60. Robey, D. "User Attitudes and MIS Use." *Academy of Management Journal, Vol. 22(3) (1979), pp. 527–538.*

61. Schultz, R. L. and Slevin, D. P. "Implementation and Organizational Validity: An Empirical Investigation," in *Implementing OR/MS*, ed. Schultz, R. L. and Slevin, D. P. (North Holland. 1975), pp. 153–182.
62. Shultz, G. and Whisler, T. (eds.). *Management, Organization and the Computer* (Free Press. New York, 1960).
63. Simon, H. A. *Administrative Behavior* (Macmillan. New York, 1947).
64. Sprague, R. H. "A Framework for the Development of Decision Support Systems." *MIS Quarterly* (December, 1980), pp. 1–26.
65. Sprague, Ralph H. and Olson, Ronald L. "The Financial Planning System at Louisiana National Bank." *MIS Quarterly,* Vol. 3(3) (1979), pp. 35–46.
66. Stern, H. "Human Relations and Information Systems." *Interfaces,* Vol. 1(2) (February, 1971).
67. Swanson, E. B. "Management Information Systems: Appreciation and Involvement." *Management Science,* Vol. 21(2) (October, 1974), pp. 178–188.
68. Welsch, Gemma M. "Successful Implementation of Decision Support Systems Preinstallation Factors, Service Characteristics, and the Role of the Information Transfer Specialist." Ph.D. Dissertation, Northwestern University, 1980.
69. Zmud, R. W. "Individual Differences and MIS success: A Review of the Empirical Literature." *Management Science,* Vol. 25(10) (1979), pp. 966–979.
70. Zmud, R. W. and Cox, J. F. "The Implementation Process: A Change Approach." *MIS Quarterly,* Vol. 3(2) (1979), pp. 35–43.

## BIBLIOGRAPHY

Churchman, C. W. *The Design of Inquiring Systems* (Basic Books, New York, 1971).

Gallagher, C. A. "Perceptions of the Value of an MIS," *Academy of Management Journal,* Vol. 17 (1), 46–55 (1974).

King, W. R. and Cleland, D. I. *Strategic Planning and Policy* (Van Nostrand Reinhold, New York, 1978).

Lientz, B. P., Swanson, E. B., and Tompkins, G. E. "Characteristics of Application Software Maintenance," *Communications of the ACM,* Vol. 21 (6), (1978).

Maish, A. M. "A User's Behavior Toward His MIS," *MIS Quarterly,* Vol. 3 (1), (1979).

Robey, D. and Zeller, R. L. "Factors Affecting the Success and Failure of an Information System for Production Quality," *Interfaces,* Vol. 8, (2), 70–75 (1978).

Schewe, C. D., "The Management Information's Systems User: An Exploratory Behavioral Analysis," *Academy of Management Journal,* Vol. 19 (4), 577–590 (1976).

# Section V
# Project Planning

This section provides "tools" that can be used in planning a project. It focuses on the various perspectives of the "owners," bidders, managers, and other project stakeholders.

In Chapter 13, David I. Cleland describes the process by which the impact of the various project stakeholders can be taken into account in developing the project plan.

Chapter 14 focuses on the work breakdown structure (WBS)—one of the most basic project planning tools. There, Gary D. Lavold discusses its use in various phases of the planning process as well as its role in project control.

The best-known tools of project planning—network plans—are reviewed by Joseph J. Moder in Chapter 15. Network plans, useful in both project planning and control, vary from the simple to the complex, with varying data requirements and ease of use. Moder reviews a wide range of these techniques from which the most appropriate one may be selected.

In Chapter 16, the linear responsibility chart (LRC)—a powerful tool of organizational planning—is discussed by David I. Cleland and William R. King. The LRC is shown to have a variety of uses in the organization as well as in the project.

In Chapter 17, Harold Kerzner discusses the pricing dimension of project planning. Pricing requires that a work breakdown structure (Chapter 14) first be developed. Then, it can be "priced out." Kerzner relates pricing to the WBS as well as to other planning tools—the linear responsibility chart (Chapter 16) and network plans (Chapter 15).

Contract development, treated in Chapter 18 by M. William Emmons, represents the transition from project planning to project implementation. Mr. Emmons presents contract development as the keystone that provides support to the overall project management process.

The view of a project from the standpoint of the bidder is treated by Hans J. Thamhain in Chapter 19—"Developing Winning Proposals."

# 13. Project Stakeholder Management*

David I. Cleland†

The management of a project's "stakeholders" means that the project is explicitly described in terms of the individuals and institutions who share a stake or an interest in the project. Thus, the project team members, subcontractors, suppliers, and customers are invariably relevant. The impact of project decisions on all of them must be considered in any rational approach to the management of a project. But management must also consider others who have an interest in the project and by definition are also stakeholders. These stakeholders are outside the authority of the project manager and often present serious management problems and challenges.

Because project stakeholder management (PSM) assumes that success depends on taking into account the potential impact of project decisions on *all* stakeholders during the entire life of the project, management faces a major challenge. For in addition to identifying and assessing the impact of project decisions on stakeholders who are subject to the authority of the management, they must consider how the achievements of the project's goals and objectives will affect or be affected by stakeholders outside their authority. Project stakeholders, often called *intervenors* in the nuclear power plant construction industry, can have a marked influence on a project. At one nuclear power plant numerous bomb threats over the

---

* Portions of this chapter have been paraphrased from David I. Cleland, "Project Stakeholder Management," *Project Management Journal,* September 1986, pp. 36–44. Used by permission.

† David I. Cleland is currently Professor of Engineering Management in the Industrial Engineering Department at the University of Pittsburgh. He is the author/editor of 15 books and has published many articles appearing in leading national and internationally distributed technological, business management, and educational periodicals. Dr. Cleland has had extensive experience in management consultation, lecturing, seminars, and research. He is the recipient of the "Distinguished Contribution to Project Management" award given by the Project Management Institute in 1983, and in May 1984, received the 1983 Institute of Industrial Engineers (IIE)-Joint Publishers Book-of-the-Year Award for the *Project Management Handbook* (with W. R. King). In 1987 Dr. Cleland was elected a Fellow of the Project Management Institute.

life of the project impacted construction schedules, shut down work on select areas, frustrated managers and professionals, and forced more intensive security provisions to include physical searches of people, equipment, and vehicles. Further impact on the project came about in the form of antinuclear blockades and demonstrations which impacted productivity. In the fall of 1981, the Abalone Alliance, an antinuclear organization, planned, organized, and attempted a blockade of the plant. The plant operating crew, management staff, draftsmen and national guard and law enforcement personnel had to be housed and fed at the plant. Total costs associated with such intervenor action cannot be calculated; schedule delays, inefficient work activity, and absenteeism occurred because of the physical threat factors.

Stakeholder management is an important part of the strategic management of organizations. There is abundant literature in the management field that establishes the need to analyze the enterprise's environment and its stakeholders as part of the strategic management of the enterprise. See, for example, F. J. Aguilar (1), W. R. Dill (2), H. Mintzberg (3), and Weiner and Brown (4).

Political, economic, social, legal, technological, and competitive "systems" affect an enterprise's ability to survive and grow in its environment. A project is also impacted by its "systems" environment. Project managers need to identify and interact with key institutions and individuals in the project's "systems" environment. For example, see Radosevich and Taylor (5), Burnett and Youker (6). An important part of the analysis of the project's "systems" environment is an organized process for identifying and managing the probable stakeholders in that environment. This management process is necessary in order to determine how the probable stakeholders are likely to react to project decisions, what influence their reaction will carry, and how the stakeholders might interact with each other and the project's managers and professionals to affect the chances for success of a proposed project strategy. Cleland and King (7), Rothschild (8), King and Cleland (9), Freeman (10), and Mendelow (11) have presented strategies for dealing with stakeholders in the corporate context. This chapter will suggest a strategy for the assessment of the influence of outside or external project stakeholders and a technique for the management of such stakeholders.

## OUTSIDE STAKEHOLDER IMPACT

Effective management cannot be carried out without considering the probable influence that key *outside* stakeholders may have on the project.

Some recent project management experiences highlight the role of these stakeholders:

- In the investigation of the management prudence of the Long Island Lighting Company (LILCO) Shoreham Project, the County of Suffolk, the New York State Consumer Protection Board, and the Long Island Citizens in Action (intervenors) argued that the project suffered from pervasive mismanagement throughout its history. The record, in the view of these intervenors, established that approximately $1.9 billion of Shoreham's cost was expended unnecessarily "as a result of LILCO's mismanagement, imprudence or gross inefficiency" (12).
- One reason that the Supersonic Transport program failed in the United States was that the managers had a narrow view of the essential players and generally dismissed the key and novel role of the environmentalists until it was too late (13).
- Some stakeholders can provide effective insight into strategic issues facing an industry. For example, in the nuclear power generation industry, the Advanced Reactor Development Subpanel of the Energy Research Advisory Board's Civilian Nuclear Power Panel submitted a report in January 1986 on the status of Advanced Reactor Development Program in the United States. This comprehensive statement reported on the three key areas on this program: (a) problem justification and current realities, (b) program redirection, and (c) Advanced reactor program recommendations. In the recommendations of this report, clear direction was given for future reactor development leading to more economical and fuel-efficient reactors. Such recommendations help to develop future strategies for other stakeholders, such as electric utilities and their suppliers, the Department of Energy, state public utility commissions, etc. More importantly, the proceedings of the Advanced Reactor Development Subpanel provide a forum for an exchange of viewpoints about nuclear power among stakeholders, some of whom may be viewed as adverse by other stakeholders. For example, the Union of Concerned Scientists, a prominent antinuclear group that represents intervenors in proceedings before the Nuclear Regulatory Commission (NRC) has been critical of a "cozy relationship" between government regulatory officials and utility officials (14).
- Public utility commissions (PUCs) are key and formidable stakeholders in the design, engineering, construction, and operation of nuclear power generating plants. In the past two years, state PUCs have

disallowed the recovery of billions of dollars in electrical rate setting. Some utilities have been penalized for imprudent spending on nuclear plants; others have been told that their plants were not needed. For example, the Pennsylvania State Public Utility Commission ruled that the Pennsylvania Power and Light Company's newly opened 945-MW $2 billion Susquehanna Unit 2 nuclear plant would provide too much generating capacity for the utility's customers. The utility was only allowed to recover taxes, depreciation, and other operating costs. The Missouri Public Service Commission recently disqualified Union Electric Company from charging ratepayers for $384 million of the $3 billion spent on the new Callaway nuclear plant in central Missouri. The commission cited high labor expenses, improper scheduling of engineering, and "inefficient, imprudent, unreasonable, or unexplained costs" during four years of delay (15).

- Diverse stakeholders, or intervenors, are taking active roles in rate-setting case hearings. For example, when the Union Electric Company of St. Louis, Missouri, instituted proceedings for authority to file tariffs increasing rates for electric service, the following parties were granted permission to intervene in the proceedings: "twenty-five cities, the State of Missouri, the Jefferson City school district, the Electric Ratepayers Protection Project, the Missouri Coalition for Environment, the Missouri Public Interest Research Group, Laclede Gas Company, Missouri Limestone Producers, Dundee Cement Company, LP Gas Association, Missouri Retailers Association, the Metropolitan St. Louis Sewer District, and the following industrial intervenors: American Can Company, Anheuser Busch, Inc., Chrysler Corporation, Ford Motor Company, General Motors Corporation, Mallinckrout, Inc., McDonnell Douglas Corporation, Monsanto Company, National Can Corporation, Nooter Corporation, PPG Industries, Inc., Pea Ridge Iron Ore Company, River Cement Company, and St. Joe Minerals Corporation (Monsanto et al.)." (16).

## MANAGEMENT RESPONSES

- Care was taken during the design and construction of the Hackensack Meadowlands sports complex to develop cooperation among each of the groups concerned with environmental impact, transportation, development, and construction.
- On the James Bay project, special effort was made to stay sensitive to social, economic, and ecological pressures (17).
- James Webb and his colleagues at NASA were adept at stakeholder

management during the Apollo program. NASA gained the support not only of the aerospace industry and related constituencies but also of the educational community, the basic sciences, and the weather forecaster profession (18).

- Obviously, in addition to special groups, the general public, often synonymous with the consumers or customers, is an important stakeholder group. At the Niagara Mohawk Power Corporation in Syracuse, New York, plans for achieving public acceptance of the atom as a source of electric power began long before the company had any specific plans for constructing its own nuclear plant. Niagara Mohawk began to inform the public of progress in using the atom for electric power generation soon after the Atomic Energy Act was signed in 1954. A full-scale successful public relations program was carried out before and continued after the initiation of the project (19). A study of quality problems in the Nuclear Power Plant Construction Industry provides insight into stakeholder management.

Action taken by two electric utilities to facilitate an effective relationship with the NRC stakeholder included the following:

—Florida Power & Light established a special office near the NRC headquarters to facilitate exchange of information during the licensing process.

—Senior management of Arizona Public Service established the following policy concerning NRC:

> Don't treat NRC as an adversary; NRC is not here to bother us—they see many more plants than the licensee sees; inform NRC of what we (APS) are doing and keep everything up front; and nuclear safety is more important than schedule. (20)

Attitudes of key managers play an important role in successfully dealing with key stakeholders. In the nuclear power generation industry, management commitment to quality and a management view that NRC requirements are not the ultimate goals for performance carry great weight. For example:

> Of the projects studied there tended to be a direct correlation between the project's success and the utility's view of NRC requirements: more successful utilities tended to view NRC requirements as minimum levels of performance, not maximum, and they strove to establish and meet increasingly higher, self-imposed goals. This attitude covered all aspects of the project, including quality and quality assurance. (21)

## PSM Justification

The principal justification for adopting a PSM perspective springs from the enormous influence that key external stakeholders can exert. Thus it can be argued that the extent to which the project achieves its goals and objective is influenced by the strategies pursued by key stakeholders. Stakeholder management leading to stakeholder cooperation enhances project objective achievement, while stakeholder neglect hinders it.

In working with project managers in the development of project strategy which encompass a PSM philosophy, a number of basic premises can serve as guides for the development of a PSM approach:

- PSM is essential for ensuring success in managing projects.
- A formal approach is required for performing a PSM process:
  - —Projects extending over multiyear life cycles are subject to so much change that informal means of PSM are inadequate.
  - —Reliance on informal or hit-or-miss methods for obtaining PSM information is ineffective for managing the issues that can come out of projects.
- PSM should provide the project team with adequate intelligence for the selection of realistic options in the management of project stakeholders.
- Information on project stakeholders can be gained from a variety of sources, some of which might superficially seem to be unprofitable.

## Objective of PSM

PSM is designed to encourage the use of proactive project management for curtailing stakeholder activities that might adversely affect the project and for facilitating the project team's ability to take advantage of opportunities to encourage stakeholder support of project purposes. These objectives can be achieved only by integrating stakeholder perspectives into the project's formulation processes and developing a PSM strategy. The project manager is then in a better position to influence the actions of the stakeholders on project outcome.

Failing to recognize or cooperate with adverse stakeholders may well hinder a successful project outcome. Indeed, strong and vociferous adverse stakeholders can force their particular interest on the project manager at some time, perhaps at a time least convenient to the project. PSM is thus a necessity, allowing the project manager to set the timetable so that he can maintain better control. A proactive PSM process is designed to help the project team develop the best possible strategies.

## PSM Process

This process consists of the execution of the management functions of planning, organizing, motivating, directing, and controlling the resources used to cope with external stakeholders' strategies. These functions are interlocked and repetitive; the emergence of new stakeholders will require the reinitiation of these functions at any time during the life cycle of the project. This management process is continuous, adaptable to new stakeholder threats and promises, and changing strategies of existing stakeholders. Putting the notion of stakeholder management on a project life-cycle basis emphasizes the need to be aware of stakeholder influence at any time.

The management process for the stakeholders consists of the phases depicted in Figure 13-1 and discussed below.

## Identification of Stakeholders

The identification of stakeholders must go beyond the internal stakeholders. Internal stakeholders must of course be taken into account in the development of project strategies. Their influence is usually supportive of project strategies since they are an integral part of the project team. A prudent project manager would ensure that these internal stakeholders play an important and supportive role in the design and development of project strategies. Such a supportive role is usually forthcoming since the project manager has some degree of authority and influence over these individuals. External stakeholders may not be supportive.

External stakeholders are not usually subject to the legal authority of the project manager; consequently, such stakeholders provide a formidable challenge to manage. A generic set of external stakeholders would include the following:

- Prime contractor.
- Subcontractors.
- Competitors.
- Suppliers.
- Financial institutions.
- Governmental agencies, commissions; judicial, legislative and executive branches.
- The general public represented through consumer, environmental, social, political, and other "intervenor" groups.
- Affected local community.

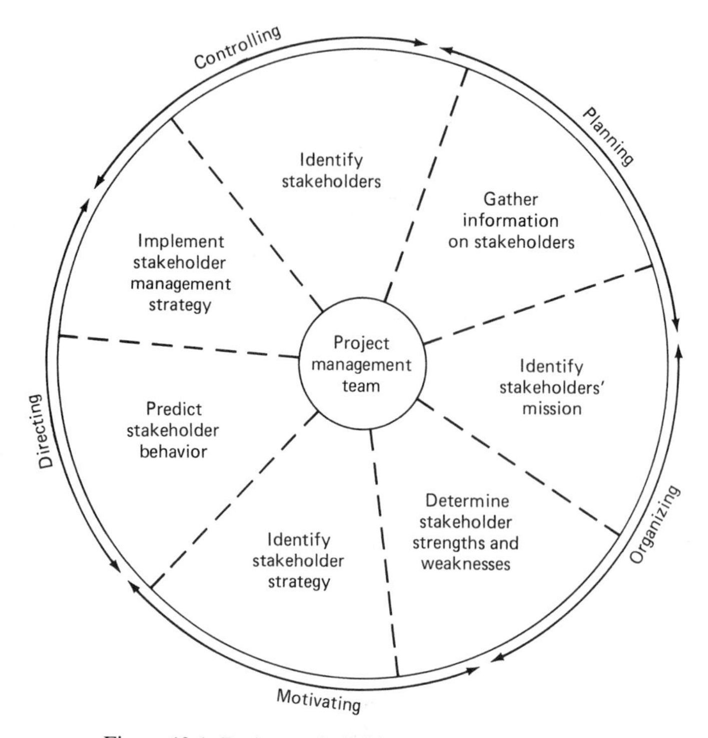


Figure 13-1. Project stakeholder management process.

Figure 13-2 depicts a typical network of project stakeholders, both internal and external.

Care must be taken to identify all of the potential stakeholders, even those whose stake may seem irrelevant at the time. Freeman points out that a historical analysis of an organization's interface with its environment is useful in identifying potential stakeholders (10). The development of a list of the "strategic issues" that currently face and have faced the parent organization and the industry over the past several years can be useful in identifying stakeholders who have been involved in these issues.

## PROJECT STRATEGIC ISSUES

A strategic issue is a condition or pressure, either internal or external, to a project that will have a significant effect on, for example, the financing, design, engineering, construction, licensing, and/or operation of a nuclear power generating plant (22). Strategic issue management is found in the

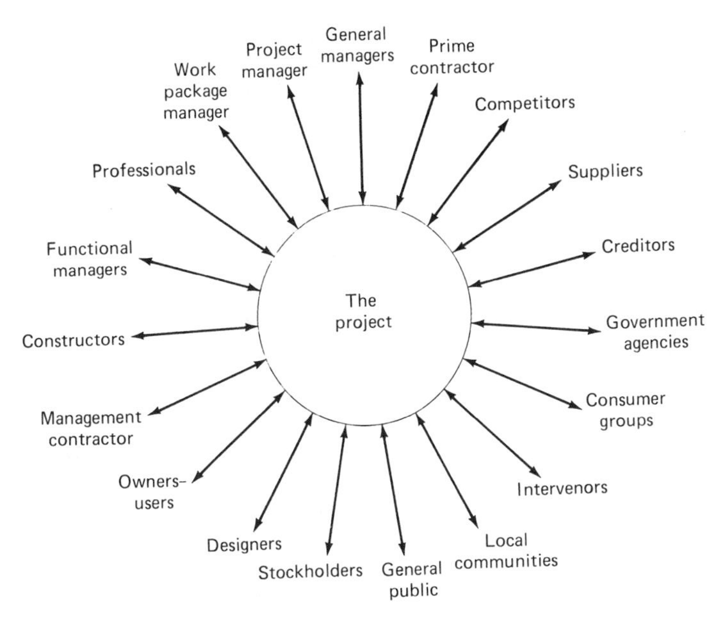


Figure 13-2. Project stakeholder network.

field of strategic planning and management (23). In project management the notion of strategic issue management is not as pervasive. Yet a project that has as long a life cycle as a nuclear power generating plant will probably be impacted during its life cycle by many issues that are truly strategic by nature. For example, both a project team and a project owner need to be aware of the following typical strategic issues that will have a significant effect on the outcome of a nuclear power generating plant project.*

- *Licensability*
  —Unless a plant can be licensed in the United States, it has little value to the utility industry since all of their plants must meet the federal codes and standards as well as the nuclear regulatory guides for the particular concept. Many of these codes, standards, and guides are not applicable to a concept if it has not been previously licensed. The first-of-a-kind of a particular concept through

---

* These strategic issues were developed during a research project by David I. Cleland and Dundar F. Kocaoglu: The Design of a Strategic Management System for Reactor Systems, Development and Technology, DOE, with the assistance of A. N. Tardiff and C. E. Klotz of the Argonne National Laboratory.

the process becomes precedent setting. As such, it will receive a commensurate amount of attention from the Nuclear Regulatory Commission (NRC) staff, so much so that joint groups will be set up with representation from the Department of Energy (DOE) and the NRC, and a bevy of consultant experts to answer the thousands of questions posed by the NRC staff and to draft appropriate revisions to the existing federal codes, regulations, and guides for future applicability to the new concept.

This strategic issue can take years to resolve when one includes the judicial and state and local hearing processes that a nuclear plant must face. The lack of firm and predictable policy emanating from the NRC at the present time adds to the risk and uncertainty involved in the management of this strategic issue. Such risks and uncertainties are reflected in the increased costs and schedules for the project. The NRC, which licenses the plant, and state and local governments who conduct hearings to ascertain the proper allocation of costs for the utility's rate base are key stakeholders.

- *Passive Safety*
  —Passive safety, as it relates to a nuclear power plant, refers to the plant's ability to take advantage of inherent, natural characteristics to move itself into a safe condition without the need to activate an automatic auxiliary safety system or to impose a set of predetermined operator procedures to do the same.

  For example, all of the commercial reactors being built and operated in the United States today require the activation, within a prescribed time period, of an auxiliary system, automatically or by an operator, to shut down.

If one allows the system to operate without adding reactivity (similar to adding coal to a fire) and assuming the cooling systems remain effective (the pumps operate, the valves open and close on cue, the heat exchangers transfer heat, etc.) the reactor will eventually shut itself down. The difficulty comes when the operating plant cannot remove reactivity from the reactor (like removing coal from the fire) and/or maintain the effectiveness of the cooling systems.

Passive safety is the dominant strategic issue facing the nuclear power generating industry today. The nuclear accidents at Three Mile Island and Chernobyl have intensified the search for a nuclear power plant that promises passive safety. Nuclear vendors and utility companies are key stakeholders keenly interested in passive safety. Indeed, all of us are stakeholders wanting economical and safe power generating capacity in the modern world.

- *Power Costs*
  —The components of the power costs are capital costs, operations and maintenance (O&M), and fuel costs. For a typical nuclear power plant, the capital cost component is four times the O&M cost, which is approximately equal to the fuel cost. Hence, it is evident that capital cost is the most significant component, and it is discussed in more detail below.
  —One of the significant factors leading to the current hiatus in orders for nuclear power plants is that these systems are extremely capital intensive and have relatively low fuel costs. Coal and oil fire plants are the opposite, that is, they have a relatively low capital cost component while their fuel costs are extremely high.
  —Construction times for many of the recent U.S. nuclear plants have exceeded ten years. The licensing and judicial processes in the United States have accounted for much of the delay, but other factors, such as utility management, have also taken their toll. Whatever the reasons, the delays have an extraordinary impact on the resultant capital investment in these plants even before they have produced one kilowatt-hour of electricity. It is not uncommon to experience the interest paid on the capital to build the plant to be greater than 50% of the capital investment in the plant. As a result, there has been an inordinate increase in the capital cost component such that nuclear power has lost its competitive power cost edge over its closest competitor, coal. Five approaches might be pursued to alleviate this problem:
    Reform the licensing process.
    Design the plants to be constructed quickly on-site.
    Build smaller plants.
    Supply turnkey plants with guarantees.
    Simplify the plants and reduce the amount of material used.
    —Utilities, nuclear reactor manufacturers, A&E firms, plant constructors, and state regulatory commissions are the principal stakeholders concerned about the power costs strategic issue.

- Reliability of Generating System
  —The reliability of a nuclear power plant must be extremely high, particularly as it relates to the safety systems and components. There are reliability differences from one concept to another, for example, one might have fewer moving parts, fewer systems, fewer components, fewer things to go wrong. Generally, a plant that has been designed, developed, tested, fabricated, constructed, operated, and maintained under a stringent quality assur-

ance (QA) umbrella will be more reliable than one that has not. Concepts that maximize factory vs. on-site fabricated and assembled systems tend to be more reliable since QA can be more easily applied at the factory. Gravity and natural circulation-dependent systems tend to be more reliable than forced circulation systems. The importance of these and more reliable approaches to a nuclear power plant cannot be overemphasized, particularly in view of Three Mile Island, Chernobyl, and public attitudes toward nuclear power. Utilities are the principal stakeholders here.

- *Nuclear Fuel Reprocessing*
  —Commercial nuclear fuel reprocessing in the United States is virtually nonexistent. Instead the U.S. government has agreed to accept for a price the spent fuel for U.S. reactors for long-term storage. Europe and Japan, however, have viable programs to recover for future use the nuclear fissionable fuel from spent fuel assemblies. Any concepts, such as breeder, which require reprocessing technology must carry the burden of developing this technology as well as the nuclear proliferation stigma attached to it. Thus any future nuclear plant ordered by authority in the U.S. may require the arrival of a liquid metal reactor technology which provides for the use of reprocessed fuel. The time frame for such fuel-reprocessing capability is circa 2040 by best current estimates. Utilities and reactor manufacturers are the principal stakeholders.
- *Waste Management*
  —Public reaction (e.g., to shipments of nuclear wastes) is becoming increasingly severe. Hence, minimum waste streams and movement of such wastes outside the plant boundaries is advisable. Poor management and cost overruns aside, one of the biggest issues for the nuclear power industry is the 1500 metric tons of lethal atomic waste that it produces each year. The waste disposal program conceived and managed by the U.S. government and the nuclear power industry to store radioactive fuel safely is troublesome (24). Utilities, states (where storage sites are located), and the general public are vested stakeholders in this strategic issue.
- *Capital Investment*
  —Closely akin to the strategic issue of power costs is the financial exposure and risks that investors of nuclear power plants have experienced over the last several years. For support of nuclear power to be resumed by the Wall Street financial houses, it is important that the current conditions change along the lines noted in the discussion of the power costs issue, that is,

Reform the licensing process.
Design the plants to be constructed quickly on-site.
Build smaller plants.
Supply turnkey plants with guarantees.
Simplify the plants and reduce the amount of material used.

—Investment agencies are the principal stakeholders along with the state public utility groups that must rule on the acceptability of a capital investment cost into the utility's rate base.

- *Public Perception*

—Table 13-1 summarizes this strategic issue quite clearly. The experts rank nuclear power twentieth in the list of high-risk items, whereas the others rank it first or close to first. Note that X-rays and nonnuclear electric power fall into the same dilemma. When the United States converted from DC to AC in the early 1920s, the same problem arose. Some extensive, innovative technical and management approaches must be successfully implemented to turn the rankings around.

—The public perception of nuclear power and its associated risks aggravated by the nuclear accidents at Three Mile Island and Chernobyl has made this strategic issue more acute.

- *Advocacy*

—Not many government interest research programs can proceed through the government bureaucracy without a strong advocate with the ability to establish and maintain a substantial support base for the program. The support must be broad and include, as is the case with the research in an advanced reactor development program, key individuals within the Department of Energy, White House, Office of Management and Budget, Congress and staff offices, the nuclear community (the stakeholders), the science community (NSF, NAS, certain universities), Wall Street, and others. With such backing the "public" generally supports the program by definition. An effective advocate(s) is an essential ingredient. Military aircraft and the aircraft carrier had Billy Mitchell; the nuclear submarine fleet had Hyman Rickover; the space program had Werner von Braun—the list of successful efforts led by able champions is long. Thus a reactor manufacturer who contemplates obtaining government funds to be added to corporate monies for the development of research in advanced nuclear reactors would be a vitally interested stakeholder to determine what advocacy existed for such research both in the government and in the corporation itself.

## Table 13-1. Risk: A Matter of Perception. Four Groups Rank What's Dangerous and What's Not[a,b]

| | EXPERTS | LEAGUE OF WOMEN VOTERS | COLLEGE STUDENTS | CIVIC CLUB MEMBERS |
|---|---|---|---|---|
| Motor vehicles | 1 | 2 | 5 | 3 |
| Smoking | 2 | 4 | 3 | 4 |
| Alcoholic beverages | 3 | 6 | 7 | 5 |
| Handguns | 4 | 3 | 2 | 1 |
| Surgery | 5 | 10 | 11 | 9 |
| Motorcycles | 6 | 5 | 6 | 2 |
| X-rays | 7 | 22 | 17 | 24 |
| Pesticides | 8 | 9 | 4 | 15 |
| Electric power (nonnuclear) | 9 | 18 | 19 | 19 |
| Swimming | 10 | 19 | 30 | 17 |
| Contraceptives | 11 | 20 | 9 | 22 |
| General (private) aviation | 12 | 7 | 15 | 11 |
| Large construction | 13 | 12 | 14 | 13 |
| Food preservatives | 14 | 25 | 12 | 28 |
| Bicycles | 15 | 16 | 24 | 14 |
| Commercial aviation | 16 | 17 | 16 | 18 |
| Police work | 17 | 8 | 8 | 7 |
| Fire fighting | 18 | 11 | 10 | 6 |
| Railroads | 19 | 24 | 23 | 20 |
| Nuclear power | 20 | 1 | 1 | 8 |
| Food coloring | 21 | 26 | 20 | 30 |
| Home appliances | 22 | 29 | 27 | 27 |
| Hunting | 23 | 13 | 18 | 10 |
| Prescription antibiotics | 24 | 28 | 21 | 26 |
| Vaccinations | 25 | 30 | 29 | 29 |
| Spray cans | 26 | 14 | 13 | 23 |
| High school and college football | 27 | 23 | 26 | 21 |
| Power mowers | 28 | 27 | 28 | 25 |
| Mountain climbing | 29 | 15 | 22 | 12 |
| Skiing | 30 | 21 | 25 | 16 |

[a] People were asked to "consider the risk of dying as a consequence of this activity or technology."

[b] *Source:* Decision Research; Eugene, Oregon (*Washington Post*, May 21, 1986).

- *Environment*
  —From an environmental viewpoint, the nuclear advocates had essentially convinced the general public that nuclear power plants were essentially environmentally benign—until the press convinced the public otherwise after the Three Mile Island incident. The Chernobyl incident reinforced this conviction. Certainly the environmental impact of the Chernobyl accident on its surrounding area appears to be serious.
  —Recovering from the image of Chernobyl will be no easy task. Much work must be done to assure that such an incident cannot occur in the United States, and this fact must be convincingly transmitted to the potential owners of nuclear power plants, the administration, the Congress, and above all the general public itself. The most environmentally benign and inherently safe nuclear plant should go a long way in settling this issue. Unfortunately, such a plant may be decades away. Environmental groups such as the Sierra Club see themselves as key stakeholders concerned about this strategic issue.
- *Safeguards*
  —One must keep fissionable material out of unauthorized hands—the objective of the nuclear safeguards activity. A nuclear plant security system that does this better than another should have an edge. For example, if throughout the fuel cycle of a particular plant the fissionable fuel avoids a plant configuration that can be used as source material for a weapon, then one could say that it is "nonproliferation proof."
  —Then the next consideration is, "Which plant configuration minimizes the exposure of the weapons-grade nuclear material during its fuel cycle operations?" A fuel cycle concept which keeps the fuel in the reactor vessel or at least on site-during its lifetime may have some significant safeguard advantages.
  —Nuclear power is an emotional issue aggravated by the recent accidents at Three Mile Island and Cernobyl. Before these accidents the public had a false sense of security; a continuing lack of education about radiation produced a fear of the unknown. Many stakeholders exist whose mission in whole or in part is directed to reshape public and legislative opinion about nuclear power. A nuclear power plant project team needs to be aware of all these potential stakeholders; there are many. For example, a partial list would include

    Advanced Reactor Development Subpanel of the Energy Research Advisory Board's Civilian Nuclear Power Panel

U.S. Committee for Energy Awareness
Institute of Nuclear Power Operations
Edison Electric Institute
American Public Power Association
Electric utilities
State public utility commissions
National Rural Electric Cooperative Association
Nuclear Regulatory Commission
American Nuclear Insurers
Mutual Atomic Energy Liability Underwriters
Oak Ridge Associated Universities
American Nuclear Energy Council
Atomic Industrial Forum (Nuclear)
Committee on Radioactive Waste Management (Nuclear)
Educational Foundation for Nuclear Science
Fusion Energy Foundation
Institute of Nuclear Materials Management
International Atomic Energy Agency
Nuclear Energy Women
Nuclear Records Management Association
Universities Research Association
Americans for Nuclear Energy
Citizens Energy Council
Clamshell Alliance
Coalition for Non-Nuclear World
Committee for Nuclear Responsibility
Concerned Citizens for the Nuclear Breeder
Environmental Coalition on Nuclear Power
League Against Nuclear Dangers
Mobilization for Survival
Musicians United for Safe Energy
National Campaign for Radioactive Waste Safety
Natural Guard Fund
Nuclear Information and Resource Service
Safe Energy Communications Council (Nuclear)
Sierra Club Radioactive Waste
Supporters of Silkwood (Nuclear)
Task Force Against Nuclear Power
Union of Concerned Scientists
Western Interstate Energy Board
World Information Service on Energy

Committee to Bridge the Gap
Constructors/A&E firms
Project management contractors
Reactor vendors

Although a historical perspective can give insight into a project's probable stakeholders, the project team should be alerted to strategic issues in the competitive and environmental systems that can change the project's future.

For example, one key strategic issue facing the U.S. nuclear power industry is to foster public acceptance of nuclear power. Recognizing this, the industry has launched a major public relations campaign to improve its image. The U.S. Committee for Energy Awareness (CEA) launched a $20 million media campaign in July 1983 to facilitate public acceptance as well as understanding that leads to agreement and support (25).

Identification of the relevant external stakeholders can be accomplished through the interaction of the project team. Through discussion and compilation of a list of some of the strategic issues facing the project, the less obvious stakeholders can be discovered. Once a list of the stakeholders has been developed, that list should become an integral part of the project plan and be reviewed along with other elements of the plan during the project's life cycle to determine if the stakeholders' perceptions or views of the project have changed. To do so will require information on the stakeholders.

## GATHERING INFORMATION

To systematize the development of the stakeholder information means that questions such as the following need to be considered:

- What needs to be known about the stakeholder?
- Where and how can the information be obtained?
- Who will have responsibility for the gathering, analysis, and interpretation of the information?
- How and to whom will the information be distributed?
- Who has responsibility for the use of the information in the decision context of the project?
- How can the information be protected from "leakage" or misuse?

Some of the information collected on the project's external stakeholders may include sensitive material. One cannot conclude that all such stakeholders will operate in an ethical fashion. Consequently, all information collected should be assumed to be sensitive until proven otherwise and protected accordingly. This suggests the need for an associated security system patterned after a company's business intelligence system. Such a system would include a classification system for some information on a "need-to-know" basis while some would be available to all interested parties.

The following precautions should be considered in planning for a PSM information system:

- One individual responsible for security.
- Internal checks and balances.
- Document classification and control such as periodic inventory, constant record of whereabouts, and prompt return.
- Locked files and desks.
- Supervised shredding or burning of documents no longer useful.
- Confidential envelopes for internal transmission of confidential documents.
- Strict security of offices containing sensitive information (26).

Information on the stakeholders is available from a wide variety of sources. When such information is obtained, the highest standards of ethical conduct should be followed. The potential sources of stakeholder information and the uses to which such information can be put are so numerous that it would not be practical to list all sources and uses here. The following sources are representative and can be augmented according to a particular project's needs:

- Project team members.
- Key managers.
- Business periodicals such as the *Wall Street Journal, Fortune, Business Week, Forbes,* and others.
- Business reference services—*Barrons, Moody's Industrial Manual,* the *Value Line Investment Survey,* etc.
- Professional associations.
- Customers/users.
- Suppliers.
- Trade associations.
- Local press.
- Trade press.
- Annual corporate reports.

- Articles, papers presented at professional meetings.
- Public meetings.
- Government sources (27).

Once the information has been collected, it must be analyzed and interpreted by the substantive experts. The project manager should draw on the company's professional personnel for help in doing this analysis. Once the analysis has been completed, the specific target of the stakeholders' mission can be determined.

### Identification of Mission

Once the stakeholders have been identified and information gathered about them, analysis is carried out to determine the nature of their mission or stake. This stake may be a key building block in the stakeholder's strategy. For example, the Nuclear Regulatory Commission manages the licensing of nuclear power plants to promote the safe and peaceful commercial use of the atom. A useful technique to better understand the nature of the *external* stakeholder's claim in the project is to categorize the stake as *supportive* or *adverse* to the project. It is in the best interest of the project manager to keep the supportive stakeholders well informed of the project's status. Care has to be taken in dealing with the potentially adversary stakeholders. Information for these stakeholders should be handled on a "need-to-know" basis because such information can be used against the project. This information should be treated as confidential. However, communication channels with these stakeholders should be kept open, for this is critical to getting the project point of view across. Adversary stakeholders will find ways to get information on the project from other sources which can be erroneous or incomplete, giving the opportunity for misunderstanding and further adversary behavior.

Once the stakeholders' mission is understood, then their strengths and weaknesses should be evaluated.

### Stakeholders' Strengths and Weaknesses

An assessment of stakeholders' strengths and weaknesses is a prerequisite to understanding their strategies. Such analysis is found in nearly all prescriptions for a strategic planning process (28). This process consists of the development of a summary of the most important strengths on which the stakeholders base their strategy and the most significant weaknesses they will probably avoid in pursuing their interests on the project.

Identifiying five or six strengths and weaknesses of a stakeholder should provide a sufficient data base on which to reach a judgment about the efficacy of a stakeholder's strategy.

An adversary stakeholder's strength may be based on such factors as

- The availability and effective use of resources.
- Political alliances.
- Public support.
- Quality of strategies.
- Dedication of members.

Accordingly, an adversary's weaknesses may emanate from

- Lack of political support.
- Disorganization.
- Lack of coherent strategy.
- Uncommitted, scattered membership.
- Unproductive use of resources.

Once these factors have been developed, they can be tested by answering questions for each proposed project strategy for coping with the stakeholders:

- Does this strategy adequately cope with a strength of the stakeholder?
- Does this strategy take advantage of an adversary stakeholder's weakness?
- What is the relative contribution of a particular stakeholder's strength in countering the project strategy?
- Does the adversary stakeholder's weakness detract from the successful implementation of the stakeholder's strategy? If so, can the project manager develop a counter strategy that will benefit the project?

For a proposed strategy to be successful, it should be built on a philosophy which recognizes the value of going through a specific strengths-weaknesses analysis and developing the project strategy to facilitate the project's success. This can be done, however, only if there is a full understanding of the stakeholder's strategy.

### Identification of Stakeholder Strategy

A stakeholder strategy is a series of prescriptions that provide the means and set the general direction for accomplishing stakeholder goals, objectives, and mission. These prescriptions stipulate

- What resource allocations are required.
- Why they are required.
- When they are required.
- Where they will be required.
- How they will be used

These resource allocations include plans for using resources, policies, and procedures to be employed, and tactics used to accomplish the stakeholder's end purposes. Once the stakeholder's strategy is understood, then the stakeholder's probable behavior can be predicted.

### Prediction of Expected Stakeholder Behavior

Based on an understanding of external stakeholder strategy, the project team can then proceed to predict stakeholder behavior in implementing strategy. How will the stakeholder use his resources to affect the project? Will an intervenor stakeholder picket the construction site or attempt to use the courts to delay or stop the project? Will a petition be circulated to stop further construction? Will an attempt be made to influence future legislation? These are the kinds of questions that, when properly asked and answered, provide a basis for the project team to develop specific countervailing strategies to deal with adversary stakeholder influence. In some cases a stakeholder will provide help to another stakeholder. For example, a group of dedicated nuclear advocates formed an industry association to assure the nuclear operating safety that the Nuclear Regulatory Commission was not able to provide. This association, the Institute of Nuclear Power Operations (INPO), is dedicated to improving the safety of nuclear plants. INPO has over 400 employees, an operating budget of $400 million, and sufficient clout to bring its 55 utility members into line. INPO sets safety standards and goals, evaluates plant safety, and provides troubleshooting assistance to its sponsors.

Currently INPO is overseeing the training of plant operators and supervisors. In its role as a stakeholder of nuclear power, INPO works closely with the Nuclear Regulatory Commission. If INPO finds areas for improvement in a utility's operation, it is the utility that alerts the Nuclear Regulatory Commission (29).

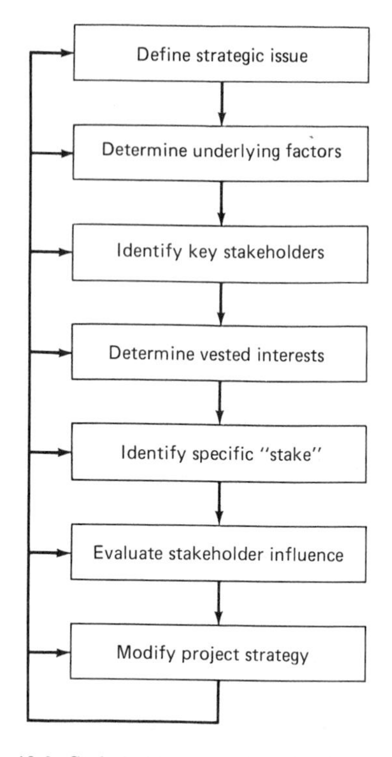


Figure 13-3. Stakeholder impact evaluation process.

The prediction of stakeholder behavior can be facilitated by the project team taking the lead in analyzing the probable impact of the stakeholder on a project. A step-by-step approach for analyzing such impact on a project would consist of the following, depicted in Figure 13-3 and described below (30):

- Identify and define each potential strategic issue in sufficient detail to ascertain its relevance for the project.
- Determine the several key factors which underlie each issue and the forces that have caused that issue to emerge. These forces can usually be categorized into *political, social, economic, technological, competitive, or legal forces*.
- Then identify the key stakeholders that have, or might feel that they have, a vested interest in the project. Remember that one strategic issue may have several different stakeholders who share a vested interest in that issue. Stakeholders usually perceive a vested interest in a strategic issue because of the following:

—*Mission Relevancy*. The issue is directly related to the mission of the group. For example, members of the Sierra Club see the potential adverse effect of a nuclear power plant project on the environment which club members vow to project.
—*Economic Interest*. The stakeholders have an economic interest in the strategic issue. A crafts union would be vitally interested in the wage rates paid at a project construction site.
—*Legal Right*. A stakeholder has a legal right in the issue such as is the case of the Nuclear Regulatory Commission's involvement in the licensing process for the operation of a nuclear generating plant.
—*Political Support*. Stakeholders see the issue as one in which they feel the need to maintain a political constituency. A state legislator would be concerned about the transportation of nuclear wastes from a power plant to a repository site within the state, or the transportation of wastes across the state.
—*Health and Safety*. The issue is related to the personal health and safety of the group. Project construction site workers are vitally interested (or should be) in the working conditions at the site.
—*Life Style*. The issue is related to the life style or values enjoyed by a group. Sportsman groups are interested in the potential pollution of industrial waste in the waterways.
—*Opportunism*. Opportunists see the issue as one around which they can rally, such as a protest meeting at a nuclear power plan construction plant.
—*Competitive Survival*. The issue is linked to the reason for existence of a group of stakeholders. For example, members of the investment community see clearly the financial risks of nuclear plant construction today, considering the uncertainty in the licensing of a nuclear power plant.

Once the stakeholders have been identified, clarify the specific stake held by each stakeholder, then reach a judgment on how much potential influence the stakeholder has on the project and its outcome. Table 13-2 can be used to summarize such interests. Development of this table should be done by the project team, who are in the best position to identify the probable impact of a stakeholder's vested interest. By perusing such a table a general manager can get a summary picture of which stakeholders should be "managed" by the project team. Stakeholders whose interest scores high on the table should be carefully studied and their strategies and actions tracked to see what potential effect such actions will have on the project's outcome. Once such potential effect is

Table 13-2. Summary of Stakeholders' Interest

| STAKEHOLDER INTEREST | STAKEHOLDERS | | | | | | | | | | |
|---|---|---|---|---|---|---|---|---|---|---|---|
| | 1 | 2 | 3 | 4 | 5 | 6 | 7 | 8 | 9 | 10 | 11 |
| Mission relevance | | | | | | | | | | | |
| Economic interest | | | | | | | | | | | |
| Legal right | | | | | | | | | | | |
| Political support | | | | | | | | | | | |
| Health and safety | | | | | | | | | | | |
| Life style | | | | | | | | | | | |
| Opportunistic competitive survival | | | | | | | | | | | |

*Vested Interest:*
High - H
Low - L
Medium - M

determined, then the project strategy should be adjusted through resource reallocation, replanning, or reprogramming to accommodate or counter the stakeholder's actions through a stakeholder management strategy.

**Project Audit**

An independent audit of the project conducted on a periodic basis will also help the project tream to get the informed and intelligent answers they need on strategic issues and stakeholder interests. Both internal and external audits performed by third parties to analyze the project's strengths, weaknesses, problems, and opportunities can shed light on how well the stakeholders are being managed. There is a symbiotic relationship between the project and its stakeholders. The project cannot exist without its stakeholders; conversely, the stakeholders rely to some extent on the project for their existence.

**Implementation of Stakeholder Management Strategy**

The final step in managing either the supportive or adverse project stakeholders is to develop implementation strategies for dealing with them. An organizational policy which stipulates that stakeholders will be actively managed is an important first step of such implementation strategies. Once this important step has been taken, then additional policies, action plans, procedures, and the suitable allocation of supporting resources can be established to make stakeholder management an ongoing activity. Once implementation strategies are operational, then the project team has to take a proactive posture in doing the following:

- Insure that the key managers and professionals fully appreciate the potential impact that both supportive and adverse stakeholders can have on the project outcome.
- Manage the project review meetings so that stakeholder assessment is an integral part of determining the project status.
- Maintain contact with key external stakeholders to improve chances of determining stakeholder perception of the project, and their probable strategies.
- Insure an explicit evaluation of probable stakeholder response to major project decisions.
- Provide an ongoing, up-to-date status report on stakeholder status to key managers and professionals for use in developing and implementing project strategy.
- Provide a suitable security system to protect sensitive project information that might be used by adverse stakeholders to the detriment of the project.

## SUMMARY

The specification of a project stakeholder management process helps to assure the timely and credible information about the capabilities and options open to each stakeholder. Once these options have been identified, the project team is in a position to predict stakeholder behavior and the effect such behavior might have on the project's outcome. Then the project team can develop its own strategies to best "manage" the stakeholders.

Attitudes play an important role in the management of a project. A positive attitude which accepts that proactive management of stakeholders can reduce the chances of the project team being surprised and unprepared for adverse stakeholder action will be a meaningful contribution to project success. The alternative is for the stakeholder to "manage" the project with the risk of an outcome detrimental to the project's best interests.

An example of project stakeholders and the "stake" such organizations have was drawn from the nuclear power plant industry, an industry that has had managerial problems and challenges on a monumental scale. In addition, an approach was suggested for analyzing the impact of stakeholders on a project.

## REFERENCES


1. Aguilar, F. J. *Scanning the Business Environment* (Macmillan. New York, 1967).
2. Dill, W. R. "Environment as an Influence on Managerial Autonomy." *Administrative Science Quarterly* (March, 1958), pp. 409–443.
3. Mintzberg, H. *The Structure of Organizations* (Prentice-Hall. Englewood Cliffs, N.J., 1979).
4. Weiner, E. and Brown, A. "Stakeholder Analysis for Effective Issues Management" *Planning Review* (May, 1986), pp. 27–31.
5. Radosevich, R. and Taylor, C. *Management of the Project Environment.* U.S. Department of Agriculture.
6. Burnett, N. R. and Youker, R. EDI Training Materials, copyright (copyright) 1980 by the International Bank for Reconstruction and Development.
7. Cleland, David I. and King, William R. *Systems Analysis and Project Management,* 3rd Ed. (McGraw-Hill. New York, 1983).
8. Rothschild, W. E. *Putting It All Together: A Guide to Strategic Thinking.* (AMACOM. 1976).
9. King, William R. and Cleland, David I. *Strategic Planning and Policy* (Van Nostrand Reinhold. New York, 1978).
10. Freeman, R. E. *Strategic Management—A Stakeholder Approach* (Pitman. Boston, 1984).
11. Mendelow, Aubrey. "Stakeholder Analysis for Strategic Planning and Implementation," in *Strategic Planning and Management Handbook* (Van Nostrand Reinhold. New York, 1986).
12. *Case 27563, Long Island Lighting Company—Shoreham Prudence Investigation,* State of New York Public Service Commission, Recommended Decision by Administrative Law Judges Wm. C. Levy and Thomas R. Matias, March 13, 1985, p. 57.
13. Horwitch, Mel. "The Convergence Factor for Successful Large-Scale Programs: The American Synfuels Experience as a Case in Point," in *Matrix Management Systems Handbook,* ed. Cleland, David I. (Van Nostrand Reinhold. New York, 1984).
14. Reported in *The Phoenix Gazette,* June 27, 1984, p. PV-12.
15. Glasgall, William. "The Utilities' Pleas Are Falling on Deaf Ears." *Business Week* (June 17, 1985), p. 113.
16. *Cases No. ER-85-160 & ED 85-17,* State of Missouri Public Service Commission, Jefferson City, March 29, 1985.
17. See Behr, Peter G. "James Bay Design and Construction Management." *ASCE Engineering Issues, Journal of Professional Activities* (April, 1978).
18. Ginsburg, E., Kuhn, J. W. and Schnee, J. *Economic Impact of Large Public Programs: The Nash Experience* (Salt Lake City. Olympus Publishing Company, 1976).
19. See Albright, Donald C. "What to Do Before the Atom Comes to Town." *Public Relations Journal* (July, 1965), pp. 16–20.
20. NUREG 1055, *Improving Quality and the Assurance of Quality in the Design and Construction of Nuclear Power Plants, A Report to Congress,* Division of Quality Assurance, Safeguards, and Inspection Programs, Office of Inspection and Enforcement, U.S. Regulatory Commission, Washington, D.C. 20555, May, 1984.

21. See Reference 20 above.
22. Definition paraphrased from Brown, J. K. "This Business of Issues: Coping with the Company's Environments." *The Conference Board Report,* No. 758, 1979.
23. King, William D. "Environmental Analysis and Forecasting: The Importance of Strategic Issues." *Journal of Business Strategy* (Winter, 1981), p. 74.
24. See Janet Novack. "Billion Dollar Shaft." *Forbes* (August 25, 1986), pp. 113–116.
25. See Reference 14 above.
26. Paraphrased from King, W. R. and Cleland, D. I. *Strategic Planning and Policy* (Van Nostrand Reinhold. New York, 1978).
27. Gathering stakeholder information is similar to gathering information on competitors. For a detailed discussion of how this can be done, see Chapter 11 in King, W. R. and Cleland, D. I. *Strategic Planning and Policy* (Van Nostrand Reinhold. New York, 1978), pp. 246–270.
28. See Reference 8 above.
29. For more on the role of INPO, see Cook, James. "INPO's Race Against Time," *Forbes* (February 24, 1986), pp. 54–55.
30. Paraphrased from Weiner, Edith and Brown, Arnold. "Stakeholder Analysis for Effective Issues Management." *Planning Review* (May, 1986), pp. 27–31.

# 14. Developing and Using the Work Breakdown Structure

Garry D. Lavold*

## INTRODUCTION

During the last two decades the emergence of projects with diverse ownership, long time spans, integral government involvement, and requirements for large quantities of diverse resources has put new strains on project management capabilities and project communication systems. The widespread usage of the personal computer (PC) has added another dimension for the requirement of an effective means of system integration utilizing a common communication language. This chapter proposes that a properly designed and implemented work breakdown structure (WBS), with associated coding structure and dictionary, forms an effective basis for project control systems, policies, and procedures for all projects. The WBS helps in organizing and planning all phases of a project.

Project management requires effective, precise information throughout all phases of the project and between all personnel involved with the project. A well-designed WBS provides the basis for the design of these project control information systems (either PC or mainframe based). The common WBS with PCs at different locations allows for easy data coordination. The definition of the WBS as supplied by the Department of Energy in its Performance Measurement Systems guidelines is as follows:

> Work Breakdown Structure. A product-oriented family tree division of hardware, software, services, and other work tasks which organizes,

* Mr. Garry Lavold holds a Bachelor of Sciences in Chemical Engineering and a Master of Business Administration from the University of Alberta. As an engineer with Gulf Oil from 1969 to 1974 he was involved in all phases of construction and start-up of an 80,000-BPD grass roots refinery. From 1974 Mr. Lavold has worked with NOVA, an Alberta Corporation, including two years as manager of project control for the prebuild of the Alaska Highway Gas Pipeline Project. At the present he is President of Novacorp Pressure Transport with responsibility for all company operations.

defines, and graphically displays the product to be produced, as well as the work to be accomplished to achieve the specified product.*

The "product to be produced" is the completion of a project within a specified time frame and budget while conforming to constraints of public interest groups and governments. Having the WBS as a discipline applied to the project ensures that all participants, both owners and contractors, are fully aware of the work required to complete the project. This utilization of the WBS as the foundation on which all estimates, schedules, and project outlines are developed ensures that the WBS will become the central medium through which all groups communicate information with one another.

Essential to the management of the project is the establishment of the WBS early in the project life. This will enable all participants to implement effective information channels at the beginning of the project life cycle. Utilizing the WBS as an information basis with outside groups, such as governmental agencies, will simplify the regulatory process in that all communication regarding the project will be via a common basis, thereby enabling both industry and government personnel to communicate on a common basis of understanding.

This chapter presents an overview of the environment within which the WBS should operate, the essential elements and concepts to be included during the design of the WBS, and one example of a WBS currently being used on a pipeline project.

## COMMUNICATION—USERS OF THE WBS

In the environment of large projects consisting of large cash expenditures, multiple owners, many contractors, and in most cases, government involvement with complex technological facilities, the requirement for information integration and communication is an order of magnitude greater than encountered in the past. These large projects deal with millions of dollars over a multiyear span; this means that as the projects proceed, the environment within which the project was conceived is quite often very different from the environment in which the project is completed. The requirement of government involvement, regulations, and monitoring necessitates that all groups have a common information basis despite changing environments. The WBS coding philosophy and methodology can also

* "mini-PMS Guide," Performance Measurement System Guidelines, Attachment 1 (Department of Energy. Washington, D.C. 1977), pp. A1–4.

be applied to small projects with each project using a common WBS on a PC-based system.

This chapter proposes that for all projects, the WBS should become the common information basis, the common language, and the device whereby diverse users can communicate back and forth from the very inception of the project to its final completion. These users include owners, project management personnel, contractors, designers, and government agencies. The integration of the users and their information is illustrated in Figure 14-1.

The left side of Figure 14-1 illustrates the functional groups (prime users), either in the owner's organization, contractor's organization, or a mixture thereof, that must perform the work required to design and construct the project. These users are responsible for the work and exchange information with each other, using the WBS as a common basis of understanding. A properly designed WBS will enable these functional groups to have a precise communication linkage by which all the data illustrated as "input data" can be gathered and distributed. Normally required key input data also are illustrated as follows:

- Budgets—which represent the expected yearly cash flows.
- Estimates—which provide the cost of the project on a facility basis.
- Productivities—the expected production rates to be achieved by design groups, drafting groups, and the construction crews.

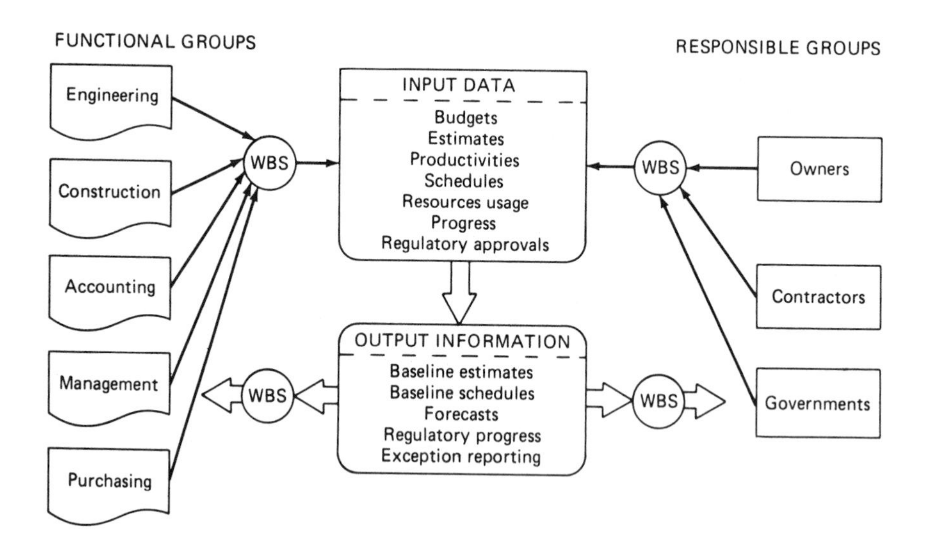


Figure 14-1. Work breakdown structure information integration.

- Schedules—the expected timing and sequence of all the activities necessary to complete the project.
- Resource usage—the expected quantities of manpower, equipment, and consumables required over the life of the project.

All the preceding information is prepared, using the WBS to define all the required elements. This use of the WBS enables all of the elements to be correlated on a common basis. The interrelating of cost, schedule, and productivity on a consistent basis is essential for accurate progress measurement and project control. Having all information collected on a common basis ensures that all work to be done is comparable to a well-defined baseline.

All input data would be collected from the functional groups, using the WBS to define all the elements. Shown in the middle section of Figure 14-1 is the "output information." The data from all the groups are consolidated to provide overall project budgets, schedules, and estimates. This provides a baseline estimate and schedule for the construction of the project, using the WBS; and as the project proceeds, forecasts against these baselines are made utilizing the WBS. The regulatory progress and exception reporting are also done, utilizing WBS as the common basis and the device by which all progress reporting and forecasting are done. Thus, the WBS is an integral part of all project reporting and project planning.

The original groups who prepare the outline and the concept of the project define it to a stage from which the WBS can be prepared. Thus, the first phase is the complete definition of a project and its associated WBS. The WBS works as an effective tool in organizing the work into logical groupings.

The next phase is to report against the baseline, and finally, to prepare a reconciliation against the estimate to measure overall performance on the project. The use of a common coding, a common structure, and a common language, from the start of the project life to the finish, enables problems and their solutions to be readily definable by the common WBS.

The right side of Figure 14-1 illustrates the responsible groups—the people who require the project, who are involved in building the project, or who are involved in approving the project. The owners, the contractors, and, on the larger projects, the government agency must be communicated with. This communication or information flow will use the WBS at a summary level, whereas the functional groups on the left side communicate normally at a detailed level of the WBS. The output information from the consolidated baseline format at a summary level of WBS will be utilized by the owners and the contractors for preparing the proposals to shareholders and/or for submissions to the government. Later, this WBS breakdown will be used in preparing the original bid documents and, as the progress proceeds, in preparing the progress to date and forecast to completion reports.

Thus, as can be seen, the WBS should be used from the start to the finish of the project for planning, tracking, and reconciliation. It is the device by which the users such as owners, contractors, and the government can organize information among themselves and with the people who are performing the work required to complete the project successfully.

## SYSTEM INTEGRATION—USING THE WBS

The user community, as described in the preceding section, communicates with each other using a common language defined by the WBS. To give the users the information they require involves support from project control systems and accounting systems. Typical project control systems include scheduling, progress and performance measurement, manpower, equipment, material tracking, cost monitoring, and forecasting systems. Project accounting systems usually include ledgers such as accounts receivable, accounts payable, capital assets, and a project cost ledger. Each one of these systems in both accounting and project control may be independent, automated, or manual, although for most large projects, these systems would be automated. Or, for smaller projects on different sites, the common WBS would be used to report to a central control center. When all the systems above are grouped as project control and/or accounting, systems information transfer between the respective systems can be analyzed.

As the explanation proceeds describing the project control and project accounting information transfer, it must be realized that this transfer would be applied to all other subsystems as well. By highlighting the two overall systems, the principles, ideas and concepts will be explained on an overall basis.

Figure 14-2 illustrates the project accounting and project control systems which both collect data using the WBS. These systems receive data on the common basis of the dictionary and code structure directly associated with the WBS. The project control systems, whether manual or automated, always serve the purpose of collecting, as precisely as possible, timely information which is current and which can be used for management reporting and forecasting. The project control system's key function is to warn management early of any impending problems which, with management's decisions, can be solved or at least have their impact re-

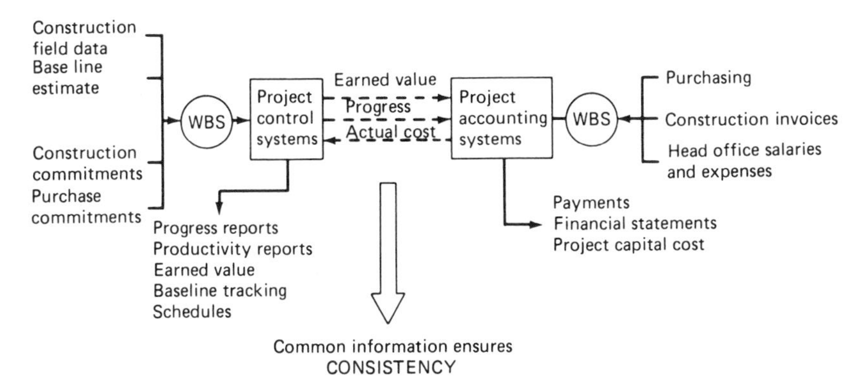


Figure 14-2. Work breakdown structure system integration.

duced. These systems do not have precise cost information but supply vital current key information to manage the project. All information is collected, sorted, and reported via the WBS code structure. The project control systems, either central or PC-based, use this code for all aspects of monitoring cost, schedule, and productivity, and future planning.

The use of the WBS code for entering all information into the project control systems ensures that all progress data collected are comparable to a baseline. Use of actual progress and resource data compared to the baseline allows forecasts of anticipated problems to be analyzed. These results from the systems can be used for control in a meaningful manner through the use of WBS, since all references to the information are made via a uniform and consistent referencing method. Thus the WBS forms an integral part of the control process.

The project accounting systems, which by design are precise but not normally as timely as the project control systems, collect the official or auditable information for cost and resource usage on the project. The resource usage or cost is collected via invoices from contractors, time sheets, personnel working on the project, and expenses of the personnel doing the work. This information is collected by the WBS, with payments being recorded for the WBS elements. Having the project accounting system able to verify the actual costs that were spent, against the estimate and against the budget, by common means allows the actual cost of the progress to date to be tracked precisely against the estimate of cost as well as against the forecasted cost to complete. The more this common tie (i.e., WBS), between the accounting and the project control systems is used, the better it is for the management of the project since it is possible to analyze results, not system discrepancies. With all information collected on a common basis, the WBS ensures that the engineer, the accountant, and management are all referring to the same information with the same meaning.

One of the problems typically present in many projects is that the accounting and project control systems are not using identical coding. Utilizing the WBS in conjunction with the existing accounting coding, or using it uniquely for the project without having to adapt to the existing accounting code, can ensure that as the project proceeds, estimates can be verified and sound projections can be made. This is essential to proper management of the project because it ensures that explanations of cost or schedule problems will be made on a organized basis. The key to providing proper explanations is the disciplined usage, by every person involved in both systems, of the WBS, which is a dictionary of definitions and a fixed coding structure that is unique for all project activities. The precision of this type of reporting ensures that, as the project proceeds, every individual involved is well aware of the project problems, their proposed solutions, the cost estimate, and the actual cost of these problems or their solutions.

The method of integration using the WBS as described above, between the project accounting and the project control systems, is also applicable to the purchasing system, which may feed either project control or project accounting, and any other systems that are utilized within the project. Each system, whether manual or automated, should include the WBS as part of its system definition. It is a discipline and an information organizer be present which should in all applicable systems to ensure that the required common linkages are available.

## WBS—RESPONSIBILITY RELATIONSHIP

The WBS provides an information organizer between both the users and the systems. This was previously illustrated in Figures 14-1 and 14-2. Communication utilizing the WBS is on a facility basis or a contract basis. The WBS defines the project in a structured format via the facilities and the items required to build the facilities, or the contracts required to complete construction of the facilities. The WBS structure should reflect as accurately as possible on paper the physical project to be completed. Utilizing the WBS in this format requires that the management structure or organization responsibility centers (which are responsible for the various components of the project) be defined separately. The relationship between the management structure and the WBS is illustrated in Figure 14-3.

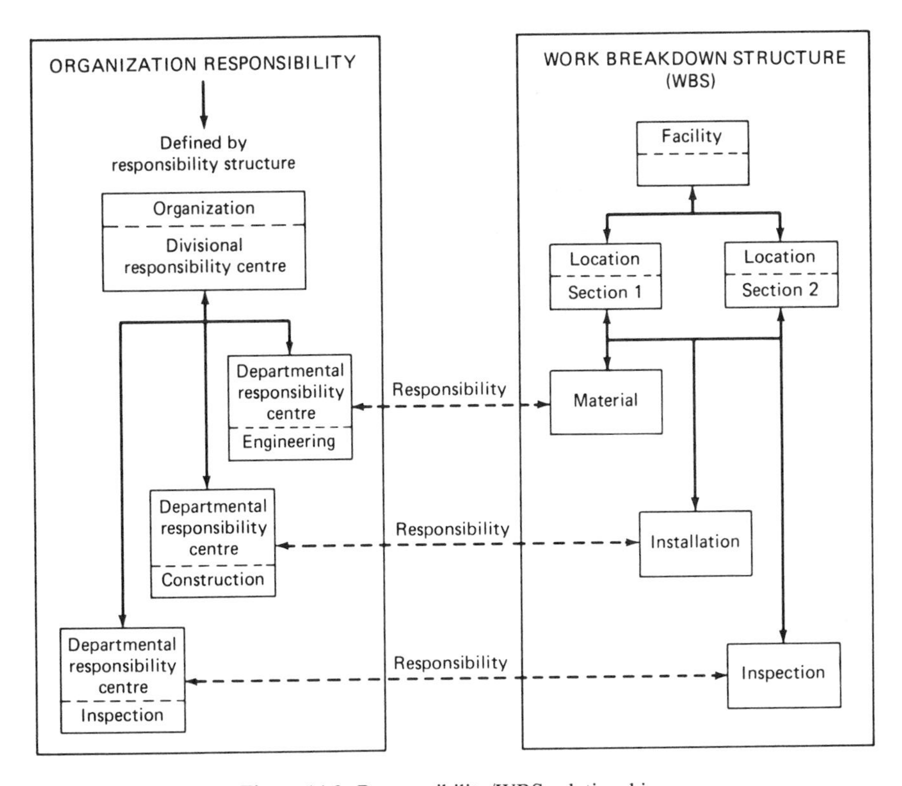


Figure 14-3. Responsibility/WBS relationship.

The right side of Figure 14-3 illustrates the WBS or the project tree which defines the project. On the left is the organizational responsibility as defined by a responsibility structure. The overlay of this organizational responsibility on the WBS creates the management matrix. This management matrix is designed specifically to allow for the case of the large project where there may be many organizations involved in the project. Usually, the basic facility definition does not change through time, whereas organizational responsibilities will change as the project progresses through the project life cycle. The WBS will be designed to provide the total precise project definition on a facility basis, with the major responsibility assignments being overlayed as required to allow assignment of responsibility centers for managing parts of the project, without upsetting the project records, the information exchanges, and systems required to manage the project.

The concept of using the management matrix is essential to effective management of projects where costs, schedules, and resource usage need to be tracked from project inception through to its completion, which can be several years later. The common fixed basis is the WBS which will not change with organizational responsibility, but can only be revised with a scope change in the project definition. At that time the WBS is normally added to with the basic structure and definitions of the WBS remaining. The WBS provides stability to the information for the cost control and accounting systems as well as to the personnel involved with the project for the life of the project. As the organization responsibility assignments change, the management responsibilities for the various segments of the project will change; but by having the WBS tied to the facilities, this changing organizational responsibility will not interfere with cost, schedule, and progress reporting. The key to this will be that the information from the WBS data base will be provided to different people at different times, but the base information remains the same and is collected on that basis.

The consistency in tracking information through the use of the WBS allows for consistent reporting to governments, senior management, owners, contractors, and other participants. This is a key factor in project control. It is therefore important in designing the WBS not to build the organization-responsibility structure as part of the WBS, because if the organization structure is built into the WBS, the end result, due to rapidly changing definitions, will be a nonuseful project control tool. Therefore, the WBS should be designed, structured, and coded to the project being performed, regardless of the organization managing it.

The organization structure should be considered in light of how the project is to be managed on an overall basis. Next must be considered the design and development of a WBS, as a key project control tool to be used between the groups responsible for the project and systems utilized to measure project progress.

## WBS DEVELOPMENT

### Overview

When the project plan is developed, the WBS must be defined with regard to all the elements that are required to make it a working entity. The base elements are

1. Structure.
2. Code.
3. Reporting.

Before integrating these elements, one must first look at each one separately, then their relationship to each other.

In the design of the project WBS, the "management philosophy" must be considered. The WBS can be a facility-oriented tree or a contract-oriented tree representing the hierarchical components of the project to be managed. The approach to the design of the WBS and its development will depend upon the management philosophy adopted. If it is contract-oriented, it must be related to the entity to be built.

**WBS Structure**

The structure of the WBS must be such that each level is significant and meaningful, both from a data collection and an overall reporting point of view. This means that every level of the structure developed has significance and relates to a facility contract actually being built or managed within the project, and can be used to generate a required or meaningful report.

The overall design of the structure of the WBS is the key to an effective working system. Therefore, it must be studied very carefully from an input and output of information point of view. Since the WBS serves as a common information exchange language, it is the language, the code, and the structure by which all information on the project is gathered, and it is also the device by which all information of the project is disseminated. Therefore, the structure will be built in a hierarchical manner, or as a tree, such that the bottom level will represent the detailed information and will be large in scope. The base level of the WBS structure is the lowest level of information required to manage the project. This is the lowest level of information at which a user can foresee a need to communicate or monitor. It is the lowest level that the line managers and the construction personnel will require to manage the project.

The next level up the structure will be narrower and will supply information to another level of users. These upper levels will supply significant information for management—significant from the point of view of providing information that is meaningful to various levels of management. This significance ensures that not too many levels are built into the structure. The structure must be designed so that it is meaningful and hierarchical. Twenty levels are too many to manage effectively. Four, five, up to six generally appears to be an adequate number of levels in a large project. In some cases, two sets of the five levels may be used—a base set of five for the detailed collection of data rolling up to a contract level or a major facility level; and five as a superstructure or overlay which ties together the larger components of the facility or the larger contracts. This would allow up to a total of ten levels, but with two distinct purposes. This double-level structure of the WBS works very well and does not restrict the WBS development.

As each level of the structure is designed, consideration must be made as to how information will flow upward to the next level. This transition from one level to another should happen in a natural manner. It should not be forced so that it is difficult or, as the information flows upward, becomes meaningless. As a new structure is being designed, it should be based on the most likely case and should have some flexibility for additions, although this flexibility for additions will come mainly from the coding once the structure has been set. A simple example of such a structure would have the facilities of the top level followed by the items to build the facility at the lower levels.

When a WBS is designed, provisions should be made so that when the structure is translated into code, the coding is meaningful to the user. This means that the user can identify the WBS as a facility tree of physical assets which he can recognize when he goes out in the field. It is of paramount importance that the usage of the WBS be designed in such a way that it becomes THE project language. Thus, in a project, items that the user understands and sees as a major physical unity become elements of the WBS.

The structure is the essential base around which the coding is built and the reporting capabilities of the WBS determined. Thus the structure design is key to an effective WBS.

### Code Design

The design of the coding is the key to establishing the WBS as the device to be used by the accounting and project control systems. An effective, meaningful code will assist the user and will complement the structure design described above. Whether the user be the field accountant, the field clerk, or senior management, the code should have common meaning to all. The top level of the code could be relevant major subsystems such as pipeline, compressor stations, and meter stations, or the process plant, buildings, and off-sites. It would be a level above this which would represent the project. The code is the ingredient which the user and the functional groups building the project must work with on a daily basis. In designing the code, one must give consideration to the information collected and the methods used to collect it. The user, who analyzes the raw data collected and puts the proper code on it so that the information can be entered into the applicable recording system via the WBS code, must understand the code.

The code design is directly related to the structure development. Each level of the structure represents a segment of the code. The code design is the assimilation of a group of digits to represent a physical facility to be built. At the top level the project does not need to be coded; at the next level the key facilities to be built are coded utilizing the first digit of the code. If the number of key facilities to be managed is nine or less, the code will typically be a one-digit code, assuming only numerics are used for coding. If alphas and numerics are used, then the level can have 35 different items. The next level below the facility in the structure represents the key items or key contracts to be utilized in building the facility. This level will typically be a two-digit code which gives the flexibility to define 99, or, if alphas are used, more than 99 different items. In the designing of the code, the level above always determines the meaning of the level below. An example of this is illustrated in Figure 14-4, which shows a WBS utilized for pipeline construction.

Figure 14-4 illustrates the code which can be used in a four-level WBS. The top level, "Pipeline Construction," is one digit (2), the next level, "Mainline Location," is three digits (212). These three digits represent

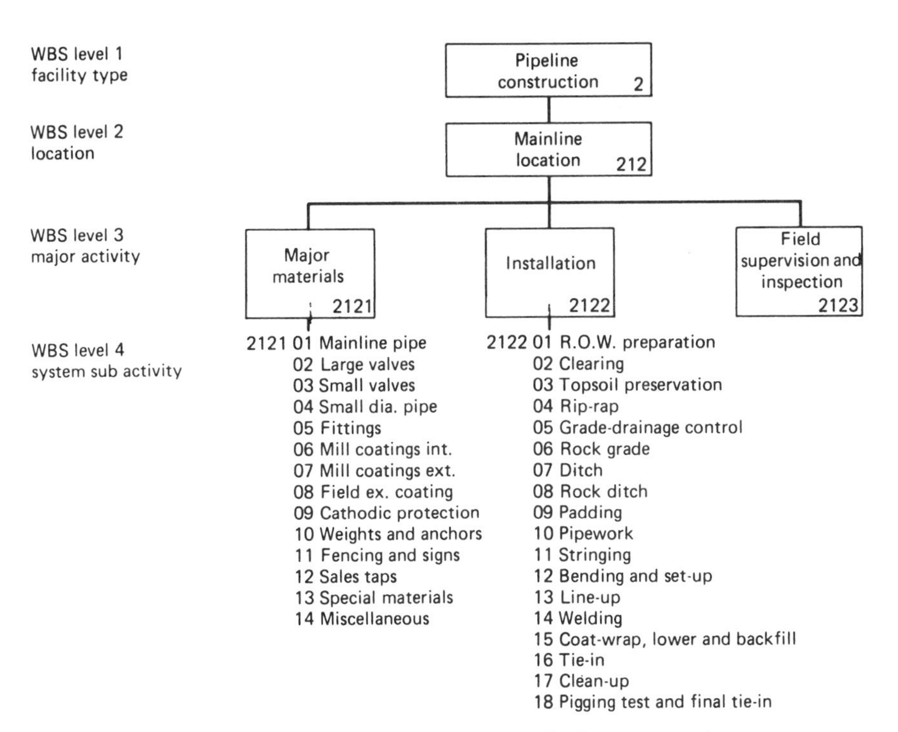


Figure 14-4. Work breakdown structure pipeline construction.

the type of facility to be built and its physical location. The next two levels, shown as 3 and 4, provide a breakdown of all the items required to build the facility. Note that each level of the code hierarchy is dependent on the levels above to determine the complete definition of a given level. This allows elements in level 4 to vary according to the type of facility to be built as defined in levels 1, 2, or 3. If at all possible, at a specific level, identical coding should reference similar information. This will facilitate a more understandable code. This ability for lower levels of the WBS to have different meanings depending on the upper levels allows for more project scope to be accommodated without adding unnecessary digits to the code. Although the code is developed in a hierarchical manner, it is desirable within the structure for the code at a given level to be the same in as many places as possible across the project. The code illustrated in Figure 14-4 always uses the same level 3 code of ___1 for Materials, ___2 for Installation, and ___3 for Inspection for all facilities on the project. This provides the capability for material, installation and/or inspection costs, schedules, or productivity information to be produced.

When the code is designed, the users must be considered. The users are the people who must code all the information that is to be utilized by the systems. Development of the code should occur in such a way that the user can understand its meaning and significance. Many companies have used alpha characters to give this meaning in a simple form. For example, they may code M for manpower, E for equipment, and C for consumables.

Integration of the code and the structure is such that every level of the structure has a specific number of digits of code assigned to it. This is where the structure hierarchy becomes important in the code design. If the structure has 20 levels, this necessitates that the coding have a minimum of 20 digits, which is too long; thus a compact structure which gives compact coding will supply a system that the people will use. It is paramount in the design of the code and the structure that usage and simplicity be kept clearly in mind. The "nice-to-have" information should be of secondary importance in the design, ensuring that essential information can be retrieved from a simple effective structure and related coding.

### Reporting Considerations

In the designing of the WBS, all levels of reporting should be looked at from senior management, or overall project management, down to the lowest level of the person recording the information and the project engineer who wants to know in detail exactly what is happening on this project. The WBS should be designed so that all reports generated from the WBS are automatic without requiring extravagant report-writing methods to extract the information collected. The reports required should be looked at and checked to see that both the WBS and the reports are meaningful and representative of what is really required. Once this has been determined, the WBS design should reflect the reports that will be produced for the various levels of management who are involved in the project. Thus, a level in the structure often becomes a level in the reporting hierarchy to the various project management personnel. When the structure, the reporting, and the usage considerations are incorporated into the coding, the first part of the code supplies the management report, and as the code expands it supplies the detailed project reports.

The consideration of the reporting requirements in the WBS design also helps to define the exact reports that will be available to management as the project proceeds. In going through this design phase, management will be able to review the reports they expect to see from the appropriate project management people, and on the detail level, the personnel doing the estimates and the schedules of the baseline work will be able to review exactly the level of detail they will see later for verifying their estimates. On the output side of the WBS, the structure and code design will be influenced by the requirements of the different groups responsible for different areas of management. This is not to be confused with the responsibility overlay matrix.

As the reports are designed, the prime requirement is to generate the applicable management information required on a facility basis and not the responsibility information required for functional or organizational responsibility reporting. It must be kept clear in the designing of the reports that they are not the departmental or functional reports but the progress reports for the progress to date for the completion or construction of a particular facility. It must be clearly stated to the users that the WBS only applies to facility-related reports.

### Coordination of Structure, Code, and Report Requirements in WBS Design

The preparation of the WBS requires the integration of the structure, code, and reporting requirements. Initially, the scope of the project is outlined through a pictorial representation of the WBS which should be prepared. This illustration, note Figure 14-4 for example, should be circulated without coding to the user community at both the worker level and the management level.

At this formative stage, it is important that much forward looking or insight into the project be considered. The structure should be simple, clear and have meaning, and then this should be linked to the code after the structure has been finalized. The coding should then be prepared. Sample reports using the WBS should be drawn up and circulated for review. These reports should be generated using test data and the proposed code.

The design of the structure, the code, and the reports should contain as much input as possible from the groups that will be using it, given whatever time, cost, or system constraints which are applied to the particular project. In many cases existing systems such as accounting, or existing policies and procedures within the company, may dictate the shape of the WBS. These constraints must be worked around, and an effort should be made to make the WBS as close as possible to the "ideal" required. These constraints should not be the primary determinants of the structure but must be considered in the design. These stumbling blocks and the hurdles must be overcome because this project control tool (WBS) is an absolute necessity in any project with multiple owners, many contractors, and/or government intervention.

When the WBS is completed, it must be presented and explained to all the users. It should be in book form, which can be readily updated. The first section in the book should illustrate the WBS structure with pictorial drawings, the next section should illustrate the code, either pictorially or graphically, and the final section should be a dictionary of definitions defining the content of each WBS element. These definitions are necessary in projects, ensuring that when a term is used, it is used as the project means it to be used, not with historical meanings which vary from group to group. A careful documentation of the meaning in the dictionary of each WBS element with regard to cost, schedule, and resource requirements for the activity ensures that all users will gather and supply information with common meaning to personnel involved with the project. The WBS manual containing coding, definitions, and explanations for usage is the last key step in the development of the WBS. A supplement to the manual may contain samples of the report formats to be utilized by the groups and illustrations as to how these tie into the WBS.

Upon completion of the manual, it should then be the responsibility of the project manager's staff to explain it to all users and personnel involved in the project. They should explain why it is necessary, and when and what it is required for. The why is to ensure that all information collected, reported, and forecast against has common meaning, regardless of the information source. It should be explained for usage in all communications (the what) on budgets, estimates, schedules, productivities, performance, and items associated with management of the project and should be done with reference to the applicable WBS element. This reference should be used and maintained from the very first day (the when) the project is clearly defined to the final reconciliation of the project. WBS should be used by the engineers, project control, and accounting. The WBS is used everywhere that information on the project's progress is collected.

## EXAMPLES OF A WBS

### Background

The previous section described the required elements for the preparation of the WBS as the foundation for establishing project control. The characteristics to be considered in WBS development and usage are

1. Management philosophy.
2. User groups.
3. System integration.
4. WBS—Responsibility Relationship.
5. WBS Components:
   - structure
   - code
   - reporting

The development process described was used for the preparation of a WBS for the Canadian section of the Alaska Highway Gas Pipeline Project (AHGPP).

Figure 14-5 illustrates the overall project, which starts from Prudhoe Bay through Alaska down through the Yukon, British Columbia, Alberta, and Saskatchewan. Thus the project covered a large geographical area. Overall, as can be seen by the map, there were many companies and governments involved. The lines from Caroline to Monchy, Saskatchewan, and from Caroline to Kingsgate, British Columbia were begun in late 1980 and completed two years later. This prebuild project of about $1 billion Canadian was completed successfully under budget and ahead of schedule. The WBS as described was used in the manner and for all the applications mentioned.

### Management Philosophy

The management philosophy for the project required a WBS designed on a facility basis.

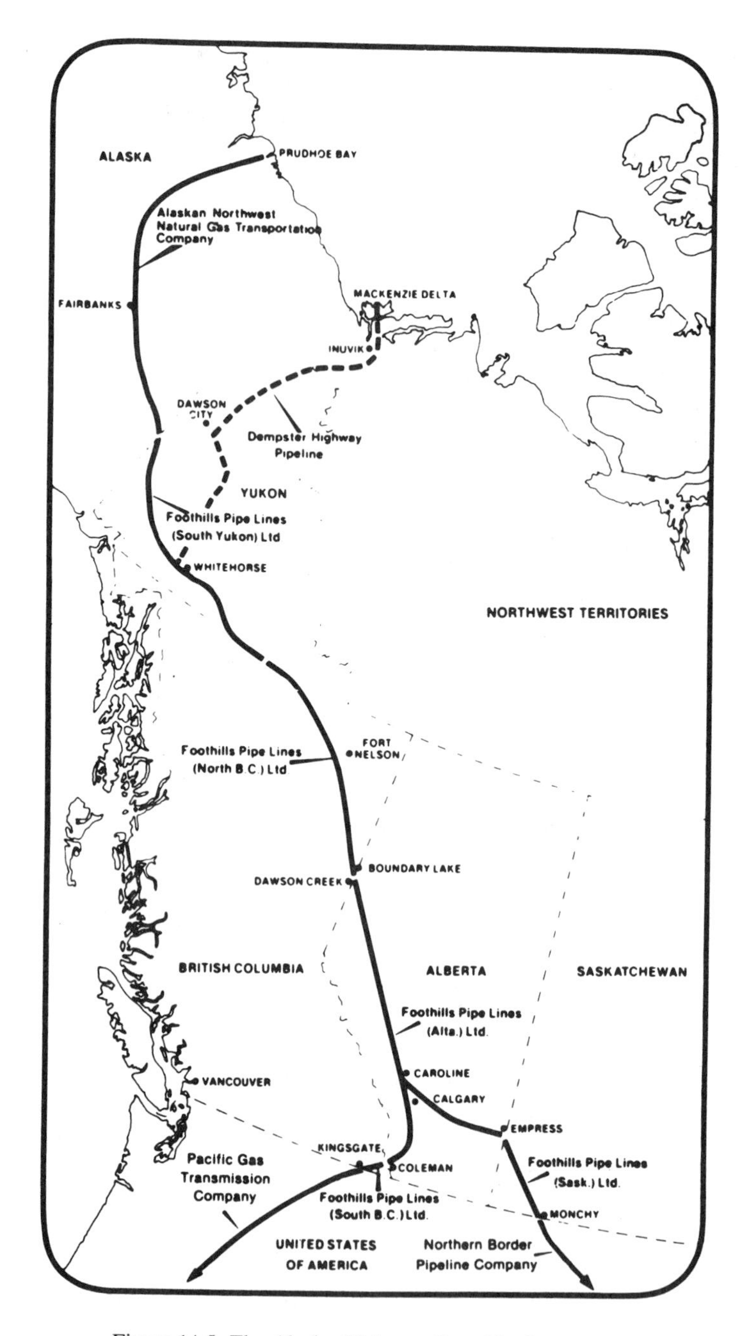


Figure 14-5. The Alaska Highway Gase Pipeline Project.

### User Groups

The planned Canadian line was over 2000 miles with pipe sizes varying from 36 to 56 inches in diameter. Along this section of the route were approximately 21 compressor stations each with over 25,000 horsepower. The user community for this project, as shown in Figure 14-1, had functional groups which were part of four companies, with the responsible groups including two owner companies, government agencies, and contractors. This diverse owner community with the multifunctional centers required a very precise WBS to ensure effective project information flow.

### Systems Integration

The functional groups in all six Canadian companies used the WBS as part of their budgeting, estimating, accounting, and scheduling systems utilizing both manual and automated systems. All information collected by these systems was summarized using the WBS and then forwarded to management on a common basis using both manual and automated interfaces. The prebuild project clearly illustrated the effective integration of systems, as the WBS was utilized by all required systems.

### WBS Development

The WBS was developed to meet the needs of all the users described previously. A segment illustrating the WBS structure developed is shown as Figure 14-4. This illustrates the different levels of the structure with their applicable coding. As can be seen from Figure 14-4, the first level is facility, the second location, the third prime activity, and the fourth describes subactivity. For each of these levels, there are specific reports. The top level—facility type—supplies cost, productivity, and schedule information on a facility basis, which is a top management report. To provide additional detail, the reports for each specific facility was generated for cost, schedule, productivity, manpower, and equipment usage for each level of the WBS shown. The final two levels of activity—the prime activity and subactivity—enable the designer, the cost engineer, and the scheduler to monitor specific items required to build a particular facility. These levels represent the engineering technical level which is required for the detailed management of the project. As can be seen in Figure 14-4, the hierarchical rollup is natural in that each level rolls to the next level in a meaningful fashion.

The code is shown in Figure 14-4 and is graphically illustrated in 14-6. On the left of Figure 14-6 is a map of Alberta showing the facilities to be

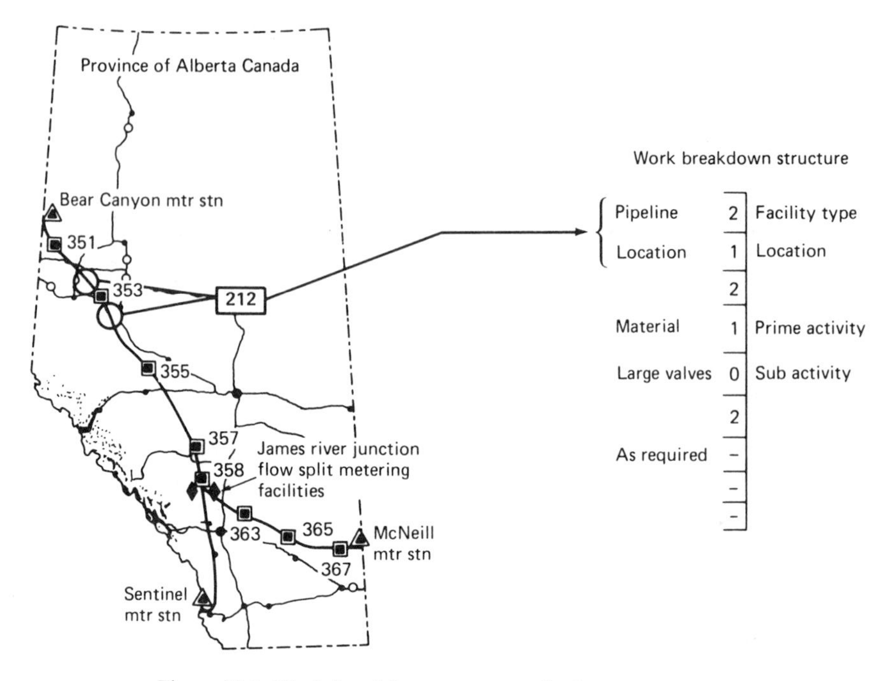


Figure 14-6. Work breakdown structure pipeline construction.

constructed. The section highlighted is a section of pipeline to be built at the particular location shown. Code 2 is always pipeline, with 212 being pipeline at location 12. Also in the drawing are locations 363, etc., which are compression facilities at designated locations. This figure illustrates a portion of the code and its meaning. This type of illustration is very useful in training staff in the usage of WBS.

An example of the usage of this coding is presented in Figure 14-7. This is the standard commitment report on a WBS basis for a particular location. The left side of the figure is the WBS code (six digits in this case). The capability of nine digits is allowed to supply future flexibility as the project proceeds. If more information is needed in a particular area, then the WBS can expand to receive the information without redesigning the code or structure, while maintaining the flow of information in a hierarchical manner. The first three digits identify the facility type and its location, as shown in Figure 14-6. Then come the materials as the next digit, which is represented by a "1." The materials code for the items required for pipeline are next. The authorized dollars illustrated are for the items authorized for purchase for the particular location. As the project pro-

| WBS Code | WBS Description | Current authorized | Incurred cost this period | Incurred cost to date | Committed cost this period | Committed cost to date | Estimate to complete | Estimated final total |
|---|---|---|---|---|---|---|---|---|
| 212100 | Unallocated | | | | | | | |
| 212101 | Mainline pipe | 37,505 | | | | | | |
| 212102 | Large valves | 420 | | | | | | |
| 212103 | Small valves | 21 | | | | | | |
| 212104 | Small diameter pipe | 2 | | | | | | |
| 212105 | Fittings | 47 | | | | | | |
| 212106 | Mill coating internal | 329 | | | | | | |
| 212107 | Mill coating external | | | | | | | |
| 212108 | Field coating external | 460 | | | | | | |
| 212109 | Cathodic protection | | | | | | | |
| 212110 | Weights and anchor | 785 | | | | | | |
| 212111 | Fencing and signs | | | | | | | |
| 212112 | Sales taps | | | | | | | |
| 212113 | Special materials – coating | 54 | | | | | | |
| 212114 | Miscellaneous | 88 | | | | | | |
| | Sub-total materials | 39,711 | | | | | | |

ceeds, the incurred costs will be recorded as well as committed costs. In addition, estimates to complete will be done so that an estimated final total cost can be determined for this part of the project. The report shown in Figure 14-7 is only one of many which can be generated within the WBS framework. This example of the pipeline project illustrates a WBS which is utilized in the multicompany environment for a large project where the WBS is a key project control tool. Thus, as illustrated, a properly designed WBS is the basis for effective project control tools.

## CONCLUSIONS

It appears that for now and the foreseeable future, projects with long time spans from the conceptual stage to operation, large capital expenditures, complex ownership, and government involvement will become more common.

In this complex environment, or with many small projects, it is absolutely essential to have a precisely defined methodology for the WBS with which all involved personnel can exchange information, plan the project, and organize reporting. With the usage of personal computers, many smaller projects at different locations can be managed using the same WBS to give overall control on a common reporting basis.

This chapter has outlined the reasons why a properly designed WBS becomes the essential tool for effective project management in a project environment. It has illustrated the components and requirements to prepare the required WBS. The prime reasons the WBS should be used are the following:

1. Developing the WBS early in the project life cycle provides a method for clear definition of the project scope, and the process of WBS development helps all participants to clearly understand the project during the initial stages.
2. The use of the WBS code for monitoring and forecasting of all cost, schedule, and productivity information ensures that project management personnel will have a baseline to which comparison can be made. When a common definition for all information on the project is established, effective and logical management decisions can be made.
3. With multiple participants and changing personnel, it is essential that all terms used mean the same to all participants. This consistency of definition is established through the development and use of the WBS with associated code and dictionary.

4. The WBS becomes the basis from which all information flow between information systems can be established, and upon which all facility-type reporting is available.

Thus a properly designed and developed WBS, with its structure, code, and dictionary, supplies the common base for project management by having cost, schedule, and productivity information all using WBS definitions forming the foundation of quality project control for any project.

# 15. Network Techniques in Project Management

Joseph J. Moder*

*Project management* involves the coordination of group activity wherein the manager plans, organizes, staffs, directs, and controls, to achieve an objective with constraints on time, cost, and performance of the end product. This chapter will deal primarily with the planning and control functions, and to some extent the organization of resources. *Planning* is the process of preparing for the commitment of resources in the most economical fashion. *Controlling* is the process of making events conform to schedules by coordinating the action of all parts of the organization according to the plan established for attaining the objective.

It can also be said that project management is a blend of art and science: the art of getting things done through and with people in formally organized groups; and the science of handling large amounts of data to plan and control so that project duration and cost are balanced, and excessive and disruptive demands on scare resources are avoided. This chapter will deal with the science of project planning and control that is based on a network representation of the project plan, also referred to as critical path methods.

It is appropriate at this point to elaborate on the term *project*. Projects may, on the one hand, involve routine procedures that are performed repetitively, such as the monthly closing of accounting books. In this case, critical path methods are useful for *detailed* analysis and optimization of the operating plan. Usually, however, these methods are applied to

---

* Joseph J. Moder is Professor in the Department of Management Science at the University of Miami, Coral Gables, Florida. He received his B.S. degree from Washington University, his Ph.D. from Northwestern University, and he did Post Doctoral work in Statistics and Operations Research at Iowa State University and Stanford University. He was a Visiting Professor of Engineering Production at the University of Birmingham, England. His research interests include applied statistics and project management methodology. He has published numerous articles and several books in these fields, and has conducted short courses and research projects in these areas.

one-time efforts; notably construction work of all kinds; maintenance operations; moving, modifying, or setting up a new factory or facility of some sort; etc. Critical path methods are applicable to projects which encompass an extremely wide range of resource requirements and duration times.

In project management, although similar work may have been done previously, it is not usually being repeated in the identical manner on a production basis. Consequently, in order to accomplish the project tasks efficiently, the project manager must plan and schedule largely on the basis of his experience with similar projects, applying his judgment to the particular conditions of the project at hand. During the course of the project he must continually replan and reschedule because of unexpected progress, delays, or technical conditions. Critical path methods are designed to facilitate this mode of operation.

## HISTORY OF THE EARLY DEVELOPMENT OF CRITICAL PATH METHODS

Until the advent of critical path methods, there was no generally accepted formal procedure to aid in the management of projects. Each manager had his own scheme which often involved the use of bar charts originally developed by Henry Gantt around 1900. Although the bar chart is still a useful tool in production management, it is inadequate as a means of describing the complex interrelationships among project activities associated with contemporary project management.

This inadequacy was overcome by the significant contribution of Karol Adamiecki in 1931 (1).* He developed a methodology in a form that he called a Harmony graph. This is essentially a bar chart, rotated 90 degrees, with a vertical time scale, a column (movable strip) for each activity in the project, and a very clever means of showing the interrelationship among project activities. This work was evidently completely overlooked by others in this field. It was not until 1957–1958 that a more formal and general approach toward a discipline of project management occurred. At this time several techniques were developed concurrently, but independently. The technique called Critical Path Method (CPM) was developed in connection with a very large project undertaken at Du Pont Corporation by Kelley and Walker (6). The objective there was to determine the optimum (minimum total cost) duration for a project whose activity durations were primarily deterministic variables.

---

* Numbered references are given at the end of this chapter.

A similar development occurred in Great Britain where the problems of overhauling an electricity generating plant were being studied (7). The principal feature of their technique was the determination of what they called the "longest irreducible sequence of events."

A somewhat different approach to the problem, called Project Evaluation and Review Technique (PERT), was developed in conjunction with the Polaris weapons system by Malcolm and others (8). The objective there was to develop an improved method of planning, scheduling, and controlling an extremely large, complicated development program in which many of the activities being conducted were at or beyond the state of the art, and hence the actual activity duration times were primarily random variables with considerable variance.

## DEVELOPMENT OF THE NETWORK PLAN CONCEPT

Although all of the above developments were conducted independently, they are essentially all based upon the important concept of a *network* representation of the project plan. The network diagram is essentially an outgrowth of the bar chart which was developed by Gantt in the context of a World War I military requirement. The bar chart, which is primarily designed to control the time element of a program, is depicted in Figure 15-1(a). Here, the bar chart lists the major activities comprising a hypothetical project, their scheduled start and finish times, and their current status. The steps followed in preparing a bar chart are as follows:

1. Analyze the project and specify the basic approach to be used.
2. Break the project down into a reasonable number of activities to be scheduled.
3. Estimate the time required to perform each activity.
4. Place the activities in sequence of time, taking into account the requirements that certain activities must be performed sequentially while others can be performed simultaneously.
5. If a completion date is specified, the diagram is adjusted until this constraint is satisfied.

The primary advantage of the bar chart is that the plan, schedule, and progress of the project can all be portrayed graphically together. Figure 15-1 shows the five-activity plan and 15-week schedule, and current status (end of third week) indicates, for example, that activity B is slightly behind schedule. In spite of this important advantage, bar charts have not been too successful on one-time-through projects with a high engineering content, or projects of large scope. The reasons for this include the fact that the simplicity of the bar chart precludes showing sufficient detail to

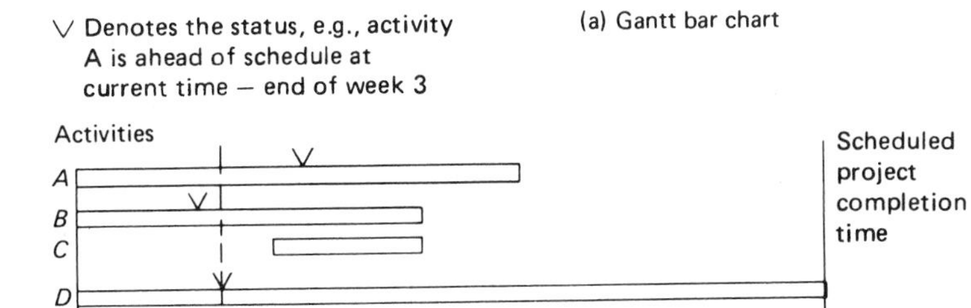


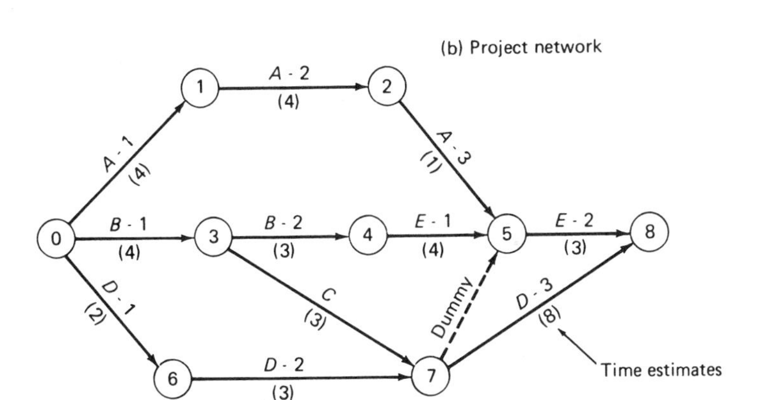


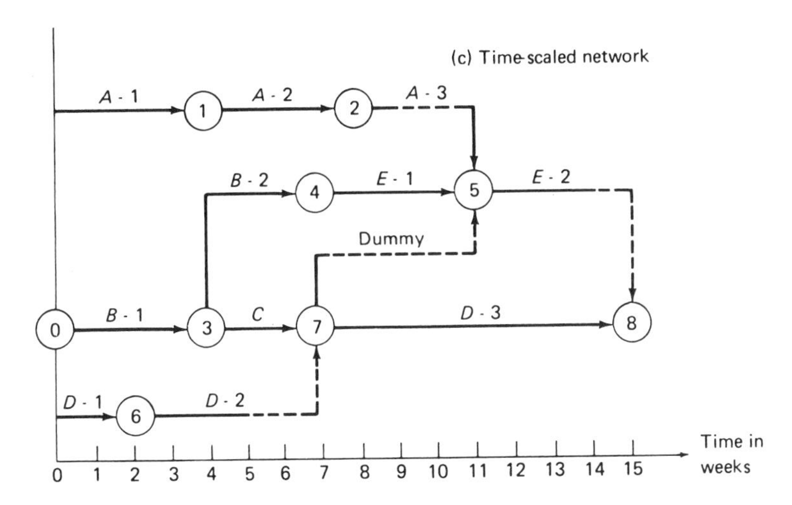


Figure 15-1(a),(b),(c). Comparison of bar chart, project network, and time-scaled network. (From *Project Management with CPM, PERT and Precedence Diagramming,* 3rd Ed., J. J. Moder, C. R. Phillips and E. W. Davis, © 1983 by Litton Educational Publishing, Inc. Reprinted by permission of Van Nostrand Reinhold Co.)

enable timely detection of schedule slippages on activities with relatively long duration times. Also, the bar chart does not show explicitly the dependency relationships among the activities. Hence, it is very difficult to impute the effects on project completion of progress delays in individual activities. Finally, the bar chart is awkward to set up and maintain for large projects, and it has a tendency to quickly become outdated and lose its usefulness. With these disadvantages in mind, along with certain events of the mid-fifties such as the emergence of large technical programs, large digital computers, general systems theory, etc., the stage was set for the development of a network-based project management methodology. Something like the critical path method literally had to emerge.

Before taking up the logic of networking, it will be useful to preview the scope of critical path methods as the basis of a dynamic network-based planning, scheduling, and control procedure, as shown in Figure 15-2.

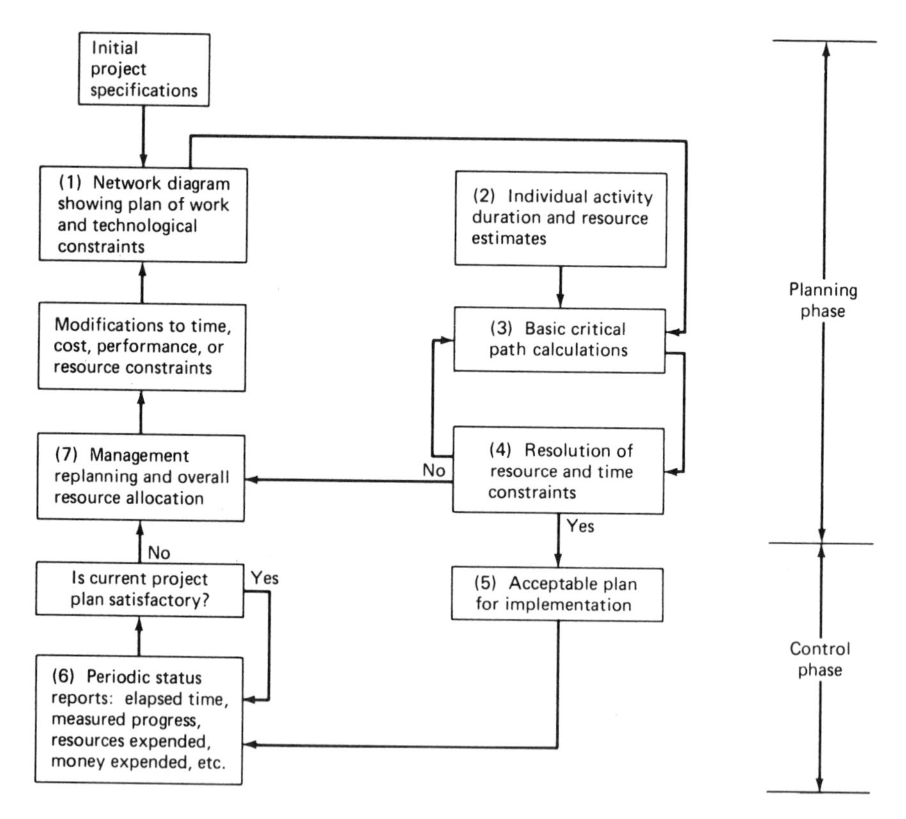


Figure 15-2. Dynamic network-based planning and control procedure.

Step 1, which is the representation of the basic project plan in the form of a network, will be treated in the next section. Steps 2 and 3 will then be considered to estimate the duration of the project plan and determine its critical path. Considered next are the techniques which comprise Step 4; they are designed to modify the initial project plan to satisfy time and resource constraints placed on the project. Finally, the control phase of project management, Step 6, will be considered.

## THE LOGIC OF NETWORKS AS MODELS FOR PROJECT PLANS

The first step in drawing a project network is to list all jobs (activities) that have to be performed to complete the project, and to put these jobs in proper technological sequence in the form of a network or arrow diagram. An aid to this process is to organize the project activities in the form of a hierarchical (tree) structure called a work breakdown structure. Such a diagram is shown in Figure 15-3, with the work broken down vertically. In addition, you can break down the organization that is to be used to carry out the project as shown horizontally at the bottom of Figure 15-3. When hierarchical codes are assigned to each of these breakdowns, it is then easy to produce specialized reports by selecting only those activities having the desired codes. For example, an organizational breakdown report could be produced to cover all Engineering activities, or Engineering-Design. Similarly, a work breakdown report could be produced to cover all Engine, or Engine-Compressor activities.

Each job in the project is indicated by an arrow, with nodes, called events, placed at each end of the arrows. Events represent points in time and are said to occur when all activities leading into the event are completed. In Figure 15-4, for example, when the two activities "select operators" and "prepare training material" are completed, the event numbered 10 is said to occur. It should be pointed out that the two predecessor activities of Event 10 need not be completed at the same time; however, when they are both completed, Event 10 occurs, and only then may the activity "train operators" begin. Similarly, when this activity is completed, Event 15 occurs, and the successor activities "test process A" and "test process B" each *may* then begin. It is important to note that the ordering of these activities is based on the "technology" of the resources being utilized.

Activities require the expenditure of time and resources to complete; eight time units and three instructors in the above example. The length of the arrow is not important, but its direction relative to other activities and events indicates the *technological constraints* on the order in which the activities making up the project may be performed.

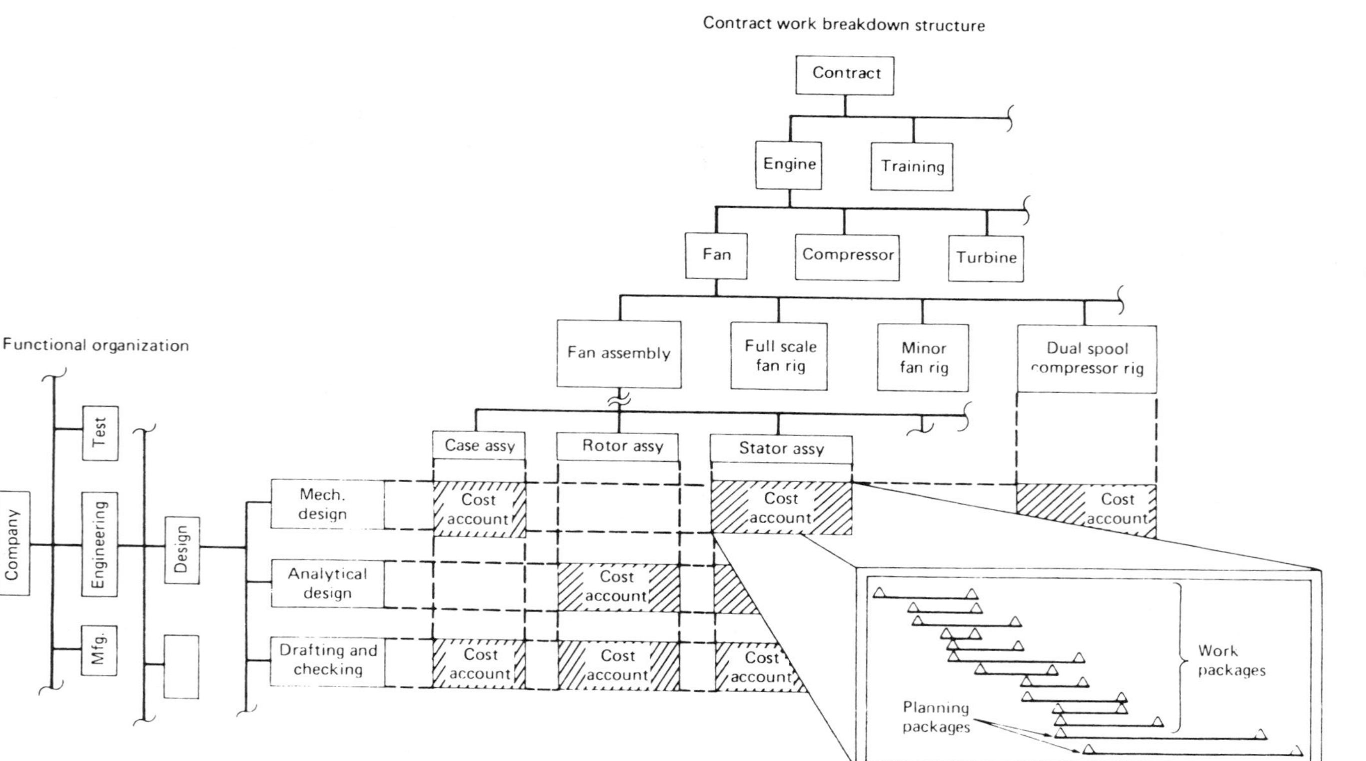


Figure 15-3. Interrelationship between work breakdown structure and functional organization. (From *Project Management with CPM, PERT and Precedence Diagramming*, 3rd Ed., J. J. Moder, C. R. Phillips, and E. W. Davis, © 1983 by Litton Educational Publishing, Inc. Reprinted by permission of Van Nostrand Reinhold Co.)

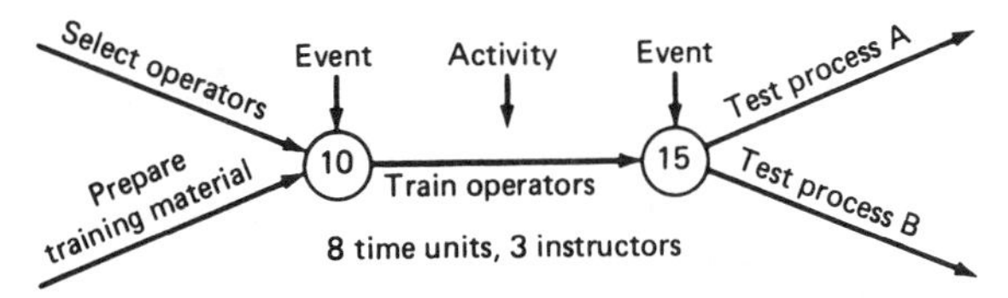


Figure 15-4. An example arrow diagram activity.

There is also a need for what is called a *dummy* activity, which requires neither time nor resources to complete. Activity 7-5 in the middle of Figure 15-1 is an example of such an activity. Its sole purpose is to show precedence relationships, that is, that activities C and D-2 must (technologically) precede activity E-2.

The project network is then constructed by starting with the initial project event which has no predecessor activities and occurs at the start of the project. From this event, activities are added to the network using the basic logic described above. This process is continued until all activities have been included in the network, the last of which merge into the project end event which has no successor activities. In carrying out this task the novice must be extremely careful to avoid the common error of ordering the activities arbitrarily according to some preconceived idea of the sequence that the activities will probably take when the project is carried out. If this error is made, the subsequent scheduling and control procedures will be unworkable. However, if the network is faithfully drawn according to technological constraints, it will be a unique project model which only changes when fundamental changes in the plan are made. It will also present maximum flexibility in subsequent scheduling of the activities to satisfy resource constraints.

The preparation of the project network presents an excellent opportunity to try out, or simulate on paper, various ways of carrying out the project, thus avoiding costly and time-consuming mistakes which might be made "in the field" during the actual conduct of the project. It also presents an opportunity to solicit inputs from project team members to gain their wisdom and their future cooperation and allegiance. At the conclusion of the planning operation, the final network presents a permanent record giving a clear expression of the way in which the project is to be carried out so that all parties involved in the project can see their involvement and responsibilities.

### The Time Element

After the planning or networking, the *average duration* of each job is estimated, based upon the job specifications and a consideration of the resources to be employed in carrying out the job. The best estimates will usually be obtained from the person(s) who will supervise the work or who has had such experience.

These time estimates are placed beside the appropriate arrows. If we were then to sum the durations of the jobs along all possible paths from the beginning to the end of the project, the longest one is called the critical path, and its length is the expected duration of the project. Any delay in the start or completion of the jobs along this path will delay completion of the whole project. The rest of the jobs are "floaters" which have a limited amount of leeway (slack) for completion without affecting the target date for the completion of the project.

These concepts are illustrated at the bottom of Figure 15-1 where the network activities have been plotted to scale on a time axis. This diagram shows the critical path quite clearly. It consists of activities B-1, C, and D-3, and has an overall duration of 15 weeks. The slack along the other network paths is shown by the dashed portion of the network arrows. For example, the path D-1 and D-2 has 2 weeks of slack, that is, 7 weeks are available to carry out these two jobs which are expected to require only 5 weeks to complete.

The above time-scaled network can be considered as a graphical solution to what is called the *basic scheduling computations*. This is not an operational procedure; it was used here primarily for illustrative purposes.

The objective of the scheduling computations is to determine the critical path(s) and its duration, and to determine the amount of slack on the remaining paths. It turns out that this can best be accomplished by computing the earliest start and finish, and latest start and finish times for each project activity.

## BASIC SCHEDULING COMPUTATIONS

A programmable algorithm for the basic scheduling computations is given by equations 1 through 7 below, in terms of the following nomenclature.

| | |
|---|---|
| $D_{ij}$ | estimate of the mean duration time for activity *i-j* |
| $E_i$ | earliest occurrence time for event *i* |
| $L_i$ | latest allowable occurrence time for event *i* |
| $ES_{ij}$ | earliest start time for activity *i-j* |
| $EF_{ij}$ | earliest finish time for activity *i-j* |
| $LS_{ij}$ | latest allowable start time for activity *i-j* |
| $LF_{ij}$ | latest allowable finish time for activity *i-j* |

| | |
|---|---|
| $S_{ij}$ | total slack (or float) time for activity *i-j* |
| $FS_{ij}$ | free slack (or float) time for activity *i-j* |
| $T_s$ | schedule time for the completion of a project or the occurrence of certain key events in a project |

*Earliest and Latest Event Times*

Assume that the events were numbered (or renumbered by a simple algorithm) so that the initial event is l, the terminal event is $t$, and all other events ($i$-$j$) are numbered so that $i < j$. Now let $E_1 = 0$ by assumption, then

$$E_j = \max_i \, (E_i + D_{ij}) \qquad 2 \leq j \leq t \tag{1}$$

$E_t$ = (expected) project duration, and
$L_t = E_t$ or $T_s$, the scheduled project completion time. Then,

$$L_i = \min_j \, (L_j - D_{ij}) \qquad 1 \leq i \leq t - 1 \tag{2}$$

*Earliest and Latest Activity Start and Finish Times and Slack*

$$ES_{ij} = E_i \qquad \text{all } ij \tag{3}$$

$$EF_{ij} = E_i + D_{ij} \qquad \text{all } ij \tag{4}$$

$$LF_{ij} = L_j \qquad \text{all } ij \tag{5}$$

$$LS_{ij} = L_j - D_{ij} \qquad \text{all } ij \tag{6}$$

$$S_{ij} = L_j - EF_{ij} \qquad \text{all } ij \tag{7}$$

The above equations embody two basic sets of calculations. First, the *forward pass calculations* are carried out to determine the earliest occurrence time for each event $j$ ($E_j$), and the earliest start and finish times for each activity $i$-$j$ ($ES_{ij}$ and $EF_{ij}$). These calculations are based on the assumption that each activity is conducted as *early* as possible, that is, they are started as soon as their predecessor event occurs. Since these calculations are initiated by equating the initial project event to time zero ($E_I \equiv 0$), the earliest time computed for the project terminal event ($E_t$) gives the expected project duration.

The second set of calculations, called the *backward pass calculations*, are carried out to determine the latest (allowable) occurrence times for each event $i$ ($L_i$), and the latest (allowable) start and finish times for each activity $i$-$j$ ($LS_{ij}$ and $LF_{ij}$). These calculations begin with the project end event by equating its latest allowable occurrence time to the scheduled project duration, if one is specified ($L_t \equiv T_s$), or by arbitrarily equating it to $E_t(L_t \equiv E_t)$ if no duration is specified. This is referred to as the "zero-slack" convention. These calculations then proceed by working backwards through the network, always assuming that each activity is conducted as *late* as possible.

To facilitate these hand calculations, *all times will be assumed to be "end-of" times*. Thus, the initial project activities, that is, those without predecessors, that follow the *initial* network event, will have an early start time of zero, which corresponds to the scheduled calendar date for the project start. A start time of zero means at the *end of* day zero, which is the same as the *start of* day one. On the other hand, computer runs give outputs transformed to calendar dates; but they also follow a different convention.

**Computer Date Convention:** Activity start times denote the *beginning of* the day (or other time period) corresponding to the given activity start date, while finish times denote the *end of* the day (or other time period) corresponding to the given finish date of the activity.

Again, to simplify the hand calculations, all times will be assumed to be *end-of* times. Hence for an activity start time of $t$, it means at the *end of* working day $t$, which is the same as the beginning of working day $t + 1$.

### Role of Hand Computation Procedure

The misuse of computers is not uncommon in the application of critical path methods. This occurs notably in making the above scheduling computations during the initial development of an acceptable project plan; an operation previously described as Steps 3 and 4 in Figure 15-2. At this stage it is important that the momentum of a project planning session must not be broken by the requirement for a computer run, and furthermore, it is more economical to perform these computations once by hand, regardless of the size of the network.

For this purpose a set of special networking symbols is useful to avoid making arithmetic errors. The key to these symbols is given in Figure 15-5, and their application is given in Figure 15-6, where the network employed is essentially the same as that used in Figure 15-1.

The start of the project at time zero is noted by setting $E_0 = 0$ in Figure 15-6. Then, equation 4 gives the early finish time for activity 0-1 as $EF_{01} = E_0 + D_{01} = 0 + 2 = 2$. Since event 1 has but one predecessor, activity 0-1, its early occurrence time is given by $E_1 = EF_{01} = 2$. The application of equation 1 occurs at all "merge" events (5, 7 and 8). For example, at event 5 the early event time $E_5 = 11$ is computed as follows:

$$E_5 = \max_{i=2,4} (E_2 + D_{25} = 6 + 1 = 7; E_4 + D_{45} = 7 + 4 = 11) = 11$$

Reading earliest expected and latest allowable activity start and finish times and float from the special symbols

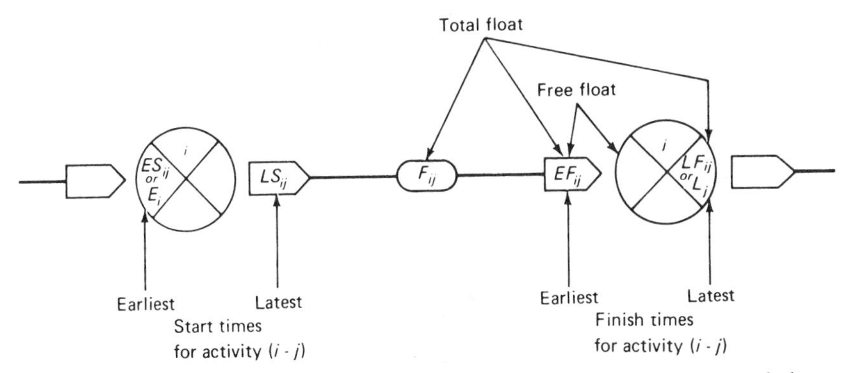


Figure 15-5a. Key to use and interpretation of special activity and event symbols.

Forward pass

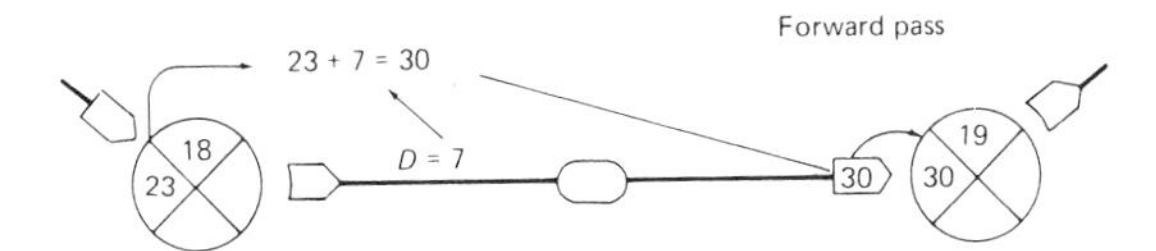


Begin with zero for the earliest start time for the initial project event and compute earliest finish times for all succeeding activities. For a typical activity, place its earliest start time (say, 23 days from project start) in the left quadrant of the event symbol. Then add its duration (7) to the earliest start time to obtain its earliest finish time (30). Write 30 in the arrow head.

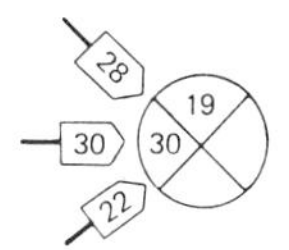


Where activities merge, insert in the left quadrant of the event symbol the largest of the earliest finish times written in the arrowheads of the merging activities.

Backward pass

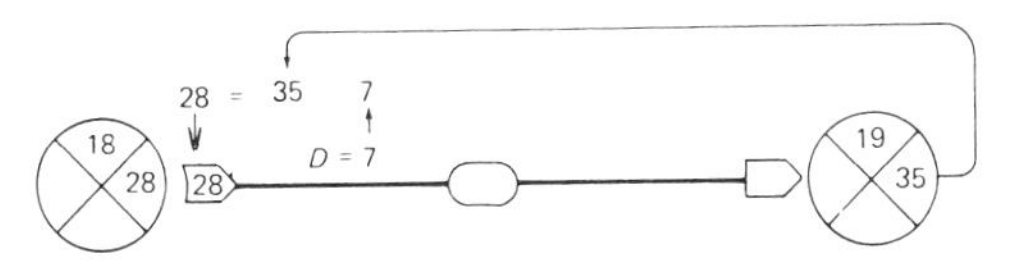


Place the scheduled completion time for the final event in the right quadrant of the project terminal event symbol. For other events, insert instead the latest allowable event occurrence time. For a typical activity, subtract its duration (7) from the latest completion time (35) to obtain the latest allowable activity start time (28). Write 28 in the arrow tail.

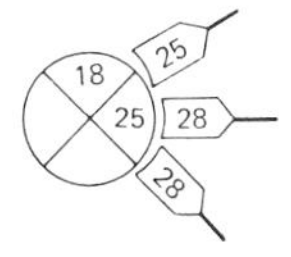


Where two or more activities "burst" from an event, insert in the right quadrant of the event symbol the smallest of the latest allowable activity start times.

Figure 15-5b. Steps in scheduling computations using special activity and event symbols. (From *Project Management with CPM, PERT and Precedence Diagramming*, 3rd Ed., J. J. Moder, C. R. Phillips and E. W. Davis, © 1983 by Litton Educational Publishing, Inc. Reprinted by permission of Van Nostrand Reinhold Co.)

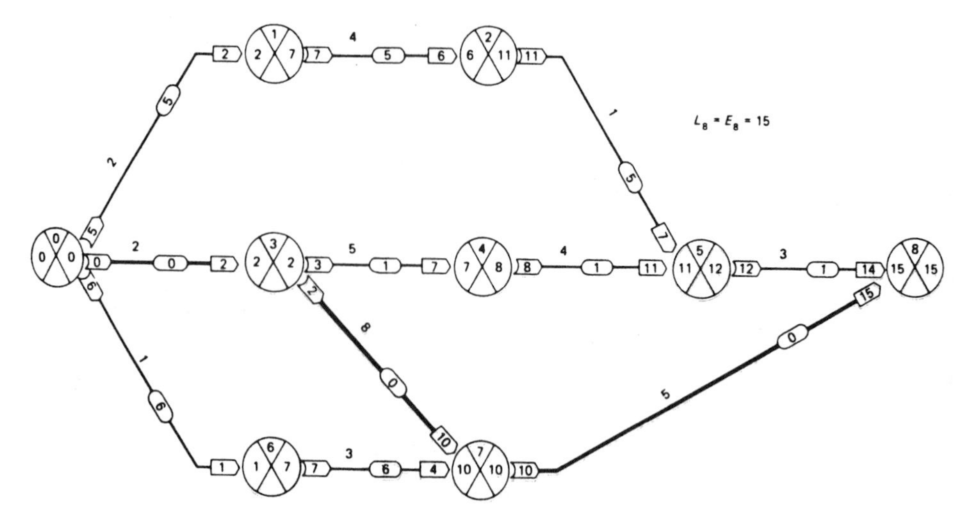


Figure 15-6. Illustrative network employing the special activity and event symbols showing completed computations. (From *Project Management with CPM, PERT, and Precedence Diagramming,* 3rd Ed., J. J. Moder, C. R. Phillips and E. W. Davis, © 1983 by Litton Educational Publishing, Inc. Reprinted by permission of Van Nostrand Reinhold Co.)

The backward pass is initiated by using the zero-slack convention, that is, letting $L_8 = E_8 = 15$. Working backwards from here, the latest start time for activity 5-8 is obtained from equation 6 as $LS_{58} = L_8 - D_{58} = 15 - 3 = 12$. Since event 5 has but one successor, activity 5-8, its latest occurrence time is given by $L_5 = LS_{58} = 12$. The application of equation 2 occurs at all "burst" events (0 and 3). For example, at event 3 the latest event time $L_3 = 2$ is computed as follows:

$$L_3 = \min_{j=4,7} (L_4 - D_{34} = 8 - 5 = 3,\ L_7 - D_{37} = 10 - 8 = 2) = 2$$

## The Critical Path and Slack Paths

Among the many types of slack defined in the literature, two are of most value and are discussed here: they are called total activity slack, or simply total slack, and activity-free slack, or simply free slack. They are also referred to as total float and free float, with the same definitions.

**Total Activity Slack.** *Definition:* Total activity slack is equal to the difference between the earliest and latest allowable start or finish times for the activity in question. Thus, for activity *i-j*, the total slack is given by

$$S_{ij} = LS_{ij} - ES_{ij} \quad \text{or} \quad LF_{ij} - EF_{ij}$$

**Activity-Free Slack.** Merge point activities, which are the last activities on slack paths, have what is called activity-free slack.

*Definition:* Activity-free slack is equal to the earliest occurrence time of the activity's successor event, minus the earliest finish time of the activity in question. Thus, for activity *i-j*, the free slack is given by

$$FS_{ij} = E_j - EF_{ij} \quad \text{or} \quad ES_{jk} - EF_{ij}$$

**Critical Path Identification.** *Definition:* The critical path is the path with the least total slack.

We will point out later that whenever scheduled times are permitted on intermediate network events, the critical path will not always be the longest path through the network. However, the above definition of the critical path always applies. For this reason, a computer run should be made before introducing intermediate scheduled times so that the longest path through the network can be determined by the computer.

If the "zero-slack" convention of letting $L_t = E_t$ for the terminal network event is followed, then the critical path will have zero slack. This situation is illustrated in Figure 15-6, where $L_8 = E_8 = 15$. However, if the latest allowable time for the terminal event is set by $T_s$, an arbitrary scheduled duration time for the completion of the project, then the slack on the critical path will be positive, zero, or negative, depending on whether $T_s > E_t$, $T_s = E_t$, or $T_s < E_t$, respectively. The last situation indicates, of course, that the completion of the project is expected to be late, that is, completion after the scheduled time, $T_s$. This is generally an unsatisfactory situation, and replanning (Steps 3 and 4 in Figure 15-2) would be required.

To carry out this replanning, it is quite helpful to be able to determine the critical path and its duration with a minimum of hand computation. This can be accomplished from the *forward pass computations alone*. Referring to Figure 15-6, start with the end event, 8, which must be on the critical path. Now trace backwards through the network along the path(s) with $EF_{ij} = E_j$. In this case we proceed to event 7 because $EF_{78} = E_8 = 15$, while $EF_{58} = 14 \neq E_8$. In like manner we proceed backwards to event 3 and then to the initial event zero. Thus the critical path is 0-3-7-8, with a duration of 15 time units, determined from the *forward pass computations alone*.

If the backward pass computations are also completed, then total slack and free slack can also be computed. For example, path 0-1-2-5 has a total slack of 5. This is the amount of time by which the actual completion time of this path can be delayed without causing the duration of the overall project to exceed its scheduled completion time. When the critical path has zero slack, as in this example, then the total slack is equal to the amount of time that the activity completion time can be delayed without affecting the earliest start time of any activity or the earliest occurrence time of any event *on the critical path*. For example, activity 0-1 has a total slack of 5 and a free slack of 0. If its completion time is delayed up to 5 time units, it will affect the early start times of the remaining activities on this slack path; however, it will not affect any event on the critical path (event 8 in this case). On the other hand, activity 2-5 has a total slack of 5 and a free slack of 4. Its completion can be delayed up to 5 time units without affecting the critical path (event 8), and it can be delayed up to 4 (free slack) without affecting *any* other event or activity in the network.

### Multiple Initial and Terminal Events, and Scheduled Dates

In most projects there will be any number of key events, called milestones, which denote the *start* or *finish* of an activity or group of activities. These milestones may be required to conform to arbitrary scheduled dates, and thus they may override the usual forward and backward calculations. These schedule constraints may be one of three different types:

- NET: *N*ot *E*arlier *T*han a specified date.
- NLT: *N*ot *L*ater *T*han a specified date (this is the usual type of schedule).
- ON: Exactly *ON* a specified date.

Conventions: A NET schedule affects the *early* time of an activity start or finish and hence it is considered in making the *forward* pass calculations. Similarly, a NLT schedule affects the *latest* time of an activity start or finish and hence it is considered in making the *backward* pass calculations. Finally an ON schedule affects both the *early* and *late* times and hence it is used in both the *forward* and *backward* pass calculations, and becomes both the early and late time for the activity in question.

These schedule times apply to all network events, including the initial and terminal network events. To illustrate the effect of a scheduled time for an intermediate network event, consider event 5 in Figure 15-6, with the schedule constraints shown in Table 15-1.

The four cases in Table 15-1 marked as no effect involve scheduled times that are less restrictive than the existing forward or backward pass calculations. For example, a NET = 10 on event 5 has no effect because activity 4-5 already has an early finish time of 11 ($EF_{45} = 11$) which is later or more constraining than NET = 10. Hence, $E_5 = 11$ remains unchanged by the imposition of this schedule constraint. This same result would also hold for any NET time less than 10. However, NET = 12 (or more) would

Table 15-1. Effects of Various Schedule Constraints on the Intermediate Event 5 in Figure 15-6.

| TYPE OF CONSTRAINT | SCHEDULE TIME | EFFECT ON FORWARD AND BACKWARD PASS CALCULATIONS |
|---|---|---|
| NET | 10 or less | None |
| NET | 12 (or more) | $E_5$ = 12 (or more) instead of 11 |
| NLT | 13 or more | None |
| NLT | 11 (or less) | $L_5$ = 11 (or less) instead of 12 |
| ON | 11 | None on forward pass |
| ON | 11 | $L_5$ = 11 (on backward pass) instead of 12 |
| ON | 12 | $E_5$ = 12 (on forward pass) instead of 11 |
| ON | 12 | None on backward pass |
| ON | 10 | $E_5$ = 10 (on forward pass) instead of 11 |
| ON | 10 | $L_5$ = 10 (on backward pass) instead of 12 |

affect the forward pass calculations because it is more constraining (later) than the early finish times of activities 2-5 and 4-5, that is, $EF_{25} = 7$ and $EF_{45} = 11$.

Table 15-1 indicates how these schedule constraints may or may not change the early/late times computed in the forward and backward pass calculations. If a NLT schedule of 10 was placed on event 5, then the latest time for this event would become $L_5 = 10$, and the critical path would become 0-3-4-5, since it would have the *least slack* of $-1$ time units ($LF_{45} - EF_{45} = 10 - 11 = -1$). Note also that this path does not go completely through the network. The longest path through the network would continue to be 0-3-7-8.

Another network complication is the occurrence of multiple initial and/or terminal events. For example, suppose there are several projects, each with their own networks, that are competing for a common set of resources. Since a number of algorithms require single initial and terminal events, a procedure is needed to combine these projects into one network with a single initial and terminal event. This can be accomplished by the use of dummy initial and terminal events to which each project connects with dummy activities. Duration times are assigned to the latter to impute the correct project start time relative to the early start time assigned to the initial dummy event, and the correct finish time relative to the late finish time assigned to the terminal dummy event.

## TIME-COST TRADE-OFF PROCEDURES

The determination of the critical path and its duration was described above. This constitutes Step 3 in Figure 15-2. Moving on to Step 4, if the earliest occurrence time for the network terminal event exceeds the scheduled project duration, then some modification of the network may be required to achieve an acceptable plan.

These modifications might take the form of a major change in the network structure. For example, changing the assumption that one set of concrete forms is available to the availability of two sets may result in a considerable change in the network and reduction in the project duration.

A different procedure that is frequently employed to handle this problem is referred to as time-cost trade-off. Referring to Figure 15-6, we might ask the question, how can we most economically reduce the duration of this project from its current level of 15 time units, say weeks, to 14 weeks? To accomplish this, the critical path, that is, 0-3, 3-7, 7-8, must be reduced by 1 week. The decision in this case would be to buy a week of time on that activity (or those activities) where it is available at the lowest additional (marginal) cost. If this turns out to be activity 3-7 or 7-8, then the resulting project will have two critical paths, each of 14 weeks duration, that is, 0-3, 3-4, 4-5, 5-8, and 0-3, 3-7, 7-8. Thus further reductions in this project duration will be more complicated because both paths must now be considered. One must also constantly consider buying back time previously bought on certain activities. This problem very rapidly reaches the point where a computer is required to obtain an optimal solution.

### The Critical Path Method (CPM)

The CPM procedure, developed by Kelley and Walker (6) to handle this problem, arises when we ask for the project schedule which minimizes *total project costs*. This is equivalent to the project activity schedule that just balances the marginal value of time saved (in completing the project a time unit early) against the marginal cost of saving it. The total project cost is made up of the indirect costs, determined by the accounting department considering normal overhead costs and the "value" of the time saved, plus the *minimum* direct project costs, determined as follows by the CPM procedure.

The CPM computational algorithm is based on an assumed linear cost versus time relationship for each activity. With this input, this problem can be formulated as a linear programming problem to minimize the total project *direct* costs, subject to constraints dictated by the activity time-cost curves, and the network logic.

Although this is an elegant algorithm, it is rarely applied today, primarily because of the unrealistic basic assumption of the unlimited availability of resources. Nevertheless, it is an important concept that is frequently applied in the simple manner illustrated at the beginning of this section.

The important consideration of limited resources is treated in the next section.

## SCHEDULING ACTIVITIES TO SATISFY TIME AND RESOURCE CONSTRAINTS

To illustrate how Figure 15-6 can be used to solve resource allocation problems, suppose that activities 1-2, 3-4, and 5-8 require the continuous use of a special piece of equipment during their performance. Can this requirement be met without causing a delay in the completion of this project?

With the aid of Figure 15-6, it is very easy to see that the answer to this question is yes, if the following schedule is used. The reasoning proceeds as follows. First, activities 1-2 and 3-4 must precede 5-8, so the first question is which of these two activities should be scheduled first. Reference to Figure 15-6 indicates that both have an early start time of 2, and since the floats are 5 and 1 for activities 1-2 and 3-4, respectively, the activity ordering of 1-2, 3-4, and 5-8 follows.

One can, of course, ask more involved questions dealing with the leveling of the demand for various personnel skills. From a computer standpoint, these questions are the most important ones involved in the use of critical path methods.

### A Heuristic Resource Scheduling Procedure

Resource allocation problems in general can be categorized as the determination of the scheduled times for project activities which do one of the following:

1. Level the resource requirements in time, subject to the constraint that the project duration will not be increased.
2. Minimize the project duration subject to constraints on the limited availabilities of resources.
3. Minimize the total cost of the resources and the penalties due to project delay—the long-range planning problem.

The combinatorial nature of this problem has prevented it from yielding to the optimal solution techniques of mathematical programming. Because of this lack of success with optimization procedures, major attention has been devoted to developing heuristic procedures which produce "good" feasible solutions. The procedures described below are an example of such heuristic procedures known as the "least slack first rule," or its equivalent name, the "minimum late start time rule." Collectively, they are essentially schemes for assigning priorities to the activities that are used in making the activity sequencing decisions required for the resolution of resource conflicts.

### Serial Versus Parallel Scheduling Procedures

A popular scheduling procedure to solve the first and second problems defined above consists of scheduling activities one day at a time, working from the first to the last day of the project. Each day, the activities that are ready to start (all predecessors complete) are ordered in a list with least slack first. Then, working through this list, as many activities as possible are scheduled (resource availability permitting). At the end of each day, the resources available are updated, as well as the early start and finish times of all delayed activities. This process is then repeated until the entire project has been scheduled. This procedure is known as *parallel scheduling*.

Another approach, known as *serial scheduling,* ranks all activities in the project only once at the start, using some heuristic such as *least slack first*. The activities are then scheduled one at a time (serially), as soon as all of their predecessors are *scheduled* (not necessarily completed), and they are considered for scheduling in the order in which they appear in the list. This procedure tends to schedule activities serially along network paths, whereas the approach described above tends to schedule activities in parallel along different paths.

An example of these two methods of scheduling is shown in Figures 15-7 and 15-8; it utilizes the network previously shown in Figure 15-6. Two resources are considered, A and B; however, the method could handle any number of resources. The schedule shown in Figure 15-7 is based on unlimited resources and shows all activities at their early start/finish times. Note that a maximum of 14 units of resource A and 8 units of resource B are required. Now consider the limited resource case where only 9 units of A and 6 units of B are available. If we apply the serial and parallel procedures described above, the order in which the activities are actually scheduled, along with their scheduled start/finish times are shown in Table 15-2, and the corresponding bar chart in Figure 15-8.

In this example, the two methods resulted in the same activity schedules as shown in Table 15-2. This will not, of course, always be the case. However, the order in which the activities were scheduled is quite different for these two methods. The activities are grouped in Table 15-2 by paths. The first three activities (0-3, 3-7, and 7-8) have zero slack and form

| Activity | Resource Req. A | Resource Req. B | D | ES | S | LS | Time 1 | 2 | 3 | 4 | 5 | 6 | 7 | 8 | 9 | 10 | 11 | 12 | 13 | 14 | 15 |
|---|---|---|---|---|---|---|---|---|---|---|---|---|---|---|---|---|---|---|---|---|---|
| 0-1 | 3 | — | 2 | 1 | 5 | 6 | x 3A | x 3A | | | | | | | | | | | | | |
| 1-2 | — | 2 | 4 | 3 | 5 | 8 | | | x 2B | x 2B | x 2B | x 2B | | | | | | | | | |
| 0-3 | 6 | — | 2 | 1 | 0 | 1 | x 6A | x 6A | | | | | | | | | | | | | |
| 3-4 | — | 2 | 5 | 3 | 1 | 4 | | | x 2B | x 2B | x 2B | x 2B | x 2B | | | | | | | | |
| 2-5 | 4 | — | 1 | 7 | 5 | 12 | | | | | | | x 4A | | | | | | | | |
| 4-5 | 2 | — | 4 | 8 | 1 | 9 | | | | | | | | x 2A | x 2A | x 2A | x 2A | | | | |
| 0-6 | 3 | — | 1 | 1 | 6 | 7 | x 3A | | | | | | | | | | | | | | |
| 3-7 | 4 | 4 | 8 | 3 | 0 | 3 | | | 4A 4B | 4A 4B | 4A 4B | 4A 4B | 4A 4B | 4A 4B | 4A 4B | 4A 4B | | | | | |
| 6-7 | 5 | — | 3 | 2 | 6 | 8 | | x 5A | x 5A | x 5A | | | | | | | | | | | |
| 5-8 | — | 5 | 3 | 12 | 1 | 13 | | | | | | | | | | | | x 5B | x 5B | x 5B | |
| 7-8 | 2 | — | 5 | 11 | 0 | 11 | | | | | | | | | | | x 2A | x 2A | x 2A | x 2A | x 2A |
| Level of resource A assigned | | | | | | | 12 | 14 | 9 | 9 | 4 | 4 | 8 | 6 | 6 | 6 | 4 | 2 | 2 | 2 | 2 |
| Level of resource B assigned | | | | | | | | | 8 | 8 | 8 | 8 | 6 | 4 | 4 | 4 | | 5 | 5 | 5 | |

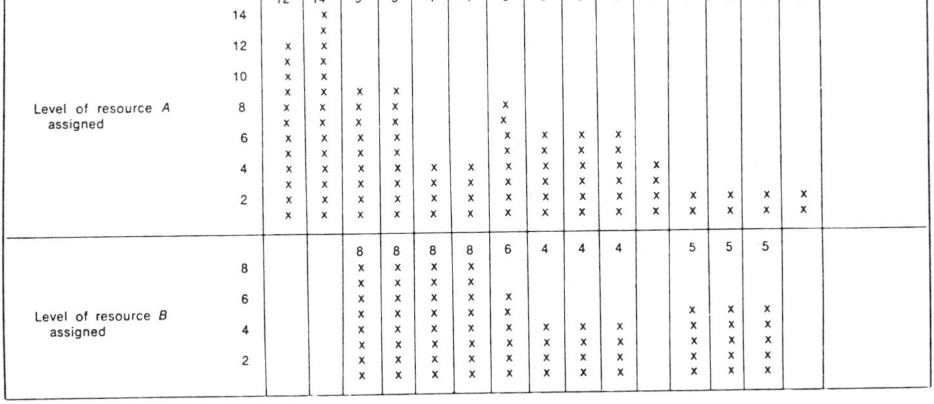

Figure 15-7. Resource loading with all activities scheduled at their early start times. (From *Project Management with CPM and PERT,* 2nd Ed., J. J. Moder and C. R. Phillips, © 1970 by Litton Educational Publishing, Inc. Reprinted by permission of Van Nostrand Reinhold Co.)

the critical path. Note how the *serial* method tends to schedule the activities sequentially along each path, whereas the *parallel* method moves back and forth among all four of the paths in this network.

## EVALUATION OF SEVERAL SCHEDULING HEURISTICS

Two categories of heuristics that have been found most effective are those incorporating some measure of time, such as activity slack or duration, and those incorporating some measure of resource usage. Davis (4) has

| Activity | Resource Req. A | Resource Req. B | D | ES | S | LS | Time 1 | 2 | 3 | 4 | 5 | 6 | 7 | 8 | 9 | 10 | 11 | 12 | 13 | 14 | 15 |
|---|---|---|---|---|---|---|---|---|---|---|---|---|---|---|---|---|---|---|---|---|---|
| 0-1 | 3 | — | 2 | 1 | 5 | 6 | 3A | 3A | | | | | | | | | | | | | |
| 1-2 | — | 2 | 4 | 3 | 5 | 8 | | | ~~2B~~ | ~~2B~~ | ~~2B~~ | ~~2B~~ | | 2B | 2B | 2B | 2B | | | | |
| 0-3 | 6 | — | 2 | 1 | 0 | 1 | 6A | 6A | | | | | | | | | | | | | |
| 3-4 | — | 2 | 5 | 3 | 1 | 4 | | | 2B | 2B | 2B | 2B | 2B | | | | | | | | |
| 2-5 | 4 | – | 1 | 7 | 5 | 12 | | | | | | | ~~4A~~ | | | | | 4A | | | |
| 4-5 | 2 | — | 4 | 8 | 1 | 9 | | | | | | | | 2A | 2A | 2A | 2A | | | | |
| 0-6 | 3 | — | 1 | 1 | 6 | 7 | ~~3A~~ | | 3A | | | | | | | | | | | | |
| 3-7 | 4 | 4 | 8 | 3 | 0 | 3 | | | 4A 4B | 4A 4B | 4A 4B | 4A 4B | 4A 4B | 4A 4B | 4A 4B | 4A 4B | | | | | |
| 6-7 | 5 | — | 3 | 2 | 6 | 8 | | ~~5A~~ | ~~5A~~ | 5A | 5A | 5A | | | | | | | | | |
| 5-8 | — | 5 | 3 | 12 | 1 | 13 | | | | | | | | | | | | ~~5B~~ | 5B | 5B | 5B |
| 7-8 | 2 | — | 5 | 11 | 0 | 11 | | | | | | | | | | | 2A | 2A | 2A | 2A | 2A |
| Level of resource A assigned (trigger level = 9) | | | | | | 8, 6, 4, 2 | ~~3~~ 9 | ~~3~~ 9 | ~~3~~ 7 | ~~4~~ 9 | ~~4~~ 9 | ~~4~~ 9 | ~~4~~ 8 | ~~2~~ 6 | ~~2~~ 6 | ~~2~~ 6 | ~~2~~ 4 | ~~4~~ 6 | 2 | 2 | 2 |
| Level of resource B assigned (trigger level = 6) | | | | | | 6, 4, 2 | | | ~~2~~ ~~4~~ ~~8~~ 6 | ~~2~~ ~~4~~ ~~8~~ 6 | ~~2~~ ~~4~~ ~~8~~ 6 | ~~2~~ ~~4~~ ~~8~~ 6 | ~~2~~ 6 | ~~4~~ 6 | ~~4~~ 6 | ~~4~~ 6 | 2 | | 5 | 5 | 5 |

Figure 15-8. Resource loading with limited resources (9 units of A and 6 units of B) using either serial or parallel scheduling. (From *Project Management with CPM and PERT*, 2nd Ed., J. J. Moder and C. R. Phillips, © 1970 by Litton Educational Publishing, Inc.

Table 15-2. Comparison of Serial Versus Parallel Scheduling for the Illustrative Network Shown in Figure 15-6.

| | | TYPE OF SCHEDULING | | | |
|---|---|---|---|---|---|
| | | PARALLEL | | SERIAL | |
| ACTIVITY | INITIAL FLOAT | ORDER OF SCHEDULING | SCHEDULE[a] | ORDER OF SCHEDULING | SCHEDULE[a] |
| 0-3 | 0 | 1 | 1-2 | 1 | 1-2 |
| 3-7 | 0 | 3 | 3-10 | 2 | 3-10 |
| 7-8 | 0 | 9 | 11-15 | 11 | 11-15 |
| 3-4 | 1 | 4 | 3-7 | 3 | 3-7 |
| 4-5 | 1 | 8 | 8-11 | 4 | 8-11 |
| 5-8 | 1 | 11 | 13-15 | 8 | 13-15 |
| 0-1 | 5 | 2 | 1-2 | 5 | 1-2 |
| 1-2 | 5 | 7 | 8-11 | 6 | 8-11 |
| 2-5 | 5 | 10 | 12 | 7 | 12 |
| 0-6 | 6 | 5 | 3 | 9 | 3 |
| 6-7 | 6 | 6 | 4-6 | 10 | 4-6 |

[a] A schedule of 3-7 for activity 3-4 means the activity is scheduled to be performed on the 3rd through the 7th days, including the 3rd day. Hence, the start time is at the *beginning of* while the finish time is at the *end of* the working day indicated.

made an extensive comparison of eight heuristics on some 83 network problems for which the optimal solutions were obtained using his bounded enumeration procedure. The rules tested included:

1. *Minimum Late Start Time* (LST)—order by increasing LST.
2. *Minimum Late Finish Time* (LFT)—order by increasing LFT.
3. *Resource Scheduling Method*—order by increasing $d_{ij}$, where $d_{ij}$ = increase in project duration resulting when activity $j$ follows $i$; = max[0; $(E_i - L_j)$], where $E_i$ and $L_j$ denote the early finish time of activity $i$ and the late start time of activity $j$, respectively. The above activity comparison is made on a pairwise basis among all activities in the eligible activity set.
4. *Shortest Imminent Operation*—order by increasing activity duration.
5. *Greatest Resource Demand*—order by decreasing total resource demand.
6. *Greatest Resource Utilization*—priority given to that combination of activities which results in maximum resource utilization in each scheduling interval; a rule which requires the use of zero–one integer programming to implement.

7. *Most Jobs Possible*—similar to Rule 6, except the number of active jobs is maximized.
8. *Select Jobs Randomly*—order the eligible activities by a random process.

The first four rules above were studied because they are very popular in the open literature on scheduling. The next three rules were included because they have been reported to be used in some of the many computer programs available for project scheduling on a commercial basis. The detailed workings of these programs have been kept secret. The last rule was included as a benchmark of human performance—presumably an experienced scheduler can outperform this rule.

The primary evaluation made in this study was based on the average percentage increase in project duration over the optimal schedule. On this basis the first three rules, having percentages of 5.6, 6.7, and 6.8, respectively, were considerably better than Rule 8, based on random selection, which had a percentage of 11.4. Also, Rules 5, 6, and 7, having percentages of 13.1, 13.1, and 16.0, respectively, gave poorer schedules than Rule 8.

While average performance is a reasonable guide in selecting scheduling rules, it should be pointed out that it is the nature of heuristics that no one rule will always give the best schedule. For this reason, one can argue that if the problem warrants a near optimal schedule, then several different heuristics should be applied. It also suggests that an important research area is to relate heuristic rule performance with simple parameters that describe the network and its resource constraints.

#### A Realistic Scheduling Procedure

Although the above heuristic scheduling procedure is oversimplified for most practical applications, it has three important properties. First, it can handle any number of resources. Second, it can handle any number of projects as long as their scheduled start and finish times are given. Finally, the procedure can be used as the basis for a more generally applicable scheduling procedure, such as that developed by Wiest (11). Some of its features are as follows:

1. Variable crew sizes are permissible.
2. Splitting or interrupting an activity is permissible.
3. Assignment of unused resources is incorporated.

The application of Wiest's procedure to solve Problem 2 cited in the section entitled A Heuristic Resource Scheduling Procedure is obvious. It can also be used to solve the long-range planning problem, 3 above, by evaluating the total cost of alternative levels of available resources and the penalties associated with delays in the completion of certain projects.

## PROBABILISTIC CONSIDERATIONS IN NETWORKING

There are two probabilistic aspects of critical path methods that are of some importance. The first involves those projects in which special milestone events occur, such as the end of test or evaluation activities. The special nature of these events is that they may have several *possible* successor activities, but only one will be selected and the others will be ignored. This situation is referred to as probabilistic branching. For example, in a space vehicle project, an evaluation activity may result in the choice of a solid or a liquid fuel engine, but not both. Also, as a result of a "failure" in some test, such projects may require recycling to an earlier network event, forming a closed loop. Neither of these situations is permissible according to the network logic assumed above.

The occurrence of these situations can be handled by drawing the network in general rather than specific terms. For example, the network plan for the above situation would be drawn up without reference to whether the engine was liquid or solid fuel. Also, the loop situation would be handled by omitting the loop and including its time effect in other network activities. Where more refined planning is required, a special simulation language called GERT (Graphical Evaluation and Review Technique) has been developed by Pritsker (10) which permits the above situations to be built into the network.

The second stochastic aspect of critical path methods deals with the fact that the actual duration of a project activity is usually a (hypothetical) random variable rather than a deterministic constant. Up to now, the effects of the variance in activity performance times on the procedures we have discussed have either been assumed to be negligible or have been neglected. The initial consideration of this problem led to the development of PERT, as cited in the opening section.

### The PERT Statistical Approach to Project Management

One of the chief concerns in the development of PERT was meeting the schedules placed on key milestone events, where considerable uncertainty in actual activity performance times existed. Because of this emphasis on events, which is a long-standing United States government practice in controlling projects by monitoring milestones, the activity labels were placed inside the event symbols. This convention, however, has no effect on the network logic described above, and thus represents a minor difference from the networking procedures described above. A major difference in procedures arises, however, from the efforts to estimate, from the project plan, the probability that the milestone schedules would be met.

The approach to this problem, which is frequently taken in developments of this type, was to collect input information on the basic elements of the system, and from it synthesize their effects on system performance. In this case the input information consisted of a measure of the uncertainty in activity duration times, and from this the uncertainty in meeting schedules was computed.

### PERT Three Time Estimates

In the PERT approach, the actual activity performance time, $t$, is assumed to have a hypothetical probability distribution with mean, $t_e$, and variance, $\sigma_t^2$. It is referred to as hypothetical because its parameters must be estimated before any actual observations are made. When the activity is finally completed, the actual time can be regarded as the first (and last) sample from this hypothetical distribution. Estimates of $t_e$ and $\sigma_t^2$ must therefore be based on someone's judgment, which in turn is based on a "sampling" of prior work experience.

The PERT activity input data is in the form of three time estimates, called $a$, $m$, and $b$. They denote the optimistic, most likely, and pessimistic estimates of $t$, respectively. Statistically, these are the zero percentile, the mode, and the 100 percentile of the hypothetical probability distribution.

A rule of thumb in statistics is that the standard deviation can be estimated roughly as 1/6 of the range of the distribution. This follows from the fact that at least 89% of any distribution lies within three standard deviations from the mean, and for the normal distribution this percentage is more than 99.7%. Thus the estimate of the variance is given by

$$\text{variance of } t \equiv \sigma_t^2 = [(b - a)/6]^2 \tag{8}$$

While the above formula is a part of the original PERT procedure, the author prefers to define $a$ and $b$ as the 5 and 95 percentiles, which in turn calls for replacing the divisor 6 in equation 8 by 3.2.

To derive an estimate of the mean requires an assumption about the shape of the probability distribution of $t$. In the development of PERT, it was assumed that a plausible (and mathematically convenient) distribution for $t$ is the Beta distribution whose standard deviation was 1/6 of its range. For this distribution, equation 9 gives a linear approximation to the true (cubic) relationship between the mean, $t_e$, and the mode, $m$:

$$\text{mean of } t \equiv t_e = (a + 4m + b)/6 \tag{9}$$

**PERT Probability Calculation**

At this point the scheduling computations described in the section entitled Basic Scheduling Computations can be carried out using only the mean values computed from equation 9 for each activity. The PERT procedure then considers the activities on the critical path(s) through the network, and ignores all others; a rather strong simplifying assumption. If there are several critical paths, then the one with the largest variance is chosen to represent the network. Assuming the actual activity performance times for these activities to be *independent random variables* with means, $t_{ei}$, and variances, $\sigma_{ti}^2$, the statistical properties of the "project" duration follow directly from the central limit theorem. Assuming the critical path consists of $N$ activities, and denoting the sum of their actual durations by $T$, this can be written as follows:

$$T = \sum_{i=1}^{N} t_i$$

$$\text{mean of } t \equiv T_e = \sum_{i=1}^{N} t_{ei} \tag{10}$$

$$\text{variance of } T \equiv \sigma_\text{T}^2 = \sum_{i=1}^{N} \sigma_{ti}^2 \tag{11}$$

shape of distribution of $T$: Normal

probability of meeting schedule $T_s$

$$= P\{T \leq T_s\} = P\left\{Z \leq \frac{T_s - T_e}{\sigma_T}\right\} \tag{12}$$

where $Z$ has a normal distribution with zero mean and unit variance, so that the last probability is read from the standard table of the cumulative normal distribution. By varying $T_s$ over a range of times of interest, one can obtain a graph giving the cumulative probability of meeting the project schedule for alternative scheduled completion times.

The basic assumption that the $t_i$'s above are independent random variables must be emphasized. Since a project manager will normally expedite a project when it falls behind schedule, the independence is violated. Hence the interpretation of the probability given by equation 12 is *the probability that the project will meet the schedule without having to be expedited.* This, of course, is very useful for planning purposes since it is computed at the outset of the project. If the calculated probability is low, say <0.75, then the project manager can anticipate the need to expedite the project and can exercise convenient or inexpensive options early in the project.

The simplifying assumption made above, that is, basing the probability computation on the critical path and ignoring all others, warrants further discussion. It is possible for a subcritical path, with a relatively high variance, to have a lower probability of meeting a schedule than the "longer" critical path. A more bothersome point is that the effect of this assumption at every network merge event is to introduce a negative bias in the estimated earliest expected time for the event. While these effects can assume practical significance, it is surprising to the author how accurate the PERT estimates are in most cases.

There are ways of estimating when the above assumption will cause a significant error. The most appropriate solution, where the refinement is called for, is to use simulation. The GERT language cited earlier is very easy to use for this purpose. The output of the simulation includes, among other things, the probability that each activity will be on the actual critical path through the network. This notion replaces the idea of a fixed critical path and slack on the remaining paths. An alternative practical solution is to use a method called PNET (12). It involves a relatively simple procedure of determining a *set* of (assumed) statistically independent paths that "represent" the project network. The probability of meeting a scheduled date is then approximated by the product of the separate probabilities that each path in the *set* will meet the schedule. This method works surprisingly well.

### Applications of PERT

PERT is much like the CPM time-cost trade-off algorithm in that it is seldom used. However, the reasons are different. It is the author's opinion that most project managers either have not learned to use PERT probabilities effectively, or they have no confidence in them. This is unfortunate because there are legitimate situations where PERT probabilities can be a useful tool, and there are also some basic advantages in the three time estimate system.

Several studies have shown that when the variance of $t$ is high, the mean activity duration time can be estimated more accurately using the three time estimate PERT procedure, than the one time estimate system which is now used quite widely. Also, a project manager's attention should be drawn to the high-variance activities as potential problem areas in the conduct of the project.

## NETWORK TIME AND COST CONTROL PROCEDURES

Having completed the presentation of planning and scheduling techniques, the attention now turns to project control as depicted by Step 6 in the dynamic project management procedure outlined in Figure 15-2. To periodically assess how well the plan is working, actual progress information regarding time and cost performance of activities is entered into the system, and the network is updated.

### Network Time Updating

Updating a network to reflect current status is similar to the problem introduced in the section entitled Multiple Initial and Terminal Events, and Scheduled Dates, in that a project under way is equivalent to a project with multiple start events. After a project has begun, varying portions of each path from the initial project event to the end event will have been completed. By establishing the status on each of these paths from progress information, the routine forward pass scheduling computations can then be made. No change in the backward pass computation procedure is necessary, since progress on a project does not affect the network terminal event(s), unless the scheduled completion date is revised.

Additional updating information is required if changes in the project plan are made which require revisions in the network or in the activity duration time estimates. Also, if an activity has not started, but its predecessor event has occurred since the last network update, then a scheduled start time or some "built-in" assumption about its start time must be entered into the system.

An update may indicate that the critical path has shifted, or more important, that the slack on the critical path has become negative. In this case, replanning will be in order to bring the project back onto schedule.

To illustrate this updating procedure, consider the network presented in Figure 15-9, which indicates an expected project duration of 15 days. Suppose we have just completed the fifth work day on this project, and the progress is as reported in Table 15-3.

The actual activity start and finish times given in Table 15-3 have been written above the arrow tails and heads, respectively, in Figure 15-9. Events that have already occurred have been cross-hatched, and activi-

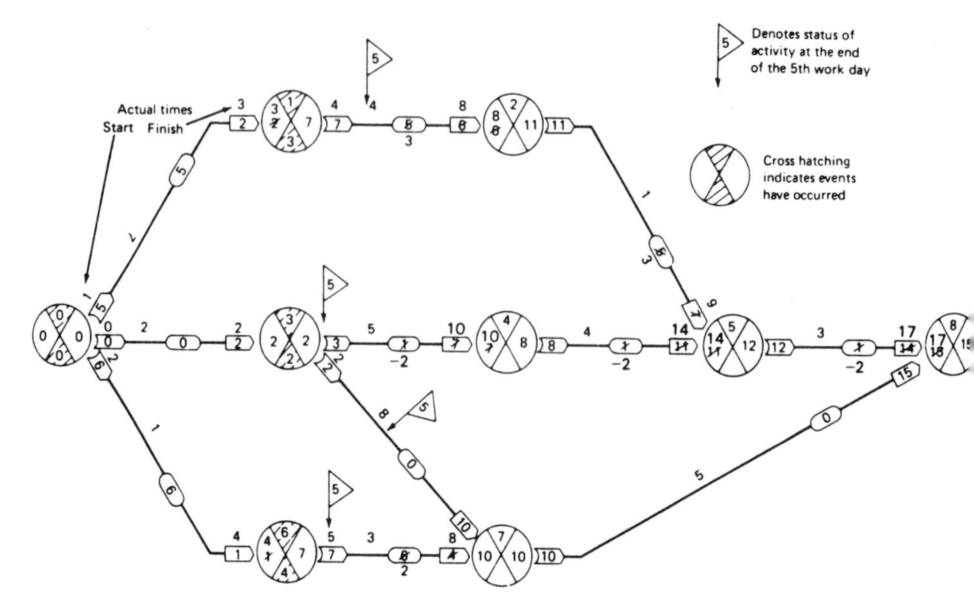


Figure 15-9. Illustrative network showing time status of project. (From *Project Management with CPM, PERT and Precedence Diagramming,* 3rd Ed., J. J. Moder, C. R. Phillips, and E. W. Davis, © 1983 by Litton Educational Publishing, Inc. Reprinted by permission of Van Nostrand Reinhold Co.)

* Moder, J. J., C. R. Phillips and E. W. Davis *Project Management with CPM, PERT and Precedence Diagramming,* 3rd Ed., (Van Nostrand Reinhold, New York, 1983), p. 92.

Table 15-3. Status of Project Activities at the End of the Fifth Working Day

| ACTIVITY | STARTED | FINISHED |
|---|---|---|
| 0–1 | 1 | 3 |
| 1–2 | 4 | — |
| 0–3 | 0 | 2 |
| 3–7 | 2 | — |
| 0–6 | 2 | 4 |
| 3–4 | 5 | — |
| 6–7 | 5 | — |

NOTE: all times given are at the *end* of the stated working day.

ties that are in progress have been so noted by a flag marked 5 to denote that the time of the update is the end of the fifth working day.

Having an actual, or assumed, start time for the "lead" activities on each path in the network, the forward pass calculations are then carried out in the usual manner. The original times are crossed out, with the new updated times written nearby. These calculations indicate that the critical path has shifted to activities 3-4-5-8, with a slack of minus two days. Assuming we were scheduled to complete the project in 15 days, the current status indicates we are now two days behind schedule.

Computerized network updating essentially follows the above procedure. However, with computerization, additional flexibility and convenience are offered. Most programs will allow any *one* or *two* of the following items as inputs for each activity in progress during the last reporting period.

Actual Activity Start Time.
Actual Activity Finish Time.
Duration Completed.
Percentage of Duration Completed.
Duration Remaining.

One other input is the "data date", that is, the date up to which the above progress is being reported. The computer will then compute appropriate values for the remaining three or four items listed above that are not given as inputs. Also, after all activity inputs are complete, it will compute updated early/late start/finish times for each activity in the project.

### Network Cost Control

Network cost control considers means of controlling the dollar expenditure as the project progresses in time and accomplishment. While network-based expenditure status reports may take many forms, they are primarily directed at the following basic questions.

1. What are the actual project costs to date?
2. How do the actual costs to date compare with planned costs to date?
3. What are the project accomplishments to date?
4. How do the actual costs of specific accomplishments compare with the planned costs of these same accomplishments?
5. By how much may the project be expected to overrun or underrun the total planned cost?
6. How do the above questions apply to various subdivisions and levels of interest within the project?

The major problem in the development of systems to answer these questions is the conflict between traditional functionally oriented accounting and a system based upon network activities. One solution to this problem is the use of groups of activities, called "work packages," in the coding of cost accounts. For example, in the construction industry a work package is often taken as a separate bid item. This, however, still does not solve all of the problems of allocating overhead and sharing various joint costs.

When used as an accounting base, network activities lend themselves to major increases in the amount of detail available to and required of the manager. This is both the promise and the inherent hazard of such systems, and it is one of the primary tasks of the system designer to achieve the level of detail that provides the greatest return on the investment in the system.

Network cost control employs an "enumerative cost model" in which activity costs are assumed to occur linearly in time. Thus, if the project budget is apportioned among the activities, cumulative cost versus time curves can be computed based on the earliest and latest allowable activity times. These two curves will bound the curve based upon the scheduled times for each activity. The latter is then taken as the plan or target against which progress is measured. Such a curve is shown as the middle curve in Figure 15-10, and is marked "Budgeted cost and work value." Using the nomenclature shown below, two important control variances can be defined.

$T_{Now}$ = Time of Update or Time Now
$T_S$ = Scheduled Project Completion Time
$T_F$ = Forecasted Project Completion Time
$ACWP$ = Actual Cost of Work in Place at $T_{Now}$
$BCWS$ = Budgeted Cost of Work Scheduled for Completion at $T_{Now}$
$BCWP$ = Budgeted Cost of Work in Place at $T_{Now}$

$$\text{Cost Variance at } T_{Now} = \left(\frac{BCWP - ACWP}{BCWP}\right) 100\% \tag{13}$$

$$\text{Schedule Variance at } T_{Now} = \left(\frac{BCWP - BCWS}{BCWP}\right) 100\% \tag{14}$$

The cost variance given in equation 13 is computed at each update time. It gives the total percent project cost over (under) run up to time $T_{Now}$, and is used to aid in forecasting the eventual total project cost.

The schedule variance given in equation 14 is used to compare planned

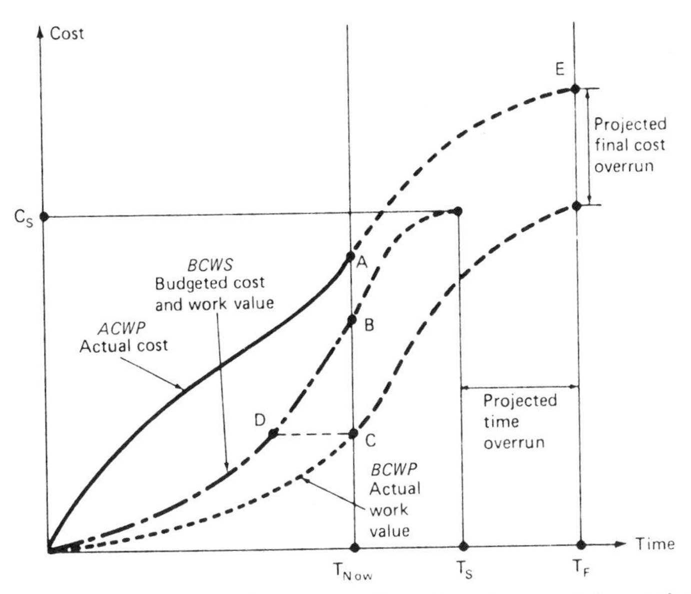


Figure 15-10. Project cost vs. time curve to illustrate cost and cash flow variances.

versus actual budgeted expenditure *rates* to aid the evaluation of *time* status. For example, a zero cost variance and a negative schedule variance would indicate project cost is currently on budget, but resources are not being applied to the project at the planned rate. This in turn may indicate that manpower limitations may not allow making this up in the future, and hence completion of the project may be delayed. The two variances in equations 13 and 14 are defined so that positive percentages are desirable (under budget and ahead of schedule) and negative percentages are undesirable (over budget and behind schedule).

To illustrate these concepts, consider the following oversimplified setting which involves only one activity rather than an entire network.

| | |
|---|---|
| Project Plan: | Construct foundations for 40 identical tract homes in one continuous period. |
| Schedule Plan: | Construct 10 foundations per week. |
| Cost Plan: | Cost (budget) per each home is $2000, for a total project cost of $80,000. |
| Progress Report: | At the end of one week 8 foundations have been completed at a total (actual) cost of $18,000. |
| Questions: | Where does this project stand with respect to cost and budget? If the current trend continues, what will be the total project cost and duration? |

From $T_{Now}$ = 1 week, and the Schedule and Cost plans, $BCWS$ is computed to be \$20,000.

$$\text{BCWS} = 10 \text{ foundations } \times \left(\frac{\$2000}{\text{foundation}}\right) = \$20{,}000.$$

From the Progress report we determine that $ACWP$ = \$18,000 and $BCWP$ = \$16,000.

$$\text{BCWP} = 8 \text{ foundations in place} \times \left(\frac{\$2000}{\text{foundation}}\right) = \$16{,}000.$$

Thus, the Cost and Schedule variances at $T_{Now}$ = 1 week are as follows:

$$\text{Cost Variance} = \left(\frac{\$16{,}000 - \$18{,}000}{\$16{,}000}\right) 100\% = -12.5\%$$

$$\text{Schedule Variance} = \left(\frac{\$16{,}000 - \$20{,}000}{\$16{,}000}\right) 100\% = -25\%$$

Thus, the project is 12.5% over budget and 25% behind schedule. If this trend continues throughout the project, the following will result:

Estimated Total Project Cost = \$80,000 × 1.125 = \$90,000.
Estimated Project Duration = 4 weeks × 1.25 = 5 weeks.

### Progress Control Signals

Another approach to the cost control problem is to use absolute dollar value variances as given by equations 15 and 16.

$$\text{Cost Variance} = BCWP - ACWP \tag{15}$$
$$\text{Schedule Variance} = BCWP - BCWS \tag{16}$$

Referring to Figure 15-10, the Cost Variance is the distance between points C and A, which in this case indicates a large negative difference or cost overrun. Similarly, the Schedule Variance is the distance between points C and B, which is also a large negative difference denoting behind schedule. The months behind could be approximated by the time difference for points C minus D.

Equations 15 and 16 are defined so that positive values are good (under budget and ahead of schedule) whereas negative values are undesirable (over budget and behind schedule). A powerful set of control signals, suggested by Brown (2), can be developed by combining this information with the project float on the critical path. The latter can be positive (ahead of schedule), or negative (behind schedule); it could also be expressed as a percentage by dividing the float by the expected project duration. Consider the following nomenclature for these signals.

C+ Positive Cost Variance (Under Budget)
C− Negative Cost Variance (Over Budget)
S+ Positive Schedule Variance (Ahead of Schedule)
S− Negative Schedule Variance (Behind Schedule)
F+ Positive Float on the critical path (Ahead of Schedule)
F− Negative Float on the critical path (Behind Schedule)

These three signals expressed in percentages (equations 13 and 14) or dollars (equations 15 and 16) are issued at each project update. They can be used to give very useful information about the status of the project. Recall that C± deals with cost management, whereas F± deals with schedule management of the critical path, and S± deals with schedule management of the entire project, not just the critical path activities.

For example, (C+, S+, F+) would indicate that project performance is good from all angles, whereas the opposite signal, (C−, S−, F−), might be explained by possible labor problems, budgets and schedules too tight, or just poor overall management. A mixed signal like (C+, S+, F−) might indicate a well-managed project (C+ and S+) that needs a recovery plan for the critical path (F−). Similarly (C+, S−, F+) might indicate a well-managed (C+ and F+) but understaffed project (S−).

Each of the major computer firms, plus a number of other corporations, have developed cost control computer packages of varying complexity. They may include the following:

1. Separate progress report outputs for three or more levels of indenture in the organization, for example, a program manager report, subproject manager reports, and finally task manager reports under each subproject.
2. Elaborate computer-printed graphical type outputs of resource requirements versus time; actual and planned expenditures versus time; bar chart type outputs showing activity early start, late finish, and scheduled times; etc.
3. Data base reports for cost estimating, labor standards, etc.
4. All too infrequently, a resource-leveling subroutine.

## OTHER NETWORKING SCHEMES

The activity-on-arrow networking logic presented above was the system utilized in the development of PERT and CPM, and is still used today. However, it is not the easiest for the novice to learn. For this reason, another scheme, called activity-on-node, has gained considerable popularity.

### The Activity-on-Node Networking Scheme

The activity-on-node system is merely the reversal of the other, that is, the nodes represent the activities and the arrows become the connectors to denote the precedence relationships. Neither of these networking schemes, however, can cope with the problem of rapidly escalating numbers of activities when two or more jobs follow each other with a lag. Since this situation occurs quite frequently, particularly in construction work, the networking scheme called precedence diagramming is gaining considerable attention.

### Precedence Diagramming

An extension to the original activity-on-node concept called precedence diagramming appeared around 1964 in the User's Manual for an IBM 1440 computer program (5). Extensive development of this procedure has since been conducted by K. C. Crandall (3). This procedure extends the PERT/CPM network logic from a single type of dependency to include three other types, illustrated in Figure 15-11. It is based on the following nomenclature.

$SS_{ij}$ denotes a start-to-start constraint, and is equal to the minimum number of time units that must be complete on the preceding activity ($i$) prior to the start of the successor ($j$).

$FF_{ij}$ denotes a finish-to-finish constraint, and is equal to the minimum number of time units that must remain to be completed on the successor ($j$) after the completion of the predecessor ($i$).

$FS_{ij}$ denotes a finish-to-start constraint, and is equal to the minimum number of time units that must transpire from the completion of the predecessor ($i$) prior to the start of the seccessor ($j$). (Note, this is the sole logic constraint used in PERT/CPM, with $FS_{ij} = 0$.)

$SF_{ij}$ denotes a start-to-finish constraint, and is equal to the minimum number of time units that must transpire from the start of the predecessor ($i$) to the completion of the successor ($j$).

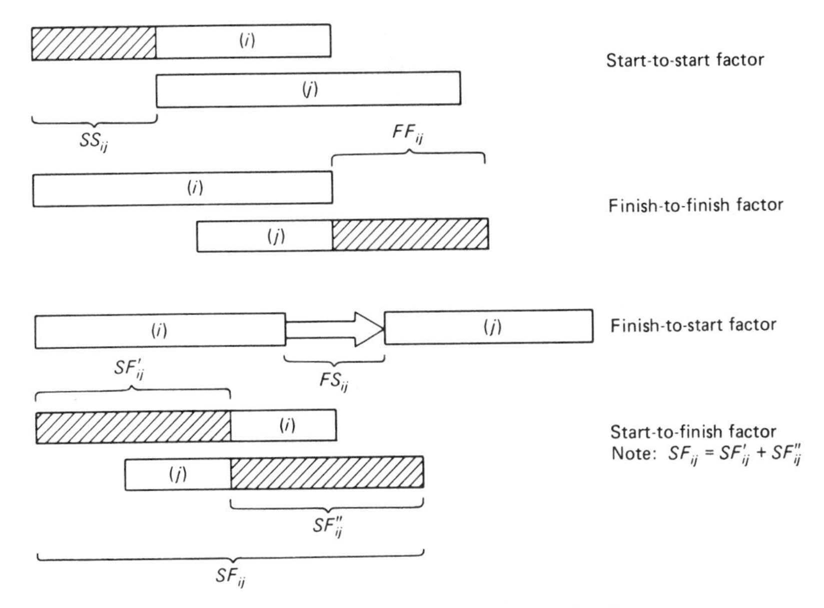


Figure 15-11. Precedence diagramming lead/lag factors.

The above constraint logic will be applied in the next section to illustrate the powerful features of precedence diagramming. It will also point up an anomaly that can occur that needs explanation. Before this is done, however, a simple example will be presented to illustrate the use of arrow versus node versus precedence diagrams.

## Comparison of Arrow, Node, and Precedence Diagrams

Consider a simple project consisting of digging and forming a foundation as shown in Figure 15-12. Each task is estimated to take three days, one day each for sections A, B, and C. Also, assume that for technical reasons, the three sections must be carried out in the order A, then B, and then C.

A strictly sequential arrow diagram plan is shown in part (a) of Figure 15-12; it would require six days to execute. To save time, suppose Forming A is started on the second day rather than the fourth day as assumed in part (a). This plan would require only four days, and is shown in part (b), again in the form of an arrow diagram. Notice that a dummy activity, 3-4, and a total of seven activities are required to diagram this project plan. An identical plan is shown in part (c) in the form of a node diagram. Note here that all arrows merely denote precedence relationships; they are all *FS*0

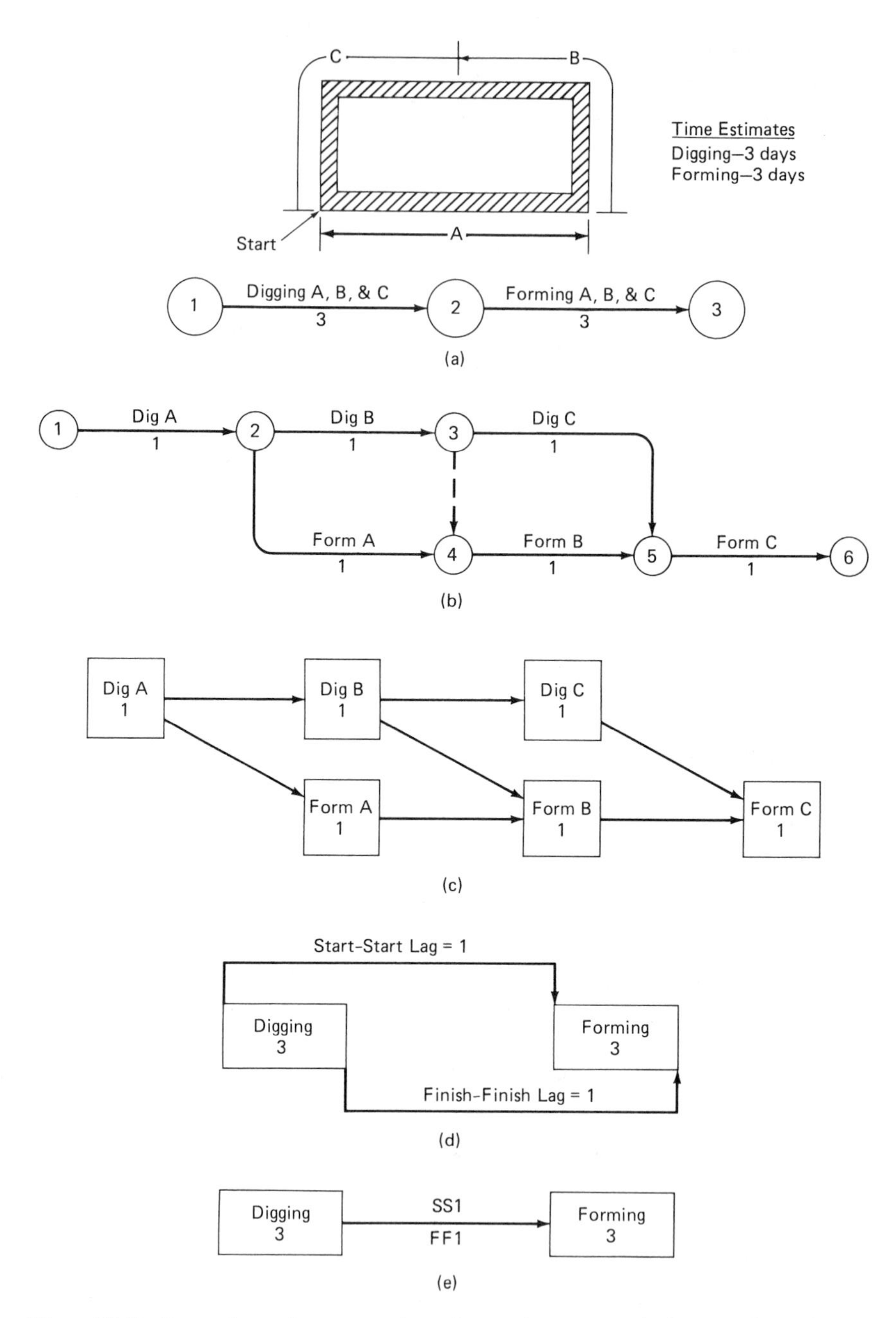


Figure 15-12. Comparison of arrow, node, and precedence network diagrams for a simple project.

constraints, the only type allowed in arrow or node diagrams. The node diagram is preferred by most users because it eliminates the need for the subtle dummy shown as activity 3-4 in part (b). Although the node diagram in part (c) is easier to draw than the arrow diagram, it still requires that Digging and Forming each be broken down into three parts. This proliferation of activities can be avoided by using precedence diagramming as shown in part (d), or the alternative form in part (e).

Because node diagrams eliminate the need for the subtle dummy, and are generally considered easier to draw, they have rapidly replaced the arrow diagram in most microcomputer project management software packages. Similarly, because precedence diagramming avoids the need to break an activity into several pieces when it is conducted concurrently with other activities as shown in Figure 15-12(b), (c), (d), and (e), this form of networking is being used in all of the better project management software packages. It should be noted that precedence diagramming restricted to *FS*0 constraints becomes a node diagram, and hence it can be used to describe a node or precedence diagram.

While the forward and backward pass computations for precedence diagrams are quite complex, it is essentially the same for arrow and node diagrams. This is illustrated in Figure 15-13, which is the node form of the arrow diagram previously presented in Figure 15-6. The early/late start/finish times are presented in a compact form in node diagrams, around the node symbol.

### Precedence Diagram Anomalies

Consider a construction subcontract consisting of *Framing* walls, placing *Electrical* conduits, and *Finishing* walls, with the duration of each task estimated to be 10 days, using standard size crews. If the plan is to perform each of these tasks sequentially, the equivalent arrow diagram in Figure 15-14(a) shows that a project duration of 30 days will result.

To reduce this time, these tasks could be carried out concurrently with a convenient lag of, say, two days between the start and finish of each activity. This plan is shown in Figure 15-14(b), in precedence diagram notation. The equivalent arrow diagram shown in Figure 15-14(c) indicates a 14-day project schedule. One important advantage of Figure 15-14(b) over 15-14(c) is that each trade is represented by a single activity instead of two or three subactivities. Also note how the $SS = 2$ and $FF = 2$ lags of Figure 15-14(b) are built into the equivalent arrow diagram in Figure 15-14(c). For example, the first two days of the Electrical task in Figure 15-14(c) must be separated from the remainder of this task to show that two days of Electrical work must be completed prior to the *start* of

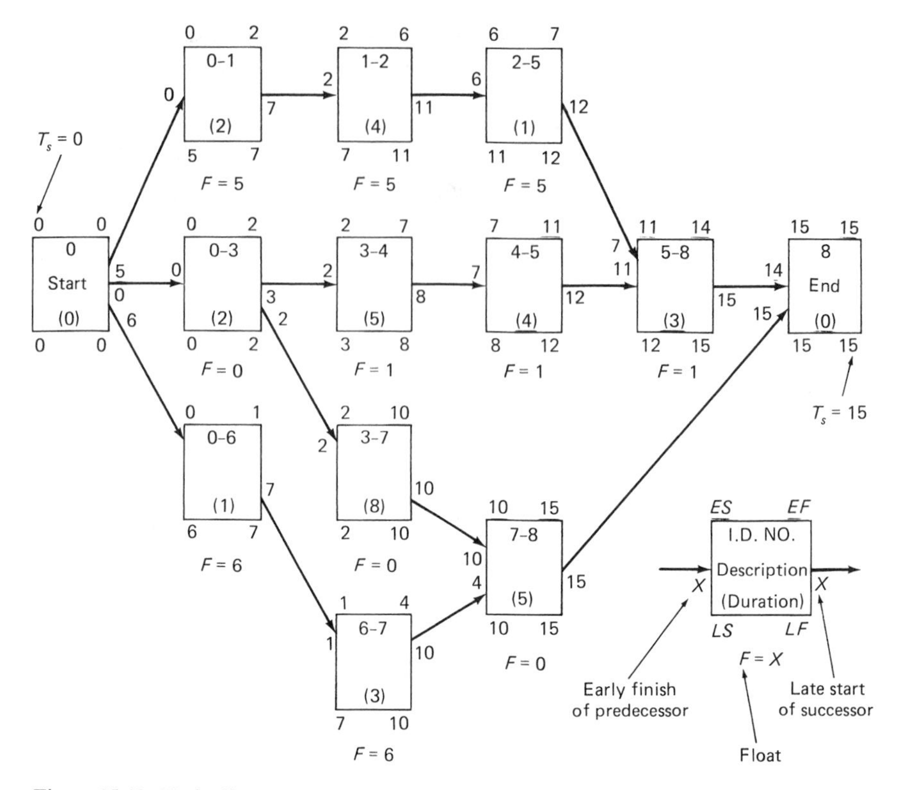


Figure 15-13. Node diagram equivalent to Figure 15-6, showing the forward and backward pass calculations and float.

the Finishing task. Similarly, the last two days of Electrical work must be separated from the remainder of this task to show that Framing must *finish* two days before Electrical is finished. Thus, the 10-day Electrical task must be broken up into three subactivities of two, six, and two days duration, respectively.

So far, precedence diagramming is easy to follow and is parsimonious with activities. But let us see what happens if the durations of the three tasks in this project are unbalanced by changing from 10, 10, 10, to 10, 5, and 15 days, respectively. These changes are incorporated in Figures 15-14(d) and 15-14(e), along with appropriate new lag times. Note that $SS = 2$ was chosen between Framing and Electrical to insure that a full day's work is ready for Electrical before this task is allowed to start. Similarly, $FF = 3$ was chosen between Electrical and Finishing because the last day of Electrical work will require three days of Finishing work to complete the project. The other lags of one day each were chosen as minimal or

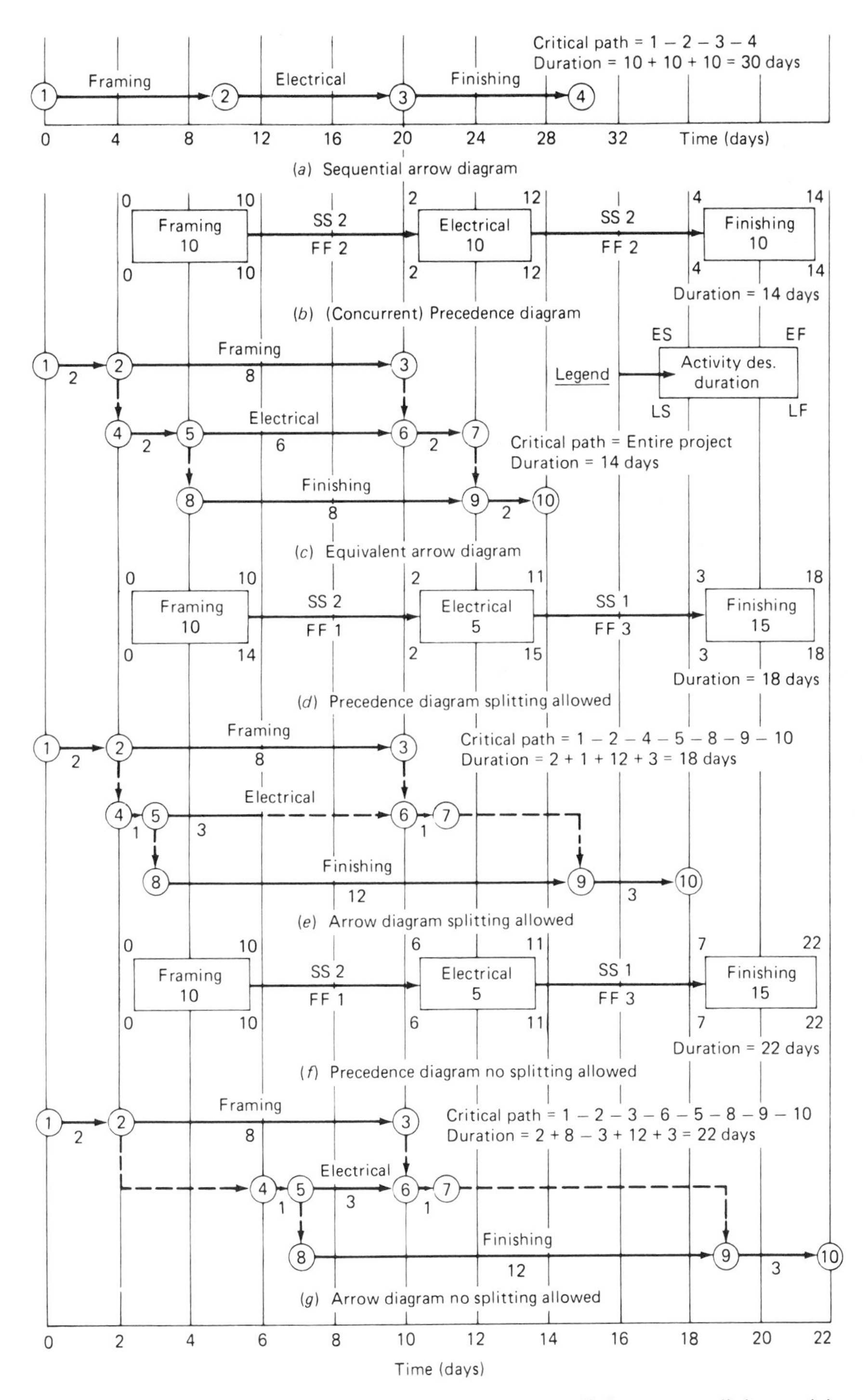


Figure 15-14. Arrow and precedence networks to illustrate splitting vs. no splitting; activities are shown at their early start times. (From *Project Management with CPM, PERT and Precedence Diagramming*, 3rd Ed., J. J. Moder, C. R. Phillips, and E. W. Davis, © 1983 by Litton Educational Publishing, Inc. Reprinted by permission of Van Nostrand Reinhold Co.)

convenience values needed in each case. These lags define the activity breakdown shown in Figure 15-14(e) where we see that the critical path is the *start* of Framing (1-2), then the *start* of Electrical (4-5), and finally the *totality* of Finishing (8-9-10). This is also shown in the precedence diagram, Figure 15-14(d), where $ES = LS = 0$ for the *start* of Framing, $ES = LS = 2$ for the *start* of Electrical, and finally $ES = LS = 3$ *and* $EF = LF = 18$ for the totality of Finishing. Since the precedence diagram shows each of these tasks in their totality, $EF \neq LF$ even though $ES = LS$ for the Framing and Electrical tasks. For Framing in Figure 15-14(d), $LF - EF = 14 - 10 = 4$ days of float, which corresponds to the 4 days of float depicted by activity 7-9 in Figure 15-14(e). Similarly, for Electrical in Figure 15-14(d), $LF - EF = 15 - 11 = 4$ days of float, which is also depicted by activity 7-9 in Figure 15-14(e). The middle Electrical activity (5-6) in Figure 15-14(e) *appears* to have an additional path float of four days, or a total of eight days. This attribute is not shown at all in Figure 15-14(d) because it depicts only the beginning and end points of each activity, but not intermediate subactivities such as 5-6. Closer examination will show, however, that any delay in the start of activity 5-6 exceeding three days would cause the Finishing crew to run out of work, and hence the critical path would be delayed. This problem is shared by both arrow and precedence diagrams, and the user should understand this. It does not, however, present a real problem in the applications since the job foreman generally has no difficulty in the day-to-day management of this type of interrelationship among concurrent activities. It is generally felt that it is not worthwhile to further complicate the networking and the computational scheme to show all interdependencies among activity segments, since these tasks can be routinely managed in the field.

A very important difference between Figures 15-14(c) and (e), other than the four-day difference in the project durations, lies in the Electrical task which is represented by three subactivities in both diagrams. In Figure 15-14(c) these three subactivities are expected to be conducted without interruption. However, in Figure 15-14(e) this is not possible. Here, the last day of the Electrical task (6-7) must follow a four-day interruption because of the combination effect of constraint $SS = 1$ depicted by activity 5-8, and constraint $FF = 1$ depicted by activity 3-6. This forced interruption will henceforth be referred to as *splitting* of the Electrical task.

If necessary, *splitting* can be avoided in several ways. First, the duration of the Electrical task could be increased from five to nine days. But this is frequently not desirable in projects such as maintenance or construction because it would decrease productivity. The second way to avoid *splitting* would be to delay the start of the Electrical task for four days, as shown in Figure 15-14(g), where it is assumed that activity splitting is not allowed. At first, it may seem that there is no difference between these two alternatives, but this is not so. Reflection on Figure 15-14(g) shows that delaying the start of the Electrical task to avoid *splitting* will delay the start of the Finish work, and hence the completion of the project is delayed by four days. But increasing the duration of the Electrical task will not have this effect. Actually, we have described an anomalous situation where an *increase* of four days in the duration of an activity on the critical path (starting four days earlier and thus running four days longer), will *decrease* the duration of the project by four days, from 22 to 18. If you are used to dealing with basic arrow diagram logic ($FS = 0$ logic only), this anomaly will take some getting used to. It results from the fact that the critical path in Figure 15-14(g) goes "backwards" through activity 5-6, and thus *subtracts* from the total duration of this path. As a result, the project duration *decreases* while the duration of an activity on the critical path is *increased*. This anomalous situation occurs whenever the critical path *enters* the *completion* of an activity through a *finish* type of constraint (*FF* or *SF*), goes backwards through the activity, and leaves through a *start* type of constraint (*SS* or *SF*).

The precedence diagram in Figure 15-14(f) shows that the entire project is critical, since $ES = LS$ and $EF = LF$ for each task. While it appears that the Electrical task has float in Figure 15-14(g), this is not true since *splitting* is not allowed. No-splitting is a constraint not explicitly incorporated in the arrow diagram logic.

### Critical Path Characteristics

Wiest (13) describes the anomalous behavior of activity 5-6 in Figure 15-14(g) picturesquely by stating that this activity is *reverse* critical. Similarly, in Figure 15-14(d) and (e) both Framing and Electrical are called *neutral* critical. They are critical because their $LS = ES$, but they are called *neutral* because their $LF > EF$, and the project duration is independent of the task duration. A task is *neutral* critical when a pair of start time constraints result in the critical path entering and exiting from the starting point of the task, or a pair of finish time constraints enter and exit from the finish point of a task. These situations could also be referred to as *start* or *finish* critical. In Figure 15-14(d) and (e), the Framing and Electrical tasks are both *start* critical, while Finishing is *normal* or *increase* critical. That is, a delay in the completion of the Finishing task will have a *normal* effect on the project duration, causing it to *increase*. Wiest

(13) suggests that precedence diagram computer outputs would be more useful if they identified the way in which tasks are critical. The author suggests that the following nomenclature be considered for this purpose:

IC— denotes an activity that is critical to an *In*crease in its duration.
DC— denotes an activity that is critical to a *De*crease in its duration.
BC— denotes an activity that is *Bi*critical, both to an *In*crease or *De*crease in its duration.
SC— denotes an activity that is critical to its *St*art Time.
FC— denotes an activity that is critical to its *Fi*nish Time.
MIC— denotes an activity whose *M*iddle portion is critical to an *In*crease in its duration.
MDC— denotes an activity whose *M*iddle portion is critical to a *De*crease in its duration.
MBC— denotes an activity whose *M*iddle portion is *Bi*critical to both an *In*crease or *De*crease in its duration.
NC— denotes an activity that is *N*oncritical.

To conclude this discussion, it should be noted that the critical path always starts with a job (or a job start), it ends with a job (or a job finish), and in between it consists of an alternating sequence of jobs and precedence arrows. Although the critical path may pass through a job in any one of the many ways listed above, it *always moves forward* through precedence constraint arrows. Hence, any *increase* in the lead-lag times associated with *SS, SF, FF,* or *FS* constraints on the critical path will always result in a corresponding *increase* in the project duration.

Following the suggestion of stating the nature of the criticality of activities on the critical path, for Figure 15-14(d) this would consist of the following alternating activities and precedence constraints: Framing (*S*tart *C*ritical—*SC*); *SS*2; Electrical (*S*tart *C*ritical—*SC*); *SS*1; Finishing (*I*ncrease *C*ritical—*IC*). Similarly, for Figure 15-14(f) it would be: Framing (*IC*); *FF*1; Electrical (*DC*); *SS*1; Finishing (*IC*). It should be noted here that Electrical is labeled decrease critical (*DC*), which puts the manager on notice that any *decrease* in the duration of this activity will increase the duration of the project. As stated above, it is *decrease* critical because its predecessor constraint is a *finish* type (*FF*1), and its successor constraint is a *start* type (*SS*1).

### Computational Procedures

Obviously the forward and backward pass computational problem becomes more complex with precedence diagramming, and it calls for establishment of somewhat arbitrary ground rules which were unnecessary with the unique nature of basic arrow diagram logic. In the computational procedures to follow, we will assume that the specified activity durations are fixed, for example, because of the productivity argument cited above. This assumption can be relaxed, of course, by varying the activity durations of interest, and repeating the calculations. Regarding task splitting, three basic cases will be treated.

Case 1: Activity splitting *is not* allowed on any activities, such as shown in Figure 15-14(g).
Case 2: Activity splitting *is* allowed on all activities, such as shown in Figure 15-14(e).
Case 3: Combination of 1 and 2; activity splitting is permitted only on designated activities.

Figures 15-14(g) and (e) represent Cases 1 and 2, respectively. The effect of not allowing splitting (of the Electrical task) is a four-day increase in the project duration. Here, the choice must be made between the (extra) cost of splitting the Electrical task, and the cost of a four-day increase in project duration. Case 3 is provided to allow the project manager to take the possible time (project duration) advantage concomitant with splitting on those activities where it can be tolerated, and to avoid splitting on those activities where it cannot be accommodated.

The computational procedure for Case 1 is reasonably simple and will be described below. The procedure for Case 2 is considerably more complex; it is given in Reference 8. The computational procedure for Case 3 merely amounts to the application of the Case 1 *or* the Case 2 procedure to each activity in turn, depending on whether the activity is designated as one where splitting *is not* allowed, or *is* allowed, respectively.

### Computational Assumptions

The computational procedure for Case 1—No Splitting Allowed, is analogous to the arrow diagram procedure described above. In making the forward pass calculations, one must consider *all* constraints leading into the activity ($j$) in question, that is, the start time constraints ($SS_{ij}$ and $FS_{ij}$) *as well as* the finish time constraints ($SF_{ij}$ and $FF_{ij}$). For *each* constraint, the early start time for activity ($j$) is computed, and the maximum (latest) of these times then becomes the early start time ($ES_j$) for activity ($j$). Because some project activities may only have finish time constraints, it would be possible for the above procedure to lead to a negative $ES_j$ time, or a time earlier than the specified project start time. For example, referring to Figure 15-15, we see that activity $D$ has no *start* time constraint. If the duration of activity $D$ was 22 (instead of 12), then its early start time would be $EF - D = ES$, or $19 - 22 = -3$ (instead of 7). This would be an erroneous negative value. To prevent the occurrence of this error, an additional time, called the INITIAL TIME, is introduced. It is usually set equal to zero, or else to an arbitrarily specified (nonzero) project scheduled start time, and it overrides the start times computed above if they are all negative, or less (earlier) than the specified project start time.

The backward pass computations follow a similar procedure to find the late finish times for each activity, working backwards along *each* constraint leaving the activity ($i$) in question. In this case, an additional time, called TERMINAL TIME, is required to prevent the occurrence of a late finish time ($LF_i$) *exceeding* the project duration, or the scheduled project completion time. As usual, the project duration is taken as the maximum (latest) of the early finish times computed for each activity in the forward

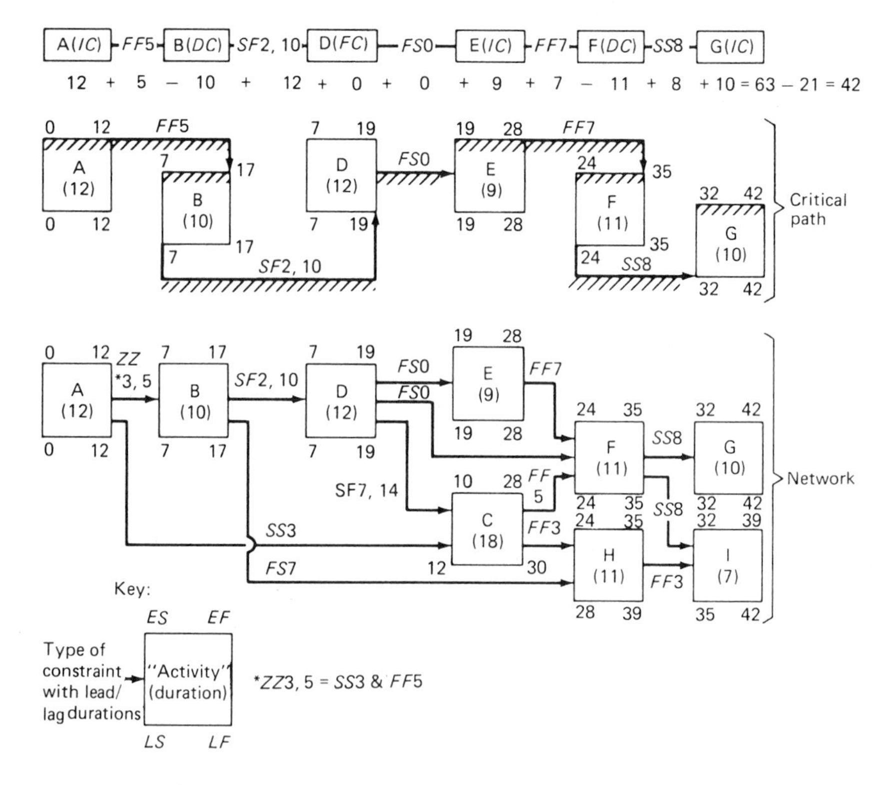


Figure 15-15. Example network with forward and backward pass times shown—no splitting allowed.

pass computations. For example, this is equal to 42 units in Figure 15-15, which is the largest of all activity early finish times.

The computational procedure given below has the same requirement that prevailed for arrow diagram computations. It requires that the activities be topologically ordered. That is, activities are arranged so that successors to any activity will *always* be found below it in the ordered list. The two-step computational procedure is then applied to each activity, working the list from the top down. When the computations are performed by hand on a network, this ordering is accomplished automatically by working one path after another, each time going as far as possible. Again, this is the same procedure required to process an arrow diagram.

### Forward Pass Computations—No Splitting Allowed

The following two steps are applied to each project activity, in topological sequence. The term called INITIAL TIME is set equal to zero, or to an arbitrarily specified project scheduled start time.

STEP 1: Compute $ES_j$, the early start time of the activity ($j$) in question. It is the maximum (latest) of the set of start times which includes the INITIAL TIME, and one start time computed from *each* constraint going to the activity ($j$) from predecessor activities indexed by ($i$).

$$ES_j = \underset{\text{all } i}{\text{MAX}} \begin{Bmatrix} \text{INITIAL TIME} \\ EF_i + FS_{ij} \\ ES_i + SS_{ij} \\ EF_i + FF_{ij} - D_j \\ ES_i + SF_{ij} - D_j \end{Bmatrix}$$

STEP 2: $EF_j = ES_j + D_j$

### Backward Pass Computations—No Splitting Allowed

The following two steps are applied to each project activity in the reverse order of the forward pass computations. The term called TERMINAL TIME is set equal to the project duration, or to an arbitrarily specified project scheduled completion time.

STEP 1: Compute $LF_i$, the late finish time of the activity ($i$) in question. It is the minimum (earliest) of the set of finish times which includes the TERMINAL TIME, and one finish time computed from *each* constraint going from activity ($i$), to successor activities indexed by ($j$).

$$LF_i = \underset{\text{all } j}{\text{MIN}} \begin{Bmatrix} \text{TERMINAL TIME} \\ LS_j + FS_{ij} \\ LF_j + FF_{ij} \\ LS_j + SS_{ij} + D_i \\ LF_j + SF_{ij} + D_i \end{Bmatrix}$$

STEP 2: $LS_i = LF_i - D_i$

**Example Problem**

To illustrate the application of the above algorithm, a small network consisting of nine activities with a variety of constraints, is shown in Figure 15-15. The forward pass calculations are as follows, based on the assumption that the project starts at time zero, that is, INITIAL TIME = 0.

*Activity A*

$$ES_A = \{\text{INITIAL TIME} = 0\} = 0$$
$$EF_A = ES_A + D_A = 0 + 12 = 12$$

*Activity B*

$$ES_B = \underset{A}{\text{MAX}} \left\{ \begin{matrix} \text{INITIAL TIME} = 0 \\ ES_A + SS_{AB} = 0 + 3 = 3 \\ EF_A + FF_{AB} - D_B = 12 + 5 - 10 = 7 \end{matrix} \right\} = 7$$
$$EF_B = ES_B + D_B = 7 + 10 = 17$$

*Activity D*

$$ES_D = \underset{B}{\text{MAX}} \left\{ \begin{matrix} \text{INITIAL TIME} = 0 \\ ES_B + SF_{BD} - D_D = 7 + (2 + 10) - 12 = 7 \end{matrix} \right\} = 7$$
$$EF_D = ES_D + D_D = 7 + 12 = 19$$

*Activity C*

$$ES_C = \underset{A,D}{\text{MAX}} \left\{ \begin{matrix} \text{INITIAL TIME} = 0 \\ ES_A + SS_{AC} = 0 + 3 = 3 \\ ES_D + SF_{DC} - D_C = 7 + (7 + 14) - 18 = 10 \end{matrix} \right\} = 0$$
$$EF_C = ES_C + D_C = 10 + 18 = 28$$

ETC.

The backward pass calculations are as follows, wherein the TERMINAL TIME is set equal to the project duration, determined from the forward pass calculations to be 42, that is, the *EF* time for the last critical path activity, G.

*Activity G*
$LF_G = \{\text{TERMINAL TIME} = 42\} = 42$
$LS_G = LF_G - D_G = 42 - 10 = 32$

*Activity I*
$LF_I = \{\text{TERMINAL TIME} = 42\} = 42$
$LS_I = LF_I = 42 - 7 = 35$

*Activity H*

$$LF_H = \underset{I}{\text{MIN}} \begin{Bmatrix} \text{TERMINAL TIME} = 42 \\ LF_I - FF_{HI} = 42 - 3 = 39 \end{Bmatrix} = 39$$

$LS_H = LF_H - D_H = 39 - 11 = 28$

*Activity F*

$$LF_F = \underset{G,I}{\text{MIN}} \begin{Bmatrix} \text{TERMINAL TIME} = 42 \\ LS_G - SS_{FG} + D_F = 32 - 8 + 11 = 35 \\ LS_I - SS_{FI} + D_F = 35 - 8 + 11 = 38 \end{Bmatrix} = 35$$

$LS_F = LF_F - D_F = 35 - 11 = 24$
ETC.

From the computational results shown in Figure 15-15, the critical path consists of activities A − B − D − E − F − G. The nature of the criticality of each activity is indicated at the top of Figure 15-15, along with the critical constraints between each pair of activities. Activities A, E, and G are *increase* (normal) critical, activities B and F *decrease* critical (noted by the reverse direction cross hatching), and activity D is only *finish* time critical. The duration of the critical path, 42, is also noted, with the net contributions of the activity durations being $(12 - 10 + 0 + 9 - 11 + 10) = 10$ and the contributions of the constraints being $(5 + 12 + 0 + 7 + 8) = 32$, for a total of 42 time units. The early/late start/finish times for each activity have the conventional interpretations. For example, for the critical activity E, both the early and late start/finish times are 19 and 28; the activity has no slack. But for activity H, the early start/finish times are 24 and 35, while the late start/finish times are 28 and 39. In this case, the activity has four units of activity slack or free slack, because the completion of activity H can be delayed up to four units without affecting the slack on its successor activity I.

## CONCLUDING REMARKS

Critical path methods represent a modern tool to aid the project manager. But they are only models of the dynamic real world interplay of money, people, materials, and machines, directed in time to accomplish a stated goal. Starting with the simple logic of the deterministic arrow diagram, they can be embellished to capture the stochastic elements of the problem, the random duration of the activity times by PERT, and the random nature of the network by GERT. More recently, precedence diagramming has been added to this array of models to depict more closely how many projects are actually conducted, without the proliferation of project activities. Finally, the role of the computer looms large when sophisticated resource allocation or general management information systems development questions are asked; or, when large projects extending over a long period require frequent updating, possibly for several levels of management, to control both time and cost. Network techniques form the vehicle for the conduct of these important management developments.

There are literally hundreds of project management microcomputer software packages on the market today.* As mentioned above, PERT probabilities and CPM time-cost trade-off techniques are not often used today. For this reason, they have only been incorporated in a very few specialized packages. Except for these procedures, programs costing less than $1000 can be found to carry out all of the *basic* procedures outlined in this chapter, including the drawing of bar charts, project networks (not time-scaled), tracking costs and resources, and in some cases allowing the use of precedence diagramming. In addition to these capabilities, programs costing between $1000 and $2000 can be found to carry out *all* of the procedures (not just the basics) outlined in this chapter, including multiproject limited-resource scheduling or resource leveling, earned-value cost analysis, and the production of a wide array of reports that can be customized by the user. Programs costing more than $2000 only provide a few additional features such as plotting time-scaled networks, the flexibility of allowing any type of networking (arrow, node, or precedence), and almost unlimited network size. Microcomputer project management software packages are usually quite "friendly," so even the "small" user can afford to adopt them. Their use will go a long way to advancing the application of the planning and control procedures outlined in this chapter.

---

* Computer software for project management is discussed in Chapter 28.

## REFERENCES


1. Adamiecki, Karol. "Harmonygraph." *Przeglad Organizacji* (Polish Journal of Organizational Review) (1931).
2. Brown, John W. "Evaluation of Projects Using Critical Path Analysis and Earned Value in Combination." *Project Management Journal* (August, 1985), pp. 59–63.
3. Crandall, Keith C. "Project Planning with Precedence Lead/Lag Factors." *Project Mngt. Quarterly,* Vol. 6(3) (1973), pp. 18–27.
4. Davis, E. W. and Patterson, J. H. "A Comparison of Heuristic and Optimum Solutions in Resource Constrained Project Scheduling." *Manage. Sci.* (1974).
5. IBM. *Project Management System, Application Description Manual* (H20-0210) (IBM. 1968).
6. Kelley, J. F. "Critical Path Planning and Scheduling: Mathematical Basis." *Oper. Res.,* Vol. 9(3) (1961), pp. 296–320. Kelley, J. and Walker, M. "Critical-path Planning and Scheduling" in *Proceedings of the Eastern Joint Computer Conference,* 1959.
7. Lockyer, K. G. *An Introduction to Critical Path Analysis,* 3rd Ed. (Pitman. London, 1969), p. 3.
8. Malcolm, D. G., Roseboom, J. H., Clark, C. E. and Fazar, W. "Applications of a Technique for R and D Program Evaluation (PERT)." *Oper. Res.,* Vol. 7(5) (1959), pp. 646–669.
9. Moder, J. J., Phillips, C. R. and Davis, E. W. *Project Management with CPM, PERT and Precedence Diagramming,* 3rd Ed. (Van Nostrand Reinhold. New York, 1983).
10. Pritsker, A. B. and Burgess, R. R. *The GERT Simulation Programs.* Department of Industrial Engineering. Virginia Polytechnic Institute, 1970. Pritsker, A. B., et al. "GERT: Graphical Evaluation and Review Techniques, Part I. Fundamentals—Part II. Probabilistic and Industrial Engineering Applications." *J. Ind. Eng.,* Vols. 17(5) and 17(6) (1966).
11. Wiest, J. D. "A Heuristic Model for Scheduling Large Projects with Limited Resources." *Manage. Sci.,* Vol. 13(6) (February, 1967), pp. B359–B377.
12. Ang, A. H-S, Abdelnour, J. and Chaker, A. A. "Analysis of Activity Networks Under Uncertainty." *J. of the Eng. Mech. Div.* (Proc. of Am. Soc. Civil Eng.), Vol. 101(EM4) (August, 1975), pp. 373–387.
13. Wiest, Jerry D. "Precedence Diagramming Methods: Some Unusual Characteristics and Their Implications for Project Managers." *Journal of Operations Management,* Vol. 1(3) (February, 1981), pp. 121–130.

# 16. Linear Responsibility Charts in Project Management*

David I. Cleland†
William R. King‡

The organizational model which is commonly called the *organization chart* is much derided in the literature and in the day-to-day discussions among organizational participants. However, organizational charts can be of great help in both the planning and implementation phases of project management.

In this chapter we shall explore a systems-oriented version of the traditional chart. Initially, we shall do this in the context of a chart which will be helpful to managers in aligning the project organization, that is, the implementation function. We shall then present an adaptation of the concept of the systems-oriented chart which has proved to be useful in the planning phase of a project.

---

* Portions of this chapter have been paraphrased from *Systems Analysis and Project Management*, 3rd Ed., by David I. Cleland and William R. King (McGraw-Hill, New York, 1983 with the permission of the publisher).

† David I. Cleland is currently Professor of Engineering Management in the Industrial Engineering Department at the University of Pittsburgh. He is the author/editor of 15 books and has published many articles appearing in leading national and internationally distributed technological, business management, and educational periodicals. Dr. Cleland has had extensive experience in management consultation, lecturing, seminars, and research. He is the recipient of the "Distinguished Contribution to Project Management" award given by the Project Management Institute in 1983, and in May 1984, received the 1983 Institute of Industrial Engineers (IIE)-Joint Publishers Book-of-the-Year Award for the *Project Management Handbook* (with W. R. King). In 1987 Dr. Cleland was elected a Fellow of the Project Management Institute.

‡ William R. King is University Professor in the Katz Graduate School of Business at the University of Pittsburgh. He is the author of more than a dozen books and 150 technical papers that have appeared in the leading journals in the fields of management science, information systems, and strategic planning. Among his major honors are the McKinsey Award (jointly with D. I. Cleland) for the "outstanding contribution to management literature" represented by their book *Systems Analysis and Project Management*, the IIE award for the first edition of this book, and designation as a fellow of the Decision Sciences Institute. Further biographical information is available in *Who's Who in the World* and *Who's Who in America*.

## THE TRADITIONAL ORGANIZATIONAL CHART

The traditional organizational chart is of the pyramidal variety; it represents, or models, the organization as it is *supposed* to exist at a given point in time.

At best, such a chart is an oversimplification of the organization and its underlying concepts which may be used as an aid in grasping the concept of the organization. Management literature indicates various feelings about the value of the chart as an organization tool. For example, Cyert and March say: (1)

> Traditionally, organizations are described by organization charts. An organization chart specifies the authority or reportorial structure of the system. Although it is subject to frequent private jokes, considerable scorn on the part of sophisticated observers, and dubious championing by archaic organizational architects, the organization chart communicates some of the most important attributes of the system. It usually errs by not reflecting the nuances of relationships within the organization: it usually deals poorly with informal control and informal authority, usually underestimates the significance of personality variables in molding the actual system, and usually exaggerates the isomorphism between the authority system and the communication system. Nevertheless, the organization chart still provides a lot of information conveniently—partly because the organization usually has come to consider relationships in terms of the dimensions of the chart.

Jasinski is critical of the traditional, pyramidal organizational chart because it fails to display the nonvertical relations between the participants in the organization. He says: (2)

> Necessary as these horizontal and diagonal relations may be to the smooth functioning of the technology or work flow, they are seldom defined or charted formally. Nonetheless, wherever or whenever modern technology does operate effectively, these relations do exist, if only on a nonformal basis.

## LINEAR RESPONSIBILITY CHARTS (LRCs)

The linear responsibility chart (LRC) goes beyond the simple display of formal lines of communication, gradations of organizational level, departmentation, and line-staff relationships. In addition to the simple display, the LRC reveals the task-job position couplings that are of an advisory, informational, technical, and specialty nature.

The LRC has been called the "linear organization chart," the "linear chart," and the "functional chart." None of these names adequately describes the device. The LRC (or the table or grid, as Janger calls it) (3) shows who participates, and to what degree, when an activity is performed or a decision made. It shows the extent or type of authority exercised by each executive in performing an activity in which two or more executives have overlapping authority and responsibility. It clarifies the authority relationships that arise when executives share common work. The need for a device to clarify the authority relationships is evident from the relative unity of the traditional pyramidal chart, which (1) is merely a simple portrayal of overall functional and authority models and (2) must be combined with detailed position descriptions and organizational manuals to delineate authority relationships and work-performance duties.

The typical pyramidal organizational chart is not adequate as a tool of organizational analysis since it does not display systems interfaces. It is because of this inadequacy that a technology of position descriptions and organizational manuals has come into being. As organizations have grown larger and larger, personnel interrelationships have increased in complexity, and job descriptions and organizational manuals have grown more detailed. Typical organizational manuals and position descriptions have become so verbose that an organizational analysis can be lost in semantics. An article in *Business Week* reflected on the problem of adequate organizational tools in this manner: (4)

> The usual way to supplement it [the pyramid organization chart] is by recourse to a voluminous organizational manual prescribing the proper relationships and responsibilities. But the manuals—cumbersome and often outdated—rarely earn much attention.

Position descriptions do serve the purpose of describing a single position, but an executive is also concerned with how the people under his jurisdiction relate to one another. On many occasions, executives are confronted with the task of examining and explaining relationships. Project management, corporate staff organization, concepts of product planning, the development of a corporate plan—all these lead to highly complex working relationships. A dynamic organization is often—even continually—redefining large numbers of positions and establishing new responsibility and authority patterns.

### Structure and Philosophy of the LRC

Typically, the LRC shows these characteristics:

1. Core information from conventional organizational charts and associated manuals displayed in a matrix format. (5)
2. A series of position titles along the top of the table (columns).
3. A listing of responsibilities, authorities, activities, functions, and projects down the side of the chart (rows).
4. An array of symbols indicating degree or extent of authority and explaining the relationship between the columns and the lines.

Such an arrangement shows in one horizontal line all persons involved in a function and the extent and nature of their involvement. Furthermore, the one vertical line shows all functions that a person is responsible for and the nature of his responsibility. A vertical line represents an individual's job description; a horizontal line shows the breakout of a function or task by job position.

One potential value of such a chart is the analysis required to create it, that is, the necessary abstracting and cross-referencing from position descriptions and related documentation manuals. The LRC in Figure 16-1 illustrates the authority interrelationships of a series of positions composing a definable unit. This chart conveys the same message by extensive organizational manuals, position descriptions, memorandums of agreement, policy letters, etc. It shows at a glance not only the individuals' responsibilities for certain functions but, what may be even more valuable, the way a given position relates to other positions within the organization.

But why not use the more conventional procedure of position analysis and position description for this sort of thing? There are two primary advantages to this mode of presentation. First, position descriptions and position guides are better at laying down responsibilities and authority patterns than at *portraying relationships*. Second, this type of charting depicts the work of top management as an *integrated system* rather than as a series of individual positions. The chart makes it easy to compare the responsibilities of related executives; in the coordination of budgets, for example, six individuals share the responsibility, ranging from "must be consulted" to "may be consulted" and "must be notified." The filled-in chart provides a quick picture of all the positions involved in the performance of a particular function.

| | Board | President | Vice-president marketing-advertising | Vice-president engineering and R&D | Vice-president finance | Director of manufacturing | Secretary-treasurer | Vice-president foreign operations |
|---|---|---|---|---|---|---|---|---|
| Establish basic policies and objectives | 2 | 1 | 3 | 3 | 3 | 3 | 3 | 3 |
| Direct operations, control and planning functions | 2 | 1 | 4 | 4 | 4 | 4 | 4 | 4 |
| Fix relationships between central office and operating divisions | 2 | 1 | 3 | 3 | 3 | 3 | 3 | 3 |
| Control expansion – merger – acquisition plans | 2 | 1 | 3 | 3 | 3 | 3 | 3 | 3 |
| Administer merger – acquisition operations | | 2 | 1 | 3 | 3 | 3 | 3 | 3 |
| Establish marketing policies and procedures | | 2 | 1 | | 4 | | | 4 |
| Coordinate sales forecasts and projections | 5 | 2 | 1 | | 5 | | | 3 |
| Coordinate advertising plans | 5 | 5 | 1 | | | | | 4 |
| Coordinate engineering, research and development | 5 | 2 | 3 | 1 | 4 | 4 | | 4 |
| Coordinate new product programs | | 2 | 3 | 1 | 4 | 3 | | 4 |
| Administer research and development center | | 3 | | 1 | | | | |
| Establish accounting policies and procedures | | 2 | | | 1 | | | |
| Administer financing, borrowing, equity | 2 | 2 | | | 1 | | 3 | |
| Coordinate budgets | 5 | 2 | 3 | 4 | 1 | 3 | | |
| Administer legal and tax matters | | 2 | | | 1 | | | |
| Utilization of manufacturing facilities | | 3 | | | | 1 | | |
| Coordinate training and safety programs | | 2 | 1 | | | 1 | | |
| Coordinate and administer capital expenditures | 2 | 2 | 4 | 4 | 3 | 1 | 3 | |
| Administer insurance plans and stockholder relations | | 2 | | | 4 | | 1 | |
| Coordinate foreign and export operations | | 2 | 4 | 4 | | | | 1 |

Code
1 Actual responsibility
2 General supervision
3 Must be consulted
4 May be consulted
5 Must be notified

Figure 16-1. Authority interrelationships in a unit. (From Allen R. Janger, "Charting Authority Relationships," *The Conference Board Record* (December, 1964).

In the words of Allen R. Janger, concerning the chart of Figure 16-1: (6)

> The top line . . . shows that the *president* is responsible for establishing basic policies and objectives. He works under the general supervision of his *board of directors*, and with the consultation of his corporate staff. Responsibility for coordinating engineering, research and development . . . is parceled out in a bit more complicated fashion. The *vice president, engineering, research, and development* carries the actual responsibility. He operates under the general supervision of the *president*, but must carry on close consultations with the *vice president, marketing and advertising* and the *director of manufacturing*. Consultation with the *vice president, finance* on R&D matters is not mandatory but may be required. It is also understood that the *board of directors* will be informed of significant developments.
>
> By reading down the chart it is possible to summarize rapidly a position's salient responsibilities. As depicted, the *president* has actual responsibility for establishing basic policies and objectives, direction of operating control and planning functions, fixing relationships between the corporate headquarters and the product divisions, and control of expansion, merger, and acquisition plans. Other top management functions are the direct responsibilities of other corporate executives, although the president generally exercises supervision over them. He must at least be consulted on the administration of the R&D center and the utilization of manufacturing facilities. The *vice president, marketing and advertising* need only notify him about advertising plans.

Staff-line and project-functional relationships pose some of the more challenging problems of project management, particularly in light of the desirability of the deliberate conflict between the functional managers and the project manager. The deliberate conflict must be planned so that respective prerogatives are recognized and protected. The use of a chart similar to that shown in Figure 16-2 can do much to define and postulate the functional-project relationships, as well as the staff-line, staff-staff interfaces in the project environment.

The chart shown in Figure 16-2 is different from the chart for top management shown in Figure 16-1 in that the starting point is different. Figure 16-2 emphasizes the purchasing function and its subfunctions. The essence of the analysis is the determination of the sphere of each executive's authority in each of the key purchasing activities and of the extent of that authority. When these facts are ascertained, the relevant positions are listed at the top of the chart, and the appropriate symbols are added.

The chart shows the roles of the various executives in manufacturing-related purchasing activities.

Figure 16-1 does more than clarify authority relationships; it can double as a collection of position guides. Its perspective is adequate to permit it to be used as an organizational chart of the top management of the organization. Figure 16-2, on the other hand, cannot serve as an organizational chart since it is not possible to get an overall picture of a position or a unit and its responsibilities from the chart. Figure 16-2 shows only the purchasing activities related to manufacturing; the nonpurchasing activities of the positions, which could be significant, are not shown.

## Limitations of LRCs

Charts such as those shown in Figures 16-1 and 16-2 are not a panacea for all organizational difficulties. The LRC is a pictorial representation, and it is subject to the characteristic limitations and shortcomings of pyramidal organizational charts. The LRC does reveal the functional breakout of the work to be done and the interrelationships between the functions and job positions; however, *it does not show how people act and interact.*

It is doubtful that any contemporary management theorists would deny that organizational effectiveness is as dependent on the informal organization of human actions and relations as it is on the structured, formal organization. The LRC, as we have so far discussed it, is limited to showing the man-job relationships that constitute the formal organization; it does not purport to reveal the infinite number of variations in human relations arising out of the informal organization. The LRC technique simply extends the scope of charting formal organizations wherever they are located in the hierarchical order. Thus, a note of caution is in order about the LRC. But, as Karger and Murdick have implied, we still must give it a vote of confidence. (7)

> Obviously, the LRC chart has weaknesses, of which one of the larger ones is that it is a mechanical aid. Just because it says something is a fact does not make it true. It is very difficult to discover, except generally, exactly what occurs in a company—and with whom. The chart tries to express in specific terms relationships that cannot always be delineated so clearly; moreover, the degree to which it can be done depends on the specific situation. This is the difference between the formal and informal organizations mentioned. Despite this, the Linear Responsibility Chart is one of the best devices for organization analysis known to the authors.

| PURCHASING ACTIVITIES | Corporate | | | | | Division | | | Plant | | | | |
|---|---|---|---|---|---|---|---|---|---|---|---|---|---|
| | Director of purchasing | V.P. manufacturing | Controller | Manager, engineering | Manager, trade relations | General manager of division | Division purchasing | Division engineer | Manager, construction purchasing | Plant manager | Plant purchasing agent | Plant controller | Plant engineer |
| PURCHASING-OPERATIONS | | | | | | | | | | | | | |
| A. Raw materials (controlled commodities) | | | | | | | | | | | | | |
| 1. Development of annual plan for purchases of major raw materials. | ○ | ▲ | | | | △ | ● | | | | | | |
| 2. Purchase or requisition of raw materials according to annual plan. | | | | | | | ○ | | | △ | ● | | |
| 3. Revisions in annual plan as to supplier, quantity, and (or) price. | ○ | ▲ | | | | △ | ● | | | ○ | ○ | | |
| 4. Selection of appropriate suppliers. | ○ | △ | | | ○ | | ● | | | ○ | ○ | | |
| 5. Conducting of any contract or other negotiations with suppliers. | ○ | ▲ | | | ○ | △ | ● | | | | | | |
| B. Maintenance contracts | | | | | | | | | | | | | |
| 1. Under $10,000 and on standard contract form. | | | | | | | | | | △ | ● | | ● |
| 2. Over $10,000 or a nonstandard contract (also approved by Legal Department and Trade Relations Department). | ○ | ▲ | | | | | △ | | | ▲ | ● | ○ | ● |
| C. Surplus disposal | | | | | | | | | | | | | |
| 1. Request for disposition of fixed assets. | ▲ | ▲ | ▲ | ▲ | | ▲ | ○ | | | ▲ | | ▲ | ● |
| 2. Disposition of surplus construction materials, equipment and supplies. | ▲ | | ▲ | | | | ▲ | | ▲ | ▲ | ● | ○ | ○ |
| 3. Disposition of self-generated scrap (other than metal), supplies, and waste materials up to $100,000. | ▲ | | ▲ | | | | ▲ | | ▲ | ▲ | ● | ○ | ○ |
| D. General stores and supplies | | | | | | | | | | | | | |
| 1. Determination of minimum inventory requirements. | | | | | | | ○ | | | △ | ● | | |
| 2. Ordering of stores and supplies. | | | | | | | ○ | | | △ | ● | | |
| E. Rental agreements – equipment (under the authorization to execute contracts and purchasing policy). | ▲ | ▲ | | | | ▲ | ▲ | | | ▲ | ● | | |
| PURCHASING-CONSTRUCTION | | | | | | | | | | | | | |
| A. Buildings and equipment | | | | | | | | | | | | | |
| 1. Development of process designs and equipment specifications. | | ○ | | △ | | | | ▲ | ○ | ○ | | | ● |
| 2. Request of and appropriation of capital funds (RFI procedure in accordance with dollar authorizations). | | ▲ | ▲ | ▲ | | ▲ | | ▲ | | ▲ | ○ | ● | ● |
| 3. Determination of the bid list – items over $10,000. | ▲ | ▲ | | | ▲ | | ▲ | ▲ | ▲ | ▲ | ● | | |
| 4. Selection of the successful bidder – items over $10,000. | ▲ | ▲ | | | ▲ | | ▲ | ▲ | ▲ | ▲ | ● | | |
| 5. Follow up on the rate and amount of expenditures of the appropriated capital funds (capital expenditures report). | | ○ | | ▲ | | ○ | | ● | | ○ | | ● | ● |

KEY:

△ Authorizes and (or) actuates
▲ Approves
○ Recommends and (or) reviews and counsels
● Does the work (personally or within the department)

Figure 16-2. Functional authority relationships. (*From Allen R. Janger,* "Charting Authority Relationships," *The Conference Board Record* (December, 1964).

### The LRC In Input-Output Terms

The LRC can be visualized as an input-output device. For example, if the job positions of the managing subsystem are considered to be the inputs, task accomplishments the outputs, and matrix symbols the specific task-to-job relationships, then the overall LRC can be looked upon as a diagram of the managing subsystem from a systems viewpoint (Figure 16-3 diagrammatically illustrates this idea).

If two additional steps are added to this charting scheme, the systems viewpoint can be made more explicit. First, if systems terminology is used to structure the LRC matrix symbols and if the personnel affected are indoctrinated in the philosophy of an LRC, then many of the facets of the informal organization (8) can be formalized and assimilated along with the formal organization into the managing-subsystem structure. The second step is to use the *systems symbols* from one row (one task) of the LRC to draw a schematic diagram, as indicated by the symbols of that row; this would show the interrelationships or intercouplings between the persons involved in accomplishing a task (see Figure 16-4).

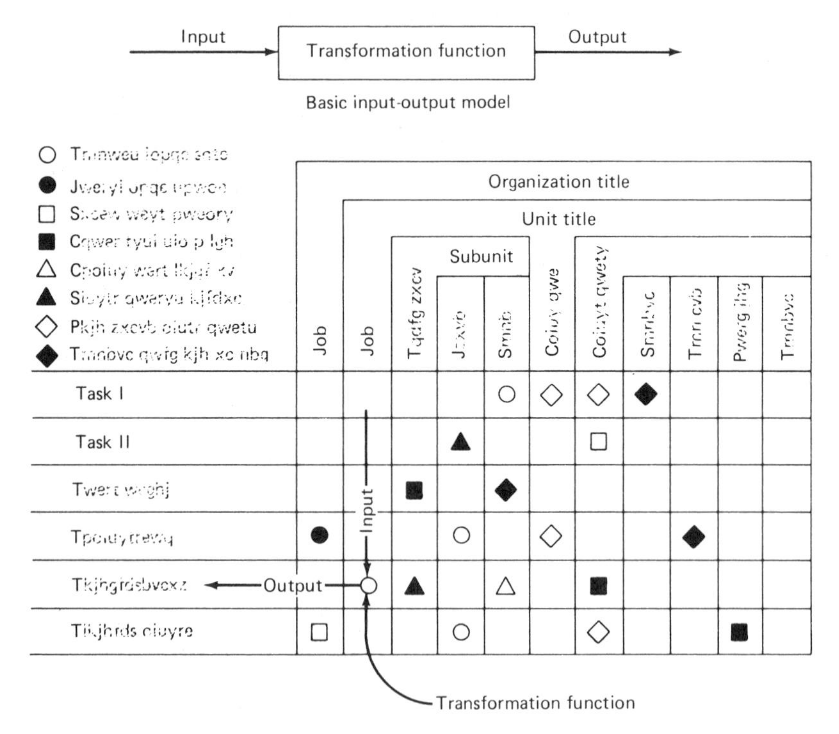


Figure 16-3. Input-output-device schematic.

| Task/Job Relationships Symbol titles | Equipment-test division | | | | | | | | | | | | | | | | | | | |
|---|---|---|---|---|---|---|---|---|---|---|---|---|---|---|---|---|---|---|---|---|
| ○ Work is done<br>● Direct supervision<br>□ General supervision<br>■ Intertask integration<br>△ Occasional intertask integration<br>▲ Intertask coordination<br>◇ Occasional intertask coordination<br>◆ Output notification mandatory | | Test management and plans branch | | | | | | Test operations branch | | | | | | | | | | | | |
| | | | | | | | | | Data A & A section | | | Elec. mach. testing section | | | | Electronic testing section | | | | |
| | Division manager | Branch manager | Test-activity integrator | Test-facility manager | Test-data planner | Test and documents coordinator | | Branch manager | Section supervisor | Instrumentation engineer | | Section supervisor | Test-equipment engineer | Test-article engineer | | Section supervisor | Electrical-systems engineer | R. F. systems engineer | Cmd and C+I systems engineer | Test-equipment engineer |
| Major functional area: test program activities | | | | | | | | | | | | | | | | | | | | |
| Approve test-program changes | ○ | ▲ | | | | | | ▲ | ◇ | | | | | | | | | | | |
| Define test objectives | ● | ○ | | | | | | ■ | ▲ | | | ▲ | | | | ▲ | | | | |
| Determine test requirements | □ | ● | ○ | ▲ | ▲ | ▲ | | ■ | ◇ | | | ◇ | | | | ◇ | | | | |
| Evaluate test-program progress | ● | ○ | ▲ | ◇ | | | | ▲ | | | | | | | | | | | | |
| Make test-program policy decisions | ○ | ▲ | | | | | | ▲ | | | | | | | | | | | | |
| Write test-program responsibility doctrine | □ | ● | ○ | | | | | ■ | | | | ◇ | | | | ◇ | | | | |
| Major functional area: integration of test-support act | | | | | | | | | | | | | | | | | | | | |
| Chair-test working group | □ | ● | ○ | | | | | ■ | | | | ◇ | | | | ◇ | | | | |
| Prepare milestone-test schedules | | ● | ○ | | | | | ■ | ◆ | | | ◆ | | | | ◆ | | | | |
| Write test directive | | ● | ○ | ▲ | ▲ | ▲ | | △ | | | | ◇ | | | | ◇ | | | | |
| Write detailed test procedures | | | ◇ | ◇ | ◆ | ◆ | | □ | ■ | △ | | ● | ■ | ○ | | ● | ○ | ○ | ○ | ■ |
| Coordinate test preparations | | | △ | ■ | ◇ | | | ● | ○ | ▲ | | ○ | ▲ | ▲ | | ○ | ▲ | ▲ | ▲ | ▲ |
| Verify test-article configuration | | | ◆ | | | ▲ | | | | ■ | | | | ○ | | | ○ | ○ | ○ | |
| | | | | | | | | | | | | | | | | | | | | |
| Major functional area: all systems test | | | | | | | | | | | | | | | | | | | | |
| Certify test readiness | ● | ■ | △ | △ | | | | ○ | ▲ | | | ▲ | | | | ▲ | | | | |
| Perform test-director function | ○ | ▲ | ▲ | ◇ | | | | ■ | | | | | | | | | | | | |
| Perform test-conductor function | ● | | | | | | | ○ | ■ | | | ■ | | | | ■ | | | | |
| Analyze test data | | | | | ▲ | ◇ | | | ■ | ■ | | ● | ○ | ○ | | ● | ○ | ○ | ○ | ○ |
| Resolve test anomalies | □ | | | | ◇ | | | ● | ▲ | ▲ | | ○ | ▲ | ■ | | ○ | ■ | ■ | ■ | ▲ |
| Prepare test report | □ | ■ | ◇ | | ◇ | ◆ | | ● | ○ | | | ○ | | | | ○ | | | | |

Figure 16-4. Systems LRC for equipment test division.

The organization's work-subsystem chart could yield another advantage if the titles for the tasks and activities of the LRC were used. If this were done, the managing-subsystem schematic could also be used for diagrammatically integrating the work subsystem and the managing subsystem. A third step would be to superimpose a string of managing-subsystem schematics on the *total* work-subsystem chart to give an overall analytical view of how the organization operates; this would show the stream of interpersonal relations that serve to control, change, and otherwise facilitate the accomplishment of the tasks essential to the realization of the organizational goals.

### The Systematized LRC

A systematized LRC could be structured to serve as the basis for drawing a managing-subsystem schematic diagram by following the three steps.

**Arrangement and Form of Inputs: Job Positions.** The job positions involved in the analysis are listed across the top of the LRC matrix. As can be seen from the sample in Figure 16-4, these job positions are arranged in such a manner that the line structure around them indicates the administrative ordering of the job positions. This method of showing job positions provides a means of integrating the pyramidal organization chart into the LRC. This can be seen by comparing the top portion of the LRC in Figure 16-4 with the corresponding traditional organizational chart in Figure 16-5. The chart should show only the jobs being analyzed. If the analysis concerns only the executives and engineers of an organizational unit, for example, nothing will be gained by including jobs such as those of secretaries, clerks, and draftsmen, even though they are vital parts of the organizational effort. The LRC, like any other chart, must be brief and simple to be effective.

**Listing Outputs: Tasks and Activities.** The tasks and activities related to the job positions are listed on the left side of the matrix. These tasks and activities should be listed in groupings and subgroupings that would facilitate the analysis and enhance the perspective view of the chart. If the tasks are extracted from the organization's work-subsystem chart, this scheme would enable the work subsystem to be integrated into the managing subsystem. Whether this method or some other is chosen, it is very important that complete, accurate, and agreed-upon descriptions of the tasks be selected. The relative ease of accomplishing the analysis and the subsequent usefulness of the LRC will depend strongly on the adequacy of the task statements.

**Description and Definition of Matrix Symbols: Task-Job Relationships.** There are a number of ways in which a person (a job position) can be related to a task. For example, he may be the person who takes some direct action concerning the task, he may be the person's supervisor, or

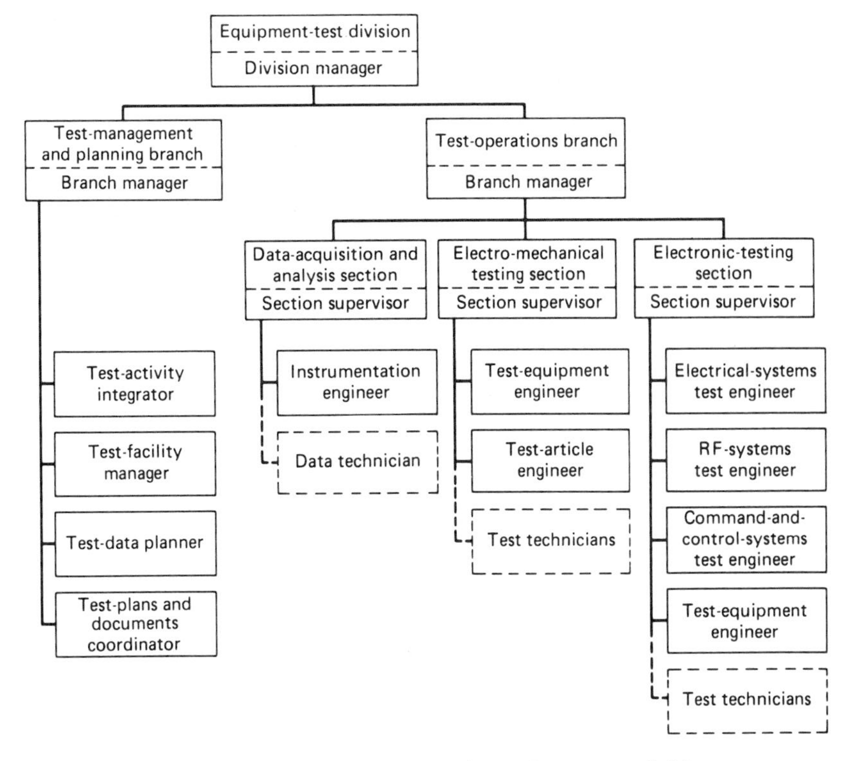


Figure 16-5. Organizational chart for equipment test division.

perhaps he is an advisor on how to do the work. Perhaps he advises as to what needs to be done from a standpoint of intertask sequencing, for example, as a production scheduler. He may even be someone in the system who only needs to be notified that an operation on the task has been completed. In each case, there exists what may be called a *task-job relationship* (TJR).

From a systems viewpoint, each TJR can be visualized as falling into one of three major categories: (1) transfer function, (2) control loop, or (3) input-output stream. The LRC shown in Figure 16-4 for the equipment test division of an electrical equipment company uses the eight TJRs defined below. The first three of these are usually found in papers and articles about the LRC; the other five have been retitled and changed so as to be more meaningful for a systems treatment.

**Transfer Function: Work Is Done/TJR (WID/TJR).** The WID/TJR is the transfer-function aspect of the managing-subsystem model. It is the actual juncture of the managing and the work subsystems. The given inputs of information, matter, and energy (i/m/e) are transformed into predetermined outputs of i/m/e in accordance with the program of instructions (policies, rules, procedures) furnished to the person for this job position.

**Control Loop: "Direct Supervision/TJR (DS/TJR).** The DS/TJR constitutes the prime operational control element in the WID/TJR control loop. The person in the DS job position is considered to be the administrative supervisor of the person in the WID job position. The DS/TJR evaluates the quantity, quality, and timeliness of the WID/TJR outputs through the use of policy guidance information, program directives, procedures, WID input-output comparisons, schedules, and other managerial feedback, measurement, and control devices. The omission of this TJR from a row (task) indicates that the WID/TJR is of such a routine, stable nature that frequent contact with the DS/TJR is not normally required in operation of the transfer function.

**General Supervision/TJR (GS/TJR).** The GS/TJR is a second-order operational-control element and a first-order or prime source of policy guidance for the WID/TJR. The person in the GS job position is the administrative supervisor of the person in the DS job position. The primary role of the GS/TJR is to furnish the DS and WID job positions with a framework of policies and guidance of a scope that permits as much *closed-loop* decision-making flexibility as possible in attainment of the desired WID/TJR outputs. The exclusion of the GS/TJR from a given WID/TJR control loop indicates that WID actions are seldom taken that involve questions of conformance to, or exceptions from, existing GS policy.

**Intertask Integration/TJR (II/TJR).** The II/TJR is placed in the WID/TJR control loop to indicate the need for consideration of functional compatibility between this WID/TJR and other WID/TJRs. The extent of the involvement of the II/TJR in the control loop is the extent to which the transfer functions of the tasks concerned are interlocked or functionally interdependent. The person in the II job position does not have an administrative role in the WID/TJR control loop.

**Occasional Intertask Integration/TJR (OII/TJR).** The OII/TJR is similar in concept and definition to the II/TJR, discussed above. The principal difference is the specialty nature of this TJR as opposed to the general or routine nature of the II/TJR. The omission of this TJR and (or) the II/TJR from a control loop indicates that the WID/TJR is, as a rule, functionally independent of, and hence decoupled from, other transfer functions in the work subsystem.

**Input-Output Stream: Intertask Coordination/TJR (IC/TJR).** The IC/TJR is an information *input* to the WID/TJR and hence does not appear in the control loop; it has nothing to do with the "shape" or "how" of the transfer function. Intertask sequencing, schedule compatibility, quantities, qualities, and other matters of a "what" and "when" nature are indicated by the use of the IC/TJR.

**Occasional Intertask Coordination/TJR (OIC/TJR).** The OIC/TJR is similar to the IC/TJR, discussed above, except that its use indicates only specialized instances of coordination instead of being a routine input.

**Output Notification Mandatory/TJR (ONM/TJR).** The ONM/TJR is placed in the *output* of the WID/TJR transfer function when it is essential or critical that the ONM job position receive some specialized, exact, or timely information concerning the WID/TJR outputs. The concept of this TJR is one of passive transmission of information only and is not coordinative.

The LRC format, as discussed to this point, consists of job positions administratively ordered, tasks and activities grouped in some meaningful way, and TJRs defined along the lines of systems terminology. The subsequent analytical process involves the findings and determinations that lead to inserting the TJRs in the boxes of the LRC matrix. If a TJR symbol does not appear in a box, the analyst has concluded that the job position is not intercoupled with the WID job position of the task.

The completed analysis is then a systematized LRC model of the managing subsystem. This model displays the following characteristics of, and information about, an organizational unit:

- The pyramidal organizational chart is incorporated into the top of the matrix through the use of lines to partition the job positions administratively.
- The tasks and activities of the work subsystem are listed according to some functional flow plan.
- The TJR symbols in the LRC matrix show the types of intercouplings between the person who offsets the work subsystem—the WID job position—and other persons with an interest in the task.
- The TJRs also serve to show how the managing and work subsystems are integrated.

The overall model yields a perspective view of how the *static* formal organization is combined with the *passive* work subsystem to become a *dynamic* functioning entity that maintains a continuous balance with its environment while transforming its information-matter-energy inputs into the outputs that satisfy the ever-changing goals and objectives of the organization.

### System Model Schematic Diagrams

One additional diagram can be constructed that will further illustrate the systems nature of the managing subsystem. If a WID/TJR and its associated task title are combined in one rectangle, and if the remaining TJRs are then intercoupled with that WID/TJR, and if each is enclosed in separate rectangles arranged about, and interconnected with, the WID in accordance with their respective TJRs, the resulting schematic diagram will convey a systems concept of the interpersonal relations involved in accomplishing a given task. Figure 16-6 shows the arrangement of TJRs in a system model schematic format. Figure 16-7 is a system model schematic of the task "Write detailed test procedures," which appears in Figure 16-4. If all tasks of the organization's work subsystems are properly analyzed and charted and if the system model schematic for each task is shown with its respective tasks and interconnections with the other schematics, the results should give an overall portrayal of the organization's integrated work subsystem and managing subsystem, such as is shown by the sketch in Figure 16-8.

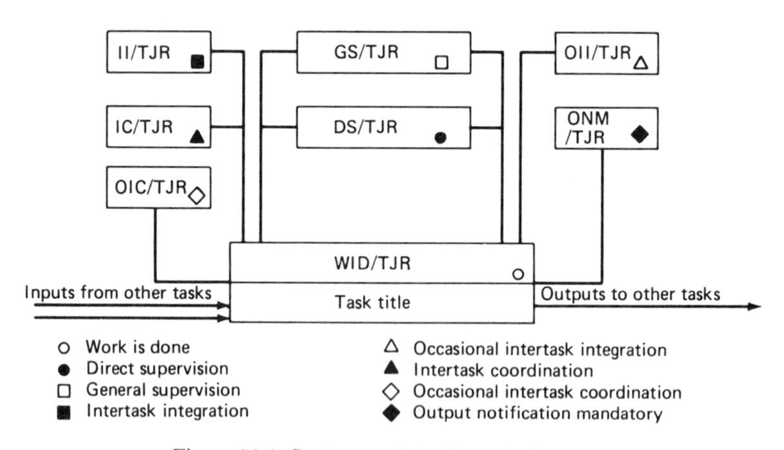


Figure 16-6. System model schematic format.

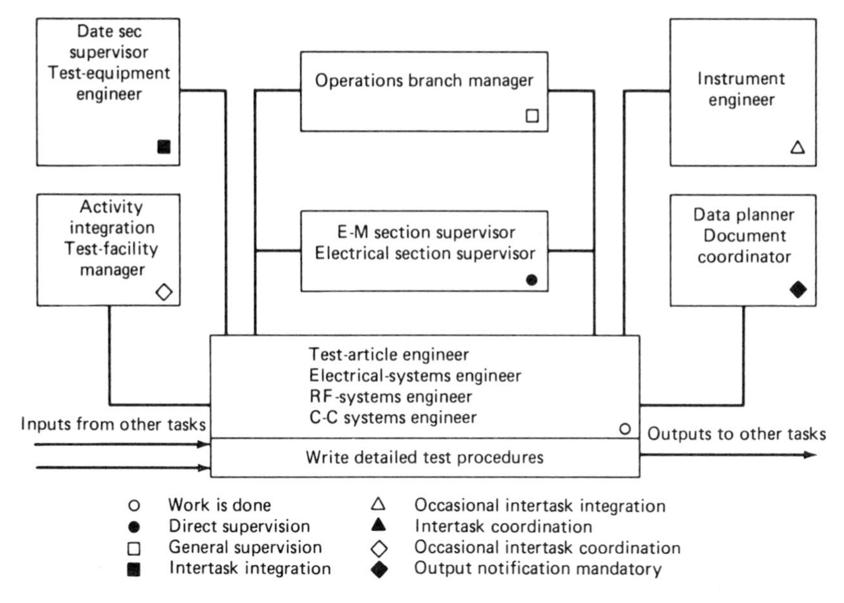


Figure 16-7. System model schematic of the task, "Write detailed test procedures."

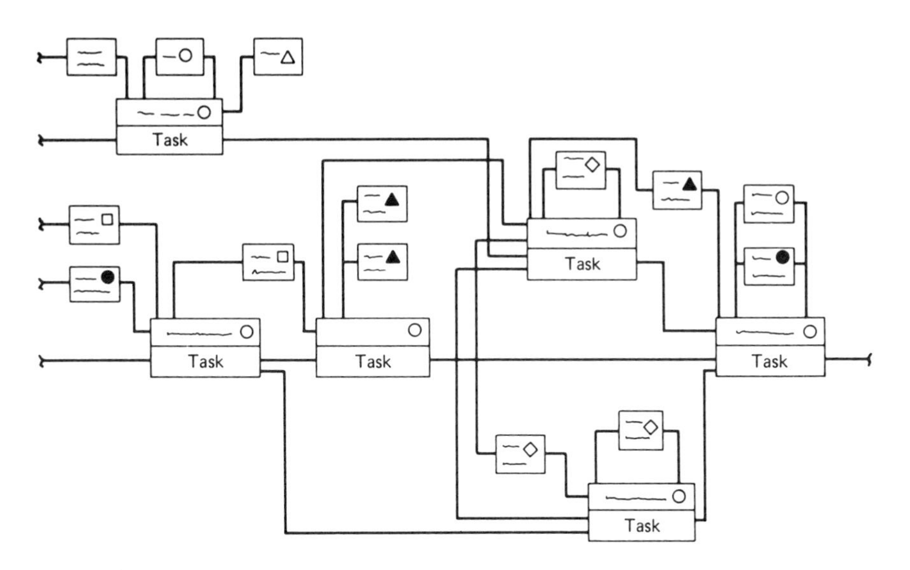


Figure 16-8. Integration of an organization's work and managing subsystems.

## THE LRC IN PLANNING

While the primary focus of the LRC has been descrptive, it may also be used in a prescriptive mode to aid in the systems design aspect of planning. This approach has been developed by the authors within the context of a Department of Justice study in the Buffalo, New York, Police Department. (9)

The approach utilizes both a descriptive and normative LRC-like model of the organization as the basis for developing a "consensus" model through negotiation.

Figure 16-9 depicts a descriptive chart which characterizes one aspect of policy planning—policy formulation.

The entries in the chart represent a number of organizational characteristics with regard to the planning decision area:

1. Authority and responsibility relationships.
2. Initiation characteristics.
3. Input-output characteristics.

The codes used to describe these characteristics for internal units are

I—Initiation
E—Execution
A—Approval
C—Consultation
S—Supervision

Subscripts on these coded symbols describe with whom the relationship exists. For instance, the simplified macrolevel chart of Figure 16-9 shows on the first row that the analysis of routine complaints (E) is handled at the police captain level under the supervision of an inspector (S) with the police commissioner having approval authority (A). In performing this function, the captain has the consultation of uniformed patrolmen ($C_7$ where the subscript 7 indicates with whom the consultation takes place). Another consultation takes place when the deputy commissioner consults with the commissioner ($C_4$) at the approval stage.

Various informational linkages with interfacing environmental organizations are also depicted in these charts. Figure 16-9 shows only one such linkage—that involving "other city departments" who both provide input (i) to and receive output (o) from the mayor (2) in his approval role.

The model of Figure 16-9 is a descriptive one in that it depicts authorities, responsibilities, initiations, inputs, and outputs *as they actually occur in the organization.*

| | 1 City council | 2 Mayor | 3 Comptroller commissioner budget director | 4 Police commissioner | 5 Deputy commissioner | 6 Inspector | 7 Captain | 8 Uniformed patrolman | 9 Police administrator | 10 Other city departments | 11 Boards and agencies | 12 Federal government |
|---|---|---|---|---|---|---|---|---|---|---|---|---|
| Routine complaints | | | | A | $C_4$ | S | E | $C_7$ | | | | |
| Observation of field practices | | | | A | S | E | | | | | | |
| Crime analysis | | | | | | | | | | | | |
| Court decisions | | | | | | | | | | | | |
| Analysis of social problems | | | | | | | | | | | | |
| New legislation | | | | A | S | E | | | | | | |
| Issue clarification definition | | | | A | S | E | | | | | | |
| Selection of alternatives | | | | A | E | $C_5$ | | | | | | |
| Obtaining relevant facts | | | | | | | | | | | | |
| Analysis of facts | | | | | | | | | | | | |
| Review | | A | | E | | | | | | $2^{1.0}$ | | |
| Formulation | | | | E | C | | | | | | | |
| Articulation | | | | A | $S_4$ | S | E | | | | | |
| Training for implementation | | | | | A | S | E | | | | | |
| Execution and control | | | | A | S | E | $C_4$ | | | | | |

Figure 16-9. Model of existing policy-making process.

A descriptive model of the organizational and environmental system such as that provided by Figure 16-9 and other associated charts is a useful "road map" for guiding information analysis. It provides insights into "who does what," the interactions among organizational units and between internal and external units, the general nature of information required, the direction of information flow, and the manner in which information requirements are generated.

However, the use of a model of this variety alone as a basis for system design would represent an abrogation of the information analyst's proper role. Instead of creating a system to serve an existing organizational system, he should attempt to restructure the decision-making process so that the system may be oriented toward the support of a more nearly "optimal" process.

To do this, the analyst must call on the best of the knowledge and theory of management to construct a normative model of the organization. For instance, a police department which is not already using a program budget structure should be aided in developing one. A *normative* for the same "policy formulation" area to which the descriptive model of Figure 16-9 applies may be developed. Most organizations will not find it desirable to directly adopt such a prescriptive model. However, an "open minded" organization will usually find some elements of the model which it wishes to adopt.

The development of a consensus model hinges on an objective comparison of a descriptive model, such as that of Figure 16-9, with a normative model. This comparison and evaluation must be done by managers, with the aid and advice of analysts.

One possible medium for this process, which has been used successfully by the authors, is that of a "participative executive development program." The program involved the system participants as "students" and the analysts as "teachers." The normative model was developed and discussed in lecture-discussion sessions. After it had been communicated fully, workshops were used to facilitate the detailed evaluation and comparison of the descriptive and normative models. Recommendations emanating from the workshops were reviewed by top management, and those which were approved were incorporated into a consensus model of the system.

## SUMMARY

This chapter demonstrates the values and limitations of traditional organizational charts and introduces a variety of charts—all based on the *linear responsibility chart* (LRC)—which can aid in both the planning and implementing phases of management.

## REFERENCES

(1) Richard M. Cyert and James G. March, *A Behavioral Theory of the Firm* (Prentice-Hall, Englewood Cliffs, N.J., 1963), p. 289.
(2) Frank J. Jasinski, "Adapting Organization to New Technology," *Harvard Business Review* (January–February, 1959), p. 80.

(3) Allen R. Janger, "Charting Authority Relationships," *The Conference Board Record* (December, 1964).
(4) "Manning the Executive Setup," *Business Week* (April 6, 1957), p. 187.
(5) For example, one writer proclaimed: "On one pocket-size chart it shows the facts buried in all the dusty organizational manuals—plus a lot more."
(6) Allen R. Janger, "Charting Authority Relationships," *The Conference Board Record* (December, 1964).
(7) Delmar W. Karger and Robert G. Murdick, *Managing Engineering and Research* (The Industrial Press, New York, 1963), p. 89.
(8) The informal organization is not what the name implies, i.e., a casual, loosely structured community of people who have similar interests. The informal organization can be most demanding on its members. Its standards of performance and loyalty and its authority patterns can be anything but loose. It can be the most powerful of alliances existing between people having vested interests.
(9) See William R. King and David I. Cleland, "The Design of Management Information Systems: An Information Analysis Approach," *Management Science,* Vol. 22(3) (November, 1975), pp. 286–297.

# 17. Pricing Out the Work

Harold Kerzner*

The first integration of the functional unit into the project environment occurs during the pricing process. The total program costs obtained by pricing out the activities over the scheduled period of performance provides management with a fundamental tool for managing the project. During the pricing activities, the functional units have the option to consult program management for possible changes to work requirements as well as for further clarification.

Activities are priced out through the lowest pricing units of the company. It is the responsibility of these pricing units, whether they be sections, departments, or divisions, to provide accurate and meaningful cost data. Under ideal conditions, the work required (i.e., man-hours) to complete a given task can be based upon historical standards. Unfortunately for many industries, projects and programs are so diversified that realistic comparison between previous activities may not be possible. The costing information obtained from each pricing unit, whether or not it is based upon historical standards, should be regarded only as an estimate. How can a company predict the salary structure three years from now? What will be the cost of raw materials two years from now? Will the business base (and therefore the overhead rates) change over the period of performance? The final response to these questions shows that costing out performance is explicitly related to an environment which cannot be predicted with any high degree of certainty.

Project management is an attempt to obtain the best utilization of resources within time, cost, and performance. Logical project estimating

---

* Dr. Harold Kerzner is Professor of Systems Management and Director of The Project/Systems Management Research Institute at Baldwin-Wallace College. Dr. Kerzner has published over 35 engineering and buisness papers and ten texts: *Project Management: A Systems Approach to Planning, Scheduling and Controlling; Project Management for Executives; Project Management for Bankers; Cases and Situations in Project/Systems Management; Operations Research; Proposal Preparation and Management; Project Management Operating Guidelines; Project/Matrix Management Policy and Strategy; A Project Management Dictionary of Terms; Engineering Team Management.*

techniques are available. The following thirteen steps provide a logical sequence in order to obtain better resource estimates. These steps may vary from company to company.

## STEP 1: PROVIDE A COMPLETE DEFINITION OF THE WORK REQUIREMENTS

Effective planning and implementation of projects cannot be accomplished without a complete definition of the requirements. For projects internal to the organization, the project manager works with the project sponsor and user (whether they be executives, functional managers, or simply employees) in order for the work to be completely defined. For these types of in-house projects, the project manager can wear multiple hats as project manager, proposal manager, and even project engineer on the same project.

For projects funded externally to the organization, the proposal manager (assisted by the project manager and possibly the contract administrator) must work with the customer to make sure that all of the work is completely defined and that there is no misinterpretation over the requirements. In many cases, the customer simply has an idea and needs assistance in establishing the requirements. The customer may hire an outside agency for assistance. If the activity is sole-source or perhaps part of an unsolicited effort, then the contractor may be asked to work with the customer in defining the requirements even before any soliciting is attempted.

A complete definition of project requirements must include

- Scope (or statement) of work.
- Specifications.
- Schedules (gross or summary).

The scope of work or statement of work (SOW) is a narrative description of all the work required to perform the project. The statement of work identifies the goals and objectives which are to be achieved. If a funding constraint exists, such as "this is a not-to-exceed effort of $250,000," this information might also appear in the SOW.

If the customer supplies a well-written statement of work, then the project and proposal managers will supply this SOW to the functional managers for dollar and man-hour estimates. Unless the customer maintains a staff of employees to provide a continuous stream of RFP/RFQs,*

* RFP (Request for Proposal); RFQ (Request for Quote).

the customers must ask potential bidders to assist them in the preparation of the SOW. As an example, Alpha Company wishes to build a multimillion-dollar chemical plant. Since Alpha does not erect such facilities on a regular basis, Alpha would send out inquiries instead of a formal RFP. These inquiries are used not only to identify potential bidders, but also to identify to potential bidders that they will have to develop an accurate SOW as part of the proposal process. This process may appear as a feasibility study. This is quite common especially on large dollar-value projects where contractors are willing to risk the additional time, cost, and effort as part of the bidding process. If the proposal is a sole-source effort, then the contractor may pass this cost on to the customer as part of the contract.

The statement of work is vital to proposal pricing and should not be taken lightly. All involved functional managers should be given the opportunity to review the SOW during the pricing process. Functional managers are the true technical experts in the company and best qualified to identify high-risk areas and prevent anything from "falling through the crack." Misinterpretations of the statement of work can lead to severe cost overruns and schedule slippages.

The statement of work might be lumped together with the contractual data as part of the terms and conditions. The proposal manager may then have to separate out the SOW data from the RFP. This is vital for the pricing effort.

This process is essential because misinterpretation of the statement of work can cause severe cost overruns. As an example, consider the following two situations:

- Acme Corporation won a Navy contract in which the Government RFP stated that "this unit must be tested in water." Acme built a large pool behind their manufacturing plant. Unfortunately, the Navy's interpretation was the Atlantic Ocean. The difference was $1 million.
- Ajax Corporation won a contract to ship sponges across the United States using aerated boxcars. The project manager leased boxcars that had doors on the top surface. The doors were left open during shipping. The train got caught in several days of torrential rainstorms and the boxcars eventually exploded, spreading sponges across the countryside. The customer wanted boxcars aerated from below.

The amount of money and time spent in rewording the technical data in the SOW for pricing is minimal compared to cost of misinterpretation.

The second major item in the definition of the requirements is the identification of the specifications, if applicable. Specifications form the basis from which man-hours, equipment, and materials are priced out. The specifications must be identified such that the customer will understand the basis for the man-hour, equipment, and materials estimates. Small changes in a specification can cause large cost overruns.

Another reason for identifying the specifications is to make sure that there will be no surprises for the customer downstream. The specifications should be the current revision. It is not uncommon for a customer to hire outside agencies to evaluate the technical proposal and to make sure that the proper specifications are being used.

Specifications are in fact standards for pricing out a proposal. If specifications either do not yet exist or are not necessary, then work standards should be included in the proposal. The work standards can also appear in the cost volume of the proposal. Labor justification backup sheets may or may not be included in the proposal, depending upon RFP/RFQ requirements.

For R&D proposals, standards may not exist and the pricing team may have to use educated guesses based upon the estimated degree of difficulty, such as:

- Task 02-15-10 is estimated to be 25% more difficult than a similar task accomplished on the Alpha Project, which required 300 man-hours. Hours needed for Task 02-15-10 are therefore 375.
- Task 03-07-02 is estimated at 450 hours. This is 20% more than the standard because of the additional reporting constraints imposed by the customer.

The standards mentioned here are usually the technical standards only.

The technical standards and specifications may be called out by the customer or, if this is a follow-on project, then the customer will expect the work to be performed within the estimate on the previous activity. If the standards or specifications will be different, then an explanation must be made or else the customer (and line managers) may feel that he has been taken for a ride. Customers have the tendency of expecting standards to be lowered on follow-on efforts because the employees are expected to be performing at an improved position on the learning curve.

The key parameter in explaining the differences in standards is the time period between the original cost estimate and the follow-on or similar cost estimate. The two most common reasons for having standards change are

- New technology requires added effort.
- Key employees with the necessary skills or expertise have either left the organization or are not available.

In either event, justifications of the changes or modifications must be made so that the new ground rules are understood by all pricing and reviewing personnel.

The third item in the identification of the requirements is the gross schedule. In summary, the gross schedule identifies the major milestones of the project and includes such items as:

Start date.
End date.
Other major milestone activities.
Data items and reports.

If possible, all gross schedules which are used for pricing guidelines should contain calendar start and end dates. Unfortunately, some projects do not have definable start and end dates and are simply identified by a time spread. Another common situation is where the end date is fixed and the pricing effort must identify the start date. This is a common occurrence because the customer may not have the expertise to accurately determine how long it will take to accomplish the effort.

Identifying major milestones can also be a tedious task for a customer. Major milestones include such activities as long-lead procurement, prototype testing, design review meetings, and any other critical decision points. The proposal manager must work closely with the customer or in-house sponsor either to verify the major milestones in the RFP or to identify additional milestones.

Major milestones are often grossly unrealistic. In-house executives of the customer and the contractor occasionally identify unrealistic end dates because either resources will be idle without the completion at this point in time, not enough money is available for a longer project, or management wants the effort completed earlier because it affects management's Christmas bonus.

All data items should be identified on the gross schedule. Data items include written contractual reports and can be extended to include handout material for customer design review meetings and technical interchange meetings. Data items are not free and should be priced out accordingly. There is nothing wrong with including in the pricing effort a separate contingency fund for "unscheduled or additional" interchange meetings.

### STEP 2: ESTABLISH A LOGIC NETWORK WITH CHECKPOINTS

Once the work requirements are outlined, the project manager must define the logical steps necessary to accomplish the effort. The logic network (or arrow diagram as it is more commonly referred to) serves as the basis for the PERT/CPM diagrams and the Work Breakdown Structure.* The arrow diagram simply shows the logical sequence of events, generally at the level which the project manager wants to control the program. Each logic diagram activity should not be restricted to specific calendar dates at this point because line managers should price out the work, initially assuming

- Unlimited resources.
- No calendar constraints.

If this is not done during the initial stages of pricing, line managers may commit to unrealistic time, cost, and performance estimates. After implementation, the project manager may find it impossible to force the line manager to meet his original estimates.

## STEP 3: DEVELOP THE WORK BREAKDOWN STRUCTURE

The simplest method for developing the work breakdown structure is to combine activities on the arrow diagram. If each activity on the arrow diagram is considered to be a task, then several tasks can be combined to form projects and the projects, when combined, will become the total program. The WBS may contain definable start and end dates in accordance with the gross schedule at this point in time, although they may have to be altered before the final WBS is firmly established. Most project managers prefer to work at the task level of the WBS (Level 3). The work is priced out at this level and costs are controlled at this level. Functional managers may have the option of structuring the work to additional levels for better estimating and control.

Often the arrow diagram and WBS are considered as part of the definition of the requirements, because the WBS is the requirement that costs be controlled at a specific level and detail.

## STEP 4: PRICE OUT THE WORK BREAKDOWN STRUCTURE

The project manager's responsibility during pricing (and even during execution, for that matter) is to establish the project requirements which identify the "What," "When," and "Why" of the project. The functional managers now price out the activities by determining the "How," "Who," and "Where" of the project. The functional managers have the right to ask the project manager to change the WBS. After all, the func-

---

* See Chapter 14 for a detailed treatment of WBSs and Chapter 15 for a discussion of PERT/CPM and network plans.

tional managers are the true technical experts and may wish to control their efforts differently.

Once the Work Breakdown Structure and activity schedules are established, the program manager calls a meeting for all organizations which will be required to submit pricing information. It is imperative that all pricing or labor costing representatives be present for the first meeting. During this "kickoff" meeting, the Work Breakdown Structure is described in depth so that each pricing unit manager will know exactly what his responsibilities are during the program. The kickoff meeting also resolves the struggle-for-power positions of several functional managers whose responsibilities may be similar or overlap on certain activities. An example of this would be quality control activities. During the research and development phase of a program, research personnel may be permitted to perform their own quality control efforts, whereas during production activities, the quality control department or division would have overall responsibility. Unfortunately, one meeting is not sufficient to clarify all problems. Follow-up or status meetings are held, normally with only those parties concerned with the problems that have arisen. Some companies prefer to have all members attend the status meetings so that all personnel will be familiar with the total effort and the associated problems. The advantage of not having all program-related personnel attend is that time is of the essence when pricing out activities. Many functional divisions carry this policy one step further by having a divisional representative together with possibly key department managers or section supervisors as the only attendees to this initial kickoff meeting. The divisional representative then assumes all responsibility for assuring that all costing data be submitted on time. This may be beneficial in that the program office need only contact one individual in the division to learn of the activity status, but may become a bottleneck if the representative fails to maintain proper communication between the functional units and the program office or if the individual simply is unfamiliar with the pricing requirements of the Work Breakdown Structure.

During proposal activities, time may be extremely important. There are many situations where a Request for Proposal (RFP) requires that all responders submit their bids no later than a specific date, say 30 days. Under a proposal environment, the activities of the program office, as well as those of the functional unit, are under a schedule set forth by the proposal manager. The proposal manager's schedule has very little, if any, flexibility and is normally under tight time constraints in order that the proposal may be typed, edited, and published prior to date of submittal. In this case, the RFP will indirectly define how much time the pricing units have to identify and justify labor costs.

The justification of the labor costs may take longer than the original cost estimates, especially if historical standards are not available. Many proposals often require that comprehensive labor justifications be submitted. Other proposals, especially those which request almost immediate response, may permit vendors to submit labor justification at a later date.

In the final analysis, it is the responsibility of the lowest pricing unit supervisor to maintain adequate standards, if possible, so that almost immediate response can be given to a pricing request from a program office.

The functional units supply their input to the program office in the form of man-hours. The input may be accompanied by labor justifications, if required. The man-hours are submitted for each task, assuming that the task is the lowest pricing element, and are time-phased per month. The man-hours per month per task are converted to dollars after multiplication by the appropriate labor rates. The labor rates are generally known with certainty over a 12-month period but from there on are only estimates. How can a company predict salary structures five years hence? If the company underestimates the salary structure, increased costs and decreased profits will occur. If the salary structure is overestimated, the company may not be competitive. If the project is government funded, then the salary structure becomes an item under contract negotiations.

The development of the labor rates to be used in the projection are based upon historical costs in business base hours and dollars for either the most recent month or quarter. Average hourly rates are determined for each labor unit by direct effort within the operations at the department level. The rates are only averages, and include both the highest-paid employees and lowest-paid employees together with the department manager and the clerical support.* These base rates are then escalated as a percentage factor based upon past experience, budget as approved by management, and the local outlook and similar industries. If the company has a predominant aerospace or defense industry business base, then these salaries are negotiated with local government agencies prior to submittal for proposals.

The labor hours submitted by the functional units are quite often overestimated for fear that management will "massage" and reduce the labor hours while attempting to maintain the same scope of effort. Many times management is forced to reduce man-hours either because of insufficient

---

* Problems can occur if the salaries of the people assigned to the program exceed the department averages. Also, in many companies department managers are included in the overhead rate structure, not direct labor, and therefore their salaries are not included as part of the department average.

funding or just to remain competitive in the environment. The reduction of man-hours often causes heated discussions between the functional and program managers. Program managers tend to think in the best interests of the program, while functional managers lean toward maintaining their present staff.

The most common solution to this conflict rests with the program manager. If the program manager selects members for the program team who are knowledgeable in man-hour standards for each of the departments, then an atmosphere of trust can develop between the program office and the functional department such that man-hours can be reduced in a manner which represents the best interests of the company. This is one of the reasons why program team members are often promoted from within the functional ranks.

The ability to estimate program costs involves more than just labor dollars and labor hours. Overhead dollars can be one of the biggest headaches in controlling program costs and must be estimated along with labor hours and dollars. Although most programs have an assistant program manager for cost whose responsibilities include monthly overhead rate analysis, the program manager can drastically increase the success of his program by insisting that each program team member understand overhead rates. For example, if overhead rates apply only to the first forty hours of work, then, depending on the overhead rate, program dollars can be saved by performing work on overtime where the increased salary is at a lower burden.

The salary structure, overhead structure, and labor hours fulfill three of four major input requirements. The fourth major input is the cost for materials and support. Six subtopics are included under materials/support: materials, purchased parts, subcontracts, freight, travel, and other. Freight and travel can be handled in one of two ways, both normally dependent on the size of the program. For small dollar-volume programs, estimates are made for travel and freight. For large dollar-volume programs, travel is normally expressed as between three and five percent of all costs for material, purchased parts, and subcontracts. The category labeled other support costs may include such topics as computer hours or special consultants.

The material costs are very time-consuming, more so than the labor hours. Material costs are submitted via a bill of materials which includes all vendors from whom purchases will be made, project costs throughout the program, scrap factors, and shelf lifetime for those products which may be perishable.

Information on labor is usually supplied to the project office in the form of man-hours/department/task/month. This provides a great degree of flexibility in analyzing total program costs and risks, and is well worth the added effort. Costs can be itemized per month, task, or even department. Computers, with forward pricing information, will convert the man-hours to dollars. Raw materials are always priced out as dollars per month with the computer providing the forward pricing information for escalation factors.

## STEP 5: REVIEW WBS COSTS WITH EACH FUNCTIONAL MANAGER

Once the input is received from each functional manager, the project team integrates all of the costs to ensure that all of the work is properly accounted for, without redundancy. An important aspect of this review is the time-phased manpower estimates. It is here where the project manager brings up the subject of limited rather than unlimited resources and asks the line managers to assess the various risks in their estimates.

As part of the review period, the project manager must ask the following questions:

- Was sufficient time allowed for estimating?
- Were the estimates based upon history or standards, or are they "best guesses"?
- Will the estimates require a continuous shifting of personnel in and out of the project?
- Will there be personnel available who have the necessary skills?

Obviously, the answers to these questions can lead into a repricing activity.

## STEP 6: DECIDE UPON THE BASIC COURSE OF ACTION

After the review with the functional managers, the project manager must decide upon the basic course of action or the base case. This is the ideal path that the project manager wishes to follow. Obviously, the decision will be based upon the risks on the project and the projected trade-offs which may have to be made downstream on time, cost, and performance.

The base case may include a high degree of risk if it is deemed necessary to satisfy contractual requirements. This base case approach and accompanying costs should be reviewed with the customer and upper-level management. There is no point in developing finalized, detailed PERT/CPM schedules and the program plan unless there is agreement on the base case.

### STEP 7: ESTABLISH REASONABLE COSTS FOR EACH WBS ELEMENT

Since the project will be controlled through the WBS, the project manager must define, with reasonable accuracy and confidence, his target costs for each WBS element, usually at Level 3. Once the project is initiated, these costs will become the basis for the project targets. The problem here is that the costs were based upon unlimited resources. Limited resources may require overtime, or perhaps the work will have to be performed during higher cost escalation periods. These factors must be accounted for.

### STEP 8: REVIEW THE BASE CASE COSTS WITH UPPER-LEVEL MANAGEMENT

Once the base case is formulated, the pricing team member, together with the other program office team members, perform perturbation analyses in order to answer any questions that may come up during the final management review. The perturbation analysis is designed as a systems approach to problem solving, where alternatives are developed in order to respond to any questions that management may wish to consider during the final review.

The base case, together with the perturbation analysis costs, are then reviewed with upper-level management in order to formulate a company position for the program as well as to take a hard look at the allocation of resources required for the program. The company position may be to cut costs, authorize work, or submit a bid. If the program is competitive, corporate approval may be required if the company's chief executive officer has a ceiling on the dollar bids he can authorize to go out of house.

If labor costs must be cut, the program manager must negotiate with the functional managers as to the size and method for the cost reductions. Otherwise, this step may simply entail the authorization for the functional managers to begin the activities or to develop detailed plans.

### STEP 9: NEGOTIATE WITH FUNCTIONAL MANAGERS FOR QUALIFIED PERSONNEL

Once the base case costs are established, the project manager must begin the tedious effort of converting all estimates to actual calendar dates and time frames based upon limited resources. Detailed schedules cannot be established without some degree of knowledge as to exactly which employees will be assigned to key activities. Highly qualified individuals may be able to accomplish the work in less time and may be able to assume added responsibilities.

Good project managers do not always negotiate for the best available resources because either the costs will be too great with those higher-paid individuals or the project priority does not justify the need for such individuals.

Accurate, detailed schedules cannot be developed without some degree of knowledge as to who will be available for the key project positions. Even on competitive bidding efforts, customers require that the resumes of the key individuals be included as part of the proposal.

### STEP 10: DEVELOP THE LINEAR RESPONSIBILITY CHART

Once the key employees are assigned to the activities, the project manager works with the functional managers in assigning project responsibilities. The project responsibilities may be assigned in accordance with assumed authority, age, experience on related efforts, maturity, and interpersonal skills.

The linear responsibility chart, if properly developed and used, is an invaluable tool not only in administering the project, but also in estimating the costs.* The linear responsibility chart permits the project manager the luxury of assigning additional work to qualified personnel, of course upon approval of the functional managers. This additional work may be assigned to lower-salaried individuals so that the final costs can come close to the departmental averages, assuming that the work was priced out in this fashion.

The linear responsibility chart development has a direct bearing upon how the costs are priced out and controlled. There are three methods for pricing out and controlling costs.

- Work is priced out at the department average and all work is charged to the project at the department average salary, regardless of who performed the work.
- Work is priced out at the department average but all work performed is billed back to the project at the actual salary of those employees who are to do the work.
- The work is priced out at the salary of those employees who will perform the work and the costs are billed back the same way.

---

* See Chapter 16 for detailed treatment of the linear responsibility chart.

Each of these methods has its advantages and disadvantages as well as a serious impact on the assignment of responsibilities.

## STEP 11: DEVELOP THE FINAL DETAILED AND PERT/CPM SCHEDULES

Work standards are generally based upon the average employee. The assignment of above or below average employees can then cause the schedules to be shifted left or right. These detailed schedules are now based upon limited resources and provide the basis for accurate cost estimating. If at all possible, "fat" and slack time should be left in the schedules so as to provide some degree of protection for the line managers. Fat and slack should be removed only as a last resort to lower costs, such as in the case of wanting to remain competitive or on buy-ins.

It should be obvious at this point that project pricing is an iterative process based upon optimization of time, cost, and performance together. After the detailed schedules are developed, the entire pricing process may have to be reaccomplished. Fortunately, the majority of the original estimates are usually salvagable and require only cosmetic modifications unless the customer provides major changes to specifications or quantity revisions because initial cost estimates were grossly unacceptable.

## STEP 12: ESTABLISH PRICING COST SUMMARY REPORTS

Although the pricing of a project is an iterative process, the project manager must still burden himself at each iteration point by developing cost summary reports so that key project decisions can be made during the planning. There are at least two times when detailed pricing summaries are needed: in preparation for the pricing review meeting with management and at pricing termination. At all other times it is possible that "simple cosmetic surgery" can be performed on previous cost summaries, such as perturbations in escalation factors and procurement cost of raw materials. The list identified below shows the typical pricing reports.

- A detailed cost breakdown for each WBS element. If the work is priced out at the task level, then there should be a cost summary sheet for each task, as well as rollup sheets for each project and the total program.
- A total program manpower curve for each department. These manpower curves show how each department has contracted with the project office to supply functional resources. If the departmental manpower curves contain several "peaks and valleys," then the project manager may have to alter some of his schedules so as to obtain

some degree of manpower smoothing. Functional managers always prefer manpower-smoothed resource allocations.

- A monthly equivalent manpower cost summary. This table normally shows the fully burdened cost for the average departmental employee carried out over the entire period of project performance. If project costs have to be reduced, the project manager performs a parametric study between this table and the manpower curve tables.
- A yearly cost distribution table. This table is broken down by WBS element and shows the yearly (or quarterly) costs that will be required. This table, in essence, is a project cash flow summary per activity.
- A functional cost and hour summary. This table provides top management with an overall description of how many hours and dollars will be spent by each major functional unit, or division. Top management would use this as part of the forward planning process to make sure that there are sufficient resources available for all projects. This also includes indirect hours and dollars.
- A monthly labor hour and dollar expenditure forecast. This table can be combined with the yearly cost distribution, except that it is broken down by month, not activity or department. In addition, this table normally includes manpower termination liability information for permature cancellation of the project by outside customers.
- A raw material and expenditure forecast. This shows the cash flow for raw materials based upon vendor lead times, payment schedules, commitments, and termination liability.
- Total program termination liability per month. This table shows the customer the monthly costs for the entire program. This is the customer's cash flow, not the contractor's. The difference is that each monthly cost contains the termination liability for man-hours and dollars, on labor and raw materials. This table is actually the monthly costs attributed to premature project termination.

These tables are used both by project managers and upper-level executives. The project managers utilize these tables as the basis for project cost control. Top level management utilizes these tables selecting, approving, and prioritizing projects, as shown in Figure 17-1.

## STEP 13: DOCUMENT THE RESULTS INTO A PROGRAM PLAN

The final step in cost estimating is to document all of the results into a project plan. The cost information will also be the basis for the cost volume of the proposal. The logical sequence of events leading up to the

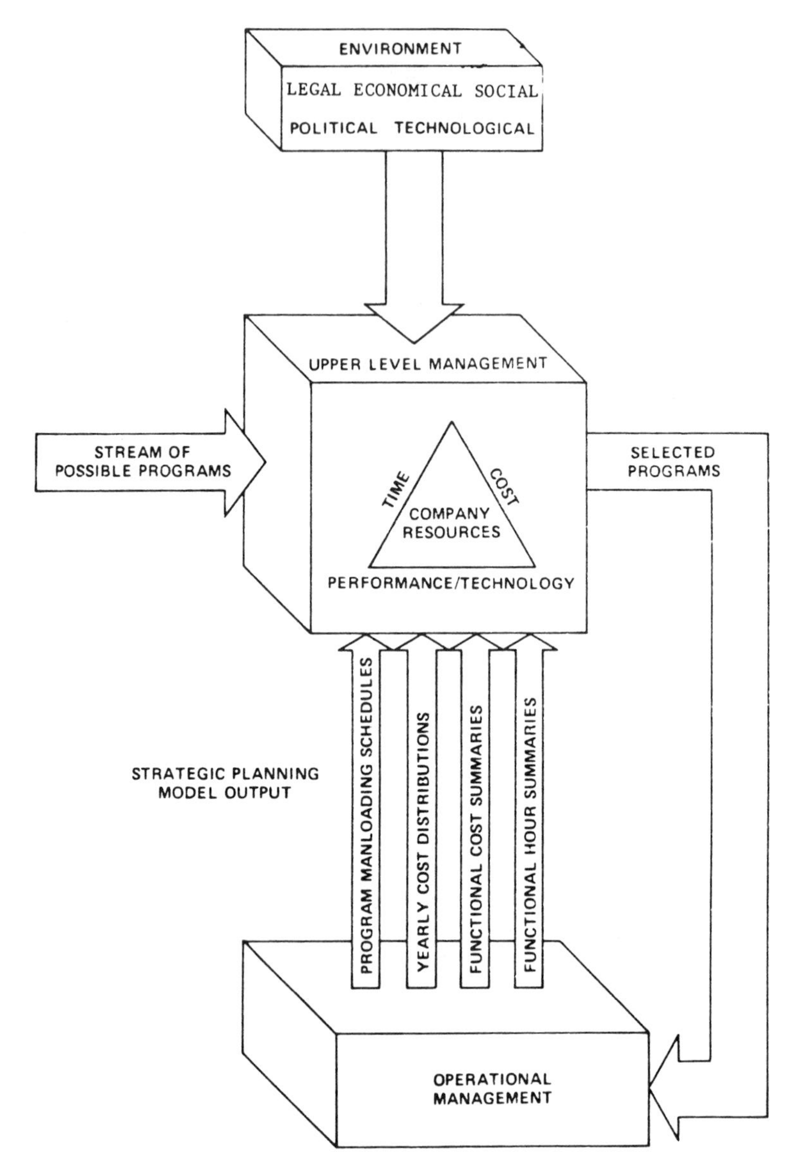


Figure 17-1. Systems approach to resource control. (From Harold Kerzner, *Project Management: A Systems Approach to Planning, Scheduling and Controlling*, Second Edition (Van Nostrand Reinhold, New York, 1984, p. 681).

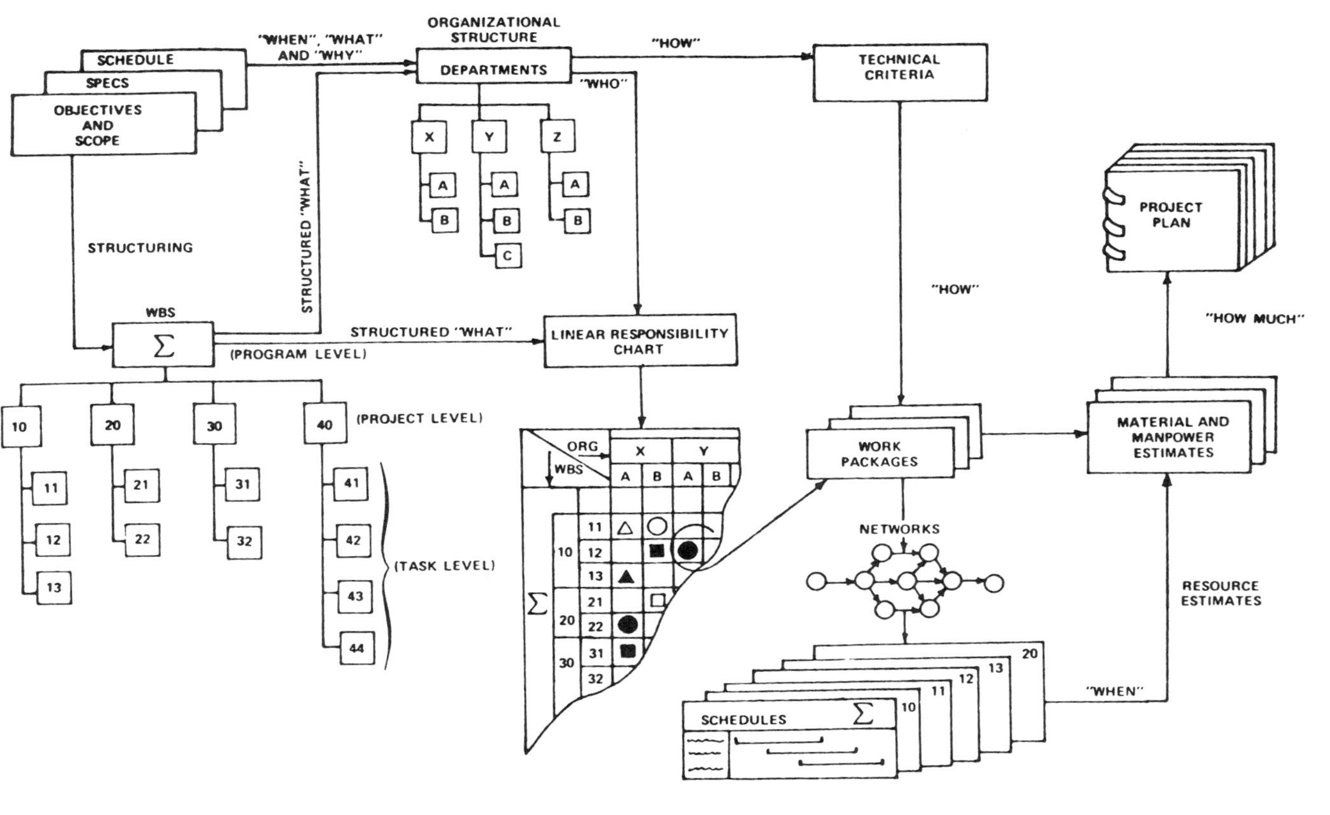


Figure 17-2. Project planning. (From Harold Kerzner, *Project Management for Executives* Second Edition (Van Nostrand Reinhold, New York, 1985, p. 585).

program plan can be summarized as in Figure 17-2. Pricing is an iterative process, at best. The exact pricing procedure will, of course, differ for projects external to the organization as opposed to internal.

Regardless of whether you are managing a large or small project, cost estimating must be accomplished in a realistic, logical manner in order to avoid continuous panics. The best approach, by far, is to try to avoid the pressures of last-minute estimating, and to maintain reasonable updated standards for future estimating. Remember, project costs and budgets are only estimates based upon the standards and expertise of the function managers.

# 18. Contracts Development—Keystone in Project Management

M. William Emmons*

A keystone is placed at the crown of an arch to provide integrity to the structure and to capitalize on the strength of each individual component. Contracts development as the keystone in project management provides the support to and brings together those activities which have been accomplished and those activities which are to be accomplished. The keystone concept is that the precontracting activities have been accomplished by the owner and the post-activities will be accomplished by the contractor; contracts development is an activity which requires a mutual accomplishment—the best efforts of both owner and contractor. Figure 18-1 depicts the project phases and activities performed in each phase.

## INTRODUCTION

### Owner and Contractor Objectives

Contracts development presents challenges to contractor and owner. The interaction is based on long-term and short-term objectives of each. In the short term, the more narrow view, the contractor is competing for the project to build its work load; the owner wants to find the best contractor to accomplish the execution successfully.

---

* M. William Emmons retired from Exxon Corporation in July, 1986, after 34 years in project management related assignments. During his last 11 years with Exxon, Mr. Emmons headed the contracting function of Exxon Chemical Company, assisting world wide affiliates in selecting contractors and negotiating contracts for a full range of petrochemical and associated facilities. Mr. Emmons is currently Principal Associate Contract Management with Pathfinder, Inc., of Cherry Hill, New Jersey, a professional consulting firm assisting owners, contractors, financial institutions, legal institutions, and insurance firms with a comprehensive scope of organizational, venture, and project management services.

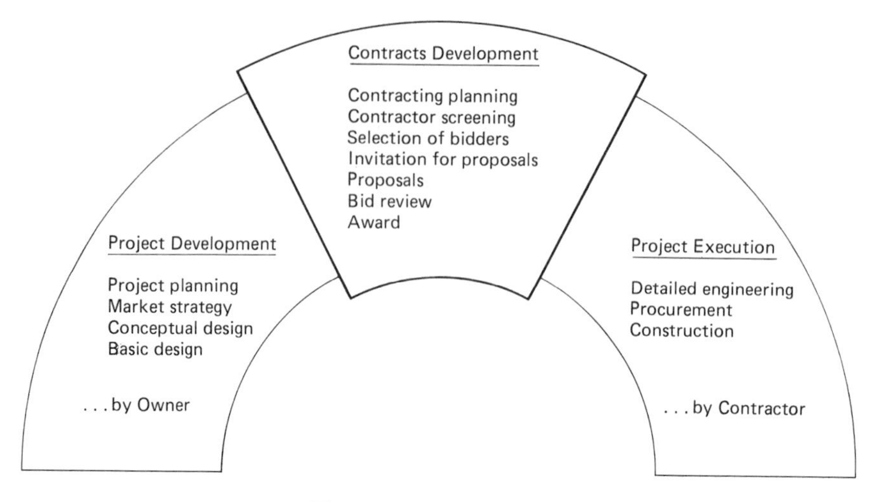


Figure 18-1. Project phases.

The longer-term objectives are more complex but also tend to become incorporated into short-term objectives. The contractor's objectives include:

- Winning a sufficient number of contracts to maintain a viable and growing organization.
- Being awarded contracts which add to its library of technology and skills.
- Providing a reasonable financial return to its owners, investors, and employees.

The owner's longer-term objectives include:

- Completing projects within anticipated/estimated cost—to meet its criteria for return on invested funds.
- Completing on or before scheduled completion—to meet its commitment to customers.
- Providing quality plants/facilities—to meet its operation/maintenance criteria and ability to produce quality products.
- Constructing facilities safely, and providing facilities which will operate safely—to meet its commitment to provide a safe and healthy environment for its employees and for the community.

With these objectives firmly in place, the owner and contractor each approach project management in general, and contracts development in particular, from its own perspective. The owner's and contractor's parallel approaches are diagrammed in Figure 18-2.

### The Art and Science of It

Defining "art" as a skill in performance acquired by experience, study, or observation, and "science" as knowledge attained through study or practice, contracting has elements of both. Both skill and knowledge are applied: the skill of understanding persons and organizations—how they act and react in specific circumstances—and of being able to synthesize scenarios to plan for reasonable and successful discussions and negotiations. The application of knowledge requires a disciplined gathering, analysis, and implementation strategy in order that each project experience provides a vernier adjustment to contracting tools—documents, procedures, and evaluation mechanisms.

### Business and Technical Linked

The design and construction of projects is, on the surface, a technical exercise. To build safely and accurately, and to provide a project/facility which will operate as planned and provide a safe and healthy environment for those who use it requires specialized technical knowledge and skill. But the other side of the coin is to do it economically. To provide such a project/facility within the budgeted funding and to operate within the cost parameters established requires specialized business knowledge and skill. Project management and contracts development need a combination of technical and business knowledge and skill to assure a successful venture.

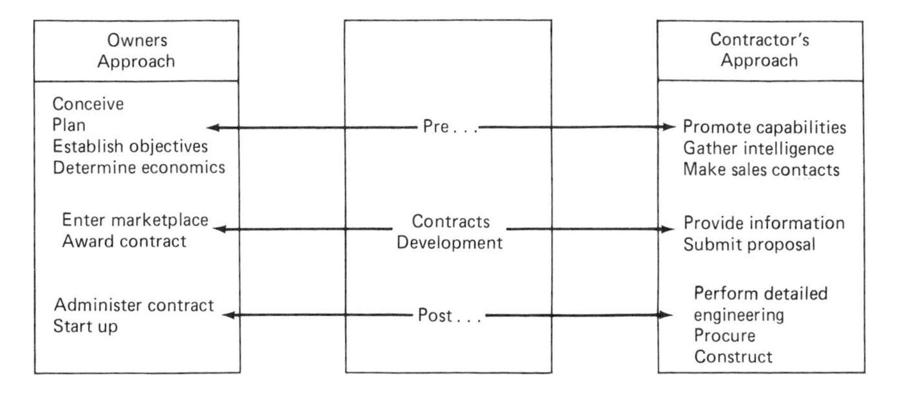


Figure 18-2. Approaches—project/contracts development.

### Human Traits Emphasis

Sensitivity, trust, integrity, and openness are active words in contracts development, as they should be in all relationships. To find the best solution, to provide the best in service to company and individuals, and to achieve professional stature are the goals. Human traits are part of the recipe; we should, as the old song says, "accentuate the positive and eliminate the negative" in our encounters.

## CONTRACTING PRINCIPLES

The basic principles which guide contracting are those which will provide for a successful venture, and will provide to owner and contractor the benefits each expect for their involvement. Only the most basic principles are cited here, since each company, whether contractor or owner, must establish its own set of principles and communicate these to its own employees and to its clients and customers.

### Competitive Bidding

The first principle is competitive bidding. There is little argument that competitive bidding, that is, awarding contracts based on the solicitation and review of proposals from several contractors, improves the industry. The owner benefits from the pressures of the marketplace in terms of economic design, procurement of equipment and materials, subcontracting, and optimized cost in general. The contractor benefits in terms of sharpening the skills and creativity of its employees, discovering techniques and methods to improve productivity, and lowering costs and increasing profits in general. However, inherent within competitive bidding as a principle is the recognition that contracting without competition may be required under certain circumstances. Such negotiated contracting is achieved by soliciting a proposal from a single selected contractor and negotiating mutually agreed terms and conditions. This is not counter to competitive bidding, it only indicates a limitation to competitive bidding. Two prime circumstances which dictate negotiation versus competition are (1) when there is a single source of technology, thus limiting the owner's choice of contractor, and (2) when the "best" contractor is already identified as the result of current work at site, duplication of facilities, or recent bidding experience. A corollary principle related to negotiated contracts is that the terms and conditions proposed by the contractor and sought by the owner should reflect the current marketplace; neither side should seek an unwarranted advantage.

### Single Responsibility

The second principle is related to the term "single responsibility." In such a circumstance the owner awards a single contract to a single contractor for all work (engineering, procurement, construction) on all facilities included in the project. Lines of communication, responsibilities, liabilities, and accountabilities are clearer, and control systems related to cost, schedule quality, and safety have the best chance to be successfully applied. However, in this less-than-simple world, a split of the work is often required to take advantage of the best available skills, technology, capabilities, and knowledge in the marketplace. Figure 18-3 indicates such vertical and horizontal splits. Alternative X is a horizontal split, alternative Y vertical, and alternative Z a mix.

### Fairness and Ethics

A very important principle to be cited here is that fair and ethical practices be established and followed by all parties and persons. It is important to create a mutual sense of trust within and between the parties. Without such trust a partnership cannot be successful, a contractual relationship will be strained, and the success of the venture may be in jeopardy. Beyond the moral obligations, fair and ethical practice makes good business sense. Written communications to document all discussions between the parties, equal treatment of all contractors by the owner, and strict sealed bid procedures all contribute to creating the desired atmosphere of trust.

| Activity | Alternative X | | Alternative Y | | Alternative Z | |
|---|---|---|---|---|---|---|
| | On-sites | Off-sites | On-sites | Off-sites | On-sites | Off-sites |
| Engineering | A | | A | B | A, B | C, D |
| Procurement | | | | | | |
| Construction | B | | | | E | |

Figure 18-3. Project execution alternatives. (*Note:* A, B, C, D, and E indicate separate contractors.)

### Management's Right To Know

A final principle may seem obvious but often is forgotten: management has the right to know. The group responsible for contracts development needs to make an intentional effort to keep management advised of progress, at times when approvals are required, of course, but also when key activities have been achieved. It is too easy to perform the tasks as planned, achieve anticipated results, and expect to move on. But such progress should be reported. This principle provides for "no surprises," and allows communication between management and the contracting group.

## SEQUENCE OF CONTRACTING ACTIVITIES

As in all undertakings, contracting follows a set sequence of activities. And each step builds on the previous step and provides the basis for the next step. Figure 18-4 sets forth the sequence of contracting activities. The elapsed time has not been shown in Figure 18-4 since each contracting effort will require a different time frame depending on type of contract, owner's requirements related to reviews and approvals, complexity of projects including number of plants, technology, grass roots versus revamp, location of facilities and contractors' offices, etc. In general, a minimum of three months should be allowed for a reimbursable cost con-

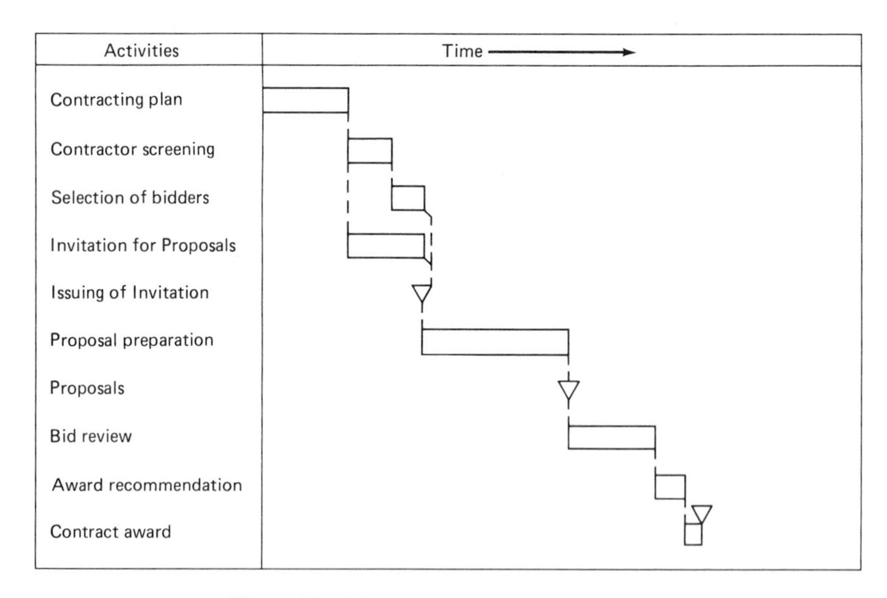


Figure 18-4. Sequence of contracting activities.

tracting exercise, and five months for lump sum. (Lump sum requires more extensive prequalification of bidders, and more time for contractors to prepare their bids.) It is important that the contracting effort start early to allow sufficient time for decisions to be made on appropriate data and study, to avoid hurried decisions and potential mistakes. And (this is very important) even if the contracting time is condensed due to market pressures, business confidentiality, or other reason, each step shown should be taken, even briefly, to assure that the best contractor is selected and an acceptable contract is developed to provide a solid basis for a successful project.

## CONTRACTING PLAN

The architect develops the plans before the cathedral is built. So it is with contracting. The development of the contracting plan is an essential ingredient to successful contracts development. Without a plan, there is no direction/guide for the participants. A poorly conceived plan will result in false starts, mistakes, and frustrations to both owner and contractors. It is not only the contracting effort that is affected, but the whole of project execution: cost, schedule, quality, and safety.

### The Planning

The contracting planning consists of (1) the gathering of all data and information related to the project which will have, or has the potential to have, an impact on the contracting effort; (2) a thorough analysis of such data and information including the effect on cost, schedule, quality, and safety aspects of the project; (3) the structuring of the potential alternative contracting strategies/approaches including qualifying and quantifying the advantages and disadvantages of each; and (4) selecting the best contracting strategy and viable alternatives.

**Information and Data Gathering.** The information and data required will cover (1) facilities (type of units, technology, size, cost); (2) location factors (existing versus new, construction mode, potential conflicts if working within an operating plant, e.g., hot work permits, emergency situations); (3) scope of services required from contractor (engineering, procurement, construction); (4) key project dates (availability of basic design specifications, turnaround/shutdown timing, completion date); (5) project and appropriation approval requirements and timing; (6) financing considerations (source, amounts, restraints, contractor involvement); (7) contracting marketplace (capacity, work load, technology, applicable experience); and (8) status of vendor and subcontractor marketplace.

**Analysis and Strategy.** All information and data are analyzed, checking the interactions among all elements, and then developing various contracting alternatives, each of which is considered applicable to the project and is judged capable of bringing the project to a successful conclusion.

After determining the advantages and disadvantages of each potential alternative, it is necessary to quantify the differences in order to select that alternative which best meets the owner's objectives—cost, schedule, quality, and safety. Viable retreat strategies also need to be identified in the event the chosen alternative cannot be achieved.

### The Contracting Plan

The results of the contracting planning are documented for presentation to, and review and approval by, the appropriate management; this document is the Contracting Plan. Included in the Contracting Plan will be the following sections:

- Executive Summary
- Purpose of Plan
- Project Background and Factors
- Contracting Principles and Policies
- Analysis of Contracting Alternatives
- Recommended Contracting Strategy
- Viable Retreat Strategies
- Detailed Schedule of Contracting Activities
- Preliminary Schedule for Project Execution
- Potential Concerns/Problems and Resolutions
- Contracting Procedures
- Contractors to Be Screened
- Draft of Screening Telex/Letter
- Decisions and Approvals Required

The Contracting Plan, after management approval, will guide the contracting effort. However, it will be subject to review at various times, for example, after screening, to verify or change the strategy. Changes may require the Plan to be revised and reissued.

## CONTRACTOR SCREENING

For most projects, the number of contractors who can perform the work, or claim such capability and experience, is large. The process of contractor screening is to start with the total known population; select those who appear to have the capability and experience required (the "long list"); contact that group to gather up-to-date information on their capacity, experience, capability, and other items; and select a limited number to submit bids (the "short list"). This process has a number of advantages: (1) by considering the total population and developing the "long list" the owner is more sure that all contractors with applicable capability are considered, and contractors are assured that efforts made to keep owners aware of their company's capabilities will be productive; (2) contacting the "long list" to gather up-to-date information assures that recent/current changes in offices, experience, projects, etc., are part of the basis for selecting bidders; and (3) limiting the number of bidders provides each contractor a fair chance in a competitive atmosphere, and conserves the owner's resources required for bid review.

The screening process is in itself an application of the fairness principle stated as one contracting principle. All potential bidders receive the same information, and the capability, experience, and interest of each is considered in selecting the bidding slate.

### The Screening Telex/Letter

The initial contact with the potential contractors is through the sending of a screening telex or letter. The use of a written document is an absolute necessity. The use of a written document ensures that all contractors receive the same information, thus avoiding misunderstandings or incomplete data which are likely to occur as a result of telephone calls and personal conversations. And it also ensures the uniformity of replies from the contractors, thus allowing for an easier and surer analysis and selection of the bidding slate. The written reply from the contractor will be an organizational response, not individual, and the owner can accept the reply as the contractor's management decision without a need for further time-consuming discussions.

The screening telex/letter is essentially in two parts. The first part provides information to the contractors regarding the facilities, contracting strategy, and key contracting-related dates. The second part requests information from the contractors related to their capability to accomplish the project. The information provided to the contractors includes the following:

- Description of facilities—nonproprietary/nonconfidential information on technical basis, capacity, product slate, special materials or features, etc.
- Proposed contracting basis and timing—type of contract, scope of

contractor's services, dates established for release of invitation for proposals, receipt of proposals, and contract award.

- Key project dates—availability of basic design specification, start of detailed engineering, start of field construction, and mechanical completion.
- Other significant factors.

The information requested from the contractors includes:

- Willingness to submit a proposal—on the contracting basis described.
- Engineering office—location, number of personnel, and current/projected work load.
- Tentative project execution plans—organization, application of experience, construction mode, and location of primary activities.
- Corporate and office experience—facilities, technology, engineering for location/country, and construction at location.
- Other significant information.

### Analysis of Contractors' Replies

The replies to the screening telex/letter will be consistent and will lend themselves to a direct line-by-line comparison and evaluation. The objective of the analysis is to identify the "short list" of contractors who are judged to be able to accomplish the project, meeting all the owner's objectives, and are judged able to submit competitive proposals in the current marketplace. The analysis normally consists of two steps: first are "musts," items with which the contractor must conform to be considered as a potential bidder; second are "wants," items which define the work required to complete the project but which need to be evaluated in order to determine the comparative levels of capabilities of the contractors.

The "musts" list should be relatively short, including (1) willingness to submit a proposal in conformance with the established basis; (2) sufficient personnel, both in skills and number, to perform the work; (3) financial stability to assure completion of work; (4) no conflict related to technology; and (5) ability to perform work at the location, including meeting any licensing, permit, or registration requirements. There may be others but the list should be kept short. The result of this step will be simply, yes, the contractor can be considered as a bidder, or no, it cannot.

The second step is to evaluate the "wants" list for each contractor which passed the "musts." The owner, for this exercise, must establish a list of items which are important to and required for the execution of the project, weighting each one as to its impact and importance. The items normally comprise the data and information requested from the contractors in the screening telex/letter including office capacity and work load, experience, and tentative execution plans. These are weighted based on the specific requirements of the project. For example, technology experience would be weighted high if a project included some highly sophisticated technology; or, in a location which had a long history of labor relation problems, knowledge and experience in construction at the location would receive a higher weighting.

Each contractor is rated for each item on the "wants" list. The scale is established by the owner depending on the degree of sophistication wanted and/or the relative/comparative differences anticipated in the contractors' performances. (A scale could be: 1 = "above average," 2 = "average," 3 = "below average"; or 1 = "oustanding" . . . 3 = "above average" . . . 5 = "average" . . . 7 = "below average" . . . 10 = "unacceptable.")

### Selection of Bidders

The selection process is now prepared. The rating given to each contractor for each item on the "wants" list is multiplied by the weighting for that item to establish the component for each line item. The sum of the components for each contractor will identify those contractors which best meet the objectives of the owner for the specific project. Sometimes the result of the analysis will confirm the judgments previously made by the owner, but often there will be surprises. In either case, the process is successful.

But the final selection of the four or five contractors for the "short list" is not complete. Beware of the "numbers game." The sum of the components will identify the best group, possibly six to eight, with small differences in the totals. Subjective factors, factors not able to be included in the "wants" list, should be considered in the final selection. However, this final step should be for the purpose of confirming the rating-times-weighting results or making changes based on real issues; it is not the time to revert to a "my-favorite-contractor" syndrome.

After the selection of the bidding slate has been approved, the selected bidders are contacted to reconfirm their willingness to bid. Only then can the Invitation for Proposals be issued.

## INVITATION FOR PROPOSALS

Since the owner has planned the venture, in the planning has established certain policies and principles which will guide its implementation, and has established economic goals, the owner must communicate these and all other aspects of contracts development and project execution to all the bidders in a consistent manner. That is the purpose of the Invitation for Proposals. The Invitation for Proposals is a set of documents, each with its own purpose, which provides a complete basis for the preparation and submission of proposals by contractors, and contains the base documents which will make up the contract to be signed by the parties at contract award. The Invitation for Proposals consists of:

- Transmittal Letter.
- Information to Bidders.
- Proposal Form.
- Agreement Form.
- Job Specification.

All the above documents are issued to all bidders; at contract award, after appropriate modifications as a result of agreements between the owner and the selected contractor, the Agreement and the Job Specification constitute the Contract for the project.

### Transmittal Letter

The Transmittal Letter is an essential part of the Invitation. It is sent jointly to all bidders; thus each will know its competition, and if the selection of the bidding slate has been done well, the degree of competition will increase immediately. The letter sets forth only the key items, for example, describes the content of the Invitation package, indicates that the Invitation documents are to be returned to owner at award, states date and time the proposals are to be submitted, and includes other critical, significant items.

### Information To Bidders

If each bidder were left on its own to prepare a response to a solicitation for a proposal, the comparison by the owner would at best be chaotic. The Information to Bidders (and the Proposal Form) are developed to avoid that situation by ensuring the consistency and comparability of all proposals. The Information to Bidders is a memorandum which describes in detail the ground rules and procedures which each contractor must follow in the preparation and submission of its proposal. It references the other documents in the Invitation package and establishes the content of the contractor's proposal, including, in certain cases, specifying the format required in submitting information and/or data. The Information to Bidders will contain as a minimum the following:

- Location and brief description of facilities.
- Type of contract (lump sum, reimbursable cost, etc.), contracting mode (competitive, negotiated), scope of services (engineering, procurement, construction).
- Content of Invitation for Proposals.
- Anticipated schedule for issuing owner-prepared specifications to contractor.
- Owner's contact(s) during bidding (name, address, telephone, etc.).
- Requirements regarding exceptions to and comments on Agreement Form.
- Requirements regarding exceptions and alternatives to Job Specification.
- Requirements and content of the technical proposal to be submitted by the contractor.
- Requirements and content of the project execution proposal to be submitted by the contractor.
- Instructions to the contractor regarding the submission of its proposal (number of copies, date and time, name, address, notifications, etc.).

Other significant items will be included which clarify the owner's policies (conflict of interest, business ethics, written documentation); emphasize owner's intent or requirements (safety, merit versus union shop, signing of contract at award, audit, etc.); or direct the contractor's attention to portions of other documents in the Invitation package (confidentiality of owner's information, risk management and insurance policy, use of Proposal Form, content of Job Specification, etc.). When the Information to Bidders is prepared, it should be remembered that the contractor's proposal can only be as good as the instructions provided. Well-developed instructions will make the contractor's work of preparation more efficient and effective, and the selection of the successful contractor for award easier.

### Proposal Form

The Proposal Form is prepared to serve as the base document used by the contractor in submitting its proposal. It is developed in parallel with the Agreement Form so that for every piece of information and data required to complete the Agreement Form and prepare it for signature is requested from the bidders. Attachment forms, request for special attachments, etc., are all included as part of the Proposal Form. A representative listing of such items requested includes the following:

- Mechanical completion date.
- List of subcontracted services or work.
- Names of key project personnel.
- Insurance coverages and deductibles.
- Lump sum prices.
- Rates for reimbursable cost contracts, including payroll burdens, departmental overheads, etc.
- Reimbursable cost fixed fees.
- Rates for computer, reproduction, and construction tools.
- Salary and wage charts.
- Estimated scope of engineering and construction management services.
- Employee policies for travel, relocation, etc.
- Exceptions and comments on Agreement Form.
- Exceptions and alternatives to Job Specification.

Thus, the Proposal Form becomes the base document in the contractor's proposal; extra copies of the Proposal Form are sent with the Invitation package and the contractor is required to fill in the form (no changes or alterations allowed), have it signed by an officer, and submit it to the owner (each contractor fills in the required number, each is signed).

### Agreement Form

The Agreement Form is a draft prepared by the owner which contains the terms and conditions which it proposes as the basis of the contractual agreement between the parties. Said terms and conditions will govern the rights, duties, obligations, responsibilities, and liabilities of the owner and contractor during the execution of the project. It is either (1) a legal document which is administered by the project teams (owner's and contractor's) during project execution, or (2) an administrative document to guide the actions of the parties during execution, which must be in conformance with law, or (3) both. In any case, it is very important that this document, after it is modified during the bidding and bid review by mutual consent of the owner and contractor, becomes part of the day-to-day tools of the project teams; it eases problems and avoids conflicts.

The owner, in preparing the document, should assess and distribute risk in accordance with the ability of each party to control or provide for in some way. The contractor, in preparing its proposal, should plan execution procedures to control its risk, provide for potential liabilities through insurance and/or risk contingencies in its quotes, and offer exceptions only when it cannot control or provide for otherwise. It is of course necessary to achieve an acceptable contractual basis for a contract to be awarded. And this must be done prior to contract award so that the contractor and owner have no extraneous issues to resolve which will interfere with the execution of the project.

The owner should have standard documents, developed with input from and approval of its law, tax, financial, risk management, and other staff functions, to use as the base document for all contracts. This will assure more efficient contracting, consistency of approach to contractors, and easier responses by a contractor from project to project.

### Job Specification

The Job Specification is a compilation of various individual documents, which when taken as a whole define (1) the administrative and procedural requirements for performing all work and services under the contract ("coordination procedure"); (2) the standards to be achieved related to operability, safety, and quality of materials and equipment, and the facilities as a whole (general and specific "standards and practices"); and (3) description of the facilities to be engineered and constructed ("design specifications").

The Job Specification will be a dynamic, evolving document. The type of contract will determine the scope of the Job Specification included in the Invitation for Proposals. For lump sum contracting, the entire Job Specification is required; for reimbursable cost, only the "coordination procedure" is required. Thus, for reimbursable cost contracting much of the Job Specification is issued to the contractor over a period of time during the early phases of project execution. Changes to the Job Specification can only be authorized by the owner, and in each event, if there is a resulting impact on the contractor's cost, schedule, and lump sum price or reimbursable cost fee, a change order is issued by the owner to adjust the respective item.

The contractor will offer exceptions and alternatives to the Job Specification as part of its proposal; these will be resolved during bid review and appropriate modifications will be made in the various documents. At contract award the Contract will consist of the Agreement and the Job Specification; the Agreement will be the precedent document.

## CONTRACTORS' PROPOSALS

The Invitation for Proposals is issued to the bidders and each contractor begins the preparation of its proposal. The key to the work of preparing a proposal, and to submitting a proposal which will be viewed favorably by the owner, is a thorough knowledge of the content of the Invitation for Proposals documents. The proposal preparation team should spend sufficient time to become familiar with all aspects of the documents. Questions should be asked of the owner to clarify any item not fully understood (assumption of meanings must be avoided). Visits to the site should normally be made to gain firsthand understanding of potential problems and possible solutions. Often the owner will require such visits, usually a combined visit by all bidders to assure consistency in information given to all bidders. The contractor should assign a leader to its proposal effort, preferably the person to be proposed as its project manager, and enlist other full-time and part-time personnel as appropriate.

The contractor's proposal must address the needs of the specific project. Standard "boilerplate" will not give the owner a satisfactory impression that the contractor understands the project requirements or the scope of work. Alternative execution schemes should be developed and evaluated. These alternatives would address the following items: project organization, key personnel, and staffing; conventional versus computer-assisted engineering and drafting; alternative sources of equipment and materials; subcontracting versus direct hire construction; type of control systems; etc. Evaluation of alternatives provides the basis from which the contractor can choose the proper ingredients which will provide the owner with the lowest-cost facility consistent with the quality and safety parameters and within the desired completion schedule.

Although the owner's Job Specification may seem specific and even appear inflexible, an owner is normally willing to consider unique concepts and innovative ideas which will enhance the project without compromising plant safety and operability. And such concepts and ideas often give the contractor a competitive edge.

The contractor's proposal will consist of three separate proposals:

- Commercial Proposal.
- Project Execution Proposal.
- Technical Proposal.

### Commercial Proposal

The Commercial Proposal will consist of the Proposal Form, with all blanks filled in as required and including all attachments, tables, and information requested. This section of the contractor's proposal contains all business terms (lump sum prices, reimbursable cost fees, fixed rates, estimates of services, etc.), plus proposed exceptions, comments, and alternatives to the Agreement Form and Job Specification. It is the most sensitive section of the proposal. Therefore, the Commercial Proposal must always be kept separate from other sections and sealed in special envelopes when the proposal is delivered to the owner. This "sealed bid" procedure enables the owner to maintain the security of the information. The content of the Commercial Proposal is limited to members of the bid review team who have a need to know; no one outside the team has access to the information.

### Project Execution Proposal

The content of the Project Execution Proposal will be that defined in the Information to Bidders; it will contain the following material:

- Corporate organization charts including reporting and financial responsibilities to parent organizations.
- Experience on similar projects and in performing engineering and/or construction work at the location.
- Project execution plans which detail how the work on the project will be accomplished including the coordination and management within the engineering office, between offices, and between engineering and field construction.
- Project organization chart showing lines of authority and communications, and job description of each key position, names of proposed key personnel (with resumes and references).
- Work load charts showing total technical personnel and the number of engineering and construction management persons available for assignment to this project.
- Preliminary work schedule reflecting as much detail as is available to the contractor during the preparation of its proposal.
- Detailed explanation of how the contractor plans to perform each function of project execution, namely: engineering, procurement, subcontracting, cost control and estimating, planning and scheduling, expediting, inspection, material control, field construction, and construction management.

- An explanation of contractor's safety practices and how they will be implemented on this project.

### Technical Proposal

The content of the Technical Proposal will vary more than any other portion, from very detailed to none required. If the proposal encompasses the provision of the technology, then of course the contractor must provide sufficient information for the owner to make a comparison of proposals and a decision. However, when a competitive proposal is submitted, the technology should be presented in two packages. One will contain nonproprietary/nonconfidential information only; this is submitted to the owner as part of the base proposal, and will be opened and evaluated as part of the initial phase of the bid review. The second package will contain proprietary/confidential information; this will be completed at the same time as the base proposal, sealed for security, and either submitted to owner for holding or held by contractor or a third party until it is needed. If the contractor's proposal remains in contention after the initial review, a confidentiality agreement can be signed and the owner can inspect the second package. This procedure relieves both the owner and contractor of the unnecessary exposure of proprietary information.

Lump sum proposals will contain certain technical details on equipment and materials to be incorporated into the project. This provides the owner a basis for evaluating the facilities to be designed by the contractor to assure itself that the plant will meet the requirements of the Job Specification.

Reimbursable cost proposals for projects which have no technology or process design component normally do not require a Technical Proposal.

## BID REVIEW

### Bid Review—General

Proposals are submitted by contractors to the person designated by the owner, at the location and time stipulated. Their work for the moment is completed; they must now wait for a response. The owner initiates the bid review phase of contracting. It is often said that contracting planning is the most important step in contracting, and it may be. However, if the owner does not properly plan and implement a thorough and effective bid review, the whole process can be "blown," and the objective of selecting the best contractor, in terms of cost, schedule, quality, and safety, may not be achieved. Bid review is the most difficult step in contracting. It requires a sensitive balancing of facts versus judgments, of objective versus subjective reasoning, of work tasks versus cost, of time versus contract price, of risks versus opportunities. Lump sum and reimbursable cost bid reviews differ considerably. Although the basic procedures apply to all bid reviews, each bid review will be different and the procedures for each effort must be tailored.

### Bid Review Practices

The bid review is a complex effort and will follow certain basic practices:

- Bid review planning.
- Dedicated bid review team.
- Sequential review.
- Reviews within bid review.
- Owner-contractor interactions.
- Contract signed at award.

**Bid Review Planning.** Continuing, progressive planning is required throughout the contracts development. Planning for the bid review ensures that it will be an efficient, effective exercise. The plan developed will require all parties involved to "buy in" and will be the only guide and basis during the actual bid review. Of course events may require some adjustments, but, since the adjustment will be made from a written plan, it will be quicker and easier, and better understood by the participants.

The planning should address all related project and contracting factors including the expected content of the contractors' proposals, bid review objectives, the bid review team's indication of time and activities for each, and a detailed explanation of activities from bid opening to contract award. A typical table of contents of an actual Bid Review Procedure Plan is as follows:

Table of Contents

Project Background
Contracting Background
  Strategy
  Screening
  Bidders
Invitation for Proposals
Contractors' Proposals
Bid Review—General
  Location

- Objectives
- Sequence
- Award

Bid Review Team and Responsibilities

Bid Review—Specifics

- Bid Opening
- Documentation/Internal Communications
- Communications with Bidders
- Initial Review/Selecting Contenders
- Visits to Contractors' Offices
- Evaluation of Key Personnel
- Selection of Contractor for Award

Contract Award

- Award Recommendation
- Contract Award/Signing

In addition to the body of the document, attachments providing relevant details where necessary to further explain the project, and forms which will be used during the bid review are included.

**Dedicated Bid Review Team.** Qualified personnel to review the various portions of the contractors' proposals must be selected, relieved of their normal duties during the bid review period, and fully informed of the project and contracting matters. A thorough review/discussion of the Bid Review Procedure Plan with the bid review team prior to receiving proposals is a prime requirement. A typical bid review team organization is shown in Figure 18-5.

**Sequential Review.** The bid review will follow the steps set out in the Bid Review Procedure Plan. There is a basic two-step sequence normally followed. The first step occurs after opening and registering all proposals. All parts of each proposal are read completely, all data and information are tabulated to check for completeness and consistency, all necessary calculations are made (for all items not quoted as fixed prices) in order to compare business terms, and the technical, project execution, contractual and other items are evaluated. With this accomplished, the contractors who offer the most complete proposals, submit competitive business terms, provide technical bases conforming to the specifications in the Invitation for Proposals, and offer satisfactory project execution plans and procedures will be identified as contenders for award. The obverse is that the contractors who submit high business terms, and whose technical

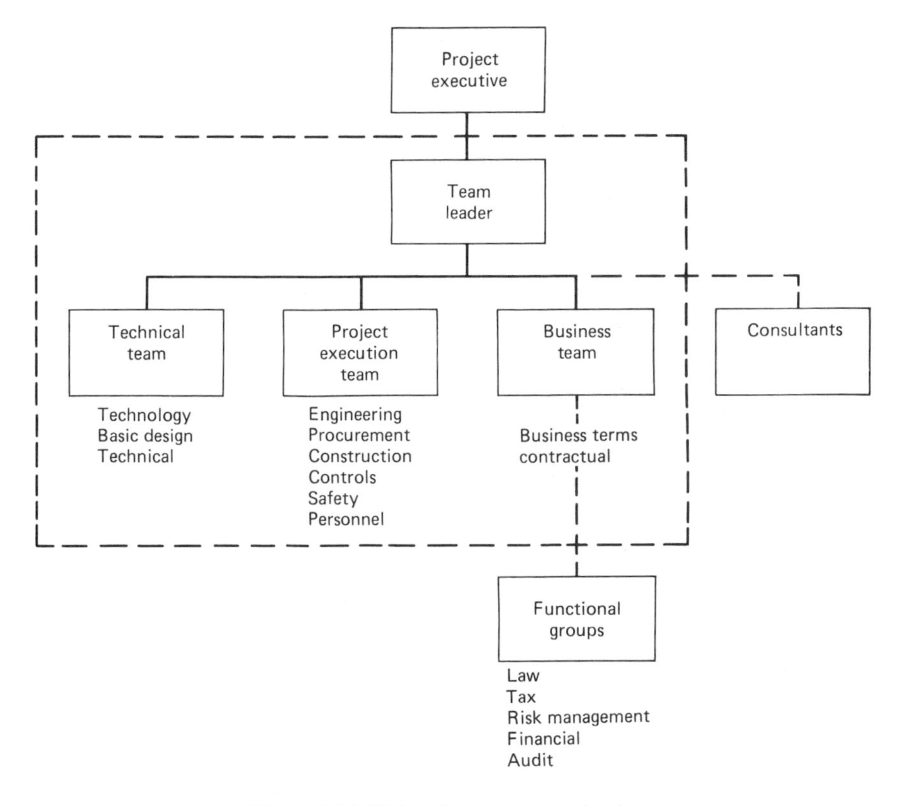


Figure 18-5. Bid review team organization.

and/or project execution plans and procedures are judged to be no stronger than the other contractors, will be dropped from contention.

The second step is then to make a thorough review of the proposals of all contractors remaining in contention. This is the time-consuming portion of the bid review. Whereas the initial step may take up to four days, this step normally takes up to four weeks—even longer in some cases.

**Reviews Within Bid Review.** Figure 18-5 (bid review team organization) provides the clue to the reviews-within-bid review concept. If the project/contracting requires technical input, then there will be three separate reviews: technical, project execution, and business. Each is conducted as a closed review, limited to those persons assigned to that area of the review. No person on an individual team will provide details of the evaluation and appraisal of their part of the review to any person on the other teams. Only the bid review team leader will have access to the entire procedures and actions. This procedure ensures an objective approach for the entire review, and helps maintain security of the proposals' content.

At the conclusion of the detailed review, the three teams share their conclusions and the melding of comparisons and evaluations leads to the selection of the contractor for contract award.

**Owner-Contractor Interactions.** There will be extensive interactions between owner and contractor during this period. Formal, written documentation will be the rule. The normal mode is owner-question/contractor-answer. It should be kept in mind that the owner's effort is aimed at clarifying and understanding the proposals, and trying to ''upgrade'' each proposal to its best possible position. Many times the owner's questions may be a way of making suggestions to strengthen the proposal. The contractor's effort is aimed at providing all necessary information, answering questions promptly, and watching for opportunities to offer technical and/or project execution alternatives which will enhance its proposal.

The bid review requires fair and impartial treatment of all proposals and all contractors. Personnel of owner and contractors must remain vigilant to assure that they are acting in the best interests of both parties. The actual advantage should be with the contractor who makes the best proposal—and only that. Contacts should always be at a business level during bid review; to vary from that can only lead to questions of propriety and perceived or actual conflicts. Being concerned for business ethics is the primary guide in owner-contractor interactions.

**Contract Signed at Award.** An important strategy/practice is that all reviews, activities, discussions, and evaluations must be completed before contract award. This includes, most importantly, the resolving of all exceptions, alternatives, and comments to the Contract documents, allowing for the signing of the Contract at contract award. This allows the work to begin immediately after contract award, and the contractor's and owner's project teams can give their undivided attention to the project. In addition, it is simple logic that to know what you have agreed to do before you do it will result in better relationships.

### Lump Sum Bid Review

The objective of a lump sum bid review is to determine which contractor offers the lowest lump sum price and provides a proposal which conforms to all technical, project execution, and contractual requirements. Normally the lowest lump sum price, after bid review, determines the successful contractor. However, such factors as plant operating costs (related to equipment selection and/or technology), extraneous owner's cost for travel or relocation costs, etc., or other items may influence the final decision if the quoted lump sum prices are close.

Figure 18-6 charts a typical lump sum bid review. Since lump sum prices are quoted, relatively little work is required in the review of business terms. The major effort is in the review of the technical proposals. Specialists in process technology and design, process equipment, materials, machinery, control systems, electrical, safety and fire protection, and environmental will review the technical proposals to ensure that the plant and facilities being proposed fully conform to the requirements of the Job Specification. Contractors will offer exceptions and alternatives which will provide a more economical plant and a more competitive proposal. These must be evaluated and accepted or rejected (usually with a corresponding adjustment to the quoted lump sum price).

The review of the project execution proposal focuses on the proposed execution plan, key project personnel, and the contractor's procedures and techniques to provide assurances that the contractor can deliver a plant as put forward in its proposal.

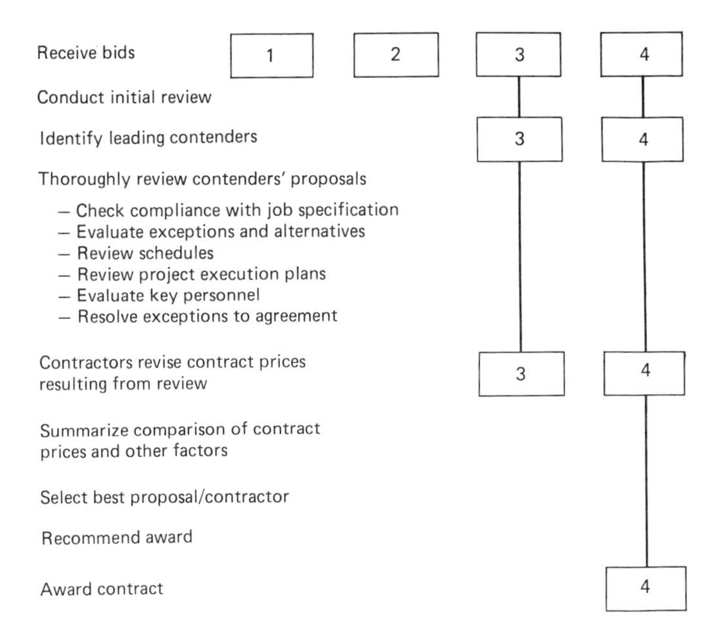


Figure 18-6. Lump sum bid review.

In order to be able to sign the contract at the time of contract award, participants must resolve all exceptions to the Agreement Form during bid review.

After all portions of the proposals have been reviewed, all exceptions, alternatives, questions, and comments have been resolved, and the bid review team is satisfied that all elements are acceptable and in conformance with the Invitation for Proposals, then the selection of the contractor for award can be made.

### Reimbursable Cost Bid Review

The objective of a reimbursable cost bid review is to determine which contractor offers the best combination of business terms, technical expertise, and project execution capability. The review is more difficult than a lump sum bid review since a firm total price is not quoted and the bid review team must evaluate the contractors' expertise and capability and quantify the differences to determine which contractor will be able to meet all the owner's technical and project execution requirements and complete the project at the lowest overall cost. This exercise requires experienced personnel with the ability to make objective and subjective evaluations and judgments and translate these into concrete impacts on project cost and project schedules.

Figure 18-7 charts a typical reimbursable cost bid review. The reimbursable cost bid review will consist of a detailed review of all portions of contractors' proposals—commercial, technical, and project execution.

**Commercial Proposal.** The commercial proposal includes business terms and contractual considerations. Business terms consist of fee, salaries and wages, payroll burdens, departmental overheads, and fixed rates for computer, reproduction, and other services.

Fee is normally quoted as a fixed amount covering all services performed by the contractor. It consists of the contractor's general corporate overhead and all profit. The contractor also may quote certain cost factors related to changes and an allowance for adjustment to the fee over the life of the project for such changes.

An estimate is made of the total cost of each contractor's engineering, procurement, and construction management services, using the contractor's business terms included in the proposal and the owner's estimate of the hours to be expended. An allowance to cover changes is added.

After all calculations are completed, checked, and tested for reasonableness, a summary of the results is prepared. Figure 18-8 shows a typical summary of business terms. The comparison of the business terms

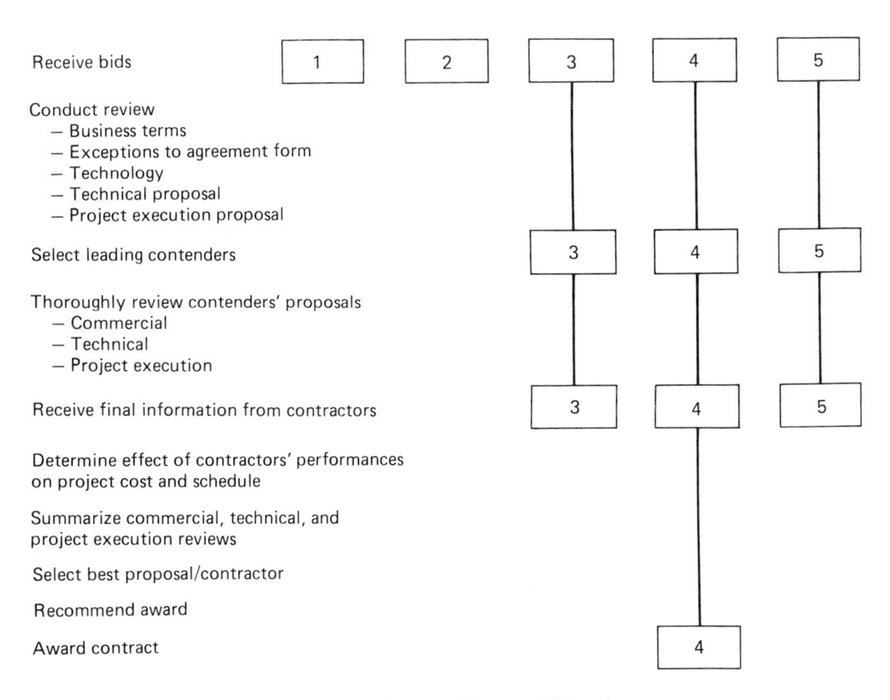


Figure 18-7. Reimbursable cost bid review.

| | Contractors | | | |
|---|---|---|---|---|
| | A | B | C | D |
| Fees | XXX | XXX | XXX | XXX |
| Engineering and procurement | | | | |
| Salaries and wages | XXX | XXX | XXX | XXX |
| Payroll burdens | XXX | XXX | XXX | XXX |
| Departmental overheads | XXX | XXX | XXX | XXX |
| Allowance for changes | XXX | XXX | XXX | XXX |
| Subtotal | XXX | XXX | XXX | XXX |
| Construction management | | | | |
| Salaries and wages | XXX | XXX | XXX | XXX |
| Payroll burden | XXX | XXX | XXX | XXX |
| Expenses | XXX | XXX | XXX | XXX |
| Allowance for changes | XXX | XXX | XXX | XXX |
| Subtotal | XXX | XXX | XXX | XXX |
| Other services (as applicable) | XXX | XXX | XXX | XXX |
| Total estimated services | XXX | XXX | XXX | XXX |
| Differential Cost | Base | + | + | + |

Figure 18-8. Summary of business terms.

is an important element in the bid review and determines the relative competitiveness of the bidders. However, it is not, as in a lump sum bid review, the primary selection criterion. Experience has shown that a contractor's performance capability can have sufficient impact on savings related to total project cost to offset higher business terms.

**Project Execution Proposal.** The project execution proposal contains details on the plans, systems, techniques, and procedures which the contractor will follow in executing the project. These are grouped into four categories:

- Detailed Engineering.
- Procurement/Purchasing.
- Project Controls.
- Field Construction.

Each category is reviewed in detail with each contractor to confirm strengths and discover weaknesses. The objective is to encourage each contractor to modify its proposal in such ways as to capitalize on the strengths and eliminate the weaknesses to bring its project execution proposal up to its best possible position.

When the proposals are analyzed and upgraded, the following items/elements within each category are addressed:

- Detailed Engineering
  —Project organization
  —Office capacity
  —Availability and quality of personnel
  —Key project personnel
  —Project team effectiveness
  —Coordination of work functions
  —Quality of drawings and specifications
  —Standards, procedures, and techniques
  —Use of models and computers
  —Experience: facilities, owner, location
  —Familiarity with codes
- Procurement/Purchasing
  —Project organization
  —Availability and quality of personnel
  —Procedures and techniques
  —Forms and paper flow
  —Experience: facilities, owner, location

—Knowledge of marketplace
—Knowledge of vendors and subcontractors
—Interaction with engineering
—Familiarity with codes

- Project Controls
  —Project organization
  —Procedures and techniques
  —Forms and paper flow
  —Use of computers
  —Availability and quality of personnel
  —Forcasting procedures
  —Reporting systems: quality and timeliness
  —Progress measurement techniques
  —Interaction with engineering and field
- Field Construction
  —Project organization
  —Quality control procedures
  —Safety awareness and procedures
  —Reporting and communications
  —Availability and quality of personnel
  —Knowledge of local labor
  —Experience: facilities, owner, location

The technique for evaluating the project execution capability and translating the evaluation into a cost impact generally is as follows.

*EVALUATING ITEMS/ELEMENTS WITHIN CATEGORIES.* Each item/element is evaluated with a judgment as to whether it is a strength in the proposal which would have a positive impact on the project cost (i.e., lower the cost), or a weakness which would have a negative impact, or a neutral item having no impact on the cost. These evaluations/judgments are made based on the proposals, question and answer exchanges, review of contractor's past experience, reference checks, and visits to the contractor's office to discuss the project execution with the contractor's personnel (in particular those who will be working on the project) and review actual details on some of the contractor's current assignments. The visit to the contractor's office is the most important.

*RATING CONTRACTOR IN EACH CATEGORY.* The strength-weakness-neutral evaluations for the items/elements within each category can be weighted by the bid review team, based on the importance of each to the project objectives, and each category can be assigned a numerical rating. The rating will reflect the anticipated performance of the contractor measured against a standard of performance experienced by an "average" contractor on the owner's previous projects. (In a rating scale of 1 to 5, 3 would represent an "average" performance, with 1 representing an anticipated outstanding performance, and 5 representing an anticipated poor performance.)

*CALCULATING IMPACT ON PROJECT COST.* The performance of the contractor within each category, or a subcategory, will impact a certain portion of the project cost. For example, a contractor's performance in the category of purchasing/procurement will impact the cost of materials, equipment, and subcontracts. The bid review team can determine which areas are impacted within each category and from the owner's cost estimate assign the cost to that area. The team also needs to determine the variation in anticipated cost from an outstanding performance (a 1 rating in the above rating scale) and a poor performance (a 5 rating). (If the team decides that the variation will be ±10%, a 1 rating will reduce the cost area effect by the category by 10%, and a 5 rating will add 10%.)

The calculation of the cost impact for each category will be (i) the ratio of assigned rating versus "average" rating ("1" assigned 100% of savings, "2" assigned 50% of savings, "3" no impact, "4" assigned 50% of added cost, "5" assigned 100% of added cost) times (ii) the cost of area affected times (iii) the anticipated percentage variation. The algebraic sum of the calculated impacts for each category indicates the relative impact which each contractor's performance will have on the project cost and identifies the best execution contractor.

**Technical Proposal.** The review of the technical proposal is not as detailed as for a lump sum proposal unless the proposal includes the provision of technology. Otherwise the review is generally integrated into the project execution review.

### Bid Review Summary

Upon completion of the three separate reviews—commercial, project execution, technical—the bid review team will combine the results and make the selection of the contractor to recommend for award.

## AWARD RECOMMENDATION

After the bid review is completed, and after the best proposal/best contractor has been identified and selected, an Award Recommendation is prepared. The purpose of the Award Recommendation is twofold: (1) to present the conclusion to management for approval, and (2) to document the results of the bid review for the records. The Award Recommendation will consist of the following:

- Introduction/Purpose.
- Project Background.
- Contracting Background.
- Award Recommendation.
- Summary of Business Terms.
- Summary of Project Execution Proposals.
- Summary of Technical Proposals.
- Status of Contract Documents.
- Significant Issues and Resolutions.
- Summary of Bid Review.
- Steps to Contract Award.

## CONTRACT AWARD

After approval by management, all open items, if any, are resolved with the selected contractor, the Contract documents are modified to reflect the owner-contractor agreements, and the Contract is awarded and signed.

## CONCLUSION

The partnership created during the contracts development phase of a project, following well-established techniques and procedures, has a firm foundation. The project can be successful; the owner can meet its objectives; the contractor can meet its objectives. But success is not assured. The project is now in the hands of the contractor's and owner's project management teams. A continuing emphasis on partnership, a continuing basis of trust and mutual respect, and the application of fairness and reasonableness by both parties will allow the proper application of engineering and construction skills and knowledge for the successful completion of the work.

# 19. Developing Winning Proposals

Hans J. Thamhain*

## MARKETING PROJECTS AND PRODUCTS

New contracts are the lifeblood for many project-oriented businesses. The techniques for winning these contracts follow established bid proposal practices which are highly specialized for each market segment. They often require intense and disciplined team efforts among all organizational functions, especially from operations and marketing. They also require significant customer involvement.

### What Makes Project Marketing Different?

Projects are different from products in many respects. Marketing projects requires the ability to identify, pursue, and capture one-of-a-kind business opportunities. The process is characterized as follows:

1. *Systematic Effort.* A systematic effort is usually required to develop a new program lead into an actual contract. The program acquisition effort is often highly integrated with ongoing programs and involves key personnel from both the potential customer and the performing organization.
2. *Custom Design.* While traditional businesses provide standard products and services for a variety of applications and customers, projects are custom-designed items to fit specific requirements of a single customer community.
3. *Project Life Cycle.* Project-oriented businesses have a beginning and an end and are not self-perpetuating. Business must be generated on

* Hans J. Thamhain is Associate Professor of Management at Bentley College in Waltham, MA. He has held engineering and project management positions with GTE, General Electric, Westinghouse, and ITT. Dr. Thamhain is well known for his research and writings in project management which include 4 books and 60 journal articles. He also conducts seminars and consults in all phases of project management for industry and government.

a project-by-project basis rather than by creating demand for a standard product or service.

4. *Marketing Phase.* Long lead times often exist between project definition, start-up, and completion.
5. *Risks.* Risks are present, especially in the research, design, and production of programs. The program manager has not only to integrate the multidisciplinary tasks and program elements within budget and schedule constraints, but also to manage inventions and technology.
6. *Technical Capability to Perform.* This capability is critical to the successful pursuit and acquisition of a new project or program.

In addition to the above differences of project versus product businesses, there is another distinction that can be drawn on the basis of external versus internal opportunities. While the remainder of this chapter focuses on developing new business from external sources, the process and challenges are very similar for pursuing company-internal business such as a new product development. Competition over scarce resources and business alternatives, as well as sound business practices, make the budget approval for a new internal development often a very intricate, involved, and highly competitive process. A brief comparison of internally versus externally funded contrast acquisitions and executions is summarized in Table 19-1.

### Selling a Project

In spite of the risks and problems, contract profits are usually very low on projects in comparison with those from commercial business practices. One may wonder why companies pursue project businesses. Clearly there are many reasons why projects and programs are good business.

1. Although immediate profits, as a percentage of sales, are usually small, the return on capital investment is often very attractive. Progress payment practices keep inventories and receivables to a minimum and enable companies to undertake programs many times larger in value than the assets of the total company.
2. Once a contract has been secured and is being managed properly, the program is of relatively low financial risk to the company. The company has little additional selling expenditure and has a predictable market over the life cycle of the program.
3. Program business must be viewed from a broader perspective than motivation for immediate profits. It provides an opportunity to de-

Table 19-1. Characteristics of Internal Versus External Contracts.

| ITEM | COMPANY-INTERNAL DEVELOPMENT SUCH AS NEW PRODUCT DEVELOPMENT | COMPANY-EXTERNAL CONTRACT VIA BID PROPOSAL |
|---|---|---|
| New product idea | Evolutionary process driven either by market needs or technological capabilities. Often the process involves personnel involvement throughout the organization. | Opportunity-driven process based on single customer needs and requirements. |
| Product (or service design) | Standard product designed for a variety of applications and customers, company-funded development. | Custom-designed item to fit specific requirement. Customer-funded development. |
| Funding | Company-internal funding through budgeting process. | Customer funding via competitive bidding process. |
| Pricing | Competitive pricing, highly dynamic, strategically influenced. | Negotiated price, often based on actual cost. |
| Engineering kickoff | Design kickoff based on general management decision after careful study. | Design kickoff. Based on customer contract award after complex bidding process. |
| Role of engineering-marketing team | Strong engineering-marketing cooperation prior to product development. Predominately marketing "effort during product introduction and beyond." | Strong engineering-marketing team effort throughout preproposal and proposed phase—then effort shifts to engineering. |

velop the company's technical capabilities and to build an experience base for future business growth.

4. Winning one program contract often provides attractive growth potential, such as (1) growth with the program via additions and changes; (2) follow-on work; (3) spare parts, maintenance, and training; and (4) the ability to compete effectively in the next program phase, such as nurturing a study program into a development and finally a production contract.

In summary, new business is the lifeblood of an organization. It is especially crucial in project-oriented businesses, which lack the ongoing nature of conventional markets.

## PROPOSAL TYPES AND FORMATS

The majority of proposals that are prepared by companies are based on inquiries received from prospective clients. The inquiry stipulates the conditions under which the clients wish the work to be done. The responses we make to inquiries received from clients are termed "proposals," even though in many cases no commitment is proposed. Accordingly, proposals can be classified broadly in two major categories: *qualification proposals* and *commercial bid proposals*.

The *qualification proposal* generally gives information about company organization, qualifications, working procedures, or information for a specific area of technology. Qualification proposals make no offer to perform services and make no commitments of a general or technical nature. These are also called *informational proposals* if the contents relate to company organization, general qualifications, and procedures. They are sometimes called *white papers, technical presentations,* or *technical volumes* if technical and economic data are provided for a specific area of technology. A special form of the qualification proposal is the presentation.

The commercial bid proposal offers a definite commitment by the company to perform specific work or services, or to provide equipment in accordance with explicit terms of compensation. A commercial bid proposal may also contain the type of information usually found in qualification proposals.

Both qualification and commercial proposals may be presented to the client in various forms under a wide variety of titles, depending on the situation, for example, the client's requirements and the firm's willingness to commit its resources under the specific circumstances involved. The most common forms are:

- Letter proposals.
- Preliminary proposals.
- Detailed proposals.
- Presentations.

There are no sharp distinctions among these on the basis of content. Differentiation is mainly by the extent of the work required to prepare them. Included in the following paragraphs are definitions of these most common forms.

*Letter proposals* are either qualification or commercial proposals. They are brief enough to be issued in letter form rather than as bound volumes.

*Preliminary proposals* are either qualification or commercial proposals and are large enough to be issued as bound volumes. They may be paid technical and/or economic studies, bids to furnish services, or other offerings of this kind.

*Detailed proposals* are most often commercial bid proposals, generally including the preparation of a detailed estimate. They are the most complex and inclusive proposals. Because of the high cost of preparation and the high stakes involved in the commitments offered, organization and contents of the documents are defined and detailed to a much greater degree than other kinds of proposals.

*Presentations* are generally in the nature of oral qualification proposals. Selected personnel, specialized in various areas, describe their subjects verbally to client representatives in time periods varying from an hour to an entire day. To aid in the success of a presentation, audio-visual aids are encouraged. Many companies maintain a library of photographic slides developed just for this purpose.

## DEFINING THE MARKET

Customers come in various forms and sizes. Particularly for large programs with multiple-user groups, customer communities can be very large, complex, and heterogeneous. Very large programs, such as military or aerospace undertakings, are often sponsored by thousands of key individuals representing the user community, procuring agencies, Congress, and interest groups. Selling to such a diversified heterogeneous customer is a true marketing challenge which requires a highly sohpisticated and disciplined approach.

The first step in a new business development effort is to define the market to be pursued. The market segment for a new program opportunity is normally in an area of relevant past experience, technical capability, and customer involvement. Good marketeers in the program business have to think as product line managers. They have to understand all dimensions of the business and be able to define and pursue market objectives consistent with the capabilities of their organizations.

### Market Predictability

Program businesses operate in an opportunity-driven market. It is a mistaken belief, however, that these markets are unpredictable and unmanageable. Market planning and strategizing are important. New program opportunities develop over periods of time, sometimes years for larger programs. These developments must be properly tracked and cultivated to form the basis for management actions such as (1) bid decisions, (2) resource commitment, (3) technical readiness, and (4) effective customer liaison.

The strategy of winning new business is supported by systematic, disciplined approaches which are illustrated in Figure 19-1 and discussed in five basic steps:

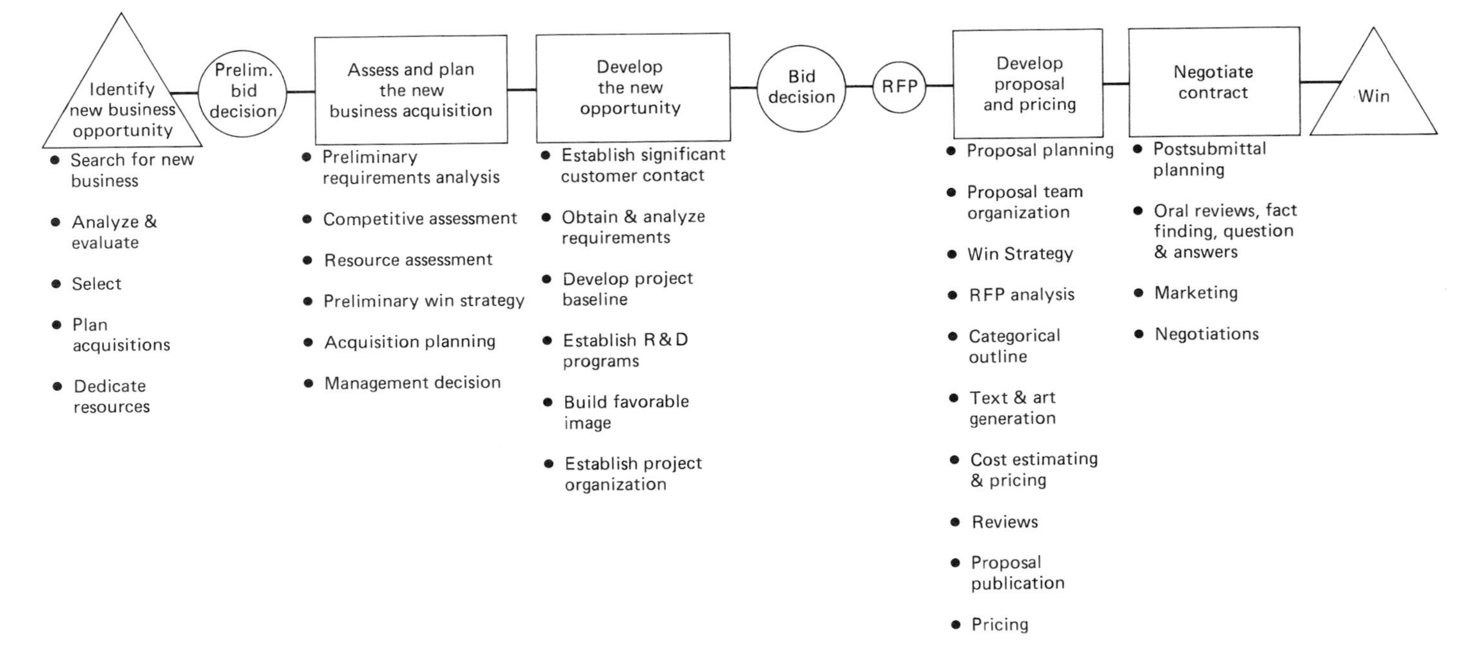


Figure 19-1. Major phases and milestones in a bid proposal life cycle.

1. Identifying new business opportunities.
2. Planning the business acquisition.
3. Developing the new contract opportunity.
4. Developing a winning proposal and pricing.
5. Negotiating and closing the contract.

## IDENTIFYING NEW BUSINESS OPPORTUNITIES

Identifying a new program opportunity is a marketing job. During the initial stages one does not evaluate or pursue the opportunity—that comes later. Furthermore, identifying new opportunities should be an ongoing activity. It involves the scanning of the relevant market sector for new business. This function should be performed by all members of the project team in addition to marketing support groups. There are many sources for identifying new business leads such as customer meetings on ongoing programs, professional meetings and conventions, trade shows, trade journals, customer service, advertising of capabilities, and personal contacts.

All one can expect at this point is to learn of an established or potential customer requirement in one of the following categories:

- Follow-on to previous or current programs.
- Next phase of program.
- Additions or changes to ongoing programs.
- New programs in your established market sector or area of technological strength.
- New programs in related markets.
- Related programs, such as training, maintenance, or spares.

For most businesses, ongoing program activities are the best source of new opportunity leads. Not only are the lines of customer communication better than in a new market but, more importantly, the image as an experienced, reliable contractor hopefully has been established, giving a clear competitive advantage in any further business pursuit.

The target result of this analysis is an acquisition plan and a bid decision. Analyzing the new opportunity and preparing the acquisition plan is an interactive effort. Often many meetings are needed between the customer and the performing organization before a clear picture emerges of both customer requirements and contractor's capabilities. A major fringe benefit of proper customer contact is the potential for building confidence and credibility with the customers. It shows that your organization understands their requirements and has the capability to fulfill them. This is a necessary prerequisite for eventually negotiating the contract. The new business identification phase concludes with a formal analysis of the new project opportunity, which is summarized in the following section.

## PLANNING THE BUSINESS ACQUISITION

The acquisition plan provides the basis for the formal bid decision and a detailed plan for the acquisition of the new business. The plan is an important management tool which provides *an assessment of the new program opportunity as a basis for appropriating resources for developing and bidding the new business.*

Typically the new business acquisition plan should include the following elements:

1. *Brief Description of New Business Opportunity.* A statement of the requirements, specifications, scope, schedule, budget, customer organization, and key decision makers.
2. *Why Should We Bid?* A perspective with regard to establishing business plans and desirable results, such as profits, markets, growth, and technology.
3. *Competitive Assessment.* A description of each competing firm with regard to their past activities in the subject area, including (a) related experiences, (b) current contracts, (c) customer interfaces, (d) specific strengths and weaknesses, and (e) potential baseline approach.
4. *Critical Win Factors.* A listing of specific factors important to winning the new program, including their rationales. (Example: low implementation risk and short schedule important to customer because of need for equipment in two years.)
5. *Ability to Write a Winning Proposal.* The specifics needed to prepare a winning proposal, including (a) availability of the right proposal personnel, (b) understanding of customer problems, (c) unique competitive advantage, (d) expected bid cost to be under customer budget, (e) special arrangements such as teaming or license model, (f) engineering readiness to write proposal, and (g) ability to price competitively.
6. *Win Strategy.* A chronological listing of critical milestones guiding the acquisition effort from its present position to winning the new program. It should show those activities critical for positioning yourself uniquely in the competitive field. This includes timing and responsible individuals for each milestone. For example, if low implementation risk and short schedules are important to the customer, the summary of the win strategy may state:

(a) Build credibility with the customer by introducing key personnel and discussing baseline prior to request for proposal, (b) stress related experiences on ABC program, (c) guarantee 100% dedicated personnel and list program personnel by name, (d) submit detailed schedule with measurable milestones and specific reviews, (e) submit XYZ module with proposal for evaluation.

7. *Capture Plan.* A detailed action plan in support of the win strategy and all business plans. This should integrate the critical win factors and specific action plan. All activities, such as timing, budgets, and responsible individuals identified, should have measurable milestones. The capture plan is a working document to map out and guide the overall acquisition effort. It is a living document which should be revised and refined as the acquisition effort progresses.
8. *Ability to Perform Under Contract.* This is often a separate document, but a summary should be included in the acquisition plan stating (a) technical requirements, (b) work force, (c) facilities, (d) teaming and subcontracting, and (e) program schedules.
9. *Problems and Risks.* A list of problems critical to the implementation of the capture plan, such as (a) risks to techniques, staffing, facilities, schedules, or procurement; (b) customer-originated risks; (c) licenses/patents/rights; and (d) contingency plan.
10. *Resource Plan.* A summary of the key personnel, support services, and other resources needed for capturing the new business. The bottom line of this plan is the total acquisition cost.

There are many ways to present the acquisition plan. However, an established format, which is accepted as a standard throughout the organization, has several advantages. It provides a unified standard format for quickly finding information during a review or analysis. Standard forms also serve as checklists. They force the planner not only to include information conveniently obtainable, but also to seek out the other data necessary for winning new business. Finally, a standard format provides a quick and easy assessment of the new opportunity for key decision makers.

### The Bid Board

Few decisions are more fundamental to new business than the bid decision. Resources for the pursuit of new business come from operating profits. These resources are scarce and should be carefully controlled.

Bid boards serve as management gates for the release and control of these resources. The bid board is an expert panel, usually convened by the general manager, which analyzes the acquisition activities to determine their status and also to assess investment versus opportunity in acquiring new business. An acquisition plan provides the major framework for the meeting of such bid boards.

Major acquisitions require a series of bid board sessions, starting as early as 12 to 18 months prior to the request for proposal. Subsequent bid boards reaffirm the bid decision and update the acquisition plans. It is the responsibility of the proposal manager to gather and present pertinent information in a manner that provides the bid board with complete information for analysis and decision. A simple form, such as shown in Table 19-2, can help in organizing and summarizing the information needed. This requires significant preparation and customer contact. A team presentation is effective as all disciplines should be involved.

Table 19-2. Request for Approval of Bid Decision.

Date: ______
No: ______

1. Client ______
2. Project ______
3. Units and Capacities ______
4. Location ______
5. Scope of Work ______
6. Value of Project ______
7. Type of Contract ______
8. Type of Proposal ______
9. Proposal Cost ______
10. Due Date of Proposal ______
11. % Probability to Go Ahead ______
12. % Probability for Award to Company ______
13. Other Factors
    a. Time of Award (if long range) ______
    b. Financing (if financing involved) ______
    c. Source of Know-how (if other than company) ______
    d. Special Contract Conditions (if not company standard) ______
    e. Manpower Availability (if not readily available) ______
    f. Secrecy Conditions (if any) ______
    g. Competition (if known) ______
    h. Unusual factors (if any) ______

Management Review Committee
Approved By: ______ Date: ______

### The Bid Decision

After the proposal inquiry is received and logged in, it should be screened as soon as possible to facilitate the bid/no-bid decision. Because most inquiries have a short response time, the sooner a decision can be made, the longer you have to prepare your proposal. As part of the screening process, it is important that the sales representative or proposal manager review the document thoroughly to determine its total value to the organization. It also allows the decision makers to determine if they have the capabilities required to bid on the job. The technical nature of the current marketplace is such that organizations must compete in specialized areas of an industry. Careful screening of the inquiry may also reveal that more information is needed to prepare a responsive proposal properly. A bid/no-bid decision is usually made based upon a set of criteria judged to be important in selecting projects that contribute to an organization's continued growth and success. A typical set of criteria to be answered is shown in Table 19-3.

Once a decision is made not to bid on a project, the customer should be notified in writing. Some organizations respond with a form letter, but such a response could cause the customer to interpret this as lack of interest and could result in loss of future work. A specific letter for that inquiry should be prepared, explaining to the customer why your organization could not respond.

## PHASES OF A TYPICAL CONTRACT DEVELOPMENT

We live in a competitive world. Winning new business requires significant homework in preparation for the bid proposal. Selling a new program is often unique and different from selling in other markets. It requires selling an organization's capability for a custom development—something that has not been done before. This is different from selling an off-the-shelf product that can be examined prior to contract. It requires establishing your credibility and building confidence in the customer's mind so that your organization is selected as the best candidate for the new program. Such a "can-do" image can be built in four phases.

### First: Significant Customer Contact

Early customer liaison is vital in learning about the customer's requirements and needs. It is necessary to define the project baseline, the potential problem areas, and the risks involved.

Table 19-3. Checklist in Support of Bid Decision.

1. Does company have capabilities and resources to perform the work?
2. Can company phase in the work to meet client schedule?
3. What is company's technical position?
4. What is company's approach to project execution?
5. Is project of special importance to client?
6. Would doing project enhance our reputation?
7. What has been our past experience and contractual relationship with client?
8. What is company's commercial approach and price strategy?
9. What are client's future capital expenditures?
10. Who is the competition and do they have any special advantages?
11. Does client have preferred contractor and if so, why?
12. What is the probability of project going ahead?
13. Does project meet the immediate or long-range objectives of the company?
14. Other work prospects for company in next six months? one year?
15. Other special factors and considerations?

Establishing meaningful customer contact is no simple task. Today's structured customer organizations involve many key decision-making personnel, conflicting requirements and needs, and biases. There rarely is only one person responsible for signing off on a major procurement. Technical and marketing involvement at all levels is necessary to reach all decision-making parties in the customer community. Your new business acquisition plan will be the road map for your marketing efforts. The benefits of this customer contact are that you:

- Learn about the specific customer requirements.
- Obtain information for refining the baseline prior to proposal.
- Participate in customer problem solving.
- Build a favorable image as a competent, credible contractor.
- Check out your baseline approach and its acceptability to the customer.
- Develop rapport and a good working relationship with the customer.

### Second: Prior Relevant Experience

Nothing is more convincing to an engineering customer than demonstrated prior performance in the same or a related area of the new program. It shows the customer that you have produced on a similar task. This reduces the perceived technical risks and associated budget and schedule uncertainties. Therefore it is of vital importance to demonstrate to the customer that your organization (1) understands their new requirements, and (2) has performed satisfactorily on similar programs. This image of an experienced contractor can be communicated in many ways:

- Field demonstration of working systems and equipment.
- Listing of previous or current customers, their equipment, and applications.
- Model demonstrations.
- Technical status presentations.
- Product promotional folders.
- Technical papers and articles.
- Trade show demonstrations and displays.
- Slide or video presentation of equipment in operation.
- Simulation of the system, equipment, or services.
- Specifications, photos, input/output simulations of the proposed equipment.
- Advertisements.

Demonstrating prior experience is integrated and interactive with the customer liaison activities. To be successful, particularly on larger programs, requires both leadership and discipline. Start with a well-defined customer contact plan as part of your overall acquisition plan. This requires well-planned involvement at all levels in order to make these contacts with relevant personnel in the customer community. One major benefit received from these intensive marketing efforts is that you create an image with the customer as an experienced, sound contractor. Second, you are learning more about the new program, its specific requirements, the risks involved, as well as the concerns and biases of the customer. This information will make it easier to respond effectively to a formal or informal request for proposal.

### Third: Readiness to Perform

Once the basic requirements and specifications of the new program are known, it is often necessary to mount a substantial technical preproposal effort to advance the baseline design to a point that permits a clear definition of the new program. These efforts may be funded by the customer or borne by the contractor. Typical efforts include (1) feasibility studies, (2) system designs, (3) simulation, (4) design and testing of certain critical elements in the new equipment or the new process, (5) prototype models, or (6) any developments necessary to bid the new job within the desired scope of technical and financial risks.

Development prior to contract is expensive, has no guarantee of return, and precludes the company from pursuing other activities. Then why do organizations spend their resources for such development? It is often an absolutely necessary cost for winning new business. These early developments reduce the implementation risks to an acceptable level for both the customer and the contractor. Further, these developments might be necessary to catch up with a competitor or to convince the customer that certain alternative approaches are preferable.

Clearly, preproposal developments are costly. Therefore they should be thorough and well detailed and approved as part of the overall acquisition plan. The plans and specific results should be accurately communicated to the customer. This will help to build quality image for your firm while giving the potential contractor additional insight into the detailed program requirements. Finally, one should not overlook two sources of funding for these activities: (1) customer funding for these advanced programs prior to contract—often the customer is willing to fund contract-definition activities because it may reduce the risks and the uncertainties of contractual performance; (2) inclusion with other ongoing developments—the program manager might find that a similar effort is under way in a corporate research department or even within the customer's organization.

#### Fourth: Establishing the Organization

Another element of credibility is the contractor's organizational readiness to perform under contract. This includes facilities, key personnel, support groups, and management structure. Reliability in this area is particularly critical in winning a large program relative to your company size. Often a contractor goes out on a limb and establishes a new program organization to satisfy specific program or customer requirements. This may require major organizational changes.

Few companies go into reorganization lightly, especially prior to contract. However, in most cases it is possible to establish all the elements of the new program organization without physically moving people or facilities, and without erecting new buildings. What is needed is an organization plan exactly detailing the procedures to be followed as soon as the contract is awarded. Further, the new program organization can be defined on paper together with its proper charter and all structural and authority relationships. This should be sufficient for customer discussion and will give a head start once the contract is received. Usually it is not the moving of partitions, people, or facilities that takes time, but determining where to move them and how to establish the necessary working relationships.

As a checklist, the following organizational components should be defined clearly and discussed with the customer prior to a major new contract:

- Organizational structure.
- Charter.
- Policy-management guidelines.
- Job description.
- Authority and responsibility relationships.
- Type and number of offices and laboratories.
- Facilities listing.
- Floor plans.
- Staffing plan.
- Milestone schedule and budget for reorganization.

A company seldom needs to reorganize completely to accommodate a new program. It requires resources and risks for both the contractor and the customer. Most likely the customer and program requirements can be accommodated within the existing organization by redefining organizational relationships, authority, and responsibility structures without physically moving people and facilities. Matrix organizations in particular have the flexibility and capacity to handle large additional program business with only minor organizational changes.

### Fifth: The Kickoff Meeting

Soon after the decision to bid has been made, a preliminary proposal kickoff meeting or proposal strategy planning meeting should be held. This meeting, chaired by the proposal manager, consists of the heads of the various contributing departments, the sales representative, and possibly senior management. Because there is a limited amount of time available for preparation of a proposal, it is mandatory that the proposal effort be planned in all aspects to make the most use of that time.

After the strategy meeting and development of a preliminary proposal plan, the proposal manager calls a kickoff meeting. This meeting is attended by all participants working on the proposal or their representatives. The proposal manager writes a kickoff memo or notice to inform the participating departments what is required to be discussed at the meeting regarding the proposal effort. The purpose of the kickoff meeting is to inform all participants of the proposal plan and objectives.

To ensure a good start, the following topics should be covered in every kickoff meeting:

- *Project scope.* The type of plant or unit, the client, the project location, the order of magnitude total installed cost (if known), the job or proposal number, and any other general identification are designated, as well as the proposal manager and other key personnel.
- *Commercial objectives.* The type of proposal required is discussed as well as the management philosophy to be incorporated into the proposal. The marketing representative can often effectively provide background information on the client's requirements, exceptions, and the competitive challenges.
- *Proposal staffing.* Three factors—the proposal schedule, man-hour budget, and the technical proposal requirements—determine which departments will participate and also set the specific proposal staffing requirements. Staffing should have been settled before the kickoff meeting.
- *Assignments of participants.* Work assignments with clear areas of responsibility will be made. Relationships between participating departments should be spelled out.
- *Proposal dates, schedules, and budget.* The key dates for issuance of proposal documents and completion of important portions of the work should be discussed and the participants advised of the available man-hour budget and deadlines.
- *Qualifications.* The proposal must be tailored to sell the client on your qualifications and capabilities. The desired strategy to accomplish this should be communicated to all involved.
- *Type of estimate.* Unless the type of estimate to be submitted with the proposal has already been decided, the proposal manager defines clearly the requirements and the type of information that must be generated by the participants in the proposal effort.
- *Final proposal contents.* Specific requirements for the contents should be brought to everyone's attention. The proposal manager develops and includes a tentative table of contents for the proposal documents under discussion. Definite assignments for preparation of the draft write-ups for each section of the proposal are also made.
- *Working information.* Copies or a summary of the inquiry documents are distributed before or during the kickoff together with any other information from the client that is to be used in preparing the proposal.

The kickoff meeting is the prefered device for starting the proposal work, especially for complex efforts. It is held shortly after work authorization has been received and sales objectives have been established. Production work on the proposal begins after the kickoff meeting.

The proposal manager sets the meeting time and place. Attendance is limited to those who actually contribute to the definition and initiation of the proposal effort, such as the following persons:

- Sales representatives responsible for the inquiry.
- Heads of departments and groups responsible for any of the proposal work.
- Key personnel from departments and groups responsible for any of the proposal work, if they have been selected.
- The general manager of operations and a member of the legal department responsible for proposal review, who are notified of the meeting so they may attend if they wish.

There are no strict rules for the agenda of a kickoff meeting because of the wide variations in the nature of the inquiries received and the proposals developed in reply to them. However, in general, the following topics should be covered in every kickoff meeting to ensure a good start:

1. Purpose. Type of proposal and estimate to be prepared.
2. Plan of approach. Explanation of methods to be used.
3. Action. Specific duties of individuals and groups.
4. Proposal schedule. Indication of key dates for internal submission of parts of the work.

In addition to covering the elements of the client's request and the proposal response to them, the kickoff meeting also serves to:

- Outline the marketing strategy behind the proposal.
- Establish management's interest in the project.
- Introduce team members to each other.
- Create interchange of ideas and suggestions early.
- Obtain overall agreement of work assignments and timing.
- Develop a winning attitude in the proposal effort.

The proposal manager is responsible for preparing the minutes of the kickoff meeting, which are issued no later than the second working day after the meeting.

## DEVELOPING A WINNING PROPOSAL

Bid proposals are payoff vehicles. They are one of the final products of your marketing effort. Whether you are bidding on a service contract or an engineering development, a government contract or a commercial program, the process is the same and, in the end, you must submit a proposal.

Yet with all due respect to the importance of the bid proposal as a marketing tool, many senior managers point out that the proposal is only one part of the total marketing effort. The proposal is usually not the vehicle that sells your program—the proposal stage may be too late. The program concept and the soundness of its approach, the alternatives, your credibility, and so on, must be established during the face-to-face discussions with the customer. So why this fuss about writing a superior proposal? Because we still live in a competitive world. Your competition is working toward the same goal of winning this program. They, too, may have sold the customer on their approaches and capabilities. Hence among the top contenders, the field is probably very close. More importantly, beating most of the competition is not good enough. Like in a poker game, there is no second place. Therefore, while it is correct that the proposal is only part of the total marketing effort, it must be a superior proposal. Proposal development is a serious business in itself. Table 19-4 shows a topical outline for a typical major bid proposal which is broken into three volumes: technical, management, and cost, a common subdivision used in formal bidding practices.

Most people hate to work on proposals. Proposal development requires hard work and long hours, often in a constantly changing work environment. Proposals are multidisciplinary efforts of a special kind. But like any other multifunctional program, they require an orderly and disciplined effort which relies on many special tools to integrate the various activities of developing a high-scoring quality proposal. This is particularly true for large program proposals which require large capital commitments. Smaller proposals often can be managed with less formality. However, at a minimum, they should include the following tasks to ensure a quality bid proposal:

- Proposal team organization.
- Proposal schedule.
- Categorical outline with writing assignments and page allocation.
- Tone and emphasis/win strategy.
- RFP analysis.
- Technical baseline review.
- Draft writing.
- Reviews.
- Art/illustration development.
- Cost estimating.
- Proposal production.
- Final management review.

Table 19-4. Bid Proposal Content: Three Volumes.

The bid package consists of three basic parts or volumes with the following subtopics

| 1. TECHNICAL PROPOSAL | 2. MANAGEMENT | 3. COST PROPOSAL |
|---|---|---|
| Introduction and background to contractor's company | Process schedule | Price for services offered in proposal |
| Organization of contractor's company | Process description | Breakdown of price (materials, labor, etc.) |
| Schedule of professional personnel | Operating requirements | Escalations (lump sum contract) |
| Resume of key personnel or a resume summary | Plot plans and elevations | Amount for subcontract work |
| Project management policy or philosophy | Process flow diagrams | Amount of off-site facilities |
| Description of contractor's engineering department | Engineering flow diagrams | Taxes |
| Description of contractor's procurement department | Utilities flow diagrams | Royalty payment |
| Description of contractor's financial controls department | Heat and material balance | Alternative systems |
| Experience list of similar plants built | Equipment list | Optional equipment |
| Experience list of large complexes built | Equipment data sheets | Prior adjustments (labor, efficiency, etc.) |
| Experience list of all plants built by contractor | General facilities, such as piping, instrumentation, electrical, civil, construction, etc. | Schedule of payments |
| Experience list of using a client's process | Contractor's or client's specifications or standards | |
| Photographs of plants built by contractor | Services provided by contractor | |
| Draft contract | Services provided by client | |
| | Model and/or rendering of proposed plant | |

For each activity or milestone the plan should define the responsible individual(s) and schedules.

### Storyboarding Facilitates Group Writing

Most bid proposals are group writing efforts. Organizing, coordinating, and integrating these team efforts can add significantly to the complexities and difficulties of managing proposal developments. Especially for the larger engineering development bids, storyboarding is a technique that facilitates the group writing process by breaking down its complexities and integrating the proposal work incrementally.

### How Does It Work?

Storyboarding is based on the idea of splitting up the proposal writing among the various contributors and then developing the text incrementally via a series of writing, editing, and review phases, typically in the following order and timing:

| | | |
|---|---|---|
| 1. | Categorical Outline | Completion at Day 1 |
| 2. | Synopsis of Approach | Day 3 |
| 3. | Roundtable Review | Day 4 |
| 4. | Topical Outline | Day 5 |
| 5. | Storyboards | Day 10 |
| 6. | Storyboard Review | Day 11 |
| 7. | Storyboard Expansion | Day 22 |
| 8. | Staff Review | Day 24 |
| 9. | Editing | Day 28 |
| 10. | Printing and Delivery | Day 30 |

The number and type of phases indicated in the above listing, together with the timing, might be typical for a major bid proposal development with a 30-day response time. However, the above listing can also serve as a guide for larger or smaller proposals by scaling the effort up or down. For larger efforts more iterations are suggested among phases 5 and 9, while smaller efforts can be scaled down to seven phases, including only phases 4 through 10.

The actual timing of proposal developments should be scheduled like any other project. A bar graph schedule is sufficient and effective for most proposal efforts. Each phase is briefly described next.

### Phase One: Categorical Outline

The first step in the storyboarding process is the development of a categorical outline. This is a listing of the major topics or chapters to be covered in the proposal. For larger proposals, the categorical outline might form the first two levels of the table of content, such as shown in Figure 19-2. The categorical outline should also show for each category (1) the responsible author, (2) a page estimate, and (3) references to related documents. The Categorical Outline can often be developed *before* the receipt of the RFP, and should be finalized at the time of proposal kickoff.

**CATEGORICAL OUTLINE**

SHEET______OF________
DATE___________________

PROGRAM:

| | PROPOSAL ADDRESS | TITLE | AUTHOR | PAGE ESTIMATE TEXT | ART | SPEC REFERENCE |
|---|---|---|---|---|---|---|
| WP# | | | | | | |

Figure 19-2. Typical Categorical Outline form for a major proposal.

## Phase Two: Synopsis of Approach

A synopsis of approach is developed for each categorical topic by each responsible author. The Synopsis is an outline of the approach which addresses three questions related to the specific topic:

1. What does the customer require?
2. How are we planning to respond?
3. Why is the approach sound and good?

The format of a typical Synopsis is shown in Figure 19-3. As an alternative, the proposal manager can prepare these forms and issue them as policy papers instead of having each author develop them.

In preparation for the review, the completed Synopsis forms and Categorical Outline can be posted on a wall, in sequential order. This method of display facilitates in a very effective way open group reviews and analysis.

# SYNOPSIS OF APPROACH

| PROPOSAL NAME: | VOL: | PROPOSAL ADDRESS | WORK PACK |
|---|---|---|---|
| | | RFP REFERENCE: | |

| AUTHOR: | EXT: |
|---|---|
| BOOK BOSS: | EXT: |

UNDERSTANDING OF REQUIREMENTS:

APPROACH & COMPLIANCE:

SOUNDNESS OF APPROACH:

RISKS:

Figure 19-3. Typical Synopsis form which is used to expand each categorical topic from Figure 19-2.

### Phase Three: Roundtable Review

During this phase, all Synopsis of Approach forms are analyzed, critiqued, augmented, and approved by the proposal team and its manager. It is the first time that, in summary form, the total proposal approach is continuously displayed. Besides the proposal team, technical resource managers, marketing managers, contract specialists, and upper management should participate in this review, which usually starts four days after the proposal kickoff.

### Phase Four: Topical Outline

After the review and approval of the Synopsis forms, the Categorical Outline is expanded into the specific topics to be addressed in the proposal. This Topical Outline forms the Table of Content for the bid proposal. Similar to the Categorical Outline, each topic defines the responsible writer, page estimates, and document references. A form example is shown in Figure 19-4.

**TOPICAL OUTLINE**

SHEET_____OF_______

DATE_______________

PROGRAM:

| PROPOSAL ADDRESS | TITLE | AUTHOR | PAGE ESTIMATE TEXT | ART | SPEC REFERENCE |
|---|---|---|---|---|---|
| | | | | | |

Figure 19-4. Topical Outline sample form, which explains the Categorical Outline of Figure 19-2.

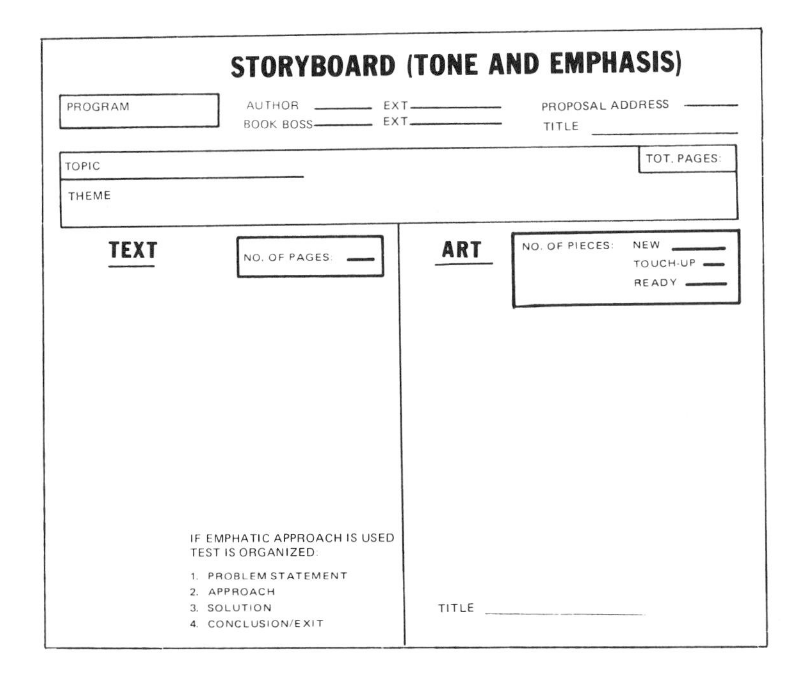
STORYBOARD (TONE AND EMPHASIS)

PROGRAM

AUTHOR ______ EXT ______
BOOK BOSS ______ EXT ______

PROPOSAL ADDRESS ______
TITLE ______

TOPIC

TOT. PAGES:

THEME

TEXT

NO. OF PAGES: ___

ART

NO. OF PIECES: NEW ___
TOUCH-UP ___
READY ___

IF EMPHATIC APPROACH IS USED
TEST IS ORGANIZED:

1. PROBLEM STATEMENT
2. APPROACH
3. SOLUTION
4. CONCLUSION/EXIT

TITLE ______

Figure 19-5. Storyboard format.

**Phase Five: Storyboard Preparation**

The Storyboard preparation is straightforward. Typically, a one-page Storyboard is prepared for each topic by the assigned writer. As shown in Figure 19-5, it represents a detailed outline of the author's approach to the writeup for that particular topic. Often the Storyboard form is divided into three parts: (1) topic and theme section, (2) text outline on the left side of the form, and (3) summary of supporting art to be prepared on the right side of the form.

The Storyboard takes a first cut at the key aspects of each topic. The topic heading, theme (if any), problem statement, and exit or conclusion must be written out in full, just the way they might appear in the final text. Expression of these key sentences is important. They must be relevant, responsive, and comprehensive.

The composition of text and art is arranged on the Storyboard in this format for convenience only. As shown in Figure 19-6, the text format can be chosen either modular or nonmodular. For the modular concept the storyboard format is copied into the proposal layout, text left, illustrations right; while in nonmodular form the art becomes an integral part of

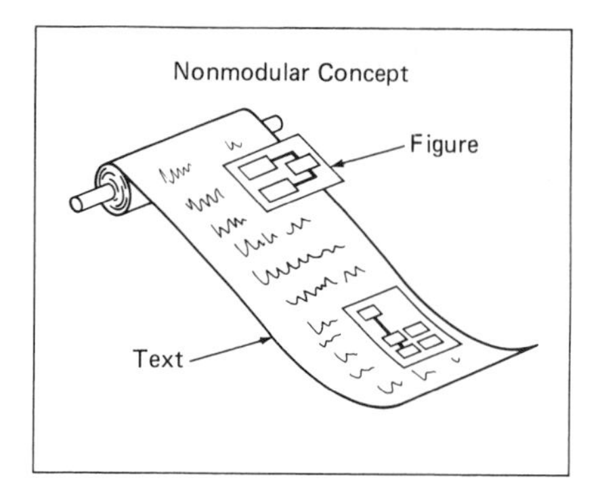


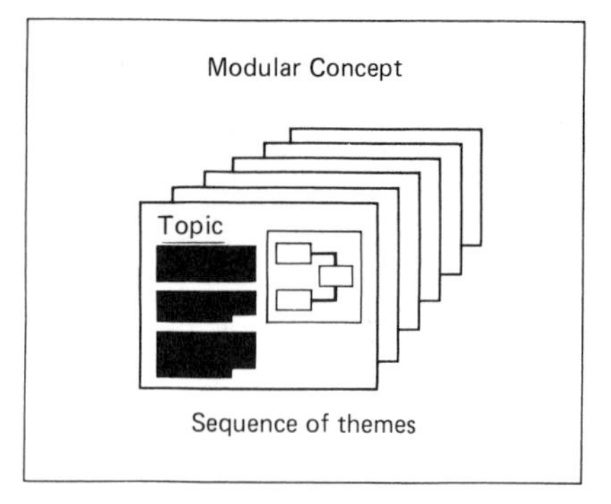


Figure 19-6. Nonmodular versus modular concept of text and art presentation.

the text. The final layout should, however, be of no concern to the authors at this point in the proposal development cycle.

Storyboards are one of the most important elements in the proposal development. They should be typed for clarity and easy comprehension during the review sessions.

### Phase Six: Storyboard Review

The completed Storyboard forms are pasted on the walls of the review room in a logical sequence, hence presenting the total story as we want to tell it to the customer.

Storyboard Reviews should start within five working days after the Synopsis Reviews. The reviews are held in a special area and attended by the author, the proposal team, and key members of the functional organization. The Storyboard Review permits a dialogue to take place between the proposal team and its management and the author. All participants of each notice of the review is given by both the daily bulletin and microschedule.

The review permits the proposal team to insert, modify or correct any approach taken by an author. An additional output of the Storyboard Review is a final Proposal Outline. This Proposal Outline includes categorical headings, topics, and art log numbers, and provides the PMT with a proposal overview and serves as a control document. The Storyboard Review provides the team with the single most important opportunity to change direction or change approaches in the proposal preparation.

Similar to the Synopsis Review, Storyboarding is an interactive process. During the reviews, a copy of the latest Storyboard should always be on display in the control room.

### Phase Seven: Final Text Generation by Expanding Storyboards

After Storyboard approval, each author prepares a Storyboard Expansion. It is the expansion of information on each topic into about a 500-word narrative. As part of the Storyboard Expansion, all authors should finalize their art work and give it to the publications specialist for processing. The completed Storyboard Expansions present the first draft of the final proposal. The material is given to the technical editor who will perform edit for clarity. Each responsible author must review and approve the final draft, which might cycle through the editing process several times.

This final text generation is the major activity in the proposal development process. All other prior activities are preparatory to this final writing assignment. If the preparations up to the Storyboard Review are done properly, writing the final text should be a logical and straightforward task without the hassles of conceptional clarification and worries about integration with other authors.

As a guideline, 10 working days out of a 30 working day RFP-response period may be a reasonable time for this final text generation. Because of its long duration, it is particularly important to set up specific milestones for measuring intermediate progress. The process of final text generation should be carefully controlled. The proposal specialist, if available, will play a key role in the integration, coordination, and controlling of this final text generation and its output.

The final text should be submitted incrementally to the publication department for editing, word processing or typing, and media storage for future retrieval.

### Phase Eight: Staff Review—A Final Check

The final proposal review is conducted by the proposal team and its management group plus selected functional managers who will later provide the contractual services to the contract. In addition, a specialty review committee may be organized to check the final draft for feasibility, rationale, and responsiveness to the RFP. For a given chapter or work package, this Staff Review is accomplished in less than a day. The comments are reviewed by the original authors for incorporation.

### Phase Nine: Final Edit

After these review comments are incorporated, the entire proposal is turned over to the publication department for final format editing, completion of art, word processing, and proofreading of the final reproducible copy. The proposal is then returned to the publication specialist or proposal specialist with all of the completed art for a final check. The authors are then given a final opportunity to look at the completed proposal. Any major flaws or technical errors that may have crept into the copy are corrected at that time.

### Phase Ten: Printing and Delivery

The proposal is now ready for paste-up, printing, and delivery to the customer.

## NEGOTIATING AND CLOSING THE CONTRACT

Sending off the bid proposal signals the start of the postsubmission phase. Regardless of the type of customer or the formalities involved, even for an oral proposal, the procurement will go through the following principal steps:

1. Bid proposals received.
2. Proposals evaluated.
3. Proposal results compared.
4. Alternatives assessed.
5. Clarifications and new information from bidders.
6. Negotiations.
7. Award.

While bidders usually have no influence on proposal evaluation or source selection process, they can certainly prepare properly for the upcoming opportunities of customer negotiations. Depending on the procurement, these opportunities for improving the competitive position come in the following forms:

Follow-up calls and visits.
Responding to customer requests for additional information.
Fact finding requested by customer.
Oral presentations.
Invitations to field visits.
Sending samples or prototypes.
Write paper.
Supportive advertising.
Contact via related contract work.

Plant or office visits.
Press releases.
Negotiations.

The bidder's objective on all postsubmission activities should be improving the competitive position. For starters, the bidder must assess the proposal relative to customer requirements and competing alternatives. In order to do this realistically the bidder needs customer contact. Any opportunity for customer contact should be utilized. Follow-up calls and visits are effective in less formal procurements, while fact finding and related contract work are often used by bidders in formal buying. These are just a few methods to open officially closed doors into the customer community.

Only through active customer contact is it possible to assess realistically the competitive situation and organize for improvement and winning negotiations. Table 19-5 provides a listing of steps for which the bidder should organize. Customer contact and interaction are often difficult to arrange, especially in more formal procurements, which may require innovative marketing approaches.

While the proposal evaluation period is being used by the customer to determine the best proposal, the bidder has the opportunity to improve his position in three principal areas: (1) clarification of proposed program scope and content, (2) image building as a sound, reliable contractor, and (3) counteracting advances made by competing bidders.

The proposal evaluation period is highly dynamic in terms of changing scores, particularly among the top contenders. The bidder who is well organized and prepared to interact with the customer community stands the best chance of being called first for negotiations, thus gaining a better basis for negotiating an equitable contract. Table 19-5 provides some guidelines for organizing and preparing for the postsubmission effort.

## RECOMMENDATIONS TO MANAGEMENT

Winning a bid proposal depends on more than the pricing or right market position, or luck. Winning a piece of business depends on many factors which must be carefully developed during the preproposal period, articulated in the bid proposal, and fine-tuned during contract negotiations. Success is geared to hard work, which starts with a proper assessment of the bid opportunity and a formal bid decision. To be sure, a price-competitive bid can help in many situations. However, it may be interesting to note some research findings. A low price bid is advantageous toward winning only in contracts with low complexity, low technical risk, and

Table 19-5. Organizing the Postsubmission Effort.

1. *Reassess your proposal.* Study your proposal and reassess (1) strengths, (2) weaknesses, and (3) compliance with customer requirements.
2. *Plan action.* Develop an action plan listing all weak points of your proposal, potential for improvement, and actions toward improvement.
3. *Open communications.* Establish and maintain communications with the customer during the proposal evaluation period. Determine the various roles people play in the customer community during the evaluation.
4. *Find your score.* Try to determine how you scored with your proposal. Find out what the customer liked, disliked, and perceived as risks; credibility problems and standing against the competition. To determine your score realistically and objectively requires communications skills, sensitivity, and usually a great deal of prior customer contact. Determining your proposal score is an important prerequisite for being able to clarify specific items and to improve your competitive position.
5. *Seek interaction.* Seek out opportunities for interacting with the customer as early as possible. Such opportunities may be presenting additional information for clarifying or enhancing specific proposed items. The meeting or presentation should be requested by the customer. It is a great opportunity for the bidder to "sell" a proposal further. This includes clarifications, modifications, options, and image building.
6. *Prepare for formal meetings.* Be sure you are well prepared for meetings, fact-finding sessions, or presentations requested by the customer. This is your opportunity not only to clarify, but also to strengthen your proposal, show additional material, and introduce new personnel if needed.
7. *Reassess cost and price.* Cost and proposed effort are often fluid during the initial program phases. Many times the discussion of the proposal narrows down the real customer requirements. This provides an opportunity to reassess and adjust the bid price.
8. *Obtain start-work order.* It is often possible to obtain a start-work order before the program is formally under contract. This provides a limited mutual commitment and saves time.
9. *Stay on top until closure.* From the time of bid submission to obtaining the final contract, the bidder must keep abreast with all developments in the customer community which affect the proposal. Try to help the customer justifying the source selection and be responsive to customer requests for additional information and meetings. Frequent interaction with the customer is pervasive.
10. *Conduct formal negotiations.* Negotiation comes in many forms. Program contract negotiations mostly center on the technical performance, schedule, and cost. They should be conducted among the technical and managerial personnel of the contracting parties. If in addition to the technicalities the contract covers extensive legal provision, terms, and conditions, the bidder should seek the interpretation and advice of legal counsel.

high competition. In most other situations price is a factor toward winning only in the context of all competitive components such as the following:

- Compliance with customer requirements.
- Best-fitting solution to customer problem.

- Real demand.
- Relevant past experience.
- Credibility.
- Long-range commitment to business segment.
- Past performance on similar programs.
- Soundness of approach.
- Cost credibility.
- Competitive price.
- Delivery.
- After-sale support.
- Logistics.

The better you understand the customer, the better you will be able to communicate the strength of your product relative to the customer requirements.

Some specific recommendations are made to help business managers responsible for winning new contracts, and professionals who must support these bid proposal efforts, to better understand the complex interrelationship among organizational, technical, marketing, and behavioral components, and to perform their difficult role more effectively.

1. *Plan Ahead.* Develop a detailed business acquisition plan which includes a realistic assessment of the new opportunity and the specific milestones for getting through the various steps needed for bidding and negotiating the contract.
2. *Involve the Right People, Early.* In order to get a realistic assessment of the new opportunity against internal capabilities and external competition, form a committee of senior personnel early in the acquisition cycle. These people should represent the key functional areas of the company and be able to make a sound judgment on the readiness and chances of their company to compete for the new business effectively.
3. *Closeness to the Customer.* Especially for the larger contracts, it is important that the bidding firm has been closely involved with the customer prior to the RFP, and if possible, during the bid preparation. A company that has been closely involved with the customer in helping to define the requirements, in conducting feasibility studies with the customer, or in executing related contract work will not only understand the customer requirements better, but also have higher trust and credibility regarding their capacity to perform by comparison to a company that just submits a bid proposal.
4. *Select Your Bid Opportunities Carefully.* Bid opportunities are usu-

ally plentiful. Only qualified bidders who can submit a competitively attractive proposal have a chance of winning, submitting "more" proposals does nothing to improve your win ratio. It only drains your resources. Each bid opportunity should be carefully assessed to determine whether you really have the ingredients to win.

5. *Make Bid Decisions Incrementally.* A formal bid decision, especially for the larger proposals, requires considerable homework and resources. By making these decisions in several steps, such as initial, preliminary, and final, management can initially quickly screen a large number of opportunities and narrow them down to a shorter list without spending a lot of time and resources. Then the available resources can be concentrated to analyze those opportunities which really seem to be most promising.
6. *Be Sure You Have the Resources to Go the Full Distance.* Many bid proposal activities require large amounts of resources and time. In addition, resources may be needed beyond the formal bidding activities for customer meetings, site visits, and negotiations. Further, the customer may extend the bid submission deadline, which will cost you more money as you continue to refine your proposal. Serious consideration should be given at the very beginning of any potential bid whether you truly have the resources and are willing to commit them. Develop a detailed cost estimate for the entire proposal effort.
7. *Obtain Commitment* from senior management to make the necessary resources available when needed. This includes personnel and facilities.
8. *Do Your Homework.* Before any proposal writing starts, you should have a clear picture of the strengths, weaknesses, and limitations of (1) the competing firms and (2) your company. In addition, you need to fully understand (3) the customer requirements, constraints, such as budgets, and biases. This requires intense customer contact and market research.
9. *Organize the Proposal Effort* prior to the RFP. Run the proposal development like any other project. You need a well-defined step-by-step action plan, schedules, budgets, team organization, and facilities. You also have to prepare for support services such as editing and printing.
10. *Grow a Proposal Specialist.* The efficiency and effectiveness of the proposal development can be greatly enhanced with a professional proposal specialist. This is an internal consultant who can lead the team through the proposal development process including providing the checks and balances via reviews and analysis.

11. *Know Your Competition.* Marketing intelligence comes in many forms. The marketeer who is in touch with the market knows his competition. Information can be gathered at trade shows, bidder's briefings, customer meetings, and professional conferences, and from the literature and via special market service firms.
12. *Develop a Win Strategy.* Define your niche or "unfair advantage" over your competition and build your win strategy around it. Only after intensive intelligence gathering from the competition and the customer and careful analysis of these market data against your strengths and weaknesses, can a meaningful win strategy be developed. Participation of key personnel from all functions of the organization is necessary to develop a meaningful and workable win strategy for your new contract acquisition.
13. *Develop the Proposal Text Incrementally.* Don't go from RFP to the first proposal draft in one step. *Use the Storyboard process.*
14. *Be fully compliant to the RFP.* Don't take exceptions to the customer requirements unless absolutely unavoidable. A formal RFP analysis listing the specific customer requirements helps to avoid unintended oversights and also helps in organizing your proposal.
15. *Demonstrate Understanding of Customer Requirements.* A summary of the requirements and brief discussions helps to instill confidence in the customer's mind that you understand the specific needs.
16. *Demonstrate Ability to Perform.* Past, related experiences will score the strongest points. But showing that your company performed on similar programs, that you have experienced personnel, and that you have done analytical homework against the requirements may rate very favorably with the customer too, especially when you have other advantages such as innovative solution, or favorable timing or pricing.
17. *Progress Reviews.* As part of the incremental proposal development, assure through reviews which check (1) compliance with customer requirements, (2) soundness of approach, (3) effective communication, and (4) proper integration of topics into one proposal.
18. *Red Team Reviews.* For "must-win" proposals it may be useful to set up a special review team that evaluates and scores the proposal, similar to the process used by the customer. Deficiencies which may otherwise remain hidden can often be identified and dealt with during the proposal development. Such a "red team review" can be conducted at various stages of the proposal development. It is important to budget enough time for revising the proposal after red team review.

19. *Use Editorial Support.* A competent editor should work side-by-side with the technical proposal writers. A good editor can take text at a rough draft stage and "finalize" it regarding logic, style, and grammar. However, the proper content has to come from the technical author. Therefore, for the process to work, text often cycles between the author and editor several times until both agree to it. The professional editor not only frees the technical writer for the crucial innovative technical proposal development, but also provides clarity and consistency to the proposal. Further, this procedure increases the total writing efficiency by a factor of 2. That is, using professional editors reduces the total proposal *writing* budget to one half of what it would cost otherwise.
20. *Price Competitively.* Pricing is a complex issue. However, for most proposals, a competitively priced bid has the winning edge. Knowing the customer's budget and some of the cost factors of the competition can help in fine-tuning the bidding price. Further, cost-plus proposals must have cost credibility, which is being built via a clearly articulated cost model and a description of its elements of cost. Pricing should start at the time of the bid decision.
21. *Prepare for Negotiations.* Immediately after proposal submission, work should start in preparation of customer inquiries and negotiations. Responses to customer inquiries, regarding clarifications on the original bid, can be used effectively to score additional technical points and build further credibility.
22. *Conduct Postmortem.* Regardless of the final outcome, a thorough review of the proposal effort should be held and the lessons learned be documented for the benefit of future proposals.

## A FINAL NOTE

Winning new contract business is a highly competitive and costly undertaking. To be successful, it requires special management skills, tools, and techniques which range from identifying new bid opportunities to bid decisions and proposal developments.

Companies that win their share of new business usually have a well-disciplined process; they also have experienced personnel who can manage the intricate process and lead a multifunctional team toward writing a unified winning proposal. Their managements use good logic and judgment in deriving their bid decisions; they also make fewer fundamental mistakes during the acquisition process. These are the managers who position their companies uniquely in the competitive field by building a quality image with the customer and by submitting a responsive bid proposal that is competitively priced. These are the business managers who target specific opportunities and demonstrate a high win ratio.

## CONTRACT INFORMATION SOURCES

### HANDBOOKS AND REFERENCE BOOKS

*Anatomy of a Win,* by Jim M. Beveridge Associates, 8448 Wagner Creek Road, Talent, OR 97540. 1978.
*Business Guide to Dealing with the Federal Government,* Drake Publishers Inc., 381 Park Avenue South, New York, NY 10016. 1973.
*Contract Planning and Organization,* United Nations Publications, United Nations, LX2300, New York, NY 10017. 1974.
*Grantmanship and Fundraising* by Armand Lauffer, Sage Publication. 1984.
*How to Create a Winning Proposal,* by J. Ammon-Wexler and Catherine ap Carmel, Mercury Communications, 730 Mission Street, Santa Cruz, CA 95060. 1977.
*Positioning to Win* by Jim M. Beveridge and E. J. Velton, J. M. Chilton, 1982.
*Selling to United States Government,* United States Small Business Administration, Washington, DC 20416. 1973.
*Source Guide to Government Technology and Financial Assistance* by Harry Greenwald et al, Prentice-Hall, 1982
*Research and Development Directory,* Government Data Publications, annually.
*Technical Marketing to the Government,* by Robert A. Rexroad, Dartness Corp., Chicago.

### PERIODICALS AND NEWSPAPERS

*Briefing Papers,* Federal Publications, 1725 K Street, NW, Washington, DC 20006. Bimonthly.
*Commerce Business Daily,* U.S. Department of Commerce, Office of Field Services, U.S. Government Printing Office, Washington, DC 20402.
*Forms of Business Agreement,* Institute of Business Planning, IPB Plaza, Englewood Cliffs, NJ 07632. Monthly.
*Government Contractor,* Federal Publications, 1725 K Street NW, Washington, DC 20006. Biweekly.
*Government Contracts Reports,* Commerce Clearance House, 4025 West Peterson Avenue, Chicago, IL 60646. Weekly.
*New Business Report* (Monthly Newsletter) by Executive Communications, Inc., New York, NY.
*NCMA Newsletter,* National Contract Management Association, 675 East Wardlow Road, Long Beach, CA 90807. Monthly.

### DIRECTORIES

*Directory of Government Production Prime Contracts,* Government Data Publication, Washington, DC. Annual.

*Government Contracts Directory,* Government Data Publications, 422 Washington Building, Washington, DC 20005. Annual.

*Government Contracts Guide,* Commerce Clearing House, 4025 West Peterson Avenue, Chicago, IL 60646.

*Selling to NASA,* U.S. Government Printing Office, Washington DC 20402.

*Selling to Navy Prime Contractors,* U.S. Government Printing Office, Washington DC 20402.

*United States Government Purchasing and Sales Directory,* U.S. Small Business Administration, U.S. Government Printing Office, Washington DC 20402. 1972.

## ON-LINE DATA BASES

*Defense Market Measurement System,* Frost and Sullivan, 109 Fulton Street, New York, NY 10038.

*Federal Register,* Capitol Services, 511 Second St. NE, Washington, DC 20002.

*U.S. Government Contract Awards,* SCD Service, System Development Corp., 2500 Colorado Ave., Santa Monica, CA 90406.

## ASSOCIATIONS AND SOCIETIES

*Electronic Industries Association* (EIA), 2001 Eye Street, NW, Washington, DC 20006.

*National Contract Management Association,* 6728 Old McLean Village Dr., McLean, VA 22101, (703) 442-0137.

*National Council of Technical Service Industries,* 1845 K Street, NW, Suite 1190, Washington, DC 20006.

---

## BIBLIOGRAPHY

Behling, John H. *Guidelines for Preparing the Research Proposal,* U. Press of America, 1984.

Edelman, F. "Art and Science of Competitive Bidding." *Harvard Business Review, 43* (July–August 1965).

Frichtl, P. "Tactics for Tactics." *Industrial Distributor:* 9 (July, 1986).

Guyton, Robert, et al. *Prerequisits for Winning Government R&D Contracts,* Universal Technology Corporation, 1983.

Hanssman, F. and Rivett, B. H. P. "*Competitive Bidding.*" *Operations Research Quarterly,* 10:49–55 (1959).

Kerzner, Harold and Loring, Roy J. *Proposal Preparation and Management Handbook,* Van Nostrand Reinhold, New York, 1982.

Holtz, Herman. *Government Contracts: Proposal-manship and Winning Strategies.* Plenum, New York, 1979.

Lehman, D. H. "A Technique for Lowering Risks During Contract Negotiations." *Transportation Engineering Managers, 33:*79–81 (May, 1986).

Merrifield, D. Bruce, *Strategic Analysis, Selection and Management of R&D Projects,* American Management Association, New York, 1977.

Newport, J. P., Jr. "Billion Dollar Bids in Sealed Envelopes." *Fortune:*42 (April, 1985).

Owens, Elizabeth, "Effective Proposals of the competitive Businessman," *Data Management, 25*:22 (January, 1987).

Park, W. R. *Construction Bidding for Profits*. Wiley, New York, 1979.

Porter-Roth, Bud. *Proposal Development—A Winning Approach*. PSI Research, 1986.

Public Management Institute Staff, *How to get Federal Contracts*, Public Management, 1980.

Robertson, J. "Government 11-th Hour Bid Cancellation Case No-Bids Team Efforts." *Electronic News, 30* (May 21, 1984).

Rugh and Manning, *Proposal Management Using the Modular Approach*, Peninsula Publication Company, 1982.

Simmonds, K. "Competitive Bidding—Deciding the Best Combination of Non-Price Features." *Operational Research Quarterly, 19*:5–15 (1968).

Steward, Rodney, D. and Steward, Ann L. *Proposal Preparation*, Wiley, New York, 1984.

Thamhain, H. J. "Marketing in Project-Oriented Business Environments." *Project Management Quarterly* (December, 1982).

Wantuck. "Bidding for Fair Play." *Nations Business, 73*:39–40 (September, 1985).

Whalen, Tim, Improved Proposal Writing Utility, Coherance, and Emphasis," *Bulletin of the Association for Business Communication*, Vol. 69 (December 1986).

# Section VI
# Project Implementation

This section focuses on a number of dimensions that are of particular importance in the successful implementation and execution of projects. As in all human endeavor, the best laid plans may come to naught if inadequate attention is paid to certain critical factors.

In Chapter 20, Jeffrey K. Pinto and Dennis P. Slevin identify the "critical success factors" for effective project management using data collected from project managers in various contexts. These factors may be taken as guides to the project manager as to whether he or she should focus attention in the various phases of the project life cycle.

John L. Heidenreich, in Chapter 21, addresses the critical quality dimension in the project management. Since quality considerations are becoming ever more important in industry, this factor warrants special attention as any project is implemented.

In Chapter 22, the legal standards for prudent and efficient project management, which must serve to guide the way in which the project is carried out, are outlined. In our litigious society, legal considerations, like quality concerns, stand out as warranting special attention in implementing a project.

# 20. Critical Success Factors in Effective Project Implementation*†

Jeffrey K. Pinto‡
Dennis P. Slevin**

## INTRODUCTION

The process of project implementation, involving the successful development and introduction of projects in the organization, presents an ongoing challenge for managers. The project implementation process is complex, usually requiring simultaneous attention to a wide variety of human, budgetary, and technical variables. As a result, the organizational project manager is faced with a difficult job characterized by role overload, frenetic activity, fragmentation, and superficiality. Often the typical project

* Portions of this chapter were adapted from Dennis P. Slevin and Jeffrey K. Pinto, Balancing Strategy and Tactics in Project Implementation', *Sloan Management Review*, Fall, 1987, pp. 33–41, and Randall L. Schultz, Dennis P. Slevin, and Jeffrey K. Pinto, 'Strategy and Tactics in a Process Model of Project Implementation'', *Interfaces*, 16:3, May-June, 1987, pp. 34–46.



** Dennis P. Slevin is an Associate Professor of Business Administration at the University of Pittsburgh's Joseph M. Katz Graduate School of Business. He holds a B.A. in Mathematics from St. Vincent College, a B.S. in Physics from M.I.T., an M.S. in Industrial Administration from Carnegie-Mellon, and a Ph.D. from Stanford University. He has had extensive experience as a line manager, including service as the CEO of four different companies, which qualified him as a member of the Young Presidents' Organization. He presently serves as a director of several corporations, and consults widely. He publishes in numerous professional journals, and is co-editor of *Implementing Operations Research Management Science; The Management of Organizational Design, Volumes I and II;* and *Producing Useful Knowledge*. He has written the pragmatic *Executive Survival Manual* for practicing managers.

‡ Jeffrey K. Pinto is Assistant Professor of Organization Theory at the College of Business Administration, University of Cincinnati. He received his B.A. in History and B.S. in Business from the University of Maryland, M.B.A. and Ph.D. from the University of Pittsburgh. He has published several papers in a variety of professional journals on such topics as project management, implementation, instrument development, and research methodology.

manager has responsibility for successful project outcomes without sufficient power, budget, or people to handle all of the elements essential for project success. In addition, projects are often initiated in the context of a turbulent, unpredictable, and dynamic environment. Consequently, the project manager would be well served by more information about those specific factors critical to project success. The project manager requires the necessary tools to help him or her focus attention on important areas and set *differential priorities* across different project elements. If it can be demonstrated that a set of factors *under the project manager's control* can have a significant impact on project implementation success, the project manager will be better able to effectively deal with the many demands created by his job, channeling his energy more efficiently in attempting to successfully implement the project under development.

This chapter reports on a program of research that has developed the following tools and/or concepts for the practicing project manager.

- A set of ten empirically derived critical project implementation success factors.
- A diagnostic instrument—the Project Implementation Profile (PIP) for measuring the ten factors.
- A ten-factor model of the project implementation process.
- Measures of the key elements of project Strategy and Tactics.
- The effect of Strategy and Tactics on project implementation success.
- The impact of the project life cycle on the relative importance of the critical success factors.

In addition, we propose that as the project moves forward through its life cycle, the project manager must be able to effectively transition from strategic to tactical issues in order to better influence project success. Implications are suggested for practicing managers along with specific approaches to managing the strategy-tactics interface.

## DEFINITIONS

Before attempting a discussion of the project implementation process, it is first important that some of the key concepts in this chapter be adequately defined, in an effort to remove some of the ambiguity from concepts which are often subject to a wide range of individual interpretations.

### What Is a Project?

While almost everyone has had experience with projects in one form or another, developing a definition of what exactly a project is is often difficult. Any definition of a project must be general enough to include examples of the wide variety of organizational activities which managers consider to be "project functions." However, the definition should be narrow enough to include only those specific activities which researchers and practitioners can meaningfully describe as "project-oriented." Two of the many definitions of projects that have been offered may be considered as follows:

> A project is an organization of people dedicated to a specific purpose or objective. Projects generally involve large, expensive, unique, or high risk undertakings which have to be completed by a certain date, for a certain amount of money, within some expected level of performance. At a minimum, all projects need to have well defined objectives and sufficient resources to carry out all the required tasks. (24, p. 498)

The second definition is offered by Cleland and Kerzner (7), in their work *A Project Management Dictionary of Terms,* and includes the following characteristics:

> [A project is] A combination of human and nonhuman resources pulled together in a temporary organization to achieve a specified purpose. (7, p. 199)

A project, then, can be defined as possessing the following characteristics:

- A defined beginning and end (specified time to completion).
- A specific, preordained goal or set of goals.
- A series of complex or interrelated activities.
- A limited budget.

### What Is Successful Project Implementation?

In addition to defining the concept of organizational projects, it is important, before attempting any discussion of the steps leading to a successful project, to describe just exactly what a "successful project" is. Project implementation success has been defined many ways to include a large variety of criteria. However, in its simplest terms, project success can be thought of as incorporating four basic facets. A project is generally considered to be successfully implemented if it

- Comes in on-schedule (time criterion).
- Comes in on-budget (monetary criterion).

- Achieves basically all the goals originally set for it (effectiveness criterion).
- Is accepted and used by the clients for whom the project is intended (client satisfaction criterion).

By its basic definition, a project comprises a defined time frame to completion, a limited budget, and a specified set of performance characteristics. Further, the project is usually targeted for use by some client, either internal or external to the organization and its project team. It seems reasonable, therefore, that any assessment of project implementation success should include these four measures.

### The Project Life Cycle

One method that has been used with some regularity in order to help managers conceptualize the work and budgetary requirements of a project is to make use of the idea of the project life cycle. The concept of the life cycle is familiar to most modern managers. Life cycles are used to explain the rise and demise of organizations, phases in the sales life of a product, etc. In a similar fashion, managers often make use of the life-cycle concept as a valuable tool for better understanding the stages in a project and the likely materials requirements for the project through each distinct phase.

Figure 20-1 shows an example of a project life cycle. This representation of the project life cycle is based on the work of Adams and Barndt (1) and King and Cleland (10). As can be seen, the project's life cycle has been divided into four distinct stages, including:

1. *Conceptualization*—The initial project stage. At this stage a project is determined as being necessary. Preliminary goals and alternatives are specified, as well as the possible means to accomplish those goals.
2. *Planning*—This stage involves the establishment of a more formalized set of plans to accomplish the initially developed goals. Among planning activities are scheduling, budgeting, and the allocation of other specific tasks and resources.
3. *Execution*—The third stage involves the actual "work" of the project. Materials and resources are procured, the project is produced, and performance capabilities are verified.
4. *Termination*—Once the project is completed, there are several final activities that must be performed. These activities usually include the release of resources and transfer of the project to the clients and, if necessary, the reassignment of project team personnel.

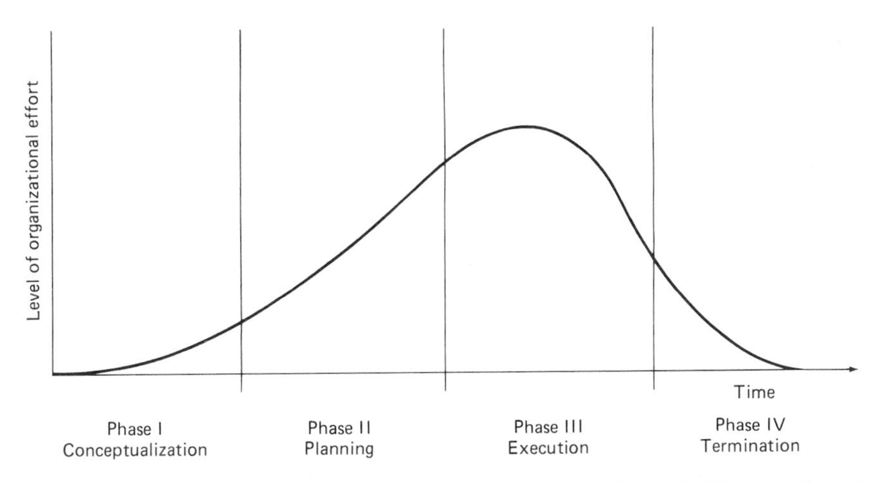


Figure 20-1. Stages in the project life cycle. (Based on Adams and Barndt; King and Cleland (10).)

As Figure 20-1 also shows, in addition to the development of four project stages, the life cycle specifies the level of organizational effort necessary to adequately perform the tasks associated with each project stage. Organizational effort can be measured using surrogates such as amount of man-hours, expenditures, assets deployed, or other measures of organizational resource utilization. As one would suspect, during the early Conceptualization and Planning stages, effort requirements are minimal, increasing rapidly during late Planning and project Execution, before diminishing again in the project's Termination. As a result, the concept of project life cycles can be quite useful to a manager, not only in terms of distinguishing among the stages in the project's life, but also through indicating likely resource requirements to be expected at each stage.

## DEVELOPMENT OF THE TEN-FACTOR MODEL OF PROJECT IMPLEMENTATION

Project information was obtained from a group of over 50 managers who had some project involvement within the last two years. Participants were asked to consider a successful project with which they had been involved and then to put themselves in the position of a project manager charged with the responsibility of successful project implementation. They were then asked to indicate things that they could do that would substantially help implementation success. This procedure, sometimes called Project Echo, was developed by Alex Bavelas (4). Responses were then sorted into categories by two experts. Both experts sorted the responses into ten categories and interrater agreement based on percentage of responses similarly sorted across the total number was 0.50, or 119 out of 236. Eliminating duplications and miscellaneous responses, a total of 94 usable responses were classified across 10 factors. These 10 factors formed the basis for the conceptual model and the diagnostic instrument for measuring relative strength of each factor.

The first factor that was developed was related to the underlying purpose for the implementation and was classified *Project Mission*. Several authors have discussed the importance of clearly defining goals at the outset of the project. Morris (15) classified the initial stage of project management as consisting of a feasibility decision. Are the goals clear and can they succeed? Bardach's (3) six-step implementation process begins with instructions to state the plan and its objectives. For both these authors and the purposes of our study, Project Mission has been found to refer to the condition where the goals of the project are clear and understood, not only by the project team involved, but by the other departments in the organization. Underlying themes of responses classified into this factor include statements concerning clarification of goals as well as belief in the likelihood of project success.

The second factor discerned was that of *Top Management Support*. As noted by Schultz and Slevin (19), management support for projects, or indeed for any implementation, has long been considered of great importance in distinguishing between their ultimate success or failure. Beck (6) sees project management as not only dependent on top management for authority, direction, and support, but as ultimately the conduit for implementing top management's plans, or goals, for the organization. Further, Manley (14) shows that the degree of management support for a project will lead to significant variations in the clients' degree of ultimate acceptance or resistance to that project or product. For the purposes of our classification, the factor Top Management Support refers to both the nature and amount of support the project manager can expect from management both for himself as leader and for the project. Management's support of the project may involve aspects such as allocation of sufficient resources (financial, manpower, time, etc.) as well as the project manager's confidence in their support in the event of crises.

The third factor to be classified was that of *Project Schedule/Plans*. Project schedule refers to the importance of developing a detailed plan of the required stages of the implementation process. Ginzberg (8) has drawn parallels between the stages of the implementation process and the Lewin (12) model of Unfreezing-Moving-Freezing, viewing planning and scheduling as the first step in the "Moving" stage. Kolb and Frohman's (11) model of the consulting process views planning as a two-directional stage, not only as necessary to the forward-going change process, but as an additional link to subsequent evaluation and possible reentry into the system. Nutt (16) further emphasizes the importance of process planning, breaking down planning into four stages: formulation, conceptualization, detailing, and evaluation. As developed in our model, Project Schedule/Plans refers to the degree to which time schedules, milestones, manpower, and equipment requirements are specified. Further, the schedule should include a satisfactory measurement system as a way of judging actual performance against budget and time allowances.

The fourth factor that was determined is labeled *Client Consultation.* The "client" is referred to here as anyone who will ultimately be making use of the result of the project, as either a customer outside the company or a department within the organization. The need for client consultation has been found to be increasingly important in attempting to successfully implement a project. Indeed, Manley (14) found that the degree to which clients are personally involved in the implementation process will cause great variation in their support for that project. Further, in the context of the consulting process, Kolb and Frohman (11) view client consultation as the first stage in a program to implement change. As this factor was derived for the model, Client Consultation expresses the necessity of taking into account the needs of the future clients, or users, of the project. It is, therefore, important to determine whether clients for the project have been identified. Once the project manager is aware of the major clients, he is better able to accurately determine if their needs are being met.

The fifth factor was concerned with *Personnel* issues, including recruitment, selection, and training. (See Table 20-1.) An important, but often overlooked, aspect of the implementation process concerns the nature of the personnel involved. In many situations, personnel for the project team are chosen with less-than-full regard for the skills necessary to actively contribute to implementation success. Some current writers on implementations are including the personnel variable in the equation for project team performance and project success. Hammond (9) has developed a contingency model of the implementation process which includes "people" as a situational variable whose knowledge, skills, goals, and personalities must be considered in assessing the environment of the organization. Only after such a diagnosis takes place can the project management team begin to set objectives and design the implementation approach. For the model, Personnel, as a factor, is concerned with developing a project team with the requisite skills to perform their function. Further, it is important to determine whether project management has built sufficient commitment toward project success on the part of team members.

The sixth factor to be discussed was labeled *Technical Tasks.* It is

Table 20-1. Factor Definitions.[a]

1. *Project Mission*—Initial clearly defined goals and general directions.
2. *Top Management Support*—Willingness of top management to provide the necessary resources and authority/power for project success.
3. *Project Schedule/Plan*—A detailed specification of the individual actions steps for project implementation.
4. *Client Consultation*—Communication, consultation, and active listening to all impacted parties.
5. *Personnel*—Recruitment, selection, and training of the necessary personnel for the project team.
6. *Technical Tasks*—Availability of the required technology and expertise to accomplish the specific technical action steps.
7. *Client Acceptance*—The act of "selling" the final project to its ultimate intended users.
8. *Monitoring and Feedback*—Timely provision of comprehensive control information at each stage in the implementation process.
9. *Communication*—The provision of an appropriate network and necessary data to all key actors in the project implementation.
10. *Troubleshooting*—Ability to handle unexpected crises and deviations from plan.

Source: Slevin and Pinto, (1986, pp. 57–58), From the article "The Project Implementation Profile: New Tool for Project Managers" which appeared in *Project Management Journal*, September, 1986.

important that the implementation be well managed by people who understand the project. In addition, there must exist adequate technology to support the project. Technical Tasks refers to the necessity of not only having the necessary personnel for the implementation team, but ensuring that they possess the necessary technical skills and have adequate technology to perform their tasks. Steven Alter (2), writing on implementation risk analysis, identifies two of the eight risk factors as being caused by technical incompatibility: the user's unfamiliarity with the systems or technology, and cost ineffectiveness.

In addition to Client Consultation at an earlier stage in the project implementation process, it remains of ultimate importance to determine whether the clients for whom the project has been initiated will accept it. *Client Acceptance* refers to the final stage in the implementation process, at which time the ultimate efficacy of the project is to be determined. Too often project managers make the mistake of believing that if they handle the other stages of the implementation process well, the client (either internal or external to the organization) will accept the resulting project. In fact, as several writers have shown, client acceptance is a stage in project implementation that must be managed like any other. As an implementation strategy, Lucas (13) discusses the importance of user participation in the early stages of system development as a way of improving the likelihood of later acceptance. Bean and Radnor (5) examine the use of "intermediaries" to act as a liaison between the designer, or implementation team, and the project's potential users as a method to aid in client acceptance.

The eighth factor to be considered is that of *Monitoring and Feedback*. Monitoring and Feedback refer to the project control processes by which at each stage of the project implementation, key personnel receive feedback on how the project is comparing to initial projections. Making allowances for adequate monitoring and feedback mechanisms gives the project manager the ability to anticipate problems, to oversee corrective measures, and to ensure that no deficiencies are overlooked. Schultz and Slevin (19) demonstrate the evolving nature of implementation and model-building paradigms to have reached the state including formal feedback channels between the model builder and the user. From a budgeting perspective, Souder et al. (23) emphasize the importance of constant monitoring and "fine-tuning" of the process of implementation. For the model, Monitoring and Feedback refers not only to project schedule and budget, but to monitoring performance of members of the project team.

The ninth factor was that of *Communication*. The need for adequate communication channels is extremely important in creating an atmosphere for successful project implementation. Communication is not only essential within the project team itself, but between the team and the rest of the organization as well as with the client. As the factor Communication has been developed for the model, it refers not only to feedback mechanisms, but the necessity of exchanging information with both clients and the rest of the organization concerning project goals, changes in policies and procedures, status reports, etc.

The tenth and final factor to emerge from classification of the model is *Trouble Shooting*. As the participants in the study often pointed out, problem areas exist in almost every implementation. Regardless of how carefully the project was initially planned, it is impossible to foresee every trouble area or problem that could possibly arise. As a result, it is important that the project manager make adequate initial arrangements for "troubleshooting" mechanisms to be included in the implementation plan. Such mechanisms make it easier not only to react to problems as they arise, but to foresee and possibly forestall potential trouble areas in the implementation process.

## THE MODEL

As Figure 20-2 shows, a framework of project implementation has been developed for heuristic purposes, based on the ten factors discovered in our analysis. Some general characteristics of the model should be noted:

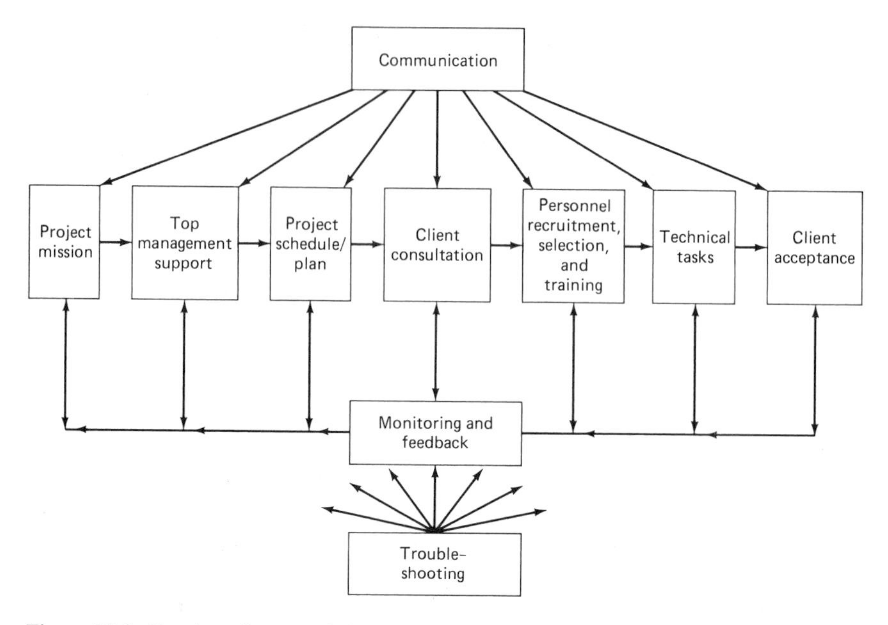


Figure 20-2. Ten key factors of the project implementation profile. (Copyright © 1984 Randall L. Schultz and Dennis P. Slevin. Used with permission.)

1. The factors appear to be both time sequenced and interdependent.

Conceptually, one could argue that the factors are sequenced to occur (or be considered) in a logical order instead of randomly or concurrently. To illustrate, consider that, according to the framework, it is first important to set the goals or define the mission and benefits of the project before seeking top management support. Furthermore, one could argue that unless consultation with the project's clients has occurred early in the process, chances of subsequent client acceptance and use, denoting successful implementation, will be negatively affected. Nonetheless, it is important to remember that in actual practice, considerable overlap and reversals can occur in the ordering of the various factors and the sequencing as suggested in the framework is not absolute.

2. The factors for a project implementation can be laid out on a critical path.

Related to the temporal aspect, the factors of project implementation can be laid out in a rough critical path, similar to the critical path methodology used to develop a new product or to determine the steps in an OR/MS project. In addition to the set of seven factors along the critical path, ranging from Project Mission to Client Acceptance, other factors such as Communication and Monitoring and Feedback are hypothesized to necessarily occur simultaneously and in harmony with the other sequential factors. As several project managers have indicated to us over the course of this research, it is important that Communication always occur or that Troubleshooting be available throughout the implementation process. It should be noted, however, that the arrows in the model represent information flows and sequences, not necessarily causal or correlational relationships.

3. The model allows the manager to actively interact with and systematically monitor his project.

The sequence of a project implementation is an important consideration for any project manager. Not only are there a prescribed set of steps to be taken in the project implementation process, but because of the order of the steps to be taken, the manager is provided with a checklist for determining the status of the project at any given stage. This monitoring capacity enables the manager to determine where the project is in terms of its life cycle and how rapidly it is moving forward. Further, the manager has the ability to determine the chances for successful implementation given attention has been paid to the proper sequencing of steps and consideration of relevant critical success factors in the implementation process.

A 100-item instrument (10 items per factor) was developed and has been used to measure the relative level of each of these critical success factors (21). This instrument was further refined and reduced to a 50-item instrument (5 items per factor) and is a useful diagnostic tool for project implementation. This instrument has been included in its entirety along with percentile norms for over 400 projects as an implementation aid for project managers.

Table 20-2 demonstrates the results of a recent study in which the ten critical factors were assessed in terms of their overall contribution to project success (17). A data base of over 400 projects were sampled in an effort toward empirical verification of the importance of each of the ten initially developed critical success factors. As can be seen, each of the ten factors was found to be significantly related to project success. Further, the cumulative $r$-square value, representing total amount of the variance explained by the ten factors, was .615. In other words, over 61% of the causes of project implementation success can be explained by the ten critical success factors.

Table 20-2. Results of Multiple Regression on the Ten Critical Success Factors.[a]

| FACTOR | BETA | $T$-VALUE | SIG. $T$ |
|---|---|---|---|
| Mission | .72 | 19.99 | $p < .001$ |
| Top Management Support | .32 | 10.60 | $p < .001$ |
| Schedule | .32 | 10.92 | $p < .001$ |
| Client Consultation | .39 | 11.86 | $p < .001$ |
| Personnel | .31 | 10.54 | $p < .001$ |
| Technical Tasks | .43 | 11.25 | $p < .001$ |
| Client Acceptance | .39 | 11.46 | $p < .001$ |
| Monitoring and Feedback | .29 | 10.89 | $p < .001$ |
| Communication | .32 | 10.38 | $p < .001$ |
| Troubleshooting | .35 | 11.15 | $p < .001$ |

[a] Total regression equation $F = 47.8$, $p < .001$. Cumulative adjusted $r$-square = .615.

## STRATEGY AND TACTICS

As one moves through the ten-factor model shown in Figure 20-1, it becomes clear that the general characteristics of the factors change. In fact, the factors can be grouped into meaningful patterns, or more general subdimensions. As Table 20-3 shows, the first three factors, (Mission, Top Management Support, and Schedule) are related to the early "planning" phase of the implementation process. The second dimension, composed of the other seven factors (Client Consultation, Personnel, etc.), may be seen as concerned with the actual process, or "action," of the implementation. These factors seem less planning in nature and more based on the operationalization of the project implementation process.

These "planning" versus "action" elements in the critical implementation success factors show significant parallels to the distinction between

Table 20-3. Strategic and Tactical Critical Success Factors.

| STRATEGY | TACTICS |
|---|---|
| Mission | Client Consultation |
| Top Management Support | Personnel |
| Schedule/Plans | Technical Tasks |
| | Client Acceptance |
| | Monitoring and Feedback |
| | Communication |
| | Troubleshooting |

strategy and tactics in the strategic management field. Strategy is often viewed as the process of deciding on overall organizational objectives as well as planning on how to achieve those goals. Tactics are seen as the deployment of a wide variety of human, technical, and financial resources to achieve those strategic plans. Strategy, then, is concerned with the up-front planning, while tactics are specifically focused on how best to operationalize, or achieve, those plans.

It is important that managers understand the differences between strategic and tactical issues. Both are vital to project success, but differentially so as the project moves forward to completion. One method for clarifying the distinction raised between strategy and tactics is through the development of a taxonomy that demonstrates the diverse nature of the two functions. This taxonomy is especially useful if applied to the project management context because it has important implications for determining the relationship between strategy and tactics and the previously mentioned planning versus action aspects of the implementation process. Table 20-4 shows a sample of ten issues which have differing implications for project implementation when approached from either a strategic or a tactical viewpoint.

From a conceptual standpoint, the first three critical success factors are primarily "strategic" in nature, while the last seven are more "tactical." Using the model and the measurement instrument (See Appendix), it is possible to monitor the level of strategy (sum of percentile scores on the first three factors) and tactics (sum of percentile scores on the last seven factors) as the project moves forward in time. In addition to showing the Project Implementation Profile, Appendix 1 also exhibits the set of percentile scores for each of the critical success factors, based on a data base of 418 projects. The manager is able to assess scores on each of the ten factors for his specific project and compare those percentile scores with this previously gathered sample of projects.

## STRATEGY-TACTICS INTERACTION

In addition to the above conceptualization regarding project implementation as a two-stage process, involving initial strategic actions and supporting tactical activities, there are further implications for project performance based on a consideration of strategic and tactical issues. Figure 20-3 shows the breakdown of strategy and tactics by high and low scores depending upon the level to which these issues were addressed in the project implementation. A high "score" on strategy would imply that the strategy is well developed and effective, as is the similar case with tactics. This value could be assessed either in a subjective or intuitive manner or

Table 20-4. Taxonomy of Strategic Versus Tactical Issues.[a]

| | STRATEGY | TACTICS |
|---|---|---|
| 1. *Level of Conduct*—Level within the organization at which project implementation activities and issues are performed or addressed. | Top management | Mid- to lower levels of management |
| 2. *Subjective/Objective Assessment*—The activities concerned with assessing project goals or status. | Greater subjectivity used at strategic level | Less use of subjective values |
| 3. *Nature of Problem*—The types of problems which arise and must be dealt with during the project implementation process. | Unstructured, one at a time | More structured and repetitive |
| 4. *Information Needs*—The determination of the types and quantity of information that is required for the project. | Large amount of information needed, much that is external | Need for internally generated, specific information |
| 5. *Time Horizons*—The scope or time frame of management's vision in implementing and evaluating the project. | Long-term, but it varies by the problem | Short-term and more constant |
| 6. *Completeness*—The degree to which the scope of the entire organization is considered. | Covers the entire scope of the organization | Concerned only with the suborganizational unit involved |
| 7. *Reference*—Involves the source, or frame of reference, of the activity to be considered. | The source of all planning in the organization is original | Done in pursuit of strategic plans |
| 8. *Detail*—Concerned with how broad or specific problems are laid out and how generally they need to be addressed. | Broad and general | Narrow and problem specific |
| 9. *Ease of Evaluation*—The ease of determining the efficiency and effectiveness of various activities involved in the implementation. | Difficult, because of generality | Easier, because of specificity |
| 10. *Point of View*—The assessment of the focus or viewpoint of the various actors involved in the project implementation. | Corporate | Functional |

"Taxonomy of Strategic vs. Tactical Issues", Source: Schultz, Slevin, and Pinto, (1987, p. 38) adapted originally from G. A. Steiner, *Top Management Planning,* MacMillan, NY, 1969.

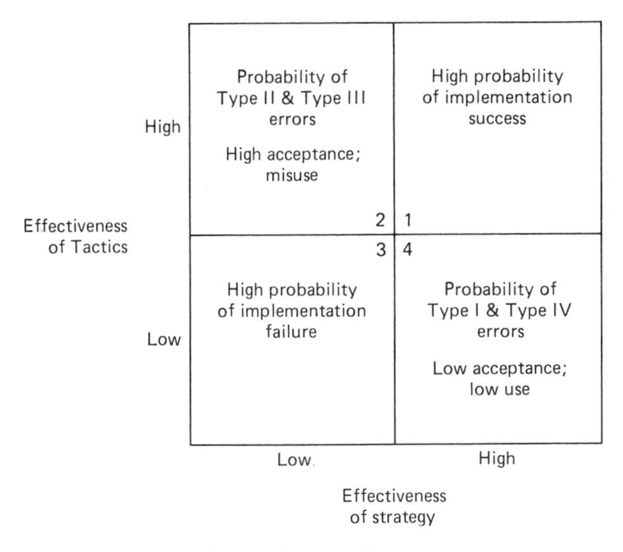


Figure 20-3. Strategy/tactics effectiveness matrix.

more systematically, through use of a project implementation assessment instrument. For example, it may be determined by the project manager that one or more factors within the strategic or tactical clusters is deficient, based on a data base of similar, successful projects. Such deficiencies could have serious implications for the resulting viability of the project under construction.

It may further be possible to speculate on some of the likely outcomes for projects being implemented, given the various combinations of strategic and tactical scores. Figure 20-3 demonstrates the four possible combinations of performance of strategic and tactical activities. It is important to note that the values "high" and "low" in Figure 20-3 are meant to imply strategic and tactical *quality,* that is, effectiveness of operations performed under the two clusters.

Four types of errors may occur in the implementation process. The first two error types were originally proposed in the context of the development of the field of statistics and statistical tests. The last two error types have been suggested as the result of research on implementation and other organizational change paradigms.

In an organizational setting, *Type I error* occurs when an action should have been taken and was not. To illustrate, consider a situation in which strategic actions have been adequately performed and suggest development and implementation of a project. Type I error will occur when little action is subsequently taken and the tactical activities are inadequate to the degree that the project is not developed.

*Type II error,* in the context of project implementation, is defined as taking an action when, in fact, none should be taken. In practical terms, Type II error would likely occur in a situation in which project strategy was ineffective, inaccurate, or poorly done. However, in spite of initial planning inadequacies, goals and schedules were operationalized during the tactical stage of the implementation.

*Type III error* may also be a consequence of low strategy effectiveness and high tactical quality. Type III error has been defined as solving the wrong problem, or "effectively" taking the wrong action. In this scenario, a problem has been identified, or a project is desired, but due to a badly performed strategic sequence, the wrong problem has been isolated and the subsequently implemented project has little value in that it does not address the intended target. Again, the implications for this error type are to develop and implement a project (tactics), often involving large expenditures of human and budgetary resources, for which inadequate or incorrect initial planning and problem recognition was done (strategy).

The final type of error that is likely to be seen in project implementation is *Type IV error*. Type IV error can be defined as taking an action which solves the right problem but the solution is not used by the organization. An example of Type IV error would occur following an effective strategy that has correctly identified the problem and proposed an effective, or "correct" solution; in this case, a project. Type IV error would result if, following the tactical operationalization, the project was not used by the clients for whom it was intended.

In addition to commenting on possible types of error which may be associated with each cell in Figure 20-3, it is important to understand some of the other aspects of likely outcomes for projects falling within each of the four cells.

### Cell 1: High Strategy—High Tactics

Quadrant 1 shows the setting for those projects which have been rated highly effective in carrying out both strategy and tactics during the implementation process. Not surprisingly, we would expect that the majority of projects corresponding to this situation would be successfully implemented. In addition to high quality strategic activities (Mission, Top Management Support, Project Schedule) these projects have also been effectively operationalized. This operationalization has taken the form of a "high" rating on tactical issues (Client Consultation, Personnel, etc.). As stated, it would be reasonable to expect resulting projects to generally show a high frequency of implementation success.

### Cell 3: Low Strategy—Low Tactics

The reciprocal of the first case is in the third quadrant and consists of a situation in which both strategic and tactical functions were inadequately performed. It would be expected that projects falling into this quadrant would have a high likelihood of implementation failure. Not only is initial strategy low, or poorly performed, but subsequent tactics are also ineffective.

### Cell 4: High Strategy—Low Tactics

While the results of projects rated as high strategy—high tactics and low strategy—low tactics may be intuitively obvious, perhaps a more intriguing question concerns the likely outcomes of projects found in the "off-diagonal" of Figure 20-3, namely, high strategy—low tactics and low strategy—high tactics. It is interesting to speculate on the result of these "mixed" scenarios in attempting to assess project success. In fact, it has been found that project implementation efforts falling within these two cells often tend to exhibit characteristics of unique, but fairly consistent, patterns.

Cell 4 refers to the situation in which the project strategy was effectively performed but subsequent tactics were rated as ineffective. As can be seen from Figure 20-3, in addition to a high likelihood of Type I and Type IV errors, one would expect projects classified in this quadrant to exhibit a strong tendency toward "errors of inaction" such as low acceptance and low use by organization members or clients for whom the project was intended. Little is done in the way of effective tactical project implementation following initial competent strategic activities. Low acceptance and use are likely outcomes because tactical duties, including Client Consultation and "selling" of the final project, are poorly performed.

### Cell 2: Low Strategy—High Tactics

The final cell represents the reverse of the previous case. In this alternative, project strategy is poorly conceived or initial planning is inadequately developed but tactical operationalization is effectively managed. One of the likely outcomes for projects classified into this cell is what are referred to as "errors of action." Because of poor strategy, a project may be initially developed and rushed into its implementation without clear ideas of its purpose. In fact, the project may not even be needed by the organization. However, tactical follow-up is well managed to the point where the inadequate or unnecessary project is implemented. This scenario represents a classic example of the "errors of action" in many areas of modern management. The mind-set is often one of "Go ahead and do it" rather than spending enough time early in the project's life to fully develop the strategy and assess whether or not the project is needed and how it should be approached.

## STRATEGY AND TACTICS OVER TIME

Strategy and tactics are both essential for successful project implementation, but differently so at various stages in the project life cycle. Strategic issues are most important at the beginning of the project. Tactical issues become more important towards the end. This is not to say that there should not be a continuous interaction and testing between the strategic and tactical factors. Strategy is not static and often changes in the dynamic corporation, making continuous monitoring essential. Nevertheless, a successful project manager must be able to transition between strategic and tactical considerations as the project moves forward.

As Figure 20-4 shows, a recent study of over 400 projects has demonstrated that strategic issues become less important and tactical issues become more important to project success over the life of a project (17). The importance value shown in Figure 20-3 has been measured by regression beta weights showing the combined relationships between strategy, tactics, and project success over the four project life-cycle stages. During the early stages, conceptualization and planning, strategy is shown to be

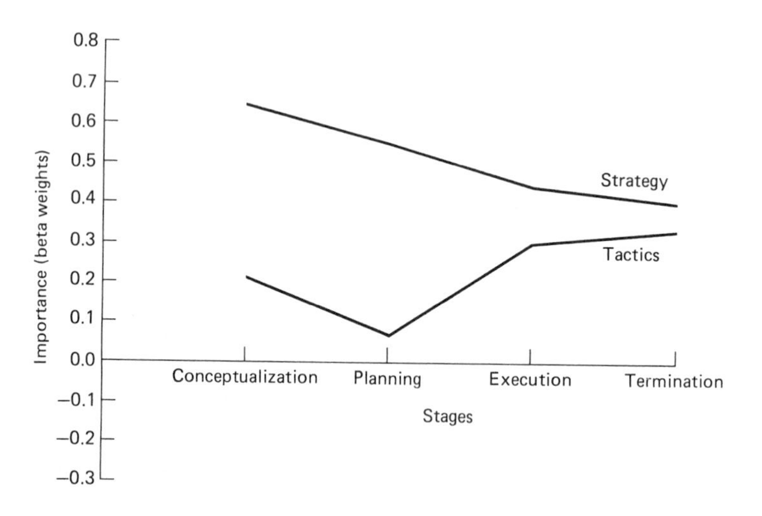


Figure 20-4. Changes in strategy and tactics across the project life cycle ($n$ = 418).

of significantly greater importance to project success than are tactics. As the project moves toward the final termination stage, project strategy and tactics achieve almost equal importance. It appears that throughout the project, initial strategies and goals continue to "drive" the project tactics. In other words, strategy continues to influence and shape tactics. At no point does strategy become unimportant to project success, while tactics increase in efforts to operationalize strategic demands.

These changes in the importance of strategy and tactics to project success have important implications for the project manager. The successful manager must be versatile and able to adapt to these changing circumstances. A project manager who is a brilliant strategist but an ineffective tactician has a strong likelihood of committing errors of Type I and Type IV as the project moves downstream. In addition, these errors may occur after substantial resources have already been expended and commitment made for the project. In contrast, the project manager who is excellent at tactical execution but weak in strategic thinking has a probability of committing errors of Type II and Type III as shown in quadrant 2, Figure 20-3. These errors will more likely occur early in the process, but perhaps stay somewhat undiscovered because of the effectiveness of the manager's execution skills.

## IMPLICATIONS FOR MANAGERS

Based on the demands facing project managers and the discussion of strategy and tactics which has been developed in this chapter, there are several conclusions which can be drawn relative to project critical success factors, along with practical implications for managers to help control the project implementation process.

### 1. Use a Multiple-Factor Model

Project management is a complex task in which the manager must attend to many variables. The more specific one can be with regard to the definition and monitoring of those variables, the more likely a successful outcome for the project will occur. Earlier in this chapter, we had listed a set of ten critical success factors which have been empirically shown to be strongly related to project success, as demonstrated by recent research. In addition to simply providing a list of factors for the project manager to consider, our research has also led to the development of a process framework of project implementation. Within this framework, the ten critical success factors are shown to contain a degree of sequentiality, in that the various factors become more critical to project success at different points in the project life cycle. As a result, it is important for the project manager to make use of a multiple-factor model, first to understand the variety of factors impacting on project success, and then to be aware of their relative importance across stages in the project implementation process.

### 2. Think Strategically Early in the Project Life Cycle

Another important implication in our discussion of project strategy and tactics is the breakdown of the ten critical factors into two distinct subdimensions, relating to the concepts of strategy and tactics. Further, it was shown that it is important to consider the "strategic" factors early in the project life cycle, during the Conceptualization and Planning stages when they become most important. As a result, it is necessary to accentuate the strategy factors (Mission, Top Management Support, and Schedule/Plans) during these early stages. It is argued that at this time, these factors are the most significant predictors of project success.

A practical suggestion for organizations implementing projects would be to bring the project manager and his team on board early in the project life cycle (preferably during the Conceptualization phase). Many managers make the mistake of not involving members of their project teams in early planning and conceptual meetings, perhaps under the assumption that the team members should only concern themselves with their specific jobs. In fact, it is very important at an early stage that both the project manager *and* the project team members "buy in" to the goals of the project and the means to achieve those goals. The more project team members are aware of these goals, the greater the likelihood of their taking active part in the monitoring and troubleshooting of the project and, consequently, the higher the quality of those activities for the project implementation.

### 3. Think More Tactically as the Project Moves Forward in Time

As Figure 20-4 shows, by the later "work" stages of execution and termination, strategy and tactics are of almost equal importance to project implementation success. Consequently, it is important that the project manager shift the emphasis in the project from "What do we want to do?" to "How do we want to do it?" The specific critical success factors associated with project tactics tend to reemphasize the importance of focusing on the "How" instead of the "What." Factors such as Personnel, Client Consultation, Communication, Monitoring, etc., are more concerned with attempts to better manage the specific action steps in the project implementation process. While we argue that it is important to bring the project team on board during the initial strategy phase in the project, it is equally important to manage their shift into a tactical, action mode in which their specific project team duties are performed to help the project toward completion.

### 4. Make Strategy and Tactics Work for You and Your Project Team

One of the points we have attempted to reinforce in this chapter is that either strong strategy or strong tactics by themselves will not ensure project success. When strategy is strong and tactics are weak, there is a great potential for creating strong, well-intended projects that never get off the ground. Cost and schedule overruns, along with general frustration, are often the side effects from projects which encounter such "errors of inaction." On the other hand, a project which starts off with a weak or poorly conceived strategy and receives strong subsequent tactical operationalization has the likelihood of being successfully implemented, but solves the wrong problem (Type III error). New York advertising agencies can tell horror stories of ad campaigns which were poorly conceived but still implemented, sometimes costing millions of dollars, and were subsequently assessed a disaster and scrubbed.

In addition to having project strategy and tactics working together, it is important to remember (again following the diagram in Figure 20-4) that initially conceived strategy should be used to "drive" tactics. Strategy and tactics are not independent of each other, but should be used together in sequence. Hence, strategy, which is developed in the earliest stages of the project, should be made known to all project team members during the entire implementation process. At no point do the strategic factors become unimportant to project success, but instead they must be continually assessed and reassessed over the life of the project. Using the example of a military scenario, tactics must be used in constant support of the overall strategy. Strategy contains the goals that were initially set and are of paramount importance to any operation.

### 5. Consciously Plan for and Manage Your Project Team's Transition from Strategy to Tactics

The project team leader needs to actively monitor his or her project through its life cycle. Important to the monitoring process is the attempt to accurately assess the position of the project in its life cycle at several different points throughout the implementation process. For the project manager, it is important to remember that the transition between strategy and tactics involves the inclusion of an additional set of critical success factors. Instead of concentrating on the set of three factors associated with project strategy, the project manager must also include the second set of factors, thus making use of all the ten factors relating to both strategy and tactics.

An important but often overlooked method to help the project leader manage the transition from strategy to tactics is to make efforts to continually communicate the changing status of the project to the other members of the project team. Communication reemphasizes the importance of a joint, team effort in implementing the project. Further, it reinforces the status of the project relative to its life cycle. The project team is kept aware of the specific stage in which the project resides as well as the degree of strategic versus tactical activities necessary to successfully sequence the project from its current stage to the next phase in its life cycle. Finally, communication helps the project manager keep track of the various activities performed by his or her project team, making it easier to verify that strategic vision is not lost in the later phases of tactical operationalization.

## CONCLUSIONS

This chapter has attempted to better define the process of project implementation through exposing the manager to a set of empirically derived factors found to be critical to project success. A ten-factor model has been presented showing these key factors and their hypothesized interrelationships. In addition, a diagnostic instrument, the Project Implementation Profile, has been presented in its entirety as a potential tool for project management and control (See Appendix). It is suggested that the PIP be used on a regular basis as a monitor of these ten key behavioral factors. It was shown that these factors may be subdimensionalized to include those activities related to initial project strategy and subsequent tactical follow-up. These dimensions of strategy and tactics are useful for the project manager in that they prescribe a two-stage process to successful project implementation.

The ability to transition successfully between early strategy and later tactics is an important characteristic for project managers to possess. Figure 20-3 showed a 2-by-2 diagram of likely outcomes for projects when strategy and/or tactics were poorly performed. Figure 20-4 demonstrated the relative importance of strategy and tactics over four distinct stages in the project life cycle, showing that strategy is of great importance initially and decreases over the life cycle while tactics steadily increase in importance. Finally, some specific suggestions were presented for project managers, in an effort to help them better manage the transition which projects go through over their life cycle.

As was stated initially, the project management process represents a complex task. The project manager is continually assaulted with a wide variety of demands on his time and resources. Because of the dynamic nature of most projects, it is becoming increasingly difficult for the project manager to keep adequate control over every aspect in the project which requiresattention. This chapter has offered some suggestions to project managers who are intent on better understanding their project during its implementation process, but are at a loss as to how to go about attempting to more adequately ensure project success.

## REFERENCES

1. Adams, J. R. and Barndt, S. E."Behavioral Implications of the Project Life Cycle," in *Project Management Handbook,* ed. Cleland, D. I. and King, W. R. (Van Nostrand Reinhold. New York, 1983), pp. 222–244.
2. Alter, S. "Implementation Risk Analysis," in *The Implementation of Management Science,* ed. Doktor, R., Schultz, R. L. and Slevin, D. P. (North-Holland. New York, 1979), pp. 103–120.
3. Bardach, E. *The Implementation Game* (MIT Press. Cambridge, Mass., 1977).
4. Bavelas, A. "Project Echo: Use of Projective Techniques to Define Reality in Different Cultures." Personal communication, Stanford University, 1968.
5. Bean, A. S. and Radnor, M. "The Role of Intermediaries in the Implementation of Management Science," in *The Implementation of Management Science,* ed. Doktor, R., Schultz, R. L. and Slevin, D. P. (North-Holland. New York, 1979), pp. 121–138.
6. Beck, D. R. "Implementing Top Management Plans Through Project Management," in *Project Management Handbook,* ed. Cleland, D. I. and King, W. R. (Van Nostrand Reinhold. New York, 1983), pp. 166–184.
7. Cleland, D. I. and Kerzner, H. *A Project Management Dictionary of Terms* (Van Nostrand Reinhold. New York, 1985).
8. M. J. Ginzberg, "A Study of the Implementation Process," in *The Implementation of Management Science,* ed. Doktor, R., Schultz, R. L. and Slevin, D. P. (North-Holland. New York, 1979), pp. 85–102.
9. Hammond, J. S. III. "A Practitioner-Oriented Framework for Implementation," in *The Implementation of Management Science,* ed. Doktor, R., Schultz, R. L. and Slevin, D. P. (North-Holland. New York, 1979), pp. 35–62.
10. King, W. R. and Cleland, D. I. "Life Cycle Management," in *Project Management Handbook,* ed. Cleland, D. I. and King, W. R. (Van Nostrand Reinhold. New York, 1983), pp. 209–221.
11. Kolb, D. A. and Frohman, A. L. "An Organizational Development Approach to Consulting." *Sloan Management Review,* Vol. 12 (1970), pp. 51–65.
12. Lewin, K. "Group Decision and Social Change," in *Readings in Social Psychology,* ed. Newcomb and Hartley (Holt, Rinehart and Winston. New York, 1952), pp. 459–473.
13. Lucas H. C. Jr. "The Implementation of an Operations Research Model in the Brokerage Industry," in *The Implementation of Management Science,* ed. Doktor, R., Schultz, R. L. and Slevin, D. P. (North-Holland. New York, 1979), pp. 139–154.
14. Manley, J. H. "Implementation Attitudes: A Model and a Measurement Methodology," in *Implementing Operating Research and Management Science,* ed. Schultz, R. L. and Slevin, D. P. (Elsevier. New York, 1975), pp. 183–202.
15. Morris, P. W. G. "Managing Project Interfaces—Key Points for Project Success," in

*Project Management Handbook,* ed. Cleland, D. I. and King, W. R. (Van Nostrand Reinhold. New York, 1983), pp. 3–36.

16. Nutt, P. C. "Implementation Approaches for Project Planning." *Academy of Management Review,* Vol. 8 (1983), pp. 600–611.
17. Pinto, J. K. "Project Implementation: A Determination of Its Critical Success Factors, Moderators, and Their Relative Importance Across the Project Life Cycle." Unpublished doctoral dissertation, University of Pittsburgh, 1986.
18. Pinto, J. K. and Slevin, D. P. "Critical Factors in Successful Project Implementation." *IEEE Transactions on Engineering Management,* Vol. EM-34 (1987) pp. 22–27.
19. Schultz, R. L. and Slevin, D. P. "Implementation and Management Innovation," in *Implementing Operations Research and Management Science,* ed. Schultz, R. L. and Slevin, D. P. (Elsevier. New York, 1975), pp. 3–22.
20. Schultz, R. L., Slevin, D. P. and Pinto, J. K. "Strategy and Tactics in a Process Model of Project Implementation." *Interfaces,* Vol. 17, May–June, 1987 pp. 34–46.
21. Slevin, D. P. and Pinto, J. K. "The Project Implementation Profile: New Tool for Project Managers." *Project Management Journal,* Vol. 18 (1986), pp. 57–71.
22. Slevin, D. P. and Pinto, J. K. "Balancing Strategy and Tactics in Project Implementation.' *Sloan Management Review,* Vol 29, No. 6, pp. 33–41.
23. Souder, W. E., Maher, P. M., Baker, N. R., Shumway, C. R. and Rubenstein, A. H. "An Organizational Intervention Approach to the Design and Implementation of R&D Project Selection Models," in *Implementing Operations Research and Management Science,* ed. Schultz, R. L. and Slevin, D. P. (Elsevier. New York, 1975), pp. 133–152.
24. Steiner, G. A. *Top Management Planning,* (MacMillian, New York, 1969).
25. Tuman, G. J. "Development and Implementation of Effective Project Management Information and Control Systems," in *Project Management Handbook,* ed. Cleland D. I. and King, W. R. (Van Nostrand Reinhold. New York, 1983), pp. 495–532.

# PROJECT IMPLEMENTATION PROFILE

Project Name: ______________________________

Project Manager: ____________________________

Profile Completed By: _______________________

Date: _______________

Briefly describe your project, giving its title and specific goals:

_______________________________________________________________

_______________________________________________________________

_______________________________________________________________

_______________________________________________________________

Think of the project implementation you have just named. Consider the statements on the following pages. Using the scale provided, please circle the number that indicated the *extent* to which you agree or disagree with the following statements as they relate to activities occurring in the project about which you are reporting.

# PROJECT IMPLEMENTATION PROFILE

## FACTOR 1 – PROJECT MISSION

| | Strongly Disagree | | | Neutral | | | Strongly Agree |
|---|---|---|---|---|---|---|---|
| 1. The goals of the project are in line with the general goals of the organization . . . . . . . . . | 1 | 2 | 3 | 4 | 5 | 6 | 7 |
| 2. The basic goals of the project are made clear to the project team . . . . . . . . . . . . . . . . . . | 1 | 2 | 3 | 4 | 5 | 6 | 7 |
| 3. The results of the project will benefit the parent organization . . . . . . . . . . . . . . . . . . . | 1 | 2 | 3 | 4 | 5 | 6 | 7 |
| 4. I am enthusiastic about the chances for success of this project. . . . . . . . . . . . . . . . . . . . . | 1 | 2 | 3 | 4 | 5 | 6 | 7 |
| 5. I am aware of and can identify the beneficial consequences to the organization of the success of this project . . . . . | 1 | 2 | 3 | 4 | 5 | 6 | 7 |

| **Factor 1 – Project Mission Total** | |
|---|---|

## FACTOR 2 – TOP MANAGEMENT SUPPORT

| | Strongly Disagree | | | Neutral | | | Strongly Agree |
|---|---|---|---|---|---|---|---|
| 1. Upper management is responsive to our requests for additional resources, if the need arises . . . . . . . . . . . . . . . . . . . . . . . . | 1 | 2 | 3 | 4 | 5 | 6 | 7 |
| 2. Upper management shares responsibility with the project team for ensuring the project's success . . . . . . . . . . . . . . . . . . . . . . . | 1 | 2 | 3 | 4 | 5 | 6 | 7 |
| 3. I agree with upper management on the degree of my authority and responsibility for the project . . . . . . . . . . . . . . . | 1 | 2 | 3 | 4 | 5 | 6 | 7 |
| 4. Upper management will support me in a crisis . . . . . . . . . . . . . . . . . . . . . . . . . . . . . . . | 1 | 2 | 3 | 4 | 5 | 6 | 7 |
| 5. Upper management has granted us the necessary authority and will support our decisions concerning the project . . . . . . . . . | 1 | 2 | 3 | 4 | 5 | 6 | 7 |

| **Factor 2 – Top Management Support Total** | |
|---|---|

## PROJECT IMPLEMENTATION PROFILE

FACTOR 3 – PROJECT SCHEDULE/PLAN

| | Strongly Disagree | | | Neutral | | | Strongly Agree |
|---|---|---|---|---|---|---|---|
| 1. We know which activities contain slack time or slack resources which can be utilized in other areas during emergencies . . . | 1 | 2 | 3 | 4 | 5 | 6 | 7 |
| 2. There is a detailed plan (including time schedules, milestones, manpower requirements, etc.) for the completion of the project . . . . . . . . . . . . . . . . . . . . . . . . | 1 | 2 | 3 | 4 | 5 | 6 | 7 |
| 3. There is a detailed budget for the project. . . . | 1 | 2 | 3 | 4 | 5 | 6 | 7 |
| 4. Key personnel needs (who, when) are specified in the project plan . . . . . . . . . . . . . . | 1 | 2 | 3 | 4 | 5 | 6 | 7 |
| 5. There are contingency plans in case the project is off schedule or off budget. . . . . . . . | 1 | 2 | 3 | 4 | 5 | 6 | 7 |

| **Factor 3 – Project Schedule/Plan Total** | |
|---|---|

FACTOR 4 – CLIENT CONSULTATION

| | Strongly Disagree | | | Neutral | | | Strongly Agree |
|---|---|---|---|---|---|---|---|
| 1. The clients were given the opportunity to provide input early in the project development stage . . . . . . . . . . . . . . . . . . . . . | 1 | 2 | 3 | 4 | 5 | 6 | 7 |
| 2. The clients (intended users) are kept informed of the project's progress . . . . . . . . . | 1 | 2 | 3 | 4 | 5 | 6 | 7 |
| 3. The value of the project has been discussed with the eventual clients . . . . . . . . . | 1 | 2 | 3 | 4 | 5 | 6 | 7 |
| 4. The limitations of the project have been discussed with the clients (what the project is *not* designed to do) . . . . . . . . . . | 1 | 2 | 3 | 4 | 5 | 6 | 7 |
| 5. The clients were told whether or not their input was assimilated into the project plan . . . . . . . . . . . . . . . . . . . . . . . . . | 1 | 2 | 3 | 4 | 5 | 6 | 7 |

| **Factor 4 – Client Consultation Total** | |
|---|---|

## PROJECT IMPLEMENTATION PROFILE

### FACTOR 5 – PERSONNEL

| | Strongly Disagree | | | Neutral | | | Strongly Agree |
|---|---|---|---|---|---|---|---|
| 1. Project team personnel understand their role on the project team | 1 | 2 | 3 | 4 | 5 | 6 | 7 |
| 2. There is sufficient manpower to complete the project | 1 | 2 | 3 | 4 | 5 | 6 | 7 |
| 3. The personnel on the project team understand how their performance will be evaluated | 1 | 2 | 3 | 4 | 5 | 6 | 7 |
| 4. Job descriptions for team members have been written and distributed and are understood | 1 | 2 | 3 | 4 | 5 | 6 | 7 |
| 5. Adequate technical and/or managerial training (and time for training) is available for members of the project team | 1 | 2 | 3 | 4 | 5 | 6 | 7 |

| **Factor 5 – Personnel Total** | |
|---|---|

### FACTOR 6 – TECHNICAL TASKS

| | Strongly Disagree | | | Neutral | | | Strongly Agree |
|---|---|---|---|---|---|---|---|
| 1. Specific project tasks are well managed | 1 | 2 | 3 | 4 | 5 | 6 | 7 |
| 2. The project engineers and other technical people are competent | 1 | 2 | 3 | 4 | 5 | 6 | 7 |
| 3. The technology that is being used to support the project works well | 1 | 2 | 3 | 4 | 5 | 6 | 7 |
| 4. The appropriate technology (equipment, training programs, etc.) has been selected for project success | 1 | 2 | 3 | 4 | 5 | 6 | 7 |
| 5. The people implementing this project understand it | 1 | 2 | 3 | 4 | 5 | 6 | 7 |

| **Factor 6 – Technical Tasks Total** | |
|---|---|

## PROJECT IMPLEMENTATION PROFILE

### FACTOR 7 – CLIENT ACCEPTANCE

| | Strongly Disagree | | | Neutral | | | Strongly Agree |
|---|---|---|---|---|---|---|---|
| 1. There is adequate documentation of the project to permit easy use by the clients (instructions, etc.) . . . . . . . . . . . . . . . . . . | 1 | 2 | 3 | 4 | 5 | 6 | 7 |
| 2. Potential clients have been contacted about the usefulness of the project. . . . . . . . . . . | 1 | 2 | 3 | 4 | 5 | 6 | 7 |
| 3. An adequate presentation of the project has been developed for clients . . . . . . . . . . . . . . | 1 | 2 | 3 | 4 | 5 | 6 | 7 |
| 4. Clients know who to contact when problems or questions arise . . . . . . . . . . . . . . . . . . . . . . . | 1 | 2 | 3 | 4 | 5 | 6 | 7 |
| 5. Adequate advanced preparation has been done to determine how best to "sell" the project to clients . . . . . . . . . . . . . . . . . . . . . . . | 1 | 2 | 3 | 4 | 5 | 6 | 7 |

| **Factor 7 – Client Acceptance Total** | |
|---|---|

### FACTOR 8 – MONITORING AND FEEDBACK

| | Strongly Disagree | | | Neutral | | | Strongly Agree |
|---|---|---|---|---|---|---|---|
| 1. All important aspects of the project are monitored, including measures that will provide a complete picture of the project's progress (adherence to budget and schedule, manpower and equipment utilization, team morale, etc.) . . . . . . . . . . . . . . . . . . . . . . . . . | 1 | 2 | 3 | 4 | 5 | 6 | 7 |
| 2. Regular meetings to monitor project progress and improve the feedback to the project team are conducted . . . . . . . . . . . . . | 1 | 2 | 3 | 4 | 5 | 6 | 7 |
| 3. Actual progress is regularly compared with the project schedule . . . . . . . . . . . . . . . . . . . . . | 1 | 2 | 3 | 4 | 5 | 6 | 7 |
| 4. The results of project reviews are regularly shared with all project personnel who have impact upon budget and schedule . . . . . . . . . . . | 1 | 2 | 3 | 4 | 5 | 6 | 7 |
| 5. When the budget or schedule requires revision, input is solicited from the project team. . . . . . . . . . . . . . . . . . . . . . . . . . | 1 | 2 | 3 | 4 | 5 | 6 | 7 |

| **Factor 8 – Monitoring and Feedback Total** | |
|---|---|

## PROJECT IMPLEMENTATION PROFILE

FACTOR 9 – COMMUNICATION

| | Strongly Disagree | | | Neutral | | | Strongly Agree |
|---|---|---|---|---|---|---|---|
| 1. The results (decisions made, information received and needed, etc.) of planning meetings are published and distributed to applicable personnel | 1 | 2 | 3 | 4 | 5 | 6 | 7 |
| 2. Individuals/groups supplying input have received feedback on the acceptance or rejection of their input | 1 | 2 | 3 | 4 | 5 | 6 | 7 |
| 3. When the budget or schedule is revised, the changes *and* the reasons for the changes are communicated to all members of the project team | 1 | 2 | 3 | 4 | 5 | 6 | 7 |
| 4. The reasons for the changes to existing policies/procedures have been explained to members of the project team, other groups affected by the changes, and upper management | 1 | 2 | 3 | 4 | 5 | 6 | 7 |
| 5. All groups affected by the project know how to make problems known to the project team | 1 | 2 | 3 | 4 | 5 | 6 | 7 |

| **Factor 9 – Communication Total** | |
|---|---|

FACTOR 10 – TROUBLESHOOTING

| | Strongly Disagree | | | Neutral | | | Strongly Agree |
|---|---|---|---|---|---|---|---|
| 1. The project leader is not hesitant to enlist the aid of personnel not involved in the project in the event of problems | 1 | 2 | 3 | 4 | 5 | 6 | 7 |
| 2. "Brainstorming" sessions are held to determine where problems are most likely to occur | 1 | 2 | 3 | 4 | 5 | 6 | 7 |
| 3. In case of project difficulties, project team members know exactly where to go for assistance | 1 | 2 | 3 | 4 | 5 | 6 | 7 |
| 4. I am confident that problems that arise can be solved completely | 1 | 2 | 3 | 4 | 5 | 6 | 7 |
| 5. Immediate action is taken when problems come to the project team's attention | 1 | 2 | 3 | 4 | 5 | 6 | 7 |

| **Factor 10 – Troubleshooting Total** | |
|---|---|

# PROJECT IMPLEMENTATION PROFILE

## PROJECT PERFORMANCE

| | Strongly Disagree | | | Neutral | | | Strongly Agree |
|---|---|---|---|---|---|---|---|
| 1. This project has/will come in on schedule . . . . . | 1 | 2 | 3 | 4 | 5 | 6 | 7 |
| 2. This project has/will come in on budget. . . . . . . | 1 | 2 | 3 | 4 | 5 | 6 | 7 |
| 3. The project that has been developed works, (or if still being developed, looks as if it will work) . . . . . . . . . . . . . . . . . . . . . . . . . | 1 | 2 | 3 | 4 | 5 | 6 | 7 |
| 4. The project will be/is used by its intended clients . . . . . . . . . . . . . . . . . . . . . . . | 1 | 2 | 3 | 4 | 5 | 6 | 7 |
| 5. This project has/will directly benefit the intended users: either through increasing efficiency or employee effectiveness . . . . . . . . . | 1 | 2 | 3 | 4 | 5 | 6 | 7 |
| 6. Given the problem for which it was developed, this project seems to do the best job of solving that problem, i.e., it was the best choice among the set of alternatives. . . . . . . . . . . . . . . . . . . . . . . . . . | 1 | 2 | 3 | 4 | 5 | 6 | 7 |
| 7. Important clients, directly affected by this project, will make use of it . . . . . . . . . . . . . | 1 | 2 | 3 | 4 | 5 | 6 | 7 |
| 8. I am/was satisfied with the process by which this project is being/was completed . . . . . . . . . . . . . . . . . . . . . . . . . . . | 1 | 2 | 3 | 4 | 5 | 6 | 7 |
| 9. We are confident that nontechnical start-up problems will be minimal, because the project will be readily accepted by its intended users . . . . . . . . . . . . . . . . . . . . . . . . | 1 | 2 | 3 | 4 | 5 | 6 | 7 |
| 10. Use of this project has led/will lead directly to improved or more effective decision making or performance for the clients. . . . . . . . . . . . . . . . . . . . . . . . . . | 1 | 2 | 3 | 4 | 5 | 6 | 7 |
| 11. This project will have a positive impact on those who make use of it. . . . . . . . . . . . . . . . | 1 | 2 | 3 | 4 | 5 | 6 | 7 |
| 12. The results of this project represent a definite improvement in performance over the way clients used to perform these activities. . . . . . . . . . . . . . . . . . . . . . . . . . . | 1 | 2 | 3 | 4 | 5 | 6 | 7 |

| PROJECT PERFORMANCE TOTAL | |
|---|---|

## PROJECT IMPLEMENTATION PROFILE

Percentile Scores
How does your project score?

Now see how your project scored in comparison to a data base of 409 projects. If you are below the 50 percentile on any factor, you may wish to devote extra attention to that factor.

| Percentile Score | Raw Score | | | | |
|---|---|---|---|---|---|
| % of individuals Scoring Lower | Factor 1 Project Mission | Factor 2 Top Management Support | Factor 3 Project Schedule/ Plan | Factor 4 Client Consultation | Factor 5 Personnel-Recruitment, Selection, Training |
| 100% | 35 | 35 | 35 | 35 | 35 |
| 90% | 34 | 34 | 33 | 34 | 32 |
| 80% | 33 | 32 | 31 | 33 | 30 |
| 70% | 32 | 30 | 30 | 32 | 28 |
| 60% | 31 | 28 | 28 | 31 | 27 |
| 50% | 30 | 27 | 27 | 30 | 24 |
| 40% | 29 | 25 | 26 | 29 | 22 |
| 30% | 28 | 23 | 24 | 27 | 20 |
| 20% | 26 | 20 | 21 | 25 | 18 |
| 10% | 25 | 17 | 16 | 22 | 14 |
| 0% | 7 | 6 | 5 | 7 | 5 |

## PROJECT IMPLEMENTATION PROFILE

After you have compared your scores, you may plot them on the next page and mark any factors that need special effort.

| Percentile Score<br>% of Individuals Scoring Lower | Raw Score | | | | | |
|---|---|---|---|---|---|---|
| | **Factor 6** Technical Tasks | **Factor 7** Client Acceptance | **Factor 8** Monitoring and Feedback | **Factor 9** Communication | **Factor 10** Trouble-shooting | Project Performance |
| 100% | 35 | 35 | 35 | 35 | 35 | 84 |
| 90% | 34 | 34 | 34 | 34 | 33 | 79 |
| 80% | 32 | 33 | 33 | 32 | 31 | 76 |
| 70% | 30 | 32 | 31 | 30 | 29 | 73 |
| 60% | 29 | 31 | 30 | 29 | 28 | 71 |
| 50% | 28 | 30 | 28 | 28 | 26 | 69 |
| 40% | 27 | 29 | 27 | 26 | 24 | 66 |
| 30% | 26 | 27 | 24 | 24 | 23 | 63 |
| 20% | 24 | 24 | 21 | 21 | 21 | 59 |
| 10% | 21 | 20 | 17 | 16 | 17 | 53 |
| 0% | 8 | 8 | 5 | 5 | 5 | 21 |

## PROJECT IMPLEMENTATION PROFILE

Tracking Critical Success Factors Grid

Percentile Rankings

| | 0% | 10% | 20% | 30% | 40% | 50% | 60% | 70% | 80% | 90% | 100% |
|---|---|---|---|---|---|---|---|---|---|---|---|
| 1. Project Mission | | | | | | | | | | | |
| 2. Top Management Support | | | | | | | | | | | |
| 3. Project Schedule | | | | | | | | | | | |
| 4. Client Consultation | | | | | | | | | | | |
| 5. Personnel | | | | | | | | | | | |
| 6. Technical Tasks | | | | | | | | | | | |
| 7. Client Acceptance | | | | | | | | | | | |
| 8. Monitoring and Feedback | | | | | | | | | | | |
| 9. Communication | | | | | | | | | | | |
| 10. Troubleshooting | | | | | | | | | | | |
| Project Performance | | | | | | | | | | | |

# 21. Quality Program Management in Project Management

John L. Heidenreich*

Completing a project on time and within cost loses importance if quality is not attained. The news media frequently report instances of inadequate quality that result in costly product recalls, construction condemnation before completion, rework and delays, accidents, and, on occasion, deaths. In addition to the adverse financial and safety effects, inadequate quality is detrimental to the image of an organization and can adversely affect future business potential. Project management includes managing for quality, as well as managing to complete a project on time and within cost.

Quality Program management is the element of project management that is designed to assure the attainment of quality. Its goal is to do the right things right the first time. It accomplishes its goal by utilizing basic elements of good management: planning, organization, implementation, feedback, and corrective action.

This chapter addresses management responsibilities and actions regarding Quality Program management in project management, including:

- Understanding quality.
- Establishing objectives, philosophy, and policies.
- Balancing schedule, cost, and quality.
- Quality Program planning.
- Organization.

---

* John L. Heidenreich has 13 years of hands-on experience in quality management and consulting. He has consulted for the U.S. Nuclear Regulatory Commission, the Chicago Operations Office of the U.S. Department of Energy, state and local government agencies, oil companies, utilities, contractors, manufacturers, and material manufacturers and suppliers. His management analysis of U.S. Nuclear Regulatory Commission programs was published in NUREG 1055. He has prepared Quality Program manuals and procedures, conducted seminars, and performed management evaluations and audits nationally and internationally. He holds a B.S. degree in Physics from Western Illinois University.

- Quality Program implementation.
- Feedback of quality-related information.
- Taking corrective action.

## UNDERSTANDING QUALITY

### Quality Achievement

What is quality? In construction and manufacturing industries, quality has generally become accepted to mean fitness for use or compliance with requirements. For an organization to prosper, fitness for use must be attained in an efficient and economical manner. Compliance to requirements must be preceded by the establishment of appropriate requirements. Quality doesn't just automatically happen. To achieve quality requires a clear definition of function, performance objectives, and design, environmental and service conditions; translation of those requirements into working documents, such as specifications, drawings, procedures, and instructions; identification of the working documents in construction and manufacturing process control documents; performance of quality-affecting activities in accordance with requirements of the working documents; and verification that the requirements have been met.

Top management and all project personnel must understand that attaining quality is the responsibility of top management and all project personnel performing quality-affecting activities, not just the Quality department. When an organization experiences financial problems, top management does not hold the Finance department solely responsible. Instead, it recognizes that the financial problems are not caused by the Finance department, but are the result of other problems within the organization. Quality problems are usually attributed to deficiencies or breakdowns in Quality Programs. Because Quality Programs are prepared by Quality departments, they are blamed and held accountable for the problems. However, the Quality department did not define function, performance objectives and design, environmental, and service conditions, did not translate those requirements into working documents, and did not manufacture, install, or construct the item. Marketing, usually with the assistance of engineering, is responsible for the clear definition of requirements to be met. Engineering is responsible for translating those requirements into specifications, drawings, procedures, and instructions. Purchasing, manufacturing, fabrication, installation, and construction are responsible for performing activities in accordance with the requirements of the working documents. Management of each function performing the above activities is responsible for assuring the activities are properly performed. Typically, the Quality department is responsible for establishing a Quality Program to control these activities and for verifying that the Quality Program is implemented and the requirements of specifications, drawings, procedures, and instructions identified on the process control documents are met. Quality problems are normally the result of breakdowns in the management and implementation of any of these activities. Quality problems can be minimized through the understanding of quality and the implementation of basic management practices.

### Quality Program

A Quality Program is a systematic approach to planning and controlling quality during a project. It improves management by establishing consistent methods of operation for all quality-affecting activities and controls changes in those methods. It results in increased productivity and a reduction in overall project costs.

Management in many organizations views a Quality Program as an unnecessary expense and an impediment to productivity. In such organizations, Quality Programs either are not established or are not effective. In regulated industries that have Quality Program requirements, management in many organizations views a Quality Program as a necessary expense and evil, but an impediment to productivity. In such organizations, Quality Programs are established, but usually are not effective. Management that views a Quality Program as an effective management tool, discovers that quality improves through focusing attention on preventing errors, productivity improves and project costs decrease through the reduction in rework and schedule delays, and the costs for the Quality Program are offset by the resulting savings in costs.

Quality Program activities of establishing documented methods of control of quality-affecting activities, training personnel in the methods, implementing the methods, verifying implementation, measuring effectiveness, and identifying and correcting problems to prevent recurrence are simply good business practices. There is no contradiction between good controls for quality and good business.

## OBJECTIVES, PHILOSOPHY, AND POLICIES

The top management of the organization responsible for a project is responsible for defining the objectives and philosophy of the organization. It is also responsible for establishing policies to assure clear understanding of those objectives and the philosophy by all managers. It is important to include quality in the objectives, philosophy, and policies in order to document top management's commitment to quality and its expectations that the entire organization considers the attainment of quality to be one of its primary goals. No one is basically opposed to quality, so top management often believes that its attainment is a naturally understood goal. Through its actions, top management establishes what is important. Failure to emphasize quality results in a perception by other management that quality is not as important as other goals, such as schedule and cost. This perception filters throughout the organization to the worker level, and quality may be sacrificed at the expense of schedule and cost. Top management inclusion of quality in its objectives, philosophy, and policies is an important initial step toward establishing an effective Quality Program and the attainment of quality.

## BALANCING SCHEDULE, COST, AND QUALITY

For a project to be successful, top management must establish a meaningful balance between schedule, cost, and quality. Emphasis on schedule can result in a project completed on time that is over cost and has unacceptable quality. Emphasis on cost can result in a project within cost that is not completed on time and that has unacceptable quality. Emphasis on quality can result in a project with the desired quality that is not completed on time and is over cost. To establish the necessary balance between schedule, cost, and quality, top management of the organization responsible for a project must include the quality function with the other organizational functions in providing input during early project planning. To maintain the balance, top management must provide for project management integration of cost, schedule and quality planning, control and performance data throughout all phases of the project. The quality function must be included with the other organizational functions in providing feedback to top management concerning project problems and the effectiveness of the Quality Program. Top management must react promptly and effectively to cause changes in the Quality Program and project management as dictated by the feedback.

## QUALITY PROGRAM PLANNING

Quality Program planning must be performed before the start of project activities and should involve all of the organizational elements responsible for project management. The early involvement of each organizational element permits integration of quality concepts into initial project management activities, such as the development of project plans, work breakdown structures, work packages, and cost accounts. It provides the opportune time to develop each organizational element's understanding of the importance of quality to the project. Through each organizational element's input, it also provides them with a sense of ownership of the Quality Program. The Quality Program does not just become a program invoked by top management which disregards the concerns of those to be held responsible for its implementation.

Most projects progress through several stages, such as conceptual, development, design, manufacture, construction, installation, operation, maintenance, and modification. The Quality Program should evolve with the project since the quality-affecting activities change as the project progresses through its stages. In addition, requirements, technology, and the working environment may change throughout the life of a project. Project personnel must accept this fact and react positively in response to such change.

The controls necessary for an organization are dependent upon the activities for which it is responsible. To be cost effective, the degree of control should be based upon the significance of activities in relation to their impact on safety, quality, cost, and schedule. Factors to consider include the consequences of malfunction or failure, complexity of the project organization and the work to be performed, accessibility for repair or replacement, degree to which compliance can be determined through inspection and tests, and quality history.

The initial step in planning is to identify the objectives to be achieved and all activities and items critical to achieving the objectives in accordance with requirements. This task can be accomplished through the project management activity of establishing project plans and work breakdown structures.

The second step is to analyze the significance of the activities and items by using various analysis techniques, such as Failure Mode and Effects Analysis and Design of Experiments. The use of these analysis techniques can be found in many publications. A brief description of these techniques follows.

### Failure Mode and Effects Analysis

Failure Mode and Effects Analysis is conducted prior to production or construction activities to prevent the first problem from occurring. It involves identifying each possible failure mode of the item, system, or structure and its probability of occurrence; the effects of failure and severity of the effects; potential causes of failure; current controls intended to prevent the cause or detect the failure; and the likelihood of the failure being detected by existing controls; and then initiating action to eliminate the potential failure modes or reduce their occurrence.

### Design of Experiments

Design of Experiments is also conducted prior to production or construction activities. It involves choosing certain factors for study, varying those factors in a controlled fashion, observing the effect of such action, and making a decision based upon the most favorable results.

The third step in planning is to establish control measures commensurate with the significance of the activities and items. This can be accomplished by classifying the activities and items into several groups based upon their significance and defining the measures of control to be applied to each group. Inspection and test plans should be prepared for each group or for specific activities and items, as applicable, which itemize inspection, surveillance, and test points in the sequence of events pertaining to the activities and items.

The fourth step is to integrate the control measures and inspection, surveillance, and test points into the overall management of the project. The control measures should be defined in a Quality Assurance Manual, and administrative and technical procedures. The inspection, surveillance, and test points should be incorporated into work package activities and administrative and technical procedures.

During Quality Program planning, consideration should be given to the need for special controls, processes, skills, and equipment, and to incorporate the appropriate features of the concepts of Quality Control, Quality Assurance, Statistical Process Control, Quality Improvement, Employee Participation Teams, and Quality Costs. Organizations have tended to utilize a specific concept as the solution to their quality problems. The results have often been disappointing. Each concept is but one aspect of a broader, comprehensive Quality Program. A description of each of these quality concepts follows.

### Quality Control

Quality Control has become accepted as the measurement of the characteristics of an item or process to determine its conformance to specified requirements and the taking of action when nonconformance exists. Quality Control is commonly considered to be inspection, examination, and testing and is an important aspect of Quality Programs. Inspection, examination, and testing of every item and activity, although often necessary based upon the significance of an item or activity, is prohibitively expensive. The factors of monotony, fatigue, and human error result in unacceptable items still being accepted. Sampling plans have been developed based upon theories of probability to provide a more economical and efficient basis for Quality Control. Samples of items and activities are inspected, examined, or tested and the results dictate acceptance or rejection of the entire lot of items or activities. Different sampling plans are applied to different characteristics following classification of the characteristics based upon their significance to safety, quality, cost, and schedule. Sample plans recognize and provide that a small percentage of items and activities will still be unacceptable. The judicious use of Quality Control is a vital part of verification of quality. Its use is normally dictated by an evaluation of the costs of Quality Control versus the affects of finding problems late in the project. It is used at the initial stages of an activity or process to verify that the activity or process is producing satisfactory results, at various stages of activities or processes during their periods of performance to verify that the activity or process continues to produce satisfactory results, and for final acceptance of completed items and structures. However, inspections, examinations, and tests are conducted after work is performed and they identify problems after they occur. When experiencing quality problems, the initial reaction of management is often to increase the number of Quality Control personnel, believing that this will solve their problems. It won't solve their problems. It will result in increased verification of quality, but does not address the more important issue of preventing unsatisfactory quality.

Organizations that base their Quality Programs solely on Quality Control do not have cost-effective programs. Quality cannot be obtained effectively and economically solely by Quality Control.

### Quality Assurance

Quality Assurance has become accepted to mean planned and systematic actions to provide confidence that specified requirements are met and that items, systems, and structures will perform satisfactorily in service. It will not improve quality when requirements are inadequately specified or design is inadequate. Quality Assurance concentrates on preventing problems in addition to identifying and correcting them. Quality Assurance includes both the functions of attaining quality and of verifying quality. It extends from the top executive to the workers and typically encompasses activities of marketing, engineering, procurement, production control, material control, manufacturing, fabrication, inspection, testing, handling, storage, shipping, receiving, erection, installation, operation, maintenance, repair, and modification. The nature and extent of actions is dependent upon the quality-affecting activities performed or contracted.

The documented description of the planned and systematic actions is the organization's Quality Assurance Program. It typically describes the activities and items to which it applies and the organizational structure, interfaces and interface responsibilities, functional responsibilities, levels of authority, lines of communication, and methods or systems of control for quality-affecting activities.

In the manufacturing and construction industries, the following activities generally are included in a Quality Assurance Program:

- Reviewing project objectives or incoming contracts or order documents to determine and define technical and quality requirements, to identify omissions and inadequate definition of requirements, and to identify and provide for any required controls, processes, equipment, and skills.
- Controlling design activities to assure that quality is designed into the product and to assure consideration of performance, reliability, maintainability, safety, producibility, standardization, interchangeability, and cost. This includes specifying design methods and design inputs; translating the inputs into design and technical documents; verifying design adequacy prior to release through design reviews, alternate calculations, or qualification tests; obtaining vendor and contractor assistance early in the design process; and controlling design interfaces, design changes, and as-constructed configurations. Design interfaces are the interfaces and controls between participating design organizations.
- Controlling procurement activities, which includes selection and qualification of vendors and contractors who have shown evidence of their capability of achieving quality; assuring that technical, quality, and Quality Program requirements are included or referenced in procurement documents; reviewing and approving vendor and contractor Quality Programs and administrative and technical procedures; assessing vendor and contractor performance against requirements; requiring corrective action for conditions adverse to quality; and requiring notification and approval of design, material, and Quality Program changes.

  Supplier and contractor performance to requirements is a key ingredient to achieving quality during the project. The value of purchased materials, items, and services is normally a major percentage of the overall project cost. Yet many organizations utilize price as the key factor in awarding contracts and purchase orders. They fail to realize that throughout all project phases, the least costly item or service is not always the one with the lowest initial price. Costs associated with the repair or rework of completed work and the replacement of items as a result of inadequate quality is more costly than savings realized through the lowest initial price. Vendor and contractor selection and continued use should be based upon a combination of cost, schedule, and quality considerations.

- Establishing documented instructions, procedures, specifications, and drawings which include or reference acceptance criteria, for performing quality-affecting activities.
- Controlling the preparation, review for adequacy, approval, issuance, and revision of documents specifying quality requirements and describing quality-affecting activities. This includes establishing measures for assuring the adequacy and control of supplier and contractor documents as well.
- Controlling purchased items and services, including evaluation of their adequacy. This is normally achieved through performing surveillances of supplier and contractor activities, performing inspection, and reviewing supplier and contractor furnished evidence of quality, such as test reports and documentation of statistical process control.
- Identifying and controlling material and items so that only accepted materials and items are used or installed.
- Controlling processes affecting quality through the use of process control documents which specify applicable instructions, procedures, specifications and drawings.
- Controlling inspections, examinations, and tests to verify conformance to specified requirements and to demonstrate that items will perform satisfactorily in service. This includes identification of status to assure that required inspections, examinations, and tests are performed.
- Controlling measuring and testing equipment to maintain its accuracy.
- Controlling handling, cleaning, preservation, storage, packaging, and shipping of items to prevent damage or loss.
- Controlling activities and items that do not conform to specified requirements.
- Identifying and promptly correcting conditions adverse to quality.
- Specifying, preparing, maintaining, and later dispositioning records that furnish evidence of quality. Records are considered a primary

means of providing objective evidence of Quality Program implementation and the attainment of quality.

- Performing audits to determine compliance with the Quality Program and its effectiveness.
- Performing trend analysis to identify repetitive conditions adverse to quality.

Quality Assurance, by being planned and systematic actions, is a management system and a key management tool for attaining quality.

### Statistical Process Control

Statistical Process Control involves using statistical data to improve processes and prevent deficiencies. It consists of performing capability studies to determine variability in a process or equipment, identifying the causes of variability, changing the process to reduce the variability, and monitoring the results of the process change. It is used to improve consistency and to prevent nonconforming work, rather than to identify nonconformances and take action based upon inspection results. The result is improved quality and productivity and reduction in inspection, rework, scrap, and process costs. Statistical Process Control is based upon mathematical theories of probability and uses tools such as frequency distributions, histograms, control charts, and problem solving techniques. However, implementation of Statistical Process Control can be accomplished by most personnel with basic training in application of the techniques and without education and training in the mathematical theories. Control charts, with control limits tighter than tolerance requirements and based upon process capability, are used to plot and trend values of characteristics to determine when something is about to go wrong with a process. As points fall outside the control limits, the cause is determined and corrective action is taken. Establishing control charts for all characteristics is not cost effective. Therefore, control charts are established for the key characteristics, and other characteristics are subject to random inspections. Numerous computer programs exist for the analysis of Statistical Process Control data, and the principles of Statistical Process Control can be applied to problem solving in many areas of project activities.

### Quality Improvement

Quality Improvement consists of actions that result in improved quality of activities, items and structures. It involves establishing and collecting measurement data, identifying problems or potential improvements, and taking action to improve performance or usability. The concept has been used to improve quality and productivity of personnel in all departments within an organization and to make error prevention an inherent part of each activity. It provides top management with a measure of effort and forces each organizational element to assess its performance and productivity and take action to improve it. Productivity is simply measured as a ratio of output to input and can be managed through measurement in nearly every activity. It is affected by facilities, equipment, materials, workers, methods, and management. It can be improved through analysis and redefinition of tasks, work flow and allocation of resources. It is not uncommon for the costs of errors, rework, and repeated operations of all departments within an organization to exceed 25% of the total sales of the organization. To be effective, improvement actions must have the support of top management. A common approach is to establish a council or committee consisting of various members from top management and the management of operational departments. The council or committee is responsible for organizing objectives, developing strategy and measurement methods, publicizing results, and providing employee recognition. Working groups established by the council or committee include a council or committee member to provide continuity and evidence of management involvement. Through the establishment of Employee Participation Teams, problems are solved at the lowest possible level. The improvement program is a continuous process throughout all phases of the project.

### Employee Participation Teams

Employee Participation Teams, also called Quality Circles, are teams of employees formed to identify and solve problems. Their success requires creation of a participative environment and an understanding of small-group dynamics, decision-making processes, and structured problem-solving techniques, such as brainstorming, cause and effect analysis, pareto analysis, and data gathering and sampling. Pareto analysis involves listing problems by their level of importance and frequency of occurrence and concentrating initial actions on correcting the most significant problems and their causes. Employee Participation Teams are an effective tool to gain employee involvement in problem identification and solving and have resulted in increased motivation of employees to attain quality.

### Quality Costs

Quality Costs are used to measure the effectiveness of Quality Programs and to control the efforts of all departments. They are based upon facts and are an effective management tool for budgeting, identifying major and costly problem areas, and allocating manpower. Their use involves determining quality-related costs and their trends and taking action to reduce the costs and improve quality. Quality Costs are an effective Quality Program management tool for presenting the need for and impact of a Quality Program to top project management.

Four categories are typically established for Quality Costs, with numerous elements in each category. The categories are prevention, costs associated with establishing and implementing a Quality Program; appraisal, costs associated with evaluating conformance to requirements; internal failure, costs associated with materials, products, and structures that fail to meet requirements and are scrapped or cause rework or repair activities during their completion; and external failures, costs associated with defective products and structures after their completion as identified by the user. The costs are normally identifiable by accounting through established accounting systems with the assistance of the Quality department. Quality Cost Reports are a financial document which should be issued to top management by accounting on a monthly or quarterly basis. The basic concept is that investments in prevention and appraisal reduce internal and external failures, resulting in lower costs and less rework. The importance of Quality Costs as a management tool has been recognized by the military, who included use of Quality Cost data in MIL-Q-9858A in 1963.

## Quality Program Preparation

Through the incorporation of appropriate features of the various quality concepts at the Quality Program planning stage, top management establishes a management system that stresses prevention of quality problems, active involvement of all project personnel, efficient and economical use of inspection, and utilization of factual performance data to continuously improve performance throughout the project.

The description of the managerial and administrative controls to be used for attaining and verifying quality during the project needs to clearly identify what is to be done, who is to do it, how it is to be done and documented, and when it is to be done. This description is the Quality Assurance Manual of the organization. Details for implementation of the controls are described in supplemental administrative and technical procedures. Administrative procedures are usually prepared by the Quality department with the assistance of other departments. Technical procedures are usually prepared by other departments in accordance with requirements of the administrative procedures and contain the specific instructions for the performance of specific activities. The Quality Assurance Manual, inspection and test plans, and supplemental procedures constitute the Quality Program and must be as simple as possible to permit their understanding and encourage their use. They must be written for the level of the user, not the writer. Complicated documents which cannot be understood will not be used. Project personnel tend to ignore such documents instead of trying to comprehend and implement them. Many organizations write their Quality Assurance Manual and supplemental procedures using vague, nonspecific terminology such as: when applicable, as appropriate, periodically, frequently, as required, the company, etc., in an attempt to provide flexibility in implementation of their program. Although permitting flexibility, such terminology does not provide a clear description of activities and leaves the interpretation of the terminology open to dispute and to each of those persons performing the activities. Such terminology is often used later as an excuse or justification for not performing an activity as intended.

## ORGANIZATION

### Organizational Structure

Top management of the organization responsible for a project has the responsibility for establishing an organizational structure which is conducive to maintaining the balance between schedule, cost, and quality. To accomplish this, it is necessary for top management to have a clear understanding of the scope and magnitude of the project and the role of each project organization. Top management should document the organizational structure in organization charts which include the job titles of managers and supervisors responsible for quality-affecting activities and for managing the project, and the major contractors for the project. The organization charts identify the project organizational elements and their basic functions and are to be included in the Quality Assurance Manual.

In establishing the organizational structure, top management is faced with the problem of how to incorporate the quality function.

One approach has been that each organizational element is responsible for attaining quality objectives as well as meeting schedule within cost. Therefore, each organizational element should do what it deems necessary to assure the quality of its work.

Another approach has been that, although each organizational element is responsible for attaining quality objectives as well as meeting schedule within cost, each organizational element is biased and quality might be sacrificed to meet the interests of the element and to maintain schedule within cost. This has often happened in organizations where top management has failed to establish and maintain a meaningful balance between schedule, cost, and quality.

Many regulations, codes, and standards contain Quality Program requirements which stipulate a separation of quality-attaining functions from quality assurance functions. The quality-attaining functions are those of defining requirements to be met, translating those requirements into working documents, and performing project activities in accordance with the requirements of the working documents. The quality assurance functions are typically those of planning a Quality Program and checking, reviewing, and verifying that the program is implemented and the desired quality is attained. Personnel performing quality assurance functions are required to have sufficient authority, organizational freedom, and independence from cost and schedule considerations to carry out their functions. Unfortunately, this has often resulted in a perception that quality is the responsibility of the quality assurance function instead of the attainers of quality. As a result, an adversarial role has often developed between personnel performing quality-affecting activities and personnel verifying attainment of quality and implementation of the Quality Program.

In view of the above, where should the quality function be located in an organization? Both approaches have been used with varying amounts of success. Where the performance and management of quality assurance functions have been delegated to each organizational element, there has been a tendency for each element to do as it pleases, resulting in inconsistent, and often inadequate, controls. To prevent this from occurring, it is necessary to coordinate the quality-related activities of each organizational element. The Quality Department accomplishes coordination of activities by establishing management and administrative control guidelines to be followed by each organizational element in establishing their controls, by reviewing the controls and procedures established by each element for their adequacy and completeness, and by performing surveillances and audits of each organizational element's implementation of their controls and procedures. The Quality Department must report to an organizational level and have sufficient independence from schedule and cost considerations which permits it to carry out these functions. Where an adversarial role has developed between personnel performing and verifying quality-affecting activities, top management must reemphasize that quality is everyone's responsibility.

The organizational structure selected and the location of the Quality Department within the organization is often dependent upon the size of the organization and its role in the project. Caution must be taken to assure that multiple layers of management does not result in the filtering of information as it is passed from one level to another, both upward and downward. Such filtering of information results in incomplete information being received by top management for decision-making purposes and by lower-level personnel for carrying out desired actions. Because many project activities involve more than one organizational element, good communication between elements, or horizontally, must also be established and maintained.

Regardless of the organizational structure selected and the location of the quality function, it must be conducive to obtaining and maintaining the balance between schedule, cost, and quality.

### Responsibilities, Authorities, and Interfaces

While establishing the organizational structure for the project, top management must clearly define and document the functional responsibilities, levels of authority, and interfaces of both the organizational elements within its organization and any key contractors. In many projects, activities are contracted to other organizations. Because of the reputation of the organization to perform the activities, top management of the project assumes that organization will properly perform their activities and produce the required quality. Top management does not require controls over the contractors to the degree that would be exercised if the activities were to be performed by its own organization. Without such controls, top management becomes aware of major problems only after they have occurred and recognizes, too late, that had they known the status of the situation earlier, they could have taken responsible action to correct the situation and minimize its effects. The organization responsible for a project is ultimately responsible for all project phases and activities, whether it performs the activities itself or has them performed by contractors or suppliers. Only by clearly defining the responsibilities, authorities, and interfaces can top management provide for adequate control over the contractor activities and stay up-to-date with the status of the project.

### Key Project Personnel

Top management of the organization responsible for a project is responsible for assuring that qualified personnel are assigned to project activities. To accomplish this task, top management must carefully select key project personnel based upon their expertise and experience. The key project personnel are responsible for assuring that their subordinates are qualified for their areas of activity based upon expertise, experience, and training. Effective selection and training of project personnel can help to minimize the levels of supervision and number of personnel required.

## QUALITY PROGRAM IMPLEMENTATION

A well-planned and documented Quality Program is useless if it is not understood and effectively implemented. Top management must provide active support and leadership to assure Quality Program understanding and implementation by project level management, first-line supervision, and all project personnel.

### Indoctrination and Training

The key to Quality Program understanding is indoctrination and training of personnel. Personnel performing quality-affecting activities must have indoctrination and training in the Quality Program and the activities they are to perform. Capabilities should be determined by evaluation of education, experience, and training, or by examination or capability demonstration. Performance should be regularly evaluated and additional training performed or reassignment made, as appropriate. Most Quality Programs require indoctrination and training schedules and records. By looking at the schedules and records, one can determine that the indoctrination and training were performed. However, this does not assure that the indoctrination and training were adequate. The effectiveness of the indoctrination, training, and motivation have major impacts on performance of an activity. Do the personnel know what is to be done and how to do it? Are they capable of doing it? Do they want to do it? It is imperative that personnel comprehend that the purpose of the Quality Program is to assure that activities are properly performed and the desired quality is attained. Unfortunately, personnel in many organizations lose sight of the purpose of the program and concentrate on complying with the words of the program. For example, the purpose of reviewing specifications, drawings, procedures, and instructions is to assure their completeness and adequacy. The purpose of signing and dating the document, or a review record for the document, is to provide objective evidence that the review was performed and the document is complete and adequate. When pressured with a backlog of documents to review, a multitude of other demands for the time of the reviewer, and expressed urgency for release of the document to production, the reviewer may sign and date the document to release it, justifying his action on either his confidence in the preparer of the document or on the urgency of the need for the document. As a result, evidence exists for the review of the document as required by the Quality Program, but the document was not properly reviewed to assure its completeness and adequacy. Project personnel must clearly understand the importance of the responsibilities of their positions.

### Encouragement of Quality

Top management can provide support and leadership by creating an atmosphere for the project that encourages quality. Emphasis should be placed on preventing, identifying, and correcting problems instead of hiding them and accepting them as a cost of doing business. Measures should be established to prevent the harassment and intimidation of personnel who identify and report problems. Top management should include quality performance in the performance evaluation of all project personnel and as a topic in all Project Review meetings.

Top management can also provide support and leadership for the Quality Program by providing clear, consistent direction to the project; establishing a firm, expeditious decision-making process; providing the necessary instructions, equipment, and materials; having regular assessments performed of the program adequacy and effectiveness; and assuring prompt and effective corrective action for problems. Failure to provide active support results in the perception by project personnel that top management is not really concerned about quality. If top management is not concerned, why should they be?

### Compliance with Quality Program

It is the responsibility of all project personnel to implement quality-affecting activities in accordance with the requirements of the Quality Assurance Manual, administrative and technical procedures, drawings, specifications, and instructions, or to cause the requirements of these documents to be reviewed and changed, if necessary. When activities cannot be performed as specified, project personnel are responsible for so notifying their supervisors, who are responsible for resolving the issue or initiating a change to the requirements. Permitting activities to be performed contrary to requirements, without causing the requirements to be changed, results in a perception by project personnel that the requirements are not important. Soon, other activities will also be performed contrary to the requirements, without the project personnel even bothering to notify their supervisors.

Top management should establish a Change Committee to review proposed changes and to minimize the impact of changes on schedule, cost, and quality.

## FEEDBACK OF QUALITY-RELATED INFORMATION

Although Quality Program planning is performed at the beginning of a project, the necessary level of control is not always appropriately assigned to each activity and item. Establishing a Quality Program and providing the necessary support for its implementation does not assure that the program will be implemented or is adequate and effective in attaining quality. Top management must establish measures for feedback of quality-related information as well as schedule and cost information so it can effectively manage, coordinate, and control the project. The information must pertain to the quality performance of each organizational element of the project, including suppliers and contractors; must be factual; and must be provided frequently. Such information can be provided through written reports to top management and through the inclusion of quality-related information as a topic at all Project Review meetings. Meeting notes must clearly identify actions to be taken and responsibility for the actions to enable follow-up at later meetings. Management of each organizational element implementing the program should also regularly assess the adequacy and effectiveness of the program as it applies to their quality-affecting activities and recommend any appropriate changes to enhance the program. Quality problems seldom happen overnight. An effective feedback system will promptly identify problems and their causes so corrective action can be taken. Audits and Trend Analysis are two methods frequently used for providing feedback of quality-related information to top management.

### Audits

Audits are an organized method of finding out how business is being conducted and comparing the results with how business should have been conducted. How business should have been conducted is described in the Quality Assurance Manual and administrative and technical procedures, specifications, drawings, and instructions identified in the process control documents. Audits determine if personnel know their responsibilities, verify whether required activities are being properly performed, and identify specific instances of deficiencies so that causes can be determined and corrective action taken. They are performed by interviewing personnel, observing operations, checking computer systems and records, and reviewing documentation, records, and completed work. Audit programs include the use of audit schedules, checklists for items and activities to be audited, and audit reports. However, the existence of such documents does not always assure that the intent of the audit was achieved. Audits should be performed by trained personnel who do not have responsibility in the areas audited. Personnel are not likely to find or report problems in their own area of responsibility. Further, audits should be performed using detailed checklists which identify specific requirements. Checklists are often not specific enough to verify compliance with requirements and to identify instances of deficiencies. Unidentified requirements are usually not audited and often show up later as major deficiencies. Two common problems with audits performed during a project are that the audit is either programmatic and based upon documentation only, or that the audit is hardware or technically oriented with a failure to recognize that the deficiencies are but symptoms of larger problems in system controls. Audits need to strike a balance between programmatic and hardware and technical orientation to get a true picture of the Quality Program implementation and effectiveness. Audits of Quality Program implementation are usually included as a method of control in the Quality Assurance Manual. Such audits are typically performed by personnel within the project organization, usually by personnel in the Quality department. In addition to such audits, top management can arrange for an evaluation or audit of the Quality Program by personnel who are independent of the project. Such evaluations and audits are commonly called management audits and eliminate organizational bias, cut through organizational barriers, and objectively report the Quality Program's deficiencies and effectiveness.

### Trend Analysis

Trend analysis involves reviewing data and determining if any trends exist of conditions adverse to quality. It is normally performed by reviewing nonconformance reports, inspection reports and logs, audit reports, corrective action reports, and quality cost data. It helps focus attention on repetitive deficiencies so that their cause can be determined and corrective action taken.

## CORRECTIVE ACTION

Top management is responsible for establishing measures for prompt and effective corrective action to project problems. The identification of problems and conditions adverse to quality normally occurs during the performance of reviews, inspections, surveillances, tests, audits, and trend analyses. Corrective action consists of four phases. The first phase, the fix, involves fixing or correcting the specific deficiency that was identified. The second phase, the purge, involves reviewing similar documents and activities to determine the extent of the deficiencies and correcting deficiencies in those documents and activities. The third phase, preventive action, is to determine the cause of the deficiencies and take corrective action to prevent recurrence of the deficiencies. The fourth phase, the close-out, is to verify that the corrective action taken was adequate and effective. Corrective action consisting of just the fix, or the fix and purge, is not effective. Without identification of the cause of the problem and the taking of action to eliminate the cause, the problem will reappear and much time and expense will be expended in repeatedly fixing the same problem. Corrective action consisting of just the fix, purge, and preventive action will also not be effective. Without verification that the corrective action taken was adequate and effective, there is no assurance that the same problem will not recur.

Top management must also assure that corrective action is taken for problems reported through the feedback of quality-related information. One approach is for top management to receive audit and trend analysis reports and to initiate corrective action through project management personnel responsible for the area of activity pertaining to the problem. Another approach is for top management to initiate corrective action during Project Review meetings and to follow up on completion of the corrective action at future meetings. Prompt and effective corrective action is a key factor in managing for quality.

## SUMMARY

Achieving quality is as important as completing a project on time and within cost and involves all project organizations and personnel performing quality-affecting activities. Quality problems are the result of breakdowns in the overall management of a project and can be prevented through the implementation of a comprehensive Quality Program. A Quality Program is a systematic approach to planning and controlling quality through the use of good business practices. Quality Program management is the element of project management that is designed to assure the attainment of quality. It accomplishes its goal by utilizing basic elements of good management: planning, organization, implementation, feedback, and corrective action. The top management of the organization responsible for a project must provide for project management integration of schedule, cost, and quality planning, control, and performance data throughout all project phases. Through incorporation and implementation of the appropriate features of quality concepts into a comprehensive Quality Program, which integrates the concepts into the activities of each organizational element, quality can be achieved, productivity can be increased, and overall project costs can be reduced. The Quality Program should stress prevention of problems, active involvement of all project personnel, efficient and effective methods of control, and the utilization of factual performance data to continually improve performance throughout the project.

# 22. The Legal Standards for Prudent and Efficient Project Management

Randall L. Speck*

In our increasingly disputatious society, the project manager has become the focus of ever-closer scrutiny and occasional opprobrium. Although most projects are completed successfully to resounding kudos, some have suffered serious calamities—skyrocketing costs, repeated delays, and unreliable quality. The most notorious examples have occurred in the construction of state-of-the-art nuclear power plants where tenfold cost increases, oft-postponed completion dates, and massive rework have been commonplace. In some of those cases, utility company stockholders have been penalized to the tune of several hundred million dollars for their project managers' apparent lapses. Every project manager is vulnerable to censure, however, whenever the project falls short of its goals and in the process injures a third party (e.g., consumers, owners, contractors, the government, or even stockholders). This chapter reviews the legal standards that have evolved over the last decade to measure project managers' performance and offers a few practical guidelines that may help to fend off unwarranted criticism.†

In the legal vernacular, a project manager's conduct is usually acceptable if his or her performance is considered "prudent" and "reasonable." Obviously, such directives make poor guideposts for the manager about to embark on a project that may be fraught with risks and uncertainties.

---

* Randall L. Speck has been lead attorney and project manager in several major legal proceedings to determine whether particular projects were managed prudently. Most notably, he represented the State of Alaska in the seven-year litigation to set the tariff for the $8 billion Trans Alaska Pipeline System. Mr. Speck is Managing Director in the law firm of Rogovin, Huge & Schiller in Washington, D.C.

† This brief primer on the putative standards for prudent project management is not intended as a substitute for legal advice that has been tailored to the circumstances of a particular project. The law varies by jurisdiction, and, as noted below, the touchstone for acceptable behavior is often elusive. It is advisable, however, to solicit anticipatory counsel rather than waiting until misfortune has become reality.

These nebulous terms have been infused with a modicum of meaning, however, based on the very large, analogous body of negligence law. Under those hoary principles, unattainable perfection is not required. The law in most state and federal jurisdictions merely prescribes the actions that a "reasonable person" would take under similar conditions, but several states have recently adopted more rigorous criteria in specified regulatory contexts. Even though those standards are not yet well defined, they seem to demand project management techniques that will produce the most efficient performance possible. Project managers should certainly be aware of the tests that will be used to evaluate their actions and adjust their behavior accordingly.

Given an environment in which day-to-day decisions made under the pressures of a dynamic project may be strictly scrutinized, the project manager should be prepared to document the reasonableness of his or her choices. Managers are normally given substantial latitude in running a project, and courts are loathe to second guess a supervisor's judgment. Nevertheless, the project manager may, as a practical matter, be forced to explain cost, schedule, or quality deviations in some detail, and the consequences of inadequate data could be extremely costly. Again, forewarned is forearmed. Thorough contemporaneous documentation (as opposed to explanations constructed after the fact) can provide a palpable defense to charges of mismanagement.

There is no facile recipe for prudent management. The other chapters in this *Handbook* cover the range of issues that are likely to arise in any challenge to management's performance, and there are a myriad of subissues that could loom large in litigation over a major project. There are three critical aspects of the project manager's responsibilities, however, that are the most likely candidates for probing review: (1) planning, (2) organization, and (3) control. A brief case study of the New York Public Service Commission's evaluation of the construction of the Shoreham Nuclear Generating Facility provides a good framework for analyzing the substantive requirements for prudent project management.

Finally, claims of unreasonable management will inevitably be predicated on some form of monetary loss—for example, excessive cost of the final product, inadequate performance that requires expensive repairs, or lost profit attributable to delays. If there is evidence of management shortcomings, the court or regulatory agency will be required to reconstruct the project as it would have unfolded if management had performed prudently. Of course, that exercise tempts the protagonists to flights of imagination and to hypotheses of an aerial project that has no relevance to reality. Judges are commonly called upon, however, to weigh models or estimates of what would have happened if the facts had been different.

Thus, with appropriate data derived from the project itself, the parties can formulate a reasonable assessment of the cost consequences of any management dereliction.

## CONTEXTS FOR CHALLENGES TO THE REASONABLENESS OF PROJECT MANAGEMENT

Almost any project may be subjected to a reasonableness review whenever a service or product is to be produced for a third party based on a specified standard, with a specified time for completion, or for a specified price. If those expectations are frustrated, the injured party is likely to seek redress, particularly where the stakes are large. The most common forum for examination of project management is in regulatory proceedings to set the rates that a utility will be allowed to charge. Very similar issues arise, however, in the execution of government contracts and in disputes between owners and contractors over performance under the terms of the contract.

### Regulation as a Substitute for Competition

It is a basic premise of most regulatory policy in the United States that the regulator serves as a surrogate for competition. In theory, the marketplace and the profit motive provide the impetus for unregulated enterprises to operate efficiently, and no external controls are necessary or desirable. For certain natural monopolies (e.g., electric utilities), however, there is no realistic opportunity for competition. Thus, a governmental watchdog is designated to act on behalf of the consumer and the public interest to monitor the company's performance against the standards that would be expected in a competitive environment. As the Federal Energy Regulatory Commission (FERC) has held:

> [m]anagement of unregulated business subject to the free interplay of competitive forces have no alternative to efficiency. If they are to remain competitive, they must constantly be on the lookout for cost economies and cost savings. Public utility management, on the other hand, does not have quite the same incentive. Regulation must make sure that the costs incurred in the rendition of service requested are necessary and prudent. [*New England Power Company,* 31 F.E.R.C. ¶ 61,047 at 61,083 (1985) (quoting *Midwestern Gas Transmission Co.*, 36 F.P.C. 61, 70 (1966), *aff'd,* 388 F.2d 444 (7th Cir.), *cert. denied,* 392 U.S. 928 (1968) (cited herein as "NEPCO")]

Similarily, the New York Public Service Commission concluded that "[t]he prudence rule is a regulatory substitute for the discipline that would be imposed by a free, competitive market economy where the penalty for mismanagement and imprudent costs is a loss of jobs, profits or business failure" (*Long Island Lighting Co.*, 71 P.U.R. 4th 262, 269 (1985) (cited herein as "LILCO"). Finally, the Iowa State Commerce Commission articulated its regulatory duty as follows:

> to maintain surveillance over costs associated with a particular decision, and in the absence of the kind of incentive provided by a competitor, the responsibility falls upon us to provide the requisite incentives. We do not believe we are unduly interfering with management prerogatives when we attempt to distinguish between reasonable and unreasonable [management] decisions. We believe such an inquiry is required of us by the legislature's directive that rates we allow be reasonable and just. [*Iowa Public Service Co.,* 46. P.U.R. 4th 339, 368 (1982)]

The basic principle of regulatory review of utility expenditures is at least 60 years old. The U.S. Supreme Court in 1923 held that a regulated company is entitled to a return on its investment that is "adequate, under efficient and economical management," to enable the utility "to raise the money necessary for the proper discharge of its public duties" (*Bluefield Waterworks & Improvement Co. v. Public Service Commission,* 262 U.S. 679, 693 (1923)). Justice Brandeis elaborated on this precept in his oft-quoted definition of "prudent investment."

> The term prudent investment is not used in a critical sense. There should not be excluded from the finding of the [capital] base [used to compute rates], investments which, under ordinary circumstances would be deemed reasonable. The term is applied for the purpose of excluding what might be found to be dishonest or obviously wasteful or imprudent expenditures. Every investment may be assumed to have been made in the exercise of reasonable judgment, unless the contrary is shown. [*Missouri ex rel. Southwestern Bell Telephone Co. v. Public Service Commission,* 262 U.S. 276, 289 (1923) (Brandeis, J. concurring)]

In recent years, regulatory bodies have used these principles to deny recovery to utilities for very significant portions of their expenditures related to major projects. For instance, the New York Public Service Commission disallowed $1.395 billion of the cost of the Shoreham Nuclear Project based on its findings of unreasonable project management,

*LILCO, supra,* and subsequently denied recovery of approximately $2 billion of the capital cost of the Nine Mile Point 2 nuclear project, *Re Nine Mile Point 2 Nuclear Generating Facility,* 78 P.U.R. 4th 23, 41 (N.Y. 1986); the Missouri Public Service Commission excluded $384 million for imprudent management in constructing the Callaway Nuclear Point, *Re Union Electric Co.,* 66 P.U.R. 4th 202, 228 (Mo. 1985); the Iowa State Commerce Commission excluded $286 million of the costs for the same plant, *Re Union Electric Co.,* 72 P.U.R. 4th 444, 454 (Iowa 1986); the Michigan Public Service Commission rejected $397 million of the Femi 2 capital expenditures as imprudent, *Re the Detroit Edison Company,* Case No. U-766 [Mich. P.S.C., April 1, 1986); the Kansas State Corporation Commission denied recovery of $244 million of the costs of the Wolf Creek Nuclear Generating Facility, *Re Wolf Creek Nuclear Generating Facility,* 70 P.U.R. 4th 475, 508 (Kan. 1985); the New Jersey Board of Public Utilities reduced the rate base for the Hope Creek nuclear plant by $432 million based on unreasonable management, *Re Public Service Electric and Gas Co.,* No. ER8512116 (N.J. Bd. of P.U., April 6, 1987); the Illinois Commerce Commission disallowed $101 million of the Byron 1 Nuclear plant costs, *Re Commonwealth Edison Co.,* 71 P.U.R. 4th 81, 98 (Ill. 1985); and the California Public Utilities Commission refused to permit Southern California Edison to include $330 million of the plant costs for San Onofre-2 and -3 because it had been imprudently spent, 80 P.U.R. 4th 148, 153 (Calif. 1986).

Most of these dramatic deductions were related to nuclear plants, but the same maxims have been used to deny cost recovery for projects involving traditional fossil power plants, a synthetic natural gas plant, oil pipelines, and the management of scheduled outages for nuclear power plants. As regulatory commissions become more comfortable with their role as sentinel against unreasonable costs, they are likely to expand their inquiries to include any significant utility expenditure that might have been controlled more effectively by management. Thus, the project manager for any regulated entity or its contractors should expect his or her decisions to be subjected to microscopic attention.

### The Prudent Project Management Standard for Contract Disputes

The measure of performance under many contracts is almost identical to the prudent management standard that has been applied in the regulatory context. In fact, by statute, contracts for some defense projects are subject to renegotiation if the contractor earns "excessive" profits, but

> [i]n determining excessive profits, favorable recognition must be given to the efficiency of the contractor or subcontractor, with particular regard to attainment of quantity and quality production, reduction of costs, and economy in the use of materials, facilities and manpower. [50 U.S.C., App. § 1213(e)]

Thus, a project manager who can demonstrate his project's efficiency may be able to retain profits that would otherwise be returned to the government.

Virtually any project-related contract could also be the subject of litigation focusing on the manager's performance. In a very typical case, a subcontractor may claim that it lost profits on a project when the prime contractor or the owner failed to integrate all of the project elements efficiently and caused a delay in the subcontractor's work. Similarly, an owner may sue its services contractor for failure to manage a project component prudently so that it would satisfy the owner's requirements (e.g., failure to debug computer software adequately or on time to permit initiation of a new manufacturing process). Occasionally (and preferably) the performance standard is spelled out in sufficient detail in the contract itself, but much more frequently, the parties are relegated to presenting evidence on the reasonableness of project management under the particular circumstances of that contract.

## STANDARDS FOR PRUDENT PROJECT MANAGEMENT

### The Reasonable Project Manager

**The Reasonable Person Test.** The Federal Energy Regulatory Commission (FERC) concluded in its seminal *New England Power Co.* decision that "the most helpful test" in resolving issues of "prudent investment" is the "reasonable person" test, which the Commission defined as follows:

> In performing our duty to determine the prudence of specific costs, the appropriate test to be used is whether they are costs which a reasonable utility management . . . would have made, in good faith, under the same circumstances, and at the relevant point in time. [*NEPCO, supra,* at 61,084]

This "reasonable person" test has been consistently applied by state public service commissions in evaluating the prudence of costs incurred by utilities under their jurisdictions and has been applied as well in other areas of law involving regulated companies, including occupational safety and health, banking, and government contracting.

The "reasonable person" is widely accepted as a standard in large part because it is an *objective* test that avoids the adverse policy implications of alternative legal criteria such as strict liability or "guilty knowledge." A strict liability approach would deem a project manager imprudent whenever a management decision produced harm significantly greater than its benefits. Under that analysis, however, regulatory bodies would undoubtedly be inundated with requests by public utilities for advance approval of projects before they undertake substantial capital investments. A "guilty knowledge" approach would consider management imprudent only when the manager acted with a conscious apprehension that his or her conduct was wrongful or otherwise faulty. That standard tends to exculpate irresponsible management, however, because it would be virtually impossible to prove that management acted with conscious knowledge of its wrongdoing. The "reasonable person" standard avoids these legal pitfalls and provides an appropriate level of regulatory or judicial scrutiny.

The "reasonable person" in the project management context draws its meaning from an extensive body of tort cases involving issues of negligence. It is clear from these well-established principles of tort law that the "reasonable person" standard is, above all, an objective standard, not dependent on individual judgment:

> The standard which the community demands must be an objective and external one, rather than that of the individual judgment, good or bad, of the particular individual. [*Restatement (Second) of Torts* § 283 comment c, at 12 (1965)]

Thus, the "reasonable person" standard does not depend on what a particular person considers reasonable under the circumstances, but rather on a standard of reasonableness imposed by the community. Indeed, the courts have gone to great lengths to emphasize the abstract and hypothetical character of the reasonable person:

> The reasonable man is a fictitious person, who is never negligent, and whose conduct is always up to standard. He is not to be identified with any real person; and in particular he is not to be identified with the members of the jury, individually or collectively. [*Restatement, supra,* § 283 comment c, at 13]

**Community Standards.** Community standards as a measure of the reasonable person's behavior may be established in a variety of ways, not the least of which are published treatises by respected project managers such as those included in this *Handbook*. Professional codes such as the Project Management Institute Code of Ethics (e.g., requiring application of state-of-the-art project management tools and techniques to ensure that schedules are met and that the project is appropriately planned and coordinated) and the Canons of Ethics of the American Society of Civil Engineers may also help define the parameters of prudent project management. In some cases there may even be a statute or regulation mandating a particular level of project management attention. For example, the Department of Interior Stipulations that governed construction of the Trans Alaska Pipeline project dictated that the owners should "manage, supervise and implement the construction . . . to the extent allowed by the state of the art and development of technology" (Agreement and Grant of Right-of-Way for Trans Alaska Pipeline, January 23, 1974). The U.S. Department of Defense also established very clear community standards for high-technology projects in its "Cost/Schedule Control Systems Criteria, C/SCSC Joint Surveillance Guide," initially issued in the late 1960s and periodically updated. Courts and regulatory agencies have used these external measures to assess project managers' performance.

**Internal Project Standards.** In some instances, however, the most relevant criteria for evaluating project management's prudence may not be set by the community, but by the managers themselves. Certainly the most applicable estimates, schedules, or quality norms are those that were tailored to the particular project at issue. Project management presumptively considered all pertinent constraints when they set those benchmarks, and it is reasonable to apply those standards to assess project execution. The New York Public Service Commission followed that approach in the Shoreham case. The company argued that it would be more appropriate to compare the procurement cycle actually achieved at Shoreham with those achieved at other nuclear construction projects. The Commission concluded, however, that:

> such a comparison would not be germane. This is because the cycles planned but unachieved at Shoreham were those that [the utility's] management considered essential if the procurement function was to succeed in supporting the engineering and construction schedules. A cycle short enough to support construction in some other plant's schedule might nevertheless have been too lengthy to achieve that same objective at Shoreham. Conversely, the failure of procurement to sup-

> port construction at another plant would not establish that Shoreham was prudently managed despite such failures. [*LILCO, supra,* at 286]

Similarly, the Missouri Public Service Commission used the definitive estimate for the Callaway Project as "the proper starting point for an investigation of cost overruns and a determination as [to] whether cost[s] incurred on the project are reasonable" (*Re Union Electric Co., supra,* at 229).

Internally approved project standards have an initial attractiveness that has seduced some fact finders to divine imprudence whenever project goals are not met. That conclusion is clearly inappropriate. Project objectives may be set for a variety of purposes—for instance, to provide an ambitious target that will always be just beyond the reach of all but the most capable managers. Moreover, the project managers may simply have erred and established standards or procedures that are impractical. Finally, circumstances may have changed so that the norms conceived at the beginning of the project no longer have any relevance. Thus, judges and commissioners should not blithely adopt the project's standards as coincident with prudent management without first testing the objective reasonableness of those criteria within the framework of the conditions that actually existed.

**Standards That Exceed Common Practice.** Even compliance with an established precedent—whether set by the community or by project management internally—may not be sufficient, however, to demonstrate prudence. The reasonable person standard applied by courts and juries reflects an observation made by Judge Learned Hand more than half a century ago in his opinion in *The T. J. Hooper:* "in most cases reasonable prudence is in fact common prudence" (60 F.2d 737, 740 (2d Cir.), *cert. denied,* 287 U.S. 662 (1932)). Thus, evidence of the usual and customary conduct of others under similar circumstances is normally relevent and admissible as an indication of what the community of project managers regards as proper.

Proof that project management practices and organizational structures consistently fell short of contemporaneous industry standards and practices serves a particularly useful function for the trier of fact:

> Proof that the defendant took less than customary care has a use different from proof that the defendant followed business usages: Conformity evidence only raises questions, but subconformity evidence tends to answer questions. If virtually all other members of the defendant's craft follow safer [or more efficient] methods, then those methods are practi-

cal; the defendant has heedlessly overlooked or consciously failed to adopt common precautions. [Morris, *Custom and Negligence,* 42 Colum. L. Rev. 1147, 1161 (1942)]

It should be emphasized that although failure to conform to industry standards establishes imprudence, proof of limited conformity to the practices of others does not carry the same weight in establishing prudence. Consequently, even if one or more specifically identifiable "real-world" project managers would have acted in a particular fashion, such evidence of limited conformity would not establish prudence. For unlike the fictional "reasonable manager" of the law, "real-world" managers, even though they are generally considered "reasonable" by their peers, sometimes act unreasonably or imprudently.

Because even people who are generally reasonable may sometimes act negligently, it is not surprising that the law refuses to allow any one individual to set the standard of prudent behavior by his or her conduct alone. Indeed, the courts have consistently held that even adherence to an industry-wide custom or practice will not insulate a defendant from liability, because an entire industry may be negligent. This principle was perhaps most eloquently articulated by Judge Hand in his oft-cited opinion in *The T. J. Hooper:*

[I]n most cases reasonable prudence is in fact common prudence; but strictly it is never its measure; a whole calling may have unduly lagged in the adoption of new and available devices. It never may set its own tests. . . . Courts must in the end say what is required; there are precautions so imperative that even their universal disregard will not excuse their omission. [*The T. J. Hooper, supra,* 60 F.2d at 740]

The New York Public Service Commission applied a similar analysis to the Shoreham nuclear project and found that "if gross inattention to cost and schedule control was typical of the industry, industry practices on their face would be unreasonable and could not excuse [the utility] from its responsibility to act reasonably" (*LILCO* at 278). Indeed, commissions have also found that utilities are not necessarily prudent simply because they produced project results that were better than the norm. The Illinois Commerce Commission found that although the Byron nuclear power plant was:

one of the cheaper plants to be built recently, that certainly does not preclude investigation into particular aspects of the project to deter-

> mine whether there were reasonably avoidable diseconomies . . . . The favorable plant cost comparisons do, however, help to prevent the Commission from inferring mismanagement simply from cost increases, or increased project ratios, or other such simple arithmetical comparisons. [*Re Commonwealth Edison Co., supra,* at 101]

**Requirement for Expert Project Management.** The common use by commissions and courts of a reasonable manager standard implies application of the qualifications required from a specialist, which differs substantially from the criteria applied to the ordinary person engaged in ordinary activities. This expert standard, again, is a familiar facet of negligence law, which has traditionally demanded more than ordinary care from those who undertake any work calling for unique skill. Specialists have always had a duty to display "that special form of competence which is not part of the ordinary equipment of the reasonable man, but which is the result of acquired learning and aptitude developed by special training and experience" (*Restatement (Second) of Torts,* § 299A comment a).

An expert generally is held to "the standard of skill and knowledge required of the actor who practices a profession or trade"—the "skill and knowledge," in other words, "which is commonly possessed by members of that profession or trade in good standing" (*Id.* § 299A comment e). Thus, as members of a particularly skilled group, project managers will normally be held to a standard based on the distinctive skill and knowledge commonly possessed by members of the profession they undertake to practice.

The level of expertise demanded by the courts will be commensurate with the complexity and challenge of the project. For instance, significantly greater talent and experience will be expected from the manager of a multibillion-dollar nuclear power plant project than from the project manager responsible for the addition to a residential home. In general, the greater the risk of calamitous outcomes (e.g., runaway costs or injury to the environment or populace from quality shortcomings), the greater the burden that will be imposed on the project manager.

### Reasonable Project Management "Under the Circumstances"

**Hindsight Prohibited.** Courts and regulatory agencies have uniformly applied the criteria for prudent project management applicable at the time decisions were made based on the facts that were available to the decision maker at that time. For instance, in adopting a reasonable utility manager standard in *NEPCO,* the FERC remarked that

> while in hindsight it may be clear that a management decision was wrong, our task is to review the prudence of the utility's actions and the costs resulting therefrom based on the particular circumstances existing . . . at the time the challenged costs were actually incurred. [*NEPCO, supra,* 31 F.E.R.C. ¶ 61,047, at 61,084]

Thus, the Commission made it clear that the standard to be used is not one of perfection, that is, judging the reasonableness of management decisions with the benefit of hindsight. Instead, management conduct must be evaluated according to the circumstances that existed at the time the relevant decision was made.

Some project managers have attempted to invoke severe time constraints as a mitigating circumstance that might justify less than optimal procedures. The courts, however, have imposed two important limitations on the rule that an actor's conduct must be evaluated according to the circumstances (including "crisis circumstances") that existed at the time of the challenged conduct. First, crisis conditions are not considered as a mitigating circumstance when the actor's own negligence creates the crisis. "The fact that the actor is not negligent after the emergency has arisen does not preclude his liability for his tortious conduct which has produced the emergency" (*Restatement, supra,* § 296(2), at 64). Thus, an individual may be held liable in a situation in which he acts "reasonably in [a] crisis which he has himself brought about" (*Id.* § 296 comment d, at 65). Second, an actor who engages in an activity in which crises arise frequently is required to anticipate and prepare for those situations. In particular, experts or professionals who perform work that is characterized by frequent crises (i.e., most project mangers) are required to have particular skill and training to deal with those situations (*Restatement, supra,* § 296 comment c, at 65).

As part of the "reasonable person" standard, the FERC has expressly held that "management must operate its systems to avoid circumstances that give rise to emergencies." In *Texas Eastern Transmission Corp.,* 2 F.E.R.C. ¶ 61,277 (1978), the FERC precluded the gas company from recovering the costs of emergency gas purchases because the commission found that the company imprudently operated its system so as to create a situation in which emergency purchases were necessary. The commission emphasized that it was not judging the company's behavior with the benefit of hindsight; rather, it found that, based on information available to the company at the time, it was imprudent in failing to take steps early in the year that would have eliminated the need for later emergency gas purchases (*Id.* ¶ 61,277, at 61,617–18).

**The "Large Complex Project."** A few commentators have argued that some projects (which they dub "large complex projects" or LCPs) are *sui generis* and that their peculiar circumstances (e.g., size, complexity, application of new technology) make it impossible to define meaningful management criteria for assessing performance. This position has been soundly rejected. In the proceeding before the New York Public Service Commission to establish the allowable costs for the Shoreham nuclear project, the utility advocated a "theory which suggests that large-scale complex projects are inherently unmanageable" and a standard of conduct that "would insulate [the utility's] management from a finding of imprudence short of outright fraud, self-dealing, blatant carelessness, or gross negligence" (*LILCO, supra*, at 269). The Commission disdained this approach because "[t]he public is entitled to expect that such undertakings by public utilities are controllable" (*Id.*) The development of the project management discipline over the past 25 years would appear to confirm the commission's judgment. Project managers are unlikely to be able to hide their failures behind rationalizations that their projects were somehow unique and not amenable to standard project management techniques.

### Project Managers' Responsibility for Agents' Actions

In most instances, the focus of any judicial inquiry will be on whether the project manager acted prudently. Of course, that investigation should include an examination of the project manager's role in selecting contractors, defining the scope of their work, supporting their efforts, and monitoring performance. Any dereliction in these duties would obviously be the project manager's direct responsibility.

According to some regulatory commissions, however, the project manager may also be vicariously liable for the imprudent conduct of his or her agents. For example, the Maine Public Utility Commission found that under its regulatory scheme, the ratepayers should "pay no more than the reasonably necessary costs to serve them. Any other reading [of the statute] is likely to lead to economic inefficiency, to excess costs, and sometimes to dubious practices between utilities and their suppliers" (*Re Seabrook Involvements by Maine Utilities,* 67 P.U.R. 4th 161, 168–69 (Me. P.U.C. 1985)). The Maine Commission held explicitly that "a supplier's unreasonable charges, even when not found to have been imprudently incurred by the utility, cannot be passed on to the utility's ratepayers" (*Id.*). Similarly, the Pennsylvania Public Utility Commission found that because the project manager, not the ratepayers, chose the contractor, the risk of performance failures should be borne by the stockholders.

> It must be recognized that the basic question is who should pay for the cost of the improper design and manufacture of the Salem 1 generator. . . . Including these costs in [the utility's rates] means that all ratepayers are charged for [the contractor's] actions. This insulated both [the utility] and [the contractor] from responsibility for the generator failure. Conversely, denying . . . recovery places the costs on the party most capable of pursuing legal remedies and negotiating future contractual protections, [the utility]. Only [the utility] can structure its operations in such a fashion as to minimize the costs of contractor error or pursue damages should errors occur. [*Re Salem Nuclear Generating Station,* 70 P.U.R. 4th 568, 606–07 (Pa. P.U.C. 1985)]

Although these cases ostensibly hold the project manager responsible for contractor negligence, regardless of his or her own fault, reality may not dictate so harsh a result. The fundamental premise underlying the commissions' holdings is an expectation that the project manager can control the contractor's performance, either through negotiation of strict contract terms that make the contractor accountable for any mismanagement or through careful monitoring and direction of the contractor's work. The project manager's reasonable steps to preclude contractor misfeasance should provide an adequate defense, particularly if the contractor withheld material information about its failures from the project manager. The manager should be penalized only if there were steps that he or she could have taken to avoid or mitigate the contractor's imprudence.

### An Alternative Standard: Efficient Project Management

In some jurisdictions, regulated companies should anticipate being held to a somewhat more rigorous performance standard if they expected to obtain full recovery for their project costs. In Texas, for instance, the legislature has determined that a utility seeking to include an allowance for construction work in process (CWIP) in its rate base must show that a project has been managed "prudently" *and* "efficiently" (Texas Public Utility Regulatory Act, § 41(a)). Under the most likely interpretation of this statute, utilities will have to show more than the reasonableness of their conduct; they will have to show that the project used "the most effective and least wasteful means of doing a task or accomplishing a purpose" and performed in "the best possible manner" (*Houston Lighting & Power,* Docket No. 5779, Examiner's Report (December 20, 1984) at 17, *aff'd* (January 11, 1985)). Similarly, the Illinois Commerce Commission has held that its statute requires a demonstration of more than mere reasonableness before a utility can recover its project costs:

> In addition to prudency, considered narrowly, the act now directs the commission to consider efficiency, economy, and timeliness, so far as they affect costs. [*Re Commonwealth Edison Co., supra,* at 94]

These seemingly broader mandates for project review have not been fully tested in the courts, but they appear to imply a greater focus on the results that are actually achieved. All projects are plagued with niggling inefficiencies, and management's task is to minimize them to the extent possible. Rigid application of the efficiency standard, however, might mean that no project, no matter how well managed, would be able to demonstrate absolute efficiency and recover 100% of its costs, at least in the context of extraordinary rate relief such as CWIP. In these cases, there will be an even greater premium on the project manager's competence.

## THE BURDEN OF PROVING PRUDENCE

Courts and regulatory agencies have long recognized the importance of giving managers relatively free reign to run projects as they see fit and to avoid second guessing managers' decisions. Thus, absent a significant showing to the contrary, they have presumed that managers act reasonably. The FERC has formulated the following general rule:

> Utilities seeking a rate increase are not required to demonstrate in their cases-in-chief that all expenditures were prudent. . . . Where some other participant in the proceeding creates a *serious doubt* as to the prudence of an expenditure, then the applicant has the burden of dispelling these doubts and proving the questioned expenditure to have been prudent. [*Minnesota Power and Light Co.,* 11 F.E.R.C. ¶ 61,312 (1980) (emphasis added)]

Several states have concluded that a "serious doubt" is raised about management's prudence whenever the final project costs materially exceed the originally estimated costs, thus shifting the burden to the company to show that it acted reasonably and that all costs were justified (*Re Union Electric Co., supra* at 212; *Houston Lighting & Power Co.,* 50 P.U.R. 4th 157, 187 (Tex. P.U.C. 1982); *Consumers Power Co.,* No. U-4717, slip op. at 8 (Mich. P.S.C. 1978)). Other factors that might create a "serious doubt" include performance that deviates significantly from the industry average, a major accident or component failure, or a fine imposed by a regulatory body for violation of statutes or regulations (e.g., an OSHA or Nuclear Regulatory Commission fine for safety infractions). If any of these tokens is present, the project manager may be called upon to marshal evidence to defend his or her administration.

This allocation of accountability is consistent with the basic legal maxim that the party having best access to the relevant information must normally carry the initial burden of proof. The project manager, with intimate knowledge of project planning, organization, and control, should be in the preeminent position to vindicate his or her prudence. Such proof should be straightforward for a meticulously run project that pays assiduous attention to documentation. Many managers have been rudely surprised, however, by their inability to demonstrate their project's good health to an outsider despite hale and hearty prognoses throughout the project's life. The project manager must be able to point to contemporaneous documentation (not "post-hoc rationalizations") to confirm management's prudence.

Two recent decisions illustrate the weight courts and commissions have given to data that was created as a routine part of the project. In *Long Island Lighting Co.*, the New York Public Service Commission was faced with a contradiction between a very critical report prepared by the architect/engineer during the project and the owners' later disclaimers. The commission found that:

> [t]he company's reconstructed version of the facts is not plausible. [The A/E's] report identified a significant problem in need of correction when the construction effort was under way and when [the A/E] was intimately familiar with the problem by virtue of its role as architect/engineer. The Judges had a choice between [the A/E's] contemporaneous analysis or an analysis prepared by [the owner] in 1984 for purposes of this proceeding. Thus, they quite reasonably attached more credence to the former. . . . [I]f the report were faulty, [the owner's] management failed to discern its alleged infirmities back in the 1970's when it was important that the design change process be carefully appraised and effectively managed. [*LILCO, supra,* at 290]

Similarly, the Missouri Public Service Commission in *Re Union Electric Co., supra,* relied on reports generated by the utility, its consultants, and its contractors during the course of the project in finding that the utility imprudently managed the Callaway nuclear project.

In evaluating after-the-fact explanations of events, courts have long recognized "the familiar phenomenon of post-hoc reconstruction." This phenomenon refers not to deliberately false testimony, but rather to "the often-encountered tendency, though unintentional, to testify to what one believes 'must have happened' and not to what did happen" (*United States ex rel. Crist v. Lane,* 577 F. Supp. 504, 511 n. 11 (N.D. Ill. 1983), *rev'd on these grounds,* 745 F.2d 476 (7th Cir. 1984), *cert. denied,* 105 S.Ct. 2146 (1985)). Because "post-hoc reconstruction" is so familiar, courts traditionally give greater weight to contemporaneous statements or documents than to later explanations of events offered by witnesses at trial. Moreover, it is well established that business records have special indicia of reliability based on the fact that managers relied on them in making decisions in the ordinary course of the project.

The prudent project manager, anticipating the possibility of a subsequent challenge, should pay particular attention to documentation as the project progresses. Ordinarily, the same records that should be used to plan and control the project will also be most useful in championing it through later trials. One of the most glaring weaknesses in the defenses mounted to date has been the project manager's inability to demonstrate the specific causes for major cost increases. The shibboleths of "regulatory interference" or "changed conditions" as justifications for broken budgets have not been adequate when millions of dollars are at stake. As knowledgeable project managers are well aware, however, a rudimentary configuration management system enhances rational decision making to accommodate change during the project and at the same time creates a concrete record that can justify cost increases in any later proceedings. Some companies (notably Southern California Edison) now prepare a "pedigree" for each design change that includes a definition of the source of the design requirements, an evaluation of alternatives, a cost/benefit analysis, and steps taken to monitor expenditures versus the estimate. This data will be invaluable for managing the project, but will also demonstrate the reasonableness of the design change should there be a future interrogation. The most crucial lesson of the last decade of prudence proceedings is the absolute necessity for comprehensive record keeping.

## SUBSTANTIVE STANDARDS FOR PRUDENT PROJECT MANAGEMENT

It is impossible to prescribe the unassailable tenets of project management that, if followed, will always invoke the imprimatur of prudence. Standards will evolve as the discipline develops, and courts may be more or less tolerant of shortcomings based on the need to meet other societal goals (e.g., an energy crisis may cause a temporary relaxation of efficiency standards in order to meet the immediate demand). Nevertheless, a few basic principles can be adduced from recent cases. This section reviews some of the illustrative holdings from the New York Public Service Commission's Shoreham prudence proceeding in three critical areas: planning, organization, and control.

### Planning

Planning is the essence of prudent management. Every element of a project demands foresight and a design for realizing goals. Planning is even more essential as the tasks become larger and more complex. Specialized expertise must be marshaled to anticipate technical specifications, lead times, potential impediments, and control requirements. Moreover, all of the parts must mesh to enable the owners and senior managers to assign priorities, assess risks, define the organization, and measure performance. Prudent planning minimizes foreseeable risks and therefore becomes more crucial as predictable risks increase. Precautions that might arguably be optional become obligatory when the peril of inaction is grave.

The New York Commission identified planning as the lynchpin for a prudent project.

> Reasonable management would have foreseen the need for a systematic approach to this large-scale construction project and would, therefore, have exercised its responsibilities by formulating a plan to achieve its objectives. [The owner] failed to commence this project with a baseline plan defining what was to be built; how the work would be performed, and by whom; how changes would be incorporated into the plans if necessary; and, most critically, how the status of the project would be monitored and kept on schedule. This failure resulted in numerous problems throughout the project's history, including an inability to perform cost and schedule monitoring; confusion over roles and areas of responsibility among the project participants; low labor productivity; and the absence of a mechanism for providing the Board of Directors with sufficient, timely information to form a basis for providing guidance and making policy decisions. Accordingly, [the utility's] planning failure not only constituted imprudence, . . . but it also had direct and foreseeable adverse consequences on the course of the project. . . . [*LILCO, supra,* at 275–76]

Any prudence review will focus first on the adequacy of project planning.

### Organization

Four basic premises guide a prudent project organization: (1) clearly defined roles and responsibilities for all of the parties at the project's inception; (2) delegated responsibility and authority based on the project's plan; (3) a relatively stable organizational structure for the project's duration; and (4) experience as the basis for building an organization. Conversely, an imprudent project organization is typically characterized by duplication, poorly defined roles, antagonism among the parties, delayed decision making, and instability.

Organizational flaws were at the heart of many of the problems pinpointed by the New York Commission on the Shoreham project.

> The Judges found that the lack of comprehensive, adequately explicit planning at the inception of the project laid the groundwork for serious conflicts and confusion over the respective roles of the various participants in the construction effort. They found that the failure to define discrete areas of responsibility led to a situation in which [the utility's] own project manager interfered with [the A/E's] managers,
>
> > producing friction, resentment and antagonism between [sic] [the A/E, utility] and the major contractors. Because of these interferences and poorly defined authority, major tasks of planning, supervision and coordination were not performed. Construction was adversely affected.
>
> . . . [T]he record evidence establishes that in 1975 [the utility's project manager] began allowing contractors to submit problems to him instead of to [the A/E]. This was interference with [the A/E's] construction management authority, which undermined [the A/E's] control over contractors and engendered antagonism between [the utility] and its architect/engineer. [The utility's] interference is also evidenced by contemporaneous documentation describing confusion over who, as between [the utility] and [the A/E], was in charge of construction management. [*LILCO, supra,* at 276]

A common problem on many projects has been the absence of a clearly designated project manager who could coordinate all elements of the project. An astonishing number of major projects in the 1970s attempted to function without a distinct focus for management authority. The result was predictable confusion and inattention to crucial problems.

**Control**

Prudent project controls must be commensurate with management's experience and the nature of the contractual relations. For instance, inexperienced managers should be given a shorter reporting leash so that corrective action can be taken at an appropriate stage before the project gets out of hand. Similarly, the owner's control under a cost-plus contract must be substantially more stringent than with fixed-price contracts where the contractor has an incentive to control its own costs. Fundamentally, controls must be tailored to fit project conditions. Regulatory bodies have expressed a strong preference, however, for formal controls over ad hoc, informal mechanisms, and some form of network analysis has become virtually *de rigueur*.

The New York Commission found that controls on the Shoreham project were sadly deficient and that the utility "never instituted a reporting system adequate to enable management to discern problem areas and to make well-informed decisions about possible corrective action" (*LILCO, supra,* at 293). The Commission pointed to evidence that there were "no genuinely informative field reports"; that due to understaffing, the reports to the project manager on construction problems lacked substance and were frequently inaccurate; that reporting problems persisted; and "generally that the reporting system was misleading and difficult to interpret" (*Id.*). The Commission concluded that the utility "failed to establish a monitoring and reporting system capable of providing the information that it needed for the purpose of making intelligent decisions about Shoreham's course and progress" (*Id.* at 294). Without these controls, the Commission found, the project could not be managed prudently.

## CALCULATION OF PRUDENT PROJECT COSTS

Once there has been a determination that a project suffered from some imprudence, the fact finder must determine whether this imprudence increased the project's costs, and if so, by how much. Causation is a factual question, and again there is no simple formula for whether particular consequences are sufficiently connected to the underlying management misconduct to warrant some form of penalty. Generally, courts and commissions have relied on expert testimony to establish a "clear causal connection" between management imprudence and resulting excess project costs (*LILCO, supra,* at 316; *Re Union Electric Co., supra,* at 228).

At first blush, the task of sorting prudent from imprudent project costs would seem quixotic. The courts have recognized, however, that reconstruction of the expenses that would have been incurred if management had acted differently is an extraordinarily imprecise art, and mathematical precision cannot be attained. This principle is not unique to the regulatory arena. Courts have long held in numerous substantive contexts that damages are at best approximations and can only be proved with whatever definiteness and accuracy the facts permit. One of the most frequently cited cases is the United States Supreme Court's decision in *Story Parchment Co. v. Paterson Parchment Paper Co.,* 282 U.S. 555, 563 (1931) (an antitrust case), in which the Court held that

> [w]here the tort itself is of such a nature as to preclude the ascertainment of the amount of damages with certainty, it would be a perversion of fundamental principles of justice to deny all relief to the injured person, and thereby relieve the wrongdoer from making any amend for his acts. In such case, while the damages may not be determined by mere speculation or guess, it will be enough if the evidence shows the extent of the damages as a matter of just and reasonable inference, although the result be only approximate. The wrongdoer is not entitled to complain that they cannot be measured with the exactness and precision that would be possible if the case, which he alone is responsible for making, were otherwise.

In three recent cases examining the prudence of nuclear power plant construction, state utility commissions have addressed the question of how to quantify the costs that should be disallowed as a result of imprudent management. In each, the commissions relied where possible on specific data tracing the cause of cost overruns. When this information was not available, the commissions adopted reasonable estimates to establish the cost of imprudence.

In *LILCO,* the New York Commission first concluded that the utility's management of the Shoreham nuclear project was imprudent in several respects. The Commission rejected the utility's argument that the staff was required to quantify the effect of each discrete instance of imprudence in order to show that alleged acts of mismanagement directly caused specific costs (*LILCO, supra,* at 316–17). The Commission concluded that the staff's methodology, which compared Shoreham cost data in four categories (engineering and construction man-hours, schedule delay costs, and diesel generator problems) with cost data of other nuclear power plant construction projects was "logical and rational" and reached "a just and reasonable result" (*Id.* at 317). Based on these approximations, the Commission disallowed $1.395 billion in costs (*Id.* at 326).

In *Re Union Electric Co.,* the Missouri Public Service Commission addressed the issue of whether the utility had prudently managed construction of the Callaway nuclear project. The Missouri Commission concluded that, as a general matter, the utility "failed to meet the prudence standard," and that this imprudence "require[d] significant disallowances in order to establish 'just and reasonable rates' " (*Re Union Electric Co., supra*).

To quantify the imprudent costs, the Missouri Commission compared the actual cost of construction with a definitive cost estimate generated by the utility during the early stages of construction. The Commission rejected the utility's argument that it should not be "held" to the definitive estimate, stating that "the definitive estimate is the proper starting point for an investigation of cost overruns and a determination as [to] whether cost[s] incurred on the project are reasonable'' (*Id.*). With respect to particular quantification methodologies, the Commission acknowledged that the staff's calculation ''represent[ed] an approximation,'' but concluded that the staff's model ''allow[ed] a reasonable estimate of these costs'' (*Id.* at 243).

Finally, the Kansas State Corporation Commission, in *Re Wolf Creek Nuclear Generating Facility,* found that:

> lack of management attention coupled with the lack of efficient effective management on the part of the owners, resulted in schedule delays and increased costs that could have been mitigated by strong management action earlier in the project. [*Re Wolf Creek Nuclear Generating Facility, supra,* at 495]

To quantify those increased costs, the Kansas Commission adopted two approaches. First, it relied on the owners' definitive estimate of costs, their reconciliation of those projections with final costs, and the staff's independent estimates of specific costs. Second, the Commission estimated the delay in the critical path that was caused by imprudent management and quantified the effect of that delay.

All of these approaches to calculating the costs of a prudently managed project have potential drawbacks or inequities. Great care must be exercised, for instance, in choosing an appropriate baseline cost and schedule for the project. As noted above, estimates and schedules may have been devised for a variety of objectives and may not accurately reflect realistic project goals. Comparisons with costs and schedules on other projects also present significant pitfalls because no two projects are exactly comparable, and it is difficult to make fair adjustments between projects. The basic admonition to project managers, however, is a familiar refrain in this chapter—be prepared with contemporaneous data to justify the costs that were incurred. Deviations from previous projections or from experience on other projects should be explained. To the extent that there are justifications for cost increases, a prudent project manager should be able to document them.

## CONCLUSION

The last decade has been both the boon and the bane of the project manager. The higher profile of megaprojects has generated intense postmortems that have focused on project managers' behavior. A shadowy profile of the ''reasonable project manager'' has emerged from these reviews, but it does not yet offer a model that can be easily emulated. In order to toe the line of prudence, managers must be constantly aware of the most sophisticated techniques and alert to document their reasonable efforts to apply those strategies in their projects. The law should serve the salutary purpose of enhancing and improving project efficiency, and the growing body of cases addressing prudent project management will almost certainly have that effect.

# Section VII
# Project Control

This section focuses on the project control process that complements project planning.

In Chapter 23, James A. Bent develops the project control concept as it relates to project planning. He emphasizes project control ideas in the construction context, but his ideas are widely applicable in other situations as well.

In Chapter 24, Mr. Bent presents 'project control basics' in the form of a briefing that could be provided to familiarize management with some of the fundamental historical statistical relationships that drive a typical project. He has selected those 'rules of thumb' which, although developed in the construction context, he believes apply to many industries.

Kenneth O. Chilstrom provides in Chapter 25 the framework for the management of audit of projects—a diagnostic tool that can serve important control objectives.

In Chapter 26, Harold Kerzner presents various techniques for assessing the performance of project personnel. Project control is thereby conceived as not only involving *project* assessments, but also assessments of people.

In Chapter 27, John Tuman, Jr., discusses the development and implementation of an effective system for the control of a project. He relates the control function to the information that is necessary if control is to be exercised and provides information flow models and modular configurations for a project management information and control system. (The systems development methodology that he prescribes is related to that described in terms of the linear responsibility chart in Chapter 16.)

'Computers in project management''—the subject of Chapter 28—ties together the planning techniques of Section V with the control ideas of this section. Most modern organizations use the computerized project planning and control software that is discussed by Harvey A. Levine to support their project management efforts.

# 23. Project Control: An Introduction

James A. Bent*

## SECTION 1.

### General

Philosophical discussions on defining "control" are never ending. There is the long-stated opinion that actual control is only exercised where the right of decision is vested—in this case, the decision making of the project manager, line supervisors, and design engineers.

It is stated that cost and schedule engineers only provide information and, therefore, have no exercise of control. This is partly true. Often, a staff function does become one of reporting and accounting. However, reporting, trending, and analysis are essential ingredients for forecasting which, in turn, is an essential ingredient of control.

It is also true that control is minimal where there is little creative analysis and only reporting and accounting.

*The fundamental elements of control are the cost estimate and project schedule.*

### Planning the Project

One of the most important functions in the life cycle of a project is project planning, especially in the preliminary phases when basic decisions are being made that will affect the entire course of the project. The purpose of project planning is to identify the work that must be done, to gain the participation of those best qualified to do the work, and to develop appro-

---

* James A. Bent has been a consultant in project management since 1980. Previously, he worked more than 30 years for owners/contractors, such as Mobil and M. W. Kellogg. Mr. Bent's experience covers both onshore plants and offshore platforms, with project assignments in the United States, United Kingdom, Australia, Netherlands, Germany, Norway, and Italy. He is an author and presents training courses (3 days–3 weeks) in project management, scheduling, estimating, cost control, contracting, claims/management/negotiating, and shutdowns/turnarounds. These training assignments take Mr. Bent to the United States, United Kingdom, Norway, Europe, Japan, South America, and South Africa.

priate project cost and schedule objectives. Sound planning will minimize lost motion and clearly define for all participants—owner, contractor, associated corporate departments, and outsiders—their role in the project. Sound planning will also provide adequate consideration of all project elements, and will ensure a proper effort to meet the completion date. Figure 23-1 illustrates the major elements of project planning.

*The project manager should personally supervise this effort with the support of business, cost, and schedule specialists.* Project planning should consider such items as organization, communication channels, personnel skills, client requirements, business-political environment, and project execution strategy, and a plan should be drawn up to set in motion these operations. The project manager should develop a *Project Coordination Procedure,* after consulting with client and others, as necessary. This document will identify all principals concerned with the project, define their functions and responsibilities, and indicate appropriate con-

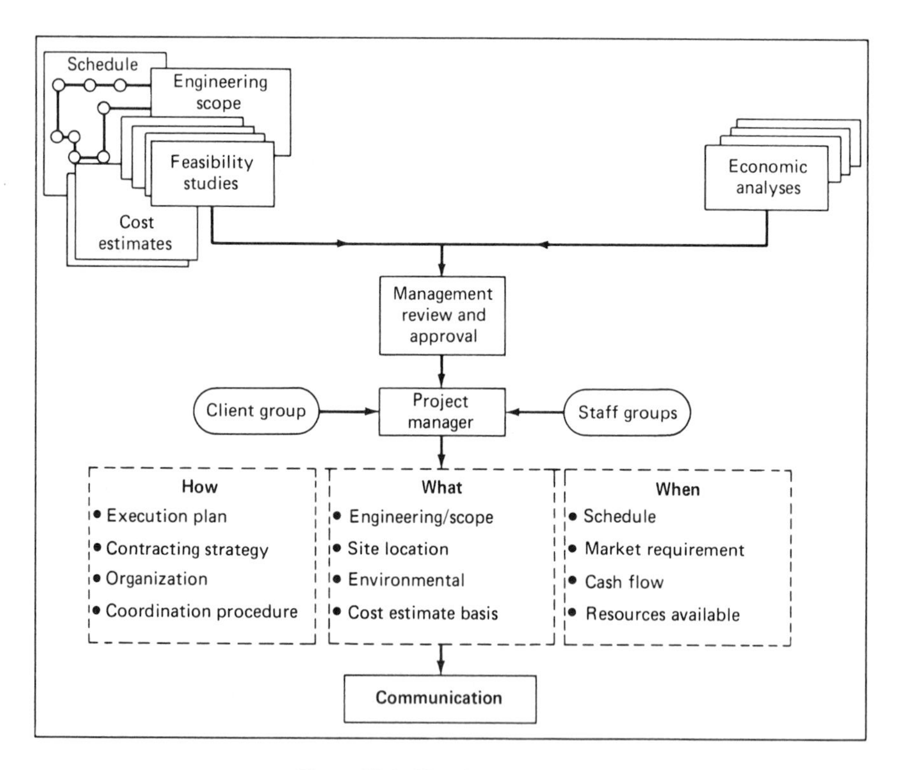


Figure 23-1. Planning the project.

tacts for each. The purpose of the document is to provide an effective basis for coordinating company activity and communications on the project, especially in the early stages of project execution when project scope and other elements are being defined. *Effective communication channels are essential for successful control.*

*Who determines the organization, control systems, and resource requirements?*

Too often, project managers will set up projects without seeking the support, advice, and assistance of staff personnel. On large projects this can be disastrous, particularly for the project control and estimating function. Resource requirements, control systems, and organizational arrangements should be matters of consultation and discussion with staff groups prior to decision by the project manager. This will also ensure that anticipated manpower requirements and resources are adequately reflected in the early conceptual estimates.

Apart from project size, the proposed execution plan and contracting strategy are the most significant elements for determining the control basis and associated organizations for the project. Figure 23-2 shows the typical phases of a project from an owner's feasibility and front-end studies to full implementation by a prime contractor. This typical life cycle is for a large process plant and shows durations of 8 months for a Phase I and 33 months for a Phase II operation. The durations for the front end vary widely.

There are many possible variations of project life cycles, this particular configuration is a typical routine of large oil corporations. Many owners use a phased approach, rather than a straight-through approach. This provides the owner with less risk on capital investment and also the ability to fully investigate the feasibility and financial viability of multiple projects at the same time.

A phased approach, particularly of large projects, also provides for more control by corporate management as the project is being developed in the feasibility, scoping, and design phase. However, it may add costs and will increase the overall project duration.

The following brief explanations cover the various phases as illustrated in Figure 23-2.

*Owner front end* is the feasibility stage when a design specification is produced by engineering, economic and market evaluations by the affiliate, and capital cost estimate and schedule by the cost group. The design specification is sometimes produced by a contractor, in greater detail than an owner-engineered design specification, but not to the detail of a Phase I operation. The control basis will be set by overall corporate objectives, mainly in the form of a development budget.

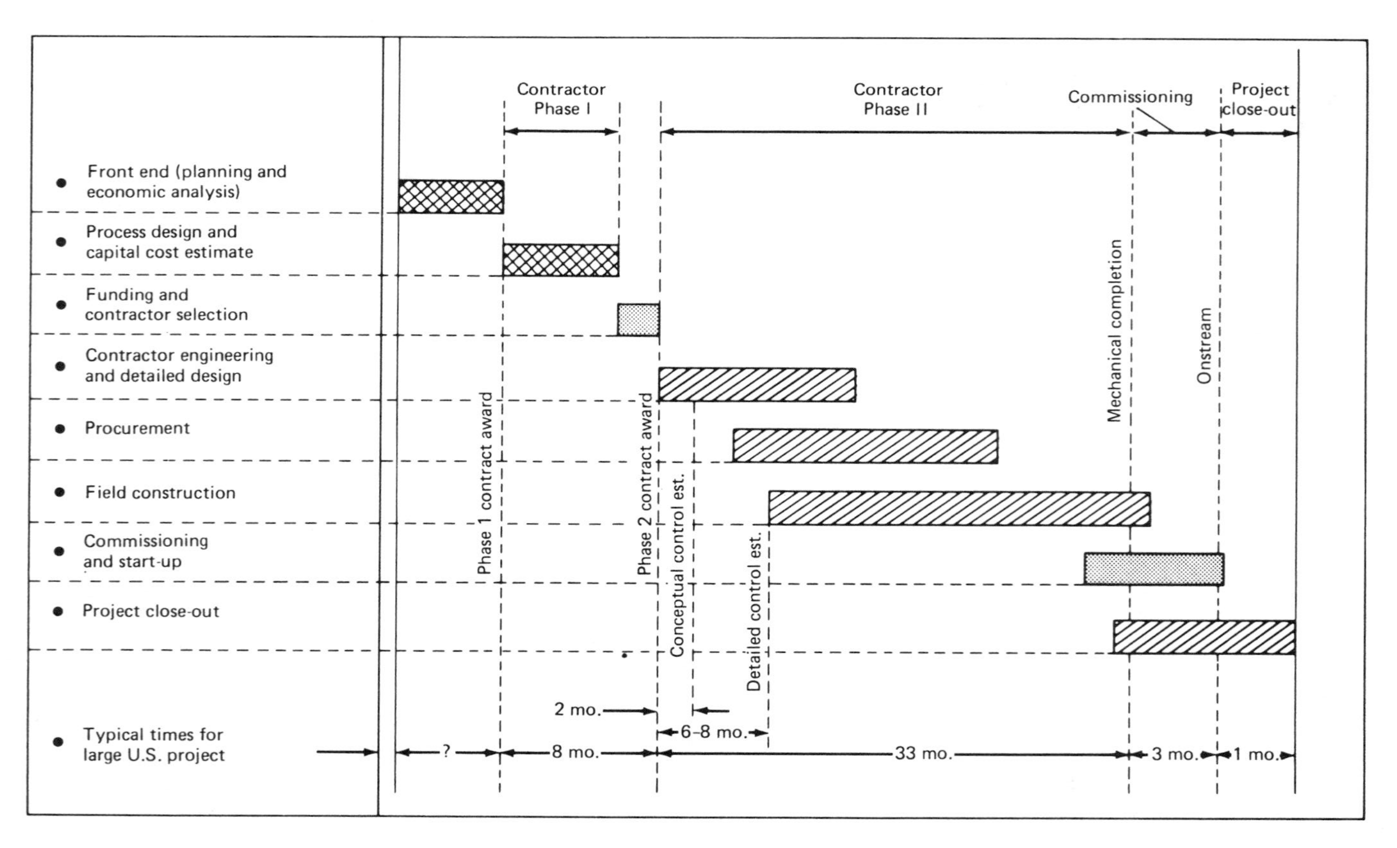


Figure 23-2. Project life cycle—typical.

*Phase I* generally covers conceptual design, process selection, optimization, upgrading of estimate/schedule, environmental/governmental studies, and finalization of the process design. The "authorization for funds" for the Phase II work is then prepared and presented, and, when approved, a contractor is selected to carry out the work. The Phase I work is carried out by a contractor, normally on a reimbursable basis with a small Owner Project Task Force (PTF) in attendance. There are two basic objectives for a Phase I operation. For large projects and revamps, it provides greater definition of scope, schedule, and cost. On small projects, it provides a design package suitable for lump sum bids. An important element of a Phase I operation is to provide an execution plan for Phase II. The control basis will be the expenditure and cost of the contractor man-hours and a milestone project master schedule.

*Phase II* is full execution of the project by a contractor. The normal project philosophy is that of a prime contractor with single responsibility for engineering, procurement, and construction. Most large projects are executed on a reimbursable basis with an Owner PTF directing/monitoring the work. This will require a complete project control system.

There are variations of a Phase II, where engineering and construction responsibilities are split and awarded to different contractors. This is the method usually adopted by utility companies where architect-engineers provide the design and construction contractors manage the field work on a subcontract basis. This approach does not provide a single responsibility and the designer and constructor can blame each other for errors of design and installation.

As outlined, the phased approach requires different control methods for each phase. A front end (feasibility study), usually carried out within the owner's organization, is authorized by an operating affiliate from its own development budget. As these budgets are developed in one-year and five-year cycles, there is rarely a need for detailed cost and schedule control at this stage. Expenditures can range from $100,000 for a small project, to $5 million for a very large project.

A contractor Phase I, on a reimbursable basis, requires a monthly monitoring of engineering man-hours and associated costs. Controls will be manual expenditure curves and progress measurement of engineering design. Expenditures can range from $1 million for small projects to $20 million for very large projects.

A contractor Phase II will require full schedule control for reimbursable and lump sum bases, but minimal cost control if on a lump sum basis.

A further variable on control requirements is the question of technology. New technology, such as synthetic fuels and offshore facilities, will generally require additional controls due to the lack of an existing data base. The past decade of the Alaska Pipeline, nuclear power plants, and North Sea platforms has clearly shown that prototype engineering, project size, hostile environments, and lack of data have produced poor cost estimates and schedules. This type of project will generally require a phased approach in order to develop data for a detailed project execution plan.

It cannot be emphasized too strongly that poor cost estimates and unrealistic project schedules can only result in an "out of control" project.

## Project Execution Plan

Figure 23-3 shows major elements of a project execution plan. This plan is developed during Phase I and covers all aspects of scope, associated services, infrastructure, approach to engineering-procurement-construction (EPC), resources, organization structure, and project control requirements.

This detailed execution plan is essential for developing a quality control estimate and project schedule. Large overseas projects with remote job sites require that the execution plan consider logistics and material han-

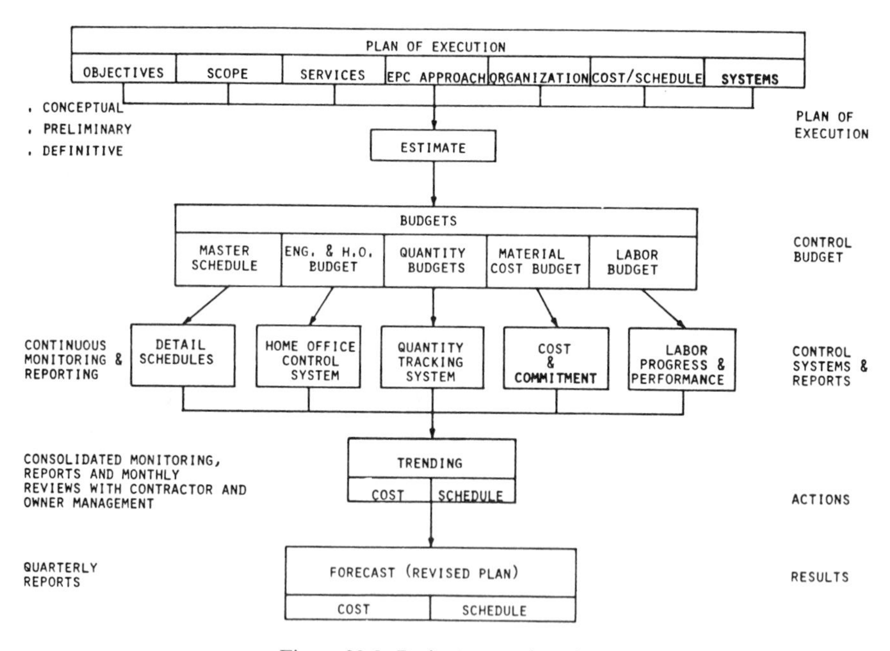


Figure 23-3. Project execution plan.

dling, local infrastructure and resources, camp facilities, training, expatriate conditions, and national and governmental requirements. A quality execution plan will provide a good estimate, control budgets, detailed schedules, and a breakdown of the project into controllable areas and cost centers. The project organization will be similarly structured, as will trending, control, and reporting systems.

*Key Items of Execution Plan* (this list is not all-inclusive):

- *Objectives*—reach agreement with owner on broad objectives.
  —National engineering and construction content.
  —Limits of authority.
  —Community responsibility/town planning.
  —Public relations. T.V., press, job site tours.
  —Contractual relationship/responsibilities.
- *Scope*
  —Process decisions/engineering specifications.
  —Capacity/feed stock and product slate.
  —Owner products for use during construction.
- *Services*—contractor and owner responsibilities.
  —Subcontracts.
  —Procurement.
  —Commissioning and start-up assistance.
  —Training. Management and craft labor.
- *Engineering-Procurement-Construction (EPC) Approach*
  —Licensors and other third parties.
  —Location of design offices.
  —Purchasing, procedures and practices.
  —Infrastructure. Local area and job site interface.
  —Project procedures.
  —Work week for engineering and construction.
  —Contractor employee conditions and procedures.
  —Preassembly/modularization.
  —Constructability analysis.
  —Labor relations and recruiting strategy.
  —Construction equipment plan/rigging studies.
  —Construction preplanning. Path of construction, field facilities.
- *Infrastructure*
  —Camp. Messing and personnel facilities.
  —Local resources. Banks, postal, religious, etc.
  —Transportation. Job site and local area.
  —Rest and recreation.
  —Security.

- *Organizations*
  - —Size and complexity. Integration and project management.
  - —Breakdown of project. Cost and management.
  - —Engineering and construction management.
  - —Third party integration.
  - —Owner organization. Relationship with contractor.
  - —Organization development (OD).
  - —Communication system.
  - —Matrix, task force, and functional considerations.
  - —Decision process. Delegation, strategic, tactical.
- *Cost and Schedule*
  - —Resource evaluation. Manpower and manufacturing.
  - —Control estimate/work breakdown structure.
  - —Project control system.
  - —Trending systems/quantity control.
  - —Schedule milestones and owner interfaces.
  - —Long-lead items.
  - —Logistics and material handling.
  - —Environmental, governmental regulations and permits.
- *Systems*
  - —Manual versus computer.
  - —Owner requirements.
  - —Level of detail and distribution.
  - —Flexibility requirements. Contraction and expansion.
  - —Frequency of reports.
- *Auditing System*
  - —Terms of reference.
  - —Evaluations and reports.
  - —Procurement and financial.
  - —Documentation.
- *Procurement*
  - —World-wide operation.
  - —National requirements.
  - —Purchasing procedures and strategy.
  - —Centralized buying/field purchasing.
  - —Owner approvals.
  - —Negotiation practices.
- *Subcontracting*
  - —Content. Work category and contract type.
  - —Organization and control requirements.
  - —Prequotation meetings.

- *Material Control*
  - —Material take off. Control and reporting.
  - —Freight consolidation.
  - —Marshalling yards.
  - —Job site controls.
  - —Weather protection and maintenance.
  - —Documentation.
- *Project Run-Down and Demobilization*
  - —What to control and at what point.
  - —Level of control and reporting.
  - —Personnel demobilization.
  - —Material surplus program.

## Contract Strategy

The current market environment plus the project cost and schedule objectives will generally determine the contracting strategy. Lump sum work is generally the most efficient method; however, a well-defined engineering package and stable market conditions are essential. There are several alternatives for the reimbursable project, and a phrased approach, though lengthy, can reduce the financial risk of a "straight-through" project.

*Lump sum* (fixed price) bids are expensive to produce and contractors are not anxious to pursue this course without a reasonable expectation of success. A poor owner definition can cause a low contractor estimate, resulting in continuous claims and extras by the contractor. It can also result in a large contingency being applied by the contractor.

Under lump sum contracts, control of time and money is the primary concern of the contractor, as his performance directly affects his profits. Here, the owner is concerned with checking contractor's compliance with project requirements, with evaluating cost extras, and with periodic analyses of the project schedule.

Under most cost-plus contracts, however, the contractor has limited incentive for controlling time and money beyond professional responsibility. In such cases, the owner is more deeply involved in the project control function than on lump sum projects. Here, owner personnel must supervise closely contractor's preparation of the definitive cost estimate and control system. This is necessary to ensure that the estimates and evaluations are prepared for facilities that are adequate for owner's needs, and to provide the owner with a better insight and understanding of the reliability and accuracy of the contractor estimates.

Target cost and schedule incentives can produce improved performance. However, the owner thereafter faces a contractor program to inflate the cost target with high estimates of engineering changes and extras.

A fixed fee, based on a percentage of the total cost, can reward poor performance. The higher the cost, the greater the fee.

Omnibus-type fees for portions of engineering and construction can result in the lack of necessary services. A fee for engineering can result in lack of optimization, poor design, overgenerous specifications, and poor equipment engineering, resulting in high-priced equipment. Material costs are reimbursable. Similarly, a fee for construction equipment can result in excessive use of labor, leading to higher labor costs and schedule extension. Labor costs are reimbursable. A fixed fee for construction management can result in lack of supervision and services, particularly if construction conditions change from those anticipated.

The above problems can be magnified with projects on a "fast-track" approach where there is a greater element of the unknown.

### The Control Estimate

Most owners develop an estimate at the front-end and feasibility stage. This conceptual estimate would generally fall in the ±30% accuracy level and would be based on cost-capacity curves or equipment and bulk ratio breakdowns.

This estimate could be updated as the design is developed, or the control could be transferred to the contractor's estimate, which is probably being developed on a different basis. Using the contractor's estimate will generally produce a greater sense of commitment and responsibility by the contractor. Whichever estimate is used to control the project costs, it is not recommended that the contractor be forced to structure his estimate to the work breakdown and account codes of the owners.

### The Project Schedule

In addition to a conceptual estimate, an overall schedule is developed by the owner at the front end of a project. This schedule is developed on a summary basis as scope and execution plans are still in a preliminary stage. As the project develops, it is recommended that daily control and detailed planning be transferred to the contractor's scheduling operation. The owner should maintain overall monitoring of the contractor's schedules and planning operation.

This early schedule provides the time basis for the estimate and presents to management an overall program showing the major decision points. At this stage, it is vital that this information be easily and clearly communicated to management.

The best format for this summary schedule is a time-scaled network. It will provide an excellent picture of time and the major phases and dependencies of the project. From a technical viewpoint, time-scaled networks are inefficient as they can require considerable rework and redrafting, but from a communication viewpoint, they are outstanding.

Figure 23-4 is a typical example of a summary schedule. This schedule, of a synthetic fuel plant, shows a phased approach, the major scope elements of a process plant and a coal mine, environmental requirements, contracting decision points, mechanical completion, and plant start-up.

With an adequate scheduling data base, the following significant information can be easily developed with this schedule:

- Escalation midpoints for material and labor.
- Progress curves for engineering and construction (Phase II).
- Manpower histograms for engineering and construction (Phase II).
- Owner manpower and project team requirements.

Activity durations are determined by judgment, past experience (data base), or a combination of both.

### A Project Control Organization

Figure 23-5 illustrates a typical organization for project control. It is recommended that the project control section be part of the project management division, whereas estimating and its associated functions can be a separate group.

The project control section would have three main project support groups and one staff support group: cost control and scheduling support groups organized on a geographic or manufacturing basis; a central group for methods development, training, manpower planning; and a specialist group to handle subcontract administration and construction management.

Rotational assignments and career development objectives should ensure the movement of personnel through the project control and project management groups. This would improve manpower utilization, provide greater training opportunities and increasing individual skill levels.

Personnel in the cost and schedule support groups should be developed to handle both cost and schedule work. Capability in both functions would be beneficial for providing home office "suitcase" services and also personnel for control manager positions.

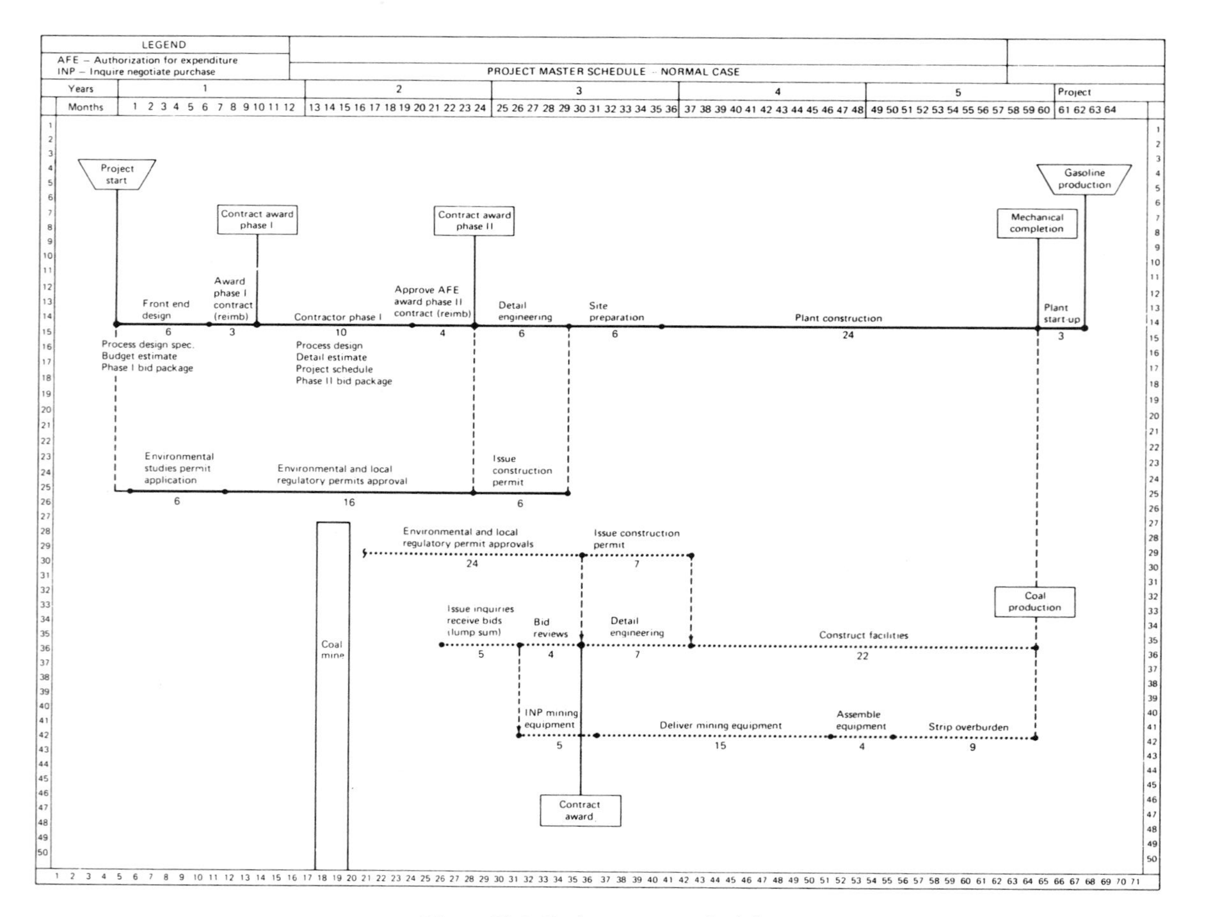


Figure 23-4. Project master schedule.

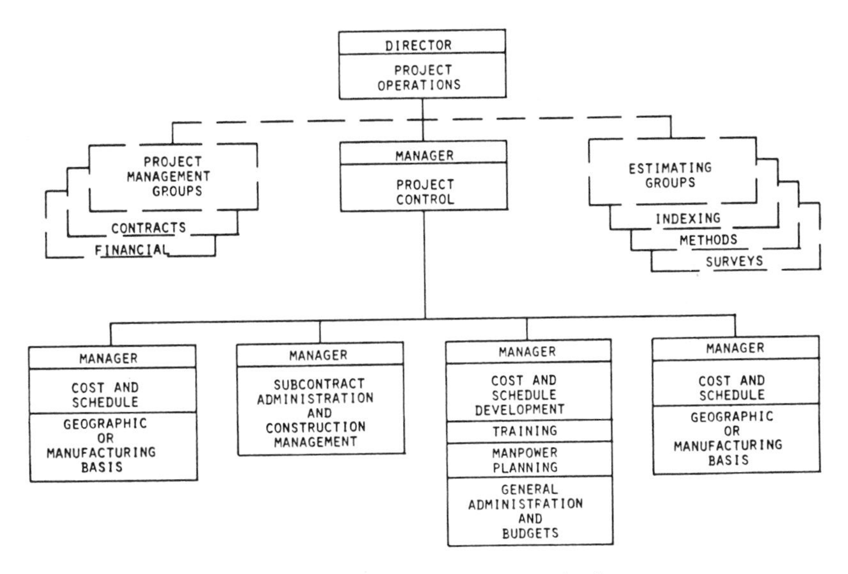


Figure 23-5. A project control organization.

Due to high work load, large projects would require separate functions of cost and scheduling.

A significant organization problem of project management and a staff project control group is the "we and they" attitude. When the project control group is part of the project management division, the "we and they" attitude is greatly reduced. In addition, the "audit image" is also reduced.

Alignment in project management divisions can sometimes stifle independent and adverse evaluations by project control personnel.

## SECTION 2. A CONTROL ENVIRONMENT

### General

Without question, *it is the project manager's responsibility to create an environment which will enable "control" to be exercised.* This means he will seek counsel, accept sound advice, and stretch control personnel to the extent of their capability.

A key element for effective control is timely evaluation of potential cost and schedule hazards and the presentation of these evaluations with recommended solutions to project management. This means that the control engineer must be a skilled technician and also be able to effectively communicate to management level. Sometimes, a skilled technician's performance is not adequate because he is a poor communicator. *Technical expertise will rarely compensate for lack of communication skills.* As in all staff functions, the ability to "sell" service can be as important as the ability to perform the service. Project teams are mostly brought together from a variety of "melting pots," and the difficulty of establishing effective and appropriate communications at all levels should not be underestimated. In this regard, the project manager is responsible for quickly establishing a positive working environment where the separate functions of design, procurement, construction, and control *are welded into a unified, cost-conscious group*. Project managers who relegate the control function to a reporting or accounting function are derelict in their duties.

Project control can be defined as the process which:

- Forecasts and evaluates potential hazards prior to occurrence so that preventive action can be taken.
- Reviews trends or actual situations to analyze their impact and, if possible, proposes action to alleviate the situation.
- Provides constant surveillance of project conditions to effectively and economically create a "no-surprise" condition.

### Task Force versus Functional Organization

The question of a functional organization versus a task force approach is a much debated subject. It is the writer's opinion that a task force approach is more efficient for large projects, whereas the functional organization can be adequate for small projects. The task force approach brings a greater concentration of resources and fewer levels of management as the reporting line to the functional departments becomes one of personnel allocation and advice, rather than direction.

Many owners now use task forces to monitor contractor performance. Some contractors are of the opinion that this approach increases schedule durations and project costs. However, in today's volatile marketplace with associated contractors' reluctance to bid on a lump sum basis, owners believe that task forces are necessary and that they make a clear, positive contribution to meeting owner objectives. In addition, owner's project control expertise, in many instances, is equal to contractor's capability.

Figure 23-6 illustrates a typical owner task force organization. This shows an owner operation with a central engineering department having responsibility for the corporation capital project program. The operating company, or client group, is responsible for funding the project and, in a

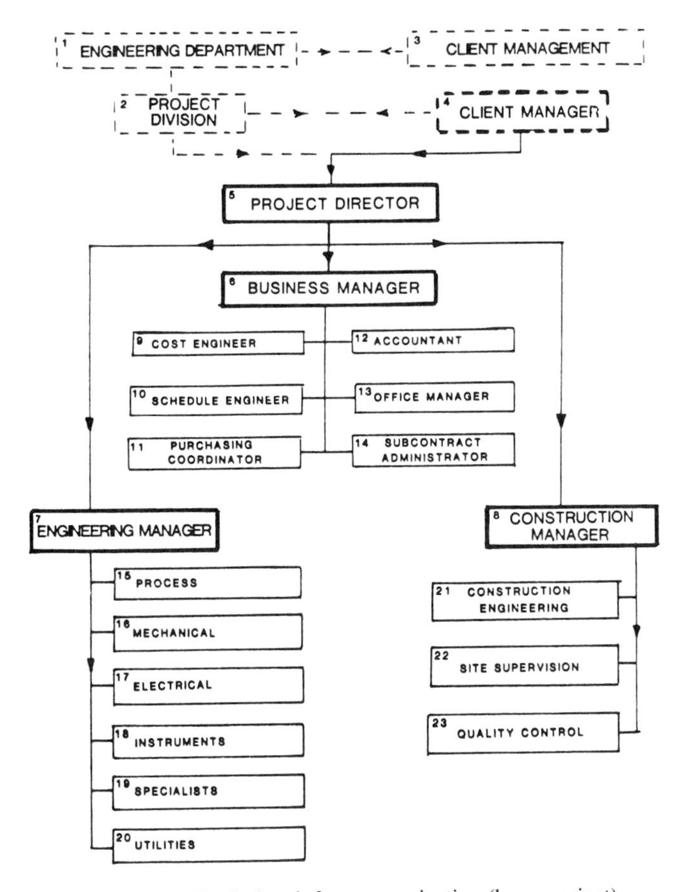


Figure 23-6. Typical task force organization (large project).

sense, hires the central engineering department to manage the project. This requires that the project director have two reporting lines: a functional line to the engineering project division and a financial line to the client manager.

The dual relationship can cause conflict. This mostly occurs when the client manager attempts to manage the project director in functional project business. The most common situation of conflict is when the client manager works directly with the contractor.

The focal point for instructions to the contractor must be through the project director and then flow from the owner task force to the contractor organization. Owner and contractor must structure their task force organizations to harmonize. The better the coordination and communication of owner and contractor personnel in this joint task force operation, the greater the prospect for successful project execution. *Systems and procedures do not build projects—people do.*

### Owner-Contractor Relationships

A significant feature of a successful project control operation is the relationship between contractor and owner personnel. One item in the initial phase of a capital project is the "screening and qualifying" of contractors prior to contract award. During this activity, owner control requirements can be clearly explained and an implementation program obtained from the contractors can be evaluated. Some owners have a formal system for evaluating contractors.

After contract award, the reality of the implementation program will be tested during detailed discussions in setting up a mutually acceptable system. These should be conducted in a spirit of equal partnership. *The owner control specification will be the basis of discussions on control organization, procedures, systems, and controls.* These early reviews can prevent later system changes, costly reorganizations, and personnel reassignments. Such discussions should be promptly followed by meetings with the contractor's engineering, procurement, construction, and project services groups to verify mutual understanding and acceptance of a common approach to planning, scheduling, and cost control. At this stage, the discussions must necessarily be brief and to the point. Everybody is busy. But they are essential to ensure that contractor's control system meets owner's requirements.

*Detailed planning, scheduling, and cost control are the contractor's responsibility, and it is his responsibility to see that they are efficient operations, effectively utilized.* This is an equal partnership operation.

Apart from estimating systems, many owners have established control data such as the following:

- Engineering man-hours per piece of equipment and man-hours per drawing.
- Construction man-hours per work category.
- Standard engineering and construction productivity profiles.
- Standard engineering and construction progress profiles.
- Overall milestone durations and dependent relationships.
- Standard procurement and subcontract relationships.
- Typical man-hour expenditure curves.
- Typical material commitment curves.
- Standard engineering discipline relationships.

- Home office and construction indirect relationships.
- Standard engineering and construction rate profiles.
- Typical breakdown of engineering by discipline and section.
- Typical breakdown of construction by craft and prime account.
- Domestic and worldwide productivity factors.
- Typical manpower buildup and rundown.
- Construction manpower density/productivity curves.
- Domestic and worldwide labor and material escalation rates.

Data, as indicated above, enables owners to check contractors' estimates and continuously monitor performance through all phases of a project. Many contractors have invested heavily in the development of PERT, CPM techniques and control systems*. In spite of this investment, and resulting sophisticated systems with their associated heavy running costs, owners continue to comment on poor execution of the contractor project control function. In turn, contractors complain that owners do not clearly identify their project objectives, change their minds on scope causing costly recycles of engineering, and are often disorganized. A major complaint by contractors is that owners monitor their activities too closely. It is essential that the owner's cost and schedule representatives refrain from continuously getting into "too much detail." This, invariably, causes an adverse relationship. Contractors should be allowed freedom of action and an occasional error.

There are two significant procedures which attempt to clearly establish the detailed working relationship of owner and contractor: the coordination procedure, outlined earlier, which covers organizational and functional relationships, and a "document action schedule" which specifies the owner involvement in all documents produced by the contractor. This covers engineering drawings, specifications, inquiry packages, bid tabulations, purchase orders, subcontracts, and all control and reporting documents.

When too tight a level of approval is imposed by the owner, it can result in additional costs and lengthening of the schedule.

A major complaint by owners is in contractor scheduling. Rarely does the owner encounter a contractor's performance where the planning, scheduling, and control of engineering, procurement, and construction phases are effectively bound into one system. Too often, rigid departmentalization of contractors has forced owners' representatives to act as catalysts and coordinators to achieve efficient execution.

---

*See Chapter 15 for a detailed discussion of PERT, CPM and other network planning techniques.

Overdepartmentalization is evident when separate groups of a contractor's organization operate to an appreciable degree to the exclusion of the interests of associated groups and departments. In particular, owners experience too many instances where engineering, design, procurement, project, and construction departments act as separate companies. Corporate politics sometimes are allowed to override project objectives and the true long-range objectives of the engineer-contractor. Unless engineering, procurement, and construction groups operate as a team, with differing functions but common objectives, project execution will be inefficient and costly.

All contractors emphasize in sales presentations the unified application of their resources to the owner's project. Departmental flexibility and coordination are stated as being strengths of the company organization. In practice, the owner too often finds that planning, scheduling, and control are exercised only within compartmented contractor departments. While it is highly desirable that individual departments and departmental sections participate in the setting of schedules, and in controlling to these schedules, overall progress scheduling and control are the owner's prime concern. For this reason, final schedule authority must rest in a strong, active project management, supported by adequate staff schedule personnel.

Alternatively, owners sometimes find scheduling operations consolidated in autonomous groups, the output of which is voluminous, but unused. If the engineer-contractor is to meet the owner's objectives, and in the long run, his own objectives, the output of planning and scheduling groups must be both usable and used by the project team.

### An Integrated System

Like any control function, effective project control requires that all efforts be fully integrated; that status be fully and accurately reported; that costs, programs, and engineering scope be compared against budget estimates, schedules, and specifications (the norms); and that the loop be closed either by modifying and correcting the control system, or by changing the control methods. This cycle of events is necessary, and should be continual for successful project execution. The owner's interest and participation in these events will vary from project to project and depend primarily on the type of project contract. In short, for effective project control, a project team (not an individual) must concentrate on anticipating and detecting deviations from project norms, and then take full and timely action to handle such deviations. Project norms should only be revised when it is absolutely certain that they are beyond achievement; *however, prompt reports should indicate deviations as they become apparent,* even though no immediate action is taken.

Figures 23-7 and 23-8 illustrate major elements of integrated scheduling and cost control systems.

Figure 23-7 is a flowchart indicating the elements necessary for an integrated schedule system. It is the writer's opinion that owners and contractors need to achieve fully integrated and coordinated control sys-

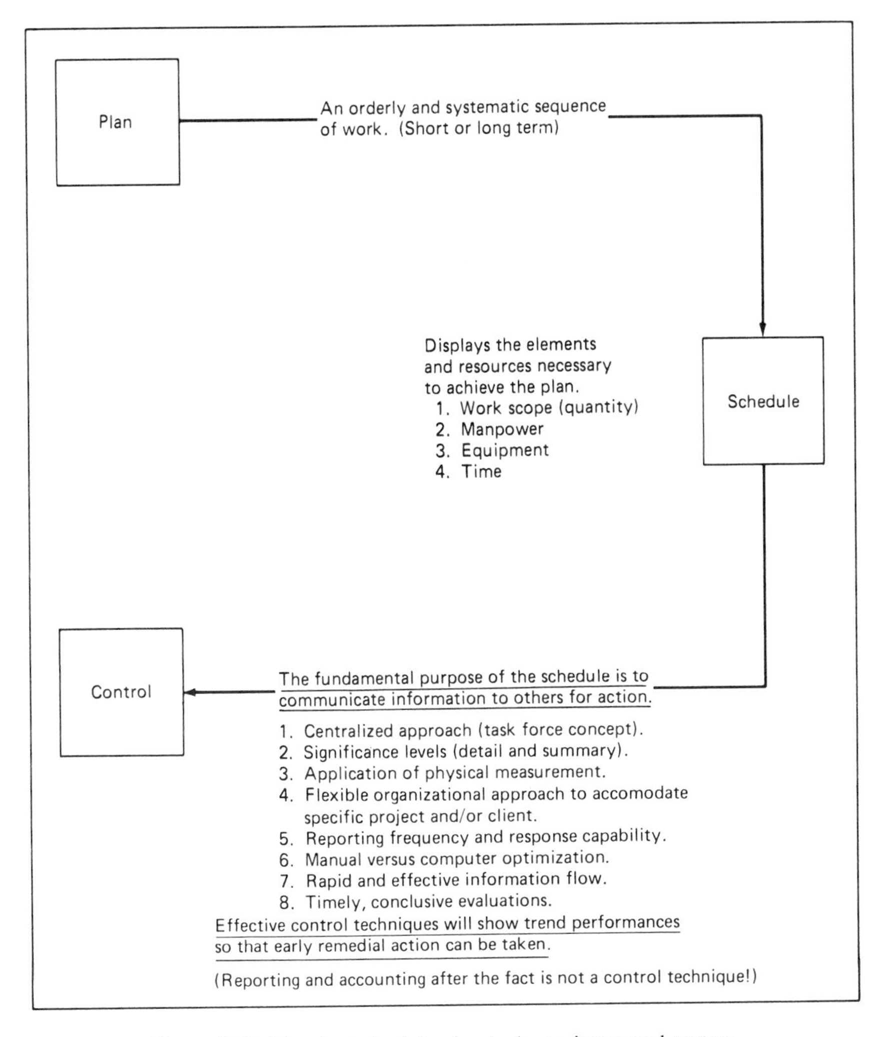


Figure 23-7. Planning/scheduling/control—an integrated system.

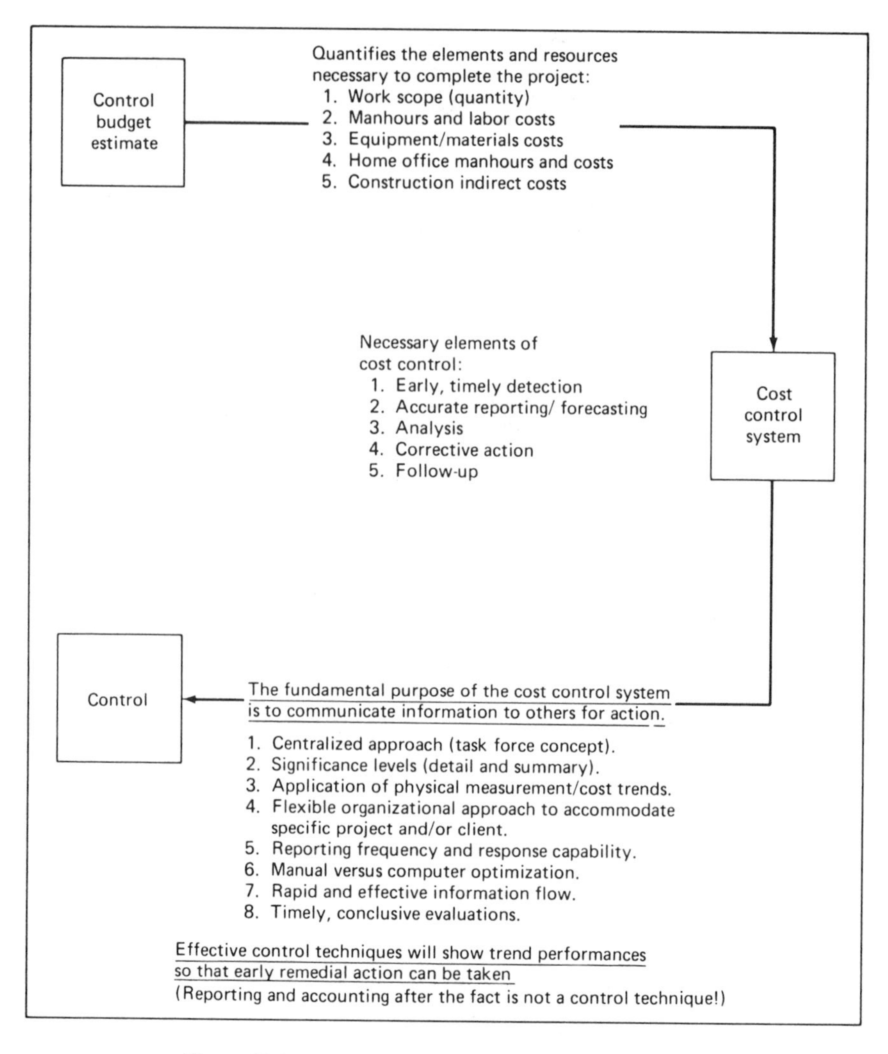


Figure 23-8. Project cost control—an integrated system.

tems along these general lines. To do so will require, in many instances, a thorough rethinking of schedule-related operations, and upgrading of personnel. In some instances, judicious "headknocking" is going to be required to call attention to outmoded practices and attitudes, and failures to conform to stated management policies.

Figure 23-8 is a flowchart indicating the elements necessary for an integrated cost control system. The major items are a quality estimate, based on quantities, an effective trending system, and qualified personnel working on a task force basis.

### Communication—Manual or Computer

Figure 23-9 is a flowchart of a typical management information system, or, in other words, the operating levels of the project control system.

Again, the key word is communication.

The project control system must generate summary and detailed information for different levels of management. Information must be current, timely, and accurate. This flowchart shows four levels of detail, which are typical for most large projects.

Information is generally a combination of computer programs and manual reports. It is difficult to conclude that computer programs are better than manual systems. There are obvious advantages with the computer, but many systems prove ineffective due to the tremendous level of detail.

Scheduling systems with tens of thousands of activities are rarely effec-

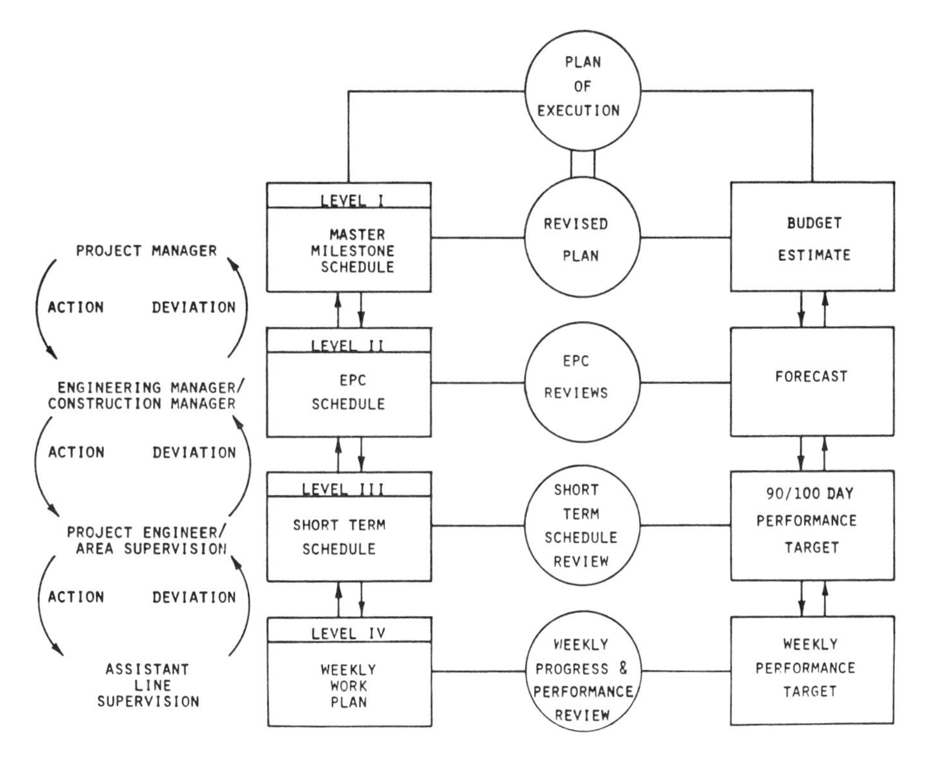


Figure 23-9. Flow of management information system.

tive. Alternatively, it is very time-consuming to produce a detailed field progress report without a computer program.

Each project and each contractor operation should be thoroughly investigated for application of a computer approach to project control. (See Chapter 28 for a discussion of the use of computer software in project management.)

As most owners work in a monitoring role, it is unlikely that owners would need their own extensive computer programs for control purposes.

### Owner Review of Contractor Control System

On large projects, early after contract award, a team should be established to review, in detail, the contractor's cost and schedule system, organization, and assigned personnel. The purpose is to recommend to the project manager a complete project control system for the project. The team should be led by a senior member of the home office control group and consist of task force and staff cost and schedule personnel. As this review will take four to six weeks, the addition of home office personnel is generally necessary as the work load of the project-assigned personnel is very heavy at this time. The team leader must be very experienced in order to understand and handle the complete range of a contractor's project control operation. Hence, a supervisor from the home office is generally required.

Personnel should be nominated by the manager of project control, and a timed execution plan presented to the project manager for approval prior to commencement of the work.

Specific objectives of this review are listed below:

- To investigate the project control systems and organization of prime contractor, joint venture or management contractors and prepare a recommended total project control system. *The investigation should be based on maximizing the use of existing contractor systems and resources. Changes should be minimal and only significant deficiencies should require modification.*

  Should a contractor system or organization have significant deficiencies, it is recommended that the contractor modify his system by supplementing it with the appropriate owner procedure and formats. However, it is important that changes be kept to a minimum and the contractor be permitted full use of procedures and methods with which he is familiar.
- To prepare a detailed report covering the investigation and recommendation.
- To prepare a schedule for the implementation of the above recom-

mendations, extending to the point where the control system is fully operational.

- To present the plan and recommendation to the owner and contractor executive management to obtain full understanding and endorsement by management at an early stage in the project.

### Implementation Schedule—Project Control System

It is essential to establish a quality Project Control System at the earliest possible date. As an aid to meeting this objective, it is recommended that a detailed "Implementation Schedule" be prepared showing the completion dates agreed to with the contractor. This schedule should be developed in summary and detailed form and will outline all facets of the proposed control system, showing deadlines for completion and personnel allocations for the work.

Contractor should list and provide "duration estimates" for all procedures, such as Schedules, Reports, Estimates, Computer Programs and Organization Charts, etc., which constitute the overall Project Control System. It is suggested that a flowchart(s) showing the major elements of the system be prepared by the contractor.

The contractor should provide schedules and details of resource for completion of the Project Control System.

This owner review and preparation of associated implementation schedule can be a frustrating time for contractors. It can be doubly so if owner personnel lack experience and the contractor has to spend considerable time in education as well as explanation. The process is time-consuming and may require the time of the key contractor control personnel who are already heavily engaged in the project.

However, this is the time for contractors to fully explore owner control requirements, provide effective and detailed explanations of their systems, accept obvious improvements, and defend "poor" programs which they believe are effective and which they have proved out on projects.

Owners should have a "minimal change" policy and contractors should encourage owner personnel to live up to this policy.

Figure 23-10 illustrates a segment of a typical implementation schedule.

This schedule should cover major categories of procedures, schedules, cost, computer, measurement, and reporting. It should be updated weekly or biweekly for progress and status.

### Jumbo Projects

A significant aspect of project work in the 1970s was the increasing size and complexity of projects.

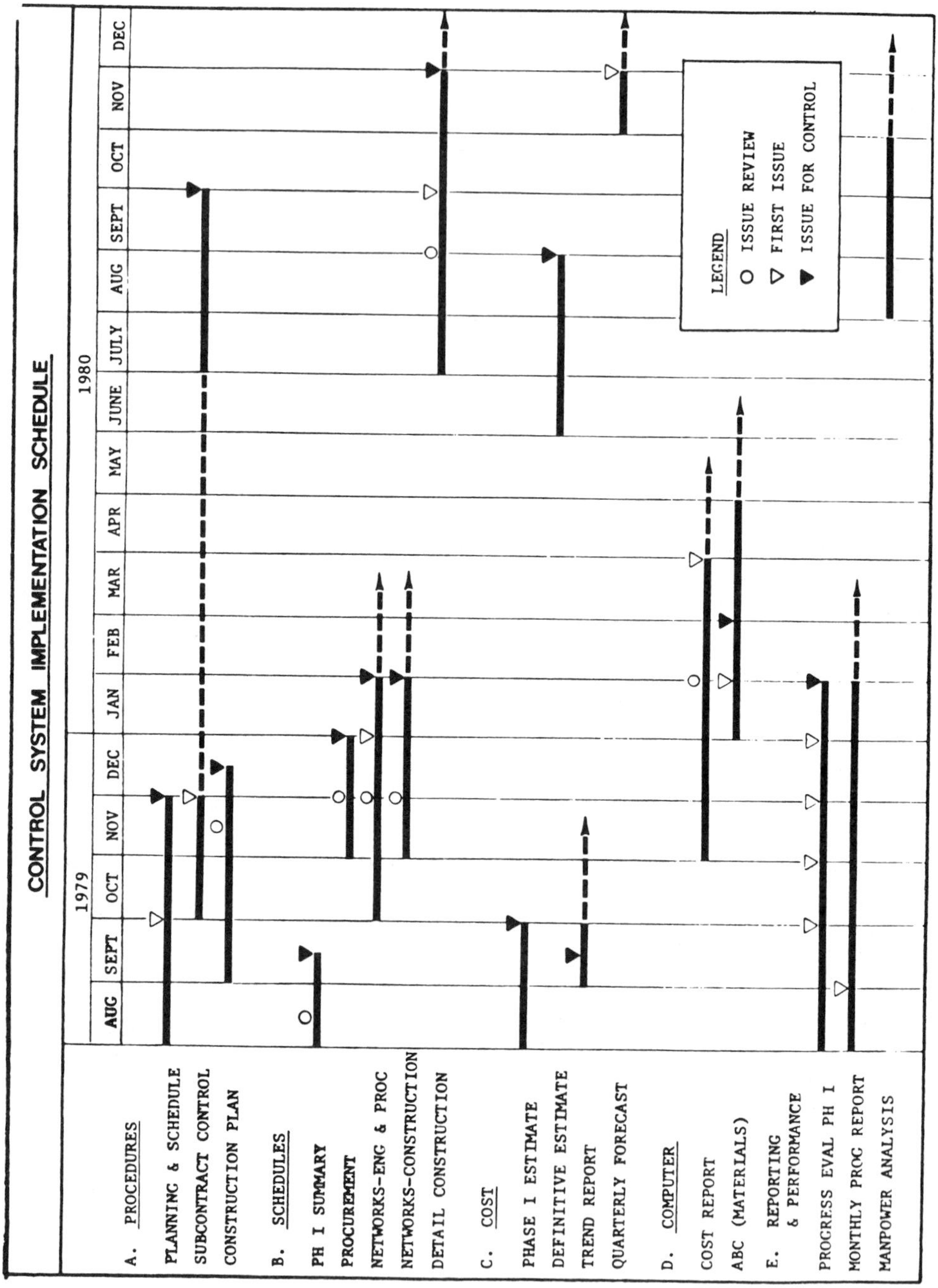

## CONTROL SYSTEM IMPLEMENTATION SCHEDULE

1979: AUG SEPT OCT NOV DEC
1980: JAN FEB MAR APR MAY JUNE JULY AUG SEPT OCT NOV DEC

A. PROCEDURES

PLANNING & SCHEDULE
SUBCONTRACT CONTROL
CONSTRUCTION PLAN

B. SCHEDULES

PH I SUMMARY
PROCUREMENT
NETWORKS-ENG & PROC
NETWORKS-CONSTRUCTION
DETAIL CONSTRUCTION

C. COST

PHASE I ESTIMATE
DEFINITIVE ESTIMATE
TREND REPORT
QUARTERLY FORECAST

D. COMPUTER

COST REPORT
ABC (MATERIALS)

E. REPORTING & PERFORMANCE

PROGRESS EVAL PH I
MONTHLY PROG REPORT
MANPOWER ANALYSIS

LEGEND
○ ISSUE REVIEW
▽ FIRST ISSUE
▼ ISSUE FOR CONTROL

Major examples are the Trans-Alaska pipeline, offshore platforms in the North Sea, gas-gathering facilities in the Middle East, the Sasol synthetic fuel plant in South Africa, and the Syncrude Tar Sands plant in Alberta. These are termed jumbo or mega projects.

New and changing technology, a hostile environment (Alaska and the North Sea), construction on a massive scale, plus the minimum of experience and data, provided the background to estimating, planning, and scheduling of these facilities.

The oil industry was breaking new technological barriers in terms of size and complexity of production facilities. The resultant first-generation jumbo projects experienced a considerable degree of last-minute innovation, and were built without full scope definition and little appreciation of offshore construction. Because of the urgent need to bring these facilities on stream, companies were tackling many of the problems during the construction and installation stages. Therefore, cost and schedule overruns were common occurrences.

It was not until about 1976 that realistic criteria and appropriate techniques had been developed to control these very large and complex projects. It was discovered that current concepts and practices of functional and task force organizations were not very effective. In particular, a task force with centralized decision making was not adequate. Management layers stretched out communication channels and decision making.

The following size parameters give a general breakdown of projects into small, medium, large, and jumbo:

| | Small | Medium | Large | Jumbo |
|---|---|---|---|---|
| (a) Engineering Manhours | 100,000 | 600,000 | 1,500,000 | 6,000,000 |
| (b) Engineering manpower | 100 | 200 | 400 | 1,000 |
| (c) Construction manhours | 500,000 | 400,000 | 8,000,000 | 50,000,000 |
| (d) Construction | 400 | 1,500 | 3,000 | 10,000 |
| (e) Construction staff | 50 | 150 | 500 | 1,000 |
| (f) Schedule (months) (Detailed engineering to completion of construction) | 25 | 30 | 35 | 50/60 |

Comparing the jumbo projects of the 1970s with conventional plants, the major lessons learned were:

- The desirability of a decentralized approach to place decision making as close to the work as possible.
- The need to combine owner and contractor project teams into one operating unit.

- The need to reduce management layers so that decentralized project teams could communicate quickly with overall project management.
- The increased effect that basic organization changes can have on a very large project.
- The importance of leadership, as opposed to managerial skills, in an effective project management organization.
- The increased importance of a quality execution plan prior to the start of detailed engineering, procurement, and construction (Phase II). The execution plan is to provide a base for the estimate, as well as a plan for executing the project.
- The significance of greatly increased influence of governmental agencies and joint venture partners.
- The inadequacy of existing data base and assumptions of size effect. It is possible that the traditional ''scale effect,'' where increased size and units reduce unit costs, does not apply on jumbo projects. Pioneer projects are likely to experience unit-cost increases as their technology advances. Extreme caution must be exercised in scaling up capacity-cost ratios of conventional plants for jumbo projects requiring new technology and prototype engineering.

The following comments further amplify a new approach to a jumbo project.

**Decentralization.** During Phase II, the project should be divided into major cost centers, to an approximate value of ±$200MM, each with its own budget, schedule, and complete project organization. Jumbo projects would then have 15–20 of such individual cost centers.

Decision making should be by the individual project organization, constrained only by its budget and schedule and objectives set by the central project management group. The central project group would be responsible for coordination of resources and common services, overall cost and schedule objectives, and interfaces with client, corporate, and government groups.

Cost, schedule, procurement, and engineering specialists of the individual project group would report directly to their project manager and functionally to the specialist manager of the central group. They would receive their day-to-day direction from their project manager and technical guidance from the functional manager of the central group.

Phase I (conceptual process design) and the commissioning and start-up phases should be organized on a central project group basis. As the major decisions of a Phase I operation are comparatively few, mainly process design and selection, execution plan, and contracting strategy, the decision-making process should be in the hands of a few people. Similarly, construction at the 95–98% point will move into the commissioning and start-up phase. This requires the reuniting of the individual projects for a common approach to start-up and operations.

**Owner-Contractor Partnership.** An adversary or stand-alone relationship between owner and contractor will add costs and extend the schedule on jumbo projects. The amounts of money are large. Decision making requires greater evaluation and analysis. Fast decision making requires that owner and contractor work as a team during the evaluation process to prevent loss of time with major reviews and presentations.

As most jumbo projects are built on a "fast-track" basis, fast decision making is essential if the schedule is to be achieved.

Continuous agreement at working levels between owner and contractor will generally require owner personnel additional to the traditional levels of the past.

Even though there will be a united team approach, it is vital that the contractor be allowed to freely operate at the daily working level.

A new concept is the completely integrated owner-contractor project team, where owner personnel may have supervisory and subordinate roles. The major problems of this approach are questions of contractor responsibility, professional pride, personnel relationships, and proprietary information.

The concept has much to offer and is one that deserves considerably more study, analysis, and development.

**Organization Changes.** The need for organizational and procedural changes can be recognized and the problem reduced with an organization development group.

**The O.D. Group.** This group would be established to unblock decision-making bottlenecks and improve inadequate procedures. Its objectives would be to constantly monitor and evaluate organization, communications, procedures, and methods. This function requires specialized personnel with experience to cover all phases and functions of the project.

Due to the wide range of experience required, it is probable that two groups will be required: one group for the home office covering engineering and procurement, the second group for the field covering construction. About four to six personnel, at peak, would be required for an effective O.D. group.

**Leadership Versus Managerial Skills.** People skills are essential in the management and control of jumbo projects. With task forces ranging in size from 500 to 1000, the importance of people skills cannot be overemphasized.

It is possible that leadership skills are more important than managerial skills. Personnel motivation is an essential ingredient of a successful project team.

**Control Estimate.** As a quality estimate is vital to the project control effort, an owner-contractor team should be established to develop the estimate. This will provide continuous working agreement on such significant elements as escalation, productivity levels, unit rates, work breakdown structure, control areas, and individual cost centers.

A detailed estimate could be produced about 12–16 months after Phase II contract award and would probably require 40 contractor and 10 owner personnel. With this approach, management review and approval could take one week instead of the months of review and reconciliation, which is the more normal case.

**Planning and Scheduling.** The size of the activity network is not the major consideration. The quality of the weekly construction program is the main concern. Construction man-hours will be in the range of 40MM–100MM. With peaks of 10,000–15,000 men, a quality weekly work program is absolutely essential.

It is likely to be a manual system and should be based on quantities, unit man-hour rates, and varying productivity adjustments and be reconciled against the objectives of the overall schedule. Productivity goals should be preplanned and then reported against on a weekly basis.

**Quantity Control.** This technique is rarely used. On jumbo projects, where the amounts of money are so large, a quantity tracking system is essential for effective cost control.

Appropriate "bulk quantities" (earth, concrete, piping, etc.) should be selected and tracked, by a random sampling technique, from the process design of Phase I through detailed engineering of Phase II.

**Rundown Control.** This method is rarely used in present-day project work. Again, due to size, this is an essential technique for jumbo projects.

As engineering and construction commence their rundown (about 80% complete), individual budgets, schedules, and manpower histograms should be developed to separately control the remaining work.

**Governmental Agencies and Joint Venture Partnerships.** Many of today's jumbo projects have governments as partners. Governmental regulations and agencies, partner and joint venture relationships add a further dimension that must be recognized by the planning, scheduling, and cost effort.

Governmental energy companies may require "preferred purchasing" (buying in the host country), extensive training programs for supervisory staff and craft labor, and the development of an infrastructure local to the project job site.

Joint venture partners require a vote in major decisions. This takes time. Major purchases can require approval of partners prior to purchase. Again, this takes time. Periodic reviews and presentations can be required by partners. This takes effort and costs money.

All of the above aspects should be carefully considered when developing the project execution plan and schedule.

Figure 23–11 vividly illustrates the effects of problems outlined in the opening paragraph of this subsection. This study (R-2481-DOE) by the Rand Corporation for the Department of Energy shows final costs versus initial feasibility estimates for many jumbo projects. As can be seen, the cost growth is 200–300%. This is caused by either bad estimates, poor performance, or a combination of both. Major changes causing significant

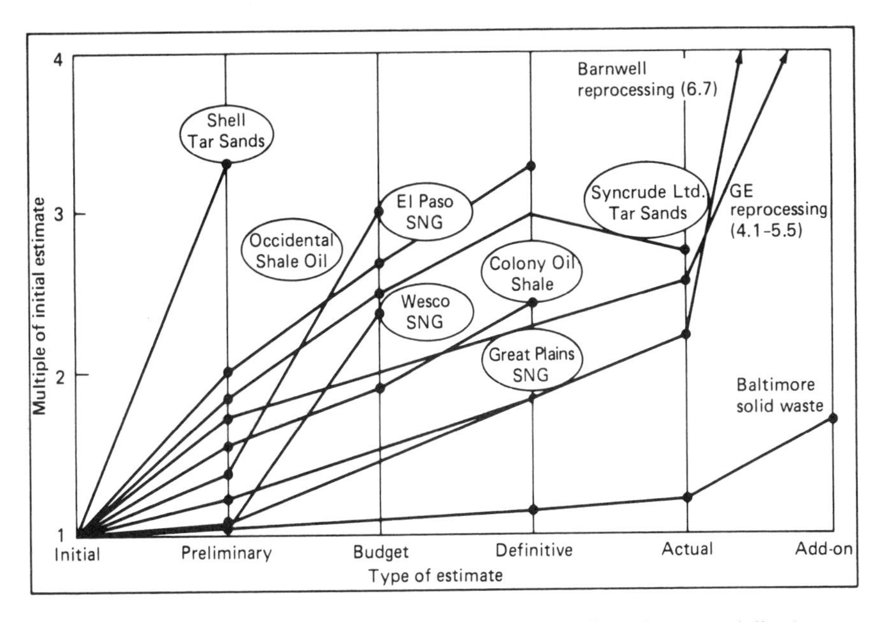


Figure 23-11. Cost growth in pioneer energy process plants (constant dollars).

cost additions can be classed as poor performance. Concern is being expressed that current estimates for jumbo synthetic fuel projects will follow the same patterns as the chart in Figure 23-11.

### Project Levels of Control and Reporting

Levels of control and reporting vary widely in the industry. They can be dependent on:

- Recognition, understanding, and need for control.
- Company commitment to control.
- Personnel resources and capability.
- Size and complexity of projects.
- Owner/contractor contractual arrangements.
- Owner/contractor control relationship/expertise.
- Acceptance of cost control.
- Cost effectiveness of control.

Most major contractors have comprehensive project control systems. However, very few owners have a similar capability, or even detailed control specifications which would enable contractors to thoroughly understand the owner's project control requirements.

As already outlined, an early, effective project control program is an essential requirement. It is difficult to achieve this objective on reimbursable projects if owners are not able to specify, in detail, their requirements. Even on lump sum projects, a similar approach is necessary, as effective planning, scheduling, and progress measurement should be an owner requirement. Apart from an adequate change order procedure, cost control reporting is the contractor's sole responsibility on lump sum projects.

Figure 23–12 lists typical project control requirements for the following project categories:

- Feasibility Study (0–10,000 engineering hours).
- Small Project (10,000–100,000 engineering hours).
- Medium Project (100,000–500,000 engineering hours).
- Large Project (500,000–1,500,000 engineering hours).

In an attempt to quantify project size, engineering man-hours have been allocated to these categories. *This can only be a guide as project size is dependent on the size of the company.*

It is generally recognized that as project size increases, additional con-

| JAMES BENT ASSOCIATES, INC.<br>CONSULTANTS IN PROJECT MANAGEMENT AND PROJECT CONTROL | | | | | DATE<br>DECEMBER 1980 |
|---|---|---|---|---|---|
| PROJECT LEVELS OF CONTROL & REPORTING – REIMBURSABLE PROJECTS<br>For the designated project size, the outlined techniques, reports and procedures are additive to the previous level lump sum projects would require most of the scheduling procedures. See project control manual for details of method and procedure. | | | | | |
| **PROJECT** | **OVERALL** | **ENGINEERING** | **PROCUREMENT** | **CONSTRUCTION** | **SUBCONTRACTS** |
| 1. FEASIBILITY (10,000 eng. hours) | Summary schedule<br>Execution plan<br>Estimate<br>Cost report<br>Monthly report | Project master schedule<br>Manhour curve<br>Manhour rate curve<br>Manpower histogram | Delivery lead times<br>Logistics evaluation | Site survey<br>Soil report | Licensor packages |
| 2. SMALL PROJECT (100,000 eng. hours) | Trend report<br>Project status report<br>● Engineering<br>● Material commitment<br>● Construction<br>Contingency rundown curve<br>Cash flow curve | Discipline schedule (milestones)<br>Engineering manhour curve<br>Home office manhour curve<br>Bid evaluation program | P.O. commitment register<br>Material status report<br>Overall commitment curve ($)<br>Vendor dwg. report | Pre-planning program<br>Manforce report<br>Three month schedules<br>Construction progress barchart (overall)<br>Manpower histogram<br>Overall manhour curve<br>Overall rate curve | Overall schedules (by subcontract)<br>Progress/status report (by subcontract)<br>Summary cost report |
| 3. MEDIUM PROJECT (500,000 eng. hours) | Contractors evaluation program<br>Detailed control specs.<br>Task force approach<br>Computer scheduling program<br>Coordination procedure<br>Extra work/change order procedure | Engineering change log<br>Material requisition curve<br>Document and action schedule<br>Progress measurement program (discipline)<br>● Quantities/hours<br>● Progress curves<br>● Manpower curves<br>● Productivity curves<br>Front end schedules (3 mo.)<br>Dwg. schedules<br>H.O. Expense expenditure curve ($) | Equipment commitment curve ($)<br>Bulk material commitment curve ($)<br>Material requisition curves<br>Inspection – expediting reports | Field estimate (quantities)<br>Construction area progress barcharts<br>Weekly work program<br>Progress measurement program<br>● Quantities/hours<br>● Progress curves<br>● Manpower curves<br>● Productivity curves<br>Status report<br>● Progress<br>● Productivity<br>● Manpower<br>Indirects expenditure curve ($)<br>Staff schedule<br>Equipment schedule<br>Backcharge register<br>Cost report | Subcontract commitment curve ($)<br>Subcontract preparation schedule<br>Unit price subcontracts<br>● Cost report<br>● Quantity report<br>● Performance evaluation<br>● Progress curve<br>● Manpower histogram |
| 4. LARGE PROJECT (1,500,000 eng. hours) | Contractors screening program<br>Project control implementation schedule<br>Weekly manforce report | Account requisition curves<br>Piping design program<br>Drawings tracking curves (P&r's) (foundations) (isometrics)<br>Quantity tracking program<br>Rundown control program<br>● Drawings<br>● Manhours<br>● Dates<br>● Manpower<br>● Progress<br>● Productivity<br>Punch lists | Account commitment curves ($)<br>Account requisition report<br>Critical purchasing list<br>Material delivery histogram<br>Surplus material report | Work lint tracking curves ● earthwork<br>● concrete<br>● piping<br>Area status reports<br>Staff manhour & rate curve<br>Equipment manhour rate profile<br>Field office expense expenditure curve ($)<br>Indirect manhour and rate curve<br>Rundown control program<br>● Manhours<br>● Manpower<br>● Progress<br>● Productivity<br>Punch lists | Independent bid analysis program<br>Sensitivity analysis<br>Performance curve and report |

Figure 23-12. Project control requirements.

trol procedures are necessary. Obviously, control for control's sake should be avoided. Typical examples of costly and inefficient control systems are very large activity network programs, duplication of effort by owner and contractor on reimbursable projects, and some governmental reporting procedures.

The outlined control procedures are divided into the major phases of a project. They are additive as the project category increases in size. A small project would require the items listed in categories 1 and 2. Similarly, a large project would require all the listed items.

### Manpower Planning—Engineering Department

One of the more difficult areas of project control is in-house company planning of engineering personnel. The major uncertainty, causing difficulty, is forecasting the amount and type of future project work. The difficulty is usually greater for owners.

The contracting industry has two major considerations: an annual estimate of the owner's capital projects program and an assessment of their ability to obtain a share of that work.

The owner's engineering department can face the following:

- Amount of feasibility studies.
- Technical service requirements.
- Methods development and technical research.
- Actuality of probable or anticipated projects.

Many owner central engineering departments act as a nonprofit service company to operating divisions of the corporation. As such, their work load is largely dependent on the capital projects program of the operating divisions. It is not too difficult to assess technical service requirements, methods development, and technical research based on past experience. But assessments of feasibility studies and capital projects depend on factors often outside the control of the engineering department:

- Quality of corporate strategic planning program.
- Corporate financing.
- Project economic viability.
- Communication channels with operating divisions.
- Relationships with operating divisions.
- "Project charter" of engineering department.
- Image/credibility/capability of engineering department.

Even though there can be many uncertainties in work load, one thing is certain: Quality evaluations of work load and associated manpower planning are essential—particularly with the typical shortfall of engineers and the industry prediction that the shortfall will increase for the long term.

The following exhibits outline a systematic approach to engineering manpower planning.

**Planning by Individual.** Individual planning is the lowest level of detail. Not only does it provide an assessment of manpower needs to meet a projected work load, it also provides a program of career development for each engineer.

Figure 23–13 illustrates a three-year plan for project services personnel (estimating, cost control, scheduling). This shows feasibility work, project assignments (home office and task force), methods development, rotational assignments, transfers, replacements, and recruiting requirements. This should be a "dynamic document" as conditions/requirements can quickly change. The control sheet should be constantly updated and issued monthly.

It is recommended that all section personnel "plans" be evaluated and summarized by the project services group into a monthly engineering department manpower report. It is probable that this would be a computer-based program so as to provide overall manpower reports by individual listing, project assignments, feasibility work, sections, etc.

As manpower plans are only as good as assessments of work load, it is vital that work projections be evaluated each month. This requires close liaison/coordination between project, engineering, construction, and project services groups to ensure that current and future work assessments are adequate.

**Planning by Project.** The following report format is mostly used by contractors as it concentrates on project manpower allocations.

Figure 23-14, usually a computer report, assesses manpower allocations and requirements based on budget man-hours, forecast, man-hours to date, schedule, and man-hour allocations for the past six weeks. Only three months of the schedule are shown and continuation sheets would provide requirements for the complete schedule. The past six weeks show current trends and also a base to assess the viability of future requirements. The computer program will take the man-hour forecast, to-date man-hours, schedule, and hourly workweek assessment and forecast the weekly scheduled manpower requirement.

The bottom two lines show men required against men available, and the difference provides the necessary recruiting program.

| COLOR CODE: | HOME OFFICE | PROJECT ASSIGNMENT | TRANSFER OUT | |
|---|---|---|---|---|
| Activity PROJECT SERVICES | Staff Assignments and Personnel Planning | | | PAGE: DATE: |
| **NAME** | **SCHEDULE** | | | **REMARKS** |
| ESTIMATING | | | | |
| 1. EVANS | GEN. / ABC PHASE I / XYZ FEASIBILITY | TASK FORCE (XYZ) COST CONTROL | GENERAL EST'G | |
| 2. DAVIES | GENERAL ESTIMATING | • BUDGETS • FEASIBILITY | | |
| 3. JONES | A.F.E. ESTIMATING | | | |
| 4. WILLIAMS | METHODS DEVELOPMENT / PRODUCTIVITY FACTORS | LABOR & MATERIAL ESCALATION | TRANSFER TO SCHEDULING | |
| COST CONTROL | | | | |
| 5. PRICE | TASK FORCE | TRANSFER TO PROJECTS | | |
| 6. BENNETT | "SUITCASE" PROJECTS 103/201/430 | XYZ PROJECT TASK FORCE | | |
| 7. GRAHAM | "SUITCASE" PROJECTS 120/150/250/310 | | | |
| SCHEDULING | | | | |
| 8. ROBERTS | FEASIBILITY SCHEDULES & "SUITCASE" PROJECTS | | TRANSFER TO DESIGN | |
| 9. JENKINS | FEASIBILITY SCHEDULES & "SUITCASE" PROJECTS | XYZ PROJECT TASK FORCE | | |
| 10. LONGDEN | METHODS DEVELOPMENT / FEASIBILITY SCHEDULES | | | |
| RECRUITING | | | | |
| 11. COST ENGINEER | RECRUIT / TRAINING | GENERAL COST CONTROL (SUITCASE) | TASK FORCE | TO REPLACE PRICE |
| 12. ESTIMATOR | | RECRUIT / TRAINING | GENERAL ESTIMATING | TO REPLACE WILLIAMS |
| 13. SCHEDULER | | RECRUIT / TRAINING | GENERAL SCHEDULING | TO REPLACE ROBERTS |
| 14. ESTIMATOR | RECRUIT / TRAINING | GENERAL ESTIMATING | | TO REPLACE EVANS |
| | JAN. FEB. MAR. APR. MAY JUN. JUL. AUG. SEPT. OCT. NOV. DEC. | JAN. FEB. MAR. APR. MAY JUN. JUL. AUG. SEPT. OCT. NOV. DEC. | JAN. FEB. MAR. APR. MAY JUN. JUL. AUG. SEPT. OCT. NOV. DEC. | |
| | 1980 | 1981 | 1982 | |

Figure 23-13. Project services—personnel.

| Engineering Department – Manpower Planning | | | | | | | | | | | | | |
|---|---|---|---|---|---|---|---|---|---|---|---|---|---|
| Project number | Manhours | | | Manhours for past six weeks | | | | | | Men weeks to go | Weekly scheduled men | | |
| | Budget | Forecast | To date | | | | | | | | Jan | Feb | Mar |
| 1 | 2 | 3 | 4 | 5 | 6 | 7 | 8 | 9 | 10 | 11 | 12 | 13 | 14 |
| | | | | | | | | | | | | | |
| | | | | | | | | | | | | | |
| | | | | | | | | | | | | | |
| | | | | | | | | | | | | | |
| | | | | | | | | | | | | | |
| | | | | | | | | | | | | | |
| | | | | | | | | | | | | | |
| | | | | | | | | | | | | | |
| | | | | | | | | | | | | | |
| | | | | | | | | | | | | | |
| | | | | | | | | | | | | | |
| Total above projects | | | | | | | | | | | | | |
| Miscellaneous projects | | | | | | | | | | | | | |
| Development work | | | | | | | | | | | | | |
| Total manhours | | | | | | | | | | | | | |
| Total men available | | | | | | | | | | | | | |
| Total men required | | | | | | | | | | | | | |

Use continuation sheet for rest of schedule →

Figure 23-14. Manpower planning.

| ENGINEERING DEPARTMENT – MANPOWER PLANNING – WORK CATEGORY | | | | | | | | | NOTES: 1. DRAFTING IS OUTSIDE CONTRACT. | | | | | | |
|---|---|---|---|---|---|---|---|---|---|---|---|---|---|---|---|
| ANNUAL PLAN | | | | | | | | | % FIGURES ARE AS OF MIDYEAR AS OF MARCH | | | | | | |
| CODE | WORK CATEGORY | % | J | F | M | A | M | J | J | A | S | O | N | D |
| | TECHNICAL PERSONNEL | | | | | | | | | | | | | |
| | CURRENT AFE PROJECTS | 48 | 188.7 | 197 | 197 | 192.1 | 188.6 | 180.4 | 184.1 | 183.3 | 181.1 | 179.4 | 159.6 | 159.6 |
| | PROBABLE PROJECTS | 5 | 0 | 0 | 0 | 1 | 3.3 | 18.6 | 26.6 | 35.3 | 41.8 | 44.3 | 55.4 | 55.6 |
| | FEASIBILITY STUDIES | 12 | 43.2 | 43 | 43 | 43.5 | 44.6 | 44.8 | 45.8 | 46 | 46.1 | 46.1 | 45.3 | 44.9 |
| | TECHNICAL SERVICE | 4 | 14.2 | 14.2 | 14.2 | 14.2 | 14.2 | 14.2 | 14.2 | 14.2 | 14.2 | 14.2 | 14.2 | 14.2 |
| | TECHNICAL METHOD DEVELOPMENT | 12 | 42.8 | 42.8 | 43.8 | 43.8 | 43.3 | 43.4 | 43.6 | 43.5 | 43.5 | 43.5 | 43 | 43 |
| | START UP/OPERATIONS | 1 | 2 | 2 | 2 | 2 | 2 | 2 | 2 | 2 | 2 | 2 | 2 | 2 |
| | SUB TOTAL TECHNICAL | 82 | 290.9 | 299 | 300 | 296.6 | 296 | 303.4 | 316.3 | 324.3 | 328.7 | 329.5 | 319.5 | 319.3 |
| | MANAGERS & SECRETARIES | 7 | 27.2 | 27.2 | 27.2 | 27.2 | 27.2 | 27.2 | 27.2 | 27.2 | 27.2 | 27.2 | 27.2 | 27.2 |
| | OTHER INDIRECTS (SERVICES ETC.) | 11 | 42.9 | 42.9 | 42.9 | 42.9 | 42.9 | 42.9 | 43 | 43 | 43 | 43 | 43 | 43 |
| | MISCELLANEOUS | 0 | 0 | 0 | 0 | 0 | 0 | 0 | 0 | 0 | 0 | 0 | 0 | 0 |
| | | | | | | | | | | | | | | |
| | | | | | | | | | | | | | | |
| | REQUIRED TOTAL | 100 | 361 | 369.1 | 370.1 | 366.7 | 366.1 | 373.5 | 386.5 | 394.5 | 398.9 | 399.7 | 389.7 | 389.5 |
| | | | | | | | | | | | | | | |
| | ACTUAL PAYROLL | | 319 | 327 | | | | | | | | | | |
| | | | | | | | | | | | | | | |

Figure 23-15. Manpower planning—work category.

| ENGINEERING DEPARTMENT – MANPOWER PLANNING – BY SECTION | | | | | | | | | NOTES: 1. DRAFTING IS OUTSIDE CONTRACT. | | | | | | |
|---|---|---|---|---|---|---|---|---|---|---|---|---|---|---|---|
| ANNUAL PLAN | | | | | | | | | % FIGURES ARE AS OF MIDYEAR AS OF MARCH | | | | | | |
| CODE | SECTION | % | J | F | M | A | M | J | J | A | S | O | N | D |
|  | EMPLOYEE RELATIONS | 1 | 4.5 | 4.5 | 4.5 | 4.5 | 4.5 | 4.5 | 4.8 | 4.9 | 4.9 | 5 | 5 | 5 |
|  | PROCUREMENT | 1 | 4.2 | 4.4 | 5 | 5.3 | 5.7 | 4.8 | 4.7 | 4.8 | 5.4 | 5 | 5.2 | 5.4 |
|  | PROJECTS – U.S. REFINING & CHEMICAL | 5 | 14.6 | 15.9 | 15.9 | 16.7 | 17.3 | 19.7 | 19.8 | 20.1 | 21.4 | 20.8 | 19.7 | 20.1 |
|  | – OVERSEAS R & C | 6 | 17.6 | 18.3 | 17.6 | 17.4 | 17.3 | 21.7 | 23.4 | 24.7 | 24 | 23.2 | 23.2 | 22.6 |
|  | – MIDDLE EAST | 3 | 9.5 | 9.5 | 9.5 | 9.5 | 9.5 | 9.5 | 9.5 | 9.5 | 9.5 | 9.5 | 9.5 | 9.5 |
|  | – OFFSHORE/SYNFUELS | 12 | 45.2 | 45.2 | 45.2 | 46 | 45.8 | 46.2 | 46.1 | 45.2 | 44.3 | 44.3 | 44.3 | 44.3 |
|  | PROJECT SERVICES (EST'G./COST & SCHED.) | 11 | 40 | 40 | 40.5 | 40.5 | 40.5 | 40.5 | 42.5 | 44.3 | 46 | 47 | 45 | 45 |
|  | CONTRACTS | 1 | 4.5 | 4.5 | 4.5 | 4.5 | 4.5 | 4.5 | 5.5 | 5.5 | 5.5 | 5.5 | 5 | 5 |
|  | GENERAL SERVICES (NON TECH.) | 5 | 22.4 | 22 | 21.9 | 21 | 21 | 20.2 | 22.4 | 23.5 | 21.6 | 21 | 21 | 21 |
|  | PROCESS ENGINEERING | 14 | 52.6 | 55.8 | 55.8 | 54.5 | 52.7 | 53.2 | 48.8 | 49.2 | 48.3 | 48.7 | 46 | 45 |
|  | FACILITIES ENGINEERING | 26 | 93.3 | 96.1 | 98.2 | 94.8 | 94.7 | 97 | 104.5 | 108.8 | 114.1 | 115.6 | 111.2 | 112.3 |
|  | OFFICE & PLANT SERVICES | 5 | 17.7 | 17.7 | 17.7 | 17.7 | 17.7 | 17.7 | 17.8 | 17.8 | 18 | 18.1 | 18.2 | 18.3 |
|  | OVERSEAS ENGINEERING OFFICES | 10 | 34.9 | 35.2 | 33.8 | 34.3 | 34.9 | 34 | 35.7 | 36.6 | 35.9 | 36 | 36.4 | 36 |
|  |  |  |  |  |  |  |  |  |  |  |  |  |  |  |
|  | REQUIRED TOTAL | 100 | 361 | 369.1 | 370.1 | 366.7 | 366.1 | 373.5 | 386.5 | 394.5 | 398.9 | 399.7 | 389.7 | 389.5 |
|  | ACTUAL PAYROLL |  | 319 | 327 |  |  |  |  |  |  |  |  |  |  |

Figure 23-16. Manpower planning—by section.

This particular report illustrates an overall engineering manpower report. A similar report could be produced for each section.

**Planning by Work Category.** Figure 23–15 illustrates a report format generally used by owners. It is similar to the previous contractor project report, but has additional categories: probable projects, feasibility studies, technical service, etc. Also, it separates technical from nontechnical and managers/secretaries. Obviously, some managers are technical. But this provides a continuous assessment of number of managers to engineers and relationship of technical to nontechnical. Both relationships need to be evaluated for an efficient operation. This report shows an annual plan. Additional years could be developed based on the quality of the individual plan cycle.

As previously stated, assessments for feasibility studies and probable projects can be difficult.

The outlined numbers illustrate a large, international operating company having a central engineering department of some 300 engineers. Evaluation of these manpower relationships should bear in mind that detail drafting and other services can be outside contracts. A typical relationship of draftsmen to engineers can be about 3.5 to 1.

This report should be issued monthly and would undoubtedly be derived from a computer program.

**Planning by Section.** Figure 23–16 is a report for the same company, as previously illustrated. Whereas the previous report showed manpower by work category, this report shows manpower by section. The construction group is part of the project management groups.

Individual section reports would clearly indicate a "shortfall" or "overmanning" of personnel by engineering classification. Adequate recruiting and training programs could be developed from this information.

Manpower requirements based on physical assessments can only be made for design groups where drawing/document take-offs and man-hour assessments can be made. Historical relationships, engineering department "character"/responsibilities, control requirements, and company policy can determine allocations of service personnel to project work.

Chapter 24 deals with control techniques that are useful to the project manager.

# 24. Project Control: Scope Recognition

James A. Bent*

## PRACTICAL PROJECT CONTROL—BASIC SCOPE APPRECIATION

The use of historical data, typical relationships, statistical correlations, and practical "rules of thumb" can greatly add to the effectiveness of a project control program. Such information can provide guidance, in:

- Developing/evaluating schedules.
- Assessing manpower requirements.
- Determining appropriate productivity levels.
- Improving cost/schedule assumptions.
- Carrying out trend analysis.
- Establishing the cost of the project.
- Evaluating the status and performance of the work.
- Recognizing the scope of work, at all times.

It is this last item that really highlights the key to effective project control. That key is *SCOPE RECOGNITION*. This equates to the ability to properly establish the scope in the first place, through a good estimate, and thereafter to constantly recognize the true scope of the work as the project develops and is executed. The "testing" and measuring of actual performance against past experience can be a valuable source of verifying status, determining trends, and making predictions. Naturally, the appli-

---

* James A. Bent has been a consultant in project management since 1980. Previously, he worked more than 30 years for owners/contractors, such as Mobil and M. W. Kellogg. Mr. Bent's experience covers both onshore plants and offshore platforms, with project assignments in the United States, United Kingdom, Australia, Netherlands, Germany, Norway, and Italy. He is an author and presents training courses (3 days–3 weeks) in project management, scheduling, estimating, cost control, contracting, claims/management/negotiating, and shutdowns/turnarounds. These training assignments take Mr. Bent to the United States, United Kingdom, Norway, Europe, Japan, South America, and South Africa.

cation of historical data to a specific project must always be carefully assessed.

The following figures represent historical and typical cost/schedule "rules of thumb" that can assist in the establishment and development of scope during the execution phase of a project. This information is especially useful at the front end of a project when engineering is at a low percentage of completion, resulting in a broad and preliminary cost estimate and overall schedule.

Figure 24-1 shows the scheduling relationship of engineering and construction for a project on a fast-track program, with the full scope of engineering, procurement, and construction (EPC). Such a project is often referred to as an EPC project. The schedule relationship/duration only covers the execution phase. This phase is sometimes referred to as Phase 2, where the earlier Phase 1 covers the conceptual design studies of process/utility alternatives and case selection.

It should also be noted that the "schedules" are illustrated as trapezoids. This concept is extremely important, as it shows that all "complex" work is executed with:

- A buildup.
- A peak period.
- A rundown.

This concept, *this fact,* then forms the basis of quality scheduling and manpower assessment. It is only "simple" work, where the task is performed by a single crew or squad, that has no buildup or rundown.

It should also be noted, and such a note is found on the figure, that the ratio breakdowns of the buildup, peak, and rundown have been rounded off to whole numbers. The historical numbers are slightly different, but at this overall level of scheduling, such minor numerical differences are of little consequence. However, further reference to the writer's historical data base would show slightly different numbers. For example, the engineering buildup, shown as 20%, is 22%, and the construction buildup, shown as 50%, is 57%.

Figure 24-2 shows the trapezoidal technique for the construction phase. The calculation procedure shows two formulas.

*Formula 1* is used to determine the peak duration, shown as X, on the basis that the following information is known, or assumed:

- Scope, in man-hours.
- Effective monthly hours per man.
- Buildup, usually developed from standard schedules.

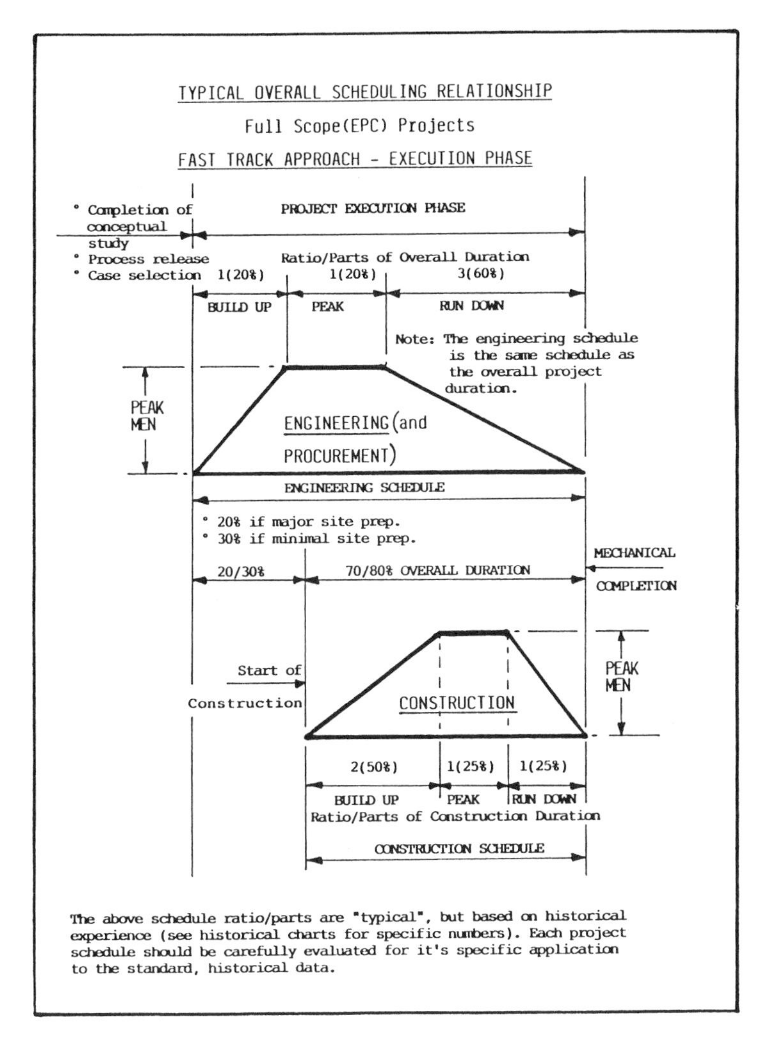


Figure 24-1. Typical Overall Scheduling Relationship.

- Rundown, half of buildup.
- Peak men, as per formula 2.

*Formula 2* covers the calculation of the peak men, if the battery limits area (plot plan) is known. By evaluating a labor density level (usually in the range of 150–300 square feet/man), one can determine the peak number of men. (See Figure 24-3 for information on labor density.)

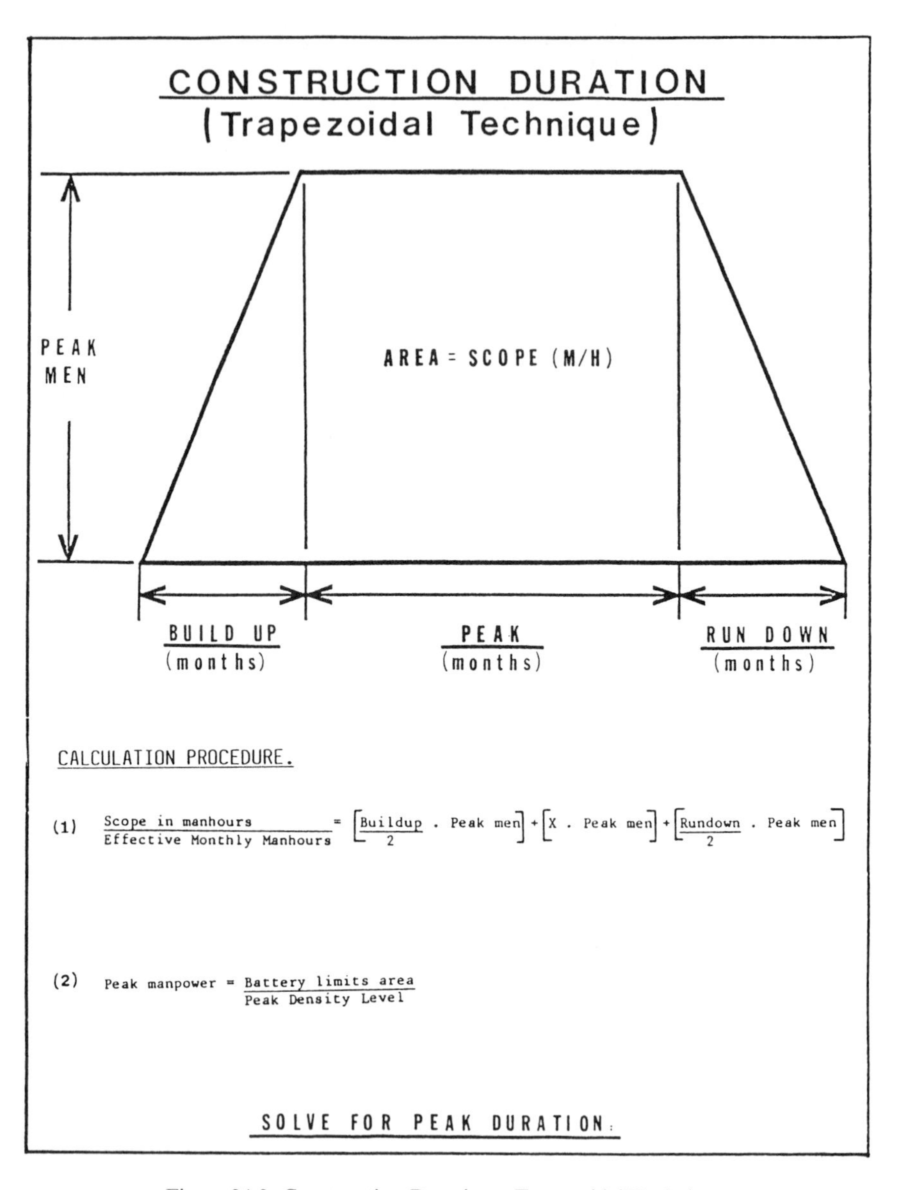


Figure 24-2. Construction Duration—Trapezoidal Technique.

This schedule/manpower evaluation technique is a very powerful program for developing or checking an overall construction duration. The key assumption, requiring good judgment, is the assessment of the labor density level. If this assessment is good, then the resulting scheduling evaluation is of a high quality.

It is emphasized that the peak manpower/density level application can only be used on single process units. The calculation process does not always work for smaller areas of work.

Figure 24-3 shows a working example of the trapezoidal density method when applied to an FCC unit. As noted in the figure, the density level of 250 square feet/man was too optimistic for such a complex unit as an FCC plant.

The first step in the calculation process is to properly assess the man-hour scope. As shown, allowances have been made for indirect labor working in the same area as direct labor and for the estimating allowance. An allowance for better subcontract productivity has been made. Assessing the density/level is extremely important. A 12% absenteeism allowance has been made.

The buildup duration was determined from a standard schedule that showed that the peak labor period would be reached at 30% of the piping duration. This activity was preceded by foundation and equipment installation work, 3 months and 2 months, respectively. The total piping duration was 15 months.

For the subcontract labor case, the man-hours are reduced to reflect a better productivity than with the direct-hire case, possibly a 10% improvement.

In the case of engineering, manpower density is not a consideration. Figure 24-4, therefore, shows the calculation of peak manpower rather than the calculation of the duration.

The same routine can be carried out for construction, if the duration is known.

This figure also shows the calculation formula for a simple piece of work, the 400-hour task, and the error of applying the "simple formula" to a complex task, the 20,000-hour task. As shown, the calculation error of 25 men, instead of 41 men, is significant.

An alternative method to the trapezoidal calculation for complex work is to use the "simple formula" and multiply the result by a "peak factor" (usually 1.6/1.7). The resulting solution is adequate.

The specific factors, based on historical data, are:

Engineering peak factor—1.65.
Construction peak factor—1.45.

**EXAMPLE**

Refinery FCC unit:

- Plot area = 320 ft × 200 ft = 64,000 ft$^2$
- Scope (direct hire) = 445,000 man-hours
- Allowance for indirect labor in area + 10% = 44,500 man-hours
- Estimating allowance + 15% = 66,700

556,200 total scope for evaluation

Consider two cases: Case 1, direct-hire labor, and case 2, subcontract labor.

ASSUMPTIONS:

- Due to "criticality", use Density of 250 (but 300 more probable)
- Allow 12% absenteeism for "effective manhours"
- "Buildup" duration from standard schedule (fdns+eqpt+piping buildup)

*Case 2: Subcontract Labor.* The project strategy, based on experience, is that local subcontractors are more productive than prime contractors (direct hire).

Scope = 556,200 man-hours for direct hire
less 10% productivity adjustment for local subcontractor labor
\- 55,600
= 500,600 man-hours

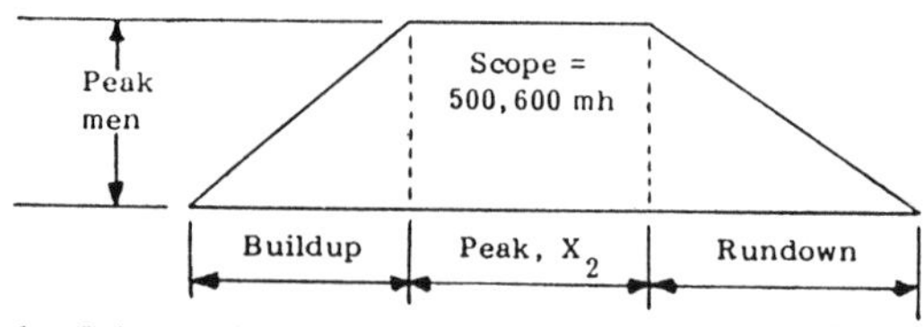


1. Labor availability: No restraint, no adjustment to manpower level.
2. Peak density level: U.S. large project, subcontract labor (from curves) = 250 ft$^2$/man.
3. Peak manpower = $\dfrac{64{,}000 \text{ ft}^2}{250 \text{ ft}^2/\text{man}}$ = 256 men.
4. Effective man-hours per man-month = 40 × 88% × 4¼ = 153 h.
5. Buildup (by judgment) = 3 + 2 + 5 = 10 months.
6. Rundown (by judgment) = 6 months.

Solve for peak, $X_2$:

$$\frac{500{,}600}{153} = \left(\frac{10}{2} \times 256\right) + \left(X_2 \times 256\right) + \left(\frac{6}{2} \times 256\right)$$

$X_2$ = 4.8 (say 5 months)

Therefore

Total construction duration (subcontract labor) = 10 + 5 + 6 = 21 months

This confirms that a subcontract operation, even though more productive, will generally take longer than one on a direct-hire basis.

Figure 24-3. Worked Example—Trapezoidal/Density Method.

## SCHEDULE BASIC

1. A SMALL ENGINEERING TASK - 400 HOURS - IS REQUIRED TO BE COMPLETED IN 2 WEEKS

HOW MANY MEN ARE REQUIRED?

2. CALCULATION METHOD IS

$$\frac{\text{MANHOURS}}{\text{MANHORS/WEEK} \times \text{No WEEKS}} = \text{No MEN}$$

$$= \frac{400}{40 \times 2} = \text{5MEN}$$

3. TASK INCREASES 50 TIMES IN SIZE -20,000HOURS - IS NOW REQUIRED TO BE COMPLETED IN 20 WEEKS

CALCULATION $\frac{20,000}{40 \times 20} = 25 \text{ MEN}$

IS THIS CORRECT ? NO

4. CALCULATION PRINCIPLE - TRAPEZOIDAL TECHNIQUE FOR SIMPLE TO COMPLEX TASKS

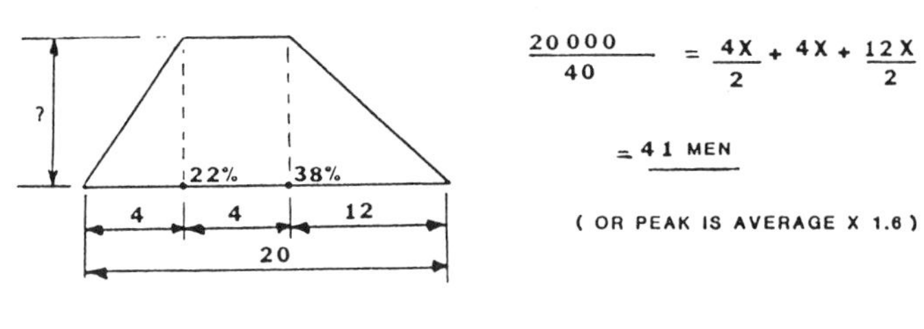


$$\frac{20\,000}{40} = \frac{4X}{2} + 4X + \frac{12X}{2}$$

$= 41$ MEN

( OR PEAK IS AVERAGE X 1.6 )

Figure 24-4. Worked Example—Engineering Trapezoid.

But the common use of 1.6/1.7, for either category of work, is a practical approach.

Figure 24-5 shows the construction indirect and direct costs in the "true" trapezoidal configuration. The "trapezoidal reality" is extremely important for the development of construction claims. As the majority of

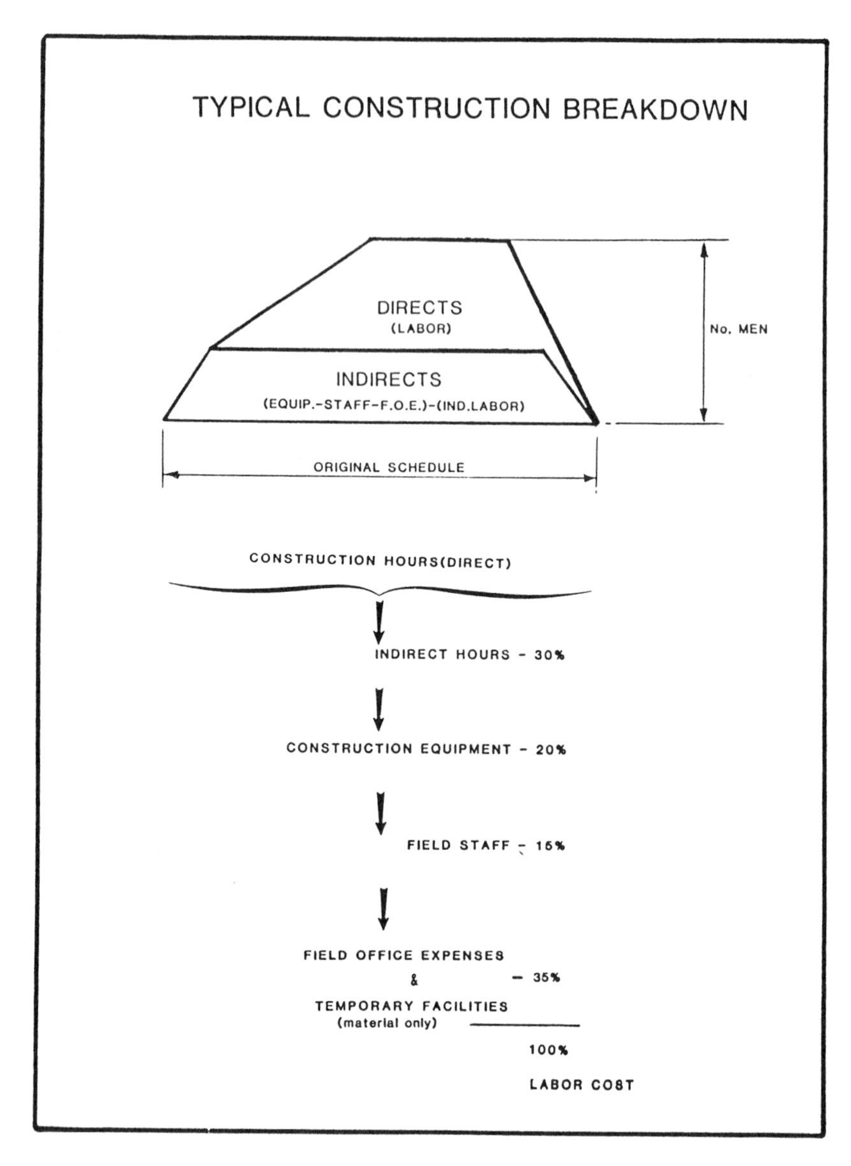


Figure 24-5. Typical Construction Cost Breakdown.

construction claims are time/schedule related, the understanding of the trapezoidal reality is vital. In fact, quality assessments of claims cannot be made without this application.

A typical breakdown of the major indirect costs is also shown. Individual companies might allocate their indirect costs slightly differently, but there is a high degree of conformity to this direct/indirect allocation within the process industry.

Figure 24-6 takes the information in Figure 24-5 a stage further, by adding some engineering information to the construction data. Again, it is emphasized that the stated information only applies to a "full EPC" project.

The emphasis of this figure is on the engineering/construction man-hour relationship. This is shown as the ratio of 1 : 6. In other words, one engineering man-hour "automatically" generates six direct construction man-hours. This is a very useful "rule of thumb" and, of course, the ratio does vary a little as the design complexity varies.

This relationship highlights the need for design engineers to *fully* realize that as they are designing, they are also generating the construction man-hours. Full realization of this fact should lead the designers to more carefully consider the question of *constructability*. This is the design process of working to construction installation considerations as well as working to standard design specifications. Constructability considerations can result in significant savings in construction labor. Such considerations are essential in the following types of construction:

- Heavy lifts.
- Prefabrication and preassembly.
- Modularization.
- Offshore hookup work.
- Site problems of limited access.

The relationship between home office support services (project management, project control, procurement, computer, clerical, etc.) and engineering is also shown, at 40%. This relationship is in man-hours.

Figure 24-7 shows a typical relationship between construction complexity and labor density. Judgment is required in assessing the appropriate density level for the specific project.

**Complexity (Man-hours/Square Foot).** Complexity is "automatically" generated by the design specifications. This statistic is based on the number of direct construction man-hours (within the plot) divided by the plot area (battery limits.) As noted, this assumes that there is no preinvestment in the design basis. Preinvestment is a fairly common practice and is

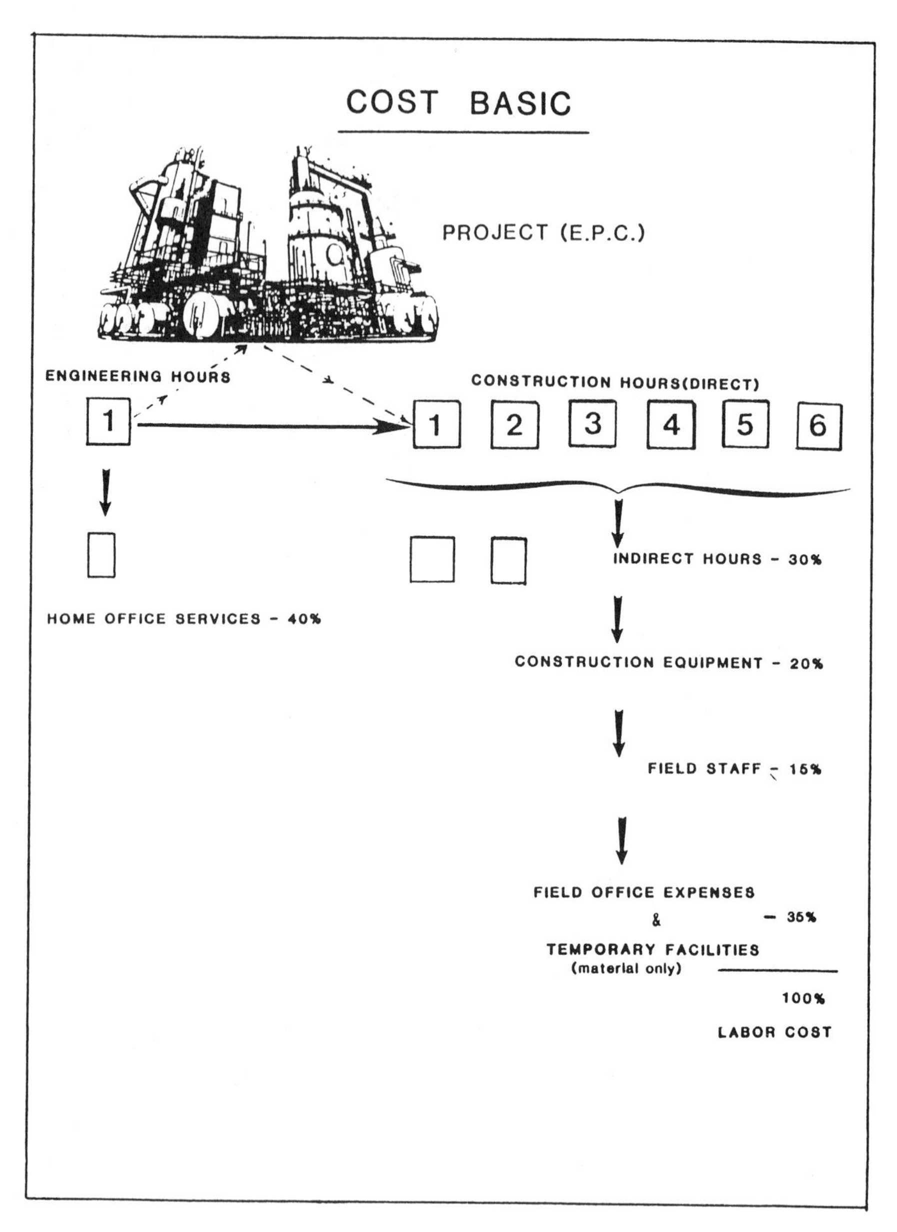


Figure 24-6. Cost Basic.

CONSTRUCTION COMPLEXITY AND LABOR DENSITY

- Only applicable to complete Process Units(small or large)
- Assumes an economic design - no "Preinvestment"
- Based on "average" US labor productivity(Calif./Union)

COMPLEXITY (direct manhours/sq.ft.)

| | Manhours/sq.ft. |
|---|---|
| ° SIMPLE UNIT | 4/5 |
| ° AVERAGE UNIT | 6/7 |
| ° COMPLEX UNIT | 8/10 |

LABOR DENSITY (sq.ft./man)

Tied to above Complexity Data.

| | Sq.ft./man |
|---|---|
| ° SIMPLE UNIT(4/5 mh./sq.ft.) | 150/180 |
| ° AVERAGE UNIT(6/7 mh./sq.ft.) | 180/250 |
| ° COMPLEX UNIT(8/10 mh./sq.ft.) | 250/300 |

Density data is based on a Prime EPC Contractor/reimbursable contract. For fixed or unit price contracts, density numbers should be increased by about 50sq.ft./man. This reflects the need for lower numbers of men to achieve higher productivity to meet "hard money" financial requirements.

Figure 24-7. Construction Complexity and Labor Density.

carried out when forward company planning has determined that the plant will need to be expanded within a few years. At the moment, the planned (design) capacity is sufficient. The design preinvestment, therefore, usually includes extra area in the plot for future installation of equipment. This "extra" area, open at the moment, would give "false" complexity numbers and density levels.

**Density (Square Feet/Man).** This statistic is based on the "economic" total number of craftsmen working at peak (supervision is not included). The data is based on historical experience and is tied to the complexity of the area. The more complex, the lower the number of men that can work in the area. In practical terms, this means that greater complexity has more equipment (more man-hours) per square foot, thus taking up space for the men to work in.

The chart in Figure 24-8 enables an overall schedule to be instantly generated, if the construction man-hours are known. The result is not of high quality, but is intended to provide a preliminary schedule during the early development or feasibility stage of a project.

The chart provides "add factors" for straight-through and overseas projects. Also shown are curves for projects that were built in the 1950s and 1960s, and for Norwegian-based projects. Curves can be developed for most overseas locations.

It is interesting to note that the project durations of the past (1950s and 1960s) cannot be repeated today. The major reasons are as follows:

- Productivity reduction—engineering and construction—(cannot always be compensated for by increased men).
- Greater complexity for same capacity (higher temperatures and pressures).
- Increased environmental/regulatory engineering.
- Increased management/approval time.
- Poor project management/poor planning.
- Fast-track approach.
- Longer equipment delivery times (1970–1983).
- Lack of manpower (1970–1983).
- Increased use of the reimbursable-type contract.

Figure 24-9 provides general guidelines for establishing a direct-labor productivity profile for the construction phase. These guidelines cover incremental and cumulative profiles. As bad weather can have a significant impact on productivity, separate guidelines are provided.

This profile should be developed as soon as the physical site conditions are known and a detailed construction schedule is available. The horizontal axis should be translated from percent complete to a calendar time frame.

As direct construction labor can be 20% of total project costs, it is important that labor productivity be tracked as early as possible. Productivity can only be properly measured if construction progress is evaluated with physical quantities and associated work measurement units.

**Application.** These guidelines show incremental productivity for the major phases of construction.

The mobilization phase (first 15%) is shown with a reduced productivity of 10% from the construction estimate. It then improves with phases of

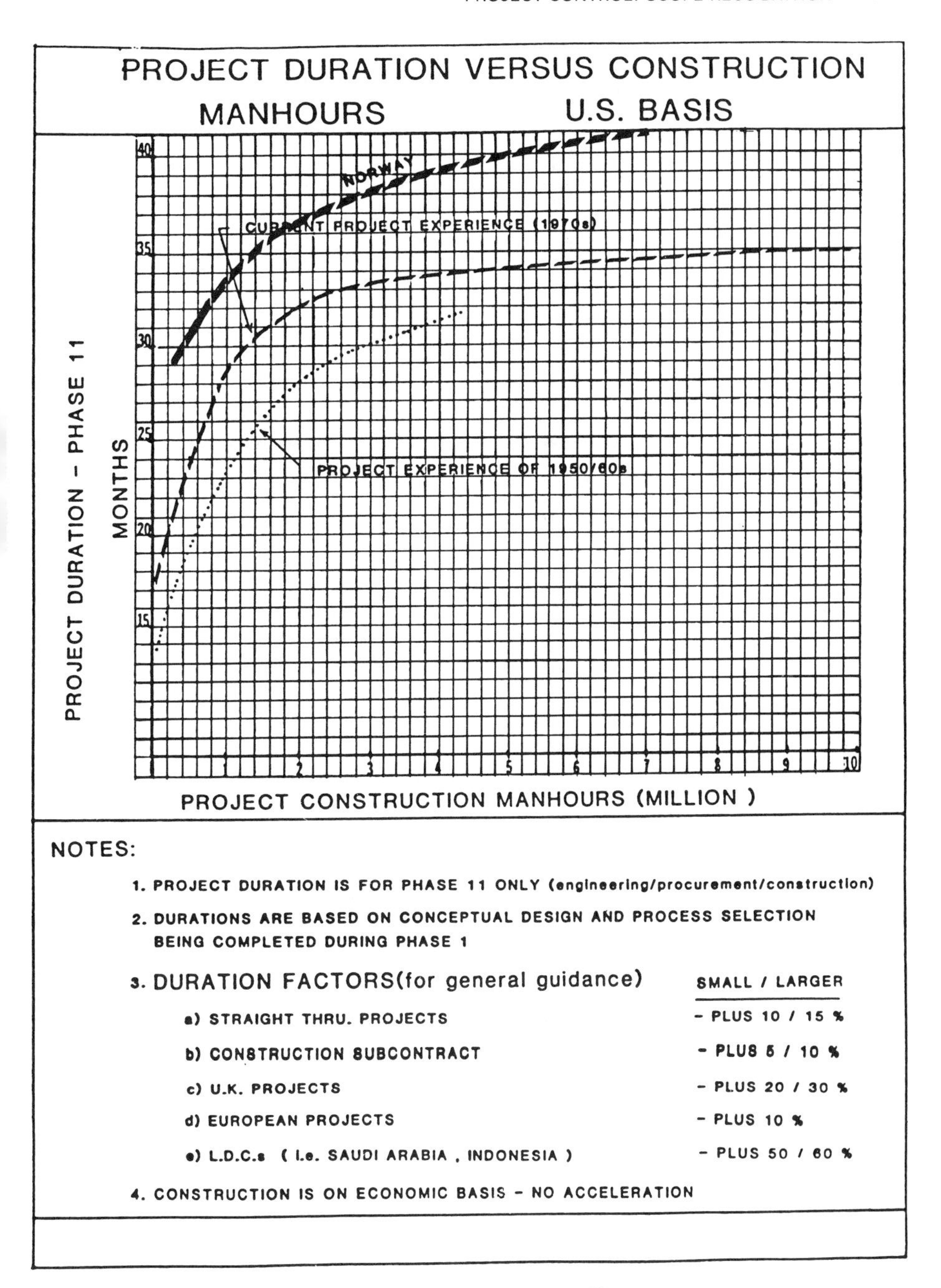


Figure 24-8. Project Duration Chart.

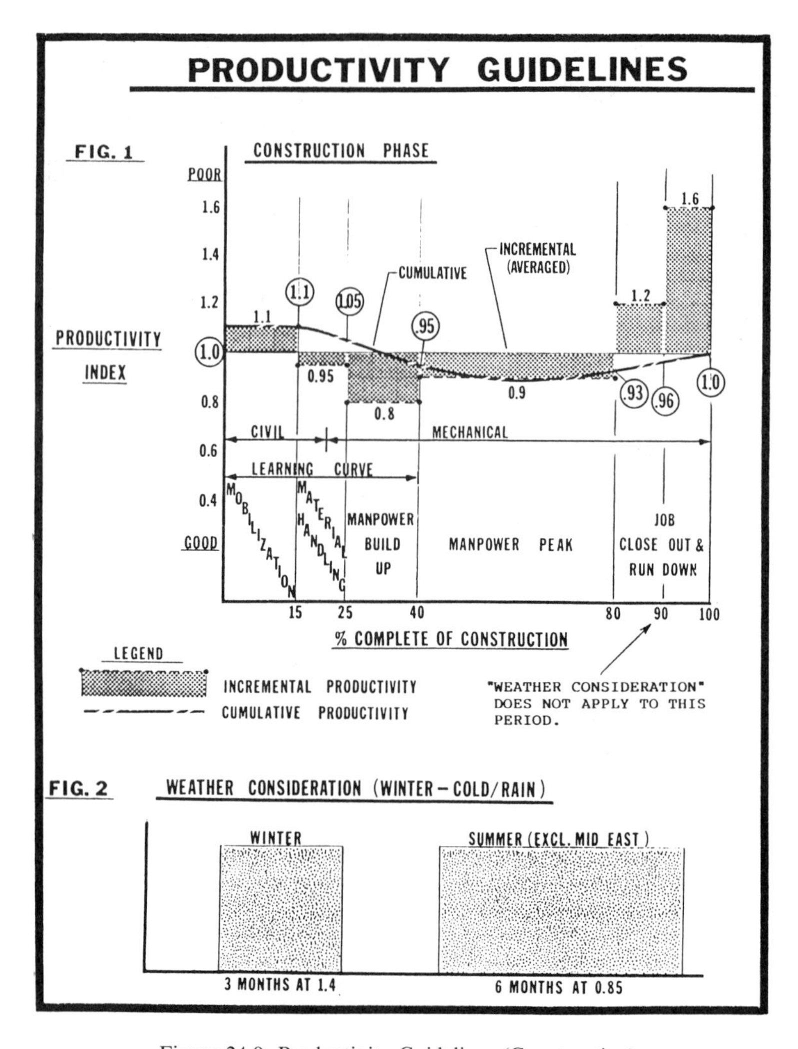


Figure 24-9. Productivity Guidelines (Construction).

5% and 20% for the material handling and manpower buildup phases. At manpower peak (40% of construction), the incremental productivity is shown to be still good at 0.9. Thereafter, for the last 20% of construction, it is shown as rapidly deteriorating.

The cumulative curve is calculated and is shown as tracking from poor to good and ending at 1.0. Additional factors for weather would be superimposed on the top profile. If the winter occurred at 40% of construction, the 0.9 could be multiplied by 1.4, resulting in a projection of 1.3 for the period. If the other periods were as shown on the chart, then the overall productivity of 1.0 would not be achieved. The "weather consideration" does not apply during the last 20% of the job.

This evaluation can be made early in the project, and these guidelines and this method can greatly assist in monitoring and forecasting productivity levels.

**The "Poor" Rundown Productivity.** Based on historical experience, the reasons for the last 20% "poor" productivity are as follows:

1. The major work during this period is punch list/check out type of work. Such work has low budget value, hence the earned value system "measures" low productivity.
2. This is also the stage of remedial work and changes, mostly required by the operational and maintenance staff. This work does not usually fall into the class of official change orders that would result in increased budget.
3. Labor has a poor attitude. The work is drawing to a close and the on-site labor may not be eager to go to a new job. In fact, there may be no other work available in the area.
4. Management/planning is poor. There is, sometimes, the tendency for the construction manager to order the work to be executed on a "crash basis" so as to finish quickly and get to the next job. This can result in significant overmanning and poor work. Costs can increase drastically.

Figure 24-10 is a typical breakdown of total home office man-hours for a "full scope" project. It is based on historical data for small-to-medium-sized projects engineered on a reimbursable basis and executed during the period 1955–1975.

**Application.** This information can be used to check an estimate or a contractor proposal of home office man-hours. It can be used for early evaluations of home office manpower and schedules when only total costs or man-hours are available.

**Example.**

1. For a typical project we can assess the percent piping man-hours. This is derived by summing the hours required for piping engineering activities (plant design, 16.4%; piping engineering, 2.1%; bill of materials, 2.1%; and model, 0.4%, for a total of 21%).
2. As a percent of engineering only, piping becomes 21%/0.67 = 32%.

Home Office Man-hour Breakdown

| Description | | % Man-hours |
|---|---|---|
| *Design & Drafting* | *Full-Scope (%)* | |
| Civil & structural | 25.00 | 10.000 |
| Vessels | 7.50 | 3.000 |
| Electrical | 15.00 | 6.000 |
| Plant design (piping) | 41.00 | 16.400 |
| Piping engineering | 5.25 | 2.100 |
| Bill of material | 5.25 | 2.100 |
| Model | 1.00 | 0.400 |
| | 100.00 | 40.000 |
| *Administration—indirect drafting* | | 4.000 |
| *Engineering* | | |
| Instrument (engineering & drafting) | | 3.000 |
| Mechanical (rotating machinery, plant utilities, metallurgy, etc.) | | 3.000 |
| Mechanical (consultants) | | 0.200 |
| Project management | | 7.500 |
| Project engineering | | 6.000 |
| Project (operating expenses, services administration) | | 0.200 |
| Process design | | 3.000 |
| Process technology services | | 0.100 |
| *Project services* | *67%* | *Engineering* |
| Estimating & cost control | | 4.000 |
| Proposals | | – |
| Computer control | | – |
| Computer systems | | 1.000 |
| Initial operations—office | | 0.200 |
| Technical information | | 0.200 |
| Scheduling | | 2.000 |
| *Procurement* | | |
| Purchasing | | 5.000 |
| Inspection and Expediting | | 5.000 |
| *General office* | | |
| Stenographic | | 4.500 |
| Accounting | | 7.000 |
| Office services | | 2.000 |
| Labor relations | | 0.100 |
| Construction (office) | | 2.000 |
| | Total | 100.000 |

Figure 24-10. Home Office Man-hour Breakdown.

As overall engineering and piping design are often on the critical path, individual evaluations are frequently required. Where information is lacking, use the following:

- Engineering man-hours as a percent of total home office: 65%.
- Piping man-hours as a percent of engineering: 35%.

Figure 24-11, an overall breakdown of project cost, is based on historical data for projects built in the United States on a prime contract basis during the period 1955–1975.

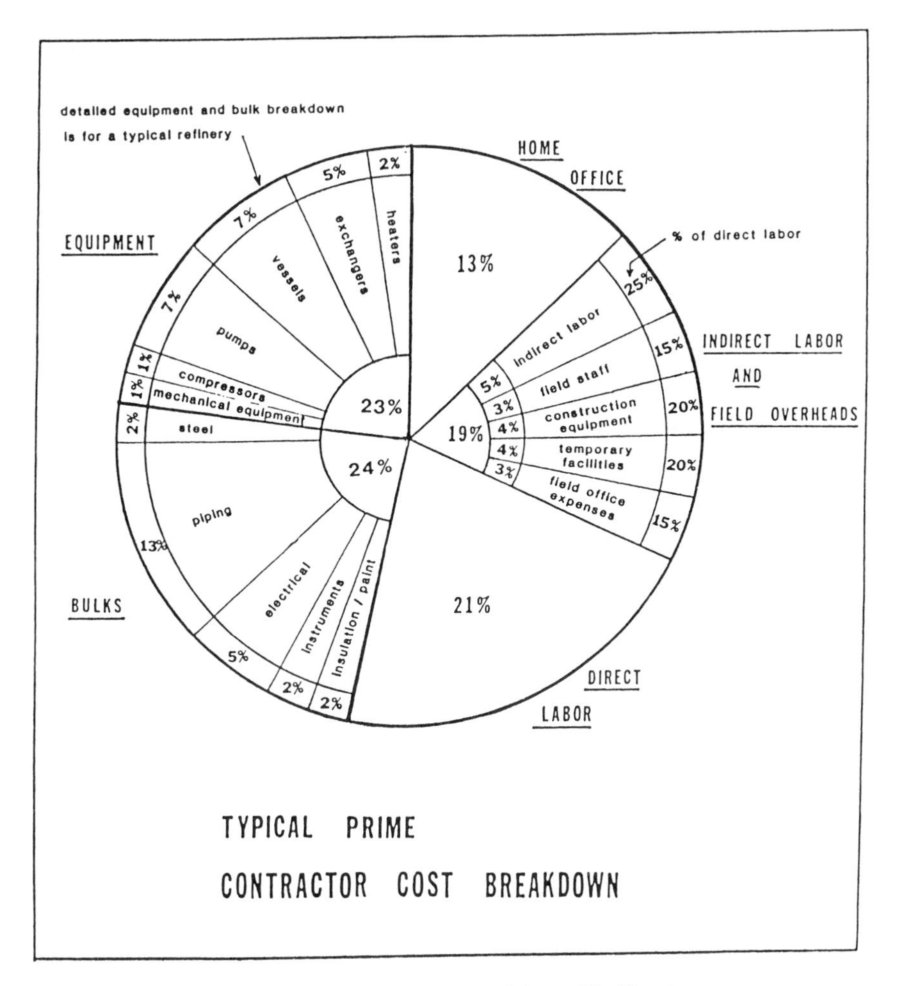


Figure 24-11. Overall Cost Breakdown (Pie Chart).

**Application.** When only an overall cost is known, this breakdown can be useful in providing overall data for a quick evaluation of engineering and construction schedules.

**Example.** Assume that a project has an estimated overall cost of $100 million.

1. From the diagram, home office costs are roughly 13%, or $13 million. By a further assumption that the contractor home office all-in cost is $30/hour, we can derive a total number of home office man-hours:

$$\text{No. of man-hours} = \frac{13{,}000{,}000}{30} = 433{,}000$$

   Thus, a gross schedule and manpower evaluation can now be made.
2. From the chart, direct field labor costs are roughly 21%, or $21 million. By a further assumption that the direct field labor payroll cost is $10/hour:

$$\text{No. of man-hours} = \frac{21{,}000{,}000}{10} = 2{,}100{,}000$$

   Applying known and historical relationships allows gross evaluations for engineering and construction durations to be made. These in turn can be used to prepare manpower histograms and progress curves.

**NOTE:** For larger projects, the percent of home office and field overheads increases.

Figure 24-12 compares the breakdowns (%) of overall costs and construction man-hours for large, grass roots projects against small, revamp projects. The data for the large, grass roots breakdown is based on historical experience, whereas the small, revamp breakdown is typical only. As there are wide variations in the EPC makeup of small projects, the "typical breakdown" should be examined very carefully for its application to a specific project.

As with the previous figure, these breakdowns/relationships, can be helpful in evaluating man-hours, schedules, and manpower requirements. The data for Figure 24-13 has been compiled from historical experience. This figure shows that the indirect curves are essentially constant throughout the execution of a project. Early buildup for field organization and installing temporary facilities is matched by a late buildup for final job cleanup and demobilization.

TYPICAL PROJECT COST BREAKDOWN

(CONTRACTOR - TOTAL SCOPE PROJECT)

| GRASS ROOTS-LARGE % | ITEM | SMALL REVAMP % |
|---|---|---|
| 10 | ENGINEERING (D&D) | 20 |
| 3 | HOME OFFICE (SUPPORT) | 5 |
| 47 | MATERIAL (DIRECT) | 40 |
| 21 | CONSTRUCTION DIRECT | 17 |
| 19 | CONSTRUCTION INDIRECT | 18 |
| 100% | | 100% |

DIRECT CONSTRUCTION LABOR HOURS

| GRASS ROOTS-LARGE % | ITEM | SMALL REVAMP % |
|---|---|---|
| 10 | SITE PREPARATION | 1 |
| 12 | FOUNDATIONS/UNDERGROUNDS | 8 |
| 7 | STRUCTURAL STEEL/BUILDINGS | 5 |
| 10 | EQUIPMENT | 12 |
| 35 | PIPING | 48(INCL.FAB) |
| 11 | ELECTRICAL | 10 |
| 6 | INSTRUMENTS | 8 |
| 4 | PAINTING | 3 |
| 4 | INSULATION | 3 |
| 1 | HVAC/FIREPROOFING | 2 |
| 100% | | 100% |

Figure 24-12. Typical Project Cost Breakdown.

The purpose of this figure is to illustrate a typical relationship between *direct work* and *indirects*.

Direct-work progress is a measure of physical quantities installed, and, as shown, the direct-work curve is identical to the historical construction curve included in the writer's data base.

Indirect construction progress cannot be assessed by measuring physical quantities and is usually measured in man-hours. This typical curve shows the rate at which these man-hours would normally be expended.

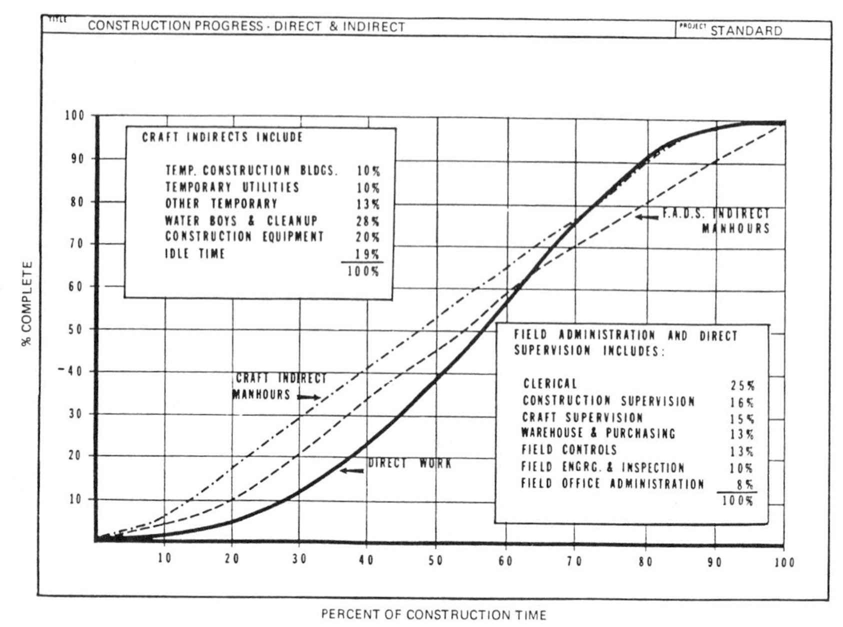


Chart of construction indirects—craft and staff.

Figure 24-13. Construction Progress—Direct and Indirect.

Indirect and direct construction curves for a project could be compared with the curves in this exhibit. During construction, actual performance should be compared with these profiles. This can provide an early warning that the expenditure of man-hours is deviating from the norm.

The percentage breakdowns for craft indirects and field administration and direct supervision can be used to check estimates and performances of individual categories.

On megaprojects, individual control curves could be developed for individual categories.

There are occasions when a project is placed on extended overtime to shorten the schedule. In many cases, productivity will be reduced and costs will increase. If this condition was not part of the original estimate, an assessment of the increased cost, as well as the schedule advantage, should be made. The schedule evaluation should recognize increased man-hours in the duration calculation. It is also possible that absenteeism will increase, sometimes to an extent that there is no schedule advantage for the increased workweek.

Figure 24-14 presents data compiled from the sources indicated on the chart. It plots labor efficiency against overtime hours worked, based on

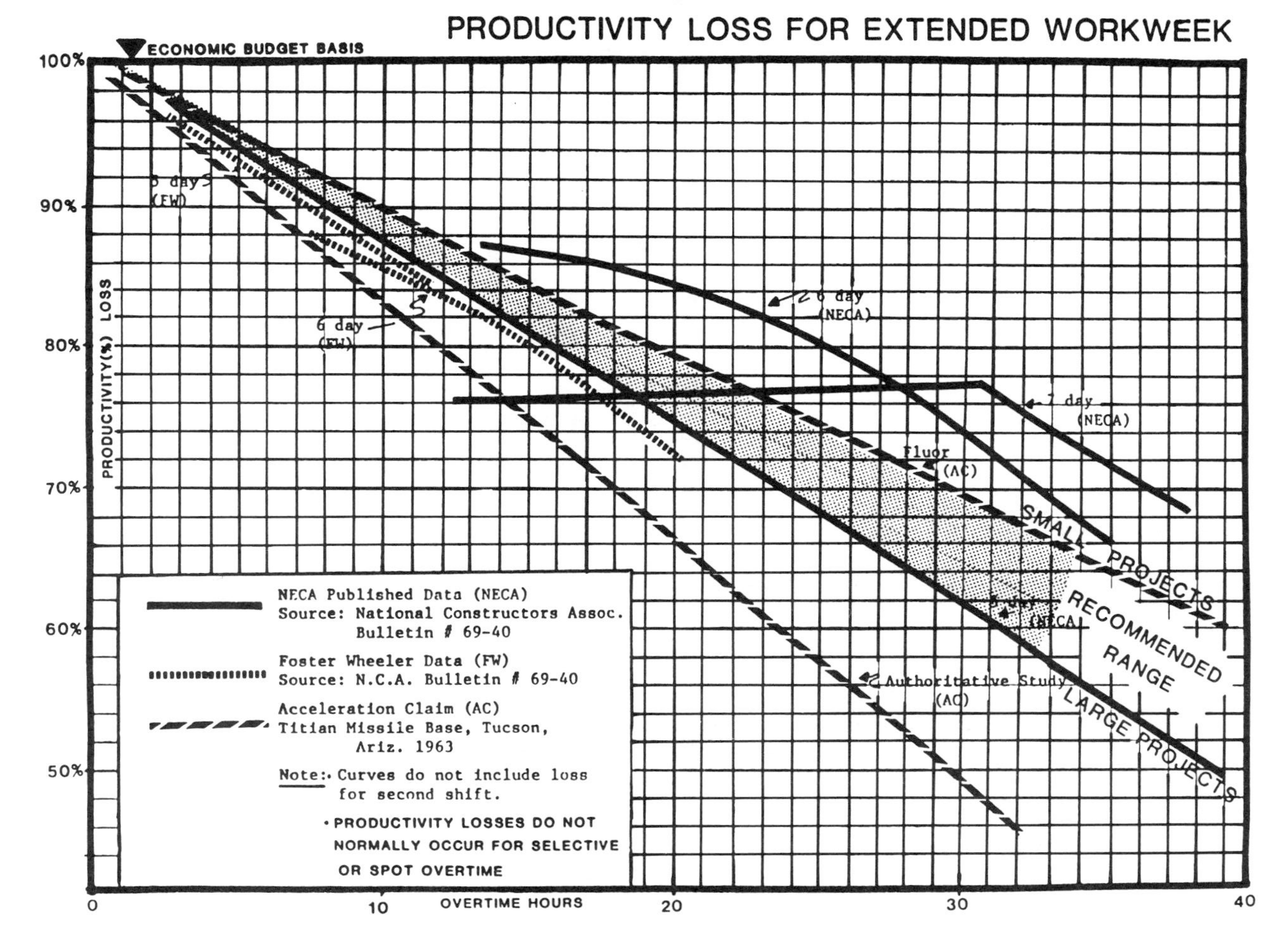


Figure 24-14. Productivity Loss for Extended Workweek.

| ACCOUNT | CODE | % | BOIL | BRICK | CARP | ELEC | LAB | % INSUL | OPER | MILL | PAINT | PIPE | IRON | TEAM |
|---|---|---|---|---|---|---|---|---|---|---|---|---|---|---|
| | | | CRAFT MIX BY ACCOUNT | | | | | | | | | | | |
| SITE PREPARATION | | 9 | | | | | | | | | | | | |
| FOUNDATIONS | | 10 | | | 37 | | 47 | | 4 | | | | 10 | 2 |
| BUILDINGS | | 3 | | 9 | 24 | 1 | 26 | | 3 | | | 8 | 27 | 2 |
| STRUCTURAL STEEL | | 5 | 27 | | 4 | | 4 | | 12 | 1 | | 4 | 47 | 1 |
| SPECIAL EQUIPMENT | | 1 | 14 | | 11 | 6 | 5 | | 6 | 25 | | 22 | 9 | 2 |
| HEATERS | | 4 | 43 | 7 | 7 | | 23 | 5 | 5 | | | 6 | 2 | 2 |
| EXCHANGERS | | 1 | 17 | | 6 | | 6 | | 14 | 5 | | 44 | 7 | 1 |
| VESSELS | | 2 | 64 | | 3 | 1 | 14 | 1 | 7 | | | 4 | 4 | 2 |
| TOWERS | | 2 | 80 | | 3 | | 4 | | 8 | | | 1 | 2 | 2 |
| TANKS | | 1 | 61 | | 5 | | 9 | | 10 | 2 | | 10 | 3 | |
| PUMPS-COMPRESSORS | | 2 | 2 | | 5 | 3 | 8 | | 4 | 56 | | 16 | 6 | |
| PIPING | | 32 | | | 2 | | 5 | | 9 | | | 82 | 1 | 1 |
| ELECTRICAL | | 7 | | | 4 | 86 | 8 | | 1 | | | | 1 | |
| INSTRUMENTS | | 6 | | | 1 | 45 | 1 | | 1 | | | 51 | 1 | |
| PAINTING-INSULATION | | 9 | 2 | | 7 | | 10 | 59 | 1 | | 16 | 2 | 2 | 1 |
| TEMPORARY CONSTRUCTION | | 5 | 3 | | 30 | 18 | 21 | | 5 | | | 20 | 3 | |

Figure 24-15. Craft Mix by Account.

5-, 6-, and 7-day workweeks. These data apply only to long-term extended workweeks. Occasional overtime can be very productive with no loss of efficiency. The exhibit shows a recommended range of productivity loss by project size (small to large.)

**Application.** This chart can be useful in an overall evaluation of the impact of overtime hours on schedule and cost. It can establish an increase in total labor hours required for a loss in efficiency due to an extended workweek. However, judgment should be used on an individual location basis. Some areas, particularly less developed countries (LDCs), work 60-hour weeks which are as productive as 40-hour weeks.

**Example.** Assume that a project has a total construction scope of 1 million man-hours and is based on a 5-day, 40-hour workweek. If the same workweek were increased by 8 hours to 48 hours, look to the chart for 8 hours of overtime and, using the (NECA) 5-day (large-project) curve, read across to an efficiency of 90%. This indicates that 10% more hours will be required to accomplish the same amount of work due to a loss in efficiency. Thus, we estimate that the total man-hours will be 1 million × 1.10 = 1,100,000. man-hours. Schedule and cost evaluations can now be made for an additional 100,000 man-hours but at an increased level of work. Obviously, there is a schedule advantage.

**NOTE:** These curves do not include efficiency losses for a second shift, which can be about 20%. However, shift work losses depend on the type of work, company organization, and experience. In the offshore industry, where shipyards traditionally work on a shift basis, losses can be minimal, or zero.

Most construction estimates are prepared on an account basis. Scheduling manpower resources at a detailed level would, therefore, require a craft breakdown of labor by account.

Figure 24-15 shows a typical breakdown of craft labor by major account for the United States. Overseas labor practices will often not conform to this mix of labor.

# 25. Project Needs and Techniques for Management Audits

Kenneth O. Chilstrom*

A periodic checkup has become routine for determining an individual's well-being and for early diagnosis of physical problems. Similarly, we wouldn't permit the operation of complex equipment such as airplanes or nuclear power stations without performing periodic inspections to ensure safe and proper performance. Applying this logic during the management of projects is just as appropriate in order to determine the health of the organization and its performance in meeting all objectives.

Management's knowledge of the actual status of a project varies with the frequency of reporting and the credibility of the data. It is also recognized that most organizations have a "velvet curtain" which frequently shields the manager's vision and hearing, thereby limiting his ability to take early corrective action. It is generally true that there is a reluctance to tell the boss the bad news, thereby delaying an early recovery plan with a minimum impact. Another factor which can be expected with a highly professional group of seasoned experts is overconfidence coupled with complacency that contributes to a false sense of security. For many people, there is a danger that experience is no more than doing today what was thought of yesterday, either in the same old way or in a slightly modified manner. The black book of failure is filled with the stories of people who have had experience. Experience should be a servant, not a handicap. Unfortunately, the aftermath of the Shuttle Challenger disaster in January 1986 indicated that conditions mentioned above were all prevalent in a very mature project.

* Kenneth O. Chilstrom's management auditing experience has been extensive in both government and industry. Early assignments in the Air Force as an experimental test pilot, R&D Staff Officer, and Program Manager led to later assignments in charge of management surveys of programs. His industrial experience includes program management surveys within the General Electric Company, and the auditing of projects and functional areas. He was an industry consultant to the GAO on a special study to determine the management "Lessons Learned" on what makes projects successful since then, he has had assignments with Science Applications, Inc., and Pratt & Whitney, United Technologies Corporation.

The periodic audit of a project is yet another tool that assists management in maintaining effective control.

## AUDIT POLICY

### Authority Established

Every business has corporate policies formally established in writing for essential business functions. It is also necessary that audit policy be established at the company level. This identifies the technique and use of audits with the boss—the President or General Manager. Unless this occurs, the likelihood of having the support required to conduct successful audits will be greatly reduced. The policy should be adequate in scope to cover responsibilities and the general approach for conduct of project audits. This would include how projects are selected for audit, who selects the projects and the audit team, the support responsibilities of the project team and functional organizations, and how the audit team will report their findings.

### Process for Selection of Project Audits

Audit policy established at the corporate level should identify the President or General Manager as the final authority in the selection of a project for audit. He may choose to make his own selection; however, the approach found most successful is when he and his immediate subordinates mutually agree as to which projects need an audit and when best to accomplish this. This procedure then commits these individuals to the support of the audit since they have agreed to its necessity. Support from the functional organization is absolutely essential to the successful conduct of an audit. When this does not exist, the individuals involved will clam up, compromising the availability of facts, and thus extend the time required to complete the audit.

## PLANNING FOR THE AUDIT

### Scope and Subject Areas Defined

At the time the decision is made to have a management audit of a project, there is usually a reason why top management has selected the particular project. This reason or purpose will most often establish the scope of the audit. Since this indicates the area of major concern, it generally is not necessary to accomplish a full-scale review of all areas of project management. However, if it is early in the schedule, such as six months from go-ahead, it is usually considered necessary to examine all areas. The most important factor is to have the areas of audit, which may limit the scope, well defined and understood at the beginning. Then there can be no misunderstandings about the objective and expectations of top management and what the audit team members will concentrate on. An example which may typify a very limited scope audit would be a project in early qualification testing of the hardware which is experiencing some failures although other areas of technical, schedule, and cost performance are in good control. Experience of many companies has shown that most new projects deserve a thorough review of all technical and nontechnical areas in their first year. It should be obvious that the early phase of a project is where weaknesses will show up in estimating, scheduling, interface relationships, and cost control. Also, the more complex the project, the more people involved and, where past experience is limited, the greater the opportunity for management problems in all areas.

### Organization of Audit Team

In a majority of companies which have 1000 employees or more, there is often a staff function of internal management specialists who form the nucleus of the audit function. In large companies there may exist a good-size office of 10 persons or more whose sole activity is operational auditing of projects, or reviews/surveys of special subject areas.

It is desirable and recommended that a professional staff be responsible for the organization and planning of audits, and then augmented with other specialists as required to conduct the audits. These full-time management auditors would be experienced specialists in functional areas as well as seasoned operational auditors, and would individually complement and supplement each other. It is the team chief/chairman who must decide on the makeup of the team. The number of persons required to cover the areas of the audit and the technical or nontechnical expertise and experience required are critical considerations bearing on the audit success. Criteria in the selection process would generally include but not be limited to the following:

- Specialist versus generalist (need good mix).
- Must have professional acceptance at all levels.
- Technical competence in specialized areas.
- Analytical mind, articulate, and personable.
- Writing ability.
- Listening ability.
- Maturity and adaptability.

- Line, staff, supervisor experience.
- No project involvement—increasing objectivity.
- Enthusiasm and support of audit assignment.

The results of the audit will always be compared to the caliber of the team and its individual members. As is true in other tasks—one should not send out a boy to do a man's job. The audit payoff will always be directly relatable to the qualifications of the audit team.

**Development of Audit Plan**

Preparing a plan for the conduct of a project audit not only assists the audit team members, but helps the project team and functional areas in knowing what the audit team needs and where the audit will concentrate its attention. The audit plan should not be a detailed voluminous coverage of how to do an audit, but rather a detailed outline of the subject areas to be covered. The audit team members who have not had recent experience in the actual conduct of the audit should be coached in the technique prior to the auditing phase by the team chief and other members who have had past experience.

In the development of an audit plan, the problems and shortcomings of other projects can provide an insight or a yardstick to judge the project performance. These "lessons learned" are worth identifying in order to determine if they are being repeated and are the source of problems or a successfully applied benefit. It is also important to identify, during the course of the audit, those activities that are performing well. Too often audits concentrate solely on the trouble areas, which may compromise the true perspective in judging the overall health of the project and the performance of the project team. However, as an example of where many projects incur pitfalls, the experience of others may identify such typical deficiencies as these:

- Techniques of estimating are poorly developed. They reflect an unvoiced assumption which is quite untrue—that all will go well.
- Estimating techniques confuse effort with progress, hiding the assumption that men and months are interchangeable.
- Schedule progress is poorly monitored.
- When schedule slippage is recognized, the natural response is to add manpower. This can make matters worse, particularly for the budget.
- Planners are optimists, so the first false assumption that underlies the scheduling of projects is *that all will go well, that each task will take only as long as it ought to take.*

The plan for a project audit that includes all areas of project management should include the following:

- Organization of project team.
- Functional support and relationships.
- Master plan.
- Contract committments.
- Work definition and assignment.
- Work progress reporting and control.
- Technical plan and capabilities.
- Manufacturing plan and capabilities.
- Product integrity/quality control.
- Logistic support plan.
- Customer relationships.
- Company/corporate policies and procedures applicable.

### Initial Data Review

Upon completion of preparing the audit plan and selection of the full audit team, the team chief should conduct an initial meeting with the project manager. This will enable the project manager to understand the scope of the project and the subject areas to be covered, and would identify the first need and request for the audit team. The team chief, with the assistance of the project manager, can then develop a listing of project data to be provided for the audit team as soon as possible. This data package, which would include such items as the contract, project operating plan, technical and management progress reports, etc., would provide the audit team background information in order to get well acquainted with the details of the project. This step is essential in order to avoid going in cold when starting the interviews with the project team. An adequate understanding about the project, its status, customer, etc., is important for the audit team member to prepare himself for the interviews and further pursuit of data. The ability to ask the right question, to find the trail, to penetrate to the right depth, is a measure of the auditor's effectiveness, and to a great extent will depend upon the amount of preparation given before the actual auditing begins.

## GAINING ACCEPTANCE FOR THE AUDIT

### Establishing the Environment

Audits of a project, or any area of an organization, can be accomplished either by an internal team or by an outside management consultant firm.

In most cases, an internal team should have advantages since they have the benefit of inside knowledge of why things are the way they are, including organization politics and personalities. However, in some situations, an outside person or team can be more effective for not having any bias and for being able to see the issues or reasons for problems in better focus. However, regardless of the use of either approach, it is essential that the organization know that top management is supportive of the audit and should make it known both in writing and vocal opportunities. When this is done, the individuals at working levels should be more willing to recognize and accept the audit in a constructive way than judging it to be an investigation or a witch hunt. When audits are seldom used, they will be more suspect and people less cooperative than if routinely used on all projects and looked upon as a normal way of business.

### Top Management Support

In those instances in which outside auditors are hired, it is usually undertaken by top management. When this is done, the reasons for the audit should be made known and may include an explanation for the advantages of having an outside group. In addition to this expressed support at the beginning, it may require frequent assessments to ascertain the progress of the audit team and those factors which may be hindering their activity. The success of an audit team can be severely handicapped if it finds an environment which is relating the way things should be done rather than what is actually being done, or only answering Yes or No without offering any information. In most cases the responsiveness of the working levels will depend upon the support that supervision and top management provide the audit on a day-by-day basis.

In the situation where a business routinely has internal auditors review all projects and functional areas for management effectiveness and efficiency, the problem of top management support is minimized. When audits have become a way of life, the reason for their being has been overcome and generally accepted. Although top management support is a continuing requirement, the communication problem is less in this circumstance.

Again experience has proven that in most cases management auditing is a tool for the boss and his top management team and requires their support.

### Team Credentials

There is a distinct difference between the known credentials of an auditor who is hired from an outside firm as compared to those persons assigned to auditing from within the organization. As expected, the outside person is only known by his written resume of experience along with the reputation of his company, and this must be sufficient to impress management However, when persons from internal resources are assigned either full-time or temporarily to an audit team, then everyone judges them on many years of personal observations and assessments of character as well as job performance It is well known that it may be easier to accept a person that you do not know too well. When this situation is recognized it, becomes most important that the best people with related experience be selected for audit team assignments. Some companies have a small permanent audit staff who provide the nucleus of the team and who are augmented on a temporary assignment with others from within the organization. It is normally the responsibility of the team chief to identify his team needs and then recommend individuals from throughout the organization who could meet the requirements. It is very important that not only the best-qualified people are identified but that they are recognized by top management as having the needed qualifications and have top management's approval.

If top management does not select and approve the audit team, then their confidence in the team's findings will not exist and the final payoff may be drastically affected. The criteria for individual selection must always recognize the need for specialists in functional or technical fields. However, other factors are equally important in selecting those who are broad thinkers, but still analytical, and who understand the human behavioral aspects of job performance. Too often, managers are guilty of giving ad hoc or other than normal assignments to those persons who are more available rather than selecting individuals on a criteria basis. There must be a general consensus that the findings of an audit team will generally reflect the qualifications of the auditors. If significant results are needed and expected, then the quality of the audit team is essential; therefore each individual's qualifications are important.

### Announcements

Communication of audit activity from the top down is an essential factor influencing the success of an audit. Usually the President or General Manager of the company should sign the internal memorandum which announces the audit, the reason for it, and the support required, and identifies the team members. In addition, verbal announcements should follow from the President to his staff, and from each level down to that of first supervision. It is hoped and desired that discussions at each organizational level will be an affirmation that all will cooperate and support the audit, and that there will not be any negative expressions which will promote withholding of information or foot dragging. People at the working levels can usually discern the true feelings of those providing direction and may then exercise a choice of full, some, or no support to the auditors.

It cannot be overemphasized that when top management decides to have an audit they select the best audit team possible, and provide evidence of their support in written and vocal form to all within the organization.

## CONDUCTING THE AUDIT

### Protocol

It is very important, as a step toward gaining further acceptance for the audit team, that normal protocol be observed in the early phase of interfacing with the project team and the supporting functional areas. The initial point of contact and arrangements for interviews must start with the project manager, and it is most appropriate to have the entire team meet with the project manager, his deputy, and other key staff members. Such a meeting should include the discussion and outline of the contact plan for the individual interviews to follow. The best procedure is to have a member of the project team assigned to the task as the interface or focal point for arranging all interviews. This has several advantages in that it allows the project team to exercise their prerogative in identifying the sequence of audit areas, provides an organized approach for each day's activities, and ensures that each individual is notified in advance that an interview has been scheduled.

Audit interviews should generally follow the organization structure, starting at the top and proceeding to lower levels. This gives those in charge the first opportunity to provide data which they view as important, and to point out areas which they judge as impacting the success or failure of project activities; therefore, talking to a subordinate should not occur before talking to the subordinate's boss. Since auditors are outsiders to the project team, but are having an opportunity to observe the inside operation, they must observe ethics and politeness. If an auditor disregards this approach and instead acts as a privileged superior who barges in whenever and wherever he pleases, then the doors will be hard to open, and data and knowledge will be difficult to obtain.

A rule for an auditor's conduct is to treat each person the way one would like to be treated if the roles were reversed.

## Team Operation

The results of a project audit are the accomplishments and product of the individual auditors; however, their performance can best be directed by the team chief/chairman. He shoulders the ultimate responsibility for the audit success; therefore his involvement in the planning and the conduct of the audit is absolutely essential. Once the scope of the audit is decided, the plan outlined, and area assignments made, then the team chief becomes a manager of the audit team to make sure that they are ready for the job at hand and that daily performance is as a team rather than a mix of individual efforts. There must be a close working relationship between audit team members during the interview phase. Often the findings of one auditor overlap into another auditor's area and may have an unforeseen impact. A sharing of knowledge and data is needed on a daily basis; therefore "end-of-day" meetings pay off and the audit team chief will ensure that communications occur for mutual benefits. As the audit progresses, the individual auditors begin to arrive at preliminary conclusions or findings. These need to be identified as the interview phase continues for each major subject area that is being reviewed. Once they surface, they need to be challenged and discussed by the audit team to ascertain the validity and sufficiency of data. This will assist the auditor in writing his portion of the report.

Since audit teams may have persons with experience in both auditing and specialized areas, it stands to reason that individual capability is enhanced. The team achieves the most benefit when these areas complement and supplement each other. At the start of an audit, it is often best to have a person who has never conducted an audit interview get his feet wet by accompanying an experienced auditor. Although the audit team chief continually works to get the working interface established early, experience has revealed that the individual auditors are so preoccupied with their own areas that during the early phase they fail to see the need for exchange. As the audit progresses, this attitude usually changes and each auditor becomes a better team player.

## Selection of Audit Areas

The development of an audit plan, regardless of the amount of detail contained in the plan, is essential at the start. The team chief will have a general agreement with those in authority who have directed that an audit be performed. The scope, along with the reason or purpose of the project audit, has been included in the initial announcement which sets the stage for outlining the audit plan. The contents of this plan should include, but need not be limited to, the following subject areas:

*Audit Plan for Project XX*

I. Purpose
II. Scope
III. Approach for Conduct
  A. Team Assignments
  B. Schedule/Itinerary
IV. Audit Areas
V. Interview Questions by Area

In the case of a new project, an audit plan is usually developed to cover all management areas that could affect the success of the project. In this event, a typical plan would include the following areas to be audited: Example:

IV—*Audit Areas for Project XX*

A. Organization
B. Policies and Procedures
C. Master Planning and Control
D. Work Authorization
E. Contract Administration
F. Engineering
G. Manufacturing
H. Quality Control
I. Test
J. Logistics Support
K. Customer Requirements
L. Vendor Support

The assignment of individual auditors to specific areas is usually done according to the background and experience of the auditor. On a large-scale project it may be necessary to have more than one auditor per area, and whenever possible this is desirable since the combined talents of two may be needed. The scope, size, and sophistication of the project will dictate the number and assignment of auditors to review the many subject areas involved.

### Interview Techniques

Interviews for the auditors should be made at least one day in advance. As discussed above, this is best accomplished by having one person arrange all interviews for the project team. Schedules always require some negotiating and last-minute changes may occur; however, it is expected that people will make themselves available when the auditors request.

The time scheduled for an interview should usually be not less than one hour and not more than two. In many cases the best practice is to have a return interview rather than extend the time to a half a day. Auditors often need time to assess the information and data provided or to confer with other members of the audit team before deciding what additional data is needed.

Prior to the actual interview, the auditor should make preparations by familiarizing himself with both background data and information on the responsibilities of the individual involved. In addition, it is necessary to be aware of the status of progress in the area as well as any known problems. The auditor, of course, will only have enough knowledge to be conversant during the early interviews; however, as time goes by, he will be continually adding to his own knowledge and his proficiency at interviewing will improve. For those auditors who are interviewing for the first time, the development and use of a list of interview questions are a must. As expected for those persons who have been auditing projects for several years, their need of a checklist of questions is more for reminders than being dependent upon the use of such a tool. The following examples of checklists of interview questions are presented only to encourage their development by the audit team members before the interviews.

Example: Interview Questions
for: *Organization and Management*

1. Request organizational charts for the project, delineating relationships between operations.
   a. Clearly indicate who reports to whom.
   b. How does structure reflect department management emphasis on the project?
   c. Where and how are project management responsibilities defined?
   d. How many people are actually managing the project effort, and what authority and responsibilities are delegated to each?
2. Does the project have its own policies and procedures for assignment of responsibilities and work accomplishment?

Example: Interview Questions
for: *Logistics Support*

1. What method is used to assure the timely delivery of spare parts?
2. What method is used to assure that instructions in manuals will allow the accurate operation and maintenance of equipment?

During the interview the auditor is a listener, asking only those questions necessary to keep the discussion on track. Requesting copies of a document needed is the best approach rather than taking time to write lengthy descriptions. The use of a list of interview questions will enable the auditor to keep the interview moving along a logical path, and will permit the taking of short notes on the replies to questions asked. In addition to the modus operandi of the auditor in pursuing the subject material, the auditor's style is important. Since he is not an investigator looking for violation of law, it is essential that the tenor of the conversation be friendly. When a healthy rapport develops between the auditor and interviewee, the likelihood of productive results increases dramatically. In contrast, an adversary situation will make it difficult to get the data needed and will also be a mutually unpleasant experience. If at all possible, it pays to be a nice guy.

### Development/Preparation of Findings

As the interviews are completed, and data are assembled and analyzed, the auditor will have reached conclusions which can then be identified as Findings. This represents the culmination of the auditor's work and requires concerted effort to ensure that each Finding is accurately stated, is fully supported, and will stand up to challenge. The format for documenting the Findings is simple and provides the framework to report the auditor's efforts. This approach and format have been used by many auditors.

*FINDINGS*

Subject: (Use a short descriptive title)
Finding No.: (Number by subject area)
State the Finding briefly but include a statement that describes the problem or outstanding condition, the cause, and the effect.
Discussion: Present as thorough and comprehensive an analysis of the condition as is necessary to prove the statements in the Finding. Include corrective actions at end.
Recommendations: State what action must be taken and by what office or position in the organization.

This phase of analyzing and writing the Findings often requires as much time as was spent in the planning, preparation, and undertaking of the interview. Seldom is an initial draft of a Finding adequate in statement of the problem or in the supporting evidence of the discussion. This is not a quick and easy task and deserves whatever time it takes to do it justice. The involvement of other team members is a good practice, for the view and perspective should help in the writing and the final acceptance by the team chief and other members.

### Validation of Findings

In the preparation of the Findings, and before they are finalized, it is good to go back and discuss them with those individuals directly involved. Most frequently the people involved and concerned the most are really the first to realize what may have gone wrong and what is necessary to correct the situation. In addition, since the individuals involved have had a hand in revealing the situation, they are more willing to accept the Findings, and may assist in determining what the recommended solutions should be. There may be times when it seems impossible to discuss the Findings without a confrontation; but this is usually the exception. In most instances, all levels of management will cooperate and appreciate a postmortem critique. This has a double effect, for if they accept the Findings, they have now become a part of the solution—which you want them to be. Taking the Findings and achieving confirmation of them on up to the project manager is the goal, and it will mean more when the complete audit results are presented to general management.

## REPORTING RESULTS OF THE AUDIT

### Report Preparation

A report is the end product of the combined efforts of the audit team. The payoff of the audit effort and the effectiveness of follow-up actions will depend upon the manner in which the report is written and presented. Findings must be clearly stated, and discussions must contain only factual information to support each Finding. Avoid lengthy philosophies, opinions, and observations. Where credit is deserved it should be recognized and receive equal treatment in comparison to problem areas and deficiencies. The team chairman should hold frequent coordination meetings to review progress and accomplish the interexchange of information. Drafts of Findings should be made as early as possible since considerable review is usually necessary to get agreement within the team, accomplish validation with those involved, and finally satisfy the team chairman. Experience has shown that preparing the audit results in report and briefing form will take as many or more hours as conducting the interviews and analysis. Another rule of thumb when writing the Findings is that most often it may take five iterations before achieving the final version for use in the report. There is a natural tendency to rush this final phase of the audit since those on temporary assignment will be anxious to return to their regular jobs. At this point, it is the team chairman who must hold to accepting nothing less than a well-expressed, accurate, and complete report that all can be proud of.

### Report Format

The results of the project audit must be presented to the person or persons who originally directed and requested its accomplishment. This usually requires a verbal briefing/presentation and a written report. Typically, the briefing would be a summary of the written report. The following outline is recommended for a project audit report:

AUDIT REPORT FORMAT

*PART I—INTRODUCTION*

*Section I—Purpose*

(Give a brief explanation of any special reasons that audit is being conducted.)

*Section II—Scope*

(Give a description of scope of audit including limitations imposed.)

*Section III—Audit Team*

(List team membership by name, title, and organization.)

*Section IV—Audit Interviews*

(List all persons interviewed, by name, title, and organization.)

*PART II—AUDIT RESULTS*

*Section I—Summary and Recommendations*

The summary of results will be a one-page abstract of the major findings and recommendations. Following each specific recommendation will be the action office responsible for that recommendation.

*Section II—Findings, Discussion, Recommendations*

This portion of the report will contain the detailed Findings, discussion, and recommendations that pertain to the program.

1. Subject: Use a short descriptive title, e.g., "Overtime."
2. Finding: The Finding should be brief but include (1) a statement that describes the condition, i.e., problem or outstanding condition; (2) the cause or reason for this condition or problem; and (3) the effect or impact resulting from the condition. *Summarized, the Finding should reflect a condition, a cause, and the effect.*

3. Discussion: Mention the pertinent factors collected during discussion with others or revealed through your personal investigations. Present as thorough and comprehensive an analysis of the condition as necessary to prove the statements in your Finding.
4. Recommendations: If corrective action(s) are suggested by the Finding, they should be recorded at the end of the discussion. Following each recommendation, note the action assignments.

*PART III—SUPPLEMENTARY DATA*

Note: The appendices listed are for guidance only and will not necessarily apply to each audit report. Conciseness should be employed. As an example, data under appendices for program history and description of system should not normally exceed one page each.

Appendix A—Project history
B—Description of system
C—Documentation and reporting
D—Programming and funding history
E—Customer organization
F—Program organization and management controls

### Briefings to Management

This is another time for respecting protocol and recognizing prerogatives. Just as it was important during the audit to start at the top and work down, it is now important to start at the bottom and work up. Early discussion of Findings with the working persons involved will establish credibility and ensure that data is accurate and complete. When the report is essentially complete and a briefing structured, there may be an opportunity to have a dry run with a second level of the project team which could provide a shakedown and then the chance for a final tune-up before a more formal review with the project manager and his staff. At this preview, the report should be 98% solid, with no holes or obvious shortcomings, and hopefully the project manager will not only endorse the report but say it's a job well done.

Since project managers seldom have all the resources under their control, the need to reach top management is absolutely necessary. It is most likely that a majority of the Findings will require decisions and actions by the functional managers. Further, it is the man at the top who can make sure it all happens—if he is convinced of the project's needs. A good briefing is the best way to get the audit results to the top management team. The team chairman may elect to do the entire briefing or to include members of the audit team to cover their specialized areas. If at all possible, members of the audit team should be included for it provides them an opportunity for recognition which is due. Copies of the briefing charts should be provided to all recipients of the report for they serve as good summaries and ready references. This last step of the audit process which requires effective written and verbal communication becomes the final measure for judging the degree of success of the entire audit effort.

## FOLLOW-UP ACTIONS

### Responsibilities

The auditor, in the recommendations for the Findings, should make every effort to determine the organization and person responsible for taking corrective action. In some instances there may be shared assignments; in others assignments may not be absolutely clear-cut. Experience has shown that final assignment of responsibilities for each and every recommendation may occur during the briefing to management. This level makes the final decisions, and their acceptance and involvement are an important step to the follow-up activity to the project audit.

It has been known that some audit operations report only the conditions and are not required to make recommendations or identify responsibilities for corrective actions. As a general practice, this is not recommended for it does not take full advantage of the audit team's capabilities and tends to shackle their initiative and limit their contribution.

### Closeout of Report

The audit team, as a part of the audit plan, should present a follow-up and closeout plan at the time of final briefings. This provides an organized means to get actions under way by the responsible individuals. When top management accepts the plan requiring a 30-day report and a 90-day final report from all involved in action assignments, then management direction has occurred and there is a control system to ensure response. At the time of these two reports to management, the audit team chairman and other appropriate team members should be present. This has several benefits, for if actions described are not adequate to correct the condition entirely, then the audit team member should have the opportunity to express his concern. Also this enables him to see the completion of his efforts, and to see a job completed is a part of the final reward.

## EVALUATION OF AUDIT FUNCTION

### Management Assessment

Since the use of the audit function of projects or other specialized subject areas is a tool that best serves the interests of management, it stands to reason that they should periodically question the value it serves. In the event that the audit results are not sufficient to warrant the use of this tool, management should determine what is required to make it more effective or do away with it. Experience has shown that the audit tool has been successful when top management has actually used it and supported it. This is also true for other internal management consultant functions which are staff support activities. A good example of the right environment was the approach used by a top executive of a high-technology firm who immediately after winning a new contract would assess past performance for needed improvements. Sometimes he would ask for an assessment by his entire team so that the next project could benefit from the lessons learned. The alternative is to continue either getting by or even repeating the same mistakes.

### Applying "Lessons Learned"

The greatest opportunity for payoff from project management auditing may often occur for the next project in order to avoid early shortcomings in applied manpower, policies, and procedures on the next project. The results of a project audit during any phase will reveal problems which, if given visibility and understanding, can assist the next project manager and top management to avoid or reduce the probability of similar deficiencies occurring again. Although there is universal lip service in recognizing the potential of lessons learned, the action needed to correct such conditions is too often lacking. Here is where top management can take direct action and change people, resources, policies and procedures, and their own involvement. For example, the lessons learned from one project audit could provide the basis for the following plan to be required on the management of new projects:

- More frequent management reviews in the first year of the project.
- More careful selection of key people with proven experience.
- Early assessment of customer satisfaction.
- Early assessment of test results.

As can be expected, people and methods are slow to change; therefore it is the responsibility of management to make changes happen. Results of each project audit will enable the project manager to better see himself and his team and determine the immediate needs and changes to be initiated. Response to such needs may work best when self-initiated, so top management should permit and encourage corrective actions whenever possible. However, there are other times when only management direction gets things done.

Whereas the impatience and daring of a new project manager are often to be admired, it is concluded that as a great philosopher—George Santayana—said, "Those who cannot remember the past are condemned to repeat it."

# 26. Evaluating the Performance of Project Personnel

Harold Kerzner*

In most traditional organizations, the need for project management is first recognized by those functional, resource, or middle managers who have identified problems in allocating and controlling resources. The next step is the tedious process of trying to convince upper-level management that such a change is necessary. Assuming that upper-level management does, in fact, react favorably toward project management, the next step becomes critical. Many upper-level managers feel that project management can be forced on lower-level subordinates through simple directives together with continuous upper-level supervision.

This turns out to be a significant turning point in the implementation phase. Upper-level management must obtain functional employee support before total implementation can be achieved. Functional employees have two concerns. Their first concern is with their evaluation. Who will evaluate them? How will they be evaluated? Against what standards will they be evaluated? Who will help them put more money into their pockets through merit increases or promotions?

The employee's second concern is centered about the resistance to change. Functional employees, especially blue-collar workers, have a strong resentment to changing their well-established occupational life styles. They must be shown enough cases (i.e., projects) in order to be convinced that the new system will work. This could easily take two to

* Dr. Harold Kerzner is Professor of Systems Management and Director of The Project/Systems Management Research Institute at Baldwin-Wallace College. Dr. Kerzner has published over 35 engineering and business papers, and ten texts: *Project Management: A Systems Approach to Planning, Scheduling and Controlling; Project Management for Executives; Project Management for Bankers; Cases and Situations in Project/Systems Management; Operations Research; Proposal Preparation and Management; Project Management Operating Guidelines; Project/Matrix Management Policy and Strategy; a Project Management Dictionary of Terms; Engineering Team Management*.

three years to accomplish. It is therefore imperative that the first few projects be successful. Most upper- and middle-level managers agree to the necessity for initially demonstrating success, but often forget about the importance of looking at the evaluation procedure problems.

## UNDERSTANDING THE NATURE OF THE PROBLEM

In most project situations, the functional employee reports to at least two bosses: a functional manager and a project manager. If the employee happens to be working on three or four projects simultaneously, then he or she can have multiple project managers to whom they must report, either formally or informally. This concept of sharing functional employees is vital if project management is to be successful because it allows better control and use of vital manpower resources by allowing key functional personnel to be shared.

In almost all cases, the relationship between the employees and their superior is a "solid" line where the manager maintains absolute employee control through the use of promotions, job assignments, merit and salary increases. The ability of the manager to motivate personnel is easily achieved through the use of the employee's purse strings.

The project manager, on the other hand, will probably be in a "dotted" line relationship and be less able to motivate temporarily assigned project personnel by using monetary rewards. Therefore, what types of interpersonal influences can a project manager use to motivate people who are assigned temporarily for the achievement of some common objective and who might never work together again? The most common interpersonal influence styles used by project managers are:

- Formal authority.
- Technical expertise.
- Work challenge.
- Friendship.
- Rewards (and punishment).

Formal authority is the ability to gain support from the functional employees because they respect the fact that the project manager has been delegated a certain amount of authority from upper-level management in order to achieve a specific objective. The amount of delegated authority may vary with the amount of risk that the project manager must take. Formal authority is not a very effective means of motivating and controlling employees because all the employees know that they have come to the project recommended by their managers.

Technical expertise is the ability to gain support because employees respect the fact that the project manager possesses skills which they lack or because he or she is a recognized expert in their field. If a project manager tries to control employees through the use of "expert power" for a prolonged period of time, conflicts can easily develop between the project and functional managers as to who is the "true" expert in the field.

Work challenge is an extremely effective means of soliciting functional support. If the employees find the work stimulating and challenging, they tend to become self-motivating with a strong desire for achievement in hopes of attaining some future rewards.

Friendship, or referent power, is a means of obtaining functional support because the employee feels personally attracted to either the project manager or the project. Examples of referent power might be when:

- The employee and the project manager have strong ties, such as being in the same foursome for golf.
- The employee likes the project manager's manner of treating people.
- The employee wants specific identification with a specific product line or project.
- The employee has personal problems and believes that he can get empathy or understanding from the project manager.
- The employee might be able to get personal favors from the project manager.
- The employee feels that the project manager is a winner and that the rewards will be passed down to the employee.

Rewards, or reward power, can be defined as the ability to gain support because the employee feels that the project manager can either directly or indirectly dispense those rewards which employees cherish. If the employee is assigned directly to the project manager, such as project office personnel, the project manager has the same direct rewarding system as does a functional manager. However, the project manager may have only indirect reward power with regard to the employees that are assigned to the project but are still attached administratively to a functional department. This chapter focuses on those problems of rewarding the temporary functional employees. The last two items under friendship are examples of reward power as well as referent power. Project managers prefer work challenge and rewards as the most comfortable means for soliciting functional support. Unfortunately, project managers are somewhat limited as to what rewards they can offer directly to the employee. What commitment should or can the project manager make in the way of:

- Salary?
- Grade?
- Responsibility?
- Evaluation for promotion?
- Bonus?
- Future work assignment?
- Paid overtime?
- Awards?
- Letters of commendation?

The major problem with project management is that, in theory, the project manager can *directly* provide only paid overtime rewards, and even this can be questionable. If the project manager cannot directly provide the necessary organizational rewards in order to motivate temporary employees, then what inducement is there for the employee to do a good job? Employees believe in the equity theory, which states that a fair day's work should receive a fair day's pay. The difficulty lies in the fact that employees occasionally perceive themselves as working for a project manager who cannot guarantee them any of these awards.

A special note need be mentioned concerning a project manager's ability or authority to provide an employee with additional responsibility. Most companies that adopt project management have rather loose company policies, procedures, rules, and guidelines. To illustrate this point, a functional employee can be performing the same task on three separate projects, yet his responsibilities might be quite different.

The problem appears when the project manager attempts to upgrade an employee. As an example, a Grade 7 employee does a good job on a Grade 7 task and receives an excellent evaluation. The project manager, having established a good working relationship with this employee, and not wanting to see it end, decides to let this employee assume the responsibilities of a Grade 8 on a follow-on task. The employee again performs above average and receives an outstanding evaluation by the project manager. The employee then demands that his manager promote him since he has now successfully performed the work of a Grade 8. The manager now becomes overly upset and claims that the project manager had no right to upgrade an employee without prior approval from the manager.

### THE INDIRECT REWARDING PROCESS

Under the definition of rewards, we stated that the project manager could either directly or indirectly dispense the valued organizational rewards.

Each project, although considered as a separate entity within the company, is still attached administratively to the company through policies and procedures. These policies and procedures dictate the means of administering the wage and salary program. It is operationally disastrous for the project manager and functional manager to be administering different wage and salary policies at the same time.

When employees are assigned to a new project, their first concern is with the identification of the mechanism by which they can be assured that their manager will be informed if they perform well on their new assignment. A good project manager will make it immediately clear to all new employees that if they perform well on this effort, then he (the project manager) will inform their manager of their progress and achievements. This assumes that the manager is not providing close supervision over the employee and is, instead, passing on some of the responsibility to the project manager. This is quite common in project management organizational structures. Obviously, if the manager has a small span of control and/or sufficient time to monitor closely the work of subordinates, then the project manager's need for indirect reward power is minimal.

Many good projects as well as project management structures have failed because of the inability of the system to properly evaluate the employee's performance. This problem is, unfortunately, one of the most often overlooked trouble spots in project management.

In a project management structure there are basically six ways that an employee can be evaluated on a project.

- *The project manager prepares a written, confidential evaluation and gives it to the functional manager.* The line managers will evaluate the validity of the project manager's comments and prepare their own evaluation of the employee. The employee will be permitted to see only the evaluation form filled out by the line manager.
- *The project manager prepares a nonconfidential evaluation and gives it to the functional manager.* The project manager prepares his own evaluation form and both evaluations are shown to the functional employee. This is the technique preferred by most project and functional managers. However, there are several major difficulties with this technique. If the employee is an average or below-average worker, and if this employee is still to be assigned to this project after the evaluation, the project manager might rate the employee as above average simply to prevent any repercussions or ill feelings downstream. In this situation the manager might want a confidential evaluation instead knowing that the employee will see both evaluation forms. Employees tend to blame the project manager if they receive a

below-average merit increase, but give credit to the manager if the increase is above average. The best bet here is for the project manager to periodically inform the employees as to how well they are doing, and to give them an honest appraisal. Of course, on large projects with vast manpower resources, this approach may not be possible. Honesty does appear to be the best policy in project management employee evaluation.

- *The project manager provides the functional manager with an oral evaluation of the employee's performance.* Although this technique is commonly used, most functional managers prefer documentation on employee progress.
- *The functional manager makes the entire evaluation without any input from the project manager.* In order for this technique to be effective, the functional manager must have sufficient time to supervise each subordinate's performance on a continual basis. Unfortunately, most functional managers do not have this opportunity because of their broad span of control.
- *The project manager makes the entire evaluation for the functional manager.* This technique can work if the employee is assigned to only the one project or if the project is physically located at a remote site where he cannot be observed by his functional manager.
- *The project and functional managers jointly evaluate all project functional employees at the same time.* This technique may be limited to small companies with less than fifty or so employees; otherwise the evaluation process might be time consuming for key personnel. A bad evaluation is known by all.

In five of the above six techniques the project manager has either a direct or indirect input into the employee's evaluation process.

## WHEN AND HOW TO EVALUATE

Since the majority of the functional managers prefer written, nonconfidential evaluations, we must determine what the evaluation forms look like and when the employee will be evaluated. The indirect evaluation form should be a relatively simple tool to use or else the indirect evaluation process will be time consuming. This is of paramount importance on large projects where the project manager may have as many as 200 employees assigned to various activities.

The evaluation forms can be filled out either when the employee is up for evaluation or after the project is completed. If the evaluation form is to be filled out when the employee is up for promotion or a merit increase, then the project manager should be willing to give an *honest* appraisal of the employee's performance. Of course, the project manager should not fill out the evaluation form if the employee has not been assigned long enough to allow a fair evaluation.

The evaluation form can be filled out at the termination of the project. One problem with this technique is that the project may end the month after the employee is up for promotion. One advantage of this technique is that the project manager may have been able to find sufficient time both to observe the employee in action and to see the complete output.

| Performance factors | Excellent (1 out of 15) | Very good (3 out of 15) | Good (8 out of 15) | Fair (2 out of 15) | Unsatisfactory (1 out of 15) |
|---|---|---|---|---|---|
| | Far exceeds job requirements | Exceeds job requirements | Meets job requirements | Needs some improvement | Does not meet minimum standards |
| Quality | Leaps tall buildings with a single bound | Must take running start to leap over tall building | Can only leap over a short building or medium one without spires | Crashes into building | Cannot recognize buildings |
| Timeliness | Is faster than a speeding bullet | Is as fast as a speeding bullet | Not quite as fast as a speeding bullet | Would you believe a slow bullet? | Wounds himself with the bullet |
| Initiative | Is stronger than a locomotive | Is stronger than a bull elephant | Is stronger than a bull | Shoots the bull | Smells like a bull |
| Adaptability | Walks on water consistently | Walks on water in emergencies | Washes with water | Drinks water | Passes water in emergencies |
| Communications | Talks with God | Talks with angels | Talks to himself | Argues with himself | Loses the argument with himself |

Figure 26-1. Guide to performance appraisal.

| EMPLOYEE'S NAME | DATE |
|---|---|
| PROJECT TITLE | JOB NUMBER |
| EMPLOYEE ASSIGNMENT | |
| EMPLOYEE'S TOTAL TIME TO DATE ON PROJECT | EMPLOYEE'S REMAINING TIME ON PROJECT |

TECHNICAL JUDGEMENT:

| | | | | |
|---|---|---|---|---|
| ☐ Quickly reaches sound conclusions | ☐ Usually makes sound conclusions | ☐ Marginal decision making ability | ☐ Needs technical assistance | ☐ Makes faulty conclusions |

WORK PLANNING:

| | | | | |
|---|---|---|---|---|
| ☐ Good planner | ☐ Plans well with help | ☐ Occasionally plans well | ☐ Needs detailed instructions | ☐ Cannot plan at all |

COMMUNICATIONS:

| | | | | |
|---|---|---|---|---|
| ☐ Always understands instructions | ☐ Sometimes needs clarification | ☐ Always needs clarifications | ☐ Needs follow-up | ☐ Needs constant instruction |

ATTITUDE:

| | | | | |
|---|---|---|---|---|
| ☐ Always job interested | ☐ Shows interest most of the time | ☐ Shows no job interest | ☐ More interested in in other activities | ☐ Does not care about job |

COOPERATION:

| | | | | |
|---|---|---|---|---|
| ☐ Always enthusiastic | ☐ Works well until job is completed | ☐ Usually works well with others | ☐ Works poorly with others | ☐ Wants it done his/her way |

WORK HABITS:

| | | | | |
|---|---|---|---|---|
| ☐ Always project oriented | ☐ Most often project oriented | ☐ Usually consistent with requests | ☐ Works poorly with others | ☐ Always works alone |

ADDITIONAL COMMENTS: ____________________

Figure 26-2. Project work assignment appraisal.

Figure 26-1 represents a rather humorous version of how project personnel perceive the evaluation form to look. Unfortunately, the evaluation process is very serious and can easily have a severe impact on an individual's career path with the company even though the final evaluation rests with the manager.

Figure 26-2 shows a simple type of evaluation form where the project manager identifies the box that best describes the employee's performance. The project manager may or may not make additional comments. This type of form is generally used whenever the employee is up for evaluation, provided that the project manager has had sufficient time to observe the employee's performance.

| EMPLOYEE'S NAME | DATE |
|---|---|
| PROJECT TITLE | JOB NUMBER |
| EMPLOYEE ASSIGNMENT | |
| EMPLOYEE'S TOTAL TIME TO DATE ON PROJECT | EMPLOYEE'S REMAINING TIME ON PROJECT |

| | Excellent | Above average | Average | Below average | Inadequate |
|---|---|---|---|---|---|
| Technical judgement | | | | | |
| Work planning | | | | | |
| Communications | | | | | |
| Attitude | | | | | |
| Cooperation | | | | | |
| Work habits | | | | | |
| Profit contribution | | | | | |

Additional comments ______________________________________________

______________________________________________

Figure 26-3. Project work assignment appraisal.

Figure 26-3 shows a typical form that can be used to evaluate an employee at project completion. In each category the employee is rated on a scale from one to five. In order to minimize time and paper work, it is also possible to have a single evaluation form at project termination for all employees (Figure 26-4). As before, all employees are rated in each category on a scale of one to five. Totals are obtained to provide a relative comparison between employees.

Even though the project manager fills out an evaluation form, there is no guarantee that the functional manager will give any credibility to the project manager's evaluation. There are always situations where the project and functional managers disagree as to either quality or direction of work. This can easily alienate the project manager into recommending either a higher or lower evaluation than the employee's work justifies. If

| EMPLOYEE'S NAME | DATE |
|---|---|
| PROJECT TITLE | JOB NUMBER |
| EMPLOYEE ASSIGNMENT | |
| EMPLOYEE'S REMAINING TIME ON PROJECT | EMPLOYEE'S TOTAL TIME TO DATE ON PROJECT |

CODE:

Excellent = 5
Above average = 4
Average = 3
Below average = 2
Inadequate = 1

| NAMES | Technical judgement | Work planning | Communications | Attitude | Cooperation | Work habits | Profit contribution | Self motivation | Total points |
|---|---|---|---|---|---|---|---|---|---|
| | | | | | | | | | |
| | | | | | | | | | |
| | | | | | | | | | |
| | | | | | | | | | |
| | | | | | | | | | |
| | | | | | | | | | |
| | | | | | | | | | |
| | | | | | | | | | |

Figure 26-4. Project work assignment appraisal.

the employee spends most of his time working alone, then the project manager may have difficulty appraising quality and give an average evaluation when in fact the employee's performance is superb or inferior. There is also the situation where the project manager knows the employee personally and may allow personal feelings to influence the evaluation.

Another problem situation is where the project manager is a "generalist," say at a Grade 7 level, and requests that the functional manager assign his best employee to the project. The functional manager agrees to the request and assigns his best employee, a Grade 10. Now, how can a Grade 7 generalist evaluate a Grade 10 specialist? The solution to this problem rests in the fact that the project manager might be able to evaluate the expert only in certain categories such as communications, work habits, problem solving, and other similar topics but not upon his technical expertise. The functional manager might be the only person qualified to evaluate personnel on technical abilities and expertise.

It has been proposed that employees should have some sort of reciprocal indirect input into a project manager's evaluation. This raises rather interesting questions as to how far we can go with the indirect evaluation procedure.

From a top management perspective, the indirect evaluation process brings with it several headaches. Wage and salary administrators readily accept the necessity for utilizing a different evaluation form for white-collar workers as opposed to blue-collar workers. But now, we have a situation in which there can be more than one type of evaluation system for white-collar workers alone. Those employees who work in project-driven functional departments will be evaluated directly and indirectly, but based upon formal procedures. Employees who charge their time to overhead accounts and nonproject-driven departments might simply be evaluated by a single, direct evaluation procedure.

Many wage and salary administrators contend that they cannot live with a dual white-collar system and therefore have tried to combine the direct and indirect evaluation forms into one, as shown in Figure 26-5. Some administrators have gone so far as to utilize a single form company-wide, regardless of whether an individual is a white- or blue-collar worker.

The last major trouble spot is the design of the employee's evaluation form. The designs must be dependent upon the evaluation method or procedure. Generally speaking, there are nine methods available for evaluating personnel:

- Essay Appraisal
- Graphic Rating Scale
- Field Review
- Forced-Choice Review
- Critical Incident Appraisal
- Management by Objectives (MBO)
- Work Standards Approach
- Ranking Methods
- Assessment Center

Descriptions of these methods can be found in almost any text on wage and salary administration. Which method is best suited for a project-driven organizational structure? To answer this question, we must analyze the characteristics of the organizational form as well as those of the

I. Employee information

1. Name ______________________ 2. Date of evaluation ______________________

3. Job assignment ______________________ 4. Date of last evaluation ______________________

5. Pay grade ______________________

6. Employee's immediate supervisor ______________________

7. Supervisor's level: ☐ Section ☐ Dept. ☐ Division ☐ Executive

II. Evaluator's information:

1. Evaluator's name ______________________

2. Evaluator's level: ☐ Section ☐ Dept. ☐ Division ☐ Executive

3. Rate the employee on the following:

| | Excellent | Very Good | Good | Fair | Poor |
|---|---|---|---|---|---|
| Ability to assume responsibility | | | | | |
| Works well with others | | | | | |
| Loyal attitude toward company | | | | | |
| Documents work well and is both cost and profit conscious | | | | | |
| Reliability to see job through | | | | | |
| Ability to accept criticism | | | | | |
| Willingness to work overtime | | | | | |
| Plans job execution carefully | | | | | |
| Technical knowledge | | | | | |
| Communicative skills | | | | | |
| Overall rating | | | | | |

4. Rate the employee in comparison to his contemporaries:

| Lower 10% | Lower 25% | Lower 40% | Midway | Upper 40% | Upper 25% | Upper 10% |
|---|---|---|---|---|---|---|
| | | | | | | |

5. Rate the employee in comparison to his contemporaries:

| Should be promoted at once | Promotable next year | Promotable along with contempories | Needs to mature in grade | Definitely not promotable |
|---|---|---|---|---|
| | | | | |

6. Evaluator's comments: ______________________

______________________

Signature ______________________

III. Concurrence section:

1. Name ______________________

2. Position: ☐ Department ☐ Division ☐ Executive

3. Concurrence ☐ Agree ☐ Disagree

4. Comments: ______________________

______________________

Signature ______________________

IV. Personnel Section: (to be completed by the Personnel Department only)

6/79
6/78
6/77
6/76
6/75
6/74
6/73
6/72
6/71
6/70

Lower 10% | Lower 25% | Lower 40% | Midway | Upper 40% | Upper 25% | Upper 10%

V. Employee's signature ______________________ Date: ______________________

Figure 26-5. Job evaluation.

personnel who must perform there. As an example, project management can be described as an arena of conflict. Which of the above nine evaluation procedures can best evaluate an employee's ability to work and progress in an atmosphere of conflict? Figure 26-6 compares the above nine evaluation procedures against the six most common project conflicts. This type of analysis must be carried out for all variables and characteristics which describe the project management environment. Many compensation managers would agree that the MBO technique offers the greatest promise for a fair and equitable evaluation of all employees. Unfortunately, MBO implies that functional employees will have a say in establishing their own goals and objectives. This might not be the case. In project management, the project manager or functional manager might set the objectives and the functional employee is told that he has to live with it. Obviously, there will be advantages and disadvantages to whatever evaluation procedures are finally selected.

Having identified the problems with employee evaluation in a project environment, we can now summarize the results and attempt to predict the future. Project managers must have some sort of either direct or indirect input into an employee's evaluation. Without this, project managers may find it difficult to adequately motivate people with no upward mobility. The question is, of course, how this input should take place. Most wage and salary administrators appear to be pushing for a single procedure to evaluate all white-collar employees. At the same time, however, administrators recognize the necessity for an indirect input by the

| | Essay appraisal | Graphic rating scale | Field review | Forced-choice review | Critical incident appraisal | Management by objectives | Work standards approach | Ranking methods | Assessment center |
|---|---|---|---|---|---|---|---|---|---|
| Conflict over schedules | • | • | | • | • | | • | • | |
| Conflict over priorities | • | • | | • | • | | • | • | |
| Conflict over technical issues | • | | | • | | | • | | |
| Conflict over administration | • | • | • | • | | | • | • | • |
| Personality conflict | • | • | | • | | | • | | |
| Conflict over cost | • | | • | • | • | | • | • | • |

Figure 26-6. Rating evaluation techniques against types of conflicts.

project manager and, therefore, are willing to let the project and functional managers (and possibly personnel) determine the exact method of input, which can be different for each employee and each project. This implies that the indirect input might be oral for one employee and written for another, with both employees reporting to the same functional manager. Although this technique may seem confusing, it may be the only viable alternative for the future. A process of good employee evaluations is essential to project success.

# 27. Development and Implementation of Project Management Systems

John Tuman, Jr.*

As we move forward toward the twenty-first century, project managers in all fields of endeavor will be called upon to accomplish their responsibilities in an increasingly complex technical, social-economic, and political environment. Thus, project managers will deal with a broad range of issues, requirements, and problems and make decisions which involve complexities and risk well beyond anything experienced in the past. Yet the project manager will still be expected to be the focal point for planning, scheduling, measuring, evaluating, informing, and directing a myriad of organizations, complex tasks, and valuable resources to accomplish some specific technological or business undertaking.

As always, the project manager's situation will be doubly demanding because he functions for the most part outside the traditional organizational structure. He must complete a difficult job by certain dates with limited resources, and in most cases he must accomplish his goals with people and organizations who do not work for him. In addition, the project manager is considered to be the fountainhead of knowledge and information about every aspect of the project. In this type of setting it goes without saying that the project manager's only hope for survival rests with some type of well-developed system for the systematic management of project information and action. Without such a system, the project

* John Tuman, Jr., is a Senior Consultant with Gilbert/Commonwealth, Inc., a major U.S. engineering and consulting firm. Mr. Tuman has been involved in all aspects of project management for more than 25 years. He has been a project manager and a program manager on several major military, aerospace, and R&D programs for the General Electric Company and the AVCO Corporation. In recent years he has directed the development of advanced computer-based project management systems as well as providing consulting expertise to major companies in the United States, the Middle East, and Asia. He has given numerous presentations and seminars on project management and has written extensively on project management methods, techniques, systems, and problems. Mr. Tuman is a registered professional engineer; he has a M.S. degree in Computer Science and B.S. degree in Mechanical Engineering.

manager and the project participants will soon be lost in a quagmire of conflicting plans, schedules, reports, activities, and priorities.

Experience has shown that as projects become large, or more numerous, the work environment becomes more complex, and the number of people, organizations, functions, and activities involved become increasingly interdependent. As this happens, the project manager tends to become further and further removed from the total day-to-day project requirements and problems, and his ability to deal with real-time issues begins to be seriously diminished. In short, the effectiveness of the project manager and the project participants tends to be inversely related to the size and/or complexities of the project. However, this process can be minimized if we provide the project manager with "systems" which are specifically designed to increase his range of control and effectiveness over project activities. Thus, in the following we will discuss why it is necessary to design unique systems for project management, how to actually design and develop these systems, and finally, how to implement these systems to ensure that they perform as intended.

## WHY DEVELOP A SYSTEM?

Before we consider the approach to defining, designing, and developing a system we must clearly establish that there is a need for such a system. In view of the large investments that have been made by corporations in the development of management information systems in recent years, one would question the need for development of yet another system within the corporate environment. And yet, this is exactly what is proposed here. The soundness of our rationale becomes evident if we examine the role of the project manager and his function within the traditional corporate hierarchical structure.

Traditionally, management information systems have been designed primarily to support functional units within the corporate structure. The computerized accounting system, the payroll system, the general ledger system, and so on provide systematized approaches to handling the corporate financial functions. In similar fashion, computer-supported systems for marketing and sales exist to aid and improve the efficiency of these functions, and personnel subsystems have been developed to aid the human resource functions of the corporation. In the operational or production areas (manufacturing, construction, services, tests, etc.) there are a myriad of computer-supported systems available to aid the product/service-producing end of the business. By and large, all of these systems have been designed to perform as efficiently as possible to collect and process data to produce information for their respective organizational functions. In more advanced management information systems designs, we see the utilization of the Data Based Management System (DBMS), which is designed to efficiently share information between functions. However, for the most part management information systems in today's corporate environment are designed to support decision making in the traditional hierarchical organizational structure.

Unfortunately, by the very nature of project management, the project manager must cut across functional organization lines to accomplish his goal of integrating and directing specific resources of the organization(s) toward a particular goal. The question naturally arises, can he do this effectively utilizing the traditional available management information resources?

In the earlier days of project management, and in fact, even today, for relatively small projects, the project manager acts either in a staff capacity to top management or in a line capacity within a functional organization. As a staff function, the project manager strives to coordinate in the name of top management certain capabilities of the functional organization. In this role, the project manager relies on the functional departments for the detailed planning, scheduling, budgeting, and control of their specific tasks. The project manager's information needs are limited and, generally, are of a summary nature. Even in the line function, the project manager is working in an environment where he can rely on the information system already in place.

However, as projects become larger and more complex, companies have established large project organizations which are functioning either deliberately or indirectly in a matrix fashion with the total organization. In this environment project managers have attempted to carry out their responsibilities utilizing the information systems already in place. However, for the most part these project managers have found that their information resources were severely limited. Some of the more typical problems encountered by project managers in trying to do their job while relying on existing or the traditional management information systems resources of the corporation are listed below:

1. *Usefulness*—Existing corporate management information systems do not generate the specific information required by the project manager and other project participants. The needed information is not generally available in a useful form, and it requires considerable time and money to revise the existing systems to get the needed data in a timely manner.
2. *Quality Versus Quantity*—Too much detailed information is generated. It is necessary to pour through reams of computer printouts to

extract the required data. It is difficult to get exception reports, especially when several functions may be involved.

3. *Integration*—There is little uniformity between corporate systems. Hence, it is difficult to develop a total project picture where several different companies are involved. Even within one organization's management information system it's difficult to reconcile information between diverse functions like finance, personnel, and operations to develop an integrated project status report.
4. *Responsiveness*—Whenever top management requests an answer to a specific question or problem, it initiates a mad scramble to obtain the required data. The existing information systems are not structured to integrate across functions to produce timely exception reports.

The essence of all of the above problems is that the traditional corporate management information systems cannot be efficiently and effectively used by the project manager simply because these systems were designed for another purpose, namely, that of enabling the functional organizations to efficiently carry out their responsibilities. This is now a fairly well recognized fact. Many companies with formalized project management organizations are now making the effort to design and implement computer-based project information and control systems specifically suited to their *unique* project management requirements.

## CHARACTERISTICS OF THE PROJECT MANAGEMENT SYSTEM

Before getting into the methodology of building a project management system, it is necessary to examine some of the basic characteristics of these systems. For, in order to design a system which is truly effective, we must have a clear idea of what these systems should do for the project manager, executive management, and the total organization at large.

Let us begin by examining the basic concept of the project. A project is an organization of people dedicated to a specific purpose or objective. Projects may be large, expensive, unique, or high-risk undertakings; however, all projects have to be completed by a certain date, for a certain amount of money, within some expected level of performance. At a minimum, all projects need to have well-defined objectives and sufficient resources to carry out all the required tasks. The important elements of a project can be depicted by the simple cybernetic diagram shown in Figure 27-1.

As shown in Figure 27-1, the project is responsible for accomplishing certain specific objectives or outputs. Those outputs can be defined in

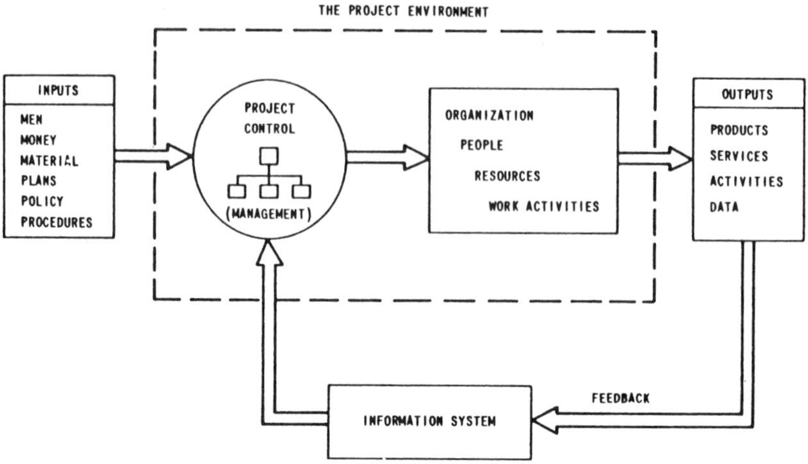

Figure 27-1. The total project management process.

terms of activities, products, services, or data generated by the project. In order to accomplish these objectives (outputs), the project needs appropriate inputs or resources. These can be defined as men, money, or material which the project will expend in the process of accomplishing its objectives.

The control function in this process is management. Management's responsibility is to allocate to the project only those resources required to do a good job, no more and no less. And, of course, management is also concerned that these resources are used in an optimum manner. The question is, how does management determine the quantity of resources to allocate to the project and decide whether or not these resources are being used effectively in terms of the project goals and accomplishments? The answer is, through the information system. The primary function of the information system is to enable management to assess how the project is performing against its established goals and thereby formulate timely decisions for the effective utilization of valuable resources.

This simplified view of the project is useful to highlight two important elements which must be carefully considered in the design of any project management system. These two elements are (1) the information system, and (2) the control system. It is particularly important to note that these two distinct and different elements are mutually related and dependent on each other. The reasons for this are obvious if we examine Figure 27-2. Note that the *information* element of the system concerns itself primarily

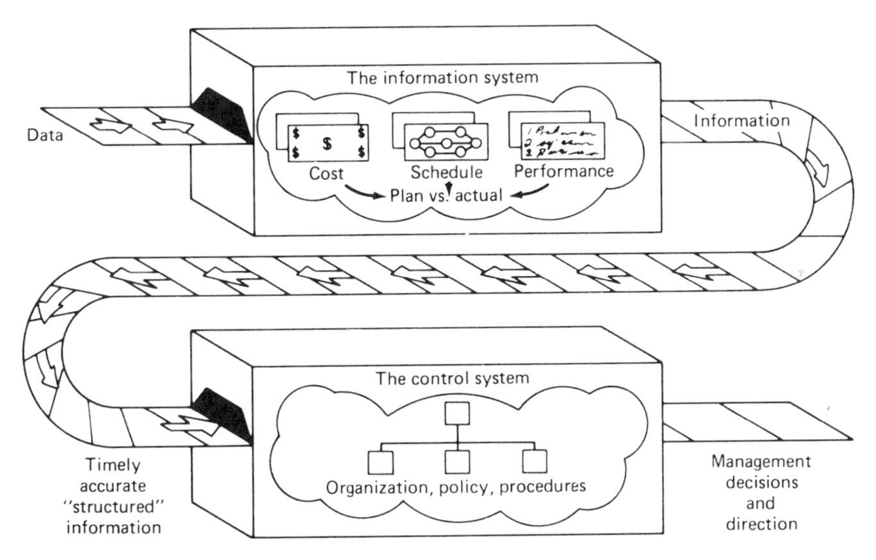


Figure 27-2. Information and control. The information system must be designed and matched to the control system such that management decisions are a natural output of the process. This means that the right kinds of information must go to the appropriate levels of management at the right time and the decision-making process must be initiated as a direct result of established procedures and routines.

with the task of processing data to produce timely, accurate, structured information regarding the cost, schedule, and performance aspects of the project. On the other hand, the *control* element of the system is concerned primarily with using the information supplied to formulate decisions and give direction relative to future utilization of resources and/or resolution of problems. Unless the control element and the information element are designed to be mutually compatible and dependent on each other, they will not function as an integrated system.

With this brief view of the "system" we can define the "project management" system as the people, policies, procedures, and systems (computerized and manual) which provide the means for planning, scheduling, budgeting, organizing, directing, and controlling the cost, schedule, and performance accomplishments of a project. Implicit in this definition is the idea that *people plan and control projects,* and *systems serve people by producing information.* The design and implementation of the procedures and methodologies which integrate people and systems into a unified whole is both an art and a science. Some of the more pragmatic aspects of these procedures and methodologies are considered next.

## GETTING STARTED

Once the need to develop a system has been established, there is generally a great temptation to make an industry-wide survey to find out what everyone else is doing. These whirlwind tours usually result in a mind-boggling collection of facts and philosophy on computer hardware, data base management systems, CPM packages,* classification schemes, programming concepts, etc. At best, a survey of other systems will give the uninitiated a feel for the magnitude and complexities of the undertaking.

The other extreme is to call in the computer or systems consultants who, in most cases, recommend the purchase of a particular set of software packages. Unfortunately, instant implementation of these packages is generally not possible. Extensive customizing may be required to enable usage of the system for a particular project. Also, software packages are designed for specific purposes and unless the buyer knows exactly what he is going to do with these packages, they are unlikely to be used to their fullest potential.

The only sure way to develop a project management system that fits a particular business environment is to first formulate a step-by-step system program plan. This master plan or program plan should fulfill two needs. First, it should be sufficiently detailed to serve as a long-range blueprint for the total program, and second, it should serve as a mechanism for obtaining continued top management support. The development effort will have a much greater chance of survival if top management has more than just a vague understanding of what the system will eventually do for the organization.

The system program plan should be a living document which is updated frequently and circulated to those who are involved in or provide support to the program. At a minimum, the system program plan should contain the following:

1. *System Objectives*—System objectives should give a concise description of what the system is supposed to accomplish, and for whom. The system objectives should define the functions, disciplines, and levels of management to be served by the system, as well as the types of information to be provided. One reason for establishing system objectives is to determine the scope and complexity of the system to be developed. It is especially important to avoid glittering generalities

* See Chapter 15 for a discussion of CPM and Chapter 28 for a discussion of computer software that is useful in project management.

such as, "The system will provide management with all the information necessary to carry out their responsibilities." In some instances it is valuable to define the areas that *will not* be served by the system. This will help avert potential misunderstandings in the future, especially with organizational entities not directly involved in the project.

2. *The System Criteria*—Fairly comprehensive criteria must be established to define the system parameters. All the disciplines to be included in the system (i.e., planning, scheduling, estimating, accounting, cost management, material management, etc.) should be defined, as well as the level of detail of information that will be addressed by these disciplines. In effect, the system criteria establish the philosophy by which the projects will be managed and define, or provides boundaries for, the information and level of control needed to effectively manage these projects. *These criteria should accurately reflect the project management environment in which the system will operate.*
3. *The Work Plan*—The basic segments of work related to the design, development, implementation, and maintenance of the system should be spelled out in broad terms. Also, the organizational groups responsible for doing the work must be identified.
   In the early stages of conceptualizing the system, development of a detailed work plan is of little value. This can be done after a comprehensive study is made to identify the system resources that currently exist in the organization and the new ones that must be developed.
4. *Schedule and Budget*—A general phasing schedule covering the major blocks of work and a gross overall budget should also be included in the system program plan. Here again, the main emphasis should focus on establishing the time and cost boundaries for the total program. Attempting to define more detailed schedules and budgets at this stage would be, for the most part, an exercise in wishful thinking.

It cannot be overstated that the most important step in the successful development of an effective information and control system is defining the nature of the system itself and the environment in which it must operate. Establishing the program plan, as outlined above, is a good start in this direction; however, the real effort required to develop a comprehensive set of system objectives and system criteria will begin following a thorough study and analysis of the organization's *existing* system resources and project management methodology. This type of study will set the stage for development of the detailed work plans, budgets, and schedules that will be used to carry out the program through actual system design, development, and implementation.

## PROGRAM SCOPE

A typical program for the development of a new or improved computer-based project management system will involve three distinct phases of work including:

- Phase I —Study and Analysis (Determine what we have now and what we need for the future.)
- Phase II —Design, Development, and Implementation (Specify the system, build it, and actually apply it to a project.)
- Phase III—Documentation, Training, Test, and Support (Ensure people know how to use the system; make it work using actual project data; and improve it as needed.)

The first phase of effort involves a study of the organization's existing information and system resources to determine what is presently available for use in building a computer-based project information system. Out of this analysis should come a list of systems and procedures that will need to be procured or developed. In addition, the project management approach or mode of operation for management of future projects should be established to identify the types of information resources that must be made available. This analysis must produce very comprehensive system criteria, as well as a preliminary description of the total project information system concept. The final output of the Phase I study and analysis effort should be the action plan which specifies how the recommendations should be carried out, by whom, when, and at what approximate cost. The flowchart given in Figure 27-3 shows the major activities and accomplishments to be realized in the three phases of the program. These will be discussed in detail later.

In the second phase of the program, efforts will concentrate on design, development, and implementation of the software, hardware, and related procedures for the total system. Typically, during this phase of the program, studies are made of commercially available software and hardware which will meet specific project requirements. Appropriate analyses and cost trade-off studies will be made to select those systems which lend themselves to the total project system concept. Or, in the situation where the management systems of several organizations or firms may be integrated to produce a composite management system, the focus of the study will be on selecting the best elements of the individual systems and integrating them in as needed. Specific software packages will be defined and the related programming and coding will be accomplished. To the maximum extent possible, new systems will be operated in parallel with exist-

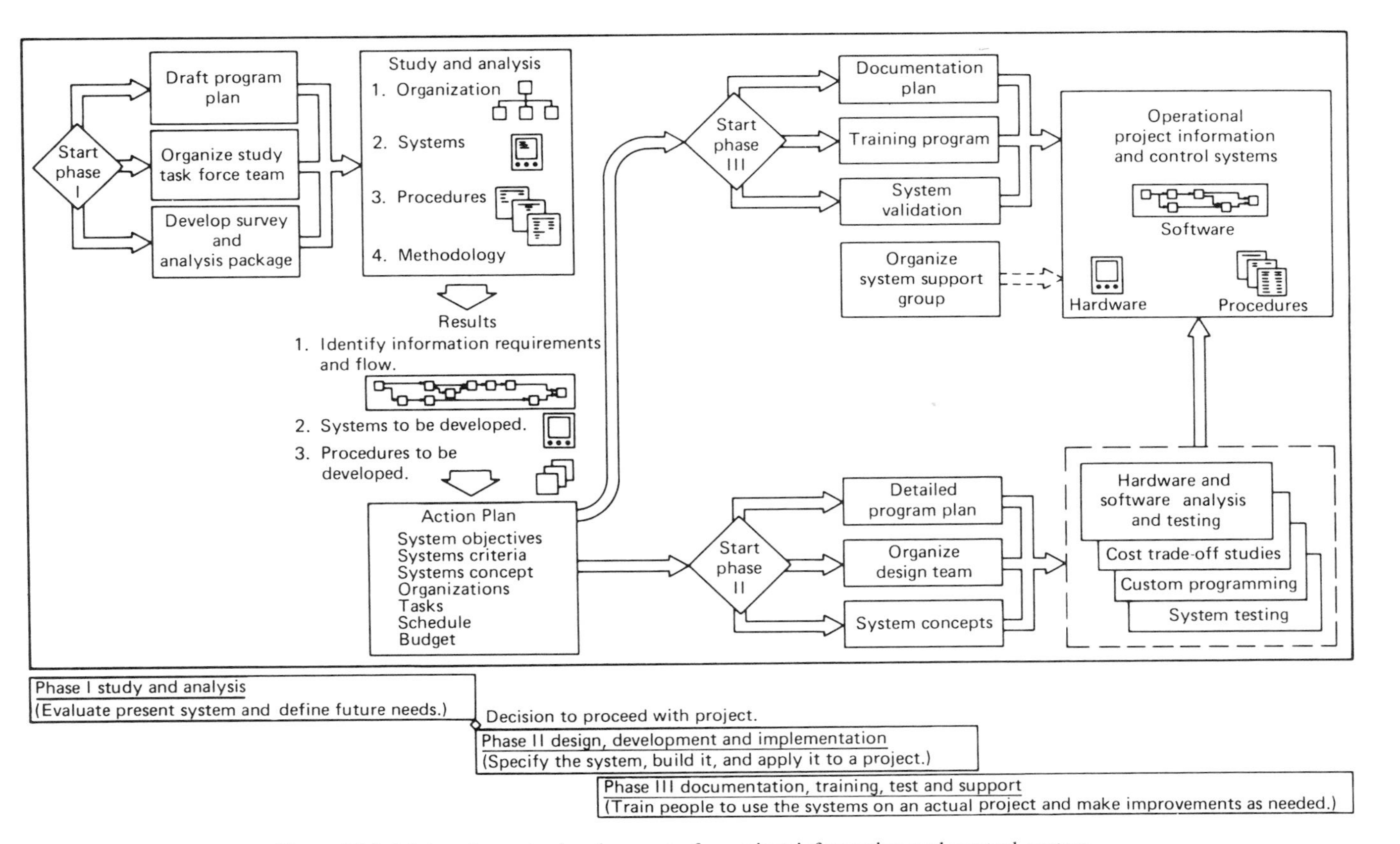


Figure 27-3. Major phases in development of a project information and control system.

ing systems in order to subject them to real-life project environments. Generally, all new systems are tested in parallel with existing systems until such time as these systems are debugged and documented and the results verified. Only then will the systems be formally turned over to the using project organizations.

The third and final phase of the program, which will overlap the second phase, involves training all personnel who will use or be served by the system. Appropriate training sessions must be organized and seminars scheduled for all levels of management. In addition, all documentation should be finalized as each element of the system completes it validation test. Once the user's organizations are satisfied that the system fulfills their needs, the development team will phase out of the program.

This is a brief overview of the total program scope; however, to get a feel for the problems that must be faced in actually carrying out such an undertaking, it will be necessary to examine each program phase in some detail.

## PHASE I—STUDY AND ANALYSIS

### The Study Team

Determining the extent and value of the organization's existing systems and defining the system requirements in detail requires that personnel be designated to organize, direct, and accomplish this effort. Obviously, a team of some type is in order. In most companies this assignment falls upon a committee which is organized for that specific purpose. This committee is most often comprised of part-time members from various departments including engineering, data processing, finance, and the project office.

Almost without exception, this committee will do a poor job, and for good reason! They have other, more immediate responsibilities, and are generally unschooled in the art of making a system survey and analysis. Experience has shown that the most successful approach is to organize a full-time task force team dedicated specifically to making a study, and providing recommendations and a proposed implementation plan. A typical team (see Figure 27-4) would include a program manager, who is specifically *responsible* for the work of the team, and several specialists with expertise in systems analysis, planning, scheduling, estimating, cost management, material management, etc. The size of the team can vary as new members are added to focus on specific topics; but, at a minimum, a core group of individuals should be identified as part of the project team until the study is completed. The team's program manager should report

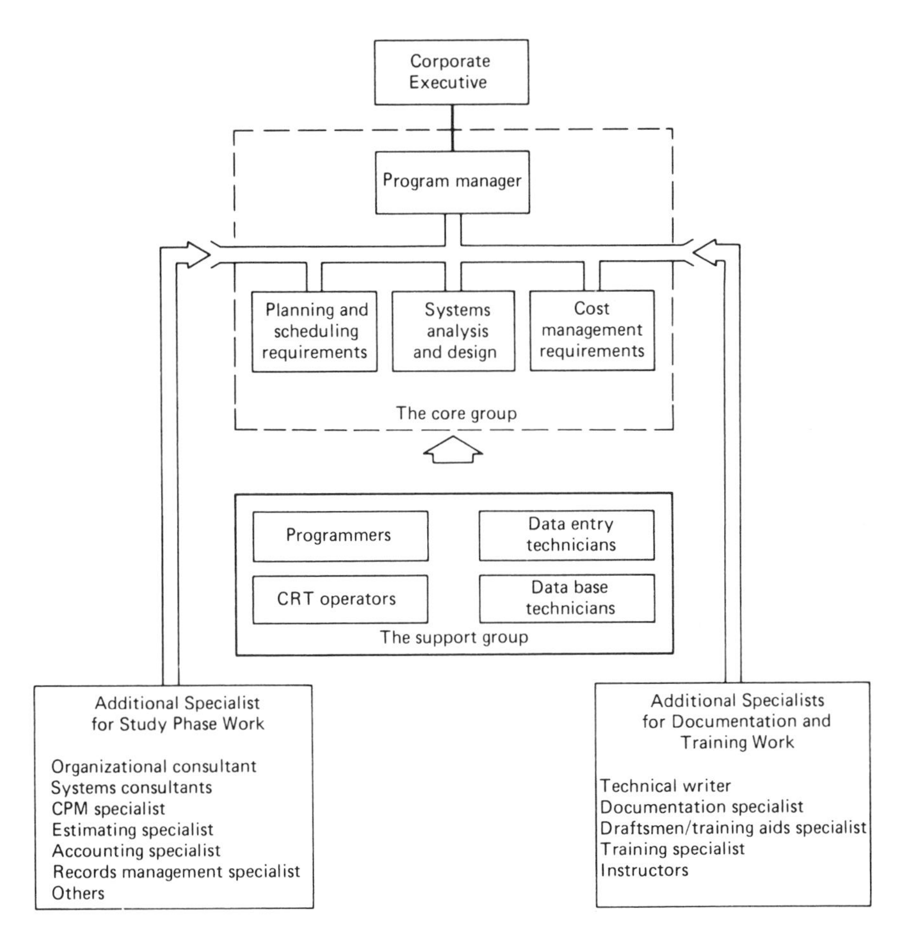


The Core Group is responsible for carrying the total program through to completion. This group is augmented by various specialists during all phases of the program. In addition, a support staff is organized to bring the system into actual operation. Selected members of the core group and the support group can be designated to maintain the system once it becomes operational.

Figure 27-4. The program team organization.

to a high level of management, in most large companies a senior vice president. The senior vice president should, in turn, be held *accountable* for the work of the study team.

### Study Methodology

Once a program team is organized and committed to making the study, preparatory work is needed to ensure that the team will make an objective evaluation and collect the required information in an organized, systematic manner. To this end, the team will need to map out the organizational elements to be included in the study, develop appropriate interview and questionnaire forms, and establish a mechanism for sorting out and evaluating all the information that will be collected. As shown in Figure 27-3, in Phase I the team's study and analysis work will focus on the organization, systems, procedures, and methodologies now in place to support project management.

A well-organized survey plan would include the following:

1. *Memo of Introduction*—A brief memo should be directed to all those to be included in the study to advise them of the purpose of the study, the topics to be covered, and the length of time required. The memo should advise all participants of the importance of the effort and should be signed by top management.
2. *Survey Questionnaire*—A well-designed questionnaire is invaluable for ensuring that all the appropriate topics are covered consistently from one interview to the next interview. The interviewers can utilize an outline or checklist (see Figure 27-5) to keep the conversation flowing along the required topics. The questionnaire can be completed after the meetings and, if necessary, follow-up discussions can be held to fill in the gaps.

   The questionnaire should be designed to capture information on the following topics from the individual being interviewed:

   - Responsibilities and functions of the unit.
   - Interfaces with the unit.
   - Primary work tasks.
   - Data inputs needed to perform these tasks.
   - Data outputs generated as a result of performing these tasks.
   - Problems, requirements, and suggestions.*

   In addition, the questionnaire should investigate at least three major aspects of the organization including: (a) information requirements and information flow, (b) methods and procedures, and (c) systems used by the organization. For all three of these major areas, specific questions should be developed. Since the project management system is concerned to a large degree with the information that a unit needs in order

---

* Chapter 16 describes the way in which a linear responsibility chart may be used as a basis for obtaining this information in the systems design process.

| INTERVIEWER | | |
|---|---|---|
| NAME ______ | | WORK ORDER ______ |
| DEPT. ______ | PROJECT ______ | FILE NO. ______ |
| DATE ______ | | PAGE ____ OF ____ |
| TIME START ______ | CLIENT ______ | |
| TIME STOP ______ | | |

SURVEY OUTLINE & INSTRUCTIONS
*(CHECK BLOCKS TO MONITOR PROGRESS OF THE DISCUSSION)*

I. ORGANIZATIONAL ANALYSIS & INFORMATION FLOW

- ☐ 1. IDENTIFY ORGANIZATIONAL LEVEL & LOCATION
- ☐ 2. PRIMARY RESPONSIBILITIES OF THIS ORGANIZATION AS UNDERSTOOD BY THE INTERVIEWEE.
- ☐ 3. STRONGEST CAPABILITY AND/OR TALENT OF THIS ORGANIZATION
- ☐ 4. INTERFACES (INTERNAL & EXTERNAL)
- ☐ 5. ROUTINE TASKS PERFORMED (LIST)
  SPECIAL ASSIGNMENTS (WHAT AND HOW OFTEN)
- ☐ 6. PROBLEMS
- ☐ 7. INPUTS (DATA/INFO) REQUIRED
- ☐ 8. OUTPUTS (DATA/INFO) PRODUCED
- ☐ 9. NEW IDEAS – PROBLEM SOLUTIONS
- ☐ 10. OTHER AREAS TO LOOK INTO

II. PROCEDURES ANALYSIS

- ☐ 1. WHAT PROCEDURES ARE USED (FORMAL AND/OR INFORMAL) HOW CLOSELY UTILIZED
- ☐ 2. EFFECTIVENESS OF THESE PROCEDURES
- ☐ 3. PROBLEMS
- ☐ 4. ADDITIONAL PROCEDURES REQUIRED (IDENTIFY)
- ☐ 5. IDEAS & SUGGESTIONS

III. SYSTEMS & METHODOLOGY

- ☐ 1. DEFINE SYSTEMS NORMALLY USED – COMPUTER BASED
  – MANUAL SYSTEMS
- ☐ 2. SYSTEMS UNDER DEVELOPMENT OR BEING CONSIDERED
- ☐ 3. MAJOR PROBLEM AREAS
- ☐ 4. IDEAS & SUGGESTIONS FOR SYSTEMS

IV. SPECIFIC QUESTIONS (SEE ATTACHED LIST)

Figure 27-5. Survey questionnaire.

to carry out its function, much effort must be devoted to mapping out this requirement. Simple input/output charts (see Figure 27-6) can be developed for each of the organizational units surveyed. These charts can then be connected (since the information outputs of one unit become the inputs of another unit) to develop a composite system information flow (see Figure 27-7). This total system information flowchart will then become an excellent tool for designing the total system logic.

3. *Survey Timetable and Score Card*—Every individual and/or area to be covered in the study should be identified and a fairly comprehensive timetable developed to be sure that the survey effort does not exceed the time allotted. It is important that the study team talk to everyone while management's interest in the project is still strong. Also, it is equally important that the results of the survey be catalogued while still fresh in the interviewers' minds. Thus, a simple matrix will not only ensure that all required areas are covered, but will also provide an evaluation of the effectiveness of the coverage.

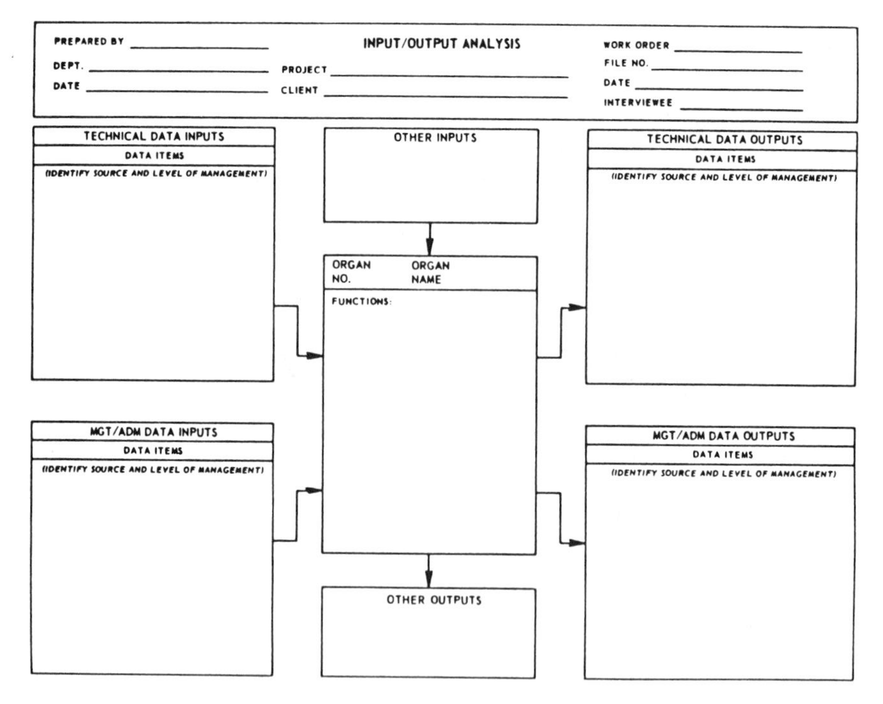


Figure 27-6. Input/output chart.

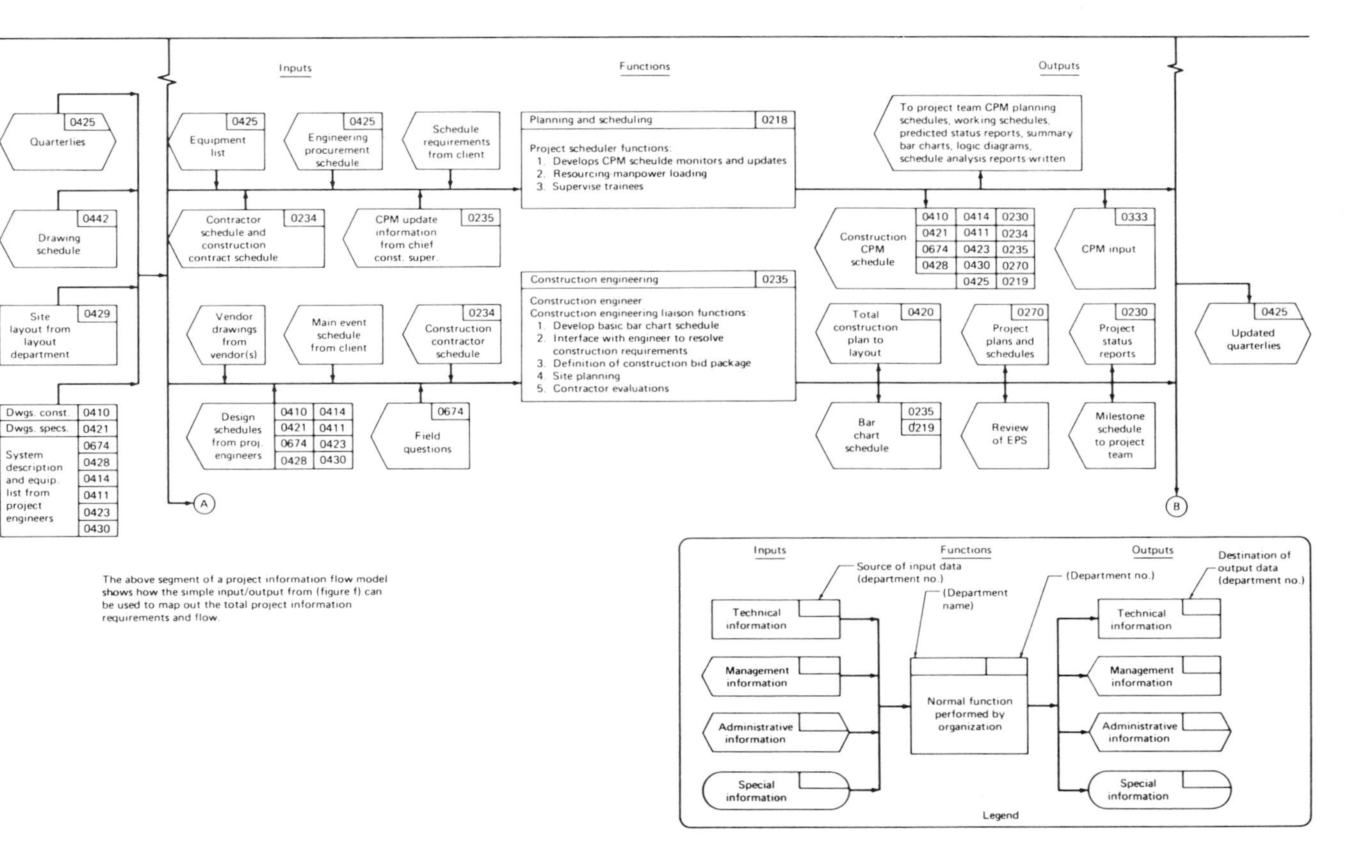


Figure 27-7. Constructing an information flow model for the project.

### Compiling the Findings

Armed with a well-organized survey plan, the study team can proceed to conduct their interviews and review the organization's current systems, procedures, and methodology for managing projects. As might be expected, information or systems studies are, at best, very subjective. The problem will be to separate facts from opinions. To have some degree of confidence in the final results, efforts must be directed at evaluating all interview data on a consistent basis; otherwise, the investigators will fall into the trap of devising systems which respond to what the interviewers think people said, and not what the people actually did say. The following approach will help to minimize the subjective influences:

1. *Compile and Structure the Findings*—To ensure a high degree of consistency in evaluating information collected, the information should first be compiled by discipline and then by level of management. Once this is done, the information requirements should be further subdivided according to technical, management, and administrative types of information needed to support each level of management and each discipline.
2. *Categorize the Results*—It is especially important to categorize the inputs obtained during the study to make a clear distinction between those things which are applicable to management information system (MIS) requirements and those which are not. The following three categories are suggested:
   a. Problems and requirements for the information system.
   b. Problems and requirements which are procedural in nature.
   c. Problems and requirements which are managerial or organizational in nature.
3. *Analysis*—After compiling the information and carefully categorizing all the facts, the study team should perform an analysis from two distinct viewpoints: first, from the viewpoint of the organizational element (and the disciplines within these elements) and second, from the viewpoint of the different levels of management. The results of these two analyses should be consolidated to identify all common requirements that must be addressed by the system to be developed. From this, a priority list can be developed to specify the system development sequence.

### Making Recommendations

If the study team has carefully organized and documented their efforts as outlined above, the team should be in a position to define the concept for a project management system which reflects the personality and unique requirements of their company. At a minimum, the study recommendations should identify:

1. The specific information resources required by
   a. Each organization or function involved in the project.
   b. Each level of management within the company that will contribute to, or be affected by, the project.
2. The new systems (hardware and software) that may have to be developed or procured.
3. The existing systems that may have to be modified.
4. The existing systems that will be utilized.
5. The organizations that must contribute to the development effort.
6. The timetable for developing the total system.
7. A budget estimate.
8. An overview (pictorial flowchart) of the system concept.

The final recommendations should include some discussion of what the system will do to increase the effectiveness of the project management organization. Often there is an attempt to identify cost savings as a means for justifying system development. However, the value of a good project management system lies in its ability to enable a small team to manage something large, complex, or very important to the organization. Unfortunately, this is extremely difficult to quantify in terms of dollar savings.

From the foregoing it should be obvious that the Phase I Study and Analysis is the key to the eventual development of a truly effective project management system. Successful development of these systems can be assured to a large degree if the study team follows a well-established approach of the type outlined here. Equally important, however, is the need to have a team composed of people who have worked in the project environment and know from experience the value and need for systems. These types of individuals, armed with a structured study and analysis methodology, should be able to produce the detailed information and plans necessary to start the actual design of the proposed project management system.

## PHASE II—DESIGN, DEVELOPMENT, AND IMPLEMENTATION

In the first phase of our program considerable effort has been expended to carefully establish parameters for a project management systems which will effectively function within a particular organizational environment. In addition, we have inventoried the existing systems and procedures and have attempted to evaluate these in terms of their effectiveness in supporting present and future project management requirements. Thus, we have developed a fairly detailed blueprint of what the future system will look like, what it will do for management, and how long and how much it will take to get there. To the system analyst, most of this work would be categorized under the heading of the functional specification. Regardless of what it is called, the purpose is the same, namely to spell out as meticulously as possible all of the user requirements prior to actually designing the system. These user requirements should identify the particular features and capabilities of the systems which must function within a given industry, organization, and management environment. However, in spite of the wide range of applications for these systems, all project management systems should have the capabilities to support the basic requirements given in Figure 27-8.

It should be noted that the requirements identified in Figure 27-8 hold true for large sophisticated computer-based systems as well as simple

| PROJECT MANAGEMENT FUNCTIONS | | BASIC ACTIVITIES INVOLVED IN THESE PROJECT MANAGEMENT FUNCTIONS |
|---|---|---|
| 1. Project Objectives | — | Define the cost, schedule and performance goals for the project. |
| 2. Work Definitions | — | Define work task to be done and the organizations responsible. |
| 3. Scheduling | — | Define the sequence for doing the work and the time constraints. |
| 4. Budgeting | — | Define the resources (men, money, material available for doing the work. |
| 5. Baseline | — | Define the parameters for measuring cost, schedule, performance accomplishments. |
| 6. Monitoring/Reporting | — | Define how progress will be tracked (the events and level of detail) and how this will be reported. |
| 7. Analysis | — | Define how and who will assess progress against plans. |
| 8. Corrective Action | — | Define who is responsible for corrective action, how it is to be implemented and when. |

The Project Management System must provide the people, policies, procedures, systems (manual and computer) to accomplish the basic task involved in each of the eight Project Management functions.

Figure 27-8. Minimum requirements for project management systems.

manual systems. Irrespective of the degree of sophistication of the system, it must be designed to support the project management process.

### Design Concepts for Project Management Information and Control Systems

As noted earlier, the "systems" discussed here involve people, procedures, and computer software and hardware integrated into a unified approach to processing data to produce information to effect a timely management decision process. Note that we do not merely talk about generating information to support the decision-making process, but rather we infer that the decision-making process must be forced to take place as a result of the systems information outputs. The implication here is quite important, because for the system concept to truly work for the project management process, the people involved in the process must be an integral part of the input, output, feedback cycle depicted in the cybernetic diagram shown back in Figure 27-1. This means that our project management system must include, in addition to the procedures and systems for generating information, procedures and systems which ensure that decisions or actions are generated as a result of the information inputs. A general concept for doing this is depicted in Figure 27-9.

The system concept for project management information and control given in Figure 27-9 utilizes a one-for-one modular concept to integrate information and management control. That is to say, for every specialized module we create (computer-based or manual) to process data and generate information relative to a specialized discipline or topic, we also create a module for project control action. These control modules identify the people (organizational functions) involved and the process that must be initiated as a result of the outputs of the information module. Thus, in designing our "systems" we must focus on the two unique requirements of information and control. Some suggestions for the methodology to follow in designing these elements of our system are discussed in the next section. However, an advantage of using a modular approach to our design is that we can build a system which can eventually support a wide range of project management requirements.

The system shown in Figure 27-10 addresses a wide range of project management requirements. These include the technical requirements (Group I Module); the usual cost, schedule, and performance requirements (Group II Module); and the predictive requirements (Group III Module) of a project. In this example we show the type of system which might need to be developed for a project environment dealing with large scale, high risk, high cost, or unusual technical or business development

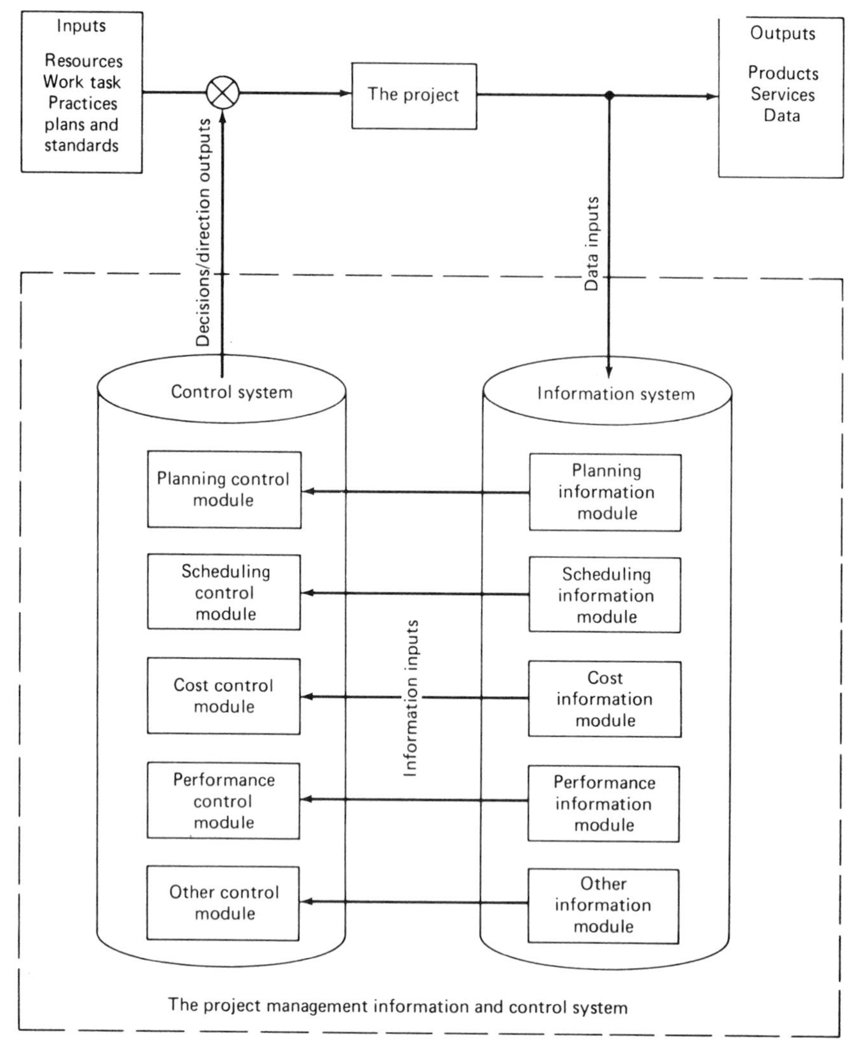


The information system receives data from the project and evalutes the plans, schedules, cost, and performance of the project against established plans and goals. Information is developed relative to the variance between what the project planned to do and what it has actually done. This information is fed to the control system. The control system evaluates the variances for various project parameters against established standards to determine if progress to date is acceptable. The outputs of the control system are decisions/directions which redirect resources, work task, practices and plans and standards of the project.

Figure 27-9. Project information flow.

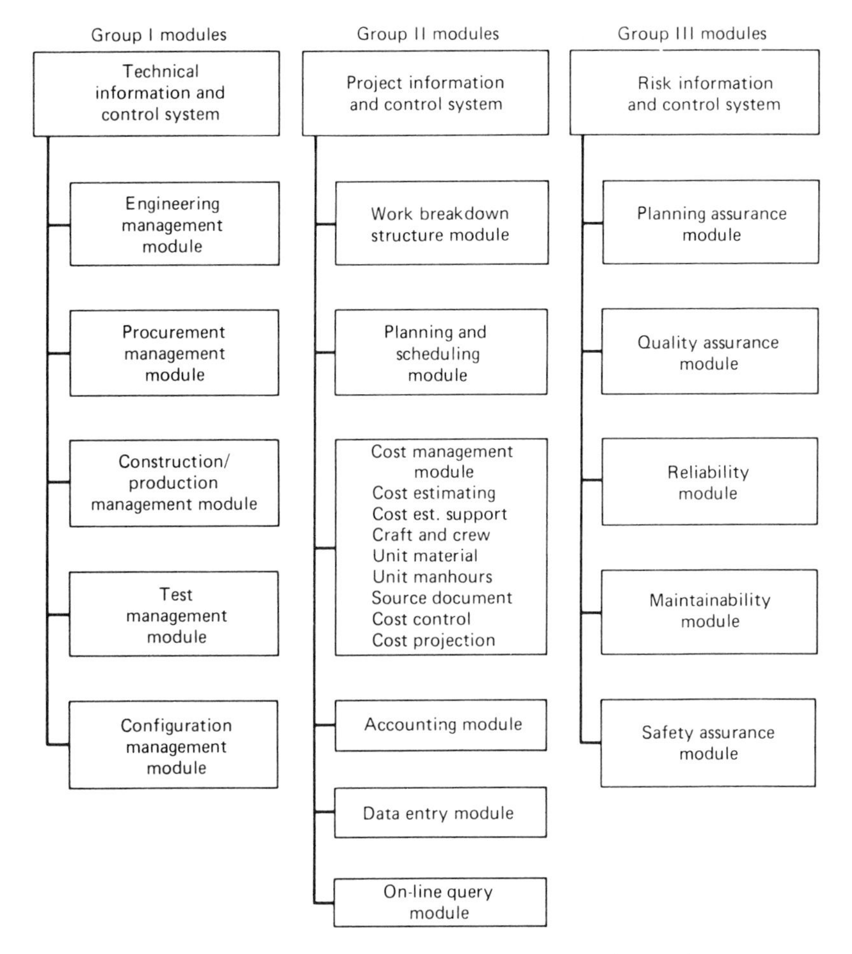


Figure 27-10. The project management information and control system.

requirements. By utilizing the modular building block approach, the design effort can focus on the priority areas and make these operational in tune with management's needs and availability of resources. The point that needs to be stressed here is that it is extremely important to conceptualize the total system in the beginning, before starting actual design; otherwise there is a high possibility that the resulting product will be a hodge-podge of poorly related systems. The design process should begin only when the objectives and the design concepts have been carefully defined, understood, and agreed to by all.

**Design of the Information System**

A modern management information system normally consists of two major elements (see Figure 27-11). The first is a data management system, which is the heart of the system, and is comprised of a series of data base-related software packages which "manage" (store and retrieve) the project's data on an integrated and logical basis. The second major element of the system consists of a series of computer software packages, or modules, which provide the means for generating information on specific functions of the project. For a project management application these modules will typically address planning, scheduling, estimating, cost management, project accounting, and so on. Additional modules can be utilized to produce information on almost any aspect of the project that management may wish to control. The primary objective of the information system module is to organize, collect, store, and process data quickly and effi-

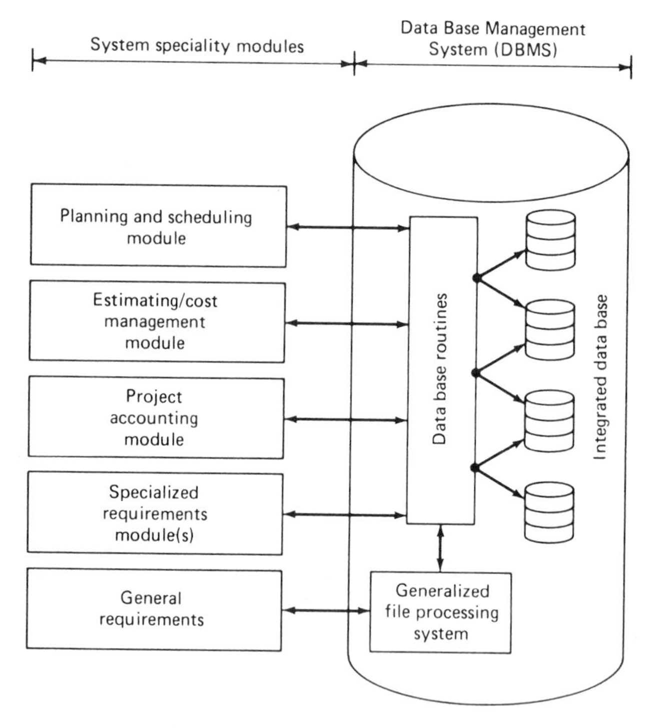


Figure 27-11. Project information system concept. The DBMS provides a convenie method for managing all project data on a logical basis. The individual application programs process the data to produce "information" or new data. This "information" is directed to management and/or to the DBMS for use by other modules.

ciently to produce meaningful information which will advise management on the project's status, trends, and potential problem areas.

### Elements of the Information System

For project management some of the basic elements of the information system that will need to be designed are described as follows:

1. *Planning and Scheduling Module*—Effective management of a large or complex project calls for a systematic method of depicting the time-sensitive relationships and the interdependency between such functions as engineering, design, procurement, operations, testing, and so on. Normally, the Planning and Schedule Module utilizes Critical Path Method (CPM)* scheduling techniques with resource leveling and target scheduling to provide the tools for planning and monitoring the project. The information generated by the project CPM also provides the keys for more detailed monitoring and analysis of subtasks within various disciplines. This is done by creating data base files for specific work tasks or requirements of individual project disciplines. Because of the integrated nature of the data base files themselves, the information can be used to support other modules. Thus, drawing lists, specification lists, purchasing schedules, and all types of cost information can be interrelated and used by all the project's organizations (engineering, purchasing, operations, tests, etc.) to manage their specific tasks as well as by the project management organization to overview the total effort.
2. *Project Accounting Module*—The project will require a formalized method to monitor, record, and report all costs, and to develop the final cost records upon completion of the project. In the Project Accounting Module this is accomplished through computer-based systems which, together with predefined operating procedures, produce periodic reports and provide for development of actual cost records as the project progresses. Typical outputs for this type of module might be
   a. *Commitments and Invoice Record of Purchase Orders and Contracts*—Used to determine the cost status of each purchase order and contract, and through appropriate coding techniques, to keep track of change orders against these orders.
   b. *Statement of Commitments and Recorded Expenditures*—Provides a quick reference to the status of all purchase orders and contracts

* See Chapter 15.

including base amounts, change orders, recorded cost, remaining commitments, and retentions.

c. *Project Ledger*—Detailed cost record for the project.

d. *Change Order Status Reports*—Provide information relative to all change orders associated with each purchase order and contract.

e. *Statement of Recorded Expenditures by Account*—A list, at the account code level, of all actual expenditures, current estimates, and current balances for the project.

3. *Estimating/Cost Management Modules**—Effective cost management for the project requires that the system provide an efficient method for making comparisons between current and budget estimates and scheduled and actual cash flow. This is accomplished through the computer-based Estimating/Cost Management Modules and the interface which this module maintains through data base with the Accounting and Planning and Scheduling Modules. Typical outputs of the Estimating/Cost Management Modules are

   a. *Detailed Estimates*—Detailed estimates in account code sequences.

   b. *Summary Estimates*—Detailed estimates summarized into work packages.

   c. *Updated Estimate/Cost Report*—Updated estimate, at the account level, providing a comparison of the new estimate with the previous estimate for each account.

   d. *Functional Cost Report*—Cost reporting by major functional categories, i.e., material purchase orders, installation purchase orders, field purchase orders, etc.

   e. *Outstanding Commitments Report*—A comprehensive profile showing the total, actual, and outstanding commitments against each purchase order.

   f. *Forecast of Cash Requirements (Summary and Detail)*—Cash forecast reports at various levels and detailed at monthly, quarterly, and yearly increments.

In addition to the above basic modules, many other modules may be developed to address specialized needs of a project, such as labor resources, materials inventory and controls, document indexing and retrieval, health, safety, and environmental records, etc. Any requirement that can be defined by tangible data elements can be designed and integrated into the total system in building block fashion. Since a data base management approach is at the heart of the system, the system designers

---

* See Chapter 19.

can address all future project needs with relative ease, once these needs are identified.

At this point, it is important to note that the amount of design and development work that will be required to develop a specific project management system will depend on the complexity of the system (which should be dictated by the needs of the project), the time and money available to develop the system, and the knowledge and experience of those responsible for getting the system up and running. Obviously, trade-offs must be made when all the factors are taken into consideration; however, three fundamental strategies and combinations thereof can be used to obtain the desired results. These are listed below:

1. *Specialized Design and Customizing*—For projects which are unique, large, or complex it may be necessary to design and program specialized modules or application programs from scratch. It may even be appropriate to procure commercially available software packages and customize them to meet specific user needs. Specialized design and customizing may represent a sizable investment; however, the investment may be well justified if the final system supports the management style of the project participants and enables them to manage a high-value or high-risk undertaking at a reasonable cost.
2. *Procurement of Project Management Packages*—Projects with routine or straightforward requirements may be satisfied by one of the many commercially available project management packages. Within recent years there has been a virtual explosion of new products for project management. Commercially available software packages range from the relatively simple and inexpensive package which is designed to run on the personal computer, very sophisticated systems which run on powerful minicomputers, and the most comprehensive project management system which takes full advantage of the speed and capacity of the large mainframe computer.

   Selection of a commercially available or canned project management package may be quite difficult and time consuming in view of the large number of systems now available. Nevertheless, the problem can be managed if the user follows the procedure prescribed in our Phase I—Study and Analysis—and uses the results to specify and evaluate the candidate project management packages.
3. *System Integration*—The strategy here is to devise a project management system by integrating major elements of the information and control systems that the project participants have in place. This strategy may be quite suitable for very complex undertakings or superprojects where the contributions of many diverse firms and highly special-

ized organizations are required. Building the actual system involves assessing the internal systems that the project participants have in place and carefully selecting and integrating elements of the system into a composite system. This approach is possible where the participants' technologies are compatible and where their respective project responsibilities can be segmented by discrete or unique information boundaries and work tasks.

Regardless of the strategy employed, the fundamental problem is to ensure that the systems which are devised support the specific project-related requirements and are compatible with the control procedures and philosophy of the management environment where they are being applied. For projects of any significance it is paramount that the system be able to collect, process, and present information in a form which will support the diverse range of organization, management, and technical needs which are unique to the project environment. Thus, for most project management system development efforts, the logical starting point will be the selection of a data base management system.

There are a number of excellent data base management systems (DMBS) currently available on the market.* The goal is to select the one most suited to the user's needs. To this end, the user will have to develop some type of evaluation criteria. At a minimum, these criteria must address in detail the following:

1. *Technical Capabilities and Requirements of the DBMS*—core requirements, interfaces, security, performance statistics, editing features, utilities available, batch/on-line, maintenance, etc.
2. *Flexibility of the DBMS*—control feature, data access, languages supported, data storage devices, linkage capabilities, search capabilities, etc.
3. *Standardization*—comply with various standards that have been developed for data bases.
4. *Resource Support Requirements*—internal and external support requirements, documentation, etc.
5. *Design Features*—data levels, indexing techniques, networking features, etc.

It is strongly recommended that the user carefully establish his criteria prior to consulting with vendors or other users. Otherwise, the user will be barraged by bewildering terminology and an array of philosophies on DBMS. It is difficult to provide specific guidelines in this area because individual needs vary so widely. However, common sense dictates that

* See Chapter 28.

the user should not purchase a DBMS more sophisticated than he needs or has the *capability* to use and maintain.

The same situation is true with respect to development of the Planning and Scheduling Modules. For large projects involving design, procurement, and construction, CPM (Critical Path Method) networks are a very popular (and very effective) way of depicting all of the project's major requirements. For large research and development projects involving activities whose outcomes are doubtful, PERT (Program Evaluation and Review Technique) can be utilized most effectively. Since PERT can establish the probability of meeting deadlines, it can be helpful in the development of alternative plans. Fortunately, there are many excellent CPM/PERT software packages on the market today to satisfy a variety of requirements*. Here again, the goal is to define the needs of the individual project, develop an appropriate criterion, and begin an investigation to choose the package which offers the features, options, and capabilities most closely suited to management's needs.

The decision to procure and modify software packages to meet a project's individual requirements versus the prospect of developing a system from scratch will depend on the uniqueness of the system requirements and the availability of system design and programming support. Certainly, it does not pay to design a DBMS or a CPM/PERT software package, in view of the number of well-designed systems currently on the market. But, by the same token, it may not be practical to try to modify someone else's estimating system to suit a particular project's cost control requirements.

### Design of the Control System

The design of the information modules for our system essentially involves the development of procedures for collecting, storing, and processing data to produce useful information in a timely manner. Many of these procedures will be instructions for the computer (programs). Hence, a good portion of our information system design efforts deal with selecting the appropriate software packages or designing new packages. Unfortunately, when it comes to the control element of our system, there are few in the way of "canned" packages that are available for our use, primarily because it is generally assumed that if managers are given the information they need, they will automatically initiate the appropriate action to "control" the situation. If we are going to devise a project management system which informs and controls as an integral part of the total project function, we cannot rely on a tacit understanding of what management is expected to do. The mechanism for control must be built into the system

* See Chapter 28.

and it must be activated automatically by the appropriate system stimuli. To establish the parameters for design of the control modules of our system, it is necessary to define exactly what we mean by control.

### Control

The purpose of control is to ensure that events conform to plan. Controlling involves locating or identifying deviations from plan and taking appropriate action to ensure desired results. Furthermore, control is concerned with the present and involves regulation of what is happening *now*. In a large measure we are concerned with regulating present activities in order to influence future outcomes.

For a project manager, the importance and the need to control are quite clear. The project manager is the one individual totally responsible for accomplishing project objectives on time and within budget. However, to be able to control, the project manager must have some frame of reference to measure against and he must have some way of determining when he deviates from this reference. This brings us to the essential elements of control.

### Elements of Control

There are four essential elements involved in control, and these provide the framework for any good project control system. These elements of control are:

1. Setting objectives.
2. Reporting.
3. Evaluating.
4. Taking corrective action.

Obviously, when we talk of controlling something we assume that we have some predefined target or goal. For a project these targets or goals are usually defined in terms of schedule, cost, and technical and quality objectives or requirements. Certainly, the project manager must know what he is trying to accomplish (i.e., get a power plant designed and constructed, design and implement an MIS system, etc.). His problem is to regulate the activities, resources, and events to accomplish the technical, cost, and schedule goals defined in the project plan. This can be done with appropriate status visibility and timely feedback. Thus, he needs an information system which reports on all the important facets of a project, which brings us to the second element of control: reporting.

Since the act of controlling is concerned with the present, the project manager needs a reporting system which is time sensitive. That is, the reporting system must identify problems and requirements in a manner which enables the project manager to make decisions and given direction while there is still time to make a positive change. If the reporting system can only provide feedback considerably after the fact, as a matter of history, then the project manager cannot control his project. Thus, at the heart of control is an information system which gives timely visibility to significant project events. This is why it is so important to have an information system specifically for the project management process.

The third essential element of control is the interpretation and evaluation of the information generated by the information system. This is extremely important because it is the basis for taking corrective action. We know from experience that problems in their early stages of development are seldom black or white. Thus, careful evaluation of indicators, or trends, in project cost, schedule, or technical parameters is extremely vital to the whole process of control. Here again a comprehensive project information system can provide the project manager with a powerful tool for spotlighting early problem areas and requirements. Of course, once having identified a problem, the project manager must take prompt corrective action. This is the fourth and final element of control.

Corrective action means that the project manager has identified a situation which is going to cause a deviation from a desired goal and does something about it. Thus, the project manager must develop a number of alternative approaches to solving the problem and he must select the best approach. In effect, the project manager will examine his options and implement a course of action that best utilizes the resources at his command.

Our interest in designing the control system is to ensure that we have established the appropriate interface between the information modules and those who control and that we have put into place the procedures which will ensure that appropriate action is taken as a result of the information process.

### Overview of the Control System

A general scheme for the control element of our system is given in Figure 27-12. The basic function of the control module is to receive inputs from the information system relative to the current status of project activities and accomplishments. Generally, these inputs will be in terms of the cost, schedule, and performance aspects of the project. Each discrete status input is measured against the previously established project goal, for that

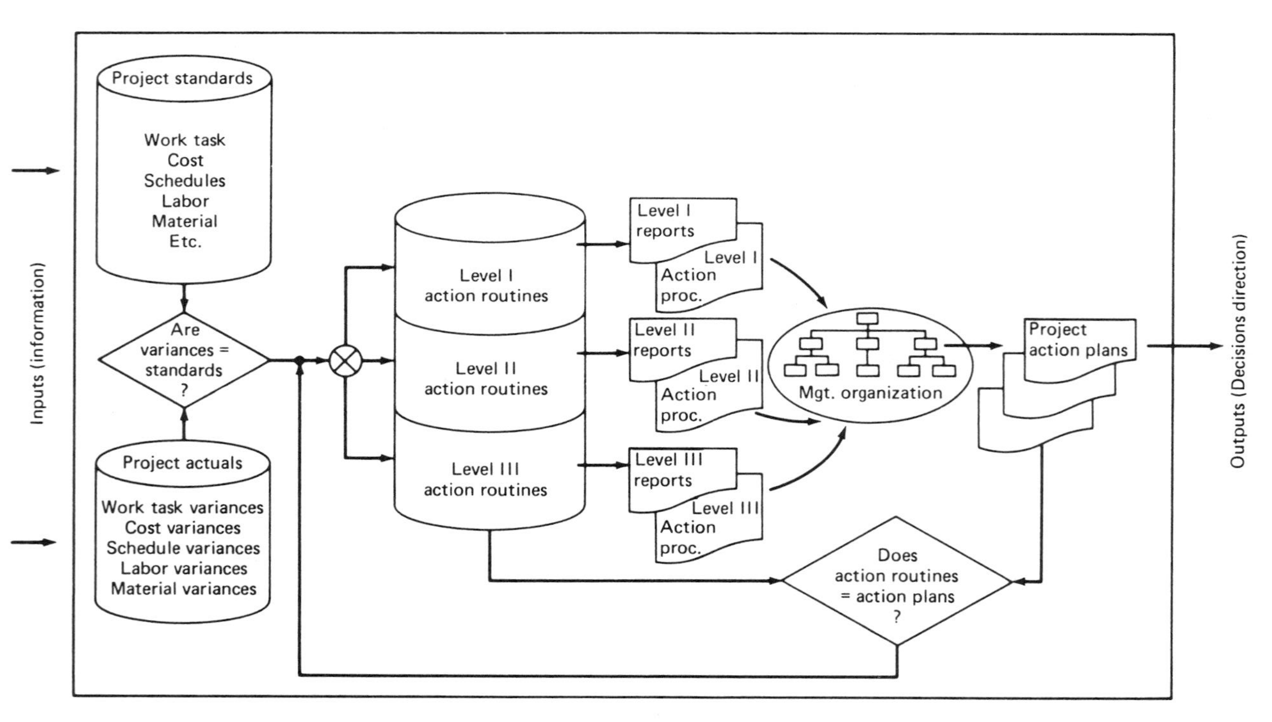


Figure 27-12. Project control system concept.

measurement period, to determine if a variance exists. In simple terms the control module is asking: "How well are we doing against what we had planned to do?" When deviations are identified they are compared against standards which have been established by management. Again, a new set of deviations are identified. We shall call these deviations "action level deviations." That is, the magnitude of the deviations from our standards will automatically identify the level of management responsible for taking corrective action and the time available for initiating this action. Routines are built into the control module so that, if action is not taken, or is not satisfactory by the next reporting period, the module automatically triggers the problem up to the next level of management. The whole process is repeated until the problems are resolved or the problem reaches the highest level of management.

The tangible actions that can be initiated by a project manager to change the course of events of his project are surprisingly few. Basically, these actions involve manipulating or controlling the following:

1. *Resources*—The allocation of resources (men, money, material, facilities, and time) to the project participants (a powerful mechanism for control).
2. *Scope of Work*—Increasing or decreasing the amount of work or the type of work to be done.
3. *Practices*—Establishing or changing the methods, techniques, policies, procedures, systems, or tools used on the project.
4. *Plans and Standards*—The degree and level of effectiveness of control is determined by the project plans and standards established. These may be changed or modified during the life of the project.

Thus, the final outputs of the control module are decisions by project mangement (at all levels) which essentially deal with one or more of the above four factors.

Once the information modules and the control modules have been designed so that they contain the systems and procedures which enable them to function as a unified system within a specific project environment, consideration must be given to the methodology for implementation.

### Implementation

Management information system textbooks offer a variety of schemes for putting a new system into operation. These include (a) running parallel systems, (b) operating a pilot system, (c) using the phase-in/phase-out technique, or (d) employing the cut-off method (burning the bridges). In most project environments, it is usually a matter of operating parallel systems until the new system proves itself. If no computer-based system presently exists, the other alternative is to run a pilot system until all the refinements are made. This then becomes the real system.

In any event, the system must be implemented in a real project environment and the test of the system effectiveness is whether or not management is able to use it to *make decisions*. Unfortunately, experience has shown that this has not been the case for most new systems. And this is why the next phase of the program is the most important of all.

## PHASE III—DOCUMENTATION, TRAINING, TEST, AND SUPPORT

### Documentation Requirements

One task generally disliked by system designers is documentation. Yet, in terms of the effective utilization of the total system, this is probably the most important task to be accomplished. It is especially important that proper system documentation be developed to meet the needs of the system users and those who will maintain and, more than likely, eventually enhance the system. In this regard, four distinct levels of documentation have been identified as follows:

1. *System Documentation*—These documents are designed to provide management with an understanding of how the computer system works at all levels. It will summarize all interfaces, files, and the logic connecting all jobs, providing an overview of the general concepts, features, capabilities, and constraints.
2. *Program Documentation*—Program Documentation provides the programmers and analysts with an understanding of the relation between their own work and the entire effort. This procedure provides a definition of the step-by-step logic developed within each program.
3. *Operations Documentation*—The Operations Documentation dictates the relationship between the functional tasks and the procedures, and establishes a time sequence. It dictates the responsibility for each task, providing a procedure to determine all action taken following a request. This procedure must contain the necessary information to process job steps.
4. *User Documentation*—This section provides a formal description of all functions necessary to input data into the system. Relevant information for the control and processing of source documents and reports should be described.

Actual development of the above required documentation should start quite early in the design effort. In fact, documentation should be a mandatory effort in parallel with system design and development. Ideally, the program manager should devise a documentation checklist which specifies the four levels of documents associated with each major system element or module and provides a timetable for the rough draft, final draft, and fully released documents. Generally speaking, the rough draft documents, which may consist of simple outlines, will suffice throughout the early design effort. However, by the time the systems are fairly well defined (during the program development and implementation stages), this documentation should begin to evolve into descriptive manuals. Once the validation and demonstration tests of the systems have been completed, the revision and publication of final system documentation manuals should be a routine, straightforward task.

A special note of caution is in order relative to documentation. In any program requiring a year or more to complete, it is highly unlikely that the original team will remain intact. Personnel turnover is inevitable; therefore, it is absolutely vital to maintain a consistent, strong documentation trail throughout the project.

### Training Considerations

In similar fashion, training should start very early in the program. A common error is to wait until all the bugs are totally out of the system before attempting to train system users. The system will have greater acceptance and be utilized more effectively if all levels of management are gradually made to understand the philosophy and mechanics of the system. This can be accomplished by handling the training program in stages as follows:

- *1st Stage Training—System Philosophy.* A series of orientation seminars should be planned to explain to management the role of computer-based information and control systems in the management of large projects. These seminars should focus on the types of information that can be provided to all levels of management and dwell on how the system is used to tie together all the project functions. It is particularly appropriate to organize workshop sessions to get the various levels of management to critique prototype reports and system approaches. This feedback can be used to enhance the actual system design work.
- *2nd Stage Training—System Capabilities.* This stage of training will

discuss the "nuts and bolts" of the system and focus on middle management and project specialists. Training will deal with specifics and should address such items as development of the CPM network schedule and the report outputs of the Planning and Scheduling Module, Estimating Module, and so on. Those organizations that will depend on the system for regular data and reports must be made to understand what type of information they can and cannot get from the system.

- *3rd Stage Training—System Operations.* Formalized training must be provided to those who will operate and maintain the system. Generally, this type of training is directed to technicians and system engineers and will include specialized courses offered by equipment manufacturers and software vendors. This portion of the training program should also focus on standard operational procedures associated with the system that may be developed by the in-house MIS organization.
- *4th Stage Training—System Utilization.* This stage of training should be a natural follow-up from the first stage or System Philosophy training discussed above. A very strong attempt should be made to get project management, selected middle management, and top management involved in seminar-type sessions to give them an opportunity to see firsthand how the system can be used to enhance their functions and capabilities. Very carefully structured "what if" type problems can be used to illustrate use of the computer-based information system in the quick evaluation of a number of alternatives or options to arrive at practical project decisions.

Obviously, the scope and level of the training to be provided will be somewhat dependent on budget restrictions and time availability. Nevertheless, money spent for training purposes will help to dispel the mystique surrounding the computer-based system and increase the likelihood that the system will be implemented and utilized successfully.

### System Validation

The last major milestone in the project information and control system development should be a validation test. This test should not be confused with the unit test that system designers or programmers will perform to check out the software packages. A validation test should cover all aspects of the system operation and utilization in a real-life project environment. This includes everything from data collection, receipt and utilization of the final output report, and subsequent management action. A formal test plan should be written and, at a minimum, this test plan must address all of the system criteria defined in the system program plan. If the program manager has been conscientiously keeping his program plan up to date, this should be a fairly straightforward effort.

Actual formal validation testing may be spread over a long period of time, especially if the system is developed and brought into actual use one module at a time. The program manager and system designers must objectively assess the results of each test and determine if they are within the established criteria. Obviously, there will always be some revisions and improvements that are desirable, but the program manager must be resolute and selective and ensure that the system is usable in a timely manner.

### System Support

A final word is in order relative to the long-term use of computer-based project information and control systems. While these systems are complex and costly to develop, this cost is small in comparison to the cost of the projects to be managed and the benefits to be derived from proper use of the systems. To ensure that the system is used effectively, management must take an additional step and provide a full-time staff of personnel dedicated totally to maintaining, supporting, improving, and constantly educating the user with regard to system capabilities. Otherwise, the system's effectiveness will diminish and reams of unread computer reports will begin to pile up on the corner of the project manager's desk.

## FUTURE DEVELOPMENTS

Truly useful information and control systems for project management have become a reality in recent years primarily because of the advancements in computer hardware and software. Computer speed and capacity have increased dramatically while cost has declined substantially. Innovative software programs have been devised while computer literacy has become commonplace. Yet we have barely tapped the potential of technology to devise more powerful tools for project management. The next generation of systems for project management will capitalize on the maturing techinques in communication, hardware for workstations, and system software for problem solving and decision making. The fundamental job of project management to plan and direct organizations and resources to accomplish a specific objective will be enhanced by tools for communication which can integrate organization and function over widely separated areas. Thus, satellite communication, electronic conferencing, local area network, electronic mail, and the automated office or workstation will be fundamental to any project management system. In addition, the ability to process information—that is, the receipt, storage, manipulation, organization, and presentation of information in a form that is useful to the user—is now possible because of advancements being made in computers, data base systems, and software. Because these systems can be integrated to form networks, we can devise project management systems in which the project participants can share information on a real-time basis and where detailed planning and control is performed at the level where the work is done (see Figure 27-13).

However, the most exciting potential for project management is the field of Artificial Intelligence (AI). Advancements in AI software and computer workstations are contributing to the development of so-called expert systems for planning, scheduling, monitoring, analysis, and decision making for a wide range of management and technical problems. The potential of AI in devising so-called expert systems is to enhance the ability of the user by providing him with tools that not only access a vast range of information quickly, but provide the means to interpret this information through some rational process much in the fashion as an expert would. The specific applications of AI are well beyond the goals of this chapter; however, the potential for providing very powerful tools for

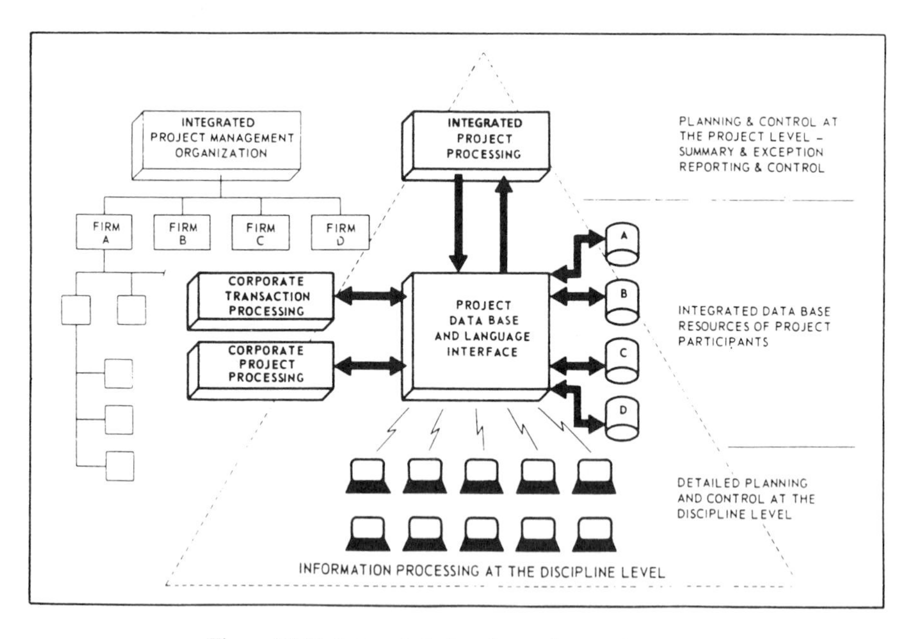


Figure 27-13. Integrated planning and control system.

project management must be recognized and accepted with an open mind. For it is only by visualizing the potential for applying these new tools to project management problems that will we be able to bring order and direction to an increasingly complex environment.

## SUMMARY

As projects become more numerous, larger, and more complex, the need increases for systems which provide for the systematic management of project information and action. These systems should be designed to function within a particular organizational environment and must reflect the unique project management requirements of that organization.

It is important to address the information system and the control system as two separate but highly dependent components of a total integrated computer-based project management system. In the design of the information system, the focus is on the procedures and techniques for collecting, storing, and processing data to produce information concerning project plans, schedules, cost, and performance parameters. In the control system the goal is the development of procedures and routines for evaluating plans against actuals to define the deviations or variances in plans, schedules, cost, and performance; the assessment of the acceptability of these variances (evaluated against some predefined standards); and the systematic development and implementation of action plans by the appropriate organizational units to produce decisions and/or directions which attempt to correct the project's deviations from plan.

Actual design, development, and implementation of an integrated system should follow a well-defined project methodolgy. The first step in this process involves making a comprehensive analysis of what is required in the way of information resources and control procedures to make project management function successfully within a particular organization environment. Once this analysis is completed, the actual design and development of the system(s) can be initiated. Design and development involves procurement of specialized software packages and/or design of new packages, as well as the development of the procedures for management action. The eventual success of the new systems will depend to a large degree on how accurately they meet the requirements specifications (as defined in the initial study), how well the systems have been documented, and how adequately the using organizations have been trained to apply these systems to the actual project environment. The true value of a project information and control system will be realized only when it enables a relatively small project management organization to successfully plan, direct, and control a complex, expensive, high-risk undertaking.

## BIBLIOGRAPHY


*Articles*

Baugh, Eddie W. and Scamell, Dr. Richard W. "Team Approach to Systems Analysis." *Journal of Systems Management,* 32–35 (April, 1975).

Brown, Foster. "The Systems Development Process." *Journal of Systems Management,* 34–39 (December, 1977).

Chapman, Charles H. et al. "Project Cost Controls for Research, Development and Demonstration Projects." *PMI Proceedings,* 53–63 (October, 1979).

Clarke, William. "The Requirements for Project Management Software: A Survey of PMI Members." *PMI Proceedings,* 71–79 (October, 1979).

Cullingford, Graham, Mawdesley, Michael J. and Chandler, Robert L. "Design and Implementation of an Integrated Cost and Schedule System for the Construction Industry." *PMI Proceedings,* 390–397 (October, 1977).

Finneran, Thomas R. "Data Base Systems Design Guidelines." *Journal of Systems Management,* 26–30 (March, 1978).

Gildersleeve, Thomas R. "Optimum Program Structure Documentation Tool." *Journal of Systems Management,* 6–11 (March, 1978).

Herzog, John P. "System Evaluation Technique for Users." *Journal of Systems Management,* 30–35 (May, 1975).

Mattiace, John M. "Applied Cybernetics Within R&D." *Journal of Systems Management,* 32–36 (December, 1972).

Miller, Earl J. "Chapter 9—Project Information Systems and Controls." In *Planning, Engineering, and Construction of Electric Power Generation Facilities,* Jack H. Willenbrock and H. Randolph Thomas, ed. Wiley, New York, 1980.

Niwa, Kiyoshi et al. "Development of a 'Risk' Alarm System for Big Construction Projects." *PMI Proceedings,* 221–229 (October, 1979).

Ramsaur, William F. and Smith, John D. "Project Management Systems Tailored for Selective Project Management Approach." *PMI Proceedings,* IV-A.1–IV-A.7 (October, 1978).

Ross, Ronad G. "Evaluating Data Base Management Systems." *Journal of Systems Management,* 30–35 (January, 1976).

Tuman, John, Jr. "The Problems and Realities Involved in Developing an Effective Project Information and Control System." *PMI Proceedings,* 279–293 (October, 1977).

Wilkinson, Joseph W. "Guidelines for Designing Systems." *Journal of Systems Management,* 36–40 (December, 1974).

*Books*

Archibald, Russel D. *Managing High-Technology Programs and Projects.* Wiley, New York, 1976.

Ashby, W. Ross. *An Introduction to Cybernetics.* Methuen, London, 1964.

Carlsen, Robert D. and Lewis, James A. *The Systems Analysis Workbook: A Complete Guide to Project Implementation and Control.* Hall, New Jersey, 1980.

Cleland, David I. and King, William R. *Systems Analysis and Project Management.* 3rd ed. McGraw-Hill, New York, 1983.

Fuchs, Walter Robert. *Cybernetics for the Modern Mind.* Translated by K. Kellner. Macmillan, New York, 1970.

Katzan, Harry, Jr. *Computer Data Management and Data Base Technology.* Van Nostrand Reinhold, New York, 1975.

Kerzner, Harold. *Project Management: A Systems Approach to Planning, Scheduling and Controlling. Van Nostrand Reinhold, New York, 1979.*

Martin, Charles C. *Project Management: How to Make It Work.* AMACOM, New York, 1976.

Murdick, Robert G. and Ross, Joel E. *Information Systems for Modern Management.* Prentice-Hall, Englewood Cliffs, N.J., 1971.

Myers, Glenford J. *Reliable Software Through Composite Design.* Petrocelli/Charter, New York, 1975.

O'Brien, James J. *Scheduling Handbook.* McGraw-Hill, New York, 1969.

Orlicky, Joseph. *The Successful Computer System Its Planning, Development, and Management in a Business Enterprise.* McGraw-Hill, New York, 1969.

Prothro, Vivian C. *Information Management Systems Data Base Primer.* Petrocelli/Charter, New York, 1976.

# 28. Computers in Project Management*

Harvey A. Levine†

Project management can be a very involved process, requiring a great deal of expertise in many disciplines. It requires that these processes be very structured and organized. It requires the development and processing of large volumes of data. It requires frequent reporting of plans and progress.

While the complete scope of project management requires much more than planning and scheduling, tracking and control, it is these specific functions that can be so effectively helped by the use of computers. It was 30 years ago that the old lumbering mainframes were put into service to support project management. For most of those years, access to computerized project management was reserved for the large organizations: those that had management information system operations, an army of dedicated project control specialists, and barrels of money to spend on hardware and software. But because of the computer technology changes during the first half of this decade, the benefits of computerized project management have been put within the reach of any potential user, for any project application.

During the past few years, the world of automation has been turned

---

* Chapter material used with permission from Osborne/McGraw-Hill, CA.

† Harvey A. Levine is the founder and principal of The Project Knowledge Group, Clifton Park, New York, a consulting firm specializing in project management software selection and evaluation, project management training, and project management using microcomputers. Mr. Levine is the author of the book *Project Management Using Microcomputers* (Osborne/McGraw-Hill), 1986. Mr. Levine provided project management applications, systems design, and consulting services to the General Electric Company for 24 years, has written several articles on project management, and has taught project management in both business and university environments. He is past Chairman of the Board of the Project Management Institute.

inside out by the fantastic success of the microcomputer and the acceptance of the microcomputer in the business community. With that acceptance has come the development of computer programs for use in solving business problems. Now, with a minimum investment, and bypassing the MIS bureaucracies, the doors to computer utilization in the business place have been opened to all of us.

Who would have believed, just a few years ago, that we would have this abundance of project management software available for the casual, as well as the serious, user, and at prices that are enticingly low. The microcomputer has given us accessibility to sophisticated programs that only recently were the private domain of the information systems gurus. The nature of project management systems, a combination of simple algorithms, calculations, and data base management, is a natural for computers. The need to do "what if" analyses, in the typical project management environment, was an additional driver of the microcomputer explosion.

These products address the entire range of the project management marketplace. There are programs for the local theater group that can help with the planning of their next production. There are programs for bankers, and programs for researchers. Programs exist at every level for the assignment and tracking of resources, and for cash flow planning and monitoring. Even formal project management organizations, with mainframe computer systems, are finding it advantageous to supplement, or even replace, their expensive batch systems with some of the very sophisticated professional-level project management software programs that are available for the microcomputer.

This chapter fully explains the planning and control process, while discussing why computers are a useful—almost necessary—tool for the performance of project management. It begins with a definition of what gets managed, and a list of typical project planning and control functions and phases.

Next, we invite you to follow the typical process, in detail, starting with establishing project objectives, defining the work, and developing the baseline schedule; then on to the development of resource plans and budgets; to the tracking of schedule, resources, and costs; and finally to a review of reporting and graphics for displaying the results of the project control efforts.

Throughout these discussions, we will define the functions themselves and then provide information relative to the support for these functions that is available through the various commercial microcomputer-based software packages, and furnish guidance on how to specify and evaluate such software for your applications and needs.

## PROJECT MANAGEMENT: A DEFINITION

It is important to understand the specific characteristics of a project if we are to understand the functions of project management. Furthermore, if we cannot recognize the differences between the management of a project and the management of the day-to-day business functions, then why should we acquire and learn new software? This is one of the fastest-growing segments of the software industry. There must be something significantly different in these two management areas, and indeed there is!

An underlying factor in general management is that one is dealing with a long-term, continuing business, with many of the measurements being associated with current performance as compared to prior performance as well as annual or other time-phased objectives.

In a nonproject environment, this management would concentrate on productivity-oriented and time-phased measurements. How many widgets did we manufacture this week? How does this compare to last week? How does this compare to the same week, last year? What were profits, or sales, or orders this week, as compared to whatever? What were our applied hours for this week versus last? Or more important, what were our unapplied hours? What were our expenditures for the last quarter versus our budget for that quarter? What is our staffing level in Department A versus our plan for this date? How is product A doing against product B? How is our product A doing against our competitor's product A?

Quality and human resource factors are, of course, also measured. What is our reject trend? Has our employee satisfaction index improved? What is the trend in customer satisfaction? Are we meeting our equal opportunity employment goals? The industry does not matter here. We can ask these questions about a construction firm. We can equally apply them to a library, a manufacturing firm, or a fast-food chain. The underlying factor is that one is dealing with a long-term, continuing business, with many of the measurements being associated with current performance as compared to prior performance, as well as annual or other time-phased objectives.

Keep this definition of general management in mind as we define the characteristics of a project and see that those characteristics demand that we use a different basis for project management.

A *project* is a group of tasks, to be performed in a definable time period, to meet a specific set of objectives. In general, a project will exhibit most of the following conditions:

- It is likely to be a unique, one-time program.
- It will have a life cycle, with a specific start and end.

- It will have a work scope that can be broken up into definable tasks.
- It will have a budget for its execution.
- It may require the utilization of multiple resources. Many of these resources may be in short supply and have to be shared with other projects.
- It may require the establishment of a special organization for its execution or require the crossing of traditional organizational boundaries.

While some of the general business measurements can apply to a projects situation, the project measurement framework will be primarily structured so that budgetary and manpower control is based on the accomplishment of the defined work rather than time frames. Also, reporting and measuring will be categorized by the various sections of the work (deliverables, physical areas, cost accounts) as well as by the traditional functional divisions (responsible areas).

It is these differences, this different measurement framework, that generated the need for the development of separate project management software for planning and control.

> Project management software supports the planning and control of such elements as the work scope, or contents, of a project; the project timing; human and nonhuman resources; budgeting and costs, and communications.

### What Gets Managed

According to a definition developed by the Project Management Institute, in the managing of projects we will normally be involved in the following:

- Scope management.
- Time management.
- Human resource management.
- Cost management.
- Quality management.
- Communication management.

For each of these functions, it is necessary to plan, organize, direct, and control. Every project must have a plan. And it is this plan that becomes the basis for control. As this chapter addresses the utilization of microcomputers in project management, we will concentrate on the planning and control aspects of project management.

## THE TYPICAL PLANNING AND CONTROL FUNCTIONS

The following is an listing of the typical functions involved in planning and controlling projects:

- Establish the project objectives.
- Define the work.
- Determine the work timing.
- Establish the resource availability and requirements.
- Establish a cost baseline.
- Evaluate the baseline plan.
- Optimize the baseline plan.
- "Freeze" the baseline plan.
- Track work progress.
- Track actual costs.
- Compare progress and costs to the baseline plan.
- Evaluate performance.
- Forecast, analyze, and recommend corrective action.

A flowchart of these functions is presented in Figure 28-1. Each of these functions will be discussed in detail later in this chapter.

## PHASES OF PROJECT MANAGEMENT

Before we leave this description of project management, let's go back to the project management definition statement again. We said that project management is the planning, organizing, directing, and controlling of resources for a specific time period to meet a specific set of one-time objectives. That specific time period generally consists of three primary phases:

- Proposal phase.
- Implementation Phase.
- Closeout phase.

The implementation phase really consists of two distinct subphases:

- Planning phase.
- Execution phase.

In Figure 28-1, we can see the functions associated with these two implementation phases. Those items leading to the development of the

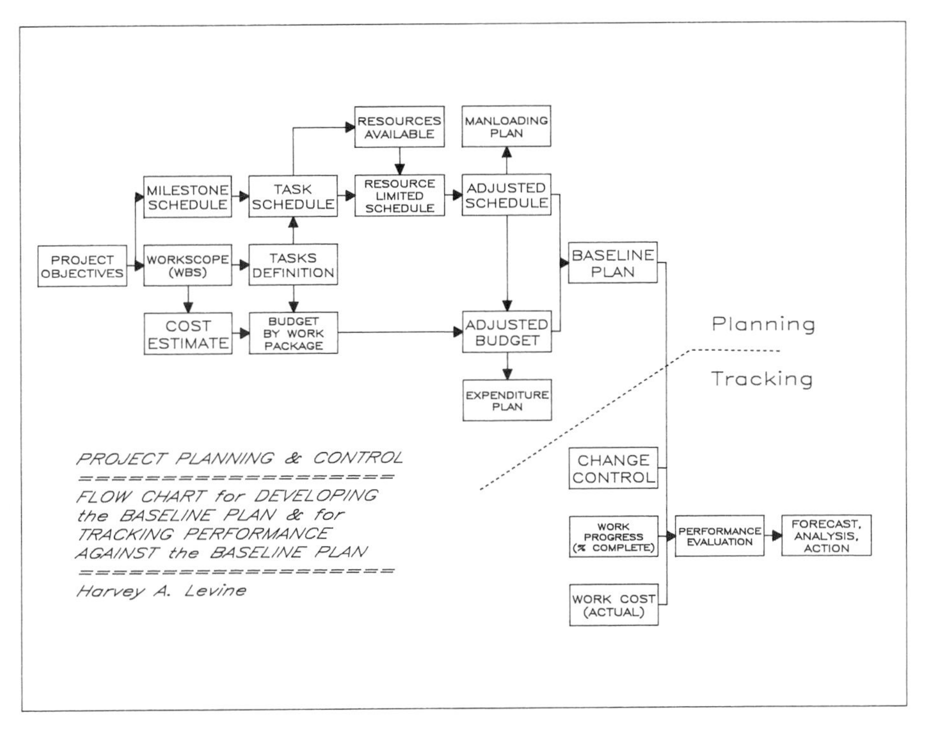


Figure 28-1. Project planning and control—flowchart.

baseline plan are associated with the planning phase. The remaining functions are part of the execution phase.

It is extremely important to recognize the impact of these phases on your choice of a planning and control system. If you only need to support the development of proposals or baseline plans, then you can utilize software that only does *planning*. On the other hand, if you are to *control* a project, you will need software that provides for progress tracking and comparison to the plan.

When many of the PC-based project management software programs were first introduced, several of them did not have the capability to retain a baseline plan. In such programs, the entry of any current data obliterated the original data, thereby prohibiting the direct use of the project data base for the comparison of actual to plan—a necessary ingredient for project control. Mention of the lack of these functions is also often missing in the magazine reviews of software. Today, there are very few programs that do not have some means of comparing the current status to the plan, but they vary considerably.

Below are listed the necessary features for project control.

**Necessary Features for Project Control**

- The ability to save a version of your data base as a baseline or target plan, or to be able to freeze a set of data within your project data base for comparison to current status and actual expenditures.
- The ability to enter actual start and finish dates without having those entries override the plan dates.
- The ability to revise task durations or remaining durations without having those entries override the plan dates.
- The ability to enter actual costs incurred, for comparison to the budget.
- The ability to record actual resource use.

## THE TYPICAL PLANNING AND CONTROL FUNCTIONS AND THE ROLE OF THE COMPUTER

Getting back to the list presented earlier in this chapter, let's complete this overview of project management with a closer look at those primary functions and a look at how the computer can be effectively used to aid the project manager in the execution of these functions.

### Establish the Project Objectives

**The Function.** It is strange, perhaps even annoying, that this first item is most often missing from the discussion of project planning and control. But isn't this the very reason for having project planning and control? The plan must be developed to support a set of objectives. These objectives should be defined in the general terms of time objectives, budget objectives, technical, etc. But, also, they should be defined, where possible, in the terms of deliverable end items. The definition of the project objectives should also be related to the overall organization objectives and to the organization's policies.

**Where Does the Computer Fit In?** This first set of project planning tasks sets the stage for work that will eventually be processed on the computer.

This setting of the objectives, and associated strategizing and organizing, cannot be helped much by the computer. Don't look to the computer to do the front-end thinking and decision making that must be the responsibility of the project manager and top management. There are several software packages that provide some guidance for this function in their users' manuals.

### Define the Work

**The Function.** To develop a baseline plan, you will push forward in three parallel paths, with considerable crosstalk between them. You will be defining the work to be done, the resources and budget for that work, and the timing for that work. Obviously, the whole thing is dependent upon the quality of the work scope definition.

A popular and useful approach toward defining the work scope is called the *Work Breakdown Structure* (WBS). In this approach, the project objectives are placed at the top of a diagram as deliverable end items. Each top-level segment of the project is then subdivided into smaller groups, in a manner similar to typical organization charts. The work breakdown continues until you have groups of activities (or tasks) that would comprise a natural set of work. These smallest groups are often called *work packages*. Several other names have been given to this approach, all of which also are descriptive of the technique. These include *tree diagram, hierarchical chart, top-down planning,* and *gozinto chart*. In all cases, the lower elements go into the next higher level. A hierarchical numbering system is generally used to facilitate the rolling-up (or summarization) of data to the various levels of the WBS. A multipart numbering system can be used to also include codes for roll-up in other categories, such as department codes, accounting codes, etc. Figure 28-2 illustrates such a WBS.

Each of the work packages is reviewed to establish a set of task definitions. There are many ways to get to this necessary list of tasks. The WBS approach has many advantages, including better organization of the work, elimination of "reinventing the wheel," and use of the WBS as a checklist. The WBS also facilitates summarization and reporting. The list of the tasks will be used as a basis for scheduling, resource planning, and budgeting.

**Where Does the Computer Fit In?** Most of this portion of the project planning process is *not* done by the computer. A few of the vendors, notably ViewPoint and Harvard Total Project Manager II, have incorporated a top-down planning or WBS feature in their software. In these

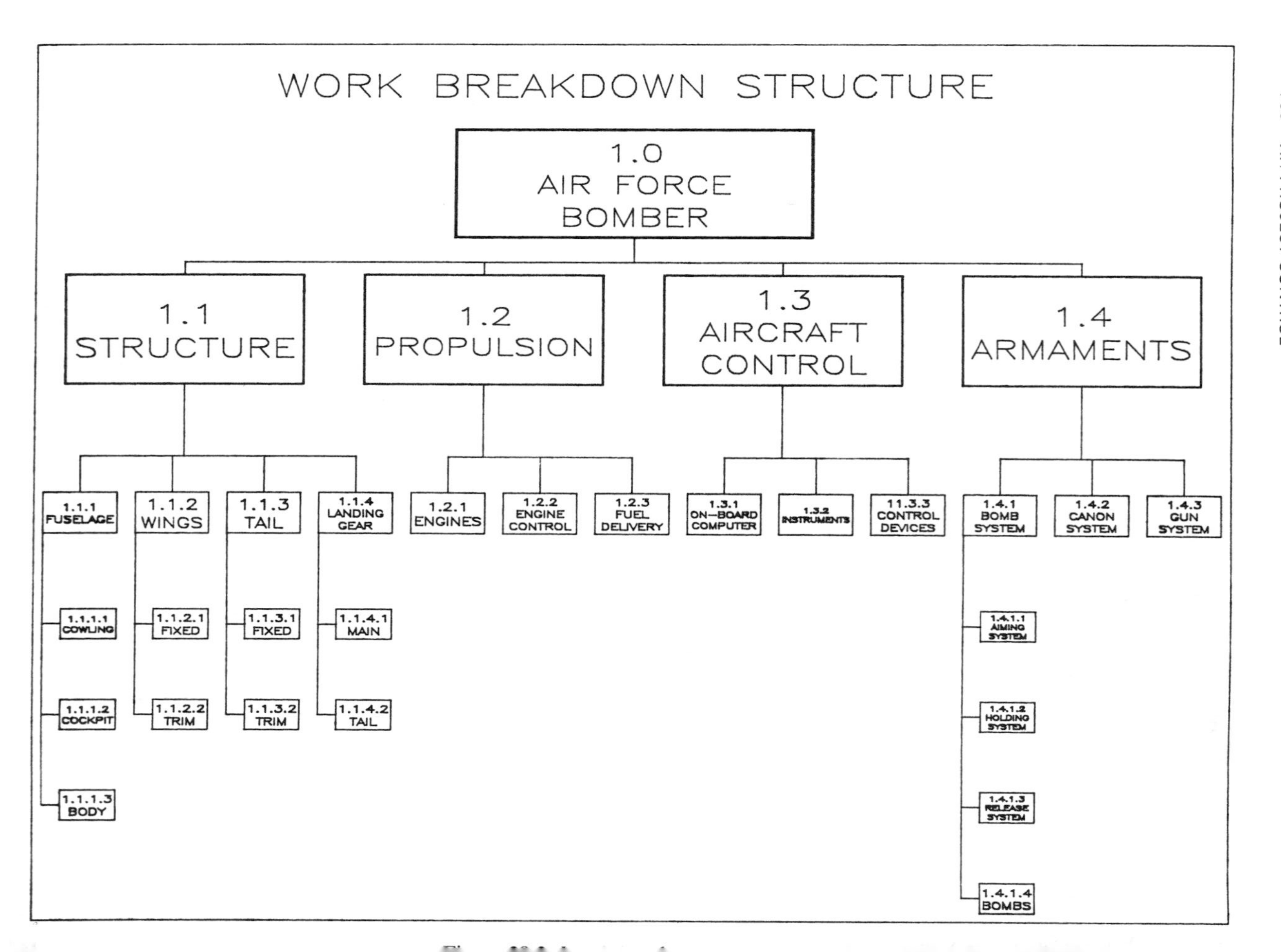
WORK BREAKDOWN STRUCTURE
1.0 AIR FORCE BOMBER
1.1 STRUCTURE
1.2 PROPULSION
1.3 AIRCRAFT CONTROL
1.4 ARMAMENTS
1.1.1 FUSELAGE
1.1.2 WINGS
1.1.3 TAIL
1.1.4 LANDING GEAR
1.2.1 ENGINES
1.2.2 ENGINE CONTROL
1.2.3 FUEL DELIVERY
1.3.1 ON-BOARD COMPUTER
1.3.2 INSTRUMENTS
11.3.3 CONTROL DEVICES
1.4.1 BOMB SYSTEM
1.4.2 CANON SYSTEM
1.4.3 GUN SYSTEM
1.1.1.1 COWLING
1.1.1.2 COCKPIT
1.1.1.3 BODY
1.1.2.1 FIXED
1.1.2.2 TRIM
1.1.3.1 FIXED
1.1.3.2 TRIM
1.1.4.1 MAIN
1.1.4.2 TAIL
1.4.1.1 AIMING SYSTEM
1.4.1.2 HOLDING SYSTEM
1.4.1.3 RELEASE SYSTEM
1.4.1.4 BOMBS

programs, the user can define the WBS to the system, which then automatically uses that definition as a hierarchical structure for the project data. While certainly useful, the important thing is that the program that you select have the facility to allow the defining of a coding structure for the WBS and other selection/summarization criteria. You should opt for at least three code fields for you to enter WBS codes, responsibilities, cost account codes, phase codes, and so forth. The system, of course, should allow you to sort and select on any of these codes.

A potentially helpful feature of the aforementioned ViewPoint program is that the plans developed for each of the WBS sections can saved as a "region" in a "library." These library regions can be recalled when a new schedule is being developed and inserted into the new network.

### Define the Work Timing

**The Function.** Part of the definition of project objectives will be the establishing of the overall timing parameters of the project. These timing objectives and assumptions should be first processed into a project milestone schedule (see Figure 28-3). The milestone schedule provides guidance by defining the time windows into which the task scheduling will attempt to fit.

Now, with the list of tasks and the milestone schedule in hand, you can take a first cut at developing the task schedule.

The detailed schedule can be either duration driven or resource driven. The more common basis is the duration-driven approach. Working with your list of tasks, you will add the expected time duration for the execution of each task. Next, identify the relationship of each task to each other task in the project, that is, develop the precedence rules for each task. Methods of doing this are numerous, although they all aim at the same objective—to enable the computation of the earliest start and finish times for each task.

The assignment of task durations and relationships would produce a schedule that assumed that there were no other factors that dictated when an activity could start or must be finished, except those two components. Yet, we know that there are times when even if an activity's predecessors are complete, the activity cannot be started until other, external conditions are satisfied. For instance, you may not want to plant grass in Wisconsin in January, even if the preceding landscaping has been completed. Or you may want to force the completion of the roof of a building before the snow falls, even though the network logic may dictate that the roof need not be completed until February to support the successor tasks.

Therefore, the last element of determining the work timing is to identify

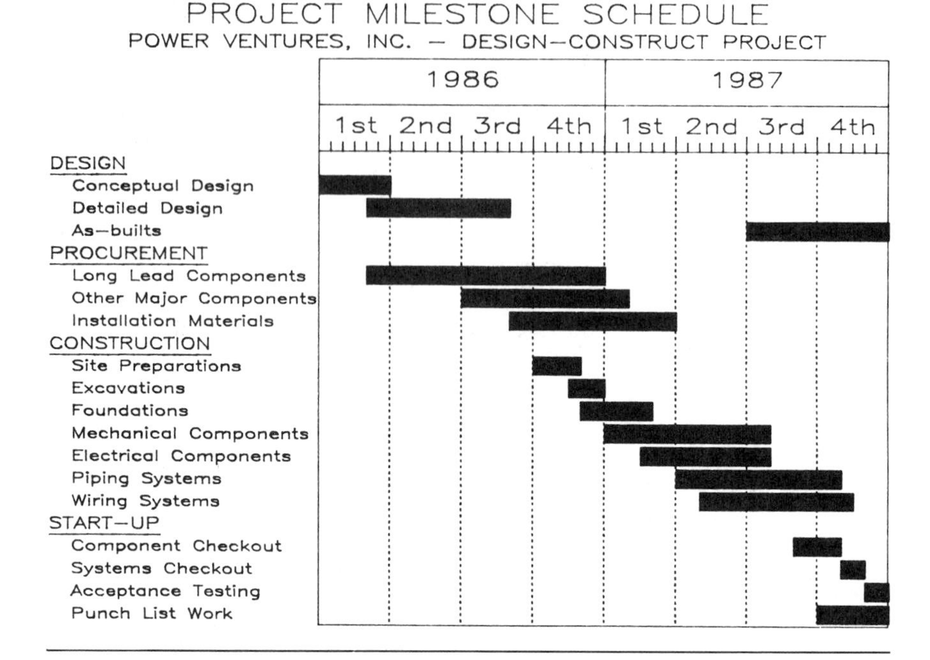


Figure 28-3. Project milestone schedule.

all activities that require imposed date constraints, often referred to as "start no earlier than" and "finish no later than" constraints.

In a resource-driven task schedule, the precedence relationships are still required. But, instead of single task durations, one identifies the labor resources required. This entails the quantification of the total man-hours (for each resource) required to do the task, and the allowable crew sizes. The scheduler (computer) can then determine the task durations. The overall result is the same as above. It is a common practice to do a duration-driven schedule first so as to get an idea of the best schedule attainable without resource limit considerations.

**Where Does the Computer Fit In?** Essentially, at this point you have built a model of the project work. This is called a *critical path network*. The computation of the early start and finish is usually performed by a computer and is referred to as the *network analysis*. Once the latest task is identified, via the forward pass of the network analysis, a backward computation can be made to determine the latest that each activity can start and finish without delaying the overall project. You will also have identified the relative time priority of each activity, because the difference between the early dates and the latest dates indicates the *float* or *slack* for each task.

The task schedule can be displayed in three typical formats: a tabular listing (see Figure 28-4), a bar (Gantt) chart (see Figure 28-5), and a network diagram. The network diagram can be presented in two formats. The simple flowchart style, as shown in Figure 28-6, depicts only the relationships between activities without a graphic reference to the time

Schedule Name: PROJECT SPLASH
Project Manager: Esther Williams
As of date: 29-May-86 1:31am Schedule File: C:\TLDATA\POOL529

| Task | How Long | Early Start | Early End | Late Total Start | Late Total End |
|---|---|---|---|---|---|
| Decide To Do | 2 days | 31-Mar-86 8:00am | 1-Apr-86 5:00pm | 31-Mar-86 8:00am | 1-Apr-86 5:00pm |
| Gather Info | 8 days | 2-Apr-86 8:00am | 11-Apr-86 5:00pm | 2-Apr-86 8:00am | 11-Apr-86 5:00pm |
| Concept | 5 days | 14-Apr-86 8:00am | 18-Apr-86 5:00pm | 14-Apr-86 8:00am | 18-Apr-86 5:00pm |
| Establish Program | 3 days | 28-Apr-86 8:00am | 30-Apr-86 5:00pm | 28-Apr-86 8:00am | 30-Apr-86 5:00pm |
| Estimate Costs | 2 days | 1-May-86 8:00am | 2-May-86 5:00pm | 1-May-86 8:00am | 2-May-86 5:00pm |
| Apply For Permit | 3 days | 1-May-86 8:00am | 5-May-86 5:00pm | 27-May-86 8:00am | 29-May-86 5:00pm |
| Issue Permit | 15 days | 6-May-86 8:00am | 26-May-86 5:00pm | 30-May-86 8:00am | 19-Jun-86 5:00pm |
| Apply For Loan | 3 days | 5-May-86 8:00am | 7-May-86 5:00pm | 7-May-86 8:00am | 9-May-86 5:00pm |
| Issue Loan | 10 days | 8-May-86 8:00am | 21-May-86 5:00pm | 12-May-86 8:00am | 23-May-86 5:00pm |
| Price Fencing | 5 days | 21-Apr-86 8:00am | 25-Apr-86 5:00pm | 21-Apr-86 8:00am | 25-Apr-86 5:00pm |
| Order Fencing | 2 days | 5-May-86 8:00am | 6-May-86 5:00pm | 16-May-86 8:00am | 19-May-86 5:00pm |
| Fencing Delivered | 15 days | 7-May-86 8:00am | 27-May-86 5:00pm | 20-May-86 8:00am | 9-Jun-86 5:00pm |
| Erect Fence | 5 days | 10-Jun-86 8:00am | 16-Jun-86 5:00pm | 10-Jun-86 8:00am | 16-Jun-86 5:00pm |
| Price Patio Kits | 5 days | 21-Apr-86 8:00am | 25-Apr-86 5:00pm | 21-Apr-86 8:00am | 25-Apr-86 5:00pm |
| Patio Detailed Plans | 5 days | 5-May-86 8:00am | 9-May-86 5:00pm | 16-May-86 8:00am | 22-May-86 5:00pm |
| Obtain Support Material | 2 days | 21-May-86 8:00am | 22-May-86 5:00pm | 23-May-86 8:00am | 26-May-86 5:00pm |
| Dig Holes | 1 day | 27-May-86 8:00am | 27-May-86 5:00pm | 27-May-86 8:00am | 27-May-86 5:00pm |
| Form | 3 days | 28-May-86 8:00am | 30-May-86 5:00pm | 28-May-86 8:00am | 30-May-86 5:00pm |
| Pour | 1 day | 2-Jun-86 8:00am | 2-Jun-86 5:00pm | 2-Jun-86 8:00am | 2-Jun-86 5:00pm |
| Cure | 3 days | 3-Jun-86 8:00am | 5-Jun-86 5:00pm | 3-Jun-86 8:00am | 5-Jun-86 5:00pm |
| Strip & Backfill | 1 day | 6-Jun-86 8:00am | 6-Jun-86 5:00pm | 6-Jun-86 8:00am | 6-Jun-86 5:00pm |
| Order Patio Kit | 2 days | 5-May-86 8:00am | 6-May-86 5:00pm | 7-May-86 8:00am | 8-May-86 5:00pm |
| Deliver Patio Kit | 14 days | 9-May-86 8:00am | 28-May-86 5:00pm | 20-May-86 8:00am | 6-Jun-86 5:00pm |
| Erect Patio Base | 4 days | 9-Jun-86 8:00am | 12-Jun-86 5:00pm | 9-Jun-86 8:00am | 12-Jun-86 5:00pm |
| Erect Patio Rails | 2 days | 13-Jun-86 8:00am | 16-Jun-86 5:00pm | 13-Jun-86 8:00am | 16-Jun-86 5:00pm |
| Finish Patio | 2 days | 17-Jun-86 8:00am | 18-Jun-86 5:00pm | 17-Jun-86 8:00am | 18-Jun-86 5:00pm |
| Get Pool Quotes | 5 days | 21-Apr-86 8:00am | 25-Apr-86 5:00pm | 21-Apr-86 8:00am | 25-Apr-86 5:00pm |
| Order Pool | 2 days | 5-May-86 8:00am | 6-May-86 5:00pm | 5-May-86 8:00am | 6-May-86 5:00pm |
| Pool Delivered | 18 days | 7-May-86 8:00am | 30-May-86 5:00pm | 7-May-86 8:00am | 30-May-86 5:00pm |
| Install Pool | 10 days | 2-Jun-86 8:00am | 13-Jun-86 5:00pm | 2-Jun-86 8:00am | 13-Jun-86 5:00pm |
| Fill Pool | 3 days | 16-Jun-86 8:00am | 18-Jun-86 5:00pm | 17-Jun-86 8:00am | 19-Jun-86 5:00pm |
| Clean-up | 2 days | 17-Jun-86 8:00am | 18-Jun-86 5:00pm | 17-Jun-86 8:00am | 18-Jun-86 5:00pm |
| Send Invites | 1 day | 16-Jun-86 8:00am | 16-Jun-86 5:00pm | 18-Jun-86 8:00am | 18-Jun-86 5:00pm |
| Get Party Supplies | 1 day | 17-Jun-86 8:00am | 17-Jun-86 5:00pm | 18-Jun-86 8:00am | 18-Jun-86 5:00pm |
| Hold Pool Party | 1 day | 19-Jun-86 8:00am | 19-Jun-86 5:00pm | 19-Jun-86 8:00am | 19-Jun-86 5:00pm |

Figure 28-4. Tabular schedule (Time Line).

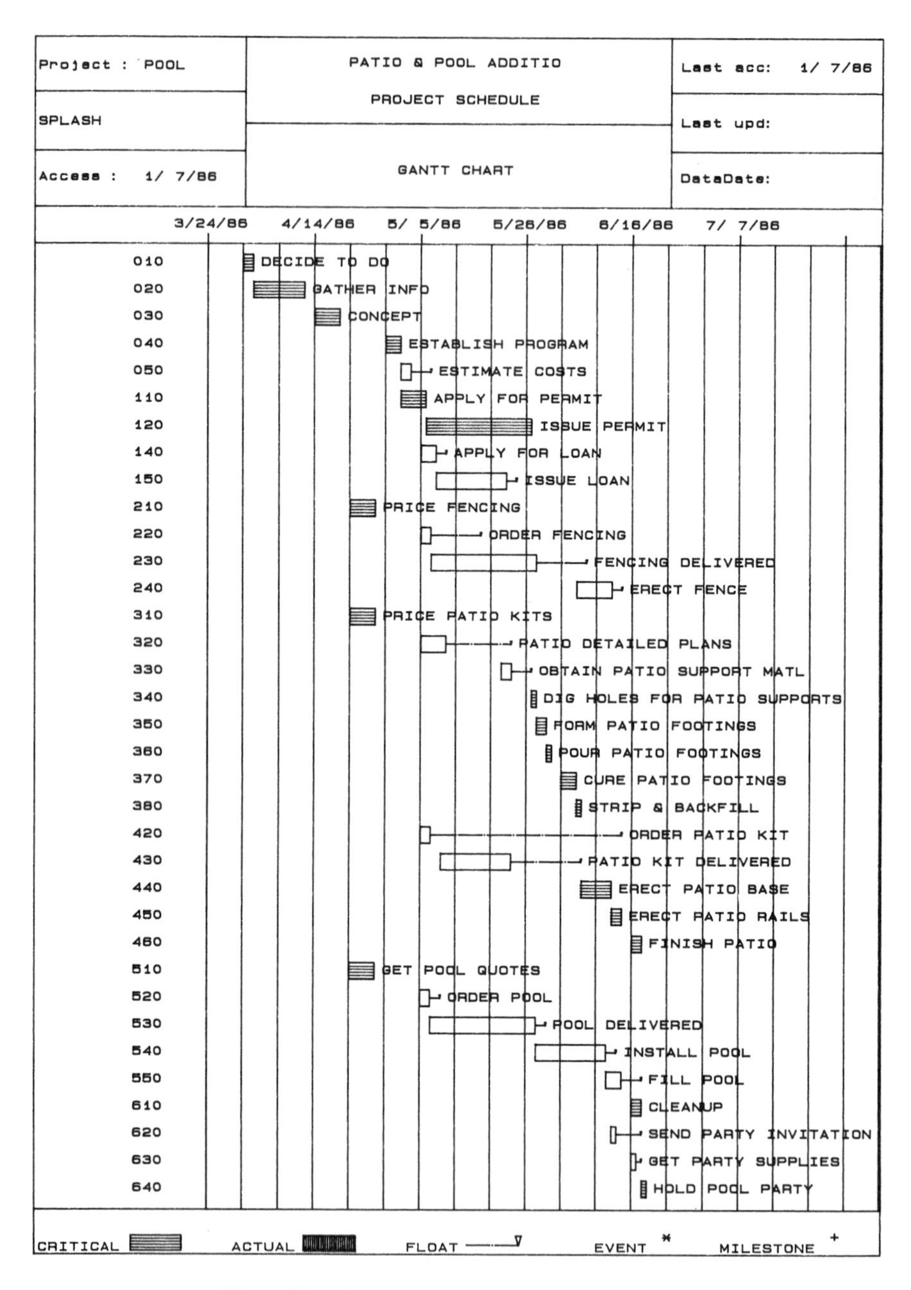


Figure 28-5. Bar (Gantt) chart schedule (PROMIS).

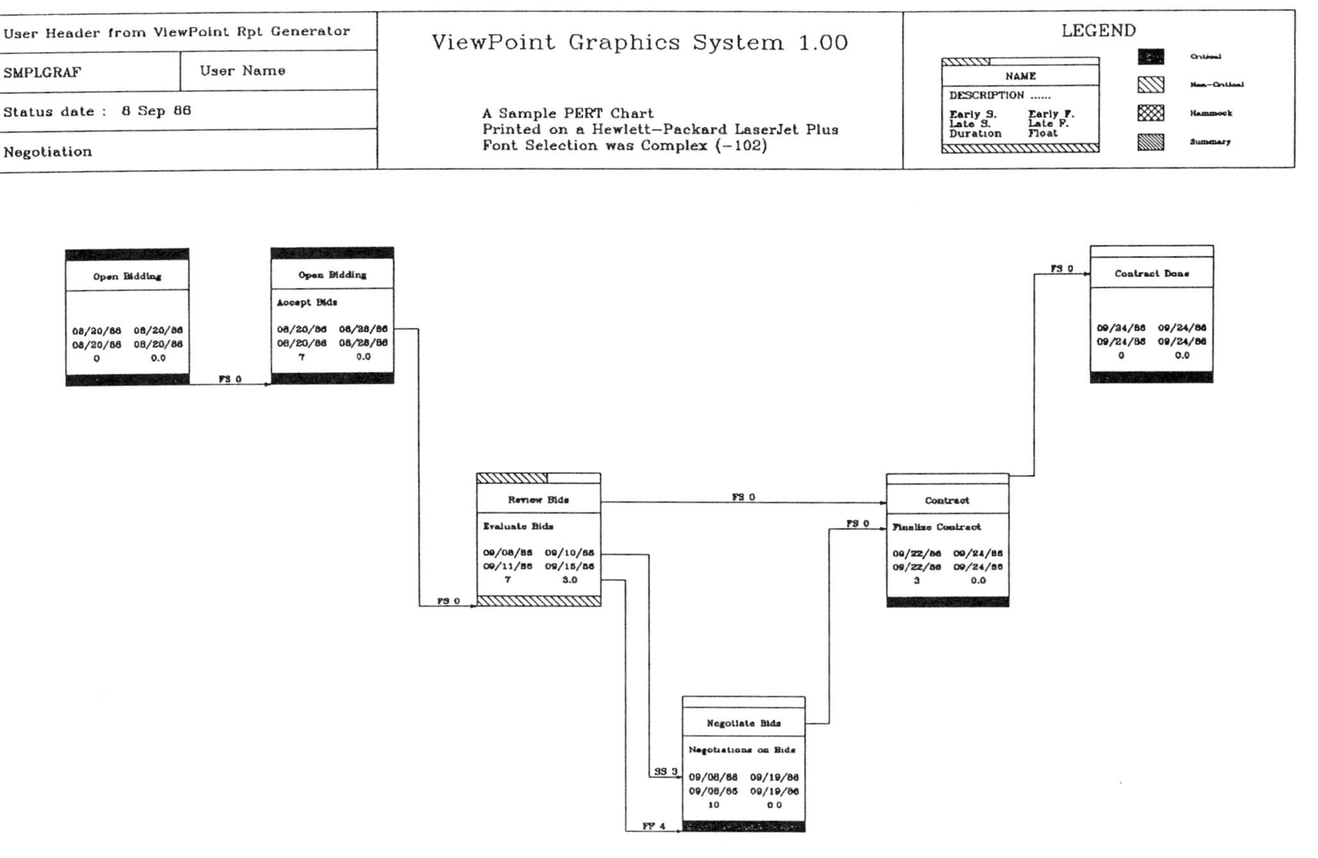


Figure 28-6. Network diagram—node format (ViewPoint).

duration of each task. Eash task is usually drawn in a box describing the task and its duration, schedule dates, and float. A more cogent format is the time-phased logic diagram (see Figure 28-7), which plots the activities against a time scale, thereby depicting both the relative time and dependencies of each task. While the latter style is more informative, we usually pay a price in loss of clarity in presentation. As task placement is now controlled by task duration and schedule, the time-phased network logic diagram can become very busy and difficult to read. This is controlled by careful selection and grouping of activities, assuming that your software will allow you to have such control over the diagramming process.

The mass market media and software vendors appear to have developed their own language to define these two network diagram formats. A common, but erroneous, term for the simple network logic diagram is "PERT chart." On occasion you will also see the time-phased logic diagram referred to as a "CPM diagram." PERT and CPM refer to the two original computerized scheduling protocols, developed in the late 1950s,

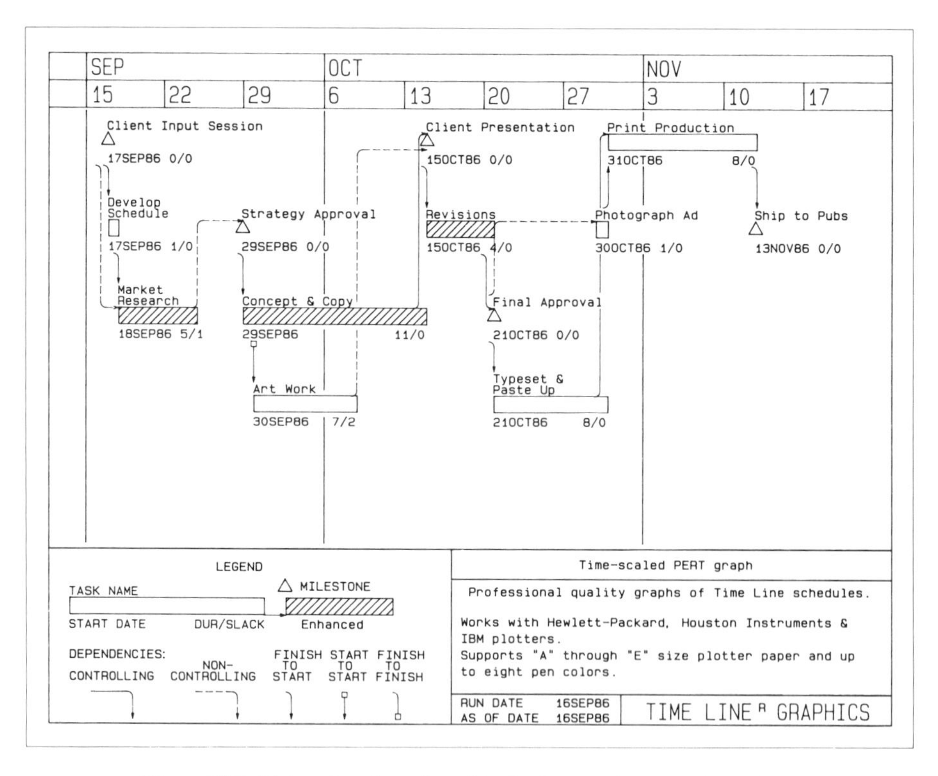


Figure 28-7. Network—time-phased logic diagram format (Time Line).

that form the basis for today's popular project management programs. Most of today's programs use the activity-on-the-node format rather than the event-oriented PERT format or the activity-on-the-arrow CPM format.

### Establish the Resource Availability and Requirements

**The Function.** Your needs in resource allocation and costing will determine how you look at the resource and cost functions of project management software. Resources include both manpower and materials. There is a tremendous range of capabilities for resource planning. It is important, then, to take a look at the levels of detail that can be addressed in resource planning and to determine what capabilities you will need for the level of planning and control that you will need to exercise on your projects. Beginning at the lowest end of the detail range, you may wish to:

- Assign task responsibility to an individual.
- Assign more than one individual to a task.
- Assign multiple resources (manloading) to a task from groups of pooled resources, without identifying a specific, named individual for a specific task.
- Define a *constant* limit for any pooled resource and have the scheduling system show the demand against the availability (resource indication).
- Define a *variable* limit for any pooled resource and have the scheduling system show the demand against the availability (resource indication).
- On a person-to-person basis, have the system reschedule tasks so that an individual is not working on more than one task at any time (simplified version of resource allocation).
- On a person-to-person basis, define the man hour availability (per day or week), and have the system reschedule tasks so that an individual is not working on a task load that exceeds that availability (another simplified version of resource allocation).
- Define a limit (single or variable) for any pooled resource and have the scheduling system reschedule tasks to hold the resource demand to that availability (resource/limited-resource allocation).
- Define a desired limit for any pooled resource, and a must-hold end date, and have the scheduling system reschedule tasks to attempt to stay within the resource limits, but only if the rescheduling does not violate the required (latest) dates (time/limited-resource allocation).

- Define a desired limit for any pooled resource, and a must-hold end date, and have the scheduling system reschedule tasks, *and availability,* to level the resource demand and hold the end date (resource leveling).

Even within these definitions, there are finite differences than can be important to the user. For instance, if you are assigning multiple resources to a task, do all of the assigned resources have to be assigned for the full duration of the task? In the real world, it is likely that some resources will only have to be applied for parts of certain tasks. But in most programs, especially at the mass market ($395–$495) level, you cannot define this criterion.

**Where Does the Computer Fit In?** As you can see, the variations on a resource theme are almost without limit and cannot be determined by a simple review or specification statement that such-and-such a program has resource leveling. Your selection of a project management software program will have to take into consideration which of the levels of detail you need, based on the preceding listing.

Regardless of the level of detail used, the process essentially calls for you to work with the list of tasks, itemizing the resources (labor, materials, services, etc.) required for each task. If you have developed a schedule for these tasks, you can now get an indication of the total resources required to execute the job on a period-by-period basis. Computers are very good at producing *resource histograms*. If you can define the desired or mandated resource availability, you should be able to note and evaluate when the required resources exceed these limits. This function (see Figure 28-8a) is called *resource aggregation* or *resource indication*.

If you require both resource allocation and leveling capabilities, how likely are you to get efficient utilization of available resources from your program? Well, that depends on the algorithms programmed into that software, and how finite the work resource parameters can be defined. A major cause of peaks and valleys (after leveling), is the inability to "split" activities. The assumption, by most programs, is that once a task has been started (or allocated resources), it must be scheduled to completion without interruption. There are a few sophisticated programs (MicroPlanner, for one) that allow the user to state whether a task can be split, and if so, what the minimum split duration is. SuperProject is another.

Once again, when it comes to evaluating programs against your resource criteria, you will first have to clarify those criteria per the above variations, and then actually use the candidate programs to see if they do what you need them to.

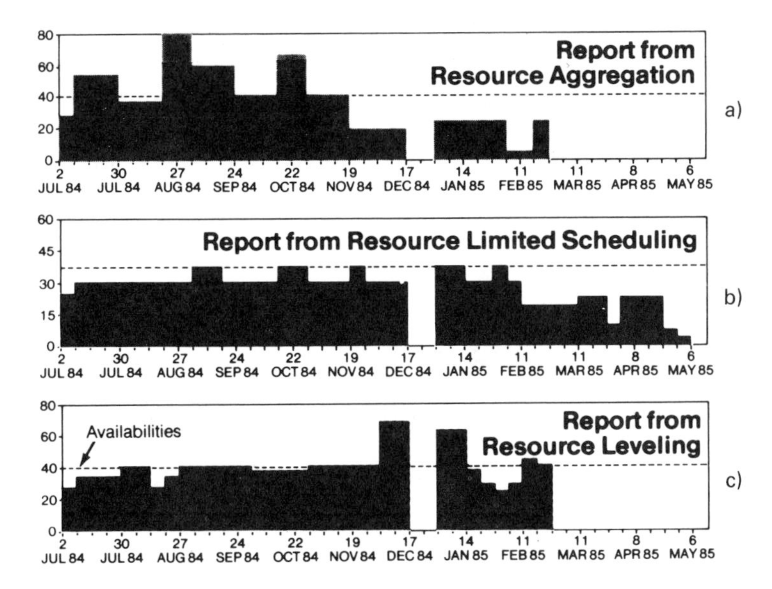


Figure 28-8. Resource histograms. (Courtesy of Computerline, Inc.)

Before we leave the area of resource planning, let's dwell a moment on the concepts of the smoothing of peaks and valleys, and staying within resource limits. While we can use the common labels of "resource leveling" and "resource allocation," the terms are not always clear, and, as usual, there can be variations to the basic concepts. Many programs give you more than one option, all of which have valid applications. These include

1. Resource-Limited Scheduling:
   Activities are scheduled when resources are available. Resource availability is not exceeded. When many activities vie for scarce resources, the resources are assigned to those activities that are more critical (or possibly by other priority schemes), and the other activities are forced to slip to later dates. The overall project schedule may slip because of resource shortages. See Figure 28-8b.
2. Time-Limited Scheduling:
   As above, activities will be scheduled according to the availability of resources. However, activities must be scheduled within the total float. Prescribed resource limits can be exceeded in order to hold the overall project schedule. See Figure 28-8c. This is often called resource leveling.

3. Time/Resource Limited Scheduling:
   This is a combination of the two above. The user can define a secondary resource level which can be invoked when the target end date cannot be held with the base resource availability. Or, the user may specify a date that the schedule can slip to if the original, non-man-loaded date cannot be met with the available resources. Optimizing the schedule/resource plan is usually achieved via trial and error computations with various resource levels and target dates.

### Establish a Cost Baseline

**The Function.** To control costs, you will want to establish a work scope-oriented budget. This can be done at the task level, or at the next higher level, the work package level. At the task level, the budget can be established by determining the resources for each task, the cost rate for each resource, the duration for each task, and the fixed costs and other expenses. Let's look at this function further as we examine the various approaches that are possible with the computer.

**Where Does the Computer Fit In?** With most computer programs, a table of resources and resource costs is established, so that much of the budget process is automatic. With the entering of task durations, resources, and fixed costs, the budget is created. As with the resource area, costing can be approached at various levels of detail.

Actually, there are two primary ways that project management programs address cost planning. We initiated this topic by describing the preferred way to deal with costs for each activity, which is on the basis of the specific resources that are used and the resource rates. This we can call resource-driven costing. If you cannot define the costs and other variables for each resource in a resource definition table or library, then you are limited to define all costs for a task as if they were fixed costs. Since, in reality, they are not fixed but affected by hourly rates, duration of the activity, and quantity of resources assigned, you will have to calculate these costs manually and input them yourself. If you do have a resource-driven program, you should be able to define both resource-generated and fixed costs. In that case, the computer will do your calculating for you, a decided advantage.

In the nonresource (fixed-cost) environment, the variations of detail could be as follows:

- Define a cost (budget) for each task.
- Establish a cost library (cost categories) and assign a series of "fixed" costs to each task.

- Assume all costs to be apportioned evenly across the duration of the task.
- Specify whether each cost item is to be allocated evenly or at the start of the task or at the end of the task.
- Report on task budgets by task.
- Report on task budgets by task and by cost categories (summary).
- Calculate and/or plot a project expenditure schedule (cash flow plan).

In the *resource-driven costing system* it gets much more involved. Here is a list of the possible variations (in addition to those above):

- Define a resource library with a single cost rate for each resource.
- Define a resource library with a multiple cost rate for each resource (standard time, overtime, etc.).
- Define a resource library with a stepped cost rate for each resource (escalation, union rate changes, etc.).
- Enter task costs by resource, by fixed cost, and by overhead or G&A categories.
- Allocate costs by four modes: even distribution, all at start, all at end, and at the resource library *rate* for the specified resource duration, for that task.

### Evaluate the Baseline Plan

**The Function.** At this time, having done all of the above, you have a baseline plan. But is it the one that you want? Does it meet the project time, cost, and resource objectives? Chances are that it will require some "adjustment."

You will want to evaluate how the schedule of the individual tasks, with their durations and relationships, correlates with the milestone schedule. This is probably the first opportunity to validate the earlier assumptions. Will you need additional resources, more work hours per week, or adjusted relationships in order to meet the overall timing objectives?

**Where Does the Computer Fit In?** The computer is an excellent tool for performing the "what if" processing required to evaluate and optimize the baseline plan. First, in its scheduling mode, the computer will trace the logic from project start to finish, calculating the earliest start and finish dates for each activity. After completing this "forward pass," the computer makes a "backward pass" to determine the latest dates that any activity can start and finish to meet the target end dates. The difference between the early dates and the late dates is the "float" or "slack" for each activity. Those that have more float are easily noted, and one's attention can be given to those activities with the least float (the most critical activities).

The other major area to evaluate will be the resource loading and distribution. It is very likely that your first cut of the schedule will produce a resource utilization profile having significant peaks and valleys, such as in Figure 28-8a. This is, at the least, an inefficient and costly condition and, in fact, may not be supportable by your labor sources. You will want to attempt to smooth the resource profiles out, which means rescheduling activities that can be moved without creating delays to the overall project milestones.

Look to your project management software to help you get at the specific information you need to evaluate the baseline plan. How well will your software allow you to sort and select records and fields from your data base? You don't want to sift through 50 pages of reports where just a page or two will focus in on the specific areas that are outside of your desired schedule and resource targets. Look for software that will allow you, the user, to pull out just the information that you need, in the formats that you want. Schedule reports should be retrievable by total float (critical path) and by key interest groups, such as area, responsibility, WBS, and so forth. Resource reports should be retrievable by specific resource and by time periods, in both resource quantity and cost formats.

### Optimize the Baseline Schedule

**The Function.** If the overall duration of the project computes to a longer—and unacceptable—time than specified, you will want to reexamine the task durations, and especially the task relationships. We often define more stringent relationships between activities than actually exist in our project. For instance, if we have three tasks, consisting of digging 300 feet of trench, laying 300 feet of pipe, and backfilling the trench, we may schedule them as three sequential activities, each with a finish-to-start relationship. Yet, in reality, we can start to lay the pipe at one end of the trench while we are still digging further along the path. Likewise, we can be backfilling at the first end while completing the pipe laying at the other end.

**Where Does the Computer Fit In?** If your software package allows the use of overlapping dependencies (start-to-start and finish-to-finish), you can redefine some of the finish-to-start relationships to shorten the schedule (see Figure 28-9). This should only be done to reflect actual overlap situations. Misuse or abuse of this feature to arbitrarily shorten the schedule will only come back to haunt you later.

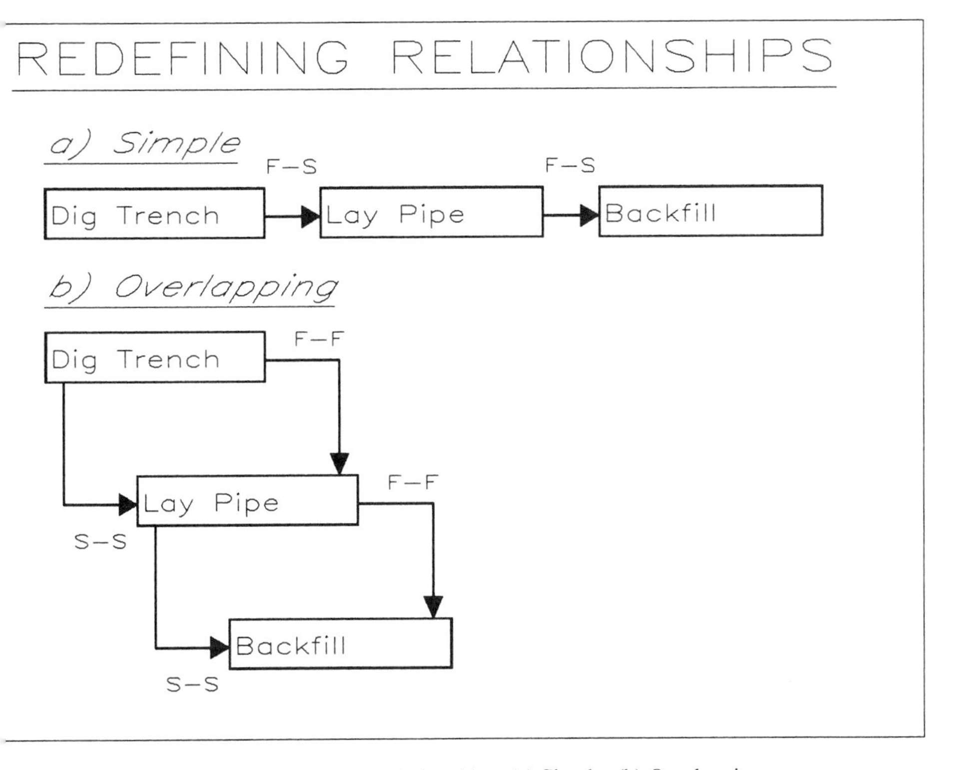


Figure 28-9. Redefining relationships. (a) Simple. (b) Overlapping.

As we stated earlier, your computer can help you in this schedule adjustment quest by organizing your data. Use your exception reporting capability to call for the longest path through the project, the "critical path," by asking the computer to sort the activities by *least float*. This will produce an analysis report (see Figure 28-10) with the activities listed in the order in which you will want to review their durations and relationships.

To the extent that you can reasonably do so, evaluate your critical activities and adjust durations and relationships. Reschedule and rerun your least float reports and continue to adjust until you have a realistic schedule that supports your objectives, if possible.

### Optimize the Resource Plan

**The Function.** If you have resource peaks and valleys, you may wish to "level" the resources, by moving the more highly loaded tasks, that have

My Humble Hacienda | PRIMAVERA PROJECT PLANNER | PROJECT SPLASH

REPORT DATE 29MAY86 RUN NO. 7 | START DATE 31MAR86 FIN DATE

TABULAR SCHEDULE REPORT - Sorted by Criticality | DATA DATE 31MAR86 PAGE NO. 1

| ACTIVITY NUMBER | ORIG DUR | REM DUR | PCT | CODE | ACTIVITY DESCRIPTION | LEVELED START | LEVELED FINISH | LATE START | LATE FINISH | TOTAL FLOAT |
|---|---|---|---|---|---|---|---|---|---|---|
| 10 | 2 | 2 | 0 | | DECIDE TO DO | 31MAR86 | 1APR86 | 31MAR86 | 1APR86 | 0 |
| 20 | 8 | 8 | 0 | | GATHER INFO | 2APR86 | 11APR86 | 2APR86 | 11APR86 | 0 |
| 30 | 5 | 5 | 0 | | CONCEPT | 14APR86 | 18APR86 | 14APR86 | 18APR86 | 0 |
| 210 | 5 | 5 | 0 | | PRICE FENCING | 21APR86 | 25APR86 | 21APR86 | 25APR86 | 0 |
| 310 | 5 | 5 | 0 | | PRICE PATIO KITS | 21APR86 | 25APR86 | 21APR86 | 25APR86 | 0 |
| 510 | 5 | 5 | 0 | | GET POOL QUOTES | 21APR86 | 25APR86 | 21APR86 | 25APR86 | 0 |
| 40 | 3 | 3 | 0 | | ESTABLISH PROGRAM | 28APR86 | 30APR86 | 28APR86 | 30APR86 | 0 |
| 110 | 3 | 3 | 0 | | APPLY FOR PERMIT | 1MAY86 | 5MAY86 | 1MAY86 | 5MAY86 | 0 |
| 120 | 15 | 15 | 0 | | ISSUE PERMIT | 6MAY86 | 26MAY86 | 6MAY86 | 26MAY86 | 0 |
| 340 | 1 | 1 | 0 | | DIG HOLES FOR PATIO SUPPORTS | 27MAY86 | 27MAY86 | 27MAY86 | 27MAY86 | 0 |
| 350 | 2 | 2 | 0 | | FORM PATIO FOOTINGS | 28MAY86 | 29MAY86 | 28MAY86 | 29MAY86 | 0 |
| 360 | 1 | 1 | 0 | | POUR PATIO FOOTINGS | 30MAY86 | 30MAY86 | 30MAY86 | 30MAY86 | 0 |
| 370 | 3 | 3 | 0 | | CURE PATIO FOOTINGS | 2JUN86 | 4JUN86 | 2JUN86 | 4JUN86 | 0 |
| 380 | 1 | 1 | 0 | | STRIP & BACKFILL | 5JUN86 | 5JUN86 | 5JUN86 | 5JUN86 | 0 |
| 440 | 4 | 4 | 0 | | ERECT PATIO BASE | 6JUN86 | 11JUN86 | 6JUN86 | 11JUN86 | 0 |
| 450 | 2 | 2 | 0 | | ERECT PATIO RAILS | 12JUN86 | 13JUN86 | 12JUN86 | 13JUN86 | 0 |
| 460 | 2 | 2 | 0 | | FINISH PATIO | 16JUN86 | 17JUN86 | 16JUN86 | 17JUN86 | 0 |
| 610 | 2 | 2 | 0 | | CLEANUP | 16JUN86 | 17JUN86 | 16JUN86 | 17JUN86 | 0 |
| 640 | 1 | 1 | 0 | | HOLD POOL PARTY | 18JUN86 | 18JUN86 | 18JUN86 | 18JUN86 | 0 |
| 240 | 5 | 5 | 0 | | ERECT FENCE | 6JUN86 | 12JUN86 | 9JUN86 | 13JUN86 | 1 |
| 630 | 1 | 1 | 0 | | GET PARTY SUPPLIES | 16JUN86 | 16JUN86 | 17JUN86 | 17JUN86 | 1 |
| 50 | 2 | 2 | 0 | | ESTIMATE COSTS | 1MAY86 | 2MAY86 | 5MAY86 | 6MAY86 | 2 |
| 520 | 2 | 2 | 0 | | ORDER POOL | 5MAY86 | 6MAY86 | 7MAY86 | 8MAY86 | 2 |
| 140 | 3 | 3 | 0 | | APPLY FOR LOAN | 5MAY86 | 7MAY86 | 7MAY86 | 9MAY86 | 2 |
| 530 | 15 | 15 | 0 | | POOL DELIVERED | 7MAY86 | 27MAY86 | 9MAY86 | 29MAY86 | 2 |
| 150 | 10 | 10 | 0 | | ISSUE LOAN | 8MAY86 | 21MAY86* | 12MAY86 | 23MAY86 | 2 |
| 330 | 2 | 2 | 0 | | OBTAIN PATIO SUPPORT MATERIAL | 21MAY86 | 22MAY86 | 23MAY86 | 26MAY86 | 2 |
| 540 | 10 | 10 | 0 | | INSTALL POOL | 28MAY86 | 10JUN86 | 30MAY86 | 12JUN86 | 2 |
| 550 | 3 | 3 | 0 | | FILL POOL | 11JUN86 | 13JUN86 | 13JUN86 | 17JUN86 | 2 |
| 620 | 1 | 1 | 0 | | SEND PARTY INVITATIONS | 12JUN86 | 12JUN86 | 17JUN86 | 17JUN86 | 3 |
| 220 | 2 | 2 | 0 | | ORDER FENCING | 5MAY86 | 6MAY86 | 15MAY86 | 16MAY86 | 8 |
| 230 | 15 | 15 | 0 | | FENCING DELIVERED | 7MAY86 | 27MAY86 | 19MAY86 | 6JUN86 | 8 |
| 320 | 5 | 5 | 0 | | PATIO DETAILED PLANS | 5MAY86 | 9MAY86 | 16MAY86 | 22MAY86 | 9 |
| 430 | 10 | 10 | 0 | | PATIO KIT DELIVERED | 9MAY86 | 22MAY86 | 23MAY86 | 5JUN86 | 10 |
| 420 | 2 | 2 | 0 | | ORDER PATIO KIT | 5MAY86 | 6MAY86 | 21MAY86 | 22MAY86 | 12 |

Figure 28-10. Critical path report (Primavera).

positive float, to periods of lesser demand for that resource. You may also wish to reset the maximum limit on certain resources and recalculate a revised schedule and project completion date.

At this point, you are involved in an iterative process, aimed at balancing the various means at hand to come as close as possible to establishing a baseline plan that meets the project overall objectives, that is consistent with organizational policies and guidelines, and that can be lived with by the project team.

**Where Does the Computer Fit In?** Obviously it would be most difficult and tedious to do this iterative process by hand. The computer becomes virtually a necessity if you are to attempt to perform this optimization. With today's programs, there should be no reason why you cannot have this capability. Most of the programs will provide some degree of resource allocation, of course, within the variations discussed earlier.

### "Freeze" the Baseline Plan

**The Function.** If you are going to exercise project control, you have to be able to compare the project progress to the plan. Therefore, once you have developed an acceptable project plan you will want to save it for further reference.

**Where Does the Computer Fit In?** Some programs allow you to label one or more "target" schedules. Others allow you to save the baseline plan and do your updates on a copy of that plan. I find the first approach to be preferable. In these programs, the target and current schedules reside in the same data base and the program provides prestructured reporting functions to allow the users to compare the two schedules. These reports may be either tabular formats with variance indication and/or Gantt charts with separate bars for each schedule. Although target schedule capabilities have been recently added to a few of the lower-priced programs that did not have that capability in earlier versions, most have still not provided formats that are as practical and readable as the programs at the higher end. In any event, as stated in the introduction, if you do not have some way to compare your current status to the baseline plan, your program will be of limited use in the execution of project control.

### Track the Work Progress

**The Function.** Now we are in the execution phase. Tracking entails the recording of how much work has been done, what resources were utilized, and what costs were incurred. In tracking the work progress, you will want to note when any task started and was finished. In addition, at any update or measurement point, usually at regular intervals, you will need to record the current status of in-progress activities. This may be noted in terms of remaining duration, expected completion, or percent complete.

Tracking the work progress also entails keeping track of the work scope changes and amending of the network plan and budgets to account for such scope changes.

**Where Does the Computer Fit In?** We've already discussed the difference between a planning-only program and a planning/tracking program. Here we will concentrate on variations within the latter group: those that allow us to enter current status for eventual comparison to the baseline. While the basic elements all fall within the three data items listed above—remaining duration, expected completion, or percent complete—they can be approached in different ways. These subtle variations can affect your satisfaction with the software.

*Remaining duration* is usually the time, from the data date, for an activity in progress to be completed. The *data date* is a very important element of the project progressing process. It is the reference point for many of the calculations, including remaining duration and earned value. When using the remaining duration technique of defining progress, you must first decide on the as-of date for your update. If you decide to change the as-of date (data date), you may have to revisit the remaining duration entries to evaluate them against the new data date. Most programs will not let you input an actual date that is later than the data date, or an imposed date that is earlier than the data date. Figure 28-11 illustrates the effect of the data date. In this example, the remaining duration of five days is figured from the data date of 6/10. The system assumes that, even though the task started a week ago, it is 38% complete (five of eight days remaining to be done).

Some programs, like QWIKNET, have an auto-progress setting. When in auto-progress mode, the system assumes that any activity that was scheduled to start prior to the data date has actually progressed as planned, although no progress has been entered. For instance (see Figure 28-12), with a data date of 4/9/86, a four-day activity that was scheduled to

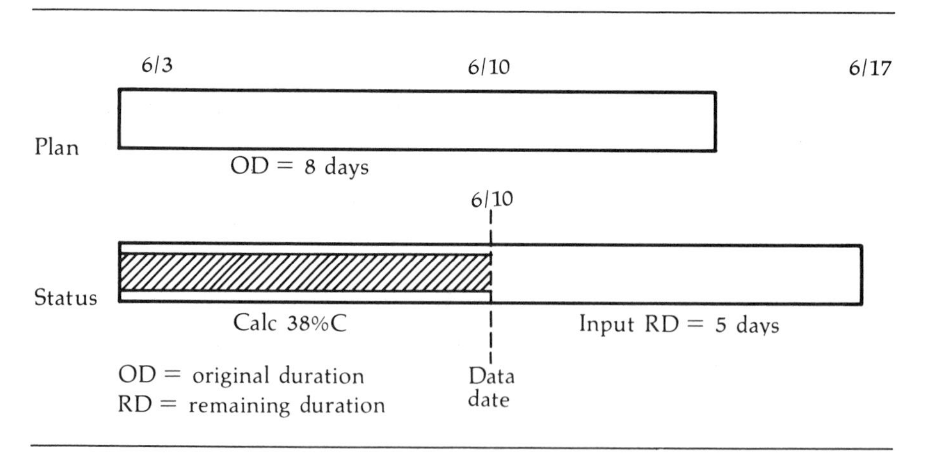


Figure 28-11. Effect of data date on remaining duration.

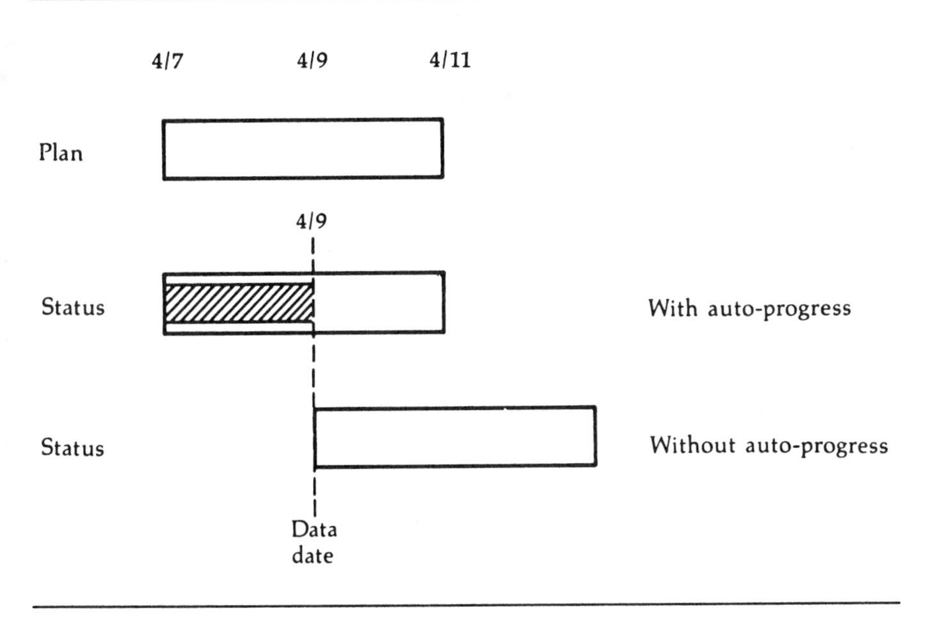


Figure 28-12. Effect of auto-progress mode on remaining duration.

start on 4/7/86 will be shown to be 50% complete and have a two-day remaining duration. Without the auto-progress mode invoked, the system will assume no progress unless reported. In this case, the preceding example would be slipped to start no earlier than the data date. The user is advised to use the auto-progress feature, if available, with judicious constraint. There is a tendency to reduce one's diligence toward quality updating when each and every in-progress activity does not have to be specifically addressed. In updating progress in ViewPoint, if you indicate that a duration-driven activity is 100% complete and do not enter actual start and finish dates, the system assumes that the task took place as scheduled, saving the user from unnecessary inputting. User override is possible, however, by just entering any dates that are different from the plan dates.

*Activity percent complete* reporting, when available, represents the percent of the activity duration that is complete. In other words, the system will calculate a remaining duration by subtracting the percent complete, times the planned duration, from that planned duration. In Figure 28-13, we show an eight-day task that is only 50% complete five days after it started. The system will assume a remaining duration of four

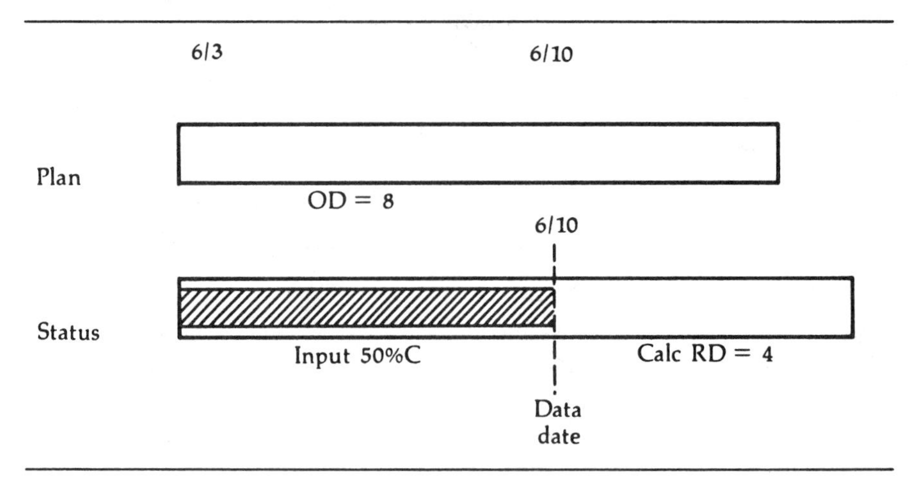


Figure 28-13. Effect of percent complete data on remaining duration.

days (50% of the eight day OD). The activity percent complete will be taken, by most systems, to also mean the resource and cost percent complete, although that may not be the specific case in all instances. Some programs will allow the user to specify resource and cost percent complete separately. The Primavera approach toward this methodology is defined later in this chapter (under resource status). Also, while some programs tie the percent complete and remaining duration figures together, interlocking those two data items, other programs provide a user option as to whether these measurements are linked or untied.

*Expected completion* is not available as an input option in all programs, and it is just as well, as many people tend to misuse this field as a scheduling option rather than for inputting status on tasks in progress. Abuse of this feature, when available, can result in degradation of the network logic. When an expected completion field is not available, the user is forced to calculate the remaining duration, so that the computer can tell you what you already know about that activity. Therefore, when used properly, the expected completion field can be helpful.

One other variable to consider is how any program deals with *out-of-sequence activity progress*. Remember that the activity logic and durations, which were defined in establishing the baseline plan, were only a best guess at how the work would be done. Remember, also, that these critical path project management techniques are used primarily for work that is unique in nature, and, therefore, the plans do not have the benefit of previous experience or hindsight. In practice, we could expect that the actual durations will vary and that some of the planned logical constraints will be violated.

There are great variations as to how the various programs deal with this phenomenon. Some will flag an out-of-sequence progress as an error and require the user to correct the status or the logic. Many will reduce the duration of the statused activity to the remaining duration, but leave it in the existing logical path (see Figure 28-14). A few may give the user an option to ignore the predecessor constraints for any activity with progress. In this "logic override" mode, the subject activity is processed as if all its predecessors had been completed, and all its successors may be scheduled to follow the new completion of that activity.

### Track Actual Resource Usage

**The Function.** In resource tracking, we will be interested in similar data as in the task tracking, except in this case we will address the status of specific resources for each task. Of course this assumes that we were able to input data for specific resources in our baseline plan. If we can, we will address the actual work performed by each resource in such quantifications as actual units used, estimated units remaining, and/or percent re-

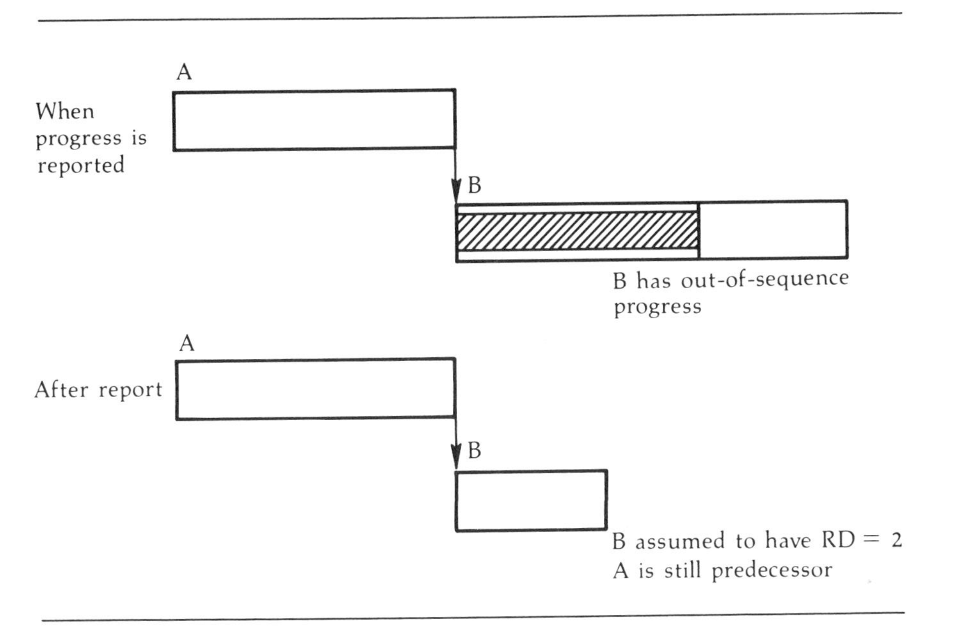


Figure 28-14. Effect of out-of-sequence progress.

source used. If your baseline plan dealt with delayed and partial duration application of the resources, then you would probably wish to further address these specifications during the project update.

If your task durations are resource driven, then you will have to reexecute your schedule computation after entering the resource usage data.

**Where Does the Computer Fit In?** Some of the software packages do not have the facility to enter actual resource usage. That is, they allow you to define resource requirements, in the planning stage, to establish resource-driven budgets, and to support resource aggregation and allocation functions. However, you may not enter data relative to the resources actually used. With these programs, you can show the effect of resource usage only by entering the *cost* of the actual resources used. This does permit analysis of the project cost performance, and may be satisfactory in many project control situations. The negative aspects are that (1) you have to figure out the costs (resource usage times resource rate); (2) if the rates have changed from the plan, you lose accuracy of any resource usage measurement; and (3) it is difficult to perform any resource productivity measurements.

Several programs allow you to enter actual resource utilization experience, for each activity, and to provide data as to the resource percent complete (which may be different from the activity percent complete) and the estimate to complete. Such programs, Primavera for example, will also usually have the facility to report these resource performance and productivity data. Figure 28-15 illustrates the entry of resource data for an activity which is in progress. Here we show that, although we are one day into this two-day task, we estimate that it will take another two days to complete. We enter the data that indicates that we have spent two man-days, to date, of the resource "Carpenter," and that we estimate that four additional man-days are required. The system calculates the forecast at completion, which is six, and the variance, which is −2.

We can observe the aggregate results of the resource statusing, in such reports as shown in the next two figures. Figure 28-16 reports the budget, actual, estimate to complete, forecast at completion, and variance for each resource, with the details for each activity for that resource. Figure 28-17 is a current versus target resource report, by week, for the resource "Carpenter."

With the rescheduling resulting from the entering of schedule progress, we may find that we no longer have a schedule that respects our resource limits. If there has been a schedule slippage, in all likelihood we have lost some of the float that was used to level critical resources. Therefore, as part of the resource tracking and analysis function, we will have to review

```
                              ACTIVITY DATA                               POOL
Activity number:     350                                                 TF:  -1
          Title:  FORM PATIO FOOTINGS                                   PCT:  33

    ES:              EF: 30MAY86  Orig. duration     2  Actual start:  28MAY86
    LS:              LF: 29MAY86  Rem.  duration     2  Actual finish:

Activity Codes:  P1
================================================================================
RESOURCE SUMMARY:        Resource 1          Resource 2          Resource 3

Resource code               C
Cost acct code/type             21L
Units per day                   2.00                0.00                0.00
Budget quantity                    4                   0                   0
Resource Lag/Duration       0      0            0      0            0      0
% Complete                         0                   0                   0
Actual qty this Period             0                   0                   0
Actual qty to date                 2                   0                   0
Quantity to complete               4                   0                   0
Quantity at completion             6                   0                   0
Variance (units)                  -2                   0                   0
================================================================================
Commands:Add Delete Edit Help More Next Return autoSort Transfer View Window
Windows :Act.codes Blank Constraints Dates Financial Log Resources Successors
```

Figure 28-15. Entering resource usage data (Primavera).

the current resource aggregation and, if unsatisfactory, rerun the resource leveling.

### Track Actual Costs

**The Function.** Once the work is initiated, you will need to track the actual costs incurred (see Figure 28-18). Ledgers should be set up both for the incurrence of cost commitments (as in purchase orders or labor contracts), and for actual (invoiced) costs. The tracking of committed costs against the budget permits an early warning system for cost overruns. Unfortunately, most systems track only actual costs, which may provide you with cost performance data too late for corrective action.

In the planning phase, we showed how you can tie the projected resource and fixed expense data to your activities and develop a time-phased budget for the project, and each of its parts. We said that by having this activity level cost plan, we would be able to measure cost performance against this plan and be able to best forecast the probable project cost performance, based on the experience to date. The activity-

My Humble Hacienda — PRIMAVERA PROJECT PLANNER — SPLASH

REPORT DATE 23NOV85 RUN NO. 38 — RESOURCE CONTROL REPORT — START DATE 31MAR86 FIN DATE 18JUN86

RESOURCE CONTROL REPORT - by RESOURCE — DATA DATE 29MAY86 PAGE NO. 1

| ACTIVITY NO | RESOURCE CODE | ACCOUNT CODE | ACCOUNT CATEGORY | UNIT MEAS | BUDGET | PCT CMP | ACTUAL TO DATE | ACTUAL THIS PERIOD | ESTIMATE TO COMPLETE | FORECAST | VARIANCE |
|---|---|---|---|---|---|---|---|---|---|---|---|
| | C | - CARPENTER | | | | | UNIT OF MEASURE = MD | | | | |
| 350 | C | 21L | LABOR | MD | 4 | 33 | 2 | 0 | 4 | 6 | -2 |
| 300 | C | 21L | LABOR | MD | 2 | 0 | 0 | 0 | 2 | 2 | 0 |
| 440 | C | 21L | LABOR | MD | 8 | 0 | 0 | 0 | 8 | 8 | 0 |
| 450 | C | 21L | LABOR | MD | 4 | 0 | 0 | 0 | 4 | 4 | 0 |
| TOTAL | C | | | | 18 | 10 | 2 | 0 | 18 | 20 | -2 |
| | L | - LABORER | | | | | UNIT OF MEASURE = MD | | | | |
| 240 | L | 22L | LABOR | MD | 15 | 0 | 0 | 0 | 15 | 15 | 0 |
| 340 | L | 21L | LABOR | MD | 2 | 100 | 2 | 0 | 0 | 2 | 0 |
| 380 | L | 21L | LABOR | MD | 2 | 0 | 0 | 0 | 2 | 2 | 0 |
| 460 | L | 21L | LABOR | MD | 2 | 0 | 0 | 0 | 2 | 2 | 0 |
| 610 | L | 1 L | LABOR | MD | 4 | 0 | 0 | 0 | 4 | 4 | 0 |
| TOTAL | L | | | | 25 | 8 | 2 | 0 | 23 | 25 | 0 |
| | M | - CONCRETE WORKER | | | | | UNIT OF MEASURE = MD | | | | |
| 360 | M | 21L | LABOR | MD | 3 | 0 | 0 | 0 | 3 | 3 | 0 |
| TOTAL | M | | | | 3 | 0 | 0 | 0 | 3 | 3 | 0 |

Figure 28-16. Resource control report (Primavera).

oriented project budget therefore was used to establish a framework for the inputting and analysis of cost performance.

In this tracking and analysis phase, we will be interested in seeing how we are doing, cost-wise, against the budgets for each activity, and, perhaps, in summarizing and evaluating cost performance by resource, cost account, or work breakdown structure grouping. We may also wish to review the cost experience against the cash flow plan, and, using a process called "earned value analysis," we can look at the cost performance against the value of the work actually performed. Next, we may want to estimate the costs for the remaining work (selective override of the uncompleted budget), generating forecast at completion data and variances from the plan. Last, having inputted the project experience to date and analyzed the results against the budget plan, we will need to evaluate

My Humble Hacienda | PRIMAVERA PROJECT PLANNER | SPLASH

REPORT DATE 23NOV85 RUN NO. 39 | RESOURCE USAGE REPORT - WEEKLY | START DATE 31MAR86 FIN DATE 18JUN86

RESOURCE USAGE REPORT | DATA DATE 29MAY86 PAGE NO. 1

| PERIOD | ---DAILY AVAIL--- | | --------EARLY SCHEDULE-------- | | | --------LATE SCHEDULE--------- | | | -------TARGET 1 SCHEDULE------ | | |
|---|---|---|---|---|---|---|---|---|---|---|---|
| ENDING | NORMAL | MAXIMUM | USAGE | AVG.DAILY | CUMULATIVE | USAGE | AVG.DAILY | CUMULATIVE | USAGE | AVG.DAILY | CUMULATIVE |
| C - CARPENTER | | | | | | UNIT OF MEASURE = MD | | | EARLY | | |
| 4APR86 | 0 | 0 | 0 | .0 | 0 | 0 | .0 | 0 | 0 | .0 | 0 |
| 11APR86 | 0 | 0 | 0 | .0 | 0 | 0 | .0 | 0 | 0 | .0 | 0 |
| 18APR86 | 0 | 0 | 0 | .0 | 0 | 0 | .0 | 0 | 0 | .0 | 0 |
| 25APR86 | 0 | 0 | 0 | .0 | 0 | 0 | .0 | 0 | 0 | .0 | 0 |
| 2MAY86 | 0 | 0 | 0 | .0 | 0 | 0 | .0 | 0 | 0 | .0 | 0 |
| 9MAY86 | 0 | 0 | 0 | .0 | 0 | 0 | .0 | 0 | 0 | .0 | 0 |
| 16MAY86 | 0 | 0 | 0 | .0 | 0 | 0 | .0 | 0 | 0 | .0 | 0 |
| 23MAY86 | 0 | 0 | 0 | .0 | 0 | 0 | .0 | 0 | 0 | .0 | 0 |
| 30MAY86 | 0 | 0 | 6 | 1.2 | 6 | 0 | .0 | 0 | 4 | .8 | 4 |
| ***DATA DATE*** | | | | | | | | | | | |
| 6JUN86 | 0 | 0 | 2 | .4 | 8 | 0 | .0 | 0 | 4 | .8 | 8 |
| 13JUN86 | 0 | 0 | 10 | 2.0 | 18 | 0 | .0 | 0 | 10 | 2.0 | 18 |
| 20JUN86 | 0 | 0 | 2 | .4 | 20 | 0 | .0 | 0 | 0 | .0 | 18 |
| 27JUN86 | 0 | 0 | 0 | .0 | 20 | 0 | .0 | 0 | 0 | .0 | 18 |
| 4JUL86 | 0 | 0 | 0 | .0 | 20 | 0 | .0 | 0 | 0 | .0 | 18 |

Figure 28-17. Resource usage report—current vs. target (Primavera).

ACTIVITY DATA | POOL

Activity number: 430 | TF:

Title: PATIO KIT DELIVERED | PCT: 90

ES: | EF: | Orig. duration 10 | Actual start: 9MAY86

LS: | LF: | Rem. duration 1 | Actual finish:

Activity Codes: P2

| FINANCIAL SUMMARY: | Resource 1 | Resource 2 | Resource 3 |
|---|---|---|---|
| Resource code | M$ | | |
| Cost acct code/type | 21M | | |
| Budgeted cost | 350.00 | 0.00 | 0.00 |
| Actual cost this period | 0.00 | 0.00 | 0.00 |
| Actual cost to date | 192.50 | 0.00 | 0.00 |
| Percent expended | 55 | 0 | 0 |
| Percent complete | 50 | 0 | 0 |
| Earned value | 175.00 | 0.00 | 0.00 |
| Cost to complete | 192.50 | 0.00 | 0.00 |
| Cost at completion | 385.00 | 0.00 | 0.00 |
| Variance | -35.00 | 0.00 | 0.00 |

Commands :Add Delete Edit Help More Next Return autoSort Transfer View Window

Windows :Act.codes Blank Constraints Dates Financial Log Resources Successors

Figure 28-18. Entering actual cost data (Primavera).

possible corrective action or changes in the remaining plans to try to achieve the original project objectives.

It's important to note that cost tracking and analysis are a crucial and often neglected part of project management. All too often, we collect tons of data, calculate, process, and massage the data until we produce more tonnage of output, and then do nothing. There are many reasons (or excuses) for this. One is the proverbial "we're too busy putting out fires to bother with planning to correct for future potential problems." Another is that to consider corrective action requires one to admit that there may, indeed, be some problems. Many project managers will wait until the problem is absolutely undeniable or uncorrectable before acceding to the long-projected facts. Yet another is that the project manager does not understand the project cost performance data reported by the system.

A final, and very prevalent, reason is that the cost-tracking framework was not set up in such a manner as to match the tracking and analysis needs for the project. This latter item, while being a major cause of project control failures, is actually the easiest to rectify. The problem is often manifested by the setting up of tracking categories (cost accounts) to one structure, and then collecting data to a totally different structure. The "baseline" is therefore no longer usable for tracking, and there is no target against which to compare the actual experience. The message here, of course, is to think very carefully about how you are going to collect your actual cost data, before you structure your cost account system.

**Where Does the Computer Fit In?** The data considerations and options for cost tracking are similar to the ones discussed under resource tracking. The most essential data item to enter is actual costs to date. If we have a resource-tracking system then this data item should be automatically calculated from the resource usage data. Of course, we will want to have a user override for those instances where the actual costs deviated from the direct resource usage times resource rate computation.

If there is one piece of information that top management most often wants to have (in addition to the project completion date), it is the project forecasted cost at completion. Many systems can provide that data for you, essentially by computing the estimated cost for the uncompleted work and adding that value to the actual costs to date. Here again, the user will want to be able to get at these data elements to override the default calculations when the current estimate to complete is no longer reflective of the baseline plan. There are several minor variations in how each program addresses these functions, and they are constantly changing as the vendors respond to techniques preferred by their users. These requests have indicated that there are multiple preferred approaches, all of which are perfectly valid. One vendor, Primavera, responded to this dilemma by providing a user-selectable resource/cost rules screen. You may work with a set of system defaults, or a set of project-specific defaults, and further selectively override the rules that you have selected, for any activity.

### Compare Progress and Costs to the Baseline Plan

**The Function.** Tracking progress and cost do not constitute control. Project control consists of many steps, leading to the taking of corrective action directed at achieving the project objectives. In this progressive action, the first step, following the collection of progress data, is to examine that progress against the target schedule. There are two measurements that should be monitored. The most obvious is the project end date that was calculated by the network critical path. If there have been any delays to activities on or near the critical path, they will generate a schedule delay. One should also observe the general progress of the noncritical activities. If enough of these fall behind, you can assume that the lower production rate will eventually impact on the critical schedule dates. Project control systems that provide measurement of production rates and earned value will facilitate this appraisal of *schedule variance*.

In the case of completed activities, it is satisfactory to compare the actual cost to the budget, for each activity. However, for in-progress tasks, the measurement of actual costs and comparison to the budget can only be effective if based on the *"earned value"* principle. The earned value, for any activity, is simply the activity budget times the activity percent complete. The biggest advantage of the earned value method is that it reduces all data to a common denominator of dollars. This facilitates the roll-up (summarization) of performance data to any level, and by any coded grouping.

**Where Does the Computer Fit In?** We've already discussed some of the attributes of computers in presenting a comparative analysis of baseline and current schedule data, namely the tabular listings and Gantt charts. For a comparison of resource and cost performance to the baseline, we will use an entirely different set of reporting structures. Let's look at these in detail as part of the next topic: performance evaluation.

### Evaluate Performance.

**The Function.** As noted earlier, there are two primary measurements that will be watched by the people whose careers may depend on the success of a project. One, of course, is the project completion date. The other is the project margin. These measurements are a composite of individual task measurements. While the specific period data is the basis for the measurement of performance, the most revealing data will generally be the *trends* that are indicated by tracking that performance over several periods. A dip in production rate, or a spike in costs, for a particular period may only be a matter of measurement fidelity or timing, and should not be a cause for alarm. But to ignore such perturbations over an extended period is certainly courting disaster.

The most important condition for performance evaluation and trend analysis is consistency. Hence, again, the computer is a major aid to that end.

In general, when evaluating resource and cost performance, we are interested in comparing the actual resource usage and costs to the budget. We may want to perform that comparison by activity, by resource, or by each resource within each activity. We may also wish to perform that comparison at the detailed level (each task and resource) or at various summary levels based on WBS, responsibility, phase, cost accounts, and what have you.

In any of these modes, comparing actual to budget is only helpful for work that has been completed. As we noted earlier, for work that is in progress, we need to compare the actuals to the value of the work that has been accomplished. This "earned value" is the percent complete of the item times the budget for the item. An alternate reporting mechanism is to compare the budget to the forecast at completion, which is the sum of the actual to date plus the estimate to complete.

Both of these approaches overlap greatly and have generated a common language and set of acronyms. They are also the basis for the two well-known government protocols. The Department of Defense uses an instruction known as DODI 7000.2, also known as Cost/Schedule Control System Criteria (C/SCSC). A similar protocol, developed by the Department of Energy, goes by the name of Performance Measurement System (PMS).

The overall objective of the performance measurement approach is to measure progress and cost against the same definition as the baseline, and to measure this performance in comparable units. Another important facet is to conduct these measurements periodically. Certainly it is essential to good project control that performance be measured while there is still time to take corrective action, and that the effect of the periodic review and corrective actions can be revealed by the trend data coming out of these periodic measurements and reviews.

In a performance measurement system, therefore, we will want to have data pertaining to the schedule and budget, both expressed in comparable units (usually dollars, and sometimes man-days). We will want to have this data for the entire duration of the project, and for specified periods during the project. We will need to be able to project (estimate or forecast) the results at the end of the project, and to express the variance of current and/or projected status against the plan.

The language in these PMS protocols is very descriptive of the data items that are used in the system. Table 28-1 lists and defines the common acronyms used in PMS.

**Where Does the Computer Fit In?** This is where the computer really earns its keep. All of the expressions described in Table 28-1 are easily programmed into the project management software systems, and many of the vendors, but not all, have chosen to do so. Hence, if the user wishes to have reports that present this kind of data, the choices are certainly there. Look for programs that provide reports that actually use this terminology. They are likely to provide the types of analytical data that you will need to evaluate and report project performance. A good example of such is the newly released QWIKNET Professional, which supplies the data item and computation formats, and allows the user to select the terms to be used for the field headers. See Figure 28-19.

Look for systems that will allow you to select, sort, and summarize based on your WBS, responsibility, phase, cost account, and other codes. Depending on your needs, ascertain whether the program can arrange these data by activity, by resource, by cost account, by time period, or any combination of these. Check into how the program deals with actuals and forecast to complete. It should either match your mode of progress tracking or permit you to set to criteria to meet your needs.

## REPORTING AND EXPORTING DATA

The real test of program functionality will be not only if the data formats match your environment and needs, but also whether the program will give you the control over the preparation of hard copy (reports and graphics) for presentation to the various project stakeholders.

Perhaps the most noticeable differences between the mass market and higher end programs will be in the reporting and graphics area. In general, the lower end programs (primarily those in the $395 to $695 price class) will emphasize on-screen graphics and ease of use, as opposed to an emphasis on functionality and hard-copy reporting, which are featured in

Table 28-1. PMS Terms and What They Mean.

| ACRONYM | STANDS FOR | MEANING |
|---|---|---|
| BAC | budget at completion | The budget for an entire item or any group of items, or roll-up (summary) of the entire project. |
| BCWS | budgeted cost of work scheduled | At any specified point in time the percent complete that the item is scheduled to be at that time, times the BAC, for that item. (Item Planned %C × BAC = BCWS) |
| BCWP | budgeted cost of work performed | At any specified point in time, the actual percent complete, times the BAC, for that item. This is the earned value of the work performed. (Item Actual %C × BAC = BCWP) |
| ACWP | actual cost of work performed | At any specified point in time the actual cost incurred for the work. The timing of the actual cost measurement should be the same as the %C progress measurement so that you can compare actual cost to earned value (BCWP). |
| ETC | estimate to complete | An estimate of the cost to be incurred to complete the remaining work. |
| EAC | estimate at completion | The sum of the actual cost to date, plus the estimate to complete. (ACWP + ETC = EAC) |
| FTC | forecast to complete | A forecast of the cost to be incurred to complete the remaining work. In some applications the FTC and FAC are synonymous with the ETC and EAC. In others the *forecast* refers to a computerized extrapolation of the performance to date, based on a built-in forecast algorithm, and *estimate* refers to a judgmental expression of the cost for the remaining work. |
| FAC | forecast at completion | The sum of the actual cost to date plus the forecast to complete. (ACWP + FTC = FAC) |
| CV | cost variance | The difference between the value of the work performed and the actual cost for that work. (BCWP − ACWP = CV) |
| SV | schedule variance | The difference between the value of the work performed, and the value of the work that had been planned to be performed, at the measurement time. The schedule variance is expressed in dollars, so that the same measurement units are used as in the cost variance and so that both the cost and schedule variance can be plotted on the same graph. (BCWP − BCWS = SV) |

My Humble Hacienda — EARNED VALUE REPORT — Page
Jose Doe — Project Start 31Mar86
Project Name POOL PROJECT SPLASH — Project Finish 19Jun86
Run Date 21Dec86 22:16 — Data Date 29May86

| Activity ID | DESCRIPTION | TO DATE | | | | | AT COMPLETION | | |
|---|---|---|---|---|---|---|---|---|---|
| | | BCWS ($$) | BCWP (EV) ($$) | ACWP ($$) | SV ($$) | CV TODATE ($$) | BAC ($$) | EAC ($$) | CV AT END ($$) |
| 230 | FENCING DELIVERED | 250.00 | 250.00 | 250.00 | 0.00 | 0.00 | 250.00 | 250.00 | 0.00 |
| 240 | ERECT FENCE | 0.00 | 0.00 | 0.00 | 0.00 | 0.00 | 1200.00 | 1200.00 | 0.00 |
| 330 | OBTN PATIO SPRT MATL | 150.00 | 150.00 | 150.00 | 0.00 | 0.00 | 150.00 | 150.00 | 0.00 |
| 340 | DIG HOLES | 160.00 | 160.00 | 160.00 | 0.00 | 0.00 | 160.00 | 160.00 | 0.00 |
| 350 | FORM | 200.00 | 133.32 | 200.00 | -66.68 | -66.68 | 400.00 | 600.00 | -200.00 |
| 360 | POUR | 0.00 | 0.00 | 0.00 | 0.00 | 0.00 | 288.00 | 288.00 | 0.00 |
| 380 | STRIP & BACKFILL | 0.00 | 0.00 | 0.00 | 0.00 | 0.00 | 360.00 | 360.00 | 0.00 |
| 430 | PATIO KIT DELIVERED | 350.00 | 350.00 | 425.00 | 0.00 | -75.00 | 350.00 | 425.00 | -75.00 |
| 440 | ERECT PATIO BASE | 0.00 | 0.00 | 0.00 | 0.00 | 0.00 | 800.00 | 800.00 | 0.00 |
| 450 | ERECT PATIO RAILS | 0.00 | 0.00 | 0.00 | 0.00 | 0.00 | 400.00 | 400.00 | 0.00 |
| 460 | FINISH PATIO | 0.00 | 0.00 | 0.00 | 0.00 | 0.00 | 160.00 | 160.00 | 0.00 |
| 530 | POOL DELIVERED | 6000.00 | 0.00 | 0.00 | -6000.00 | 0.00 | 6000.00 | 6000.00 | 0.00 |
| 540 | INSTALL POOL | 0.00 | 0.00 | 0.00 | 0.00 | 0.00 | 4000.00 | 4000.00 | 0.00 |
| 550 | FILL POOL | 0.00 | 0.00 | 0.00 | 0.00 | 0.00 | 75.00 | 75.00 | 0.00 |
| 610 | CLEAN-UP | 0.00 | 0.00 | 0.00 | 0.00 | 0.00 | 320.00 | 320.00 | 0.00 |
| TOTALS | | 7110.00 | 1043.32 | 1185.00 | -6066.68 | -141.68 | 14913.00 | 15188.00 | -275.00 |

Figure 28-19. Earned value report with user-selected headers (QWIKNET Professional).

most of the higher end programs ($1495 and up). Given my choice, and assuming that I need and can operate the more functional systems, I would opt for the latter group.

Look for your system to provide schedule reports in three formats, tabular, bar charts, and network diagrams. The bar charts and network diagrams, as produced on most printers, are of dubious quality. Look for those products that include or offer optional graphics functions to produce bar charts and network diagrams on plotters or laser printers. A compromise capability, to print these diagrams on a printer in a graphics mode, is a questionable compromise because of the usual slowness of that process. Plotter graphics capabilities are included in Scitor Project Scheduler Network, SuperProject EXPERT, Open Plan, PROMIS, Plantrac, ARTEMIS Project, Project Workbench (partial, and ViewPoint. Optional

My Humble Hacienda — PRIMAVERA PROJECT PLANNER — SPLASH

REPORT DATE 14DEC85 RUN NO. 47 — EARNED VALUE REPORT - COST — START DATE 31MAR86 FIN DATE 18JUN86

EARNED VALUE REPORT - by RESOURCE — DATA DATE 29MAY86 PAGE NO. 1

| ACTIVITY NO | RESOURCE CODE | ACCOUNT CODE | ACCOUNT CATEGORY | UNIT MEAS | BUDGET | PCT CMP | ACTUAL TO DATE | EARNED VALUE | VARIANCE |
|---|---|---|---|---|---|---|---|---|---|
| | L | - LABORER | | | | | UNIT OF MEASURE = MD | | |
| 240 | L | 22L | LABOR | MD | 1200.00 | 0 | .00 | .00 | .00 |
| 340 | L | 21L | LABOR | MD | 160.00 | 100 | 160.00 | 160.00 | .00 |
| 380 | L | 21L | LABOR | MD | 160.00 | 0 | .00 | .00 | .00 |
| 460 | L | 21L | LABOR | MD | 160.00 | 0 | .00 | .00 | .00 |
| 610 | L | 1 L | LABOR | MD | 320.00 | 0 | .00 | .00 | .00 |
| TOTAL | L | | | | 2000.00 | 8 | 160.00 | 160.00 | .00 |
| | C | - CARPENTER | | | | | UNIT OF MEASURE = MD | | |
| 350 | C | 21L | LABOR | MD | 400.00 | 33 | 200.00 | 132.00 | -68.00 |
| 380 | C | 21L | LABOR | MD | 200.00 | 0 | .00 | .00 | .00 |
| 440 | C | 21L | LABOR | MD | 800.00 | 0 | .00 | .00 | .00 |
| 450 | C | 21L | LABOR | MD | 400.00 | 0 | .00 | .00 | .00 |
| TOTAL | C | | | | 1800.00 | 10 | 200.00 | 132.00 | -68.00 |
| | M | - CONCRETE WORKER | | | | | UNIT OF MEASURE = MD | | |
| 360 | M | 21L | LABOR | MD | 288.00 | 0 | .00 | .00 | .00 |
| TOTAL | M | | | | 288.00 | 0 | .00 | .00 | .00 |
| | M$ | - MATERIAL COST | | | | | UNIT OF MEASURE = $ | | |
| 230 | M$ | 22M | MATERIAL | $ | 250.00 | 100 | 250.00 | 250.00 | .00 |
| 330 | M$ | 21M | MATERIAL | $ | 150.00 | 100 | 150.00 | 150.00 | .00 |
| 430 | M$ | 21M | MATERIAL | $ | 350.00 | 100 | 425.00 | 350.00 | -75.00 |
| 530 | M$ | 1 M | MATERIAL | $ | 6000.00 | 0 | .00 | .00 | .00 |
| TOTAL | M$ | | | | 6750.00 | 12 | 825.00 | 750.00 | -75.00 |
| | S$ | - SUBCONTRACTOR COST | | | | | UNIT OF MEASURE = $ | | |
| 540 | S$ | 1 S | SUBCONTR | $ | 4000.00 | 0 | .00 | .00 | .00 |
| 550 | S$ | 1 S | SUBCONTR | $ | 75.00 | 0 | .00 | .00 | .00 |
| TOTAL | S$ | | | | 4075.00 | 0 | .00 | .00 | .00 |
| | | | REPORT COST TOTALS | | 14913.00 | 8 | 1185.00 | 1042.00 | -143.00 |

Figure 28-20. Earned value report—by resource by activity (Primavera).

plotter graphics programs are offered with Time Line, MicroTrak, and Primavera.

Look for your system to provide resource and cost performance data in several ways. Graphically you will want to see incremental resource and cost histograms, and cumulative resource and cost curves. While these diagrams are very graphic, they do not provide specific quantities. These are best obtained from a tabular presentation of the resource and cost data.

The most revealing information relative to resource and cost performance is presented in the aforementioned earned value type reports. A good example of these resource and cost reports can be seen in Figures 28-20 and 28-21, featuring Primavera's approach. Similar reports can be

My Humble Hacienda — PRIMAVERA PROJECT PLANNER — SPLASH

REPORT DATE 29MAY86 RUN NO. 15 — COST CONTROL ACTIVITY REPORT — START DATE 31MAR86 FIN DATE 18JUN86

COST CONTROL REPORT - by ACTIVITY by Resource — DATA DATE 29MAY86 PAGE NO. 1

| ACTIVITY NO | RESOURCE CODE | ACCOUNT CODE | ACCOUNT CATEGORY | UNIT MEAS | BUDGET | PCT CMP | ACTUAL TO DATE | ACTUAL THIS PERIOD | ESTIMATE TO COMPLETE | FORECAST | VARIANCE |
|---|---|---|---|---|---|---|---|---|---|---|---|
| 330 OBTAIN PATIO SUPPORT MATERIAL<br>RD 0 AS 21MAY86 AF 22MAY86 | | | | | | | | | | | |
| | M$ | 21M | MATERIAL | $ | 150.00 | 100 | 150.00 | .00 | .00 | 150.00 | .00 |
| | | | | | 150.00 | 100 | 150.00 | .00 | .00 | 150.00 | .00 |
| 340 DIG HOLES FOR PATIO SUPPORTS<br>RD 0 AS 27MAY86 AF 27MAY86 | | | | | | | | | | | |
| | L | 21L | LABOR | MD | 160.00 | 100 | 160.00 | .00 | .00 | 160.00 | .00 |
| | | | | | 160.00 | 100 | 160.00 | .00 | .00 | 160.00 | .00 |
| 350 FORM PATIO FOOTINGS<br>RD 2 AS 28MAY86 EF 30MAY86 LF 29MAY86 TF -1 | | | | | | | | | | | |
| | C | 21L | LABOR | MD | 400.00 | 33 | 200.00 | .00 | 400.00 | 600.00 | -200.00 |
| | | | | | 400.00 | 50 | 200.00 | .00 | 400.00 | 600.00 | -200.00 |
| 360 POUR PATIO FOOTINGS<br>RD 1 ES 2JUN86 EF 2JUN86 LS 30MAY86 LF 30MAY86 TF -1 | | | | | | | | | | | |
| | M | 21L | LABOR | MD | 288.00 | 0 | .00 | .00 | 288.00 | 288.00 | .00 |
| | | | | | 288.00 | 0 | .00 | .00 | 288.00 | 288.00 | .00 |
| 380 STRIP & BACKFILL<br>RD 1 ES 6JUN86 EF 6JUN86 LS 5JUN86 LF 5JUN86 TF -1 | | | | | | | | | | | |
| | C | 21L | LABOR | MD | 200.00 | 0 | .00 | .00 | 200.00 | 200.00 | .00 |
| | L | 21L | LABOR | MD | 160.00 | 0 | .00 | .00 | 160.00 | 160.00 | .00 |
| | | | | | 360.00 | 0 | .00 | .00 | 360.00 | 360.00 | .00 |
| REPORT COST TOTALS | | | | | 1358.00 | 38 | 510.00 | .00 | 1048.00 | 1558.00 | -200.00 |

Figure 28-21. Cost control report—by activity by resource (Primavera).

produced by programs such as PROMIS, QWIKNET Professional, View-Point, and PMS-II. Some of the other programs can be made to produce such reports, but only by using supplementary programs or report writers. Some of the mass market programs can also produce earned value and PMS type reports, but with a lesser degree of detail and user control. Time Line, Microsoft Project, and QWIKNET Professional also offer a matrix type of reporting (as illustrated in Figures 28-22 through 28-24), a different and interesting way of presenting resource and cost data.

Regardless of how varied and thorough a package's reporting capability may be, many users will want to do more with the data than can be provided in a general-purpose program. One need may be to develop one's own computation algorithms for forecasting. Another may be to feed selected data to other data bases. Many vendors are providing export capabilities. These fall into three general categories: export to a mainframe, export to other non-project management programs on the micro, and, lately, export to other microcomputer-based project management programs. Export to the mainframe can be product specific, such as PS-

Schedule Name: PROJECT SPLASH
Project Manager: Esther Williams
As of date: 29-May-86 12:31am Schedule File: C:\TLDATA\POOL529

This is a selective report. All items shown
(Additionally, some tasks were manually selected or excluded.)

| TASK | C | L | M | M$ | S$ | TOTAL |
|---|---|---|---|---|---|---|
| Fencing Delivered | | | | 250 | | 250 |
| Erect Fence | | 1,200 | | | | 1,200 |
| Obtain Support Material | | | | 150 | | 150 |
| Dig Holes | | 160 | | | | 160 |
| Form | 600 | | | | | 600 |
| Pour | | | 288 | | | 288 |
| Strip & Backfill | 200 | 160 | | | | 360 |
| Deliver Patio Kit | | | | 425 | | 425 |
| Erect Patio Base | 800 | | | | | 800 |
| Erect Patio Rails | 400 | | | | | 400 |
| Finish Patio | | 160 | | | | 160 |
| Pool Delivered | | | | 6,000 | | 6,000 |
| Install Pool | | | | | 4,000 | 4,000 |
| Fill Pool | | | | | 75 | 75 |
| Clean-up | | 320 | | | | 320 |
| TOTALS | 2,000 | 2,000 | 288 | 6,825 | 4,075 | 15,188 |

Figure 28-22. Matrix type cost report—activities vs. resource (Time Line).

PROJECT SPLASH

Project: POOLMP  Date: Dec 15, 1985 12:11 PM
Timescale: Month

| Period ending | May 1, 1986 | June 1, 1986 | July 1, 1986 |
|---|---|---|---|
| 1 LABORER | $0.00 | $160.00 | $1840.00 |
| 2 CARPENTER | $0.00 | $400.00 | $1400.00 |
| 3 CONCRETE WORKER | $0.00 | $288.00 | $0.00 |
| 4 MATL-PAT/FNC | $0.00 | $750.00 | $0.00 |
| 5 MATL-POOL | $0.00 | $6000.00 | $0.00 |
| 6 SUBCON-POOL | $0.00 | $4000.00 | $75.00 |
| Total: | $0.00 | $11598.00 | $3315.00 |

Figure 28-23. Matrix type cost report—resource vs. time (Microsoft Project).

J. L. GRAPHICS and PRINTING  ACTIVITIES VS. TIME REPORT  Page 1.1
Prj Mgr: Joseph Leahy  Units ($$)  Project Start 15Jun87
Project Name BROCHURE Brochure section of CURE-ALL product  Project Finish 20ct87
Run Date 30-Jul-86 15:05  Data Date 15Jul87

| Activity ID / Project ID | 15-jun-87 To 29-jun-87 | 29-jun-87 To 13-jul-87 | 13-jul-87 To 27-jul-87 | 27-jul-87 To 10-aug-87 | 10-aug-87 To 24-aug-87 | 24-aug-87 To 7-sep-87 | Totals |
|---|---|---|---|---|---|---|---|
| ART BROCHURE | 0.00 | 0.00 | 0.00 | 0.00 | 0.00 | 0.00 | 0.00 |
| ART-PREP BROCHURE | 0.00 | 2049.75 | 4571.75 | 0.00 | 0.00 | 0.00 | 6621.50 |
| PASTEUP BROCHURE | 0.00 | 0.00 | 0.00 | 0.00 | 0.00 | 0.00 | 0.00 |
| PLAN-ART BROCHURE | 2763.63 | 363.63 | 872.72 | 0.00 | 0.00 | 0.00 | 3999.99 |
| OTHER BROCHURE | 0.00 | 0.00 | 0.00 | 0.00 | 0.00 | 0.00 | 0.00 |
| SETCOPY BROCHURE | 0.00 | 0.00 | 0.00 | 0.00 | 0.00 | 0.00 | 0.00 |
| SCHED-PROJ BROCHURE | 1200.00 | 150.00 | 0.00 | 0.00 | 0.00 | 0.00 | 1350.00 |
| PRINTING BROCHURE | 0.00 | 0.00 | 0.00 | 0.00 | 0.00 | 0.00 | 0.00 |
| WRITE BROCHURE | 0.00 | 0.00 | 0.00 | 0.00 | 0.00 | 0.00 | 0.00 |
| REVIEW2 BROCHURE | 0.00 | 0.00 | 0.00 | 0.00 | 0.00 | 1475.00 | 1475.00 |
| DRAFTFINAL BROCHURE | 0.00 | 0.00 | 0.00 | 0.00 | 0.00 | 1156.50 | 1156.50 |
| DRAFT2 BROCHURE | 0.00 | 0.00 | 0.00 | 1156.50 | 2313.00 | 385.50 | 3855.00 |
| DRAFT1 BROCHURE | 0.00 | 2056.50 | 3384.00 | 1342.00 | 0.00 | 0.00 | 6782.50 |
| OUTLINE BROCHURE | 1217.65 | 1685.31 | 168.53 | 0.00 | 0.00 | 0.00 | 3071.50 |
| REVIEW1 BROCHURE | 0.00 | 0.00 | 0.00 | 3277.50 | 0.00 | 0.00 | 3277.50 |
| TOTALS ($$) | 5181.29 | 6305.19 | 8997.00 | 5776.00 | 2313.00 | 3017.00 | 31589.49 |

Figure 28-24. Matrix type cost report—activities vs. time (QWIKNET Professional).

DI's QWIKNET to PROJECT/2, or can be in ASCII. The most popular export function is to produce files, containing selected project data, in formats that can be accessed by other microcomputer programs. Initial capabilities centered on the more universal ASCII, DIF, and CSV formats. While these files could eventually be read by most other programs, an intermediary conversion step was necessary. The trend lately is to write files directly in spreadsheet and data base formats, with .WKS (Lotus 1-2-3) and .DBF (dBASE) formats being the most popular. If you have an extensive need to add supplementary data items to your project data base, you will want to look into one of the programs that is written in an easily user-accessable data base language such as Open Plan (dBASE III) or the new Artemis Project (Artemis 2000). These programs actually allow the user to modify the project data base structure, as well as to add extensions to that structure.

## CONCLUDING COMMENTS

All of the functions described in the past several pages are of little value unless they contribute to the attainment of the project objectives. Good plan development, coupled with accurate, dependable progress tracking, and followed up by truthful forecasting and analysis are the necessary ingredients to a successful project planning and control endeavor. Being able to develop alternatives for corrective action, and being able to back up any recommendations with a presentable project story, are equally essential. It has been generally found that the discipline of a computer-based project planning and control system leads to more cohesive and supportable project management and reporting. The general acknowledgment of this condition has helped project managers to be more respected and, in turn, to get better support from their contributors and management.

## TRADEMARKS

The following names are trademarked products of the corresponding companies.

| | |
|---|---|
| Artemis 2000® and Artemis Project® | Metier Management Systems Limited |
| dBASE II® | Ashton-Tate |
| dBASE III® | Ashton-Tate |
| Harvard™ | Software Publishing Corporation |
| Lotus® | Lotus Development Corporation |

| | |
|---|---|
| Micro Planner® | Micro Planning Software, USA |
| Microsoft® | Microsoft Corporation |
| MicroTrak® | SofTrak Systems, Inc. |
| 1-2-3® | Lotus Development Corporation |
| Open Plan™ | Welcom Software Technology |
| PLANTRAC® | Computerline, Inc. |
| PMS-II | North America Mica, Inc. |
| Primavera Project Planner® | Primavera Systems, Inc. |
| Project Scheduler Network™ | SCITOR Corporation |
| PROJECT/2® | Project Software & Development, Inc. |
| Project Workbench® | Applied Business Technology Corporation |
| PROMIS™ | Strategic Software Planning Corporation |
| QWIKNET® | Project Software & Development, Inc. |
| QWIKNET Professional | Project Software & Development, Inc. |
| SuperProject® EXPERT | Computer Associates International |
| Time Line® | Breakthrough Software Corporation |
| ViewPoint | Computer Aided Management, Inc. |

# Section VIII

# Behavioral Dimensions and Teamwork in Project Management

In the first chapter in this section (29) Dennis Slevin and Jeffrey K. Pinto discuss the important intangible elements of the project manager's job—leadership and motivation. They include diagnostic tools that project managers, or aspiring project managers, will find to be useful in assessing themselves for this role.

Teams and teamwork are so important to project management that three chapters are devoted to these topics in this section.

In Chapter 30, Raymond E. Hill and Trudy Somers treat the management of the social conflict that inherently arises in the project context. They prescribe how, in the team context, this conflict may be managed constructively.

Teamwork is also the key to Chapter 31 by Thomas E. Miller. He shows how failure in teamwork creates resistance to change using a clinical case study from a large urban fire department to illustrate and underscore his diagnosis.

Hans J. Thamhain, in Chapter 32, characterizes an effective team and delineates barriers and drivers of effective team performance. He discusses the organization of the project team, examines team building as an ongoing process, and presents recommendations for effective team management.

In Chapter 33, David L. Wilemon and Bruce N. Baker discuss some major research findings regarding the human element in project management. This chapter is particularly useful to the project manager who is experiencing some problems managing this aspect of the project.

# 29. Leadership, Motivation, and the Project Manager*

Dennis P. Slevin†
Jeffrey K. Pinto‡

## INTRODUCTION

The project manager typically works through a project team consisting of individuals with diverse backgrounds, education, experiences, and interests. One secret to successful project implementation is the project manager's ability to get this diverse set of actors performing at maximal effectiveness. Consequently, the project manager must be both a *leader* and a *motivator* of the members of the project team. He or she is often required to work through others while possessing minimal legitimate line authority over their actions. Consequently, the techniques of effective leadership and motivation become very important in the project management context.

---

* Portions of this chapter are adapted from *Executive Survival Manual,* by Dennis P. Slevin, Innodyne: Pittsburgh, 1985. The authors are indebted to Dr. Thomas V. Bonoma for the development of the Bonoma/Slevin Leadership Model and to Dr. S. Lee Jerrell for the development of the Jerrell/Slevin Management Instrument.



† Dennis P. Slevin is an Associate Professor of Business Administration at the University of Pittsburgh's Joseph M. Katz Graduate School of Business. He holds a B.A. in Mathematics from St. Vincent College, a B.S. in Physics from M.I.T., an M.S. in Industrial Administration from Carnegie-Mellon, and a Ph.D. from Stanford University. He has had extensive experience as a line manager, including service as the CEO of four different companies, which qualified him as a member of the Young Presidents' Organization. He presently serves as a director of several corporations, and consults widely. He publishes in numerous professional journals, and is co-editor of *Implementing Operations Research Management Science, The Management of Organizational Design, Volumes I and II,* and *Producing Useful Knowledge*. He has written the pragmatic *Executive Survival Manual* for practicing managers.

‡ Jeffrey K. Pinto is Assistant Professor of Organization Theory at the College of Business Administration, University of Cincinnati. He received his B.A. in History and B.S. in Business from the University of Maryland, M.B.A. and Ph.D. from the University of Pittsburgh. He has published several papers in a variety of professional journals on such topics as project management, implementation, instrument development, and research methodology.

This chapter is intended to assist the project manager in these two important areas. First, a contingency model of leadership is presented along with an instrument to aid the project manager in diagnosing his or her leadership style. The emphasis is on the conscious selection of a leadership style contingent upon the varying parameters of the project situation. Second, the issue of motivation is addressed both from the conceptual standpoint of motivating others and from the standpoint of diagnosing one's internal motivational structure and determining if it is appropriate for the project management situation. This chapter is intended to give the reader both a conceptual grounding in leadership and motivation issues along with diagnostic techniques to determine appropriate actions with regard to these concepts in varying project management situations.

## LEADERSHIP DEFINED

Leadership is a complex process that is crucial to successful management. Behavioral scientists have been studying the leadership problem over the past half century in an attempt to better understand the process and to come up with prescriptive recommendations concerning effective leader behaviors. Years of careful research have generated a variety of findings that at best are sometimes confusing to the practicing manager and at worst are at times internally inconsistent. For example:

- Leaders will be most effective when they show a high level of concern for both the task and the employees (1).
- Leaders should be task-oriented under conditions where they have either high or low control over the group. When there is moderate control over the group, leaders should be people-oriented (4).
- Leaders will be effective to the extent that they assess the required quality of the decision to be made in relation to the required level of acceptance by subordinates and seek participation by subordinates in decision making accordingly (25).
- Leaders will be accepted and will motivate employees to the extent that their behavior helps employees progress toward valued goals and provides guidance or clarification not already present in the work situation (11).
- A participative approach to leadership will positively influence employees' morale and increase their commitment to the organization (5).
- Under stressful conditions (e.g., time pressure) an autocratic leadership style will lead to greater productivity (6). A minimum level of friendliness or warmth will enhance this effect (24).

- Even when a leader has all the needed information to make a decision and the problem is clearly structured, employees prefer a participative approach to decision making (8).
- Effective leaders tend to exhibit high levels of intelligence, initiative, personal needs for occupational achievement, and self-confidence (21, 7).
- Effective leadership behavior may be constrained by the organizational setting so that similar behaviors may have different consequences in different settings (2, 20).
- Leadership behavior may not be important if subordinates are skilled, if tasks are structured, if technology dictates actions, and if the administrative climate is supportive and fair (12).
- Leadership behavior can be divided into task behavior (one-way communication) and relationship behavior (two-way communication). An effective leadership style using these behaviors can be selected according to the maturity level of subordinates relative to accomplishing the task (9).

These principles of leadership present a variety of sometimes conflicting premises which make it difficult to select appropriate behaviors in practice. Some of the work suggests that a participatory style is best in all situations, while other work suggests that a participatory style is more effective in some situations than in others. Some of the work suggests that the traits of the leader determine effectiveness, while other work shows that the dynamics of the interaction of the leader with subordinates or the interdependency with other organizational factors determines effectiveness. Still other work suggests that, under some conditions, leadership style is not important at all.

What is needed for the practicing executive is a conceptual model that clarifies some aspects of the concept of leadership, suggests alternative leadership styles that may be used, and provides normative recommendations on conditions under which alternative styles might be used. One of the purposes of this chapter is to describe a *cognitive* approach to leadership that will help the practicing executive consciously select the leadership style correct for alternative situations.

## A DIAGNOSTIC

Before continuing, the reader may wish to develop a diagnostic on his or her leadership style without being biased by the conceptual discussion. Respond to the questions in the Jerrell/Slevin management instrument in Figures 29-1, 29-2, and 29-3. This instrument has been used effectively

LEADERSHIP 1

## JERRELL/SLEVIN MANAGEMENT INSTRUMENT

Complete this instrument by circling the number for each item that represents your best estimate.

| | Strongly Disagree | Disagree | Neutral | Agree | Strongly Agree |
|---|---|---|---|---|---|
| 1. I don't like it when others disagree with me. | 1 | 2 | 3 | 4 | 5 |
| 2. I like quick results. | 1 | 2 | 3 | 4 | 5 |
| 3. I find it hard to accept others decisions. | 1 | 2 | 3 | 4 | 5 |
| 4. I have a strong ego. | 1 | 2 | 3 | 4 | 5 |
| 5. Once I make up my mind I stick to it. | 1 | 2 | 3 | 4 | 5 |
| 6. I enjoy giving orders. | 1 | 2 | 3 | 4 | 5 |
| 7. The work group should determine its own vacation schedule. | 5 | 4 | 3 | 2 | 1 |
| 8. The work group should determine its own work schedule. | 5 | 4 | 3 | 2 | 1 |
| 9. I feel comfortable being placed in a powerful position. | 1 | 2 | 3 | 4 | 5 |
| 10. I like working in a group situation. | 5 | 4 | 3 | 2 | 1 |
| Total your D score (Items 1–10 above) | | D = ______ | | | |
| 11. It is easier to make a decision in a group. | 1 | 2 | 3 | 4 | 5 |
| 12. Groups usually take up more time than they are worth. | 5 | 4 | 3 | 2 | 1 |
| 13. I often ask for information from subordinates. | 1 | 2 | 3 | 4 | 5 |
| 14. Groups give a deeper analysis of a problem. | 1 | 2 | 3 | 4 | 5 |
| 15. I often use what subordinates have to say. | 1 | 2 | 3 | 4 | 5 |
| 16. No one else can know as much about the problem as I do. | 5 | 4 | 3 | 2 | 1 |
| 17. I usually make my decision before calling a staff meeting. | 5 | 4 | 3 | 2 | 1 |
| 18. Better decisions are made in group situations. | 1 | 2 | 3 | 4 | 5 |
| 19. A group is no better than its best member. | 5 | 4 | 3 | 2 | 1 |
| 20. Group decisions are the best. | 1 | 2 | 3 | 4 | 5 |
| Total your I score (Items 11–20 above) | | I = ______ | | | |



Figure 29-1. A leadership assessment instrument.

LEADERSHIP 1

## JERRELL/SLEVIN MANAGEMENT INSTRUMENT
(Continued)

Scoring Instructions

1. Record your D score (the sum of the answers to items 1-10 on the previous page)
   D = ______
2. Record your I score (the sum of the answers to items 11-20 on the previous page)
   I = ______
3. Determine your percentile score from the table below

| D | |
|---|---|
| Raw Score | Percentile |
| 19 | 1 |
| 20 | 1 |
| 21 | 1 |
| 22 | 3 |
| 23 | 5 |
| 24 | 6 |
| 25 | 9 |
| 26 | 12 |
| 27 | 15 |
| 28 | 22 |
| 29 | 27 |
| 30 | 37 |
| 31 | 42 |
| 32 | 53 |
| 33 | 64 |
| 34 | 72 |
| 35 | 81 |
| 36 | 85 |
| 37 | 91 |
| 38 | 94 |
| 39 | 97 |
| 40 | 98 |
| 41 | 99 |
| 42 | 99 |
| 43 | 100 |

| I | |
|---|---|
| Raw Score | Percentile |
| 22 | 1 |
| 23 | 1 |
| 24 | 1 |
| 25 | 2 |
| 26 | 2 |
| 27 | 4 |
| 28 | 6 |
| 29 | 7 |
| 30 | 8 |
| 31 | 15 |
| 32 | 18 |
| 33 | 26 |
| 34 | 39 |
| 35 | 48 |
| 36 | 56 |
| 37 | 69 |
| 38 | 78 |
| 39 | 84 |
| 40 | 87 |
| 41 | 92 |
| 42 | 96 |
| 43 | 98 |
| 44 | 99 |
| 45 | 99 |
| 46 | 100 |

4. Plot yourself on the grid on the following page

   Percentiles are estimates based on data collected from 191 American managers.

Figure 29-2. A leadership assessment instrument. (*continued*)

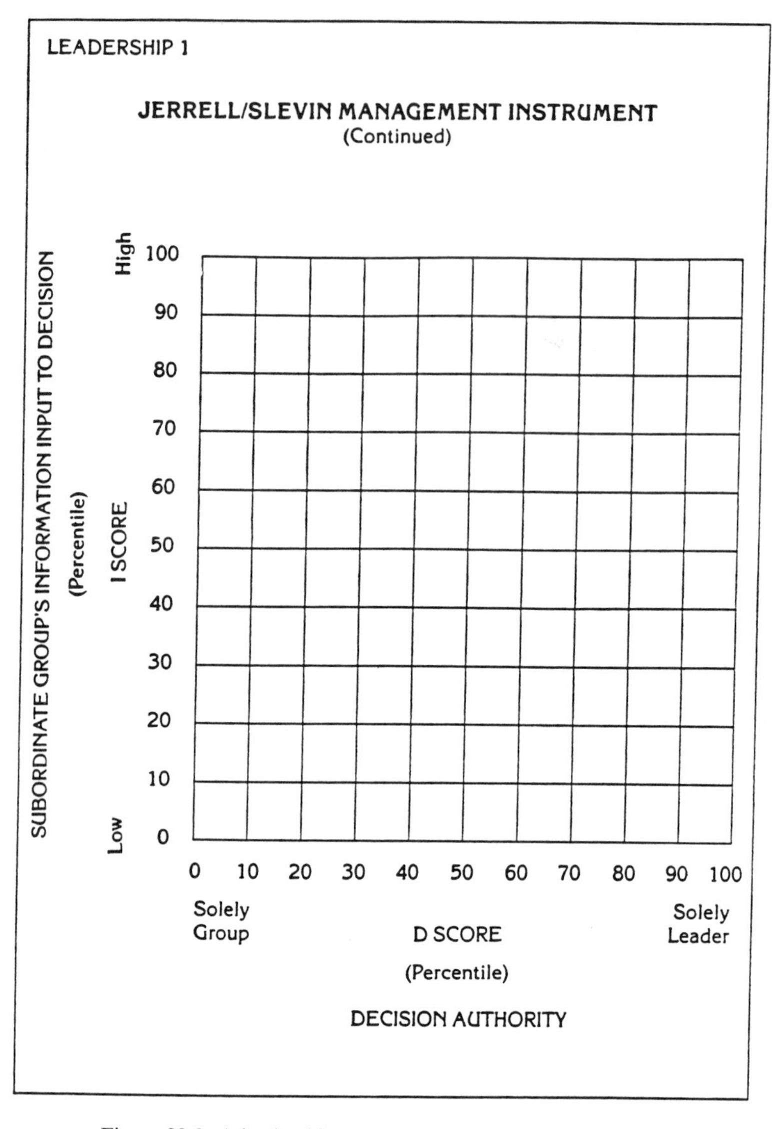


Figure 29-3. A leadership assessment instrument. *(continued)*

with thousands of managers in both explaining the theory and providing them with a diagnostic on their particular style.

## THE BONOMA/SLEVIN LEADERSHIP MODEL

Suppose that you are the leader of a group of five immediate subordinates. You are faced with a problem that is complex and yet a decision must be made. From the standpoint of leadership, before you make the decision, you must answer two "predecisional" questions:

1. *Where will you get the information input?* (Whom do you ask?)
2. *Where should you place the decision authority for this problem?* (Who makes the decision?)

The first question asks which members of the group you head will furnish information about a particular decision. The second asks to what extent you maintain all the decision authority and make the decision yourself or to what extent you "share" your decision authority with members of your group and have them make the decision in more or less democratic fashion.

The first dimension is one of information, the second one of decision authority. These two critical dimensions are essential for effective leadership and they have been plotted on the graph in Figure 29-4. As a leader you may request large amounts of subordinate information input into a decision or very small amounts of input. This is the vertical axis—information input. As a leader you may make the decision entirely yourself or you may share power with the group and have the decision made entirely as a group decision. This is the horizontal axis—decision authority. If we use percentile scores, any leadership style may be plotted in the two-dimensional space. For convenience we will refer to the horizontal axis (decision authority) first, and the vertical axis (information) second in discussing scores.

## FOUR LEADERSHIP STYLES

Using this plotting system, we can describe almost any leadership style. Refer to Figure 29-4. The four extremes of leaders you have known (depicted in the four corners of the grid) are the following:

1. *Autocrat (100, 0).* Such managers solicit little or no information input from their group and make the managerial decision solely by themselves.

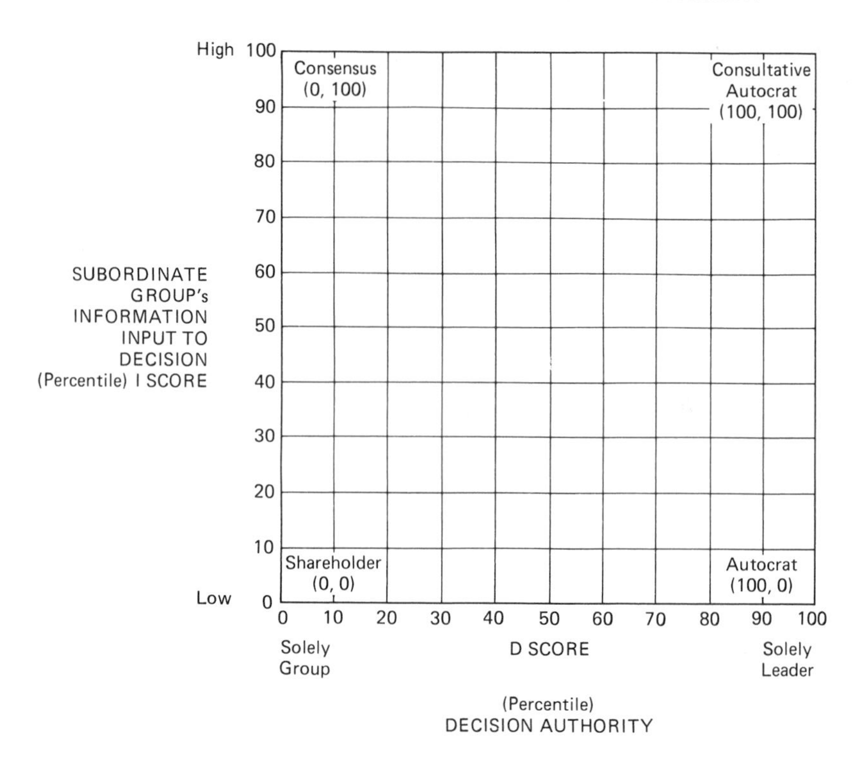


Figure 29-4. Bonoma/Slevin leadership model.

2. *Consultative Autocrat (100, 100).* In this managerial style intensive information input is elicited from the members, but such formal leaders keep all substantive decision making authority to themselves.
3. *Consensus Manager (0, 100).* Purely consensual managers throw open the problem to the group for discussion (information input) and simultaneously allow or encourage the entire group to make the relevant decision.
4. *Shareholder Manager (0, 0).* This position is literally poor management. Little or no information input and exchange takes place within the group context, while the group itself is provided ultimate authority for the final decision.

The advantages of the leadership model, apart from its practical simplicity, become apparent when we consider three traditional areas of leadership and managerial decision style:

- Participative management.
- Delegation.
- Personal and organizational pressures affecting leadership.

## PARTICIPATIVE MANAGEMENT

The concept of "participation" in management is a complex one with different meanings for different individuals. When the construct of participation is discussed with practicing managers in a consulting setting, the following response is typical:

"Oh, I participated with my subordinates on that decision—I asked each of them what they thought before I made the decision."

To the practicing manager, participation is often an *informational* construct—permitting sufficient subordinate input to be made before the hierarchical decision is handed down.

When the construct of participation is discussed with academics, the following response is typical:

"Managers should use participation management more often—they should involve their subordinates in consensual group processes."

To the academic, the concept of participation is often an issue of *power,* that is, a moving to the left on the Bonoma/Slevin model such that decision authority is shared with the group.

In actuality, participation is a *two-dimensional* construct. It involves both the solicitation of information and the sharing of power or decision authority. Those familiar with the Vroom-Yetton model might be interested that this conjecture is reinforced by their five positions as well. The first four positions of the Vroom-Yetton model (A1, A11, C1, C11) can be plotted vertically on the Bonoma/Slevin model as one moves from autocrat (100, 0) to consultative autocrat (100, 100), as shown in Figure 29-5 and Table 29-1. The final Vroom-Yetton position (G11) can be consigned to the consensus manager position (0, 100) on the Bonoma/Slevin model. The authors have found in a number of consulting situations that the Vroom-Yetton style selected by practitioners for a given situation is quite consistent with the position taken for the same situation on the Bonoma/Slevin model.

## DELEGATION

A good manager delegates effectively. In doing so he negotiates some sort of compromise between the extreme of "abdication"—letting subordinates do everything—and "autocratic management"—doing everything himself. When you have a task that might be delegated, you may wish to

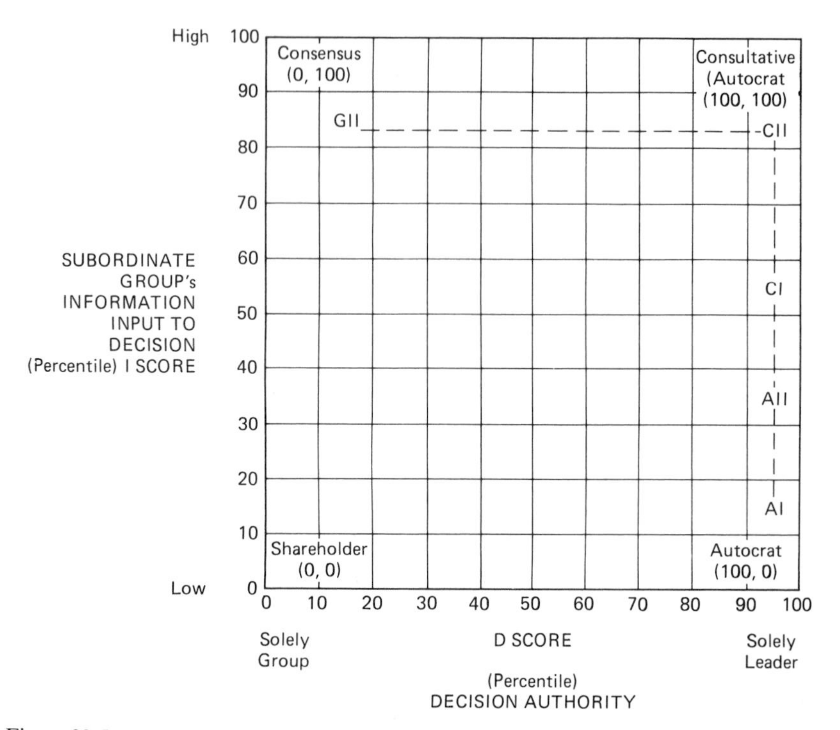


Figure 29-5. Bonoma/Slevin leadership model showing the Vroom-Yetton leadership decision styles.

ask yourself these questions which recent research has shown to relate to delegation (13):

1. How capable are my subordinates?
   (Capable subordinates make delegation more likely.)
2. How important is the decision?
   (Important decisions are less likely to be delegated.)
3. How large is my work load?
   (High supervisor work load increases delegation.)

Once you have decided to delegate a task, you may wish to use the leadership model explicitly. We have found the model useful for managers as they delegate work to their subordinates. After exposure to the model, managers are more likely to be more explicit about the informational and decision authority requirements of a delegated task. For example, the practitioner might say to his subordinate, "Get the information that you need from my files and also check with marketing [information]. You

Table 29-1. Decision Styles For Leadership: Individuals and Groups.[a]

A1. You solve the problem or make the decision yourself, using information available to you at that time.

A11. You obtain any necessary information from subordinates, then decide on the solution to the problem yourself. You may or may not tell the subordinates what the problem is in getting the information from them. The role played by your subordinates in making the decision is clearly one of providing specific information that you request, rather than generating or evaluating solutions.

C1. You share the problem with the relevant subordinates individually, getting their ideas and suggestions without bringing them together as a group. Then *you* make the decision. This decision may or may not reflect your subordinates' influence.

C11. You share the problem with your subordinates in a group meeting. In this meeting, you obtain their ideas and suggestions. Then *you* make the decision, which may or may not reflect your subordinates' influence.

G11. You share the problem with your subordinates as a group. Together, you generate and evaluate alternatives and attempt to reach agreement (consensus) on a solution. Your role is much like that of chairman, coordinating the discussion, keeping it focused on the problem, and making sure that the critical issues are discussed. You do not try to influence the group to adopt "your" solution, and you are willing to accept and implement any solution that has the support of the entire group.

[a] From Gibson, James L., John M. Ivancevich, James H. Donnelly, Jr., *Organizations, Behavior, Structure, Processes,* Business Publications, Inc. Plano, Texas, 1985 Fifth Edition. Adapted from Vroom, Victor H. and Philip W. Yetton, *Leadership and Decision Making,* University of Pittsburgh Press, 1973.

make the decision yourself [decision authority], but notify me in writing as soon as you have made it so that I am kept informed." Thus the subordinate understands both the information and decision authority aspects of the delegated tasks. Delegation often fails when the communication is unclear on either or both of these dimensions.

## PRESSURES AFFECTING LEADERSHIP

As a manager, you may act differently as a leader under different conditions, depending on three kinds of pressure:

1. Problem attributes pressures.
2. Leader personality pressures.
3. Organization/group pressures.

Think of the leadership model in terms of a map of the United States. Table 29-2 summarizes these pressures on leadership style in terms of geographical direction (for example, a movement "north" is a movement upward on the vertical axis of the leadership model, and so on).

Table 29-2. Three Leadership Style Pressures.

| 1. PROBLEM ATTRIBUTE PRESSURES | DIRECTION OF PRESSURE ON LEADERSHIP MODEL |
|---|---|
| • Leader lacks relevant information; problem is ambiguous | North: more information needed |
| • Leader lacks enough time to make decision adequately | South and east: consensus and information collection take time |
| • Decision is important or critical to organization | North: information search maximized<br>West: if implementation is crucial |
| • Decision is personally important to leader | North and east: personal control and information maximized |
| • Problem is structured or routine | South and east: as little time as possible spent on decision |
| • Decision implementation by subordinates is critical to success | West and north: input and consensus required |
| **2. LEADER PERSONALITY PRESSURES** | **DIRECTION OF PRESSURE ON LEADERSHIP MODEL** |
| • Leader has high need for power | East: personal control maximized |
| • Leader is high in need for affiliation; is "people oriented" | North and west: contact with people maximized |
| • Leader is highly intelligent | East: personal competence demonstrated<br>South: if leader lacks trust in subordinates |
| • Leader has high need for achievement | East: personal contribution maximized |
| **3. ORGANIZATION/GROUP PRESSURES** | **DIRECTION OF PRESSURE ON LEADERSHIP MODEL** |
| • Conflict is likely to result from the decision | North and west: participative aspects of decision making maximized |
| • Good leader/group relations exist | North and west: group contact maximized |
| • Centrality, formalization of organization is high | South and east: organization style matched |
| • Highly participative culture—strong norm for power sharing | North and west: consensus maximized |
| • High level of intergroup conflict | South and west: Communication between groups break down<br>(North and west preferred as a solution to conflict) |

### Problem Attributes

Problem attribute pressures generate eastward forces in the leader. This is especially true when problems are characterized as:

- Time-bound.
- Important.

- Personal.
- Structured and routine.

In such cases, it is very tempting to take "control" over the decisional process personally and "get the job done."

However, northward pressures can occur too, given:

- Important decisions.
- Ones in which you as the leader lack the resources to make the decision yourself.
- Problems in which subordinate implementation is critical to success.

In these cases, information input and exchange will be maximized.

### Leader Personality

Some managers tend to be inflexible in their leadership style because of who they are and how they like to manage. Such managers have at least one of the following characteristics:

- Have a high need for power.
- Are task-oriented.
- Are highly intelligent.

They will make many decisions themselves that might otherwise be left to subordinates, and they also may make decisions without acquiring information from the group. These managers typically see the sharing of the decision-making authority with subordinates as an abdication of power, or "selling out." Further, if they are highly task-oriented, their motivation is simply to get the job done or make the decision, with little thought given to the feelings of subordinates. A third characteristic, high intelligence, can often have the effect of influencing more autocratic decisions in the belief that they are more intellectually capable of arriving at the "best" decision.

People-oriented leaders, on the other hand, will act to maximize information inputs from their subordinates and to share their decision authority as well. Both activities are people processes.

### Organizational/Group Pressures

If conflict is likely to result from any decision made, effective managers are most likely to want their subordinates as involved as possible in both the input (northward) and authority (westward) dimensions, so as to help manage the potential conflict. Should good leader/group relations exist, pressure northward (but *not* necessarily westward) will be felt. The leader will feel great pressures to fit into the "culture" of the organization. The R&D lab expects a consensual approach; most business groups expect a consultative autocrat approach; the factory floor may expect an autocratic approach. It is important to match your style to the norms, needs, and expectations of your subordinates.

### Flexibility

The key message here is that this is a contingency model of leadership. No one style is best for all situations. Rather the successful manager is flexible and moves around the model as appropriate to the given situation. If time pressures are overwhelming, autocrat decisions may be appropriate. For a majority of management decision making, the consultative autocrat may be appropriate. And in dealing with high-tech professionals, the manager may choose to use a more consensual style. The key to success is *to match your leadership style to the situation and people involved.*

## FINDINGS FROM THE FIELD

Based on the presentation and discussion of this model with thousands of practicing managers, we would like to share with you our conclusions concerning important principles.

1. *You're more autocratic than you think.*
   In the eyes of your subordinates you are probably closer to the autocrat on the graph than you are in your own eyes. Why? Because you're the boss and they are the subordinates. No matter how easy-going, friendly, participative, and supportive you are, you are still the boss. There is almost always difference in leadership style as perceived by supervisor and subordinates.
2. *But it's O.K.*
   Often, when you ask subordinates where they would like their boss to move, they respond, "Things are O.K. as they are." Even though there are perceptual discrepancies concerning leadership style, there may not necessarily be a felt need or pressure for change. The status quo may be O.K.
3. *It's easy to move north and south.*
   It's easy to move vertically on the graph. Why? Because management is a job of communications. It's easy to collect more informa-

tion or less information before you make the decision. The information dimension is the least resistant to change.

4. *It's hard to move west.*
   Most managers of our experience find it quite threatening to move westward too quickly. Why? Because a westward movement upsets the basic power realities in the organization. If your head is in the noose, if things don't work out, then it is hard to turn the decisions totally over to your subordinates.
5. *If subordinates' expectations are not met, morale can suffer.*
   What would you expect to be the consequences if your subordinates expected you to use a (50, 90) process and instead you made the decision using a (90, 10) style? Very likely the result would be dissatisfaction and morale problems. As mentioned before, decision process can be as important as decision outcome, especially from the standpoint of motivating subordinates.
6. *Be flexible.*
   A successful manager is autocratic when he needs to be, consultative when necessary, and consensual when the situation calls for it. Move around the leadership space to fit the needs of the situation. Unsuccessful managers are inflexible and try the same style in all situations. Most managers feel that their score on the Jerrell/Slevin management instrument is a function of the particular job they have been in over the last few months. Be flexible and match your leadership style to the job.

## LEADERSHIP CHECKLIST

Let us now see how well your leadership style squares with your organization.

| | Yes | No |
|---|---|---|
| 1. *Dominant Leadership Style* | | |
| Should I try to change my own style? | ☐ | ☐ |
| Should I become more autocratic? | ☐ | ☐ |
| Or more consensual? | ☐ | ☐ |
| 2. *Fit* | | |
| Is my leadership style inappropriate for my organization? | ☐ | ☐ |
| Am I an autocratic peg in a participative hole? | ☐ | ☐ |
| 3. *Flexibility* | | |
| Do I fail to move my leadership style around the graph to match the different problems that I face? | ☐ | ☐ |
| Am I sometimes too inflexible for the responsibilities that I have? | ☐ | ☐ |

4. *Subordinate Fit*
   Do my subordinates have expectations and needs that are not in line with my leadership style? ☐ ☐
   If so, do I need new subordinates? ☐ ☐
5. *Information*
   Do I receive insufficient information for decision making? ☐ ☐
   Do my subordinates fail to regularly send me meaningful reports? ☐ ☐
6. *Decision Authority*
   Do I fail to share decision authority appropriately? ☐ ☐
   Am I stifling my subordinates by not letting them participate in decisions affecting them? ☐ ☐

If you answered "no" to all of the questions, you're O.K. in all six areas and are in excellent shape. If you answered "yes" to any questions, these are your problem areas and you should examine them. For example, if your subordinates don't seem to fit your own leadership style, maybe you need some new subordinates. Or you need to change. Or, if you just don't seem flexible enough to take on the broad range of problems you face, maybe you need to reexamine your leadership style. If you have three or more problem areas, it's undoubtedly time to give that corporate headhunter a call. (Or maybe you'd better engage in a serious high-priority program of leadership therapy.)

## AN EXAMPLE

Does the leadership model really work? We have applied it in a number of research, training, and field management settings. Here's an example of the sorts of results that might be forthcoming.

### Production Supervisor, Polyurethane Division, Large Chemical Corporation

This manager kept a running diary over a three-week period of 41 decision situations he faced on the job. Later he plotted them on the model. He found a strong divergence between the way he thought he behaved and the way he actually behaved in most leadership situations. It became obvious to him that his dominant leadership style was not that of a consensus manager but more that of a consultative autocrat, for when he plotted his flexibility space on the basis of actual behavior, he found that the only direction in which he was flexible was vertical (that of allowing increased information input from subordinates). In no case did he behaviorally allow any decision authority sharing, as he had previously supposed he did. He concluded:

The merit of the Bonoma-Slevin Leadership Model was in its simplicity, since it uses only two dimensions to describe alternative leadership styles. The model confirmed that my actual leadership style was inconsistent with my preconceived image of leadership style. It presented a visual conception of changes necessary for me to alter my present leadership style as that of a Consultative Autocrat to become a Consensus Manager.

My recommendation for anyone wanting to use the Leadership Model for self analysis would be to keep a diary of decisional situations faced over a period of time.

### Vice President, Division of International Manufacturing Firm

This manager supervises approximately five other managers and five people not holding managerial status. He asked each of his subordinates, first, to estimate his dominant decision style over the last 12 months, and, second, to recommend how he should change. As a result, he learned that his subordinates rated him approximately 20% more autocratic than he had rated himself. However, he also found that his subordinates were quite happy with his leadership style, and desired only slightly less autocratic decisions and slightly more information input.

## CONCLUSION: ORGANIZATIONAL IMPLICATIONS OF LEADERSHIP

The leadership framework presented in our model forces the manager to ask two key questions concerning decision making:

- Whom do I ask?
- Who makes the decision?

Obtaining accurate information input from one's subordinate group is crucial to effective management. Similarly, decision-making authority must be located in the right place vis-à-vis the leader's group.

In addition, this model might be broadened to apply to even more fundamental questions of your managerial job. Forgetting for a moment that you are a leader of a subordinate group, answer the following questions by considering both vertical and horizontal relationships in your organization:

1. *Where do I get my information?*

Every manager needs accurate information to make effective decisions. Do you get sufficient information:

- From your boss?
- From your peers?
- From the formal information system?
- From all sources in the organization that can aid you in your job?

2. *Who makes the decisions?*

Is the decision authority vested in the right people, not just concerning your subordinates but also upward and laterally? Should other departments be making decisions that you now make or vice versa?

The answers to these questions have broad implications for the management information system and the power structure of your organization. If you are not getting necessary information, perhaps the management information system should be modified to provide it. If you have a problem of misplaced decision authority (downward or upward), this problem should be addressed. If the answers to these fundamental questions are satisfactory, then only one important question remains, which cannot be addressed here: are the decisions good ones?

## MOTIVATIONAL ISSUES

Motivation is important to the project manager from two perspectives. First, the individual must be motivated to *be* a project manager. If one does not have sufficient intrinsic motivation to take the types of managerial steps required, then one is not likely to succeed at the project management task. Second, the project manager must be able to motivate others. For this it is crucial that the project manager have an adequate understanding of motivation and techniques for motivating others. This section is intended to provide the reader with the ability to diagnose his or her intrinsic motivation to manage and also the ability to understand better the motivational structures that others bring to their work settings and ways to utilize them for an effective performance.

## MOTIVATION TO MANAGE DEFINED

Do you have the necessary internal motivation to be an effective project manager? Many people who find themselves in project management situations are technically skilled and analytically oriented. Consequently, they often feel uncomfortable in assuming the managerial role which requires numerous interventions into the work of other people. The construct of Motivation to Manage (17) can be a useful tool for the current or prospective project manager.

According to research conducted over the past two decades, the Motivation to Manage and its associated attitudes and motives are likely to cause one to:

- Choose a managerial career.
- Be successful in a managerial position.
- Move rapidly up the managerial ladder.

In other words, the Motivation to Manage is essential for managerial success. The six components included in the Motivation to Manage, outlined by John B. Miner, follow.

The manager should have:

1. Favorable Attitude Toward Authority—Managers are expected to behave in ways that do not provoke negative reactions from their superiors; ideally, they elicit positive responses. Equally, a manager must be able to represent his group upward in the organization and to obtain support for his actions at higher levels. This requires a good relationship between a manager and his superior.
2. Desire to Compete—There is, at least insofar as peers are concerned, a strong competitive element built into managerial work; a manager must compete for the available rewards, both for himself and for his group. Certainly, without competitive behavior, rapid promotion is improbable. On occasion a challenge may come from below, even from among a manager's own subordinates.
3. Assertive Motivation—There is a marked parallel between the requirements of the managerial role and the traditional assertive requirements of the masculine role as defined in our society. Although the behaviors expected of a father and those expected of a manager are by no means identical, there are many similarities: both are supposed to take charge, to make decisions, to take such disciplinary action as may be necessary to protect the other members of their group. . . . when women are appointed to managerial positions, they are expected to follow an essentially assertive behavior pattern, at least during the hours spent in the work situation.
4. Desire to Exercise Power—A manager must exercise power over his subordinates and direct their behavior in a manner consistent with organizational (and presumably his own) objectives. He must tell others what to do when this becomes necessary and enforce his words through positive and negative sanctions. The individual who finds such behavior difficult and emotionally disturbing, who does not wish to impose his wishes on others or believes it wrong to do so, probably cannot be expected to meet this requirement.

5. Desire for a Distinctive Position—The managerial job tends to require a person to behave differently from the ways his subordinates behave toward each other. He must be willing to take the center of the stage and assume a position of high visibility; he must be willing to do things that invite attention, discussion, and perhaps criticism from those reporting to him; and he must accept a position of considerable importance in relation to the motives and emotions of others.
6. A Sense of Responsibility—The managerial job requires getting the work out and staying on top of routine demands. The things that have to be done must actually be done; constructing budget estimates, serving on committees, talking on the telephone, filling out employee rating forms, making salary recommendations, and so on. To meet these requirements, a manager must be capable of dealing with this type of routine and, ideally, of gaining some satisfaction from it.*

## RESEARCH RESULTS

Can you predict your degree of success and job satisfaction by knowing your Motivation to Manage? It appears that you can. Research has compared the initial Motivation to Manage with subsequent promotions over a five-year period in two departments. The results are presented below:

Motivation to Manage

Initial Motivation to Manage and Number of Subsequent Promotions over Next 5 Years in the R&D Department of an Oil Company

| Motivation to Manage | No Promotions | One Promotion | Two or More Promotions |
|---|---|---|---|
| High | 3(25%) | 6(38%) | 15(71%) |
| Low | 9(75%) | 10(62%) | 6(29%) |

Initial Motivation to Manage and Rate of Subsequent Promotion over Next 5 Years in the Marketing Department of an Oil Company

| Motivation to Manage | No Promotions | Slow Promotion Rate | Fast Promotion Rate |
|---|---|---|---|
| High | 13(32%) | 14(64%) | 16(89%) |
| Low | 28(68%) | 8(36%) | 2(11%) |

*Source:* Miner (17), pp. 28–29.

* (Copyright © 1973 by the President and Fellows of Harvard College; all rights reserved. Reprinted by permission of *Harvard Business Review*. "The Real Crunch in Managerial Manpower," by John B. Miner, Vol. 51, No. 6, 1973, p. 153.

Numerous other results have suggested the basic premise:

- To be a successful manager you must have the Motivation to Manage.

What is the relevance of this for you? Many project managers evolved into their managerial positions out of areas of technical expertise such as accounting, finance, marketing, and engineering. It is not uncommon to have highly qualified individual contributors promoted to the managerial ranks. The change is dramatic. In the previous roles they communicated primarily with things. They now must communicate primarily with people. Previously they could spend long uninterrupted hours reading, reflecting, collecting data, and working on technical projects. They now must spend large amounts of time in group meetings, talking on the telephone, being interrupted by subordinates, and doing the other frenetic activities that encompass a typical manager's day. Before, they could self-actualize by coming up with good technical solutions. Now they must engage in the sometimes unpleasant process of exercising power over others and engaging in competition for key resources.

Management is recognized as a step up in the career ladder. It is noted as a logical progression in one's professional advancement. Nevertheless, for the talented specialist with a low Motivation to Manage, it may be a traumatic metamorphosis indeed. A crucial question every manager should ask himself is: "Do I have sufficient Motivation to Manage to ensure that I will be satisfied and successful in this position?"

## DIAGNOSING YOUR MOTIVATION TO MANAGE

Now it's time to give you an opportunity to see where you stand on Motivation to Manage (MTM). Complete the Motivation to Manage Audit (Motivation 1, Figure 29-6). This is a subjective assessment and you may have some error, but at least it will sensitize you to the key variables and encourage you to think about these issues.

## MTM AND PREFERRED JOB CHARACTERISTICS

Activities that are preferred and enjoyed are very different for the person with a high versus a low Motivation to Manage. Jobs with given characteristics are better and less well suited to each managerial type. Table 29-3 indicates those job characteristics that are best matched with people with high MTM and low MTM.

MOTIVATION 1

## MOTIVATION TO MANAGE AUDIT

How motivated to manage are you?

Complete this instrument by circling the number for each item that represents your best estimate of your current level.

| | Well Below Average | | | | | Average | | | | | Well Above Average |
|---|---|---|---|---|---|---|---|---|---|---|---|
| 1. Favorable attitude toward authority | 0 | 1 | 2 | 3 | 4 | 5 | 6 | 7 | 8 | 9 | 10 |
| 2. Desire to compete | 0 | 1 | 2 | 3 | 4 | 5 | 6 | 7 | 8 | 9 | 10 |
| 3. Assertive motivation | 0 | 1 | 2 | 3 | 4 | 5 | 6 | 7 | 8 | 9 | 10 |
| 4. Desire to exercise power | 0 | 1 | 2 | 3 | 4 | 5 | 6 | 7 | 8 | 9 | 10 |
| 5. Desire for a distinctive position | 0 | 1 | 2 | 3 | 4 | 5 | 6 | 7 | 8 | 9 | 10 |
| 6. A sense of responsibility | 0 | 1 | 2 | 3 | 4 | 5 | 6 | 7 | 8 | 9 | 10 |

TOTAL MOTIVATION TO MANAGE = ______

Now place a check mark next to the number for each item that represents your best estimate of where you would *like to be*. The difference between your *desired* and *actual* score for each item is your Motivation to Manage deficit on that factor. It will be used in completing your Motivation to Manage Action Plan (Motivation 2).

Figure 29-6. Instrument for assessing the motivation to manage.

Table 29-3. Job Characteristics Associated with Motivation to Manage (MTM).

| LOW MTM | HIGH MTM |
|---|---|
| Relatively small span of control | Large span of control |
| Small number of subordinates | Large number of subordinates |
| High technical/engineering component | High people/budgetary component |
| Maintain "hands-on" expertise | Surround oneself with technical experts |
| Limited number of activities per day | As many as 200 activities per day |
| Few interruptions | Many interruptions |
| Time for reading, analyzing | Time for interactions |
| Serve as facilitator to staff | Serve as "boss" to staff |
| Career progression = increase in technical expertise | Career progression = managerial advancement |
| Little exercise of power required | Much intervention in lives of others |
| Lower stress position | Higher stress position |

## MTM/JOB FIT

In order for you to be satisfied and successful in an organizational role, you must first attempt to match your basic motivation to the job. Does your job fit your Motivation to Manage? Different jobs require different levels of MTM. Look at Table 29-3 and try to determine if you are in a high or low MTM job. Suppose that you have just concluded that you have low Motivation to Manage? What should you do?

You have two alternatives:

1. Select jobs that are more appropriate to your MTM.
2. Attempt to change your MTM.

In order for you to be happy, fulfilled, and successful in your job, you must fit the job requirements. There must be a match between your motivation and the characteristics of the job, the activities you like to perform and those demanded by the job. Thorough and accurate self-assessment regarding your Motivation to Manage can present you with very important information as you make career choices.

## CHANGING YOUR MOTIVATION TO MANAGE

It is generally accepted in psychology that a certain amount of motivation is learned. McClelland (15) has claimed the ability to teach people to increase their need for achievement. He has also concluded that successful managers have a higher need for power than their need for affiliation

(16). This seems compatible with the MTM in that one must be prepared to exercise power over others in order to succeed as a manager.

Is it possible to increase a person's Motivation to Manage? We think so. Little research has been done in this area. There are no figures to cite. However, look at the six components of Motivation to Manage. They are learned motives. Therefore one should be able to change them. Is it possible to change your Motivation to Manage? Definitely! If you want to.

Do you want to change your Motivation to Manage? If yes, you will need to formulate an action plan for changing each of the components of managerial motivation.

Go back to the Motivation to Manage Audit (Figure 29-6). Look at your desired level for each of the factors in the Motivation to Manage in your present position. The difference between the desired level and your actual level provides a managerial motivation deficit for each factor. List on the Motivation to Manage Action Plan (Motivation 2, Figure 29-7) the specific steps that you might take to increase your MTM on each factor to remove the deficit.

## MOTIVATING OTHERS

So far we have focused on what motivates you, on diagnosing your internal motivational structure. As a successful manager you must be able to effectively motivate others. We have attempted here to provide you with a pragmatic approach to the problem of motivating others. In the old days this was easier, or it at least appears to have been from today's perspective. To dramatize this, let's go back to the Bethlehem Steel Company labor yard in Bethlehem, Pennsylvania, in the spring of 1899. Frederick Winslow Taylor, the father of scientific management, comes over to the laborer Schmidt whose job it is to load pig iron onto gondola cars. (His real name was Henry Noll, but Taylor thought "Schmidt" sounded better in his historical record.) Schmidt's job was quite simple: pick up a 91-pound pig of iron, walk horizontally across the yard with it, up an inclined ramp, and deposit it in the gondola car. He then returned to the pile for another pig of iron and continued to do this throughout the day. During a typical day, Schmidt would load between 12 and 13 tons (long tons = 2240 pounds each) of pig iron per day. For doing this he earned his daily wage of $1.15. Taylor studied him "scientifically." He carefully timed how long it took to pick up a pig, the speed with which one could walk horizontally up the incline with a load, back down the ramp unloaded, etc. He then made Schmidt a proposition: "Follow my instructions and increase your output and pay." Schmidt agreed, since he was put on a piece-rate system under which he could now earn $1.85 per day provided he reached his

MOTIVATION 2

## MOTIVATION TO MANAGE ACTION PLAN

Record your Motivation to Manage deficit (desired - actual) for each factor below. Then specify appropriate action steps you might take to increase your Motivation to Manage on each factor and remove the deficit.

1. Favorable Attitude Toward Authority Deficit: ________
   Action Plan: ________________________________________
   ____________________________________________________
   ____________________________________________________
   ____________________ Probability of Success: ________

2. Desire to Compete Deficit: ________
   Action Plan: ________________________________________
   ____________________________________________________
   ____________________________________________________
   ____________________ Probability of Success: ________

3. Assertive Motivation Deficit: ________
   Action Plan: ________________________________________
   ____________________________________________________
   ____________________________________________________
   ____________________ Probability of Success: ________

4. Desire to Exercise Power Deficit: ________
   Action Plan: ________________________________________
   ____________________________________________________
   ____________________________________________________
   ____________________ Probability of Success: ________

5. Desire for Distinctive Position Deficit: ________
   Action Plan: ________________________________________
   ____________________________________________________
   ____________________________________________________
   ____________________ Probability of Success: ________

6. A Sense of Responsibility Deficit: ________
   Action Plan: ________________________________________
   ____________________________________________________
   ____________________________________________________
   ____________________ Probability of Success: ________

Figure 29-7. Motivation to manage action plan.

target. His target amounted to 45–48 tons of pig iron per day. According to Taylor (22), Schmidt reached his target and continued to perform at this rate on a regular basis.

The enormity of his task is hard to contemplate in today's world. To accomplish his 45 tons per day, Schmidt had to walk the equivalent of 8 miles each day with a 91-pound pig of iron in his arms. He then had to run 8 miles back to the pile (23). According to Taylor's reports, Schmidt was not a large man, weighing about 130 pounds, but was particularly suited to his work. Based on discussions that we have had with the Human Energy laboratory at the University of Pittsburgh, one can conclude that Schmidt's caloric energy output amounted to at least 5000–6000 calories per day. The poor man would have had to spend much of his waking hours eating just to keep from slowly disappearing over time.

As an interesting human interest side of this story, although Taylor reported Schmidt as happy with his work, later reports indicate that Henry Noll became an excessive drinker and lost both his home and his job. Also, although Taylor's self-reports seem to indicate great success and relative ease of implementation, such was not the case. The excellent history written by Daniel Nelson (19) demonstrates some extremely difficult problems such as worker resistance, lack of cooperation from top management, political infighting, and other problems reminiscent of the difficulties of implementation of modern-day organizational change.

The moral of this story is that it is much easier to motivate an extremely deprived and hungry worker. Taylor's writings include stories of tremendous accomplishments with his "first-class men," individuals who were willing to work at extremely high physical rates. He was able to accomplish these feats because the workers were at near subsistence levels and were willing to work quite hard to get that potential 60% increase in pay. Imagine the difficulties that Taylor might have in a modern-day steel yard!

## MASLOW'S HIERARCHY OF NEEDS

One very useful model for explaining the changes that have occurred in human motivation over the years is that developed by Abraham Maslow (14). Maslow's hierarchy argues that man's needs come in an ordered sequence that is arranged in the following five need categories:

1. Physical needs: the need for food, water, air.
2. Safety needs: the need for security, stability, and freedom from threat to physical safety.
3. Love needs: the need for friends with whom one may affiliate.
4. Esteem needs: the need for self-respect and esteem of others. This includes recognition, attention, and appreciation from others.

5. Self-actualization needs: the need for self-fulfillment to be able to grow and learn.

Maslow argued that these needs were arranged on a "hierarchy of prepotency." In other words, they must be fulfilled in sequential fashion starting with the lower-order needs first and progressing up the need hierarchy. One who is dying of dehydration in the desert is not interested in esteem needs; one who is being threatened by a criminal is not interested in self-actualization, etc. Substantial follow-up research (and even Maslow's original speculations) indicate that the needs do not always have to be fulfilled in a lock-step fashion. At times the artist may be willing to starve in order to create, etc. But, in general it is a useful managerial model when we consider the problem of how to motivate workers. Over the eight decades since Taylor's initial success, society at large has moved dramatically up the need hierarchy. It would be quite difficult today to get a worker to triple output in return for a 60% increase in wages.

The manager of today must be able to assess where each of his subordinates and co-workers are on the hierarchy and attempt to appeal to the appropriate needs. Some people crave status and recognition. Others want strongly to be a member of a cohesive team and "to belong." Others have a tremendous need to be creative, innovative, and learn new skills. If you have a motivational problem with a worker, attempt to answer these three questions:

1. Where is he or she on Maslow's Need Hierarchy?
2. What needs will motivate him or her?
3. How can I help him or her to satisfy these needs?

The key point for you as manager and potential motivator of your subordinates is this: individuals are driven by differing sets of needs and, consequently, must be motivated individually. In other words, the manager who supposes that one motivation style will work for all employees (e.g., Taylor, who thought all workers were motivated solely by the need for more money) greatly oversimplifies his or her job in attempting to motivate subordinates. Adequate time must be spent up front with subordinates, in getting to know them and understand those drives on which their needs are based.

## HERZBERG'S MOTIVATION HYGIENE THEORY

Herzberg (10) has suggested that there are two types of motivational factors: hygiene factors and motivators. He suggests that the hygiene factors are necessary conditions for a satisfied worker, but do not guarantee satisfaction. If they are absent, you will have an unhappy worker; but their presence does not guarantee contentment. The hygiene factors include:

- Company policy and administration.
- Supervision.
- Relationship with supervisor.
- Working conditions.
- Salary.
- Relationship with peers.
- Personal life.
- Relationship with subordinates.
- Status.
- Security.

Table 29-4. Herzberg and Maslow Compared.[a]

| NEED THEORY | | TWO-FACTOR THEORY |
|---|---|---|
| Self-actualization and fulfillment | Motivational | Work itself<br>Achievement<br>Possibility of growth |
| Esteem and status | Motivational (Advancement, Recognition) / Maintenance (Status) | Advancement<br>Recognition<br>Status |
| Belonging & social needs | Maintenance | Relations with supervisors<br>Peer relations<br>Relations with subordinates<br>Quality of supervision |
| Safety & security | Maintenance | Company policy and administration<br>Job security<br>Working conditions |
| Physiological needs | Maintenance | Pay |

[a] Keith Davis and John W. Newstrom, "Human Behavior at Work," *Organizational Behavior,* New York; McGraw-Hill, 1985, p. 77. Used with permission.

In other words, the hygiene factors satisfy the lower-level Maslow needs. On the other hand, there are motivators which are factors that account for satisfaction in the worker. The motivators include:

- Achievement.
- Recognition.
- Work itself.
- Responsibility.
- Advancement.
- Growth.

In other words, the motivators are found at the higher levels of Maslow's Hierarchy (see Table 29-4).

Herzberg's model has been challenged on both empirical and conceptual grounds. However, it has been demonstrated to be quite powerful as a general guide to the process of "job enrichment." The basic philosophy is that you can enrich someone's job and induce that individual to work harder by making the work more interesting and satisfying.

If you would like to explore the possibility of job enrichment, perform the following steps:

1. Remove serious dissatisfiers. For example:
    - Make sure working conditions are pleasant:
      Be sure pay is equitable, make supervision fair.
2. Vertically load the job content on the satisfiers. For example:
    - Push responsibility downward,
    - Push planning upward,
    - Provide meaningful modules of work.
    - Increase job freedom.
    - Introduce new and difficult tasks.

It is well known that individuals tend to do what is satisfying to them. They are more likely to repeat behaviors that result in rewards and to not repeat behaviors that do not. Consequently, if you can design a work environment in which individuals are reinforced by the work itself, you will experience much greater effectiveness as a manager.

## IN CONCLUSION: MOTIVATING YOURSELF AND OTHERS

You have now had an opportunity to assess in a personal way your Motivation to Manage. The logical steps in this assessment are portrayed in the flowchart shown in Figure 29-8. Try to accomplish this in as perceptive a way as possible. It's fun to consider your own personal motivational structures and to talk to others about career, job, and personal needs. If you can better understand where you are concerning your Motivation to Manage, you will be in a better position to perform your job at peak efficiency. If your Motivation to Manage is insufficient for your current or future job prospects, then you must seriously consider chang-

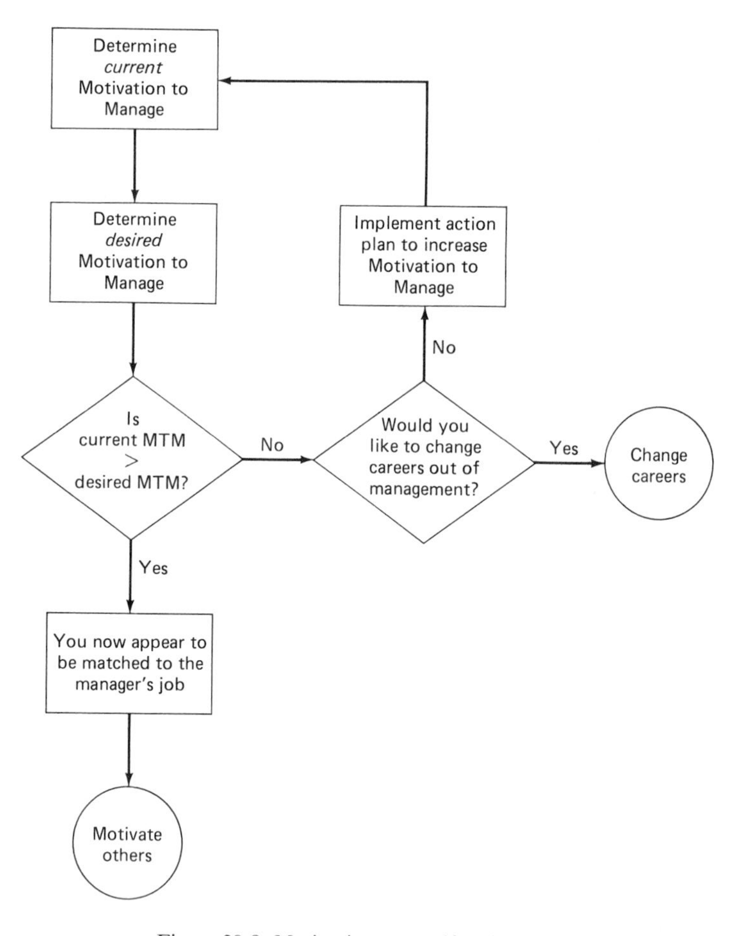


Figure 29-8. Motivating yourself and others.

ing these needs or changing your career. Millions of people get matched to millions of jobs through ad hoc and almost accidental sequences of events. In this chapter you are provided with a framework for consciously and analytically attempting to assess the match between your motivation structure and the manager's job.

The second point of this section consists of attempting to motivate others. Remember that different people have different motivational needs. You may have to respond in a very contingent manner to individuals on your team. Entire books have been written on this topic, and human motivation is indeed a complex area. In brief, you should try to assume the role of a facilitator that links up need satisfactions with desired job performances. If you can structure an environment in which diverse individual needs are being met through job performance, you will find yourself with a more cohesive and dedicated work group. If you can master these techniques, you will be well on your way to becoming a highly effective and successful project manager.

## SUMMARY

A good project manager is a capable leader. You must be able to lead the project team—often under conditions of minimum legitimate power. The leadership model presented here should enable the project manager to select leadership styles consciously, based on the parameters of the situation. A good project manager similarly must be able to motivate the project team. The models presented here should enable the project manager to assess his or her own internal motivation and also help in the selection of strategies for the motivation of others. These two skills, leadership and motivation, are important tools for the successful project manager.

## REFERENCES

1. Blake, R. R. and Mouton, J. *The Managerial Grid* (Gulf Publishing. Houston, 1964).
2. Dansereau, F., Green, G. and Haga, B. "A Vertical Dyad Linkage Approach to Leadership Within Formal Organizations: A Longitudinal Investigation of the Role Making Process." *Organizational Behavior and Human Performance,* Vol. 13 (1975), pp. 45–78.
3. Davis, K. and Newstrom, J. *Human Behavior at Work,* 7th Ed. (McGraw-Hill. New York, 1985).
4. Fiedler, F. E. "Contingency Model and the Leadership Process," in *Advances in Experimental Social Psychology,* Vol. 11, ed. Berkovitz, L. (Academic Press. New York, 1978).
5. Fleishman, E. A. 1973, "Twenty Years of Consideration and Structure," in *Current Developments in the Study of Leadership,* ed. Fleishman, E. A. and Hunt, J. G. (Southern Illinois University Press. Carbondale, 1973).

6. Fodor, E. M. "Group Stress, Authoritarian Style of Control, and Use of Power." *Journal of Applied Psychology,* Vol. 61 (1976), pp. 313–318.
7. Ghiselli, E. *Exploration in Managerial Talent* (Goodyear. Santa Monica, Calif., 1971).
8. Heilman, M. E., Hornstein, H. A., Cage, J. H. and Herschlag, J. K. "Reactions to Prescribed Leader Behavior as a Function of Role Perspective: the Case of the Vroom-Yetton Model." *Journal of Applied Psychology,* Vol. 69 (1984), pp. 50–60.
9. Hersey, P. and Blanchard, K. H. *Management of Organizational Behavior: Utilizing Human Resources,* 3rd Ed. (Prentice-Hall. Englewood Cliffs, N.J., 1977).
10. Herzberg, Frederick. "One More Time: How Do You Motivate Employees?" *Harvard Business Review,* Vol. 46(1) (1968), pp. 53–62.
11. House, R. J. "A Path-Goal Theory of Leader Effectiveness." *Administrative Science Quarterly,* Vol. 16 (1971), pp. 321–333.
12. Kerr, S. and Jermier, J. M. "Substitutes for Leadership: Their Meaning and Measurement." *Organizational Behavior and Human Performance,* Vol. 22 (1978), pp. 375–403.
13. Leana, Carrie R. "Predictors and Consequences of Delegation." *Academy of Management Journal,* Vol. 29(4) (1986), pp. 754–774.
14. Maslow, A. H. "A Theory of Human Motive Acquisition." *Psychological Review,* Vol. 1 (1943), pp. 370–396.
15. McClelland, David C. "Toward a Theory of Motive Acquisition." *American Psychologist,* Vol. 20 (1965), pp. 321–333.
16. McClelland, David C. and Burnham, David H. "Power Is the Great Motivator." *Harvard Business Review* (March–April, 1976).
17. Miner, John B. *The Human Constraint* (The Bureau of National Affairs, Inc. Washington, D.C., 1974).
18. Miner, John B. "The Real Crunch in Managerial Manpower." *Harvard Business Review,* Vol. 51(6) (1973), pp. 146–158.
19. Nelson, Daniel. *Frederick W. Taylor and the Rise of Scientific Management,* (University of Wisconsin Press. Madison, 1980).
20. Salancik, G. R., Calder, B. J., Rowland, K. M., Leblebici, H. and Conway, M. "Leadership as an Outcome of Social Structure and Process: A Multi-Dimensional Analysis," in *Leadership Frontiers,* ed. Hunt, J. G. and Larson, L. L. (Kent State University, Kent, Ohio, 1975).
21. Stogdill, R. *Handbook of Leadership* (Free Press. New York, 1974).
22. Taylor, Frederick W. "Time Study, Piece Work and the First-Class Man." *Transactions of the American Society of Mechanical Engineers,* Vol. 24 (1903), pp. 1356–1364, Reprinted in Merrill, Harwood F. *Classics in Management* (American Management Associates. New York, 1960).
23. Taylor, Frederick. *The Principles of Scientific Management* Harper & Row. New York, 1911).
24. Tjosvold, D. "Effects of Leader Warmth and Directiveness on Subordinate Performance on a Subsequent Task." *Journal of Applied Psychology,* Vol. 69 (1984), pp. 422–427.
25. Vroom, V. H. and Yetton, P. W. *Leadership and Decision Making* (University of Pittsburgh Press. Pittsburgh, 1973).

# 30. Project Teams and the Human Group

Raymond E. Hill* and Trudy L. Somers†

A central theme in managing the human side of project teams is the management of social conflict. Conflict in organizations is pervasive, inevitable, and ubiquitous; organizations develop specialized differentiated subunits which then obey many of the principles of general systems theory, not the least of which is the emergence of opponent processes among the differentiated parts. Project teams, by the very act of bringing together representatives of the specialized subunits, become a microcosm of the larger organizational dynamics.

There are a variety of issues around which conflict arises. Thamhain and Wilemon (1)‡ have reduced these issues to seven fundamental areas which include the following: project priorities, administrative procedures, technical opinions and performance trade-offs, manpower resources, cost estimates, scheduling and sequencing of work, and personality conflict. Thamhain and Wilemon found significant variation in intensity of the seven conflict types over the life cycle of a project. Data from one hundred project managers indicated considerable variation over time in the intensity of conflict from almost all sources except personality clashes. Personality conflicts were relatively constant during all phases of a project life cycle including project formation, buildup, main program phase,

---

* Raymond E. Hill is Associate Professor of Organizational Behavior in the Graduate School of Business Administration, The University of Michigan. He is a member of the American Psychological Association and the Academy of Management. His research interests are focused on project management, with recent interests centered around career choice and development processes, particularly of systems and computer personnel. He has published several articles in professional journals, and has edited a book of readings on Matrix Organization.

† Trudy L. Somers is a Ph.D. student of Organizational Behavior in the Graduate School of Business Administration, The University of Michigan. She is a member of the Academy of Management and the American Psychological Association. Her current research interests arise from years of working experience in computer-related firms and center in strategic human resource management, with special interest in careers.

‡ Numbered references are given at the end of this chapter.

and phaseout. In discussing personality clashes as a source of conflict, Thamhain and Wilemon suggest that while it is not as intense as some other conflict types, it is nevertheless problematical. In particular, they suggest, "Project managers emphasized that personality conflicts are particularly difficult to handle. Even apparently small and infrequent personality conflicts might be more disruptive and detrimental to overall program effectiveness than intense conflict over nonpersonal issues, which can often be handled on a more rational basis" (2).

It is the purpose of this chapter to relate two frameworks for conceptualizing personality conflict, both of which can be diagnosed and assessed using standardized questionnaires, to discuss the relationship of these frameworks to group performance, to report on how differences can be managed for more effective project performance, and to identify additional references for the interested reader. The first framework (MBTI) deals with different problem-solving strategies determined by how the individual (1) gathers and (2) processes information. The second framework (FIRO-B) focuses on emotional dynamics expressed through interpersonal relational styles.

## PROBLEM-SOLVING STYLES: THE MBTI FRAMEWORK

This section explores a framework for understanding the cognitive or problem-solving styles of group members, and how these styles influence team functioning. Problem solving can be viewed as involving two dimensions: the process of gathering information, and the process of making evaluations, decisions, or judgments based on that information. Information gathering involves such tasks as finding out about a problem, interpreting the world, and inferring meaning from all the complexity which usually surrounds a given event. As Hellriegel and Slocum (3) note, information gathering also involves rejecting some data and reducing much of it to a manageable, comprehensive form. Judgment, the process by which information is evaluated and decisions are made, is the second important operation in problem solving.

A framework for conceptualizing and assessing problem-solving styles has been developed by Isabel Briggs-Myers based on Jung's theory of psychological types (4). The basic dimension of information gathering is referred to as perception and takes the two opposing forms of either sensing or intuition. Information evaluation, or judgment, takes the opposing forms of either thinking or feeling. The Myers-Briggs Type Indicator (MBTI) is a questionnaire that assesses which type of perception and judgment an individual prefers to use in problem solving. An initial categorization of problem-solving styles can be depicted as the four quadrants of Figure 30-1 (adapted from Hellriegel and Slocum (5)).

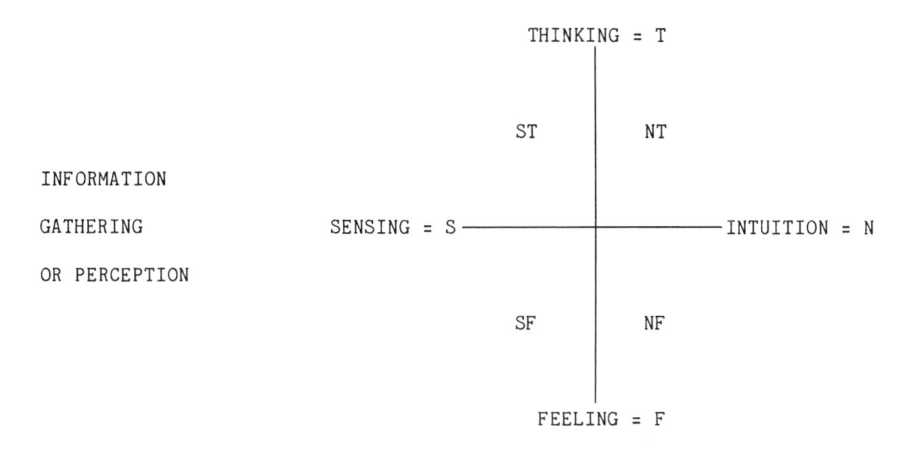


Figure 30-1. Problem-solving styles. (Adapted from Hellriegel, D. and Slocum, J. W., ''Preferred Organizational Designs and Problem Solving Styles: Interesting Companions,'' *Human Systems Management,* 1: 151–158 (1980).

### Two Ways of Perceiving: Sensing and Intuition

Sensing types (S) look at the world and see the hard data that is directly available through the senses (i.e., seeing, hearing, touching.) They tend to be realistic, pragmatic, and focused in the here and now. They prefer problems which lend themselves to standardized solution procedures. They prefer routine work and enjoy working in structured settings. They dislike unstructured problems, high degrees of uncertainty and change, and working with ''soft data.''

Intuitive (N), types on the other hand, look at the world and see not what is ''just there,'' but rather what might be. They see beyond the here and now into the realm of possibilities, implicit meanings, and potential relationships among events. Intuitives often use ''soft data,'' hunches and possibilities which come to them rather spontaneously. They also tend to see the ''big picture'' or totality rather than the fine print or details, as sensing types do. Intuitive types also enjoy complexity, unstructured problems, and newness, and are burdened by too much routine. Mary McCaulley, president of the Center for Applications of Psychological Type (CAPT, Inc.), who oversees applications and development of the MBTI, suggests that ''In a new and complex venture, a team needs more intuitive types; in a venture which requires careful management of many details, a team needs more sensing types'' (6).

### Two Types of Judgment: Thinking and Feeling

The second dimension in the model involves making decisions, and again there are two opposing approaches. Those who prefer thinking as their favorite judgment process like putting things into a structured format and using cause-effect logic. Thinking types (T) tend to be uninterested in people's feelings and seem "hard-headed" (or hard-hearted) and are able to fire or reprimand people when necessary. They enjoy applying data and formulas to problems, and are particularly attracted to the contributions made by operations research and management science to decisional problems. They like to think of management as a science, not an art. As Hellriegel and Slocum (7) note, there is considerable similarity between thinking types, the scientific method, and work in bureaucratic organizations which emphasize rationality and impersonality in human relations.

Feeling types (F) also make decisions by a rational process. As McCaulley notes, "Feeling does not refer to emotions, but rather to the process of setting priorities in terms of values, weighing their greater or lesser importance to oneself and others" (8). They tend to become sympathetic, are concerned about their own and other's reactions in decision situations, and reflect a concern for the establishment of harmonious interpersonal relations. They do not particularly like to tell others unpleasant things, and often gravitate toward mediator and conciliator roles in team situations. Whereas thinking types approach decisions with abstract true or false judgments and formal logic rules, the feeling type invokes more personalistic judgments of good or bad, pleasant or unpleasant, and like or dislike.

### The Composite Styles

Each type of perception can team up with each type of judgment, so that the quadrants of Figure 30-1 represent four basic problem-solving styles. The sensing thinkers (STs) are the traditionalists of organizational life. They can absorb, remember, and manage a large number of facts and details. They strive to create and maintain order, structure, and control; sensing thinkers are realistic, pragmatic, economically motivated. They follow through on assignments and generally represent the "quintessential organization man," in the positive sense of the term. They are good task leaders.

Sensing-feeling types (SFs) also prefer structured organizations and work, but could be characterized as "loyalists," and gravitate toward social-emotional leadership. They tend to accept people as they are, know a lot of personal things about the people with whom they work, and are sensitive to the fit between the person and the job.

Intuitive thinkers (NTs) are visionaries. They are innovative and enjoy creating a new system, rather than maintaining or running an established program. They need to be supported by persons who are good at implementation and maintenance of new programs. They feel burdened by overly structured organizations and jobs, and function best in looser organization designs.

Intuitive-feeling types (NFs) could be considered catalysts in a team setting. They focus on possibilities, and can often get other people excited about new projects, or spark new enthusiasm in the face of obstacles or current problems. They prefer democratic, less structured organizations, and are particularly interested in the quality of working life experienced by both themselves and others. Other people are likely to attribute personal charisma to NFs, because they are popular and tend to become personally involved with those around them. Sometimes, this involvement makes them vulnerable since they prefer positive harmonious relations with others. These four composite styles are shown in Table 30-1, which has been reproduced here from *Introduction to Type* (9).

**Other Aspects of the MBTI**

The Myers-Briggs Type Indicator output is somewhat more complex than the two-dimensional model shown in Figure 30-1, but the present discussion explicates a central part of the framework. In addition to providing dimensionality to problem-solving strategies, the perception and judging categories are considered as definitions of basic mental functions or processes. The functions serve to provide distinct goals for conscious mental activity. "Sensing (S) seeks the fullest possible experience of what is immediate and real. Intuition (N) seeks the broadest view of what is possible and insightful. Thinking (T) seeks rational order and plan according to impersonal logic. Feeling (F) seeks rational order according to harmony among subjective values" (10). Because these functions aim in different directions, one of them must be dominant for purposeful action by the individual, although all of them are present in any individual to some degree. The dynamic interaction of the functions provides additional substance for the MBTI framework.

Two more bipolar scales are part of the MBTI dimensions. The first is an extraversion-introversion* scale which measures complementary atti-

* Extraversion is a Jungian term describing a preference for the outer world as an attitude toward life. It is frequently equated by the layman with the similar concept of extroversion, meaning "not shy." The spelling in this chapter will be extraversion, because most of the references are in the context of the MBTI framework, and therefore refer to the Jungian scale.

Table 30-1. Composite Styles: Effects of the Combinations of Perception and Judgment.[a]

| PEOPLE WHO PREFER | ST SENSING + THINKING | SF SENSING + FEELING | NF INTUITION + FEELING | NT INTUITION + THINKING |
|---|---|---|---|---|
| focus their attention on | Facts | Facts | Possibilities | Possibilities |
| and handle these with: | Impersonal analysis | Personal warmth | Personal warmth | Impersonal analysis |
| Thus they tend to become: | Practical and and matter-of-fact | Sympathetic and friendly | Enthusiastic and insightful | Logical and ingenious |
| and find scope for their abilities in: | Technical skills with facts and objects | Practical help and services for people | Understanding and communicating with people | Theoretical and technical developments |
| for example: | Applied science<br>Business<br>Production<br>Construction<br>Etc. | Patient care<br>Community service<br>Sales<br>Teaching<br>Etc. | Behavioral science<br>Research<br>Teaching<br>Etc. | Physical science<br>Research<br>Management<br>Forecasts, analysis<br>Etc. |

[a] Adapted from Myers, Isabel B. *Introduction to Type,* Gainesville, Florida: Center for Application of Psychological Type, 1976, 2nd ed. and reproduced by special permission of the publisher, Consulting Psychologists Press, Inc., 577 College Ave., Palo Alto, Calif. 94306.

tudes toward life. An extraverted (E) individual prefers to operate primarily in the outer world of people, action, and objects. The following characteristics are typical of the extraverted attitude: "awareness and reliance on the environment for stimulation and guidance; an action-oriented, sometimes impulsive way of meeting life; frankness; ease of communication; or sociability" (11). On the other hand, the introvert (I) operates in a more contemplative, inner world of ideas and concepts. The extrovert experiences an attention flow from the individual to the environment. The intravert concentrates energy from the environment. An introverted attitude is frequently characterized by the following attributes: "interest in the clairy of concepts and ideas; reliance on enduring concepts more than on transitory external events; a thoughtful, contemplative detachment; and enjoyment of solitude and privacy" (12).

The final scale reflects whether a judging or perceiving attitude will be taken toward the outer world. Judging types (J) "tend to live in a planned, decided, orderly way, aiming to regulate life and control it. Perceptive types (P) . . . tend to live in a flexible, spontaneous way, aiming to understand life and adapt to it" (13). (For more detailed descriptions of the MBTI, the reader is referred to References 4, 8, 9, 10, 15, and 16.)

#### Implications for Team Functioning

Effective team functioning requires that the psychological and intellectual resources of all individuals be utilized. One of the greatest barriers to effectiveness is the truncation of one person's resources by another simply because they perceive and react to problems in quite different ways. Sensing-thinking types, for instance, are inclined to want the facts and an explicit statement of the problem before considering possible solutions. Intuitive-feeling types (the opposite of STs) must first be inspired by interesting possibilities before they will pay much attention to facts. And even then the "facts" seen by the NF will more likely appear as "soft" impressionistic data to the ST.

As Myers (14) notes, problem solving generally could be viewed as a sequential use of the various perception and judgment functions: sensing establishes the facts, intuition suggests possible solutions, thinking establishes the probable consequences of different courses of action, and feeling weighs the likelihood of acceptance among relevant persons. Thus all the psychological functions should ideally be brought to bear on problems, and well-balanced teams which allow broad participation should have greater resources than even the best balanced individual decision maker. In fact, leadership in a team could be defined as insuring that the appropriate functions are brought to bear on problems at the appropriate stage.

The formation of interdepartmental project teams is likely to bring together different types because there is some evidence to suggest type is related to occupational specialization (15,16). Engineering is dominated by thinking types, although the theoretical specialties tend to be more thinking oriented than the applied specialties. McCaulley notes that two important implications flow from this trend (17). First, a team heavily weighted with similar types can develop group "blind spots" characteristic of the dominant type. Thinking types often neglect the human aspect of project work because technical work is more appealing to them. Second, if feeling types are in a minority, they may be undervalued and criticized, which causes them to withdraw their vitality and energy from the team effort. Third, thinking types may be inclined to take a laissez-faire attitude toward persuading others to accept their work, believing, often erroneously, that "the logic will speak for itself," and that high-quality technical work will unquestioningly be accepted by project clients. Fortunately, as McCaulley notes, the engineering professions contain a large number of feeling types who are of great value in gaining acceptance of project work with the client-users. The leadership challenge, however, is to create a climate where thinking judgment can predominate during planning stages and feeling judgment during project implementation.

The type situation does vary quite a bit across the specialties. The *Atlas of Type Tables* (18), compiled from over 60,000 of the records submitted to CAPT for scoring from 1971 through 1982, includes classifications for many standard occupational codes. (See Table 30-2.) Within any engineering project team, and especially for teams that combine specialties, there may be a wide range of types, as indicated in the prior discussion. For the planning stage of a project, it is conceivable to combine engineers from, for instance, mining and mechanical specialties. The typical profile for each field is quite different—in fact, opposite. For mining engineers, the predominant type is ESFP. Mechanical engineers tend to be INTJ. Even within a specialty, there may be various combinations of types of

Table 30-2. Distribution of Type Among Engineers[a]

| | PREFERENCE TYPE[b] | | | | | | | |
|---|---|---|---|---|---|---|---|---|
| FIELD | E | I | S | N | T | F | J | P |
| Engineers | 47.7% | 52.3% | 52.6% | 47.4% | 63.6% | 36.4% | 60.5% | 39.5% |
| Engineering Specialty | | | | | | | | |
| Aeronautical | 50.0% | 50.0% | 42.6% | 57.4% | 40.7% | 59.3% | 53.7% | 46.3% |
| Chemical | 50.0% | 50.0% | 53.9% | 46.1% | 71.2% | 28.8% | 78.9% | 21.1% |
| Electrical/ Electronic | 37.0% | 63.0% | 51.9% | 48.1% | 66.7% | 33.3% | 63.0% | 37.0% |
| Mechanical | 46.8% | 53.2% | 58.4% | 41.6% | 70.1% | 29.9% | 62.3% | 37.7% |
| Mining | 53.2% | 46.8% | 54.2% | 45.8% | 25.8% | 74.2% | 44.2% | 55.8% |
| Engineering and Science Technicians | | | | | | | | |
| General | 48.0% | 52.0% | 55.4% | 44.6% | 57.1% | 42.9% | 56.0% | 44.0% |
| Electrical/ Electronic | 42.1% | 57.9% | 56.1% | 43.9% | 68.4% | 31.6% | 63.2% | 36.8% |

[a] Adapted and compiled from Macdaid, Gerald P., McCaulley, Mary M., and Kainz, Richard I., *Atlas of Type Tables*, Gainesville, Florida: Center for Application of Psychological Type © 1986 (1987) and reproduced by special permission of the publisher, Center for Application of Psychological Type, P. O. Box 13807, Gainesville, Florida 32604.

[b] E = extraversion — I = introversion
S = sensing — N = intuitive
T = thinking — F = feeling
J = judging — P = perceiving

individuals. For example, in an engineering project, there may be all engineers in the planning stage of a project. When the project moves to the implementation stage, technicians may be also a part of the staff. In electrical engineering, these profiles are quite similar. Both groups exhibit dominant preferences for ISTJ. However, for the field of engineering in general, the profile preferences indicate that engineers exhibit a preference for Introversion, while engineering technicians exhibit a preference for Extraversion. This may cause a change in the composition of the group, and alter the ability of the group to function at all. The relationship of type to group development is discussed in a later section of this chapter.

For the past six years, McCaulley and her associates at CAPT have participated in a longitudinal study of a consortium of engineering schools concerned with the constructive uses of individual differences in the educational setting for engineers. Various institutions have different "typical" student profiles. Recent analysis of the panel data indicates that some types of students who initially majored in engineering either transferred to another major or left the institution. McCaulley comments in the six year follow-up:

> Engineering has and will continue to have a challenge to engage students whose minds work in a linear fashion (S) and those who see patterns more easily. You seem to be losing substantial numbers of the types who can do engineering well—the practical, hands-on linear thinkers. . . . Many of [the fast-moving extraverts] are transferring out of engineering. . . . Engineering . . . attracts large numbers of hands-on practitioners and theoretical visionaries. It is essential . . . to develop mutual respect and communication so that the skills of both—who see their world so differently—can be used constructively. (19)

The challenge to institutions which prepare students for engineering continues for the industries in which they will subsequently work. (Additional information about the types in engineering can be found in references 20 through 26.)

In summary, our purpose in this section has been to introduce the MBTI framework in its simpler form and to suggest how it can be useful in understanding conflict between team members. Individual team members must come to see others as sources of differing resources, or "differing gifts," rather than as sources of antagonism. Table 30-3 illustrates how complementary functions can be supplied by opposite types to the benefit of both (adapted from *Introduction to Type*) (27). This process could be greatly facilitated by having either an external or internal organization

Table 30-3. Mutual Usefulness of Opposite Types[a]

| INTUITIVE NEEDS A SENSING TYPE | SENSING NEEDS AN INTUITIVE TYPE |
|---|---|
| To bring up pertinent facts | To bring up new possibilities |
| To apply experience to problems | To supply ingenuity on problems |
| To read the fine print in a contract | To read the signs of coming change |
| To have patience | To have enthusiasm |
| To keep track of essential details | To watch for new essentials |
| To face difficulties with realism | To tackle difficulties with zest |
| To remind that the jobs of the present are important | To show that the joys of the future are worth working for |
| **FEELING NEEDS A THINKING TYPE** | **THINKING NEEDS A FEELING TYPE** |
| To analyze | To persuade |
| To organize | To conciliate |
| To find the flaws in advance | To forecast how others will feel |
| To reform what needs reforming | To arouse enthusiasms |
| To hold consistently to a policy | To teach |
| To weight "the law and the evidence" | To sell |
| To fire people when necessary | To advertise |
| To stand firm against opposition | To appreciate the thinker |

[a] Adapted from Myers, Isabel B., *Introduction to Type,* Gainesville, Florida: Center for Application of Psychological Type, 1976, 2nd ed. and reproduced by special permission of the publisher, Consulting Psychologists Press, Inc., 577 College Ave., Palo Alto, Calif. 94306.

development specialist administer and conduct educational seminars for teams using the MBTI. It is an instrument which is gaining increasing application in the behavioral sciences to a variety of management problems.

## EMOTIONAL STYLES: THE FIRO FRAMEWORK

Whereas the prior section examined a framework for understanding individual problem-solving styles, this section presents a framework for understanding emotional styles in a group. It is important to recognize that the location of the problem in most so-called personality conflicts does not reside solely in one person. Most social conflicts are inherently relational; that is, a problem does not exist until two or more persons have to work together, or live together, etc. The dysfunction then is typically not located in one person or the other, but rather is located in their relationship. Thus any framework which systematically attempts to explain personality conflict must in fact be a theory of interpersonal relationships. The framework used in the present study was developed by Schutz (28), and is both concise and operational in the sense that it provides for a method of assessing the degree of potential interpersonal strife or incompatibility in any relationship.

### Three Basic Interpersonal Needs

In the following discussion, interpersonal incompatibility will be used as synonymous with "personality conflict," whereas interpersonal compatibility suggests harmony and lack of conflict. The basis of Schutz's theory is the individual's fundamental interpersonal relations orientation, or FIRO as it is usually abbreviated. One's FIRO is an "interpersonal style" which is hypothesized to be rather stable and to have developed from psychological forces in the person's childhood and developmental history. It reflects a person's central emotional position with respect to other people generally. That is, people learn a way of relating to others along certain dimensions, and they tend to carry that style around with them as a rather stable aspect of their personality which affects their work and social relations. The FIRO is in fact a set of three basic interpersonal needs which are common to all persons in greater or lesser degrees. These three needs (inclusion, control, and affection) are considered to be predictive in a general sense of the fundamental behavior that occurs interpersonally. Inclusion refers to the need to be included in other people's activities, or to include others in one's own activities, and is analogous to the extraversion-introversion dimension of other authors, or to sociability. It entails moving toward or away from people psychologically. Control refers to the need to give and receive structure, directions, influence, power, authority, and responsibility, to the need to engage in leadership or followership. Affection is concerned with emotional closeness to others, friendship, liking, or disliking, and refers to the need to act close or distant toward others.

There are two aspects to each of the three interpersonal needs. One is what we do or have a need to express toward others. The second is how we want others to behave toward us. This is shown schematically in Table 30-4. That is, people have a need to both give and receive in each need area and this forms the basis for interpersonal harmony or strife. Harmony (compatibility) results when one party has a need to give (or express) what the other party is interested in getting (or wants).

### Scaling Group Interpersonal Incompatibility

If we symbolize the need to express behavior as "e," the need to receive from others as "w," and the three need areas of inclusion, control, and

Table 30-4. The FIRO Framework.

| | NEED AREA | | |
|---|---|---|---|
| | INCLUSION | CONTROL | AFFECTION |
| e[a] | eI<br>Need to initiate interaction with others—need to reach out and include others in one's activity | eC<br>Need to assume leadership, responsibility, control and exert influence | eA<br>Need to act close and personal toward others—express friendship |
| w[b] | wI<br>Need to be invited to join others—need to be included in interaction | wC<br>Need to receive directions, guidance, assume followership roles, receive influence | wA<br>Need to be on the receiving end of friendship and personal closeness |

[a] What I need to express to others is symbolized by e.
[b] What I need or want from others is symbolized by w.

affection as I, C, and A, then any individual can be characterized by the six scales: eI, wI, eC, wC, eA, wA. Schutz has developed a questionnaire, referred to as FIRO-B, which is designed to measure an individual's need levels in each of the six categories (the "B" indicates the questionnaire is designed to predict behavior). The six categories are measured on a scale from a low of zero to a high of nine. This scaling provides a way of assessing the potential conflict or incompatibility in an interpersonal relationship.

Schutz has developed methods for scaling the degree of potential interpersonal incompatibility in a group by combining the FIRO-B scores of all individuals in the group according to certain formulas. The control dimension will be used to explain these formulas since control is often at the heart of difficulties in work settings. The examples shown below are based on Pfeiffer and Heslin's (29) work with the FIRO-B instrument in human relations training.*

| | Supervisor | Subordinates |
|---|---|---|
| Need to express (eC) | low score | low score |
| Need to receive (wC) | low score | high score |

* Used with permission and reprinted from J. W. Pfeiffer, R. Heslin and J. E. Jones, *Instrumentation in Human Relations Training* (2nd Ed.) San Diego, Calif.: University Associates, 1976.

Subordinates: "Boss, we're ready to go, tell us what you want us to do."

Supervisor: "Look over the situation, and do whatever you think is best."

Pfeiffer and Heslin report a situation where the physical education department of a college had this constellation of scores. The employees would say, "Fred, what do you think we should do with the intramural program this year?" Fred would usually respond, "I don't care fellows. Do whatever you want." Needless to say, this caused employee frustration and low performance since the employees wanted (and needed) influence from others to be effective.

The reverse case also occurs:*

| | Supervisor | Subordinates |
|---|---|---|
| Need to express (eC) | high score | high score |
| Need to receive (wC) | low score | low score |

Subordinates: "Boss, let us do it our way and we'll give you the best sales department in the country."

Supervisor: "You'll do things the way I say to do them."

This relational constellation represents competitive, aggressive incompatibility wherein it is difficult for the parties to share control and leadership. Since the supervisor has formal authority and can invoke various forms of punishment for noncompliance, the subordinates often lose interest in work and withdraw their vitality from the job. On the other hand, these subordinates may actively express their anger in the form of sabotage and other counterproductive maneuvers.

Schutz's concept of originator compatibility reflects the degree to which one person's excess of a need to express or receive in a given need area is balanced by the other person's excess in the reverse direction; for example, if one party's e score is greater than the w score, the other party's w score should be greater than the e score so that one difference counterbalances the other. The originator compatibility formula symbolized as OK for person i and person j is given in equation 1.

$$\text{Originator compatibility: } OK_{ij} = (e_i - w_i) + (e_j - w_j) \qquad (1)$$

The score, therefore, would range from 0 to 18. For example, returning to the extreme case illustrated earlier of a competitive, incompatible relationship around control, the two persons, i and j, would be

* Ibid.

| | i | j |
|---|---|---|
| need to express (eC) | 9 | 9 |
| need to receive (wC) | 0 | 0 |

$$OK_{ij} = (9 - 0) + (9 - 0) = 18, \text{ perfect incompatibility}$$

Two persons who are compatible around initiation-reception might be

| | i | j |
|---|---|---|
| need to express (eC) | 7 | 2 |
| need to receive (wC) | 2 | 7 |

$$OK_{ij} = (7 - 2) + (2 - 7) = 0, \text{ perfect compatibility}$$

Person i's propensity to initiate control is directly counterbalanced by person j's propensity to receive control, and hence they would be compatible with i initiating and j following. There must be some complementarity, reciprocity, or "oppositeness" in the relationship for originator compatibility to exist. Schutz (30) suggests this aspect of relationships reflects the old maxim "opposites attract."

Another maxim seemingly contradictory to the first about interpersonal relationships is that "birds of a feather flock together," or similarity attracts. The second type of compatibility can be used to assess this dimension of relationships. It is called interchange compatibility (symbolized as IK) and refers to whether two persons have a need to be similarly active in a given need area. Incompatibility results where two persons are very different in how much interchange (activity) they prefer in the various need areas.

For inclusion, the typical conflict is between the joiner or participator (high interchange) who likes to be surrounded by people doing things together, and the more withdrawn person who prefers to work alone (low interchange). In the control area, the conflict is between those who like to create a system of rules and clearly delineated roles (high interchange) versus those who prefer to "live and let live" in a more permissive, unstructured atmosphere (low interchange). For affection, interchange incompatibility reflects a conflict between those who want to be close, personal, and confiding versus those who prefer cooler, more distant, and nondisclosing relationships. Interchange compatibility between persons i and j is given by the following formula

$$\text{Interchange compatibility: } IK_{ij} = [(e_i + w_i) - (e_j + w_j)] \qquad (2)$$

Interchange compatibility also ranges from 0 (most compatible) to 18 (least compatible). In our three prior examples, all pairs would be interchange compatible in the sense that within each pair there would be an interest in the same amount of "general control activity" even though not all cases were originator compatible. Thus people may be compatible in one sense, but not the other. Below is an example that would be originator compatible, but not interchange compatible:

| | i | j |
|---|---|---|
| need to express (eC) | 9 | 1 |
| need to receive (wC) | 8 | 2 |

$$OKij = (e_i - w_i) + (e_j - w_j) = (9 - 8) + (1 - 2) = 0, \text{ compatible}$$

However,

$$IKij = (e_i + w_i) - (e_j + w_j) = 17 - 3 = 15, \text{ incompatible}$$

Thus, person j would usually want to work in an unstructured fashion, whereas person i would be very concerned about establishing order, roles, rules, and leader-follower behavior. Control would be a source of friction for this pair. However, if person j could accept the necessity for leader-follower behavior he or she would likely be the follower in relation to person i. It is no paradox, incidentally, to be high in both expressed and wanted control. The ideal military officer, for instance, must take orders from above (be high on wanted control) and turn around and give orders to those below (be high on expressed control). In short, individual i in the last example would be comfortable both as a leader and as a follower.

There is another type of compatibility, referred to as reciprocal compatibility, which is rather similar to originator compatibility and will not be discussed here. Interested readers are referred to Schutz (31), Pfeiffer and Heslin (32), and Hill (33) for further information on reciprocal compatibility.

The typical data for a group is displayed in the format shown in Table 30-5, where each compatibility type is shown for each need area. In addition, the rows and columns can be averaged to illustrate compatibility by need area or type. Finally, the entire matrix can be averaged to assess the total group compatibility. Inspection of this matrix can signal the source and nature of potential conflicts in a group. Schutz indicates that the most frequently occurring types of incompatibility which affect work group functioning usually center around inclusion interchange, control interchange, and control origination.

Table 30-5. Compatibility Types and Areas.

| | | AREAS OF COMPATIBILITY | | | |
|---|---|---|---|---|---|
| | | INCLUSION | CONTROL | AFFECTION | ROW AVERAGES |
| Type of compatibility | O[a] | OK(I) | OK(C) | OK(A) | Overall OK |
| | I | IK(I) | IK(C) | IK(A) | Overall IK |
| | Column average | Overall K, Inclusion | Overall K, Control | Overall K, Affection | Average K, entire matrix |

[a] O = originator compatibility.
I = interchange compatibility.

Inclusion interchange problems in project teams are usually expressed as conflicts between members who prefer to work in groups (high-inclusion activity) versus those who prefer to work more individualistically (low-inclusion activity). The differences manifest themselves in the group setting, because, whenever a decision must be made, those high on inclusion prefer working it out together, whereas those low on inclusion attempt to go off on their own in pursuit of the decision.

Control interchange problems tend to pervasively affect a group. While high-control group members attempt to create a group structure and definite roles and responsibilities, those low in the control area will be resisting the high-control persons and attempting to create an unstructured group with freedom to play different roles at different times as they see fit.

Control originator compatibility problems, like many interpersonal conflicts, are also expressed through the task all too often. If someone makes a suggestion in a group meeting, it is improbable that someone else would say, "Your suggestion is okay, but I want mine accepted because I want to have the most influence here." The person is more likely to say, "Your idea is okay, but these are its disadvantages, and I think we should do so and so." The merits of suggestions are often secondary to the deeper agenda of who will have how much control over the decision process. This competitive incompatibility often produces a general feeling of struggle and conflict in the group. Apathetic incompatibility, on the other hand, usually produces feelings of emptiness and boredom with the group process.

Affectional conflicts are perhaps less frequently felt in work settings. However, a high-interchange person in a low-interchange group would probably complain of an impersonal group in which there was not enough encouragement and support.

### Using the FIRO-B to Enhance Team Performance

The FIRO system then can be used to conceptualize interpersonal strife in any work group or team which must accomplish some objectives. Probably its greatest value lies in diagnosing and defining the kind of conflict likely to occur in a group. If the FIRO-B questionnaire is administered to a project team, a profile can be obtained of the potential points of friction, and the manager as well as the entire team can then be more informed as to the likely origin and nature of conflict in the team. This procedure would probably best be carried out with the assistance of a staff specialist, and in an open manner wherein the results were fed back to the entire team and the meaning and nature of the FIRO system fully explained. In short, all of the usual organizational development group rules regarding survey feedback (i.e., voluntary participation, disclosure of results) would ideally be adhered to, and the feedback effort itself would become an intervention to facilitate team development. The FIRO system has been used to select submarine crews and police teams, and for personnel placement in various industries.

If the FIRO-B instrument is used to help a team see its conflict areas more clearly, and deal with them more constructively, there are several uses which Pfeiffer and Heslin (34) suggest as relevant for group and individual development. These uses are reproduced below:

> Generating a Personal Agenda. Giving the FIRO-B scale early in a training session can provide participants with insights into their inclusion, control, and affection desires and behavior which they may wish to modify or change by trying out new behaviors within the group setting.
>
> Sensitization to Interpersonal Dimensions. Scoring and discussing the FIRO-B can make participants aware of dimensions of interpersonal relations with which they will be dealing during a training session. It introduces terminology for understanding inclusion, control, and affection problems.
>
> Checking Self Understanding. Administering the FIRO-B can be preceded by asking members to estimate how they expect to score (high, medium, or low) on each scale. If the group has been in existence long enough, members can also predict how they expect the other members to score on the instrument. FIRO-B is not a deceptively-worded instrument, so pre-awareness should have little effect on the respondent's scores.
>
> Individual Interpretations. The FIRO-B can be given in a group followed by a general discussion of the subscales. Later the facilitator can

meet with members individually to interpret each person's pattern of scores in detail and discuss how this feedback affects the individual's understanding of his or her past and future group behavior.

The upshot of most group interventions with the FIRO-B is that it serves to sensitize the members to, and to educate them about, the nature of interpersonal conflict. It also serves as a stimulus for an open discussion of problem areas within the group, which, when attended by a competent facilitator, can enable a group to work through the problem of interpersonal agendas contaminating the productive efforts of the group.

## RELATIONSHIPS BETWEEN THEORETICAL FRAMEWORKS

The discussion so far has presented the FIRO-B and the MBTI as separate theoretical frameworks for assessing differences in the human personality, but the two are not unrelated. Indeed, many of the theoretical frameworks used to measure personality differences are related in some ways to each other. They do not measure the same constructs entirely, but the correlated measures, which indicate some overlap, help to illuminate the complexities of the human personality and relationships in the human group. Table 30-6 relates the MBTI and the FIRO-B frameworks to each other. Not surprisingly, the emotional stance measures of the FIRO-B are strongly related to the Feeling preference of the TF dimension of the MBTI, the measure of which kind of judgment to rely upon when making a decision. Also, the high Inclusion scores of the FIRO-B are associated with the Extraversion preference of the MBTI.

Multiple pictures of individuals' personality characteristics can be used together to help group functioning. Obviously, the process is complicated by using multiple frameworks, but the additional information is potentially of great value. The nature of the project and the project team would affect a firm's decision to make the time and money commitment to employ the exploration of multiple personality frameworks to enhance group performance. Such factors as sensitivity to production delays, anticipated disastrous consequences if an intact project team disbands prematurely, or the need for a project planning task force to begin work quickly could make such an investment attractive to a firm. Again, using multiple instruments can be enhanced by having either an external or internal organization development specialist administer and conduct educational seminars for teams using these frameworks. Additional frameworks, some of which require the intervention of a trained psychologist, that might be of use are presented here. Table 30-7 provides correlations between some of them and the MBTI.

Table 30-6. Relationship of MBTI Continuous Scores with the FIRO-B.[a]

| FIRO-B SCALE | EI | SN | TF | JP |
|---|---|---|---|---|
| Expressed inclusion | E[b] | | | |
| Wanted inclusion | E | | F | |
| Expressed affection | E | | F | |
| Wanted affection | E | | F | |
| Expressed control | | | T | |
| Wanted control | | S | | |

[a] Adapted and compiled from Myers, Isabel B. and McCaulley, Mary H., *Manual: A Guide to the Development and Use of the Myers-Briggs Type Indicator*. Gainesville, Florida (1985) and reproduced by special permission of the publisher, Consulting Psychologists Press, Inc., 577 College Ave., Palo Alto, Calif. 94306.

[b] The values in the table indicate that the scales are correlated at beyond the .001 level. Since the Jungian preferences are bidirectional, a strong negative correlation is indication of relationship to one preference, a strong positive correlation is indication of relationship to the other. For example, in the FIGO-B comparison, the Expressed inclusion scale is correlated significantly with the MBTI Extraversion scale. Individuals who tend to score high on one of the scales also tend to score high on the other.

*Study of Values* (35) rank orders the importance of six categories of values (Theoretical, Economic, Aesthetic, Social, Political, and Religious) for the individual. The relationship of this measure to the MBTI is included in Table 30-7.

*Edwards Personality Preference Schedule* (36) includes measures for such categories as Achievement, Order, and Affiliation.

*Vocational Preference Inventory* (37) measures six basic vocational interest scales: Realistic, Investigative, Artistic, Social, Enterprising, Conventional. The relationship of selected scales in this measure to the MBTI is included in Table 30-7.

*Strong Campbell Interest Inventory* (38) provides scales for Holland's scales, 23 general occupational themes and over 200 specific occupations. The relationship of selected scales in this measure to the MBTI is included in Table 30-7.

*SYMLOG* (39, 40, 41) measures overt group actions as well as individual values along three dimensions: Friendly versus Unfriendly, Emotionally expressive versus Instrumentally controlled, and Dominant versus Submissive.

Table 30-7. Relationship of MBTI Continuous Scores with Other Scales[a]

| STUDY OF VALUES (35) | EI | SN | TF | JP |
|---|---|---|---|---|
| Theoretical | | N[b] | T | |
| Economic | | S | T | |
| Aesthetic | I | N | | P |
| Social | | | F | |
| Political | E | S | | |
| Religious | | | F | |
| **VOCATIONAL PREFERENCE INVENTORY (37) (SELECTED OCCUPATIONAL THEMES)** | **EI** | **SN** | **TF** | **JP** |
| Investigative | | N | | |
| Artistic | | N | | |
| **STRONG CAMPBELL INTEREST INVENTORY (38) (SELECTED BASIC INTEREST SCALES)** | **EI** | **SN** | **TF** | **JP** |
| Mechanical Activities | | | T | |
| Business Management | E | | | |
| Art | | N | | |

[a] Adapted and compiled from Myers, Isabel B. and McCaulley, Mary H., *Manual: A Guide to the Development and Use of the Myers-Briggs Type Indicator.* Gainesville, Florida (1985) and reproduced by special permission of the publisher, Consulting Psychologists Press, Inc., 577 College Ave., Palo Alto, Calif. 94306.

[b] The values in the table indicate that the scales are correlated at beyond the .001 level. Since the Jungian preferences are bidirectional, a strong negative correlation is indication of relationship to one preference, a strong positive correlation is indication of relationship to the other. For example, in the Vocational Preference Inventory comparison, individuals who tend to score high on the Investigative scale also tend to score high on the N dimension of the MBTI scale.

*Kolb Learning Style Inventory* (42) provides scores along three dimensions of preferred style of learning: Concrete versus Abstract, Experiential versus Conceptual, Reflective versus Active.

*California Psychological Inventory* (43) measures such traits as Dominance, Tolerance, Flexibility, and Extraversion-introversion.

*Omnibus Personality Inventory* (44) measures attributes such as Artistic sensitivity, Likes being with people, Impulsive or imaginative, and Likes applied, concrete activities.

*Comrey Personality Scales* (45) include such measures as Extraversion versus Introversion and Orderliness versus Lack of compulsion.

## IMPLICATIONS FOR TEAM BUILDING AND DEVELOPMENT

The discussion thus far has assumed that a group exists, and that understanding the individual differences in a group, the potential for conflict, will help the group to function more efficiently and effectively. Certainly, in industry, the extant group is the norm, but in firms which deal with turbulent environments or deal with rapidly changing production needs or planning duties, the project team may be an ad hoc phenomenon. Further, it may be subject to a matrix organization. These groups need not only to function, but also first to become groups instead of a collection of individuals if they are to perform most effectively and efficiently. As a group moves through various stages of development, from inception through productive functioning to disbandment, different aspects of the individuals involved in the group may variously cause or ease conflict in stage transitions.

As a work group or project team develops, it progresses through several stages: forming, storming, norming, and performing (46, 47, 48, 49). Each stage is typified by a major dilemma to resolve. Appropriate (successful, in some sense) resolution of the dilemma is necessary before moving to the next stage. It is possible for a group to remain plateaued at any of the stages. Not surprisingly, a major factor in group development is the ability to handle conflict generated by the dilemmas. This section discusses each stage and suggests some of the characteristics which might be most relevant, the ones to which a group member or group leader should be most sensitive.

Forming, the initial group stage, includes resolving the conflicting demands of the individual's self versus the external pressures. In the case of the project group, these external demands are often such things as project schedules and deadlines or organizational goals and standards. This dilemma is labeled inclusion and consists of setting group boundaries and goals (50). In this stage, it is common for group members to exchange their opinions of what is expected of them. In the MBTI framework, for example, STs will be impatient to get on with the task in this stage, while NFs will want to explore all the possibilities open to them before taking any action at all. In the FIRO-B framework, as another example, members high on wanted affection will be concerned with first impressions and how people react to each other. On the other hand, project members low on wanted affection will be concerned with getting to the task at hand. See the potential for conflict? Further, if group members who need to have possibilities explored and feelings affirmed are not vocal or strong enough to have these personal needs met, they are unable to progress to the next stage. If the "group" progresses without dealing with these needs, the group is fragmented into subgroups, and effective functioning is impaired.

When the group proceeds to the storming stage, new tasks bring out new aspects of the individuals who make up the group. Determining the authority structure within the group is the major task at this stage (51). If the project team is an ad hoc group, perhaps drawn from different functional specialties, different people will be working from different agendas. In the framework of the FIRO-B, individuals high on expressed control and high on wanted control will want to be in charge, or have the vice president of their own division acknowledged as the real "final authority" on the project. Those who are high on wanted control and low on expressed control will be content as long as somebody is designated "head." In the MBTI framework, SFs are likely to consider the way it was done the last time, while NFs will want to find a new, better way to resolve this authority situation.

Norming, the third stage of group development, is concerned with agreeing on an appropriate level of intimacy for the project group (52). There must be mutual agreement on issues such as the shared level of trust, the mix of personal and professional issues, and whether contributions are to consist of individual effort or of consensualized group work, if the project team is to perform optimally. In this stage, if the norm is set for impersonal, work-only items, individuals who are F in the MBTI preference for TF may experience isolation and complain of a group that is not sensitive to personal needs. On the other hand, T-preference individuals will be delighted with the resolution of the stage dilemma in this "no nonsense" way. The potential for conflict looms large as one remembers that engineering groups are typically split along this dimension, with a mix of T and F types. A similar scenario develops for the FIRO-B scale of expressed affection.

Once these stages have been successfully negotiated, the group is considered a performing group. In the real world, many groups never acknowledge or resolve these early developmental stages, and, consequently, try to perform without successful completion of a prior stage. The project team output may be good, but it could probably be better if these group stage needs were successfully met. The literature suggests a final group stage, that of separation and termination, to provide some sort of closure to the life of the group (53). Without this, some workers are not able to release an involvement in prior group experience and to proceed to the next project group situation. A nagging sense of something left undone may interfere with the ability to engage in the next project for the MBTI J type. Because workers tend to anticipate future group experiences based on their prior project team efforts, an experience in which the stages are not successfully negotiated may make a good engineer reluctant to contribute further in a project team setting.

## MANAGING CONFLICT CONSTRUCTIVELY*

The previous discussion of differences in the individuals who work in project teams suggests that moderate levels of conflict are inevitable, and desirable, particularly if people are able to constructively utilize conflict. The key is effective conflict management by supervisors and members of the team. Note the key word here is conflict management rather than resolution. Probably the majority of social conflicts are not amenable to complete resolution, and it is probably more realistic to think in terms of managing conflict for productive purposes rather than resolving it. Hill (54) studied the characteristics of project team leaders employed in a large oil company who were selected by their organization as being outstanding project managers, and compared them to a sample of average managers in terms of the leadership practices which distinguished the two samples. The descriptions below (55) borrow heavily from the results of these interviews.

There seem immediately to be two general aspects in which high-performing managers differed from the lower-performing in terms of responding to internal team conflict. First, the high performers reflected a much larger repertoire of responses. They simply had more ideas and choices about how to deal with conflict generally. Second, they seemed much less afraid of disagreements, and intimated much more willingness to approach conflict rather than avoid it. This latter point is a common theme in management literature and has been noted by other authors as a preference for confrontation rather than withdrawal as a conflict-handling mode. The lower-producing managers had a more prevalent feeling that conflict would "go away" if left unattended. With these general differences in mind, the next question became, "What specific behavior did the higher-performing managers report which distinguished them from their lower-performing counterparts?"

---

* Parts of this section are reprinted from "Managing Interpersonal Conflict in Project Teams" by Raymond E. Hill, *Sloan Management Review,* Vol. 18, No. 2, by permission of the publisher. 

## Personal Absorption of Aggression

Being willing to hear subordinates out when they are particularly disturbed by a peer was a common theme. One manager described a situation in which two subordinates were making life rather miserable for a third and had essentially rejected this third subordinate. When confronted about their behavior, one of the two team members launched a brief personal attack on the manager. Instead of counterattacking, the manager simply asked the subordinate involved what was really wrong, as it appeared that some hidden agenda was more responsible for the anger. The subordinate declined to answer, and abruptly walked away. At this point, the manager felt some concern about losing the respect of the two subordinates. However, the next day, they both came to the manager's office, apologized, and explained their feelings that the third subordinate did not take enough initiative and do an appropriate share of work, which subsequently left them carrying most of the load. Whereas the manager had originally feared loss of respect, it now appeared that perhaps the reverse was true. Equally important, the manager had started a process of owning up to interpersonal antagonisms which could then be worked on with future benefits to team functioning.

If a manager does not flinch in the face of negative interpersonal feelings, and accepts them as a normal part of working life, differences between people are viewed as legitimate and their expressions are not inhibited. This is closely related to the next differentiating characteristic.

## Encouraging Openness and Emotional Expression

Interpersonal relationships as well as leadership behavior have long been characterized by at least two fundamental dimensions: instrumental and expressive behavior. The higher-performing managers seemed more concerned with how their subordinates felt about work, the organization, their peers, etc., and reported more initiative in attempting to allow expression of those feelings. More of the high-performing managers claimed to have an "open-door" policy in which subordinates were free to speak with them anytime. However, there was much more to the picture than just a manager sitting passively in his or her office with the door open. One manager, who initiated conversation with subordinates "anytime" and frequently, commented, "I don't like to be in the dark about what's going on out there or what people are thinking."

In addition to encouraging expression and being employee-centered directly, a more subtle difference seemed to characterize the high-performing managers as a group. It seemed that they simply enjoyed social interaction more than their less effective counterparts. Although there were exceptions, as a group they talked more enthusiastically, spontaneously, and longer during the interviews. Comparison of the magnitude of the total interpersonal needs on the FIRO-B scale for the high- and low-performing project managers showed that the higher-performing managers averaged 27.2 whereas the lower group averaged 22.9. The scale, a combination of the six subscales, would run from a low of 0 to a high of 54. This difference was not statistically significant at the usual 5% level but was in the direction expected from clinical observation—the higher performers reflected a greater propensity for interpersonal activity.

### Norm Setting, Role Modeling, and Counseling

One of the most fascinating aspects of the study involved managers who, in essence, "taught" their subordinates how to cope with interpersonal conflict in productive ways. Several of the high-performing managers felt it was important for them to "set an example" when it came to reacting to personality clashes. They felt it was more legitimate for them to urge a subordinate to listen to his or her emotional rival with more understanding if they in fact did that themselves. One manager noted that "Effective supervisors teach others to listen by doing it themselves. Some analysts have trouble listening . . . they keep talking when it is inappropriate. A good boss will be emulated, though, and find that is one of the best ways to get across an idea on how to behave."

An interesting correlate of this process was the observation that often a peer would intercede and act out a third-party conciliation role much like the manager might normally perform. Thus two parties in conflict would find themselves the target of peer pressure to live up to a norm which involved at least trying to understand the other party's point of view. At the same time, each party would also be likely to find other peers who tried to be impartial, but reassuring that it was okay to feel hostility. The norm seemed to be one of acceptance of conflict rather than suppression, and was apparently felt by members of high-producing groups more often than lower-producing teams.

Other comments which reinforced the idea that supervisors served as role models included the observation that managers set the "climate" in the group, and that if conflict was handled poorly in a group, it was usually because people did not feel free to "open up" in front of the supervisor. In fact, one manager observed that many groups seem to take on the personality characteristics of the supervisors. Of course, it is not the "group" which takes on the manager's characteristics, but rather the individuals who comprise it. Lower-producing managers seemed to verbally encourage openness with admonishments about the value of keeping people informed, but did not report as many instances where they actually practiced it themselves or taught it by example.

### Awareness of the Utility of Conflict

The higher-producing managers seemed to more frequently evidence the attitude that conflict could be harnessed for productive ends. One high-producing manager noted that "You have to break people in to the idea that conflict does not have to be personally destructive, but can be important toward task accomplishment. . . . I try to encourage freedom of expression, and consensus on issues with my team." On the other hand, the lower-producing managers seemed to speak more frequently of the disruptive effects of conflict.

### Pacing and Control of Potential Conflict

While the prior factors suggested a pattern of high-producing managers confronting differences, they also intimated a sense of when to do just the opposite. There were cases when high-producing managers delayed face-to-face group meetings because they felt two rival members were on the edge of acrimonious outbursts. The higher-producing managers seemed more willing to stop work and to socialize with two or three persons over coffee, and on occasion would take the entire team out to lunch as a way of getting away from work pressures. In fact, it seemed that informal work stoppages were more frequent during periods of high work stress such as deadlines and project phaseout. Sometimes, however, the process was more formal and involved allowing team members time off from work (with no pay penalty) because they had recently put in a large amount of overtime. People were becoming exhausted and tempers were getting short. The extreme of this general containment strategy involved removing people from teams; only one high-producing manager had actually done this, although others reportedly threatened it on rare occasions.

The important aspect of pacing and control of conflict as a coping strategy was that high-producing managers seemed to be in close enough touch with team members that they could judge whether it was appropriate to approach or to avoid conflict. The lower-producing managers did not exude the same sense of relatedness to subordinates and interpersonal sensitivity.

### Summary of the Effective Project Manager

The composite picture of a high-producing manager which emerged was one who "came on straight" with subordinates and who was open in dealing with their conflicts, who also encouraged subordinates to express their problems, and signaled to all concerned a tolerance of negative and hostile feelings. High-producing managers also "taught" their team members through example as well as direct counseling how to respond to conflict. This appeared to be a critical phenomenon since it apparently expanded the conflict-managing capacity of the entire team. While exuding a belief in approaching conflict, the high-producing manager also had enough knowledge of subordinates to know when to avoid conflict, and postponed meetings or confrontations when necessary. The higher-producing managers tended to play out a third-party conciliation and interpersonal peacemaking role. It is important to reemphasize, however, that these actions were taken primarily in response to what was perceived to be personality clashes rather than disagreement over substantive issues (even though they are often difficult to separate). As Walton notes:

> The distinction between substantive and emotional issues is important because the substantive conflict requires bargaining and problem solving between the principals and mediative interventions by the third party, whereas emotional conflict requires a restructuring of a person's perceptions and the working through of feelings between the principals, as well as conciliative interventions by the third party. The former processes are basically cognitive; the later processes more affective.(56)

The particular kinds of third-party conciliative roles involved several actions. First, empathic support and reassurances that hostile feelings are accepted in someone's eyes is important in getting parties to express real differences between themselves. Second, helping parties express their differences by patient listening is crucial to the management or resolution of them. As Walton (57) suggests, differentiation puts a certain reality and authenticity into the relationship of the principals to the conflict. In addition, it provides information as to opinions and attitudes in the relationship which can be checked and corrected as to accuracy. In short, an expressional function is critical to interpersonal conflict because a person cannot begin productive resolution of differences until he or she is clear what the real differences in fact are. In addition, under stress one usually has to be emotional before being rational. Third, superior knowledge of the principals' situation and feelings helped the higher-producing managers pace the confrontation of conflict. Confrontation per se is not universally a panacea for conflict management, but rather confrontations in which the principals can exhibit a modicum of rationality and problem-solving behavior are what is needed. Fourth, so-called counseling tended to place the manager in the role of an interpersonal process consultant.

There are some crucial limitations on the effectiveness of organizational superiors as third-party conciliators of subordinate conflict. Walton (58) suggests that effective third-party consultants should not have power over the fate of the principals, and should also be neutral as to the substantive outcome. This is rarely if ever approximated in most organizational settings. However, the fascinating aspect of the study results suggested that team member peers often acted out third-party conciliator roles by modeling and identification with their manager. Peers usually have no formal power over the fate of the principals to the conflict, and are potentially able to be more neutral as to the outcome. Thus, peer members of a conflict pair often supplied a third-party influence which the manager could not. This phenomenon, however, appeared to depend critically on whether subordinates identified with the manager. By creating a more open interpersonal climate, the high-performing managers apparently leveraged their ability to manage personality clashes by stimulating resolution responses from the conflicted parties' peers. This is similar to Likert's (59) observation that participative management systems stimulate leadership behavior from subordinates themselves, or "peer leadership" as he calls it.

A more subtle process may also have been operating through the mechanism of identification with the superior. Heider (60) proposes a "balance theory" of interpersonal conflict which suggests that two parties find it more difficult to maintain negative feelings toward each other when they both feel positively toward a third party. Thus the higher-producing managers who created positive subordinate relations may have ameliorated conflict largely by an unconscious process. Levinson (61) expands on the dynamics of the process by saying that "A generalized process of learning how to behave and what to become occurs through identification. . . . By acting as the focal point of unity—the ego ideal of the group or organization—the leader serves as a device for knitting people together into a social system."

## CONCLUSION

If this discussion has served to pique interest in the use of theoretical personality frameworks among project teams, our purpose will have been served. Carl Jung suggested that individual development and maturity involves an integration over time of the four psychological functions of sensing, intuition, thinking, and feeling. A team can similarly develop, given democratic leadership and norms which value the uniqueness and individuality of its members. Orchestrating these unique contributors can provide results beyond that of any single individual contribution.

## REFERENCES


1. Thamhain, H. J. and Wilemon, D. L. "Conflict Management in Project Life Cycles." *Sloan Management Review* (Spring, 1975), pp. 31–49.
2. Op. cit., p. 39.
3. Hellriegel, D. and Slocum, J. W. Jr. "Preferred Organizational Designs and Problem Solving Styles: Interesting Companions." *Human Systems Management,* Vol. 1 (1980), pp. 151–158.
4. Myers, Isabel B. *Manual: The Myers-Briggs Type Indicator* (Consulting Psychologist Press. Palo Alto, Calif., 1962, 1975).
5. Op. cit.
6. McCaulley, M. "How Individual Differences Affect Health Care Teams." *Health Team News,* Vol. 1(8) (1975).
7. Op. cit.
8. McCaulley, M. "Psychological Types in Engineering: Implications for Teaching." *Engineering Education,* Vol. 66 (1976), pp. 729–736, p. 732.
9. Myers, Isabel B. *Introduction to Type,* 2nd ed. (Center for Application of Psychological Type. Gainesville, Fla., 1976), p. 3.
10. Myers, Isabel B. and McCaulley, Mary H. *Manual: A Guide to the Development and Use of the Myers-Briggs Type Indicator.* (Consulting Psychologists Press. Palo Alto, Calif., 1985), p. 13.
11. Ibid.
12. Ibid.
13. Myers, Isabel B. *Type and Teamwork* (Center for Application of Psychological Type. Gainesville, Fla., 1974), p. 2.
14. Op. cit., p. 3.
15. Macdaid, Gerald P., McCaulley, Mary H. and Kainz, Richard I. *Atlas of Type Tables* (Center for Application of Psychological Type. Gainesville, Fla., 1987).
16. McCaulley, Mary H. *Personality Variables: Model Profiles that Characterize Various Fields of Science* (Center for the Application of Psychological Type. Gainesville, Fla., 1976).
17. McCaulley, M. "Psychological Types in Engineering: Implications for Teaching," op. cit., p. 732.
18. Macdaid, op. cit.
19. McCaulley, M. H. and Macdaid, G. P. "Results of a Six-Year Study of Retention at Eight Engineering Schools." Paper presented at the 94th Annual Conference of the American Society for Engineering Education, Cincinnati, Ohio (1986), p. 16.
20. McCaulley, M. H., Godleski, E. S., Yokomoto, C. G., Harrisberger, L. and Sloan, E. D. "Applications of Psychological Type in Engineering Education." *Engineering Education,* Vol. 73 (1983), pp. 394–400.
21. McCaulley, M. H., Macdaid, G. P. and Granade, J. G. "ASEE-MBTI Engineering Consortium: Report of the First Five Years." Paper presented at the 93rd Annual

Conference of the American Society for Engineering Education, Atlanta, Georgia (1985).

22. Myers, Isabel B. with Myers, Peter B. *Gifts Differing* (Consulting Psychologists Press. Palo Alto, Calif., 1985).
23. Sloan, E. D. "An Experiential Design Course in Groups." *Chemical Engineering Education,* Vol. 16 (1982), pp. 38–41.
24. Thomas, C. R. "Personality in Engineering Technology." *Journal of Engineering Technology* (March, 1984), pp. 33–36.
25. Thomas, C. R. "Results of an MBTI Utilization in Engineering Technology." *Journal of Psychological Type,* Vol. 8 (1985), pp. 42–44.
26. Yokomoto, C. F. and Ware, J. R. "Individual Differences in Cognitive Tasks." In *Proceedings: 1984 Frontiers in Education Conference* (1984).
27. Op. cit., p 5.
28. Schutz, William C. *The Interpersonal Underworld,* Reprint ed. Science and Behavior Books, Palo Alto, Calif., 1966.
29. Pfeiffer, J. William and Heslin, Richard. *Instrumentation in Human Relations Training* (University Associates. Iowa City, 1973), p. 144.
30. Op. cit., p 118.
31. Op. cit.
32. Op. cit.
33. Hill, R. E. "Interpersonal Compatibility and Workgroup Performance." *The Journal of Applied Behavior Science,* Vol. 11 (1975), pp. 210–219.
34. Op. cit., p. 140.
35. Allport, G. W., Vernon, P. E. and Lindzey, G. *Study of Values: A Scale for Measuring the Dominant Interests in Personality,* 3rd ed. (Houghton Mifflin. Boston, 1960).
36. Edwards, A. L. *Manual for the EPPS* (The Psychological Corporation. New York, 1954).
37. Holland, John L. *Vocational Preference Inventory: Manual* (Consulting Psychologists Press. Palo Alto, Calif., 1978).
38. Campbell, David P. and Hansen, Jo Ida C. *Manual for the SVIB-SCII Strong-Campbell Interest Inventory,* 3rd ed. (Stanford University Press. Stanford, Calif., 1981).
39. Bales, Robert F. and Cohen, Stephen P., with Williamson, Stephen A. *SYMLOG: A System for the Multiple Level Observation of Groups* (The Free Press. New York, 1979).
40. Bales, Robert F. *Personality and Interpersonal Behavior* (Holt, Rinehart and Winston. New York, 1970).
41. Bales, Robert F. *Interaction Process Analysis* (University of Chicago Press. Chicago, 1976).
42. Kolb, David A. *Learning Style Inventory: Technical Manual* (McBer and Company. Boston, 1976).
43. Gough, H. G. *Manual for the California Psychological Inventory* (Consulting Psychologists Press. Palo Alto, Calif., 1975).
44. Heist, P. A., McConnell, T. R., Webster, H. and Yonge, G. D. *Omnibus Personality Inventory* (The Psychological Corporation. New York, 1963).
45. Comrey, A. L. *Manual for the Comrey Personality Scales* (Educational and Industrial Testing Service. San Diego, Calif., 1970).
46. Bennis, Warren F. and Shepard, H. A. "A Theory of Group Development." *Human Relations,* Vol. 9 (1956), pp. 415–437.
47. Davies, D. E. and Kuypers, H. C. "Group Development and Interpersonal Feedback." *Group and Organization Studies,* Vol. 10(2) (1985).

48. Lacoursiere, R. B. *The Life Cycle of Groups: Group Developmental Stage Theory* (Human Sciences Press. New York, 1980).
49. Tuckman, B. W. "Developmental Sequence in Small Groups." *Psychological Bulletin,* Vol. 63 (1965), pp. 384–399.
50. Bennis and Shepard, op. cit.
51. Ibid.
52. Ibid.
53. Ibid.
54. Hill, R. E. "Managing Interpersonal Conflict in Project Teams." *Sloan Management Review,* Vol. 18 (1977), pp. 45–62.
55. Ibid.
56. Walton, Richard E. *Interpersonal Peacemaking: Confrontations and Third Party Consultation* (Addison-Wesley. Reading, Mass., 1969), p. 75.
57. Op. cit.
58. Ibid.
59. Likert, Rensis A. *New Ways of Managing Conflict* (McGraw-Hill. New York, 1976).
60. Heider, F. *The Psychology of Interpersonal Relations* (Wiley. New York, 1958).
61. Levinson, Harry *The Exceptional Executive: A Psychological Conception* (Mentor. New York, 1968), pp. 163–164.

# 31. Teamwork—Key to Managing Change

Thomas E. Miller*

This new work schedule dreamed up by the experts in city hall is for the birds! The men don't like it and are laying down on the job. A lot of them aren't showing up or they're late, or calling in sick. Injuries are up, equipment gets lost or it's in the repair shop half the time, and we battalion chiefs are spending God knows how many hours detailing men from one fire station to another. We've told city hall about our morale problem, but they say if the men refuse to fulfill the contract agreement, we have no recourse but to reprimand, suspend, and in chronic cases, fire them. But, hell, if any of these steps are taken, we just aggravate the manpower shortage. If you were a seasoned firefighter, how would you like to put your life on the line with a raw young recruit next to you?

I'm swamped with paperwork and I have to go to a training course that's supposed to tell me how to handle the men, but I spend most of my time listening to their complaints. I don't have any time for training, equipment inspection, or fire prevention as the plan says I should. I'm lucky to make it to a fire since I'm usually in a meeting trying to iron out the kinks in this schedule. We chiefs have all complained to city hall, but they argue that the new work schedule is a success. The computer says so. Now, if they can only get the computer to fight the fires, we can all go home.

---

* Dr. Thomas E. Miller is Professor of Administration and Human Relations at the University of Missouri-Kansas City, where he teaches courses in human relations, organizational behavior, and theory of communication. He also taught at Northwestern University and the University of Kansas, and was a training fellow and instructor in human relations at the Harvard Business School. Dr. Miller has been a trainer in the private as well as the public sector, and has authored several case studies and articles on change, teamwork, and management. His current research interests focus on career development, promotions, and leadership, particularly in federal, law enforcement, and fire suppression organizations.

These remarks were made by a middle manager, a battalion chief, some months after a new work schedule had been introduced in his fire department.

Any reader of the organizational behavior literature will quickly recognize in this excerpt many of the familiar "resistance-to-change symptoms," such as restriction of output, sick-outs, work slowdowns as well as some more subtle symptoms—passive indifference, hostility and "rationalizations" on why the changes won't work.*

The "change-resistance syndrome"† has been well documented in the literature.‡ Roethlisberger believes the syndrome has the properties of reciprocalness, similar to the reciprocal relations encountered in mathematical relations.** The reciprocal pattern begins when management introduces a change. Immediately, the workers counter with resistance. Anticipating this resistance, managers resort to strategems designed to overcome the workers' objections. These approaches can take many forms such as (1) giving elaborate logical explanations as to why the change is necessary; (2) hiring outside consultants to "justify" the change or to help "facilitate" its acceptance; (3) using "feedback" sessions involving managers, union, and/or staff personnel to alert all parties to the "realities" behind the change; and sometimes, as in this case study, (4) sending middle managers, who usually experience the brunt of the change, to training programs to acquire techniques on how to handle resistance.

When all else fails, management proposes *new* (and different) changes to cure the illnesses caused by the original change. Frequently, these new proposals are also resisted, thus reactivating the reciprocal cycle. In frustration and anger, management then forces the change on the workers by

---

* Paul R. Lawrence, "How to Deal with Resistance to Change," *Harvard Business Review*, Vol. 32 (1954), p. 49.

† The phrase is the author's.

‡ For a sampling of classic studies see: Fritz J. Roethlisberger and William J. Dickson, *Management and the Worker* (Cambridge: Harvard University Press, 1949), pp. 657–668 and 579–580; Kurt Lewin, "Group Decisions and Social Change," in G. E. Swanson, T. M. Newcomb, and E. L. Hartley (Eds.), *Readings in Social Psychology*, Rev. ed. (New York: Holt, 1952), pp. 459–473; Lester Coch and John R. P. French, "Overcoming Resistance to Change," *Human Relations*, Vol. 1(4) (1948), pp. 512–532; Alvin Zander, "Resistance to Change—Its Analysis and Prevention," *Advanced Management*, Vol. 15 (1950), pp. 9–11; David Klein, "Some Notes on the Dynamics of Resistance to Change: The Defender Role," in W. G. Bennis, K. D. Benne, and R. Chin (Eds.), *The Planning of Change* (New York: Holt, Rinehart and Winston, 1976), pp. 117–124; Gary Powell and Barry Z. Posner, "Resistance to Change Reconsidered: Implications for Managers," *Human Resource Management*, Vol. 17(1) (1978), pp. 29–34.

** Fritz J. Roethlisberger, *The Elusive Phenomena* (Boston, Mass.: Division of Research, Harvard Graduate School of Business Administration, 1977), pp. 169–170.

utilizing its positional authority. As resistance persists, management punishes the holdouts and rewards the collaborators. This reward-punishment pattern drives wedges among the organizational social groupings and the battle continues until the ringleaders quit or the workers passively accept the change, or management decides to chuck the whole idea. At this point, stability in the organization is restored, except for one major difference: often bitterness and resentment linger between the two parties and make the next innovation and its acceptance that much more difficult. Thus the changes that should be accomplished during the normal day-to-day work routines are avoided, brushed aside, and minimized by management and the worker until external or internal pressures force them to face the change process again.*

In spite of all of our managerial knowledge on how to handle change, the change-resistance syndrome still appears and appears frequently in modern-day organizations.† Some writers believe that the symptom-by-symptom attack that management is prone to take in overcoming resistance does not get below the surface to the human factors involved. By failing to recognize the hydra-headed nature of the social situation with which it is faced when introducing technical innovation, management will cut off one head, only to have two new ones appear.‡ Other writers argue that the whole idea of resistance to change needs reconceptualization since management's anticipation of resistance may lead them to the creation of resistance—a sort of self-fulfilling prophecy.** Whatever the basis, most writers concede that there is a gap between our existing knowledge (theory) and managerial practice (skill) when introducing change.

The purpose of this chapter is to examine once again the causes of the change-resistance syndrome by reporting a clinical case study of change

---

* Larry E. Greiner, "Patterns of Organizational Change," *Harvard Business Review*, Vol. 45 (1967), pp. 119–130.

† Recent publications indicate the "change-resistance syndrome" is still very much with us. See, for example: John P. Kotter and Leonard A. Schlesinger, "Choosing Strategies for Change," *Harvard Business Review* (March–April, 1979), pp. 106–114; Jay W. Lorsch, "Managing Change," in Paul R. Lawrence, Louis B. Barnes, and Jay W. Lorsch, (Eds.), *Organizational Behavior and Administration* (Homewood, Ill.: Irwin, 1976); Gerald Zaltman and Robert Duncan, *Strategies for Planned Change* (New York: Wiley, 1977); Michael Beer, *Organization Change and Development* (Pacific Palisades, Calif.: Goodyear Publ. Co., 1980); Douglas Basil, *The Management of Change* (New York: McGraw-Hill, 1974); E. F. Huse, *Organization Development and Change* (New York: West, 2nd ed., 1980); Rosabeth Moss Kanter, *The Change Masters* (New York: Simon & Schuster, 1983); Arnold Brown and Edith Weiner, *Supermanaging: How to Harness Change for Personal and Organizational Success* (New York: McGraw-Hill, 1984).

‡ Paraphrased from Fritz J. Roethlisberger, "The Foreman: Master and Victim of Double Talk," in *Man-In-Organization: Essays of F. J. Roethlisberger* (Cambridge: Belknap Press of Harvard University Press, 1968), p. 36.

** Powell and Posner, "Resistance to Change Reconsidered," p. 29.

in a large urban fire department. The thesis is that resistance to change is a *failure in teamwork.* Where teamwork exists resistance does not emerge, or if it does, the proactive collaboration between those affected—both managers and workers alike—allows them to develop technical and social skills in handling both their logical and social involvement in the change process. Teamwork gets to the core of resistance by facilitating the cooperative interactions between people, irrespective of their status, in the accomplishment of a common task. This is the basic insight from the relay assembly room of the Western Electric Studies: where teamwork existed, no special resistance was encountered between workers and investigators in spite of the fact that several major technical changes were introduced. "What actually happened," wrote Mayo, "was that six individuals became a team and the team gave itself wholeheartedly and spontaneously to cooperation in the experiment."* The team process held the human situation steady and permitted the participants to develop both technical and social skills in relation to change phenomena.† The workers did not view or experience change as something externally imposed by management; rather, change became a way for them to adapt and experience social complication in their work. Satisfaction came from meeting and coping with the challenge.

More often than not, when teamwork is left to chance, managers do not get teamwork; they get instead the kind of resistance behavior which was described by the battalion chief in the opening paragraphs of this chapter. Resistance is management's external evaluation of the workers' behavior when they do not accept a technical change; opposition to management's change is the workers' internal response in protecting the social character of their world. Teamwork bridges the gap between these internal and external processes during change. However, unless both managers and workers diagnostically understand this process and can behaviorally respond to each other in such fashion as to create cooperation, then teamwork will elude both.‡ Before we explore further the worth of this idea

---

* Elton Mayo, *The Social Problems of an Industrial Civilization* (Boston, Mass.: Division of Research, Harvard Graduate School of Business Administration, 1945), p. 72.

† Ibid.

‡ A recent and widely adopted organizational approach to encourage worker participation is known as quality circles. Many American managers see their utilization as a major breakthrough for encouraging cooperation and shared decision making in a common task. Although teamwork is not specifically mentioned in definitions of quality circles, teamwork would certainly be a major prerequisite to their success. The following are some recent sources on quality circles: Robert I. Patchin, *The Management and Maintenance of Quality Circles* (Homewood, Ill.: Dow Jones-Irwin, 1983); Sud and Nima Ingle, *Quality Circles in Service Industries* (Englewood Cliffs, N.J.: Prentice-Hall, 1983); Laurie Fitzgerald and Joseph Murphy, *Installing Quality Circles: A Strategic Approach* (San Diego, Calif.: University Associates, Inc., 1982).

and its implications, let us give a brief description of the changes so colorfully alluded to in the battalion chief's remarks.

### Introduction of a Change in a Fire Department

In a major move from conventional scheduling, the management of a large urban city placed its firefighting personnel on a 40-hour workweek consisting of three eight-hour work shifts. This contrasted with the department's former 48-hour a week schedule of 24 hours on duty and 48 hours off duty. The primary objective of the plan was to maximize the on-duty time of the firefighting personnel, thus providing a better level of fire service to the citizens and taxpayers of the city.

The eight-hour shift was just one component of the city's comprehensive fire protection plan, but was viewed by management as central to the plan and an innovative step forward in fire service. Many advantages seemed to be inherent in the plan: fresh personnel every 8 hours rather than every 24 should reduce life risks due to fatigue; an increase in on-duty productivity and communications should result because the men would work their full shifts instead of sleeping part of them; and last, round-the-clock attention could be given to fire prevention, training, and equipment maintenance. Also management believed that by putting the men on an eight-hour schedule it would be easier to justify raising their pay commensurate with other protective service personnel in the city. The eight-hour shift was thus negotiated as part of a pay package requested by the local firefighters' union. Pay increases in the contract were tied to a "good faith, best effort" on the firefighters' part in carrying out management's plan.

During the three years that the eight-hour shift was in operation, the department was in almost constant turmoil because of the firefighters' steadily mounting resistance to the changes in the work schedule. Management pointed to the many tactics the firefighters were using to resist the plan: slowdowns, work stoppages, sick-outs, and damaging and losing equipment. On their part, the firefighters felt that the city was squandering the taxpayers' money on what was a totally unworkable plan, designed only to "punish" them for their pay demands.

The fire chief and his technical assistants held many meetings with the battalion chiefs in an effort to work out "the bugs" in the plan. Several adjustments were made to the plan as a result of these meetings: 313 persons were promoted; relief companies were formed to provide backup manpower; and a special task force consisting of several senior battalion chiefs was created to "track" the progress of the fire protection plan. But in spite of all of management's efforts, few of the chiefs and even fewer of the rank-and-file firefighters "accepted" the new work schedule. A six-months' evaluation of the plan prepared by the special task force concluded that the eight-hour work shift had failed to achieve management's objective of greater productivity and had been devastating to the morale of the firefighters.

When contract negotiations came due, communication relations between city management and the union were at the boiling point. The union was determined to change the eight-hour workday; management was equally determined to keep the plan intact, but was open to any internal modification to make the work schedule more flexible. When negotiations broke down, the union initiated a massive work slowdown. Management responded by discharging some 40 firefighters for contract violations. Fearing that the discharged firefighters would not be rehired because of the slowdown, the men returned to their jobs and negotiations resumed. When it then became apparent to the union that the city did not intend to rehire the discharged firefighters, the union called a strike, leaving the stations virtually unmanned except for the battalion chiefs. In short order, the city manager called in the police and the national guard to man the fire stations and their presence naturally inflamed the situation. Before the paralyzing strike was over, an international firefighters' union, the governor of the state, the city council, the mayor, and many prominent citizens had become involved in trying to resolve the crisis. Public opinion was divided. Some citizens believed that the firefighters had failed to honor their agreement with the city and were attempting to usurp the city's authority to run the department. Others felt that the city had "forced" an impractical plan on the firefighters and that the latter were justified in their counteractions.

After the strike was settled, the eight-hour shift was replaced by a 10/14-day workweek, a schedule regarded by the firefighters as better than the eight-hour shift but not as desirable as the original 24/48 workweek. The outcome of the situation seemed to be losses for all concerned: the city had spent a great deal of time and money on the innovative plan which they were finally forced to abandon largely due to outside pressures; the firefighters settled for a longer workweek with the same pay and few additional benefits; the fire chief with over 30 years of service resigned; the union president was passed over for promotion although he was more than qualified; and many outstanding firefighters left the department either through early retirement or by seeking employment in other fire departments. These were the immediate and apparent consequences—hundreds of thousands of dollars lost, turnover, lowered morale and motivation, and no noticeable improvement in fire service.

Briefly, then, these are the facts of the implementation of a change in a fire department. We might now ask: how did management and the firefighters evaluate their experiences in the presence of the same concrete phenomena? Needless to say, both managers and firefighters pointed to the technical and organizational alterations as the single cause of their problems, although each group drew very different conclusions from the same events because of their dissimilar organizational and group roles and individual personal needs. However, as we saw it, both top management and the firefighters responded to the change process in the *same* fashion. In other words, both the city manager and his technical support staff as well as the rank-and-file firefighters made the assumption (and this was clearly evident in their behavior) that what was happening in the fire department was strictly a rational-logical experience. Let us explore further the way managers and firefighters alike reasoned about the change.

### How Management Reasoned

To management, obviously, there was nothing basically wrong with the plan and any number of adjustments could be made to eliminate the "bugs." The real problem, they concluded, was the firefighters' refusal to carry out the plan in "good faith." Being good logicians, they were at first reluctant to draw this inference (although they felt it in their guts) until more evidence was in. So, to be consistent at all costs, upper management spent innumerable hours and considerable sums of money getting the "bugs" out of the work schedule, in order to allay the complaints of the battalion chiefs and the firefighters. In particular, technical specialists, both inside and outside the department, along with a specially appointed associate city manager and the fire chief, spent endless hours discussing the battalion chiefs' and firefighters' objections to the schedule and making appropriate modifications only to find that each modification required another and yet another until at the end of the first six months, a major adjustment was made approximately once a week: management seemed to be operating on the assumption that the better the firefighters and, particularly, the battalion chiefs, understood the logics of the plan, the more likely it was that they would support it. When this assumption did not come true, upper management then drew the inference that the battalion chiefs and the firefighters were "unreasonable" because of their stubborn refusal to face the facts.

### How the Firefighters and Battalion Chiefs Reasoned

In the presence of the same happenings and utilizing the same rational approach, the firefighters, along with the union officers and middle-management battalion chiefs, drew just the opposite inference from that of top management: "The plan is not working because it is not suitable to our work situation, and no amount of propping it up can save it." The battalion chiefs and the union countered logic with, from their point of view, better logic. Like upper management, they produced lengthy reports—filled with facts and figures—to demonstrate how the plan, not they, had failed. As the firefighters never tired of pointing out to city management, "a firefighter's job is unique and will not fit into a nine-to-five work schedule." The battalion chiefs and the union argued that other cities had tried the eight-hour work schedule and had eventually abandoned it as impracticable. The more they talked and tried to explain their position (just like top management), the more they tended to evaluate management as being unreasonable and resisting logic.

### A Stalemate Results from the Rational Approach

Because of the circular nature of the rational approach, it is difficult for its practitioner to realize consciously what he is doing at any given point in the reasoning process. Within this framework, the practitioner evaluates as follows: change → causes consequences (both expected and unexpected) → leading to the need for correction. Since the "unexpected" consequences are by definition not anticipated (and therefore not desirable), someone or something is to *blame*. Either the designer of the schedule did not set it up properly, the supervisors did not effectively implement it, or the firefighters did not do what they were told.

In meeting after meeting both city managers and firefighters pointed to the external, technical phenomena and blamed the other party for its failure to comprehend the "facts." Both assumed they were talking only about the technical shortcomings of the plan, when in reality they were responding also to their uncomfortable feelings and disturbed social relationships. Both firefighters and battalion chiefs tried to convey to top management the social confusion which had been brought into their lives by the change, but in management's presence their explanations always came out as logical rationalizations, which management, with a superior set of logic, always succeeded in beating down. Likewise, the chiefs never understood nor were able to accept management's emotional involvement in the change. As one chief put it: "It's the city's plan; let them make it work!" So, in time the positions froze because of a basic misevaluation: that social evaluations and behavior are not involved in technical change.

The conflict between management and firefighters becomes more phenomenologically understandable when viewed through the membership commitments each had to their respective social groups. Social membership roles were manifested in their meetings and discussions, but neither management nor the men understood how they were linked to the success or failure of the work schedule change. It was essentially their *social* world which was threatened by the change, and this factor needs to be placed in perspective along with the technical aspects of the change. Let us now turn first to a description of the fire department as a social system and, second, to an analysis of the four major subgroupings which determined in a large degree the social memberships of both firefighters and battalion chiefs.

### A Description of the Fire Department as a Social System

As the writers on social systems have stressed, people who work together over a period of time begin to form into collective configurations or groupings. Because of the formal and logical divisions established in an organization, certain people are brought together more frequently than others. Initially, people interact with one another in certain prescribed ways dictated by the wider society and the formal requirements of the work. But, in time, these prescribed patterns are modified, changed, or adapted to accommodate to individual and group differences and needs. And, gradually, groups form among those who share the same values, the same sentiments, and similar needs. Out of these shared interactions, sentiments, and activities, norms of behavior develop. Group members as a whole develop certain ideas about how they should be treated, what their contributions are worth, and what are proper and improper ways of behaving according to their status and job roles. It is these group processes and evaluations which emerge and feedback on the purposive organization that we point to when we speak of social phenomena.

Although social phenomena are related to technical phenomena, they are also different. Technical phenomena often are created, ordered, modified, and even eliminated without consideration of the feelings of people. Technical phenomena can be talked about, pointed to, diagrammed, and manipulated far more easily than social phenomena. Social phenomena exist at lower levels of abstraction; they are "relations of interconnectedness which exist among persons" and are part and parcel of concrete natural systems.* These phenomena are more difficult to point to, to talk about, and to manipulate. More important, they are *naturally ordered* by individuals and groups to bring stability and meaning to their lives. Technical phenomena are logically ordered to attain the purposes of the organi-

* Roethlisberger, *The Elusive Phenomena,* p. 144.

zation. Social phenomena are nonlogical in origin and seldom can be modified by logic alone.

The social interconnectedness among the members of the fire department was "tight." Firefighting is the most dangerous of occupations, and firefighters literally depend upon each other for survival. This fact alone would create the need for cooperation among them. In addition, living together "around-the-clock" in the fire stations increased opportunities for social interactions. The firefighters—from raw recruits to senior battalion chiefs—ate, slept, and fought fires together; occasionally, some died together. They thought of their stations as home; many brought television sets and furniture to make the stations more livable, more like home. The family atmosphere and group loyalty was expressed in the elaborate parties given for those who had retired, the large attendance at funerals for "old-timers," and contributions made toward gifts to fellow firefighters at times of noteworthy occasions in their lives.

To illustrate the deep sense of dedication, loyalty, and friendship that bound them together, here is a typical statement from the casewriter's data:

> Our job is different, it isn't just a nine-to-five kind of job. We actually spend more time with each other than with our families. We have trust because of our work. Remember, when you go out on a run, that man next to you can save your life or let you burn. We are all brothers. This feeling is hard to explain to someone who has never been a firefighter. If you make friends with a firefighter, you make it for life.

These sentiments were expressed by a senior battalion chief nearing the end of his career. Most firefighters would share this viewpoint. However, agreement with these sentiments did not mean that as a consequence all firefighters thought alike, behaved alike, shared the same values, or followed the same norms of social behavior. Depending on their background values, personal needs, and social ranking, some firefighters regarded work as more important than friendship, status more important than group respect, competence more important than seniority, and pleasing the boss more important than conforming to his co-workers' expectations.

### Four Natural Groups

The different values brought to the work situation and the values which emerged on the job while they interacted resulted in the formation of several social groups in the fire department. We will briefly describe the four major groups which we delineated from the data, keeping in mind, of course, that not every individual fitted neatly into any one group.*

**Technical-Specialist Organizational.** As the name suggests, these people are most comfortable relating to technical phenomena and organizational authority. They tend to be "standoffish" in social relations and they are usually regarded by other group members, particularly the social regulars, as isolates, although they may be admired for their technical competence. Their isolation is often self-imposed. If they rise in the management hierarchy—many do because of their performance—their social isolation is reinforced by their becoming organizationals, namely, "boss oriented." Their values: "Technical competence is foremost. My job is to get the work done regardless of whose toes I step on." Technical specialists can be high producers.

**Social-Specialist Regular.** Feeling lonely, misunderstood, or disliked is the worst thing that can happen to a social regular. Satisfaction is secured not exclusively from doing a good job, but is only secondary to being accepted by one's colleagues. Leaders of this group have great influence with their peers and subordinates because they are people, rather than power, oriented. Their values: "We stick together, and that's how we get the job done. Keep confidences, never hurt a brother." Their production can be on the line or high (never low) depending on how established their group membership is and the way their performance is regarded by their colleagues, not by top management.

**Underchosen.** These men may truly be unhappy isolates, not self-chosen like the technical specialists. They are underchosen by both the technicals and the socials because of some critical value out-of-lineness such as age, competence, ethnicity, education, or personality. If they rise in the hierarchy, they are seen as sycophants by their colleagues because of their subservience to management. Around their fellow workers they often behave like "good Joes," but their social influence is limited. By "working both sides of the street," they hope to gain acceptance and status. Their values: "Whatever you desire me, I'll be. I've paid my dues and I want what's coming to me." Their production is often minimal and they usually keep a low profile when conflict emerges because of their unstable organizational status and group membership.

---

* For some of the ideas and the distinctions in the social groupings, the author is indebted to David Moment and Abraham Zaleznik, *Role Development and Interpersonal Competence*, Division of Research, Harvard Graduate School of Business Administration, Boston, pp. 122–125.

**Power Specialist.** These people seek recognition and acceptance by exercising power. To enhance their standing with the social regulars, with whom they ultimately identify, they openly confront hierarchical authority in the name of a "good cause." Both their organizational status and group membership can be threatened because of their aggressive activities. On the one hand, they are admired by social regulars—never organizationals or the underchosen—for their accomplishments, but on the other hand, they are feared because of their reckless actions. They often have enormous group influence and power, particularly at times of crisis. Their values: "Go for broke, rock the boat, the cause is everything." Their production is variable like their status and group membership. They have many labels—politicos, power-seekers, do-gooders, troublemakers—depending on where you and they identify in the social system.

### The Impact of the Social Groups in the Change Process

The delineation of social groupings could be extended or modified in terms of the many existing social patterns in the fire department. These groups were chosen because they provided a comprehensible framework for viewing and understanding the departmental members' behavior during the change.

As the tensions mounted over the work schedule, group membership became a major determinant of the firefighters' evaluations. The technical-organizational group, composed of the fire chief, an associate city manager, and their assistants, clearly identified with the values of upper management. These members experienced no role conflict within their own group; their frustrations and conflict resulted from their inability to enlist many followers. Since upper management had originally entrusted them with the creation and implementation of the plan, these men had to gain the men's acceptance of the work schedule. But because they were identified with upper management's values and goals, they were regarded with suspicion by other group members and their effectiveness was limited. The target of the technical organizationals was, of course, the chiefs who had strong social ties with both the social regular firefighters and the power specialist union members. In time, the only people the fire chief and his group influenced significantly were a handful of the department's staff personnel and some underchosen low producers and "good Joes."

Explicitly opposed to the work schedule was the power specialist group. Headed by the union officers, this group urged all firefighters to resist the work schedule at all costs or run the risk of forfeiting their "social" membership. Paying the price could mean accusations of group disloyalty and weakness at best or group exclusion and loneliness at worst. The pressure of this group was felt by both the social regulars and the underchosen.

Caught between these two groups, which were both vying for their support, were the critical management line supervisors, the battalion chiefs. Over the years most of the chiefs had identified with the values of the social regulars, but a few had split loyalties, partly to upper management and partly to the union of which they had historically been members.

Their organizational role placed them right in the middle of the conflict. On the one hand, the chiefs had to enforce the logics of the change to keep their supervisory positions, while on the other hand, they needed to maintain their group standing in order to sustain the firefighters' cooperation. A decision for management could mean rejection by the men; a decision for the men could mean organizational exclusion by upper management.

The chiefs' role conflict erupted into the open at the time of the six-months' evaluational report presentation. Intentionally or not, upper management evoked the incident when they selected the members of a steering committee to "track" the results of the change and write up a report. Management "bent over backwards" to appoint a fair and equitable representation of the chiefs, knowing full well that many were not sympathetic to the new schedule. In preparing the written report, the committee solicited reactions from all the chiefs and added their own inputs. The report, when completed, was signed by all battalion chiefs.

#### The Six-Months' Evaluational Report

On the day of the presentation, the committee, chaired by a senior battalion chief, met with the fire chief and his assistants. The chairman and most of the chiefs were social regulars and carried the brunt of the interactions; a few were "underchosens" and remained largely silent during the meeting. The power specialists were present as "spectators" along with several members of the news media. (News had leaked out to the press that the report on the work schedule was unfavorable.)

In a tense atmosphere, the chairman opened the meeting:

| | |
|---|---|
| Chairman: | The purpose of this meeting is to review the findings of the six months' evaluation of the plan. (He passes out copies of the report.) |
| Fire Chief: | (Thumbing through his copy.) This is the first time I've seen this report. |
| Chairman: | No one has seen it but the chiefs. Duplication was just finished yesterday. However, it isn't that long. We can all go through it together. |

| | |
|---|---|
| Fire Chief: | All the chiefs have seen the report since they all signed it. All officers have read the report except the chief officer of this department. Is that correct? |
| Chairman: | Well, Chief, you were away and. . . . |
| Associate City Manager: | The report was requested by the city manager's office. Your job was to make an objective appraisal. Just glancing through the report, it seems to contain a number of subjective statements and not much hard data. You must fold into your report the computer data on response times and allocations of equipment and manpower. Your statistics and conclusions must be carefully meshed. We must be guided by the facts in our evaluations. |
| An Assistant: | Who wrote the report? |
| Senior Battalion Chief: | We all wrote it. It bears no malice to anyone. We report the concerns of all the chiefs. |
| Chairman: | The report isn't all that bad. It's written from an operational point of view, it isn't very long, and it has several good things to say. |
| Fire Chief: | I will make no comments until I have thoroughly studied the report. |
| Chairman: | (Pause) I accept your decision, of course, but I think it's a sad day when we can't discuss our mutual problems openly. |
| Firefighter: | (Speaking from the back of the room) I think this report should be gotten out and not buried somewhere. The public should be informed about what kind of fire protection they are or are not getting. |
| Associate City Manager: | Of course, when everyone has had time to read the report, we'll certainly go through it carefully and consider every recommendation. |

At this point, the meeting turned to other matters, and shortly afterwards was adjourned.

Ostensibly, the committee had met to discuss the report, but intergroup relationships and feelings were such that an open discussion was impossible. The report was a "plea for understanding and help," although this was probably an unconscious wish on the part of the chiefs; however, management perceived the report as a "slap in the face." Management reacted only to the "slap" and "slapped back." In fact, all committee participants evaluated what happened in the meeting only from their respective group referents.

These were the reactions after the meeting:

| | |
|---|---|
| Technical Organizational: | The chiefs are just too close to the men. They should support the city. I told the city manager and fire chief they'd get a bad report from this committee. (He had not read the report.) |
| Social Regular: | Had the report been given to the city first, all the findings unfavorable to them would just have been buried. There's no way in good conscience we chiefs can ignore the men. |
| Underchosen: | I didn't want to sign the report, but what else could I do? I was on the spot. |
| Power Specialist: | After this, maybe the chiefs will come over to our side where they belong. |

Subsequent communications within each group and between groups only reinforced these perceptions. Upper management immediately came to the defense of the technical organizationals by verbally reprimanding the chiefs for publicly airing departmental differences. (They believed the chiefs were responsible for the leak to the press.) The city officials felt strongly and, understandably, that they had been "set up" to look bad in the public's eye. The chiefs felt misunderstood, but admitted they had mishandled the presentation of the report. At the next meeting, the committee members apologized to upper management; management reciprocated. Following this exchange, a technical organizational commented: "It's good we got this off our chests; now, we can turn to the real work, the facts of the report." In short, he was saying that now they could dismiss the uncomfortable social dimensions of the situation and return to the only feasible task at hand, the logical implementation of the work schedule change.

In the months ahead, staff meetings were concerned exclusively with the technical problems caused by the work schedule. Upper management met less and less with the battalion chiefs until finally all meetings between them ceased. As time went on, management tended to exclude the chiefs from all major decisions, and thereby, consciously or not, reduced their status and authority. This exclusion by upper management, which was exploited by the power specialists, eventually forced the chiefs to make a choice. In time, they joined the social regulars and the power specialists in their overt opposition to upper management and the technical organizationals. By joining with these groups, the chiefs became ineffective in working out a managerial compromise between the city and the firefighters' union.

The technical organizationals totally understimated the strength of intrinsic group membership values and never really understood how these values were linked to success in achieving extrinsic goals. At least, if they did understand, they were never able to act upon their insights in any effective way. The power specialists learned only to exploit the weaknesses and strengths of the various groups for their own purposes. The social system of the fire department, for all practical purposes, became frozen and the change-resistance syndrome intensified between upper management and the union. At contract negotiation time, when the union was free of legal obligations and all communications had broken down between the city and the firefighters, the union called a strike.

A few months after the strike, the fire chief resigned. In subsequent months, his resignation was followed by that of the associate city manager, all technical specialists who had been connected with the fire protection plan, and eventually the city manager himself. Although these personnel changes cannot entirely be attributed to the upheaval in the fire department, certainly the failure of the fire protection plan was in the picture. In 1985, the union successfully negotiated with the new city manager a return to the *original* 24/48-hour work shift. In addition, by this time, most of the technical innovations introduced with the fire protection plan had been abandoned.

### Summary

This case study has illustrated once again a breakdown in communications and cooperation between management and workers during a change process. In the fire department, both upper management and the firefighters were quickly swept up in the vicious cycle of the change-resistance syndrome. Both parties evaluated the technical consequences of the change as the problem; both failed to grasp the impact that the technical change had on their interactions and interpersonal feelings. In their meetings and discussions, they responded to breakdowns in communication as though they were only technical misunderstandings which could be corrected through logical coordination and formal rules. Because only the logical impact of the change was addressed, the vital social group processes which were disturbed by the change went unattended.

This occurred not because one or both parties were power-oriented, stupid, or illogical. Both intuitively understood and responded to their own needs and their own group involvement; less clearly did each understand the impact that their behavior had on each other as they went about satisfying their needs for security, belonging, and status. The fire department situation could be best described not as a battle for power between opposing forces—as both management and the firefighters evaluated it—but rather as a struggle between the logics of change and the sentiments of group membership.

When introducing change, the manager's role becomes crucial. On the one hand, he must determine what changes are necessary for the effective survival of the organization, and on the other, he must secure and maintain the effective cooperation of individuals and groups to insure their acceptance. How is he to achieve this?

### Implications for Practice

**Diagnostic Skills.** It may seem strange today to say that we need to improve our understanding of the dynamics of organizations, but all our clinical research points to this need.* A manager is a practitioner and requires diagnostic skills similar to those of the clinician. Before a physician can act, he must diagnose the sort of illness he is handling—whether it be a case of measles, mumps, or cancer. Managers often behave as if they require no such equivalent skills in diagnosis. Too often, if we are not careful, we try to solve our problems of teamwork and change intellectually, abstractly, and analytically rather than through the diagnostic identification and sizing up of individual values and group norms in relation to organizational requirements.

To become better diagnosticians requires that we understand our own involvement and the impact of our behavior on others. This is particularly true during an introduction of change when actions on the part of both managers and workers are so critical and subject to such easy misunderstanding. A manager is an involved participant and member of his social system, and from this involvement there is no escape. He is both an instrument of change and a recipient of that change; his dilemma is: how can he (a part of a system) effect a change in other parts of the system as a whole of which he is an involved member without destroying cooperative relations in the process? A manager is the bridge between organizational requirements and interpersonal cooperative processes, and how well he understands social processes will largely determine whether his change is accepted or rejected.

Here is an example of a simple social pattern which was unrecognized by upper management. When it was altered, the change led to much bitterness and deepened the chasm between management and firefighters.

---

* For several ideas in this section, the author is indebted to Roethlisberger, *Man-In-Organization*, pp. 65, 139, 169.

> The social custom of the "cook shack" was a long established practice in the fire stations under the former work schedule. The fire companies took enormous pride in their food preparation and there was considerable vying among the various stations for a reputation as "best house." Deputies and battalion chiefs were often invited as special guests to share a meal and pass judgment on the "table."
>
> On these occasions, the senior ranking officers sat at the head of the table, while the less senior in rank and age took their respective places around the sides. The conversation was not limited to social matters; often chiefs, captains, and experienced firefighters shared mutual problems about particular fires, equipment, etc., while the younger men listened and added their comments.

According to the "logics" of the new work schedule, firefighters were discouraged from cooking in the stations. They were expected to "brown bag" it or eat at nearby restaurants. Neither the city nor the firefighters properly understood how this alteration in eating habits with the new work schedule impacted upon morale. "Eating together" was a social process which paid off handsomely at the fire site later in terms of strengthening the patterns of company and battalion teamwork.

**Distinguishing Fact from Sentiment.** Diagnostic understandng and intuitive familiarity with social uniformities deters a skillful manager from making a quick size-up when introducing change. He begins not with a theory or technique but rather, with the slow, laborious task of observing social behavior in relation to technical behavior. By practicing a diagnostic skill, a manager can begin to improve his communications with others; he can become gradually more proficient at promoting cooperation and participation during the change process.*

For example, a skillful manager realizes he cannot express his own feelings without regard to the impact they will have on others, particularly his subordinates, who cannot be expected to distinguish between his expression of feelings and his statements of fact. He must help them to make that distinction. This confusion between fact and sentiment was at the core of the conflict between managers and firefighters.

Basically, the confusion arises when one participant makes a value judgment of another's behavior from the former's point of view. Not only do people make value judgments, but they also insist that the other person

---

* Fritz J. Roethlisberger, "Conversation," in William Dowling, Ed., *Effective Management and the Behavioral Sciences* (New York: AMACOM, A Division of American Management Association, 1978), p. 207.

accept these judgments as correct, factual, and true. When this pattern occurred in the case study, all cooperation ceased. A manager needs to practice a skill of communication which distinguishes between fact and sentiment in his dealings with others. He must learn to identify the values, sentiments, and norms which are important for him and others to maintain in their interactions. Simply put, when people are not kept busy defending their personal feelings, they have more time and *desire* to cooperate.

**Utilizing Listening Skills to Facilitate Teamwork.** No group was more critical to the successful introduction of change in the fire department than the middle-management battalion chiefs. Both upper management and the union knew this. As we have stated, the chiefs were caught in a role conflict aggravated by a change which they intellectually understood but emotionally found difficult to handle. Their job organizationally, as seen by upper management, was simply to enforce the change. This placed the chiefs, as they saw it, in the role of being "watchdogs," a role they found personally uncomfortable and virtually impossible to carry out and at the same time maintain their image in the eyes of the rank-and-file firefighters. Time and time again they had to do things that were "disloyal" from upper management's point of view in order to get the work done. On the other hand, with the firefighters in the field, the chiefs frequently had to "distort reality," plead ignorance, or be vague about some happening or regulation in order to maintain their social group standing.

Had upper management recognized the chiefs' role dilemma and responded to it more skillfully by listening, two important processes would have been set in motion. First, upper management would have heard not only the chiefs' *words,* but more important, what these words *pointed to*—the chiefs' role conflict. Listening implies we attend to (1) what people want to tell you, (2) what they don't want to tell you, and (3) what they cannot tell you without help. Listening reduces social distance and softens social status differences between groups. It is a facilitator to understanding and cooperation. By being listened to, the chiefs would have felt more accepted and less defensive in their communications with upper management.

Secondly, the systematic practice of listening gives the manager more phenomenologically relevant data upon which to take action, since such data come to him from an internal as opposed to an external point of view. Had upper management understood that the chiefs' role conflict sprang not from a lack of "management knowledge about logics" but rather from the very nature of the concrete social system of which they were a part, they might have made a more proper evaluation of the chiefs' "double talk" during their meetings. With this insight, management's actions could have been made consistent with the chiefs' internal experiences. This is one of the most important lessons we need to learn when introducing change—that a little difference, such as listening, can make a big difference among organizational members.*

**Listening and Teamwork Go Hand in Hand.** Although many managers would readily concede the importance of listening, particularly as a tool in creating teamwork, the skill is not widely practiced in organizations, certainly not in the fire department under consideration in this chapter. Generally stated, most of us don't listen because we honestly believe it is a waste of time. We already have our minds made up and all we want from our subordinates and colleagues is acceptance of our point of view. It is much easier, neater, and quicker just to tell people what is expected of them logically and hope for the best. As the city manager remarked in the early stages of the resistance, "In time the firefighters will come to accept the change because they will see the logic in it."

At a deeper and more personal level, the resistance to listening is subtle and difficult to cope with. To this writer, the major barrier to the practice of listening arises from the uncomfortable and insecure feelings which the practice sets up in the practitioner.† These uncomfortable feelings are aroused in the practitioner because listening forces him to be more conscious of his own behavior—his attitudes and feelings—and take responsibility for the consequences of his behavior in relation to others. Listening requires a willingness to see, appreciate, and even accept points of view different from his own.

For most of us these ideas are difficult to accept emotionally. It is not difficult for us to understand intellectually that people are motivated more by feelings and sentiments than by facts and logic. It is not difficult to appreciate that people are members of groups and act in accordance with the sentiments of these groups. Furthermore, it is even obvious to most of us that we do not all perceive the world in the same way. What is important to one person is not important to another. This is a way of saying that we do not respond to reality as it actually is, but rather as we perceive it to be.‡ These matters are all easy to understand intellectually, but to practice this understanding by listening to others is quite another matter.

Time after time, city managers and firefighters ignored nonlogical behavior in their dealings with each other. As a consequence the resistance

---

* Ibid., p. 217.

† For certain ideas in this section, the author is indebted to Roethlisberger, *Man-In-Organization,* pp. 105–106.

‡ For a recent statement on a relativistic view of "reality," see James Burke, *The Day the Universe Changed,* (Boston,: Little, Brown, 1985) pp. 336–337.

between them intensified. Listening requires that we address individual and group sentiments and evaluations in face-to-face interactions, and failure to do so means that teamwork will languish or even worse, atrophy, and the change-resistance syndrome will persist.

Breakdowns in cooperation and communications between management and workers continue to be commonplace. Disillusionment with interpersonal skills such as listening and understanding is also rising among managers, possibly because they expect too much from them too soon. It is only human to want the "quick fix," particularly when we introduce change. There is no magic nor any guarantees of success in the listening process. However, management's recent shift back to reliance on organizational structure as the exclusive change variable may prove, as it did in this case study, costly, disruptive, and self-defeating. Perhaps it is time we take Mayo's admonition of nearly 40 years ago seriously: Teamwork ". . . is the problem we face in the . . . twentieth century. There is no 'ism' that will help us to solution; we must be content to return to the patient, pedestrian work at the wholly neglected problem of the determinants of spontaneous participation.*

* Mayo, *The Social Problems,* p. xvi.

# 32. Team Building in Project Management

Hans J. Thamhain*

Team building is important for many activities. It is especially crucial in a project-oriented work environment where complex multidisciplinary activities require the integration of many functional specialties and support groups. To manage these multifunctional activities it is necessary for the managers and their task leaders to cross organizational lines and deal with resource personnel over whom they have little or no formal authority. Yet another set of challenges is the contemporary nature of project organizations with their horizontal and vertial lines of communication and control, their resource sharings among projects and task teams, multiple reporting relationships to several bosses, and dual accountabilities.

To manage projects effectively in such a dynamic environment, task leaders must understand the interaction of organizational and behavioral variables in order to foster a climate conducive to multidisciplinary team building. Such a team must have the capacity of innovatively transforming a set of technical objectives and requirements into specific products or services that compete favorably against other available alternatives in the marketplace.

## A New Management Focus

Building effective task teams is one of the prime responsibilities of project managers. Team building involves a whole spectrum of management skills required to identify, commit, and integrate various task groups from traditional functional organizations into a multidisciplinary task/management

---

* Hans J. Thamhain is Associate Professor of Management at Bentley College in Waltheim, Massachusetts. He has held engineering and project management positions with GTE, General Electric, Westinghouse, and ITT. Dr. Thamhain is well known for his research and writings in project management, which include 4 books and 60 journal articles. He also conducts seminars and consults in all phases of project management for industry and government.

system. This process has been known for centuries. However, it becomes more complex and requires more specialized management skills as bureaucratic hierarchies decline and horizontally oriented teams and work units evolve. Starting with the evolution of formal project organizations in the 1960s, managers in various organizational settings have expressed increasing concern and interest in the concepts and practices of multidisciplinary team building. Responding to this interest many field studies* have been conducted, investigating work group dynamics and criteria for building effective, high-performing project teams. These studies have contributed to the theoretical and practical understanding of team building and form the basis for the discussion of the fundamental concepts in this chapter.

### The Process of Team Building

Team building is an ongoing process that requires leadership skills and an understanding of the organization, its interfaces, authority and power structures, and motivational factors. It is a process particularly critical in certain project situations, such as:

- Establishing a new program.
- Improving project-client relationships.
- Organizing for a build proposal.
- Integrating new project personnel.
- Resolving interfunctional problems.
- Working toward major milestones.
- Transitioning the project into a new activity phase.

*Team Building is defined* as the process of taking a collection of individuals with different needs, backrounds, and expertise and transforming them into an integrated, effective work unit. In this transformation process, the goals and energies of individual contributors merge and support the objectives of the team.

Today, team building is considered by many management practitioners and researchers as one of the most critical leadership qualities that determines the performance and success of multidisciplinary efforts. The outcome of these projects critically depends on carefully orchestrated group efforts, requiring the coordination and integration of many task specialists in a dynamic work environment with complex organizational interfaces. Therefore, it is not surprising to find a strong emphasis on teamwork and

---

* See bibliography numbers 3, 4, 5, 8, 17, 18, 22, 24, and 49

team building practice among today's managers, a trend which, we expect, will continue and most likely intensify for years to come.

## A SIMPLE MODEL

The characteristics of a project team and its ultimate performance depend on many factors. Using a systems approach, Figure 32-1 provides a simple model for organizing and analyzing these factors. It defines three sets of variables which influence the team characteristics and its ultimate performance: (a) environmental factors, such as working conditions, job content, resources, and organizational support factors; (b) leadership style; and (c) specific drivers and barriers toward desirable team characteristics and performance. All of these variables are likely to be interrelated in a complex, intricate form. However, using the systems approach allows researchers and management practitioners to break down the complexity of the process and to analyze its components. It can further help in identifying the drivers and barriers toward transforming resources into specific results under the influence of managerial, organizational, and other environmental factors.

## CHARACTERISTICS OF AN EFFECTIVE PROJECT TEAM

Obviously, each organization has its own way to measure and express performance of a project team. However, in spite of the existing cultural and philosophical differences, there seems to be a general agreement among managers on certain factors which are included in the characteristics of a successful project team:*

1. Technical project success according to agreed-on objectives.
2. On-time performance.
3. On-budget/within-resource performance.

Further, over 60% of the managers who identified these three measures ranked them in the above order.

When describing the characteristics of an effective, high-performing project team, managers point at the factors summarized in Figure 32-2. These managers stress consistently that a high-performing team not only produces technical results on time and on budget but is also characterized by specific job- and people-related qualities as shown in Table 32-1.

---

* In fact, over 90% of the project managers interviewed during a recent survey (46) mentioned the measures among the most important criteria of team performance.

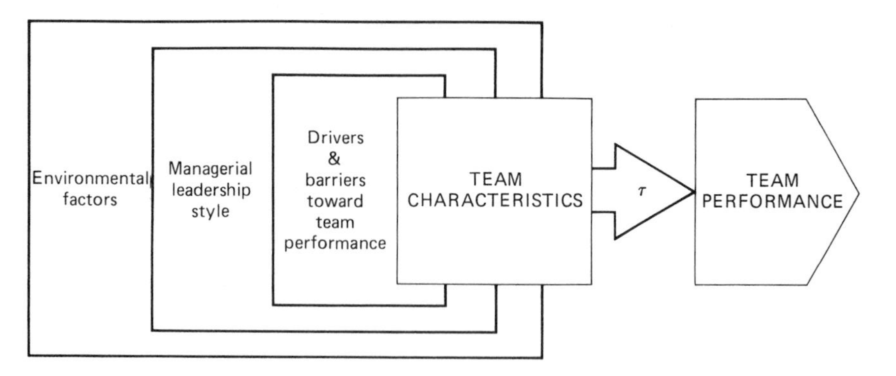


Figure 32-1. A simple model for analyzing project team performance.

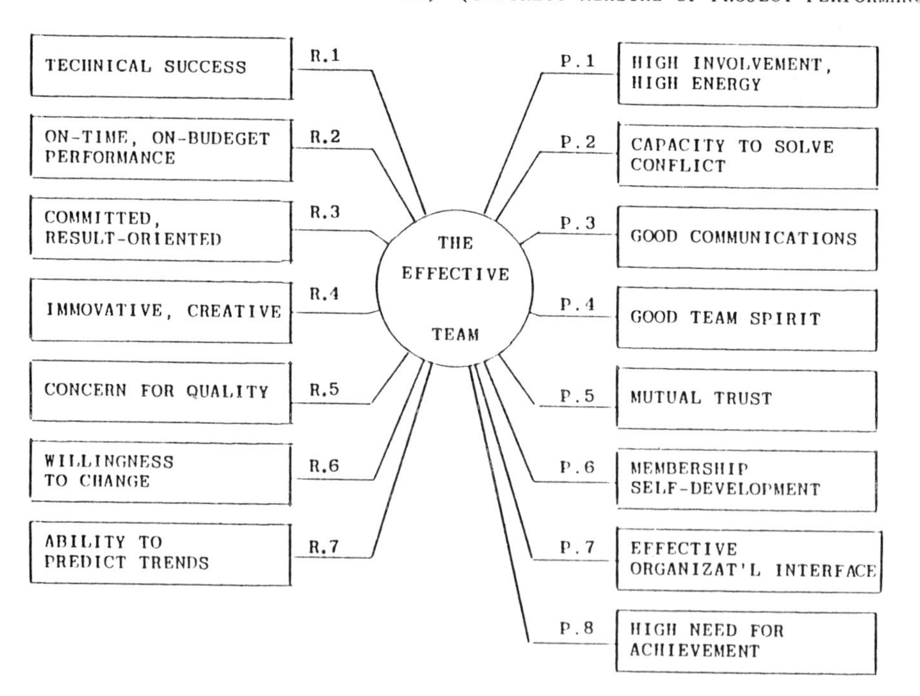


Figure 32-2. Characteristics of an effective project team.

Table 32-1. Qualities Associated with High-Performing Teams.

| TASK-RELATED QUALITIES | PEOPLE RELATED QUALITIES |
|---|---|
| • Committed to the project | • High involvement, work interest, and energy |
| • Result-oriented attitude | • Capacity to solve conflict |
| • Innovative and creative | • Good communication |
| • Willingness to change | • High need for achievement |
| • Concern for quality | • Good team spirit |
| • Ability to predict trends | • Mutual trust |
| | • Self-development of team members |
| | • Effective organizational interfacing |

In fact, field research shows (41) a statistically significant association between the above team qualities and team performance at a confidence level of $p = 95\%$ or better.*

The significance of determining team performance characteristics is in two areas. First, it offers some clues as to what an effective team environment looks like. This can stimulate management thoughts and activities for effective team building. Second, the results allow us to define measures and characteristics of an effective team environment for further research on organization development efforts, such as defining drivers and barriers toward team performance.

## DRIVERS AND BARRIERS OF HIGH TEAM PERFORMANCE

Additonal management insight has been gained by investigating drivers and barriers to high performance.† Drivers are factors associated with the project environment, such as interesting work and good project direction. These factors are preceived as enhancing team effectiveness and correlate *positively* to team performance. Barriers are factors, such as unclear objectives and insufficient resources, that are perceived as impeding team performance and statistically correlate *negatively* to performance.

Studies by Gemmill, Thamhain, and Wilemon (39, 41, 42) into work group dynamics clearly show significant correlations and interdependencies among work-environmental factors and team performance. These studies indicate that high team performance involves four primary fac-

---

* Specifically, a Kendall-Tau rank-order correlation model was used. These measures yielded an average association of $\tau = .37$. Moreover, there appears to be a strong agreement between managers and project team members on the importance of these characteristics, as measured via a Kruskal-Wallis analysis of arience at a confidence level of $p = 95\%$.

† The Kendall-Tau Rank-Order correlation was used to measure to association between these variables. Statistical significance was defined at a confidence level of 95% or better.

tors: (a) managerial leadership, (b) job content, (c) personal goals and objectives, and (d) work environment and organizational support. The actual correlation of 60 influence factors to the project team characteristics and performance provided some interesting insight into the strength and effect of these factors. One of the important findings was that only 12 of the 60 influence factors were found to be statistically significant.* All other factors seem to be much less important to high team performance. Specifically, the *six drivers which have the strongest positive association to project team performance are:*

1+ Professionally interesting and stimulating work.
2+ Recognition of accomplishment.
3+ Experienced engineering management personnel.
4+ Proper technical direction and leadership.
5+ Qualified project team personnel.
6+ Professional growth potential.

The *strongest barriers to project team performance are:*

1- Unclear project objectives and directions.
2- Insufficient resources.
3- Power struggle and conflict.
4- Uninvolved, disintegrated upper management.
5- Poor job security.
6- Shifting goals and priorities.

It is furthermore interesting to note that the six drivers not only correlated favorably to the direct measures of high project team performance, such as the technical success and on-time/on-budget performance, but also were positively associated with the 13 indirect measures of team performance, ranging from commitment to creativity, quality, change-orientation, and needs for achievement. The complete listing of the 16 performance measures is shown in Figure 32-3. The six barriers have exactly the opposite effect. These conclusions provide consistently and measurably support to the findings from a variety of field studies (4, 8, 9, 16, 21, 24, 28, 38, 41).

What we find consistently is that successful organizations pay attention to the human side. They seem to be effective in fostering a work environment conducive to innovative creative work, where people find the as-

---

* The research on team characteristics and drivers versus barriers is based on a field study by H. Thamhain and D. Wilemon, published as "A High-Performing Engineering Project Team," *IEEE Transactions on Engineering Management,* August, 1987.

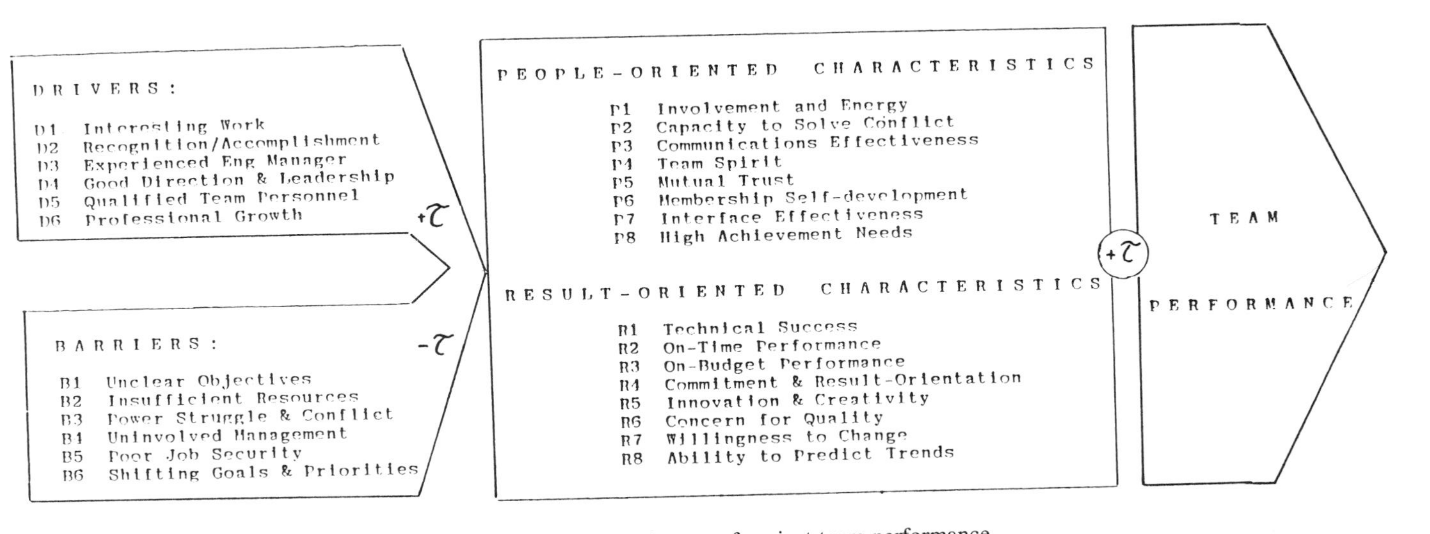


Figure 32-3. Influence factors of project team performance.

signments challenging, leading to recognition and professional growth. Such a professionally stimulating environment also seems to lower communication barriers and conflict, and enhances the desire of personnel to succeed. Further, this seems to increase organizational awareness as well as the ability to respond to changing project requirements.

Secondly, a winning team appears to have good leadership. That is, management understands the factors crucial to success. Managers are action oriented, provide the needed resources, properly direct the implementation of the project plan, and help in the identification and resolution of problems in their early stages. Taken together, the findings support three propositions:

P1 : The degree of project success seems to be primarily determined by the strength of six driving forces and six barriers which are related to (a) Leadership, (b) job content, (c) personal needs, and (4) the general work environment.

P2 : The strongest driver toward project success is a professionally stimulating team environment, characterized by (a) interesting, challenging work; (b) visibility and recognition for achievements; (c) growth potential; and (d) good project leadership.

P3 : A professionally stimulating team environment also leads to low perceived conflict, high commitment, highly involved personnel, good communications, change orientation, innovation, and on-time/on-budget performance

To be effective in organizing and directing a project team, the leader must not only recognize the potential drivers and barriers but also know when in the life cycle of the project they are most likely to occur. The effective project leader takes preventive actions early in the project life cycle and fosters a work environment that is conducive to team building as an ongoing process.

The effective team builder is usually a social architect who understands the interaction of organizational and behavioral variables and can foster a climate of active participation and minimal dysfunctional conflict. This requires carefully developed skills in leadership, administration, organization, and technical expertise. It further requires the project leader's ability to involve top management, to assure organizational visibility, resource availability, and overall support for the new project throughout its life cycle.

It is this organizational culture which adds yet another challenge to project team building. The new team members are usually selected from hierarchically organized functional resource departments led by strong individuals who often foster internal competition rather than cooperation. In fact, even at the contributor level, many of the highly innovative and creative people are highly individualistically oriented and often admit their aversion to cooperation. The challenge to the project manager is to integrate these individuals into a team that can produce innovative results in a systematic, coordinated, and integrated way to an overall project plan. Many of the problems that occur during the formation of the new project team or during its life cycle are normal and often predictable. However, they present barriers to effective team performance. They must be quickly identified and dealt with.

## ORGANIZING THE NEW PROJECT TEAM

Too often the project manager, under pressure to start producing, rushes into organizing the project team without establishing the proper organizational framework. While initially the prime focus is on staffing, the program manager cannot effectively attract and hold quality people until certain organizational pillars are in place. At a minimum, the basic project organization and various tasks must be defined before the recruiting effort can start.

These pillars are not only necessary to communicate the project requirements, responsibilities, and relationships to team members, but also to manage the anxiety which usually develops during the team formation. This anxiety is normal and predictable. It is a barrier, however, to getting the team quickly focused on the task.

This anxiety may come from several sources. For example, if the team members have never worked with the project leader, they may be concerned with his leadership style and its effect on them. In a different vein, team members may be concerned about the nature of the project and whether it will match their professional interests and capabilities. Other team members may be concerned whether the project will be helpful to their career aspirations. Furthermore, team members can be anxious about lifeor work style disruptions. As one project manager remarked: "Moving a team member's desk from one side of the room to the other can sometimes be just as traumatic as moving someone from Chicago to Manila to build a power plant" (41).

As the quote suggests, seemingly minor changes can result in sudden anxiety among team members.

Another common concern among newly formed teams is whether or not there will be an equitable distribution of work load among team members and whether each member is capable of pulling his or her own weight. In some newly formed teams, members not only have to do their own work, but they must also train others. Within reason this is bearable, necessary, and often expected. However, when it becomes excessive, anxiety increases and morale can fall.

### Make Functional Ties Work for You

It is a mistaken belief that strong ties of team members to the functional organization are bad for effective program management and should be eliminated. To the contrary, loyalty to both the project and the functional organization is a natural, desirable, and often very necessary conditon for project success. For example, in the most common of all project organizations, the matrix, the program office gives operational directions to the program personnel and is responsible for the budget and schedule, while the functional organization provides technical guidance and personnel administration. Both the program manager and the functional managers must understand this process and perform accordingly or severe jurisdictional conflicts can develop.

### Structure Your Organization

The key to successfully building a new project organization is clearly defined and communicated responsibilities and organizational relationships. The tools for systematically describing the project organization come, in fact, from conventional management practices:

1. *Charter of the Program or Project Organization*. The charter of the program office clearly describes the business mission and scope, broad responsibilities, authorities, the organizational structure, interfaces, and the reporting relationship of the program organization. The charter should be revised for each new program. For small projects a simple one-half page statement may be sufficient, while larger programs usually require a formal charter supported by standardized company policies on project management.

2. *Project Organization Chart*. Regardless of the specific organizational structure and its terminology used, a simple organizational chart shall define the major reporting and authority relationships. These relationships can be further clarified in a policy directive. (See Chapter ).

3. *Responsibility Matrix*. This chart defines the interdisciplinary task responsibilities: who is responsible for what. The responsbility matix not only covers activities within the project organization, but also defines the functional relationship with support units, subcontractors, and committees. In a simpler format, a task roster can be used to just list project tasks and corresponding responsible personnel.

4. *Job Description*. If not already in existence, a job description shall be developed for all key project personnel, such as the program managers, system project managers, hardware project managers, task managers, project engineers, plan coordinators, and so on. The job descriptions are usually generic and hence portable from one project to the next. Job descriptions are modular building blocks which form the framework for staffing a project organization. The job description includes (a) the reporting relationship, (b) responsibilities, (c) duties, and (d) typical qualifications.

**Define Your Project**

In dealing with engineering personnel, there is seldom a problem in finding project managers capable of defining the technical components of the project. The challenge is to convince these project leaders that all four segments of the project management system must be defined, at least in principle, before staffing can begin:

Segment 1. *The Work:*
- Overall specifications
- Requirements document
- Statement of work
- System block diagram
- Work breakdown structure
- List of deliverables

Segment 2. *Timing:*
- Master schedule
- Milestone chart
- Interdependencies of interfaces

Segment 3. *Resources:*
- Budget
- Resource plan

Segment 4. *Responsibilities:*
- Task matrix
- Task rostes
- Project charter
- Work packages

Regardless of how vague and preliminary these project segments are at the beginning, the initial description will help in recruiting the appropriate personnel and eliciting commitment to the preestablished parameters of technical performance, schedule, and budget. The core team should be formed prior to finalizing the project plan and contractual arrangements. This will provide the project management team with the opportunity to participate in trade-off discussions and customer negotiations leading to technical confidence and commitment of all parties involved.

## Staff Your Project

Staffing the project organization is the first major milestone during the project formation phase. Because of the pressures on the project manager to produce, staffing is often done hastily and without properly defining the basic project work to be performed.

The results are often personnel poorly matched to the job requirements, conflict, low morale, suboptimum decision making, and, in the end, poor project performance.

The comment of a project section leader who was pressed into quick staffing actions is indicative of these potential problems: "How can you interview task managers when you cannot show them what the job involves and how their responsibilities tie in with the rest of the project?"

Therefore, only after the project organization and the tasks are defined in principle can project leaders at various levels start to interview candidates. These interviews should always be conducted one to one. *The interview process* normally has five facets which are often interrelated:

1. *Informing the candidate about the assignments.*
   What are the objectives for the project?
   Who will be involved and why?
   Structure of the project organization and its interfaces.
   Importance of the project to the overall organization or work unit; short- or long-range impact.
   Why was the team member selected and assigned to the project? What role will he or she perform?
   Specific responsibilities and expectations.
   What rewards might be forthcoming if the project is completed successfully?
   A candid appraisal of the problems and constraints which are likely to be encountered.
   What are the rules of the road that will be followed in managing the project, such as regulate status review meetings?
   The challenges and recognition the project is likely to provide.
   Why is the team concept important to success and how should it work?
2. *Determining skills and expertise.*
   Probe related experience; expand from resume.
   Probe candidate's aptitude relevant to your project environment: technology involved, engineering tools and techniques, markets and customer involvement, and product applications.
   Probe into the program management skills needed. Use current project examples: "How would you handle this situation . . .?"

Probe leadership, technical expertise, planning and control, administrative skills, and so on.

3. *Determine interests and team compatibility.*
   What are the professional interests and objectives of this candidate?
   How does the candidate manage and work with others?
   How does the candidate feel about sharing authority, working for two bosses, or dealing with personnel across functional lines with little or no formal authority?
   What suggestions does the candidate have for achieving success?
4. *Persuading to join project team.*
   Explain specific rewards for joining the team, such as financial, professional growth, recognition, visibility, work challenge, and potential for advancement.
5. *Negotiating terms and commitments.*
   Check candidate's willingness to join team.
   Negotiate conditions for joining: salary, hired, signed, or transferred, performance reviews and criteria.
   Elicit candidate's commitment to established project objectives and modus operandi.
   Secure final agreement.

### Suggestions for Handling the Newly Formed Project Team

During its formation, the project group represents just a collection of individuals who have been selected for their skills and capabilities as collectively needed to perform the upcoming project task. However, to be successful the individual efforts must be integrated. Even more demanding, these individuals have to work together as a team to produce innovative results which fit together to form an integrated new system as conceptualized in the project plan.

Initially, there are many problems which prevent the project group from performing as a team. While these problems are normal and often predictable, they present barriers to effective team performance. The problems therefore must be quickly identified and dealt with. The following list presents typical *problems which occur during a project team formation:*

- Confusion.
- Responsibilities unclear.
- Channels of authority unclear.
- Work distribution load uneven.
- Assignment unclear.
- Communication channels unclear.

- Overall project goals unclear.
- Mistrust.
- Personal objectives unrelated to project.
- Measures of personal performance unclear.
- Commitment to project plan lacking.
- Team spirit lacking.
- Project direction and leadership insufficient.
- Power struggle and conflict.

Certain steps taken early in the life of the team can help the project leader in identifying the specific problems and dealing with them effectively. These steps may also provide preventive measures which eliminate the potential for these problems to develop in the first place. *Specific suggestions* are made below.

1. *Make Sure the Assignment Is Clear.* Although the overall task assignment, its scope, and objectives might have been discussed during the initial sign-on of the person to the project, it takes additional effort and involvement for new team members to feel comfortable with the assignment. The thorough understanding of the task requirements comes usually with the intense personal involvement of the new members with the project team. Such involvement can be enhanced by assigning the new member to an action-oriented task that requires team involvement and creates visibility, such as a requirements analysis, an interface specification, or producibility study. In addition, any committee-type activity, presentation, or data gathering will help to involve the new team member and to give him or her a better understanding of the specific task and his or her role in the overall team effort.

2. *Make New Team Members Feel Professionally Comfortable.* The initial anxieties, the lack of trust and confidence, are serious barriers to team performance. New team members should be properly introduced to the group, and their roles, strengths, and criticallity to the project explained. Providing opportunities for early results allows the leader to give recognition for professional accomplishments which will stimulate the individual's desire for the project work and build confidence, trust, and credibility within the group.

3. *Be Certain Team Organization Is Clear.* Project team structures are often considered very "organic" and inconsistent with formal chain-of-command principles. However, individual task responsibility, accountability, and organizational interface relations should be clearly defined to all team members. A simple work breakdown structure (WBS) or task matrix, together with some discussion, can facilitate a clear understanding of the team structure, even with a highly unconventional format.

4. *Locate Team Members in One Place.* Members of the newly formed team should be closely located to facilitate communications and the development of a team spirit. Locating the project team in one office area is the ideal situation. However, this may be impractical, especially if team members share their time with other projects or if the assignment is only for a short period of time. Regularly scheduled meetings are recommended as soon as the new project team is being formed. These meetings are particularly important where team members are geographically separated and do not see each other on a day-to-day basis.

5. *Provide a Proper Team Environment.* It is critical for management to provide the proper environment for the project to function effectively. Here the project leader needs to tell the management at the outset of the program what resources are needed. The project manager's relationship with senior management support is critically affected by his or her credibility and the visibility and priority of the project.

6. *Manage.* Especially during the initial stages of team formation, it is important for the project leader to keep a close eye on the team and its activities to detect problems early and to correct them. The project manager can also influence the climate of the work environment by his of her own actions. The manager's concern for project team members, ability to integrate personal goals and needs of project personnel with the project objectives, and ability to create personal enthusiasm for the work itself can foster a climate which is high on motivation, work involvement, and resulting project performance.

## TEAM BUILDING AS AN ONGOING PROCESS

While proper attention to team building is critical during the early phases of a project, it is a never-ending process. The project manager is continually monitoring team functioning and performance to see what corrective action may be needed to prevent or correct various team problems. Several barometers provide good clues of potential team dysfunctioning. First, noticeable changes in performance levels for the team and/or for individual team members should always be followed up. Such changes can be symptomatic of more serious problems, such as conflict, lack of work integration, communication problems, and unclear objectives. Second, the project leader and team members want to be aware of the changing energy level in various team members. This, too, may signal more serious problems or indicate that the team is tired and stressed. Sometimes changing the work pace, taking time off, or selling short-term targets can serve as a means to reenergize team members. More serious cases, however, can call for more drastic measures, such as reappraising project objectives and/or the means to achieve them. Third, verbal and nonverbal clues from team members may be a source of information on team functioning. It is important to hear their needs and concerns (verbal clues) and to observe how they act in carrying out their responsibilities (nonverbal clues). Finally, detrimental behavior of one team member toward another can be a signal that a problem within the team warrants action.

It is highly recommended that project leaders hold regular meetings to evaluate overall team performance and deal with team functioning problems. The focus of these meetings can be directed toward "What are we doing well as a team?" and "What areas need our team's attention?" This approach often brings positive surprises in that the total team will be informed on progress in diverse project areas such as a breakthrough in the technology department, a subsystem schedule met ahead of the original target, or a positive change in the client's behavior toward the project. After the positive issues have been discussed, attention should be devoted toward areas needing team attention. The purpose of this part of the review session is to focus on real or potential problem areas. The meeting leader should ask each team member for his or her observations on these issues. Then an open discussion should be held to ascertain how significant the problems really are. Assumptions should, of course, be separated from the facts of each situation. Next, assignments should be agreed upon how to best handle these problems. Finally, a plan for follow-up should be developed. The process should result in better overall performance and promote a feeling of team participation and high morale.

Over the life of a project the problems encountered by the project team are likely to change, and as the old problems are identified and solved, new ones will emerge.

In summary, effective building is a critical determinant of project success. While the process of team building can entail frustrations and require energy on the part of all concerned, the rewards can be great.

Social scientists generally agree that there are several indicators of effective and ineffective teams. At any point in the life of a team, the project manager should be aware of certain effectiveness-ineffectiveness indicators, which are summarized in Table 32-2.

As we go through the next decade, we anticipate important developments in team building. As shown in Figure 32-3, these developments will lead to higher performance levels, increased morale, and a pervasive commitment to final results that can withstand almost any kind of adversity.

Table 32-2. Project Team Characteristics: Effective Versus Ineffective.

| LIKELY CHARACTERISTICS OF EFFECTIVE TEAM | LIKELY CHARACTERISTICS OF INEFFECTIVE TEAM |
|---|---|
| + High performance and task efficiency | – Low performance |
| + Innovative/creative behavior | – Activity-oriented |
| + Committed, results oriented | – Low level of involvement and enthusiasm |
| + Professional objectives of team members coincide with project requirements | – Low commitment to project objectives |
| + Technically successful | – Unclear project objectives and fluid commitment levels from key participants |
| + On-time/on-budget performance | – Schedule and budget slips |
| + Team members highly interdependent, interface effectively | – Uninvolved management |
| + Capacity for conflict resolution, but conflict encouraged when it can lead to beneficial results | – Anxieties and insecurities |
| + Communicates effectively | – Unproductive gamesmanship, manipulation of others, hidden feelings, conflict avoided at all costs |
| + High trust levels | – Confusion, conflict, inefficiency |
| + High achievement needs | – Subtle sabotage, fear, disinterest, or foot-dragging |
| + Result-oriented | – Frequent surprises |
| + Interested in membership self-development | – Quality problems |
| + High energy levels and enthusiasm | – Cliques, collusion, isolating members |
| + High morale | – Image problems (credibility) |
| + Change-oriented | – Lethargic/unresponsive |

## RECOMMENDATIONS FOR EFFECTIVE TEAM MANAGEMENT

Taken together, the project leader must foster an environment where team members are professionally satisfied, are involved, and have mutual trust. As shown in Figure 32-2, the more effective the project leader is in stimulating the drivers and minimizing the barriers, the more effective the manager can be in developing team membership and the higher the quality of information contributed by team members, including their willingness and candor in sharing ideas and approaches. By contrast, when a team member does not feel part of the team and does not trust others, information will not be shared willingly or openly. One project leader emphasized the point: "There's nothing worse than being on a team where no one trusts anyone else. Such situations lead to gamesmanship and a lot of watching what you say because you don't want your own words to bounce back in your own face. . . ."

Furthermore, the greater the team spirit, trust, and quality of information exchange among team members, the more likely the team will be able to develop effective decision-making processes, make individual and group commitment, focus on problem solving, and develop self-forcing, self-correcting project controls. As summarized in Figure 32-2, these are the characteristics of an effective and productive project team. *A number of specific recommendations* are provided for project leaders and managers responsible for the integration of multidisiplinary tasks to help in their complex efforts of building high-performing project teams.

1. *Barriers.* Project managers must understand the various barriers to team development and build a work environment conducive to the team's motivational needs. Specifically, management should watch out for the following barriers: (a) unclear objectives, (b) insufficient resources and unclear funding, (c) role conflict and power struggle, (d) uninvolved and unsupportive management, (e) poor job security, and (f) shifting goals and priorities.
2. *The Project Objectives* and their importance to the organization should be clear to all personnel who get involved with the project. Senior management can help develop a "priority image" and communicate the basic project parameters and management guidelines.
3. *Management Commitment.* Project managers must continuously update and involve their managements to refuel their interests and commitments to the new project. Breaking the project into smaller phases and being able to produce short-range results frequently seem to be important to this refueling process.
4. *Image Building.* Building a favorable image for the project, in terms of high priority, interesting work, importance to the organization, high visibility, and potential for professional rewards is crucial to the ability to attract and hold high-quality people. It is also a pervasive processs which fosters a climate of active participation at all levels; it helps to unify the new project team and minimizes dysfunctional conflict.
5. *Leadership Positions* should be carefully defined and staffed at the beginning of a new program. Key project personnel selection is the joint responsibility of the project manager and functional management. The credibility of project leaders among team members, with senior management, and with the program sponsor is crucial to the leader's ability to manage the multidisciplinary activities effectively across functional lines. One-on-one interviews are recommended for explaining the scope and project requirements, as well as the management philosophy, organizational structure, and rewards.

6. *Effective Planning* early in the project life cycle will have a favorable impact on the work environment and team effectiveness. This is especially so because project managers have to integrate various tasks across many functional lines. Proper planning, however, means more than just generating the required pieces of paper. It requires the participation of the entire project team, including support departments, subcontractors, and management. These planning activities, which can be performed in a special project phase such as Requirements Analysis, Product Feasibility Assessment, or Product/Project Definition, usually have a number of side benefits besides generating a comprehensive road map for the upcoming program.
7. *Involvement.* One of the side benefits of proper project planning is the involvement of personnel at all organizational levels. Project managers should drive such an involvement, at least with their key personnel, especially during the project definition phases. This involvement will lead to a better understanding of the task requirements, stimulate interest, help unify the team, and ultimately lead to commitment to the project plan regarding technical performance, timing, and budgets.
8. *Project Staffing.* All project assignments should be negotiated individually with each prospective team member. Each task leader should be responsible for staffing his or her own task team. Where dual-reporting relationships are involved, staffing should be conducted jointly by the two managers. The assignment interview should include a clear discussion of the specific task, the outcome, timing, responsibilities, reporting relation, potential rewards, and importance of the project to the company. Task assignments should be made only if the candidate's ability is a reasonable match to the position requirements and the candidate shows a healthy degree of interest in the project.
9. *Team Structure.* Management must define the basic team structure and operating concepts early during the project formation phase. The project plan, task matrix, project charter, and policy are the principal tools. It is the responsibility of the project manager to communicate the organizational design and to assure that all parties understand the overall and interdisciplinary project objectives. Clear and frequent communication with senior management and the new project sponsor becomes critically important. Status review meetings can be used for feedback.
10. *Team-Building Sessions* should be conducted by the project

manager throughout the project life cycle. An especially intense effort might be needed during the team formation stage. The team is being brought together in a relaxed atmosphere to discuss such questions as, How are we operating as a team? What is our strength? Where can we improve? What steps are needed to initiate the desired change? What problems and issues are we likely to face in the future? Which of these can be avoided by taking appropriate action now? How can we "danger-proof" the team?

11. *Team Commitment.* Project managers should determine lack of team member commitment early in the life of the project and attempt to change possible negative views toward the project. Since insecurity is often a major reason for lacking commitment, managers should try to determine why insecurity exist, then work on reducing the team members' fears. Conflict with other team members may be another reason for lack of commitment. It is important for the project leader to intervene and mediate the conflict quickly. Finally, if a team member's professional interests may lie elsewhere, the project leader should examine ways to satisfy part of the team member's interests by bringing personal and project goals into perspective.
12. *Senior Management Support.* It is critically important for senior management to provide the proper environment for the project team to function effectively. Here the project leader needs to tell management at the outset of the program what resources are needed. The project manager's relationship with senior management and ability to develop senior management support is critically affected by his or her credibility, visibility, and priority image of the project.
13. *Organization Development Specialists.* Project leaders should watch for changes in performance on an ongoing basis. If performance problems are observed, they should be dealt with quickly. If the project manager has access to internal or external organization development specialists, they can help diagnose team problems and assist the team in dealing with the identified problems. These specialists can also bring fresh ideas and perspectives to difficult and sometimes emotionally complex situations.
14. *Problem Avoidance.* Project leaders should focus their efforts on problem avoidance. That is, the project leader, through experience, should recognize potential problems and conflicts at their onset and deal with them before they become big and their resolutions consume a large amount of time and effort.

## A FINAL NOTE

In summary, effective team building is a critical determinant of project success. Building the engineering team for a new technical project is one of the prime responsibilities of the program leader. Team building involves a whole spectrum of management skills to identify, commit, and integrate the various personnel from different functional organizations into a single task group. In many project-oriented engineering organizations, team building is a shared responsibility between the functional engineering managers and the project manager, who often reports to a different organization with a different superior.

To be effective, the project manager must provide an atmosphere conducive to teamwork. Four major considerations are involved in the integration of people from many disciplines into an effective team: (a) creating a professionally stimulating work environment, (b) providing good program leadership, (c) providing qualified personnel, and (d) providing a technically and organizationally stable environment. The project leader must foster an environment where the new project team members are professionally satisfied, involved, and have mutual trust. The more effectively project leaders develop team membership, the higher is the quality of information exchanged, including the candor of sharing ideas and approaches. It is this professionally stimulating involvement that also has a pervasive effect on the team's ability to cope with change and conflict, and leads to innovative performance. By contrast, when a member does not feel part of the team and does not trust others, information will not be shared willingly or openly.

Furthermore, the greater the team spirit, trust, and quality of information exchange among team members, the more likely the team will be able to develop effective decision-making processes, make individual and group commitment, focus on problem solving, and develop self-forcing, self-correcting project controls. These are the characteristics of an effective and productive project team.

The potential gains from increased engineering productivity are great for individuals, organizations, and society as a whole. Such gains are possible only if we can utilize our engineering resources effectively. One such improvement is through effective team building.

Over the next decade we anticipate important developments in team building which will lead to higher performance levels, increased morale, and a pervasive commitment to final results. This chapter should help both the professional in the field of engineering management as well as the scholar who studies contemporary organizational concepts to understand the intricate relationships between organizational and behavioral elements by providing a conceptual framework for specific situational analysis of engineering team-building practices.

## BIBLIOGRAPHY


1. Adams, John R. and Kirchof, Nicki S. "A Training Technique for Developing Project Managers." *Project Management Quarterly* (March, 1983).
2. Altier, William J., "Task Forces—An Effective Management Tool" *Sloan Management Review,* Spring 1986
3. Aquilino, J. J. "Multi-Skilled Work Teams: Productivity Benefits." *California Management Review* (Summer, 1977).
4. Aram, J. D. and Morgan, C. P. "Role of Project Team Collaboration in R&D Performance." *Management Science* (June, 1976).
5. Atkins, S. and Katcher, A. "Getting Your Team In Tune" *Nation's Business* (March, 1975).
6. Baler, Kent H. "The Hows and Whys of Teambuilding," *Engineering Management Review* (December, 1985).
7. Barkman, Donald F., "Team Discipline: Put Performance on the Line," *Personnel Journal,* Vol 66, March 1987, p. 58
8. Benningson, Lawrence. "The Team Approach to Project Management." *Management Review, 61*:48–52 (January, 1972).
9. Carzo, R., Jr. "Some Effects of Organization Structure on Group Effectiveness." *Administrative Science Quarterly* (March, 1963).
10. Conover, W. J. *Practical Nonparametric Statistics*. Wiley, New York, 1971.
11. Diliddo, Bart A., James, Paul, C. and Dietrich, Harry J. "Managing R&D Creatively: B. F. Goodrich's Approach." *Management Review* (July, 1981).
12. Ely, D. D. "Team Building for Creativity." *Personnel Journal* (April, 1975).
13. Foster, Richard N. "A Call for Vision in Managing Technology." *McKinsy Quarterly* (Summer, 1982).
14. Galagan, Patricia, "Work Teams at Work", *Training and Development Journal,* Vol 40, November 1986, p. 33
15. Gray, James, "Team Building: Transforming Individuals into Work Groups", *Executive,* Vol 26, Winter 1986, p. 24
16. Harris, Philip R. "Building a High-Performance Team." *Training and Development Journal* (April, 1986).
17. Hayes, J. L. "Teamwork." *Management Review* (September, 1975).
18. Hopkins, D. S. "Roles of Project Teams and Venture Groups in New Product Development." *Research Management* (January, 1975).
19. Howe, R. J. "Building Teams for Increased Productivity." *Personnel Journal* (January, 1977).
20. Huesing, S. A. "Team Approach and Computer Development." *Journal of Systems Management* (September, 1977).
21. Jewkes, John, Sawers, David and Stillerman, Richard. *The Sources of Innovation*. Macmillan, New York, 1962.
22. Katz, F. E. "Explaining Informal Work Groups in Complex Organizations." *Administrative Science Quarterly* (10) (1965).
23. Kidder, John Tracy. *The Soul of a New Machine*. Avon Books, Hearst Corporation, New York, 1982.

24. Likert, Rensis. "Improving Cost-Performance with Cross-Functional Teams." *Management Review* (March, 1976).
25. Maister, D. H. "The One-Firm: What Makes It Successful." *Sloan Management Review* (Fall, 1985).
26. Miller, Donald Britten, "Understanding the R&D Culture," *Management Review,* Vol 75, December 1986, p. 34
27. Pincus, Claudio. "An Approach to Plan Development and Team Formation." *Project Management Quarterly* (December, 1982).
28. Quinn, James Brian. "Technological Innovation, Entrepreneurship and Strategy." *Sloan Management Review* (Spring, 1979).
29. Rantfl, R. M. *R&D Productivity, A Study Report.* Hughes Aircraft Company (1978).
30. Raudsepp, Eugene. "Motivating Engineers." *Engineering Management Review* (March, 1986).
31. Reich, Robert, "Entrepreneurship Reconsidered: The Team as Hero", *Harvard Business Review,* Vol 65, May-June 1987, p. 77
32. Rigby, Malcom J., "The Challenge of Multinational Team Development," *Journal of Management Development,* Vol 6, Fall 1987, p. 65
33. Rogers, L. A. "Guidelines for Project Management Teams." *Industrial Engineering* (December, 1974).
34. Salomon, B. A. "A Plant that Proves that Team Management Works" (Digital). *Personnel* (June, 1985).
35. Senia, Al, "Hewlett-Packards Team Approach beats back the Competition" *Production,* Vol 97, May 1986, p. 89
36. Shea, Gregory P. and Guzzo, Richard A., "Group Effectiveness: What really matters", *Sloan Management Review,* Vol 76, Spring 1987
37. Thamhain, Hans J., *Team Building in Technology-based Organizations,* Addison-Wesley, 1988
38. Thamhain, Hans J. and Wilemon, David L. "Building High Performing Engineering Project Teams." *IEEE Transactions on Engineering Management* (August, 1987).
39. Thamhain, Hans J. "Managing Engineers Effectively." *IEEE Transactions on Engineering Management* (August, 1983).
40. Thamhain, H. and Wilemon, D. "Skill Requirements of Engineering Program Manager." *Proceedings of the 26th Engineering Management Conference* (1978).
41. Thamhain, Hans J. and Wilemon, David L. "Anatomy of a High Performing New Product Team." *Convention Record, 16th Annual Symposium of the Project Management Institute.*
42. Thamhain, Hans J. and Gemmill, Gary R. "Influence Styles of Project Managers: Some Project Performance Correlates." *Academy of Management Journal* (June, 1974).
43. Tichy, N. "Analysis of Clique Formation and Structure in Organizations." *Administrative Science Quarterly* (June, 1973).
44. Watson, D. J. H. "Structure of Project Teams Facing Differentiated Environments: An Exploratory Study in Public Accounting Firms." *Accounting Review* (April, 1975).
45. Ward, Brian K. and Hardaker Maurice, "How to make a Team Work", *Harvard Business Review,* Vol 65, November-December 1987
46. Wilemon, David L. and Thamhain, Hans J. "Team Building in Project Management." *Project Management Quarterly* (July, 1983).
47. Wilemon, D. L., et al. "Managing Conflict on Project Teams." *Management Journal* (1974).

48. Wilemon, David L. and Thamhain, Hans J. "A Model for Developing High-Performance Teams." *Proceedings of the Annual Symposium of the Project Management Institute,* Houston, (1983).
49. Zenger, J. H. and Miller, D. E. "Building Effective Teams." *Personnel* (March, 1974).
50. Ziller, R. C. "Newcomer's Acceptance in Open and Closed Groups." *Personnel Administration* (September, 1962).

# 33. Some Major Research Findings Regarding the Human Element in Project Management

David L. Wilemon*
Bruce N. Baker†

One of the most significant developments in management practice during the past two decades has been the accelerated emphasis on project management in administering complex tasks. Project management is a widely utilized management system. The early project management literature tended to be oriented around the development and explanation of the tools and techniques of the project manager (1).‡ More recently, however, increased research attention has been placed on the behavioral and organizational dimensions of project management. This research has resulted in a growing body of knowledge which helps explain the myriad of complex human factors which contribute to project management effectiveness. This chapter summarizes the mainstream of research in the human factors of project management. This review should help users better understand the numerous interpersonal forces found in project organizations.

Although the authors reviewed dozens of articles, some important research may have been omitted. Space limitations preclude a complete coverage of all the relevant research. In some areas, pertinent research dealing with general management problems has been cited to further con-

---

* Dr. David Wilemon is a professor and director of the Innovation Management Program in the Graduate School of Management at Syracuse University. He is widely recognized for his work on conflict management, team building, and leadership skills in project-oriented work environments. He has studied various kinds of project management systems in the United States and in several foreign countries.

† Dr. Bruce N. Baker is Program Manager of Information and Computer Security at SRI International. He conducts project management seminars and is a frequent speaker on the topics of improving teamwork and success in project management environments. He received his A.B. degree from Princeton University, his M.B.A. degree from Stanford University, and his D.P.A. degree from The George Washington University.

‡ Numbered references are given at the end of this chapter.

tribute to the understanding of the interpersonal dimensions of project management.

Five major areas were selected for a review of key research contributions. These areas include (a) leadership styles/interpersonal skills; (b) conflict management; (c) decision-making styles and team-building skills; (d) organizational design and project manager authority relationships; (e) communications in project management; and (f) project team relationships with the parent, client, and other external organizations.

## LEADERSHIP STYLES/INTERPERSONAL SKILLS

The leadership abilities and interpersonal skills of the project manager are critical to effective project management performance. While there has been much discussion on the role of leadership in project management, only recently has there been a growing interest in empirical investigations of some of the determinants of effective project management leadership.

Lawerence and Lorsch investigated the differences between effective and ineffective integrators (managerial positions like project managers) in terms of their behavioral styles in dealing with others in their organizations (2). Ten integrators were rated as "effective" and ten were evaluated as "less effective" (superiors' ratings were utilized). It was found that:

- Effective integrators had a significantly higher need for affiliation than the integrators rated as less effective. Differently put, the effective integrators had higher needs for interpersonal involvement, interactions, and demonstrated empathy in dealing with others.
- No statistically significant findings were found between the effective and less effective integrators in their need for achievement. A tendency, however, did emerge which seemed to indicate that the more effective integrators had a lower need for achievement than their counterparts.
- The need for power was rated approximately the same for the effective and the less effective integrators.
- Integrators rated as effective "prefer to take significantly more initiative and leadership, they are aggressive, confident, persuasive, and verbally fluent. In contrast, less effective integrators avoid situations that involve tension and decisions"(3).
- Effective integrators also were more ambitious, forceful, and effective in communications than those rated as less effective.

Hodgetts empirically addressed the means of overcoming the "authority gap" in project management (4). Researching project management in aerospace, construction, chemicals, and state government environments, he found the following:

- Negotiation skills were important in aerospace and construction project environments.
- Personality and/or persuasive ability was considered important in all the project management situations.
- The project manager's competence was considered important in aerospace, construction, and chemicals.
- Reciprocal favors were noted as important as a surrogate for authority in aerospace and construction.
- The combined sample of firms (aerospace, construction, chemicals, and state government) rated the four authority supplements as "very important" or "not important." The following represents the significance of each technique as rated by the project managers in overcoming authority deficiencies. (Percentages are for those authority surrogates rated as either very important or important.)

| | |
|---|---|
| —Competence | 98% |
| —Personality and/or Persuasive Ability | 96% |
| —Negotiation Ability | 92% |
| —Reciprocal Favors | 47% |

Gemmill and Wilemon's exploratory research on 45 project managers and supporting project team members focused on identifying several influence bases utilized by project managers in eliciting support (5). Their research suggested the following:

- Authority, reward, punishment. expertise, and referent power are sources of influence frequently utilized by project managers in gaining support. Each influence mode can have different effects on the climate of the project organization.
- Two fundamental management styles used by project managers were identified. The first style relied primarily on the project manager's authority, ability to reward, and ability to "punish" those who did not furnish needed support. The second relied on an expert and referent power influence style.

Gemmill and Thamhain's empirical research of 22 project managers and 66 project support personnel addressed the relationship of the project manager's utilization of interpersonal influence and project performance (6). Their research revealed the following:

- Support project personnel ranked the eight influence methods as follows (1 is most important, 8 is least important):

| *Influence Method* | *Mean* |
|---|---|
| Authority | 3.0 |
| Work Challenge | 3.2 |
| Expertise | 3.3 |
| Future Work Assignments | 4.6 |
| Salary | 4.6 |
| Promotion | 4.8 |
| Friendship | 6.2 |
| Coercion | 7.8 |

- Project managers who were perceived to utilize expertise and work challenge as influence modes experienced higher levels of project performance.
- Project performance was positively associated with high degrees of support, open communication among project participants, and task involvement by those supporting the project manager.
- The use of authority by project managers as means to influence support personnel led to lower levels of project performance.

The work of Fiedler has been a catalyst to the research and literature concerning effective leadership styles under various levels of authority and for various task situations (7). Space limitations preclude describing his model which supports a contingency-oriented approach to leadership, but one of his major findings is that:

> Both the directive managing, task-oriented leaders and the nondirective, human relations-oriented leaders were successful under some conditions. Which leadership style is the best depends on the favorableness of the particular situation and the leader. In very favorable or in very unfavorable situations for getting a task accomplished by group effort, the autocratic, task-controlling, managing leadership works best. In situations intermediate in difficulty, the non-directive, permissive leader is more successful [8]. This corresponds well with our everyday experience. For instance:
>
> - Where the situation is very favorable, the group expects and wants the leader to give directions. We neither expect nor want

the trusted airline pilot to turn to his crew and ask, "What do you think we ought to check before takeoff?"

- If the disliked chairman of a volunteer committee asks his group what to do, he may be told that everybody ought to go home.
- The well-liked chairman of a planning group or research team must be nondirective and permissive in order to get full participation from team members. The directive, managing leader will tend to be more critical and to cut discussion short; hence he or she will not get the full benefit of the potential contributions of group members.

The varying requirements of leadership styles are readily apparent in organizations experiencing dramatic changes in operating procedures. For example:

- The manager or supervisor of a routinely operating organization is expected to provide direction and supervision for subordinates to follow. However, in a crisis the routine is no longer adequate, and the task becomes more ambiguous and unstructured. The typical manager tends to respond in such instances by calling the group together for a conference. In other words, the effective leader changes his or her behavior from a directive to a permissive, nondirective style until the operation again reverts to routine conditions.
- In the case of a research planning group, the human relations-oriented and permissive leader provides a climate in which everybody is free to speak up, to suggest, and to criticize. Brainstorming can help institutionalize these procedures. However, after the research plan has been completed, the situation becomes more structured. The director now prescribes the task in detail, and he specifies the means for accomplishing it. Woe betide the assistant who decides to be creative by changing the research instructions! [9].

Other research findings also support a contingency-based view of project management organization design.

## CONFLICT MANAGEMENT

It is widely accepted that project environments can produce intense conflict situations (10). The ability of project managers to handle conflict is a determinant of successful project performance. Researchers have addressed the causes of disagreements in project management as well as the means by which conflict is managed.

### Determinants of Conflict

Wilemon's study on delineating fundamental causes of conflict in project management revealed that (11):

- The greater the diversity of expertise among the project team members, the greater the potential for conflict to develop.
- The lower the project manager's power to reward and punish, the greater the potential for conflict.
- The less the specific objectives of a project are understood by project team members, the more likely that conflict will occur.
- The greater the ambiguity of roles among the project team members, the more likely conflict will develop.
- The greater the agreement on superordinate goals (top management objectives), the lower the potential for detrimental conflict.
- The lower the project manager's formal authority over supporting functional and staff units, the higher the probability of conflict.

Butler also developed a number of propositions on the primary causes of conflict in project management (12). Many of his propositions are supported by prior research on conflict in various organizational settings—not exclusively project management. A few of the propositions advanced by Butler may be summarized as follows:

- Conflict may be either functional (beneficial) or dysfunctional (detrimental).
- Conflict is often caused by the revised interaction patterns of team members in project organizations.
- Conflict can develop as a result of the difficulties of team members adapting their professional objectives to project work situations and requirements.
- Conflict often is the result of the difficulties of diverse professionals working together in a project team situation where there is pressure for consensus.
- Role ambiguity and stress by the project managers and supporting functional personnel are more likely to occur when project authority is not clearly defined.
- Competition over functional resources, especially functional personnel, is likely to produce conflict.
- Conflict may develop over the lack of professional incentives derived from functional specialists participating in project-oriented work.

Thamhain and Wilemon's research focused on the causes and intensity of various conflict sources (13). Utilizing a sample of 100 project managers, their study measured the degree of conflict experienced from several variables common to project environments which were thought particularly conducive to the generation of conflict situations.

- The potential sources of conflict researched revealed the following rank-order for conflict experienced by project managers:
  1. schedules
  2. project priorities
  3. manpower resources
  4. technical conflicts
  5. administrative procedures
  6. cost objectives
  7. personality conflicts
- The most intense conflicts occurred with the supporting functional departments, followed by conflict with personnel assigned to the project team from functional departments.
- The lowest degree (intensity) of conflict occurred between the project manager and the immediate subordinates.

Thamhain and Wilemon followed their 1974 research with a study focused on measuring the degree of conflict experienced in each of the four generally accepted project life-cycle phases, namely, project formation, buildup, main program, and phaseout (14). Results reported from this research include the following:

- Disagreements over schedules result in the most intense conflict situations over the entire life cycle of a project.
- The mean conflict intensities over the four life-cycle stages reveal the following rank order:

  *Project Formation*
  1. Project priorities.
  2. Administrative procedures.
  3. Schedules.
  4. Manpower resources.
  5. Cost.
  6. Technical conflicts.
  7. Personality.

  *Buildup Phase*
  1. Project priorities.
  2. Schedules.
  3. Administrative procedures.

4. Technical conflicts.
5. Manpower resources.
6. Personality.
7. Cost.

*Main Program Phase*

1. Schedules.
2. Technical conflicts.
3. Manpower resources.
4. Project priorities.
5. Administrative procedures, cost, personality.

*Phaseout*

1. Schedules.
2. Personality.
3. Manpower resources.
4. Project priorities.
5. Cost.
6. Technical conflicts.
7. Administrative procedures.

### Conflict-handling Methods

If recognizing some of the primary determinants of conflict is a first step in effective conflict management, the second step is understanding how conflictful situations are managed in the project environment. Lawrence and Lorsch examined the methods that "integrators" used in handling conflicts (15). The following items from their study are considered pertinent:

- The uses of three conflict-handling modes were examined, namely, the confrontation or problem-solving mode, the smoothing approach, and the forcing mode. The utilization of forcing often results in a win-lose situation.
- The most effective integrators relied most heavily on the confrontation approach.
- Functional managers supporting the integrators in the most effective organizations also relied more on the confrontation approach than the other two modes.
- Functional managers in the highly integrated organizations employed "more forcing, and/or less smoothing behavior" than their counterparts in less effective organizations.

Building on the methodologies of Lawrence and Lorsch (16), Blake and Mouton (17), and Burke (18), Thamhain and Wilemon examined the effects of five conflict-handling modes (forcing, confrontation, compromising, smoothing, and withdrawal) on the intensity of conflict experienced (19). They found:

- When interacting with personnel assigned from functional organizations, the forcing and withdrawal methods were most often associated with increased conflict in the project management environments.
- Project managers experienced more conflict when they utilized the forcing and confrontation modes with functional support departments.
- The utilization of the confrontation, compromise, and smoothing approaches by project managers were often associated with reduced degrees of conflict in dealing with assigned personnel.
- The withdrawal approach was associated with lower degrees of conflict. (This of course may be detrimental to overall project performance.)

To determine the actual conflict-handling styles utilized by project managers, research was conducted by Thamhain and Wilemon in conjunction with their study on conflict in project life cycles (20). The results reported included:

- The problem-solving or confrontation mode was the most frequently utilized mode of project managers (70%).
- The compromising approach ranked second, with the smoothing approach ranking third. The forcing and withdrawal approaches ranked fourth and fifth.
- Project managers often use the full spectrum of conflict-handling modes in managing diverse personalities and various conflict situations.

Several suggestions for minimizing or preventing detrimental conflict were also provided by the study.

First, conflict with supporting functional departments is a major concern for project managers. Within the various categories of common conflict sources (schedules, project priorities, manpower resources, technical opinions, administrative procedures, and cost objectives), the highest conflict intensity occurs with functional support departments. The project manager frequently has less control over supporting functional departments than over his assigned personnel or immediate team members, which contributes to conflict. Moreover, conflict often develops due to the functional department's own priorities which can have impact on any of the conflict categories, that is, manpower resources and schedules.

Minimizing conflict requires careful planning. Effective planning early in the life cycle of the project can assist in forecasting and perhaps minimizing a number of potential problem areas likely to produce conflict in subsequent project phases. Consequently, contingency plans should be developed early in the life of a project. Senior management involvement in and commitment to the project may also help reduce some of the conflicts over project priorities and needed manpower resources and administrative procedures. In the excitement and haste of launching a new project, good planning by project managers is often insufficient.

Second, since there are a number of key participants in a project, it is important that major decisions affecting the project be communicated to all project related personnel. When project objectives are openly communicated there is a higher potential for minimizing detrimental, unproductive conflict. Regularly scheduled status review meetings, for example are for communicating important project-related issues.

Third, project managers need to be aware of their conflict resolution styles and their potential effect on key interfaces. Forcing and withdrawal modes appear to increase conflict with functional support departments and assigned personnel, while confrontation (problem-solving) and compromise can reduce conflict. Again, it is important for project managers to know when conflict should be minimized and when it should be induced. In some instances project managers may deliberately create conflict to gain new information and provoke constructive dialogue. Creating an open dialogue can produce positive results for the decision-making process.

Fourth, a definite relationship appears to exist between the specific influence mode of project managers and the intensity of conflicts experienced with interfaces. For example, the greater the work challenge provided by a project manager, the less conflict experienced with assigned project personnel. Thus, project managers need to consider the importance of work challenge not only in eliciting support but also in assisting in the minimization of conflict. One approach is to stimulate interest in the project and to match the professional needs of supporting personnel with the task requirements of the project.

Conflict with functional departments also can develop if the project manager overly relies on penalties and authority. The overuse of these power sources can have a negative effect in establishing a climate of support, cooperation, and respect.

Thus, project managers not only must be aware of the approaches they use in eliciting support but also of the effect of the conflict resolution approaches they employ. For the project leader each set of skills is critical for effective performance. If a project manager, for example, is initially skillful in gaining support but cannot manage the inevitable conflict situations which develop in the course of a project, then his or her effectiveness as a manager will erode.

## DECISION-MAKING STYLES AND TEAM-BUILDING SKILLS

The degree of participative decision making and esprit de corps have considerable impact upon not only the human aspects of the project environment but also upon the perceived success of projects.

Baker, Murphy, and Fisher, in their study of over 650 projects, including over 200 variables, found that certain variables were significantly associated with the perceived failure of projects, others were significantly associated with the perceived success of projects, and still others were linearly related to failure/success (21), for example:

- Lack of project team participation in decision making and problem solving, lack of team spirit, lack of sense of mission within the project team, job insecurity, and insufficient influence of the project manager were variables significantly associated with perceived project failure.
- In contrast, project team participation in setting schedules and budgets was significantly related to perceived success.
- The relative degree of goal commitment of the project team and the degree to which task orientation (with a backup of social orientation) was employed as a means of conflict resolution were linearly related to project success.*

Kloman's study contrasting NASA's Surveyor and Lunar Orbiter projects revealed that several elements contributed to the higher levels of actual and perceived success associated with the Lunar Orbiter project:

- Lunar Orbiter benefited from a strong sense of teamwork within both the customer and contractor organizations and in their relations with each other. Surveyor was handicapped by a lack of teamwork, particularly in the early years of the program.
- Senior management was committed to full support of the Lunar Orbiter project and was personally involved in overall direction at both

* These findings are discussed in detail in Chapter 35.

the NASA field center and in the prime contractor's organization. There was far less support and involvement in the case of Surveyor (22).

There has been an accelerated use of team building in project management in the last few years. Varney suggests that there are three primary reasons for the increasing interest in team building (23). First, there are more specialists/experts within organizations whose talents need to be focused and integrated into the requirements of the larger task. Second, many organizational members want to become more involved in the overall project rather than just perform narrowly defined roles. Third, there is ample evidence that people working well together can create synergy and high levels of creativity and job satisfaction.

In a recent exploratory research study with over 90 project managers, Thamhain and Wilemon identified some of the major barriers project leaders face in their attempts to build effective teams (24). The results of the exploratory field probe revealed the following barriers to team building:

- Differing priorities, interests and judgments of team members
  A major barrier is that team members can have different professional objectives and interests. Yet project accomplishment can require team members to place "what's good for the project" above their own interests. When team members are reluctant to do so, severe problems can develop in building an effective team. This problem is further compounded when the team relies on support groups which have widely different interests and priorities.
- Role conflicts
  Team development efforts are thwarted when role conflicts exist among the team members. Role conflicts are most likely to occur when there is ambiguity over who does what within the project team and between the team and external support groups. Overlapping and ambiguous responsibilities are also major contributors to role conflicts.
- Lack of team member commitment
  Lack of commitment to the project was cited as one of the most common barriers. Lack of commitment can come from several sources, such as: the team members' professional interests lying elsewhere; the feeling of insecurity sometimes associated with projects; the unclear nature of the rewards which may be forthcoming upon successful project completion; and intense interpersonal conflicts within the team.

  Other issues which can result in uncommitted team members are suspicious attitudes which may exist between the project leader and a functional support manager or between two team members from two warring functional departments. Finally, it was found that low commitment levels were likely to occur when a "star" on a team "demanded" too much deference or too much pampering from the team leader.

- Communication problems
  Not surprisingly, poor communication was a major barrier to effective team development efforts. The research findings revealed that communication breakdowns could occur among the members of a team as well as between the project leader and the team members. Often the problem was caused by team members simply not keeping others informed on key project developments. The "whys" of poor communication patterns are far more difficult to determine than the effects of poor communication. Poor communication can result from low motivation levels, poor morale, or simply carelessness. It was also found that poor communication patterns between the team and support groups could result in severe team-building problems, as did poor communication with the client. Poor communication practices often led to unclear objectives, poor project control and coordination, and uneven work flow.
- Project objectives/outcomes not clear
  One of the most frequently cited team-building barriers was unclear project objectives. As one project leader in the study remarked:

  How can you implement a team building program if you're not clear on what the objectives of the project are? Let's face it, many teams are muddling along on fifty percent of their potential because no one is really clear on where the project should be headed.

  Thus, if objectives are not explicit, it becomes difficult, if not impossible, to clearly define roles and responsibilities.

- Dynamic project environments
  A characteristic of many projects is that the environments in which they operate are in a continual state of change. For example, senior management may keep changing the project's scope, objectives, and resource base. In other situations, regulatory changes or client demands for new and different specifications can drastically affect the internal operations of a project team. Finally, the rate by which a team "builds up" to its full manpower base may present team-building barriers, e.g., not fully sharing "the vision" of the project.

- Credibility of the project manager
  Team-building efforts also are hampered when the project leader suffers from poor credibility within the team or with important managers external to the team. In such cases, team members are often reluctant to make a commitment to the project or the leader. Credibility problems may come from poor performance skills, poor technical judgments, or lack of experience relevant to the project.
- Lack of team definition and structure
  One of the most frequently mentioned barriers was the lack of a clearly delineated team. The study found this barrier to be most likely to occur among computer system managers and R&D project leaders. A common pattern was that a work unit (not a project team) would be charged with a task but no one leader or team member was clearly delegated the responsibility. As a consequence, some work unit members would be working on the project but not be clear on the extent of their responsibilities.

  In other cases, a poorly defined team will result when a project is supported by several departments but no one person in these departments is designated as a departmental coordinator. Such an approach results in the project leader being unclear on whom to count on for support. This often occurs, for example, when a computer systems project leader must rely on a "programming pool."
- Competition over team leadership
  This barrier was most likely to occur in the early phases of a project or if the project ran into severe problems and the quality of team leadership came into question. Obviously, both cases of leadership challenge can result in barriers (if only temporary) to team building. These challenges were often covert attacks on the project leader's managerial capability.
- Project team member selection
  This barrier centered on how team members were selected. In some cases, project personnel were assigned to the teams by functional managers, and the project manager had little or no input into the selection process. This, of course, can impede team development efforts especially when the project leader is given "available personnel" versus the best, hand-picked team members. The assignment of "available personnel" can result in several problems, for example, low motivation levels, discontentment, and uncommitted team members. As a rule, the more influence the project leader has over the selection of his or her team members, the more likely team-building efforts will be fruitful.

## ORGANIZATIONAL DESIGN CONSIDERATIONS IN PROJECT MANAGEMENT

Several research studies have investigated the impact of organizational arrangements and the authority of the project manager. Baker, Murphy, and Fisher found that with respect to organizational and authority arrangements:

- Excessive structuring within the project team and insufficient project manager authority were significantly related to perceived project failure.
- Adequate and appropriate organizational structures and effective planning and control mechanisms were significantly related to perceived project success. (Note that no particular type of organizational structure or particular type of planning and control mechanism was associated with success. This finding supports the contingency theory of management.)
- Degree of bureaucracy and degree of spatial distance between the project manager and the project site were linearly related to success/failure; that is, the greater the bureaucracy and the greater the spatial distance, the more likely the project was perceived as a failure (25).*

Marquis and Straight studied approximately 100 R&D projects (mostly under one million dollars) and found that:

- Projects in which administrative personnel report to the project manager are less likely to have cost or schedule overruns (26).
- Projects organized on a functional basis produce better technical results.
- Matrix organizations in which there is a small project team and more than half of the technical personnel remain in their functional departments are more likely to achieve technical excellence and, at the same time, to meet cost and schedule deadlines, than purely functional or totally projectized organizations (27).

Baker, Fisher, and Murphy also found that insufficient project manager authority and influence were significantly related to cost and schedule overrun. Chapman found that:

---

* These findings are discussed in greater detail in Chapter 35.

- A matrix structure works best for (a) small, in-house projects, where project duration is two years or less; (b) where assignments to technical divisions are minimal; and (c) where a field installation has substantial fluctuation in the amount of project activity it is handling.
- A matrix structure begins to lose its flexibility on large, long-duration projects, and therefore a more fully projectized structure is appropriate in these circumstances (28).

In contrasting functional organizations with project organizations, Reeser found some unique human problems associated with projectized organizations:

- Insecurity about possible unemployment, career retardation, and personal development is felt by subordinates in project organizations to be significantly more of a problem than by subordinates in functional organizations.
- Project subordinates are more frustrated by "make-work" assignments, ambiguity and conflict in the work environment, and multiple levels of management than functional subordinates.
- Project subordinates seem to feel less loyal to their organization than functional subordinates (29).

## COMMUNICATIONS IN PROJECT MANAGEMENT

Increasingly, effective interpersonal communication is being recognized as a critical ingredient for project success. A study by Tushman, for example, clearly illustrates the role and managerial consequences of effective communication networks within R&D-oriented project work environments (30). He notes that for complex problem solving, "verbal communication is a more efficient information medium than written or more formal media (e.g., management information systems)."

Tushman found the following communication patterns existing for complex research projects conducted in a large corporate laboratory:

- There were high degrees of problem solving and administrative communication within the high-performing teams. Further, the frequency of these two types of communication were positively associated with performance.
- The high-performing project teams relied more on peer decision-making interaction than on supervisory direction.
- Communication to provide feedback and technical evaluation to areas outside the project but within the host organization tended to be

highly specialized for the more effectively managed research projects.
- The high-performing research teams made effective use of "gatekeepers" to link with expertise external to the project team, for example, universities and professional societies.

From his major findings Tushman developed an information-processing model to help plan and manage communication requirements for complex projects and programs.

## RELATIONSHIPS OF THE PROJECT TEAM WITH THE PARENT ORGANIZATION, THE CLIENT, AND THE EXTERNAL WORLD

The patterns of relationships among the project team, the parent, the client, and other external organizations are highly important to the perceived success of projects. Baker, Murphy, and Fisher found that:

- Coordination and relation patterns explained 77% of the variance of perceived project success. (Stepwise multiple regression analysis with perceived success as the dependent variable; perceived success factor included satisfaction of all parties concerned and technical performance.)
- Success criteria salience and consensus among the project team, the parent, and the client also significantly contributed to perceived project success (second heaviest factor in the regression equation).
- Frequent feedback (but *not* meddling or interference) from the parent and the client, a flexible parent organization, lack of legal encumbrances or governmental red tape, and a minimal number of public governmental agencies involved with the project were pertinent variables significantly related to perceived project success (31).*

These findings supported Kloman's earlier study:

- From a management viewpoint, the greatest contrast between the Surveyor and Lunar Orbiter projects was the nature of the relationships of participating organizations, or what might be called the institutional environment. For Surveyor, there was an unusual degree of conflict and friction between Headquarters, JPL, and the prime contractor. For Lunar Orbiter, harmony and teamwork prevailed. Institutions and people worked together in a spirit of mutual respect (32).

---

* These findings are discussed in Chapter 35.

## CONCLUSIONS

Research regarding the human element in project management has enabled practitioners to formulate strategies which can not only improve the behavioral aspects of project management (the climate) but which also results in more effective project performance. The many research projects are relatively consistent with each other. Some of the principal findings which should be consistently stressed are:

- There is no single panacea in project management; some factors work well in one environment while other factors work well in other environments.
- It is important to vest a project manager with sufficient authority; once vested with authority, the project manager is well advised to also utilize his expertise and work challenge as major influence modes.
- The problem-solving approach is more successful than the smoothing or the forcing mode of conflict resolution.
- Participative decision-making styles are generally more successful than other styles; commitment, teamwork, and a sense of mission are important areas of attention in project management.
- Project organizational design must be tailored to the specific task and the environment, but higher degrees of projectization and higher levels of authority for the project manager result in less probability of cost and schedule overruns.
- To attain high levels of perceived success (including not only adequate technical performance but also satisfaction of the client, the parent, and the project team), effective coordination and relations patterns are important; also, success criteria salience and consensus among the client, the parent, and the project team are crucial.

Fully understanding the complexity of the interpersonal network in project management requires an ongoing research effort. We hope that research on this crucial area of project management will continue to produce new knowledge in the future.

## REFERENCES

1. Such a focus is a natural development in the life cycle of many management concepts. In the area of systems analysis, for example, the early literature centered on the hardware, software, and technical information handling processes. An earlier version of this paper appeared in the *Project Management Quarterly* (March, 1977), pp. 34–40.

2. Lawrence, Paul R. and Lorsch, Jay W. "New Management Job: The Integrator." *Harvard Business Review* (November-December, 1967), pp. 142–151.
3. Ibid., p. 150.
4. Hodgetts, Richard M. "Leadership Techniques in the Project Organization." *Academy of Management Journal,* Vol. 11 (1968), pp. 211–219.
5. Gemmill, Gary R. and Wilemon, David L. "The Power Spectrum in Project Management." *Sloan Management Review* (Fall, 1970), pp. 15–25.
6. Gemmill, Gary R. and Thamhain, Hans J. "Influence Styles of Project Managers: Some Project Performance Correlates." *Academy of Management Journal* (June, 1974), pp. 216–224. Also see Gemmill, Gary R. and Thamhain, Hans J. "The Effectiveness of Different Powerstyles of Project Managers in Gaining Project Support." *IEEE Transactions on Engineering Management* (May, 1973), pp. 38–43.
7. Reprinted by permission of the *Harvard Business Review*. Excerpt from "Engineer The Job To Fit The Manager" by Fred E. Fiedler (September/October 1965). p. 119 Copyright 1965 by the President and Fellows of Harvard College; all rights reserved.
8. Ibid.
9. Ibid.
10. Gaddis, Paul O. "The Project Manager." *Harvard Business Review* (May-June, 1959), pp. 89–97; Goodman, Richard M. "Ambiguous Authority Definition in Project Management." *Academy of Management Journal* (December, 1967), pp. 395–407; Steward, John M. "Making Project Management Work." *Business Horizons* (Spring, 1967), pp. 63–70.
11. Wilemon, David L. "Project Management Conflict: A View from Apollo." *Proceedings of the Project Management Institute* (1971).
12. Butler, Arthur G. "Project Management: A Study in Organizational Conflict." *Academy of Management Journal* (March, 1973), pp. 84–101.
13. Thamhain, Hans J. and Wilemon, David L. "Conflict Management in Project-Oriented Work Environments." *Proceedings of the Project Management Institute* (1974).
14. Thamhain, Hans J. and Wilemon, David L. "Conflict Management in Project Life Cycles." *Sloan Management Review* (Summer, 1975), pp. 31–50.
15. Lawrence, Paul R. and Lorsch, Jay W. (same reference as footnote 2), pp. 148–149.
16. Lawrence, Paul R. and Lorsch, Jay W. (same reference as footnote 2).
17. Blake, R. R. and Mouton, Jane S. *The Managerial Grid* (Gulf Publishing Company. Houston, 1964).
18. Burke, Ron J. "Methods of Resolving Interpersonal Conflict." *Personal Administration* (July-August, 1969), pp. 48–55.
19. Thamhain, Hans J. and Wilemon, David L. (same reference as footnote 13).
20. Thamhain, Hans J. and Wilemon, David L. (same reference as footnote 13).
21. Murphy, David C., Baker, Bruce N. and Fisher, Dalmar. *Determinants of Project Success.* Springfield, Va. 22151; National Technical Information Services, Accession number: N-74-30392, 1974, pp. 60–69.
22. Kloman, Erasmus H. *Unmanned Space Project Management—Surveyor and Lunar Orbiter,* a Report Prepared by the National Academy of Public Administration and sponsored by the National Aeronautics and Space Administration, Washington, D.C.; U.S. Government Printing Office, 1972, p. 14.
23. Varney, Glenn H. *Organization Development for Managers* (Addison-Wesley. Reading, Mass., 1977), p. 151.
24. Thamhain, Hans J. and Wilemon, David L. "Team Building in Project Management." *Proceedings of the Project Management Institute,* Atlanta (1979).
25. Murphy, Baker, and Fisher (same reference as footnote 21).

26. Marquis, Donald G. and Straight, David M. *Organizational Factors in Project Performance*. Washington, D.C.: National Aeronautics and Space Administration, July 25, 1965.
27. Marquis, Donald G. "A Project Team + PERT = Success or Does It?" *Innovation* (5) (1969), pp. 26–33.
28. Chapman, Richard L. *Project Management in NASA*, a report of the National Academy of Public Administration Foundation (January, 1973).
29. Reeser, Clayton. "Some Potential Human Problems of the Project Form of Organization." *Academy of Management Journal* (December, 1969), p. 467.
30. Tushman, Michael L. "Managing Communication Networks in R & D Laboratories." *Sloan Management Review* (Winter, 1979), pp. 37–49.
31. Murphy, Baker, and Fisher (same reference as footnote 21).
32. Kloman (same reference as footnote 22), p. 17.

# Section IX

# The Successful Application of Project Management

This section deals with successful applications of project management in the sense of both contextual applications and "success factors" that contribute to project success.

In Chapter 34, Michael K. Gouse and Frank A. Stickney provide an overview of a wide range of project management applications. Their annotated bibliography is an excellent guide to the broader literature describing various such applications.

In Chapter 35, Bruce N. Baker, David C. Murphy, and Dalmar Fisher discuss "project success and failure factors"—those factors that have been demonstrated to affect the perceived success or failure of projects. Such factors can provide the project manager with insight into those characteristics that might influence the eventual degree of success that his project is perceived to have achieved.

In Chapter 36, Baker, Fisher, and Murphy present a comparative analysis of public- and private-sector projects in terms of their success and failure patterns.

In Chapter 37, Laura C. Leviton and Gordon K. MacLeod discuss the successful application of project management in the health care sector.

In the concluding Chapter 38, David I. Cleland discusses the cultural ambience of the matrix organization—those "climatic" or cultural factors that appear to be associated with successful matrix organizations.

# 34. Overview of Project Management Applications

Michael K. Gouse*
Frank A. Stickney†

This chapter updates material included in the first edition of *Project Management Handbook*. It also identifies and discusses more recent applications of project management in specific contexts including health care management, research and development, government/defense, construction, and business and industry. The continued and increased use of the project management concept further convinces the authors that project management is the most effective approach to the management of multiple tasks, projects, programs, and products in many contemporary organizational situations. A major factor contributing to the increased use of project management has been the rapid development of computer software packages devoted to simplifying the process of implementing and executing a project management program. These recent innovations in computer applications to project management and their impact will also be analyzed.

Over the past 25 years, many contemporary organizations have adopted and implemented the project management concept for improving planning and control of their multiple tasks, projects, programs, and products. Project management should not, however, be considered a panacea for every organization encountering problems in multiple task manage-

---

* Michael K. Gouse is pursuing a Masters of Business Administration degree from Wright State University, Dayton, Ohio. His field of specialization is financial management.
† Frank A. Stickney is a Professor of Management at Wright State University. He received his Ph.D. from The Ohio State University in 1969. In addition to his academic research activity and consulting, he has extensive work experience in project management and related activities. His primary areas of research interest are organizational design, job satisfaction, the supervisory process, and strategic management.

ment. Project management is not relevant to all organizational situations because the implementation and execution of project management requires the use of additional resources and creates strains on the traditional roles and relationships within an organization. It should be considered only when the benefits derived from its use outweigh these additional costs. The authors propose that any organization considering the adoption and implementation of project management should follow the contingency approach; that is, if any organization is designed and operating under the traditional functional structure, without project management, and is achieving its objectives satisfactorily, then the adoption of project management may be unnecessary.

Project management received its strongest and earliest stimulus in organizations functioning in the aerospace, communications, and electronics technologies. The technologies involved in these industries are very dynamic and are characterized by rapid and continuous change. However, these conditions of technological change are rapidly spreading to organizations with historically stable technological environments. Organizational growth, complexity, and change in this environment are evident in the various applications presented in this chapter. These organizational changes, coupled with the impact of recent innovations in computer applications on the project management process, are also discussed.

The following structure will be used to present and review these applications:

1. A definition of project management and the circumstances under which it should be used.
2. A brief review and analysis of project management applications cited in the first edition of the handbook.
3. Description and analysis of recent project management applications.
4. The impact and use of the computer in project management.
5. Summary and conclusions.

## A DEFINITION OF PROJECT MANAGEMENT AND THE CIRCUMSTANCES UNDER WHICH IT SHOULD BE USED

Project management is the application of the systems approach to the management of technologically complex tasks or projects whose objectives are explicitly stated in terms of time, cost, and performance parameters. A project is, in reality, one of several subsystems in an organization. All of these subsystems must be managed in an integrative manner for the effective and efficient accomplishment of organizational objectives. Project management provides an interfunctional structure for a specified project and has a primary management orientation. It involves a project manager who, through funding allocation and control, obtains the required resources from the various functional departments: engineering, marketing, production, etc. This project manager should be proactive on tasks, as compared with the functional manager who is proactive on functional specialization, but reactive on tasks. Through the process of planning, organizing, directing, and controlling, the project manager coordinates the application of these resources to a given project with specific objectives providing the primary focus. The project manager should be evaluated on the successful completion of the project in terms of timeliness, budgetary constraints, and task performance.

The project organization is the focal point for all activities on a given project and provides total project visibility. It is integrative in nature and becomes the hub of all activities, both internal and external, which affect the project. It is a management mechanism and supplements, rather than replaces, the functional activities of the various departments. It is again emphasized that project management is not applicable to every task, project, product, or program. It is not an organizational panacea.

Many organizations employing project management use a "team or task-force approach," "matrix structure," "program structure," or "product structure." Not all of these approaches/structures fully qualify as pure project management applications, but all contain some characteristics of the project management concept. Specifically, the differences among these recognized subsets of the general project management approach are simply a matter of emphasis (1).* For example, the task-force approach is interdisciplinary, but does not have a finite termination date. The term "program" structure is usually used to define a multi-"project" structure (basically similar, but more complex).†

The authors consider project management a means of managing technological change when organizations are engaged in multiple tasks. It provides the organizational structure necessary for the successful management of these efforts relative to organizational and functional as well as task/project objective achievement. Several criteria have been developed to be used in evaluating the applicability of project management. The

---

* Numbered references are given in a bibliography at the end of the chapter.

† For a more thorough discussion of the forms of project management, please refer to Adams, John R., Barndt, Stephen E. and Martin, Martin D., *Managing by Project Management* (Dayton, Ohio: Universal Technology Corp., 1979) pp. 1–14.

following organizational conditions should exist when considering the applicability of project management:

- An objective-oriented work allocation structure is required.
- Identifiable product and/or task with a distinct life cycle (as well as performance, time, and cost objectives) are the primary outputs.
- The environment is highly uncertain and dynamic.
- Complex short- and long-run products are the organization's primary output.
- Complex and functionally integrated decisions are required with both innovation and timely completion.
- Several kinds of sophisticated skills are required in designing, producing, testing, and marketing the product.
- A rapidly changing marketplace requires significant changes in products between the time they are conceived and delivered.
- There is a stream of projects, some of which are not achieving project objectives.
- A task/project or other organizational activity requires simultaneous coordination of two or more departments (interdisciplinary) (11).

The above are the major factors which an organization should consider in determining whether to implement project management. The organization should ensure that the benefits of better project planning, control, and performance outweigh the additional costs required by the implementation of project management. The examples cited in the first edition of the handbook and the more recent project management applications reviewed in this chapter appear to have benefits which exceed additional costs.

Figure 34-1 is a model illustrating the interaction between the organizational (system) objectives, project/product objectives, and functional objectives. Notice that a system of "checks and balances" must be in effect to preclude the enhancement of one set of objectives to the detriment of the others. This figure is a conceptual model introduced to analyze and understand the project and functional management relationships and does not represent an actual organizational design.

## A REVIEW AND ANALYSIS OF PROJECT MANAGEMENT APPLICATIONS CITED IN THE FIRST EDITION OF THE HANDBOOK

These previously cited project management applications are presented by type of technology. The references are indexed to the bibliography for more complete information.

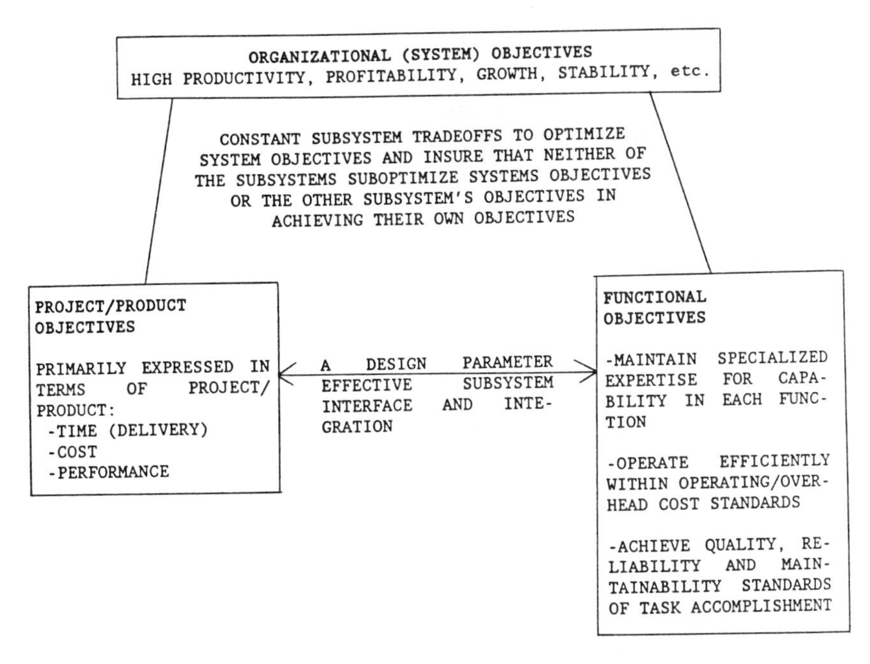


Figure 34-1. System and subsystem objectives. (From Frank A. Stickney and William R. Johnston, "Communication: The Key to Integration," *1980 Proceedings of the Project Management Institute* (The Project Management Institute. Drexel, Pa., 1980), p. I-A.7.

## Research and Development

The Searle Laboratories Division of G. D. Searle and Company, Chicago, Illinois, is involved in drug development. As such, it operates in a complex, ever-changing environment requiring a dynamic system of operations. To meet this challenge, Searle has utilized a "collateral" project structure for achievement of both its research and organizational objectives. This structure includes the following groups:

1. A corporate *Pharmaceutical R and D Group* has goal-setting responsibilities and interfaces primarily with the Action Council in a strategic administrative capacity.
2. The *Action Council* is an interfunctional group employing worldwide technical personnel; this group has responsibility for product selection and prioritization and for selection and evaluation of project leaders.
3. *Core Committees* are interfunctional groups interfacing with the Action Council, project teams, and project task forces; they are

responsible for needed resources, and project sites, and for determining the nature of worldwide technical inputs.

4. *Interfunctional Task Forces* consist primarily of members from the Core Committees; they are responsible for project planning.
5. *Project Leaders* are responsible for managing the projects and *Project Management Consultants* are responsible for monitoring the project progress. Project leaders have personnel assigned based on skill assessment requirements determined by the interfunctional task forces; this comprises the basic management team.

All of the above groups are in continuous interaction by organizational/system plan, and their specific roles and relationships are explicitly defined. The task of these groups is made extremely difficult by government regulation and monitoring, by frequent changes and modifications, and by the nature of basic research. The project structure, although seemingly complex, appears to be conducive to achievement of objectives in these types of environment (22).

Battelle Pacific N. W. Laboratories (BNW), a division of Battelle Memorial Institute, has project contracts from both government and industry (national and international) which are extremely broad in scope, with their size and complexity ranging from the small single-discipline type to the large multimillion-dollar, multidisciplinary type. A Research Project Management System (RPMS) was established which supports a work environment conducive to the creative and innovative output expected from researchers with functional/specified discipline orientations. It also provides for effective performance of the firms' many projects.

BNW maintains a "functional" structure with an implied "project" structure. Project managers report to functional managers but maintain the flexibility for temporary assignment to other functional managers within the organization. The project organization is not reflected in the formal organizational chart but is created when needed (35).

### Construction

The James H. Campbell Plant #3 is a coal-fired power plant designed and constructed near Grand Haven, Michigan. This project was under the direction of Consumers Power Company (CPCo), plus two separate organizations, Townsend and Bottam Engineers and Contractors, the contractor, and Gilbert/Commonwealth, the engineer. These two autonomous organizations joined together in providing the single responsibility and interdisciplinary focus CPCo required. This effort was carried out by integrated project teams (utilizing the expertise of both companies) with a single project manager as the individual with key responsibility to CPCo. Complex channels of communication, distance, and a lack of understanding among the different groups of the companies were dealt with through weekly management meetings. Progress reviews were held first for each functional discipline, and then followed by a project group meeting where open communication and constructive criticism was the mode. In addition, "Team Building Sessions" were held to promote cooperation between and among all groups. The success of this project is attributable to the project management approach and the resulting integration of effort (40).

Another construction organization where project management was applied is the Los Angeles County Flood Control District, a system of flood control and water conservation facilities (i.e., dams, major channel improvements, spreading grounds, etc.) for Los Angeles County. In 1966, a small program management effort was adopted and implemented consisting of two people and a secretary, with mainly a facilitating responsibility for recommending change. The program management function grew from that state, continuing to solve problems despite encountering resistance to change. Several years later, program/project managers were directly assigned to projects and a management Systems Division was formed. This change in organizational structure is believed by many within the Flood Control District to be an appropriate vehicle for dealing with complex management problems. It clearly indicates the use of project management as an integrating mechanism for diverse activities required for achievement of common objectives (17).

### Government/Defense

NASA's space shuttle program is an immense undertaking. The complexity of this program necessitates a tremendously dynamic system structure. A structure has been developed consisting of the following four levels:

- Level I: NASA Headquarters in Washington, D.C., which interfaces directly with Congress and other government agencies. It is responsible for provision of major program direction and planning, basic performance requirements, and control of funding allocations to NASA field centers.
- Level II: the Space Shuttle Program Office, which has major interfunctional project management responsibilities.
- Level III: the actual projects, which are managed through project offices at NASA.

- Level IV: the level where contractors carry out operational duties (37).

This structure facilitates smooth communication flows and processing of information to key decision centers within the constraints of the program.

The Naval Air Systems Command uses project management in its defense weapon systems acquisition process. Project offices are formed for the weapon systems requiring large outlays of R&D and/or production dollars, and project managers are assigned overall project responsibility. The technical and administrative experts from the functional departments are assigned to projects to provide the specialized expertise on an as-needed basis. The Command's contractors (suppliers) also use the project management concept for managing their portion of the project and for interfacing with the project manager in the Command. With the complexities and interdependencies inherent in major weapon systems development and acquisition, the project management structure utilized in this situation has worked well (38).

### Business and Industry

By the end of 1973, the business environment in the automobile industry had become increasingly complex and uncertain. General Motors management determined that conditions were appropriate for project management, and in 1974 the project center concept was implemented to coordinate the efforts of GM's five automobile divisions. These project centers were temporary in nature and were formed whenever a major effort requiring interfunctional specialties was required. Work in the project centers focused on problems common to all divisions (i.e., frames, electrical systems, steering gear, and brakes). The implementation of this concept speeded new technological ideas onto the production line and utilized resources to their best advantage, eliminating much redundancy in efforts.(7)*

At the Inland Division of General Motors, a "team approach" was implemented to minimize the suboptimization of organizational objectives that often occur in complex organizations. This approach consists of interfunctional teams working on one or more major projects, rotating team chiefs with differing technical specialties, and a "board" to oversee the

---

* See the Business and Industry Section, page 32, Recent Project Management Applications, for more recent information on G.M.

team operations and to provide strategic management direction. Although the situation does not meet the exact criteria for a project office, it is another illustration showing positive results of the project management approach (12).

The management of Standard Steel Company of Burnham and Latrobe, Pennsylvania, a traditionally functional organization, made the decision to implement project management. This structure was integrated with the functional structure, and teams of specialists (production, finance, metallurgy, etc.) jointly managed a product line or geographic area. The approach worked well at Standard Steel, although other companies in the metals industry may not find this application of project management applicable to their situations (9).

### Health Care Management

The British National Health Service was reorganized in 1974, involving the overlapping of the functional hierarchical organization with various lateral relationships, some permanent, some ad hoc. The lateral groupings were tasked with a variety of responsibilities ranging from direct provision of services by health service professionals to developing policies and/or plans for use by both in-house personnel and non-health agency personnel. In addition, this organizational form has a synergistic effect upon the health service. When those with technical expertise are brought together to solve problems, learning from the knowledge and experience of one another is inherent in the interactions, resulting in greater optimization of organizational goals (16).

In 1976, the Westside Community Mental Health Center was a traditional functional organization. Problems existed because a limited number of services were offered and service areas exhibited varying degrees of productivity. Coordination among service areas was increasingly complex, communications were poor. Often, patients were discharged from the Inpatient Department with no follow-up arrangements made with the Outpatient Department; a systems perspective among personnel was almost nonexistent. The conditions were conducive for change. A matrix project management structure was adopted and major improvements in operations were observable within two years. In 1978, there were five additional service programs offered employing seven less staff personnel than in 1976. The adoption and implementation of the project management organizational form was highly conducive to increased communication, coordination, control, and achievement of organizational objectives (44).

## DESCRIPTION AND ANALYSIS OF RECENT PROJECT MANAGEMENT APPLICATIONS

### Research and Development

Groups involved in research and development are basically concerned with turning intangible concepts, ideas, and theories into specifications, tangible products, and processes. Consequently, these groups are different from other subsystems within an organization and often require a different structure (24).

Several recent drug development successes in the Medical Research Division (MRD) of American Cyanamid are credited to a systems and project management organizational perspective. There are two main factors in this R&D success story: one, participative management in the form of interdisciplinary teams focused on project planning at the mid-management level of detail; and, two, the teams continuously evaluated project specifications which are constantly changing in the R&D projects.

Cynamide utilizes a ''platoon'' approach project structure for achievement of its research and organizational objectives. Project managers are selected from the ranks of the functional organizations, spending about 20% of their time as the project team leader. It is the project manager's responsibility to lead the project team members in preparation of a project plan, and the tracking, reporting, and adjusting of that plan through project conclusion. Representatives from each of the functional groups are on the project teams. The marketing representative has been identified as one of the most critical on the team. Constant assessment of any changes in project specifications must be made vis-à-vis market potential at projected product launch time.

Three basic guidelines for project management success have been derived from this experience:

- *Focus:* Because project objectives will change, focus on, and continuous reassessment of, these objectives is imperative. Exceptions and changes to the project are analyzed and coordinated at status review meetings.
- *Communication:* The project reporting system should be kept relatively simple; it should be objective-oriented. Members of the R&D functional organizations, such as the Regulatory and Marketing departments, should be on the project team. A formal and timely mechanism must be available to distribute information to those who need to know.

- *Team Leadership:* The team leader should be both a good manager and a good scientist because the project decisions need to be based on both scientific and managerial factors.

Basic project management concepts have been applied to the pharmaceutical R&D environment, and success in these endeavors is attributed to utilization of these concepts. Clearly defined project specifications and the availability of automated project management systems for project planning, evaluation, and control are required (23).

The Tri University Meson Facility (TRIUMF), located at Vancouver, B.C., Canada, is the only meson factory in Canada and is one of three intermediate-energy nuclear physics laboratories in the world. More than 500 people from different disciplines, countries, and cultures conduct both basic and applied research at this facility. On the average, 10 to 20 projects are active at any one time, each involving 10 to 30 people and total costs ranging from $500,000 to $2,500,000 (Cdn.). Two typical projects were both completed within two to three weeks of the projected project duration of approximately 15 months and within 10% of the budget. Such notable success is attributed to utilization of CAMS (Computer Aided Management System), timely project progress reviews, and remarkable team effort (43).

In such a dynamic, complex environment, the planning process is extraordinarily important. In 1978, the National Research Council (NRC) of Canada asked TRIUMF to submit its future plans and related budgetary proposals. The plans and budgets were divided into three categories consistent with TRIUMF organizational objectives:

- *Basic Support* covering the maintenance, repair, and operation of all commissioned facilities, general and technical services, and administration.
- *Facility Development* involving the requirement to increase the scope and capabilities of TRIUMF, including the upgrading and/or extension of existing facilities and construction of new facilities.
- *Experimental Support* including site standard instrumentation, services, and equipment required to carry out the experiments.

The above reflects the functional activities/capabilities required to support TRIUMF's projects. Each major project has a project description serving as a guideline to conceptually define and plan the project. Both the total project cost and estimated cash flows are defined. These estimates are improved as the project progresses. One of the effective keys to successful project management at TRIUMF is an effective review process which continues throughout the project life cycle. A well-designed review process is important because of the need for interactive communication and interdisciplinary expertise. At TRIUMF, the regular weekly or bi-weekly meeting could be classified as one of information, analysis, evaluation, design, and/or decision making. However, the amount of time and effort expended on reviews should not outweigh their potential benefits.

At TRIUMF, successful project management is based on user participation, feedback, and coordinated team efforts. This is an outstanding example of project management implemented with creativity in team formulation, resulting in more effective communication, integration, and interface management between project and functional support activities (43).

Goodyear Tire and Rubber Company, Akron, Ohio, has an R&D philosophy based on the premise that innovation involves taking an idea and translating it into a more efficient manufacturing process and/or a more marketable product. Basic research is constantly monitored to ensure that it supports other key activities of the organization such as distribution, marketing, and manufacturing (6).

Goodyear utilizes a project matrix management structure that works closely with R&D organizations for project planning and control. Project managers organize teams that are interfunctional and interdisciplinary in nature. A typical project team might consist of a research scientist, a development engineer, and a marketing specialist. One of the key responsibilities of these project management teams is to terminate a project when it is no longer useful. These teams have proven to be very effective in expediting a project toward either completion or termination. Departing from the traditional approach to product design and manufacture, the corporation now utilizes a systems approach that emphasizes basic scientific knowledge by simultaneously looking at all elements involved and their interdependencies, that is, the product itself, materials, and relevant manufacturing processes, etc. Goodyear has found project management an effective approach to accomplishment of its organizational objectives (6).

Stuart Pharmaceuticals is the U.S.-based operation for the research-intensive pharmaceuticals division of Imperial Chemical Industries PLC of London, England. Since 1977, the company has introduced several new products as a result of extensive research. In order to attain organizational objectives, the company has implemented project management in two areas: research development and commercial development. The following are the major project management activities in their research function:

- *Predevelopment activities* culminate with identification of the best chemical compound which has the potential to become a drug. The project manager begins coordinating the compound through the rigors of predevelopment to compound proposal and selection.
- *Compound selection and development strategy* involves the project manager editing the scientific information into a document. This document is used by an international research management team to determine whether funds are allocated for further development of this compound. If the project is funded, the project manager selects appropriate team members and convenes team meetings to ensure completion of documents.
- During the *progressive development* phase, a large percentage of the project manager's time is devoted to managing the preclinical and early clinical aspects of development with computerized monitoring of these developmental activities. The project manager summarizes team proposals and recommendations for international research management team consideration. The resulting decision is communicated to the team by the project manager, who coordinates the decision and its implementation.

After the drug is submitted for approval, research development is replaced by commercial development and the research team manages these activities to the market.

Project management at Stuart blends proven managerial techniques with the technical expertise necessary in dynamic research processes. Stuart has discovered that project management focuses upon the task/project at hand, de-emphasizing the "star" syndrome and potentially destructive infighting. The sources of project management at Stuart Pharmaceuticals are one of the primary reasons Stuart is a leader in the industry (8).

Inherent in these studies of project management application in research and development environments is the strong project/team leader. This individual must possess excellent managerial skills combined with technical expertise. The findings of Katz and Allen suggest a separation of roles between project and functional managers of R&D professionals, although these roles can never be completely separate (25).* A project management structure has been identified in this research which utilizes functional managers as project leaders on a part-time basis. This, however,

---

* For a more thorough discussion, please refer to Ralph Katzand Thomas J. Allen, "Project Performance and the Locus of Influence in the R&D Matrix," *Academy of Management Journal* (March 1985), pp. 67–87.

may be the exception rather than the rule. Management must consider the specific aspects of the organizational environment as a prerequisite to project management implementation and execution. R&D organizations that survive and prosper in today's dynamic, high-tech environment will be those that successfully combine functional expertise and the strategic and operational management of research projects.

### Construction

Construction projects requiring interdisciplinary efforts from two or more separate technologically oriented organizations by nature necessitate a project structure. An example is the expansion of Texaco's Louisiana refinery, located at Convent, about 50 miles north of New Orleans on the east bank of the Mississippi River. The overall success of the project was attributed to careful selection of contractors. Texaco contracted with Fluor Engineers, Inc., for engineering and procurement, and with Brown and Root, Inc., for construction. This was the first time Texaco had utilized separate contractors for engineering and construction. The intent was to allow Texaco to concentrate its efforts in other critical areas and to allow the contractors to work in their areas of expertise with minor Texaco input (4).

To satisfy organizational objectives in engineering, procurement, and construction, Texaco determined that an integrated schedule was necessary. To accomplish this, Texaco and contractor personnel developed an overall critical path method (CPM) schedule*. Contractor personnel reported directly to the Texaco project manager, but were never considered Texaco personnel by either contractor. Additionally, Texaco used its own personnel to consolidate cost information from both contractors to develop the current control budget.

In any project of this magnitude, communication is a critical and monumental task. At its peak, over 5000 people were working on the project and 12 groups had to interface with each other. Each company had a staff at the construction site as well as at each organization's project offices. Information between the three companies was expedited more quickly and reliably and problems were identified and solved faster. In an effort to reduce paperwork on the project, meetings were held on an "as-needed" basis, and the co-location of all three organizations' personnel facilitated timely integration (4).

Another application was the CRS Group of Houston, Texas, who are specialists in project design, development, and construction. They were faced with the formidable task to plan and implement a massive irrigation project in the remote deserts of Saudia Arabia. Using project manage-

---

* See Chapter 15.

ment, CRS completed the project on time. A key to the successful completion of this project was breaking it down from a large, unwieldy entity into smaller, more manageable parts. CRS had been hired by Lindsay International, a Houston manufacturer of irrigation equipment, to build and install systems at four locations within eight months. The four sites to be irrigated were at different locations in Saudia Arabia and each site had unique problems. CRS turned each site into a separate subproject, with Riyadh, the Saudi capital, as the coordination point for the four subprojects. With one microcomputer being used for each project, the subprojects were compared with one another on an ongoing basis. CRS credits use of the computerized project management packages with the success of the project (42).*

The Piper Jaffray Tower, a 42-story office building in downtown Minneapolis, is a $90 million project completed within the time schedule and budget that was established before final design was started. The success of this endeavor is attributed to the project manager's striving for excellence in teamwork with open and continuous communication. The project was entirely a local effort: the general partner and developer, the architect and engineers, and the construction manager/general contractor are all local Minneapolis/St. Paul firms. Most of the subcontractors that were used on the project were also local.

On-site project managers were designated and involved in all phases of the project, including these key phases:

*Postdesign Development*

- Preconstruction and major construction phases scheduled.
- Bid packages selected.
- Monthly construction phase expenditures estimated.
- All involved participants, including owners, lenders, and potential subcontractors, made aware of the overall schedule.

*Prebid*

- General strategy and construction sequence finalized.
- Comprehensive plan established to be used as a benchmark for controlling progress.
- Subcontractors encouraged to participate in plan refinement.

*Preconstruction*

- Teamwork attitude developed.
- Schedule plan refined with subcontractor assistance.

---

* See Chapter 28.

*Construction*

- Attainable short-term progress goals established.
- Plans established for smooth transition from one construction phase to another.
- Schedule problems resolved upon occurrence.
- Schedule modified as necessary.

These areas were integrated by the master project office. Emphasis was placed on all phases of construction activity with timely information flow and communication between all project participants. It is this constant coordination and project commitment by the projects' managers that enabled the project to be completed on schedule and within budget (27).

In the past three years, Paragon Engineering Services, Inc., has had the project engineering and construction supervision responsibility for at least six separate shallow-water oil and gas facilities. Project management techniques which were successfully employed onshore or in water depths greater than 50 feet had to be altered for these shallow-water projects. Paragon developed an approach which has enabled each of these projects to be built in three to six months from well completion and at cost 20% to 40% less than identical "turnkey" projects. A turnkey project is one where the contractor completes all building and installation to the point of readiness for operations, at which time it is sold to the customer at a prearranged price. This "modified turnkey" approach to project organization maintains operator control over quality while providing the shortest project schedule at least cost (5).

A project schedule or plan of project execution is prepared to monitor progress. When this schedule is set, the separate contract or engineering activities whose progress is necessary to assure start-up are identified. Progress toward attaining preset intermediate goals is reviewed weekly to identify problem areas. In the "modified turnkey" approach to project management, the activities of the operator and all vendors are continuously coordinated. Inspection and expediting are greatly simplified compared to other project organizations where the contractor has many subcontractors. This approach allowed major problems to be identified early so that they would not affect the overall project schedule (5).

One of the major considerations in any construction project is the transition period which involves training the operators and maintenance personnel so that the new facility start-up and operation is safe and efficient. Project managers from Standard Oil Company faced such a task in the planning, development, and construction of the Trans Alaska Pipeline (TAPS). The design of this project extended over a long time period because of the size of the project and environmental problems. The owner companies formed a separate organization (corporation) for the construction and operation of the pipeline. In essence, this organization had two major subsytems (groups) whose activities were integrated on a continuous basis: one for construction and one for operation.

The smooth transition of TAPS from construction to operation resulted from the formation of these two primary groups. The following are the critical elements in the operating subsystem which directly contributed to this success:

- It was established early in the design phase of the project.
- It functioned independently and had responsibility for commissioning the new facility.
- It and the project management subsystem worked on an integrated basis with continuous communication.

Separate but integrated organizations for construction and facility operations are recommended so that the demands of construction do not dilute the preparations necessary for operations. The project organization was the primary integrator between these two major functions.

The first person assigned to the operating organization should be the one who will manage the facility after start-up. This person's responsibility is to plan and to organize both the start-up and the permanent operations. The organization of the start-up manager has several overall responsibilities which include reviewing design plans to ensure the facility can be operated, preparing operational manuals and start-up procedures, planning the start-up and training personnel, and commencing operations with timely resolution of all start-up problems.

The major phases of activity for the operating group include design review, start-up planning, and start-up. During these phases, the start-up manager chairs a committee that includes all operating managers (operations, maintenance, technical, administrative) and a high-level manager from the construction project management group. One of the primary functions of this committee is to provide a formal interface between the two groups.

### Health Care Management

It has often been suggested that hospitals can be more effectively administered by a product management structure. Rather than operating hospital activities primarily along traditional functional departments such as laboratory, radiology, internal medicine, etc., specific product categories focus on specific areas such as orthopedics, obstetrics, neurosciences, urology, etc. With the advent of diagnosis-related group (DRG) prospective payment systems, discussions of product-managed administration have dramatically increased. With DRG, payments are made based on the number of patients in specific diagnostic/treatment categories rather than on allocated costs from functional departments. DRG is forcing hospitals to think in terms of product line management (30).

Organizational objectives for Miami Valley Hospital (MVH) in Dayton, Ohio include cost management and increased market share. Hospital administration felt that product line management would afford MVH an opportunity to meet those objectives by more appropriately matching products to target market segments (46).

The key to successfully implementing this structure is the selection of an appropriate product manager who possesses skills that are interdisciplinary in nature. The role of this person is significantly different from that of the functional manager. It is critical, however, to recognize that both types of managers are necessary. The organization must gradually transition into the situation where the relationship between the product and functional managers is accepted. The issue of "turf" is a difficult obstacle to overcome in product line management, but the organization must make a concerted effort to effectively organize teams at all levels for product groups.

Product line management is also utilized at Health and Hospital Services in Bellevue, Washington. Management has identified three basic elements in the administration of health care project management; they are *planning, cost control,* and *marketing*. The *planning* stage involves making decisions about what services to offer, how, and to whom; and it is based on hospital activities in terms of its products and product lines. Effective planning is reflected in end results; how well do the sets of products which the hospital offers enable the hospital to achieve its objectives? Product line *cost control* works primarily by influencing how physicians manage their patients; this technique is unlike traditional management efforts to control costs by controlling functional departments. Successful product line cost control management requires the hospital to organize, direct, and control its operations by product line categories. The product line *marketing* effort is concerned with identifying, achieving, and sustaining a specific market position for each product line. The objective should be a position in which a significant number of end-users (physicians, patients, employers) consider the hospital to be the best facility for all products in a specific product line. Marketing effectiveness can be measured by such criteria as patient volumes and mix and their effects on productivity, efficiency, and financial performance (30).

A survey was conducted of U.S. hospitals by Swedish Health Services in August, 1985, to determine which hospitals were utilizing product lines and product line management. The results were significant. Of 47 respondent hospitals, 26 (55%) were using some type of product line management, with length of use up to ten years. Six other hospitals (13%) are considering implementing the method. Some hospitals started with just one product line and added slowly, while others started with as many as 16. Product line managers share areas of responsibilities with other functional segments of the organization. These responsibilities include budgeting, staff supervision, equipment acquisition, market research, and promotional activities. The ultimate organizational objective for product line managers is for them to have complete responsibility for specified product lines (14).

Health care service institutions are complex social and technological organizations existing in a continuously changing and now highly competitive environment. As such, they require a dynamic system of operations to survive and prosper. Product line management, with its myriad uses, purposes, and design is one structure appropriate for dealing with this health care environment.

### Government/Defense

The Department of Defense (DOD) has used program/project management for the acquisition of large defense weapon systems for many years. The Polaris submarine project was managed and acquired by a formal project management office. Other major programs such as the B-1 bomber, the Phoenix missile, the F-14, F-15, F-16, and F-18 aircraft, the Apache, Cheyenne, and Blackhawk helicopters, the Trident submarine, and the M-1 tank have all been acquired through the process of project management. Smaller systems and subsytems are also routinely acquired utilizing program/project management. In fiscal year 1983, DOD spent $140.5 billion for the acquisition of defense weapon systems, much of it managed by project managers (26).

In an analysis of the successful completion of a number of these projects, several key factors were identified. These factors for success differ from project to project, but there are some basic themes. Project managers interviewed cited these success factors most often: good staffs, good contractor project managers, realistic requirements, a good contractor or manufacturer, and stability factors (personnel, funding, and product stability). The skill and ability of the project manager is also a vital factor for success. The operational background, leadership ability, education, and project management experience are critical factors. Project managers must be able to communicate well with all types of audiences (26).

In every project analyzed, the project manager was clearly the driving force; personal attributes were clearly important to the success of the project. The project staff gradually assumed the personality of the project manager; enthusiastic and hard-charging, low key, etc. The project became the project manager. The following are several of the key personal attributes which were identified:

- *Experience* in acquisition; decisions are based on judgment gained through experience.
- *Leadership* which instills loyalty and motivates team members.
- Ability to *communicate* with all segments impacting upon the project.
- Willingness to *accept responsibility* and *execute authority*.

These are general attributes which correspond to civilian projects as well as military (26).

Costs overruns still plague the defense procurement system even with attempts by Congress and the Defense Department to improve project management. Much of this problem is traced to military and civilian project managers who have limited training or experience in running large programs that deal with the research, development, and production of defense weapon systems. These people are highly capable, highly motivated individuals trained in the government system of procedures and regulations. However, their formal training often does not equip them with the negotiating skills necessary to deal with contractors. More effort is being made to develop career project managers through training and experience. Outstanding military and civilian personnel should be offered project management as an attractive career opportunity. Project managers who complete acquisition programs on time and within budget should be rewarded with promotion, recognition, and more responsible future assignments. Project management is working in the Defense Department, but improvement continues to be the challenge (19).

### Business and Industry

For an organization to survive, prosper, and grow, it must continuously be in interaction with its environment; the technological, political, social, and economic forces are vital factors in decision making. Those organizations which have not maintained this interaction have suffered. Implementation of the project management structure has allowed many organizations to work effectively within their environment.

In a 1984 reorganization, General Motors replaced its decades-old structure of seven divisions and a major subdivision with two super-groups—Buick-Oldsmobile-Cadillac (BOC) and Chevrolet-Pontiac-GM of Canada (CPC). While the five auto divisions continue to exist, they essentially have become marketing arms of the two supergroups. This move involved more than an organizational change; GM also implemented a strategy matrix form of project management. This form of project management was introduced to alleviate the frustrations traditional functional organizations experience in handling interdependent business segments. GM's decision to implement project management was prompted by increasing difficulty in planning, organizing, and controlling the traditional organizational structure. This highly centralized system couples functional lines with matrix management to provide open lines of communication between functions (45).

Another effective project management application is illustrated by an approach taken by BellSouth, Inc., one of the seven regional holding companies formed by the divestiture of American Telephone and Telegraph (AT&T). BellSouth is a holding company that owns two operating companies—South Central Bell and Southern Bell. These two operating companies jointly own BellSouth Services Company, which provides services to both operating companies. Information System Services (ISS) coordinates project management systems development for BellSouth, Inc. ISS utilizes what is termed "centralized project management," a form of project management that is a significant departure from textbook organizational designs. At ISS this approach explicitly recognizes the central role of the project manager in the success of a project throughout its life cycle.

Centralized project management is not suited for organizations with occasional demands for information systems; however, it is a good model for organizations that face continuing demands for the development, implementation, and maintenance of changing information systems. In the ISS approach, a specialized and permanent organization is established that is dedicated to systems planning, design, application, and operation. A significant part of the responsibility of three of the six vice presidents within ISS is project management, an indiciation of the organizational commitment.

The following are the benefits of centralized project management practiced by ISS:

- Projects are managed by professional, experienced project managers, eliminating the beginning-of-project lag that often occurs because a project manager selected from the user organization does not have the proper skills and experience.

- Standards for the development and maintenance of documentation and system interface are determined.
- The integration of information systems throughout the organization is guaranteed; system priorities and possible overlaps are more readily recognized; and work duplication is eliminated.
- There is no split loyalty exhibited between a project and a functional unit, which can occur with project managers in a decentralized approach.

The implementation of centralized project management represents a significant commitment of organizational resources. However, the reduction in the time and expense of developing large-scale systems under this project management structure is more that adequate reward for this commitment (15).

In 1968 Dow-Corning introduced a strategy matrix project management structure in an effort to resolve two specific issues: one, the multidimensional nature of the company's planning, budgeting, and implementation activities, and two, the extreme interdependence of the company's product divisions. This strategy matrix project management structure facilitates the involvement of both business (project) and functional managers in the formulation of organizational objectives and in the implementation of actions for their achievement. Unlike typical hierarchical organizations, decision making is deeply decentralized in the organization, allowing more people to participate in the process. A style of management is encouraged that is characterized by cooperation, trust, and team building among managers (34).

Each business within Dow-Corning is a profit center, although they are not operated as "stand-alone" divisions. A business is managed like a project. The functional managers control the company's scarce resources, while the businesses compete for a share of these pooled resources. Each profit center has its own full-time manager responsible for short-term profitability and long-term performance. Among other responsibilities of these business managers is assessment of product line, development of five-year objectives and strategies, and proposal of specific operating plans and budgets for implementing strategies.

The planning process at Dow-Corning involves the following three major activities within the strategy matrix:

1. The business managers develop one-year business targets, along with five-year objectives.
2. Functional managers allocate resources to the business plans based on the objectives.

3. Detailed cost-center budgets for the coming year are developed from these plans.

Dow-Corning's executive committee carefully reviews these activities at each stage of the planning process. These reviews serve to reconcile the business and integrate functional dimensions.

Since Dow-Corning's successful implementation of strategy matrix project management in 1968, several other major companies have introduced the concept. These companies include IBM and Shell Oil (1972), Velsicol Chemical and Westinghouse (1979), and Citibank and Federal Express (1984) (34).

Although not as dynamic as organizations in rapidly changing technological environments, the banking industry also faces environmental changes in technology, regulation, and competition. In order to adapt to and manage these external forces, banks are using project management more frequently as an administrative tool. The objectives of project management in banking are to provide senior management visibility of major projects and to assign designated project managers responsibility for project cost, time, and performance objectives (18).

Security Pacific Bank of Los Angeles effectively used project management in its automated loan collection system. Under the old system, debt collections were handled manually at each of Security Pacific's 600 loan offices and at collection centers. The bank decided to centralize all collections to six regional adjustment centers in northern and southern California and to a charge card center. Utilizing project management, the $2.3 million development project was completed on time and within budget. Security Pacific estimated it would save $4 million by the end of the first full year of the system's operation (18).

The project management structure implemented was conventional in nature. A project manager was selected and made responsible for overall results and performance. The project manager selected a team comprised of people from the different functional departments involved. This approach provided not only interdisciplinary expertise, but also made better decisions possible when time, cost, or performance trade-offs were necessary. The project team formulated its objectives, financial plan, and schedule. The team met on a regular basis until successful completion of the project. Several distinct phases in the evolution of this project were recognized:

- *Concept phase:* identification of the need was established.
- *Preliminary study:* technical, operational, and economic bases for the project were developed through feasibility studies.

- *In-depth study:* project characteristics were defined. This includes extensive study and analysis and a detailed design and evaluation of alternative solutions.
- *Project development and implementation:* development, procurement, and deployment of all required resources was made.
- *Postproject evaluation:* a written comparison between planned and actual results was made.

Each phase had a distinct beginning and end, but realistically they frequently overlapped. The more they overlapped, the less likely the project objectives were met. Because of the importance of these events in a project life cycle, it is important that senior management formally review each phase upon completion (18).

## FACILITATION OF PROJECT MANAGEMENT VIA COMPUTER APPLICATIONS*

The use of computer software to effectively implement and operate project management of complex tasks is commonplace at this time. The software aids project managers in focusing on and controlling criticial segments of the project. Project management software applications range from maintenance of simple bookkeeping systems to multimillion-dollar defense projects. Time, money, and other resource savings may be significant by using project management software.

One reason for this transition from traditional to computerized project management procedures is the increasing power of microcomputers which caused the recent proliferation of project management software packages for personal computers. This trend toward the smaller computer systems facilitates integration from project concept to completion. As these small computers become more powerful and programmers more sophisticated, a full range of project management software will be available for microcomputers (39). Additionally, costs for these software packages continue to decrease due to user demand and intense vendor competition. Original mainframe versions are being scaled down for use on smaller computers, and current packages are offered with more functionality than previously available. Although project managers using these more powerful project management packages may have to invest time in learning the new process, the result is a more effective project management operation (33).

Computer applications are possible in six generic areas in the basic scope and responsibility of a typical project manager. The following are these responsibilities and project management software applications:

* See Chapter 28 for a detailed discussion of computer software for project management.

1. *Financial Analysis* - involves planning a project budget, including cost estimates, and resource alloction during the life cycle of the project. Accounting and forecasting programs may be utilized, along with more generalized spreadsheets and data base management messages.
2. *Production Management*—includes planning the production effort of the project, scheduling project milestones, and controlling inventories by determining what resources are needed and scheduling deliveries to interact effectively with the project schedule. Specialized software programs and data base management will meet this requirement.
3. *Forecasting*—involves predicting the future events, requirements, and activities of the project with some degree of accuracy. The project manager will extensively use electronic spreadsheets and computer programs that perform mathematical calculations based on quantitative data.
4. *Report Preparation*—includes all documentation, studies,and reports needed during the project life cycle. Word processing and graphics will be used in this area.
5. *Personal Schedule*—involves the project manager effectively managing his or her time during the project. Specialized programs can be developed to manage the daily and routine matters, while word processing and data base management programs are used to generate necessary lists and documents.
6. *Office Administration*—involves normal office tasks including secretarial work such as typing, generating mailing lists, and personal administration. Word processing and data base management programs will meet these needs (2).

Project managers face myriad responsibilities and details in managing a project to its successful completion. The vast amount of project management software packages available today allows the project manager to fit programs to his or her specific needs. The successful project manager of the present and future must use the computer to the fullest extent possible (2).

In each of the application areas discussed earlier in the chapter (R&D, construction, health care, government/defense, and business and industry) project management is being coupled with computer applications. As software packages become more affordable and projects become more complex in nature, project managers will increasingly adopt and apply computer-aided project management techniques.

In R&D, Goodyear is successfully using a computer-aided design linked to computer-aided manufacturing (CAD-CAM) system. In the technical departments, every professional is given a terminal and access to necessary data bases, which enables them to achieve the highest degree of technological proficiency. CAD-CAM, finite element analysis, and statistical process control are tools Goodyear uses to help reduce testing requirements. These computer uses allow projects to be completed in a shorter life cycle and with less budgetary requirements (6).

One of the earliest uses of project management software was in the construction industry. Computer-aided construction management (CM) was born in the late 1960s when large firms began experimenting with mainframe-based CPM software and automated accounting procedures. Many CM packages are now on microcomputer versions and include scheduling, estimating, job costing, materials handling, equipment inventory, and accounting functions (13).

Management of the Red River Construction Co., Dallas, Texas, measures success by being awarded a bid and then making a profit from the contract. Since adding a computer to its project estimating procedures in 1984, Red River has had a 40% to 50% increase in productivity. In making the estimate, the computer is given the following essential information: manpower requirements, current labor rates, materials required and material costs, subcontract work requirements, and total man-hours anticipated. The company plans to expand the system to include word processing and documentation/report development (3).

Construction firms involved in offshore or arctic activities often face uncertainties due to unpredictable weather conditions. Conventional project management tools are often inadequate in managing weather-sensitive projects. Important issues such as weather downtime and the risk of not completing the project in the limited weather window often cannot be addressed satisfactorily. However, an offshore project simulation program is now available for microcomputers. This program answers "what if" questions in a fast turnaround time, allowing project managers to supplement their own expertise with computer output to make more rational decisions; the results are increased productivity and successibility to weather (10).

When TRW initiated plans to construct a new corporate headquarters in 1979, they utilized computerized facilities design to develop a detailed program which explored several growth scenarios over the anticipated 50-year life of the facility. The needs of the functions to be housed in the corporate headquarters were clearly defined prior to designing the shell of the building. The factors evaluated were:

- The building's design concept and unique design aspects.
- The ongoing assets management needs.
- The facility management process, including planning and design.

The design of the building constructed will allow TRW flexibility at its headquarters for decades (32).

Hospitals using a product management structure are finding they must put a priority on the quality of medical records and the coding of diagnoses and procedures. Administrators making decisions on expanded product lines and services must have access to timely patient data base information. Successful hospital product and data management requires the following:

- Satisfying the information needs for all staff and functional departments.
- Implementing procedures to collect, process, store, retrieve, collate, and distribute the data.
- Evaluating data quality (input and output).
- Assisting data end-users in obtaining and analyzing reports.
- Ensuring that system design objectives are being met.

The basic tenets of hospital record keeping have not changed; they have been improved and expanded by computer technology to enhance product management administration (20).

The Westinghouse Marine Division, a prime contractor to the U.S. Navy, has applied project management microcomputer technology to the Trident II Missile Launching and Handling (ML&H) system being developed for the government. Westinghouse has found the microcomputer useful for many aspects of this project. More daily uses are found as the staff becomes more familiar with the system. Administrative functions are performed in a more efficient manner, and communications have been enhanced. Specific areas of project management that have been improved in this application include planning and scheduling, organizational information lists, budget sheets, scope of work documents, and manpower planning (29).

The military is currently using computerized project management in some areas. For example, the U.S. Marine Corps' Central Design and Programming Activity in Albany, Georgia, has been effectively using project management software for years. However, this use is not mandatory on all military hardware projects. Mandatory computerized project management would provide Congress and other interested agencies a review process to ensure that projects/programs are proceeding according to budget and schedule. Pentagon officials would be more accountable on a "real-time" basis for these multibillion-dollar budgets and projects. Cost overruns and scheduling problems could be identified earlier and resolved more quickly (31).

Iowa Beef Processing, Inc. (IBP) has in-house construction and engineering departments for building and maintaining their plants and offices. These functional departments fluctuate between 100 and 300 employees and are accountable for $40 million to $80 million a year in capital expenditures. Large-volume materials as well as specialized equipment are purchased. To keep track of the many ongoing projects, IBP uses a Capital Project Management System (CPMS) software package from Data Design Associates. In order to get the right information to the right people on a timely basis, CPMS uses a three-tier, hierarchical reporting system. At the project level, reports provide an overview for top-level management and others not involved in the daily operation. At the work unit level, reports are geared to the project managers in the field. Finally, the detail level reports are used by functional specialists such as engineers, accountants, and others responsible for the daily operation of the Engineering and Construction groups. These reports ensure that projects are proceeding in a timely and cost-effective manner, as well as providing needed accountability (28).

The fifth-generation computer systems now being developed will have capabilities from those now used. These Knowledge Information Processing Systems (KIPS) will select, interpret, and adapt huge amounts of information, giving the system the ability to reason. Their artificial intelligence will enable them to learn, make inferences, and reach decisions. The impact of these systems will be significant in many areas, but especially so in project management which involves the interaction of several different disciplines simultaneously. This vast storage of information, combined with superfast processing speeds and reasoning ability, will allow these machines to operate like thousands of experts functioning as one entity. Should only a part of the promise of these machines become reality, project management will experience revolutionary changes (35).

## SUMMARY AND CONCLUSIONS

This chapter has reviewed several project management applications and the characteristics of organizations that have utilized project management systems. The ultimate product of these organizations may be an advanced weapon system, automobiles, an offshore oil rig, steel, national defense, or improved health care. The major determinants in considering the implementation and operation of a project management system include such factors as a technologically dynamic and highly uncertain environment, an organizational structure enabling rapid response to change, identifiable performance and cost objectives, and simultaneous coordination of two or more functional departments (interdisciplinary tasks).

In order to evaluate an organization from a system perspective, one is required to consider the organizational structure as a totality with input, transformation, and output processes functioning within greater environmental systems. Two major points must be understood: one is the nature of the interface of the system with its environmental systems; the other is the nature of the complex interdependencies of parts within the system itself. Although organizations vary greatly in their inputs, transformation mechanisms, and output components, they are strikingly similar with respect to the nature of the external environmental forces impacting on them. The organizational structure chosen to manage such diverse projects, tasks, programs, and products must not only be adaptable to internal and external environmental forces, but also provide for a degree of stability. The project management applications reviewed illustrate that it is meeting these requirements.

It is the authors' belief that organizations with varying technologies and differing organizational objectives may still benefit from the implementation of project management. The many well-documented cases cited here and in other literature reinforce this contention. However, organizations contemplating the adoption of project management must consider the specific aspects of its internal and external environments previously described. It has been observed that coordination costs increase as organizations strive to increase specialization to cope with rapidly advancing technology. This implies that the strategic and competitive health of an organization is contingent upon a systems perspective. This perspective allows the organizational design to be an effective trade-off between the cost and benefit factors inherent in the processes of growing specialization and more complex coordination requirements.

Project management was determined to have wide application to research and development organizations because of structural similarities: technologically complex environments, large specialized staffs requiring interfunctional integration, and differentiated outputs. Resistance to change in organizations can be a common problem among specialized professionals in these organizations. Competition among R&D groups in similar industries is critical in the contemporary environment, and those who become most successful will do so by efficiently managing their resources and evolving new products needed and wanted by their customers. Project management has proven to be effective in these roles.

The implementation of project management in the construction field resulted from the need for communication and information flows, interdisciplinary efforts, cost containment, time management, and complexities involved in the materials ordering and handling process. The construction projects cited are similar in that they all involve large numbers of employees with a wide variety of expertise and specialization. They are subject to varying environmental pressures requiring interfunctional integration for specific output with time, cost, and performance parameters inherent in the process.

Organizations in highly uncertain, dynamic environments, which must function for rapid response to internal and external changes, and those organizations with tasks requiring a high degree of reciprocal interfunctional coordination are also prime candidates for project management. Because their services and/or products require an integrated interfunctional orientation, more contempory service organizations are adopting and implementing project, program, team/task force, or product management because the project structure is appropriate and contributes to the more effective achievement of organizational goals and objectives.

### Conclusions

This chapter provides a working definition of project management including general guidelines for use in evaluating its applicability; it also familiarizes the reader with recent project management applications and pertinent special factors involved. The bibliography enables the reader to go directly to the source should more in-depth information be required.

The emergence of computer applications to the field of project management has had far-reaching implications. With these innovations, more and more contemporary organizations are determining that some form of project management is applicable to their situation. As these programs become more sophisticated, they will become more readily available to smaller-size organizations; indeed, these programs are being increasingly designed for microcomputers. Computer-managed project management applications will make project management more attractive as implementation and maintenance costs decrease. Organizations already utilizing project management will also benefit from this technology because it helps managers to perform more accurate and reliable planning, coordination, and control of their project/task.

Project management has the potential for contributing to more effective achievement of organizational objectives if internal and external environmental forces justify its use. However, project management cannot, and should not, be considered a panacea for all organizational tasks. It is the authors' contention that the increasing complexity and dynamism of today's organizational environment reflected in this review, coupled with innovative computer technology, enable growing applications of project management. The "networking" of microcomputers with each other and with mainframes will give the project and functional managers previously unavailable capability to achieve the required interfaces and integration necessary for project, interfunctional, and interorganizational task success.

The applications discussed and reviewed in this chapter illustrate that there is no one best way to apply and implement project management. Inherent in the concept of project management is the capability to be adaptive to varying organizational and environmental situations; it clearly indicates the essence of the contingency approach to management. The authors forecast that the applications of project management, undoubtedly with many modifications from the traditional applications, will increase significantly since the power and capability of the microcomputer (personal computer) will be readily available at the operating management levels in most organizations.

## BIBLIOGRAPHY


1. Adams, John, R. Barndt, Stephen E. and Martin D. (eds.). *Managing by Project Management* (Universal Technology Corp. Dayton, Ohio, 1979), pp. 1–14.
2. Adams, John R. and Morris, Jackelyn Kelley. "Microcomputer Usage for Project Management." *1984 Proceedings of the Project Management Institute* (The Project Management Institute. Drexel Hill, Pa., 1984), pp. 296–304.
3. Adams, John R. and Morris, Jackelyn Kelley. "A Farewell Bid." *Infosystems* (January, 1985), p. 102.
4. Alvi, Akhtar A. and Methven, Andrea L. "Project Management of a Billion Dollar Refinery Expansion." *1985 Proceedings of the Project Management Institute* (The Project Management Institute. Drexel Hill, Pa., 1985), Track 1.
5. Arnold, Kenneth E. "Modified Turnkey Approach Can Cut Cost of Marsh and Shallow-water Projects." *Oil and Gas Journal* (December 17, 1984), pp. 71–77.
6. Barrett, Tom H. "Research and Manufacturing Share a Common World." *New Management* (March–April, 1986), pp. 23–25.
7. Burck, Charles G. "How G. M. Turned Itself Around." *Fortune* (January 16, 1978), pp. 87–96.
8. Byers, Luanne. "The U.S. Pharmaceutical Industry—A Standard for Success." *1985 Proceedings of the Project Management Institute* (The Project Management Institute, Drexel Hill, Pa., 1985), Track 3.
9. Cathey, Paul. "How Metals Industry Uses Management Tools." *Iron Age* (November 20, 1978), pp. 38–41.
10. Chen, Henry. "Computer Program Uses Simulation Method to Help Manage Weather-Sensitive Projects." *Oil and Gas Journal* (June 24, 1985), pp. 80–86.
11. Cleland, David I. and King, William R. *Systems Analysis and Project Management,* 3rd Ed. (McGraw-Hill. New York, 1983), pp. 187–265.

12. Cobbs, John L. (ed). "G.M.'s Test of Participating." *Business Week* (February 23, 1976), pp. 89–90.
13. Cobbs, John L. "Contractors Wrangle with Planning." *Engineering-News Record* (May 31, 1984), pp. 54–58.
14. Craig, Carol. "Hospital Product Line Survey." Swedish Health Systems, August, 1985.
15. Dilworth, James B., et al. "Centralized Project Management." *Journal of Systems Management* (August, 1985), pp. 30–35.
16. Dixon, Maureen. "Matrix Organization in Health Services," in *Matrix Management—A Cross-Functional Approach to Organization,* ed. Knight, Kenneth (Gower Press. Great Britain, 1977), pp. 82–90.
17. Easton, James. "Long-Term Effects of Project and Project Management on a Large Public Works Organization." *1978 Proceedings of the Project Management Institute* (The Project Management Institute. Drexel Hill, Pa., 1978), pp. IVG.1IV19.
18. Einstein, Harold B. "Project Management: A Banking Case Study." *Magazine of Bank Administration* (April, 1982), pp. 36–40.
19. Fox, J. Ronald. "Revamping the Business of National Defense", *Harvard Business Review* (September-October, 1984), pp. 63–70.
20. Fox, Leslie Ann and Tucker, Jeanne. "Product Management Spurs Emphasis on Medical Records." *Hospitals* (March 1, 1985), pp. 92–94.
21. Fraylick, J. R. "How to Start Up Major Facilities (Trans Alaska Pipeline)." *Oil and Gas Journal* (December 3, 1984), pp. 112–115 and (December 10, 1984), pp. 92–94.
22. Gallagher, Susan C. "The Management of World-Wide Pharmaceutical Development Utilizing Geographically Spread Resources." *1975 Proceedings of the Project Management Institute* (The Project Management Institute. Drexel Hill, Pa., 1975), pp. 143–145.
23. Grudzinshas, Charles V. "Change + Communication = Challenge (Management of Pharmaceutical R & D Projects)." *1985 Proceedings of the Project Management Institute* (The Project Management Institute. Drexel Hill, Pa., 1985), Track 3.
24. Gung, H. P. and Pearson, A. W. "Matrix Organization in Research and Development," in *Matrix Management—A Cross-Functional Approach to Organization,* ed. Knight, Kenneth (Gower Press. Great Britain, 1977), pp. 23–44.
25. Katz, Ralph and Allan, Thomas. "Project Performance and the Locus of Influence in the R & D Matrix." *Academy of Management Journal* (March, 1985), pp. 67–87.
26. Kelley, P. A. "Success for Defense Projects." *1985 Proceedings of the Project Management Institute* (The Project Management Institute. Drexel Hill, Pa., 1985), Track 4.
27. Knudson, Robert E. "Case Study of the Successful Development of a Real Estate Project." *1985 Proceedings of the Project Management Institute* (The Project Management Institute. Drexel Hill, Pa., 1985), Track 2.
28. Kohlowski, David. "Capital Project Management System." *Management Accounting* (March, 1986), pp. 69–70.
29. Kohrs, Robert H. "Implementation of the Microcomputer in a Project Management Environment." *1984 Proceedings of the Project Management Institute* (The Project Management Institute. Drexel Hill, Pa., 1984), pp. 311–318.
30. MacStravic, Robin Scott. "Product-Line Administration in Hospitals." *HCM Review* (Spring, 1968), pp. 35–43.
31. Mandrell, Mel. "Budget-Balancing Computers." *Computer Decisions* (February 25, 1986), p. 12.
32. Mayne, Alfred P., Jr. and Wade, A. Dale. "Computerizing Facilities Design." *Industrial Development* (July/August, 1984), pp. 4–8.
33. Morrison, David. "High-Power Project Management." *Computer Decisions* (January 14, 1986), pp. 39–42.

34. Naylor, Thomas H. "The Case for the Strategy Matrix." *New Management* (Summer, 1986), pp. 38–42.
35. Pandia, Rafeev M. "Impact of Super-Computers on Project Management." *1984 Proceedings of the Project Management Institute* (The Project Management Institute. Drexel Hill, Pa., 1984), pp. 374–383.
36. Patrick, Miles G. "Implementing a Project Management System in a Research Laboratory." *1979 Proceedings of the Project Management Institute* (The Project Management Institute. Drexel Hill, Pa., 1979), pp. 243–252.
37. Peters, Frederick. "NASA Management of the Space Shuttle Program." *1975 Proceeding of the Project Management Institute* (The Project Management Institute. Drexel Hill, Pa., 1975), p. 154.
38. Robinson, Clarence A., Jr. (ed.) "Matrix System Enhances Management." *Aviation Week and Space Technology* (January 31, 1977), pp. 41–58.
39. Robbins, Clarence A., Jr. "Scouting the Management Trail." *Engineering-News Record* (May 31, 1984), pp. 45–53.
40. Schrontz, M. P., Porter, G. M. and Scott, N. C. "Organization and Management of a Multi-Organizational Single Responsibility Project—The James H. Campbell Power Plant—Unit #3." *1977 Proceedings of the Project Management Institute* (The Project Management Institute. Drexel Hill, Pa., 1977), pp. 258–264.
41. Stickney, Frank A. and Johnson, William R. "Communication: The Key to Integration." *1980 Proceedings of the Project Management Institute* (The Project Management Institute. Drexel Hill, Pa., 1980), p. I-A.7.
42. Trembly, Ara C. "Getting More From Your Key Resources." *Computer Decisions* (September 15, 1984), pp. 89–94.
43. Verma, Vijay, K. "Achieving Excellence Through Project Management in R&D Environments." *1985 Proceedings of the Project Management Institute* (The Project Management Institute. Drexel Hill, Pa., 1985), Track 3.
44. White, Stephen L. "The Community Mental Health Center as a Matrix Organization.'" *Administration in Mental Health* (Winter, 1978), pp. 99–106.
45. Whiteside, David E. "Roger Smith's Campaign to Change the F.M. Culture." *Business Week* (April 7, 1986), pp. 84–85.
46. Wood, William R., et al. "Product-Line Management: Final Report and Recommendations." Miami Valley Hospital, April, 1985.

# 35. Factors Affecting Project Success*

Bruce N. Baker†
David C. Murphy
Dalmar Fisher

Why are some projects perceived as failures when they have met all the objective standards of success:

- —completed on time,
- —completed within budget,
- —all technical specifications met?

On the other hand, why are some projects perceived as successful when they have failed to meet two important objective standards associated with success:

- —not completed on time,
- —not completed within budget?

---

* The study reported in this chapter was conducted under the sponsorship of the National Aeronautics and Space Administration, NGR 22-03-028. The complete report is entitled, *Determinants of Project Success,* by David C. Murphy, Bruce N. Baker, and Dalmar Fisher. It may be obtained from the National Technical Information Services, Springfield, Va. 22151, by referencing the title and the Accession number: N-74-30392, September 15, 1974.

† Dr. Bruce N. Baker is Program Manager of Information and Computer Security at SRI International. He conducts project management seminars and is a frequent speaker on the topics of improving teamwork and success in project management environments. He received his A.B. degree from Princeton University, his M.B.A. degree from Stanford University, and his D.P.A. degree from the George Washington University.

David C. Murphy is an Associate Professor at the Boston College School of Management. His research and publications have been concerned with project and program management, strategy and policy formulation, environmental analysis, and organizational decentralization. He has served as editor of *Project Management Quarterly,* and is an active member of several professional societies including the Project Management Institute. He received the D.B.A. degree from Indiana University.

Dalmar Fisher is Associate Professor of Organizational Studies at the Boston College School of Management, where he teaches courses in organizational behavior and administrative strategy. He has authored several articles and books in the areas of organizational communication, project management, and managerial behavior, and has served as associate editor of *Project Management Quarterly.* He received his D.B.A. degree from Harvard Business School.

## WHAT CONSTITUTES SUCCESS FOR A PROJECT?

If project success cannot be considered simply a matter of completing the project on schedule, staying within the budget constraints, and meeting the technical performance criteria, then how should project success be defined?

The research conducted by the authors on some 650 projects supports the following definition of success:

> If the project meets the technical performance specifications and/or mission to be performed, and if there is a high level of satisfaction concerning the project outcome among key people in the parent organization, key people in the client organization, key people on the project team, and key users or clientele of the project effort, the project is considered an overall success.

Perceptions play a strong role in this definition. Therefore, the definition is more appropriately termed, "perceived success of a project." What types of variables contribute to perceptions of success and failure? One would certainly assume that good schedule performance and good cost performance would be key ingredients of the perceptions of success and failure. But note that schedule and cost performance are not included in the above definition.

How do cost and schedule performance relate to the perceived failure and success of projects? It was found that cost and schedule overruns were not included in a list of 29 project management characteristics significantly related to perceived project failure. See Table 35-1. Conversely, good cost and schedule performance were not included in a list of 23 project management characteristics significantly related to perceived success, Table 35-2. Nor were cost and schedule performance included in the list of ten project management characteristics found to be linearly related to both perceived success and perceived failure, Table 35-3. If the study had been conducted solely on aerospace projects, this might not have been too surprising, but aerospace projects represented less than 20% of the responses. For project managers and project personnel who have constantly lived with heavy emphasis upon meeting schedules and staying within budgets, this finding may be difficult to swallow. A partial explanation may be as follows: the survey was concerned only with *completed* projects. As perspective is developed on a project, the ultimate satisfaction of the parent, the client, the ultimate users, and the project team is most closely related to whether the project end-item is performing as desired. A schedule delay and a budget overrun may seem somewhat

Table 35-1. Project Management Characteristics Which Strongly Affect the Perceived Failure of Projects (The absence of these characteristics does not ensure perceived success).

- Insufficient use of status/progress reports.
- Use of superficial status/progress reports.
- Inadequate project manager administrative skills.
- Inadequate project manager human skills.
- Inadequate project manager technical skills.
- Insufficient project manager influence.
- Insufficient project manager authority.
- Insufficient client influence.
- Poor coordination with client.
- Lack of rapport with client.
- Client disinterest in budget criteria.
- Lack of project team participation in decision-making.
- Lack of project team participation in major problem solving.
- Excessive structuring within the project team.
- Job insecurity within the project team.
- Lack of team spirit and sense of mission within project team.
- Parent organization stable, non-dynamic, lacking strategic change.
- Poor coordination with parent organization.
- Lack of rapport with parent organization.
- Poor relations within the parent organization.
- New "type" of project.
- Project more complex than the parent has completed previously.
- Initial under-funding.
- Inability to freeze design early.
- Inability to close-out the effort.
- Unrealistic project schedules.
- Inadequate change procedures.
- Poor relations with public officials.
- Unfavorable public opinion.

The lists in Tables 35-1, 2, and 3 are based on statistical tests in which data about each project management characteristic were grouped according to whether the project's success was rated in the upper third (successful), middle third, or lower third (unsuccessful).

unimportant as time goes on, in the face of a high degree of satisfaction and a sound foundation for future relationships. Conversely, few can legitimately claim that "the operation was a success but the patient died." If the survey had been conducted on current, ongoing projects only, the management emphasis upon meeting schedules and staying within budgets would undoubtedly have been reflected more heavily in the research results. Moreover, good cost and schedule performance *were* correlated with success but to a lesser degree than the items listed in Table 35-2.

Table 35-2. Project Management Characteristics Stongly Associated with Perceived Success. (The following were found to be necessary, but not sufficient conditions for perceived success.)

- Frequent feedback from the parent organization.
- Frequent feedback from the client.
- Judicious use of networking techniques.
- Availability of back-up strategies.
- Organization structure suited to the project team.
- Adequate control procedures, especially for dealing with changes.
- Project team participation in determining schedules and budgets.
- Flexible parent organization.
- Parent commitment to established schedules.
- Parent enthusiasm.
- Parent commitment to established budget.
- Parent commitment to technical performance goals.
- Parent desire to build-up internal capabilities.
- Project manager commitment to established schedules.
- Project manager commitment to established budget.
- Project manager commitment to technical performance goals.
- Client commitment to established schedules.
- Client commitment to established budget.
- Client commitment to technical performance goals.
- Enthusiastic public support.
- Lack of legal encumbrances.
- Lack of excessive government red tape.
- Minimized number of public/government agencies involved.

Table 35-3. Project Management Characteristics Strongly Linearly Related to Both Perceived Success and Perceived Failure. (The presence of these characteristics tends to improve perceived success while their absence contributes to perceived failure.)

- Goal commitment of project team.
- Accurate initial cost estimates.
- Adequate project team capability.
- Adequate funding to completion.
- Adequate planning and control techniques.
- Minimal start-up difficulties.
- Task (vs. social) orientation.
- Absence of bureaucracy.
- On-site project manager.
- Clearly established success criteria.

## ANALYSIS OF VARIABLES ASSOCIATED WITH PERCEIVED SUCCESS AND VARIABLES ASSOCIATED WITH PERCEIVED FAILURE

The listings of variables associated with perceived success and failure, Tables 35-1, 35-2, and 35-3, are much lengthier than anticipated. For a project to be perceived as successful, many, if not most, of the variables associated with success must be present. The absence of even one factor or inattention to one factor can be sufficient to result in perceived project failure. Similarly, most, if not all, of the variables associated with perceived failure must be absent. To add to the fragility of perceived success, the variables must be present or absent in the right degree. For example, project management is closely associated with the use of PERT and CPM networking systems.* So much so, that many managers consider project management and networking systems as synonymous terms.

*Is the use of PERT-CPM systems the most important factor contributing to project success?*

No. PERT-CPM systems *do* contribute to project success, especially when initial overoptimism and/or a "buy-in" strategy has prevailed in the securing of the contract, but the importance of PERT-CPM is far outweighed by a host of other factors including the use of project tools known as "systems management concepts." These include work breakdown structures, life-cycle planning, systems engineering, configuration management, and status reports. The overuse of PERT-CPM systems was found to hamper success. It was the *judicious* use of PERT-CPM which was associated with success. An important military satellite program was actually hampered by early reliance upon a network that covered four walls of a large conference room. The tool was too cumbersome and consumed too much time to maintain it. Fortunately, someone decided that the network was a classified document and ordered curtains to be placed over the walls. Once the curtains were up, they were never drawn again and the project proceeded as planned. More often than not, however, networking *does* contribute to better cost and schedule performance (but not necessarily to better technical performance).

## GENERAL STRATEGIES FOR DIRECTING PROJECTS

Based upon the factors associated with success and the factors associated with failure, a set of general strategies can be developed for directing projects. Some of the strategies tend to be counterintuitive or counter to traditional practice. The somewhat controversial general strategies are

---

* See Chapter 15.

presented in the form of statements which the reader is asked to indicate as true or false.

*A matrix form of project organizational structure is the least disruptive to traditional functional organizational patterns and is also most likely to result in project success. True or false?*

False. Although there are no clear definitions of the different forms of project organizational structures which have attained widespread acceptance, there are some terms which imply certain patterns. The matrix form of organization is well understood by experienced project management personnel, but the authority which goes with such a matrix form of structure varies considerably. In order to provide a spectrum of choices which attempted to avoid preconception of terms, the following organizational patterns were presented for describing the organizational structure of the project team as it existed during the peak activity period of the project:

- Pure Functional—Project manager, if any, was merely the focal point for communications; he had no authority to direct people other than by persuasion or reporting to his own superior.
- Weak Matrix—Project manager was the focal point for controls; he did not actively direct the work of others.
- Strong Matrix or Partially Projectized—Project manager was the focal point for directions and controls; he may have had some engineering and control personnel reporting to him on a line basis, while remainder of the project team was located administratively in other departments.
- Projectized—Project manager had most of the essential elements of the project team under him.
- Fully Projectized—Project manager had almost all of the employees who were on the project team under him.

Each of these organizational arrangements was associated with perceived success in certain situations, but an F-test of these different forms of organizational structure compared with perceived project success revealed that the projectized form of organizational structure is most often associated with perceived success. In general, it is important for the project manager of a large, long-duration project to have key functions of the project team under him.

In the early days of the Ranger and Surveyor lunar research programs, the project managers had only a handful of people reporting to them on a line basis. Both of these programs were relatively unsuccessful as compared to the Lunar Orbiter program, which employed a projectized organization from the start.*

The question remains, however, how should the decision-making authority of the project manager relate to the decision-making authority of the client organization (the organization which sponsored, approved, and funded the effort), and the parent organization (the organization structure above the level of the project manager but within the same overall organization)?

*When a project is critical to the overall success of a company and/or it is critical to the client organization, the parent organization and/or the client organization should take a strong and active role in internal project decision making. True or false?*

False. It is important for the client organization to establish definitive goals for a project. Similarly, especially for in-house projects, the parent organization must also establish clear and definitive goals for the project. When there is a good consensus among the client organization, the parent organization, and the project team with respect to the goals of a project, then success is more readily achieved. A path analysis revealed that success criteria salience and consensus are especially important for:

- Projects with complex legal/political environments.
- Projects which are relatively large.
- Projects undertaken within a parent organization undergoing considerable change.

Once success criteria have been clarified and agreed upon by the principal parties involved with a project, that is, the client, the parent, and the project team, then it is essential to permit the project team to "carry the ball" with respect to internal decisions.

Because some decisions require the approval of the client organization, it was found that the authority of the client contact should be commensurate with the authority of the project manager. Projects characterized by strong project manager authority and influence and strong client contact authority and influence were strongly associated with success. Unfortunately, many client organizations and parent organizations tend to believe that the more closely they monitor a project and the more intimately they enter into the internal project decision process, the more likely the project

* Many comparisons between the Surveyor program and the Lunar Orbiter program support the findings of this chapter. See Erasmus H. Kloman, *Unmanned Space Project Management—Surveyor and Lunar Orbiter,* a report prepared by the National Academy of Public Administration and sponsored by the National Aeronautics and Space Administration, Washington, D.C.: U.S. Government Printing Office (1972).

is to be successful. Close coordination and good relations patterns were found to be the most important factors contributing to perceived project success. Nonethelesss, there is a very important distinction between "close" and "meddling" and there is just as important a distinction between "supportive" and "interfering" relationships. Many factors and relationships pointed to the need for the client and the parent organization to develop close and supportive working relationships with the project team but to avoid meddling or interfering with the project team's decision-making processes. The lesson is clear: the project manager should be delegated sufficient authority to make important project decisions and sufficient authority to direct the project team. In the case of the Polaris program, for example, the head of the Special Projects Office of the U.S. Navy had extensive authority with respect to contracting arrangements. This level of authority, combined with strong levels of authority for the project managers in the contractors' plants, was a major factor contributing to the success of that program. Once given this authority, how should the project manager arrive at decisions and solve problems?

*Because participative decision making and problem solving can tend to slow up the decision-making and problem-solving processes, these approaches should not be employed on complex projects having tight schedules. True or false?*

False. First of all, participative decision making and problem solving within the project team was highly correlated with success for the total sample of projects. Second, a path analysis* revealed that under some conditions of adversity, such as a highly complex project, or one where initial overoptimism prevailed regarding the time and cost for completing the project, it was especially important to employ participative approaches to overcome these adversities.

If this pattern is successful, should the public also participate in project decisions affecting the public interest?

*Public participation is an essential ingredient of success for projects affecting the public interest. True or false?*

Mostly false. Although the trend of the past two decades has certainly been in this direction, that is, to encourage, or at least to facilitate, public participation in the decision-making process for public projects, and although value judgments may lean heavily toward this approach, the facts are that public participation often delays and hampers projects and reduces the probability of success.

Therefore, from a management standpoint (not from a value judgment standpoint), public participation should be minimized, avoided, or cir-

---

* A statistical procedure. Path analysis is explained on pp. 928–929.

cumvented as much as possible. Public participation is, of course, a legal requirement for most public projects but there seems to be little reason for overdoing it.

If too much public participation hampers success, can the cooperation and participation of several agencies help to safeguard the public interest and result in a more successful overall effort than a project undertaken by a single agency?

*Public projects involving the cooperation, funding, and participation of several governmental agencies are more likely to be successful than projects undertaken by a single agency. True or false?*

False. Again, the trend is certainly in this direction. There has been a great deal of emphasis upon:

- Interagency cooperative efforts, for example, Departments of Labor, Commerce, and Transportation.
- Intergovernmental cooperative efforts, for example, federal, state, and local jointly funded efforts.
- The creation of new, integrative agencies, for example, regional commissions combining the efforts of several states, counties, or cities to attack common problems.

Although the creation of these jointly funded, jointly managed organizational mechanisms may be desirable from the standpoint of integration of efforts, they tend to result in less successful projects as compared to projects undertaken by a single source of funding and authority. Such cooperative efforts often result in the creation of elaborate bureaucratic structures, decision delays, red tape, and relatively diminished success. The New England Regional Commission is an example of an agency which consumed millions of dollars for its own bureaucracy but failed to accomplish much of anything for New England.

Many discussions of project management focus upon qualities of an ideal project manager.

*It is much more important for a project manager to be an effective administrator than to be a competent technical person or to possess good human relations skills. True or false?*

Mostly false. All three types of skills (technical skills, human skills, and administrative skills) were found to be important but technical skills were found to be most important, followed by human skills, and then by administrative skills.

It is true that technically oriented scientists and engineers who are placed into project manager positions often perform poorly from an administrative and human relations standpoint but, on the other hand, some of the most costly blunders have been made by administrators of proven competence who ventured into unfamiliar areas. During the past two decades, much progress has been made in training technical people to acquire effective human relations and administrative skills.

Leadership style has been the subject of a great deal of research.

*The most effective project managers are nondirective, human relations-oriented leaders as opposed to directive, managing, task-oriented leaders. True or false?*

Mostly false. Fiedler, for one, conducted extensive research on this subject, finding that, "In very favorable or in very unfavorable situations for getting a task accomplished by group effort, the autocratic, task-controlling, managing leadership works best. In situations intermediate in difficulty, the nondirective, permissive leader is most successful."*

The research described in this chapter supports the concept of a leader who is task-oriented with a backup social orientation for *most* project efforts. Does this contradict Fiedler's research and the previous statement that project team participation in decision making and problem solving is important to project success? The authors believe that there is no contradiction. An effective project manager is generally one who is committed to the goals of the project and constantly stresses the importance of meeting those project goals. Yet, he calls upon key project team members to assist with problem solving and decision making. In *some* very straightforward or very chaotic settings, a project manager may find an autocratic style to be most effective. And, as Fiedler's research suggests, a project manager may need to employ different leadership styles at different times during the project effort.

More recent writers on leadership have argued that the effective leader not only responds to settings and situations that present themselves, but shapes and transforms the circumstances by reframing problems and goals in insightful ways that sharpen task definitions and arouse heightened levels of commitment by team members†. Our data support this concept of the effective project manager as one who can respond to realities "incrementally," but who can also define general strategies and spark others' commitment to them.

A comprehensive list of general strategies is shown in Table 35-4. Strategy guidelines are presented for the client organization, the parent organi-

---

* Fred E. Fiedler, "Engineer the Job to Fit the Manager," *Harvard Business Review* (September–October, 1965), p. 18.

† Christ Argyris and Donald A. Schon, *Organizational Learning* (Reading, Mass.: Addison-Wesley, 1978); and Bernard M. Bass, *Leadership and Performance Beyond Expectations* (New York: Free Press, 1985).

Table 35-4. General Strategies for Directing Projects.

| | CONCEPTUAL PHASE (BEFORE THE INVITATIONS FOR BID) | BID, PROPOSAL, CONTRACT DEFINITION, AND NEGOTIATION PHASE (BEFORE CONTRACT AWARD OR GO-AHEAD) | IMPLEMENTATION PHASE (AFTER CONTRACT AWARD OR GO-AHEAD) |
|---|---|---|---|
| The Client Organization and/or Principal Client Contact | Encourage openness & honesty from the start from all participants. | | Develop close, but not meddling, work relationships with project participants. |
| | Create an atmosphere that encourages healthy, but not cutthroat, competition or "liars' contests." Plan for adequate funding to complete the entire project. → | Reject "buy-ins." → Make prompt decisions regarding contract award or go-ahead. | Avoid arms-length relationships. |
| | | | Do not insist upon excessive reporting schemes. |
| | Develop clear understandings of the relative importance of cost, schedule, and technical performance goals. | | |
| | Seek to minimize public participation and involvement. Develop short and informal lines of communication and flat organizational structures. → | → | → |
| | Delegate sufficient authority to the principal client contact and let him promptly approve or reject important project decisions. → | → | → |

| | | |
|---|---|---|
| The Parent Organization and/or Principal Parent Contact | Select, at an early point, a project manager with a proven track record of technical skills, human skills, & administrative skills (in that order) to lead the project team. | Do not exert excessive pressure on the project manager to win the contract. |
| | Develop clear and workable guidelines for your project manager. | Do not slash or balloon the project team's cost estimates. |
| | | Avoid "buy-ins." |
| | Delegate sufficient authority to your project manager and let him make important decisions in conjunction with his key project team members. → | Develop close, but not meddling, working relationships with the principal client contact and the project manager → |
| | Demonstrate enthusiasm for and commitment to the project and the project team. Develop and maintain short and informal lines of communication with the project manager. → | |
| | Insist upon the right to select your own key project team members | Call upon key project team members to assist in decision-making and problem solving. → Employ a workable and candid set of project planning and control tools. |

Table 35-4. General Strategies for Directing Projects (*continued*)

| | CONCEPTUAL PHASE (BEFORE THE INVITATIONS FOR BID) | BID, PROPOSAL, CONTRACT DEFINITION, AND NEGOTIATION PHASE (BEFORE CONTRACT AWARD OR GO-AHEAD) | IMPLEMENTATION PHASE (AFTER CONTRACT AWARD OR GO-AHEAD) |
| --- | --- | --- | --- |
| The Project Manager and/or the Project Team | Select key project team members with proven track records in their area of expertise. | Develop realistic cost, schedule, and technical performance estimates & goals. | |
| | Develop commitment and a sense of mission from the outset among project team members. | Develop back-up strategies and systems in anticipation of potential problems. → | |
| | | | Avoid preoccupation with, or over-reliance upon, one type of project control tool. |
| | Seek sufficient authority and a projectized form of organizational structure. | Develop an appropriate, yet flexible and flat, project team organization structure. | Constantly stress the importance of meeting cost, schedule and technical performance goals. |
| | Coordinate frequently and constantly reinforce good relationships with the client, the parent, and your team. | Seek to maintain your influence over people and key decisions even though your formal authority may not be sufficient. | Generally, give highest priority to achieving the technical performance mission or function to be performed by the project end-item. |
| | Seek to enhance the public' image of the project. → | | Keep changes under control. |
| | | | Seek to find ways of assuring the job security of effective project team members. |
| | | | Plan for an orderly phase-out of the project |

zation, and the project team for three distinct phases of a project. It is important to note (1) the interlocking and interdependent relationships among the three organizational groups involved, and (2) that two of the three phases leading to overall perceived success occur before contract award or go-ahead. Although different combinations are needed for success in various situations and environments, these general strategies seem to apply to most situations.

## KEY FACTORS TO MAXIMIZE POTENTIAL OF PERCEIVED PROJECT SUCCESS

Up to this point, the ingredients to assure success and to avoid failure have been somewhat overwhelming. This portion of the chapter will attempt to focus in on the key factors which appear to be most important for achieving high levels of perceived success.

In reexamining Table 35-1, one can see that a large number of the variables associated with perceived failure center about poor coordination and human relations patterns. Therefore, in order to minimize the chances of perceived failure, project managers are well advised to put heavy emphasis on establishing good, effective patterns of coordination and human relations. Such emphasis may eliminate failure but may not necessarily promote success. Table 35-2 sheds light on the need for good, tight controls and commitment to the goals that have been established for a project in order to achieve high levels of perceived success.

Tables 35-1, 35-2, 35-3, and 35-4 also point to another important strategy: effective project planning is absolutely essential to project success. Of the 29 items listed in Table 35-1, over half the variables associated with perceived failure can be avoided through effective project planning. The role of project planning is even more apparent in Table 35-2. Almost every one of the variables associated with success is determined by, or can be significantly influenced by, effective project planning. Finally, every one of the items listed in Table 35-3 is intimately related to the project planning process. As stated previously, two of the three phases of strategies shown in Table 35-4 occur before actual work on the project end-item begins. Therefore, effective project planning is very important to project success.

In addition to the analyses summarized to this point, stepwise multiple regression analysis was conducted to determine the independent contribution of some 32 factors to Perceived Success. It should be re-emphasized that technical performance was integrally associated with success and was part of Perceived Success itself. Beyond technical performance, however, what are the principal factors contributing to project success?

Table 35.5 The Relative Importance of the Factors Contributing to Perceived Project Success.

| DETERMINING FACTORS | STANDARDIZED REGRESSION COEFFICIENT | SIGNIFICANCE | CUMULATIVE $R^2$ |
|---|---|---|---|
| Coordination and Relations | +.347 | $p<.001$ | .773 |
| Adequacy of Project Structure and Control | +.187 | $p<.001$ | .830 |
| Project Uniqueness, Importance, and Public Exposure | +.145 | $p<.001$ | .877 |
| Success Criteria Salience and Consensus | +.254 | $p<.001$ | .886 |
| Competitive and Budgetary Pressure | −.153 | $p<.001$ | .897 |
| Initial Over-Optimism, Conceptual Difficulty | −.215 | $p<.001$ | .905 |
| Internal Capabilities Buildup | +.084 | $p<.001$ | .911 |

Table 35-5 shows that the strongest seven of the determining factors explained 91% of the variance in Perceived Success. The makeup of these seven factors is shown in Table 35-6. Note the extremely important impact of coordination and relations patterns (77% of the variance). Success Criteria Salience and Consensus and avoidance of Initial Over-Optimism, Conceptual Difficulty were the next two heaviest weighted factors in the regression equation. Note also that although the factor, Adequacy of Project Structure and Control, is included in the seven principal factors contributing to Perceived Success, no particular tool, as such, is included in the factor. In other words, PERT and CPM are *not* the be-all and end-all of project management.*

Occasionally, project management personnel adopt a defeatist attitude about a project. One hears such comments as, "The project was doomed to failure from the start," or "There was no way we could make them happy on that project." Table 35-5 does not lend credence to such an attitude. An analysis of Table 35-5 reveals that a very high proportion of the key factors associated with success are within the control of the project manager and the project team. The project manager *can* help to achieve effective coordination and relations; the project manager *can*

* See Chapter 15.

Table 35-6. Items Included in the Seven Factors of Table 35-5.

Coordination & Relations Factor.
- Unity between project manager and contributing department managers.
- Project team spirit.
- Project team sense of mission.
- Project team goal commitment.
- Project team capability.
- Unity between project manager and public officials.
- Unity between project manager and client contact.
- Unity between project manager and his superior.
- Project manager's human skills.
- Realistic progress reports.
- Project manager's administrative skills.
- Supportive informal relations of team members.
- Authority of project manager.
- Adequacy of change procedures.
- Job security of project team.
- Project team participation in decision making.
- Project team participation in major problem solving.
- Parent enthusiasm.
- Availability of back-up strategies.

Adequacy of Project Structure and Control Factor.
- Project manager's satisfaction with planning and control.
- Team's satisfaction with organization structure.

Project Uniqueness, Importance and Public Exposure Factor.
- Extent of public enthusiasm.
- Project larger in scale than most.
- Initial importance of state-of-art advancement.
- Project was different than most.
- Parent experience with similar project scope.
- Favorability of media coverage.

Success Criteria Salience and Consensus Factor.
- Importance to project manager—budget.
- Importance to project manager—schedule.
- Importance to parent—budget.
- Importance to parent—schedule.
- Importance to client—budget.
- Importance to client—schedule.
- Importance to client—technical performance.
- Importance to parent—technical performance.
- Importance to project manager—technical performance.

Competitive and Budgetary Pressure Factor (Negative Impact).
- Fixed price (as opposed to cost reimbursement) type of contract.
- Highly competitive environment.
- Parent heavy emphasis upon staying within the budget.
- Project manager heavy emphasis upon staying within the budget.
- Client heavy emphasis upon staying within the budget.

Table 35-6. (continued)

| |
|---|
| Initial Over-Optimism, Conceptual Difficulty Factor (Negative Impact). |
| Difficulty in meeting project schedules. |
| Difficulty of staying within original budget. |
| Original cost estimates too optimistic. |
| Difficulty in meeting technical requirements. |
| Project was more complex than initially conceived. |
| Schedule overrun. |
| Difficulty in freezing design. |
| Unrealistic schedules. |
| Project was different than most. |
| Internal Capabilities Build-up Factor. |
| Extent to which project built up parent capabilities. |
| Original total budget. |
| Total cost of project. |

make certain that there are adequate project structure and control systems; the project manager *can* help to achieve success criteria salience and consensus; the project manager *can* help to avoid initial overoptimism and conceptual difficulty; and, the project manager *can* have some impact upon internal capabilities buildup, the atmosphere of competitive and budgetary pressure, and the project uniqueness, importance, and public exposure.

Therefore, the project manager *can* control the destiny of the project and the perceptions others will have of him or her. Even under extremely adverse circumstances, a project manager can be perceived as doing the best job possible under the circumstances.

## CONCLUSIONS

The following conclusions seem to be warranted from the research:

1. Project success cannot be adequately defined as
   - Completing the project on schedule.
   - Staying within the budget.
   - Meeting the technical performance specifications and/or mission to be performed.
2. Perceived success of a project can best be defined as
   - Meeting the project technical specifications and/or project mission to be performed.
   - Attaining high levels of satisfaction from:
     - The parent.

- The client.
- The users or clientele.
- The project team itself.

3. Technical performance is integrally associated with perceived success of a project, whereas cost and schedule performance are somewhat less intimately associated with perceived success.
4. In the long run, what really matters is whether the parties associated with, and affected by, a project are satisified. Good schedule and cost performance mean very little in the face of a poorly performing end product.
5. Next to technical performance and satisfaction of those associated with, and affected by, a project, effective coordination and relation patterns are the most important contributors to perceived project success.
6. Project managers can attain high levels of perceived project success even under adverse circumstances.

# 36. Project Management in the Public Sector: Success and Failure Patterns Compared to Private Sector Projects*

Bruce N. Baker†
Dalmar Fisher
David C. Murphy

## INTRODUCTION

How do public sector projects differ from private sector projects? Most people have definite preconceptions about the two. Some of these preconceptions may be summarized by the types of contrasting characteristics listed in Table 36-1. A number of studies have been made of public sector projects which tend to support some of these types of preconceptions. For example, a number of studies regarding cost growth and cost overrun

---

*The study reported in this chapter was conducted under the sponsorship of the National Aeronautics and Space Administration, NGR 22-03-028. The complete report is entitled, *Determinants of Project Success,* by David C. Murphy, Bruce N. Baker, and Dalmar Fisher. It may be obtained from the National Technical Information Services, Springfield, Va. 22151, by referencing the title and the Accession number: N-74-30392, September 15, 1974.

† Dr. Bruce N. Baker is Program Manager of Information and Computer Security at SRI International. He conducts project management seminars and is a frequent speaker on the topics of improving teamwork and success in project management environments. He received his A.B. degrees from Princeton University, his M.B.A. degree from Stanford University, and his D.P.A. degree from George Washington University.

Dalmar Fisher is Associate Professor of Organizational Studies at the Boston College School of Management, where he teaches courses in organizational behavior and administrative strategy. He has authored several articles and books in the areas of organizational communication, project management, and managerial behavior, and has served as associate editor of *Project Management Quarterly*. He received his D.B.A. degree from Harvard Business School.

David C. Murphy is an Associate Professor at the Boston College School of Management. His research and publications have been concerned with project and program management, strategy and policy formulation, environmental analysis, and organizational decentralization. He has served as editor of *Project Management Quarterly,* and is an active member of several professional societies including the Project Management Institute. He received the D.B.A. degree from Indiana University.

Table 36-1. Some Preconceptions Regarding Private Sector Projects vs. Public Sector Projects.

| *Private* | *Public* |
|---|---|
| Efficient | Inefficient |
| Effective | Ineffective |
| On schedule | Behind Schedule |
| Within budget | Overrun of budget |
| Well planned | Poorly planned |
| Competitive | Non-Competitive |
| Capable managers | Incapable managers |
| Competent workers | Incompetent workers |
| Free of Politics | Encumbered by politics |
| The end-product "works" | The end-product doesn't "work" |
| Minimum paperwork | Excessive paperwork |
| Definitive goals | Nebulous goals |
| Feelings of satisfaction | Feelings of dissatisfaction |
| People seem to care | People don't seem to care |
| Good team spirit | Lack of team spirit |
| Incompetent people are fired | Incompetent people can't be fired |
| Good performance is rewarded | Good performance is not rewarded |

of federal government projects have been conducted during the past two decades.

The most sophisticated studies of actual cost performance on Department of Defense programs as compared to original cost estimates were the Merton J. Peck and Frederic M. Scherer studies* and several Rand Corporation studies.

Peck and Scherer analyzed 12 typical weapon systems programs of the 1950s. All 12 systems employed cost-plus-fixed-fee contracts. The average cost growth was found to be 220% beyond original target cost.†

Almost identical results came from a later study of 22 Air Force weapon systems programs involving 68 estimates. The study, entitled *Strategy for R&D: Studies in the Microeconomics of Development,* by Thomas Marschak, Thomas K. Glennan, Jr., and Robert Summers of Rand Corporation, showed an average cost growth of 226% beyond original estimated cost.‡ These programs involved mainly the cost-plus-fee contracts of the late 1950s.

* Merton J. Peck and Frederic M. Scherer, *The Weapons Acquisition Process—An Economic Analysis* (Boston: Graduate School of Business, Harvard University, 1962).
† Ibid., p. 429.
‡ (New York: Springer-Verlag New York Inc., 1967), p. 152.

In the 1960s, incentive contracts, rather than cost-plus-fixed-fee contracts, were used for most engineering development efforts in the Department of Defense. One might therefore expect actual program costs to be closer to original cost estimates. Two such studies of the 1960s were undertaken by Rand personnel.

Robert Perry et al. reported in a study of 21 Army, Navy and Air Force system acquisition programs that " . . . [O]n average, cost estimates for the 1960s were about 25% less optimistic than those for programs for the 1950s. Thus, if reduction in bias (or reduced optimism) is a realistic index of 'better,' there is evidence of improvement in the acquisition process."* Even such a statement as this must be hedged considerably as Perry et al. were careful to do: "Still, the model has little explanatory power (in a statistical sense), and it does not indicate *why* improvements have occurred.†

Rand studies of defense procurement in the 1970s showed less cost and schedule overrun than in the 1960s, and studies of the 1980s conducted to date show improvement over the 1970s. Procurement of weapons systems that has gone through one or more production runs have shown minimal cost overrun, except those contracts which were subjected to stretch-out of deliveries.‡ All in all, cost and schedule performance on procurement of weapons systems has improved steadily from the 1950s to the 1980s, if one defines improvement in terms of final cost and schedule performance as compared to original cost and schedule estimates.§

The problems of cost and schedule overruns have not been unique to the Defense Department within the federal sector. Environmental projects, transportation projects, federal housing projects, and space projects, to name a few, have all experienced overrun problems. In many cases, the interactions of federal, state, and local governmental agencies create their own set of inherent problems in addition to the obstacles and setbacks encountered on the actual project work itself. Nuclear plant construction, freeway construction, subway systems, and environmental cleanup efforts entail interaction of many governmental agencies, many types of

---

* *System Acquisition Experience,* Memorandum RM-6072-OR prepared for United States Air Force Project Rand (Santa Monica: The Rand Corporation, November, 1969), p. 6.

† Ibid.

‡ For an excellent analysis of the program stretch-out problem, see Jacques S. Gansler, "Defense Program Instability: Causes, Costs, and Cures," *Defense Management Journal,* Vol. 22, No. 2, Second Quarter, 1986, pp. 3–11.

§ E. Dews, G. K. Smith, A. A. Barbour, E. D. Harris, M.A. Hesse, *Acquisition Policy Effectiveness: Department of Defense Experience in the 1970s,* Report R-2516-DRE (Santa Monica: The Rand Corporation, October, 1979); M. D. Rich, E. Dews, C. L. Batten, *Improving the Military Acquisition Process: Lessons from Rand Research,* Report R-3373-AF/RC (Santa Monica: The Rand Corporation, February, 1986).

business entities, and numerous public interest groups in ways that create a nightmare from a management standpoint. The public has been demanding more of a voice in all of these areas. Housing and Urban Development, for example, does not "control" housing as much as it tries to cope with the special interests it encounters wherever it turns. Projects in some of these areas encounter obstacles and setbacks that the Defense Department never dreamed of.

Personal experience often reinforces the preconceptions of public sector efforts through one's dealings with the U.S. Postal Service, governmental social service agencies, regulatory agencies, etc.

The United States does not have a monopoly on public sector project difficulties. A.P. Martin notes that Canada has experienced a plague of failures of large-scale public sector projects to reach their targets.* He cites the Panartics, James Bay, the Gentilly nuclear reactors, the NORAD defense network renewal, and the Montreal Olympics.

Very few studies have been conducted of private sector projects. For example, cost overrun data is generally not available or at least not publicized by independent sources. Yet, the cost overrun records and fiascos of some major private sector projects are comparable to many public sector projects. For example, cost overruns of pioneer process plants have averaged about 200%.† A list of private sector fiascos is shown in Table 36-2.

Although good data may not be available for a comprehensive comparison of actual cost overrun and actual schedule overrun for private sector projects versus public sector projects, the authors conducted a study which compares these dimensions as well as overall perceived project success for the two sectors. The study was designed to detail the relationships among situational, structural, and process variables as they related to project effectiveness.

The study is believed to be the largest and most comprehensive investigation to date on the subject of project management effectiveness. A sample of 646 responses to a 17-page questionnaire represented a variety of industries (34% manufacturing, 22% construction, 17% government, and 27% services, transportation, and others). Most of the respondents themselves had been directly involved in the particular project they chose to describe in their questionnaire. Of the total sample, 50% had been the project manager, 31% had been in other positions on the project team, and

* A. P. Martin, "Project Management Requires Transorganizational Standards," *Project Management Quarterly,* Vol. X, No. 3, 1979, pp. 41–44.

† E. Merrow, K. E. Phillips, C. W. Myers, *Understanding Cost Growth and Performance Shortfalls in Pioneer Process Plants,* Report R-2569-DOE (Santa Monica: The Rand Corporation, September, 1981).

Table 36-2. Some Notable Failures Among Private Sector Projects.

| | |
|---|---|
| Ford | The Edsel |
| Proctor and Gamble | Rely Brand Tampons |
| General Dynamics | Convair 880 and 990 |
| Lockheed | L-1011 Airbus |
| Four Seasons | Chain of Nursing Homes |
| John Hancock Mutual | Windows in Boston Office Building |
| Polaroid Corporation | Polavision |
| Firestone | Radial Tires |
| Dupont | Korfam |
| Gillette | Digital Wristwatches |
| Dansk Designs, Ltd. | Gourmet Product Line |
| BIC Pen Corporation | Fannyhose |
| General Foods | Burger Chef Restaurants |
| A & P | WEO Price Reduction Program |
| National Semiconductor | Consumer Products |

another 10% had been the project manager's direct superior. About one-third of the projects were described as being public in nature, the remaining two-thirds being in the private sector. The types of contracts or agreements involved included cost plus fixed fee (32%), in-house work orders (28%), fixed price (21%), and fixed price with incentives (14%). The major activity or end product involved in the projects included construction (43%), hardware or equipment (22%), new process or software (14%), and studies, services and tests (11%).

## DETERMINANTS OF COST AND SCHEDULE OVERRUN

The study revealed the principal determinants of cost and schedule overrun for both public and private sector projects. Cost overruns were found to be highly correlated with the size of the project and the difficulty of meeting technical specifications. However, schedule difficulties and resulting schedule overruns were the primary causal factors leading to cost overruns. Schedule overruns were, in turn, caused by the variables listed on Table 36-3.

In order to prevent schedule and cost overruns, or to minimize the amount of schedule and cost overrun when initial overoptimism or a "buy-in"* has occurred, the research points to the need for employing

* A "buy-in" is an intentional underestimation of costs in order to obtain a contract or to obtain approval to proceed on an effort with the hope that follow-on contracts, changes, or additional funding will compensate for the original low estimate.

Table 36-3. Determinants of Cost and Schedule Overruns.

- Cost underestimates.
- Use of "Buy-in" strategies.
- Lack of alternative backup strategies.
- Lack of project-team goal commitment.
- Functional, rather than projectized, project organization.
- Lack of project team participation in setting schedules.
- Lack of team spirit, sense of mission.
- Inadequate control procedures.
- Insufficient use of networking techniques.
- Insufficient use of progress/status reports.
- Over-optimistic status reports.
- Decision delays.
- Inadequate change procedures.
- Insufficient project manager authority and influence.
- Lack of commitment to budget and schedule.
- Overall lack of similar experience.

networking techniques, systems management approaches, participative approaches to decision making within the project team, and a task-oriented style of leadership, with a backup relationship-orientated style.

## COMPARISONS BETWEEN PRIVATE SECTOR AND PUBLIC SECTOR PROJECTS

The comparisons between the private sector projects (2/3 of the sample) and public sector projects (1/3 of the sample) were extremely intriguing. Although many of the characteristics may seem intuitively obvious and may concide with our preconceptions, some of the findings appear to be counterintuitive. Moreover, some of the characteristics commonly attributed to public sector projects do not appear on the listings of items that statistically differentiate public from private sector projects. The variables which were found to be highly related to public sector projects are shown in Table 36-4.

Of course, the bulk of the items coincide with our preconceptions and experiences. Such items as red tape, overcontrol, overinvolvement of the public, politics, paperwork, the ratio of parent managers to total employees, the number of staff-type project team members, etc., coincide well with our beliefs and experiences.

As indicated by the asterisks, however, there are at least 13 variables on the list which may not concide with our preconceptions or our intuition. We may tend to believe that the private sector has greater latitude in replacing project managers, but the survey clearly shows that project

## Table 36-4. Variables Significantly Associated with Public Sector Projects.†

| | |
|---|---|
| Delays caused by governmental red tape | (P < .001) |
| Government overcontrol | (P < .001) |
| Difficulty in obtaining funding to complete the project | (P < .001) |
| Length of project | (P < .001) |
| Scheduled length of project | (P < .001) |
| Multi-funding | (P < .001) |
| Percent of R&D budget to the total parent budget | (P < .001) |
| *Number of times the project manager was replaced | (P < .001) |
| The extent of use of work breakdown structures | (P < .001) |
| The extent of use of systems management concepts | (P < .001) |
| The extent of use of operations research techniques | (P < .001) |
| *The project manager's authority over merit raises | (P < .001) |
| *The client contact's influence in relaxing specifications | (P < .001) |
| *The client contact's authority in relaxing specifications | (P < .002) |
| *The job insecurity of project team members | (P < .002) |
| The number of governmental agencies involved with the project | (P < .002) |
| *The legal restrictions encumbering the project | (P < .002) |
| The need for new forms of government-industry cooperation | (P < .002) |
| Overinvolvement of the public with the project | (P < .003) |
| Total project team personnel | (P < .004) |
| The value of systems management concepts | (P < .004) |
| *The project manager's authority to select project team personnel | (P < .005) |
| The amount of politics involved in the contract award | (P < .006) |
| *The importance to the client of staying within the budget | (P < .006) |
| The importance to the project manager of obtaining follow-on work | (P < .007) |
| The difficulty of obtaining funding from the client | (P < .008) |
| The excessive volume of paperwork | (P < .009) |
| *The availability of back-up strategies | (P < .010) |
| The travel time between the project manager and the client | (P < .011) |
| The importance of state-of-the-art advancement | (P < .012) |
| The difficulty of keeping competent project team members | (P < .015) |
| *The degree to which competition was considered cutthroat | (P < .018) |
| The value of operations research techniques | (P < .020) |
| The extent to which problems arose because the project was different | (P < .024) |
| The ratio of the number of parent managers to total employees | (P < .024) |
| The number of staff-type project team members | (P < .027) |
| *The influence of the project manager over the selection of project team personnel | (P < .029) |
| *The client satisfaction with the outcome of the project | (P < .037) |
| The extent to which bar charts or milestone charts were used | (P < .041) |
| *The extent to which the project team participated in major problem solving | (P < .043) |

* Indicates probable counter-intuitive relationships.

† This list is based on statistical tests in which data about each project were grouped according to whether the project was a private or a public sector project.

managers on public sector projects are replaced much more often than their private sector counterparts. We may tend to believe that project managers of private sector projects have much greater authority and influence in selecting project team personnel and in determining their raises, but the study shows just the opposite. We may tend to believe that client contacts for private sector projects have greater influence and authority in relaxing specifications, and greater satisfaction with the outcome of the project, but the study shows just the opposite. We may tend to believe that the legal restrictions resulting from OSHA, EPA regulations, etc., result in greater legal encumbrances over private sector projects as compared to public sector projects, but the study shows just the opposite. We may tend to believe that greater emphasis is placed on the availability of backup strategies for private sector projects, but the study shows just the opposite. And we may tend to believe that cutthroat competition is more prevalent among private sector projects but, again, the study shows just the opposite.

Table 36-5 indicates the variables which are significantly uncharacteristic of public sector projects. As might be expected, unity between the project manager and public officials involved with the public sector effort is generally not high, and the parent organization places little importance upon achieving the technical performance goals of the project (as opposed to staying within the budget and meeting the schedule). Also, the project manager's influence in selecting subcontractors and his authority to permit subcontractors to exceed original budgets or schedules are very low.

Of greater interest than the preceding lists are the characteristics which did *not* show up as significantly different between public sector and private sector projects. Table 36-6 lists some items which did not show statistically significant differentiation. *Note particularly that actual cost and schedule overrun were not found to be significantly different.* Also, satisfaction of the parent organization, the project team, and the ultimate

Table 36-5. Variables Significantly Uncharacteristic of Public Sector Projects.

|  |  |
|---|---|
| The degree of unity between the project manager and the principal public officials involved with the effort | ($P < .001$) |
| The project manager's authority to permit subcontractors to exceed original budgets or schedules | ($P < .009$) |
| The importance to the parent organization of achieving the specified technical performance goals | ($P < .016$) |
| *The difficulty in freezing the design | ($P < .025$) |
| The project manager's influence in selecting subcontractors | ($P < .042$) |

*This characteristic may be considered counter-intuitive.

Table 36-6. Some Variables Which Did Not Differ Significantly Between Private Sector and Public Sector Projects.

| |
|---|
| Actual cost overrun. |
| Actual schedule overrun. |
| Extent of use of networking systems. |
| Advance in state-of-the-art required. |
| Difficulty in defining the goals of the project. |
| Difficulty in meeting the technical requirements of the project. |
| Satisfaction of the ultimate users, recipients, or clientele with the outcome of the project. |
| Satisfaction of the parent organization with the outcome of the project. |
| Satisfaction of the project team with the outcome of the project. |

users did not differ significantly. In fact, client satisfaction tends to be *higher* for public sector projects, as indicated in Table 36-4.

*In general, the comparisons between private sector projects and public sector projects do not support many of our preconceptions*. Public sector projects certainly have their share of problems, but they have been maligned more than the evidence of this study warrants.

## STRATEGIES FOR OVERCOMING SOME OF THE PROBLEMS FACING PUBLIC SECTOR PROJECTS

Many of the characteristics which distinguish public sector projects from private sector projects can be considered adverse in nature. These adverse conditions make the probability of success less likely for public sector projects as compared to private sector projects. Should those involved with public sector projects therefore accept their fate and be content with very low levels of relative success? The findings of this study do not support such a defeatist approach to the management of public sector projects.

Instead, several strategies have been derived from the research findings which can maximize the success potential of public sector projects. Even when combinations of adversities exist, moderate success levels can be achieved if heavy emphasis is placed upon appropriate strategies for the situation and the environment as well as upon diligent pursuit of the project goals. The strategies which follow are based upon a path analysis diagram which was derived from a series of multiple regressions. Path analysis is a relatively new analytic technique and is not to be confused with networking techniques such as PERT and CPM. The result of a path analysis is a model which explains the interaction of a large number of variables. Such a model illustrates the causal relationships contained in a series of relationships. The strengths of these relationships are measured by path coefficients. These coefficients are standardized measures which can be compared to determine the relative predictive power of each independent variable with the effects of the other variables held constant. The particular value of path analysis is that it illustrates the working relationships of many variables in a network of relative predictive powers, thus allowing one to understand the relationships among variables in a systematic manner. The strategies derived from the path analysis are summarized in Figure 36-1, contingent strategies for successful projects.

The most significant conclusion to be derived from Figure 36-1 is that a *project manager faced with one or more adversities need not and should not adopt a defeatist approach to the management of the project.* Even when combinations of adversities exist, moderate success levels can be achieved if heavy emphasis is placed upon appropriate strategies for the situation and the environment as well as upon diligent pursuit of the project goals.

A project manager can thus use Figure 36-1 as a basis for developing contingent strategies to overcome or circumvent certain adversities. The path analysis diagram was derived from the complete sample of private and public sector projects, but two of the adversities which often face managers of public projects will be analyzed:

- Legal-political difficulties.
- Large projects.

The reader can examine Figure 36-1 to see the basis for the strategies designed to overcome these adversities. Although most of these strategies can be considered general strategies, they should receive added emphasis for public projects facing one or both of these adversities. In situations where these adversities do not exist, these strategies can be played down.

### Strategies for Overcoming Legal-Political Difficulties

1. *Encourage openness and honesty from the start from all project participants and specifically seek to avoid and reject "buy-ins."*

When legal-political difficulties surround a project, these difficulties can only be compounded in the long run by permitting "buy-ins" to occur. In the short run, it may appear advantageous to secure initial program funding and initial contracts in order to enable "the camel's nose to enter the

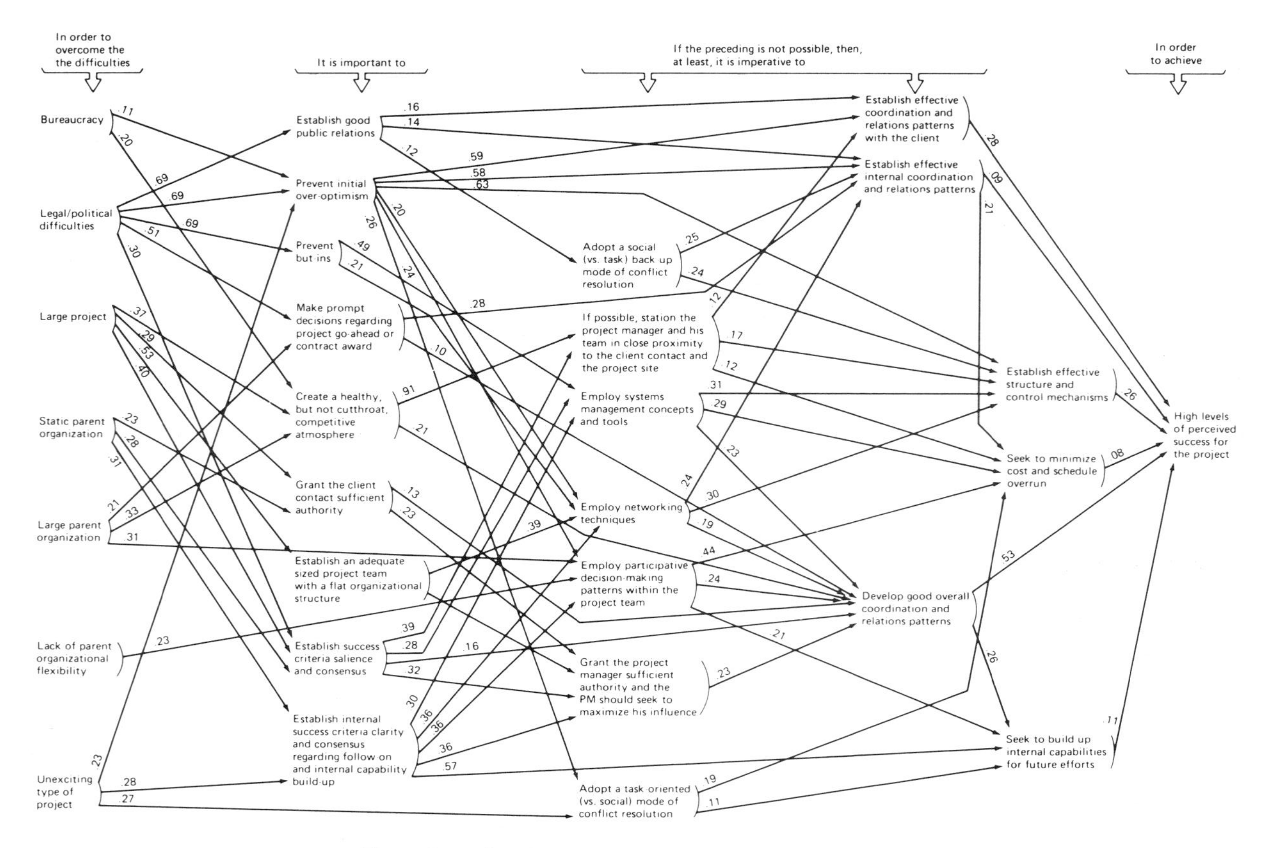


Figure 36-1. Contingent strategies for successful projects.

tent,'' but in the long run such a strategy and/or acquiescence to such strategy results in:

- Inefficient use of resources.
- Panic reprogramming of public funds.
- Diminished reputation of the agencies and contractors involved with the project.
- Loss of credibility regarding future efforts.
- Poor relations with legislative bodies.
- Poor relations with the public.

In view of these factors, the recommended strategy seems best suited to public sector projects in both the short run and the long run. Such a strategy entails planning for and securing adequate funding commitments to complete the project. If the funding is considered excessive in relation to other deserving projects, then the project may be shelved or rejected, but is not this also the best strategy and the fairest in the long run for all competing interests? The parent organization must also refrain from placing excessive pressure on the project manager to win the contract and avoid slashing or ballooning the project team's cost estimates.

2. *Develop realistic cost, schedule, and technical performance estimates and goals.*

This strategy is closely related to the first strategy. It is sometimes difficult to distinguish when intentional overoptimism (or buy-in) has occurred, rather than unintentional overoptimism. Buy-in (or intentional optimism) and unintentional overoptimism were two of the biggest factors contributing to project failure. Cost realism can best be determined by means of an independent cost estimate conducted by a truly independent organization. In most cases, so-called independent government cost estimates can hardly be considered independent because of the vested interests of the agencies involved. In practice, the ''independent'' government cost estimate may be known by the contractors, and conversely, earlier contractor estimates may be known by the independent cost-estimating team. When they are not known, the government cost estimates may vary over 100% higher or 50% lower than the dollar figures proposed by contractors for cost reimbursement contracts. As a result, contracting officers usually ignore the independent government cost estimates. Even if a truly independent cost estimate could be performed by a disinterested party, the question remains how accurate is such an estimate and how far can a contractor's estimate vary from such an estimate to be considered ''realistic''?

The answers do not rest in some rules-of-thumb or complicated formulas. The answers rest in the creation of an appropriate atmosphere and checkpoints to catch unintentional overoptimism and the creation of severe sanctions for intentional overoptimism (buy-in).

3. *Seek to enhance the public's image of the project.*

Project initiators generally are required and/or encouraged to obtain public participation during the planning and approval phases of most types of projects affecting the public interest. Some people believe that the more the public participates in these processes, the more successful the project will be. Although this concept may be appealing in the context of one's value judgments, it is not conducive to project success in a management context. High levels of public participation often result in a "tempest in a teapot." Project delays, poor public relations, and diminished project success are the results of excessive public participation. The most successful strategy from a project management success perspective is to create a good public relations image with only the minimal required levels of public participation. This is not an opinion. It is derived from the study of the hundreds of projects described in this chapter.

4. *Make prompt decisions regarding project go-ahead or contract award.*

This strategy is obviously directed to the client organization. There is nothing wrong with providing sufficient time for studies, planning, definition, etc., before seeking bids on a contract effort. But, once bids are sought, the schedule of contract award or go-ahead should be adhered to. Otherwise morale of the project team cadre deteriorates, to say nothing of the added costs of the delays.

5. *Seek to establish definitive goals for the project and seek to establish a clear understanding and consensus among the principal project participants (client organization, parent organization, and project team) regarding the relative importance of these goals.*

Although this factor was not quite as important as the previous four in overcoming legal-political difficulties, it was, nonetheless, an important factor on the road to project success.

### Strategies for Overcoming the Disadvantages of Large Projects

1. *Establish a project team of adequate size but with a flexible and flat organizational structure.*

Obviously, the larger the project, the larger the project team must be, but this does not necessarily entail the creation of excessive levels of organizational structure. Flexible and flat organizational structures were found to be an essential ingredient for project success.

2. *Seek to establish definitive goals for the project and seek to establish*

*a clear understanding and consensus among the principal project participants regarding the relative importance of these goals.*

Success criteria salience and consensus was found to be an important factor on both paths discussed here, but it was especially important for large projects.

3. *Create an atmosphere that encourages healthy, but not cutthroat, competition or "liars' contests."*

The larger the project, the more likely there will be many contractors who will be anxious to obtain the contract. It is sometimes tempting for a public agency to take advantage of this situation by creating a cutthroat competitive atmosphere, or even a subtle form of auction. In the long run, such a strategy works to the detriment of all parties concerned. A public agency and public officials must be especially careful to support the healthy aspects of the free enterprise system.

4. *Delegate sufficient authority to the principal client contract and let him or her promptly approve or reject important project decisions.*

The best way to overcome the sluggishness of a large organization and the traditional red tape associated with public projects is to delegate a high level of authority to the principal client contact.

### Difficulties that Come in Bunches

Unfortunately, when trouble occurs, it does not occur in just one dimension. The expression "a bag of snakes" is commonly heard in meetings involving public projects. In many cases, as Figure 36-1 allows, a project manager is faced not only with legal-political difficulties but also with a buy-in situation, initial overoptimism, poor public relations, delays in contract award go-ahead, and/or lack of success criteria clarity and consensus. Similarly, a project manager confronting the inherent difficulties of a large project cannot always avoid additional major obstacles such as an inadequately sized project team, heavy budgetary pressure, etc. Figure 36-1 points the way to strategies for overcoming such combinations of adversities. For each combination of adversitites toward the left of the diagram, it is possible to achieve improved levels of perceived success by placing heavy emphasis on the related strategies to the right. The reader may undertake similar analyses with the aid of Figure 36-1 for the other major adversities, such as a static or large parent organization, which may face a public sector project.

## CONCLUSIONS

In general, the comparisons derived from the research of public sector projects and private sector projects do not support many of our preconceptions regarding public sector projects. For example, the study revealed no significant difference between private sector projects and public sector projects with respect to cost and schedule overrun. It is not only possible, but also very common, to attain high levels of perceived success on public sector projects.

In order to prevent schedule and cost overruns, or to minimize the amount of schedule and cost overrun when initial overoptimism or a "buy-in" has occurred, the research points to the need for employing networking techniques, systems management approaches, participative approaches to decision making within the project team, and a task-oriented style of leadership with a backup relationship-oriented style.

Adverse environmental or "given" conditions do not necessarily affect project success directly, but often may be seen as affecting success through their influence on other intervening conditions and management processes. An adverse environmental or given condition can therefore be avoided or overcome through astute identification of those factors which it tends to affect directly, and through effective management action on those factors.

A project manager cannot afford to set his sights solely on objectively oriented targets, that is, meet the schedule, stay within the budget, and meet the technical requirements. Perceived success is in the eyes of all participants and the parties affected by the project. In the long run, perceived success appears to be more important than the traditional objective measures of success, provided that the project meets the technical performance specifications or mission to be performed.

Although many general strategy guidelines, based upon the study of hundreds of projects, have been developed to assist with improving project management, these guidelines must be tailored to the situation and the environment. Certain strategies must be given added emphasis in order to overcome specific adversities, and some strategies must be played down in order to meet the demands of other environments. Overemphasis or underemphasis of just one strategy can lead to failure. Combinations of multiple strategies can afford more frequent project successes in the public sector.

If public and private officials would pay more attention to what the project management research is trying to tell them, more frequent and higher levels of success could be attained on project efforts. There is room for improvement in both the public and private sectors in the way projects are managed.

# 37. Health Project Management in an Occupational Setting*

Laura C. Leviton, Ph.D.† Gordon K. MacLeod, M.D.‡

## EMPLOYEE HEALTH PROJECTS

Good employee health is a major asset for any organization. The protection and preservation of an employee's health reflect many important objectives for an organization. Yet, assuring good health is no easy matter. The project management approach can assist in this process. A health project involves the development or transformation of worksite health programs. After a general description of worksite health programs, we apply the project management approach to four health projects: (a) modification of health insurance benefits; (b) development of cost-effective health promotion programs for employees; (c) development of Employee Assistance Programs (EAPs); and (d) development of occupational health and safety programs.

### Definitions of Health Projects

**Modification of Health Insurance Benefits.** The traditional way in which health benefits are covered through employer-funded programs is through

---

* We gratefully acknowledge the comments and suggestions of several individuals in their review of a draft of this chapter: E. C. Curtis, M.D., Corporate Medical Director, Westinghouse Electric Corporation; Bertram D. Dinman, M.D., then Vice President, Health and Safety, Aluminum Company of America; R. L. Gibson, M.D.; Beaufort B. Longest, Ph.D.; William McClellan, M.D., former Medical Director, Gulf Oil Corporation; and Joseph J. Schwerha, M.D., Director, Industrial Medicine, United States Steel Division, U.S.X.
† Laura C. Leviton, Ph.D., is Assistant Professor of Health Services Administration at the University of Pittsburgh. She has written extensively in the area of work-site health programs. She is a former W.K. Kellogg Foundation Fellow and a former policy analyst of the Health Policy Institute of Pittsburgh.
‡ Gordon K. MacLeod, M.D., is Professor of Health Services Administration and Clinical Professor of Medicine at the University of Pittsburgh. He is a former industrial engineer at the Procter & Gamble Company who later served as the founding Director of the HMO Service for the federal government, and as Secretary of Health for the Commonwealth of Pennsylvania.

a health insurance plan that pays costs or charges to health care providers on a per diem, per case, or fee for service reimbursement basis. That is, a hospital, physician or other therapist submits a bill to the insurance carrier which is then paid, no questions asked. This traditional coverage through reimbursement for costs or changes is one of the main factors that has contributed to huge increases in the cost of medical care over the past several decades. One goal of a health benefits project, therefore, is to stimulate competition among alternative provider systems, limiting increases in medical care costs, whether by improving efficiency or by limiting utilization deemed to be unnecessary. Cost containment can be achieved either through the traditional provider system or through an alternative to that system.

**Development of Cost-effective Health Promotion Programs.** Health promotion involves actions aimed at improving health and the quality of life. Health promotion is therefore aimed at already healthy employees as well as those at risk of a specific illness. The key to health promotion is to provide opportunities for improving health, and to ensure that a high number of employees take advantage of those opportunities. Health promotion is a broader concept than health education because it includes not just information, but also changes in the environment and organizational climate that ultimately facilitate changes in health behavior (25). Health promotion can also include "wellness" activities that are aimed at more than the absence of disease. For the organization, "wellness" means high levels of productivity and morale, while for the employee wellness means minimizing physical limitations as well as enhancing subjective well-being.

**Development of Employee Assistance Programs (EAPs).** EAPs provide counseling and arrange for treatment and referral of troubled employees to deal with a variety of counseling needs. Alcohol treatment programs organized through work settings have existed for some time, but organizations now tend to expand the focus of those programs to deal with multiple problems. While EAPs are aimed at facilitating rehabilitation or improving the functioning of the employee, they also hope to reduce absenteeism and sick days, to contain utilization of health care benefits, to reduce the incidence of grievances, and thus to increase productivity.

**Development of Occupational Health and Safety Programs.** These programs focus on the reduction of injuries and on correcting work conditions conducive to injury and illness, such as exposures to hazardous substances. Programs may also include employee health services on site.

Depending on the organization, they may also involve screening for conditions related to occupational exposures, or preemployment screening to facilitate assignment of individuals to tasks most consistent with their capabilities and physical capacities.

### Background

At the present time, employee health benefits and programs are undergoing significant change. Guidance and direction are needed to keep health programs and benefits in line with long-term organizational objectives. Some background is essential in order for the reader to avoid many past pitfalls. A central problem in health project management is that the manager needs to exercise control over costs, utilization, and time lost from productive employment, but also needs to avoid the appearance of paternalism. The history of employer-provided health care shows why this is important (55). In the late nineteenth century, the high incidence of industrial injuries motivated companies to hire physicians to treat cases—and to document the injury, often serving as expert witnesses for the company in lawsuits. Some companies provided medical care, employed physicians, and in some cases even built their own hospitals. Many employees resented mandatory payroll deductions for company-provided care and wanted to choose their own doctor. In the early years of this century, occupational physicians were often held in low esteem by workers as well as by their own professional colleagues (18, 55).

Initially, growth of health insurance was stimulated during World War II by the federal government's decision to exclude health insurance premiums from taxation as employee income (10). Provision of such fringe benefits offset to some extent employee dissatisfaction with government-mandated wage and price controls intended to combat wartime inflation. After the war such benefits packages became an important part of collective bargaining. At the same time, the medical profession endorsed hospital and medical care insurance for the first time, so long as no one interfered with the actual practice of medicine or with the free choice of a physician (55). Insurance obtained through employment has become so widespread that in 1981, employers and unions paid the entire premium for 42% of privately insured people, and helped pay it for another 52% (57). Moreover, retirees frequently receive company coverage for health care in addition to Medicare.

However, health insurance with fee for service payments to physicians and lack of controls on hospitals produced increases in the cost of care, most dramatically after the passage of Medicare and Medicaid. Expenditures on health care reached 10.7% of the GNP in 1985 (58). Employers in the 1980s began to take steps to contain costs. Many providers, insurers, employers, and employee groups are concerned that freedom of choice and quality of care may suffer as a consequence.

The areas of EAP, health promotion, and occupational health are changing no less rapidly than is health insurance. EAP providers are in a state of tremendous flux as concepts such as brief psychotherapy come into ascendance (8). Both EAP and health promotion services are being aggressively sold with unprecedented marketing slickness. In the occupational health area, the model of a corporate medical director is being transfigured by several forces such as decentralization of programs, evolution in the area of toxic torts, and demands that government take over some functions that are currently provided by companies. In summary, all these areas may require health project management to meet the challenge of changing times.

## DEVELOPING A HEALTH PROJECT

This section outlines the approach to a generic health project for an organization, from planning to implementation, from monitoring to adjustment, and from quality assessment to corrective action.

### The Planning Process

The need for short- and long-term planning for health programs and projects flows from an overall statement of organizational philosophy and objectives. Specific goals and strategies are formulated to achieve desired changes. Beginning even before this formulation, consultation with the many stakeholders of such programs and projects is needed, and may have to be repeated several times during the planning process. Through such consultation, the organization will gain an ability to maximize the value and relevance of health programs.

**Objectives.** Following the format of this handbook a health project begins with a statement of the overall objectives of an organization. It is important for project managers to clarify these broad objectives at the outset and specify how they relate to specific goals for health programs (19, 27). The stated objectives should include those that benefit the employee and those that benefit the employer. Frequently, employee and employer objectives coincide, especially for preventive programs. Objectives that focus primarily on the employee include:

- Demonstrating commitment to the health of the employee.
- Demonstrating commitment to the quality of work life.

- Aiming at increasing the life span of the employee.
- Maintaining the employee's physical functioning.
- Improving the retention of employees in the organization.
- Improving employee morale and satisfaction.
- Providing a variety of employee benefits.

Objectives that focus primarily on the organization include:

- Increasing productivity.
- Containing health, life, and disability insurance costs.
- Complying with federal and state "right-to-know" laws.
- Recruiting desirable employees with attractive programs.
- Decreasing absenteeism and workdays lost.
- Saving short-term sick pay costs.
- Reducing Workers' Compensation costs.
- Limiting lawsuits about exposure to hazardous substances.
- Improving organizational image.

An additional objective of many companies needs to be discussed here. Decision makers should at least be aware of the possible impacts of their decisions on their communities. On the one hand, employer-sponsored disease prevention programs are good for the community as well as the organization, because they set a good example for other organizations. On the other hand, experience rating of a company's own health insurance impinges upon the societal good of shared risk in insurance. Although companies save money by insuring young, healthy employees, older and sicker community members must then pay more.

**Consultation with Stakeholders.** Stakeholders for health programs are all those individuals and groups, both in and outside of the organization, that have a stake in the number and types of programs and the ways in which they are offered. Involving stakeholders can facilitate acceptance of change. Among the many stakeholders for health programs are managers, the work force, dependents and retirees, and providers. These groups will not necessarily have uniform views about health programs. For example, top managers themselves can and do disagree about the importance of offering certain health programs, and unionized employees may differ from those not unionized in terms of some goals.

Several advantages can result from consulting various stakeholder groups about health objectives and goals. First, stakeholders may challenge assumptions about health programs that turn out to be false on closer examination. First-dollar coverage, for example, may be less important than management believes. The statement of a health project's purpose or even objectives may have to be reworked when such assumptions are challenged. Another good example is that prevention and rehabilitation activities can really benefit from employee input. With such participation, EAP, health promotion, and safety and health protection programs will get better compliance from employees.

**Needs Assessment.** Supplying good information on needs is a first step toward modifying health programs to be more rational and to suit employees, based on prevailing health issues and problems for the work force. Data bases to measure need vary by the type of health project described below.

**Specification of Goals.** Once an overall statement of the purpose of health programs has been formulated and discrepancies have been identified between that statement and the existing programs, the next step is to relate specific activities to the overall statement of purpose, and to develop both short- and long-range specific goals and objectives.

A principal goal for the short term in planning for any health project is to enter the networks of employers who are bringing about changes in health care for employees. Business coalitions in many communities are accomplishing this change. A major coordinating body for those coalitions is the Washington Business Group on Health, which publishes the periodicals *Business and Health* and *Corporate Commentary*. Smaller organizations can affiliate with such coalitions, through associations of smaller businesses. These coalitions have had little direct involvement with occupational health, however. Although knowledge in this area is specialized, the employer can gain useful information from the national and regional Safety Councils and from the American Occupational Medical Association.

Establishing long-term goals requires strategic planning about the future of the company and the health system. Assessing potential changes in public programs is essential for all these health projects. For example, if catastrophic health coverage for all Americans is increased, major medical plans will be affected and stop-loss coverage will become less important. On the other hand, keeping the status quo in public medical care programs will mean that an even larger proportion of retirees' health care must be paid by some other source than Medicare or Medicaid.

Insurance practices may also change. For example, insurers could probably increase their profit margin by reimbursement for preventive services. Should the organization therefore aim toward expansion of prevention? Insurers may eventually reimburse new high-technology interventions such as liver transplants or they may perform careful analysis of cost-effectiveness first. What should the organization's response be?

Several worksite policies may impact on both health and legal matters. For example, AIDS and other community health problems should be followed closely, and policy should be discussed today concerning possible workplace problems later. Drug testing may be required in some organizations—but the legal status of such testing is still under debate. Communicating with employees about health risks is a very complex issue in light of right-to-know laws and the uncertain state of toxic torts.

**Organizational Constraints on Planning.** So far we have discussed health projects and the objectives and goals they address as though there were no organizational constraints on those projects. Yet most projects are embedded in organizations with characteristics that may pose barriers to achieving rational objectives. Most companies have a past history of providing certain benefits, and change does not come easily. Changes in benefits and programs may require protracted periods. However, the cost crisis in health insurance has created an environment in which such changes can occur more quickly, because the need is more apparent to stakeholders.

The role of top management is crucial in each of the four project areas. The high cost of care has made many businesses fearful that a new benefit could result in the same overutilization they feel has occurred in hospitals. For example, some business leaders are open to inclusion of a mental health benefit, except that they see no effective controls on utilization. Also, a firm commitment from top management is essential to make prevention activities work. Without a commitment, safety and health will be relegated to a marginal status, and health promotion activities will wither (4, 40).

Unions have frequently supported cost containment efforts, unless the efforts represent give-backs. Unions understandably tend to be much more concerned with occupational hazards than with health risks based on individual lifestyle. They also tend to be wary of alcohol, drug abuse, and mental health programs offered by management, because of perceived dangers of paternalism. Cooperation with unions usually produces valuable and effective prevention programs. For example, hypertension control programs have been jointly sponsored by Ford and the United Auto Workers, and by the Storeworkers Union and Gimbel's in New York. These programs are exemplary models for such cooperation (2, 11).

### Implementing Health Projects

**Organization.** One issue in health projects involves whether health programs should be centralized in one department of an organization or decentralized into several departments. Frequently, responsibilities for health programs are divided among such departments as personnel benefits (for health insurance), human resources (for EAP), and a medical department (for occupational safety and health). Health promotion is sort of a free-floater—it is sometimes implemented by a medical department, but just as often is a sideline for a middle-level manager or committee who merely have an interest in health and fitness.

Yet for effective project management, coordination among the various departments is essential. In fact, we would argue for centralized planning and implementation of all health-related programs. Figure 37-1 presents an organizational chart for an ideal configuration, for very large organizations. In smaller organizations, much of the service delivery could be provided by outside consultants, but overall administration might be the same. A vice president for health programs can supervise both a corporate medical director and a manager of occupational health and safety. Health data management is a separate independent office that assists the vice president in policy and planning decisions. Laboratory services are in a swing position, relating to both employee health and occupational health and safety.

This type of organization is not common. It is more common, in fact, for several of these functions to be subsumed under a personnel or human resources department. Even where our ideal configuration is not possible, however, it may be feasible for conducting a time-limited health project. Our proposed organization has some advantages over others for coherently planning health programs and projects.

This configuration permits health specialists to complement each other. A medical director may deal primarily with an employee health clinic, but

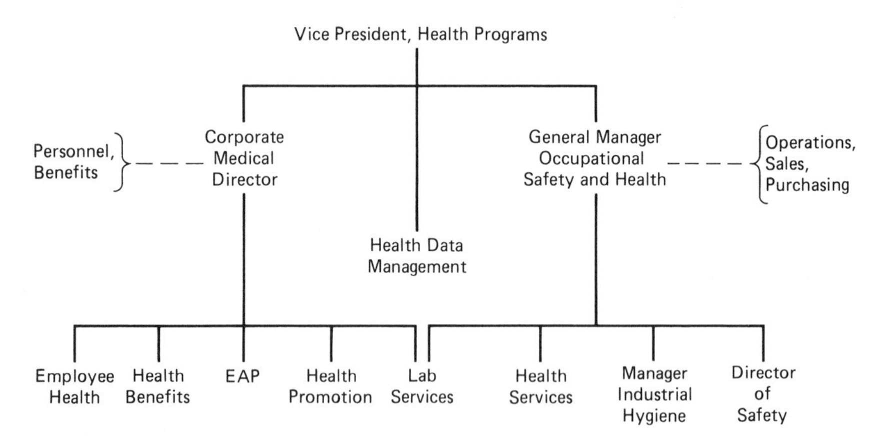


Figure 37-1. Model table of organization.

clinical training also puts him or her in a unique position to identify an individual employee's need for EAP, to suggest health promotion activities designed to lower risk of disease, and to advise health benefits managers about quality assurance in community health care programs. An EAP director frequently implements health promotion activities, and when referring employees for long-term treatment, needs the same kind of utilization review system that the benefits office may be implementing for other health problems. Occupational health nurses see a variety of substance abuse and lifestyle problems. Centralized planning and implementation of health project management are more likely to retain their focus upon and relation to organizational objectives than are fragmented efforts.

**Staffing.** Selection of personnel to administer health programs for employees will depend on the size of the organization and on the extent to which activities are conducted in-house. The staffing of benefits departments outlined by Griffes (19) applies to some extent: in small organizations, there will usually be a single manager of health benefits, who will also probably be the manager of other insurance benefits as well. In small organizations, this manager is usually a generalist with a background in benefits administration, personnel, and human resource issues. Some familiarity with EAP, health promotion, and occupational health and safety issues is also desirable. He or she is dependent on other departments or on outside consultants for medical, legal, financial, actuarial, investment, communications, accounting, and record-keeping tasks, as well as services in the areas of insurance and occupational health and safety. Health promotion and EAP will usually be provided to individual employees by outside professionals and voluntary organizations as needed.

The health project of a larger organization can administer more services and functions in-house, involving specialists in record keeping, financial management, communications, and planning and design. A part-time medical consultant, an occupational nurse, or some other in-house health-related professional may be a useful addition to administer occupational health and safety, health promotion, and EAP referrals.

The largest organizations have the fewest outside dependencies. If they self-insure for medical care, they may even choose to administer the program in-house, as opposed to giving its administration to an insurer. A medical director or vice president for medical or health affairs is highly desirable for large organizations, since he or she can relate to multiple health programs. Other health-related professionals are also desirable, such as occupational health nurses, industrial hygiene specialists and safety professionals to monitor workplace hazards, psychologists and social workers to administer EAP programs, and exercise physiologists to launch health promotion activities and maintain employee interest in those activities.

The importance of relying on professionals for the provision of health services and programs cannot be overemphasized. Today's market for health care services is extremely competitive. A department which administers health benefits and programs has an obligation both to the organization of which it is a part and to employees to choose services wisely.

**Choice of Cost-effective Services.** Cost effectiveness is defined as meeting objectives to the extent possible within the limits of the budget for a health program. There is more than one way to achieve organizational objectives, and they are not all equally effective nor equally costly. Decision makers frequently exhibit a bias that more expensive programs will produce better results. This is not always the case for health benefits, especially for prevention programs. Effectiveness and cost-effectiveness data are not always available, but as organizations gain experience, health projects can be monitored as shown below.

**Contracting with Providers.** The choice of services, providers, and benefits should reflect organizational objectives, needs of the employee population, and stakeholder positions. Services to be delivered must be specified. Moreover, contractors should provide some assurances to the organization about performance and safeguards regarding patient care.

Outside consultants may not have the same sensitivity to organizational goals that an insider would have. They may apply the "one size fits all" approach to employees in widely different organizations. This may or may not make a difference, but the situation should be monitored.

It is reasonable to hold service providers and consultants to standards of performance. Cost containment consultants should be able to describe their standards for utilization review and preadmission certification. In health promotion and EAP, providers should be able to specify expected outcomes. Consultants on safety should be able to explain the rationale for suggested changes in the workplace.

For all types of health services, assuring confidentiality is a major concern. Service providers must give assurances that employee names will not be associated with any reports to the organization about alcohol, drug abuse and mental health services. The perception of anonymity is essential to participation in many health programs, especially those that the employer recognizes as essential to health, well-being, and continued productivity on the job.

To the extent possible, providers must guarantee both a high level of quality and a high degree of access. Quality is somewhat difficult to assure directly, since it is difficult to define and monitor. However, the choice of reputable providers should contribute to quality. Access can be assured through rapid feedback on complaints, satisfaction, and utilization of health programs.

**Employee Communications.** Health program changes or additions may cause some confusion. Surveying a sample of employees from time to time may help reveal the extent to which employees understand their benefits and programs. Communications should be relatively simple and easy to understand (19). Information should be disseminated to employees through multiple channels such as pamphlets, newsletters, bulletin boards, in pay envelopes, and any other means available. In many cases, however, there is no substitute for personal contact when employees have questions.

### Monitoring Health Projects and Programs

**Use of Data.** Once a health project or program is in place, the project manager will want to monitor its progress in two ways. First, he or she will look for any overall changes in trends compared to the past experience of the organization. Second, the manager will want to examine data on individual services to spot significant deviations from the norm that need closer inspection. If, for example, the project manager spots excessive utilization of services, then he or she may want to examine utilization by those providers more closely. This second method of monitoring acts as a filter to direct the attention of the manager toward potential sources of trouble.

**Corrective Actions.** These occur in several stages. If the employer sees potential problems or deviations from the norm, then the first stage of action is verification of a problem through more intensive study. If problems do emerge, then corrective actions may include negotiating with the existing provider, choosing another provider or system, or better communication with employees.

## A PROJECT TO MODIFY HEALTH INSURANCE BENEFITS

### Planning

**Objectives.** Decision makers most frequently say that the objectives of revising the health benefits package are to contain the cost of medical care while continuing to assure quality of care and access. In addition, an attractive benefits package could assist in recruiting desirable employees into the organization (19).

**Consultation with Stakeholders.** Not all employees will share the same view about the importance of particular goals of the health benefits package. For example, unmarried employees may have different views from those who are married and have to pay an additional fee for family coverage. Also, those who are young and generally healthy may have different opinions from older employees about the scope of coverage, deductible, and coinsurance. Views on cost sharing by retirees may differ depending on whether they are recently retired and relatively healthy, or beginning to experience more serious chronic ailments.

When choosing among methods of containing costs, the employer will want to know about the acceptability of cost containment approaches to employees. For example, the general trend toward encouraging outpatient procedures has not been popular with surveyed employees (20). Both management and unions have expressed reservations about second surgical opinion programs (20). Unions also oppose increases in cost sharing by employees (5). Access to high-quality tertiary care providers must be considered. In some communities alternative provider systems are fragmented or inaccessible, because complex health care centers are competing to be the sole providers of care (32).

**Needs Assessment.** Employee demographics are helpful to the health project manager in assessing the need for the form of the health benefits package. Younger workers are less likely to have serious and costly debilitating illnesses than older workers. Health care utilization experience of the employee group helps to describe health needs. Physician practice patterns, when subjected to small-area analysis (61), can tell the health benefits manager whether to focus first on unnecessary surgery, hospital admissions, or length of stay.

Cost containment methods should be geared to problems revealed in utilization data and surveys of employee attitudes. Are there too many hospital admissions, suggesting the need for preadmission screening? Or is length of stay too long in comparison to the national average, requiring concurrent review? Likewise, the level of employee knowledge of changing patterns of health insurance coverage should be considered. How difficult is it to explain the rules of a second surgical opinion program? How many mistakes are likely to be made by employees? What will be the level of employee dissatisfaction over the consequences?

**Specification of Goals.** In the short term, managers of health projects should specify how they intend to deal with changes in governmental programs that will affect employees. For example, Medicare has never covered the full cost of medical care for the elderly (62). A goal for the short term is therefore to establish whether and how an organization will contribute to the ever-widening medigap coverage for retirees. Trends and directions of new governmental programs, such as catastrophic care coverage, warrant consideration.

Changes in the demographic makeup of the country dictate some of the long-term goals that will be set by health benefits planners. With the graying of the population, a vastly greater proportion of retirees will be supported by a reciprocally smaller work force. The retirement age is likely to increase, especially as debilitating chronic conditions in the elderly are delayed or prevented (14, 19). Creation of long-term care alternatives and effective means of coping with the cost of medical care will assist the demographic transition for both the organization and retirees.

### Implementation

**Organization and Staffing.** Staff should have substantial familiarity with the methods of cost containment described below. Consultants or staff are needed for tax, accounting and auditing, and legal services, as well as direct administration of benefits. If outside consultants are permitted a free hand with limited oversight by the organization itself, there might be negative consequences for both employee and employer. Decisions about admission, length of stay, and surgery are not always clear-cut; if they were, then cost containment might have happened years ago. The organization, not the consultant, has the responsibility to determine how far to pursue cost containment goals.

**Choice of Cost-effective Services.** Not much is known yet about the cost-effectiveness of some methods. Even for established methods, the scene is changing so rapidly that information is outdated. What follows is the best available assessment of cost effectiveness of provider systems.

*FEE-FOR-SERVICE WITH INDEMNITY INSURANCE.* Even with the traditional reimbursement system for health care services, the health benefits manager has some tools to limit unnecessary utilization and therefore limit costs. By requiring deductibles and copayments from employees, a health insurance plan can limit unnecessary utilization and thus control costs. A variety of studies, including a large-scale experiment conducted by the Rand Corporation, have demonstrated that increasing the cost sharing of patients from 0% to 25% can bring about a reduction in medical care expenditures of between 17% and 24% (17, 37). The Rand study also found that quality does not suffer, but this finding is somewhat controversial (20).

Self-insuring employers can also hire consultants or insurers to perform several tasks designed to limit unnecesary utilization of hospitals. One such task is preadmission review of hospital admissions, in which qualified professionals review patient records to determine whether admission is needed. Concurrent review of length of stay is a second task that contributes to hospital cost control; professionals determine whether the patient needs to remain in the hospital beyond a certain time period. Retrospective or claims review of hospital stays is a third task, primarily used to detect and correct accounting errors, rather than to encourage efficiency in hospital care (20). Second surgical opinions provide a fourth approach for controlling utilization and therefore costs. In such programs, employees are required to seek a second opinion for certain elective surgical procedures—the second, and even a third, opinion are reimbursed by the insurance plan, and in some cases employee travel time and time lost from work may be reimbursed as well. While mandating second opinions may well be cost effective (e.g., reference 43), the evidence is still controversial; large employers who have tried the method express the least satisfaction with it compared to other strategies (20).

Case management is a newer concept intended to affect both utilization and costs. It involves arranging for the organization and sequence of medical and social services for an individual. Case management may include elements of utilization review, but also involves a variety of steps to ensure good-quality and appropriate care. Case management is especially appropriate when a health condition is complicated and requires the services of several specialists, or when social and rehabilitative needs also exist, as in the case of the elderly or persons with AIDS.

*HEALTH MAINTENANCE ORGANIZATIONS (HMOs).* The search for responsible cost containment strategies has given rise to several alternative provider systems. Health maintenance organizations (HMOs) are in fact as old as Blue Cross and Blue Shield plans, but the early 1970s saw federal support for expansion and creation of new HMOs (30, 31, 55). An HMO is a direct service plan which combines per capita prepayment for hospital care along with physician services at a minimum to a voluntarily enrolled population on a contractual basis (31). Mental, dental, drug, and other services are often included. The feature of prepayment alters the usual incentives to the provider. Where fee for service practice may encourage overuse of hospital and medical services, physicians in an HMO need to minimize care that is not needed (17). Preventive medicine, which is seldom reimbursed by insurers under a fee for service system, is a central focus, though often not fully achieved in HMOs (31). Although patient waiting time for a nonurgent appointment may be used as a means to ration services in HMOs, the patient may utilize services as often as required for no additional or only minimal registration fees (17). Thus delays in seeking needed care should be minimized.

Two models of HMO, the prepaid group practice (PGP), and the individual practice association (IPA), are easily distinguishable. In the PGP model, physicians are either employees (staff model), or members of a partnership or corporation and related to the plan through an exclusive contractual arrangement (group model). In the IPA model, physicians and physician groups in independent practice agree to form an association to share financial risk in providing prepaid care to a voluntarily enrolled population, usually on a fee for service basis or on a capitated basis (network model). The IPA service agreement may specify fee levels and degree of risk sharing (30).

PGPs studied in the 1960s and 1970s have been found to result in 10% to 40% lower health care costs in terms of premiums plus out-of-pocket expenses, compared to the traditional fee for service care and indemnity insurance coverage (30). There is no evidence that quality of care in HMOs is different from that of fee for service care (30). Although PGP patients may be somewhat different from the general population in terms of health status or condition, the difference is probably not responsible for the superior performance of PGPs in age, sex, and occupationally adjusted populations (17). Relatively few IPA model HMOs have been studied, but existing IPA's are less effective in lowering health costs, compared to PGP's (30).

*PREFERRED PROVIDER ORGANIZATIONS (PPOs).* The expansion of another alternative provider system, the PPO, started up in the mid-1980s (46). PPOs are contractual arrangements with a defined group of providers, usually calling for discounted medical care services to be offered to a firm's employees or to retirees. Although patients are not required to use designated providers, incentives are provided for them to do so. These incentives are usually a lower deductible or copayment for medical care (46). PPOs must generally employ all the cost containment strategies outlined for fee for service practice, unless providers are exposed to financial risk factors similar to those used in HMOs. Although charges are discounted, providers have an incentive to contract with PPOs because they hope to get a substantial number of patients in a competitive market for health care. Also, PPOs usually promise rapid response to submitted claims. PPOs may be formed and marketed by insurers, employers, community groups, independent entrepreneurs, or providers themselves, often in a medical staff and hospital (MESH) joint venture (22).

In the mid-1980s PPOs vary greatly, but the majority share certain characteristics. In a survey conducted by Rice et al. (46), discounts by PPOs average 10% to 15% less than the usual charges by hospitals. Physicians are paid on a fee schedule or their usual fees are discounted, often as much as 20%. PPOs employ utilization review as the major way in which costs are controlled within a hospital. Unlike HMOs, relatively few PPOs utilize the case management approach or use networks of primary care physicians as gatekeepers for medical care. In addition to savings in out-of-pocket costs, incentives to use the PPO often include the provision of additional benefits, such as well-baby care. The organization of PPOs may produce some inherent conflicts that may not be to the advantage of employers or employees (46). PPOs formed by hospitals themselves face a conflict between health care cost containment opposed to the need to fill their beds. Thus discounts for services may be swallowed up in extra utilization. Also, a recent survey showed that insurers tend to form PPOs in order to increase their market share, and are not selective in terms of the physicians included in the contractual arrangement (46). This is also said to be true for IPAs. Until PPOs have established a track record that can be demonstrated to purchasers of care, uncertainties about quality and cost-effective performance are likely to persist.

Distinctions among HMOs, PPOs, and any other alternative may become less sharp in future, as principles behind cost-effective care become more prevalent and familiar to purchasers of medical care. With likely deregulation of the health care industry in the future, it is possible that marketplace competition may be used to further contain costs.

*NEW DEVELOPMENTS*. Two of these are worth mentioning. First, "cafeteria" benefits plans, also known as "flexible spending arrangements," are becoming more common. Under cafeteria plans, employees can choose from among several possible benefits at different levels of coverage. The IRS has ruled that these plans qualify for exclusion of income from taxation. Such plans permit employees to choose how much of their fringe benefits they wish to allocate to health, life, and disability insurance, for example.

Enthoven (10) has been a consistent advocate for a voucher system as an alternative to this development. In a voucher system, individual patients could seek their own source of care, within the limits of the voucher, or could supplement vouchers with their own funds when desired. For this arrangement to work well, consumers must be informed enough to make good choices—and frequently they are not well informed.

**Contracting with Providers and Consultants.** Providers and consultants for health benefit programs must be chosen with care. For example, consultants may or may not be vigorous in reviewing claims, certifying admissions, and monitoring length of stay. Even if they are vigorous, they may not give sufficient attention to access and quality of care. Likewise, many new HMOs are being formed. Yet not all these organizations will be federally qualified. Many HMOs claim to provide utilization controls, yet show no evidence of having such a system (20). Similarly, all PPOs will not be equally discriminating about the choice of practitioners to be involved, nor about quality control, nor about the kinds of discounts they will give (13).

It is said that the only thing worse than a bad doctor is a bad group of doctors. Organizations must become sophisticated in selecting alternative provider systems with good management and financial skills. The aggressive marketer and deep discounter may not be providing really adequate care, and complaints may surface. In general, discounts will probably turn out to be less important than controls on excess utilization. The early experience of some IPAs as well as prepaid group practices provides evidence of deficiencies in such skills (13).

The health benefits manager will also want to know the HMO's or PPO's track record on quality of care. Satisfactory information is generally not available from hospitals, and even less so from physicians. Hospital quality assurance programs vary in part due to the types of patients seen and the severity of their illnesses. The federal government has developed a system for comingling cost and quality assessment which is monitored by Peer Review Organizations (PROs) for the Medicare population. In some states this process is now spreading to all hospitalized patients. Information on physicians' practice patterns can be obtained through utilization profiles developed by third-party payors for the Medicare program. Thus some data are available for analysis of the numbers and costs of certain procedures attributed to individual hospitals and to groups of physicians.

Contracts for the delivery of hospital and physician care should specify which party will bear the risk of large medical bills. Some PPOs are now more akin to hybrid HMOs in that they receive prepayment and bear the risk of utilization that exceeds the prepayment amount. The contract should specify the basis for charges—will they be based on diagnosis related groups (DRGs)? A discount might be specified, but some analysts have warned that providers might "churn" an excessive number of patients through their facilities, or raise their prices and then offer a discount that is meaningless (13). The contract should guarantee a discount for at least a set period of time. It should also specify which forms of utilization review will be used, and the qualifications of the personnel doing the review. Ready access to second surgical opinion programs means that less travel time to designated physicians is required.

**Employee Communications.** Employees can misunderstand their benefits, so that medical services rendered to them will not be reimbursed. This is especially the case when there is no gatekeeper mechanism to control the use of medical care. In particular, mandatory second surgical opinion programs run the risk of employees' misunderstanding of the procedures and failing to obtain the required opinion prior to elective surgery.

A benefits information center may be desirable in larger organizations (19). Employees can then direct their questions to a centralized location whose sole function is to answer questions clearly and in a helpful tone. This should be advantageous to benefits managers as well, since their duties will not be interrupted by constant inquiries. In smaller organizations, a single individual might be designated to answer such questions, or the firm might contract with providers for this feature. Some hospitals, for example, now have specialists whose duty is to contact patients prior to admission and to explain what their benefits packages cover.

### Monitoring

**Use of Data: Utilization and Charges.** Information about utilization and charges should be available from the insurer for aggregates of employees. In the future, it will be available from PPOs and HMOs. In smaller employee groups, however, the manager should be aware that the overall level of these indicators is likely to fluctuate from year to year, since the medical care experience of only a few seriously ill employees may skew utilization data and charges for any given year. For this reason, it is important that the employer compare charges and utilization to age-adjusted charges and utilization of the population in the region as a whole, as well as to the past experience of the employee group.

Providers may attempt to reclassify patients into categories of care for which utilization and charges are higher. The Medicare program has developed some methods for dealing with such potential overall changes in classification, through the use of the PROs.

The benefits manager usually does not have access to the same resources as the federal government, but there are still steps that he or she can take to assure that providers do not reclassify of patients. First, the manager should track the regulatory activities of the PRO. Second, if the PRO determines that some providers are changing the classification of patients toward diagnoses that allow more liberal utilization and charges, then the manager might have to consider canceling contracts with those providers or else negotiate new health benefit arrangements with insurers or alternative provider systems.

**Use of Data: Quality and Access.** The manager who is concerned about medical care quality should keep his or her eye on certain indicators. Specifically, the manager will want to know about provider licensure, reports by the Joint Commission on Accreditation of Health Care Organizations (JCAHO), and, for HMOs, whether they are federally qualified. The manager may also want to track the proportion of employees admitted to the hospital subsequent to outpatient surgery. If this proportion is too high in comparison with past experience or with the overall experience of the region, then decisions of preadmission assessors may be overly biased against admission for procedures. This is not to say that numerous mistakes are occurring or that quality of care *is* deteriorating, but rather, that a closer look at aggregate data over time is in order. The confidentiality of individual records must be respected by the organization itself. Finally, the manager can rely on decisions made by the PROs regarding quality, although in most states these organizations monitor Medicare patients exclusively.

Access to care, patient satisfaction, continuity of care, and perceived quality should continue to be of concern to management and employees alike. Lower performance on these indicators will defeat the purposes of business in offering health benefits. In the past, the employer had to rely on the marketplace to dictate satisfaction and quality. Under new arrangements such as PPOs, the marketplace may not function so well, or so quickly. Periodic sample surveys of employees and monitoring of complaints by those employees should provide indicators of access, satisfaction, and perceived quality. An appeals process is highly desirable, to minimize actual mistakes in denial of benefits and to provide employees with opportunity to air their complaints in an appropriate manner.

## DEVELOPMENT OF COST-EFFECTIVE HEALTH PROMOTION PROGRAMS

### Planning

**Objectives.** Health promotion and disease prevention programs have emerged as a means to improve health of employees; at the same time it is hoped that these programs will help employers meet their own objectives for cost containment, increased productivity, and reduced absenteeism, among others. In this section we explore the extent to which employee and employer objectives are achieved by these programs.

Some health promotion activities have been clearly linked to improving quality of life, health, and increased life span of the employee. Others have not, although they may be regarded as "promising". Still others have not been demonstrated to be beneficial for health although they may be useful in meeting other objectives. But part of health maintenance for the employees is, in principle, to provide guidance concerning effective preventive efforts and protection from marketers of useless fads.

Primary, secondary, and tertiary disease prevention have important financial implications for businesses. Primary prevention by immunization and lifestyle modification activities avoid illness before it begins. They are relatively low-cost, potentially high-yield activities. Use of seat belts does a great deal to prevent later physical impairment (35). Flu shots and pneumovax are particularly beneficial for older employees and those at high risk of respiratory infections. Secondary prevention activities help avoid serious consequences of a condition. Activities such as smoking cessation, hypertension control, and cholesterol reduction can mitigate or reverse the progression of heart disease; in their absence, the employee may become disabled (39, 60). Premenstrual stress is a major cause of work loss; many cases now respond to short-term therapy. Tertiary prevention is primarily a medical matter intended to avoid chronic complications or limit disability from existing diseases. Handicapped employees can benefit greatly from tertiary prevention. Diabetics who learn important self-care skills suffer fewer complications, such as blindness, and require less medical care for problems such as amputation (56).

These activities could have positive effects on health and life insurance premiums, especially as the retirement age increases. Lowering long term disability costs may be especially important, for example, in avoidance of back injuries by matching employees to jobs or by redesigning the work. Savings should be greatest in companies in which the turnover rate is low and tenure long (40). However, savings from these health promotion and disease prevention activities are likely to be modest in most cases (28). Moreover, illnesses avoided will not result in reduced health insurance premiums unless a company self-insures, unless its own experience rating is used in setting premiums, or unless insurers begin to give discounts to companies where employees are willing to participate.

Achievement of higher productivity has not yet been well measured in the literature on health promotion. At present, it can only be inferred that productivity would increase if employee behavior were more consistent with a healthful lifestyle.

Health promotion can potentially reduce liability from occupational exposures, because those linked to cancer are often made worse by smoking. Diseases from tobacco along with exposure to other carcinogens will not become evident for many years. The employer can be held liable for the effects of both smoking and worksite exposures.

Recruitment of desirable employees, increase in morale, and improvement of company image are also frequently mentioned as objectives of health promotion. Indeed, recruitment appears to motivate many companies to provide the trendier health benefits, such as stress reduction, exercise, and weight loss programs (59). Some organizations use health promotion programs to enhance their identification with health, sports, or innovation. Also, managers may see the focus on health as fitting into their overall philosophy or style. Team activities such as bowling and softball are loosely health-related and also promote such an image.

**Consultation with Stakeholders.** In health promotion programs it is essential to consult all stakeholders in an organization. Employee interest must be ascertained in order to gain good participation. Employee suspicions about confidentiality must be allayed for activities such as hypertension counseling. At present, health promotion is the least accepted of the health projects described in this chapter. As such, it may have different meanings and relate to different objectives among various stakeholders at the workplace.

**Needs Assessment.** Demographic information assists in the planning of health promotion activities, as well. Younger workers are raising families, often with both parents in the work force. They may need information on parenting and preventive health care for children. Also, younger workers are more prone to experience automobile accidents, and seat belt campaigns are important interventions for them. Older workers are more prone to cardiovascular ailments; they may be nearing retirement, and need to know about self-care for the elderly. Women are entering the work force in increasing numbers and are being subjected to many of the same stresses once thought to be exclusively a male domain. This is likely to increase the importance of interventions to prevent cardiovascular disease for both sexes.

Another way to get information on needs for health promotion in the work force is to conduct health screenings and administer health risk appraisals (2, 40). In addition, screening is useful to the employer who wants to be protected from allegations of problems due to worksite exposures to hazardous substances. Health risk appraisals are not just a data collection tool. By themselves, they can be used to educate and motivate employees to make changes in their lifestyle (47).

**Specification of Goals.** For the short term, health promotion efforts should be aimed at maximizing participation by employees at risk of health problems that can be prevented or improved, minimizing dropouts from any treatment or intervention, and assisting the maximum number of participants to adhere to lifestyle changes. It is only by these three indices that an organization can hope to improve the health of entire work forces, and only in this way can intervention be made truly cost effective. A special goal for the short term is to ensure that activities are targeted especially to those who can benefit: for example, exercise programs for sedentary people, first, and then perhaps expansion to everyone.

Statements about short-term goals should use action verbs, specify a time frame for accomplishment, and specify outcomes in measurable terms. A very immediate goal for hypertension control might be: "by one year from today, the health promotion program will identify the majority of hypertensives in the work force by screening at least 75% of employees for high blood pressure." Another, longer-term goal might be the following: "by two years from today, the health promotion program will recruit and retain at least 80% of known hypertensives into a worksite hypertension counceling program." Longer-term goals relate to maintaining positive health behavior and gaining a substantial cumulative impact on employees' health.

## Implementation

**Organization and Staffing.** Two models of organization for health promotion are frequently seen. In one model, health professionals such as the medical director take a leadership role in programs. This model has the advantage that health professionals are familiar with the idea of needs assessment and are likely to focus programming onto those activities for which a health benefit can reasonably be expected. In the other model, leadership comes from a committee made up of participants from various parts of the organization. This model has the advantage that organizational members feel ownership of the program; the belief is that they are therefore more likely to participate. However, such committees are sometimes more likely to choose activities without respect to health or other needs, and without a critical appraisal of likely benefits for the organization or for the work force. Combining the two models may incorporate their best elements. Staffing of health promotion efforts does not have to be expensive, since outside consultants and volunteer organizations provide a rich resource. Occupational nurses, health educators, and exercise physiologists frequently supply health promotion services in-house.

**Choice of Cost-effective Services.** In the area of health promotion, it is fairly clear that some activities (outlined above) are more effective than others in lowering the risk of disease. Some health promotion activities simply do not produce permanent lifestyle changes. Weight reduction for obese employees is not effective because the average reduction across many studies in all settings is only 10 pounds (6). Stress management is useful for specific problems, but when provided to all employees without regard to specified conditions, participation is low and people fail to practice stress-reducing exercises on a routine basis. When stress management is targeted specifically to those who can be expected to benefit, participation and routine practice are much higher. Exercise is linked to lower cardiovascular risk. There is also suggestive evidence of a link to lower absenteeism and higher productivity (53). However, the highest participation rate in worksite exercise to date is 25% (7). Moreover, exercise participation rates frequently fail to take into account people who exercise away from the work setting.

Screening and follow up for hypertension, high cholesterol, and breast cancer are all worthwhile where community services do not pick up these conditions. However, not all screening programs are worthwhile. Also, even good screening programs must be linked to a system to get people under a physician's care. If people are not aggressively guided to a physician's care for problems found in screening, they may take on the "sick role," losing even more workdays and experiencing more anxiety and depression than before they were notified about their condition (44). Excellent health screening programs exist, and they have succeeded in getting many employees under medical care (11). However, many well-intentioned organizations do not understand the importance of aggressively referring people to medical care, and may promote screenings for unselected groups of people without a good understanding of the prevalence of underlying problems, the nature of the test, or the role that the screening information might play in the medical care system.

Even for health promotion activities that are effective in improving health, there are more and less costly approaches. Smoking offers the best example. Most people want to quit on their own (42), and self-help materials are not costly. Even though the percentage of participants who quit is not very great, participation can be so great that the yield is large. Behavioral medicine and commercially marketed programs can be very costly. Although the percent of participants who quit might be higher, such methods might not be warranted until people have tried and perhaps failed through cheaper methods.

**Contracting with Providers.** Health promotion providers often promise the purchaser more than they can deliver. These providers should be able to present several kinds of information. First, they should be able to tell the purchaser about the extent of participation they can achieve in interventions for smoking, hypertension control, weight reduction, etc. Second, they should project the likely dropout rates from these interventions and explain how they hope to prevent dropouts. Finally, they should project the likely percentage of employees who will sustain behavior changes conducive to health for a year or more. And, they should be able to describe how they plan to help sustain those changes through environmental supports in the workplace, continued contact with participants, etc.

**Employee Communications.** Access to health promotion activities means that activities are held at times and in locations that employees deem to be convenient, as ascertained through an employee interest survey. The best means of doing so is to launch an awareness campaign at several levels. Announcements of clinics or classes should appear in multiple locations and through many sources. Feedback to employees about their own Health Risk Appraisals can motivate further interest in participation for at least some activities (47). Several programs have gained good participation by giving small incentives for participation, such as T-shirts, coffee mugs, or apples. Larger incentives for lifestyle change get good participation, but it is still not clear whether lifestyle changes are permanent (28). Changes in employee perception of the corporate culture can also facilitate participation, since employees should perceive that the administration looks with favor on participation and on lifestyle changes.

### Monitoring

**Use of Data.** To measure the impact of health promotion programs, the employer can use three short-term indicators of lifestyle change: participation in clinics or activities, retention of employees in those activities, and long-term changes in the targeted outcome. For example, it should be relatively easy to track the proportion of smokers who attend a first meeting of a smoking cessation clinic, as well as the number who continue to attend most of the meetings of that clinic. Participants can be contacted one year later and their smoking and health status verified. The use of these three indicators together should give the manager an idea of the performance of the health promotion program. The average experience across many worksites is now available, and the manager can compare the experience of his or her organization to that of many other organizations (28). In general, managers may have an unrealistic idea of what health promotion can accomplish. Only a portion of the at-risk population is likely to participate in health promotion programs, only a portion of those participants will remain in a class or clinic, and only a portion of those retained employees will actually change their behavior. When compared to the results from other companies, however, an organization's health promotion effort may appear quite productive.

**Corrective Action.** If difficulties do emerge, another survey should be conducted with the object of locating access or interest problems. Participation in health promotion programs can be increased by further attention to employee interest surveys—activities could be made more convenient, occur on company time, with time sharing, or with cost-sharing by the company (28). Also, dropouts could be interviewed confidentially to determine their reasons for quitting the health promotion activity.

## A PROJECT TO DEVELOP AN EMPLOYEE ASSISTANCE PROGRAM

### Planning

**Objectives.** EAPs are mostly aimed at helping employees solve personal problems so that productivity can rise or be maintained. Recent data suggest that this is indeed the case, (29). Of special note, alcoholics detected and treated through a worksite program have a much better prognosis than do alcoholics in the community, so limitation in work performance due to alcohol abuse can be reduced or reversed.

EAPs are also aimed at helping employees solve other personal problems, as indicated by the extent to which many EAPs permit self-referrals. Indeed EAPs can be expected to achieve this goal, since counseling is generally found to alleviate anxiety and depression (54).

As a spinoff from the immediate objective, EAPs can also help alleviate several other organizational problems. Since blue-collar workers with alcohol problems are absent from work more frequently than others, EAPs have been demonstrated to lower both absenteeism and workdays lost (51). Alcohol abuse contributes to a variety of health problems, such that rehabilitation should reduce health care utilization (38). Also, there is a consistent relationship found between mental health counseling and lower utilization of medical care services (50). Finally, EAPs give an incentive to employees to deal with drug dependencies, which can prevent involvement of employees with the legal system and help the employer deal intelligently with the issue of drug testing on the job.

Since EAPs are often presented as the only alternative to terminating the employee, they may assist in preventing turnover. However, termination may be more cost-beneficial, depending on the number of years of employment ahead, and the investment in employee training (52).

**Consultation with Stakeholders.** Some of the decisions made about EAPs really involve organizational values as well as information about program effectiveness. For example, to what extent should self-referrals be permitted? Self-referrals may or may not directly assist the organization in terms of productivity or absenteeism, while referrals from superiors should. Nevertheless, self-referrals may play an important role in prevention of problems, and certainly are in line with providing a benefit to the employee. Along the same lines, counseling for troubles other than substance abuse and mental health may not be essential for productivity or other employee-focused goals. However, such counseling could be of great use to the employee and may benefit the employer indirectly. Organizations must also develop a policy about drug abusers. Is the goal of EAP to detect such drug abusers for possible dismissal? Or is the goal to provide rehabilitation and a warning? What about the role of security and prevention of thefts by hard drug users?

**Needs Assessment.** The voluntary nature of data collection for preventive activities is especially problematic for alcohol, drug abuse, and mental health programs. The prevalence of substance abuse and emotional problems can never be completely gauged (29). Instead, employers might rely on national data estimates, performance data, health care utilization data, absenteeism data, and other sources that would supply indirect evidence of a general need for an EAP. Some employees may not even admit to themselves that they have a substance abuse problem, and may require constructive confrontation by a supervisor before they are willing to seek treatment. There is often much more substance abuse and emotional disturbance in the work force than people believe.

**Specification of Goals.** The penetration rate of EAP services is the percentage of troubled employees who receive counseling by the EAP in any given year. Because the prevalence of problems is not known, staff will have to choose an estimate from the population as a whole or from similar organizations in order to plan. For example, they may project a certain number of likely troubled employees based on such estimates, and then set a short-term goal concerning what they hope is a realistic penetration rate. The organization may also wish to set short-term goals for rates of detection and referral by supervisors, as well as rates of acceptance of counseling by employees who are referred. These rates could be based on past experience or the experience of other organizations; some rates are available in the literature (29, 33).

Longer-term goals concerning productivity, absenteeism, and medical care utilization are somewhat easier to quantify, in that information concerning present status of the organization and decrements in performance of individuals can in fact be monitored. Realistic goals for the organization as a whole or for referred individuals can be set in advance.

### Implementation

**Organization and Staffing.** The choice of providers under an EAP, whether full- or part-time physicians, psychologists, social workers, or nurses, will be limited to some degree by the extent to which such services are provided in-house. In-house staff will determine whether employees require referrals for long-term treatment; in this way, EAP staff act in the same way as case managers or "channeling" staff in cost containment strategies for medical care. EAP administrators should be clinically oriented and have good familiarity with the network of providers in a city or region and their practice patterns. A danger in EAP staffing is that low-cost providers with inadequate training will provide the actual counseling. The project manager should make sure that EAP staff can recognize conditions they cannot treat and can make appropriate referrals.

**Choice of Cost-effective Services.** Services can include treatment for substance abuse, mental health problems, domestic conflict, legal and financial difficulties, etc. They can also include stress management, assistance with housing and daycare problems, gambling, disability counseling, retirement planning, job counseling, assistance in work problems, and counseling on bereavement. Planners for EAPs and similar efforts should ascertain which of the many possible services under an EAP will truly help to achieve organizational objectives. Alcohol abuse programs have been shown to be cost effective, and the likelihood of recovery by employed alcoholics is very high compared to the rest of the population (38). Less information is available concerning drug abuse programs for the working population. Psychotherapy and psychological counseling generally do improve people's self-esteem, decrease anxiety and depression, and permit better job performance (54). For the employer, however, it is important to translate these benefits into quantifiable terms—how much can the person expect to benefit in terms of improved job performance, for example? The majority of troubled employees can benefit from brief psychotherapy and need not undergo long-term treatment (8). Many health benefits managers worry that mental health utilization will expand greatly if benefits are liberalized (8). However, employers should be aware that the EAP itself constitutes a means of referring employees to cost-effective sources of care.

**Contracting with Providers.** Outside providers should be able to demonstrate their qualifications. They should also be able to cite average utilization and the types of treatment offered to EAP clients. In addition, they should spell out in writing their means of maintaining confidentiality of records and of reporting progress back to the in-house staff. Access to an EAP would require that employees could refer themselves for counseling, as well as being referred by their supervisors if necessary. In general, case-finding procedures for EAP should be spelled out in writing.

**Employee Communications.** For EAP the goal is primarily to motivate at-risk groups to utilize a needed service. For EAP utilization to occur in the manner intended, supervisors must be trained to recognize and refer employees whose performance decrements suggest possible substance abuse or emotional difficulty. Supervisors should not attempt to begin counseling and should limit their confrontation to observed and documented evidence of poor or declining performance. They should leave anything else to the EAP professionals. Many businesses have elected to permit self-referrals to EAP as well. Announcements should be posted with either the phone number or location of the EAP clearly marked. A convenient but private location will help to reduce the stigma of visiting the EAP. Also, the EAP can reduce fears of a stigma by offering other activities, such as smoking cessation or stress management. The employee would not then be identified as visiting the EAP strictly for substance abuse treatment or emotional help.

### Monitoring

**Use of Data.** Tracking the performance of an EAP is possible at several levels. The manager will want to know about the rate of case finding, types of referrals, utilization, and charges, as well as the proportion of substance-abusing employees who are rehabilitated (33). It will not be possible to know the proportion of troubled employees who are reached by the EAP, however, since all troubled employees are not known to the employer. However, other indicators may be proxies for this aspect of performance, such as the overall rate of absenteeism, performance, accident rates, or utilization of physical health services (29).

**Corrective Measures.** Improper EAP utilization may require further training or supervising, or new methods of case finding, such as inspection of long-term performance or absenteeism records. Supervisors may need retraining on how to confront employees and how to refer them to the EAP. If a great many EAP clients fail to improve in terms of productivity or absenteeism, decision makers may wish to reconsider whether they have chosen appropriate kinds of people for EAP referral.

## A PROJECT TO DEVELOP OCCUPATIONAL HEALTH AND SAFETY PROGRAMS

The motivation for organizations to take action in this area is usually furnished by law, regulation, and economics. Government has mandated much activity, such as the workplace health regulations enforced by the Occupational Safety and Health Administration (OSHA). State and federal "right-to-know" laws require that a company both *obtain* information for its employees concerning the substances with which they come in contact, and must *produce* information for its customers. The danger of liability is very great for any organization that produces materials that could present a health hazard. The importance of high-quality occupational health and safety programs is therefore strongly linked to risk management.

In all probability, more has been written on the subject of occupational health and safety than on our other three projects combined. The reader is referred to texts such as Patty's *Industrial Hygiene and Toxicology* (41), Allen, Ells and Hart (3), Gardner (16), or Hunter (21). We should say at the outset of this brief overview of occupational health and safety that data collection and control are best left to the professionals described below. Nevertheless, the decision to expand, build, or reorient a health and safety program does lend itself to the project management approach, and priorities will still be set by top managers in conjunction with health and safety professionals.

### A. Planning

**Objectives.** Occupational safety and health protection programs are intended to prevent occurrence of work-related injuries and illnesses. The employee benefits from these programs because they improve the work environment, and the employer benefits because these programs first lower premiums on and subsequently control costs for Workers' Compensation insurance. They can also prevent other costs to an organization, such as health care, long-term disability, life insurance, and replacement costs. Although occupational diseases are covered by Workers' Compensation in most states for at least some period of time, lawsuits related to occupational disease that bypass Workers' Compensation are on the increase. Employers who fail to inform workers of hazards are held liable for negligence by the courts. This liability can be avoided if the employer shows that all possible steps were taken to avoid the hazard. Right-to-know laws at federal and state levels make it even more important to demonstrate that the employer has informed its workers.

**Consultation with Stakeholders.** All literature on occupational safety and health indicates that top management support is crucial to success (4, 36). Top managers set the tone for the rest of the organization and determine whether supervisors and workers will take health and safety seriously. A health and safety committee might be especially helpful for stakeholder consultation. The committee should be made up of representatives from all departments of the organization, and should have rotating membership so that many more individuals can be exposed to the safety and health message (36). Employee health nurses, if present, should sit on this committee, since they are well qualified to speak about the human element in injury or health hazards. Supervisors have the greatest influence over the safety behavior of workers, and should therefore be consulted about any proposed changes in occupational health programs in order to gain their complete support (12). Workers are more likely to cooperate with safety and health measures when they have been consulted first. Also, employees frequently know more about the actual production processes and may have information that challenges preconceptions about problems. Protective equipment can be pilot tested using the committee's judgments concerning comfort and convenience (4).

Certain departments in the workplace should have a special relationship with the occupational health and safety program. The personnel department has a special relevance to occupational safety and health, because it is frequently involved in both training the employee and in employee relations (4). Purchasing should be consulted frequently and the health and safety manager should have input into equipment and chemical product purchasing decisions, to assure that equipment will meet standards and also to assist the organization in meeting "right-to-know" requirements (4, 36). The legal department should consult with the occupational health project manager regarding OSHA rules and possible litigation over worksite exposures. Close collaboration with the operating departments is needed to assure training required by "right-to-know" laws and to plan for the prevention of hazardous exposures and safety risks. When the organization makes certain products, the sales department must also be involved, to assure that valid and accessible "right-to-know" information goes to customers.

**Needs Assessment.** The nature of the organization and the presence or absence of particular kinds of hazards will dictate certain data needs. Workplaces must keep records of accidents under the Occupational Safety and Health Act (OSHA). OSHA inspections, levels of Workers' Compensation claims, enforcement of "right-to-know" laws, and possible litigation over workplace exposures to toxic substances all could motivate management and employees to cooperate with such a health project.

Beyond routine data collection, industrial hygienists, safety engineers, and toxicologists can provide more detailed analyses of hazardous conditions. Textbooks on occupational safety and health stress the fact that priorities for solving safety and health problems must be set—resources are finite and some hazards are more likely than others to be serious or to cause harm. These priorities must be set by decision makers in consultation with occupational health and safety professionals.

**Specification of Goals.** Record keeping permits organizations to set realistic goals for any reduction in illness and injuries. Traditionally organizations have set goals for employee compliance with health and safety rules. Setting compliance goals requires a survey of current compliance levels, plus informed judgments about the extent to which compliance can feasibly be increased. Goals can also be set for employee knowledge about hazardous materials in the worksite, so that "right-to-know" requirements can be met. Current knowledge might be assessed via a survey; goals could be expressed in relative terms, such as increases in knowledge, or in absolute terms, such as saying that the "right-to-know" requirements have or have not been met (49).

For the longer term, it is important to recognize that the face of occupational health and safety is changing a great deal. Both large and small organizations are decentralizing programs and are contracting out much of the medical care that was formerly provided by company physicians. Organizations must be aware of the potential for a dangerous fragmented approach, the very opposite of the team approach called for by right-to-know laws, product liability, toxic torts, and other new challenges. Some experts believe that the future holds one of two models for safety and health: either a hospital-based model, or a government-supported model such as those mandated in Europe.

## Implementation

**Organization and Staffing.** Larger organizations are well advised to have a medical director, while smaller ones with potential safety or health hazards should have a part-time consulting physician with expertise in occupational medicine. Any manager of occupational health and safety should know a great deal about product liability, consumer protection, community protection, toxic torts, and health legislation and regulation.

Other staff are desirable if the size of the organization warrants them. Occupational health nurses are likely to be close to the human element in the analysis of health exposures and safety problems. Safety directors are likely to have some management training and experience-based knowledge, while industrial hygienists have received specialized training in their subject. These two types of specialists are likely to have different viewpoints about the dangers present in the workplace, and these should be resolved (4).

The increasing role of engineering management to control hazardous substances suggests linkage with line and staff operating departments. Some worksites locate occupational health and safety in the legal department, because of the litigation issue. This is a totally inappropriate place for effective protection of employees (4).

**Choice of Cost-effective Services.** The major activities of a program include careful analysis of health protection and safety hazards, careful design of equipment and work settings to prevent disease and injury, monitoring of hazards and of compliance with health and safety procedures, and education and training (12). Worksites vary greatly in terms of hazards and feasible solutions. Knowledge in the area of risk assessment is full of gaps (9, 34).

**Employee Communications.** Employee education on hazards is required by hazard communication regulations. However, education is not a substitute for careful analysis or for solving a problem through redesign of the work setting (12). Since educational intervention is required, behavioral science offers some means of making it more effective. Rewarding specific behaviors that are conducive to safety seems to make a difference in several diverse settings (12). Although injury events or near misses may be dramatic events that point to the importance of safety, they are also rather infrequent events. It is better to modify behavior by rewarding more frequently occuring behavior that can then be maintained. The role of the supervisor in assisting this process cannot be overstated (12).

Development of educational materials for safety and for employee "right-to-know" purposes does not have to be inordinately expensive. Materials can be purchased from other companies that have already developed them, or they can be developed in-house for the purpose of marketing for use by other organizations (49).

### Monitoring

After implementing a project in occupational health and safety, one would expect the rate of accidents to decline or the extent of occupational exposures to be affected. In addition, the organization may wish to monitor employee compliance with safety and health protection measures. Trouble spots could then be reassessed for corrective action, in terms of engineering solutions, employee education solutions, or enforcement by the supervisor.

## CONCLUSION

In this chapter we have adapted the project management strategy to four types of health projects normally encountered in organizations. The project management approach is emminently suited to development and change in programs designed to maintain the health of organizational members.

## REFERENCES


1. Abraham, J. "A Union's View of Employee Assistance Programs." *Corporate Commentary,* Vol. 1(3) (March 1985) pp. 40–43.
2. Alderman, M. H. and Davis, T. K. "Blood Pressure Control Programs On and Off the Worksite." *Journal of Occupational Medicine,* Vol. 22 (1980), pp. 167–170.
3. Allen, R. W., Ells, M. D., and Hart, A. W. *Industrial Hygiene* (Prentice-Hall. Englewood Cliffs, N.J., 1976).
4. Asfahl, C. R. *Industrial Safety and Health Management* (Prentice-Hall. Englewood Cliffs, N.J., 1984).
5. Bieber, O. F. "Bargaining for Equitable, Cost-effective Health Care." *Business and Health,* Vol. 2(5) (1985), pp. 20–24.
6. Brownell, K. D. "Obesity: Understanding and Treating a Serious, Prevalent and Refractory Disorder." *Journal of Consulting and Clinical Psychology,* Vol. 50 (1982), pp. 820–840.
7. Colacino, D. L. and Gulbronson, C. R. "New Perspectives on Pepsico's Fitness Participation." *Corporate Commentary,* Vol. 1 (1984), p. 36.
8. Cummings, N. A. "The Dismantling of Our Health System." *American Psychologist,* Vol. 41 (1986), pp. 426–431.
9. Dinman, B. "Occupational Health and the Reality of Risk—An Eternal Dilemma of Tragic Choices." *Journal of Occupational Medicine,* Vol. 22 (1980), pp. 153–157.
10. Enthoven, A. "Health Tax Policy Mismatch." *Health Affairs,* Vol. 4(4) (Winter, 1985), pp. 6–13.

11. Erfurt, J. C. and Foote, A. *Hypertension Control in the Work Setting: The University of Michigan–Ford Motor Company Demonstration Program* (University of Michigan. Ann Arbor, 1982) (NIH Publication No. 83-2013).
12. Everly, G. S. and Feldman, R. H. L. *Occupational Health Promotion* (Wiley. New York, 1985).
13. Fox, P. D. and Spies, J. J. "Alternative Delivery Systems: What Are the Risks?" *Business and Health,* Vol. 1(3) (January/February, 1984), pp. 5–10.
14. Fries, J. F. "Aging, Natural Death and the Compression of Morbidity." *New England Journal of Medicine,* Vol. 303 (1980), pp. 130–135.
15. Gallo, P. S. "Meta-analysis—A mixed meta-phor?" *American Psychologist,* Vol. 33 (1978), pp. 515–517.
16. Gardner, A. W. (ed.). *Current Approaches to Occupational Health* (Wright PSG. Boston, 1982).
17. General Accounting Office. *A Primer on Competitive Strategies for Containing Health Care Costs* (GAO. Washington, D.C., 1982) (GAO/HRD-82-92).
18. Goldsmith, F, and Kerr, L. E. *Occupational Safety and Health* (Human Sciences Press. New York, 1982).
19. Griffes, E. J. E. *Employee Benefits Programs: Management, Planning, and Control* (Dow Jones-Irwin. Homewood, Ill., 1983).
20. Herzlinger, R. E. "How Companies Tackle Health Care Costs: Part II." *Harvard Business Review,* Vol. 86 (1986), pp. 108–120.
21. Hunter, D. *Diseases of Occupations* (English Universities Press. London, 1975).
22. Jacobs, P. *The Economics of Health and Medical Care,* 2nd ed. (Aspen. 1987).
23. Jameson, J., Shuman, L. and Young, W. "The Effects of Outpatient Psychiatric Utilization on the Costs of Providing Third Party Coverage." *Medical Care,* Vol. 16 (1978), p. 383.
24. Kannel, W. B. and Gordon, T. *The Framingham Study: An Epidemiological Investigation of Cardiovascular Disease* U.S. Department of Health Education and Welfare, NIH. Washington, D.C., 1974) (NIH) 74-475.
25. Kellogg Foundation. *Viewpoint: Toward a Healthier America* (W. K. Kellogg Foundation. Battle Creek, Mich., January, 1980).
26. Kiefhaber, A. K. and Goldbeck, W. B. "Worksite Wellness." Prepared for the Office of the Assistant Secretary for Planning and Evaluation, U.S. DHHS, in conjunction with the Private Sector Health Care Initiatives Study, contract #100-82-31, n.d.
27. King, W. R. "The Role of Projects in the Implementation of Business Strategy," in *Project Management Handbook,* ed. Cleland, D. I. and King, W. R. (Van Nostrand Reinhold. New York, 1983).
28. Leviton, L. C. "Can Organizations Benefit from Worksite Health Promotion?" *Health Services Research* (in press)
29. Leviton, L. C. "Diversity and Uncertainty in Employee Assistance Programs," in *Mental Health Policy: Patterns and Trends,* ed. Rich R. (Sage, Newbury Park, CA, in press).
30. Luft, H. S. *The Operations and Performance of Health Maintenance Organizations* (NCHSR. Washington, D.C., 1981).
31. MacLeod, G. K. "Health Maintenance Organizations in the United States." *International Labour Review,* Vol. 110(4) (1974), pp. 335–350.
32. MacLeod, G. K. and Schwarz, M. R. "Faculty Practice Plans: Profile and Critique." *Journal of the American Medical Association,* Vol. 256 (July 4, 1986), pp. 58–62.
33. Masi, D. A. *Designing Employee Assistance Programs* (American Management Associations. New York, 1984).

34. Michaels, D. and Zoloth, S. "Occupational Safety: Why Do Accidents Happen?" *Occupational Health Nursing* (October, 1982), pp. 12–16.
35. National Highway Traffic Safety Administration. *Final Regulatory Impact Analysis: Passenger Car Front Seat Occupant Protection* (U.S. Department of Transportation. Washington, D.C., 1984).
36. National Safety Council. *Handbook of Occupational Safety and Health* (National Safety Council. Chicago, 1975).
37. Newhouse, J. P. et al. *Some Interim Results from a Controlled Trial of Cost Sharing in Health Insurance* (Rand Corporation. Santa Monica, Calif., January, 1982).
38. Office of Technology Assessment. *The Effectiveness and Costs of Alcoholism Treatment* (OTA. Washington, D.C., 1983).
39. Oster, G., Colditz, G. A. and Kelly, N. L. *The Economic Costs of Smoking and Benefits of Quitting* (Lexington Books. Lexington, Mass., 1984).
40. Parkinson, R. S. and Associates. *Managing Health Promotion in the Workplace: Guidelines for Implementation and Evaluation* (Mayfield. Palo Alto, Calif., 1982).
41. Patty, F. A. *Patty's Industrial Hygiene and Toxicology,* 2nd ed. (Wiley. New York, 1985).
42. Pechacek, T. F. "Modification of Smoking Behavior," in *The Behavioral Aspects of Smoking,* ed. Krasnegor, N. NIDA Research Monograph 26, Washington, D.C. 1979. DHEW Publication No. (ADM) 79-882.
43. Poggio, E. et al. *Second Surgical Opinion Programs: An Investigation of Mandatory and Voluntary Alternatives* (Abt Associates. Cambridge, Mass., 1982).
44. Polk, B. F. et al. "Disability Days and Treatment in a Hypertension Control Program. *American Journal of Epidemiology,* Vol. 119 (1984), pp. 44–53.
45. Price, D. N. "Income Replacement During Sickness, 1948–1978." *Social Security Bulletin,* Vol. 44 (1981), pp. 18–32.
46. Rice, T., deLissovay, G., Gabel, J. and Erman, D. "The State of PPOs: Results from a National Survey." *Health Affairs,* Vol. 4(4) (Winter, 1985), pp. 26–39.
47. Rodnick, J. E. "Health Behavior Changes Associated with Health Hazard Appraisal Counseling in an Occupational Setting." *Preventive Medicine,* Vol. 11 (1982), pp. 583–594.
48. Russell, L. *Is Prevention Better than Cure?* (Brookings Institution. Washington, D.C., 1985).
49. Samways, M. C. "Cost-effective Occupational Health and Safety Training." *American Industrial Hygiene Association Journal,* Vol. 44 (1983), pp. A-6 to A-9.
50. Schlesinger, H. J., Mumford, E. and Glass, G. V. "Mental Health Services and Medical Utilization," in *Psychotherapy: From Practice to Research to Policy,* ed. Vandenbos, G. (Sage. Beverly Hills, 1980).
51. Schramm, C. J. "Measuring the Return on Program Costs: Evaluation of a Multi-employer Alcoholism Treatment Program." *American Journal of Public Health,* Vol. 7(1) (1977), pp. 50–51.
52. Schramm, C. J. "Evaluating Industrial Alcoholism Programs: A Human-Capital Approach." *Journal of Studies on Alcohol,* Vol. 41(7) (1980), pp. 702–713.
53. Shephard, R. J., Corey, P., Renzland, P. and Cox, M. "The Influence of an Employee Fitness and Lifestyle Modification Program upon Medical Care Costs." *Canadian Journal of Public Health,* Vol. 73 (1982), pp. 259–263.
54. Smith, M. L. and Glass, G. V. "Meta-analysis of Psychotherapy Outcome Studies." *American Psychologist,* Vol. 32 (1977), pp. 752–760.
55. Starr, P. *The Social Transformation of American Medicine* (Basic Books. New York, 1982).

56. Steiner, G. and Lawrence, P., eds. *Educating Diabetic Patients*. (Springer. New York, 1981).
57. U.S. Department of Commerce. *Statistical Abstract of the United States, 1984, 104th Edition* (Washington, D.C., 1983).
58. Waldo, D. R., Levit, K. R., and Lazenby, H. "National Health Expenditures, 1985." *Health Care Financing Review,* Vol. 8(1) (1986), pp. 1–22.
59. Warner, K. E. and Murt, H. A. "Economic Incentives for Health." *Annual Review of Public Health,* Vol. 5 (1984), pp. 107–133.
60. Weinstein, M. C. and Stason, W. B. *Hypertension: A Policy Perspective* (Harvard University Press. Cambridge, Mass., 1976).
61. Wennberg, J. E. "Dealing with Medical Practice Variations: A Proposal for Action." *Health Affairs,* Vol. 3 (1984), pp. 7–32.
62. Wilson, F. A. and Neuhauser, D. *Health Services in the United States* 2nd ed. (Ballinger. Cambridge, Mass., 1985).
63. World Health Organization. "Health Promotion: A Discussion Document on the Concept and Principles." (WHO. Copenhagen, 1984).

# 38. The Cultural Ambience of the Matrix Organization*

David I. Cleland†

## INTRODUCTION

The concept of matrix management has grown beyond the project management context first introduced by John Mee in 1964.‡ Project management has been the precursor of today's matrix management approach found in diverse uses today. In the multinational corporation product, functional, and geographic managers work in a sharing mode of matrix management. Matrix management is found in a wide variety of other contexts: product management, task force management, production teams, new business development teams, to name a few.

In this chapter, I describe the cultural ambience of the project-driven matrix organization.

Culture is a set of refined behaviors that people have and strive towards in their society. Culture encompasses the complex whole of a society which includes knowledge, beliefs, art, ethics, morals, law, custom, and other habits and attitudes acquired by the individual as a member of society. Anthropologists have used the concept of culture in describing primitive societies. Modern-day sociologists have borrowed this anthropoligical concept of culture and used it to describe a way of life of a

---



† David I. Cleland is currently Professor of Engineering Management in the Industrial Engineering Department at the University of Pittsburgh. He is the author/editor of 15 books and has published many articles appearing in leading national and internationally distributed technological, business management, and educational periodicals. Dr. Cleland has had extensive experience in management consultation, lecturing, seminars, and research. He is the recipient of the "Distinguished Contribution to Project Management" award given by the Project Management Institute in 1983, and in May 1984, received the 1983 Institute of Industrial Engineers (IIE)-Joint Publishers Book-of-the-Year Award for the *Project Management Handbook* (with W. R. King). In 1987 Dr. Cleland was elected a fellow of the Project Management Institute.

‡ John F. Mee, "Matrix Organization," *Business Horizons* (Summer, 1964).

people. I borrow from the sociologists and use the term culture to describe the synergistic set of shared ideas and beliefs that is associated with a way of life in an organization.

### Nature of a Business Culture

The word culture is being used more and more in the lexicon of management to describe the ambience of a business organization. The culture associated with each organization has several distinctive characteristics that differentiate the company from others. In the IBM Corporation the simple precept, "IBM means service" sets the tone for the entire organization, infusing all aspects of its environment and generating its distinctive culture. At 3M the simple motto "Never kill a new product idea" creates an organizational atmosphere of inventiveness and creativity. In some large corporations such as Hewlett-Packard, General Electric, and Johnson & Johnson, the crucial parts of the organization are kept small to encourage a local culture which encourages a personal touch in the context of a motivated, entrepreneurial spirit of teamwork.

Understanding the culture of the organization is a prerequisite to introducing project management. An organization's culture reflects the composite management style of its executives, a style that has much to do with the organization's ability to adapt to such a change as the introduction of a project management system.

## THE ROOTS OF THE MATRIX ORGANIZATION

The cultural ambience of the project-driven matrix organization is unique in many respects. But it should not be strange to us since our first organization, the family, has key features of the matrix design. In the traditional family unit, the child is responsible to, and has authority exercised over him by, two superiors (parents). A perceptive child soon learns that he must work out major decisions and such matters with both his bosses. If his parents have agreed on a "work breakdown structure" where each will exercise authority and assume responsibility over a particular aspect of raising the child, it will make it easier for him to get along with them and his peer group. A child may have to find ways of collaborating with both parents as well as his siblings and peers, adjusting to all groups.

When the child goes to school, another similar matrix design is found. The student is placed in a "home room" with a teacher whose main business is administration, "logistic support," and discipline. The student is taught by other teachers who are the "functional specialists"; thus the student has several more "superiors" to whom he is accountable as well as a larger peer group. If the student is active in extracurricular activities, still more bosses come into his life. Success and acceptance in these activities generally require peer acceptance, teamwork, and an ability to communicate with his "superiors" and his peers.

When the student leaves school and seeks employment, he may find more of a hierarchical structure, yet the new matrix is in many ways similar to those already experienced. If his initial job is on a production line, the production foreman becomes boss number one; yet the quality control specialist can shut down the production line. The perceptive individual finds that certain staff specialists (personnel, finance, maintenance, wage and salary) and even the informal leaders in the peer groups temper the "sovereign domain" of the foreman. He soon finds that certain people in the organization exercise power simply because they have control of information (such as the production control specialist) or have become experts in some areas. People look to the expert to make decisions or take a leadership role in certain matters. The role of the union steward as a tempering influence on production techniques and policies soon becomes obvious to him. If he is active in community affairs he finds many other "bosses" telling (or suggesting to) him what to do. Who's really the boss? Well, it depends on the situation—as it does in the matrix organization.

The sharing context of project management should not be foreign to any of us. Our family life, education, and work experience have given us ample exposure to working for and satisfying several bosses and of learning to communicate and work with peers as well. Then why such resistance to matrix design? I believe the resistance has its roots in several cultural factors. First is the concept that authority flows from the top of the organization down through a chain of command. The foundation of this belief springs from the "divine right" of the king, who is delegated to rule the kingdom by a deity. Historically most social institutions have had a strong vertical structure—a chain of command. Ecclesiastical organizations have contributed much to organization theory; many of these contributions have reinforced the vertical structure. Have we not always assumed that "heaven"—by whatever name it is called—is a higher place or state? The Bible speaks of ascending into heaven. (Why not moving to heaven on a lateral basis?)

A good friend of mine who is a competent minister once delivered a sermon on the theme that "Hell was a state of mind, not a place." After the sermon was over I asked him this question: If hell is a state of mind and not a place, then it follows that heaven is a state of mind and not a place. He said: "Perhaps, but we are not ready for that yet!"

Perhaps, like heaven, the matrix organization is more a state of mind that anything else!

No one would doubt the strong influence the Bible has had on our thinking. Indeed, the words of St. Matthew are familiar to everyone: "No man can serve two masters: for either he will hate the one, and love the other; or else he will hold to the one, and despise the other" (Matthew 6:24).

Part of the rationalization for the principle of "unity of command" may well be traced back to this verse. In managerial theory, this principle and its corollaries "parity of authority and responsibility," "compulsory staff advice," "line commands, staff advises," "span of control," etc., provide the conceptual foundation of the hierarchical organizational form. Indeed, many times managers and professionals have asked, "How can I work for two bosses?"

Yet Matthew also provides us with the basis for doing so: "Render therefore unto Caesar the things which are Caesar's; and unto God the things that are God's (Matthew 22:21).

I contend that in the light of both pragmatic and cultural experience there is as much a basis for the matrix design with its multidimensional sharing of authority, responsibility, accountability, and results, as there is for the hierarchical style of management.

## THE MATRIX ORGANIZATIONAL DESIGN

I use the concept of organizational design in a broad context to include organizational structure, management systems and processes, formal and informal interpersonal relationships, and motivational patterns. The matrix design is a compromise between a bureaucratic approach that is too inflexible and a simple unit structure that is too centralized. The design is fluid: personnel assignments, authority bases, and interpersonal relationships are constantly shifting. It combines a sense of democracy within a bureaucratic context.

From an organizational design viewpoint, the entire organization must be psychologically tuned to results: the accomplishments within the organization that support higher level organizational objectives, goals, and strategies. The purpose of a matrix design is not only to get the best from its strong project approach and strong functional approach but to complement these by a strong unity of command at the senior level to insure that the balance of power is maintained in the organization. In some companies only one or a few divisions might require a realignment to the project-driven matrix form; the others might be left in the pyramidal, hierarchical form. Indeed a single organizational chart cannot realistically portray the maze of relationships that exist inside a large organization because some elements select project management, others opt for the conventional line-staff design, and still others choose some hybrid form.

### The Design Is Result-Oriented

The matrix design is result-oriented and information-related. The very design itself says that there is need for someone who can manage a process of cutting across the line functions. A compromise results through the bipolarity of functional specialization and project integration. Out of the lateral relations—direct contact, liaison roles, and integration—comes a faculty to make and implement decisions and to process information without forcing an overburden on the hierarchical communication channels. It is the need to reduce the decision process on the hierarchical channels that motivates the formal undertaking of lateral relations through establishing a design which is bilateral:

1. *Project managers* who are responsible for results.
2. *Functional managers* who are responsible for providing resources to attain results.

When implementing a matrix design in the early stages, a poor harmony will usually exist between the behavioral reality and its structural form. It is at this stage that the process of integration become important and a series of critical actions must be initiated and monitored by senior management. Superior-subordinate relationships need to be modified; individual self-motivation leading to peer acceptance becomes critical. The development of strategies for dealing with conflict, the encouragement of participation techniques, and the delineation of expected authority and responsibility patterns are crucial. the complexity of the resulting organizational design, described by Peter Drucker as "fiendishly difficult," reminds us that the matrix design should only be used when there is no suitable alternative. The design lacks the simple model of the conventional hierarchy. The nature of projects each in various stages of its life cycle creates a lack of stability. Key people on the project teams must *not only know* their specialty, but also *how* the specialty contributes to the whole.

The emphasis is on flexibility, peer informality, and minimization of hierarchy. To change an existing design to a fully functioning matrix form takes time—perhaps several years.

The matrix organizational design is the most complex form of organizational alignment that can be used. The integration of specialists along with supporting staff into a project team requires strong and continuing collaborative effort. And the coordination of effort in this kind of design requires a continuing integration of the mutual efforts of the team members. Authority (and consequently power) tends to flow to the individual who has the information that is needed and whose particular skills and knowledge are necessary to make a decision. Many managers are found in the matrix design: project managers, functional managers, work package managers, general managers, staff managers. The greater the number of project teams, the greater the number of managers that will be used. As a result, the management costs in such an organizational approach are increased.

The introduction of project management into an organization tends to change established management practice with respect to such matters as authority and responsibility, procedural arrangements, support systems, department specialization, span of control, resource-allocation patterns, establishment of priorities, evaluation of people, etc. Performance goals within such organizations tend to be assigned in terms of the interfunctional flow of work needed to support a project. In so doing, established work groups within the functional agencies are often disrupted. In addition, there is a potential for the staffing pattern to involve duplication. The functional manager previously had the freedom to manage the organization in a relatively unilateral fashion, for he carried out integration himself or a higher authority handled it. Now he is forced into an interface in an environment which places a premium on the integration of resources through a project team consensus in order to accomplish project results. He must learn to work with a vocal and demanding horizontal organization.

A cultural characteristic of the matrix design causes two key attitudes to emerge: the manager who realizes that authority has its limits and the professional who recognizes that authority has its place.

## THE CULTURAL AMBIENCE

In its organizational context, a cultural ambience for matrix management deals with the social expression manifest within the organization when engaged in managing projects. A cultural system emerges which reflects certain behavioral patterns characteristic of all the members of that organization. This system influences the skills, knowledge, and value systems of the people who are the primary organizational clientele. The clientele are a "team" of people who have some vested outcome in the success of an organizational effort.

Thus, project clientele include those in the organizational society who are the managers and professionals collectively sharing the authority and responsibility for completing a project on time and within budget. Superiors, subordinates, peers, and associates are the primary project clientele who work together to bring the project to a successful completion. The cultural ambience that ultimately emerges is dependent on the way these primary clientele feel and act in their professional roles both on the project team and within the larger organizational context. The integration of these clientele results in an ambience which has the following characteristics: Organizational Openness; Participation; Increased Human Problems; Consensus Decision Making; Objective Merit Evaluation; New Criteria for Wage and Salary Classification; New Career Paths; Acceptable Adversary Roles; Organizational Flexibility; Improvement in Productivity; Increased Innovation; Realignment of Supporting Systems; and Development of General Manager Attitudes. These characteristics are discussed below.

### Organizational Openness

A propensity toward organizational openness is one of the most characteristic attributes of the matrix design. This openness is demonstrated through a receptiveness to new ideas, a sharing of information and problems by the peer group. Newcomers to a matrix organization are typically accepted without any concern. There is a willingness to share organizational challenges and frustrations with the newcomer. This openness characteristic of project team management is described in one company as "no place to hide in the organization."*

### Participation

Participation in the project-driven matrix organization calls for new behavior, attitudes, skills, and knowledge. The demands of working successfully in the matrix design create opportunities for the people as well as for the organization. For the people, there are more opportunities to attract attention and to try one's mettle as a potential future manager. Because matrix management increases the amount and the pattern of recurring contacts between individuals, communication is more intense. The resolution of conflicts is also of a more intense nature than in the traditional organization where conflict can be resolved by talking to the functional boss. In a matrix design, at least two bosses have to become involved—the manager providing the resources and the manager held accountable for results. These two managers, locked in a conflict, may appeal as a last resort to the common line supervisor for resolution. Matrix management demands higher levels of collaboration. But in order

* "Texas Instruments Shows U.S. Business How To Survive in the 1980's," *Business Week* (September 18, 1978).

to have collaboration, trust and commitment are needed on the part of the individuals. In order to be committed and to maintain trust, the individuals in the organization must take personal risks in sharing information and revealing their own views, attitudes, and feelings.

There is growing evidence that individuals today wish to influence their work situation and to create a democratic environment at their place of work. People expect variety in their life-style in the organization as well as in their private lives. The flexibility and openness of the matrix design can accommodate these demands.

The degree to which people are committed to participate openly and fully in matrix organization effort can influence results. Murphy, Baker, and Fisher, in a study of over 650 projects including 200 variables, found that certain variables were associated with the perceived failure of projects. Lack of team participation in decision making and problem solving was one important variable associated with perceived project failure. In contrast, project team participation in setting schedules and budgets was significantly related to project success.*

## Increased Human Problems

Reeser conducted research to examine the question as to whether project organization might not have a built-in capacity of causing some real human problems of its own. This research was conducted at several aerospace companies. Reeser's research findings suggested that insecurity about possible unemployment, career retardation, and personal development is felt by subordinates in project organizations to be significantly more of a problem than by subordinates in functional organizations. Reeser notes that project subordinates can easily be frustrated by "make-work" assignments, ambiguity, and conflict in the work environment. Project subordinates tend to have less loyalty to the organization. There are frustrations because of having more than one boss. The central implication of Reeser's findings is that although there may be persuasive justifications for using a matrix design, relief from human problems is not one of them.†

Even with formal definition of organizational roles, the shifting of people between the projects does have some noticeable effects. For example, people may feel insecure if they are not provided with ongoing career

---

* D. C. Murphy, Bruce N. Baker, and Delmar Fisher "Determinants of Project Success," Springfield, Va. 22151, National Technical Information Services, Accession No. N-74-30392, 1974, p. 60669. See Chapters 35 and 36.

† Calyton Reeser, "Some Potential Human Problems of the Project Form or Organization," *Academy of Management Journal*, Vol. 12 (December, 1979).

counseling. In addition, the shifting of people from project to project may interfere with some of the basic training of employees and the executive development of salaried personnel. This neglect can hinder the growth and development of people in their respective fields.

### Consensus Decision Making

Many people are involved in the making decisions. Members of the matrix team actively contribute in defining the question or problem as well as in designing courses of action to resolve problems and opportunities in the management of the effort. Professionals who become members of a matrix team gain added influence in the organization as they become associated with important decisions supporting an effort. They tend to become more closely associated with the decision makers both within the organization and outside it. Perceptive professionals readily recognize how their professional lives are broadened.

A series of documents which describe the formal authority and responsibility for decision making of key project clientele should be developed for the organization. If a manger is used to a clear line of authority to make unilateral decisions, the participation of team members in the project decision process makes management more complex. However, the result is worth the effort for the decisions tend to be of a higher quality. Also, by participating in the decision process people have a high degree of commitment toward carrying out the decision in an effective manner.

### Objective Merit Evaluation

This is an important area of concern to the individual in the matrix design. If the individual finds himself working for two bosses (the functional manager or work package manager and the project manager) chances are good that both will evaluate his merit and promotion fitness. Usually the functional manager initiates the evaluation; then the project manager will concur in the evaluation with a suitable endorsement. If the two evaluators are unable to agree on the evaluation it can be referred to a third party for resolution. For the most part, individuals who are so rated favor such a procedure as it reinforces their membership on the project team as well as ensures that an equitable evaluation is given. A project team member who has been assigned to the project team from a functional organization may find himself away from the daily supervision of his functional supervisor. Under such circumstances a fair and objective evaluation might not be feasible. By having the project manager participate in the evaluation, objectivity and equity are maintained.

### New Criteria for Wage and Salary Classification

The executive rank and salary classification of a project manager will vary depending on the requirements of his position, the importance of the project to the company, etc. Most organizations adopt a policy of paying competitive salaries. However, the typical salary classification schema is based on the number and grade of managers and professionals that the executive supervises. In the management of a project, although the project manager may only supervise two or three people on his staff, he is still responsible for bringing the project in on time, at the budgeted cost. In so doing, he is responsible for managing the efforts of many others who do not report to him in the traditional sense. Therefore, new criteria for determining the salary level of a project manager are required. Organizations with successful salary classification schema for project manager's salary have utilitzed criteria such as the following:

- Duration of project.
- Importance of project to company.
- Importance of customer.
- Annual project dollars.
- Payroll and level of people who report directly to project manager (staff).
- Payroll and level of people whom project manager must interface directly on a continuing base with project manager.
- Complexity of project requirments.
- Complexity of project.
- Complexity of project interfaces.
- Payroll and rank of individual to whom project manager reports.
- Potential payoff of project.
- Pressures project manager is expected to face.

In many companies the use of project management is still in its adolesence, and suitable salary criteria have not been determined. In such cases it is not uncommon to find individuals designated as project managers who are not coded as managerial personnel in the salary classification and executive rank criteria. Word of this will get around and the individual's authority may be compromised. The author has found this situation arising usually because of a failure of the wage salary staff specialists to develop suitable criteria for adjusting the salary grade of the project managers. This problem is not as significant in those industries where project management is a way of life, such as in aerospace and construction.

### New Career Paths

The aspiring individual typically has two career paths open to him: to remain as a manager in his technical field or to seek a general manager position. Or he may prefer to remain a professional in his field and become a senior advisor, for example, a senior engineer. Project management opens up a third career field in management. The individual who is motivated to enter management ranks can seek a position as a project manager of a small project and use this as a stepping stone to higher-level management positions. It is an excellent way to learn the job of a general manager since the job of project manager is much like that of a general manager except that the project manager usually does not have the formal legal authority of the general manager. This should not deter the project manager; it should motivate him to develop his persuasive and other interpersonal and negotiation skills—necessary skills for success in general management!

### Acceptable Adversary Roles

An adversary role emerges in project management as the primary project clientele find that participation in the key decisions and problems is socially acceptable. An adversary role may be assumed by any of the project clientele who sense that something is wrong in the management of the project. Such an adversary role questions goals, strategies, and objectives and asks the tough questions that have to be asked. Such a spontaneous adversary role provides a valuable check to guard against decisions which are unrealistic or overly optimistic. A socially acceptable adversary role facilitates the rigorous and objective development of data bases on which decisions are made. But the prevailing culture in an organization may discourage the individual from playing the adversary role that will help management to comprehend the reality of a situation. This situation is possible in all institutions of a hierarchical character.

An adversary role presumes that communication of ideas and concerns upward is encouraged. As people actively participate in the project deliberations, they are quick to suggest innovative ideas for improving the project or to sound the alarm when things do not seem to be going as they should. If the adversary role is not present, perhaps because its emergence is inhibited by the management style of the principal managers, information concerning potential organizational failures will not surface. An example of the stifling of an adversary role is found in the case of a company in the management of an urban transportation project.

In the late 1960s this company attempted to grow from a $250-million-a-year subcontractor in the aerospace industry into a producer of ground transit equipment. In pursuing this strategy, it won prime contracts for two large urban rail systems. Heavy losses in its rail programs put the company into financial difficulties. What went wrong?

The company got into difficulty in part because the chief executive dominated the other company executives even though he was unable to face overriding practical considerations. When major projects in the rail systems business were in difficulty, the unrealistic optimism demonstrated by the chief executive prevented any executive from doing much about the difficulties. In the daily staff meetings that were held, the executives quickly learned that any negative or pessimistic report on a project would provoke open and sharp criticism from the chief executive. Project managers quickly learned that in the existing cultural ambience the bringing of bad news would not be tolerated. Consequently, they glossed over problem areas and emphasized the positive in order to please the chief executive.

On one of their large contracts they submitted a bid that was 23% below the customer's own estimate, and $11 million under the next lowest competitive bid. The project manager had felt that this estimate was too low but had not argued against it because, "I didn't want to express a sorehead minority view when I was in charge of the program." The cultural ambience within this company during this period might be summarized as follows: Don't tell the boss any bad news, only report good news—if you bring bad news, you run the risk of being sharply criticized.

Members of a project team need to feel free to ask tough questions during the life of the project. When the members of a team can play an adversary role, a valuable check and balance mechanism exists to guard against decisions which are unrealistic. Within Texas Instruments a cultural ambience exists where an adversary role can emerge. Consequently, "It is impossible to bury a mistake in this company. The grass roots of the corporation are visible from the top . . . the people work in teams and that results in a lot of peer pressure and peer recognition."*

### Organizational Flexibility

The lines of authority and responsibility defining the structure tend to be flexible in the matrix organization. There is much give and take across these lines with people assuming an organizational role that the situation warrants rather than what the position description says should be done.

---

* "Texas Instruments Shows U.S. Business How To Survive in the 1980's," *Business Week* (September 18, 1978), pp. 66–92.

Authority in such an organizational context gravitates to the person who has the best credentials to make the judgment that is required.

The matrix design provides a vehicle for maximum organizational flexibility; no one has "tenure" on a matrix team. There are variable tasks that people perform, a change in the type of situations they may be working on, and an ebb and flow of work loads as the work of the organization fluctuates. When an individual's skills are no longer needed on a team, that person can be assigned back to his or her permanent functional home.

There are some inherent problems in the flexibility of the matrix organization. The need for staffing tends to be more variable. Both the quantity of personnel and the quality needed are difficult to estimate because of the various projects that are going on in the organization. For example, a structural design group may have a surplus of design engineers at a particular time who are not assigned to any one project. The manpower estimates for oncoming projects, however, may indicate that in several months these professionals will be needed for project assignments. A functional structural design manager has the decision of whether to release the men and reduce his overhead or to assign them to "make-work" for the period and forgo the future costs of recruitment, selection, and training. The same manager may anticipate assigning these professionals to an emerging project yet, if the emerging project is delayed or even canceled, the project manager may not need these people for some time.

As the work effort nears its end, and perceptive individuals begin to look for other jobs, there can be a reduction in their output level. This reduction can damage the efficacy with which the project is being managed. Paradoxically, although morale takes on added significance in the matrix team, the design itself may result in lowering it.

The organizational flexibility of project management does, therefore, create some problems as well as opportunities.

### Improvement in Productivity

Texas Instruments credits the use of project teams for productivity improvement in the company. Its productivity improvement over a period of years has slightly more than offset the combined impact of its wage and benefit increases (average 9.2% annually) and its price decreases (averaging 6.4% per unit).*

At Texas Instruments more than 83% of all employees are organized into "people involvement teams" seeking ways to improve their own productivity. The company views its people as interchangeable—"kind

* Ibid.

of like auto parts." The culture there is much like the Japanese—a strong spirit of belonging, a strong work ethic, competitive zeal, company loyalty, and rational decision making. The culture of Texas Instruments ". . . has its roots buried deep in a soil of Texas pioneer work ethic, dedication, toughness and tenacity—it (the culture) is a religion. The climate polarizes people—either you are incorporated into the culture or rejected."*

The experience of Litton Industries in its Microwave Cooking Division shows that the use of project teams in the manufacturing function has increased productivity. Since the manufacturing organization was grouped into team units, production increased fourfold in 15 months. Product quality has increased, 1000 new production workers have been added to a base of 400 people, and unit production costs have declined 10–15%.† Some other claims of productivity increases that have come to the author's attention are as follows:‡

- A steel company chief executive states: "Properly applied, 'matrix' management improves profitability because it allows progress to be made on a broader front; with a given staff size, i.e., more programs and projects simultaneously pursued (including those concerning productivity)."
- The chief executive of a company in the microprocessor industry declares that the company's success (15% of the microprocessor market, $1.8 million in revenue, 18.1% ROI) would not be possible without matrix management.
- A chemical company president claims: "Matrix management improves people productivity."

The experiences of these companies suggest that project management techniques can assist in raising productivity.

### Increased Innovation

In the private sector in those industries where a fast-changing state-of-the-art exists, product innovation is critical for survival. There is evidence that the use of project teams has helped to further innovation within such organizations. For example, the teams are successfully used in the

---

* Ibid.

† William W. Grove, "Task Teams for Rapid Growth," *Harvard Business Review* (March–April, 1977), p. 71.

‡ These are productivity claims cited to the author in correspondence.

aerospace industry where the ability to innovate is essential, particularly in the development and production of sophisticated weaponry.

Why does the project team seem to foster innovation in organizations? Innovation comes about because an individual has an idea, a technological or market idea, and surrounds himself with some people who believe in the idea and are committed to it. A small team of people is formed, who become advocates and missionaries for the idea. The team of people represents a diverse set of disciplines who view the idea differently. It's difficult to hide anything in such an environment. The openness, the freedom of expression, the need to demonstrate personal effectiveness, all seem to be conducive to the creativity necessary to innovate. Within such organizations, decision making tends to be of a consensus type. An element of esprit de corps exists. The creativity and the innovative characteristics of small teams can be illustrated by the Texas Instruments situation.*

Texas Instruments has been extrememly successful with the use of teams in over 200 product-consumer centers (PCC). In each of these centers, the manager runs a small business of his or her own with responsibilities that include both long-term and short-term considerations. These managers have access to functional groups and are able to utilize the enormous resources that the functional organizations can offer. Indeed, what Texas Instruments has done is to create an organization in which the entrepreneur—the innovator—can flourish by making available to him or her the technical resources that are needed to do the job.

Project teams used effectively can take advantage of the scale economics of large organizations and, by their team nature, the flexibility of a small innovative organization is realized. An early research effort in the use of program (project) organizations noted that such organizations seemed to have been more successful in developing and introducing new products than businesses without program organizations.†

L. W. Ellis, Director of Research, International Telephone and Telegraph Corporation, claims that temporary groups (project teams) that are well organized and have controlled autonomy can stimulate innovation by overcoming resistance to change. Cross-functional and diagonal communication within the project team and with outside interested parties helps to reduce resistance to change.‡

---

* See "Texas Instruments Shows U.S. Business How to Survive in the 1980's," *Business Week* (September 18, 1978), pp. 66–92.

† E. R. Corey and S. H. Star, *Organization Strategy: A Marketing Approach*, Chapter 6, Division of Research, Harvard Business School, Boston, Mass., 1970.

‡ L. W. Ellis, "Effective Use of Temporary Groups for New Product Development," *Research Management* (January, 1979), pp. 31–34.

Jermakowicz found that the matrix design was most effective of three major organizational forms he studied in ensuring the implementation or introduction of new projects, while a "pure" project organization produced the most creative solutions.*

Kolodny, reporting on a study of his own and citing some other studies as well, comments on the effect that matrix organizations have on new product innovation.† Kolodny cites Davis and Lawrence, who point to an apparent high correlation between matrix organization designs and very high rates of new product innovation.‡ In his summary Kolodny notes that there is an apparent relationship between high rates of new product innovation, as measured by the successful introduction of new products, and matrix organizational designs.§

There is no question that an organization whose business involves the work of temporary projects is more anxiety-ridden, tension-filled, and demanding of personal competence and equilibrium than a stable, conventionally organized one. The matrix design is complex, yet its successful operation reflects a complementary mode of collaborative relationships in an open ambience. It is an adaptive, rapidly changing temporary management system that can favorably impact on organizational innovation.

### Realignment of Supporting Systems

As the use of project management grows in an organization it soon becomes apparent that many of the systems that have been organized on a traditional hierarchical basis need to be realigned to support the project team. What initially appears to be only an organizational change soon becomes something larger. Effective project management requires timely and relevant information on the project; accordingly, the information systems have to be modified to accommodate the project manager's needs. Financial and accounting systems, project planning and control techniques, personnel evaluations and other supporting systems require adjustment to meet project management needs.

---

* Wladyslaw Jermakowicz, "Organizational Structures in the R & D Sphere," *R & D Management,* No. 8, Special Issue (1978), pp. 107–113v as cited in Kolodny below.

† Harvey F. Kolodny, "Matrix Organization Designs and New Product Success," *Research Management* (September, 1980), pp. 29–33.

‡ Stanley M. Davis and Paul R. Lawrence, *Matrix* (Addison-Wesley Publishing Company, Reading, Mass.), 1977.

§ Kolodny, op. cit., p. 32.

### Development of General Manager Attitudes

An organizational culture is in a sense the aggregate of individual values, attitudes, beliefs, prejudices, and social standards. A change for these individuals means cultural changes. The matrix design, when properly applied, tends to provide more opportunity to more people to act in a general manager mode. With this kind of general manager thinking, the individual is able to contribute more to organizational decision making and information processing.

The matrix design with its openness and demands for persuasive skills provides an especially good environment for the manager-to-be to test his ability to make things happen by the strength of persuasive and negotiative powers. A perceptive general manager knows that there is little he accomplishes solely by virtue of his hierarchical position; so much depends on his ability to persuade others to this way of thinking. Exposure to the workings and ambience of the matrix culture brings this point home clearly and succinctly.

Effective collaboration on a project team requires plenty of a needed ingredient—trust. To develop this trust, individuals must be prepared to take personal risks in sharing resources, information, views, prejudices, attitudes, and feelings to act in a democratic mode. Not everyone can do that, yet executives in successful companies are able to do so. For example, in the Digital Equipment Corporation where a matrix environment prevails, the ambience is described as "incredibly democratic" and not for everyone. Lots of technical people can't stand the lack of structure and indefiniteness. In such an organization bargaining skills are essential to survival.* The matrix design is permanent—the deployment of people is changing constantly. In such a transitional situation the only thing that prevents breakdown is the personal relationships as conflicts are resolved and personnel assignments are changed. Communication is continually needed to maintain the interpersonal relationships and to stimulate people to contribute to the project team efforts.

## SOME CAVEATS

The matrix organizational design is hard to get started and challenging to operate. The more conventional the culture has been, the more challenges will emerge in moving to the matrix form. A few caveats are in order for those who plan to initiate and use the project-driven matrix design:

---

* Harold Sneker, "If You Gotta Borrow Money, You Gotta," *Forbes* (April 28, 1980), pp. 116–120.

1. Realize that patience is absolutely necessary. It takes time to change the systems and people who make the matrix work.
2. Promote by word and example an open and flexible attitude in the organization. Encourage the notion that change is inevitable, and that a free exchange of ideas is necessary to make project management work.
3. Develop a scheme for organizational objectives, goals, and strategies that will provide the framework for an emerging project management culture.
4. Accept the idea that some people may never be able to adjust to the unstructured, democratic ambience of the matrix culture. For these people a place in the organization must be assured where they can remain insulated from the "fiendishly difficult" surroundings of the matrix organization.
5. Be mindful of the tremendous importance that the team commitment plays in managing the project activities. Make use of the winning football team analogy where the commitment to win is an absolute prerequisite to becoming a championship team.
6. Provide for a forum whereby conflict can be resolved before the conflict deteriorates into interpersonal strife.
7. Realize that project management is not a panacea for organic organizational maladies. In fact, the implementing of a project management system will bring to light many organizational problems and opportunities that have remained hidden in the conventional line and staff organization.
8. Be aware that the particular route that an organization follows in its journey to the matrix design must evolve out of the existing culture.
9. Recognize that senior management support and commitment are essential to success.
10. Work for communication within the company that is uninhibited, thorough, and complete. Information requirements for project management require definition. Those individuals who have a need to know must have access to the information to do their job. Those in key positions have to understand and use the project-generated information.
11. Be aware that shifting to a matrix form is easier for the younger organization. For large well-established companies where a rigid bureaucracy endures, the shift will be quite formidable.
12. Institute a strong educational effort to acquaint key managers and professionals with the theory and practice of project management. Time should be taken to do this at the beginning using the existing culture as a point of departure.

## SUMMARY

The real culture of project management refers to actual behavior—those things and events that really exist in the life of an organization. The introduction of project management into an existing culture will set into motion a "system of effects" which changes attitudes, values, beliefs, management systems to a participative, democratic mode. Thus, a new cultural context for the sharing of decisions, results, rewards, and accountability will ultimately emerge as an organization matures in the use of project management.

## BIBLIOGRAPHY

Cleland, D. I. and King, W. R. *Systems Analysis and Project Management,* 3rd ed. McGraw-Hill, New York, 1983.

Corey, E. R. and Star, S. H. *Organization Strategy: A Marketing Approach,* Chapter 6. Division of Research, Harvard Business School, Boston, 1970.

Davis, Stanley M. and Lawrence, Paul R. *Matrix.* Addison-Wesley, Reading, Mass., p. 19.

Ellis, L. W. "Effective Use of Temporary Groups for New Product Development." *Research Management* (January, 1979):31–34.

George, William W. "Task Teams for Rapid Growth." *Harvard Business Review* (March-April, 1977):78.

Jermakovicz, Wladyslaw. "Organizational Structure in the R & D Sphere." *R&D Management,* No. 8, Special Issue (1978):107–113v.

Kolodny, Harvey F. "Matrix Organization Designs and New Product Success." *Research Management* (September, 1980):29–33.

Mee, John F. "Matrix Organization." *Business Horizons* (Summer, 1964).

Murphy, D. C., Baker, Bruce N. and Fisher, Delmar. "Determinants of Project Success." National Technical Information Services, Accession No. N-74-30392 (1974), Springfield, Va. 22151, p. 60669.

Reeser, Clayton. "Some Potential Human Problems of the Project Form of Organization." *Academy of Management Journal, 12* (December, 1979).

Seneker, Harold. "If You Gotta Borrow Money, You Gotta." *Forbes* (April 28, 1980):116–120.

"Texas Instruments Shows U.S. Business How to Survive in the 1980s" *Business Week* (September 18, 1978).

# Index

# THE INSTRUMENTS OF AMERICA'S FOREIGN POLICY

# THE INSTRUMENTS OF AMERICA'S FOREIGN POLICY

H. BRADFORD WESTERFIELD

YALE UNIVERSITY

Thomas Y. Crowell Company, New York

ESTABLISHED 1834

Library of Congress Catalog Card Number: 63-7119
DESIGNED BY LAUREL WAGNER
MAPS BY THEODORE MILLER
Manufactured in the United States of America
by Vail-Ballou Press, Inc., Binghamton, N.Y.

*For Carolyn*

# PREFACE

CONTEMPORARY AMERICA has a distinct fundamental foreign policy, one that is often challenged, periodically restyled, occasionally even subjected to major deviation, and frequently misunderstood by both its practitioners and its critics, yet one that has persisted at the core of the conduct of United States foreign relations since 1947. It is the policy of containment directed against totalitarian expansionism for the principal purpose of securing America's survival in freedom. In pursuit of containment, the country has been learning to utilize a variety of methods and instruments, including military, economic, informational, and undercover interventionist practices. This book is centrally concerned with analyzing these techniques of struggle as America wages the ominous conflict of competitive coexistence with the Communist powers.

Of course, United States foreign policy is not and should not be entirely preoccupied with the cold war. The country has additional interests and purposes in using its available instruments in world affairs. This book deals also with many of these other aspects, but its overall focus, which emerges most clearly as the initial historical chapters proceed beyond World War II, is upon the modes of coexistential competition as America has been experiencing them.

The interpretation begins with a survey of the fundamental historical traditions of the United States in her approach to international involvements, with emphasis on her unfamiliarity with protracted conflict; soon the study leads into exploration of each of the principal instruments of contemporary American policy abroad. Thus the organizing principle of most of the book is *functional analysis*, directing attention to the ways and means by which the United States can exercise her influence overseas, and using examples of recent historical episodes in widely scattered geographical areas for comparisons and judgments about the utility of the various policy instruments available. This functional approach is the most distinctive characteristic of the treatment of American foreign relations in this volume. Another distinctive element is the lengthy, respectful attention given to undercover techniques of United States influence upon the internal politics of foreign lands. I am convinced that the serious study of contemporary American foreign affairs must attempt much more earnestly than has been customary to penetrate the mysteries of that obscure but vital aspect of our international relations.

I have also tried to give ample coverage to the military, economic, and propaganda aspects of United States overseas involvements since World War II. Inevitably, such a broad reconnaissance must be somewhat preliminary and inconclusive; much that is important cannot yet be ascertained with confidence. And I have made deliberate omissions that may bother some readers. I have not, for example, attempted to treat *diplomacy* as an independent instrument of policy, since the diplomat's traditional roles of negotiator, reporter, and exemplary representative of his people and government seem to me largely participant functions involved in the direction and implementation of the other more distinct instruments of policy. If I were making an examination of administrative structures and domestic political arrangements that affect the conduct of American foreign relations, the role of diplomats would engage my attention further, but my intention in this book has not been to analyze institutions. Nor have I attempted to cover very thoroughly the behavior of the foreign contestants in the cold war. This book is principally about the external conduct of the United States in that struggle—a scope more modest than an exhaustive appraisal of the entire conflict, yet still so ambitious that perhaps at times its broad design may appear rather indistinct among the parts. Hopefully, despite the limitations, the reader will find here some useful clarification of what the United States has in fact been doing and will acquire a firmer foundation for judging what she might better be doing within the rather narrow range of discretion realistically open to the American government in the contemporary international arena.

Probably the widest use for this book will be found in college courses on recent American foreign relations and on contemporary problems in international affairs. I hope also that well-informed citizens outside the classroom may find the analysis enlightening and that professional specialists on United States foreign policy may discern a considerable amount that is new in it, particularly in some of the chapters on military affairs and on interventionist activities. Although the experts may regret my decision to omit bibliographical citation of most of the narrow primary sources, articles, and foreign-language books I used for my research, I decided to adopt as the principle of selection for the bibliographies their utility for the general informed reader, who may benefit most from competent secondary works and some carefully picked primary source books.

My indebtedness to other researchers is only part of what I owe to all the friends who aided me in the preparation of this volume to an extent that can be acknowledged but that can never be fully repaid: the specialists who read parts of the manuscript; the Yale undergraduates on whom I tried it out in lectures, especially two of them, Emanuel Demos and John Jeavons, who cheerfully fulfilled important editorial responsibilities; my faithful typists and our department secretary, Veronica O'Neill, who cordially eased my path in manuscript preparation; the administrators of the Henry L. Stimson fund, which contributed support; Herman Makler who diligently edited it with felicity and understanding; Martha Arnott who prepared the index and Theodore Miller who rendered the maps, both of them with resourceful craftsmanship; John T. Hawes, in my publisher's office, whose trust and tolerance of my delays was vitally reassuring; and my family with its fortifying confidence and assistance—especially my wife Carolyn, who has endured this book in manuscript since first we met and without whose patience and buoyant spirit it might never have reached publication.

The contribution that friends have made is best appreciated by me as author; the final product, which remains my responsibility, is best judged by the reader. I can only hope that most of you who peruse the book to the end will be willing to conclude with me that the initiation, since 1940, of reluctant, immature America into an agonizing adulthood as the principal guardian of Western civilization has indeed evoked a potent resourcefulness in the United States—such that despite her many mistakes and failings there are sober grounds for persevering, without becoming despondent or desperate, in the protracted perilous struggle for survival in freedom.

H. B. W.

*New Haven, Connecticut*
*June 1962*

# CONTENTS

## Part Two: Foreign Policy and Military Strategy since 1940

## Part Four: Economic Relations and Foreign Policy

## MAPS

# INTRODUCTION

"BETTER RED THAN DEAD."

Ominously, the infectious defeatism drifts across the Atlantic and begins to insinuate itself into the mind of America.

Less than a generation ago many responded with a similar kind of resignation to the claim that "you cannot sustain the Past against the wave of the Future." World War II proved the contrary with regard to Fascist totalitarianism. But the war itself and its outcome, both in political geography and military technology, prepared the way for a titanic confrontation of Western constitutional democracy and Communist totalitarianism—a mounting struggle that by now has truly set at stake the future of civilization and the physical existence of most of mankind.

And by an irony of history the only country that emerged from the war strong enough—unscathed enough—to be the champion of antitotalitarianism in the new contest was the one which of all the major powers had probably been the least prepared by her historical experience to understand the requirements of any kind of protracted conflict: the United States of America. A century of virtual invulnerability before 1940 had left her with illusions ill suited to the competitive environment of most of the world. The other major powers, lacking America's unique good fortune, had had to become accustomed to the idea that at least their borderlands would usually be violable and insecure. The United States had not. Of course, the residuum of that same invulnerability was what left her so unscathed by World War II that she was strong enough to undertake the resistance to Communism thereafter. But then she found herself in an environment that had become much more relentless even than what the other major powers had previously experienced. America was constrained to learn almost simultaneously to play a persistent leading role in the world, to accustom herself to the protracted vulnerability of borderlands that she would have an interest in protecting, and—the dimension that had no modern precedent—to strive unflaggingly to avert real dangers of her own physical annihilation or of her total surrender.

This was an environment in which even peoples more accustomed than Americans to living with danger would tend to shrink and to pose the alternatives "Better Red than dead." The United States, buoyed by a congenital optimism derived more from her past good fortune than from logical calculation of her developing prospects, was bound to be slow to digest the perils and to conceive of any destiny so bleak. But if a full rational awareness of the massively impending dangers ever broke through the shell of overconfidence, this sanguineness might prove to be brittle. More and more Americans might conclude in panic that resistance was hopeless, and that it would be better to submit and be swept along with the new wave of the future.

One must anticipate that there will be increasingly keen debate in the United States on whether it would be better to be Red than dead, that is, whether it would be better for America to submit to Communism than to engage in a war of mutual annihilation with Russia, or perhaps even than to run very great risks of such a war. It is easy to respond that these are not likely to be the real alternatives. American society, particularly the leadership strata in every walk of life, might find itself both "Red" in the sense of having surrendered and then also soon dead, executed because not regarded as trustworthy by the Communist rulers. Conversely, there is certainly at least a theoretical possibility of remaining neither Red nor dead—a possibility of surviving in freedom.

One can assume that the latter would continue to be to some degree the preference of nearly all Americans, if it is indeed a realistic possibility. So the question takes on a fuller contour: how much of a preference, set against how small a chance of achieving it.

It is probably true that most Americans, perhaps even those in positions of leadership, would greatly improve their chances of physically surviving in future years if they submitted to totalitarianization, relinquishing the attributes of an open society and a government that can be held responsible. But perhaps they would also be able to improve their present prospects of surviving—this time in freedom—if they could manage to improve their country's performance in the competition of "peaceful coexistence." And perhaps they can continue to care so much about freedom as to be willing to take enormous chances in the marginal gap of danger which no foreseeable improvement in performance would suffice to close.

What then is this freedom that is to be treasured? Freedom may be defined as the existence of real opportunities for making uncoerced choices among significantly different alternatives. The choices may be made by governments of nations, by leaders of groups, and by families and individuals.

In the first sense, the demand for freedom is equivalent in the modern world to the demand for national independence, for decolonization, for the removal at least of the most direct kind of coercion, that is, military occupation by foreign troops.

Pursuing freedom further in the sense of choices exercised by groups, families, and individuals, the free society learns to put special emphasis on the autonomy of the groups, because the individual regularly needs the support of a group in whatever course he has chosen, or might in the future choose by switching his group identifications; he cannot be expected to be so utterly self-reliant as to be able to persist in nonconformity if he finds himself as merely one of many socially isolated individuals confronting a single center of authority, the state regime. Thus meaningfully free choices, both those made by leaders of groups and those made by individuals among group identifications, depend on the existence of a multiplicity of relatively autonomous associations in a society, depend in other words on the existence of a pluralistic society, a society that is not monistic, not totalitarian, not all-embracing in the range of the active concern and penetration its government exercises within it.

There is truly some meaning then in the West's slogan "the free world" even when it is used to embrace almost all the diverse states that are not Communist. Even the military dictatorships are usually pluralistic. The free world may be thought of as a world of relatively independent, nontotalitarian states.

But why desire this freedom?

Of course, not all the leaders of free-world countries do desire it (not to mention the opposition elements within those countries), and certainly those that do do not all desire it for the same reasons for which the Western democratic nations have come to value it. In the West the ideal grew out of themes in the merging Judeo-Christian–Greco-Roman tradition that tended to issue together in an emphasis on the self-realization of the individual, the unfolding of his own unique personality, with his own unique potentialities for self-improvement through experience, possibilities that were not to be contained within anyone's foreordained precise prescription for the Good Life. And for society as a collectivity, the emphasis came to be placed on the virtues of creative innovation, sheer spontaneity, the element of the unplanned (as also for the individual) and with it the element of the reversible. Neither of these—the unplanned spontaneity or the undictated reversibility—is compatible with the essence of full totalitarianism.

The value of reversibility, in particular, is one that should be impressive to very large segments of the existing free world, not only to those prosperous industrialized democracies that can associate material well-being with freedom and thus find additional reasons to treasure it. The choice of totalitarianism in the modern world, once made in any country, has been found to be virtually irreversible. The advanced technology of repression that in the twentieth century for the first time in human history has become available to rulers who are determined to remold personality and society throughout their territories is so pervasive that one should realistically recognize it has become almost impossible to destroy an established totalitarian regime except by external war; once consolidated, the control will endure, at least for a great many years, until perhaps its zeal and repressiveness mellow from within.

A country cannot experiment in choosing totalitarianism. The choice once made, whether by free elections, or civil war, or revolt, or *coup d'état,* will almost certainly not be reversed: not by free elections, or civil war, or revolt, or *coup d'état.* And very probably no nontotalitarian outside power will venture deliberately to reverse that rooted decision by initiating an invasion.

Free-world peoples who may differ on nearly everything else by the very essence of their pluralism may perhaps be persuaded that it remains desirable to avoid any irreversible choice among potential regimes—that their existing freedom of choice as groups and governments, however small or large it may be, should not be used to favor totalitarianism. Freedom, if ever surrendered to Communists on the bare possibility that their tyranny would be mild, is not soon likely to be recovered if it turns out to be oppressive. Shades of "redness" in any subjugated country, even America, would be mostly within the discretion of the new masters; the color would not be erasable, and it would take many years to fade.

But, of course, the rejoinder may be made that "deadness" too is irreversible, and that though an individual may choose it for himself, he is doubtfully entitled to choose it for most of mankind. So in proportion as the chance of preserving freedom, while still enabling the vast majority of mankind to survive physically, is judged to be smaller and smaller, more and more of those persons whose attachment to freedom is not very zealous may be expected to favor making the choice of submitting to Communism on the gamble that it will not be too oppressive.

This is the state of mind that must be forestalled. But it is not likely to be averted fully by programs of public education and indoctrination about the virtues of freedom and the evils of totalitarianism, if only because in a free society the program of instruction itself will always be bound to be the subject of confusing controversies. What is most needed is some improvement—or at least some halt in the decline—of the reasonable prospects that survival in freedom can continue, that the chances for it are good enough to be worth the enormous exertions and dangers it requires, and that those chances can be improved, or at least be prevented from getting much worse, by means of the exertions. And this need, in turn, means a need for improvement in the effectiveness of the West's instruments of policy in the interim contest of "peaceful coexistence," with a better understood comprehensive strategy for their coordinated use and with a quicker process of learning and applying the lessons of experience in their tactical utilization. In capsule, to put up a better fight in the cold war in order to avert a total hot war or surrender.

But a better fight does not mean simply a fiercer fight and a bigger fight, with more and more of everything: more missiles, more bombers, more megatons, more submarines, more soldiers, more airlift and sealift, more trade and more aid (military and economic), more propaganda and more undercover operations, more aggressiveness and more implacability. All of these instruments and attitudes may in part and at times be required, indeed, perhaps more and more. None is beyond the existing pale of the coexistential competition. But the implications of all of them for the course of that competition need continuous study and review. And that is what this book attempts to give, beginning with an interpretation of the traditional pattern of Americans' responses to international entanglements, the time-worn attitudes inculcated by this country's privileged historical experience that still largely shape her conduct in today's less congenial world.

**PART ONE**

# THE HISTORICAL UNPREPAREDNESS OF THE UNITED STATES FOR INTERNATIONAL LEADERSHIP

# CHAPTER 1

# THE HERITAGE

In the modern era with its constantly accelerating technological change and its economic and social dislocations, the characteristic experience of those men and women who chose to become Americans was *escape.* Emigrating from Europe, they postponed or abandoned a struggle for physical well-being or spiritual betterment in the land of their birth. They faced the primitive hardships of an ocean crossing and the cold inhospitality of alien cities or empty prairies. But they decided not to take their stand to the bitter end at home, in the strife of classes and nationalities, of religions and ideologies, by which centuries-old societies were painfully reconstructing themselves to utilize the unfolding opportunities of discovery and invention. Rather, the new Americans, generation after generation, quit Europe to seek the opportunities for themselves abroad, by paths which however long seemed less obstructive to the pursuit of happiness than those at home, and through gateways which however far away seemed wider open.

## Escape from Europe

Northern and western Europe passed first through the stage of transformation which replaced the feudalistic peasant village in favor of the commercial tenant farm, and for the handicrafts of the artisan's shop substituted the standardized products of the factory. This upheaval took place to the accompaniment of an astounding increase in total European population, and the uprooted casualties numbered millions. They began to surge across the Atlantic in large numbers when a relatively peaceful international environment succeeded the Napoleonic wars. The century which followed saw ebbs and flows in the tide, but never was it reduced to a trickle except when America or Europe was fighting a major war—until the immigration acts of the 1920's finally erected an almost insurmountable and indestructible dam. The figures are impressive. In the decade 1847–57, when the United States population totalled about 25 million, more than 200,000 immigrants reached these shores every year, 400,000 in 1854 alone. Before the end of the century that figure of 400,000 was surpassed many times, reaching almost 800,000 during 1882. In 1900, when the total population had reached 76 million, the census takers calculated that fully a third of the nation consisted of foreign-born persons or their children, reflecting the movement which had brought almost 20 million newcomers to America during the previous eight decades. The annual number of immigrants continued to increase, sometimes passing a million a year shortly before the outbreak of World War I (1,285,000 in 1907). After the war the totals rose rapidly again; more than 800,000 immigrants in 1921 mingled with the 105 million Americans the census had counted the year before. Then, however, restrictive legislation rapidly began to take effect and by the end of the decade annual transatlantic immigration was down below 200,000. Never again would it rise much higher than this, although the population of the country as a whole has since grown by more than half.

Almost to our own day, then, it is clear that the experience of escape from Europe was of profound importance in the outlook of a very large proportion of our people. The later generations who were not touched by it in their own immediate families were nonetheless affected in complex fashion through their reaction to the presence of the newcomers.

Typically, the adults in an immigrant family clung for many years to a hope of preserving in America as much as possible of the culture of the communities they had known in Europe—perhaps even returning there if savings could ever be accumulated in the new country. Their rejection of Europe was far from complete, and they felt no inner compulsion to conform to all American ways. Of central importance was the maintenance of the old religion with its peculiar local forms. With this went a desire to be able to continue to communicate effectively in a familiar tongue. The foreign-language press and the development of nationality distinctions among churches met some of these needs. Actually, these institutions often strengthened or even created a sense of identification with a particular European nationality, which many of the immigrants had never felt before they reached America. Since most of the immigration from central and southeastern Europe took place before modern nationalism had swept the masses of those countries, their sense of identity had been with a village or a province. In America such splintered divisions were too small to recreate. Besides, older Americans could not understand such loyalties, whereas everyone would understand nationalism. Its stirring among the immigrants was thus in large measure a defense mechanism, giving a comforting sense of group identity in face of the multiple challenges of the alien environment and, one could still hope, helping the foreign-born to salvage something of the ways of their old community.

The intensification of nationalism in Europe itself during the late nineteenth and early twentieth centuries echoed across the seas among increasingly self-conscious nationality groups; they split and realigned themselves in accordance with changes, real or hoped for, in the map of Europe. Leadership in such groupings was typically in the hands of the sons or perhaps the grandsons of the immigrants themselves; they had learned enough of American ways to be able to manage organizations. It was a way for them to get ahead and make a name for themselves. For many of them it also seemed a way to facilitate for others the adjustment they themselves were particularly eager to make—between the immigrant ways they knew at home and the new world they found at school; wasn't the multiplicity of group organization already a fundamental characteristic of the native American scene? Among the various language groupings, however, such organization tended to heighten divisive rivalries, especially as these were compounded by echoes of nationalist struggles in Europe, with each nationality there enlisting the support of its "compatriots" here and seeking through them a benevolent attitude in United States diplomacy. The resulting pressures upon American foreign policy have always been very important.

Another effect of this situation was to raise disturbing, though generally unjustified, doubts in the minds of "older" American families as to where the ultimate loyalties of the newcomers really lay. The doubts were heightened in the context of World War I, when a quest for solidaristic uniformity in the United States ran athwart the aspirations of some ethnic organizations that were concerned about the future map of Europe. There was also a tendency for many of the sons and most of the grandsons of each successive wave of immigrants to identify themselves eagerly with prevailing American patterns, in conscious rejection of the ways of their forebears. Here the sense of "escape" was most profound. These second and third generation Americans would often outdo the established families in resentment toward the "un-American" behavior of newer arrivals from different national origins; besides their unfamiliar appearance and speech, the newcomers' competition for jobs seemed to depress wages for all.

Till the Civil War era the cost of an ocean voyage was such that most of those who could afford to escape to America were farmers and artisans from above the poorest levels of European society. It was not very difficult for such persons to accept the prevailing American creed of individualistic self-reliance in an intensely competitive society, where there was fairly easy mobility up the socioeconomic scale—but also down it—across huge geographical areas. Wealthy European aristocrats, on the other hand, rarely felt an incentive to emigrate to America and never succeeded in transplanting a durable feudalistic hierarchy or its principles to the United States. This twofold sifting-out of the poorest and richest classes in the transatlantic migration helped to produce in this country an extraordinary unanimity of basic "middle-class" ideology, which was further universalized by the destruction of the deviant slave society through the Civil War. In later decades, however, with the great cheapening of steamboat fares, a more poverty-stricken class of immigrants surged across the ocean, especially from southern and eastern Europe, and there was rising fear among native Americans that these families would not adopt the traditional ways of the United States as readily as those who had come over in an earlier era. The cry went up that the "melting pot" would never be able to melt down the differences and create a homogeneous nation if new types of peoples were constantly being added in large numbers. Critics also reiterated that ethnic loyalty conflicts were jeopardizing the pursuit of America's own national interest in foreign affairs. The solution they urged was one which came readily to mind in the United States—a strategy of isolation. If the cherished American society had been so largely shaped by the fact of its remoteness from European peoples, then to keep it safe, the remoteness should be re-established as much as possible by the imposition of drastic legal restrictions on immigration. The old escape from Europe and Europeans was to be reinforced, indeed, redoubled. During the 1920's and thereafter enough citizens were persuaded of the need and feasibility of such a project to permit the passage of the severely restrictive laws which still form the basis of our immigration policy. America would develop her unique society unaffected by any further substantial direct admixture of differing cultural patterns from abroad.

## Isolation and the Balance of Power

Many other peoples have tried to preserve their own culture and escape the difficulties arising from connections with foreign nations by simply severing those connections. Even Britons in their nineteenth-century epoch of world power gloried in "splendid isolation." But few nations in modern history—and no great power except perhaps Russia—have been so favorably situated as the United States to effectuate the inclination to live apart from alien influences. "God takes care of drunks, fools, and Americans," snorted Kaiser Wilhelm; the country has always been strongly aware of its extraordinary good fortune.

The most obvious element and the one most often emphasized is the fact of geographical isolation, thousands of miles of blue water between the United States and the great powers of Europe. It was on this consideration that George Washington rested the famous injunction in his Farewell Address:

> Europe has a set of primary interests which to us have none or a very remote relation. Hence she must be engaged in frequent controversies, the causes of which are essentially foreign to our concerns. Hence, therefore, it must be unwise in us to implicate ourselves by artificial ties in the ordinary vicissitudes of her politics or the ordinary combinations and collisions of her friendships or enmities. *Our detached and distant situation invites and enables us to pursue a different course.* If we remain one people, under an efficient government, the period is not far off when we may defy material injury from external annoyance. . . . *Why forego the advantages of so peculiar a situation?*

To most Americans the question seemed unanswerable, at least as long as we could (under Washington's prescription) "safely trust to temporary alliances for extraordinary emergencies," steering "clear of permanent alliances," and yet still be able to "defy material injury from external annoyance." There has been one major line of contrary argument—that such a policy even if safe was excessively selfish. But for this view Washington also had an answer, which has echoed throughout our history: "It will be worthy of a free, enlightened, and at no distant period a great nation to give mankind the magnanimous and too novel example of a people always guided by an exalted justice and benevolence." To be part of a people whose life would be a novel example of goodness—this was the invitation which America's distant location extended. And the same distant location would enable her to pursue such a goal independently.

Geographical isolation was not, however, actually so complete in itself as to provide a shield for the republic. Obviously the defensive usefulness of geography was a function of the existing state of naval (and later aerial) technology; protection would decline if there were any marked advance of offensive over defensive systems of weapons and means of delivering them across the oceans. Washington emphasized that we ought to take "care always to keep ourselves by suitable establishments in a respectable defensive posture." He was well aware too (though unfortunately the Farewell Address did not stress the point) that the "frequent controversies" of Europe, "the causes of which are essentially foreign to our concerns," could in their consequences be exceedingly advantageous to the United States. The American colonies had resented the hardships caused by their being drawn repeatedly into general European wars in the seventeenth and eighteenth centuries; this was a prime reason for desiring isolation. But independence itself had finally been won only because of the age-old rivalry between Britain and France, and independence could most easily be preserved if, keeping her hands as free as possible, the United States was able to take advantage of similar European rivalries in the future. Indeed, except for such rivalries there was practically no hope for the young republic. Three thousand miles of ocean would have been no defense against a united Europe bent on subduing the States and carving them up, or even against a single great power if that power could have been freed of the restraining fear of being stabbed in the back when it became occupied with transatlantic operations. It follows that the ability of America to go her own way safely in relative isolation was actually dependent upon the continued existence of a division and balance of power in Europe. Otherwise, the United States would have had to rely on some super-power's dubious willingness to exercise permanent self-restraint. In short, the very struggles Americans wanted to escape made that escape feasible.*

In the first decades of the republic, when it was obviously weak, many American statesmen showed sophisticated sensitivity to shifts in European power politics, and often exhibited a capacity to utilize adventitious opportunities for the promotion of American interest; a notable example was Jefferson's acquisition of the Louisiana Territory from

* Compare how the remoteness of America helped to sift her early immigrants to produce a fairly homogeneous population with a distinctive way of life which their descendants wished to preserve inviolate, and then how that same remoteness helped them in efforts to do so.

Napoleon. But the Jacksonian Revolution ushered in nearly a century in which Americans felt so securely self-sufficient that they generally forgot the importance for them of the balance of power in Europe. The maintenance of a balance was, in fact, exceptionally easy in that period. Britain was effectively performing the role of chief balancer, but actually no great power was determined anyway to seek the full destruction of another. National rivalries were for limited objectives, and therefore the occasional wars could be kept limited with relative ease. The world economy was rapidly expanding, and whole continents in the Eastern Hemisphere were available for fresh colonization and partition by the European powers. Empire-builders could expend their aggressive energies for the most part in Asia and Africa. Fortunately for the United States, in the Western Hemisphere Great Britain had decided to content herself with economic penetration; she did not intend to expand her areas of direct political control—and correspondingly she did not wish to allow other powers to expand theirs, to the possible detriment of British economic interests. If only the markets of the Western Hemisphere could be kept open to all, Britain's industrial superiority would assure her an advantageously profitable position there; at the same time she would be able to avoid the costs and the hazardous balance-of-power vulnerability which might result from any political and military overextension in the New World. Thus the country which was in the best position to overcome the geographical defenses of the Americas—thanks to her pre-eminent sea power—was in effect willing to underwrite the Monroe Doctrine itself against other European powers, and even to make concessions when the Doctrine was asserted against her own aims by bumptious Americans.

Britain was actually providing vital protection for the United States. On the surface, however, there still remained enough friction between the two countries to keep most Americans from becoming conscious of that fact. They simply ascribed their safety to geography and did not look beyond the oceans for the power factors which temporarily made remoteness an effective defense. Consequently, at the turn of the century the United States was extremely slow to recognize how her own security was affected by the rise of Imperial Germany, which was becoming a unified industrial power, potentially strong enough to dominate the Continent to a greater degree than any country had done since Napoleon was destroyed—and hence, perhaps, was ultimately likely to win for herself a free hand to move across water even against British and American interests. Conservative Imperial Germany might show self-restraint in the use of her power overseas, but could the policy of the United

Kingdom and the United States be safely based on confidence that she would do so?

Confronted with this situation, English statesmen had but to renew an immemorial prescription of national policy: to shore up the weaker Continental powers and thus try to maintain a balance. But even the English statesmen were afraid that masses of their own people did not recognize the need for such a policy and would not willingly support it at the cost of war; consequently, many of them felt considerably relieved when Germany violated Belgian neutrality at the outbreak of World War I and thus provided Britishers with a rousing moral justification for joining the Continental struggle. As for the United States, its traditional foreign policy had never recognized even in principle the need to help maintain a balance of power in Europe. Consequently, it was even more difficult for Americans than for Britons to take themselves into war for such a purpose. In the end Americans did not do so consciously. They went to war in 1917 for legalistic and idealistic principles and emerged with almost as little understanding as when they went in of the importance of the balance of power for their national security.

## Territorial and Commercial Expansion

Geographical remoteness plus a strong navy (how strong was debatable): these were still seen as the basic defenses which would permit America's unique society to avoid the entanglements with Europe which it so strongly desired to escape. The policy of noninvolvement had been extraordinarily successful, for whatever reason. Not only had independence been preserved, but an empire had been won, stretching across a continent and an ocean halfway around the world—all this gain at a cost so slight as to leave Americans understandably doubtful that there was any necessary correlation between diplomatic success and military power. Some adroit fishing in the troubled waters of European politics, two cheap "mopping-up" campaigns against minor powers (Mexico and Spain), some spasmodic guerrilla warfare against primitive natives so few and far between that even their virtual annihilation was scarcely noticed to be an act of imperial conquest—these (plus the much more serious but largely forgotten War of 1812) were the means whereby the United States established her vast territorial possessions. The power factor was almost invisible. On the North American continent there was no rival of equal power potential; and after 1815, for nearly a century Great Britain's enlightened self-interest tacitly provided that the United

States would have no dangerous rival from across the seas. The continent was almost a power vacuum for Americans to expand in; decade after decade they recognized their "manifest destiny" to do so.

Specific motives for expansion included commerce and navigation, frontier security, a quest for living space for future generations of Americans, and a desire to spread democratic culture. Such considerations may be recognized as local variations of justifications for imperialism which were also common in Europe. The great difference was that the lack of external opposition usually made it possible for Americans to pursue these objectives without becoming really conscious that they were acting imperialistically. In attitude they were merely exercising "squatters' rights" all across an "empty" continent. It was possible for three and one-half million square miles of North America to be acquired by a people which still remained anti-imperialist in principle and continually reminded itself of its anticolonial heritage. The inconsistency could partly be reconciled by the doctrine that United States continental acquisitions were to be established as self-governing territories and ultimately would be admitted to equal statehood. Historically, too, American anticolonialism had been bound up with the idea of separation from Europe; the westward movement (though some might well describe it as a "course of Empire") was in one sense part of the same escape. Therefore, the resemblance to imperialism tended to be blurred in Americans' minds.

The Spanish-American War, however, for the first time left the United States with an overseas empire of the European type, in scattered and distant islands which were obviously not "empty"—where there were, in fact, large populations with cultures so different as to raise grave doubts whether they would ever be assimilated to full equality in American statehood. Furthermore, possessions as distant as the Philippines would not promote continental security as previous acquisitions had done, but rather would create new security problems and tend to draw the United States into entanglements like those it had so long and so successfully sought to avoid. The world was round, and one could not "escape" westward indefinitely without landing back in the toils of Eurasian controversies. In this situation Americans had to face up to some of the inconsistencies in their expansionism. Was it their "manifest destiny" only to fill the vacuum of a fabulously safe and wealthy continent, to develop there a unique society almost self-sufficient and invulnerable? Or should they enter upon the responsibilities and opportunities of world power and take up the "white man's burden" of administering and enlightening less fortunate peoples? The missionary impulse was strong, and American attitudes toward involvement in Asia were more open-minded than toward European entanglements. There was less emotional fixation that perilous embroilments and contamination would surely result, at home and abroad (very few Americans had emigrated from Asia). Nevertheless, the decision to keep the Philippines was taken only over great opposition and was never generally accepted as a permanent arrangement. The incompatibility with traditional principles of isolationism was too disturbing.

Yet it took many years to prepare the Islands for self-government; and meanwhile the United States did become, to a varying degree, a conscious participant in that regional balance of power and did partially and indirectly learn to accept a definition of vital national interest which would prevent any single power from dominating all of East Asia and the Far Pacific. Limited American commitments led ultimately to unlimited war, and to the elimination of the Japanese menace—but only at great cost in blood and treasure. It was always tempting for Americans to look back nostalgically on the halcyon days when they had been able to master a continent almost without a fight and had reaped the benefits of a favorable balance of world power without having to throw even their material abundance into the scales.

There was another privilege which United States citizens could utilize in the nineteenth century without having to fight very hard for it. That was access for her traders to most of the markets of the world. American faith in the virtues of private enterprise was indestructible, and the endorsement was extended almost indiscriminately to overseas commerce and navigation. Private business was not regarded as dangerously "entangling." In the Farewell Address, Washington himself defined the accepted distinction: "The great rule of conduct for us in regard to foreign nations is, *in extending our commercial relations,* to have with them as little *political* connection as possible." From his day forward, one of the foremost objectives of American diplomacy, usually ranking second only to national security considerations, was the protection and promotion of the business ventures of United States citizens abroad. This took the form of widespread representation by American consular staffs in foreign countries; the negotiation of reciprocity treaties securing privileges for Americans in a foreign country in return for equivalent benefits for that country's nationals in the United States; if possible, the universalization of such advantages under the "most-favored nation" principle, whereby each country would pledge to extend to the other any additional privileges which it might later decide to grant to any outside nation. Occasionally American trade was also promoted by armed force. With rare exceptions no questions, testing the value of a particular business enterprise by any criterion of high national interest, were asked, and until the twentieth century few efforts were made to induce companies to undertake special new projects abroad for the sake of a national policy objective. The mere existence of private initiative was sufficient justification for its protection, along its own spontaneous course of development. For this purpose, for example, Jefferson and Madison became involved in minor undeclared wars with the Barbary pirates of North Africa, and in later decades disciplinary campaigns were undertaken against unruly elements in the Caribbean.

In general, however, American overseas commerce was remarkably free to expand in the nineteenth century without requiring the application of much United States force to smooth the way for traders and investors. Great Britain was continuing to "paint the map red," and her own advantages in competitive economic power were still so great that she could afford to leave all her vast domains open on a relatively nondiscriminatory basis to businessmen from all countries, including Americans. Her naval power also helped to keep Latin-American commerce open to all comers. Under such leadership trade barriers shrank throughout the world: even "protectionist" countries were "free traders" by our present-day standards. Differentials of price and quality were allowed for the most part to determine the movement of goods and services from country to country and from continent to continent, without much political interference, even when the resulting adjustments were painful. International trade expanded enormously, and Americans had a sizable share of the business.

In time of war between other countries, however, it was understandably less easy for the United States to insist upon the freedom of her citizens to do business anywhere they pleased, even with belligerents. The doctrines of freedom of the seas and traditional neutral rights could not in fact be reconciled with military necessity as seen by the warring powers in a major conflict. Both in the case of the Napoleonic wars and in World War I, American efforts to go on doing business-as-usual contributed mightily to ultimate embroilment in the fighting, and at the same time drew attention away from considerations of national security and the balance of power which would have more far-reaching significance.

One other major policy, which was largely designed to promote American business abroad but which led to serious political involvements and ultimately contributed to war, was the principle of the Open Door in China. This objective was adopted at the time of the acquisition of the Philippines, and both moves were strongly influenced by grossly exaggerated visions of the profits that Americans could win in oriental trade if only the great powers could be prevented from carving up China into discriminatory imperial preserves and the door be kept open to all traders and investors equally.

Business realities were soon disillusioning; United States trade with China was never more than a small fraction of American trade with Japan. But once involved, the United States found new reasons for continuing to espouse the "administrative and territorial integrity of China." There was a vague awareness of the importance of a balance of power in East Asia for the defense of the Philippines, plus an increasing sentimental regard for the long-suffering Chinese people, among whom many American missionaries were striving. Even after 1911, when "China" became more a geographer's label for a region of anarchy than the name of any effectively governed nation, the United States continued to contend that the great powers should keep hands off, for decades if need be, until the waves of revolution finally beat themselves out and a new regime emerged dominant nationwide. But American business prospects in China itself had ceased to furnish more than a small stimulus to the persistent vitality of this doctrine. United States traders and investors preferred to take advantage of opportunities in stabler countries, and the world was full of them—for example, Japan. A market that conspicuously could probably be held only by military force was not attractive to private businessmen from a country that had learned to believe that business growth should be achieved with a minimum of political entanglements, abroad as well as at home. In the international environment of the nineteenth century this approach had been extraordinarily productive. It was naturally very difficult for Americans to bring themselves to recognize how exceptional those favorable circumstances had been, and to make the painful adjustments needed for economic and political survival in the more crowded and fiercely competitive world of the twentieth century.

## The Absence of Class Conflict

The remarkable degree to which the United States had been able to grow and prosper in the nineteenth century while keeping detached from the struggles of Europe left yet another area of political innocence which ill-fitted her to understand the deeper tensions of the new age. America had never experienced a great social revolution. Her physical environment permitted continuous expansion and abundance. The early waves of settlers who established the nation were not drawn from the most impoverished strata of European society: their dream was not to transform an existing society but to move away and help found a new one. The American Revolution was a War of Independence, a political separation, rather than a social upheaval like that of France and other countries. The cleaning out of the Tories was so easy and complete that it left almost no scars in American minds. From this experience, recounted in United States history books, our people could not acquire lasting sympathy for social revolutionaries, only for national independence movements. American sympathies were lastingly fixed in favor of the latter, sometimes being blinded thereby to other vital considerations, for example in fostering the multiplication of tiny weak nations in eastern Europe after World War I. But the "virtues" of violent social revolution, which have appealed to tens of millions of people in almost every other part of the world, have remained beyond the ken of Americans. The closest the United States came was in the Civil War, with the extirpation of slavery in the South. But for most of the country this regional overthrow represented little more than a universalization of the status quo. It was halted far short of a complete transformation even of Southern ways of life. Moreover, the issues were envisaged essentially as national unity and Negro emancipation, not as class conflict between whites. The persistent race issue has certainly helped to sensitize Americans to ethnic conflict elsewhere in the world, but the idea of class struggle has always remained for nearly all of them a suspect bit of rhetoric rather than the profoundly stirring image of reality it has been in other countries. Correspondingly, the ideologies of Left and Right which grew up around true class conflicts in the rest of the world have been practically incomprehensible to Americans, and have therefore generally been dismissed as ridiculous or shunned as contaminating. As long as the United States could escape from them, she did not need to understand them and did not try to. Indeed, it was scarcely necessary for native Americans even to be able to defend their own way of life articulately and comprehendingly since they almost never heard it basically challenged. "We hold these truths to be self-evident." They must be; who at home was arguing against them?

The time would come when the United States could no longer stay out of earshot of the clangor rising across the seas from people who had not been so blessed with the absence of conflict of class and ideology. Could Americans, then, ever really acquire sympathetic understanding of experiences they had escaped from sharing? Could the other peoples ever really understand Americans? Was true communication possible?

If ideological persuasiveness in the outside world came to be essential for the very security of the United States, would she be doomed to sound inarticulate and incomprehensible? These were troubling questions with which American policymakers did not need to concern themselves until the nineteenth-century Age of Innocence was over.

### Idealism and Apathy

By the beginning of the twentieth century the United States had become both a great power and a satiated power. Her economic growth and social harmony were phenomenal; she had rounded off her continental and mid-Pacific empire and become extremely wary of further expansion in the Far East. Naturally she was disposed to look with favor, if not with deep understanding, on the broad framework of international relations that had permitted her to develop so remarkably without involvement in a major foreign war for many decades. Impressive also was the absence of general conflict even in Europe itself since 1815. The vital workings of the balance of power and the new dislocations being caused by the rise of Imperial Germany were obscured for Americans by their own noninvolvement, by the benign effects of British policy, by the rich unhampered flow of private international commerce, and by the spread of democracy which would tend to limit the waging of deliberate aggressive war. It was easy to believe that other powers could now be as satisfied with the essential status quo as the United States was; that they could be content to live in peace with one another, continuing to substitute private economic competition for national power struggles, and increasingly accepting the application of agreed principles as rules for the settlement of international disputes rather than resorting to the arbitrament of force—at least not to general multilateral war.

The United States was not, of course, the first great power in history to reach the point of satiation and then become enamored of international law and peaceful change. But other great powers have at least had to fight hard to achieve that exalted status in the first place, and may therefore have been somewhat less inclined to believe seriously that they could expect to keep it without continuing to fight hard against new challengers. At any rate, from the beginning of the twentieth century forward, American foreign policymakers and the segments of the public which paid attention to diplomacy were notably preoccupied with a succession of projects to systematize the existing international order by various formal codes and institutions, provided always that the likely cost to the United States could be kept at bargain-basement levels. The attempted devices included numerous international treaties for the conciliation and arbitration of disputes, frequent prolonged negotiations for mutual reduction of armaments, the Kellogg-Briand promise on the part of all nations in 1928 never to resort to war, and schemes of international organization with differing degrees of military commitment on the part of the United States. Of the projects which came anywhere near fruition before World War II, only the League of Nations would have been potentially costly to Americans; they rejected it. The value of the other attempts was magnified by their supporters out of all proportion to what could reasonably be expected to be achieved by such inexpensive means in dealing with countries that were not so satisfied with the international status quo as the United States was.

With the possible exception of the League, practically all these projects placed excessive confidence in the potentialities of international law and international organization to be somehow self-enforcing. Insufficient attention was given to the ultimate need for military sanctions. Verbalizations, especially high moral affirmations or condemnations, tended to be regarded as self-sustaining. The optimism was related to a high degree of faith in the power of world public opinion to sway recalcitrant governments. But even when that faith was lacking, if the United States could not summon up the will to act, it was comforting at least to sound righteous—and the words might yet lead to action at some future time. There was also hope that governments with their power politics could be increasingly by-passed in international relations through the spread of networks of "people-to-people" contacts. Such an approach was likely to be especially appealing to Americans; it fitted well with their long tradition at home and abroad of preferring to rely heavily on private charity and missionary humanitarian enterprises as the means to alleviate hardship without encouraging government action. This suspicion of powerful governments also contributed to a special fondness for small nations, which was grounded in the old belief in colonial emancipation and national independence and was reinforced through continuous agitation by ethnic minorities in the United States, as well as by common acceptance of the success of the federal pattern of government with its states' rights protected by equal representation in the Senate. Americans thus felt tempted to romanticize small nations and their leaders, even to invert the maxim that "power corrupts" into a dubious notion that weakness somehow purifies. There was a continuing nostalgia for the days when the United States was also weak and (nicely sheltered) was able to pronounce judgment on the world's problems with sturdy strict morality, uninhibited by much sense of responsibility for positive follow-up action.

Thus, when many Americans found simple isolationism no longer satisfactory in the first decades of the twentieth century, they turned their attention outward far enough to tinker hopefully with the forms but not to grapple painfully with the substance of international relations. It is easy to contend that United States policy was thereby distracted from fruitful courses of action. On some occasions it probably was—and when the organizational emphasis was continued into World War II for American postwar planning and diplomacy, the consequences were clearly injurious. However, the formalism should be assessed in the light of the dominant tradition of isolationism. Attempts at institutionalization were partly efforts to perfect a framework within which the United States could more safely keep a basically free hand; otherwise they would never have been acceptable. Yet, at the same time, as the all-out isolationists constantly protested, these moves did imply and foster some American concern for international developments and to that extent helped to reorient opinion away from simple escape from the world's problems. The new orientation was misleading, but less so at least than the old one was coming to be.

There was an enormous void of public apathy to be bridged. A century of easy escape from painful struggles for international survival and growth had left its mark. Immersed in private competitive concerns, the overwhelming majority of Americans took no interest in foreign affairs, or at least were incapable of sustaining interest over any length of time. Even politically active citizens shunned international problems. Right up until 1939 there were only a very few brief periods when foreign affairs intruded themselves on the public consciousness enough to be adopted by politicians as serious issues for national debate. The pre-eminence of domestic issues was not even tempered by any widespread belief that their solution might depend upon some active foreign policy. The day-to-day conduct of United States foreign relations could be left to a small group of experts on whom the spotlight of publicity rarely shone because their activities seemed of such slight importance. The problems with which they dealt were technical, bailing out some American businessman or drafting some paper-peace project; they had little occasion to recommend subtle or vigorous foreign policies. Regarded objectively, policy was often "insolvent," in the sense that the publicly endorsed objectives and apparent commitments were out of balance with the power that was readily available to sustain them. But except in time of recognized crisis the typical American way to restore the balance was to let the commitment lapse in actual practice while continuing to support it with words. If this were only a strategy consciously adopted in the hope of concealing American weakness and irresolution from foreign eyes, the device might easily be defended. But in any democracy such a course necessarily also involves a large measure of *self*-deception; and in the United States it not only confused public understanding of the immediate problem (for example, China), but also helped to perpetuate the prevailing naïveté about what could be accomplished in general by mere verbalization in international politics. Why should an American bother to try to figure out the world's problems as long as he felt able to talk his way out of them or hide from them in blissful ignorance?

Public apathy may be dispelled for a while by a sufficiently strong shock. But the ignorance bred of apathy requires a long time to overcome, and unless the shock is sustained or frequently repeated, apathy soon returns and the incentive to learn is likely to be dissipated. This kind of fluctuation has been conspicuous in America's groping response to international change in the twentieth century. Except in time of crisis, the attitude of the public and most of its leaders has tended toward a complacent mood lacking clear structure in the sense that little effort has been made to define goals and rank them or to evaluate realistically the available means. Positive action is not immediately imperative; so planning is lax. Then comes the crisis; action is demanded—and there is a tendency to leap to adopt some ill-considered panacea. Tremendous energies can temporarily be mobilized in support of the chosen solution, and great hopes are raised. But when the "solution" fails to meet the expectations, overoptimism soon shifts to a mood of frustration and pessimism; and rather than persevere toward the partial success which might still be achieved with the chosen policy, Americans impatiently drop it in favor of some new expedient or else relax all their efforts in disillusionment. Any emergence of a new international equilibrium facilitates the return to apathy.

The classic example was America's participation in World War I.

## The Impact of World War I

The advent of the war was unforeseen in 1914. So were its duration and the tremendous cost to the belligerents which would be produced by the new technology of warfare. So too were the decisive victories which both sides would feel driven to demand in order to justify such costs to their own people. Few Americans at first could foresee that the war could not be brought to an end without producing a decisive shift in the balance of European power—unless the United States, as the one great outsider, should conceivably be willing to dedicate herself to imposing a compromise peace, for the sake of the long-run stability of the world order under which she thrived, and unless the United States were well-enough armed to do so effectively. Of course, no support existed for such an interventionist balance-of-power policy, although President Wilson did seek by diplomacy to promote a "peace without victory." Gradual recognition of the possibility of Allied defeat, which would leave Imperial Germany preponderant on the Continent and strong on the seas, produced increasing apprehension and larger defense expenditures in the United States. But Allied strength continued to be greatly overestimated, and there was no deep sense of danger to the Western Hemisphere. That threat would materialize too far in the future, if at all.

Commerce with areas under British control was much more important to the United States even in normal times than was her trade with Central Europe. Many decades of English leadership in world economic development made that situation inescapable; and now in time of war the growing demands of Allied mobilization were bringing fresh prosperity to American farms and factories. There can be no doubt that this situation made it awkward for the United States government to challenge Britain as firmly as it challenged Germany when they both interfered with American commerce. But as a reason for actually going to war against Germany, the economic stake was little more heeded than the long-run national security stake. What counted for Americans was the unprecedented method of Germany's interference with merchant shipping—unrestricted submarine warfare.

Underwater commerce raiders were a new contribution to the technology of warfare; they simply could not be assimilated to the established rules of international law for humane conduct on the part of belligerents. The rules had been designed to deal with naval surface vessels, of which, of course, England had a relative plenty but Germany did not. A basic requirement was that a raider must see to the safety of the passengers of any merchant ship before he was entitled to sink her. Ordinarily this was physically impossible for a submarine raider. If there were any danger that the merchant ship was armed with even a light naval gun for defense, the thin-skinned submarine could not afford to take the risk of giving warning before launching its torpedoes; in fact, a submarine was even vulnerable to being rammed by an unarmed vessel. But American opinion had not been schooled in the imperatives of naval technology. When merchant ships were attacked without warning, waves of genuine horror swept the country with an intensity that is almost inconceivable today. Where human lives were thus involved, the Wilson Administration took a righteously firm position in defense of the traditional rules of warfare, even insisting on throwing a protective mantle over *Allied* vessels on which Americans might happen to wish to travel.

In angry reaction to a number of incidents involving the loss of American lives, nearly all of them on foreign ships, public and governmental protests mounted spasmodically till at last an ultimatum was delivered on April 18, 1916. Unless there was an end to Germany's "present methods of submarine warfare against passenger and freight-carrying vessels," the United States would sever diplomatic relations; probably war would follow. For a time the Kaiser's government bowed. But the Americans had now committed their prestige in the gravest fashion, and concepts of national honor and self-respect would make it extremely difficult to avoid fighting if Germany should decide later that she simply had to make effective use of the submarine weapon against Britain. Within nine months she did. Faced with the continuing drain of a stalemate war and the British blockade, and failing to recognize the imminence of Russian collapse and French weakening, the German government resolved to make an unlimited effort to starve Britain out of the war, even if it meant bringing America in. German naval calculations were that submarine victory could not be accomplished unless all ships trading with Britain—armed and unarmed, belligerent and neutral—were subject to attack without restriction; United States intervention was assumed, but it would be too little and too late to save Britain, and then America supposedly would not insist on prosecuting a transatlantic war indefinitely. This estimate helped remove any chance that Germany would merely probe experimentally at the scope of the previous American ultimatum—torpedo, say, armed merchant ships but not unarmed, at least not those flying the United States flag. Such tactics would have made it more difficult for Americans to go to war wholeheartedly. Even Wilson was beginning to doubt that the national honor should be involved to the point of war over the right of Americans to go touring on armed British liners. But the broader right of neutral United States ships to carry harmless goods to belligerent ports could not presumably be sacrificed. The essential fabric of international law and morality seemed at stake; the honor of the nation was committed.

However, even those high objectives might not have had sufficient appeal to a nation as isolationist as the United States, in the absence of much sense that national security or even economic prosperity was yet at stake. The security argument became slightly more pointed, though still unconvincing, after the American severance of relations with Germany, when the British let it be known that Berlin was foolishly trying to entice Mexico to join in war to recover "lost territory" northward if the United States should abandon neutrality.

But Wilson like other Americans was still looking for a more soul-satisfying justification for becoming involved in Europe's conflict. His own spirit was beginning to burn with the contagious fire of a vision whose grandeur would be great enough to consume even the vast energies of the American people, now already kindled by the abrasive impact of unrestricted U-boat attacks. The United States would fight the war to make the world safe for democracy and crown the victory with a league of nations! The collapse of Czarist autocracy actually seemed to brighten the prospect, and in the peroration of his war message Wilson raised high the torch:

> It is a fearful thing to lead this great peaceful people into war, into the most terrible and disastrous of all wars, civilization itself seeming to be in the balance. But the right is more precious than peace, and we shall fight for the things we have always carried nearest in our hearts,—for democracy, for the right of those who submit to authority to have a voice in their own Governments, for the rights and liberties of small nations, for a universal dominion of right by such a concert of free peoples as shall bring peace and safety to all nations and make the world itself at last free. To such a task we can dedicate our lives and our fortunes, everything that we are and everything that we have, with the pride of those who know that the day has come when America is privileged to spend her blood and her might for the principles that gave her birth and happiness and the peace which she has treasured. God helping her, she can do no other.

This was a far, far cry from the trenches of Flanders, where every month tens of thousands spent their blood and strength and seared their spirits for a few hundred yards of devastated earth. To raise millennial hopes out of involvement in such brutality was to invite disillusionment soon afterward. Yet, could Americans possibly be brought to enter the war soberly? Was not emotional intoxication necessary, and therefore the later hangover unavoidable? Considering the sheltered and unsophisticated background of the American people in world affairs, the honest answer unfortunately seems to be yes.

Moreover, Wilson's prescription was in fact helpful medicine, even if not the panacea it was presented to be. The cost of a major war in the twentieth century is so great that democracies cannot fight it efficiently unless much more than a simple majority of the public favors participation; Wilson's tonic contributed strongly to the remarkable degree of solidarity with which traditionally isolationist America fought the distant struggle. Besides strengthening the American war effort, his formula also helped to undermine the enemy's will to resist, by seeming to promise a peace without vengeance. And after the Armistice, the League of Nations organization, Wilson's central ingredient, was in fact capable of being adapted to mitigate the hardships and imbalances with which wartime bitterness did sear the final peace settlement despite his moderating efforts. Even though the United States stayed out, the institutions established at Versailles would be adequate for Britain and France to maintain their essential position in Europe for many years if they had the foresight to remain resolute and united. Moreover, it was only by a narrow Senate margin that postwar America wearied of her sovereign remedy and refused to join the League. In all probability, if Wilson had been willing to make more concessions to the Senate opposition—which in turn the Allies would probably have found endurable—the treaty would have been ratified and the United States would have been in the League. In 1920 the inevitable strains of postwar disillusionment had not yet brought to a peak the reaction against wartime ideals. Later, when the peak came, a demand for withdrawal would have encountered the inertia of actual participation and might well have failed. Even half-hearted sharing in the day-to-day work of a League with limited enforcement powers would have had the merit of helping to educate the American public to a concern for international developments—and in a setting which was at least more attuned to power realities than were most of the other projects to which internationalists tried to direct attention. Woodrow Wilson, however, was not content to hope for long-range benefits from a half-strength prescription. If the American people no longer found the League formula a rousing stimulant at full strength, then they were not fit to share it at all. He instructed his Senate supporters to kill the treaty themselves after Senator Lodge had watered it down. Apathy and disillusion spread unchecked. America returned to isolationism.

## SELECTED BIBLIOGRAPHY

ALMOND, GABRIEL A. *The American People and Foreign Policy*, rev. ed. New York: Praeger, 1960.

BAILEY, THOMAS A. *The Man in the Street: The Impact of American Public Opinion on Foreign Policy.* New York: Macmillan, 1948.

COOK, THOMAS I., and MALCOLM MOOS. *Power through Purpose: The Realism of Idealism as a Basis for Foreign Policy.* Baltimore: Johns Hopkins Press, 1954.

HALLE, LOUIS J. *Dream and Reality: Aspects of American Foreign Policy.* New York: Harper, 1959.

HANDLIN, OSCAR. *The American People in the Twentieth Century.* Cambridge: Harvard University Press, 1954.

———. *The Uprooted: The Epic Story of the Great Migrations That Made the American People.* Boston: Little, Brown, 1952.

HARTZ, LOUIS. *The Liberal Tradition in America: An Interpretation of American Political Thought since the Revolution.* New York: Harcourt, Brace, 1955.

KENNAN, GEORGE F. *American Diplomacy, 1900–1951.* Chicago: University of Chicago Press, 1951.

———. *Realities of American Foreign Policy.* Princeton: Princeton University Press, 1954.

LIPPMANN, WALTER. *Isolation and Alliances: An American Speaks to the British.* Boston: Little, Brown, 1952.

MORGENTHAU, HANS. *In Defense of the National Interest: A Critical Examination of American Foreign Policy.* New York: Knopf, 1951.

MORISON, ELTING, ed. *The American Style: Essays in Value and Performance.* New York: Harper, 1958.

OSGOOD, ROBERT E. *Ideals and Self-Interest in America's Foreign Relations: The Great Transformation of the Twentieth Century.* Chicago: University of Chicago Press, 1953.

PERKINS, DEXTER. *The American Approach to Foreign Policy.* Cambridge: Harvard University Press, 1952.

POTTER, DAVID M. *People of Plenty: Economic Abundance and the American Character.* Chicago: University of Chicago Press, 1954.

TANNENBAUM, FRANK. *The American Tradition in Foreign Policy.* Norman: University of Oklahoma Press, 1955.

WEINBERG, ALBERT K. *Manifest Destiny: A Study of Nationalist Expansionism in American History.* Baltimore: Johns Hopkins Press, 1935.

# CHAPTER 2

# ISOLATION BY LEGISLATION?

THE NEW ISOLATIONISM of the twenties and particularly of the thirties was distinguished from the old by its self-conscious intensity. No longer could isolation simply be taken for granted as it had been by most Americans for nearly a century, almost as a law of nature decreed by geography. Now they heard it constantly challenged by many of their own fellow Americans. Worse yet, it had just recently been overcome by a fit of moral indignation and injured pride and had been supplanted for a time by a popular crusade to end all wars everywhere and create a democratic world order. Compared with such a goal, intervention had clearly "solved nothing." It must never be allowed to happen again. But what had happened once could happen again unless something positive and new were done to prevent it. The old natural separation from Europe was evidently no longer enough. It would have to be artificially buttressed, whatever the resistance of those who protested that events had proved it was no longer feasible to escape from Europe.

## The Bitter Legacy of World War I

One far-reaching new shield for isolation after World War I has already been discussed—the drastic restriction of immigration, which involved a deliberate effort to reinforce American separation from Europe by reducing as far as possible the number of Americans who would be likely to feel close emotional ties to a foreign nation. The redoubled determination to develop a unique society unaffected by further foreign cultural admixture suggested a long-run political and moral problem which was scarcely confronted: How long could America expect her vast continental possessions to seem justifiable to crowded peoples elsewhere in the world after she had ceased to constitute herself a haven of refuge? Already in the 1920's, Japan and Italy were bitterly antagonized by the new American immigration policy. Such considerations, however, carried little weight in the United States. The war in Europe had stimulated ethnic clamor in America to a disturbing degree; many believed it contributed to entanglement and endangered national unity and strength. Economic isolationism was also prevalent. The standard of living of the native American worker was not to be undermined by further competition of "cheap foreign labor" for jobs in the United States. And sharp tariff increases in 1922 and 1930 would help protect him against competition from goods produced abroad by foreigners who now could not migrate even if they wished.

The tariff increases made it even more unlikely than otherwise that the former allies would ever be able to earn enough to repay very much of the multibillion-dollar loans the American government had extended to them to help fight the war in 1917–19. Congress consented to scale down the debts on fairly liberal terms, but was not willing to write them off completely as part of a joint effort. The result was an economically absurd situation which irritated all the countries concerned. The Allies grudgingly paid some of their debts to the United States government but themselves insisted on collecting practically all the money in reparations from Germany. Germans in turn borrowed the money from private investors in the United States, who were encouraged by the American government to make loans abroad in order to promote exports. Thus, in a few years about $2 billion went circling around Europe from United States citizens to the United States Treasury. Most Americans came to believe more than ever that in succoring Europe in 1917–19 they had been acting like suckers and that somehow they still were. The United States government was willing to listen to European protests and tried to ameliorate the economic dislocations. But the American public was becoming increasingly suspicious that every time an international conference met "to get at the bottom of things, one of the things is Uncle Sam's pocket."*

* Quoted in T. A. Bailey, *The Man in the Street* (New York: Macmillan, 1948), p. 183.

When the bubble burst in 1931 and intergovernmental payments were suspended altogether, the sense of being "taken" became even stronger. Americans had always been vaguely afraid that their sheltered background had left them unsophisticated about international politics. Finding themselves citizens of a world power, they felt the psychological insecurity of parvenus. They suspected their diplomats of seeking to acquire an alien respectability by aping the Europeans. City slickers, they knew, were always lying in wait for honest country boys. The good-hearted fellow was sure to be snared by knavish trickery if he failed to keep his distance. Most of these anxieties were a natural reaction to the world situation in which Americans found themselves, but the shrinking fear of being a sucker was especially exacerbated after 1919 by a dogmatic popular belief in capitalist economic orthodoxy. The war debt issue was salt constantly rubbed in a tender wound, till ultimately the scars were so distracting that in early 1940 the safety of little Finland seemed more important to many Americans than that of Britain and France. Finland paid her debts.

This sensitive obsession with war debts was, like the simultaneous drives to raise tariff and immigration barriers, a powerful economic and cultural factor in the heightened isolationism of the 1920's. But more important yet was the force of antimilitarism and even pacifism. The sheer brutality of World War I had produced a traumatic shock. Compared with the European countries that had staggered back and forth from one blood-soaked trench to another for more than fifty months, the United States had suffered very little in the war. But compared with the results achieved, as viewed through the wrong end of Wilson's telescope, America's losses seemed tragically great. Besides, if she ever fought again might she too not pay the price that France had paid? Antimilitarism became a keynote of the new isolationism.

The impact of this development can be pinpointed if first a differentiation is made between four types of orientation in foreign policy: *unilateralism, multilateralism, interventionism,* and *noninterventionism.* Unilateralism is the policy of a state to keep a "free hand" and "go it alone." Multilateralism involves a program of systematic cooperation with certain other states, accepting some degree of regularized limitation on one's own freedom of maneuver. Interventionism and noninterventionism concern the disposition to take an active part in world conflicts, or to steer clear of them as far as possible. Commonly, these two sets of orientation are combined. Thus an imperialist policy is an extreme form of "unilateral interventionism," whereas a forceful collective-security program is a kind of "multilateral interventionism." "Unilateral noninterventionism" is strict isolationism; and "multilateral noninterventionism" can describe many devices and gestures of international cooperation lacking firm supporting sanctions.

From this perspective, the antimilitarist reaction from the horrors of World War I tended to split American multilateralists (the "internationalists") and weaken their interventionist (collective-security, League-of-Nations) wing. More than ever the latter group turned its energies away from programs which might involve war toward support of the verbalistic paper-peace projects of their multilateral *non*interventionist brethren, who, as has previously been indicated, usually tended anyway to be the major faction among the Americans who had abandoned traditional isolationism. Thus it was possible for former supporters of the League of Nations to herald the Kellogg-Briand Pact as though it were really a major contribution to world peace. If even collective-security advocates could be so optimistic about an "international kiss," why should the apathetic public rouse itself from complacency to face the realities of international politics? The postwar equilibrium seemed comfortingly safe and sound in 1928, and the horrors of war a nightmare to be put out of mind.

While multilateral interventionism was thus severely undermined by postwar antimilitarism, *uni*lateral interventionism was largely destroyed. Indeed its obliteration was so nearly total that the very fact was not much noticed. Theodore Roosevelt went out of fashion much more completely than did Woodrow Wilson. The nationalist expansionism of "Manifest Destiny," which had helped to carry Americans westward in a series of enthusiastic waves for many decades, from the Atlantic seaboard to the very shores of Asia, had already begun to pall by 1905. By the late twenties the reaction had gone so far that the United States was even getting ready to pull her troops out of the Caribbean and explicitly cancel the Roosevelt Corollary to the Monroe Doctrine, thereby abandoning her "right" to intervene. Gone too was the martial vigor of the charge up San Juan Hill and the militant righteousness of the stand at Armageddon. They were matters for ridicule and caricature. Farewell to arms. But not so much farewell to nationalism. The old nationalist enthusiasms which had always sought a free hand for America to go it alone in pursuit of her own destiny, but which had been somewhat divided at the time of the Mexican War and even more after 1898 between unilateral interventionism in quest of empire and unilateral noninterventionism in quest of insulation, now could all surge solidly in the noninterventionist direction. To assert national strength abroad might mean war. Therefore, now to want to go alone was to want to stay at home. This fusion of outlooks came to be taken so generally for granted, even by opponents, that as late as the 1950's, when unilateral interventionism had in fact re-emerged, many analysts found it very difficult to conceive of men like Senator Knowland as anything but "isolationists." Actually, over the course of United States history there has often been an influential and partly self-conscious body of American unilateral interventionist opinion, usually shunning Europe but pressing firmly westward. In the twenties and thirties, however, this movement was nearly eclipsed by antimilitarism, and as multilateral interventionism was similarly weakened, there remained extraordinarily little opposition of any kind to noninterventionism.

The increasing moral relativism of the interwar years was another key factor tending to undermine resistance to isolationism. The utopianism of the wartime era was turning sour in disillusionment. Cynicism was rife, much of it embittered. One found it increasingly difficult to believe that there could be any special righteousness in an American "national mission." More than ever before, general mockery greeted the old gospel that the United States had a responsibility to spread abroad her unique virtues for the regeneration of mankind. Unilateral interventionism had fed on this creed; now men said the creed was only a rationalization for the interventionism. Thus both were undermined. Debunkers also corroded the faith of the multilateral interventionists in a similar fashion, by raising doubts about the justice of the international order. The evils of the Versailles settlement were magnified by disillusioned Utopians to the point literally of moral paralysis where Germany was concerned. Conscience-stricken multilateralists in America (and Britain) could scarcely justify to themselves the containment of Germany in any boundaries smaller than her prewar Imperial frontiers. After all, the historical record was now being opened to reveal that the Kaiser's war guilt was only partial and that Allied diplomacy both before and during the war was strongly shaped by imperialist aims. The neat black and white of America's 1917–18 picture had turned to dirty shades of gray. How was a moral man to choose between them? Never again.

### The Nye Committee

After the onset of the Great Depression in 1929 the intellectual attack on American participation in World War I adopted a new emphasis on economics, in keeping with the universal concern about the miserable condition of the economy. Historical scholars and publicists re-examined the record of the years before 1918 from a viewpoint resting ultimately on a belief in economic determinism (though usually not consciously derived from Marxism). Many "revisionists" contended that World War I had resulted from a rivalry between Germany and England caused fundamentally by competition for worldwide markets and investment opportunities; this was no concern of the United States, but she had intervened anyway because American private business interests had overextended themselves financially in assistance to the Allies and needed to have their government help bail them out by assuring Allied victory. Such distorted oversimplifications and misrepresentations gained easy credence, even in intellectual circles, at a time when depression hardships were causing Americans to search for villains and find them in the shape of "economic royalists."

The increased preoccupation with domestic economic problems also deepened the desire of liberals and conservatives alike to seek solutions independent of influences from the outside world. Economic isolationism battened on the struggle for recovery. From the Right came demands for increased tariff protection to help restore confidence in business as usual. From the Left came far-reaching reconstruction projects which would be more manageable if they could be insulated from the world markets. Reaction had set in against the Republican efforts of the 1920's to improve prosperity by promoting American sales abroad and by encouraging United States investors to make foreign loans. That bubble had burst; loans were defaulted, and exports declined in tariff wars. Critics saw new cause to look back suspiciously at American economic involvement with Europe in 1914–17, and more reason to concentrate their reform efforts on expanding the domestic market in relative self-sufficiency.

Several years later, President Roosevelt took occasion to explain that it would be necessary for a while to replace "Dr. New Deal" with "Dr. Win-the-War." By the time of these remarks almost all New Dealers had already become caught up in the enthusiasms of World War II, and the change of "specialist consultants" caused little protest. But during the 1930's most New Dealers had been very much afraid of just that kind of future development, and many of them had warned vehemently that any involvement in war would put an end to great reforms and might well straitjacket the nation in militarist dictatorship. This anxiety on the part of liberals gave powerful reinforcement to noninterventionism in the first years of the Roosevelt Administration. It is worth noting too that another characteristic of liberal thought in that period helped to influence the form in which the new isolationism expressed itself. There was a prevalent spirit of "legislative solutionism." ("Have a problem? Well pass a law. That will take care of it.") One problem was to extend the New Deal by staying out of war. Why not try a law for that purpose too?

The momentous catalyst for all these currents of thinking was the Nye Committee investigation of 1934–36. It was sparked by an anonymous feature article in *Fortune* magazine, March 1934, entitled "Arms and the Men," about the worldwide munitions industry. Senator Nye and others were scandalized by the revelations. He introduced a resolution calling for the establishment of a special Munitions Investigating Committee. Democrats controlled the Senate; Nye was a Republican and an isolationist, coming from North Dakota where German-American and Scandinavian antipathy to the Allies had always been powerful. On the other hand he was an ardent New Dealer, and Democratic Senator Pittman, chairman of the Foreign Relations Committee, was lukewarm toward internationalism; so Pittman offered no objection to Nye's becoming head of the investigation, under the common practice that a Senator who initiates a demand for a special committee and can persuade his colleagues to establish it is then entitled to become the chairman. The committee staff, which soon included Alger Hiss, began its work in April 1934, and by September, Nye was ready to start a series of public hearings. They were resumed periodically for a year and a half and certainly had more far-reaching effects on American foreign relations than any other congressional investigation in recent decades, much more even than Senator McCarran's probe of the Institute of Pacific Relations (1951–52) or the various McCarthy investigations (1950–54).

The course of the hearings was not clearly mapped out in advance; they wandered widely. But their essential function was to publicize and popularize the revisionist critique of American participation in World War I—especially the various economic-determinist versions of that critique. Through the Nye Committee these arguments acquired common currency. In the first phase, the investigation centered on the munitions makers and armaments salesmen and dramatized the wicked machinations of "merchants of death," who bore such villainous-sounding names as Sir Basil Zaharoff. The American public was duly shocked; attention was focused; and in 1935 the committee's melodrama became slightly more sophisticated. The search for bloodsuckers shifted from commercial and industrial capitalists to finance capitalists. It was now the international bankers who had got us into war in 1917, just as it was they who had got us into depression in 1929. The theme was very popular, and by the beginning of 1936 Nye felt strong enough to dare impugn the personal integrity and honesty of President Woodrow Wilson himself. This was the final straw for Wilson's onetime Treasury Secretary, Carter Glass; as Senate Appropriations Committee Chairman, Glass cut off Nye's funds and brought the investigation finally to a close in the winter of 1936.

While the American public had sat eagerly absorbing Senator Nye's horror story, events in the outside world had been drifting rapidly toward catastrophe. The balance of power in Europe was shifting swiftly in favor of Germany under the intransigent new leadership of Adolf Hitler. America, of course, had still not learned to look at the world in terms of imbalances. Great Britain had, but her calculations failed to recognize the debility of postwar France. The British, too, were inhibited by pacifism and doubts about the justice of the Versailles treaty. And the Western democracies were all alike in their preoccupation with the internal strains of economic depression. Thus the years 1934 and 1935 slipped by; Germany rearmed, and England and France lost the power to contain her by themselves except at the cost of a ruinous war. The cooperation of Italy, Russia, or the United States would now also be needed. However, there was nothing to hope for from the United States except perhaps matériel, as in 1914–16, and the Nye Committee's intensification of isolationism was making even that unlikely. Communist Russia was untrustworthy and seemed weak. Fascist Italy was not strong either, or trustworthy, and would exact a stiff price for any cooperation against Germany or even for failing to give Germany outright support. France did ally herself loosely with Russia in 1935, but for the most part the capitalist democracies of West Europe found the effort to conciliate Mussolini much more to their taste. Italy took advantage of this attitude to seize Ethiopia and help put the Fascist Franco in power in Spain, but Rome did very little in return to put any rein on Hitler. German military strength mounted precipitously. By 1938, when the Nazi regime felt prepared to begin its open external aggression, the British and French governments almost desperately believed that their only hope lay in further appeasement.

### The Neutrality Acts

The United States was showing herself more determined than ever not to become involved, in any serious fashion, economic or political. With the aid of the Nye Committee hearings, noninterventionists were spurring a chorus of public demand for new formal legislative prescriptions against the contagion of war, though they disagreed widely among themselves regarding most of the specifics. If there was any basic lesson to be drawn from Nye's disclosures regarding World War I, it seemed to be that in any great war a high degree of American economic isolation would be required in order to supplement and preserve the presumably desired political and military isolation. But just how high a degree? How costly to what segments of the United States economy? American ships might by law be kept out of combat zones, so there would be no sinkings and consequent popular indignation—but the restrictions would tend to impoverish the merchant marine. Americans could be prevented altogether from traveling in or to belligerent countries, so they would not be killed and the public would not get angered, but the restrictions might tend to impair important business and charity ventures. Exports of arms to warring countries could be blocked—at considerable cost to the companies and workers who produced them. More companies and more workers would be financially injured by every single addition to a government list of embargoed commodities. And if loans were cut off from belligerents, not only the bankers would suffer, but also more and more American producers in farms and factories that depended on markets in the countries which had gone to war, including producers of nonstrategic goods. The United States economy was not self-sufficient. Could it really be forced to become so in order to avoid all risk of military entanglements? And if autarky were not adopted as the permanent peacetime policy (as it virtually was in the special case of immigration), what would be the shock effects of an effort to impose it suddenly at the outbreak of a major foreign war? Who would allocate the costs? In particular, how much discretion was the President to have, case by case?

In any new neutrality legislation, the Roosevelt Administration wanted to preserve for the President a maximum range of discretion when applying restrictions at the outbreak of foreign war—including notably the right to discriminate against the particular belligerent he found to be the aggressor in the conflict. Thus America would be able to continue to furnish supplies to the victims of aggression, while reducing her trade with the attackers. But this plan implied that the United States had reason to be concerned to cooperate more with one side than another in a foreign war, and that the President could be trusted to make the right distinctions. The noninterventionists protested that such favoritism between belligerents would lead America down a new road to war like 1914–17. Congress therefore refused to grant the President any authority to discriminate against an aggressor. Nevertheless, Roosevelt did not hesitate to approve the first two Neutrality Acts (1935 and 1936); they were temporary measures with short time limits and in actual operation would mainly hinder an aggressor power, Italy, in her war against Ethiopia.

The imminence of the Italian invasion was clearly foreseen when the 1935 act was signed into law. When the assault actually began a few months later, the Administration was prompt and eager to apply the mandatory restrictions of the neutrality legislation and even tried to go beyond them. As required, an embargo was imposed on sales of American "arms, ammunition, or implements of war" to either belligerent (only Italy could have afforded to buy much anyway); American ships were barred from carrying any munitions to either belligerent. The law also gave the President some discretionary authority over travel by United States citizens, and Roosevelt used it to discourage the use of Italian (or Ethiopian) vessels. Congress had not been prepared to go any further as yet in authorizing restrictions on America's wartime commerce, but Roosevelt plunged on in efforts to use the pressure of publicity as far as possible to discourage American exporters from shipping other kinds of strategic goods to Italy, including oil. The Administration was tacitly cooperating with the official League of Nations policy of sanctions against the aggressor. But the United States would obviously not go on to furnish positive military assistance to Britain or France if Italy should resort to war in Europe in order to break an oil embargo, the one she would find most costly. All the casualties and matériel losses would be British and French. Moreover, the United States seemed even less likely than Italy to be an available makeweight against the ominous rearmament of Germany in the quavering balance of world power. Britain and France needed Italian cooperation too badly to throw it away for the sake of Ethiopia. Oil sanctions in Europe never went into effect. Mussolini won his empire within a few months.

Even in America, the President's "moral embargo" was only partially effective. Roosevelt, however, had continued to favor a wide range of sanctions; and when the six-month limit of the 1935 Neutrality Act ran out early in 1936, the Administration sought new legal authority for tighter embargoes. But key congressmen were not willing to give the President authority to pick and choose among commodities when undertaking to ban American exports. The second Neutrality Act, as passed, re-enacted for another year the essentials of its predecessor, while giving the President a little new discretion to decide whether hostilities constituted "war" and therefore required an arms embargo; a ban was added against American loans to belligerents. (The Nye Committee had been turning its lurid spotlight on Wall Street.)

In 1937 the crucial test of the Roosevelt Administration's attitude toward neutrality legislation came, for Congress passed a "permanent" act. The previous enactments had been scheduled to expire after relatively brief periods and could not have been extended except by two-thirds majorities in both houses if world developments persuaded the President to fight the extensions with his veto power. Thus he would need the cooperation of only one-third of either house to get rid of restrictions in fairly short order when they ceased to seem as useful to him as they had in the case of Italy. But if permanent legislation were allowed to reach the statute books, it could later be amended only by majorities of both houses of Congress, no matter how different and urgent a future international situation might look to the White House. Nevertheless, Roosevelt concentrated his attention in 1937 on driving additional New Deal domestic legislation through Congress and overcoming the Supreme Court's opposition. For these objectives, especially in the court-packing fight, he needed every vote he could find. He could not afford to alienate liberals who happened to be isolationists; so the Administration did not fight the Neutrality Act of 1937. The enactment confirmed, "permanently," the previous mandatory arms embargo, with no authority to discriminate between belligerents; similarly the ban on loans to them was mandatory, and new provisions required a prohibition against Americans' traveling on their ships. The American merchant marine would still be permitted to carry most nonmunitions cargoes to belligerent ports; but for the limited period of two years the President was authorized to list other materials besides munitions and forbid the United States ships to transport them to warring countries. The belligerents would have to send their own ships to America to obtain these goods, "cash and carry."

This legislation had become law by the time general hostilities broke out between Japan and China in the summer of 1937. But Japan chose to minimize her aggressive campaign as a mere "China Incident." War was not declared, and, therefore, Roosevelt, to his satisfaction, was not obliged to find that any state of war existed in the Far East. Unlike Ethiopia, China could afford to purchase some matériel for her defense. In this new aggression, not only the attacker, but also the victim would get substantial benefit from permission to continue trading with the United States. Since vital provisions of the Neutrality Act could not be invoked in such a way as to discriminate against a particular belligerent, the President now preferred not to invoke the measure at all in this "incident." Roosevelt's own mildly interventionist motivation was much the same as when he had gladly imposed embargoes in the Italo-Ethiopian conflict. It was the circumstances of the war that were different. And just as Roosevelt had tried to hurt Italy by sponsoring a "moral embargo" beyond the scope of the law, so now he seemed to be preparing to move toward special restrictions against Japan without closing off channels to China. On October 5, 1937, the President raised a trial balloon in a speech in Chicago:

> It seems to be unfortunately true that the epidemic of world lawlessness is spreading. When an epidemic of physical disease starts to spread, the community approves and joins in a quarantine of the patients in order to protect the health of the community against the spread of the disease.

This image in the Quarantine Speech was certainly not self-explanatory, and the public reaction was so negative that Roosevelt never did venture to set forth a positive plan; but some form of diplomatic or economic sanctions was clearly implied. The American people, however, were vociferous in their opposition. It was not until the following summer that the first in a series of widening moral embargoes was officially proclaimed against Japan.

Meanwhile, isolationism reached its high-water mark.

Congress was called into special session late in 1937 to accelerate New Deal domestic legislation. Safely bottled up in the House Judiciary Committee was a constitutional amendment sponsored by Louis Ludlow, a Democrat from Indiana. Under its provisions, unless American territory were directly invaded, the United States could not go to war at all until a congressional declaration of war had been confirmed by a nationwide popular referendum. The delays and uncertainties inherent in such a process would render American diplomacy even less influential on the world scene than it had already become under the Neutrality Acts. Then on December 12 Japanese forces bombed and sank an American gunboat, the *Panay,* which the United States was operating under treaty rights on the Yangtze River; survivors were machine-gunned. The action was probably unauthorized by Tokyo, and prompt restitution was made. But even before it was known that the Mikado's government regretted the attack, the dominant American public reaction was not outrage at Japan but misgivings about American units' being in China at all. The effect in Congress was to produce within two days the remaining margin of signatures necessary to force the Ludlow Resolution out of its pigeonhole and onto the floor of the House. More than an absolute majority of the Representatives had thus signified their apparent approval of the constitutional amendment. The turn of events finally roused the Administration to summon all its political resources to resist this most extreme isolationist measure. Democratic party pressure was strenuously applied, and enough congressmen switched their positions to defeat the Ludlow Resolution on the floor, January 10, 1938, but only by a margin of 209–188. Noninterventionism had reached a limit. Yet it had demonstrated such great strength that Roosevelt did not venture for more than a year thereafter to lead any drive to alter the Neutrality Act, even though during that period Austria and Czechoslovakia were both swallowed up by Hitler and the West European democracies stood repeatedly poised on the brink of war—a war for which they would need arms from the United States that they could not possibly obtain under the Neutrality Act.

This was the era of Munich, and the bleak prospects regarding help from America were certainly one factor in the Anglo-French appeasement policy. In turn, the irresolution of Britain and France was a factor in the failure of interventionists in America to make a fight for a change in neutrality policy. Disunited, the Atlantic democracies thus tended to devitalize each other.

Until the very eve of Munich in September 1938, the American government was not even concentrating its attention on the ominous march of events abroad. The major preoccupations were a new economic recession in the United States and a provocative experiment by Roosevelt, seeking to purge some uncooperative congressmen out of the Democratic party in the primary elections. The shock of Munich finally shifted the focus of Administration concern outward. Plans were slowly formulated to seek congressional repeal of the arms embargo, thus putting Hitler on notice that in the event of a European war Britain and France, commanding the seas, would have access to American factories for the production of armaments. Resistance to such a move was known to be great in Congress, and the Administration was very wary of the adverse psychological effect of a public defeat on this issue. Supporters on Capitol Hill were encouraged to carry the ball for repeal in the first few months of 1939 without much open assistance from the White House until May. By that time all the free remnants of Czechoslovakia had been wiped off the map. Poland seemed next in line, and Britain had finally given her a guarantee that would mean general war if Hitler pressed his demands. Nevertheless, the House of Representatives in late June was unwilling to repeal the arms embargo; instead an ambiguous modification was adopted, 159–157, leaving the embargo essentially intact; the narrow vote was confirmed 214–173. The Administration turned to the Senate. In a famous evening conference on July 18, 1939, the President and Secretary of State pleaded with Senate leaders to permit repeal—without success. Republican leader Borah preferred his own sources of information to those of the State Department and announced that Hitler would not fight in 1939; the arms embargo question could be postponed till Congress reconvened in January 1940. There was nothing else to do. In retrospect some may be able to take dubious consolation from the later discovery that American assistance to Britain had already been discounted so much in Berlin, as a result of previous evidence of isolationism, that this new development could scarcely lower Hitler's estimate any further. His decisive concern was to neutralize Russia, as he did by the startling Nonaggression Pact in August. The United States had essentially thrown away her own influence on the disastrous course of events in Europe; it was now too late to regain it before the actual outbreak of war, even if the arms embargo had been repealed.

### The End of Invulnerability

When the war began a few weeks later, the legal situation was that belligerents could not purchase American "arms, munitions, or implements of war"—but they could purchase anything else, and American ships could transport it to their shores. The shock of the new events produced a situation in which the Administration could find sufficient votes to work out a deal with Congress, which was called back into special session. The White House gave full support to new legislation to ban United States vessels from carrying any goods whatever to belligerent ports, or even sailing at all through "combat zones." Thus, at considerable cost to the government-subsidized merchant marine, there would be minimum danger of provocative incidents involving American ships; this expensive restriction was one which had long been vainly sought by many noninterventionists reacting to the 1914–17 experience. In return for this concession to their wishes, the Administration wanted to have the arms embargo repealed and all goods made available to belligerents on a cash-and-carry basis. Under the spur of events in Europe, Congress enacted this arrangement within a few weeks in the fall of 1939. Thereafter, the Allies would be able to get whatever matériel was available for sale in the United States—as long as their money held out to pay for it and their ships to transport it and convoy it.

The opening of the American munitions market did not, however, result in any immediate flood of Allied orders to buy. Britain and France were well aware of their financial limitations and did not want to spend precious dollars for goods that could be produced at home. Less excusable was the continued complacency which surrounded their outlook toward the war. Their lack of a sense of urgency was shared by the Roosevelt Administration and even by professional American Army and Navy planners. Excessive confidence was placed in the Maginot Line and the British Navy. Behind these defensive bulwarks the Allies were supposed to be able to outlast Hitler in a long war of economic attrition, finally starving him out. Precious months were lost in a mood of sedentary "phony war," in which only Finland and Russia seemed to be really fighting. The *Blitzkrieg* had been replaced, one heard, by the *Sitzkrieg.*

Until April 9, 1940. On that date German and Allied forces clashed as each sought to forestall the other from gaining control of strategic Norway. Norwegian neutrality was deliberately violated by both sides, but the German expedition was the better prepared and the Allies failed miserably. Britain and France were still reeling from the Scandinavian defeat when on May 10 they were struck by the Nazi onslaught in the Low Countries. The new blow was incredibly disastrous. Within six weeks the military power of France was destroyed, Italy joined the war on the side of Germany, and Britain found her home islands beleaguered and largely disarmed. The balance of power in Europe was utterly shattered.

With Germany and Russia in partnership and Britain very likely doomed, the United States almost overnight confronted the prospect that all of Europe and the eastern Atlantic would soon be united under the domination of a single combination of great revolutionary powers. If, as was probable, Japan consolidated her position in the Axis alliance, the anti-status-quo partnership would embrace nearly all the resources of Eurasia. If the British fleet also fell into Nazi hands, perhaps as ransom for millions of English lives, the Western Hemisphere itself would very quickly be vulnerable to direct attack. Even if the British fleet were scuttled to keep it out of German control, the shipbuilding capacity of all Europe plus Japan would pose a grave threat within a few years to American naval supremacy in Western Hemisphere waters. The United States would be compelled to devote her energies to a vast armaments race which might change her democracy into a kind of armed camp. Her export trade with Europe, so essential to prosperity (especially of agriculture), would be conducted only at the sufferance and on the terms of politically-motivated Nazi state-trading cartels. In 1940 at the end of ten years of depression, one could readily doubt that even the the American economy was strong enough and adaptable enough to withstand such pressure. Vastly more vulnerable still would be the shaky economies of Latin America, which were enormously dependent on European trade. Would the United States be able to solve their export-import problems in addition to her own, and thus prevent their yielding to German demands? Worse yet was the tremendous danger that fascism, victorious in Europe, would look like the "wave of the future" to peoples south of the Rio Grande, already imbued with resentment against the "Colossus of the North." Strong political factions among them would be glad to try to ride the wave, cooperating with Germany and Italy. Once the familiar Latin coups and palace revolutions had put them in power, it would be extremely difficult for the United States to eject them without appearing to act imperialistically—thus alienating additional South Americans and making them more amenable to other Nazi schemes. Already in the late thirties there was very disturbing evidence of German economic and political penetration of Latin America.

Postwar disclosures from Nazi records make it appear that the United States commonly exaggerated the seriousness of Hitler's intention ever to invade the Western Hemisphere by actual force of arms, although ultimately, if unchecked, his grandiose ambition might perhaps have embraced the attempt, and American policy needed to take precautions against that possibility. But the graver, though more subtle, threat would be the sheer magnetism of Nazi power entrenched in Europe, applying its attractive influence through various agencies upon disgruntled particles of South American leadership. This kind of danger resembles that which Communism too would pose if it won control of all the Eastern Hemisphere. The Atlantic could be crossed without a battle, and Fortress America whittled down to Fortress USA.

## SELECTED BIBLIOGRAPHY

ADLER, SELIG. *The Isolationist Impulse: Its Twentieth-Century Reaction.* New York: Abelard-Schuman, 1957.

BLUM, JOHN M. *From the Morgenthau Diaries: Years of Crisis, 1928–1938.* Boston: Houghton Mifflin, 1959.

DRUMMOND, DONALD F. *The Passing of American Neutrality, 1937–1941.* Ann Arbor: University of Michigan Press, 1955.

HULL, CORDELL. *The Memoirs of Cordell Hull,* vol. 1. New York: Macmillan, 1948.

ICKES, HAROLD. *The Secret Diary of Harold L. Ickes,* 3 vols. New York: Simon and Schuster, 1953–54.

LANGER, WILLIAM L., and EVERETT S. GLEASON. *The World Crisis and American Foreign Policy: The Challenge to Isolation, 1937–1940.* New York: Harper, 1952.

———. *The World Crisis and American Foreign Policy: The Undeclared War, 1940–1941.* New York: Harper, 1953.

OSGOOD, ROBERT E. *Ideals and Self-Interest in America's Foreign Relations: The Great Transformation of the Twentieth Century.* Chicago: University of Chicago Press, 1953.

# PART TWO

## FOREIGN POLICY AND MILITARY STRATEGY SINCE 1940

# CHAPTER 3

## ENTRYWAYS TO WAR

HITLER'S CONQUESTS in the spring of 1940 came too suddenly for their full implications to be assimilated promptly in American public opinion. Generations of citizens had been accustomed to regard the oceans themselves as the basic guarantee of national security; the related indispensability of British power could not become sweepingly clear overnight, even after the shock of the destruction of France. Britain was staggering, but time was still required to convince the American people of the stark realities of their situation. On another plane of thought which they tended to find more congenial than the balance of power, they were being profoundly affected by the sheer ruthlessness of Hitler's trampling on small, would-be neutral nations. Clearly, the Nazis were not just Imperial Germans seeking to rectify the injustices of Versailles. Increasingly, American liberals were persuaded that immediate risks of "militarism" at home would be a lesser danger to American democracy than would Hitler's final victory abroad. Multilateral interventionism revived steadily. This development was also related to the course of policy toward Japan in the Pacific. But for purposes of analysis it seems useful to postpone scrutiny of the Far Eastern "back door to war" until after a brief survey of America's approach to the "front door."

### The Battle against the U-Boats

In her final dying hours France had sent desperate appeals for an immediate declaration of war by the United States. But, of course, at that stage American opinion was utterly unprepared for such action. There is no evidence that President Roosevelt even remotely considered undertaking it. He himself was more confident than a realistic appraisal seemed to justify concerning Britain's capacity to hold out alone. Incidentally also, the presidential nomination conventions of both parties were about to convene, and any open moves toward war would disrupt their proceedings and might well be checkmated through their choice of platforms and candidates. Congress did show alarm by voting heavy new appropriations for national defense, but Roosevelt allowed actual production to continue seriously disorganized, on the implicit assumption that the country did not yet have to make painful choices between civilian and military goods; Americans could have both "guns and butter."

On the other hand, Roosevelt's refusal to despair of Britain led him to respond boldly, almost recklessly, when Churchill dispatched urgent appeals for matériel that was available only in United States government stockpiles. The American Army and Navy were told to scrape the bottom of the barrel of their own current stores of guns and ammunition and sell them to private companies for resale to the British. After this move, however, Roosevelt held back and let others undertake the initial burden of leading public opinion to accept further measures during the pre-election period. He connived at a drive to enact selective service, but did not himself push for it strongly and publicly until after it had already won the endorsement of the Republican candidate, Wendell Willkie. The other major summer project was designed to meet England's desperate need for antisubmarine vessels by providing fifty American Navy destroyers, in return for bases in British Western Hemisphere possessions. The groundwork for this exchange was laid for the most part outside the government, by private American citizens anxious to find devices for aiding Britain; they secured Willkie's tacit endorsement. Again, of course, as in the case of selective service, Roosevelt's decisive action was needed to put the plan through in September. But he would go no further than these steps till after the election.

As the campaign drew to a close there was increasing pressure on both Roosevelt and Willkie to give assurances of peace. The Gallup poll in mid-October showed 83 per cent opposed to an immediate declaration of war, and an almost even split as to whether it would ultimately be more important to save Britain than to stay out of the conflict. Bidding for the noninterventionist vote, Willkie predicted the United States would be at war by April if Roosevelt won, and declared that American boys were practically on the transports already. Roosevelt replied: "I have said this before, and I shall say it again and again and again. Your boys are not going to be sent into any foreign wars." The effect of such statements was to confuse and mislead public opinion about the possibility of saving Britain by "all aid short of war"; Americans were encouraged to believe they could have their cake and eat it too—as their historical good fortune encouraged them to suppose anyway. Complacency was prolonged. Willkie explained later: "In moments of oratory in campaigns we all expand a little bit." Roosevelt, for his part, had honestly not yet made up his mind at the end of 1940 how far it would be necessary for the United States to move toward war, and partisanship in Congress was so pronounced on issues of preparedness and aid to Britain that he knew he would need every possible Democratic vote there if his third administration were to function effectively. His final 1940 election margin (55 per cent of the popular vote) was so great as to suggest that he himself could have been re-elected even while speaking more frankly on foreign policy; but any lessening of his own popularity would have cost Democratic seats in Congress. As it was, party strength in the new House and Senate was almost the same as in the previous session—and every Democratic vote was needed in crucial tests later in 1941. Nevertheless, believers in constitutional democracy are likely to find the 1940 campaign promises of both candidates a disturbing ethical problem.

In more immediate concrete terms, Roosevelt's temporizing during the election campaign meant there was about three months' delay in accumulating momentum for the drive to aid England. Obviously the lag was not fatal—thanks to a few hundred men in the Royal Air Force for whom those months meant the desperate challenge of the Battle of Britain. Yet, even after the election it took Roosevelt several more weeks to decide how to relieve the financial handicaps which confronted the British war effort. Then he found the solution in Lend-Lease. The United States government would pay for matériel and lend it to the Allies, as one would lend a garden hose to a neighbor whose house was on fire. Roosevelt explained that he was trying to get rid of that "silly, foolish, old dollar sign" and substitute a "gentleman's obligation to repay in kind." There would be no new accumulation of enormous war debts to cause postwar recrimination and bitterness. The plan was adopted by the House of Representatives on February 8, 1941, by a practically straight party-line vote; in the Senate a few weeks later Willkie's endorsement helped to encourage one-third of the Republicans to cooperate in approving the measure.

The next problem was how to deliver the goods, for the Battle of the Atlantic was still going very badly. In getting congressional approval of the Lend-Lease bill, the Administration had had to fight hard to prevent its being amended specifically to prohibit American convoys; a substitute amendment was adopted declaring that the bill was not to be construed to *authorize* them. Yet, during March and April executive officials privately debated whether to convoy Allied merchant ships halfway across the Atlantic and let the Royal Navy take over from there, thus conserving its resources. While these policy discussions were in progress, in mid-April a United States destroyer, the *Niblack,* on a secret reconnaissance mission in Icelandic waters, encountered a German submarine and depth-charged her. Roosevelt drew back from the appearance of provoking a war and compromised upon a policy of "patrolling," rather than convoying. American warships in the western Atlantic were to search for Axis submarines and broadcast the position of any U-boat to the British, so that the Allies could come and do the actual depth-charging.

On June 22, 1941, Germany turned against Russia and thereby assured the Atlantic powers of a few weeks' respite. (They did not expect the Soviets to survive much longer than that.) American public opinion was still deeply divided, as was dramatized on August 12 when only a one-member plurality could be obtained in the House of Representatives to pass a bill holding draftees in service beyond their regular one-year tour of duty. Without such an extension the newly created army would disintegrate; nevertheless, the Republicans were solid in opposition, joined by many Democrats. Roosevelt became more convinced than ever that he must not run any unnecessary risk of defeat in Congress, lest the prestige of his leadership be so impaired that a reaction would set in against all interventionist moves.

While the House was voting on the draft, the President was meeting secretly with Churchill. Their conference was of symbolic importance in that Roosevelt endorsed an "Atlantic Charter" which declared the ideal objectives of Anglo-American policy "after the final destruction of the Nazi tyranny." Personal contacts were established and greater understanding reached between civilian and military officials of both countries. Also, the United States finally agreed to undertake full convoying as far as Iceland. (The new "shoot on sight" policy was announced to the American public a few weeks later in the context of excitement over hostilities between a U-boat and a "patrolling" American destroyer which was trailing it near Iceland.)

Roosevelt then finally resolved to push American convoys all the way through to England and to Russia. The Administration wanted Congress to eliminate from the Neutrality Act the prohibitions on arming American merchant ships and sending them into combat zones. Knowing that such a measure would encounter greater resistance in the House than in the Senate, the Administration persuaded the House to pass a bill merely permitting the arming of merchantmen. Then the Senate consented to tack on an abolition of the combat-zone restrictions, and the House was asked to swallow the combined program in a conference committee report. It did so by a margin of only twenty votes. Almost no Republicans in either House gave support to the combat-zone abolition; a quarter of the House Democrats deserted the Administration on this provision. The date was November 13, 1941—just twenty-four days before Pearl Harbor.

Thus, by narrow margins in highly partisan roll calls Congress had in effect approved an undeclared naval war against Germany in the Atlantic. Nothing more, however. The program would be enough to keep England alive, but certainly not enough to accomplish "the final destruction of Nazi tyranny." Public opinion polls still revealed heavy majorities for two propositions: (1) do not go to war right away; (2) do go to war if there is no other way to defeat Hitler. Unfortunately it was still not certain which of these two sentiments was really the more fundamental. The President had not informed the public that naval warfare would not be sufficient. Perhaps he had not yet decided so himself. In the early fall of 1941 all of Roosevelt's military advisers were telling him so, but even they wanted outright American intervention delayed, because of the continued unpreparedness of United States forces. The President himself in late September was actually looking for ways to reduce the Army in order to furnish more of its equipment to Britain. He decided against this move, but the fact that he considered it at all suggests strongly that he did not yet envisage any early American involvement in all-out war.

In review, the available evidence of the years 1939–41 indicates that President Roosevelt was never, even in his own mind, more than a very few months ahead of public opinion and congressional opinion on the road to war. His leadership was not heroically individualistic like Winston Churchill's. Roosevelt shared the prevailing American reluctance to become involved in another bloodbath, and also the persistent overoptimism that somehow it could be avoided—at least in the extreme form of total war. (After Pearl Harbor a similar attitude was reflected in his eagerness to get the Russians to help in the war against Japan.) Roosevelt also shared the characteristic American reluctance to look very far into the future in foreign affairs; this attitude also was again revealed in his wartime diplomacy toward Russia. With such propensities in

1939–41, the President utilized a leadership tactic of artful maneuver to get adequate support for each successive stage of intervention, at the possible expense of public understanding and acceptance of additional steps which might become necessary in the future. Such tactics may have contributed to the likelihood of an eventual impasse—an inability, perhaps, ever to arouse the public to a sense of need for bringing the war to an end by sending an American army to Europe.

Irrespective of Roosevelt's leadership, such an impasse would have been the logical outcome of decades of isolationism and antimilitarism in the United States. With such a tradition to overcome, probably no kind of leadership could have taken a united America into total war in the absence of a direct enemy attack. And if a democracy is not united, there is good reason to doubt that it is capable of waging a total war effectively (consider the example of France and Britain in the first half of 1940).

One new factor was entering the picture in late 1941, already of some importance and sure to become more so if all-out American participation had been delayed much longer; Soviet Russia was showing surprising ability to hold out against the Nazi invaders. Study of German records after the war has revealed that a very significant relationship existed between Russia's survival capacity and the nature of the U-boat war in the Atlantic, the fighting which was most likely to provoke an American declaration of war if anything could. Hitler's top naval commander was pleading for permission to attack American vessels in retaliation for their patrolling and convoying activities, but the Fuehrer repeatedly insisted that extreme forbearance must be shown toward the United States in the Atlantic until after the collapse of Russia. Thus the continued Soviet resistance meant that, had it not been for events in the Far East, the impasse of a mere naval war in the West might have continued for a long time. Many observers have argued that this outcome would actually have been highly desirable. As long as Russia survived, the Continental balance of power—smashed in 1940 with the destruction of France while Stalin stood aside—would now be re-established in a new form between the two dictatorships. Ideally they might even go on fighting each other to the point of mutual exhaustion. Meanwhile, American material and naval aid would suffice to keep Britain alive indefinitely in a defense posture. This view of the possibilities of the situation is very attractive, and presumably it will always have advocates, but one must remember that it rests on the assumption that Stalin would be willing and able to hold out indefinitely. If Russia had actually ever become so weak as to constitute no serious threat to Hitler's rear, there is reason for grave doubt that the United States and Britain could have mustered sufficient strength in time to withstand him at the English Channel. American mobilization was still inadequate both in pace and resolution for such a venture. In late 1941, despite Stalin's unexpected tenacity, time did seem to be running out on the Eastern Front.

## Attrition of Japan

Time was also running out in the Far East. American responsibility for that situation was so marked as to lend credence to the later charge that Roosevelt had deliberately led the country into war through the back door after having reached an impasse at the front. How much support does the presently available record give to such allegations? How was noninterventionism finally overcome?

In the late thirties, despite the traditional affirmation of the "Open Door in China," American noninterventionism with regard to the Far East was actually not much less pronounced than with regard to Europe. Public and congressional reaction to the *Panay* episode was good evidence of a determination to avoid military involvement. The primary United States defense responsibility in the area was the Philippines, and even that commitment might well come to seem less compelling when the Islands received their scheduled independence in 1946. Top-ranking American Army officers were already less Pacific-minded than Europe-minded in their strategic concerns, an attitude which has been of continuing importance ever since. When, for example, Congress refused to grant Navy appropriations for defense improvements on Guam in February 1939, the War Department approved the decision, objecting to any measure that might involve the United States in major operations in the Far Pacific. The American Navy had traditionally been Pacific-minded, anticipating war against Japan, but during 1939–41 even the Navy's focus of attention was shifting to the Atlantic. United States diplomats, for their part, were divided regarding the Sino-Japanese War in its early years. Generally, they were aware of the critical economic problems at the root of Japanese expansionism, but their personal sympathies lay with the underdog China; they tended to count on the ability of Chinese manpower to achieve ultimate success through a process of attrition. Meanwhile, United States pressure was confined to a refusal to grant diplomatic recognition of Japan's conquests, plus a moral embargo on sales to her of airplanes and bombs. After the beginning of the war in Europe a supplementary embargo was placed against sales of a few strategic raw materials. Then the fall of France brought new export-controls legislation in the United States. Restrictions were imposed on top-grade scrap iron and on aviation fuel; but Japan could still buy other grades of scrap (amounting to 85 per cent of her previous purchase selections) and also gasoline which she herself could reprocess for airplanes.

A great turning point in American-Japanese relations finally came in September 1940. Tokyo moved armed forces into northern Indochina and signed a Tripartite Pact with Germany and Italy. The major provisions of the Axis alliance were brief:

> *Article 1:* Japan recognizes and respects the leadership of Germany and Italy in the establishment of a new order in Europe.
>
> *Article 2:* Germany and Italy recognize and respect the leadership of Japan in the establishment of a new order in Greater East Asia.
>
> *Article 3:* Japan, Germany, and Italy agree to cooperate in their efforts on the aforesaid lines. They further undertake to assist one another with all political, economic and military means when one of the three Contracting Parties is attacked by a power at present not involved in the European War or in the Sino-Japanese conflict.

Article 5 indicated that the treaty did not refer to Russia; the United States rightly assumed that she herself was the target, but of what exactly? In many months of tortuous negotiations American diplomats struggled to determine just how far Japan was committed to war against the United States if the United States became increasingly involved in war against Germany. Actually, a secret exchange of notes between Japan and Germany at the time of the signing promised that the applicability of the pact to any particular "attack" would be decided by mutual consultation of Tokyo and Berlin. Thus, Japan's obligation was truly vague, but she was never willing to cancel it entirely. She hoped that uncertainty would deter the United States from taking steps in the Pacific to salvage the imperial remains of France, Britain, and the Netherlands. She was also eager in September 1940 to get Hitler's treaty guarantee that Japan, not Germany, would fall heir to the imperial spoils in East Asia, although it was Germany that was destroying the colonial powers themselves. Hitler's reward for this prospective generosity was Japan's persistence in maintaining before the United States the specter of a two-front war.

When Japan thus chose to throw her weight into the European balance, the American government's reaction was prompt and severe. An instant embargo was placed on *all* grades of scrap iron. Since a third of all the scrap iron used in Japan each year had come from the United States, the impact of this embargo was punishing. It was followed by a steady extension of other export restrictions during the winter of 1940–41, but the most unbearable potential embargo, on oil, was not imposed. During the winter of 1941, American and British military staffs held secret conversations in Washington regarding the allocation of responsibilities *if* the United States should become a full participant in the war. But the American government would not anticipate involvement by moving part of its fleet to Singapore to undertake the defense of the British base; and Roosevelt sought to emphasize the contingent character of United States commitments by giving only tacit approval to the secret war plans.

It is important to recognize that successive American moves against Japan, step by step in the form of export restrictions and military staff talks, were provoked by anxiety more for the fate of the British Empire and the tin and rubber of Southeast Asia than by a pressing concern to save China. However, the nonrecognition policy still stood. Tokyo found that once American trade restrictions had actually been imposed, for whatever reason, they acquired fixity, and Japan's self-restraint in Southeast Asia alone would not suffice thereafter to get them removed. Nor would even her willingness to minimize her obligations under the Axis alliance. Japan would have to reach a general settlement of outstanding issues with the United States. In particular, since the United States had never recognized Japan's conquests in China, she would now have to give those up also, and withdraw all her forces from China on a schedule agreeable to Washington.

By failing to content herself with a stalemate war in China, Japan was antagonizing the American government so much that finally by late 1941 the Administration was willing to risk war on behalf of China herself.

The decisive turning point came after the German invasion of Russia, when Japan decided to begin picking up the imperial spoils in Southeast Asia. During July 1941, she moved into southern Indochina. The previous occupation of northern Indochina could be rationalized as merely an encircling operation directed against China. But in the south, Japan was unmistakably on the march toward wider realms. American retaliation was harsh. All Japanese assets in the United States were subjected to license controls so that they could be used only for purchases approved by Washington; no licenses were in fact granted for the purchase of the indispensable commodity—oil. The British and Dutch followed suit. The effect was to cut off 90 per cent of Japan's current oil supply at a time when her stockpiles would suffice to meet her minimum wartime needs for only two years.

Japan's southward aggression had thus provoked the United States to a point where American policy had acquired most of the character of an ultimatum. Imprecisely, but effectively, a time limit had been established. The Japanese warrior was bleeding steadily in China; yet he would get no more blood transfusions until he abandoned the fight entirely. He could refuse to yield, and resolve instead to risk seizing the vital fluid in the Indies. But in that case he would have to allow himself several months to conquer and restore the distant wells. Subtracted from the scant two years of existing supplies, this interval meant that any successful aggression would have to begin within a year; seasonal weather conditions would also affect the necessary timing. The United States government could not foresee exactly how soon Japan would have to decide whether to fight for Malaya and the Indies or to back down altogether; and no official American communication was phrased as an ultimatum to Japan. But an early decision in Tokyo was as indispensable as gasoline engines, if American policy remained adamant. Conceivably, of course, the policy might yet be softened. In September 1941, however, this possibility became more than ever unlikely when Roosevelt refused to meet personally with Japanese Premier Konoye to arrange a settlement, unless Konoye would give satisfactory advance assurances of withdrawal from China as well as from Indochina. The Premier felt unable to do so.

Japan's final offer came on November 20, 1941—a *modus vivendi,* or way of getting along temporarily. Its essence was that Japan would make no further advance in Southeast Asia and would withdraw promptly from southern Indochina—if the United States would make concessions on its embargoes, would persuade the Dutch to do likewise, and would cut off economic and diplomatic support from Nationalist China. The Tripartite Pact was still not abandoned, but the Japanese were willing to minimize its obligations. In effect, the United States was being offered a chance to buy time at the expense of Chinese morale.

The War and Navy departments had recently wanted more time. In the Pacific, time seemed to be running in favor of the United States; Japanese oil reserves were draining away, and a promising new American program was under way to reinforce the Philippines. Even in Europe, Russia's endurance made the passage of time seem less harmful than heretofore. To be sure, the Japanese might break a truce—but so could the United States, indirectly, if circumstances warranted.

However, China was not to be explicitly abandoned, even temporarily; the State Department was particularly vehement on this principle. Yet the President and Secretary of State wavered. They encouraged the department to give very serious consideration to putting American support of China back on the basis which had existed prior to the invasion of southern Indochina (when United States pressure on Japan had been mildly coercive, involving the early embargoes, but did not yet include the all-out economic sanctions which began strangling Japan on July 26 and were being used to save China as well as Indochina). A new American draft of a *modus vivendi,* designed to re-establish roughly the status quo of early July for a three-month period, was prepared. In its specific terms, the main difference between the Japanese offer and the American proposal was that the latter avoided all reference to aid to China, meaning that United States support for her would continue, but without the recent reinforcement formed by ruinous economic sanctions against Japan.

The American government never went through with this offer. After wavering, Roosevelt and Secretary of State Hull rejected the plan. In effect they insisted not only on continuing to give China diplomatic support and a little material aid, but also on continuing the system of extreme coercive pressure on Japan, very largely for the sake of China's liberation. The President's motivation on November 26, the day of decision, is still unclear. Chinese and, to a lesser extent, British influence was being brought to bear against the suggested American offer to Japan. Secretly decoded Japanese messages entitled Roosevelt to presume that the American draft would not be acceptable to Tokyo as it stood and that no time would be allowed for further negotiations to narrow the gap. If so, the United States offer might look like futile appeasement. If by chance the Japanese did prove willing to accept the proposal, it would still be condemned as appeasement by large segments of the public, for Americans had finally become aroused about Japan's aggressiveness in Asia—and at the same time were wildly overconfident about her vulnerability. Polls showed that the body of opinion willing to risk war rather than see Japan grow stronger had reached 70 per cent (close to the anti-German total at that time).

Instead of a *modus vivendi,* Japan was handed a ten-point program which epitomized America's maximum demands. Thus the record was officially "clarified" for the future. In form, there had still been no ultimatum issued, but in substance and effect one was operating. The Administration knew there was practically certain to be fighting soon somewhere in Southeast Asia, but to what extent the United States would be involved was still not certain. And whether events in the Far East would produce full-scale war with Germany was even less certain. Perhaps the handling of the *modus vivendi* did constitute a "back door to war," but if so the door was surely very poorly lighted and very hesitantly opened.

The Administration clearly expected the Japanese to attack British and Dutch but not American outposts, thus avoiding extreme provocation toward the United States. Even so, the American Far Eastern fleet would almost certainly have become involved at once, under existing understandings. But would the larger American fleet at Hawaii also have gone out promptly into Far Pacific action, recognizing that the Japanese Navy was already stronger than all other Pacific fleets combined? In those circumstances American opinion would certainly have been divided. What kind of war would the United States have fought against Japan? Would the total liberation of China have been maintained as a fundamental war aim of the American people?

Additional imponderables appear where Hitler is concerned. In the last days before Pearl Harbor the Germans actually reconsidered briefly whether it was really desirable for them to get the United States fully into war on the Atlantic as well as on the Pacific front. On December 5 they finally arranged with the Italians to go to war against the United States as soon as Japan did. But suppose the Germans had held back and left their Far Eastern ally in the lurch? What then would have happened to the Atlantic-first policy on which the Administration had grounded its actions? Would an American declaration of war against Germany still have been forthcoming from a Congress preoccupied with naval warfare off Malaya, or with avenging Pearl Harbor? If not, a "back door" would have been opened on "the wrong war in the wrong place at the wrong time," at least from the point of view of the Administration. This possibility was sufficiently clear, even before the President's *modus vivendi* decision, to make a deliberate "back door" strategy seem improbable. (Of course, from the perspective of those who would rely on a Hitler-Stalin stalemate, it might well seem desirable for America to concentrate on war with Japan. Unfortunately, however, any reliance on Russia was bound to be insecure.)

The Japanese felt they were trapped in November 1941. But in another sense, United States leaders were also trapped—headed in the Atlantic into the possible *cul de sac* of indecisive naval war as a result of the inescapable heritage of isolationism; headed in the Pacific into the distracting effort of another naval war because of a widespread sense of moral obligation to China.

Actually, though, it was the Japanese and the Germans who rescued American leadership from its fundamental handicap by suddenly providing a public united for total war on both fronts. The Japanese did so by attacking Pearl Harbor instead of only Singapore, the Germans by declaring war on America without waiting for America to declare war on them. The enemy powers thus suddenly smothered the chronic seductions of isolation as no American leadership by itself could have done, thereby releasing at last the full energies of the whole nation for total victorious war.

## SELECTED BIBLIOGRAPHY

CHURCHILL, WINSTON S. *The Second World War,* vols. 1–3. Boston: Houghton Mifflin, 1948–50.

DRUMMOND, DONALD F. *The Passing of American Neutrality, 1937–1941.* Ann Arbor: University of Michigan Press, 1955.

FEIS, HERBERT. *The Road to Pearl Harbor: The Coming of the War between the United States and Japan.* Princeton: Princeton University Press, 1950.

GREW, JOSEPH C. *Turbulent Era: A Diplomatic Record of Forty Years,* vol. 2. Boston: Houghton Mifflin, 1952.

HULL, CORDELL. *The Memoirs of Cordell Hull,* 2 vols. New York: Macmillan, 1948.

ICKES, HAROLD L. *The Secret Diary of Harold L. Ickes,* 3 vols. New York: Simon and Schuster, 1953–54.

KING, ERNEST J., and WALTER M. WHITEHILL. *Fleet Admiral King: A Naval Record.* New York: Norton, 1952.

LANGER, WILLIAM L., and EVERETT S. GLEASON. *The World Crisis and American Foreign Policy: The Challenge to Isolation, 1937–1940.* New York: Harper, 1953.

———. *The World Crisis and American Foreign Policy: The Undeclared War, 1940–1941.* New York: Harper, 1952.

MATLOFF, MAURICE, and EDWIN M. SNELL. *The United States Army in World War II: Strategic Planning for Coalition Warfare, 1941–1942.* Washington, D.C.: Department of the Army, Office of the Chief of Military History, 1953.

SCHROEDER, PAUL W. *The Axis Alliance and Japanese-American Relations, 1941.* Ithaca: Cornell University Press, 1958.

SHERWOOD, ROBERT E. *Roosevelt and Hopkins: An Intimate History.* New York: Harper, 1948.

STIMSON, HENRY L., and MC GEORGE BUNDY. *On Active Service in Peace and War.* New York: Harper, 1947.

TANSILL, CHARLES C. *Back Door to War: The Roosevelt Foreign Policy, 1933–1941.* Chicago: Regnery, 1952.

# CHAPTER 4

# TOTAL WAR AND LIMITED DIPLOMACY

IN THE SECRET Anglo-American military staff conferences early in 1941, it had been agreed, "should the United States be compelled to resort to war," that "since Germany is the preponderant member of the Axis Powers, the Atlantic and European area is considered to be the decisive theatre. The principal United States military effort will be exerted in that theatre, and operations of United States forces in other theatres will be conducted in such a manner as to facilitate that effort." Promptly after Pearl Harbor, the American military chiefs reaffirmed the doctrine: "Much has happened since February last, but notwithstanding the entry of Japan into the War, our view remains that Germany is still the prime enemy and her defeat the key to victory. Once Germany is defeated, the collapse of Italy and the defeat of Japan must follow."

## Concentration on Europe

The Europe-first policy was never seriously reconsidered except once, briefly, in the summer of 1942. Fundamentally, American policy-makers simply cared more about what was happening to Europeans under Hitler than about Asians, but the bald military arguments also seemed decisive. (1) Germany's war potential was undeniably much greater than Japan's. To be sure, it was being drained in Russia, but

Stalin might not hold out indefinitely—and even in a stalemate, Germany's scientific resources were so great that she was much more likely than Japan to be able to take advantage of time to develop decisive new weapons. (To an important degree she actually did succeed in doing so with jet fighters, rockets, and snorkel U-boats, but mercifully not with atomic bombs.) (2) An offensive across the Pacific would inevitably be subject to many months of delay while American naval strength was being constructed. Only in the European theater could large American land forces be brought into action without experiencing months of demoralizing idleness. (3) British and Russian strength was necessarily concentrated against Hitler. Until he was defeated, against him only could the United States hope to get the advantage of much active assistance from its major allies. And prolonged attrition of their military capacity would reduce their contribution to any ultimate offensive against Hitler, even if America should choose to get along without their assistance in destroying Japan first.

In retrospect, was it a mistake to relegate the Pacific theater to second place? This question may most succinctly be answered by reference to the one major consequence of the Far Eastern war as a whole which seems to have been fundamentally harmful to the interests of the United States—the creation of a power vacuum for Communism by the excessive weakening of Nationalist China and Japan. Reconsider the strategy from this point of view. A Pacific-first program would certainly not have been more likely than was the Atlantic-first policy to produce a compromise peace with Japan which would have left her effective herself in the Far Eastern balance of power. But after Pearl Harbor the acceptability of such an outcome was so slight anyway that it is more important to recognize that a Pacific-first policy would not have assured Nationalist China either of the strength to overcome the Communists after liberation from Tokyo. Japanese aggression, not Chinese Communism, was the enemy recognized in Washington; consequently, any American strengthening of Chiang Kai-shek's armies for their eventual use against the Communists would be quite unintentional. Very likely, to be sure, an American policy aimed at defeating Japan first would have incidentally left Chiang with more weapons and more American-trained troops than he actually accumulated; conceivably such a policy would have liberated China so much sooner as to make a lasting difference in the regime's morale. But these would have been accidental by-products of strategies that the United States would have consciously been directing only against Japan, not against Communism. Therefore, the likelihood that there would have been sufficient unintended "bonuses" to

Chiang to enable him, on his own, to block postwar Communist expansion does not seem great—not enough to offset, in retrospect, the surer advantages of an Atlantic-first policy in World War II. Besides, even within the adopted framework of a Hitler-first policy, there was room for lesser strategic variations which could substantially affect Chiang's strength vis-à-vis the Communists; even in a second-ranked theater such opportunities could have been readily utilized if American policy for the Far East had been less exclusively focused on operations against Japan.

This preoccupation with the conquest of Japan was central—parallel to and subordinate only to the determination to destroy Hitler. The Atlantic theater might outrank the Pacific, but within each respective area a speedy total victory over the Axis enemy was the overriding American objective. The characteristics and effects of this attitude were similar in both spheres, but in the Far East it was superimposed upon the special complexities of a vast long-term national-social revolution. Consequently, rather than tell the story twice, now and later, of United States wartime relations with China, the record there will be examined in Chapter 6 in connection with a discussion of the rise of Chinese Communism. The general problem of American pursuit of total victory can be satisfactorily considered in the light of the European phase alone in World War II, even though the perspective does detract somewhat from a full global view of the diplomatic and military operations.

## The Anglo-American Debate over Strategy

The kinds of problems which beset American strategy in Europe may be approached by noting that the Atlantic-Pacific argument can take the quite different tack that the Atlantic phase was not considered primary enough, that too many forces were actually diverted to the secondary Pacific theater in order to launch offensives which could better have been delayed till after Germany's defeat. In rebuttal, it may be observed that a prolonged time lag would have allowed Japan to consolidate her initial gains in Southeast Asia; the ultimate American roll-back campaigns might well have been more painful; and Russian intervention—at a price—might well have seemed even more essential than it actually did. At any rate, the public determination to "Remember Pearl Harbor!" would hardly have permitted much lower priorities than were actually given to the war against Japan. The more substantial arguments revolve around the allocation of fighting strength within the European theater itself, between the Mediterranean and the English Channel as avenues of approach to the conquest of Germany. British and American strategists repeatedly clashed over this issue during the war, and the nature of their dispute reveals much about the attitude of both countries toward force as an instrument of national policy.

The final push which had made the United States a full participant in the war was so infuriating that the American public was quite ready to accept a very high degree of mobilization of the nation's resources, both human and material. Potentially available were manpower almost equal to that of Germany and Japan combined, steel production twice that of the two countries with all their possessions at the height of their power, and vast amounts of exportable food. Americans had a deep sense of industrial superiority to any other power on earth, and they were eager to make the fullest possible use of their productive might in the struggle in which they found themselves. They felt quite able to expend material resources lavishly; blood was much more precious, but it too seemed more expendable to American strategists than to British.

In any war, the passage of time seems psychologically much less bearable than in conditions of peace. In peacetime a few months more or less of delay in the adoption or execution of a particular government policy rarely encounters anywhere near the degree of public concern which is stirred in the emergency mood of war. Most peoples experience this change of perspective; the United States seems especially prone to it. Her normal good fortune makes the unaccustomed hardships of war particularly painful. Quite apart from the added physical dangers, any time spent in military pursuits has seemed to most Americans to be time virtually subtracted from life, not really part of life. Even a slight prolongation of this purgatory is deeply resented. "Hurry up and get it over with, so the boys can come home!" The typical impatience can make its impact effective through elections, which continue to be held on schedule even in wartime.

This almost universal demand for speed in reaching victory and the recognition that America's unique strength lay in her farms and factories were powerful underlying motives for the choice of European strategy made by the American Army planners in World War II. It was clear that the continued success of Hitler's U-boat campaign, together with the requirements of the Pacific war, made the shortage of shipping the major bottleneck against any speedy application of American arms and manpower to the defeat of Germany. The limited number of available ships could carry more troops and more cargo to England and thereafter to northern France faster than to any other European gateway. Moreover, a concentration of forces in England for a cross-Channel invasion of the Continent would simultaneously serve the other essential purpose of providing defenses for Britain against air attack and a possible German invasion. Any other strategy would require a dispersion of Allied forces, since large numbers would inevitably have to be kept in England whatever was done elsewhere. To American strategists the direct route across the Channel through northern France to the heart of Germany seemed the shortest road to victory and the route which the Allies should concentrate their forces to traverse.

British strategy differed. To Churchill and his military chiefs the cross-Channel invasion was an undertaking which should be postponed as a kind of *coup de grâce* for the final defeat of Germany after she had already been decisively weakened by other means. They emphasized that the Reich's interior position gave her substantial advantages in shorter lines of communication and in mobility of ground forces from one front to another. To counter these enemy conveniences the British favored a strategy designed to keep German strength dispersed, so that it could not be shifted readily from one threatened front to another. For this purpose the Western Allies would take advantage of their special strength on the sea and in the air. Small-scale amphibious commando raids would be made at scattered points; guerrilla warfare would be fostered within Occupied Europe on the part of local resistance movements with Allied advice and air-dropped supplies; heavy reliance would be placed on indiscriminate mass bombing to weaken the German population's will to resist; the Nazi empire would be invaded from a variety of positions around its periphery; Germany would be prevented from concentrating her strength for a defense of the Channel coast.

In large part, these projects were advocated by American strategists also. United States planners always laid great stress on the need to keep Russia on her feet against Germany if at all possible, in order that German strength would be divided between an East and a West front. But the American Army chiefs balked repeatedly at the British view that attrition in Russia, plus bombing, subversion, and commando raids, would not suffice to soften Germany for a successful cross-Channel invasion—that diversionary Anglo-American campaigns in the Mediterranean area would also be required. Thus Army Chief of Staff George Marshall steadily resisted while Churchill successively pressed for drives in French North Africa, in Italy, and in the Balkans, eagerly improvising new projects to take advantage of sudden shifts in the constellation of forces.

Only in the Balkans did Churchill ultimately fail to get a major offensive mounted, and in that case mainly because neutral Turkey refused to run the risk of joining the Allies. Elsewhere in the Mediterranean, considerable Anglo-American resources were gradually put at the disposal of British strategies seeking to outmaneuver the enemy, despite the continuing United States propensities simply to outproduce him and to rely on overpowering him directly.

Great Britain had been in a position for many generations where her worldwide interests were far greater than her capacity to maintain them by the application of force on the ground. Unlike the United States she could not traditionally hope to ride roughshod over all who might rise against her. Her reliance would have to be placed in skillful maneuver based on flexible sea power and the political-military principle of "divide and rule." This habit of mind was clearly evident in the Anglo-American strategic disputes in World War II, but there was also a more particular motive for British hesitation to invade across the Channel. English leaders, military as well as civilian, were haunted by memories of the futile carnage of trench warfare in World War I; it must never, never be repeated. Americans had experienced only a very few months of such fighting; during most of their 1918 participation the front had been in motion, not stalemated, and they were confident that with new armored vehicles and air weapons a new Western Front, once opened, could be kept in motion also. But the British, seared by 1914–18, could not escape the insistent fear that static warfare would re-emerge if approximately equal armies ever clashed again on a limited fortified front like the French channel coast or the line of the Rhine. Actually, in World War II this narrow balance developed only in Italy, and there there was near-stalemate; so the British fears were not without some justification. They were accentuated by an awareness that England's population was aging and her birth rate falling; the long-term effects upon Britain of any renewal of trench warfare in World War II might be as debilitating as it had been for France in World War I. Moreover, English leaders could not be entirely confident of the morale of their troops or of the home front if the stalemate situation, which so many millions remembered so bitterly, reappeared. American planners were not impelled to take account of so universal a revulsion from the sheer slogging unimaginativeness, the "stupidity," of Allied strategy on the Western Front a generation earlier. The English had reason to set a high premium on intellectual creativity, opportunistic improvisation in military planning. Churchill himself was the very embodiment of this attitude; even in World War I he had been prime sponsor of a peripheral Dardanelles strategy to circumvent the deadlock in France. The Americans in their vast self-confidence were trying to stick firmly to a fixed plan to accumulate overpowering forces for a direct smashing blow across the Channel into the heart of Germany, but this reminded the British too much of World War I with its reciprocal pyramiding of greater and greater forces by both sides—all to no avail. English leaders preferred to try to reduce German channel strength by diverting it to the Mediterranean, even if a diversion of Anglo-American forces was also required to accomplish the purpose.

It is important to recognize that American leaders were usually not willing to accept these British strategic arguments at face value as grounded on military considerations. They always tended to ascribe "imperialistic," postwar power-politics objectives to the British. And this suspected twisting of military arguments for political purposes deeply offended the American commanders' sense of responsibility for upholding what they thought to be the established standards of the specialized profession of arms in a democratic society. Soldiers, they emphasized, should not be politicians. The constituted civilian authority in America was the President. He was deliberately postponing awkward decisions regarding postwar territorial settlements. Clearly, the overriding objective of his policy was the speedy defeat of the Axis powers; his "unconditional surrender" formula had the formal endorsement of Churchill and the British War Cabinet. Accordingly, top United States officers felt that their judgments and recommendations as military men had better be guided almost exclusively by considerations of technical efficiency in the rapid pursuit of the desired victory through appropriate strategy, tactics, and logistics. In principle, of course, the President remained free to overrule such advice and to redefine political objectives, but until he did so American generals would avoid intruding on that area of high policy. This kind of civil-military relationship, in the dominant Pentagon view, embodied the essence of professionalism; accordingly, the Americans were inclined to resent the somewhat different pattern which prevailed among their British counterparts. Actually the Americans were not meeting their own full professional responsibility for pressing upon the President the arguments for altering some of his broadest wartime policies in order to reinforce the postwar security of the nation. In fact the military planners themselves were absorbing the Administration's single-minded preoccupation with victory. Thus, the potential advantages of a true military professionalism were dissipated in the President's failure to formulate precise foreign policy requirements for the military planners, and in their own failure to press him to do so realistically.

Churchill in England was much more willing than Roosevelt to intrude his day-to-day views directly upon his military advisers, and the influence of diplomats was vastly more effective in London than in

Washington. Consequently, the political motivation in British military planning was much more clearly defined than in America. The United States has had little experience with the continuous use of force as an instrument of national policy. If war is viewed simply as an unavoidable means of punishing some particular, intolerably provoking nation—or else as a way of constructing a new world order essentially free of power politics—then precise political direction is likely to be actually suspect as a harmful nuisance; it might obstruct the military experts and delay the victory. This popular American attitude of deference to the particular recommendations of professional military leaders in wartime was shared to a significant degree by President Roosevelt himself, especially because he recognized that the developing Pentagon attitude toward the war as a whole was close to his own. In contrast, the professional diplomats in the State Department and Foreign Service were objects of suspicion to Roosevelt, as they traditionally were to the public at large. Consequently, the President showed extreme reluctance to overrule the congenial military advice he continually received from the Joint Chiefs of Staff, or to formulate more precise political directives for their planning—much less look to the State Department for broad strategic guidance. Yet even in England, where political motivation was clearer, it was probably not until mid-1943 that the particular focus of British concern became farsightedly anti-Kremlin to anywhere near the degree that Americans had suspiciously supposed it to be all along.

Since the war, with the change in attitudes toward the USSR, the anti-Soviet political motivation of Churchill's strategy—still somewhat exaggerated in the United States—has been transformed from a source of suspicion to one of admiration. In large measure this praise is justified. A deliberate effort to look ahead and take precautions against new imbalances of power was clearly desirable in the cold light of previous experience with Russian expansion and Communism. But it is worth noting in rebuttal that Churchill's foresight did not prevent him from putting heavy reliance on British support to Communist-led guerrilla movements, thus favoring the immediate objective of killing Germans over the postwar viability of independent governments in some countries. Furthermore, Churchill's repeated efforts to create the missing link in Mediterranean strategy by persuading Turkey to enter the war seem highly questionable in the light of the need for a strong postwar Turkey. Turkey could have exhausted herself fighting Germany just as Nationalist China exhausted herself fighting Japan, and might similarly have become a postwar victim of Communism, unless great Anglo-American resources were thrown into the local struggle both during and after the war.

Another argument deserves attention; if valid it would be much the most important. Suppose that by ignoring the Mediterranean it had actually been possible to cross the Channel in 1943 and defeat Germany by early 1944, instead of a year later. The war might well have ended with Russian forces not yet much beyond their own borders. Conceivably then a firm diplomacy might have saved much of Eastern Europe from Communism. Thus an avowedly nonpolitical American Army strategy, if pursued without any distraction at all, might perhaps have produced more useful anti-Russian political consequences than the political-military proposals Churchill advocated. There are so many ifs in this argument that it is not very persuasive, but it does point up one fact which always deserves consideration: simply that time is truly very precious in war, not just for a sense of professional efficiency or for the morale of those who want to get it all over with, but for the enduring social-political effects of the war itself upon any attainable peace settlement. Thus the passage of time is properly a factor to be weighed by top civilian leaders in giving overall political guidance to military planners, though it should not always be the single overriding factor that Americans tend to make it. This consideration together with other aspects of British and American thinking should be kept in mind in the following survey of the evolution of grand strategy for the defeat of Hitler.

## The Postponement of Settlements for Postwar Europe

The first important controversy revolved around American and Russian efforts to induce the British (who had the largest uncommitted forces at hand) to play the major role in a cross-Channel invasion attempt in 1942. It was recognized that this project, code-named "Sledgehammer," might well be a sacrifice operation. At the very best only a small beachhead could be established and held through the winter till a major invasion effort followed in 1943. More likely the Germans would be able to force a complete abandonment of the position, inflicting heavy losses on the Allies. But at least German airpower would be drawn away from the Eastern Front, thereby improving what Americans thought were the poor odds that Russia would fully survive Hitler's second summer offensive. Soviet survival till 1943 would be almost indispensable to prevent German concentration in the West against the major invasion scheduled for that year. Therefore a cross-Channel expedition in 1942, though admittedly premature, seemed justifiable to American Army planners and to the President. In May and June, Soviet Minister Molotov visited London and Washington and was repeatedly told by Roosevelt and other Americans that the United States fully expected to invade France in 1942; the British, however, insisted that no promise was being made. Churchill preferred landings in French North Africa, designed to be linked with an overland drive from Egypt in an effort to clear the enemy from the southern Mediterranean. American Army planners, however, had strong, well-grounded fears that such operations, once begun, would progressively lead to a permanent diversion of strength from the main cross-Channel effort, very likely postponing it till 1944. "Sledgehammer," even if costly, would be much less distracting in the long run. For a few weeks in early July 1942 top American officials even weighed seriously the idea of switching major efforts to the Pacific war rather than sink forces into peripheral operations around Europe. Roosevelt was determined that large American units should see action somewhere before the end of the year; he did not want a "phony war" atmosphere to develop at home while troops grew stale in England waiting for a 1943 invasion. And he, like many others in England and America, felt a sense of shame, not satisfaction, that the Russians were doing almost all the fighting in the war. After five days of discussion in July he reaffirmed the Atlantic-first policy and made the fateful decision that if the British would not cross the Channel, there was nowhere else to go in 1942 but into French North Africa or the Middle East. Agreement was finally reached on a North African campaign, which was launched in November.

The Russians, even more than the Americans, regarded this Mediterranean expedition as distinctly second best. The Soviet Union was being left to bear the brunt of German offensive power a thousand miles inside her borders. The fact that she did hold out at Stalingrad in December 1942 should not obscure the risk the Western Allies were taking in avoiding a sacrifice play in France. Fortunately the outcome was successful, and the United States and Britain were not left alone to face Germany in the following years. But there may have been a genuine sense of betrayal in Moscow. At any rate, American leaders themselves for a long time felt ashamed of their misleading second-front statements. Partly for this reason, renewed costly efforts were made to get United States supplies to the Red Army; by 1943 major shipments were at the front lines and contributing heavily to the mobility of Russian forces as they began their counteroffensives.

Another decision of far-reaching importance was made by the President at the beginning of 1943, partly in order to reassure the Soviets.

With the consent of the British Cabinet, Roosevelt announced at Casablanca that the price of peace was the Unconditional Surrender of the Axis powers: "It does not mean the destruction of the population of Germany, Italy, or Japan, but it does mean the destruction of the philosophies in those countries which are based on conquest and the subjugation of other people." Such an objective barred acceptance of a stalemate war or a separate peace which would leave Germany free to fight Communism. This fact must have gratified the Russians, though they often indicated that they regretted the effect that the publicity surrounding the Western declaration would have upon Germany's will to resist. The Soviets seemed to doubt the wisdom of advertising in advance the ruthlessness of one's intentions. The American leaders, however, were determined to make it emphatically clear that they would not settle for any regimes in Germany, Italy, and Japan which would allow the perpetuation of a military tradition; the ouster of Hitler, Mussolini, and Tojo would not be nearly enough. The enemy population must taste defeat and undergo social transformation so that never again would a demagogue be able to convince them, as Hitler did after 1918, that they had been betrayed from within, seduced by the false promises of a Woodrow Wilson to lay down their arms voluntarily. The American and British people seemed to want reassurance; publicists and legislators were volubly resentful over recent American military deals with Germany's former collaborators in French North Africa.

Criticism of the Unconditional Surrender formula has been voluminous. Fundamentally it rests on the balance-of-power concept that the destruction of Germany's military capacity was bound merely to substitute domination of the Continent by one expansionist power, Russia, in place of another, Germany, and thus leave the position of Britain and America as hazardous as before; the wise objective should have been to try to create a new balance within the Continent itself between Germany and the Soviet Union. However, the tragic fact was that Adolf Hitler's nihilistic ambitions could not possibly be contained in any such conveniently balanced framework. He would not give up his personal power, and no trust whatever could be placed in peace terms he might be driven to accept. His actions had proved that the utter destruction of his Nazi regime was the prerequisite of any security in Western Europe—regardless of how disastrous the consequences of that destruction might later be for the long term balance of power in Europe. What reliable security for Britain and America could be based upon a Continental balance between two mighty totalitarian states, both fervently detesting Western democracy and both driven toward insatiable aggression by their ideologies and by the internal compulsions of their repressive regimes? They had already linked arms against the West once (1939–41). Even today in retrospect it seems true that the elimination of German national power as a barrier against Communism was less of a danger than the survival of Nazism would have been. One of the two totalitarian governments had to be eradicated before there could be any real hope of reconstituting its nation as a nontotalitarian barrier to help contain the other—and of the two governments the Nazi was the more immediate threat to the very existence of Western democracy. Elimination of Nazism had to have priority at least.

Consequently, the critics of Unconditional Surrender must show that, except for the threat of that formula, a *non-Nazi* regime (probably drawn from the professional officer corps) would have been able to win power in Germany and make a compromise peace. The evidence on this point is very inconclusive. On the one hand, in spite of the formula, there did develop a high-level conspiracy with repeated attempts on Hitler's life, one almost successful; yet the plotters were finally rooted out by the secret police. On the other hand, one can always argue in rebuttal that conspiracies would have been more numerous and ultimately would have been successful if non-Nazi Germans, especially army officers, had been offered an alternative more hopeful than Unconditional Surrender.

Actually the worst consequences of the formula probably came in the last eight months of the war; then its bare public outlines (which in theory could have merely duplicated the generosity of Grant at Appomattox) were instead filled in with horrifying items from Secretary of the Treasury Morgenthau's plan to reduce Germany to an agrarian state. These details made it much easier for the Nazi chieftains to persuade millions of prospectively "surplus" Germans that it would be better to die fighting than to slave in Siberia or starve in a smokeless Ruhr. But the Unconditional Surrender formula had really not been designed as a cover for a Morgenthau Plan—or for any specific postwar arrangements for Germany beyond an aim to eliminate Nazism and militarism. Quite the contrary. Its essential purpose was simply to *postpone* all such controversial questions as long as possible and thus to keep the attention of a very divided coalition of peoples focused together on the immediate needs of a great war effort. To be realistically specific about war aims would only provoke disagreement, dissipate enthusiasm, and perhaps prolong the war itself. The idealistic generalities of the Atlantic Charter (August 1941) and the United Nations Declaration (January 1942) were one technique for providing rousing, unifying war cries. Another technique, tuned to a growing vengeful spirit bred by the war itself, was to raise the slogan of Unconditional Surrender. Both techniques were essentially means of postponing embarrassing issues, nor were they wholly inconsistent, though they differed in spirit. The progress of the war itself was given loose rein to determine which outlook would tend to prevail.

How clear were the issues that Roosevelt was avoiding?

In November and December 1941, when the Germans were hammering at the very gates of Moscow at the climax of their first offensive, Stalin had shown himself determined to discuss with Britain the details of postwar boundary settlements in Europe. Foreign Secretary Anthony Eden was in Moscow seeking a formal treaty of alliance with the Soviet Union in order to minimize the danger that Stalin would make a separate peace with Hitler. But the Russian asking price was British recognition of postwar boundaries roughly the same as those Stalin had acquired in his former deal with Hitler. Estonia, Latvia, Lithuania, nearly half of Poland, and parts of Finland and Rumania—all lost in 1917–21 and reconquered in 1939–40—would have to be restored to Russia. In 1939 Britain's refusal to make a similar deal against the will of the peoples concerned had been a major reason for Stalin's bargaining with Hitler, with all the disastrous consequences for Western Europe. Now in January 1942 Churchill was extremely reluctant to make these concessions, especially since at this time he shared the common Anglo-American view that Russia would not emerge from the war strong enough to impose the demands at a later date.

But Stalin continued during the winter of 1942 to use his very weakness as a diplomatic asset by throwing out broad public hints of readiness to negotiate a separate peace with Germany. By March the British government was willing to compromise to get a formal alliance. The Baltic countries would be sacrificed—but not Poland unless her own government-in-exile could be persuaded to yield up the territories Russia was demanding. Most of that eastern Polish area was land which the British in 1919 had determined to be ethnically non-Polish anyway (beyond the proposed "Curzon Line" frontier). On the other hand, in 1939 England had formally declared war in defense of Poland against Germany; therefore there was a continuing sense of moral obligation not to use outright force in inducing the Poles to come to terms with Russia in the East. Rather, Britain offered them the hope that quick territorial concessions would satisfy Stalin's appetite and persuade him to leave the remainder of Poland fully independent even if Russia emerged from the war very powerful. The Polish government-in-exile was unwilling to take such a chance. The Roosevelt Administration emphatically threatened to dissociate itself explicitly from any Anglo-Russian treaty which prejudged postwar boundaries at all; such controversial issues supposedly ought to be postponed. Partly as a substitute tactic for keeping Russia in the war against Germany, the Americans made their half-promises of a second front. Then as it became clear that Hitler was mounting a new summer offensive against Russia—not negotiating for a separate peace—the British themselves were emboldened once more to press for an Anglo-Russian alliance without any territorial clauses at all, and this time the Soviets accepted it. The treaty was published June 12, 1942. Postwar problems would be held in abeyance for several more months at least, but Russia's minimum postwar objectives were explicitly clear. If the continued fighting enhanced rather than reduced her relative power, she might well be in a position to insist upon her demands or perhaps expand them at a later date.

The turning point in the war actually came at the end of 1942 and the beginning of 1943. The Russians held at Stalingrad and began their counteroffensives. Allied shipbuilding began to outpace the tonnage destroyed by U-boats; the Battle of the Atlantic was being won. Anglo-American forces established themselves in French North Africa, although unforeseen delays dragged out that campaign until May 1943 and eliminated—as the Americans had feared—any chance of a cross-Channel invasion until 1944.

When the Russians managed to hold at Stalingrad at the end of 1942, Prime Minister Churchill began to show concern for the postwar position of the USSR, and increasingly there was an element of anti-Soviet motivation in his advocacy of expanding the Mediterranean operations. At the beginning of February 1943 a private memorandum of his own views still contained a prediction that the great powers, fearing the devastation of World War III, would make "the most intense effort . . . to prolong their honourable association . . . by sacrifice and self-restraint"; but he also anticipated that "Great Britain will certainly do her utmost to organize a coalition resistance to any act of aggression committed by any Power, and it is believed that the United States will cooperate with her, and even possibly take the lead of the world, on account of her numbers and strength, in the good work of preventing such tendencies to aggression before they break into open war." * During the course of the year 1943 Churchill pressed publicly for the closest possible postwar union between Britain and the United States, and in March, Foreign Secretary Eden took the initiative in exploratory Anglo-

* Winston S. Churchill, *The Second World War*, vol. 4 (Boston: Houghton Mifflin, 1950), pp. 711–12.

American conversations about the shape of the ultimate peace settlements. But the basic Roosevelt Administration policy of postponement still stood. There would be no attempt to pin Stalin down while vast areas of his country were still occupied by powerful German armies.

There was a significant difference in attitude between the United States and Britain which was not made explicit. The British public showed noticeably greater warmth toward Russia than did the American, and the British government was more flexible than Washington in its reaction to Soviet territorial demands at the expense of Estonia, Latvia, Lithuania, and Poland. But the American government was generally more indulgent than London toward the Soviets on such current questions as matériel shipments. The explanation lay very largely in the growing desire of the British government to *bargain* with Stalin in the granting of wartime assistance and the delimitation of spheres of influence, whereas Roosevelt and his Secretary of State preferred to think in terms of idealistic general principles and to seek by indulgent treatment to get Stalin committed to them also, overcoming his distrust of the West.

Probably the most important reason for this American attitude was the preoccupation with the war against Germany and Japan and the feeling that Russia's actual contributions against Hitler and prospective contributions against Tojo were themselves worthy of important concessions on the part of the Western Allies, quite apart from any eventual bargaining about the postwar world. Sustained Soviet pressure on Germany's Eastern Front would be required till almost the very end of the European war. And the skill and fanaticism of Japanese resistance in distant island bases presaged a protracted war of annihilation in the Pacific, which was expected to last from twenty-four to thirty-six months after Hitler's defeat; even after the home islands had been conquered, the self-supporting Japanese armies in Manchuria and North China might well fight on to the death, and Russian forces could be enormously helpful in containing them on the mainland and destroying them. Until the spring of 1945 the American air force also believed it would require the use of bases in Siberia to mount a sustained bombardment of Japan. Until then also, strategic estimates took almost no account of any potential capacity of the atomic bomb to produce surrender. Partly, this mistake was the result of extreme secrecy that excessively restricted the number of military planners who had any awareness of the development of the weapon; partly, it was the result of the facts that until July 1945 no one could be absolutely certain that the bomb would really explode or that it would be authorized for actual use. Moreover, for many additional months, production would be so slow that the few available bombs (about a dozen before 1946) could be decisive, if at all, only because of their psychological effects. Germany's demonstrated capacity to withstand mass bombing, coupled with evidence of extraordinary Japanese endurance elsewhere, seemed to indicate that ground operations would be required in addition to bombing and naval blockade, or the war would drag on for years. Therefore, Roosevelt was under constant pressure from his Chiefs of Staff to secure Russian intervention against Japan at an early enough date to prevent Tokyo from shifting its armies out of Manchuria to reinforce some other position against American troops. Speedy defeat of Japan was so much more important to Americans than to Britons that, when bargaining with the Russians, Roosevelt could not jump so readily from preoccupation with joint defeat of Hitler to an emphasis upon postwar relationships.

Anticipation of a major campaign against Japan, plus recognition of the public desire to get the boys home, further inhibited Roosevelt from wartime bargaining about postwar boundaries and spheres of influence in Europe, because he did not expect that American troops would ever remain on the scene to enforce the agreements. At Teheran in November 1943 he personally explained to Stalin that the United States would provide only air and naval forces to resist any future aggressor; land forces would have to come from Britain and Russia. Again, at Yalta in February 1945, Roosevelt twice told Stalin and Churchill that he did not expect the American people to be willing to keep an army in Europe for more than two years after the defeat of Germany, although perhaps participation in a United Nations organization might change their attitude. Given this assumption it was easy to conclude that wartime bargaining with Russia about postwar political geography, even if successful in reaching agreement on the map, would compromise the United Nations ideals of democracy and national self-determination without assuring that Stalin's future appetite would be either satiated or restricted. Stalin would have to be persuaded to restrain himself; there was no other hope. To this end, the American government bent its major efforts, struggling to win Soviet confidence and convince the eternally suspicious Communist regime that Russian national security would be adequately assured through the framework of a general United Nations organization; it would not be necessary for Stalin to establish a buffer zone of new Soviet republics and satellite states to ward off capitalist aggression.

The American reliance on international organization as a near panacea grew steadily as the war continued. This emphasis was related directly to the similar tendency in Wilsonianism, 1917–20. Wilson's strongest supporters now felt themselves justified by the supposed results of America's failure to join the League of Nations, while former opponents of the League lost their self-confidence. In postwar planning the American people and government became preoccupied with avoiding the mistakes of 1919–20. The sheer mechanics of international organization became almost the exclusive focus of public debate upon foreign policy in 1944. The central goal of United States policy for the postwar world was to institutionalize a security structure to which the great powers would all belong. Russia was thought to have made very valuable concessions when she accepted, whether freely or through bargaining, nearly all of the American proposals concerning the forms and official objectives of the United Nations. What was Stalin's own attitude? A clue may lie in the fact that he apologized at the Yalta conference for not having had an opportunity to study the subject in detail; he said he had been busy on other matters. One might suspect that this remark was mere diplomatic fencing except that he actually went on to make comments about international organization which sounded unrehearsed and inconsistent—hardly Stalin's wont when discussing matters of great importance to himself.

Clearly, the American government was grossly overoptimistic about the significance of Stalin's willingness to join the United Nations Organization. As a bridge between East and West its institutional structure did assure it of greater durability than the various verbal pledges of high principles which repeatedly were used to paper over the cracks in the coalition. And American membership did represent a major step away from traditional isolationism. But the organization could not really become the painless cure-all which Americans eagerly persuaded themselves it might be. Despite all the attention that was lavished upon it, the creation of the UN was actually a device of procrastination—somewhat like the Unconditional Surrender formula—too simple to solve the global complications being caused by World War II. To preserve Russian cooperation against the Axis, formal agreements seemed essential, and perhaps beneath the shelter of formal agreement true and durable cooperation might take root—but the chances were slim and the hopes excessive. The onrushing tide of battle was itself more likely to fill the empty forms with tension-laden content, regardless of American efforts to postpone unpleasant specifics for the sake of harmony and speedy victory.

## Military Operations and the Spheres of Influence

In mid-November 1942, as the battles for Stalingrad and Tunisia were just beginning, Winston Churchill embarked upon a major effort to induce Turkey to enter the war on the United Nations side and thus open a front in the Balkans. Interestingly, then and for more than a year thereafter, the Soviet government also showed itself anxious to get Turkey to fight. Both the British and the Russians wanted German troops to be diverted against the Turks, and each of them could see possible postwar advantages from the development of a Balkan front; the British could hope that their forces would become lodged in the region before the Red Army arrived; Stalin, for his part, could anticipate the exhaustion of Turkey, making her vulnerable to postwar pressure from Moscow. Only the Turks and the Americans did not wish to make this gamble —the Turks shunning the role of pawn, the Americans resisting all Mediterranean entanglements. But in January 1943 the United States government actually gave Churchill leave to woo Ankara; it was Turkey which, then and later, vetoed the affair and thus virtually assured that Soviet power would reach the eastern Balkans before British or American forces could do so.

Churchill turned his attention to utilizing the Anglo-American forces in North Africa for an invasion of Italy with the possibility of a later move into the southwestern Balkans. The Tunisian campaign dragged finally to a victorious close in May 1943, more than four months overdue. Sicily was next on the schedule, the objective being to open the Mediterranean to Allied shipping. The invasion came on July 10; but not until Mussolini was actually overthrown by a *coup d'état* on July 25 did the Anglo-American Combined Chiefs of Staff make a final determination of forces to be allotted to a campaign on the Italian mainland. Once again, the fears of American military strategists were largely justified. Despite the Italian government's defection, the Germans held on in Sicily long enough to enable them to occupy most of Italy, and became well entrenched—as previously in Tunisia—before the Allied invasion near Naples on September 8 could acquire momentum. A long and costly campaign would ensue, diverting Allied as well as German resources from the prospective main Western Front in France. American military leaders felt annoyed and resentful, but they undertook to minimize the diversion by securing British agreement that strength from the Mediterranean would be landed in southern France in the summer of 1944 to reinforce the climactic invasion of northern France; they were determined that no major movement be made into the Balkans. Right up until the very week of the Riviera landing on August 15, 1944, the British and American governments continued to dispute this point. In the end the United States view prevailed, and the troops and equipment that could be spared from Italy went into southern France. The Western Front was strengthened, and most of the Balkans were left to Russia.

Meanwhile on the Eastern Front, thanks very largely to the American trucks, shoes, and food which began to arrive in vast quantities by the beginning of 1943, Soviet armies acquired the mobility and resources to launch a series of offensives which drove Axis troops out of nearly all of prewar Russia by the spring of 1944. During the course of 1943 also, Stalin established a "Union of Polish Patriots" as the possible nucleus of a future satellite government, and strengthened his relations with the Czechoslovakian government-in-exile, and with Tito's partisans as an alternative to the Yugoslav government-in-exile.

The first Big Three conference at Teheran in December 1943 postponed once again the major problems of Eastern Europe, but there was a significant agreement in principle with regard to Germany. She was to be dismembered into two or more separate states. This concept continued in the forefront of Allied postwar thinking until the spring of 1945 when Stalin officially dropped the idea. Evidently he was by that time satisfied to find himself in physical occupation of a large part of Germany. Actually the zonal boundaries for occupation had been proposed by a British Cabinet Committee under the chairmanship of Labor leader Attlee in mid-1943, and the limits of the Eastern zone had been approved without significant change by Russian and American representatives early in 1944. In the prevalent anticipation of German dismemberment, this decision had never been given the critical attention it deserved in the highest Anglo-American circles. Clearly, in retrospect this was a case in which it would be vastly more advantageous to have adhered fully to the general policy of postponement and thus have allowed the fortunes of war to determine the size of occupation zones in Germany. Churchill himself evidently regarded the zonal agreement as breakable; but the Americans were less cynical, and because they had made it they retreated in 1945 and surrendered to Russia almost half of what is today the Communist-dominated republic of East Germany.

There was another series of specific big-power agreements made in 1944 which overshadowed the general policy of postponement and fixed the structure of postwar Europe—but in this case they did little more than register the actual power which Russia had already been allowed to acquire in 1942 and 1943 for combatting the Nazis. The British government had come to feel it was essential to attempt to pin down the Soviets to some territorial limits before their power grew still greater. The new agreements reached in 1944 defined specific spheres of influence for Britain and the Soviet Union, with the temporary consent of President Roosevelt.

The stimulus came from a strikingly coordinated series of Communist maneuvers in the first days of spring 1944. At that time, Soviet armies in Rumania reached and crossed the Pruth River, and Molotov announced that the USSR had no intention of annexing any more Rumanian territory or "changing the existing social order of Rumania." Almost simultaneously, Palmiro Togliatti arrived in liberated Italy from Moscow to take charge of the local Communist party, and he swiftly reversed its position from opposition to support of the British-favored monarchist government. At the same time in North Africa, French Communists joined in De Gaulle's British-supported Committee of National Liberation, and a few weeks later De Gaulle felt strong enough to declare his group the Provisional Government of the Republic of France. However, in Greece, the Communist-dominated guerrilla movement, EAM-ELAS, meanwhile declared itself to be a provisional government, thus implicitly challenging the legitimacy of the British-backed Greek government-in-exile; and when the English asked for Russia's help in pacifying that situation (as in the French and Italian arrangements) the Soviets declined to cooperate. Foreign Secretary Eden was not slow to take the hint. On May 5, 1944, he suggested a deal to the Russian ambassador, and it was soon agreed that Rumania and Bulgaria would be considered as in the Soviet "sphere of operations" and Greece and Yugoslavia (where Churchill was still hoping to land troops) would be in the British sphere. Stalin, however, wanted the United States to concur in the arrangement, and Churchill reluctantly undertook the difficult task of winning Washington's consent. Secretary of State Hull resisted stoutly, but Roosevelt finally decided to bypass him and give consent for a limited trial period of three months, which was formalized on July 14, 1944. The Russians thereupon persuaded the EAM to enter a coalition cabinet of the Greek government-in-exile, and the British in return rejected appeals from Rumania and Bulgaria to intervene on their behalf in their armistice negotiations with the USSR.

The armies marched on. After August 1944 it was clear that no very large Western forces would ever reach the northern Balkans, although Churchill was still hoping that more rapid progress in Italy would render some Allied troops surplus, available to be moved across the Adriatic into northwestern Yugoslavia. Soviet forces conquered Rumania and Bulgaria in September and linked themselves with Tito's partisans in Yugoslavia at the beginning of October. Plainly Yugoslavia was slipping out of the British "sphere of operations" despite the Anglo-Soviet bargain. Churchill moved quickly to protect his position at least in Greece by getting small detachments of British troops ashore there on October 5, but he could not secure adequate forces for major operations if the Communist-dominated Greek guerrillas should cease to cooperate. Immediately Churchill flew to Moscow to try to nail the Russians down to a new bargain. He offered Stalin 90 per cent predominance in Rumania, 75 per cent in Bulgaria, and a new 50-50 split of influence over Yugoslavia and Hungary, in return for 90 per cent British predominance in Greece. Stalin accepted without hesitation. The American government also was committed to the limited extent of having an official observer present and making no official protest. But the extraordinary durability of these power relationships (except for Hungary) was quite unforeseen and unintended by the United States government. Only in retrospect is it clear that these agreements in 1944—together with the helpful restraint which Moscow apparently imposed upon the Communist revolutionary potential in France and Italy—actually fixed the permanent political structure of Southern and Western Europe. They fitted the realities of military power on the scene. As for Hungary, the last slim chance that Western troops would ever reach her border died at the beginning of February 1945 when the surplus divisions from Italy were moved to France at American and Canadian insistence, not across the Adriatic.

The Polish question remained. At the beginning of 1944 the Red Army had entered prewar Poland. Despite British urging, the government-in-exile and the main guerrilla movement (the Polish Home Army) were still refusing to accept Russian demands for a "Curzon Line" frontier. The British government in 1919 had approved the Curzon Line as the proper ethnic frontier between Poles and Ukrainians, and it was considerably more advantageous to Poland than the Hitler-Stalin boundary of 1939–41; but it would require Poland to abandon a full third of her 1920–39 territory (with some westward compensation at the expense of Germany). The Poles remained intransigent. When the Red Army reached the outskirts of Warsaw, the local Polish Home Army units rose in revolt against the Germans on August 1, 1944, hoping to capture the city themselves a few hours before Russian occupation began and thus to improve their bargaining position against Moscow. Stalin resolved upon ruthless measures. Remembering, perhaps, his tactic of murdering several thousand Polish officers in prison camps in 1940, he decided to permit a further decimation of the potential leaders of a free Poland. He held the Red Army back from Warsaw for several weeks while the Germans crushed the Home Army revolt and systematically wiped out several tens of thousands of Polish fighters. Stalin even refused to allow British and American planes to land in Russian territory after they had already risked 800 miles of German air defenses to drop supplies to the embattled Poles.

In Britain and America the impact of Stalin's ruthlessness was sharp, but appeasement seemed essential. More than ever, Churchill made urgent efforts to persuade the Polish government-in-exile to come to terms on the Curzon Line before Moscow proceeded to settle the matter by establishing a puppet government in "liberated" Poland. The Poles remained adamant. A Big Three conference was scheduled to convene within a few weeks (at Yalta) to discuss the Polish situation, but Stalin was now determined to achieve a *fait accompli.* In early December he received De Gaulle in Moscow and signed a permanent treaty of alliance with France, thus helpfully reinforcing the subordination of French Communists to De Gaulle. Then on December 31, 1945, Stalin's nucleus puppet government of Poland declared itself to be the Provisional Government of the country, and a few days later received full Soviet recognition as such. The implied deal, France for Poland, was clear, but Britain and America never approved it. At the Yalta Conference and thereafter until June of 1945, the Western powers negotiated strenuously to replace the Communist-dominated Provisional Government with a broader coalition of Polish leaders, who would be willing to accept the Curzon Line and cooperate with Russia. Limited success was achieved in the re-establishment of a genuine opposition party in Poland. But meanwhile, Stalin further decimated its potential leadership by luring key figures from the underground out into the open for negotiations and then suddenly jailing them on March 28, 1945. Virtual Soviet occupation and Communist domination remained effective in Poland, and the main opposition party was subjected to extreme terrorism long before the "free" elections, guaranteed at Yalta "as soon as possible," were finally held in January 1947.

At Yalta on February 9, 1945, President Roosevelt had remarked: "I want *this* election in Poland to be the first one beyond question. It should be like Caesar's wife. I did not know her but they said she was pure." Stalin interjected: "They said that about her but in fact she had her sins." The point was not pursued, but it was revealing, for just as the

Westerners were bitterly to denounce the sins of the 1947 elections, on the other hand Stalin was bound to regard any genuinely free democratic election as an evil because in Poland it could not reasonably be expected to produce a government truly friendly to Communist Russia. This was a dilemma which America's addiction to high principles tended to conceal from her own negotiators. In most of Eastern Europe it was a delusion to imagine that democratically-elected governments would not be anti-Russian. Stalin must have supposed that the West had fully digested this fact, and, therefore, when the unrealistic principles of the Atlantic Charter and the United Nations Declaration were temporarily supplanted by the spheres-of-influence deals of 1944, he must have interpreted the development as evidence of a growing frankness on the part of Western diplomats. Then at Yalta, a verbalistic smoke screen was thrown around the practical Anglo-American acceptance of Communist predominance in Poland. Stalin may have been misled into believing that the United States would not give grave attention to future violations of a new proclamation of additional high principles, the Declaration on Liberated Europe.* The Soviet leaders endorsed this exalted pronouncement with remarkably little hesitation; presumably they felt reassured because no special action could be taken under the terms of the Declaration except with unanimous consent of the Big Three powers. But actually, in the minds of most American leaders the Declaration was of real importance, for they intended that it should supersede the spheres-of-influence arrangements and bring practical policy into line with the old ideals of the Atlantic Charter.

However, the quest for verbal harmony was still the overriding consideration at Yalta, and so the American interpretation was not made

* February 11, 1945. "The establishment of order in Europe and the rebuilding of national economic life must be achieved by processes which will enable the liberated peoples to destroy the last vestiges of Nazism and Fascism and to create democratic institutions of their own choice. . . .

"To foster the conditions in which the liberated peoples may exercise these rights, the three governments [United States, United Kingdom, and USSR] will jointly assist the people in any European liberated state or former Axis satellite state in Europe where in their judgement conditions require (a) to establish conditions of internal peace; (b) to carry out emergency measures for the relief of distressed peoples; (c) to form interim governmental authorities broadly representative of all the democratic elements in the population and pledged to the earliest possible establishment through free elections of governments responsible to the will of the people; and (d) to facilitate where necessary the holding of such elections. . . .

"When, in the opinion of the three governments, conditions . . . make such action necessary, they will immediately consult together on the measures necessary to discharge the joint responsibilities set forth in this declaration."

emphatically clear. Indeed, in the weeks immediately following the conference both the Russian and British governments continued to act as though the former lines of demarcation were still applicable. Churchill (as he had personally told Stalin at Yalta) was still "very much obliged to Marshall Stalin for not having taken too great an interest in Greek affairs"; the Soviets did not lift a hand in opposition when British troops massed in Greece and crushed the uncooperative Communist-led guerrillas. Correspondingly, a few weeks later in Rumania, when Russia forced a Communist-dominated government into power on March 6, 1945, the British government dragged its feet about supporting American protests based on the Declaration on Liberated Europe.

But as Britain's anticommunist position in Greece grew quickly more secure, whereas in Yugoslavia Churchill's hopes dwindled of ever achieving even the 50 per cent influence for which he had settled with Stalin, the British government like the American began to place a strong legalistic and propagandistic reliance on the Declaration; it was almost the only basis still available on which to make claims for Western influence in Eastern Europe outside Greece. During most of 1945 the Soviet government seems to have been taken somewhat aback by this renewed serious emphasis on high democratic principles. From the Soviet viewpoint, it may well have seemed that the Western powers, having gained the advantage of Communist self-restraint in Greece, Italy, France, and Belgium, were now launching a surprise diplomatic offensive in an effort to deprive Russia of her winnings in the East. Nevertheless, Stalin could be confident that the actual location of armies would still remain the determining factor in European political alignments. The progress of Russian soldiers in filling the vacuum of German defeat was not to be reversed by mere appeals to ambiguous paper promises like the Declaration on Liberated Europe.

If Churchill had had his way, the Western Allies at the end of the war would have acted on a similar cold calculation. The German armies were holding ground much more firmly on the Eastern Front than on the Western after January 1945. It began to appear that an Anglo-American drive aimed directly at Berlin would capture the German capital before Nazi forces would allow the Russians to enter the city. Churchill strongly advocated this move, but General Eisenhower was preoccupied with a campaign to prevent the remnants of Hitler's troops from withdrawing into the mountainous region of southern Germany and Austria for a suicide stand. Eisenhower's concentration on this nonpolitical military objective was fully supported by American Chief of Staff George Marshall; President Roosevelt was too sick to intervene. Churchill soon abandoned hope of beating the Russians to Berlin, but he continued to work desperately to push Western armies as far eastward in Europe as possible so that they could then hold on to advanced positions *beyond* the predetermined occupation-zone boundaries. Roosevelt died. Churchill appealed to the new President. "After examining the situation, I [Harry Truman] could see no valid reason for questioning an agreement on which we were so clearly committed, nor could I see any useful purpose in interfering with successful military operations. The only practical thing to do was to stick carefully to our agreement and to try our best to make the Russians carry out their agreements." * With Truman's authorization, Eisenhower thereupon called a halt to the advance on the central Western Front and even declined an easy opportunity to liberate the capital of Czechoslovakia, lest there be any clash between his forward units and those of the Red Army.

Berlin, Prague, and Vienna were therefore all in Russian hands as the European war ended on May 8, 1945. But a very large belt of territory in the prospective Soviet zone of occupation had come into Anglo-American possession before the Eisenhower stop-order. Churchill continued to strive at least to prevent a Western retreat from these acquisitions, which amounted to almost half of the present Communist state of East Germany.

During May a major crisis arose in the region of Trieste in northeastern Italy. In violation of previous agreements, Tito's Yugoslav partisans occupied that area before a Western force could arrive, and thereafter refused to subordinate themselves to British authority. Churchill was relieved to find that Truman was willing to take a strong stand in this situation, including a virtual Anglo-American ultimatum and show of force. Under such pressure Tito finally yielded in early June to the extent of accepting a compromise boundary between the occupation zones around Trieste. But, with this outcome, the Trieste crisis did not alter the Truman Administration's determination to adhere faithfully to the previous zonal boundaries in Germany and to hasten the withdrawal of American troops from Europe for the prospective invasion of Japan. On June 21 the President finally rejected the repeated appeals Churchill was making, and refused to delay the retirement, even for a few weeks, for use in bargaining at the forthcoming Big Three conference (Potsdam, July 1945). On July 1 Western armies began to transfer almost 15,000 square miles of German territory into Russian hands. It was a supreme demonstration of faithfulness to agreements, even when such

* Harry S. Truman, *Year of Decisions: Memoirs,* vol. 1 (Garden City: Doubleday, 1955), p. 214.

agreements had been made prematurely in wartime, in contrast to the general policy of postponement, and had proved to be disadvantageous. At least it would be very difficult thereafter for anyone to contend that the Truman Administration had not made great efforts toward harmony with Russia.

## The Postponement Policy in Retrospect

There has been much argument in the years since World War II as to the wisdom of the postponement policy as against a deliberate effort to pin Stalin down to specific political-territorial arrangements for the postwar world at an early stage when he was still fighting with his back to the wall.

Actually it still seems possible that Stalin would have been provoked by such pressure into seriously seeking a separate peace with Hitler or, failing that, into resorting to a purely defensive "stand-by" war in the Ural Mountain region (much as Chiang Kai-shek wished to do in western China), leaving Britain and America to defeat the Axis. Even if Stalin had been willing to sign territorial agreements with the Western powers under the conditions of extreme duress, 1941–43, there does not seem to be much reason to believe that he would thereby have felt restrained after the war in areas where Russian rather than Anglo-American troops emerged in physical control; surely the record of Soviet-British relations in Yugoslavia, Hungary, and Poland, 1944–46, indicates the contrary. Only a basic reallocation of Anglo-American strength during the war itself, sufficient for the Western powers to conduct major Balkan operations, would have had much chance of keeping the region southwest of Poland and Rumania out of Communist control. But the corollary result of such a strategy would probably have been still further delay in the conquest of Germany herself, with the constant danger that the resourceful enemy might revitalize his strength through new weapon developments. Inasmuch as a compromise peace with a Germany under Nazi control was impossible and a successful anti-Nazi *coup d'état* unlikely, it was probably indeed necessary for Western planning to be directed toward the total destruction of the capacity of the German state to offer organized resistance anywhere. The consequences for the European balance of power were bound to be catastrophic; nevertheless, total victory through total war was probably required, and the total mobilization needed for it had been made possible through the initiative of the Axis itself in declaring war upon the United States.

Given this almost inescapable concentration upon the German enemy, with the related need to save precious time by minimizing Mediterranean operations, the postponement of controversies with Russia can be justified. It took a variety of forms: the United Nations Declaration and the Anglo-Soviet Alliance (1942); Unconditional Surrender (1943); the formation of the UN (1944–45); and the Declaration on Liberated Europe (1945). These diplomatic moves did help to preserve the anti-Nazi coalition, while providing an available framework for any possible cooperative attitude in postwar Russia, and also a loose foundation in law and propaganda for the West to demonstrate to the postwar world which side had tried hardest to reach agreements and to keep them.

If the postponement policy was thus able to salvage something useful out of an inherently bad situation, would the West not have done better to stick to it rigidly, completely avoiding all political-territorial commitments? That point of view has strong defenders now, as it did during World War II. Applied to Europe, the criticism is directed mainly at the Anglo-Soviet spheres-of-influence bargain and the zonal boundaries arrangement for Germany.

On closer examination, however, there may be a useful distinction to be drawn between these two sets of agreements. If either side benefited from the spheres-of-influence deal, it was in retrospect the British more than the Russians. Under the pressure of American strategists, Britain's Mediterranean forces were stretched so thin that she might never have been able to establish herself in Greece had Stalin not agreed temporarily to call off the Communist dogs; even in France and Italy the restraint shown by Communists obeying the Moscow line was extremely helpful in permitting the reconstruction of local anticommunist strength during a perilous period. Stalin, for his part, found he had gained nothing from the arrangement in the Balkans except what he was in a military position to take anyway without encountering effective Western resistance. Moreover, the ambiguous relationship of the United States to the Anglo-Soviet bargaining left her basically free to reassert her ideal principles very soon in a legalistic-propagandistic countermove against the *faits accomplis* in Eastern Europe. In retrospect, the result—albeit unplanned—of this two-way policy on the part of the British and American governments was to salvage a little something for the West in a marginal situation which had in fact been rendered essentially unmanageable by the basic concentration against Germany; the real cost to the West was little more than a heightening of Soviet suspicion of Anglo-American policies.

With regard to Germany, however, the ultimate cost did become very great when the general policy of postponement was modified by the premature establishment of occupation-zone boundaries. There was good reason for avoiding any territorial commitment of any kind in the case of Germany (just as a precise definition of the common notion of "dismemberment" was again and again postponed). In the eyes of American strategists, Germany had never been a marginal theater of war like southeastern Europe. Germany was their central concern, their main target for speedy invasion by vast land armies. At the end of the war those armies were almost certain to be on the ground somewhere in western Europe. Many factors might delay their advance, but at least there was no reason to anticipate that the essence of the strategy itself (as in the Balkans) would prevent Anglo-American forces from occupying nearly all of the enemy's territory before the unreliable Russian ally arrived on the scene also. In the case of Germany, unlike southeastern Europe, there was no desperate pressure upon the Western Allies to take the dubious chance of salvaging a little something by bargaining in advance with the Russians. The balance of military force in northern Europe was such as to give ample warrant for an American extension of the postponement policy to delay the demarcation of occupation zones until after the verdict of battle itself. The chance would have been worth taking.

The original decision not to do so was made with tragically little scrutiny on the part of top officials either in Britain or the United States. In Washington, the policymakers were even influenced by a positive desire to minimize American occupation responsibilities in Germany so that more troops could be moved out to the Far East. Thus, the narrow military emphasis of United States policy, which hastened the Anglo-American conquest of Germany and thereby provided rich though unpremeditated political opportunities, also tended to dissipate those opportunities in advance by accelerating a technically efficient reconcentration of forces against Japan.

Nevertheless, one should also recognize that even if Anglo-American forces had been free to speed eastward across Germany unencumbered by predefined occupation zones, they could only have acquired a north-south belt running about 150 miles farther east than the present boundary between East and West Germany. The Russian colossus would still straddle Eastern Europe; indeed if the Western Allies had seized a larger share of Germany, that very move might well have caused the USSR to hasten still further its ruthless solidification of power in Poland and the Balkans. Inevitably, the destruction of German authority was bound to leave a vacuum throughout the continent of Europe. Lines of force would flow inward from East and West. Somewhere in central Europe—a few score miles to the East or to the West—the powers would meet and begin to probe each other's perseverance. Britain and America had long delayed these disconcerting explorations, for the sake of total war. Now the West would come to learn how bitterly inconclusive the fruits of total victory can be.

## SELECTED BIBLIOGRAPHY

ARMSTRONG, ANNE. *Unconditional Surrender.* New Brunswick: Rutgers University Press, 1961.

CHURCHILL, WINSTON S. *The Second World War,* vols. 4–6. Boston: Houghton Mifflin, 1950–53.

DEANE, JOHN R. *The Strange Alliance: The Story of Our Efforts at Wartime Cooperation with Russia.* New York: Viking, 1946.

EHRMAN, JOHN. *History of the Second World War: Grand Strategy,* vols. 5 and 6. London: Stationery Office, 1956.

FEIS, HERBERT. *Between War and Peace: The Potsdam Conference.* Princeton: Princeton University Press, 1960.

———. *Churchill, Roosevelt, Stalin: The War They Waged and the Peace They Sought.* Princeton: Princeton University Press, 1957.

HULL, CORDELL. *The Memoirs of Cordell Hull,* 2 vols. New York: Macmillan, 1948.

HUNTINGTON, SAMUEL P. *The Soldier and the State: The Theory and Politics of Civil-Military Relations.* Cambridge: Harvard University Press, 1957.

KECSKEMETI, PAUL. *Strategic Surrender: The Politics of Victory and Defeat.* Stanford: Stanford University Press, 1958.

KING, ERNEST J., and WALTER M. WHITEHILL. *Fleet Admiral King: A Naval Record.* New York: Norton, 1952.

LEAHY, WILLIAM D. *I Was There: The Personal Story of the Chief of Staff to Presidents Roosevelt and Truman.* New York: Whittlesey, 1950.

MC NEILL, WILLIAM HARDY. *America, Britain, and Russia: Their Cooperation and Conflict, 1941–1946.* New York: Oxford University Press, 1953.

MATLOFF, MAURICE, and EDWIN M. SNELL. *The United States Army in World War II: Strategic Planning for Coalition Warfare, 1941–1944,* 2 vols. Washington, D.C.: Department of the Army, Office of the Chief of Military History, 1953, 1959.

MORGENTHAU, HENRY, JR. *Germany Is Our Problem.* New York: Harper, 1945.

NEUMANN, WILLIAM L. *Making the Peace, 1941–1945: The Diplomacy of the Wartime Conferences.* Washington, D.C.: Foundation for Foreign Affairs, 1950.

POGUE, FORREST C. *United States Army in World War II: The European Theater of Operations: The Supreme Command.* Washington, D.C.: Department of the Army, Office of the Chief of Military History, 1954.

RUSSELL, RUTH B. *A History of the United Nations Charter: The Role of the United States, 1940–1945.* Washington, D.C.: Brookings Institution, 1958.

SHERWOOD, ROBERT E. *Roosevelt and Hopkins: An Intimate History.* New York: Harper, 1948.

SNELL, JOHN L., ed. *The Meaning of Yalta: Big Three Diplomacy and the New Balance of Power.* Baton Rouge: Louisiana State University Press, 1956.

———. *Wartime Origins of the East-West Dilemma over Germany.* New Orleans: Hauser Press, 1959.

STETTINIUS, EDWARD R. *Roosevelt and the Russians: The Yalta Conference.* Garden City: Doubleday, 1949.

STIMSON, HENRY L., and MC GEORGE BUNDY. *On Active Service in Peace and War.* New York: Harper, 1947.

TOYNBEE, ARNOLD and VERONICA M., eds. *The Realignment of Europe.* New York: Oxford University Press, 1955.

TRUMAN, HARRY S. *Year of Decisions: Memoirs,* vol. 1. Garden City: Doubleday, 1955.

UNITED STATES DEPARTMENT OF STATE. *Foreign Relations of the United States: Diplomatic Papers: The Conference of Berlin (the Potsdam Conference), 1945,* 2 vols. Washington, D.C.: Government Printing Office, 1960.

———. *Foreign Relations of the United States: Diplomatic Papers: The Conferences at Malta and Yalta, 1945.* Washington, D.C.: Government Printing Office, 1955.

———. *Postwar Foreign Policy Preparation.* Washington, D.C.: Government Printing Office, 1949.

WEDEMEYER, ALBERT C. *Wedemeyer Reports!* New York: Holt, 1958.

WELLES, SUMNER. *Seven Decisions That Shaped History.* New York: Harper, 1951.

WESTERFIELD, H. BRADFORD. *Foreign Policy and Party Politics: Pearl Harbor to Korea.* New Haven: Yale University Press, 1955.

WILMOT, CHESTER. *The Struggle for Europe.* New York: Harper, 1952.

## CHAPTER 5

# THE BEGINNINGS OF CONTAINMENT

MANY MONTHS passed before there was full consolidation of Communist power in the Soviet sphere of military operations. The process of reconstructing the various East European states was not one of steamroller uniformity, nor did the USSR press farther outward in every direction at once. Even where Red Army troops were stationed, the pace of communization was varied in different countries. There were general pauses and occasional slight retreats between gains, partly in reaction to countermoves on the part of the United States and Britain. To some extent, the Kremlin was presumably still feeling its way, uncertain in advance how much it could get away with without arousing dangerously great antagonism in the West. Excessive tensions would be inexpedient. Some of the most pressing goals of Soviet policy at the end of the war were beyond the limits of the Red Army's advance: a major share in the control of Japan and of the Ruhr in Germany, a military foothold at the Dardanelles and in the former Italian colonies in Africa, a multibillion-dollar reconstruction loan from the United States. As long as hope remained of achieving these goals by diplomacy and limited pressures, the Soviet Union might choose not to accelerate its "revolution from above" in Eastern Europe. Even where the Soviets already enjoyed the advantages of full or partial military occupation, they must have felt some concern in the first few months after Hitler's defeat lest the West be provoked into moving its still massive armies against the

Russian outposts in Europe. Those mass armies, as long as they still existed, also tended somewhat to stiffen the backs of anticommunists in the political struggle in Eastern Europe.

Soviet hopes of winning additional major objectives by superficially conciliatory diplomacy declined in the year after Nazi surrender—but so also did the basis for Soviet fears of Western armed might. The American government did become less free-handed about concessions to the Russians; yet the United States military forces melted away. The method and speed of demobilization left the Administration without means to exercise much influence inside the Soviet sphere at the time when alarm was finally aroused in Washington. By 1946, even outside the Iron Curtain the American capacity to deter Stalin from taking by force what he was increasingly unable to obtain by diplomacy depended mainly on Stalin's fears of the ultimate outcome of a long general war, coming too soon after Hitler's ravaging of Russia. The beginnings of such a war would surely have registered huge Soviet conquests. Soviet leaders themselves now take credit for peaceful intentions in the self-restraint they say they showed in not having seized all of Europe in the immediate postwar years; they observe that then there was nothing to stop the Red Army. The short-run opportunities they saw open to them in 1946 and 1947 must indeed have been tempting. They could see the effects of the American system of demobilization (releasing individuals in order of relative personal hardship rather than holding each military unit together until its release as a whole); the procedure was stripping all units of key personnel and rendering the forces useless for battle. Furthermore, thanks to its spies, the Kremlin evidently knew how small and indecisive the American stockpile of atomic bombs actually was until 1948 or later; the Russians probably had a better conception of the limited dimensions of this weapons system than did most Western leaders. Insofar as atomic weapons were actually a primary deterrent to Soviet aggression in 1946 and 1947, the aspect which must have been giving the Kremlin pause was their probable ultimate effects in a prolonged war of attrition, which would cause the small American production of bombs and bombers to be accelerated. In the initial stages of a war begun in those years no decisive air attacks could probably have been mounted; but gradually, as the war lengthened, American technological superiority would begin to register, even if Europe had meanwhile already been overrun by the Red Army. The long-run risk for the USSR was great.

Meanwhile, there were attractive opportunities open to Russia for political expansion and consolidation of previous gains by means unlikely to provoke the demobilizing West to undertake a general war. The Kremlin chose to feel its way ahead rather cautiously, but persistently.

## The Soviet Tide Encounters Resistance

At first, immediately after the end of the war in Europe, the Soviet government moved firmly to strengthen Russia's position. Southward, pressure was directed against Turkey and Iran. Turkey was asked to grant bases to the Russians on her strategic territory at the Dardanelles, and to return to the USSR the eastern provinces which the Czars had once conquered but which Turkey had recovered after World War I. Iran was confronted with a Soviet-sponsored separatist "Azerbaijan National Committee of Liberation" in her northern province, one which bordered on Russia and which the Red Army still occupied in the aftermath of World War II.

Westward, the USSR proceeded to annex the easternmost tenth of Czechoslovakia and transfer to Polish jurisdiction the German territories which were supposed to compensate Poland for her losses to Russia. In Yugoslavia, the native Communists with their own armed forces were somewhat independent and impatient; they endured external restraint at Trieste, but internally the drive toward Red dictatorship was relentlessly rapid. There and in Bulgaria, where it was possible to capitalize to some extent on traditional friendship for Russia, elections could be held as early as November 1945 and produce the superficially overwhelming popular endorsement which characterizes dictatorial balloting.

In Austria and Hungary, on the other hand, Communist tactics were much more restrained. In these two countries the Russians allowed generally free elections to be held in November 1945 even though the results put the Communists decidedly in a minority. The presence of Western troops in Austria is a partial explanation. And perhaps in Hungary Stalin was still adhering to his 50-50 deal with Churchill, part of the spheres-of-influence arrangement which had thus far in effect been broken only in Yugoslavia (and there more because of Tito's intransigence than through Russian initiative). Another factor may have been a desire on the part of the Russians to mollify the Western powers somewhat (for the first time the West had allowed a Big Three conference, in September 1945, to end in complete failure rather than make further concessions to Moscow). Free elections in Austria and Hungary in November did help to ease resentment over the repressive campaign tactics the Reds were using in Bulgaria and Yugoslavia that same month. The new Austrian and Hungarian governments were promptly recognized; and in December a new Big Three conference showed that the United States, which won confirmation for its own unfettered authority in Japan, was now willing to take a chance even in Rumania and Bulgaria on only a very slight broadening of the Communist-dominated governments, together with another promise of free elections.

The chance was slim, and the promise was never truly implemented. Even in Hungary, where noncommunists had been allowed to win a parliamentary majority, they were gradually outmaneuvered. The local Reds played upon the majority's fear of trying to govern without their cooperation in a country which the Soviet Army still occupied; as the price of remaining in a coalition government, Communists exacted for themselves the control over the police and armed forces. Their ultimate victory in Hungary was thereby virtually assured, but they moved toward it cautiously over a period of many months.

In Poland also, the country to which American eyes were chiefly directed in Eastern Europe, the fiercest Communist fire was not turned against the organized opposition until the end of 1946. But actually an image of Soviet obstructionism and Red imperialism had already become well established in the American mind several months before that time. The crisis which probably contributed most to this new awareness in America did not occur in Europe at all, but in Iran, and its special impact derived very largely from the glaring spotlight of publicity which the fledgling United Nations Organization turned more in that direction than toward Eastern Europe.

Late in 1945 the Soviet-sponsored Azerbaijan separatist movement staged a local revolt; Russian troops intervened to prevent the central Iranian government from subduing the rebels. The separatist demand was for autonomy, not independence, for Moscow wanted her local puppets to be able to elect representatives to the national parliament in Teheran and thus to provide additional leverage for Soviet objectives, particularly to help Russia secure oil concessions in northern Iran. The Iranian government appealed to the United Nations Security Council against this interference in the country's internal affairs. In the UN debate which ensued during January 1946, the American delegate was not active, but the British spokesman lent substantial support to Iran's case against Moscow. However, a change of administration in Teheran soon provided a convenient excuse to adjourn the debate while Iran and the USSR resumed direct negotiations. Iran's chances seemed poor.

During February the American government stiffened its position.

Charges of "appeasement" of Russia were being directed at the State Department from both political parties more widely than before. Secretary of State Byrnes reacted vigorously. One significant development was the department's request to George Kennan, American chargé d'affaires in Moscow, to set forth at length his general views on Soviet policy, which were known to be highly critical; Kennan proceeded to cable an 8000-word essay, a powerfully argued document which in various adaptations has had great and lasting influence (see pp. 165–73). Byrnes's own conclusions emerged in a firm policy speech on February 28 which dealt explicitly with the possibility of a new war, in defense of the United Nations Charter against aggression "accomplished by coercion or pressure or by subterfuge such as political infiltration." The battleship *Missouri* was dispatched to the Dardanelles (ostensibly to bear home in honor the body of the Turkish ambassador who had recently died in Washington).

At the beginning of March 1946, former Prime Minister Churchill, visiting America on a much heralded lecture tour, further highlighted the East-West conflicts. The American government formally protested Russia's failure to withdraw her troops from Iran as promised. The Iranian premier broke off negotiations in Moscow, refusing to accept the Soviet demands. A new appeal went to the UN. Then suddenly, just before the UN meeting, the Russians reopened negotiations at Teheran on the basis of modified demands and announced an "understanding" that Soviet troops would be withdrawn within six weeks "unless unforeseen circumstances arise." The new terms for evacuation proved acceptable to the Iranian premier, and the Red Army did in fact get out. The United Nations Organization was credited with a major triumph, and anti-Soviet sentiment relaxed again in the West.

In Germany, however, May 1946 brought a Russian-American cleavage which was of even deeper significance, though less publicized, than the Iranian affair. Ever since the Yalta Conference the central immediate preoccupation of Soviet diplomacy regarding Germany had been the extraction of reparations for the USSR of $10 billion in capital equipment and in currently produced goods, together with the forced labor of hundreds of thousands of German war prisoners in Russia. Indeed, the extraordinary concern the Kremlin showed for reparations at the end of World War II highlights the economic weakness of the Soviet Union at that time and helps to explain the considerable caution that was being shown in her political expansionism. The rebuilding of the devastated homeland also called for heavy assistance from the United States, if obtainable, in the triple form of Lend-Lease shipments, United Nations

Relief and Rehabilitation Administration grants, and (hopefully) a multibillion-dollar American loan. Then, when Lend-Lease was ended unexpectedly soon and the loan failed to materialize, German reparations may have seemed increasingly essential to the Kremlin. But actually, there was no possibility of extracting reparations from Germany at a rate fast enough to satisfy Moscow unless standards of living were allowed to fall to a level which the Western conscience found intolerable (and unsafe for the occupation troops), or unless American and British taxpayers made up the difference in Germany by paying vastly increased occupation costs. This was the lesson the four powers (America, Britain, Russia, and France) learned through painful experience and continual friction among themselves, during the first year of their control of Germany.

Finally, on May 3, 1946, the American Military Governor put a halt to most deliveries of reparations from the United States zone. The Russian reaction was vehement. Both sides were unyielding. Within three months the break was crystallized in an Anglo-American decision to merge and rehabilitate their West German zones regardless of Soviet opposition.

Only the East Zone remained for full Russian exploitation; Moscow's economic and political planning was severely frustrated at home and abroad. It was probably not coincidental that Communists renewed their guerrilla warfare in Greece during May 1946 with extensive support from the Red regimes on Greece's northern borders. Similarly, there was a renewal of Soviet pressure on Iran and Turkey. The puppet administration in Azerbaijan, now free from UN publicity, successfully insisted on a large measure of autonomy and on greatly increased representation in the Iranian parliament. In July 1946 the Communist-dominated Tudeh party led a very violent general strike in the vital British oil concession area of southwestern Iran, and on August 2 three Tudeh members were admitted to the cabinet in Teheran, one as vice-premier. Simultaneously, Russian demands on Turkey were sharpened in notes to Ankara, London, and Washington, calling for a joint defense of the Dardanelles by the USSR and Turkey, an arrangement that would mean in effect a partial Soviet occupation of Turkish territory.

The United States government reacted firmly to the renewed threat to the Dardanelles. A full naval task force was sent to the eastern Mediterranean, and President Truman, after consultation among the State, War, and Navy departments, decided on August 15 to oppose Moscow's "joint defense" plans, even at the privately explicit risk of war. Coincidentally, the American rejection of Russia's proposal was highlighted by a flare-up of very sharp tension between the United States and Yugoslavia; Tito's forces had begun shooting down American planes flying over his territory between Italy and Austria. Under pressure Tito backed down. On Turkey the Russians bided their time.

The British supported the American action on the Dardanelles; simultaneously, in Iran they took matters into their own hands to salvage something from the disintegrating political situation by tactics similar to those which Russia had been using in Azerbaijan. British troops were rushed to the Iraqi-Iranian border, and Iraqi politicians were encouraged to agitate for a union of the neighboring southwest Iranian region with Iraq. A revolt was organized in southern Iran, demanding local autonomy and increased representation in the national parliament. The country seemed on the verge of partition into Russian and British spheres. But in October, with the active encouragement of the American ambassador, the premier decided to test his strength against the Communists in an attempt to reunify the country. The three Tudeh ministers were dropped from the cabinet, and by mid-December loyal government troops had reoccupied Azerbaijan. Russia abstained from intervention as the puppet provincial government collapsed. Thereafter, for many months the Soviet threat to Iran was in abeyance.

Communist gains in Greece, however, were reaching crisis proportions. The economic, financial, and administrative structure of the country, feeble and venal, was being strained beyond endurance by the need to maintain an army of 100,000 and police forces of 50,000 to combat the Red guerrillas. Steady improvement in the numbers, tactics, and organization of the Communists, thanks to aid from Bulgaria, Yugoslavia, and Albania in the north, meant that even larger government defense expenditures would be required in 1947. The resulting enormous deficit and skyrocketing inflation were almost certain to demoralize resistance to the Reds. A final blow came in February 1947 when the British government decided that its own postwar financial weakness had become so perilous that it could no longer afford to continue the heavy subsidies on which Athens was utterly dependent; even the thousands of British troops that were helping the Greek government to organize its own forces would now be withdrawn in order to reduce England's expenses. In effect, Britain was preparing to abandon the last toehold in the Balkans that she had acquired through Churchill's obsolescent spheres-of-influence deal with Stalin. Or rather, she was taking the chance that the United States would assume the burden rather than see the area entirely lost.

The imminent British withdrawal was formally announced to the American government on February 21, 1947. London provided estimates of Greece's survival requirements and the equally vital, though less urgent, price for enabling beleaguered Turkey to support a modern military establishment. Reworked by American experts the figures came to $400 million in the first year of what would be a continuing program. The money would not be spent effectively unless the United States provided extensive advisory missions in most spheres of military, economic, and political life. The American involvement in Greece's internal affairs, with the consent of the Greek government, would have to be great.

Yet the Truman Administration had moved so far in an anti-Soviet direction during nearly two years of cumulative friction that extraordinarily little hesitation was now shown in taking up the new challenge. The main question was how to present the issue to Congress and the public, whether to treat it as a relatively simple transfer from Britain to the United States of financial and advisory responsibilities toward two suffering mutual friends, or to emphasize coolly the need for American police action to preserve the precarious balance of power in the eastern Mediterranean, or else to dramatize the program as an heroic defense against worldwide Communist imperialism. Acheson and Truman chose the latter technique, and encouraged the President's speech-writers to pull out all the stops. On March 12 Truman went before the Congress to appeal for the aid program and proclaimed:

> At the present moment in world history nearly every nation must choose between alternative ways of life. The choice is too often not a free one.
>
> One way of life is based upon the will of the majority, and is distinguished by free institutions, representative government, free elections, guarantees of individual liberty, freedom of speech and religion, and freedom from political oppression.
>
> The second way of life is based upon the will of a minority forcibly imposed upon the majority. It relies upon terror and oppression, a controlled press and radio, fixed elections, and the suppression of personal freedoms.
>
> *I believe that it must be the policy of the United States to support free peoples who are resisting attempted subjugation by armed minorities or by outside pressures.*
>
> I believe that we must assist free peoples to work out their own destinies in their own way.

> I believe that our help should be primarily through economic and financial aid which is essential to economic stability and orderly political processes.

This declaration, in particular the sentence which has been italicized above, was promptly dubbed by the press the "Truman Doctrine." It has come to be regarded as the basic proclamation of what is known as the "containment policy," which has formed the thread of continuity in American foreign relations since that time. The "containment" label came from George Kennan's formulation (published a few months after Truman's speech) that likened the Kremlin's "political action [to a] fluid stream which moves constantly [to fill] every nook and cranny available to it in the basin of world power"; Kennan pursued this image by prescribing a United States policy of keeping that basin impermeable through "a long-term, patient but firm and vigilant containment of Russian expansive tendencies." * The broad policy has been condemned in principle and has not always been followed in practice, but the deviations have been spasmodic and for the most part verbal and have rarely altered the basic concept of the policy. Its adoption came gradually in the specific series of crises in Soviet-American relations, 1945–47, which have been outlined above. The effort to conciliate Russia by continuing concessions was being abandoned under frustrating pressure in one location after another. The overall rationale for this practice had been formulated in the State Department by Kennan and others but had not yet been brought to public attention in a major presidential statement. The "Truman Doctrine" speech met this need in part, for it dramatized the wide-ranging threat posed by "aggressive movements that seek to impose totalitarian regimes." But the speech did not undertake to explain the relationship of the Soviet Union to this pattern of strife in Greece or elsewhere, not even in the case of Turkey. And although he made repeated references to the armed force that was being utilized for totalitarian subversion, the President's emphasis throughout was on economic techniques for combatting it. In short, the vision of world struggle which Truman projected in this speech was highly ideological and relied heavily, on the American side, on the vast material wealth of the United States, which could be used for economic aid abroad. The military aspects of the struggle were soft-pedaled. In this respect the speech was an appropriate indicator of the general spirit of the "containment policy" until the outbreak of the Korean war in 1950. In Greece, actually the American military involvement through matériel and advisors increased steadily in 1948 and 1949; but in the other theater of civil war between Reds and

* Mr. X, "The Sources of Soviet Conduct," *Foreign Affairs*, July, 1947, p. 575.

anti-Reds—China—the Truman Administration strongly resisted American commitments. Meanwhile, in places where the immediate danger was mainly economic—Western Europe, Japan, and the Philippines—United States assistance for their recovery mounted sharply.

## Repairing the Dike in Western Europe

The first massive enterprise in the implementation of containment was the European recovery program, which followed rapidly upon the inauguration of Greek-Turkish aid in the spring of 1947. Encouraged by the sharpening of American policy, both the French and Italian governments rid themselves in May of participation by Communists and fellow-travelers in their cabinets. Eastward, both the Polish and Czechoslovakian regimes indicated a desire to participate in the program for pan-European cooperative recovery which Secretary of State Marshall called into being with an offer of American assistance in June. His offer embraced even the Soviet Union, but there was small chance that the USSR would accept at the standard price of having its projected internal economic requirements laid open for evaluation by the Western powers. Kremlin secretiveness could not endure such an arrangement for itself; and the rooted Communist suspicion of the West would not tolerate the risk that any smaller state within the Russian orbit might be lured away from Soviet economic and political ties by the temptations of Wall Street gold. Moscow's reaction was swift and determined. The bonds must be tightened inextricably in Eastern Europe, while in Western Europe wherever Communists were strong they must throw their physical strength against governmental decisions to participate in the Marshall Plan.

Death, exile, or life imprisonment was the fate of the head of the main opposition party in Poland, Hungary, Rumania, and Bulgaria—all within the space of five months, May through October 1947. In Czechoslovakia, Communists speeded their infiltration of key organizations, particularly the police force, to such a degree that when anticommunist ministers tried to stop the process in February 1948, it was too late. Rather than risk civil war under circumstances where the Red Army stood ready on his borders and the Western powers were ill-prepared to intervene, President Beneš yielded to a Communist *coup d'état* (proof, the Kremlin explains, that Communism can now come to power without any violent revolution). Yugoslavia, too, during 1947, was subjected to the heightened pressure to conform to Kremlin wishes—but here alone in the Communist sphere Moscow met resistance which could not be overcome, except by military invasion with an unacceptable risk of world war.

Communism was given a reinforced organizational structure in September 1947 with the establishment of an "information bureau," the Cominform. During the remainder of the autumn, France and Italy, where the Reds were strongest in Western Europe, were wracked by Communist-led general strikes of extreme severity and violence. In large measure Moscow turned loose the revolutionary potential that had been leashed there as long as Communists were in the cabinets (and feared the proximity of British and American occupation troops in Germany and Italy). The strikes may not have represented the maximum effort of which the Reds were still capable even in 1947 after three years of stabilization; the violence did fall short of insurrection. But it might well have been enough to topple the fragile cabinets of France and Italy had not American determination been signalized by the calling of a special session of Congress in November to vote a half-billion dollars of emergency aid to those countries; this sum was designed to carry them through the interim period before the more massive Marshall Plan allotments began in 1948.

An essential element in any speedy achievement of prosperity in Europe would be the output of a reviving Germany. This proposition was accepted as fundamental in the transatlantic planning which engrossed American diplomacy in the summer of 1947. Russia, on the other hand, was still demanding vast reparations from Germany. The gulf proved to be unbridgeable at a four-power conference in November and December. Representatives of the Western powers thereupon agreed to meet early in 1948 to move toward the establishment of a West German government. The American-British-French talks began in February 1948 and continued intermittently through the spring; French obstruction was substantial, but was gradually overcome. The reaction of the Soviet government to the evidence of successive stages of progress in the Western negotiations was to impose steadily increasing harassments upon transportation between Berlin and the West. The slowly graduated character of this developing blockade suggests that it was intended as much to deter the three-power conversations as to complete the consolidation of Eastern Europe by driving Westerners out of Berlin. But failing in the first purpose it might still serve the second. The final stage of total surface blockade went into effect on June 24, 1948, the same day that the United States, Britain, and France extended to their sectors of Berlin the complete and drastic currency reform which they had just put into effect in

West Germany. Western-occupied areas would henceforth have a separate currency from the rest of Germany, and their economic recovery would be vastly accelerated. At least, Moscow reasoned, the Western "show-window" in Berlin should finally be closed out—if general war would not thereby be provoked.

The result was a nerve tingling exercise in the limited use of force, Russia versus the West, continuing for a period of nearly eleven months. Neither side wanted general war. In order to minimize the risk, President Truman rejected a venturesome recommendation of General Lucius Clay, the American commander in Germany, and resolved to rely merely on an airlift and counterblockade to maintain access to Berlin; he would not try to send an armed convoy through to the city by land. The Russians, for their part, did not shoot down the planes. Perhaps Stalin did not believe that the Western governments and the inhabitants of West Berlin would have the will to bear the burden of circumventing the blockade. (After all, to the West, Berlin was important mainly as a symbol, signifying no further retreat; to hold the city as a base for promoting unrest in Communist Eastern Europe would not yet have seemed worth the effort in 1948.) Perhaps on the other hand, Stalin was content thus to divert to central Europe the bulk of America's slim air transport arm and the special attention of the nation—during the months while in China the Communists mounted the offensives which conquered the mainland. The Berlin airlift was indeed supremely successful; so, on the other hand, was the headlong triumph of the Red Army in China, where the United States government was unwilling to pay a much greater price for containment. By the spring of 1949, the overall pattern was clear enough to induce the Russians, under real pressure themselves from West Germany's counterblockade against all East Germany, to decide to lift the barriers.

Moscow had failed to eliminate the outpost in Berlin or to halt the revival of West Germany; indeed, the American action in saving West Berliners from starvation or Communism tended mightily to overcome the hatred of Germans for their Western conquerors. The possibility of a really close association between America and a revived democratic West Germany, strengthening the containment of Soviet influence, grew. During 1949 also, the friction between Yugoslavia and Moscow, caused by the Kremlin's efforts to tighten its grip on Eastern Europe, enabled the Greek government to make decisive gains in its civil war. The Greek Communist leadership splintered. Tito's suspected sympathizers were removed, and the Greek rebel radio criticized him. In July 1949 the Yugoslav leader closed his southern boundary, leaving the guerrillas dependent on inadequate support from Albania and Bulgaria. Within a few months the guerrillas were routed. Thus the setback Stalin had suffered in Yugoslavia was compounded in Greece.

Russia's forceful moves in 1948 to counter Western European recovery, notably the Czech coup and Berlin blockade, removed any possibility that the military aspects of the containment doctrine would not be extended northward from Greece to fit some of the defense needs of other countries. The United States prepared in the spring of 1948 to begin negotiation of some form of Atlantic alliance and to provide considerable military matériel for reconstituted European forces. At the onset of the full Berlin blockade, the British government welcomed the basing of American heavy bombers in England. It was generally presumed that in case of general war they would carry atomic explosives to Russia, but American military policy was still so unclear that Defense Secretary Forrestal took the occasion to secure a considered private declaration by President Truman that the use of the A-bomb would indeed be authorized if the need arose. Actually the United States had scarcely any other readily available defense force, and even for A-bombs the adequacy of the delivery system was dubious. The number of long-range bombers (and probably the number of bombs) was pitifully small. Army ground strength at the beginning of 1948 was such that no more than one division could have been put into action without requiring partial mobilization at home. The shock of the Czech coup and tension in Germany caused a brief reimposition of the draft for a few months in 1948, but not for long enough to produce much benefit. From the onset of the containment effort in 1946 until the Korean war, the Adminstration's prevailing attitude was the one well summarized by Secretary Forrestal in a private letter, December 8, 1947:

> At the present time we are keeping our military expenditures below the levels which our military leaders must in good conscience estimate as the minimum which would in themselves ensure national security. By doing so we are able to assist in European recovery.

Forrestal might well have added that containment in the form of adequate defense in addition to European recovery could be achieved by raising taxes, running a big deficit, or cutting welfare programs; but these possibilities were ruled out by domestic politics and the prevailing Administration view that the peacetime economy could not stand a total budget much above $40 billion.

> . . . In other words we're taking a calculated risk in order to follow a course which offers a prospect of eventually achieving national se-

> curity and also long-term world stability. . . . As long as we can outproduce the world, can control the sea and can strike inland with the atomic bomb, we can assume certain risks otherwise unacceptable in an effort to restore world trade, to restore the balance of power—military power—and to eliminate some of the conditions which breed war. The years before any possible power can achieve the capability effectively to attack us with weapons of mass destruction are our years of opportunity.*

In Europe after 1946 the opportunity was well utilized; in Asia, not. But in both areas the years proved to be fewer than anyone in authority had foreseen, for late in August 1949 the Soviet Union exploded an atomic device.

The Administration's reaction was threefold: to minimize its concern in public, to outbid the Soviet Union by speeding development of a much bigger bomb (the hydrogen bomb), and to re-examine its world strategy in the light of the nuclear race and the Communist conquest of China. The re-examination produced a major secret document of the National Security Council, NSC 68, calling for an extensive build-up of American armed forces together with other costly programs of foreign aid and propaganda. Its general principles were privately approved by President Truman in April 1950, but his military budget requests to Congress were still only about $13 billion, and it was by no means clear that any major effort would be made in the following months to enlist public support for the expanded programs. American armed strength stood at ten ground divisions, eight of them undermanned, and forty-eight air wings, only three of them heavy bombers.

On June 24, 1950, Communist troops struck across the 38th parallel to invade South Korea in an act of virtually naked aggression.

## SELECTED BIBLIOGRAPHY

BYRNES, JAMES F. *Speaking Frankly.* New York: Harper, 1947.

CLAY, LUCIUS D. *Decision in Germany.* Garden City: Doubleday, 1950.

DAVISON, W. PHILLIPS. *The Berlin Blockade: A Study in Cold War Politics.* Princeton: Princeton University Press, 1958.

JONES, JOSEPH M. *The Fifteen Weeks: February 21–June 5, 1947.* New York: Viking, 1955.

MC NEILL, WILLIAM HARDY. *America, Britain, and Russia: Their Cooperation and Conflict, 1941–1946.* New York: Oxford University Press, 1953.

* Walter Millis, ed., *The Forrestal Diaries* (New York: Viking, 1951), pp. 351–52.

MILLIS, WALTER, ed. *The Forrestal Diaries.* New York: Viking, 1951.

OPIE, REDVERS, *et al. The Search for Peace Settlements.* Washington: Brookings Institution. 1951.

REITZEL, WILLIAM, MORTON A. KAPLAN, and CONSTANCE G. COBLENZ. *United States Foreign Policy: 1945–1955.* Washington, D.C.: Brookings Institution, 1956.

SETON-WATSON, HUGH. *The East-European Revolution.* New York: Praeger, 1956.

TOYNBEE, ARNOLD and VERONICA M., eds. *The Realignment of Europe.* New York: Oxford University Press, 1955.

TRUMAN, HARRY S. *Years of Trial and Hope, 1946–1952: Memoirs,* vol. 2. Garden City: Doubleday, 1956.

VANDENBERG, ARTHUR H., JR., ed. *The Private Papers of Senator Vandenberg.* Boston: Houghton Mifflin, 1952.

WESTERFIELD, H. BRADFORD. *Foreign Policy and Party Politics: Pearl Harbor to Korea.* New Haven: Yale University Press, 1955.

## CHAPTER 6

# CHINA: WORLD WAR AND CIVIL WAR

THE KOREAN WAR turned the spotlight of public and, to a much smaller extent, of official attention upon Asia in a way that had not occurred since World War II, indeed that had not occurred even during World War II except for an occasional few months of the struggle against Japan.

Korea also dramatized with shocking impact the fact that a reversal of alliances had been taking place in the Far East during the decade 1941–51. Japan had been forced into almost unconditional surrender and then had been transformed into a potential ally of the United States, while in that same period mainland China had been rescued from Japanese occupation only to become more hostile to America than any other power in Asia and, later, than any major country in the world.

The transformation of China into potentially the most threatening power that America confronts on the world scene has been so breathtakingly sudden that its causes and implications are still obscure and highly controversial. But no deep consciousness of the magnitude of the challenge that Communist power has presented to the West since 1950 is possible without some attempt at understanding the "great leaps forward" which were made in China in the 1940's during World War II and its aftermath, when Communism swept into control of a half-billion people, before the United States became sufficiently provoked to resolve to contain it, as in Europe, by using all available means, even American military manpower.

## The Vengeful Crippling of Japan

A first step in appraising the Far East policies that the United States pursued in the decade after Pearl Harbor is to focus sharply upon the ultimate reversal of alliances and to inquire whether Americans could have been brought to appreciate Japan's role in the Asian balance much sooner than they were, and thus have fought the Pacific war for limited goals that would not have left a vacuum of power into which Chinese Communism would be free to flow. Specifically, it was American policy before Pearl Harbor to demand an eventual Japanese withdrawal from China on a timetable acceptable to America, but the anticipation was that the timetable would extend many months at least; could the United States have adhered to that position in a "compromise" peace with Japan after Pearl Harbor and have allowed her to remain in charge of parts of northeast China and Manchuria until the time when Nationalist, not Communist, Chinese would truly be fit (not just willing) to take over the territory and could administer it effectively?

This notion becomes very tempting when one recognizes that the outlines of the Far Eastern situation that emerged to view after 1950 have been remarkably and frustratingly parallel to those that existed before Pearl Harbor, despite all the intervening costs in American lives and treasure. An Anglo-American coalition is once again struggling to keep Southeast Asia out of the control of the most powerful Asian country (formerly Japan, now China). As before, the Anglo-American coalition is linked to the second most powerful Asian country (then China, now Japan), but cannot expect to get much assistance from it in protecting Southeast Asia. As before, the most powerful Asian country is allied to the most powerful European enemy of the Anglo-American coalition (then Germany, now Russia), and can rely on that ally to distract much of the attention of the Anglo-American coalition away from the Far East. As before, the most powerful Asian country also knows it could find many willing collaborators among the native populations of Southeast Asia. And as before, the most powerful Asian country increasingly needs Southeast Asia's rich resources: food, oil, and minerals. A violent clash again seems likely—as before.

Could this situation have been avoided by not having destroyed Japan? The question is akin to the one concerning the unconditional surrender formula for Nazi Germany, which has been discussed (see pp. 74–75). In the case of Japan there is considerably less objective certainty than for wartime Germany that only a government that was ultimately transformed under the supervision of Allied military government could become a reasonably safe participant in world affairs for the preservation of the interests of the United States. Japan's regime was never as totalitarian as Germany's, and her aggressiveness was more opportunistic. Yet the very lack of durable central control would have produced its own dangers of a renewal of militarism and expansionism in a Japan which escaped total defeat. During the 1930's insubordination on the part of militant underlings had been so severe that they had regularly used direct action to force the hands of their more moderate superiors by assassinations at home and by provocative acts on the Asian mainland when they commanded units of Japanese occupation troops there. Consequently, one could never be sure that a cabinet in Tokyo would really be able to restrain its own forces and survive in office to effectuate its own policies. Furthermore, there would have been some possibility, perhaps remote, that Japanese who had lost a war but who had been allowed to remain strong in Asia would have continued to nurse bitterness toward the United States and to cherish vengeful aspirations, and would hence have found congenial associations with Moscow in even greater numbers than, mainly with other motives, they have in fact done since 1945. (Worthy of note in this connection is the fact that much of the insubordinate radical militancy in the Japanese Army in the 1930's was already pro-Soviet.)

Of course the attitude of the Chinese government was a principal reason for the failure of Americans to give much thought in World War II to the possible future uses of a strong Japan. The principal *raison d'être* of Chiang Kai-shek's Nationalist regime since its founding had been the unification of China and the elimination of foreign enclaves. After a generation of fighting the Japanese and their puppets, more or less continually, there was little if any sign that Nationalist morale would be able to adjust to the idea of Japanese participation in joint campaigns against the Chinese Communists. On the contrary, the American government believed that promises to remove the Japanese wholesale from China promptly after Tokyo's eventual surrender would help to buck up Nationalist morale, in the low periods of World War II when the United States found it could give China little more than encouraging words.

Besides, these sweeping retributive threats comforted the hurt pride of the Americans themselves, bitterly humiliated by Pearl Harbor. Here lies the heart of the answer to any speculation about a compromise peace in the Pacific. From December 7, 1941, Americans evinced a true hatred of Japanese which was vastly greater than their antipathy toward Germans in World War II. "Remember Pearl Harbor," the "sneak attack" whereby a nation of whom Americans had been inclined to be contemptuous had managed to immobilize the United States Pacific fleet—such villainous treachery had to be avenged! Sheer vindictiveness was a mark of the Pacific war much more than on the Western Front in Europe. (The bitterness, for example, partly accounted for the failure of Americans to take many Japanese prisoners.) Policy that was colored by that spirit could scarcely have settled for a compromise peace which left Japan with any military power. Subjectively there was even less basis for not demanding a virtually unconditional surrender from Japan than there was from Germany, even if objectively there may have been a little more basis for not doing so than in the case of the Reich.

## Conflicting American Strategies and the Attrition of Nationalist China

Assuming then that no Japanese power would be allowed to remain on the mainland of Asia for any considerable length of time after surrender, a vacuum would be created that was bound to be filled by the nearby Russian and/or Chinese Communists, unless the Chinese Nationalists were somehow made strong enough to occupy the territory promptly when the Japanese were removed and to administer it effectively, and unless the Russians and the Chinese Communists were somehow induced to permit the Nationalists to exercise such authority.

Theoretically, force, or concessions, or some combination of both might enable the Nationalists to establish this authority over all of a liberated China.

Force would obviously be out of the question for China vis-à-vis Russia; negotiation would be required and the outcome would very largely depend on the United States and its attitude toward Russian intervention in China during and after the war against Japan. But vis-à-vis the indigenous Communists, the Nationalists might consider relying more heavily on force and less on negotiated concessions—if Chiang's regime at Chungking could emerge from the Pacific war with a preponderance of strength in comparison with the Communists at Yenan.

By 1942 such an outcome had become unlikely. For a score of years the Nationalist regime, its armies, and its official party, the Kuomintang, had been waging almost continuous warfare against one opponent or another to achieve constructive ambitions that were so far-reaching that perhaps no single Chinese regime could realistically have hoped to accomplish them by itself; perhaps in laying the groundwork the Kuomintang was bound to wear itself out, behave antagonizingly, and be superseded before its task was complete by a new regime which would be fresher, look more promising, and be able to build more swiftly and radically upon the foundations that the Kuomintang had been laying.

The goals that the Nationalists were pursuing in the generation after World War I were to unite China, overcoming all the many autonomous local authorities and armies; to eliminate all the many enclaves of foreign influence and control; if then the foreigners fought back to protect their privileges (as the Japanese did), to fight until they ultimately withdrew; to carry out the economic modernization of the country; and to accomplish some degree of social revolution. By the late twenties the Kuomintang had achieved such remarkable gains that the Japanese had been impelled to resort to general military occupation to prevent Manchuria from adhering to the new China, and in the middle thirties Kuomintang successes elsewhere in China had been so substantial as to arouse militant Japanese elements to new counterencroachments. The war which had followed demonstrated that the Nationalist regime had acquired sufficient support to maintain its continuous existence and its loose framework of government over large parts of China, and to keep armies in the field that would fight hard enough to prevent even the modern forces of Japan from ever imposing peace terms and ending the "China Incident."

On the other hand, the Japanese in the early years of the war had pushed the Kuomintang southward and westward away from the developed urban areas of East China where its progressive elements had gained strength through the adherence of businessmen and intellectuals; in the backward rural areas of western China, to which the regime was driven, its survival depended more on adapting to the traditionalist village society and particularly upon conciliating the unprogressive landlord class—unless the Kuomintang had ventured, as the Communists were doing in their much smaller area of responsibility in North China, to push a social revolution accentuating the role of the peasantry, even while simultaneously warring against Japan. Understandably, Chiang chose not to make that attempt. The war brought special burdens upon the peasantry in the forms of heavy taxation and conscription, which were not ameliorated by much reform. At the same time the best educated and most forward-looking of the Kuomintang's former adherents and sympathizers were finding little outlet for their talents and hopes in the primitive areas to which the war had driven them. They were becoming disillusioned. Meanwhile, the Japanese had seized and occupied the best transport and communications facilities and had reduced the economic base of the Nationalist war effort so drastically that the regime felt obliged to resort to extreme deficit financing, which produced immense inflation. The professional and administrative classes—the Kuomintang's best hope for the future—were again among the hardest hit, since they depended on relatively fixed salaries. Disillusioned and squeezed, they became increasingly corrupt and embittered. The prolongation of the war, year after year, was rotting away the vitality of Chiang Kai-shek's regime and probably dooming it.

The overall situation was extremely ominous. Superficially, one might conclude, as some Americans did, that the sooner Japan was defeated, no matter how, the better for Chiang, and that the harder he could be induced and enabled to fight the Japanese the sooner Japan would be defeated. Chiang could reach a contrary conclusion that his regime, already suffering desperately from the attrition of war, would only suffer more if it fought harder. He might well judge that his regime had already done its share against Japan in the long years before Pearl Harbor, that now the Japanese were almost certain to be defeated by the United States eventually, no matter how little more contribution the Chinese might make to the fighting, and that therefore his principal need was to conserve and accumulate strength as much as possible for the showdown with the Communists that was sure to come after Japan's removal from China. He might gladly stockpile as many supplies as the Americans could be cajoled into giving him and might welcome United States officers to train his armies, but do his best indirectly to avoid actually using the supplies and troops in actions against Japan that would wear them down before they could be used against the Communists. If there had to be expanded military operations against Japanese in China, he could press that they be by air and on a scale so small as not to provoke the Japanese to attempt seizing the airfields in offensive action greater than Chinese troops could readily handle. (Thus Chiang was most favorable to General Clair Chennault's fighter-bomber campaigns.)

Knowledgeable Americans during World War II very commonly suspected Chiang's regime of harboring these attitudes. Their general reaction was disgust at the supposed selfishness of his calculations; rarely did they seriously weigh the possibility that very solicitous attention to the relative growth of Chiang's strength vis-à-vis the Communists ought to characterize the wartime policy of the United States also, with less single-minded concentration on the defeat of Japan. Even when Americans did manage to bring themselves to sympathize with Chiang's anxiety over the future Communist danger, they promptly assimilated this concern to their own eagerness to get on with the job of beating Japan, by contending that the Nationalist structure would only rot faster if it were allowed to be inactive and to try to recuperate for a postwar showdown with Yenan. The Kuomintang's troops should be trained and armed gradually with what little could be spared from higher priority operations in Europe and the Pacific and should then be sent into action in distant Burma in order to clear a road to India there, in order to get more supplies through to West China, in order then to engage in offensive operations against the Japanese in East China!

Chiang had tried one offensive on his own initiative to help try to keep the Burma Road open in early 1942. His fingers were badly burned; he lost one-third of the strategic reserve force that comprised the best divisions under his direct control. He had no enthusiasm for another Burma campaign. But the United States Army had a strategic concept for winning the Pacific war by establishing a link-up between, from one direction, American divisions, which would move up the island chain of the Southwest Pacific (Guadalcanal to New Guinea to the Philippines to Formosa to Canton), and, from the other direction, well-armed and trained Chinese divisions, which would move eastward in China; finally, together they would all march northward and push the Japanese off the mainland of Asia. This gigantic enterprise would require that the Chinese troops be readied for the link-up; airlift alone, over the Himalayas from India, could not suffice to equip large forces—so the airlift should be used to build a Chinese army strong enough to reopen the Burma Road; thereafter, supplies could flow into West China by land, and the grand strategy could be effectuated. Frequent calculations were that Japan would not be defeated until 1947 by this strategy, but the immensely influential Army Chief of Staff George Marshall and his friend General Joseph Stilwell, who had been appointed Chief of Staff to Chiang Kai-shek, doggedly promoted the Burma phase despite repeated delays that were caused by Chiang's resistance and by the diversion of the tiny available resources to other strategies that were simultaneously being pursued by other services.

Chennault, with Chiang's encouragement, wanted to use more airlifted material for fighter-bomber harassment of Japanese troops in China.

The principal commanders of the United States Air Force—heavy-bomber-minded as usual—wanted to construct big air bases in West China and use airlift to supply the bases, from which Japan was to be attacked directly. Air Force enthusiasm for this scheme gradually shifted to enthusiasm for getting Russia into the war, so that bases in Kamchatka could be used. But meanwhile, the United States Air Force bases in China were built, and, as Chiang had feared, their existence was so provocative to Japan that her army launched a decimating offensive late in the war—at a time when the crack new Chinese divisions, which had finally been formed, were far away in Burma struggling to open the supply route there to effectuate the United States Army's strategy.

While all these strategies were competing for air-lifted supplies and otherwise interfering with one another, the United States Navy was pursuing its own approach of island-hopping across the central Pacific—a concept that was so much more successful than the skeptical Army and Air Force had anticipated that the Navy strategy gave the Air Force island bases within range of Japan (making United States bases in Russia no more necessary than those in China), and also gave the Army island bases for a potential direct invasion of Japan (making the ground operations on the mainland of Asia unnecessary for the conquest of the Japanese home islands). But the Air Force was slow to recognize that it would not need Russian bases, and the Army was so preoccupied with the idea of mainland operations that it had been developing the theory that the Japanese would be likely to hold out around Manchuria even after the home islands were conquered—so the Allies would still need powerful mainland armies. Only now the prospect eagerly grasped was that they would be Russian armies, not Chinese and American. It would even be worth paying a price for Russian participation, at the expense of China.

Actually, the Navy's direct cross-Pacific strategy (coupled with Air Force heavy bombing from Navy-secured bases and perhaps also coupled with MacArthur's similar island-hopping in the Southwest Pacific) provided a means whereby Japan could probably have been defeated about as rapidly as she was without any exhausting operations in China or Burma, and without any Soviet contribution.

But the need to rebuild the United States Navy after Pearl Harbor had meant that this strategy could not begin to show any convincing successes until late in 1943—so late a time that all the other strategies * were already under way at their various peripheral points around the Japanese orbit, gathering such momentum that they could not then be cancelled as rapidly as they obsolesced—at least not unless the matter received more persistent and determined attention than the topmost officials were willing to spare from their preoccupation with Germany.

The basic reasons for the "Europe first" emphasis have already been reviewed here and appraised as reasonable. (See pp. 64–66.) But the

* There was also a British variation: to by-pass Burma by sea and to island-hop eastward through the Indian Ocean to Singapore and northward through the South China Sea to Hong Kong and Canton.

cubbyhole to which the China theater was relegated can surely be seen in retrospect to have been excessively unattended at the White House and 10 Downing Street. The whole relationship of the way the Pacific war was being fought to the future Japan-China-Russia-America balance in the Far East went unexamined at the highest levels of Anglo-American wartime command except fragmentarily or spasmodically. The will to require that the pieces be made to fit together was not sustained; attention habitually, almost automatically, shifted back to Europe.

The outcome underscores the need in a world struggle, where priorities are of course necessary, at least not to let any sector be so completely overshadowed. The conduct of American affairs concerning China in World War II was such as to arouse in a re-examiner a sense of slightly delirious fantasy akin to *Alice in Wonderland.* There appears a dizzying interplay of courses of action consistently pursued, utterly inconsistent with each other; marvelously logical, yet together thoroughly illogical; on all sides American Don Quixotes tilting at different windmills—courageous, idealistic, dedicated, but characterized by varying degrees of ignorance of Asia, ardently pursuing unobtainable objectives, at odds with each other—in a decidedly secondary theater of war.

In effect, the conduct of the Pacific war by Americans partook of a kind of opportunistic peripheralism and pluralism of forces and strategies such as they were inclined to disparage when promoted by the British for the European war. Indeed the American Pacific pattern, being less deliberate (in a secondary theater), emerged in some respects as a near caricature of peripheralism. Yet it did succeed in wearing down the Japanese at the fringes of their empire while numerous possible jugular thrusts were made ready for a *coup de grâce.* (The actual final *coup de grâce,* the atomic bomb, was not anticipated or incorporated in any of the principal rival strategies till almost the end of the war; see pp. 78–79.)

One net effect of the mixture of American strategies was to aggravate the postwar position of the Chinese Nationalist regime. Chiang's regime, gravely weakened anyway by the long years of struggle against Japan, was probably further injured in falling between two stools of the American conduct of the war. Chiang would probably have emerged relatively stronger if the United States Army's mainland-oriented strategy had never been undertaken at all (and he had been allowed to conserve his strength and wait out the war while it was won in the Pacific Ocean), or else if the United States Army strategy had been followed through completely no matter how long it took and Chiang had emerged with modernized armies marching side by side with Americans all the way north into

Manchuria, temporarily conciliating the Communists but overshadowing them for the showdown thereafter. The way the war was actually fought and finally won had the effect of increasing the attrition upon the Nationalists, and then not giving them the time and the means to recuperate before confronting the Communists.

## The Mirage of a Peaceful Unification of China

Furthermore, the exercises in partial peripheralism that were begun and then not fully implemented played some part in leading Americans into peculiar new diplomatic and political paths concerning Chinese affairs, to the further impairment of Nationalist interests.

Even for Europe, it will be recalled, the Americans were eager supporters of one principal aspect of peripheralism: the preservation of an active Russian front to divert most German troops away from the West. It is hardly surprising that a corresponding eagerness was felt for the establishment of a Russian front against Japan as soon as there was any hint that Moscow could be persuaded to oblige; and the willingness to pay a price for it followed shortly.

Another aspect of peripheralism to which the Americans had not objected in Europe was co-belligerency with Communists: the arming of Communist-led guerrillas (by air-drops where necessary), their affiliation with regular armies when Allied liberators arrived in force, and the formation of governing coalitions between their leaders and noncommunists. It is thus also not surprising that this desire to take the most effective advantage of Communist activism to destroy fascism—together with a rather wistful hope of meanwhile somewhat taming the Communists—appeared also in China. In 1943, when the American Army was pursuing its grandiose plans for China-Burma operations, it developed a keen interest in establishing ties with the Chinese Communists—just as in that same year the British, in pursuing their Mediterranean-Balkan ambitions, began shifting support in Yugoslavia to Josip Broz Tito, away from the less activist guerrilla forces of the prewar anticommunist government.

It can of course be argued that this procommunist dabbling, in pursuit of peripheralism, would have done no irreversible injury if the peripheralism had been fully implemented and American and/or British armies had finally gone in force through the doors that they had been intending to open for themselves by means of conciliating the Communists. The experience in Italy and France, even in Greece—when contrasted with that in Yugoslavia and China—would lend weight to this argument. In the cases of both Yugoslavia and China, Britons and Americans never arrived in force before the war was finally won (along other routes to the enemy homeland that were direct, not peripheral). Those Americans and Britons who had promoted the collaboration with Communists from motives of wartime opportunism did not get the chance to offset the harmful side-effects of their actions by means of the almost automatic advantage that local noncommunists would have eventually obtained (as they did in France, Italy, and Greece) from the arrival of the large British or American forces, which had been anticipated initially at the time when the expediential collaboration policy had been set.

But it would be a mistake in the case of China to imply that the American policy of conciliating Russia and the Chinese Communists was solely a diplomatic-political consequence of abortive peripheral military strategies. Much more, it was the product of a will on the part of Americans to avert future civil war in China—and to do so not so much because of anxiety about what China would be like if the Communists won, as because of sentimental concern about the suffering that the Chinese people would endure in any renewed fighting (following upon the weary years of warfare against Japan), and because of fear that the United States and the Soviet Union would find themselves drawn into war against each other if they remained too closely affiliated with the respective Chinese factions. The tolerant view of Yenan was possible because the Communists there were managing to appear much more progressive than the wartime Kuomintang and no more repressive, and because the doctrinaire Leninism of Yenan's publications was not taken seriously. Besides, of course, it was an era when the United States was getting along with Russia too. From 1943 onward American policy acquired a fixed central purpose: to promote reform in the Kuomintang and also a close affiliation between it and the Communists.

United States officials never did override Chiang's objections and actually arm Yenan's troops to fight the Japanese more effectively and jointly; and in the last year of the war the sense of urgency about getting any major Chinese military contribution declined among Americans. But their suasion upon Chiang for rapprochement with the Communists continued. Indeed, the principal new reason for not needing any Chinese so much against Japan—at least as seen by the United States Army and Air Force—was that Russia was becoming available; and the prospective Russian participation shed a red light upon the political future of Manchuria and North China that made the territorial unification of China under one government seem more than ever unlikely unless Chiang were persuaded to make major concessions in negotiations with Moscow and Yenan.

The United States government forced his hand, by a bargain that was suggested at Teheran in late 1943 and made final at Yalta early in 1945 designed (1) to get the Russians into the Pacific war; (2) to limit their Far Eastern gains to what the czars had formerly held, or perhaps a little less; and (3) to get Russian support for Chiang Kai-shek, or at least for a coalition in which the Kuomintang would preponderate over the Chinese Communists. The second point, of course, was in conflict with the uncompromising nationalism of the Kuomintang; its war-shaken morale was bound to be newly weakened by this Yalta obligation to concede the Russians a virtual sphere of influence in Manchuria and Mongolia. But point three might have offered great compensation, except that it was not delineated with sufficient specificity in the Yalta bargain and, more important, that the war thereafter ended so unexpectedly soon that the Kuomintang forces had not arrived on the North China scene at all—much less arrived in the company of huge American armies. And the Kremlin could scarcely resist the temptation to give major assistance to the Chinese Communists, turning over to them captured stores of Japanese arms and permitting them to become entrenched in the Manchurian countryside (as they were already entrenched in rural North China) while the arrival of the Nationalists was being delayed.

There are parallels here to the fate of the "postponement policy" regarding Europe. Unlike the premature delineation of occupation zonal boundaries for Germany, the United States in effect continued indefinitely to postpone consideration of an occupation zone for Russia in Japan—and thanks to the course of military operations never had to permit one. But somewhat like Churchill dealing about spheres of influence in the Balkans, trying to salvage something in a region where the only major armies would be Russian, Roosevelt found himself drawn into a wartime bargain with Stalin about mainland Asia. Yet there was a major difference in that, unlike Churchill pondering the fate of the Balkans, Roosevelt, ill advised, positively wanted Russians to enter Manchuria in order that the Western Allies would not have to go there; hence he was less able than Churchill and he were in Balkan affairs to hedge the Far Eastern bargain with other agreements (like the Declaration on Liberated Europe), which could be used later to throw back on Russia all the blame for her aggrandizement.

Of course, the Yalta provisions calling for Russian support of "the National Government of China" did constitute something of a hedge.

However, the Russians were careful to adhere to them in form and not to violate them grossly in practice except during the Soviet military occupation of Manchuria, which ended in April 1946. Russia's policy, as in Europe and the Middle East, was cautious; for the time being she seemed content with effectuating her "czarist" Yalta gains and with giving her fellow Communists in China an interim boost, while leaving the United States to play the most conspicuously interventionist role in most of China.

But the Americans' approach when intervening was still not wholeheartedly pro-Nationalist. So in the Far East the limited scale of Soviet violations of the Yalta understandings did not arouse much American protest. Chiang's regime was expected to absorb the morale damage of the "czarist" Yalta concessions, plus the partial failure of the Russians to adhere to their ambiguous promise to support the National Government, and get on with the task of cementing an affiliation with the Communists to avert civil war. The latter remained the overriding concern of United States policy.

The lack of a top-priority American commitment to the success of the Nationalist regime was no new departure when viewed in the light of a half-century of sentimental verbalization in the United States about the "territorial and administrative integrity of China." Despite all their protestations, Americans had never been committed to the unification of China under any particular regime to the point of running much risk of war on that account, with the possible exception of the months before Pearl Harbor. But even then, as has been noted, the American involvement had originated more in anxiety about the fate of the British Empire and the rubber and tin of Southeast Asia than in a pressing concern to save China or its current regime, however much admired. Then in the years after Pearl Harbor the decaying Nationalist regime had been losing even the admiration of many Americans, particularly some who knew it best and were inclined to grasp rather blindly for any alternative. Not surprising, then, was the unwillingness of the United States to do for Chiang Kai-shek after the war what the Europe-first priorities plus the indecisiveness about Far Eastern strategy plus the unexpectedly sudden end of the war had prevented the United States from doing incidentally for him during the war, in the course of operations against the Japanese: namely to use American military power to establish Chiang's forces in control of all China and Manchuria. Instead, United States support for the Nationalists partook of the same old low-priority, half-hearted spirit as before, yet without ever turning into total abandonment, as it hadn't before.

From the viewpoint of the Kuomintang, nursing the wounds of war on the morrow of the victory over Japan, there were three principal alternative policies. One would be to settle temporarily for control of only part of China, tacitly accepting for an indefinite period a partition of the country that would leave Russian and Chinese Communists to share among themselves the domination of Manchuria and some of North China and allow the Kuomintang to concentrate on entrenching itself in Central and South China, closer to its wartime bases of strength; a long rivalry between the two Chinas would then ensue to develop their respective regions before each would ultimately seek mastery of the whole country. A second possible policy for the Kuomintang would be to gather all the forces it could muster, weak though they were, and drive north for a prompt military showdown with the Communist armies, gambling on the vast Nationalist superiority in numbers, if they were fully concentrated, to overcome the probable Red superiority in morale and tactics. Such a policy would be certain to encounter strong American objections, which would further reduce its chance of success; so a third alternative policy would be to occupy whatever parts of China could be obtained with American cooperation and assistance—and then, if necessary, undertake to bargain with the Communists about the rest.

The partition policy might well have been acceptable to the Communists for a considerable period. They would stand to gain the most developed part of the country, particularly Manchuria, with a potential for further industrialization that would be likely to overshadow the progress of the Kuomintang in the south. Furthermore, the ideological cement on which the Communists largely depended to hold their adherents was not so centrally nationalistic as that of the Kuomintang; Communists would not be so likely to become disillusioned if the unification of China were much longer delayed. Moreover, the passage of time, even in war, had been adding to the Reds' strength and enabling them gradually to extend their limited area of control, rather than wearing them out like the widely extended Kuomintang; the same pattern might well be allowed to continue.

The converse of each of these arguments gave the Nationalists a reason for opposing even a temporary partition, although for a time in the late summer and early fall of 1945 they were in effect counseled to do so as the lesser of evils by General Albert Wedemeyer, the American who had succeeded Stilwell as Chiang's American Chief of Staff and who was decidedly pro-Nationalist.

Potential Communist gains from a partition were indeed ominous, but the overall position of the Kuomintang was so disadvantageous that there were strong reasons to believe that other policies would be even more hazardous. For example, it could and should have been foreseen that the Kuomintang would have great difficulty in promptly winning sufficient local support and providing enough competent and honest administrators to establish effective government in North China and Manchuria, even if Nationalist armies could be transported there. Never in history had Manchuria really been governed by the Kuomintang; as for most of North China, the Kuomintang had been out of control for about a decade after having been in for only about a decade. A large proportion of the administrators who had retreated southwest with the government had degenerated under the inflationary squeeze and lack of opportunities; back in North China and Manchuria they might behave like carpetbaggers. (Yet the administrators who had stayed there during the war were so tainted through collaboration with the Japanese that a Nationalist government would hesitate to employ them.) If misgovernment ensued and the local population were alienated, the Nationalist armies might be unable to supply themselves locally and would be dependent upon long lines of communication stretching back to Central China. To guard those lines from Communist strength in the countryside, the armies would have to be strung out thinly and rather ineffectively. Attrition would set in on them as it had on the Japanese, but they would be not nearly so well fitted as the Japanese to hold out.

That much could have been foreseen, and in part it was—but it was offset by anxiety over long-term gains that might well accrue to the Communists from any partition. What could not have been foreseen, in light of the Yalta promises, was the considerable degree to which the Russians during their nine months' occupation of Manchuria assisted the Chinese Communists and impeded the Nationalists and the extraordinary extent to which Moscow looted the province's industry, making it after all a much less valuable prize, at least in the short run, than Nationalist aspirations had assumed.

On the other hand, the Nationalists had meanwhile acquired a bonus inducement to reach for Manchuria, one that had not been present in Wedemeyer's initial cautionary weighing of the arguments: the Administration in Washington decided late in 1945 to cooperate with Chiang's Manchurian ambitions and to provide air and sea transportation to move much of his army there, bypassing Communist-held areas of North China. This gesture minimized the possibility that the Nationalists would attempt to concentrate their forces for a showdown with the Communists, and it improved the chance of persuading Chiang to resume negotiations for some kind of coalition with Yenan—two principal American objectives at the time.

The ultimate effect was, as Wedemeyer had feared, to overextend the Nationalist armies, guarding railroads and garrisoning cities; initiative drained away from even the few élite divisions whose training the United States Army, pursuing its mainland strategy, had completed before the sudden end of World War II—including some that had managed to survive the related campaign to reopen Burma. The Nationalist armies were not maneuverable enough to bring their countrywide numerical superiority to bear locally in 1947 when the Communists finally concentrated against them in Manchuria; the rail lines were cut, and the isolated Nationalist garrisons were destroyed one by one.

However, the fatal move into Manchuria in 1945 was only partly the fault of the misguided American gesture in providing transportation. Actually, most of the Kuomintang and its armies were so defensive-minded and so accustomed to wars of attrition after the long years against Japan that they themselves found the Manchurian occupation much more congenial than massing for an all-out campaign against Yenan. In fact, many Americans were more conscious than they of the danger that protracted fighting would produce a war-weariness that would make any alternative, even the Communists, welcome to the public. On the other hand, the Americans were naïve in concluding that the answer was not a quick showdown either, but a coalition government of some kind.

The coalition mirage continued to be pursued faithfully for most of 1946 under the auspices of a special United States presidential mission headed by the potent wartime Chief of Staff General Marshall. Americans were very slow to reconcile themselves to the fact that neither the Nationalists nor the Communists could possibly trust each other enough to give up control of their respective armies to a government in which the other would or even might come to preponderate. The parallels that could superficially be drawn with postwar Europe were not really pertinent. There it had been possible temporarily to form coalitions between Communists and noncommunists (though each such government eventually swung in the direction of the closest major foreign army). But in China the problem was not simply one of bringing hostile parties together in a single government, or even of bringing hostile guerrilla units together in a single army, but one of bringing hostile governments with million-man armies together into some kind of union, a thorough impossibility.

Yet American pressure to continue negotiations was an additional reason for the Nationalists to limit their offensive action in 1946 to the occasional grab of an additional city, thus tying down more troops in garrisons, instead of launching a concentrated drive using at least the troops that had not gone to Manchuria. Even the spasmodic Kuomintang thrusts upset the American mediators so much that an embargo prohibiting all United States shipments of arms to China was put into effect by the Administration from August 1946 until May 1947. And finally in late 1946, when General Marshall began to despair of achieving a government that would embrace the Communists as well as the Kuomintang, he shifted his emphasis not to support of the Kuomintang but to encouragement of an idealistic "Third Force" of Chinese moderates and liberals.

## No American Troops to Block Communist Victory in China

Understandably, Americans found these men more attractive than the contending armed parties, but the essence of the whole situation was that the two principal parties in conflict were heavily armed. The Third Force was not. And those who had arms were not disposed to admit the newcomers to power, unless, possibly, the American government had shown itself willing at last to offer massive support to the Kuomintang and thus in effect to purchase the admission of the Third Force to major influence.*

If the United States had been participating strongly in a joint effort in China to overcome the Communists, or at least to contain them, the Americans might conceivably have been able to induce important reforms in Nationalist policy and administration. But arms-length prodding and criticism—however well-meant and even if wise (which it often was not)—is not likely to carry much weight when the adviser shows no sense of responsibility himself for a final outcome that will be favorable to those who take the advice.

In China the only realistic alternative to the Communists was the Kuomintang. Its hopes for survival depended on internal reform as well as upon military assistance. But to have much hope of inducing the indispensable reforms, the United States would have to commit herself

* The conceivable alternative that America might side so heavily with the Chinese Communists as to reorient them was unreal, both because of opinion and politics in the United States and because Yenan's antipathy fixed a gulf that was even wider than the most pessimistic Americans at that time recognized. See pp. 128–29, 500–51.

much more deeply to the regime's military survival—and any further delay in doing so would relatively strengthen the Communists.

Yet the cost of effective aid was bound to be enormous. The real, painful demands of the China situation were finally delineated most clearly in a report which General Wedemeyer submitted in the late summer of 1947. Wedemeyer had returned to Washington from duty as Chiang's Chief of Staff early in 1946. He was known as a wartime protégé of Marshall as well as a supporter of Chiang, and was selected to go back to China as head of an investigatory mission the following year. Among the recommendations in his report were these three: (1) heavy American military matériel assistance to Chiang; (2) American training and advisory missions for Kuomintang troops down to the regimental level, but outside actual combat zones, requiring an estimated 10,000 American officers and men; and (3) a "Big Five Manchurian Guardianship" or, should one of the great powers refuse to participate in it, a formal trusteeship under the UN to save Manchuria from Communist domination; the American manpower cost of this arrangement was not strictly calculable (it would depend on the Russian and Chinese Communist response), but in light of later events in Korea, one must suppose that there might very well have been local resistance requiring the use of many tens of thousands of American troops, even if only a "holding operation" were attempted against the Reds, as the Wedemeyer Report suggested.

It seems reasonable to conclude that these aspects of Wedemeyer's proposals constituted the bare minimum of the military requirements for the survival of the Nationalist regime in control of most of China after 1946. If there was any margin of error in Wedemeyer's calculations, he was figuring the scale of necessary American involvement too low. To have succeeded, the commitment might also have had to begin sooner and then have lasted many years.

Wedemeyer's recommendations were rejected out of hand. The State Department, where General Marshall had become Secretary, showed little interest in the report, attempted to obtain Wedemeyer's signature upon an expurgated version and, failing that, suppressed the entire document as secret. The formal justification for doing so was that the recommendation concerning Manchuria would be a blow to Chinese nationalism that would also antagonize Chiang Kai-shek—as indeed it might, being an infringement upon Chinese sovereignty and a dramatization of his regime's incompetence to govern all its own territory and of its dependence upon foreign support. Many Chinese were sure to be gravely offended. But Chiang at that stage of the struggle might well have accepted the arrangement if it had been formally offered to him; his freedom of choice had become very much constricted.

Somewhat more substantial arguments for not following Wedemeyer's recommendations were a fear that the relationship he was advocating between the United States and Nationalist China would be construed as "semi-colonial," not only by many Chinese, but also by other Asians, and that they would thereby be alienated from America—and also a fear that the Russian reaction would be to give aid to the Chinese Communists, step by step, matching any increased aid from America to the Nationalists. The consequent heightening of East-West tension might lead to war; at least there would be grave weakening of the UN as the bridge in which Americans still reposed high hopes.

However, the fundamental objections of the Administration to Wedemeyer's proposals centered upon their costs to the United States in manpower and economic resources, mostly unpredictable both in quantity and in duration.

Concerning manpower, the United States had demobilized to a point where implementation of Wedemeyer's recommendations for training and advisory missions alone would probably have required the recall to active service of some hundreds or thousands of reserve officer veterans of World War II; the additional effectuation of a trusteeship over Manchuria (if it did encounter Communist resistance) would have required a reimposition of selective service. Domestic political resistance would surely have been very difficult to overcome.

As for the drain on American economic resources, their availability was bound to be limited if the Republicans, then in control of Congress, insisted (as they did) on cutting taxes and avoiding government controls over the economy, while the Administration insisted (as it did) on massive Marshall Plan aid to Europe. With an habitual Europe-first perspective upon American foreign relations, the Administration feared, logically, that if it sought large amounts of assistance for China, Congress would accordingly reduce the aid for Europe, and, in consequence, would fail to appropriate enough for success in either area. Feeling a need to choose, the Administration characteristically preferred to concentrate on Europe.

Finally, the former image of Japan bogged down in China haunted some American policymakers. They feared that the drain upon American manpower and economic resources would continue indefinitely, and that once the commitment was undertaken, withdrawal would be practically impossible, regardless of the developing consequences. Therefore, a tentative, experimental approach to intervention was also ruled out, perhaps too hastily, too much so because the potential costs to the

United States of a failure to attempt an intervention in China were not confronted by the Administration with the tough-mindedness that was applied to costing a possible decision to intervene. To be sure, the costs of not intervening were not so conspicuous as the costs of intervening; yet the apparently gross underestimation of the price of inaction can hardly be justified.

In the first place, there was consistent underestimation, at least until the middle of 1948, of the potential speed with which the Communists would conquer China. For example, in March 1947, Dean Acheson, then the Under Secretary of State, declared: "The Chinese government is not in the position at the present time that the Greek government is in. It is not approaching collapse. It is not threatened by defeat by the Communists. The war with the Communists is going on much as it has for the last 20 years." *

Second, there was underestimation of the potential speed with which a Communist regime would be able to consolidate its rule over China even if the Kuomintang did collapse. After all, no government had been able to master all of China in the previous half-century; one could easily suppose that the Communists would not be able to do so soon either, in view of the immensity of the country's area and population.

Third, and partly in consequence of the foregoing, there was underestimation of the strength with which a Communist China would be able to threaten Western interests elsewhere on the mainland of Asia, particularly in the southeast region over which the United States had previously clashed with Japan. Even the Wedemeyer Report, otherwise prescient, did not foresee the potential independent aggressiveness of a Communist China; the fears expressed were principally of Russian forces that would be enabled to use bases in China, if the United States allowed the Communists to defeat Chiang Kai-shek. After a century of patronizing China themselves, even the most far-sighted Americans were much more anxious about the new patrons that China might acquire than about what China might do on her own.

Significantly, this attitude characterized both the group of Americans that favored heavy support for Chiang Kai-shek and also the opposite group, gradually becoming preponderant in the Administration, that wanted to cut loose from him and seek some accommodation with the Chinese Communists. Both groups shared an underestimation (the fourth for many among them) of the depth of Red Chinese hatred of the West and of the degree to which a Communist regime would propagate

* Hearings on assistance to Greece and Turkey, before the House Foreign Affairs Committee, March 1947, p. 17.

the hatred to elicit mass sacrifices for the country's economic development,* and also of the degree to which the hostility would be perpetuated in Southeast Asia by China's unsatiated economic needs and political ambitions there, especially if the Kremlin were shrewd enough to yield back its northerly Yalta gains to a new, cooperative Communist regime in China. The two groups of Americans disputed whether these antipathies were so great as to bar all prospect that Yenan might be divided from Moscow if Washington decided to be conciliatory toward Yenan. But they shared an assumption that whatever anti-Westernism did exist in Yenan would make a split between the two Communist centers more (not less) difficult to achieve; that if somehow a split did come, the Communist country less hostile to America would be China. Neither group envisaged a China so utterly alienated from the United States that she would become more hostile to America than Russia was, nor a time when the Kremlin would have to restrain Peiping's bellicosity and when Communist Chinese publications would imply that if any split was developing it was because Russia was turning "Titoist."

From this final gloomy perspective, which was clouded from view until the middle and late fifties, the hopelessness of any American policy for the forties, except possibly one of paying whatever price was required to keep the Nationalist regime in control of part or all of China, appears even more clearly than it did to Wedemeyer and others who shared his views during the months of final decision.

The State Department and Secretary of State Marshall confronted the costs of action frankly, and understandably they were appalled. But they did not appreciate the costs of inaction even to the degree that Wedemeyer did. The balance of the Administration's calculations was thus overwhelmingly against further commitments to Chiang and even in favor of reducing those that already existed. The weight of the arguments seemed so clear-cut to those in authority that they decided it was more important to avoid new commitments than to have the conceivable alternatives for China fully debated in public as they had been discussed in private; hence Wedemeyer's report was suppressed in order to deprive the domestic opposition of ammunition for debate. When Congress nonetheless enacted a small, presumably inadequate military supply program for Nationalist China in 1948, the Administration did not expedite its implementation. By early 1949 the Kuomintang's position on the mainland of China had become irretrievable except possibly by massive American armies, which of course were not forthcoming.

In the autumn the Nationalist regime fled to the island of Formosa,

* See the analysis on p. 167.

130 miles off the southeast coast of the mainland. As long as Chiang might manage to remain there, his supporters in the United States would probably be influential enough to prevent the Administration from disavowing him completely and pursuing an accommodation with the new Communist government. Besides, that new government was already behaving so harshly to American diplomats that some illusions were being shaken in Washington. But in any case, the fall of Formosa to the Communists was expected to occur some time during 1950. The Administration explicitly and emphatically refused to use the limited available United States forces to defend the island, lest American involvement in China's civil war be prolonged any further.

After the anticipated fall of Formosa would come the 1950 congressional elections, and thereafter the Administration would probably begin to try seriously to come to terms with the New China.

However, the Kremlin was planning a different sequence of victories. Instead of Formosa, the first blow was to come in Korea. But there, surprisingly, the Administration was willing to respond with force—and even, thus provoked, decided to extend protection to the Nationalist regime in its remote refuge (another interesting parallel to 1941). The ensuing conflict revealed and heightened Red China's enmity for America, and the United States found herself expending the manpower and resources to save half of tiny Korea that she had been unwilling to spend when such assistance might have saved half or more of mighty China.

## SELECTED BIBLIOGRAPHY

CHIANG KAI-SHEK. *Soviet Russia in China: A Summing-Up at Seventy.* New York: Farrar, Straus, and Cudahy, 1957.

CHURCHILL, WINSTON S. *The Second World War,* vols. 4–6. Boston: Houghton Mifflin, 1950–53.

DEANE, JOHN R. *The Strange Alliance: The Story of Our Efforts at Wartime Cooperation with Russia.* New York: Viking, 1946.

EHRMAN, JOHN. *History of the Second World War: Grand Strategy,* vols. 5 and 6. London: Stationery Office, 1956.

FAIRBANK, JOHN K. *The United States and China,* rev. ed. Cambridge: Harvard University Press, 1958.

FEIS, HERBERT. *The China Tangle: The American Effort in China from Pearl Harbor to the Marshall Mission.* Princeton: Princeton University Press, 1953.

———. *Japan Subdued: The Atomic Bomb and the End of the War in the Pacific.* Princeton: Princeton University Press, 1961.

JONES, F. C., HUGH BORTON, and B. R. PEARN. *The Far East, 1942–1946.* New York: Oxford University Press, 1955.

KECSKEMETI, PAUL. *Strategic Surrender: The Politics of Victory and Defeat.* Stanford: Stanford University Press, 1958.

KING, ERNEST J., and WALTER M. WHITEHILL. *Fleet Admiral King: A Naval Record.* New York: Norton, 1952.

LATOURETTE, KENNETH S. *The American Record in the Far East: 1945–1950.* New York: Macmillan, 1952.

LEAHY, WILLIAM D. *I Was There: The Personal Story of the Chief of Staff to Presidents Roosevelt and Truman.* New York: Whittlesey, 1950.

MATLOFF, MAURICE, and EDWIN M. SNELL. *The United States Army in World War II: Strategic Planning for Coalition Warfare, 1941–1944,* 2 vols. Washington, D.C.: Department of the Army, Office of the Chief of Military History, 1953, 1959.

NEUMANN, WILLIAM L. *Making the Peace, 1941–1945: The Diplomacy of the Wartime Conferences.* Washington, D.C.: Foundation for Foreign Affairs, 1950.

ROMANUS, CHARLES F., and RILEY SUNDERLAND. *United States Army in World War II: The China, Burma, India Theater,* vols. 1–3. Washington, D.C.: Department of the Army, Office of the Chief of Military History, 1953–59.

ROSTOW, W. W. *The Prospects for Communist China.* New York: Wiley, 1954.

SHERWOOD, ROBERT E. *Roosevelt and Hopkins: An Intimate History.* New York: Harper, 1948.

SNELL, JOHN L., ed., *The Meaning of Yalta: Big Three Diplomacy and the New Balance of Power.* Baton Rouge: Louisiana State University Press, 1956.

STETTINIUS, EDWARD R. *Roosevelt and the Russians: The Yalta Conference.* Garden City: Doubleday, 1949.

STIMSON, HENRY L., and MC GEORGE BUNDY. *On Active Service in Peace and War.* New York: Harper, 1947.

TRUMAN, HARRY S. *Year of Decisions: Memoirs,* vol. 1. Garden City: Doubleday, 1955.

———. *Years of Trial and Hope, 1946–1952: Memoirs,* vol. 2. Garden City: Doubleday, 1956.

UNITED STATES CONGRESS SENATE COMMITTEE ON FOREIGN RELATIONS. *Nomination of Philip C. Jessup: Hearings.* Washington, D.C.: Government Printing Office, 1951.

———. *State Department Employee Loyalty Investigation: Hearings* and *Report.* Washington, D.C.: Government Printing Office, 1950.

UNITED STATES CONGRESS SENATE COMMITTEE ON THE JUDICIARY. *Institute of Pacific Relations: Hearings* and *Report.* Washington, D.C.: Government Printing Office, 1951–53.

UNITED STATES CONGRESS SENATE COMMITTEES ON ARMED SERVICES AND FOREIGN RELATIONS. *Military Situation in the Far East: Hearings.* Washington, D.C.: Government Printing Office, 1951.

UNITED STATES DEPARTMENT OF STATE. *Foreign Relations of the United States: Diplomatic Papers: The Conference of Berlin (the Potsdam Conference), 1945,* 2 vols. Washington, D.C.: Government Printing Office, 1960.

———. *Foreign Relations of the United States: Diplomatic Papers: The Conferences at Malta and Yalta, 1945.* Washington, D.C.: Government Printing Office, 1955.

———. *Postwar Foreign Policy Preparation.* Washington, D.C.: Government Printing Office, 1949.

———. *United States Relations with China, with Special Reference to the Period 1944–1949.* Washington, D.C.: Government Printing Office, 1949.

VANDENBERG, ARTHUR H., JR., ed. *The Private Papers of Senator Vandenberg.* Boston: Houghton Mifflin, 1952.

WEDEMEYER, ALBERT C. *Wedemeyer Reports!* New York: Holt, 1958.

WELLES, SUMNER. *Seven Decisions That Shaped History.* New York: Harper, 1951.

WESTERFIELD, H. BRADFORD. *Foreign Policy and Party Politics: Pearl Harbor to Korea.* New Haven: Yale University Press, 1955.

CHAPTER 7

# THE DIPLOMACY OF LIMITED WAR: KOREA

THE INVASION of South Korea was a profoundly shocking surprise to the American government. Numerous reports had come of a massive military build-up in North Korea, but these had been largely discounted as South Korean exaggerations, direct or indirect, designed to induce Washington to arm the southerners on such a heavy scale that they would be able to invade the North; the American government and the United Nations General Assembly supported a unification of Korea, but only if accomplished by peaceful means. The prevailing view in Washington was still that the Communist powers would avoid resorting to outright military invasions across internationally established frontiers, and would continue to restrict themselves to subversion and economic pressures in their drive for expansion. Yet if an invasion should occur after all, other places seemed more likely targets than Korea—notably Yugoslavia.

Secretary of State Acheson had given the most formal public pronouncement concerning the Administration's Far Eastern policy on January 12, 1950, when he described a "defensive perimeter" which "runs along the Aleutians to Japan and then goes to the Ryukyus" (Okinawa) and "from the Ryukyus to the Philippine Islands." About the United States commitment to defend these positions Acheson was emphatic: The "defense [of Japan] must and shall be maintained. . . . [The Ryukyus] are essential parts of the defensive perimeter of the Pacific, and they must and will be held. . . . An attack on the Philippines could not and would not be tolerated by the United States." But elsewhere in East Asia, Acheson seemed to be hedging a little: "So far as the military security of other areas in the Pacific is concerned, [should a military attack occur] the initial reliance must be on the people attacked to resist it and then upon the commitments of the entire civilized world under the Charter of the United Nations which so far has not proved a weak reed to lean on by any people who are determined to protect their independence against outside aggression." *

Actually, the American government did give sufficient stimulus to the UN to make the organization's response effective in South Korea. But in January 1950 there was good reason to doubt that anyone would intervene to save that country from invasion. The Security Council was the only organ of the UN which seemed to have the authority to initiate military sanctions—and there the Soviet Union would be able to exercise a veto. Acheson himself was known to be rather skeptical of the utility of the United Nations in the cold war situation. Most important of all, there was not in progress any substantial build-up of the woefully weak American forces in the Far East, nor a build-up of South Korean forces, except for purposes of internal security against insurrection. Thanks to new American aid programs, the power of the government of the Republic of Korea was becoming sufficiently firmly established in the South, with tolerably promising economic prospects, so that its overthrow by procommunists from within would be increasingly unlikely. Military invasion from outside would be necessary if South Korea was to be incorporated into the Red bloc in the near future. These developments, coupled with Secretary of State Acheson's questionable reliance on the UN, must have produced strong temptations for Stalin to take a chance on rapid military conquest. Indeed, the absence of the Soviet delegate from sessions of the Security Council (in protest against the failure to seat Red China there in place of Nationalist China) continued for weeks after the outbreak of war in Korea, even though this absence permitted the United States to get Security Council backing for a military defense of South Korea unhampered by a Soviet veto. The Russian boycott had been initiated in January 1950 in the context of negotiations for a general treaty of alliance between the USSR and Red China; perhaps a commitment that could not be readily reversed had been given to continue the boycott. But it seems unlikely that such a potentially awkward and cumbersome arrangement would have been maintained throughout the spring, during preparations for the North

* *Department of State Bulletin,* January 23, 1950, pp. 115–16.

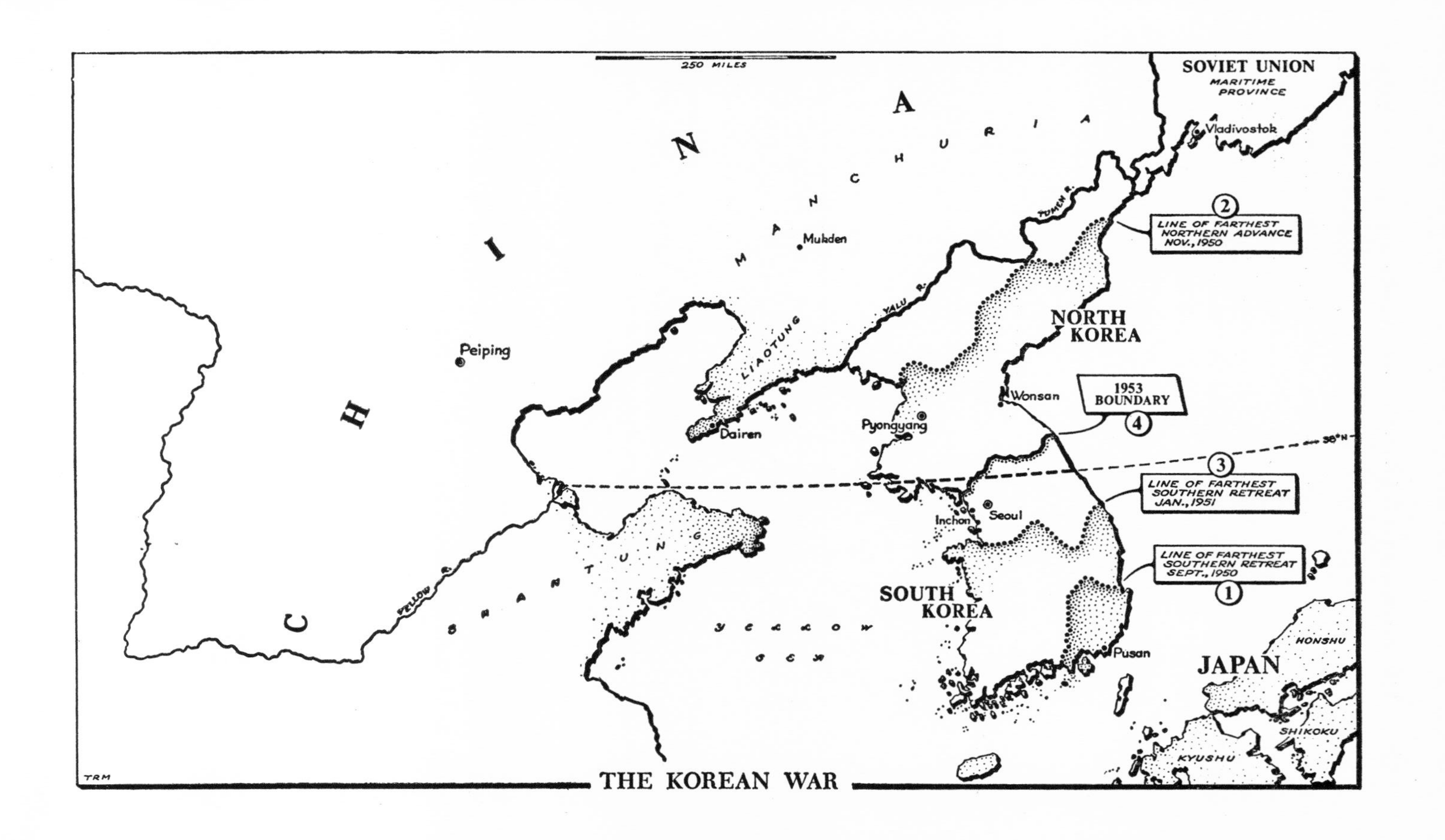


THE KOREAN WAR

Korean attack, if Stalin had seriously anticipated that the United States would act through the Security Council with vigor sufficient to defeat the invasion. On the other hand, it is possible that Stalin was experimenting, creating a test case for his future policy, by permitting a situation to arise in which he anticipated that some U.S.–UN action would be feasible and even likely, in order to see whether it actually would take place and in what form. If the action came, it would probably be too late to be effective, and then the UN and the United States would both be discredited; and even if, in the end, their defense of South Korea should happen to be successful, Stalin could reason that that particular bit of territory was not of central importance to his ultimate ambitions anyway.

### The American Military Intervention

Whatever the Kremlin may have calculated in June 1950 when it unleashed the attack by the North Korean government (which had been formed under the close supervision of Soviet occupation forces in 1945–48), the actual response in Washington was rapid and decisive. The solidarity of the Administration's swift determination to resist militarily was heightened by several considerations:

1. The North Korean action appeared as blatant massive armed aggression across an established frontier, if not across a formally recognized international boundary line; the cold or tepid war between East and West was suddenly and deliberately being turned boiling hot.

2. If unresisted, this invasion would be likely to spur additional Communist incursions elsewhere and cause demoralization in other countries outside the existing Soviet sphere. A parallel with the Japanese seizure of Manchuria and with Hitler's early conquests was in the forefront of the minds of American policymakers. Communist expansion must not be allowed to come to appear as another irresistible "wave of the future" on which all prudent men would choose to ride along rather than to stand helplessly and drown. Dramatic counteraction seemed necessary to block the formation of a defeatist "bandwagon" psychology; moreover, such action would also help to reverse the widespread incipient wavering that even the merely subversive tactics of previous postwar Communist expansion had already engendered.

3. In particular, it became apparent soon, if not at first, that an actual fighting war in progress in Korea could furnish the stimulus to induce the masses in America and in Europe to accept the sacrifices required for the worldwide rearmament previously recommended by Ad-

ministration officials in the National Security Council in the spring of 1950; * prior to the outbreak of the Korean war such efforts to rearm had appeared doomed by public complacency. With a war on, mobilization would be feasible, but the Administration never had any intention of proceeding toward a level high enough to encourage the launching of a "preventive war" against the Soviet Union. Public opinion in the United States might well have been aroused to endorse such a rash course, but national policymaking was heavily under the influence of men who were somewhat disillusioned with the fruits of total war and determined to do their utmost to avoid it, short of surrendering additional territory to Communist military aggression.

4. The United States and also the United Nations Organization had had a particularly deep formal involvement in the establishment of the Republic of Korea in the south, and consequently seemed to have acquired a special responsibility for preserving its existence.

5. Moreover, somewhat surprisingly, both the United States and the UN were in fact able to act—the United States because the invasion took place so close to the only major concentration of American troops in the Far East (the half-trained occupation forces in Japan), the UN because Russia was still boycotting the Security Council.

Spurred by these considerations, the United States, only twenty-four hours after the North Koreans crossed the frontier, obtained the adoption of a Security Council resolution, calling for immediate cessation of hostilities and withdrawal of forces to the 38th parallel boundary and appealing to UN members to "render every assistance to the United Nations in the execution of this resolution." After about thirty more hours of deliberation, the Administration determined upon air and naval intervention as its form of assistance, without waiting for authorization by Congress. Accordingly, the following day the United States delegate on the Security Council obtained passage of a resolution recommending that the UN "furnish such assistance to the Republic of Korea as may be necessary [Would land forces be included? The Administration was awaiting General MacArthur's on-the-spot recommendation, which came two days later and was promptly accepted in Washington.] to repel the armed attack and to restore international peace and security in the area." Did the phraseology about restoring security constitute an authorization to use force to do more than re-establish the status quo before the war, that is, Korea divided at the 38th parallel? Actually, the vague language was simply borrowed from Chapter VII of the United Nations

* NSC 68, in the wake of the Soviet A-bomb explosion and the Communist conquest of China.

Charter, and Security Council discussion, when the resolution was adopted, gave little clarification. But two days later Secretary Acheson explained that United States action "pursuant to the Security Council resolution" was "solely for the purpose of restoring the Republic of Korea to its status prior to the invasion from the north and of re-establishing the peace broken by that invasion." * Very little if any ambiguity remained; the American government's objective at the onset of the Korean war was evidently not to conquer North Korea nor to unify the country by force. The United States military intervention was rather narrowly defensive in purpose—a fitting extension of the means but not the ends of the containment policy.

Behind this choice to limit the objectives lay the Administration's own peaceable intent, which was rooted in the traditional American aversion to foreign wars, but also related to some recent disillusionment with the fruits of total wars fought for total victories, and confirmed by the even less bellicose attitudes of respected allies whose coöperation was wanted in other cold war theaters that seemed more important than Korea. In particular, even though the actual fighting was in the Far East, the Administration's primary concern, as usual, was Europe. The construction at last of an adequate military defense to shield the rich human, industrial, and cultural resources of the Continent would require close collaboration in the North Atlantic Treaty Organization; a bellicose American policy anywhere would impede it. Moreover, the same Administration concern for rearmament in Europe meant that the United States' own forces, though expanding rapidly, would continue to be spread dangerously thin around the world; they would not be concentrated in Korea. For these various reasons the Administration would attempt to keep the Korean war limited—in area, in weapons, in manpower, and in objectives.

### The Red Chinese Intervention

Even the Administration's new decision to protect Formosa from the Communist Chinese was principally a limitation device. Adopted simultaneously with the initial policy of air and naval intervention in Korea, its major purpose was to prevent Communist forces from getting any closer to the Philippine Republic and thus menacing it, in conjunction with formidable local guerrillas, at a time when the slim American forces in the Far East would have to be concentrated around Korea.

* *Department of State Bulletin,* July 10, 1950, p. 46.

Therefore, Formosa was officially "neutralized." The cordon of the Seventh Fleet was thrown around Chiang Kai-shek's regime to protect the island from the Chinese Communists. In form, the Communists on the mainland were also being protected from Chiang, but of course they were bound to regard the arrangement as provocative, especially since the terms of the American declaration failed to give assurance that Red China would ever in the future be allowed to overrun Formosa.

For several months previously United States policy had frankly accepted the prospective fall of the island. The apparent reversal of that line produced such a negative reaction among other members of the UN, who were reluctant to antagonize the Peiping regime, that the American government made no effort to obtain UN endorsement for its unilateral action in the Formosa Straits. The Formosa decision did help to conciliate Chiang's supporters in the United States and thus to unify American opinion for the Korean venture. The "neutralization" continued. But during the summer of 1950 the State Department sought strenuously to minimize any new provocations which might induce Russia or Red China to enter the war. American bombers were generally kept away from the northernmost fringes of North Korea where they might stray over Manchuria or the USSR. Repeated pronouncements at the highest levels of the American government affirmed a peaceful intent toward Red China. Chiang Kai-shek's offer of 33,000 Nationalist troops to support South Korea was rejected, partly because of transport shortages, doubts of their fighting quality, and the desire to have them available if needed to defend Formosa itself, but largely in order to avoid giving the Communist Chinese an excuse to intervene in Korea on the other side. And when General MacArthur twice during the summer made conspicuous gestures of enthusiastic support for Formosa, he was publicly rebuffed by the Administration.

But the American fence around Formosa remained; to the Chinese Communists the Seventh Fleet's "neutralizing" role constituted "aggression," shutting them off from part of their own country. In addition, their attitude was affected by the degree to which the prestige of international Communism as a whole was at stake in the North Korean venture; after the successful American holding action in July at the Pusan beachhead in southeast Korea, that prestige was badly in need of restoration. Perhaps also there was direct Russian pressure upon China to enter the war to redress the balance as Americans gained momentum rolling back the North Koreans. However, even without such prodding the Peiping leaders had further reasons of their own for going to war. Intervention would give the New China a chance to exhibit her strength, to win for herself prestige in Asia and in the world at large, perhaps even to secure greater respect from Moscow. Korea, an ancient dependency of the Chinese Empire, might be brought in whole or in part under Peiping's control. And even if Chinese forces were not so successful on the battlefield, the mere existence of the war would give Mao Tse-tung's regime useful excuses to consolidate its power more forcefully over the whole of mainland China.

Then came a final incentive for intervention: the rout of North Korean forces after mid-September and the UN advance up to and beyond the 38th parallel, posing apparently a new, direct threat to the Yalu River boundary of China herself.

Why did UN forces (mainly American and South Korean) cross the 38th parallel in defiance of Peiping's specific threats to intervene if they did so? Partly, the purpose was narrowly military: to clean up the remnants of North Korean armies which had eluded the trap between UN forces pushing north from Pusan and those fanning out eastward from Inchon. But further, there was the political aim to punish aggression; though involving elements of sheer vengefulness, this aim could be justified as a deterrent to future aggression by insuring that the guilty party would lose not only the fruits of its initial successes but also so much more of its original pre-invasion possessions that in comparison with the victims it would clearly be the greater sufferer from the entire venture. If such a precedent were set in this instance, it probably would strongly inhibit future Communist aggression around the world. At the very least, the North Korean regime itself would be prevented from reconstituting itself as a threat to the security of the Republic of Korea. Otherwise, without unification, the menace would return. Indeed, Syngman Rhee's South Korean government might well send its own troops on into North Korea to forestall such an eventuality whatever the United States did; but then would Americans really be able to keep out of the fighting if Rhee's forces were defeated in the north? Besides, the United States and the United Nations General Assembly had for years been committed to the objective of uniting the country. Were they now to give up the golden opportunity to do so, by the use of force, just because Peiping was issuing threats? In the sudden momentum of victory, unification seemed too easy to be waived. This attitude was also fostered by the general Western underestimation of China, of the physical strength and the aggressive purposefulness she had newly acquired under Communist rule.

In some Western minds also, including apparently General MacArthur's at this stage, the contemptuous attitude toward Chinese threats was related to an overestimation of what the Administration in Washington would be willing to do in response to Chinese intervention if the threatened incursions actually did begin. When North Korea was the only attacker, she was allowed to be bombed; there was a superficial logic in supposing that if China also became an attacker she too would be bombed (at least in Manchuria), and that knowing this, she therefore would never attack at all. Actually, taking a broad world view and recognizing how thin Western forces were stretched and how reluctant they would be to start *atomic* bombing in a secondary theater like Korea-Manchuria, the Communists could well have reached a shrewder advance estimate than MacArthur of how limited China's risks would really be if she did intervene. Possibly, also, the Red Chinese knew by espionage that they could expect to get some advance warning of any American decision to strike across the Yalu, even if they provoked that decision themselves.* In short, the Western policymakers who discounted the likelihood of Chinese intervention were underrating the power and resolution of the New China and were sometimes inclined to imagine China to be overrating the actual power and resolution of the West. Small wonder that unresisted unification of Korea seemed within the UN grasp.

The General Assembly (the Security Council being once again stymied by Russia's return) hastily approved forcible unification on October 7, 1950, in a resolution formally recommending that "all appropriate steps be taken to ensure conditions of stability throughout Korea" and that there be established under UN auspices a "unified, independent and democratic government in the sovereign State of Korea." UN forces (mainly American) were to leave any part of Korea where they ceased to be required for achieving these objectives, but this gesture toward Chinese sensibilities was not sufficient to appease Peiping. Armies from Manchuria began crossing the Yalu in mid-October about a week after Americans crossed the 38th parallel on the way north. Battles began at the end of the month between UN and Chinese forces.

A strange month ensued in which Chinese troops streamed across the Yalu bridges and massed south of the river, but did so without making

* During the summer the Administration had given a "commitment to the British not to take action which might involve attacks on the Manchurian side of the river without consultation with them." Harry S. Truman, *Years of Trial and Hope: Memoirs*, vol. 2 (Garden City: Doubleday, 1956), p. 374. This commitment might have been communicated to the Communist powers by their strategically placed secret sympathizers in the British Foreign Service, Burgess and MacLean, who might also have been expected to pass along word of the special consultation if it did come.

their military objectives clear. They may indeed not yet have been clear even to Peiping. The initial attacks in late October were followed by a Chinese disengagement to concealed positions, with an accompanying lull in the fighting, which gave Peiping time to test the scale of the American response to the new intervention. If the response were a ruinous bombardment of Manchuria—or alternatively, if it were a withdrawal to the "narrow neck" of Korea, leaving a North Korean government intact in the far north—the potential massive Chinese onslaught might be countermanded. Meanwhile, as preparation in case the offensive was finally ordered, tens of thousands of troops would move south from the Yalu in remarkable night marches that would enable them to get into position—unobserved by UN airmen—for a possible future sudden exploitation of the wide dispersion of UN army units in North Korea.

Unaware of Peiping's massing of forces, but very much worried about the presence of any Chinese, the Administration still hoped to avoid general war against them. Decisions were almost immediately made in Washington that must have relieved some remaining anxiety in Peiping about the scale of UN response to the fact of intervention. American planes were forbidden to fly more than halfway across the Yalu, even though this rule meant that they could not effectively knock out the bridges nor even fully pursue Red fighters which flew missions over Korea from bases in Manchuria.* On the other hand, there was no UN withdrawal to the "narrow neck," much less to the 38th parallel. Some serious consideration was given to offering to establish a narrow demilitarized buffer zone along the Yalu, and Red China was invited to send emissaries to state her case before the UN in New York. A regroupment of United Nations forces was tentatively explored; they were fanning out to occupy all of North Korea, and it seemed possibly desirable to concentrate them instead for defensive action against the Chinese. But reconcentration back south at the "neck," which the British wanted, encountered shrill rejection as a "Munich," particularly from General MacArthur, who claimed furthermore that his forces were too few (even if reconcentrated) to hold any line in North Korea against an all-out Chinese offensive, and therefore preferred to take a chance that no such unrestricted Chinese effort would occur.†

* The part of this restriction which forbad "hot pursuit" into Manchurian air space was established largely in response to the influence of hypercautious allies; but even on its own, the Administration was disinclined to risk the other aspect, that is, bombing the wrong bank of the river by mistake. Both limitations surprised and shocked MacArthur.

† Courtney Whitney, *MacArthur: His Rendezvous with History* (New York: Knopf, 1955), pp. 413–14.

In order to dispel the uncertainty, MacArthur launched a general offensive on November 24, in part as a "reconnaissance in force" designed to determine how many Chinese troops had entered North Korea and whether they were bluffing, or "reconnoitering in force" themselves, or methodically organizing to attack. MacArthur's headquarters had been somewhat lulled (by the weeks of Chinese disengagement in mid-November and by the absence of air reconnaissance reports of enemy deployments, which went unobserved) into believing that massive intervention had probably been deterred or blocked, particularly by the bombing of communications, which was still permissible inside North Korea. But now MacArthur quickly discovered the Chinese hordes, and as his troops reeled back he was severely criticized for having precipitated the final showdown; yet his move is understandable in view of the supposed unreliability of the Korean and Nationalist Chinese sources of intelligence data on which he otherwise would have been heavily dependent. Less excusable perhaps was the doubtful adequacy of the preparations that had been made for an orderly retreat if the Red Chinese did happen to be flushed out of cover en masse, as they were.*

## Keeping the War Limited

The war against China was now on in earnest in Korea, and the question to the fore of the minds of American policymakers was whether to extend air and naval action to Chinese territory. The Administration in Washington would not give favorable consideration to such an expansion of the war unless the UN forces were pushed back far south of the 38th parallel; leading allies—most important, the British—were very strongly opposed to expansion. From the point of view of most Europeans, the United States had already proved her support for collective security under the UN Charter by defending South Korea; the precedent they wanted was thus already established, and further risks should now be minimized. Moreover, West Europeans widely held the view that Peiping had genuine grievances against the United States (with regard to Formosa, recognition, a seat in the UN, and the advance to the Yalu), and that a general war against Red China would assume the aspect of a "race war," causing revulsion around the world. American policymakers in part shared this attitude, and insofar as they were being influenced by

* Also questionable is the timing of the offensive to begin on the very day the Red Chinese emissaries arrived in New York, since it destroyed whatever slim chance may have remained that serious negotiations would ensue in the UN.

opinion in the allied countries, were showing respect mainly for particular allies, only secondarily for the UN as an institution; the UN formed a convenient mantle of righteousness without inhibiting the Administration much more than it would have been inhibited by its regard for the British alone.

Yet the primary factors determining the limitations the Administration continued to impose upon the conduct of the war in Asia were its own calculations of the national security interests of the United States—calculations that were heavily affected by the persistent scarcity of available armed forces and the customary concentration upon building defenses for Europe.

In the months of decision at the turn of the year 1950–51, the United States had only one uncommitted, trained division; a few more would become available, but not until spring. The Administration was unwilling to embark upon a scale of mobilization that would subject World War II veterans to selective service again. Yet because the pool of manpower that had come of military age in the years 1946–50 was limited, the armed forces could not hope to expand to more than about one-fourth the World War II peak of mobilization by relying mainly on it. Governor Dewey of New York was prominent among those who urged a further increase in forces to about half of that 1944 "full mobilization" level. Such an expansion would have made feasible some very useful flexibility in American world strategy in 1951 and 1952. One possibility, for example, would have been the reconquest of North Korea on the ground, defeating China there without the necessity of bombing Manchuria or of holding troops back from Europe's defense preparations. Full mobilization may be incompatible with a limited war, but *semi*-mobilization in principle is not, and would have made practicable a "less limited" war. The very low level of partial mobilization on which the Administration resolved, in contrast, kept the range of possible uses of the American Army extremely limited, especially since so many of the available troops were earmarked for Europe.

As for the Air Force, it had more than seventy wings in late 1950, but only a few of those comprised long-range bombing planes; the Air Chief, Hoyt Vandenberg, felt that these should be conserved for retaliation against a possible Russian attack. As yet, no hydrogen bomb existed, and the supply of A-bombs was still too small to tempt statesmen to "waste" them against North Korean and Chinese troops in the field or against the extremely primitive, easily reparable communications system on which the enemy was able to rely; because these targets seemed inappropriate, nuclear weapons would probably not have been used even if there had not been international political considerations militating against their employment—Europe's fear of the new Russian A-bombs and Asia's resentful suspicion that white men would use nuclear weapons only against colored men, for example, the Japanese.

The supposed need to conserve air power for the potential defense of Europe was matched by the demand to proceed rapidly with construction of a ground defense there. The Administration was haunted by the notion that the Kremlin might be plotting to lure the bulk of the growing American army into East Asia and then to pounce upon relatively undefended Western Europe. Such fears probably were somewhat exaggerated—a reaction to the upsetting discovery in Korea that the Communist powers were willing to resort to outright invasion; containment would have to be military as well as economic and countersubversive. This principle would have to be applied to Europe as well as to East Asia, but Western policymakers may have been too quickly inclined to forget how relatively cautious Communist policy in Europe in the postwar years, contrasted with Asia, had been. Communist willingness to take a chance on invading South Korea did not betoken a willingness to take the same chance in Europe: the odds favoring victory ought probably to be higher to warrant invasion of an area which the United States would presumably fight much more punishingly to defend. The Administration's calculations regarding the imminent likelihood of an invasion of Europe may have been too pessimistic, excessively colored by residues from the total-war preconceptions of the early 1940's.

Even in the Far East, however, the United States was still ill-prepared—either physically or, perhaps as important, in the breadth of strategic doctrines then regarded as respectable—to fight a "somewhat less limited" war. There were evident advantages to be derived from successfully inflicting some lasting punishment upon the aggressors, to deter them from future invasions by depriving them of part of their preinvasion possessions. In a sense (little noted), the forcible partition of Formosa from China did come to operate as one retaliatory deprivation punishing the Communists for having started the Korean war. Red China paid this price; North Korea might have been made to pay a similar, more painful price if she had been deprived of her territory south of the Pyongyang-Wonsan "narrow neck," or some other line deep inside her territory. But to achieve such an objective against Chinese troops on the ground would require a far larger commitment of American troops than seemed justifiable to the Administration. And to seek this kind of victory (less than total, but more than a draw) by expanding the area of the fighting, by bombing and blockading part or all of China, required the improvisation of a new strategic concept in the midst of war as well as the successful communication of it to the enemy. The concept would have to be established that limitations on the war—on its objectives, its weapons, and its locale—were being relaxed but not abandoned; that a new set of limitations would be developed (involving, for example, nonatomic bombing of factories and communications in Manchuria only, in order to induce the Chinese to withdraw as far as the "narrow neck" only); and that the new limitations would be respected and great restraint be shown to prevent this partially graduated expansion of the conflict from "snowballing" into total war.

This strategic concept was uncongenial to most groups of Americans who were in positions to influence policy. Their preferences among objectives differed, but they tended to agree in envisaging the alternative goals as sharply defined: (a) the status quo before the war, (b) total victory over North Korea, (c) total victory over Red China, (d) total victory over Russia. After China entered the war, it was very difficult to suppose that a total victory could be won over one Communist power without provoking its next largest partner to offer determined military resistance, and requiring additional expansion of the war; yet preoccupation with total victory, derived from World War II, still distracted the minds of policymakers from devising schemes of partial victory the next largest Communist partner might be induced to tolerate. Thus the concept of "limited war" became identified with the extremely restrictive limitations on the conduct of the war already in force before the Chinese entered. The Chinese intervention was taken to mean that in order to preserve the restrictions it would be necessary to revert to the limited objectives for which they had originally been designed (that is, mere restoration of the prewar status quo), abandoning the wider objective of forcible Korean unification which had temporarily seemed attainable in the early autumn. Insistence on objectives beyond the status quo situation was envisaged as involving a cumulative relaxation of restrictions on the conduct of the war, which might lead to adoption of yet more grandiose objectives, and so on toward worldwide total war. Perhaps it is true that the war could never have been reconfined, if it had been extended at all beyond the precisely recognizable geographical limits of Korea, largely because of the difficulty of settling upon new limits the enemy would believe durable. But with the shortage of UN forces and the preoccupation with Europe, the Administration probably did not fully explore the possibilities of extension.

One other major set of considerations militated against any partial expansion of the war: the "privileged sanctuary" which the UN forces enjoyed in Japan. Naval and air bases and staging areas for ground troops, immune from enemy attack in the islands, were all very usefully available. Would the immunity survive a UN attack on the Communists' "privileged sanctuary" in Manchuria? Also, UN shipping was not subject to naval attack (except in the form of mines). Great havoc could be wrought by Communist submarines if any should go into action. A particularly ominous possibility, in light of the probable absence of sufficient, trained Chinese, would be the participation of Russian submarines manned by Russian crews; American forces might not be able to identify them positively as Russian—or at least might have to pretend not to, rather than undertake wider hostilities against the USSR. Numerous Russian "volunteers" for the ground and air forces in Korea were another ominous possibility. Any American move toward expanding the war might well provoke retaliatory Communist moves of these kinds, sufficient to cancel whatever advantage the United States might expect to derive from the initial partial relaxation of restrictions.

For all these reasons, the American government's policy, decided in a frantic atmosphere in Washington in December 1950, was to try to continue to localize the war, while seeking formal General Assembly recognition that China was now participating in the aggression and should be forcibly resisted "in Korea."

Compromise efforts by other powers in the UN ran afoul of Red China's minimum demands—the UN seat, Formosa, and withdrawal of all foreign forces from Korea; the United States would not "reward aggression."

But a policy of continuing the war in Korea required that there be some UN foothold in Korea from which to continue the war. Yet Chinese victories were mounting to the point that briefly, in mid-January, General MacArthur, the Joint Chiefs of Staff, and President Truman all anticipated that UN troops would be forced to withdraw from Korea. If so, or if confined to a narrow beachhead, the Americans would have been authorized, under a tentative JCS plan, to carry on the war against China herself, by blockade, air reconnaissance (but no bombing), and by supporting attacks by Chiang Kai-shek against the mainland. In short, the war would have been expanded for want of any other way to fight it. But just at this moment the tide was turning in Korea. A few days later the prospects were good for holding most of South Korea, and as the winter weeks passed American forces found themselves grinding northward again, cautiously this time, uncertain how far to try to go in this limited war, and concentrating meanwhile on slaughtering China's massive manpower in the field (for example, "Operation Killer").

Should we cross the 38th parallel again? How far? With what commitment of force? Additional ground divisions were becoming available in America, but they were thought to be needed in Europe. If air power were to continue to be confined within the boundaries of Korea, any northward movement of UN troops would narrow the area in which air attrition could be applied against Communist supply lines. The various other arguments for strictly limiting the war were rehearsed in Administration circles and among the major allies. The decision reached was that the UN military position had become respectable enough to provide a decent basis for undertaking a diplomatic initiative to inform the Communists that a cease-fire in the general area of the 38th parallel would be acceptable, that the UN would pursue its other Korean objectives nonforcibly, by negotiation. The Communists would even be given some reason to hope that if they negotiated a settlement for all Korea that was acceptable to the West, they would be paving the way for attractive agreements on other Far Eastern questions of special concern to them (for example, Formosa and the UN seat). No specific commitments on the latter points would be offered.

There was little expectation in Administration circles that Red China would yet be ready to accept these proposals, which would come close to restoring the prewar status quo, requiring both sides to give up trying to unify Korea by force. But at least the overture would help to clarify the objectives of the war in a fashion appealing to allied and neutral sentiment—and ultimately the Chinese might prove amenable. The plan was for Truman himself to present the offer with the authority and dignity of a formal presidential declaration. But General MacArthur upset the arrangements by publicly offering to do the negotiating himself and by strongly implying that there would be no related conversations concerning Formosa or the UN seat; moreover, MacArthur's offer was colored by an implied threat to expand the war into the interior of China. This pronouncement—unauthorized by the Administration, contrary to its war policy, and provocative to the Chinese—was repudiated in Washington, but Truman was unable to proceed with his own more moderate declaration in such a context. The war went on. UN forces crossed the 38th parallel for a short distance and braced to withstand a series of massive Red counteroffensives during late April and May. But this time the Chinese were unable to make major advances, despite their willingness to endure enormous losses of manpower. By early June the UN Command was forging ahead with gathering momentum, and Red China's leaders may have felt that the scale of the war was becoming too great a strain.

Meanwhile, however, the American war effort was being subjected to new and severe pressures on the home front. In April, because the Administration could no longer endure General MacArthur's willingness to allow his dissatisfaction with its limited-war policy to become public in repeated episodes, he had been unceremoniously removed from command. The brusque dismissal, however, precipitated such fierce resentment among the American people, who felt increasingly frustrated by the apparent aimlessness of the war, that some modifications of Administration policy were hastened. Substantial military supplies were sent to Chiang on Formosa, raising the possibility that his forces might ultimately be used. Negotiations with the British produced agreement that air bases inside Red China might be bombed if they were used for attacks on UN ground forces in Korea, not only, as agreed before, for attacks outside Korea (for example, on Japan). There was some remaining ambiguity as to whether the American government would have to engage in further consultation with the British before unleashing bombers in this eventuality, but the effect of this agreement, much of which was probably communicated to the Chinese, was to help extend the UN "privileged sanctuary" from Japan to Korea itself: if airfields in North Korea could be kept knocked out, as they could, UN forces on the ground need not fear bombing or strafing anywhere. This comfortable situation persisted till the end of the war. (The development showed also that the rules of limited war could be somewhat altered even in the course of hostilities without becoming perilously unsettled.)

Another UN move in the spring of 1951, partly spurred by Administration concern over pro-MacArthur sentiment, was a formal tightening of trade restrictions against Red China. More pressing efforts than before were also made to secure a larger proportion of non-American troops for the UN forces in Korea, but these efforts were never really successful. The French emphasized that they were fighting Communists in Indochina; the British that they were fighting Communists in Malaya; and all the Western Allies that they were being pressed by America to build maximum forces in their own respective areas for defense of the free world as a whole. The Administration was inclined to agree that logistically it made little sense to transport more Americans to Europe so that more Belgians or Danes could be spared for Korea, but such calculations were hard to explain to a United States public which had its attention focused on comparative casualty rates. The non-American, non-Korean forces fighting in the name of the UN never exceeded approximately 33,000 soldiers and 11,000 air, naval, and service personnel; they came from fifteen countries, while five others sent medical units.

The American contingents, in contrast, approached a half-million men; the South Korean army was ultimately even larger.

### Truce Talks and Stalemate

The most important of the Administration's moves in the spring of 1951 that were partly designed to cope with the MacArthur-removal furor was the decision to proceed in the first week of June with a public presentation of the proposal for an armistice at the 38th parallel. The war would no longer appear so aimless and interminable as before. Acheson, not Truman, was the initial spokesman; his cautious, undramatic peace feeler did elicit favorable response.

Further feelers went out from both sides during June, often including specific reference to the 38th parallel. Meanwhile in Korea, however, the UN army was grinding ahead on a line which extended as far as thirty miles above the parallel. Would it retreat from this territory in the event of an armistice? The Communists would be certain to cast an embarrassing light upon any such withdrawal. Moreover, the incentive for UN troops to continue to apply northward pressure on the battleline would decline sharply if they knew in advance that any gains they made would probably be only temporary. And the diagonal northeasterly line that the UN was on, approximately, was the shortest, most defensible line anywhere across the Korean peninsula (with the possible exception of the Pyongyang-Wonsan line a hundred miles further north).

The Administration decided to press for a cease-fire along the battleline wherever it might be when the truce was signed—not exactly on the 38th parallel despite the implications of the previous peace feelers. This policy became clear, to the Communists' dismay, when formal armistice negotiations began in Korea in early July. For six weeks the North Korean and Chinese delegates stalled the talks, manufactured and dramatized "incidents" designed to impair the prestige and impugn the motives of the American negotiators, and then finally broke off the meetings altogether. Perhaps the Communists believed that they could push the battleline back to the 38th parallel by a new offensive of their own. (UN military pressure had been relaxing lest further "provocative" advances upset the truce talks.) Perhaps also, the Communists hoped to be able to use the well-publicized setting of the San Francisco conference (where the Japanese peace treaty was to be signed in early September) to score a propaganda victory over the United States for supposedly delaying peace in Korea. But after the conference was over, it was the UN army that was advancing.

Once again the American government had to decide whether to seek negotiations. Again the offer was made—somewhat safeguarded this time by insistence on a new site and rules for the talks which would minimize the possibility of "incidents," and by a private agreement with the British that if the Communists refused to renew negotiations and instead resumed large-scale fighting, the United States would be free to relax certain restrictions on its Korean war effort without further consultation with London.*

The Communists consented to the new truce talks, since they were losing ground on the battlefront and faced possible expansion of the war. They were now also willing to accept a truce demarcation line at the battleline, provided it be the existing battleline, not the final one at the time the entire armistice agreement was complete. This proposal presented the American government with a crucially important choice. If accepted, it would destroy most of the incentive for further advances in Korea. During the course of the remaining truce negotiations the Communists would know that they were immune not only from attack outside Korea, as hitherto, but also immune, in effect, from heavy ground attacks inside Korea; the only important military pressure that the UN would be able to apply to win acceptable agreements would be air raids over North Korea. The American high command in the Far East was strongly opposed to such a situation. However, doubt remained that the Administration would ever actually choose to commit enough additional ground forces to Korea to maintain very much battlefront pressure anyway. And, of course, the farther north one advanced the less room would remain south of the Yalu for attrition by airpower. Also, there was a problem about the order of items on the agenda of the truce talks. Shortsightedly, in July the American negotiators had accepted a Communist-proposed agenda which did contain the appropriate items, but in the wrong order, as it later appeared from the Western viewpoint. The Communists now refused to proceed to the other items until the first, the armistice line, had been settled. Unless this item were compromised, there would be no way of knowing whether a basis for peace could be found with regard to the others.

The Administration was finally moved to offer a compromise about

* When this agreement was announced in London the following year, the specific American action involved was kept secret (*Parliamentary Debates*, February 26, 1952, pp. 970–78). It did not automatically include bombing of China.

the armistice line—a compromise which amounted to near surrender on this item. During a thirty-day period, if a complete armistice agreement were reached, the final demarcation line would be the battleline as of the beginning of that period, regardless of any troop movements during the thirty days. After that period, if the armistice remained unsigned, the course of battle would once more be allowed to determine the line for any eventual cease-fire. The Communists accepted this arrangement. For thirty days as anticipated, there was no incentive for the UN to advance; fighting nearly ceased. The Chinese and North Koreans used the respite to construct extraordinarily powerful fortifications along the existing line, vastly strengthening their defensive position. And when the thirty days were up after Christmas 1951, with the armistice still far from settled, there was less chance than ever that the UN forces in Korea would be strong enough to maintain enough offensive pressure on the ground to induce the Communist negotiators to make concessions. Battle positions were essentially frozen. But the war—and the truce talks—continued with ever-deepening frustration.

The item concerning provisions for safeguarding the armistice did prove to be manageable—mainly because the American negotiators were willing, step by step, to abandon insistence on detailed controls, putting their reliance on other arrangements not requiring Communist consent. The basic Administration concept was that the armistice should not result in any change in the relative strength of the two sides such that, if hostilities were later resumed, either side would have benefited much more than the other from the intervening armistice. This objective could be reached by two different approaches: The first, insisting that the same relative balance be preserved inside Korea, would require elaborate restrictions on increases in force levels and complex arrangements for inspection to detect violations; the whole scheme would be extremely difficult and time-consuming to negotiate and doubtfully reliable even if finally established. Adopting the other approach, one would be content to insure the maintenance of a balance in the Far East as a whole. Thus, if there were a build-up of Communist military strength in North Korea, violating the armistice restrictions and perhaps remaining undetected, and if this development led to a renewal of hostilities inside Korea, the UN could undertake to retaliate, no longer only inside, but also outside Korea; the "privileged sanctuary" need not again be respected. Consequently, any new Korean war would probably be a bigger war. The newly elected Conservative cabinet in Britain agreed late in 1951 to accept this arrangement, and in January 1952 the concept was endorsed by all the UN powers that had troops in Korea. It did serve its purpose of speeding up the truce talks regarding provisions for safeguarding the armistice; the American negotiators could consciously accept lame guarantees from the Communists. But the armistice as a whole was not hastened after all, for the remaining issue proved unexpectedly stubborn: What was to be done with the prisoners of war?

Actually, the war was only half over; it would go on for eighteen months more, and in that period six thousand Americans would be killed. Certainly no such cost was anticipated when the American government adopted its policy late in 1951—and possibly the cost would have had to be paid regardless of what policy was adopted on prisoners, because the Communists may have thought it advantageous to prolong the stalemate war on any excuse. But officially the issue was the principle of "no forced repatriation" of prisoners, and apparently the war would have ended much sooner if the Administration had never adopted this principle. It had not been established in the laws of war that a prisoner could refuse return to the jurisdiction of the government to which he formally owed allegiance. Indeed, a literal interpretation of the relevant international treaty, the Geneva Convention of 1949, appeared to require repatriation "without delay after the cessation of hostilities" (Article 118), though whether and how this was to be accomplished against the will of a prisoner was left unspecified. The Administration seized upon this ambiguity to insist that no prisoner in UN hands would be forced unwillingly to go back to his Communist homeland.

The reasons for this insistence were various. One was the ugly image of the period after World War II in Europe when the Western Allies had cooperated with the Soviet government in uncovering tens of thousands of Russians in their territory and forcing them to return to Soviet control; many of the victims had committed suicide rather than risk repatriation. The Korean war was unpopular enough already in the Western world without letting it end on a similarly ugly note. Matters of appearance came to seem yet more important after February 1952, when the Communists were beginning to make propaganda capital of the charge that UN forces used germ warfare in Korea. In rebuttal the UN seemed to need some glowing moral principle to uphold. Besides, there was some positive hope (probably exaggerated) that the guarantee against forcible repatriation, if once established in Korea, would be useful in encouraging persons to defect from the Communist cause in any future conflict. Furthermore, there was the problem of preserving tolerably good relations with Syngman Rhee. The Korean president was vehemently opposed anyway to a truce that would leave his country divided. How much more bitter he would be—and how disruptively he might behave—if the armistice agreement also required him to force Koreans to return to Communist jurisdiction. This problem might have been avoided if the South Koreans had been allowed to release their cooperative North Korean prisoners unobtrusively—perhaps to draft them into the South Korean Army, as the North Koreans were drafting prisoners from the South. But such a tactic had been blocked by Americans, who feared that reprisals would be taken against Americans in Communist camps; then came the spotlight of publicity, assuring that any such move in South Korea would provoke an international crisis.

No such degree of publicity had been anticipated, largely because the American policymakers were unaware of the great number of North Korean prisoners involved. The United States was not really in control of the prison camps; to conserve combat personnel most of the direct policing had been turned over to South Korean guards—but the Americans distrusted their reports about the prevalence of anticommunism among the prisoners. Actually, there was in progress a large-scale infiltration of the camps by agents of Peiping and North Korea on the one side and of Chiang Kai-shek and South Korea on the other—all posing as prisoners and fighting by speeches and violence to persuade the ordinary captives to demonstrate loyalty to the respective regimes. Ultimately this "battle of the prison camps" came shockingly to the attention of the world in a series of bloody incidents during the spring of 1952, climaxed by the kidnapping and ransom of the American brigadier general who commanded the main camp. But long before these spectacular developments the situation had been emerging among the prisoners, unrecognized by American leaders, which would make it extremely embarrassing for the Communist powers ever to agree to accept "no forced repatriation."

The Americans had anticipated having to endure a temporary interval of refusal; the interval seemed to be reaching an end satisfactorily when the Communist negotiators asked for specific figures on how many prisoners would refuse repatriation. The UN Command agreed to conduct a preliminary screening in the camps during March and April 1952, and thereby discovered, to the surprise of both sides, that about 35,000 North Koreans and 15,000 Chinese would forcibly resist repatriation. The Chinese contingent constituted two-thirds of all the Chinese being held—all, of course, supposedly "volunteers" for the Korean war. Suddenly the prisoners-of-war issue revealed itself as a huge problem. Now the unexpectedly large stakes, human and symbolic, engaged the prestige of both sides so deeply that it was too late for either to yield without great humiliation.

The "war over prisoners" continued for another fourteen months. No longer was it possible for the UN Command to exert much additional pressure on the battlefront or in the air unless there were a major commitment of larger forces to Korea, or a major change in the rules which limited the war, or both. The truce talks seemed to be getting nowhere, and were frequently suspended for long intervals. But American policy-makers were never so convinced that a truce would be unattainable as to feel impelled to reverse the Administration's established policy and begin to wage the expanded Korean war they had so long worked to avoid. Besides, there was now yet another reason for not taking the plunge: a presidential election year had arrived, with all of its customary distractions and the further contention that only a President with a fresh mandate from the people should make drastic changes in the war policy.

Also, there was a growing tendency at the Pentagon, below the top echelon, to make a virtue of necessity. One could observe that the continuance of a small-scale fighting war in Korea did furnish a psychological prod for the public in America and Europe to accept the sacrifices required for an efficiently steady, planned, general rearmament program. (The slackening the Korean truce later brought was precisely what had, from this point of view, been feared.) Similarly, attention was directed to the deep commitment of Red Chinese armies to Korea. They were being "tied down" there at least as much as the Americans were, and probably being "drained" more severely; even Chinese manpower was limited where trained, equipped soldiers were required. But if the fighting stopped in Korea, these armies, or at least their supply lines, could be rerouted to Indochina to help defeat the French, as indeed did later happen in part.

Finally, some Western officers managed to derive some satisfaction from the fact that a large and growing number of young American men were acquiring combat experience in Korea, which would stand their country in good stead in a possible war in the near future. The Russian Army might be bigger, but America possessed—or would soon possess, if the Korean war continued—even larger battle-trained reserves than Moscow could mobilize. In one sense, the war itself was turning into a colossal training exercise. Ammunition was live, and Americans were killed, but at a rate in the last year of the war much smaller than the fatality rate on American highways. The standard tour of duty for enlisted men in Korea was nine months; after that period they were "rotated" home and soon mustered out (available for recall in any new war); during the nine months they were regularly rotated out of the front lines for "rest and recreation" in South Korea or Japan; at the front lines, now stabilized, they were provided with increasing physical comforts unique in the history of warfare. Many draftees were able to acquire a kind of professional attitude that the duration or outcome of the war itself was not their concern; their personal concern was to survive nine months while putting on a soldierly performance satisfactory to their buddies in a platoon. Such an attitude on the part of the troops made indefinite stalemate a practicable mode of warfare, but also made it a progressively paralyzing one. Now the basic spirit of the army would have to be changed—and the popular rotation system altered—if the Administration's limited war strategy were reversed, an additional reason not to reverse it.

Nevertheless, however the staff officers in the Pentagon or the enlisted men in the field might manage to rationalize the endlessness of the war, the American friends and relatives of men in Korea were unreconciled and restless. One way to quiet the mounting frustration would be to take further steps to reduce United States casualties: limit offensive operations still further and replace Americans with South Koreans on the battleline. Yet the Administration was slow to avail itself of the full potentialities of Republic of Korea manpower, mainly because, as in other cases, equipment was scarce and seemed needed in Europe. Even after the war had been on for two years, so much further expansion of R.O.K. forces was still feasible that General Eisenhower was able to make effective presidential campaign propaganda out of the issue in 1952—supported by the American commander in Korea himself.

Eisenhower managed to avoid making promises to end the war by any particular method, but certainly the belief that he would "do something" contributed powerfully to his heavy electoral victory. He did proceed as President to hasten the modernization of the South Korean Army. Feasible withdrawals of American troops, to supporting roles behind the front lines, would mean that the Communists would be less able than before to count the attrition of United States forces as an advantage to be gained from prolonging the war. But this troop shift was made in the same spirit as the previous Truman Administration efforts to hold down America's military commitment in Korea. The major problem confronting the new Administration was whether to try to end the war by expanding it, and, if so, how.

Eisenhower was not in haste to take the offensive. He did publicize a policy of letting Chiang Kai-shek attack the mainland, but Chiang was not provided with the wherewithal to do it on any large scale. Trade controls were further tightened against Red China. It seems likely that eventually some more drastic action would have been taken. But luck was with the Eisenhower Administration. Stalin suddenly died (on March 5, 1953, according to the Kremlin), and his death was followed immediately by a determined intensification of the Soviet "peace offensive," which had perhaps been tentatively foreshadowed in the final months of his life. Negotiations were resumed in Korea on a newly hopeful basis. Sick and wounded prisoners were actually exchanged. Nonforcible repatriation was still a sticking point, but a compromise procedure, which would preserve the UN principle, seemed feasible. And this time the Eisenhower Administration was determined to convince Peiping that if American hopes of peace were dashed again the war would be expanded. In particular, Secretary of State Dulles himself went to India in the third week of May to persuade Prime Minister Nehru of the likelihood that selected targets in Manchuria would be bombed and that tactical atomic weapons, now at last available in large quantities, would be used in Korea. The threat was designed to be communicated to Peiping. If, conceivably, the move was only a bluff, it was fortunately never called. Communist negotiators instead hastened to accept a complex set of arrangements for the prisoner transfer which were satisfactory to the Americans.

Then, for a few perilous days in June it appeared that the nearly complete agreement would be destroyed by South Korean President Syngman Rhee. He did what he could have done quietly many months before: his men released most of the North Korean prisoners from the camps in the South. Perhaps he did not want to take a chance that any large number would choose to be repatriated to the North; certainly he did not want official North Korean agents to be admitted to South Korea to interview the men in the camps, as provided in the armistice draft. Probably also he hoped to create an incident grave enough to block the truce altogether, in order that the war would continue until Korea was unified. But in fact the Communists did not end the negotiations, and Syngman Rhee was finally pacified by major American promises of post-armistice assistance.* And at long last, on July 27, 1953, the Korean war ended.

Six months later some 14,500 Chinese and 7,500 North Korean prisoners who still refused repatriation but had not previously been released in the South were granted freedom under the UN Command's inter-

* Rhee was promised sufficient military assistance to build twenty South Korean divisions, together with vast economic aid, and a treaty with the United States which, as finally ratified, guaranteed the security of the Republic of Korea within its existing territorial jurisdiction plus any additional territory the United States in the future might decide had "lawfully" come under R.O.K. jurisdiction.

pretation of the armistice agreement; most of the Chinese were sent to Formosa. On the other side, twenty-one Americans were among the 350 men who insisted on remaining in Communist hands.

## The Results of the War

What was accomplished by the Korean war, from the viewpoint of the United States? Clearly there was no total victory or total defeat. The human and material costs of the war were roughly calculable: 23,300 Americans killed, 105,785 wounded, mostly in the first year of the conflict; and an estimated $18 billion in military expenditures directly attributable to the fighting (exclusive of many billions of servicemen's pay and other defense costs for the general rearmament effort stimulated by the war).* What did this effort achieve?

1. Aggression was successfully repelled, and somewhat punished. The American action under UN auspices managed to restore the Republic of Korea, which otherwise would certainly have been overrun, and provided it with vastly increased defensive capabilities for the future. Indeed, South Korea has become one of the most highly militarized countries in the world, with huge forces battle-proven in modern (but nonnuclear) warfare. The rival North Korean territory suffered ruinous attrition during the conflict that its regime had initiated; and even Red China, though not under direct attack, was subjected to some temporary reduction through casualties in her supply of trained military manpower, in consequence of having joined the North Korean onslaught. (On the other hand, her officers and men who survived acquired an education in modern war that may stand them in good stead in a future conflict.) A full reckoning of the punishments for aggression may also include the deprivations of territory which resulted from the Communist initiation of the struggle. North Korea lost on balance some 1,500 square miles of territory, a wedge rising northeastward above the 38th parallel, and Red China, in effect, lost Formosa, which otherwise she would have been allowed to capture. Peiping also suffered some inconvenience from the embargoes applied against her trade, from the widespread rebuff of her claims to diplomatic recognition, and from her exclusion from

* Casualties of other UN members were approximately 14,000. South Korean troops were estimated to have suffered 300,000 killed, wounded, and missing; civilian lives lost in the South may have totaled a million—out of a population of about 20 million. North Korea suffered even more heavily—20 per cent or more of its population dead or missing. Several hundred thousand Chinese were also war casualties.

UN membership. This set of punishments, though burdensome, is not acute. The resulting deterrent effect against future aggression might well have been heightened by additional deprivations—for example, conquest as far as the "narrow neck"—though any such effort would have run substantial risk of causing the Communists to refuse to cease-fire in the existing war, thereby creating further incentives for largely incalculable expansion of the conflict.

2. The American action in Korea did provide encouragement and reassurance to allies and potential allies everywhere in the world that the United States would in fact fight to defend them against outright military invasion. American promises on this subject were shown to be more than mere scraps of paper. Indeed, the circumstances of the Korean intervention indicated that the United States would feel bound to heed a government's appeal to join the resistance to invasion even in the absence of treaty guarantees referring explicitly to the particular country. The universalistic language of the United Nations Charter would be affirmed and its malleable procedures manipulated, if necessary and if possible, to confirm American defensive action. Advance treaties of alliance could make the defense more reliably efficient, but the Communist expansionists and would-be independent neutrals alike were on notice that such treaties were not indispensable to produce United States intervention on call of a victim in sudden emergency. The Korean precedent, however, was of course not irreversible, and did not clearly apply to a situation where the United States might be unable to utilize or decently circumvent the United Nations Organization.

3. In the course of fighting the Korean war, the Western world rearmed. Europe in particular would not again be so relatively defenseless on the ground as she had been since 1946. And the normally accepted levels of manpower and of expenditures for defense were raised in the United States to a new plateau persisting after the Korean war at levels more than double those of the years before 1950; a similar situation obtained in America's major ally, Great Britain. Even greater percentage increases were registered by Continental countries that had been occupied and disarmed during World War II. Germany's restoration to power in the Western family of nations was also accelerated by many months, and her defensive rearmament, though still delayed for several years longer, was put high on the international agenda, as an immediate consequence of the invasion of South Korea.

4. An Asian peninsula roughly bounded on the north by the 38th parallel was preserved from Communist control. In narrow military strategic terms, this meant that the West would have a large foothold on

the mainland of Asia, close to nearly all the centers of industrial power in the area ranging from central China to the Russian maritime provinces, in case new developments in weapons technology should ever make such a central location useful as a base for actual operations or for deterrent threats. And Communist military power would be held at a substantial distance from southern Japan. On the other hand, of course, the mainland foothold remained vulnerable and might well prove unrewardingly expensive to defend in some kinds of potential wars.

5. Finally, some propaganda, useful for a time, was made of the prisoners question—the fight for the principle of nonforcible repatriation and the conspicuous numbers of prisoners who shunned Communist rule. Other idealistic aspects of the United Nations cause were also widely publicized. The lasting impact is not calculable, but it seems fair to observe that memories seem short, at least concerning the prisoners.

To recapitulate the five points in brief: The war not only strengthened the worldwide military position of the West, but succeeded, to some extent, in communicating an awareness of this resurgence, which heightened the will to resist in the lands outside the Iron Curtain, a development apparently recognized by leaders inside the Soviet sphere. On the other hand, the war did not cause much weakening of the Communist countries: indeed, in most respects it strengthened them militarily and industrially by encouraging their leaders to extort greater sacrifices from their populations. However, this Communist build-up would much more likely have come anyway at a slightly slower pace than would Western rearmament—in the absence of outright hostilities. To this extent, the West gained on balance from the experience of the war.

## Some Lessons from Korea

The years of experience with limited war in Korea also served to accustom American strategists to the idea of this mode of warfare and to some of its problems. Nagging doubts might remain in their minds as to whether the general public would ever support another similar limited war. But if, as bitter memories faded, the political inhibitions could be overcome, the policymakers might be emboldened increasingly to apply what they had learned to better prepare for and fight this admittedly frustrating kind of conflict.

One clear lesson learned from the vicissitudes of UN forces in Korea was that the ability of the United States to fight a successful limited war outside Europe depends very largely on the availability of sufficient armed forces in existence with their equipment and rapid transport, particularly airlift facilities, to get them to the scene of action before it is overrun. Korea, though so close to the occupation troops in Japan, was nearly lost in the first weeks. Later, in 1956, when the British attacked Nasser at Suez, chances of success were reduced even more than they had been for Americans in Korea by the lack of appropriate forces and facilities. And as late as 1958, in Lebanon, when the United States intervened again in what appeared to be an incipient conflict, the speed of troop concentration was not wholly reassuring if resistance had had to be overcome and if operations in Iraq had also been necessary.

The argument has often been made that the United States cannot hope to plug the dike at every point around the Communist periphery. Usually this contention assumes that the Communist powers can shift their forces to attack heavily at many different points. But actually, the ruggedness of terrain and the primitiveness of communications severely restrict the number of points at which Communist military force can be utilized on any very large scale. If ever truly reconciled to having to fight limited wars, the West should be quite able to build the air and sea transport facilities and maintain the ready troop contingents necessary to insure that a mere shortage of forces would not of itself preclude an attempt to make a locally confined defense in case of Communist attack in southern and eastern Asia. The prevailing opinion among American Army experts, however, has been that this capability has not in fact been maintained since the end of the fighting in Korea, at least until 1962.

Another set of lessons concerning limited war that could be drawn from the 1950–53 experience and that has only in part been followed since then concerns the questionable desirability of seeking the participation of numerous allies if a war occurs outside Europe. Multilateral alliances do raise special problems for the feasibility of keeping a war limited and at the same time fighting it effectively.

An increase in the number of countries committed to fight in response to an attack upon one of them may enlarge the area of hostilities in a manner that is on balance disadvantageous, unless it succeeds in deterring the war altogether. "All-for-one-and-one-for-all" defense arrangements covering numerous countries may manage to convey an intimidating impression of massed power with a prestigeful luster of common purpose, but it is also possible that the luster grows dim as cracks become evident, and the massed power may be stymied by the hesitations of members having little to gain and much to lose by supporting action in a particular crisis situation.

Both the utility and the handicaps of such coalition enterprises were illustrated in part in the experience of the Korean war. If the United States had been better armed worldwide, the Truman Administration might have felt itself in a position (as apparently the Eisenhower Administration ultimately did feel in 1953) to wage a somewhat less limited war for somewhat more ambitious objectives than those that were finally sought in Korea after the Chinese intervention. If so, the protests of distant allies who contributed little but prestige to the fighting might have seriously frustrated the eager South Koreans and Americans in the effective conduct of an expanded but still limited war. On the other hand, the presence of mainly symbolic quotas of troops from Britain, for example, in such an expanded operation might well have provoked fatal Communist assaults upon, say, Hongkong, which could not be defended but would otherwise have been left unmolested; similarly other countries with token contingents in Korea might have been subjected to intensified retaliatory pressures, political or military, which would weaken their capacity to contribute to the containment of Communist expansion in their own areas, and thus impose upon the worldwide efforts of the United States an additional strain out of proportion to the advantages the United States would receive in manpower and prestige from their joint participation in Korea. In short, too many cooks may spoil the broth, not only because their varying recipes undo each other, but because their joint stirring makes the pot spill over and scald them, resulting in great aggravation and divergence of effort.

Actually, the system of alliances the United States has created is predominantly regional in its explicit terms, and the treaties, except perhaps for NATO, are so drafted as to accord the American government some legal discretion to determine how best to meet its joint defense obligation. The framework, therefore, does exist for responses to aggression (occurring outside Europe), which could involve different numbers of powers fighting together with the United States, according to what the objectives of the immediate situation—including the intention of trying to keep a war limited—might seem to require. Even inside Europe, NATO's war plans in case of attack upon Turkey or Scandinavia (possibly also Greece and Yugoslavia) could be devised to permit a local defense that would not compel the waging of even limited war in Germany, much less intercontinental atomic bombing. The relative tightness of NATO commitments might impede but need not block such a strategy of localization if it became feasible and acceptable in other respects. It is probably important that the formal flexibility be preserved, lest any apparent tendency for the United States to renege on a supposed pledge that it would unleash total war in retaliation for a Communist incursion be taken as evidence that all obligations to contribute to a joint defense effort were lapsing.

Yet, of course, flexibility must not be so increased as to seem to nullify in advance all sense of commitment—and with it the utility of a whole alliance as a deterrent to potential aggression. In 1949 and early 1950 the American relationship to the Republic of Korea seemed to acquire this extreme aura of uncertainty. Ultimately, the lesson of the Korean war most unequivocally taken to heart by United States policymakers in the 1950's was the judgment that it was this American disengagement, tempting the Communist bloc to invade, that actually produced the fighting there. Western statesmen were never able to determine definitely whether Stalin did anticipate any strong resistance in Korea; but the view that he did not was so persuasive as to lead many, notably John Foster Dulles, to conclude that America's commitment to defend particular territories adjacent to the Communist bloc must be made as specific and formal as possible, throughout the Soviet sphere, lest there be any vague "no man's land" into which Red armies might again probe. Such commitments, of course, tie American hands and can be criticized as reducing United States freedom of action, inhibiting the government from deciding in the concrete circumstances of a future attack that a particular territory is no longer worth defending by force. But at least the danger of war's arising through Soviet miscalculation is lessened. The line of containment is etched more deeply on the minds of responsible men everywhere.

## SELECTED BIBLIOGRAPHY

APPLEMAN, ROY E. *United States Army in the Korean War: South to the Naktong, North to the Yalu (June–November 1950)*. Washington, D.C.: Department of the Army, Office of the Chief of Military History, 1961.

BEAL, JOHN ROBINSON. *John Foster Dulles: 1888–1959*. New York: Harper, 1959.

BERGER, CARL. *The Korea Knot: A Military-Political History*. Philadelphia: University of Pennsylvania Press, 1957.

CLARK, MARK W. *From the Danube to the Yalu*. New York: Harper, 1954.

DILLE, JOHN. *Substitute for Victory*. Garden City: Doubleday, 1954.

DONOVAN, ROBERT J. *Eisenhower: The Inside Story*. New York: Harper, 1956.

EDEN, ANTHONY. *Full Circle: The Memoirs of Anthony Eden*. Boston: Houghton Mifflin, 1960.

GOODRICH, LELAND M. *Korea: A Study of U.S. Policy in the United Nations.* New York: Council on Foreign Relations, 1956.

HIGGINS, TRUMBULL. *Korea and the Fall of MacArthur: A Précis in Limited War.* New York: Oxford University Press, 1960.

HUNTINGTON, SAMUEL P. *The Soldier and the State: The Theory and Politics of Civil-Military Relations.* Cambridge: Harvard University Press, 1957.

JOY, C. TURNER. *How Communists Negotiate.* New York: Macmillan, 1955.

MARSHALL, S. L. A. *Pork Chop Hill: The American Fighting Man in Action, Korea, Spring, 1953.* New York: Morrow, 1956.

———. *The River and the Gauntlet: Defeat of the Eighth Army by the Chinese Communist Forces, November 1950.* New York: Morrow, 1953.

NEUSTADT, RICHARD E. *Presidential Power: The Politics of Leadership.* New York: Wiley, 1960.

OSGOOD, ROBERT E. *Limited War: The Challenge to American Strategy.* Chicago: University of Chicago Press, 1957.

POATS, RUTHERFORD M. *Decision in Korea.* New York: McBride, 1954.

REITZEL, WILLIAM, MORTON A. KAPLAN, and CONSTANCE G. COBLENZ. *United States Foreign Policy: 1945–1955.* Washington, D.C.: Brookings Institution, 1956.

RIDGWAY, MATTHEW B. *Soldier: The Memoirs of Matthew B. Ridgway.* New York: Harper, 1956.

SPANIER, JOHN W. *The Truman-MacArthur Controversy and the Korean War.* Cambridge: Harvard University Press, 1959.

TRUMAN, HARRY S. *Years of Trial and Hope, 1946–1952: Memoirs,* vol. 2. Garden City: Doubleday, 1956.

TURNER, GORDON B., and RICHARD D. CHALLENER, eds. *National Security in the Nuclear Age.* New York: Praeger, 1960.

UNITED STATES CONGRESS SENATE COMMITTEE ON THE JUDICIARY. *The Korean War and Related Matters: Hearings* and *Report.* Washington, D.C.: Government Printing Office, 1954–55.

UNITED STATES CONGRESS SENATE COMMITTEES ON ARMED SERVICES AND FOREIGN RELATIONS. *Military Situation in the Far East: Hearings.* Washington, D.C.: Government Printing Office, 1949.

VATCHER, WILLIAM H., JR. *Panmunjom: The Story of the Korean Military Armistice Negotiations.* New York: Praeger, 1958.

WHITING, ALLEN S. *China Crosses the Yalu: The Decision to Enter the Korean War.* New York: Macmillan, 1960.

WHITNEY, COURTNEY. *MacArthur: His Rendezvous with History.* New York: Knopf, 1956.

WILLOUGHBY, CHARLES A., and JOHN CHAMBERLAIN. *MacArthur: 1941–1951.* New York: McGraw-Hill, 1954.

# CHAPTER 8

# KOREA AND THE MILITARIZATION OF CONTAINMENT

THE MOST LASTING consequences of the Korean conflict were the militarization of the containment policy and its application to China—an intensification of the cold war for the indefinite future, with no end anywhere in sight.

## George Kennan and the Basic Western Doctrine on Communist Foreign Policy

It will be recalled that the State Department in 1946 and early 1947 had been developing a rationale for an American foreign policy of active support for what President Truman called "free peoples who are resisting attempted subjugation by armed minorities or by outside pressures." (See pp. 98, 101–102.) George Kennan, Foreign Service career officer specializing in Russian affairs, crystallized a doctrine of "The Sources of Soviet Conduct" that became, more literally and more lastingly than Kennan himself desired, the Bible of Western foreign policy in the mid-twentieth century. It is essential to understand this prevailing interpretation and prescription and to recognize how the Korean war gave a fixity to certain emphases—notably concerning military power and the treatment of China—that Kennan himself had never intended. He had been slow to consider the applicability of his theory about Soviet Russia to a potential Communist regime in China; not till after 1950 were the parallels widely noted by others. And although he used the term "counterforce" in his prescription, he was much more at home with the nonmilitary Marshall Plan than with any rearmament. He shied away from what others, under the impact of the North Korean invasion, considered the logic of his own doctrine. But despite Kennan's personal defection, the doctrine as others understood it became entrenched at the very core of Western foreign policy—and there it remains.

One should notice first that the theory implicitly rejected the view that was common during World War II and that substantially influenced Roosevelt's thinking: namely, that the objectives of the Soviet Union are "merely" the traditional strategic security objectives of Czarist Russia—at the minimum to recover the 1904 boundaries, to dominate the Balkans, and then to seek access to "warm water" in the Mediterranean, the Persian Gulf, and the Yellow Sea through secure, reliable control of lines of communication through Manchuria, Iran, and the Dardanelles. Except in Iran and at the Dardanelles (and to some extent in Finland), the Soviet regime had essentially managed to win these traditional Russian objectives by 1946. The Kremlin's continuing pressures were particularly sharp in Iran and around the Dardanelles. Many Western students of international affairs still contended at that time that these places could and should be conceded, that Russian ambitions would thus be satisfied and the basis laid for peaceful relations between the USSR and America. A related argument usually was that the United States could safely afford to make these particular concessions to Russia because resistance to them was only a tradition of the British Empire, which London had recently misled America into underwriting.

From Kennan's contrary perspective, which became generally accepted, it would be impossible thus to "buy off" the Kremlin because the Soviet Union under Communist control was essentially insatiable. Like a "fluid stream . . . its main concern is to make sure that it has filled every nook and cranny available to it in the basin of world power." This for two main reasons, which will here be elaborated beyond the compressed formulation given in Kennan's classic article.*

One motive that he emphasized was ideology; the other was a need to create enemies in order to justify dictatorial controls. One should notice that Kennan's argument on the latter score—and with it some of

* Mr. X, "The Sources of Soviet Conduct," *Foreign Affairs*, July 1947, pp. 566–82.

the reasons for degrees of Communist ideological fervor—relates particularly to a period when a Red regime is still young and its country is still backward; hence, during the 1950's, the argument may have ceased being so valid for Russia as it formerly had been, and as it had become for the New China.

Here is the argument: From motives of national security and idealization of the industrial working class, the rulers are determined to build a strong industrialized state with extreme rapidity, and therefore must exact great sacrifices from their people. Propagandistic appeals to the public to overtake and surpass the productivity of capitalist countries do not suffice. The fervor must be supplemented by repressive violence, some of it directed at particular individuals and groups whose devotion is actually suspect, some of it an almost random terrorization to produce generalized insecurity as a prod to everyone to outdo himself in seeking to follow the current line. Yet, for public justification at least, all the violence must be said to be directed against internal enemies who gravely endanger the achievement of socialist construction, and obviously the state is so strong that they would scarcely be able to do so without assistance from abroad. So the regime has to frighten its people with foreign bogeymen and must usually behave as though the outside world were irreconcilably hostile. Such behavior, of course, actually provokes some genuine opposition abroad, which "proves" the rulers were right, that Russia (later Red China) must become even stronger; so further repression is needed, further exaggeration of foreign hostility, and further provocativeness. A vicious circle is created, and the tensions are exacerbated by the increasing importance of the "specialists in violence," the armed forces and particularly the secret police, whose personal careers depend on the maintenance and, to some extent, the growth of dissension which justifies expansion of their forces.

This pattern of repression and aggression draws vital inspiration from Communist ideology, and in turn, the development creates a compelling psychological need on the part of the leaders to hold fast to faith in their creed, since so much suffering has already been caused in its name.

The ideology itself is mainly that of Lenin, who deviated markedly from the progenitor, Karl Marx. The main concepts can be summarized as follows. It had become clear in the twentieth century that in the countries where capitalism was most fully developed the standard of living of the industrial workers was not becoming increasingly miserable, as Marx had predicted, but on the contrary was rising, unevenly but perceptibly. Lenin found a superficially plausible explanation for this phenomenon. Capitalists in the advanced industrial countries had extended their operations to the backward areas of the world, where they had succeeded in acquiring control of even cheaper labor than at home; they could exploit this labor to produce raw materials and some finished goods. These products could be handled by the capitalists so very profitably that they could then afford to use some of the exploitative profits from the underdeveloped regions to "buy off" the workers in the home country, granting sufficient increases in living standards to postpone the Marxist revolution at home. Thus, Lenin contended, a very large part of the population of whole countries like Britain and France is able to live tolerably well on the exploited labor of distant peoples. The most advanced capitalist countries have become "usurer states." To a large extent, therefore, the class conflict, which Marx insisted is the essence of history, was seen by Lenin as developing into a conflict of nations, the industrially backward peoples seeking liberation from colonial or semi-colonial exploitation by the great capitalist powers. Meanwhile, since capitalist development by its very nature lacks overall central planning, the economy grows at different rates in the different advanced countries; consequently, likewise does the urgency of the need they each feel to find backward peoples to exploit in order to buy off the opposition at home. In this changing situation the national power wielded by the capitalists of a particular country and the urgency of their need to expand abroad can only temporarily coincide with their opportunities to do so by peaceful means. Therefore, there must, said Lenin, be periodic "wars of redivision" of the world between the capitalist powers. But as they fight one another, they all weaken and become less able than before to control the backward peoples about whom they fight. One of these peoples may prove to be the "weak link" in the worldwide system of exploitation; local Communists may be able to break it loose from capitalist control outside and inside and assume control themselves. According to the Leninist interpretation, the Communists would be entitled to do so in the name of the exploited population of the area as a whole, even though the local proletariat, the urban working class, is only a tiny minority.

Communists seizing power in such a country (Russia in the aftermath of the first great capitalist war of redivision, China in the aftermath of the second) obviously find a setting for revolution vastly different from what Marx anticipated. He assumed that capitalism would previously have industrialized the country, have provided elementary education for the whole population, and probably have established some of the institutions and habits of bourgeois democracy. Under such circumstances the Communist "dictatorship" would not need to be permanently ruthless. But none of this preparation exists in the backward countries which Lenin found to be the weak links of world capitalism. Finding no vast proletariat available there to be led, and unwilling to take the chance of waiting for one to grow, a Communist party inspired by Lenin has to prepare itself as a band of professional revolutionaries to find whatever support it can anywhere, in any class and any outside country, to seize power at any opportune moment—and then to accomplish the belated industrialization, education, and training, for the ultimate ideal of administration by the masses themselves.

The stupendous task calls for men capable of sustained ruthlessness. Leninism has provided them with their self-justification. And Westerners have ruefully been driven to recognize that particularly in the backward areas of the world, men flock to the cause that gives such a succinct explanation of the unremitting pressure of the age-old burdens of poverty, ignorance, and overpopulation. The identification of the villains as Westerners—in a system of thought developed by Westerners themselves—does fit well with the resentful natives' own ambivalent admiration and jealousy of the West; also, for the educated, the planned revolutionary transformation offers alluring prospects of rapid personal advancement.

Even in the advanced industrial countries, say the Leninists, social conflict is bound to grow, for the capitalists' expedient of exporting the crisis situation overseas can only temporarily delay its return home. More and more backward areas will achieve national liberation and turn Communist, and less and less of the world will be available for capitalist exploitation. With a smaller margin of overseas profit to distribute at home, the standard of living of the home-country workers will finally cease to rise. The huge proletariats of these highly industrialized states will then fully accept Communist leadership in revolution.

After Lenin's death, however, the rise of fascism in some such countries showed, according to the Kremlin's readaptation of his doctrine, that the capitalists have discovered one more expedient to postpone their doom. At high levels of employment, the industrial overproduction that cannot be distributed profitably to the workers can be utilized for armaments, and, if capitalist regimes can cause enough international tension, their populations will for a time be willing to endure taxation and inflation for this purpose. Thus capitalism in its final stages requires militarism, and to suppress growing popular resentment will eventually produce some form of fascist dictatorship. Thus the Communists contend that capitalism can keep control of its own people internally only if it conducts a cold war externally—much as Kennan and other Westerners have contended that Communism can keep control of its own people internally only if it conducts a cold war externally.

By the logic of this aspect of both theories, cold war is likely to seem more useful than hot war to those in power in the enemy country. Cold war keeps the population whipped up for economic development of the system without subjecting the system itself to the hazards of total war. However, each side has also emphasized that the other has additional reasons to be aggressive besides the aim of keeping the home front urgently producing: Westerners point out that the ultimate revolutionary aims of Communists embrace the whole earth; Leninists direct attention to the "inevitable" need of capitalists to try to conquer new areas for exploitation. Moreover, even if there were a desire on the part of the opposition to keep the war "cold," each side has had to recognize, especially since Korea, that cold war tensions may tend to be cumulative; so there is always imminent danger of drift into all-out war. The situation at best is fearfully unstable. And, therefore, each side has anticipated that the other may choose to precipitate a deliberate showdown in order to put an end to the special burdens of the protracted struggle by destroying the center of infection in the heartland of the opposing social system. For all these reasons, each side has been extremely hesitant to abandon the notion that the opposition will initiate some form of hot war, will not content itself indefinitely with the "advantages" of cold war.

The likelihood of attack would increase sharply if either side felt that its situation was so desperate that there was no other hope. But here, Kennan emphasized, Marxist theory plays an important role; for the Kremlin inherited from Marx a sublime confidence in the scientific certainty that the world will ultimately go Communist, but at the same time a conviction that this inevitable final victory will be won only through a "dialectical" process of progress through conflict in which apparent setbacks will resolve themselves advantageously for the cause. Thus any reversal can be temporarily endured, for it bears within itself the seeds of certain success. The forward march against capitalism is a "zig-zag," Lenin said, "two steps forward, one step backward." No time schedule needs to be met if it would risk losing most of what has already been won. The record of Red advance in the last half-century gives the Communist sufficient confirmation for his faith in the ultimate triumph—and shows him that victory can indeed be snatched from the jaws of defeat. For him the lesson is to press on and on unrelentingly, but without foolhardiness. As he sees contemporary history, it is the declining capitalist powers who are most likely to initiate the total war of desperation.

George Kennan, for his part, felt "reasonable confidence" in 1947 that the West was still strong enough to resist this continuing challenge with patient perseverance. His prescription was phrased with classic simplicity:

> In these circumstances it is clear that the main element of any United States policy toward the Soviet Union must be that of a long-term, patient but firm and vigilant containment of Russian expansive tendencies. . . . The Soviet pressure against the free institutions of the Western world is something that can be contained by the adroit and vigilant application of counterforce at a series of constantly shifting political and geographical points, corresponding to the shifts and maneuvers of Soviet policy. . . . The United States has it in its power to increase enormously the strains under which Soviet policy must operate, to force upon the Kremlin a far greater degree of moderation and circumspection than it has had to observe in recent years [1947], and in this way to promote tendencies which must eventually find their outlet in either the breakup or the gradual mellowing of Soviet power.

The Kremlin's own policy would be patient, Kennan argued, and flexible, but not forever. If its external expansionism were frustrated by firm counterforce for many years, the internal tensions which fostered the expansionism would be bound to alter. In the long run Kennan was optimistic.

The outcome could be "either the break-up or the gradual mellowing of Soviet power." These words of Kennan were likely to be read with the emphasis on "break-up" until after 1953 when the Kremlin showed itself able to accomplish without internecine struggles a succession to Stalin's rule of Russia. Since that time the more characteristic goal of Western policy has been a "mellowing" of Communist aggressiveness.

The vision has been one of a Soviet Union—and ultimately, but much less confidently, of a Communist China—that has become prosperous enough so that its leaders no longer feel a need to generate insecurity at home and abroad in order to force the pace of industrialization—a society that would probably continue to call itself Communist and to verbalize the old Leninist shibboleths, but that would have become so stabilized and stratified that men of influence could afford to be preoccupied with conserving their acquisitions and with passing on their privileges to their children, and would shy away from jeopardizing these gains in really dangerous foreign adventures on behalf of the worldwide Communist cause. In short, though rarely so bluntly expressed, the Communist world and the Western world would be genuinely at peace because the major Communist powers would have become almost as satiated and complacent as the West itself.

The internal developments in the Soviet Union after Stalin's death lent support to the view that this Western vision of Russia's ultimate future is not an absurdity and that a policy of containing her expansionism while awaiting such a transformation is probably not unreasonable.

But the Korean war dramatically added China to the list of powers to be contained, and any prospect that her appalling problems could be solved to the point of complacency there must be regarded as exceedingly remote. And the Korean war emphasized what Kennan personally preferred to minimize: that Communist expansion may be by military force, at least if the military weakness of noncommunists is extreme and tempting. Even Khrushchev, who has been at pains to dissociate himself from much of Stalin's statecraft, has never so much as pretended to grant that there was anything blameworthy in Soviet policy connected with the North Korean invasion. Always remembering this precedent and the Kremlin's unwillingness to disavow it, the West has not dared to avoid preparing military as well as nonmilitary instruments for containment, in order to establish an adequate defensive posture, not just against "immature" Communist China, but also against relatively "mature" Communist Russia, even though the military means of containment themselves have to be recognized as extra causes of tension that delay the process of mellowing in Russia, which remains the chief hope of the policy.

Such is the great dilemma of containment. Kennan himself would escape it by having the West run great risks by refusing to regard Communist actions as sufficient provocation for large military preparations or military action. But since Korea the prevailing Western interpretation of the requirements of containment has included a military dimension so pronounced as admittedly to impede any prospective mellowing.

The West, and in another sense the Soviet Union, would each like to lull the other to the "brink of complacency" in hope that it would fall over and have to endure a transformation before any resurrection. But each side has imagined the other to be straining toward the brink of war. And for both, the Korean conflict—the Communist invasion and the Western intervention—provided traumatic evidence to confirm their suspicions and prejudices. Hence each camp has felt a special need to take up positions near the brink of war. In the post-Korea "peaceful competitive coexistence" of the Khrushchev era, the coexistence has remained so militarily competitive that its peacefulness is highly tenuous. And any possible mood of complacency on both sides, which might permit a truly peaceful relationship without the overthrow of either, recedes to a distant future.

### The Preventive War Alternative

There is no end anywhere in sight. Some have even speculated that the United States may need—and be able to get—the help of a mellowed Russia in order to continue to contain a still unmellowed China two decades hence. And will the following adversary be African? And who will the next one be?

America's traditional unfamiliarity with a world of continuing struggle and her anxious awareness of a gradual shift in the balance of world power toward the advantage of the Communists has naturally sufficed to keep alive in the United States the argument that the Korean war period was the "now or never" time for a showdown, whatever the probable cost, not only with China, but also with the Soviet Union.

The early 1950's did indeed present what was probably the last chance for the United States to put an end to the military danger inherent in rising Soviet power and manifested in the North Korean and Chinese aggression, by mobilizing fully and undertaking to wage preventive total war, in reasonable confidence that surrender and disarmament could ultimately be imposed upon the major Communist powers at a cost which might be acceptable to the United States, in the sense that most of America's urban civilization would remain undestroyed and that only millions, not tens of millions, of her citizens would be killed.

Evidently, no such opportunity can arise again except for the side that may some day manage to perfect an antimissile and antifall-out defense for its own centers of population, thus readying itself to absorb the retaliatory capabilities of the other side, while retaining its own offensive capability—and to do so ahead of the enemy's development of comparable defenses by an interval of time (presumably several months at least) which would permit reliable calculations to be made that a knock-out blow could be inflicted before the enemy could bring any secret weapons into effective counteraction. The technological development which might make such a choice reasonable is now nowhere in sight.

But a comparable choice did conceivably exist for Americans in the early 1950's, because the Soviet capacity to atom-bomb the United States was then still small. It is important to recognize, however, that Western Europe would probably have had to be sacrificed temporarily to the vengeful ravages of Soviet A-bombs and military occupation; and it seems unlikely that the social structure and pattern of values—to say nothing of the material heritage—which constitute the very fabric of European civilization could have survived exposure to such systematic brutalization as the doomed Soviet leaders in their final months would probably have inflicted on America's allies. Europe was pitifully vulnerable, even if yet the United States was not.

Would American leaders have been justified in precipitating disaster for their allies and death for millions of their own people in order to forestall the further growth of military power within the Communist states? A series of deliberately magnified "incidents" might well have made such a course politically feasible in the United States during the Korean war, but the Truman Administration never even regarded it as an open alternative despite the existence then of considerable preventive-war agitation in the country. Total war for total victory had twice been a souring experience—after 1918 and 1945. New enemies arose to replace the old; the same could happen again. Victory in another total war would leave the United States to police and probably rehabilitate half the world. Such an enterprise of pacification would be impossible for an America herself probably suffering from atomic attack, unless the country were transformed by severe regimentation for a world-imperial mission. The prospect was so grim as to be rejected out of hand, if the alternative policies—variations of containment—still offered hope of averting a final conquest of the United States. To most American foreign affairs specialists in the early 1950's, those policies did seem to hold promise. To most of them they still do. But regrets about supposed opportunities that were lost in the early 1950's are not likely to be entirely stilled.

## SELECTED BIBLIOGRAPHY

BARGHOORN, FREDERICK C. *Soviet Russian Nationalism.* New York: Oxford University Press, 1956.

BARNETT, A. DOAK. *Communist China and Asia: Challenge to American Policy.* New York: Harper, 1960.

BRZEZINSKI, ZBIGNIEW K. *The Soviet Bloc: Unity and Conflict.* Cambridge: Harvard University Press, 1960.

CONQUEST, ROBERT. *Power and Policy in the U.S.S.R.: The Study of Soviet Dynastics.* New York: St. Martin's Press, 1961.

DALLIN, DAVID J. *Soviet Foreign Policy after Stalin.* Philadelphia: Lippincott, 1960.

DEUTSCHER, ISAAC. *The Great Contest: Russia and the West.* New York: Oxford University Press, 1960.

GOODMAN, ELLIOTT R. *The Soviet Design for a World State.* New York: Columbia University Press, 1960.

GRULIOW, LEO, ed. *Current Soviet Policies–II: The Documentary Record of the Twentieth Party Congress and Its Aftermath, 1956.* New York: Praeger, 1957.

KENNAN, GEORGE F. *American Diplomacy, 1900–1951.* Chicago: University of Chicago Press, 1951.

———. *Realities of American Foreign Policy.* Princeton: Princeton University Press, 1954.

———. *Russia and the West under Lenin and Stalin.* Boston: Little, Brown, 1961.

KULSKI, WLADYSLAW W. *Peaceful Co-existence: An Analysis of Soviet Foreign Policy.* Chicago: Regnery, 1959.

NIEMEYER, GERHART, and JOHN S. RESHETAR, JR. *An Inquiry into Soviet Mentality.* New York: Praeger, 1956.

TANG, PETER S. H. *Communist China Today.* New York: Praeger, 1957.

ULAM, ADAM B. *The Unfinished Revolution: An Essay on the Sources of Influence of Marxism and Communism.* New York: Random House, 1960.

# CHAPTER 9

# THE INTERCONTINENTAL WEAPONS RACE

THE MOST DRAMATIC change in the relative power position of East and West in the years since the Korean war has been the rapid growth in Soviet capability to devastate the United States in the event of total war. Russian development of the hydrogen bomb and delivery systems for it (long-range planes and missiles) have appeared to overtake the American lead in the air power that can be directed at the heartland of the enemy. The United States has striven to keep ahead—or at least to keep up—as the arms race has accelerated to unprecedented dimensions, involving enormous expenditures on weapons for which the rate of innovation and obsolescence is so swift as to strain severely the imagination of those who have to consider their implications for diplomacy.

Some sense of the transformations, together with an awareness of the persistence of strategic preconceptions, may be garnered from a survey of the evolution of "the great deterrent" in the decade after 1948, the year the American capacity to plant atomic explosives upon Russian cities was first openly dramatized as a means of deterring Soviet aggression in Europe.

### Bombs and Bombers

American B-29's took up positions on British airfields in the context of the Berlin blockade. Combat-proven against Japan in World War II, their range from Britain would permit round-trip missions to western Russia; in case of war, it was commonly presumed that they would carry atomic explosives. In the expanding jargon of strategy, they would engage in a "first strike," in retaliation against an invasion of Europe.

Their capability for successful bombing missions would be reduced, however, as the Soviet Union in the course of time improved its own defensive and offensive air capabilities: "active" defenses comprising better radar warning fences, interceptor planes, and antiaircraft guns and missiles; "passive" defenses, shelters; as well as offensive "strike" capabilities including better bombs, bombers, and ultimately missiles, for use in a Soviet "first strike" or retaliatory "second strike" to destroy the bases from which Western air-atomic power could operate against Russia.

Of these potentialities (offensive and defensive improvements), the offensive or "strike" capability of the Soviet air force attracted the major concern of most United States air officers. Deeply inculcated upon them was the strategic doctrine of the "knock-out" blow, the "atomic Pearl Harbor"—a concept that air power ought to be used directly upon the enemy's homeland in order to destroy his will to resist, that the correct strategy (later called "counterforce") would set as its primary target the enemy's own air force bases (after their destruction, his other facilities could be devastated with relative impunity in "city-busting" attacks), and that, as the Japanese surprise attack epitomized, dictatorial states were willing to make the most unrestrainedly effective use of air power's potentialities by striking without warning. From this doctrine, it soon followed that United States air forces, which were not likely to be allowed to launch a preventive war themselves, would probably be unable to unleash any devastating atomic first strike in retaliation against a Soviet invasion of Europe. Therefore, the entire weapons system—British based, nuclear armed B-29's (modified to become B-50's)—would have only a limited "strike back" ("second strike") capacity to deter Russian aggression.

To secure immunity for the retaliatory force, free from hazardous dependence on overseas bases, American air officers were anxious to form a substantial striking force of intercontinental bombers which would operate from United States home bases, safely beyond the range of destruction in a Soviet surprise attack, though subject to some damage in one-way suicide strikes by Soviet adaptations of the B-29. The American force would be composed of B-36's. The B-36 had been developed during World War II but had not gone into quantity production because its vast range was not needed. It was capable of being improved to reach a top "spurt" speed of 435 miles per hour and to fly nonstop 10,000 miles. The round-trip radius of 5,000 miles could blanket major Russian centers from the continental United States, providing at least a safe "second strike" capability to help deter any major Soviet blow in Europe. The Air Force wanted a few hundred of these planes, but only several score were made available before the Korean war.

The B-36's were bitterly criticized by American Navy officers (anxious to have the money spent on aircraft carriers) who emphasized the vulnerability of the piston-driven bombers to Russian air defenses that would use improved jet fighter planes. The Air Force, in short, could not escape vulnerability to Soviet offensive capabilities by using the B-36, without remaining vulnerable to rising Soviet defensive capabilities. American air chiefs, however, with their eyes fixed on striking power and on budgetary limitations, were somewhat reluctant to accept the logic of this situation until the Korean war demonstrated forcefully the availability to the Soviets of the MIG jet fighters, while at the same time sharp increases became permissible in the overall United States defense budget. The Administration's new concern with the military aspects of containment was intense. The decision reached was to proceed with mass production of the best available jet bomber, which was capable of speeds in excess of 600 miles per hour, even though its range was not much more than 3,000 miles (on the order of the B-29 and B-50). This B-47 would, it was thought, be able to cope effectively with the improved Russian defenses. Efforts would be made to reduce the vulnerability of its air bases to Russian offensive action by scattering them more widely. Strategic Air Command overseas bases were accordingly established in Iceland, Britain, Morocco, Spain, Saudi Arabia, Okinawa, and the Philippines; landing fields were also available in other allied countries. The justification that gradually gained currency was that Soviet bombers would be virtually unable to emerge undetected over all of SAC's overseas bases at the same time. If the Russians synchronized operations so as to arrive over all bases simultaneously, they would cross the advance radar warning nets at different intervals, giving SAC bombers at some bases time to get off the ground; if they synchronized so as to penetrate the estimated radar fences simultaneously, they would arrive over some bases later than others, giving time for some retaliatory bombers to become airborne. And from bases in America—anticipated to be still almost invulnerable—would come hundreds of slow B-36's, bravely ready to run the gauntlet of MIG's over Russia. Thus somewhere much of the dispersed force would survive even a Soviet first strike that was ruinous elsewhere; enough would survive to effectuate for the United States a capability for second strike "counterforce" blows so speedy and so massive as even to "blunt" the ongoing Soviet first strike. (Indeed as speed and massiveness of feasible retaliation quickened, one might not choose to wait for more than minimum evidence that the enemy had yet launched a first strike at all. What was supposedly a retaliatory second strike would "pre-empt" the field: "pre-emptive attack" could become the precipitate version of preventive war. But, of course, this was not the Administration's purpose.)

Urgent efforts were pressed to acquire a jet bomber with intercontinental range to replace the B-36. The B-52 was the plane, but it would not be operational in large numbers until the mid-1950's. Meanwhile, the B-47 would have to be the backbone of SAC's retaliatory power in the event of a Soviet invasion of Western Europe or one-way raids against the United States, or conceivably an assault upon some other position which the President might decide warranted total war. Very significantly, this reliance on the medium-range B-47 as the means of retaliation gave strong reinforcement to the American commitment to the objective of defending Western Europe, and other peripheral positions around the Soviet Union, from invasion. More than 1,400 B-47's were rushed into operation. Such a stake was bound to militate against the propensity of air power enthusiasts otherwise to prefer to operate from secure bases in the United States and to disengage themselves from dependence on overseas facilities. "Fortress America" was not yet, in this military technological sense, a self-sufficient base of operations for the most modern air warfare.

Meanwhile, of course, the Soviet offensive and defensive air capabilities continued to grow—and at rates much faster than Western policymakers anticipated. Correspondingly, the time was passing, unused and generally unregretted, when the huge new air armada could conceivably be launched offensively in a preventive-war assault upon the Soviet Union, which would then be able to retaliate massively against Europe only, not yet against the United States. Russian military progress was startling. The Soviet atomic bomb explosion in the summer of 1949, four years after the corresponding American achievement, was followed by the Soviet detonation of a thermonuclear (hydrogen) device in the summer of 1953, a few months behind or perhaps ahead of the American accomplishment (depending on still secret estimates of the comparative "packaging" of the two countries' weapons at that time); and by the appearance in 1955 of operational numbers of an intercontinental jet bomber comparable in speed and range to the new American B-52, which was also just coming into quantity production.

The B-52 and its Russian counterpart, which Americans labeled the "Bison," both would require aerial refueling for full coverage of attractive targets in the enemy's homeland on round-trip missions. But the availability of the B-52 did mean that SAC's dependence on overseas bases would be sharply reduced, while the "Bison" on the other hand, if it were produced in quantity, meant that SAC's overall retaliatory power would have to serve the primary function of deterring a direct Soviet attack (with "Bisons") upon the continental United States; deterring an attack upon Europe might come to seem secondary and even be tacitly abandoned. For the first time, however reluctantly, American policymakers would have to begin to think in terms of a possible lack of will on the part of themselves or their successors to actually precipitate what would amount to collective suicide for tens of millions of Americans as well as Russians, if the United States persisted in treating a Soviet assault on Europe as sufficient reason for a massive American retaliatory blow against Russia, for she would then be certain to retaliate in turn massively against the United States. As full awareness of the Soviet intercontinental thermonuclear capability gradually sank in, the temptation would increase in Washington to let even Europe go under rather than to precipitate deliberate mutual devastation. "Fortress America" might attempt to stand alone. At least the threat that the United States would in fact retaliate massively against anything less than a direct attack on her own territory would come to sound increasingly unconvincing to the Russian leaders who were supposed to be deterred by it and to America's allies who depended on it.

### Missiles

Very much would then depend on the degree of caution in the Kremlin itself, confronted with the horrendous stakes of even a barely possible thermonuclear war and aware, from the experience of Washington's reaction to the North Korean invasion, that American policy in a showdown is somewhat uncalculated and rash. Also, not relying primarily on Soviet caution, the Eisenhower Administration took measures of its own to help restore some of the previously favorable balance that was lost when the Russians acquired the H-bomb and the "Bison."

1. Additional efforts were made to reduce the degree to which Western Europe depended upon American bombers to deter Russian invasion. The method, as under the Truman Administration after 1949, was to create the basis for a *local ground defense* sufficiently formidable to constitute some deterrent itself against invasion. Determined emphasis continued to be put upon the need for a German contribution of troops for this army. After 1954 attention shifted to the potentialities of nuclear weapons for the ground defense, particularly in preventing Soviet reinforcements from reaching the front lines. However, the Western troop build-up was repeatedly delayed, and never did reach the point where it could hope to stop a determined Russian offensive without calling upon SAC. Quite the contrary. Probably the vulnerable ground force, including 250,000 Americans, was simply large enough to continue to convince some skeptical European and Soviet leaders that SAC would in fact be used for massive retaliation: whatever the consequences for New York and Chicago, 250,000 American troops (not to mention their dependents and the tourists) would probably not be left alone with their European allies to fight a losing battle on the ground. In a sense, the American army in Germany remained a willing hostage to Western Europe, insuring SAC retaliation on Europe's behalf, even if the B-52 replaced the B-47 and ultimately were to make overseas air bases nonessential.

2. Somewhat more successful efforts were made by the Eisenhower Administration to improve the defensive capability of the continental United States, which would now be subject to direct Soviet attack either at the onset of a war or in retaliation against a "first" or "second" SAC strike on Russia which the President might launch (determining that the ruinous decision had become ineluctable). The increased expenditures for continental defense, though substantial, went primarily to improve the protection accorded to SAC's retaliatory capability so that it would largely survive any Soviet air attack. Increasingly distant radar warning nets were strung across Canada, and interceptor planes and missiles were accumulated, but the defense arrangements for urban population centers remained primitive. Defense was to be "active" by interception, not "passive" by shelter.

3. For the retaliation, the American government, confronted with the improved Soviet offensive and defensive capabilities, sought once more to provide superior offensive weapons for its own arsenal. The new emphasis was upon missiles. Their speed would render most defenses obsolete. But their accuracy at ranges of many hundreds or even thousands of miles had seemed so dubious that efforts to develop them

for use against Russian cities had been delayed until a compact and cataclysmic warhead, a thermonuclear device, seemed likely to be available. (Development of this hydrogen bomb, in turn, had been delayed until it was known that the Russians had exploded an A-bomb.) When a belated effort was made to accelerate missile development, the Eisenhower Administration could get early production only of liquid-fueled 1500-mile Intermediate Range Ballistic Missiles (IRBM's), which would require launching sites on land on the periphery of the Soviet sphere. Additional years would be required to develop Intercontinental Ballistic Missiles (ICBM's), which could be launched directly from the United States, and solid-fueled IRBM's, which could be fired from ships at sea.

The emphasis on the liquid-fueled IRBM, it will be observed, was akin to the previous emphasis on the B-47. Once again it meant that the United States Air Force would not be able to escape dependence on overseas bases. B-29 (B-50), B-36, B-47, B-52, IRBM—the overlapping sequence was indicative. Each time it began to look as though peripheral positions were becoming nonessential, an advance in Soviet capabilities required a hasty advance in American retaliatory weaponry—one which could not, however, be perfected for intercontinental ranges at first; the new weapons system would have to be designed to operate from European bases for a few years at least. Thus the operational requirements of the means of retaliation—the IRBM, like the B-47 and the B-29—would give strong reinforcement to the American commitment to the objective of defending Western Europe from invasion, not just the aim of deterring direct attack upon the United States.

But the Soviet Union had pressed its own missile-development program much more persistently than had America since World War II, and consequently, was actually ahead. Radar in Turkey disclosed regular frequent test firings of 800-mile Soviet missiles as early as 1956, and at the end of that year, multi-stage missile flights began. From that time on, the Soviet missile capability began gradually to overshadow the peripheral SAC bases, which would be within range and defenseless against attack. The weapons system built around B-47's was therefore becoming obsolete, except for those planes that could be kept airborne, the few others that might be able to get off the ground in the few minutes warning time available before a radar-detected missile arrived, and those that could rely on aerial refueling to operate from distant bases. The ominous development also meant that even the American IRBM's, when eventually perfected and installed, might not have time enough to get off the ground in the event of a surprise Soviet attack unless they were protected ("hardened") by massive concrete shelters.

Nevertheless, American policymakers could still take some comfort from the fact that a huge fleet of 500 B-52's would be operating from bases in the United States by 1959, somewhat vulnerable to the Soviet long-range bombers but protected by elaborate continental defense facilities and presumably immunized by distance from ground-to-ground missiles. Even this prospect, however, began to darken in 1957, when Russia demonstrated its possession of a multi-stage ballistic missile with intercontinental range. As these Soviet ICBM's became operational in substantial numbers, they would begin to overshadow the B-52 bases in the United States just as Soviet IRBM's were already beginning to overshadow the B-47 bases on the periphery. There was no known defense against the missiles. Many B-52's as well as B-47's could, of course, be kept in the air, and some would have enough warning time to get off the ground; even with missile blows sufficient to wipe out all of SAC's major bases everywhere, Russia could not count on preventing final "city-busting" retaliation—unless Soviet defenses against jet bombers were also vastly improved, perhaps to the point (technically conceivable) where SAC crews were intimidated. If the combination of novel offensive and defensive capabilities actually were achieved by Russia, SAC's retaliatory bombing fleets would cease to constitute a reliable deterrent even to direct attack upon the United States. (They had already been losing their reliability as a deterrent to attack on Europe since 1955 when the advent of the "Bison" bomber exposed Americans to severe retaliation if they attacked Russia in defense of Europe.) Urgent need was recognized for incorporating in the retaliatory forces numerous ballistic missiles capable of penetrating the Soviet Union and well scattered so that they could not all be destroyed on the ground in a surprise attack. But only the IRBM's—no ICBM's—could be operational for the United States before the 1960's. European countries would have to be persuaded to provide launching sites within 1,500 miles of the major Russian cities.

Reliance on this means of retaliation from these sites would, of course, reconfirm America's intention to defend Europe. Superficially, the parallel with the B-29 and B-47 experiences was simply continuing, but actually the Soviet ICBM was transforming the structure of relationships. Previously the United States commitment to the defense of Europe—when examined from the selfish viewpoint of America's own military security, apart from considerations of principle and sentiment—had always involved a notion of sequence: Europe's fall would precede America's. And in undertaking to resist the fall of Europe in order to forestall the later fall of America, the United States government had always been able, at least until 1955, to suppose that the initial joint defense effort would not be suicidal for the United States—although in an age of hydrogen bombs one could hardly deny that it might be suicidal for the European countries concerned. Even the 1955 "Bison" with its thermonuclear weapons did not appear to be so decisive a weapon as to tempt Russia to bypass Europe and attack only the United States directly; too many "Bisons" could be intercepted. But the ICBM, if it became available to the Russians in quantity before the United States devised a defense, could conceivably induce them to destroy America first, leaving Europe to "wither on the vine," to surrender with her industrial wealth intact, if she ejected American bases from her territory. In such a situation, lacking an ICBM of its own and unwilling to put all its reliance for deterrence upon the power of its B-52's to penetrate Russia's defenses all the way from America, the United States government was impelled to clutch at its peripheral alliances with a new urgency. In hastening to station its IRBM's overseas, the Administration might appear only to be reconfirming the massive retaliation threat which had helped for years to deter invasion of Europe as well as attack on America. But the move was made with a vitally new concern that otherwise America might be not only devastated but conquered, and with a grim awareness that Americans would now share as much as Europeans and Russians in the mutual devastation if the deterrent ever failed.

Other programs of the United States government, however, would be likely to make this extreme dependence on Europe a passing phase. The American ICBM was expected to become operational by the early 1960's. Its availability, like that of the B-36 and the B-52 in earlier phases of the arms race, would tend to restore some freedom of choice to the United States government whether to honor its alliances (at the cost of devastation) in case an ally, not America, were attacked. This freedom of choice would be enhanced when the United States perfected submarines—and perhaps fleets of atomic-powered planes capable of remaining airborne for days at a time—as launching platforms for intermediate range missiles. Sites on allied territory would tend to become nonessential. How much so would depend largely on whatever success the Russians might have in devising antimissile defenses for Soviet territory, sufficiently effective to require the West to maintain huge numbers of very widely scattered Western missiles in order to insure a deterrent degree of penetrability of the USSR. Conversely, any success the United States might have in developing antimissile defenses for herself would tend to relieve her statesmen somewhat of dependence on overseas retaliatory sites for deterring direct catastrophic attacks upon America.

Thus the arms race continued to reveal a pattern whereby the interacting technological developments on both sides recurrently opened the prospect of a fairly free hand for the makers of American military defense strategy—but never actually gave it to them. The recurrent glimmer could provide sharp incentives for Western Europe to build greater deterrent forces of her own—not to continue to rely on Soviet caution and American nerve. To some extent this incentive did operate: Britain bore the cost of constructing her own retaliatory jet bomber force armed with atomic and hydrogen weapons, and then began to develop her own deterrent intermediate range missiles; under De Gaulle, France evinced similar aspirations. On the other hand, the possibility that Americans might decide that their own homeland would remain safe enough, that they could afford not to retaliate massively just for the sake of Europe, also provided a contrary incentive for worried Europeans to seek to keep hundreds of thousands of United States troops on the Continent where their sufferings in combat would be expected to excite a desperate retaliation from North America.

American policymakers, taking note of the pattern of the arms race, might well conclude in any case that the prospects for a disengagement which would be reasonably safe even for the United States were in fact only recurrent glimmers; that it would be prudent to assume that future developments in the arms race would endow positions on the periphery of the Soviet Union with new uses for the defense of the United States, as had happened before. The long-term safety of the nation could be seen to be linked to the preservation of these territories from Communist expansion, even if temporary phases in weapons technology seemed to deny the indispensability of such areas. Far-sighted military calculations thus would help to confirm the ties of sentiment and principle which bind America to the defense of Europe. But no one could be sure that the men of decision in the United States in a future crisis would be so resolute as to convince the Kremlin of their willingness to accept the immediate costs of a thermonuclear holocaust in order to implement that joint defense.

## SELECTED BIBLIOGRAPHY

BRODIE, BERNARD. *Strategy in the Missile Age.* Princeton: Princeton University Press, 1959.

DINERSTEIN, H. S. *War and the Soviet Union: Nuclear Weapons and the Revolution in Soviet Military and Political Thinking.* New York: Praeger, 1959.

GARTHOFF, RAYMOND L. *The Soviet Image of Future War.* Washington, D.C.: Public Affairs Press, 1959.

———. *Soviet Strategy in the Nuclear Age.* New York: Praeger, 1958.

HITCH, CHARLES J., and ROLAND N. MC KEAN. *The Economics of Defense in the Nuclear Age.* Cambridge: Harvard University Press, 1960.

HUNTINGTON, SAMUEL P. *The Common Defense: Strategic Programs in National Politics.* New York: Columbia University Press, 1961.

KAHN, HERMAN. *On Thermonuclear War.* Princeton: Princeton University Press, 1960.

KAUFMANN, WILLIAM W., ed. *Military Policy and National Security.* Princeton: Princeton University Press, 1956.

KISSINGER, HENRY A. *The Necessity for Choice: Prospects of American Foreign Policy.* New York: Harper, 1961.

———. *Nuclear Weapons and Foreign Policy.* New York: Harper, 1957.

KNORR, KLAUS, ed. *NATO and American Security.* Princeton: Princeton University Press, 1959.

MILLIS, WALTER, ed. *The Forrestal Diaries.* New York: Viking, 1951.

MORGENSTERN, OSKAR. *The Question of National Defense.* New York: Random House, 1959.

ROWEN, HENRY. *National Security and the American Economy in the 1960's* (Study Paper #18, U.S. Congress Joint Economic Committee, January 1960). Washington, D.C.: Government Printing Office, 1960.

SCHELLING, THOMAS C. *The Strategy of Conflict.* Cambridge: Harvard University Press, 1960.

SNYDER, GLENN H. *Deterrence and Defense: Toward a Theory of National Security.* Princeton: Princeton University Press, 1961.

TURNER, GORDON B., and RICHARD D. CHALLENER, eds. *National Security in the Nuclear Age.* New York: Praeger, 1960.

WASHINGTON CENTER OF FOREIGN POLICY RESEARCH (Johns Hopkins University). *United States Foreign Policy: Developments in Military Technology and Their Impact on United States Strategy and Foreign Policy* (Study #8, U.S. Senate Committee on Foreign Relations, December 1959). Washington, D.C.: Government Printing Office, 1959.

# CHAPTER 10

# NATO AND CONTEMPORARY WAR

ALTHOUGH THE MILITARY security of Western Europe continued all during the 1950's to depend primarily upon American strategic air power and the Kremlin's sense of caution, a substantial strengthening of local defense facilities was gradually achieved under the auspices of the North Atlantic Treaty Organization. An understanding of the experience of this joint armament effort is, like an acquaintance with the evolution of the Great Deterrent, an important foundation for any consideration of the future prospects for a coalition defense of the noncommunist world.

## The Initial Rearmament Effort for a Conventional Defense of the Continent

The extreme weakness of Western Europe's military defenses before the Korean war can scarcely be overemphasized. Sample bits of concrete evidence include the following. In May 1950 the members of the Atlantic alliance (not yet including Greece and Turkey) had troops on the Continent equivalent to about fourteen understrength divisions, supported by less than 1,000 aircraft. No central command existed; in case of attack, improvisation would be necessary. The main lines of communication ran from north to south across West Germany; they could be cut by Soviet thrusts, and allied troops could not fall back to rear positions along the supply lines. Troops were deployed in scattered units for occupation duty; they were not grouped to attempt a serious ground defense against Russian invasion. One extreme example of the unpreparedness was the fact that the British infantry was stationed behind British armored units, which in turn were stationed behind their main base—"behind" when viewed from the perspective of a war against Russia; that perspective had of course not been in view when they were put there.

The overall weakness was so disturbing that the shock of the North Korean invasion raised a sudden specter in the West that mere satellite troops from Eastern Europe—armed and advised but not reinforced by Russians—would be able to overrun much of the Continent. In terms of relative numbers and training this possibility actually seemed quite real; not until 1956, after observing the behavior of the Hungarian Army during the revolt in that country, could one blandly discount the willingness of East European satellite armies to fight against the West. Questions were raised in 1950 whether, if they did fight, the United States would attempt to counter their aggression by atom-bombing Russia. Or would America bomb the satellite populations, which were presumably not hostile to the West, or decline to engage in retaliatory bombing at all under such ambiguous provocation? But if bombing was out, where were the troops to come from in time to mount an adequate local defense? Would the satellite invasion be allowed to succeed?

An alternative possible Kremlin strategy, conceived in the somewhat heated imaginations of worried Europeans during 1950–51, might be a sudden invasion coup by the Russian forces already present in Poland and East Germany; moving with massive swiftness to overrun the flimsy ground defenses, they could then occupy major Western European population centers and hold them hostage, with the result that retaliatory bombing from Britain would be deterred before it ever got started, especially in view of the Russians' ability to threaten counterretaliation upon Britain herself with their new A-bombs.

Spurred by such urgent fears, transatlantic military planning moved at a hectic pace in the summer of 1950. The Continental powers, especially the Low Countries, wanted new defense preparations to be based on a "forward strategy" of local ground defense, which meant that the main battlefield would be Germany, not themselves. Local ground defense would require additional American divisions beyond those currently on occupation duty. (At least this requirement was urgent if one assumed that it would be desirable not to rely altogether on nuclear deterrence, and that full-scale rearmament of West Germany was out of the question.) By September 1950 Secretary of State Acheson informed NATO that the United States would indeed be willing "to participate in the immediate establishment of an integrated force in Europe, within the framework of the North Atlantic Treaty, adequate to insure the successful defense of Western Europe, including West Germany, against possible aggression." The proffered American participation would include troops and an American commander to assume responsibility for directing the integrated allied forces.

United States military forces of all branches increased to 400,000 in Europe by 1952. Other NATO powers also multiplied their available divisions. The most ambitious of the early plans (Lisbon, February 1952) envisaged accumulating trained manpower equivalent, as a defensive force, to that which the Soviet Union could throw into an all-out, nonnuclear assault on the West. The ultimate goal was nearly 100 divisions (twelve of them German) ready and assigned, or trained and earmarked for mobilization on short notice—all for deployment under NATO command in Western Europe, Greece, and Turkey, with additional forces under national command elsewhere in the world.

In terms of relative populations and physical productive capacity, it was feasible for NATO to prepare thus to repel Russia on the ground with forces conventional to World War II. But in terms of the willingness of democratic electorates to endure sacrifices in peacetime, such a strategy proved to be unacceptable even before the end of the Korean war. The massive physical facilities were successfully created: hundreds of airfields, many thousands of miles of telephone and telegraph lines, submarine cables, radio relay circuits, and oil pipelines. They were built, and in many instances had to be rebuilt later, as new developments in weapons technology made them obsolete for war. But not even temporarily was anything resembling the Lisbon manpower target achieved.

One major obstacle concerned the Germans. From the beginning of September 1950 the Administration in Washington had been insisting that some German troop contribution be made and accepted. The demand for German troops arose partly from the view that the American public could not be induced to give the support required for a semipermanent stationing of very large American draft armies in Germany unless the Germans themselves were obviously ready also to join in combat for their own defense against Communist aggression. The other major consideration militating in favor of the German contribution was the difficulty of calculating what kind of Communist effort the West needed to be prepared to meet, and hence how large the total land forces would need to be in Western Europe.

If there was an error in the calculated requirement for German soldiers, the error would supposedly be on the safe side from a narrowly military viewpoint. Yet even here one must note that the USSR increased its numbers of troops during the early years of NATO by at least as many men as did the Western powers, thus maintaining its manpower edge while racing to expand its nuclear air and missile striking forces. In addition, the USSR also squeezed its satellites with increasing severity to contribute their resources and manpower to the joint rearmament effort. Possibly these exertions were exacerbated by genuine anxiety about the rearmament more of Germany than of the rest of Western Europe.

Conversely, and somewhat ironically, the specter of a new Wehrmacht so much overshadowed that of the Red Army in the fears of millions of Western Europeans, especially Frenchmen, that they actually blocked the formation of German forces until after 1954. Often in the nightmares of NATO countries in the early years, the dreaded eastward apparitions had fused to foreshadow some new "Hitler-Stalin pact" against the West. The strain upon the worn fabric of political democracy was severe in France, especially when superimposed upon the burdens of war in Indochina and class conflict at home. French cabinets sponsored compromise plans for integrating any German army units inextricably with those of other countries willing to participate in multinational forces. Some Frenchmen saw this concept as a leash upon German rearmament; others as a hopeful step toward European federation; still others opposed it for enabling Germans to rearm at all, and others for possibly leashing France herself to the rearmed Germans. The American and British governments would not commit their own forces in such a sacrifice of national identity, but they were willing to encourage France, West Germany, Italy, and the Low Countries to proceed with an arrangement of this kind (modified for military efficiency) if there were no other way of getting a German troop contribution. And soon the view even became general in London and Washington that this "European Defense Community" (EDC) was, after all, the best, not the second-best, pattern for rearmament, because of its anticipated by-products in Franco-German harmony.

Hope for it was sustained for almost four full years. The Korean war and then the Indochina war ended with scarcely a German yet in uniform. Finally, in the latter half of 1954, the French National Assembly rejected EDC, and arrangements were made instead for West Germany to rearm as a full partner in the North Atlantic Treaty Organization, with her own distinct national forces on a basis of almost complete equality with France and the other members. The French had proved unwilling to go through with the scheme that might link their own army divisions inextricably with those of a Germany that now seemed potentially stronger than France. But the German rearmament itself would still take place, having been long postponed but not finally prevented by French political strife. Nevertheless, more years were still to pass before America's long-sought West German army became a military reality. Not till near the end of the decade was a sizable number of German divisions actually ready for combat, and by that time they served mainly as substitutes for French troops that had meanwhile become unavailable.

France had her own special reasons for failing to furnish the ground forces which NATO's commanders felt were needed from her. She fought a war in Indochina until July 1954, and promptly thereafter found herself with another war on her hands in Algeria. The drain in Algeria was harassing. Technically Algeria fell within the NATO area, and France therefore felt justified in using her NATO contingents in increasing numbers to fight to hold that territory against the rebels. But the consequence was that on the Continent, German troops eventually came to outnumber the French, despite the long delays in Germany's rearmament. NATO never obtained large forces from France and Germany simultaneously.

The protracted delays tended to be self-confirming, since obviously Moscow was not taking military advantage in Europe of the relative Western weakness on land. An additional incentive for procrastination was the spirit of "relaxation of tensions" which the Communist leaders everywhere were actively promoting after Stalin's death in 1953. The climax of this "peace offensive" came with the Geneva "summit" conference of heads of government of the Big Four powers in 1955. In such a soothing international climate the accumulation of army divisions in Western Europe seemed a matter of diminishing urgency, indeed a risky irritant to the apparently mellowing enemy, although skeptics might continue to argue that a determined forging of Western "positions of strength" would be the surest inducement for the Kremlin to continue to negotiate in earnest. That argument, however, was unconvincing to many who avowedly or tacitly had come to believe that the only realistic positions of strength anyway were SAC's air bases in America (and possibly those in Britain), that the Great Deterrent was sufficient protection for Western Europe.

## The Reliance on Nuclear Weapons

Throughout the decade the strategic concept that appeared really operative—while NATO waited half-heartedly for the elusive capacity to defend the Continent on the ground—was the notion of the "trip-wire," or "plate-glass window," which if broken would "trigger-off" massive retaliation, trusting a secure American capability to launch a first strike, or even an overpowering second strike, from dispersed bases. By this doctrine enough troops were needed on the ground to put up a local defense long enough to prove that aggression had taken place and to give Western leaders time to decide to strike massively with SAC. Accordingly, it was preferable that a large number of those troops be Americans and Englishmen, so that SAC would be most likely to be used to avenge their deaths (whatever the cost to civilians in the United States and Britain), and especially so that the Kremlin would anticipate this vengeful reaction and not initiate a ground attack at all. A corollary argument might be that no Germans were needed for the army, but NATO's top commanders were never willing to accept this conclusion. They insisted that the central front required twenty ready divisions (including Germans), plus eight available for swift mobilization, in order to insure (1) that the "trip-wire" or "plate-glass window" would be absolutely continuous across Central Europe and that no point existed which a Russian thrust could occupy without encountering heavy local military resistance; otherwise the Soviets might suppose that the West would allow them to seize and hold it, "temporarily," rather than precipitate thermonuclear war; (2) that the resistance would be strong enough to compel a pause in any Soviet probing action and induce the Kremlin to reconsider whether to continue the incursion in view of the increasing likelihood that it would soon turn thermonuclear and total; and (3) that enough Western troops existed for some to survive even an exchange of hydrogen blows and be on the scene to fight the (perhaps primitive) "broken-backed" warfare that would still be needed to occupy the ruins against any Russian survivors.

The forces NATO actually managed to accumulate were subject to severe criticism as being too numerous for a mere trip-wire but too few for a limited-war defense against any deliberate Russian incursion, even a rather small one. As for the requirements of "broken-backed" warfare, who could really tell? The calculations attempted on these points were of course confused by the uncertain availability of the

French contingents, a complexity comparable to the previously uncertain availability of the German contingents, and also by the emergence of "nuclear plenty" and the idea of "tactical nuclear war."

Near the end of the Korean war there had been developing a large enough supply of fissionable material to make credible its possible use in diversified nuclear weapons, not just against major strategic targets, but against lines of communication and troops on battlefields. The warheads might be delivered by artillery and by tactical aircraft and missiles of short range. The early prevailing theory about their use was that they would reduce the number of troops needed to defend a position; firepower could be substituted for manpower; any large concentrations of soldiers would merely constitute dangerously attractive targets for enemy atomic attack. This theory was always challenged by an important body of expert army opinion, which contended in brief that any possible reduction of troops on the battlefront would be offset by increased requirements for replacements and supporting units stationed in the rear of the immediate zone of combat. When the Russians also acquired a capability for tactical nuclear war, the likelihood that a good small nuclear army would be able to block the advance of a good big nuclear army seemed increasingly doubtful. Yet there was a will to believe in the possibility, mainly because it would permit the United States and Britain and perhaps other powers to economize on ground troops. The burdens of the draft and heavy taxation could be somewhat lightened in response to the domestic political pressures that bore upon the governments of all Western countries in the wake of the Korean armistice.

Another reason for opting for "nuclearization" was a spreading belief that it would soon become inescapable for any effective ground defense of Europe. The weapons themselves, it was argued, would virtually dictate their own employment if war came; so preparations must be made accordingly.

The development of a full tactical nuclear capability is not simply a matter of supplying soldiers with an improved form of artillery shell or a new rocket projectile. The entire organization of the army and its basic concept of land warfare must be revised drastically. Troops should be scattered in small, self-reliant units—rather like guerrilla bands—ranging with maximum mobility over a very wide combat zone, many of them possessing atomic weapons in diverse compact forms with destructive power equivalent to a few hundred or many thousands of tons of TNT—while overhead short-range planes and missiles rain down other nuclear explosives. To escape annihilation, both sides would have to avoid concentrating their troops; hence the situation would invite wide maneuver and extensive infiltration—so much so that a force deployed for this kind of warfare, relying on the broad destructiveness of its weapons to fill the gaps between manned strongpoints, could easily be overrun by conventional (World War II) forces if it were denied the use of its atomic weapons. Conversely, it has been strongly argued that if in a conflict between two nuclear powers there is any substantial prospect at all that atomic weapons may be used directly on the battlefield—not just on lines of communication—the atomic weapons will in fact have to be so used if the combat ever begins. The reasoning is this: (1) The defensive side must disperse its forces before it is attacked, because it cannot be sure that the potential aggressor will not use atomic weapons, and, if he does, any conventionally concentrated defense force would be annihilated. (2) But having dispersed itself, the defensive side will have to use atomic weapons, because its scattered positions could otherwise be overrun by the aggressor even if he tried not to use atomic weapons. (His conventional forces could prudently be kept somewhat scattered and still maintain, with the advantage of the initiative, a sufficient degree of concentration to overrun the dispersed defenses.) (3) But knowing that the defenders will therefore have to resort to atomic weapons, even if he at first does not, the aggressor also will in fact have to intend to use them (if his thrust is more determined than a mere experimental probe), although he might choose to hold off briefly from using them at the onset, in order to put the onus on the defense for initiating atomic warfare.

If this argument is not invalidated but only somewhat modified by further military staff analyses and field exercises, the conclusions are that the same forces cannot simultaneously be deployed with great effectiveness for both tactical nuclear and conventional combat against a nuclear power; that if a conventional force is trained and deployed for local defense under such circumstances it should be supported by a tactical nuclear force available in the rear to be thrown into action if the enemy does initiate atomic combat (this kind of dual capability being very costly and almost inconceivable for NATO, which has never fully achieved either capability); and finally, that tactical nuclear weapons are so likely to be used that the training and to a considerable degree the organization and equipment of troops for duty in Europe should be directed toward preparedness for local atomic combat, whether or not efforts are also made to give them a conventional-war capability.

This logic, however, ran athwart congressional legislation restricting the sharing of atomic information with other countries, and also athwart the Administration's own doubts about the desirability of multiplying the number of powers possessing actual or virtual nuclear capabilities, who might be suspected of an inclination to employ them without due circumspection. Some skeptics questioned the prudence of the French; many more hesitated to put atomic weapons into the hands of Germans, or even into German army depots watched by a few American guards. There was long procrastination in truly "nuclearizing" any ground troops except some of the Americans.

The atomic capabilities would probably have spread less sluggishly if the Western Europeans themselves had not been so reluctant to envisage their countries as potential battlefields for tactical nuclear war.

The prospect is indeed appalling. One must have the most serious doubts that in an area as congested as Europe between Poland and the English Channel any kind of war could be fought with battlefield atomic weapons, however "small" they might supposedly be, which would not have cumulative consequences as suicidal for France, Germany, and the Low Countries as massive retaliation would have for the United States, Britain, and Russia. Moreover, for the Western Europeans on the Continent, the threat to commit suicide by waging such a war would be unilateral, not collectively dooming the civilian population of Russia also, only some of her satellites, unless the tactical nuclear bombardment of troops and lines of communication were sure to be supplemented by SAC's strategic retaliation. Admittedly, the attrition of Soviet forces and the repulsion of their aggression might be made to seem so forbidding to Moscow that the attack would be deterred even though the Russian homeland, unlike Western Europe, was expected to emerge relatively unscathed. But suppose this threat of limited nuclear war failed to deter? Small wonder that if threats of suicide are the means of deterrence, many Western Europeans have wanted the United States to be also committed inescapably to the suicide pact and have regarded the continuing insufficiency of the local defense forces, even with several American divisions included, as an assurance of the interdependence, a guarantee that SAC will in fact be used whatever the consequences.

From this point of view, it is possible to contemplate the vulnerability of the NATO armies on the German front, still equipped and deployed mainly for conventional war, as a not unmitigated misfortune. By persuading the Kremlin that SAC would have to be used, the weakness on land may contribute to complete deterrence of any Soviet military probing of Europe. This means of deterrence, while admittedly perilous, may at least seem preferable to a deliberate reliance on a self-sufficient tactical nuclear defense capability, because the latter might be more devastating to the Continent if deterrence were ultimately to fail.

One modification of this pattern of thinking would contend that there has been such a decline in SAC's capability to deter a massive direct attack upon the United States, if Americans followed through on their threats to retaliate directly upon Russia in response to a mere invasion of Europe, that the losses Americans would incur by defeat in a conventional war in Germany would not be horrifying enough to rouse SAC to strike—or at least that the Kremlin would so interpret the situation and would cease to be deterred from the invasion. Perhaps then, a tactical nuclear defense capability, if not self-sufficient, would appear to the Russians to raise the scale of devastation that might occur in Europe to a point where once again they would have to take seriously the prospect that Washington would be provoked to unleash SAC no matter what the Soviet counterblows would do to American territory. The upward spiraling tendency of a partial tactical nuclear defense might thus deter initial invasion much more surely than would a conventional defense—if neither kind of capability was itself full and self-sufficient for holding the line in Germany.

This kind of argument has appealed to many Europeans and Americans and may constitute an acceptable rationalization of the mixed and limited structure of forces the West achieved for Continental defense by the late 1950's. Emphasis, whether explicit or implicit, has thus been put upon a need to reinforce the anticipated automaticity of SAC's response—an automaticity that had once seemed so dependable, at least for Europe, in the early years of the decade.

However, the American superiority in intercontinental striking forces does in fact continue to decline. And unless subjective Soviet cautiousness happens to be rising under Khrushchev in some close proportion to the objective decline in relative American retaliatory capability, the Kremlin is increasingly unlikely to feel deterred from invading Europe, and correspondingly the Western Europeans themselves will be increasingly tempted to seek a prior accommodation with Moscow on whatever terms they can get. American threats against Russia on Europe's behalf will simply sound unconvincing.

To recapitulate NATO's situation briefly: For more than a decade, reliance for the defense of Western Europe has actually been upon the threat of American massive retaliation to deter any large-scale Communist attack from starting at all; indeed, anything much above the level of a border skirmish could be stopped only if it were thus prevented (or else perhaps if it were thus punished until any advancing Russian troops withdrew again). Western local ground defense forces have, to be sure, been raised, but only to a level adequate to put up a fight in Germany long enough and conspicuous enough to give assurance that America would finally resolve to launch massive retaliation—a level that would thus simply reinforce the thermonuclear deterrent in the mind of the Kremlin and therewith the confidence of the Western Europeans. But as massive retaliation comes to be a Russian capability as much or more than it is an American one, how long can it possibly be made to sound convincing with regard to Europe? How much automaticity of response is it conceivably practicable to establish and maintain—for how many years—in order to preserve the persuasiveness of an American determination to resort ultimately to collective suicide rather than allow Russians to dominate Europe, especially if the Soviet encroachments are made skillfully gradual and ambiguous? Is there no more satisfactory alternative than making increased efforts—for example, partial tactical nuclearization—to increase the persuasive automaticity of the thermonuclear response?

The principal alternatives that might be found are ways of fighting a large-scale but less than total-nuclear war inside Europe in case of Soviet invasion. One strategy would be to establish huge conventional forces; another would be to construct a genuine dual capability; and a third would be to go all-out for a tactical nuclear defense. In common among all three strategies would be a policy of avoiding the need to resort to massive atomic bombing of the Soviet Union in order to ward off invasion.

1. The idea of developing a self-sufficient conventional defense would amount to reverting to the never-effectuated doctrine with which NATO began in the Korean war period: The Western Allies would raise whatever manpower may be required for stopping any scale of conventional attack by the Red Army. In order to deter the Russians from then using tactical nuclear weapons to break up the conventionally massed defense formations, threats of American massive retaliation (intercontinental thermonuclear war) would be issued, inasmuch as the West would not attempt, under this strategy, to maintain also a substantial tactical nuclear capability of its own.

This strategy obviously confronts enormous doubts as to the political feasibility of greatly expanded draft-calls in Western democracies. And also, one may be unwilling to rely on the massive retaliation threat to be sufficiently persuasive to deter the Russians from "escalating" a conventional war to a higher level of violence by making a tactical nuclear attack—if one is already becoming unwilling to rely on a massive retaliation threat to deter the Russians from making any kind of attack at all.

2. The notion of a true dual capability would meet the latter objection, but at great additional cost in manpower and resources. The capability would have to be genuinely dual if it were to have much chance of restricting the actual fighting to the conventional means, by keeping stand-by tactical nuclear forces in the rear, which would be so strong as to deter the Kremlin from escalating the conventional war, without the West's having to rely for that deterrence on threats of intercontinental war. Conceivably the stand-by nuclearized forces in the rear might be entirely American (and perhaps British) while the conventional troops up front would be mostly German and French. Many varied combinations of national forces are imaginable.

All of these assignments would raise grave questions of political acceptability in the NATO countries, internationally and domestically. And all of them may seem impracticable because of the magnitude of the total burden: conventional forces sufficient by themselves to ward off a Soviet conventional attack; tactical nuclear forces sufficient by themselves to ward off a Soviet tactical nuclear attack; and, of course, intercontinental strike forces sufficient to ward off a Soviet intercontinental attack; plus adequate nonmilitary instruments of policy (like foreign aid programs) and a corresponding set of defenses against Communist China. To be sure, there would naturally be some overlap. A particular instrument may serve for more than one level of East-West confrontation, perhaps because the instrument could really be employed in action in more than one kind of conflict; perhaps because the mere availability of the instrument increases the chances of escalation so much that the resulting risks may be even more likely to deter Moscow from starting a lower-level attack at all than they are to paralyze Washington from meeting the attack with whatever instruments are necessary. Nevertheless, these overlaps may still leave a total burden so great as to be unacceptable to Western countries.

3. If a choice must therefore be made between a self-sufficient conventional defense and a self-sufficient tactical nuclear defense, the latter may be judged the less dispensable of the two, the one more likely to be self-sufficient in a deterrent as well as a combat sense, the one less likely to depend for reinforcement upon discountable threats of massive retaliation, but also the one likely to be vastly more destructive to continental Europe. So much so that its acceptability as basic strategy would seem to depend upon the development of some pattern of limitations that

both sides would be willing to respect, even while fighting a nuclear war.

### Locally Limited Nuclear War

The essence of any limited war is to take advantage of the use of force while preserving calculability. In order to prevent the means of warfare from producing uncontrollable consequences, it is necessary to restrict the kinds of means used and their targets. But between relatively equal adversaries such restrictions are feasible only if both sides are known to be willing to abide by them. Such mutual restraint, in turn, is unlikely unless both sides understand the objectives of the war to be such as not to be worth for either side the hazards of exceeding the limits on the means. It is important, therefore, to formulate concrete and feasible goals for the fighting and to communicate them convincingly to the enemy without excessively provocative bellicosity—and then to stick to them consistently, not raising falsely exaggerated expectations or altering the scope of the objectives up and down in response to the shifting fortunes of battle. Clearly, there are many chances in a limited war for either side to seek advantages by expanding it a little, allowing it gradually to "snowball" on the theory that the enemy will not ultimately dare to remove all limits and therefore will accept severe local defeat. In a war situation constantly subject to changes, some of them drastic, miscalculation of the degree of provocation the enemy will endure and of which rules must be observed is easy.

The example of the Korean war illustrated some of the difficulties that can arise in limited war if the impression develops that a government's behavior is reckless or capricious or unrelated to its policy pronouncements; for example, that it will not in fact content itself with achieving its originally stated, relatively modest objective, but will press on for more. Yet the Korean war also indicated that the graduated application of force can aim at some change in the status quo if the nation possesses an abundance of diverse capabilities which can be applied in combinations that are adjusted with discretion to the circumstances of the situation (the United States lacked such worldwide capabilities at the time of Korea); but the scope of the change that could be achieved by any limited war could not be indeterminate and would have to be one acceptable to the enemy when he was confronted with unlimited thermonuclear war as the alternative to accepting such a concrete limited defeat.

In a war fought for this restricted range of objectives, the mode of limitation placed on means of combat, in order to preserve calculability, involves *weapons, theater* (locale), and *targets.* (See also pp. 161–63 on some aspects of the significance of the number of participating powers in the war.)

With regard to *weapons,* the most obvious, clearly detectable line of discrimination is that between nuclear and nonnuclear weapons. Thus nuclear weapons of all types and sizes may be explicitly or tacitly proscribed (as in Korea and Indochina), largely because of the belief that the use of any atomic explosives will spiral through retaliation into the use of larger and larger types, with consequences so devastating for the battle area as to produce irresistible incentives for their use over larger areas also. Rules can be imagined, directed particularly toward checking the explosive force of the weapons, the height at which they are detonated, and the nature of the surface (land or water) over which they are used; such restrictions would seek particularly to limit fallout, which would otherwise spread to civilian centers beyond the battle area. But the difficulties of maintaining such limitations in the heat of atomic battle would be immense. In the midst of showers of nuclear explosions and atomic fall-out in a restricted battle area, would even the most competent teams of inspectors (perhaps drawn from neutral countries) be able to discern the point at which a particular combatant power began using explosives that were more destructive than was permitted by the rules for the limited war? It would be vastly more difficult to detect and allocate responsibility for violations of stipulated "ceilings" on atomic destructiveness than to fix the blame for initiating use of A-bombs at all. Even assuming that each side would recognize the preferability of adhering to the restrictions if the other would also do so, how could a system of communications be preserved during combat, which would provide satisfactory assurance to each suspicious adversary that the other was in fact still sticking to the rules? In some areas of the world—Turkey perhaps, or Iran, or Korea in another "round"—the stake between Russians and Americans might well seem high enough to warrant their using nuclear weapons in a locally limited war, but not so high as to preclude allowing some margin of tolerance for possible errors when the adversary's weapons seemed rather oversized. But in Europe—with its concentrated population, cultural affinities, and industrial potential—hypersensitivity, not tolerance, would be the likely response to suspected violations of the rules.

Nevertheless, despite the bleak prospects confronting the endeavor, great efforts need to be made to see if any reasonably reliable techniques can possibly be devised for restricting the degrees of destructiveness of the explosives used in an atomic war, because of the even greater unlikelihood that their use can be avoided altogether in a war fought between nuclear powers. Any scheme that could be devised for manageable restraints upon a nuclear war in Europe would contribute substantially to the willingness of the populations there to persevere with local defense preparations, and thus at least might slightly reduce the degree of reliance on an American policy of massive retaliation which may come to seem too suicidal to be a convincing deterrent.

Outside Europe the possibilities of confining a war, even one fought with nuclear weapons, seem more realistic. The Korean war demonstrated that geographical limitations on the *theater* of operations can be made to stick (and can even be changed during the course of hostilities by a cautious process of reciprocally graduated pressure), at least outside Europe in a situation where nuclear weapons are taboo. Probably the weapons ban would not be indispensable. In the extreme case, a set of reciprocally adjustable geographic theater limitations could logically suffice to prevent the war from spreading even if there were no supplementary restrictions on weapons or targets. Such a case would be an area that both major contenders were willing to see virtually obliterated rather than risk retaliating beyond the clearly discernible geographical boundaries. (One contender might ultimately win most of the territory at stake in this limited war, but he would at least have been deprived of most of the value of the "desert" he conquered.)

Europe would appear to be the least probable theater for such a struggle. Therefore, any planning for possible limited nuclear war involving Europe—and most plans for such a war in the other areas of the world where it is more likely to be attempted—must examine the feasibility of devising and maintaining restrictions on the *targets* subject to attack within the theater of operations, as well as the geographical delimitation of the theater as a whole.

With regard to the feasibility of theater limitation, the general prerequisite, of course, is that the area be one in which both sides regard a "fenced in" conduct of military operations as less disadvantageous than expansion would probably be. Correspondingly, when limitation of war requires supplementary restrictions on targets inside the theater, the general prerequisite is that both sides regard a "fenced out" conduct of military operations as less disadvantageous than intrusion would probably be.

The concept is an old one in the law of war—"open cities." The underlying attitude is that civilization is urban, that the great cities of the world are most worthy of being spared the ravages of war. At one time, such immunity could be granted without great fear of strengthening the enemy, until World War I demonstrated the importance of industrial production for victory in a total war fought with modern conventional weapons. In World War II strategic air power made it possible also to attempt to destroy the enemy's will to resist directly, by "area bombing" his civilian population in homes as well as factories. But in the new age of nuclear warfare, with its probable premium on quick decisive blows struck by forces already in being at the outset, industrial production in the actual theater of operations during the course of hostilities would probably make much less difference to the outcome of the war than was the case in 1914–18 and 1939–45; in this respect there would be less reason for the enemy to insist on destroying cities, and more chance of reverting to the restraint of an earlier age. On the other hand, the destructiveness of the new nuclear weapons makes accidental intrusions much more devastating than would ever previously have been the case in an "open city," and also increases enormously the temptation for the enemy to seek victory by the threat or the reality of annihilation. In this situation the difficulty of devising an acceptable pattern of target restrictions remains very great.

The whole problem of such restraints requires most careful exploration by military experts. Possible approaches include declaring "off bounds" the following kinds of urban centers: every city over a certain size; or every such city unless in use for air bases or missile-launching sites (thus actually participating in current military operations); or every city over a certain size unless it contains facilities *potentially available* for actual military operations, the list of such prohibited facilities including perhaps specific kinds of military communications and transport centers, supply depots, troop concentrations, and so on. Perhaps it would even be possible to restrict attacks *within* cities to neighborhoods containing certain kinds of military facilities.

If the use being made of the cities, rather than their mere size, is a necessary element in an acceptable pattern of target restrictions, there would probably have to be devised some system of inspection to operate even during hostilities, implemented by neutrals or even by officially designated enemy agents who would be granted immunity of person and of communication with their home governments.

No informed person will deny the extraordinary difficulty of creating such a system for regulating nuclear war. It is further complicated by the rapidly changing technology of war, which outdates the strategic and tactical assumptions on which any particular pattern might be based

—outdates it indeed before it is fully devised. Perhaps only the simplest of the conceivable formulations is worth attempting; a pattern so detailed and narrowly restrictive as to be in some aspects intolerable in the heat of war might be as bad as no restrictions at all, because when it inevitably began to break down, each side would be inclined to suspect the other of lack of will to maintain any restraints, and, consequently, perhaps no new rules would be accepted.

There would inevitably be mistakes. Most essential would be the effort of each side to give such convincing earnest of its intention to try to preserve restraints that the other side would then accept mistakes as such even though the accidents may involve atomic bombs falling on his allies' cities and possibly on his own. These prerequisites may be attainable, given the even more ghastly alternatives. The great powers have shown remarkable restraint about "accidental" shooting down of their planes and imprisonment of their nationals, but restrained reaction to the deaths of tens of thousands of civilians has not been tested. The margin for tolerable error seems most likely to be lacking in Europe; it is probably much more present in some other places where nuclear weapons may be used locally, for example, Turkey, Iran, and Korea.

The chances of getting some agreement on at least a bare framework of limitations for nuclear war may be improved by the grim fact that the margin for tolerable deviations seems likely to rise as both the United States and the USSR tend to find themselves without a counterforce capability, neither power daring to retaliate massively against the other despite almost any kind of provocation, because each will have become unable to forestall reciprocal retaliation that would be so massive as to be unacceptable. Each side would possess hundreds, perhaps thousands, of launching platforms, most of them mobile or heavily hardened, on land, sea, and in the air, each of them containing one or more hydrogen-warhead missiles which could be lobbed in less than a half-hour across continents and oceans to almost any spot on earth. The indestructibility of each force as a whole would result from its dispersed deployment, from the relative inaccuracy of all missiles (when, for example, aimed to destroy enemy missiles on the ground before their launching), and from the rather "low yield" of explosive force and consequently the confined destructive radius of the kind of missiles that would themselves be the hardest for the enemy to catch on the ground —the mobile, rapid count-down, solid-fuel missiles. (The latter might each be only about thirty times as powerful as the Hiroshima bomb, in contrast to stationary, slow count-down, liquid-fueled missiles, which if launched successfully could be much more than one hundred times as powerful as the first atomic weapon, but which would themselves be highly vulnerable before take-off unless heavily hardened.) With further improvements in missile accuracy and weight-lifting capacity, and perhaps with further compression of warheads through nuclear testing, the dispersal and the hardening would come to be the principal assurance for Russia and America that each could maintain a safely protected second-strike capability of devastating magnitude. But as long as any true "balance of terror" could be maintained, each side would have to recognize that neither possessed a decisive first-strike capability; hence, that neither side could afford to respond to provocations by attacking totally, and, correspondingly, that each side could afford to wait to see how much other injury it would be forced to endure (and could inflict in reprisals) short of total war, because it would know that it still possessed intact its ultimate capability to mutualize any utter devastation and thus presumably to deter it.

In such a situation, threats of massive retaliation would be discounted; the powers could experiment with all kinds of more or less limited wars, attacks, and pressures without being greatly inhibited by the fear that conflicts would get completely out of hand.

## Deliberate Total War

To be sure, one would have to recognize that the balance of terror might never be fully dependable in stability and durability. There would remain at least three kinds of situations under which a power might by deliberate calculation undertake to unleash intercontinental hydrogen war.

One would resemble the final lunge of a dying man, victim of wounds which cannot be healed, determined to take his tormenters to the grave with him. The great power that found itself losing a series of limited wars and suffering piecemeal political reverses might decide that mutual devastation would at least be preferable to surrender, and incidentally conclude that it might leave the leadership of the world (what was left of it) in the hands of "third parties" (for example, Indians or Brazilians) less disagreeable than the major adversary. This policy logic is, of course, supremely pessimistic: my side cannot win, therefore both sides, with most of their inhabitants, must die.

The other logical paths toward a deliberate calculation to launch total nuclear war would be more sanguine. They would aim for victory. One side might become convinced that its own new defensive capabilities could temporarily (several months at least) reduce enemy retaliation to acceptable proportions—acceptability depending in part on the policymakers' view of the value of the lives of their own citizens and of any allies or neutrals on whom the enemy might wreak his frustrated vengeance—while its own offensive power remained devastatingly capable of penetrating the less well-developed defenses of the enemy. One side could get through to the enemy; the enemy could not get through to it. Alternatively, one side might become convinced that it was capable of launching a perfectly synchronized attack in secret, using planes, missiles (some launched from submarines and disguised merchant ships), and sabotage (for example, atomic explosives in luggage compartments of cars parked near air bases)—all destroying the enemy's means of retaliation at so nearly the same instant that no substantial counterblow could ever be struck. The chances of human failure at any point in so vastly complicated a scheme of synchronization are so great that the enterprise seems exceedingly unlikely, except perhaps as the last hope of an otherwise doomed power. (The image here is that of the trapped fugitive from a death sentence who elects to attempt to shoot it out with the surrounding posse; the odds against him are staggering, but he might conceivably succeed.)

Theoretical studies in the wake of the development of the Soviet ICBM concluded that the USSR had acquired the power to nullify America's retaliatory capacity if the Communist plans were ingeniously constructed and flawlessly executed. Such perfection is scarcely realistic, and the difficulties tend to get progressively greater as nuclear retaliatory capability is more and more widely dispersed, on missile-launching submarines, for example. The element of uncertainty about the true variety and location of secret weapons available to the enemy in a world of swiftly changing military technology will always constitute a powerful argument against taking advantage of any apparent margin of temporary superiority to launch a preventive total war unless decisive defeat seems otherwise imminent.

However, one would have to recognize that processes of rational calculation where the stakes are so titanic might be subject to peculiar distortions. In past eras when technology raised less cataclysmic alternatives no great power was content indefinitely to pursue interests based solely upon coldblooded computation of material advantages and disadvantages. The values of honor, prestige, and principle have always been major aims in international policy decisions. Surely there is a skyrocketing inflation of the price level for such immaterial goods in the hydrogen age, but have they really been priced right out of the foreign policy market? Will they not still seem important even under the pressure of nuclear threats? How much so? The materialistic calculus itself is erratic when made possibly in panic reaction to the loss of a series of marginal, distant positions with an importance mainly symbolic. In the protracted stress of the balance of terror, immature, nonrational patterns of motivation might stir decision-makers, perhaps under popular pressure, to set in motion fatal policies—possibly even a "premature" preventive total war.

## Limited Retaliation

The strategic concept of "limited retaliation" as a variety of limited war provides a good illustration of the unprecedented, hazardous potentialities for provocations on the one hand, and for deterrent punishments on the other, that may emerge within a conceivable framework of limited war if intercontinental mutual deterrence comes to seem dependable (as is likely at least for a while in the late 1960's).*

The idea of limited retaliation, when used as the strategy of a defensive power, would be to hit the aggressor with a graduated series of punishing blows that would become more and more severe until he withdrew from the territory he initially had been able to seize. Not much effort would be made to keep the blows confined to the neighborhood of the aggression. For example, an early reprisal might be to sink some of the enemy's ships at sea. Then, if necessary, to bomb conventionally a remote spot in his homeland. Then perhaps to atom-bomb a sparsely inhabited bit of his country. Then perhaps a small city, giving warning time for it to be evacuated. Then perhaps a city-busting blow that would produce many casualties. Next??

The initial invader might well decline to withdraw until he had at least inflicted some comparable counterreprisals on the defenders, but the hope would be that the resulting war of nerves could be carried on in a sort of slow-motion, with intervals for calculation, threats, and negotiation, so that a bargain could finally be struck, acceptable to both sides as the preferable alternative to their continuing to endure the reprisals and the risk that these would get completely out of hand. Of course the aggressor might still show such determination that he would

* The concept of "limited retaliation" has been most fully elaborated by Morton A. Kaplan in *The Strategy of Limited Retaliation* (Princeton: Princeton University Center of International Studies, 1959). That study is the source of many of the ideas that are presented in this section of this book.

be allowed to retain some of his initial gains. Nor is there even anything in the nature of limited retaliation that would assure that it would be normally less profitable for aggressors than for defenders. In particular, democratic governments may find it more difficult than totalitarian regimes to summon up the degree of cold-bloodedness required to initiate and later, if necessary, to escalate each exchange of blows. Isolated positions like West Berlin might in theory be more defensible by a deterrent threat of graduated limited retaliation than by conventional war or by locally limited tactical nuclear war or by threats of massive retaliation. But in fact, if for example West Berlin were seized, democratic publics outside Germany might well recoil from any nonlocal retaliation, especially if it would inflict any civilian casualties.

Moral inhibitions against "punitive raiding" would surely be one factor involved, though reprisals do have a long tradition in international law. One may question whether the decimation of North Korea in the regular course of locally limited conventional war operations was less immoral than a cold-blooded bombing of Vladivostok would have been, and one may imagine that publics that have since become inured to the idea of massive retaliation can be taught to tolerate almost any form of punishment for aggression. But fears of escalation may be more difficult to cope with than moral inhibitions, when there is time to think and argue in the intervals between graduated retaliation strikes. On the other hand, the pace of an exchange might be so fast as to be psychologically unsettling to the point where there would be pressures for vengeance that would put greater and greater strain on the mutual deterrence (the balance of terror), which was still needed to keep the war from becoming utterly total. With secure second-strike capabilities on both sides the tolerability of provocations and punishments could rise and should rise on both sides, but would it rise enough? Would nerves hold firm enough on both sides so that such a struggle—and even a series of such struggles—would not result in piecemeal Communist acquisitions or in mutual devastation from the launching of a total first strike by either side?

The uncertainties have been so great that the concept of graduated retaliation has never achieved the attention and support in the West that has been accorded to other forms of limited war. But it does bear a significant relationship to the rationale of Great Britain's policy throughout the 1950's and France's policy under De Gaulle of establishing, each for herself, an independent national nuclear striking force: Britain's "V" bombers (with atomic and eventually hydrogen weapons) which were to be superseded by her own IRBM's (the "Blue Streak"); France's fancied nuclear capability, comprising warheads, planes, and missiles, to be ready by the mid-1960's. To head off similar aspirations on the part of Germany, the notion developed of a joint deterrent force in which the Germans would share command with other Continental powers and perhaps also with Britain and the NATO headquarters, supposedly somehow apart from the United States government. Each of these forces—the British, the French, the joint—has been conceived as deterring Soviet aggression by potential blows that would, from the point of view of a receiving country as vast and mighty as Russia, amount to "limited" retaliation, although they might represent the maximum effort of the Western Europeans and might induce Moscow to inflict total devastation upon them in return. At least the Russians in their own homeland would be made to pay some bitter price for the eventual conquest of Europe, and hence might not attempt it—even though the democracies would continue to avoid heavy draft-calls (thus preventing a conventional defense), and even though the potent credibility of American massive retaliation would continue to decline. If the remaining alternative would be to undertake a tactical nuclear defense it might be almost as devastating for the Continent as any Russian reprisals for retaliatory strikes against Soviet cities, and the latter might be more likely to deter the initial invasion. Indeed, one could further argue that a nuclear war between Russians and Western Europeans, which was very likely to be graduated up to attacks, say, on Moscow and London, would be more likely to trigger off a desperate American strike than would a conventional or even a tactical nuclear war confined to Central Europe; fear of this escalation would also help to deter any Russian incursion. The conclusion to be drawn would be that ground defense forces could be kept conventional and perhaps even be reduced below present levels, because their function would simply be to put up a fight long enough and conspicuous enough to prove that there had been a Communist incursion and to prevent the Soviet troops from seizing too much territory before they were painfully pressed to give it back under the suasion of reprisals. Lest the Kremlin feel constrained to "pre-empt" by striking first at the relatively vulnerable retaliatory capability of the middle-sized Western country or coalition, a case could be made for stocking "dirty," heavy fall-out weapons of low accuracy that would be useful for city-busting, but would not be able to destroy the missiles on which the Kremlin would continue to feel dependent for deterring an American blow.

Such thinking has not yet actually made much headway, partly because of a fear in the West that Russia would find it provocative, but mostly because of the expense, which even the British government finally decided it could not bear when the need arose to substitute missiles for bombers, and which the American government has never been willing to underwrite even for an ally, at least until, like Britain, she has done most of the work on her own.

If the retaliatory force were built jointly by a number of European countries they would be able to spread the cost. And if they each retained an effective veto on its employment the potential provocation to Russia would be reduced—but so also would be its deterrent effect. Projects for joint forces have repeatedly foundered in this dilemma, without much encouragement from the United States.

The principal motive for the prolonged hesitation of the American government about engaging in much "nuclear sharing" on any basis—bilateral or multilateral—has been its anxiety about various problems of control: first of vital information, which might leak to other countries, including Russia, and later of the use of the weapon systems themselves. Moscow's own nuclear advances probably outdated Washington's concern about Soviet espionage in this field sooner than Americans recognized. But there were other countries to be worried about besides Russia. Any multiplication of nuclear powers—even current allies—came to be viewed with growing anxiety as increasing the possibility that the United States and the USSR would become embroiled in war without premeditation.

## Accidental or Anonymous Nuclear War

More and more vigorously in many countries, East and West, the argument has been pressed that there is much less likelihood of a thermonuclear holocaust's being started through a deliberate calculation made in Moscow, Washington, or London, whether rational or irrational, than of its arising accidentally through miscalculation of the consequences of a dispersion of nuclear capabilities. Imagine Israel and Nasser's Egypt both armed with nuclear weapons. A considerable number of potential atomic powers are countries with little or no recent tradition of stable government, or with rulers who are inexperienced in the responsibilities attendant upon the possession of substantial international power. They may be more susceptible than statesmen in the United States, Britain, and Russia to impulses of doubtful rationality, with grave consequences. There is a resulting possibility that "small" wars will become more numerous, or bigger, or both, than would be the case if the great powers alone possessed nuclear capabilities. And this means more occasions for possible big-power involvement with further chances for all-out expansion of the conflicts.

The entanglements, of course, could be accidental, but the possibility also exists that the lesser power is being incited by a great power experimenting with a "proxy attack" as a tactic of harassment to cripple its cold-war enemy. Intercontinental mutual devastation would probably occur if the decision were to retaliate at all heavily against the major adversary (for example, Russia), on a theory that it must be made to accept responsibility for holding in check every attack against the United States, including attacks by lesser powers, because otherwise it might be able secretly to encourage such an attack and then itself avoid punishment. Yet there may be some genuine uncertainty about which of the lesser powers was the immediate source of the blow, and hence about which one to punish when prudence dictates avoiding the awkward possibility that the Soviet Union was ultimately responsible. In a world of widely dispersed nuclear capabilities, the victim of a sudden atomic attack would (unless extraordinary new detection devices have meanwhile been developed) have grave difficulty in determining where the plane or missile came from—especially if, as seems likely, the great powers by that time are maintaining (in continuous secret patrol operations) sea and air launching platforms for atomic-warhead missiles able to attack from many different corners of the globe. Consequently, a lesser atomic power, harboring various resentments, might feel able to launch a secret attack on a great power, not sufficient, of course, to destroy it but to wound it severely (a few million dead), expecting that the government of the great power would not know against whom to retaliate. To retaliate against one or more lesser nuclear powers on the theory that they might be responsible would tend to precipitate some further suicide attacks from them and might well incur the onus of "preventive" nuclear war against "little" nations.

How, then, are the interested nuclear powers each going to calculate who is responsible for an anonymous nuclear attack and estimate the net value of paying the price for attempting to avenge the injury?

Suppose further that arms-race tension between the great powers has mounted to a point where they may be inclined to suspect the worst of each other and to be provoked, under intolerable frustration, into acting on those suspicions, even though the cost would be nearly suicidal to them both. Then the rulers of some lesser powers (if outside the probable fall-out zones) might see opportunities to enhance their relative status in the world by launching anonymous nuclear attacks upon one or more of the great powers, with the expectation that the United States and Russia would retaliate against each other. Imagine Red China in this role, precipitating a "catalytic war." There would presumably be more urgent incentives for the leaders, or individual officials, of the lesser nuclear power to do their utmost to get the great powers totally embroiled with each other if the latter were already in the process of using the territory of the lesser power for a limited nuclear war. Misery loves company.

Admittedly, the policy calculations surrounding an "anonymous nuclear war" would be fraught with gigantic uncertainty and danger for all the governments concerned. This fact need not, however, reduce the possibilities of the occurrence of a war through a miscalculation somewhere. And, as has been noted, even if none of the lesser atomic powers makes a deliberate effort to embroil the great powers in total war, the "juniors" may grow "trigger-happy" in their relations with each other and thus multiply catastrophically the occasions for mistakes and accidents.

### The Dispersal of Nuclear Weapons

Nevertheless, any projected defense of Western Europe by tactical nuclear or limited retaliation capabilities makes nuclear sharing, with some "Nth country," very difficult to avoid. A range of mutual control techniques is conceivable; for example, atomic warheads with American custodians scattered around Germany, and next door to each of them a European crew with the missile or projectile into which the warhead is to be fitted before firing. But credulity boggles at the notion that control techniques of this kind would suffice indefinitely to preserve an effective veto for the American government over every unauthorized firing. In any case, the policy on which the American and British governments have already embarked (with Russian acceptance and UN endorsement) of selling or giving away peaceful atomic reactors will provide an increasing number of underdeveloped countries with facilities which produce fissionable plutonium as a by-product. Elaborate international inspection arrangements have been imposed, but they cannot with absolute reliability prevent these countries from diverting some of this explosive to secret stockpiles. Meanwhile, their scientists will be receiving from the advanced nations a training in nuclear technology that may lead them to discover how to produce the bomb with their hidden plutonium. Even without this kind of assistance, according to expert American estimates, any country with a fair degree of scientific competence and the equivalent of a half-billion dollars to spend for a few years could acquire nuclear weapons. The list of such countries is growing longer, at least a dozen already. Some may develop the ambition to become the "Nth" atomic powers, although the burdensome cost of also constructing and modernizing a substantial delivery system for the weapons must be reckoned as an additional major handicap for any serious new competitor in the arms race. Also, an international ban on the testing of atomic weapons might be successful in preventing the bomb-builders from proving to themselves and others that their devices would actually explode. But the likelihood that they would, and the harassing potentialities of even relatively primitive delivery systems when bearing atomic warheads would at least sharply increase the range of possible diplomatic miscalculations, thereby magnifying the risk of world war. Such is the "Nth country" problem.

At least as ominous as the dispersal of atomic capabilities among nations—and superimposed upon it—is the dispersal of these mighty weapons, in practice, among small teams of individuals. Originally, according to the intent of the United States Atomic Energy Act of 1946, American bombs (then the only ones in the world) were to be kept in the possession of the Atomic Energy Commission, not the armed services; their use by the armed services in war could be authorized only by the President himself. The arrangement was admirably designed to prevent any chance of hot-headed firing across international frontiers. But it was slow. The growing danger of surprise attacks' nullifying America's retaliatory capability has pushed the country further and further away from the original safeguards against error. By 1955 atomic weapons had been dispersed to numerous pick-up points, where bomber crews on patrol could quickly get them from AEC personnel in case a retaliatory strike were ordered. Now the bombs are actually carried inside many of the patrolling bombers so that they cannot all be caught on the ground. As a surviving safeguard, the bombs are not to be wholly assembled in the air unless the crew receives a radio code word which shall have been authorized, at least officially, by the President himself (and also, for SAC bombers based in Britain, by the responsible British authorities). Upon receipt of this message, the crew will "arm" the bomb (assemble it for explosion) and proceed to target. By other related arrangements, called the "fail safe" system, the patrolling bomber may be directed through receipt of one code signal to head for its prearranged Communist target, but then must turn back at a specified point short of its goal unless it meanwhile receives a second code message (officially the President's personal command) ordering it to go all the way and drop its bombload. In the interim between the first signal and the possible follow-up, while the American bombers save time and fuel by heading toward Russia, commanders on the ground can double-check their preliminary radar reports that a Soviet surprise attack is in progress and that retaliation is truly warranted.

This system depends upon the obedient restraint of every crew; any one of the many hundreds of them might decide that the time had come for it to precipitate war between America and Russia. If it could succeed in penetrating Soviet anti-aircraft defenses and drop its bomb on an important target, the war would probably begin; perhaps even a blocked attack would be regarded by Moscow as an intolerable provocation requiring retaliation which would not remain limited. Nonetheless, the prospect for manned bombers is that there will be increasing numbers of them airborne at all times with their hydrogen bombs. Can they otherwise be securely immunized from sudden attack? If an atom-powered plane is developed, its main purpose would be to remain aloft as an "untraceable" launching platform for nuclear-warhead, air-to-ground guided missiles. Even when fired from conventional jet bombers, these missiles, with potential ranges of a few hundred miles, save the air crews themselves from running enormous risks to penetrate the inner core of Soviet defenses. Again the device is clearly necessary if retaliatory capability is to be preserved, but its existence may increase the temptation for some safe jet crew to launch its missile without authorization.

Americans may have some confidence that their fellow countrymen in uniform will not act so recklessly, but are we so sure of the Russians? How much devolution of direct control over atomic weapons has taken place on the Communist side? Probably somewhat less as yet than in the West, since the essential nature of the Soviet governmental system is likely to offer greater resistance to decentralization in any matter of such importance.*

However, the advent of ballistic missiles speeds up potential attacks by a factor of at least ten. Correspondingly, the incentive for devolution of responsibility for the separate retaliatory nuclear weapons may be magnified on both sides, if effective retaliation is to be possible at all.

Even with the use of new radar ranging thousands of miles, warning time has decreased to the point where it is much less than the time now

* It may also be argued that there is less need for devolution on the Soviet side because they may have less fear of our launching a surprise attack and wiping out their retaliatory forces than we have of their doing so. However, Communist ideology and pervasive suspicion of the West are likely to keep their fears very much alive, providing incentives for them to disperse their nuclear strength.

required to make ready ("count-down") and launch a retaliatory missile. Under such circumstances there is scarcely any possibility of a supposedly retaliatory missile's being launched by mistake before the fact of the enemy's attack is confirmed by radio reports or by actual sightings of nuclear blasts (only missiles on untouched sites can finally get off in counterattack). But within a very few years, when solid-fuel propellants are effective for long-range missiles, the "count-down" interval may be reduced to the point where great incentive will exist to launch all possible missiles while they are still usable: that is, in the five or twenty minutes that will elapse in Western Europe or America, respectively, between the first radar sighting of an apparent flight of enemy missiles and their arrival, which might be expected to destroy most of the retaliatory missile sites. Mysterious spots appear on a radar screen: Are they sunspots? Or a strayed air training mission? Or an enemy reconnaissance effort? Or the surprise attack? There would be scarcely any time for central government clearance of a series of local decisions to launch the reprisal missiles before they were destroyed—and probably no possibility of stopping them once launched by any sort of "fail-safe" system. (Even an electronic signaling device to explode them in the upper atmosphere, if the retaliatory blow proved to be unwarranted, would furnish only a few more minutes for decision-making on the ground.) Very likely the danger of war's resulting in this situation from human or mechanical error will be so great that the standing orders will prohibit long-range missile "counterattacks" until after the enemy's missiles have actually and indubitably landed, the defense thus taking a chance that some of its retaliatory capacity somewhere will survive the first attack. But even if top policymakers select this dreadful risk as the lesser danger to the nation, will the operators of individual missile sites, sensing their own doom, or disagreeing with the overall policy, not decide against orders to shoot their bolts while there is yet time to do so? Even assuming sanity under stress on the part of all concerned, and the most careful patterns of indoctrination, there will be terrible danger of human failure and also of malfunction in the mechanical equipment. Preferring to restrict the exercise of discretion as much as possible, the Administration could decide to reinforce its prohibitory orders by installing devices that would make it mechanically impossible to launch a missile without an affirmative signal from central headquarters. However, such security against accidents might be at the expense of insecurity about the ability of the top commanders to make and communicate go-ahead decisions under conditions of enemy attack. To the extent that Moscow recognized that deficiency, deterrence might be diminished. In such a situation there is compelling need to strive constantly to improve the selective screening and indoctrination of the personnel who have direct access to nuclear weapons, and to substitute new machines for human brains wherever the danger of error can thereby be reduced—all without jeopardizing deterrence.*

Further, the recognition that all the hazards of dispersion of firepower to individuals and groups are compounded by any dispersion of nuclear capabilities among nations heightens the anxiety that centers on the "Nth country" problem. One consequence is to reinforce demands for a ban on atomic weapons tests, for the purpose of restricting the multiplication of nuclear powers.

## SELECTED BIBLIOGRAPHY

BRODIE, BERNARD. *Strategy in the Missile Age.* Princeton: Princeton University Press, 1959.

BUCHAN, ALASTAIR. *NATO in the 1960's: The Implications of Interdependence.* New York: Praeger, 1960.

DEUTSCH, KARL W., and LEWIS J. EDINGER. *Germany Rejoins the Powers: Mass Opinion, Interest Groups, and Elites in Contemporary German Foreign Policy.* Stanford: Stanford University Press, 1959.

DINERSTEIN, H. S. *War and the Soviet Union: Nuclear Weapons and the Revolution in Soviet Military and Political Thinking.* New York: Praeger, 1959.

GARTHOFF, RAYMOND L. *The Soviet Image of Future War.* Washington, D.C.: Public Affairs Press, 1959.

———. *Soviet Strategy in the Nuclear Age.* New York: Praeger, 1958.

GAVIN, JAMES M. *War and Peace in the Space Age.* New York: Harper, 1958.

HUNTINGTON, SAMUEL P. *The Common Defense: Strategic Programs in National Politics.* New York: Columbia University Press, 1961.

KAHN, HERMAN. *On Thermonuclear War.* Princeton: Princeton University Press, 1960.

KAPLAN, MORTON A. *The Strategy of Limited Retaliation.* Princeton: Princeton University Center of International Studies, 1959.

KAUFMANN, WILLIAM W., ed. *Military Policy and National Security.* Princeton: Princeton University Press, 1956.

KISSINGER, HENRY A. *The Necessity for Choice: Prospects of American Foreign Policy.* New York: Harper, 1961.

* For example, a likely compromise on the firing of well-hardened, solid-fuel intercontinental missiles would leave it possible for the individual crew to take the initiative, but then would prevent the vehicle from actually taking off until after a fixed interval, say one hour, during which an automatic communications network, if still operative, would alert the top commanders to the incipient launching and enable them to stop it if they chose.

———. *Nuclear Weapons and Foreign Policy*. New York: Harper, 1957.

KNORR, KLAUS, ed. *NATO and American Security*. Princeton: Princeton University Press, 1959.

LERNER, DANIEL, and RAYMOND ARON. *France Defeats EDC*. New York: Praeger, 1957.

MOORE, BEN T. *NATO and the Future of Europe*. New York: Harper, 1958.

MORGENSTERN, OSKAR. *The Question of National Defense*. New York: Random House, 1959.

OSGOOD, ROBERT E. *Limited War: The Challenge to American Strategy*. Chicago: University of Chicago Press, 1957.

ROWEN, HENRY. *National Security and the American Economy in the 1960's* (Study Paper #18, U.S. Congress Joint Economic Committee, January 1960). Washington, D.C.: Government Printing Office, 1960.

SCHELLING, THOMAS C. *The Strategy of Conflict*. Cambridge: Harvard University Press, 1960.

SNYDER, GLENN H. *Deterrence and Defense: Toward a Theory of National Security*. Princeton: Princeton University Press, 1961.

TAYLOR, MAXWELL D. *The Uncertain Trumpet*. New York: Harper, 1960.

TURNER, GORDON B., and RICHARD D. CHALLENER, eds. *National Security in the Nuclear Age*. New York: Praeger, 1960.

WASHINGTON CENTER OF FOREIGN POLICY RESEARCH (Johns Hopkins University). *United States Foreign Policy: Developments in Military Technology and Their Impact on United States Strategy and Foreign Policy* (Study #8, U.S. Senate Committee on Foreign Relations, December 1959). Washington, D.C.: Government Printing Office, 1959.

WOLFERS, ARNOLD, ed. *Alliance Policy in the Cold War*. Baltimore: Johns Hopkins Press, 1959.

## CHAPTER 11

# CONTROL OF THE ARMS RACE

SINCE WORLD WAR II the intermittent negotiations for control and reduction of armaments have functioned almost exclusively as propaganda channels for the contending powers—as elements in the struggle for some growing sense of alignment on the part of uncommitted peoples and for the preservation of existing alliances and affinities, by reinforcing a power's reputation for pacific intentions.

Even from this cold-war perspective, the disarmament schemes and discussions have been more important than the United States government usually appeared to regard them in the first dozen years after World War II. It was tempting for weary American diplomats, long since disillusioned, to seek to withdraw from the endless harangues. Downgrading of disarmament talk was especially attractive to United States policymakers because it seemed almost inherent in the nature of the situation that the Soviet Union would on balance gain somewhat from any highlighting of the subject of arms limitation. If disarmament talk should happen to raise hopes sufficient to produce considerable international relaxation, the détente would be likely to impede the armament principally of the democratic powers, which are more susceptible to public pressures. If, however, the level of international tension remained fairly high—as seemed certain—the Kremlin's internal control apparatus would remain strong enough to ensure a continued development of military skills and material.

Increasingly in the late fifties, however, it was recognized in Washington that anxiety about attempting to prevent disarmament discussion from raising excessively demoralizing false expectations could not be sufficient reason for American diplomats simply to avoid the subject; such a tactic would cost more than could probably be lost in discussions —as long as a reputation for pacific intent is as widely attractive as it is, and as long as the technological pace of the arms race continued to hasten the day when hundreds of individual men in numerous countries would each possess the power to spark off a nuclear conflagration. The terrors of the impending situation were bound to win widening audiences for all schemes of arms limitation. The United States, like the Soviet Union, would have to continue to devise her own schemes, and to revise them in the light of technological and political change (taking care to try to protect minimum foreseeable defense needs in the remote contingency that one was accepted), and to argue for them vigorously, as though she really meant it. Indeed, as a goal, she would mean it.

## The Varied Significance of a Nuclear Test Ban

Nevertheless, the ban on nuclear weapons tests continued to be the only new measure of arms limitation that has come anywhere near acceptance since the onset of the cold war. What were the special attractions of this modest program for arms control that kept it in the forefront of attention while other proposals lay dormant?

One attraction, of course, was the very fact that the ban measure was so modest and so tentative that it could be conceived as a mere experiment in the restraint of arms racing. Even if it were to fail, there was reason to believe that neither side would have suffered great disadvantage relative to the other.

Yet the proposal did offer substantial promise of reducing the dimensions of the "Nth country" problem, which had become of increasing worldwide concern. Specifically, if there were "N" nuclear powers at the time a test ban agreement was put into effect, "N + 1" powers would likely never emerge unless the "1" obtained the prerequisites from one of the "N" powers, because probably no potential "1" would consider it worthwhile to develop nuclear explosives on its own if it could not prove to itself or to any other power that the weapons really worked. The further hope was emphasized that none of the "N" existing nuclear powers would engage in any further sharing, even with one of its allies;

admittedly, if there were sharing the test ban could not by itself stop wider dispersion.

In another vein, the idea of the ban was promoted as a kind of world health regulation, akin to prohibitions on narcotics, because it would prevent atomic fall-out in peacetime. If the latter objective had been the only one sought, however, a ban on tests in the atmosphere only, or perhaps even a mere quota system for each nation's annual permissible production of fall-out, would have sufficed. Tests underground and in outer space could have continued. The principal purpose of a total ban was to impede very severely the acquisition of atomic capabilities by "N plus" countries, at least by those that all of the existing nuclear powers distrusted somewhat. The other main purpose of a total ban was to put a halt to drastic technological innovation in at least one sphere of the arms race.

The technological race has acquired a cumulative momentum of its own that enormously exacerbates the tensions of the cold war from which it drew its early impetus. The discussion in Chapter 9 of the evolution of nuclear deterrence is one example of this grave tendency. East-West competition in the military application of rapid advances in science and technology, conducted largely in secret, is creating a succession of "generations" of weapons systems that are superseded so swiftly that often they are obsolescent before they are even in full production and may be virtually obsolete before they are efficiently usable. Unpredictability, incomprehensibility, instability, tension—these are the bewildering and provocative frustrations the mounting war of the laboratories now feverishly aggravates.

The roots of the unpredictability lie in factors of secrecy, time, institutionalization, and dispersion.

Secrecy confuses both sides, not only each with regard to what the other is doing, but also with regard to what its own units are doing.

The time factor involves, on all sides, the uncertainties of the rapid pace of innovation on the drawing boards and the not-so-rapid pace of the production, installation, personnel training, and effective operability of the new weapons systems. How much allowance must be made for "lead time" at each stage in calculating what such-and-such an invention would mean by the time it could actually be put in operation?

Institutional inertia and favoritism among particular components of the military-industrial complex in various countries add another dimension of uncertainty as to which weapons will be expedited and which ones retarded.

The disturbing hazards of dispersion, which have already been explored (pp. 209–15), compound the unpredictability that results from the factors of secrecy, time, and institutionalization. And with such myriad unpredictability comes virtual incomprehensibility to top policymakers who must struggle somehow to calculate for months and years ahead just what the implications are for international politics, diplomacy, and strategy of particular aspects of the accelerating technological competition.

Incomprehensibility is not too strong a word. In the bewildering unpredictability of the race few elements are certain except the instability itself: in the most extreme sense, the possibility is always felt that behind the veil of secrecy the enemy is pulling ahead, developing some fearful new weapon which, when unveiled, will outclass one's own weapons for a period just long enough to enable him to put an end to the almost intolerable frustrations of the arms race by going to war and winning. The dread of such a decisive leap tingles the nerves of the competitors. Lesser instabilities also keep them anxious, for the arms race lurches ahead unevenly in partly hidden jumps and spurts; it is rarely a neck and neck marathon.

Many observers have concluded that the unpredictability, incomprehensibility, and instability are bound to produce so much frustration, anxiety, and provocation among decisionmakers in the great powers that the temptation to take the chance of going to war when one imagines one is slightly ahead is likely to become irrepressible. A minority opinion, on the other hand, is that these same uncertainties will function, on balance, more to perpetuate inhibitions: no government will ever feel quite sure enough that it is really ahead to take the chance of initiating a major war. Indeed, the very pace of expensive innovation and obsolescence may already be tending to induce Russia and America not to produce a large enough quantity of any particular weapons "generation" to be very much tempted to use it before technology supersedes it with the next generation. Thus the arms race becomes more a qualitative than a quantitative competition, and the mere image that can be disseminated of a country's relative position in the technological contest becomes a kind of substitute for massive stockpiles and for actual lethal combat in helping to determine its capacity to influence events to its advantage in the world.

Such an interpretation implies that the countries immediately involved can become psychologically conditioned and economically and socially adjusted to an indefinitely protracted "war of the laboratories," deriving some fitful sense of security from the mutual incomprehensibility of the struggle in which they are engaged and finding some reassuring stability in its very instability.

Would it then be possible to confine yet a little more the realm of instability? Anyone who recognizes the dangers of the present accelerated arms race with its cumulative momentum, but who feels some confidence that the world can learn to live with a technological race of some kind as long as the cold war requires it, is likely to look for a means of channeling the competition along somewhat less perilous paths. Even those who insist that only a very drastic reduction of armaments can lower the dangers to acceptable levels may look hopefully at "arms control" measures as first steps toward "disarmament."

It is in this light that the enthusiasm for a nuclear weapons test ban should be understood. Not just fall-out and not just the "Nth country" problem have concerned its supporters. They have also aimed to halt innovation in nuclear weapons technology and thus to contribute to a stabilization of one major aspect of the arms race.

Between the fall of 1958 and the summer of 1961 the United States, Britain, and the Soviet Union did not engage in detectable nuclear testing, that is, they did not explode nuclear devices in the atmosphere, at least none larger than one thousand tons of TNT-equivalent (one "kiloton"). Furthermore, they said they were keeping a promise not to conduct tests anywhere at all—underground, in the ocean, or on missiles in outer space—but no inspection system existed to supply evidence (except about what was unilaterally detectable, the kiloton-plus atmospheric testing, and perhaps, through espionage, some underground experiments).

It seems unlikely that tests to increase the destructive force of the largest nuclear weapons, those with the power of more than a couple of hundreds of thousands of tons of TNT, could be conducted below the surface at all. Hence, avoiding the atmosphere, such explosions could probably take place only in outer space and thus would be enormously expensive if practicable. In principle, space tests could be detected by unilaterally controlled satellites and other devices about as well as by an international inspection system. It probably follows, therefore, that the only tests that would have been undetectable without some new system of joint inspection would have been those of relatively small nuclear weapons, ranging from a few hundred to some tens of thousands of tons of TNT-equivalent, not many times more powerful than the Hiroshima bomb. The implication is that even without international inspection the abstention of America, Britain, and Russia from unilaterally detectable tests would by itself gradually have tended to confine any innovation in nuclear weapons technology to the smaller devices which would be of use principally in limited tactical nuclear wars. If one did not value an improved tactical nuclear capability very highly—preferring to rely on improved conventional and strategic retaliatory capabilities both for deterring and if necessary for waging war, whether or not the enemy were secretly improving his own tactical nuclear capability—then one probably would not have needed to insist on much if any joint inspection, or even on a formal international treaty at all. The objective of putting a curb on fall-out, on "N + 1" countries, and on some of the technological race did not depend on multilateral inspection arrangements or formal agreements.

It was somewhat sounder logic, therefore, to base any demand for elaborate inspection upon a further insistence that tactical nuclear innovations on either side must be restrained, in much the way that disturbing innovations in strategic weapons technology were temporarily hindered on both sides, during 1958–61, by mutual abstention from atmospheric tests. However, no amount of inspection considered to be even conceivably acceptable to the USSR would have sufficed to discover every test of every very small atomic weapon. Hence degrees of proposed inspection amounted to degrees of proposed restriction upon nuclear technology; none could have provided assurance that innovation in the atomic weapons dimension of the arms race had been virtually ended. Some unsettling elements of instability and unpredictability would have remained whatever the scale of feasible inspection. How important these matters of degree were estimated to be was one measure of the importance of insisting on inspection systems.

Another measure was the estimated value of inspection as a good in itself and as an important precedent to be applied in wider spheres of arms control where close inspection would be much less dispensable.

If the United States government had agreed to some highly unreliable joint detection system for underground nuclear tests—perhaps partly because the dangers of technological development in this sphere might have come to be minimized in Washington—the precedent might have been hard to overcome later if opportunities happened to arise for possible extension of international controls over additional elements of the arms race that might seem more crucial.

Nevertheless, provided that a partial inspection system for nuclear tests was even marginally reliable, it might on balance have been considered in the West to be of value: principally for its use in accustoming the secretive Russians to the presence of foreigners exercising some kind of legitimate surveillance, in the hope that the Kremlin, to its own surprise, would have come to find it tolerable and allowed it to be extended. Any possible extension great enough to cover adequately the other spheres of the arms race could have been an objective worthy to be prized as being likely, not only to relieve the extra tensions with which the unpredictable momentum of secret technology exacerbates competitive coexistence, but also to increase the range of general contacts between Soviet officials and noncommunist foreigners—so much as to contribute significantly to the "mellowing" tendencies within the USSR that are the chief hope of the West for any ultimate end to the cold war.

Large-scale inspection in this sense would indeed amount to "cultural penetration," as the Kremlin has charged; multiplication of foreign inspectors would be resisted in Moscow for that very reason even if there were no other incentives for secretiveness. But the opening of the cultural Iron Curtain and the fostering of status-quo-mindedness in Communist countries is so important in the West's long-run hopes for genuine peace that arms inspection arrangements are often tacitly promoted as successive moves, necessarily rather devious, toward a distant goal when the arms race could really end; meanwhile, by the same moves it may be somewhat channeled, stabilized, and moderated.

## Potential Rechanneling of Competitive Innovation in Armaments

However, success in impeding the arms race from being run fast—or even from being run at all—in a particular qualitative channel, for example, nuclear weapons technology, cannot assure that the impetus toward limitation will carry on to block other qualitative or quantitative channels of the armaments competition.

With a major retardation of qualitative arms-racing might come an increased incentive for a country to endure the expense of going into mass production of existing models, since they would not soon be outmoded. To be sure, quantitative racing would at least be more calculable than is qualitative racing. Less uncertainty might mean less tension, hence less danger of unpremeditated attacks. On the other hand, there might be increased danger of deliberately calculated attacks—more certainty, for example, that one really was ahead in the race and could afford the risk of going to war.

Ideally, of course, successful restraints on qualitative racing would improve the climate for quantitative limits also. Thus a nuclear test ban would lead to an adequately inspected agreement on halting further production of fissionable material, and then to efforts to reduce the existing stockpiles. But each step becomes physically more difficult. Ultimately, no one knows a way to uncover all the existing atomic bombs that might be hidden away, and few can foresee agreement on how to maintain world peace if all armed forces were really reduced to the levels of internal police, as Khrushchev has proposed.

The qualitative and quantitative races might simply find new channels in which to run. There can be little assurance at the outset of a major attempt at arms control that the new combinations would be any less unsettling, on balance, than the old. But an awareness of the problem may help to minimize the risks of the probable rechanneling of the contest.

Two examples may help to illustrate the potentialities. Qualitative channels in which the arms race has not yet run very much but where it could readily be activated include civil defense arrangements (notably shelters and evacuation training) and the development of chemical, biological, and radiological weapons.

Reportedly, the Russians are considerably ahead of the United States in civil defense preparations, particularly in organization and training. But a relative lead in this sphere of the competition has been maintained without the acquisition of much absolute capability on the part of the Russians either; their civil defense looks superior only in comparison with the American effort, which has simply been negligible, at least until 1961. In allocating their resources for national security both countries have opted overwhelmingly for strike capabilities as the means of deterrence or of possible pre-emption, thus aiming to prevent the enemy from attacking heavily at all and thereby to protect their populations. Even the less ambiguously defensive kinds of capabilities have been preponderantly "active"—radar, interceptor planes, and ground-to-air missiles, intended to prevent attackers from ever arriving at strategic targets, indeed principally from catching the retaliatory strike force on the ground. Elaborate "passive" defenses, particularly shelters to enable populations to survive actual nuclear explosions, have been omitted from budgets on both sides of the Iron Curtain. In effect the American and Russian governments, at least until 1961, each showed willingness to leave its own population vulnerable as long as the population of the other was also being left vulnerable.

The logic lies partly in mutual deterrence. A highly effective shelter program, great enough to immunize nearly all of a country's population from attack, would constitute a powerful incentive, to the country that had it, to launch preventive war secure from fear of retaliation. If the United States were thus protected, she would be back in the relatively favorable strategic situation of the early 1950's. But even with that opportunity, would Americans be much more likely than they were before to actually wage the preventive war? Very likely they would still be paralyzed by moral and political inhibitions, including concern for the retribution Moscow would still be able to wreak upon America's unsheltered allies. Thus the United States might very well not exploit its advantage offensively, but the Kremlin might be equally likely not to accept this risk and might desperately attempt a preventive war itself, before America's shelters were complete. Of course other elements in the strategic balance would be likely to deter Moscow from such a fearsome gamble. More likely, the Russians would be aroused to undertake a rapid shelter-building program to protect themselves like the Americans. But if the USSR did thereupon succeed in immunizing herself, would not the Kremlin be likely to exploit its new secure position aggressively, more than Washington would be? In short, would not the United States have precipitated a channeling of the arms race along a course from which ultimately greater insecurity, not less, would flow for her and particularly for her allies?

Better perhaps to let sleeping dogs lie? Better especially in view of the enormous expense of a highly protective shelter program (tens of billions of dollars) and the possibility of having to dig deeper and deeper if the enemy responds with threats of bigger and bigger bombs? Better also in view of the related economic and social disruption and the unpredictable effects upon public morale? Some experts have forebodings of a resulting "Maginot Line" complex of neo-isolationist defensive-mindedness, or even of panic defeatism, if the public mind were ever really focused on the realities of thermonuclear war, as a massive shelter-building program would focus it. Other more probable kinds of adverse morale consequences would come from the aroused public awareness that some parts of the population, favorably located, were being protected much more thoroughly than others. Some discrimination would be practically inevitable no matter how elaborate the program might be made. (The possible inequities could only partly be concealed by such tactics as universalizing fall-out shelters while omitting blast shelters; apparently everyone would then be receiving the same treatment, the expense would be tolerable, but probably only the nonmetropolitan population would really be much improving its chances to survive attack.)

All these lines of argument help to explain why neither side in the cold war has felt impelled to precipitate an intense East-West competition in shelter-building, and why a major effort in this sphere may continue to seem worth avoiding. However, there are also strong contrary humanitarian considerations tending to induce some considerable expansion of shelter programs above their current negligible levels: deterrence may fail, and is it not then kinder to save perhaps three-quarters of one's population than one-quarter? Also the credibility of one's will to retaliate against almost any kind of an enemy first strike may be greatly diminished if massive retaliation would provoke an enemy second strike that could surely annihilate one's population, because it was still virtually without shelter.

Imagine Russia and America continuing essentially without shelter; Russia strikes first, concentrating on American bases and avoiding enormous slaughter of civilians; the United States strike capability has been impaired by the Russian attack, but if much of what is left is used to inflict punitive casualties on the unsheltered Russians (who may indeed have even acquired some protection by preliminary evacuation of their cities), then the Soviet return attack will be upon American civilians and will be annihilating; yet the American retaliatory capability may have been excessively impaired for effective counter*force* blows; how then to retaliate at all? Of course this drama would not actually have to be acted out; it would emerge in the tacit threats and counterthreats of mutual deterrence. And it may mean that unless there is some increase in sheltering, the deterrence will cease to be mutual.

Therefore, a moderate expansion of the American shelter program seems indispensable, and the Russians are likely to find similar reasons for at least keeping abreast in this new competition. Once it has begun, neither side can afford to remain much more vulnerable than the other. Hopefully, however, such a race could be conducted as if in slow motion, both sides substantially reducing their vulnerability but neither side becoming (or appearing to become) so completely invulnerable or even so very much more invulnerable than the other that the superiority would itself rouse incentives or provocations for a preventive attack.

Suppose finally, in this developing situation, that innovations in nuclear warheads were virtually halted by an effective formal or informal ban on testing. It might then appear that massive shelter-building could be undertaken with greater confidence because the edifices would less likely have to be reconstructed every time the enemy developed a more efficient warhead. The calculability of shelter adequacy would of course be even surer if some other kinds of arms control agreements put known curbs upon the enemy's multiplication of his existing types of warheads and his delivery systems for them. But in any case, the possibility emerges clearly that weapon limitations that have the effect of curbing the arms race in the channels in which it has been running can produce incentives for accelerating it in other channels—specifically in this example, that restrictions like a nuclear test ban that tend to confine strike capabilities can produce incentives for improving passive defense capabilities like shelters.

Whether the net effect is stabilizing or destabilizing is itself extremely difficult to estimate. It depends very largely on the effect an agreement has on the climate of opinion and the impetus it contributes toward effectuation of other related agreements. The point to be recognized, therefore, is that arms control arrangements, which are largely designed to make the competition less frantic by making it more intelligible, are themselves beset by most of the same incalculabilities as the rest of the arms race. It is scarcely less difficult to estimate the effects of trying to get along without a weapon than it is to estimate the effects of having it.

Attention deserves to be focused on one other important example of a hitherto dormant sphere of competition in weapons development which might be activated at any time, possibly through a rechanneling of the arms race away from the nuclear sphere. Development in chemical, biological, and radiological warfare (CBR) has apparently remained relatively stagnant in the atomic age. It may yet come into its own.

In the sense of CBR "toxic warfare," the chemical warfare is what is commonly known as "gas warfare"; the biological, "germ warfare"; and the radiological, the deliberate scattering of radioactive material (for example, dust) on a specific target area, achieving the lethal and debilitating "fall-out" effects of an atomic explosion without the accompanying destruction of useful property.

In general, this preservation of physical facilities is, for an attacker, one of the most attractive aspects of all three forms of toxic warfare, the other attractions being the incapacity of human sensory organs ordinarily to detect its use at all; the compactness and diversity of possible delivery systems to disseminate the chemical and biological agents (gases and viruses); the flexibility with which the toxic agents may be employed selectively to injure or destroy crops, or animals, or people, over varying areas; and the persistence of some of the contaminating poisons after dissemination.

Probably if an amount of research effort, remotely comparable to the actual expenditure upon nuclear progress, were devoted to developing these techniques, they could be refined into highly precise, diversified, and reliable methods of conflict that would unveil startling vistas for the imaginative conduct of war. In particular, the potentialities revealed by the limited research that proceeds in the Western world suggest extraordinary uses for bacteriological agents in sabotage operations; for example, killing or debilitating large populations by injecting viruses into food-processing plants or into the ventilating systems of office buildings. In a cold war situation a considerable amount of this kind of harassment might well be executed without its being traced to "germ warfare" at all.* In a pre-atomic-attack situation, the aggressor might use such agents to soften up his victims in certain crucial localities. And in the context of actual atomic war, some kinds of gases or viruses might be useful supplements to the regular fall-out. Moreover, CBR capability need not be a monopoly in the hands of the great powers. Production facilities for many kinds of vastly injurious chemicals and germ cultures could be established by those lesser powers that have enough open space to conduct the necessary field tests. The expense would be much less than would be necessary for them to acquire nuclear capability, and probably the requirements for the training of their scientific personnel would be less exacting. To be sure, the smaller nation might feel more vulnerable to CBR retaliation than the large. But again the possibility of "anonymous attack" arises—anonymous germ war, probably easier to wage than anonymous nuclear war. The consequent multiplication of occasions for international conflict is potentially perilous.

Not just perilous, some will say, but unspeakably horrifying. Even to discuss the subject is repulsive to many. The whole concept of "germ warfare" has been buried from view almost as completely as the canisters themselves containing radiological "wastes" from the atomic plants.

The Communist powers have contributed feverishly to a worldwide propaganda campaign with an apparent purpose of making CBR unthinkable as an instrument of national policy. Allegations that Americans did in fact use "germ warfare" in Korea were a central Red theme for more than a year. This propaganda campaign is subject to various interpretations. It may have been a rebuttal to "no forced repatriation"; or, in an oriental setting of natural epidemics, a combination of "Blame the Americans" and "Improve our hygiene"; or, to rally Asians and other colored peoples, "See how the white imperialists first drop atom bombs on Japanese and now drop germs on Chinese." Or the propaganda cam-

* The "Asian flu" epidemic of 1957 was elaborately studied by American, and perhaps also by Soviet, CBR specialists as a model for the possible future effects of a deliberate enemy campaign of debilitation by biological warfare.

paign may have been designed to upset any possible American intention of using atomic bombs (which could be made to appear embarrassingly similar to germ warfare), without implying any Communist inferiority by directly admitting anxiety about the nuclear weapons themselves; or truly to remove CBR methods entirely from the range of instruments of East-West conflict, thus facilitating concentration upon other modes of competition; or to paralyze the will of the West to develop any substantial CBR capability, while the Communist powers themselves proceeded in secret to create one for offensive use if propitious circumstances should arise. This range of possible interpretations leaves the Communist intent with regard to CBR warfare unclear. Soviet statements in other contexts make it clear that Russians regard the use of CBR agents as likely if major war breaks out again, but their massive propaganda efforts on this subject, as well as on atomic bombs, have consistently functioned to deepen moral inhibitions against the new instruments of warfare and thus to reduce the likelihood of their use—at least by the West but perhaps also by the Kremlin itself.

Indeed, if the atomic bomb had not actually been exploded in the passionate heat of war in 1945, there is little reason to suppose that it would ever have circumvented those inhibitions and become more widely tolerated than is CBR. Intrinsically, CBR technique is no more inhumane than the hydrogen bomb with, say, a fifteen-mile radius of blast damage and hundreds of square miles of fall-out, and potentially CBR technique is at least as precise and flexible as the "family" of tactical nuclear weapons on which many Western strategists have wanted to rely (perhaps at greater financial cost than a comparable CBR capability would have required). If the war against Japan had lasted another year, the American forces might have experimented with available biological agents to destroy the Japanese rice crop and thus hastened surrender by starvation. With such a recent precedent, the moral inhibitions against CBR techniques would probably never have grown much greater than those against atomic bombs. But in fact, the bomb was used and the germs were not, and Western preparedness since that time (or at least since active military preparedness efforts began at all in 1948) has been based on the assumption that nuclear explosions are a much more acceptable means of warfare—even for mutual annihilation—than are gases and germs. Campaigns to "ban the bomb" and to eliminate germ warfare did not succeed with regard to the bomb, but in the absence of actual precedents for the extensive use of CBR methods since World War I, germ warfare may have been successfully precluded, in this sense: CBR agents will not be used by either side except in retaliation against the use of CBR agents; they will not be used in retaliation against other, nontoxic modes of conflict; and correspondingly, unlike nuclear weapons, the threat of initiating their use cannot be utilized to deter other, nontoxic modes of conflict.

This policy, although unpublicized, seems inherent in the small-scale efforts of the American government to develop some CBR capabilities (with defenses). The USSR seems to be proceeding on a similarly modest level, but overoptimism on this score may perhaps be unwarranted and dangerous. American intelligence may be inclined to direct its attention to Soviet developments which parallel our own major weapons systems, and fail to note ominous offensive capabilities growing in novel and supposedly disreputable fields of activity. Nevertheless, if the most careful intelligence probes practicable disclose no evidence of a very heavy Communist effort far outstripping our own in the CBR field, there are some marked advantages for the United States in continuing generally to let the sleeping dog lie. The arms race based on nuclear weapons, with possible comparatively small admixtures of CBR capabilities possessed by various powers, is already straining the imagination of strategists and diplomats almost beyond mental capacity, without the addition of another wholly new dimension: uninhibited, massive investment in many countries of brains, money, and material in a race to develop whole "families" of new chemical, biological, and radiological weapons and the corresponding defenses against each of them. No great power with an established position of superiority in one kind of weapons system has much reason to encourage the rapid development of a very different kind in which its head-start will be less pronounced. Both the United States and the Soviet Union share this status vis-à-vis the lesser powers. If they do not come to feel that they would gain vis-à-vis each other by unleashing a CBR arms race, the existing inhibitions may be allowed to continue to confine the struggle to methods the great powers find somewhat more familiar and more manageable than toxic weapons. The "if," however, is a very big one. Each side is likely to suspect—perhaps correctly—that it is depriving itself of potentially great advantages over the other by failing to press CBR development urgently. And some of the lesser powers may see less reason for inhibition in their own weapons technology. If the general arms race continues much longer, the inattention paid to toxic warfare will probably come to depend very largely on the sheer bureaucratic inertia associated with the existing commitment of human and material resources to preparation for nuclear warfare.

Therefore, the operation of any atomic test ban that might show promise of being durable, whether by formal agreement or by mutual abstention, should induce greater alertness on all sides about whether the adversary's CBR preparations were being intensified. If it should ever prove to be possible to move beyond mutual abstention to adequate joint inspection in the matter of nuclear tests, the feasibility of a parallel move in the CBR field should be carefully explored and, if practicable, negotiated.

One might suppose that it would be easier to establish a pattern of controls to keep a relatively dormant competition like CBR quiescent than to bring the dynamic nuclear development under some restraint. Inspection problems would differ greatly, however. CBR production facilities, being small, would be very much harder to uncover than those for nuclear uses; on the other hand, hidden stockpiles would be a lesser problem for a CBR than for a nuclear control system—probably neither kind of stockpile could be found, but at least the CBR materials would tend to lose their potency faster. Testing would probably be the most practicable stage of a CBR development process to inspect, as it is with nuclear weapons. If relatively wide open spaces are really needed for CBR experiments they may be conspicuous enough to be discovered, but no externally detectable phenomena like seismic disturbances would be there to alert an inspectorate to explore one sparsely inhabited area as a possible testing ground instead of another. If, as seems likely, a larger number of foreign inspectors would be required for CBR than for nuclear development controls in order to achieve a similar degree of reliability that a ban is effective, then the antithetical East-West attitudes toward inspection itself would intensify the difficulty of imposing restraints.

These problems have received very little attention in the West as yet. They are mentioned here as indicative of the kinds of difficulties that confront partial schemes for the restriction of armaments. The ramifications of any particular scheme throughout the shape and movement of the arms race need to be studied with the utmost care before the particular scheme can satisfactorily be evaluated. Yet, typically, in the time required to accomplish such research, digest it in particular governments, and establish a mutual understanding among them concerning its implications, the race speeds on to a new weapon generation, and objectively the implications of the scheme itself have in the meantime changed.

## Redeployment, Disengagement, and Disarmament

This transitoriness is seized upon as justification, on the one hand, for abandoning the effort to negotiate partial controls and for embarking instead upon rapid, total disarmament. Or, on the other hand, accepting the growth of armaments as unavoidable even while continuing to attempt to restrain the technological race, it is considered a justification for putting principal emphasis upon preventing a major war from arising inadvertently, out of misunderstandings and provocations resulting from the particular mode of peacetime deployment of forces or from the way they are used in a limited war.

The expression "arms control" is used to denote peacetime deployment restrictions, and restraints upon the arms race of the kinds that have been discussed in this chapter. "Arms control" often also covers curbs upon a limited war, of the kinds that were considered in Chapter 10. Increasingly, in sober analytical usage, the term "disarmament" is reserved to denote only very drastic reductions in absolute levels of forces. Before commenting on disarmament in this sense, a fuller understanding of arms control requires a closer look at the notion of deployment restrictions.

Mutual redeployment is, like control measures for the qualitative and quantitative arms race, designed to reduce the exacerbating risks associated with considerations of time, dispersion, institutionalization, and secrecy.

The purposes and concerns of the form of arms control that concentrates on deployment are therefore as follows: to bring and keep under the reins of effective central command the dispersed units of armed force, each of which may possess power to provoke major war by a deliberate or accidental action of its own; to stretch back the shrinking span of reaction-time that would be available to the central commanders to decide with what level of force they should cope with a particular provocative act apparently committed by the enemy; to blow away on both sides the clouds of secrecy that surround the nature and purpose of a provocative occurrence and the intended response to it, lest either side be panicked into launching a pre-emptive total attack; to forestall a self-perpetuating and self-aggrandizing institutionalization of certain armed forces which would later be able to block redeployments that might otherwise come to be regarded as mutually advantageous.

For example: The device of keeping a substantial proportion of a strategic air force constantly in the air carrying hydrogen bombs, so that it could not be caught on the ground in an enemy first strike and would remain reliably available for a retaliatory second strike, could have the potential effect also of enabling this "airborne alert" force to proceed quite far toward enemy territory on a surprise first strike of its own before the adversary would know that something more than a routine flight was afoot; thus airborne alerts may shave the adversary's reaction-time in a way that is perilous to both sides if the routine flights regularly go near his territory. Furthermore, the dispersed crews with their bomb loads may be difficult to keep reliably under the reins of central control, whether because of human obtuseness or recalcitrance, or because of mechanical failure in radio communications.

In this situation, both sides (or one side with or without some other equivalent concession by the other) might agree to keep their airborne alert forces flying along regular routes that were known to the adversary, that could be followed on his radar, and that were far removed from his territory. Then if any planes should stray off course, toward his territory, there would be a few hours' time in which each side could attempt to clarify for the other how deliberate the incursion was and what series of mutual responses would result. A special intergovernmental communications channel and even a network of attachés and authorized observers from the adversary's camp might have been established for such an emergency in order to overcome the usual secrecy and enable both sides to know better what kind of struggle they really confronted, and to give them a little time to decide whether and how it could be prevented from becoming total.

Such an arms control scheme for aerial redeployment with some inspection would, like proposals for curbing the qualitative and quantitative arms race, change the pattern of incentives for the development and deployment of other kinds of armed force, and would have to be evaluated partly in that light. For example, by conspicuously reducing the availability of the SAC airborne alert force for a possible first strike, the scheme would add to all the other difficulties Washington has in making any persuasively deterrent threat that SAC would retaliate massively against Russia in case she attacks Europe and not (initially) the United States. An implication might be that if the airborne alert were being thus restricted, much more should be done than has been in the past to improve Western Europe's local defense forces so that they would be strong enough on their own to deter a Soviet attack.

It is with regard to Europe itself that the concept of arms control through redeployment has been most thoroughly explored and strenuously debated for several years. In this context, its usual label is "disengagement," and the variety of possibilities is immense. Common to all of them has been an effort to put some considerable distance between the main armed forces of the Soviet Union and the Western powers as they confront each other, particularly in Europe. Distance would increase reaction time in case of an incursion. It also would reduce the risk of provocative incidents arising from undisciplined actions of local units dispersed along the front. Coupled with inspection and communications facilities to dispel secrecy, it would improve the opportunities of both sides to make informed and calculated responses in emergencies.

Disengagement proposals have also put emphasis upon partial or total local "denuclearization." The relation of this objective to the effort to avoid the risks of dispersion is obvious. Most especially, the forestalling of nuclear armament of Germans is a central purpose of nearly all schemes for disengagement. Fear of the provocativeness of a people who followed Hitler and who today, divided, are the least satiated nation in Europe stands out in the thinking of supporters of disengagement; but they do differ among themselves on how dangerous Germans would be if they were allowed conventional arms, while the nuclear powers proceeded to withdraw their own forces ("disengaged") more or less distantly to the rear. Even conventional armament of Germans has been resisted as institutionalizing a new Wehrmacht whose interests become entrenched and expand to erode year by year the feasibility of many conceivable forms of disengagement.

Yet these fears and purposes that motivate the quest for arms control through redeployment in Central Europe have run athwart the NATO objective of confronting the Red Army with a very high risk of total war in case of any substantial incursion in Europe, and thus to deter it. There are also tacit fears that an "unoccupied" Germany would be so unreliable, from the point of view of either side, that the dangers it would pose outweigh the existing risks of the armed confrontation of Russian and American troops on German soil. Also, there would be immense physical difficulties, as well as great expense, in redeploying large forces, finding space for them, and reconstructing their facilities. And there is great inertia for prolonging the status quo.

Yet the analysis in Chapter 10 showed why few experts are really convinced that the status quo of the 1950's provides a defense reliable enough to be prolonged without major changes. And some of the new strategic emphases proposed would probably be compatible with some forms of disengagement. Thus the problem of devising a viable arrangement for European defense (cf. Chapter 10) may lead one to give serious attention to certain disengagement suggestions, just as in this chapter the quest for arms control has been seen to lead to consideration of deployments that would minimize provocativeness, and hence, consideration of alternative defense arrangements for Europe.

The amount of detailed examination of particular proposals necessary to evaluate them properly is beyond the scope of this book. What is important here is to emphasize that "arms control" is inextricably interrelated with military strategy and assumes that armed force will continue to constitute a critically important dimension of the cold war; the control concentrates on efforts to diminish the risks, without any early expectation of ending them. However, the intricacies of the military relationships are so baffling that a cry naturally goes up for governments to abandon the perhaps impossible calculations and plump instead for rapid total ("general and complete") disarmament.

Soviet policy and propaganda have seized upon this panacea and have thereby accentuated the sense of futility that has engulfed nearly all negotiations for lesser measures of arms control. Whether or not Russia would really be willing to disarm totally within a very few years, down to the level of internal police forces as Khrushchev in 1959 urged the world to do, her willingness to espouse such a program provided her with talking points that have a dual appeal, both to many of the most simplistic pacifists and also to some of the most sophisticated students of the arms race and of arms control, who feel almost overwhelmed by the fearful mysteries of their subject.

In 1961 the Kennedy Administration felt impelled to espouse a much more concrete project than had Eisenhower for negotiations designed to proceed by stages to ultimate general and complete disarmament, but still to do so very cautiously.

Yet if the incomprehensibility and instability of the existing weapons competition defy mitigation, if particular controls and cautiously balanced East-West reductions in forces have so many ramifications and would allow so much rechanneling as to be virtually unassessable and hence are never actually negotiable, if the bewilderment exacerbates frustrations and tensions—yet also tends somewhat to paralyze aggressive action because no government is ever sure enough that it is ahead—then perhaps, one argument runs, an extremely rapid and utterly complete worldwide disarmament program would abruptly halt the race and all its perils, and could do so so suddenly that, while in the process of jamming on the brakes, both sides would continue to be so bewildered about their relative strength at any particular moment that neither would use its partly disarmed forces to attack its partly disarmed foe. By the time all armed forces had been dismantled everywhere, there would be as much inspection as the West could desire (thus runs Khrushchev's offer). So if thereafter there were significant violations, they would be detected, and if necessary the arms race might be started again; but even then, at least it could not quickly reacquire the dimensions of utter doom with which the present competition overhangs mankind.

Confronted with the manifest horrors of the arms race currently in prospect, this logic of desperation, whatever its loopholes, would probably get more favorable attention in the West, and Khrushchev's seriousness in proposing rapid total disarmament would be probed more positively in negotiations if he were also offering any hope that the remaining nonmilitary struggle could be ended short of the Communization of the world.

The arms race certainly does now have its own momentum which fiercely exacerbates the other dimensions of East-West competition, such as political subversion and economic warfare. Total disarmament derives its attractiveness in the West from the conclusion that the only promising way to reduce that exacerbation very much is to confine international competition entirely to nonmilitary instruments. But if the nonmilitary struggle ("peaceful competitive coexistence") is itself to continue to be a struggle over the very existence of the Western pattern of civilization, the men who hold positions of responsibility for its future must weigh very gravely whether they can really afford to deny themselves and their successors the physical capacity to use military force as a last resort in the cold war—for example, in order to prevent an uncommitted country from becoming committed to the Communist bloc or to prevent a Western-oriented country—say, in Latin America—from changing sides. If "coexistence" must continue to be "competitive" to the end that a whole political-economic order of civilization be "buried," can its leaders, however much they now abjure the use of force (sincerely in private as well as in public), make it impossible for themselves to change their minds in a future emergency, when facing some crucial defeat? Could they even put their whole trust in a new international police force, which might be subverted by the adversary?

The Kremlin may also have doubts about its own future in a disarmed world, but its leaders can adopt propaganda positions and then later discard them more easily than can their Western counterparts who are driven to explain themselves more frankly. Hence, Moscow can go on appealing for total disarmament and require the West to decide whether to pursue it seriously.

Whether the West ought to pursue total disarmament is by no means clear. One's judgment must depend in part on the appraisal one finally makes of all the risks of the developing military situation explored in the preceding chapters of this book. That appraisal would also affect one's judgment on the crucially related problem of how much to insist on detailed negotiation of a series of stages of arms reduction with accompanying inspection systems, even if endless disagreement over these details were very likely to doom total disarmament. Would the hazardous uncertainties of a period during which both sides were supposed to be rapidly dismantling all their armed forces—but in the absence of adequate inspection might actually not be doing so—outweigh the hazards of the existing arms race, in light of the further possibility that in the dismantlement period the novel uncertainties might be as likely to stimulate or provoke aggressive action as to paralyze it? These highly speculative judgments should also be supplemented by as sober and analytical an appraisal as can be made of the possibility after disarmament of rearming in time to prevent total defeat, if the struggle were being decisively lost when confined to nomilitary means. And clearly of crucial importance in any judgment on total disarmament is one's estimate of the potential effectiveness with which the West can learn to use the nonmilitary means.

This final problem is the subject of the remainder of this book.

## SELECTED BIBLIOGRAPHY

BARNET, RICHARD J. *Who Wants Disarmament?* Boston: Beacon Press, 1960.

BECHHOEFER, BERNHARD G. *Postwar Negotiations for Arms Control.* Washington, D.C.: Brookings Institution, 1961.

BRENNAN, DONALD G., ed. *Arms Control, Disarmament, and Security.* New York: Braziller, 1961.

BULL, HEDLEY. *The Control of the Arms Race: Disarmament and Arms Control in the Missile Age.* New York: Praeger, 1961.

HADLEY, ARTHUR T. *The Nation's Safety and Arms Control.* New York: Viking, 1961.

HENKIN, LOUIS, ed. *Arms Control: Issues for the Public.* Englewood Cliffs, N.J.: Prentice-Hall, 1961.

HINTERHOFF, EUGÈNE. *Disengagement.* London: Stevens, 1959.

HITCH, CHARLES J., and ROLAND N. MC KEAN. *The Economics of Defense in the Nuclear Age.* Cambridge: Harvard University Press, 1960.

KAHN, HERMAN. *On Thermonuclear War.* Princeton: Princeton University Press, 1960.

KING-HALL, STEPHEN. *Defense in the Nuclear Age.* Nyack, N.Y.: Fellowship Publications, 1959.

KISSINGER, HENRY A. *The Necessity for Choice: Prospects for American Foreign Policy.* New York: Harper, 1961.

KNORR, KLAUS, ed. *NATO and American Security.* Princeton: Princeton University Press, 1959.

MELMAN, SEYMOUR, ed. *Inspection for Disarmament.* New York: Columbia University Press, 1958.

MORGENSTERN, OSKAR. *The Question of National Defense.* New York: Random House, 1959.

NOEL-BAKER, PHILIP. *The Arms Race: A Programme for World Disarmament.* New York: Oceana Publications, 1958.

NOGEE, JOSEPH. *The Diplomacy of Disarmament.* New York: Carnegie Endowment for International Peace, 1960.

NUTTING, ANTHONY. *Disarmament: An Outline of the Negotiations.* New York: Oxford University Press, 1959.

ROWEN, HENRY. *National Security and the American Economy in the 1960's* (Study Paper #18, U.S. Congress Joint Economic Committee, January 1960). Washington, D.C.: Government Printing Office, 1960.

SCHELLING, THOMAS C. *The Strategy of Conflict.* Cambridge: Harvard University Press, 1960.

———, and MORTON H. HALPERIN. *Strategy and Arms Control.* New York: Twentieth Century Fund, 1961.

TUCKER, ROBERT W. *The Just War: A Study in Contemporary American Doctrine.* Baltimore: Johns Hopkins Press, 1960.

WASHINGTON CENTER OF FOREIGN POLICY RESEARCH (Johns Hopkins University). *United States Foreign Policy: Developments in Military Technology and Their Impact on United States Strategy and Foreign Policy* (Study #8, U.S. Senate Committee on Foreign Relations, December 1959). Washington, D.C.: Government Printing Office, 1959.

# PART THREE

# THE INFORMATIONAL ARM OF POLICY

# CHAPTER 12

## ON BEING SEEN

## AND ALSO HEARD

ADHERENTS OF the basic American policy that Communism must be contained everywhere it appears in the world often find the reasons for their all-embracing concern difficult to define in particular instances. Some countries do lack resources or strategic location of a kind that can be shown to be of specific military or economic importance to either of the main contending powers. But containment insists that in the stress of cold war one should always assume that some specific use for any such country may yet be discovered, and that, in any case, every country, no matter how apparently unimportant, has symbolic significance; its loss may weaken the will to resist of more obviously valuable partners.

It is impossible to define at what point an erosion of the less-developed countries away from cooperative relations with the West would really produce economic, social, and political changes in Western Europe or America that would destroy even the Anglo-American will to resist—or would facilitate new weapons systems that would make the defenders' efforts so unilaterally suicidal as to be a matter of cool indifference to a Communist aggressor. Perhaps never. Perhaps even an utterly marooned Fortress North America could preserve its independence. Too much depends on morale and technology for the possibility to be nonsensical. But if "America" means a way of life, not just sovereign independence, then in that sense "America" would certainly be lost. Likewise Western Europe, and presumably sooner, because economically and socially more vulnerable. Therefore, the policy of containment, not knowing where ultimately the line would really have to be drawn to prevent completely fatal consequences, tries always and everywhere at least to hold the line where it is against Communism.

### Inarticulateness about Basic Policy

The development and use of various forms of American military force for coercion, intimidation, or reassurance and of economic assistance for rewards, penalties, or amelioration can be understood as means whereby the international persuasiveness of the United States is enhanced and exercised. But an ability to persuade, by whatever means, is largely meaningless unless there are goals toward which the means of persuasion can be directed, relevantly and effectively, and unless those goals or the desired steps toward them, or both, can somehow be actually communicated to those who should be persuaded. Yet clearly, containment, when it is frankly articulated as the underlying principle of contemporary American foreign policy, does have a negative tone that is unappealing, even to many Americans, and that is still less congenial to the emerging peoples of the underdeveloped world with whom America needs to try to communicate—if only in order to effectuate the containment policy itself. Thus, ambivalently and clumsily the United States tends to be inarticulate about her actual operative fundamental policy, while she makes consoling noises and gestures to herself and others, which ambiguously imply that she has other, more "positive" basic policies.

America's limited ability to communicate articulately her working principles in foreign affairs to others derives largely from a limited ability to grasp them coherently and concretely herself. Partly, the causes are the sheer size and heterogeneity of her population and the pluralism of her political institutions. But in a more fundamental sense, the cause is the lack of fundamental discontent among Americans toward the international status quo that prevailed in the twentieth century until World War II, which usually makes the "negativism" of the containment policy more congenial in actual substance to most of them than any particular "positive" alternative, when the relative costs are assessed, while nonetheless leaving Americans reluctant to contemplate that image of themselves in a mirror.

The ambivalence produces widespread confusion at home and abroad. The self-image of Americans—particularly those who work with words rather than things—has a broad streak of crusading idealism that would disown any national selfishness. But the comforts of an era of Western hegemony over the rest of the world were—and would continue to be—undeniably pleasurable for the United States; and some inclination to restore as much of them as possible can be detected in a very large part of American policy since World War II. Of course, few persons in the United States have actually contended soberly that the world order that was shaken in the 'teens and shattered in the forties could ever be fully reconstructed with its essential forms and balances. Restoration as an objective has been dismissed as a mirage. It has to be. Yet like a mirage it tends to recur to Americans as they struggle to hold their own in the wearying trackless foreign-affairs desert of erosion and drift where no other discernible goal—even frozen containment—is likely to appear quite so comforting to them.

Whatever crusading ideals they may like to proclaim, the world has simply been too good to most Americans for them to press urgently for a fundamental overturn in the established order, at home or abroad. Most Americans are likely to detect burdensome consequences for the United States in almost any basic change in the international environment. Change is inevitable in the revolutionary age of the mid-twentieth century, and the burdens may differ enormously in degree; but most Americans probably cannot realistically be expected to become positively enthusiastic about them, or honestly manage to adopt as their own personal aspirations the aspirations for drastic upheaval that fire peoples outside the North Atlantic area, however much the international image of the United States might be improved if they did.

In most of the world, where the majority of politically conscious persons probably desires some such overturn, the international status-quo-mindedness of the United States can have only limited appeal if it be articulated with frankness. If, truly, America's nostalgic Utopia is simply that she be let alone to stay at home to prosper in comfortable immunity and pontificate self-righteously about less fortunate peoples, and if any other more positive sense of national purpose in world affairs—even containment—can become acceptable to most Americans only as the lesser of the evils among which realistic choices unfortunately have to be made, then any very explicit articulation of the country's mood is likely to generate international embarrassment. It is possible to give some attractive emphasis abroad to the nonaggressiveness inherent in these essentially isolationist propensities, but in most other aspects they have too strong an aura of complacency to be very appealing.

Fortunately, however, American negativism about remaking the world (except for those adjustments that become necessary to preserve the opportunity for the country to continue along its own accustomed course of development at home) has not been duplicated in any comparable lack of a sense of direction and advance in its domestic affairs. Quite the contrary. The popular consensus that confirms the optimistic American Way of Life within the United States has been pervasive and powerful. Yet it too poses a problem in external communication, one that is somewhat different from the problem of articulating an appealing national purpose in foreign affairs. The difficulty is largely that the agreement on positive domestic objectives is so widespread that Americans can be satisfied to have them merely signaled by clichés and embodied in pragmatic arrangements; rarely is the content of this consensus systematically elucidated and intellectually comprehended as a complete, coherent creed. The task of ideologizing the American domestic *status quo* and its future promise is unnecessary at home and perhaps even unsettling there.

## Government Support for Private Contacts and for Unofficial Communications

Thus, fittingly, for attracting admirers the United States has traditionally preferred to rely more on being observed than on being explicated. Washington's Farewell Address, it will be remembered, predicted that his country, in the "worthy" role of "a great nation," would "give mankind the magnanimous and too novel example of a people always guided by an exalted justice and benevolence." An example to be imitated—not a blueprint to be followed. That was what America offered to other peoples, with a distinct impression also that those who would not immigrate but instead remained in their less fortunately situated homelands would be sadly handicapped in any efforts at imitation they might make.

If the American Way of Life was a worthy example of goodness that would generally have to be experienced directly to be understood, its persuasiveness abroad was bound to depend mainly upon the constant flow of persons to United States shores, some to observe and then return to their own countries, others to remain and continuously to send back bits and pieces of an image of America to their old acquaintances across the seas. The flow of persons did accelerate until after World War I, and the image glittered in most parts of the world. But when mass immigration ended and was soon followed by the travel barriers of World War II and the postwar economic stringency in Europe and Asia, the opportunity to experience the United States at first hand or even at second hand became relatively restricted. Millions more Americans were abroad after World War II, in civilian and military capacities, than ever before in peacetime. But an impressive flow of foreigners to the United States, to develop a feeling for American life in its native habitat, was bound to require some special governmental assistance for surmounting the obstacles. The consequence was a great expansion in organized public and private programs for the "exchange of persons" with foreign countries.

No other technique of international persuasion could be so fundamentally congenial to the American mind. In one sense, the predilection can be ascribed to a logical conclusion that the best way to convey an idea is to embody it in a human being and send him to the place where the idea should be implanted, or to bring someone from that place to watch the idea embodied in action. But there are overtones also of a relative indifference on the part of Americans toward controversial ideas as such and toward communicating them. Instead there is a warmhearted confidence in the possibility of establishing international friendship by personal contacts in which disturbing ideas simply tend to be set aside. The notion that they can be bypassed comes easily to Americans, who are accustomed to ideological consensus, not to deep-rooted ideological conflict. Correspondingly, the exchange-of-persons programs have put great emphasis on providing opportunities for observation, travel, and association, even when the formal purpose of a grant may be academic research or instruction or technical training or consultation or whatever else. Contacts become almost an end in themselves, the faith being that Americans will appear in an attractive posture to foreigners if only enough of them are enabled to establish some sort of contact.

The enthusiasm for exchange programs reflects also the traditional American faith in the virtues of individual and group initiative, an attitude that applies even to the government sponsored and subsidized exchanges. The solicitation and screening of applicants by the sponsoring agencies and the guidance given to those who are selected—even if, as in some cases, it is extensive—still leave the travelers much more independent of official direction than are the persons engaged in almost any other deliberately organized American technique of international persuasion. The individuals may be chosen partly because they seem unlikely to say or do things that the sponsors would consider harmful; but what they actually say or do, once the grants have been bestowed, is mostly of their own choice. (Contrast the position of a scriptwriter for the "Voice of America" or a press relations officer in a United States embassy.)

Key congressmen prefer the exchange programs over other forms of propaganda so much that they require that at least the formal educational exchanges (notably the scholarship program associated with the name of Senator William Fulbright) be administered separately from other government informational activity, lest the exchanges appear tainted by it.

Generally, there is full recognition that the impact of such subsidized contacts between individual Americans and foreigners can for the most part be effective only in the long run in enhancing the persuasiveness of the United States among other nations. Especially when the contacts are with young students from industrialized countries, the investment in good will cannot be expected to pay early dividends. Some other exchanges, notably the provision of guided tours of America for community and group leaders from foreign countries, have hopes of implanting ideas and images that will function promptly to further particular United States government policies. But in essence the function the government is performing in promoting exchanges can be understood to resemble official efforts to further the flow of private investment abroad (see Chapter 17). (1) The flow is considered to deserve government encouragement if only because it is regarded as a good in itself by a society that exalts private and corporate initiative. (2) Yet, in addition, the flow has consequences abroad that can be advantageous in the long run to the position of the United States in world affairs. (3) Extra subsidies, therefore, seem warranted, and with them come a few controls. (4) Ultimately, however, in a cold-war world even these devices have to be supplemented by more direct government activities.

As with economic foreign relations, so also with the organized efforts at international persuasion commonly called propaganda. The American approach is at first private, accepts increasing supplementary activity by government, yet never ceases (except in some dealings with Iron Curtain countries) to preserve intimate interrelationships between the private and the governmental undertakings. (And probably in cultural activities and the flow of nontechnical information there are not nearly so many places in the world where the role of the United States government is preponderant over that of private Americans as there are places where this is the case in economic relationships.)

In addition to the close public-private links in the exchange-of-persons programs, a large part of the American government's overseas informational effort takes the form of facilitating the circulation of newspapers, magazines, and books from the United States. Subsidization of translations and of mailing costs; purchases of publications to be given free to influential persons; arrangements for American publishers to unload their foreign currency earnings upon the American government in return for dollars when local exchange controls would prevent convertibility; maintenance of American libraries (rather like small-town libraries in the United States) in a great many cities throughout the noncommunist world: these are typical of the devices whereby publications from private American sources are utilized and supported by the American government for international persuasion.

Books, of course, can be given an official screening, individually (yet even this process was haphazard until the middle fifties). Periodicals must be selected almost entirely on the basis of past performance; once a subscription has been donated for a fixed term of months or a currency-convertibility guarantee has been provided, it can scarcely be cancelled no matter what embarrassing material later issues of the journal may contain. Nor can either the books or the periodicals be censored, though some probably unwarranted allegations have been made that dependence on renewable subsidies, even in this limited form, may tend to inhibit American journalists from attacking their government's policies. If there is any such effect, one might of course still defend it as appropriate to the jointness of public and private efforts that characterizes United States activities abroad. Fundamentally, however, the pattern of promoting the United States by promoting American publications reflects again the high confidence that Americans place in private enterprise as such. How, it is felt, could a government newssheet hope to match the New York *Times* International Edition—and why should it, when even back in Washington the officials themselves are more inclined to rely on the *Times* than on the daily intelligence digests that come from CIA? This at least is a typical American attitude.

It has been overcome, however, with regard to radio and television. Cold-war competitiveness required the United States to establish permanent arrangements for worldwide mass communication through these media after World War II, and the cost could not possibly be met from principally private sources, either through advertising or donations. Like the other major powers the United States acquired its official radio "Voice." But the American predilection for privateness re-emerged vigorously in the Korean war period when the heightening of the East-West struggle led to a demand for a more hard-hitting kind of propaganda. "Radio Free Europe" and "Radio Liberation" were then established as sounding boards for émigrés from the East European satellites and from the Soviet Union, respectively. Ostensibly the financing and direction of these broadcasting services was almost wholly nongovernmental. Actually the government's covert role was important, but, significantly, it did remain considerably limited, at least in the early years. Apart from a desire to save the government money by soliciting voluntary donations from the American public for the "Crusade for Freedom," the motivation for this arrangement reflected again an eagerness to enlist private initiative, dedication, and drive, largely freed from the traditional constraints of diplomatic etiquette and governmental red tape.

## Sloganeering: "Liberation" and the Hungarian Tragedy

Interestingly, the high hopes that were held for the radio voices—the official Voice of America and the unofficial Radio Free Europe and Radio Liberation—were in line with another recurrent strand of United States behavior in international affairs. When Americans manage partly to overcome their traditional inarticulateness and their preference for being seen accomplishing rather than explaining national purposes, and for doing so under mainly private auspices, they have often been inclined to use words clamorously in the international arena as though talk could function as a self-sustaining substitute for some costly action. High moral affirmations or condemnations, without the backing of adequate force, have been used again and again in United States history in sanguine efforts to sway recalcitrant foreign governments by rallying world public opinion. Even when hope has faded for a time, there has been comfort for Americans themselves in simply sounding righteous. Of course the words that have kept alive an idea of action may in fact eventually produce the action itself, but commonly the principal effect of the pronouncements has been self-deception in the United States, both as to the nature of the immediate international problem and as to the general efficacy of mere verbalization in world politics. The periodic twentieth-century American policies of nonrecognition of violent seizure of territory or government power have, for example, usually been nonproductive gestures, except in some cases where they were reinforced by some concrete form of sanctions. With this perspective, there appears a certain consistency in the fact that the most ambitious objective Americans have yet set before themselves in the cold-war years for international persuasion by mass communications has been the goal of "liberation" of the Soviet satellites whose domination by Moscow is considered by Westerners to be a brutal violation or perversion of wartime agreements among the great powers.

The notion that some meaningful form of liberation of Eastern Europe could be achieved in the near future and without force was given wide currency in the United States during the Korean war by important leaders of opinion, principally Republicans. Very many Americans found such propositions so congenial and so plausible as to be attracted to Eisenhower's presidential candidacy by its slogan of "Liberation!" Those who considered such talk simply demagogic were probably a much smaller minority. The previous Truman Administration itself had envisaged its containment policy as leading eventually to the break-up or mellowing of Soviet power; the East-West frontiers of early 1950 had never been accepted as permanent by the American government. And the "hard" line of propaganda broadcasting to Eastern Europe by official and unofficial American-sponsored voices was well developed before Eisenhower and Dulles took office. It was designed to accentuate dissatisfaction and keep alive hope among the captive peoples. As incitement to action these broadcasts went no further than to discourage collaboration with the Communist overlords. Such a line could at most produce only slowdowns and sabotage—unless an implication were allowed to develop that the West would provide concrete assistance in case of an insurrectionary crisis. That impression did begin to form in Eastern and Western Europe in 1952 largely as a consequence of the greatly increased vociferousness of discussion within the United States concerning the goals of "liberation" and of a "rollback" of Communism. The verbal emphasis upon such themes in the Republican election campaign, followed by GOP victory, tended to add revolutionary overtones to much of what was being broadcast—at least in the minds of the listeners—even though the actual contents of the scripts and the availability of forces for material assistance in a showdown were not changing very much.

The new Secretary of State, John Foster Dulles, allowed the illusions to mount, partly because of his fondness for bluffing abroad and for borrowing the language of oppositionists at home, and partly from an idealistic unwillingness to espouse a compromise alternative like "national Communism" that would reduce Moscow's domination over the satellites and some of the extremes of internal repressiveness but would leave local Communists in ultimate control over their peoples. Under such a formula the maximum "liberation" achievable would be a form of Yugoslav "Titoism." The Eisenhower-Dulles Administration, like its predecessor, welcomed and assisted Tito as the lesser of evils, but feared that forthright advocacy of "national Communism" by the United States would compromise Western principles and dash the hopes of anticommunists in Eastern Europe without improving—perhaps even reducing

—the likelihood that Moscow would risk allowing the satellite Communist regimes to have some independence. Yet the failure to give clear approval to a moderate and relatively realistic objective like "national Communism" tended to lead satellite populations, stirred by the talk of rollback and liberation, to expect concrete American assistance in their pursuit of forms of emancipation that could only be achieved by revolutionary violence, and not to wait indefinitely for evolutionary mellowing. And if assistance to violence might lead to war, the Eisenhower Administration was actually no more willing to attempt it than Truman's had been; the war would be too cruelly costly to those who were to be liberated as well as to the West. Containment remained the fundamental operative policy of the United States.

These partly calculated ambiguities—deceptive to the American people and to their allies and the neutrals as well as to Eastern Europeans—were somewhat resolved in June of 1953, a few months after Eisenhower's inauguration and Stalin's death, when the United States government failed to take any vigorous action to prevent the suppression of a wave of anti-Soviet riots in East Germany. But in the ensuing months, as the Russians gradually relaxed their stranglehold on Eastern Europe and Khrushchev attempted to lure Tito back into a newly loosened Communist fold, the Western powers did not clarify their immediate objectives for Eastern Europe in light of the altered situation. The need for clarification did not seem urgent; Americans were inclined to act as though the decline in their own excitement about rollback talk, which coincided generally with the ending of the military conflict between themselves and the Communists (Korea, July 1953), was also being shared by the Eastern Europeans. Later, by 1956, another election campaign was in the offing in the United States—further incentive to continue the ambiguities about "peaceful liberation."

If they were doing much harm, it was not obvious, even as late as October 22, 1956. In Poland a movement toward liberalization gathered momentum swiftly during that summer and early fall. It obtained the support of the established Communist heads, Premier Cyrankiewicz and First Secretary of the Party Ochab. They turned the top leadership over to former Party Secretary Wladyslaw Gomulka who had previously been punished by imprisonment and torture for alleged Titoist and nationalist deviations. Together with him, Cyrankiewicz and Ochab formed a joint front that defied the violent threats of the foremost rulers of Russia, who hastened to the scene.

American-sponsored broadcasts, official and unofficial, were encouraging the Polish people to support Gomulka's moves despite his Communist background, conveying the impression that this was a stage toward liberation that Washington welcomed, and that violence leading to war must be avoided.

But this clarification of United States purposes, tending, though still somewhat fuzzily, toward an explicit partial endorsement of "national Communism," came too late to forestall an onward surge of events in neighboring Hungary that went much further—far beyond what Moscow could be induced to tolerate without resorting to Red Army violence.

The events in Poland had an exhilarating impact in Hungary. But Hungarians were not Poles; they were much less fearful that Germany would be aggressive again if the Russians were entirely out, and perhaps the Hungarians had learned fewer lessons from the past about Russian power and determination. Also the new First Secretary of the Hungarian Communist party, Ernö Gërö, who had replaced an arch-Stalinist, Mátyás Rákosi, in a Soviet-endorsed move toward partial liberalization, was no Cyrankiewicz willing to stand up to the Russians in support of other much more moderate Communist leaders. And the key figure among those moderates, Imre Nagy, was no Gomulka, shrewd enough to know what the limit of Russian tolerance would be and strong enough to stop the revolutionary surge of his own people from going beyond that point. Nor was the direction of American propaganda broadcasting as temperate and realistic for Hungary as for Poland—partly because the Polish "national Communists" were able to accomplish their coup and stabilize their power so swiftly that the situation there did not begin to get out of hand so much as it did in Hungary, producing agonizing temptations for the West to encourage an ongoing complete popular revolution against Communism itself.

The crisis began on October 23, 1956, when Party Secretary Gërö returned to Budapest from a ten-day visit to Yugoslavia, during which he had presumably been somewhat out of touch with the degree to which the Polish coup had suddenly fired the confidence and determination of anti-Stalinists in Hungary. Street demonstrations were mounting against the government. Gërö adopted a fairly tough line in a broadcast speech attempting to quell them. The angered crowds soon provoked the political security police (known as the AVH) into firing. In the ensuing skirmishes the populace suddenly discovered that units of the regular police and of the army would support them against the hated AVH. The facade of AVH invincibility was destroyed. Khrushchev's right-hand man, Mikoyan, came immediately by plane from Moscow, and agreement was reached with the top Hungarian Communists to throw Russian troops into action at once on the side of the AVH (the Russian troops were already stationed in Hungary, officially as part of the Soviet bloc's defense forces).

Imre Nagy, who as Premier in the immediate post-Stalin years 1953–55 had been identified with a policy of liberalization, was formally reinstated in the premiership, but was kept under wraps; Gërö retained the key post of Party Secretary. Late in the morning of October 24, Nagy made a radio appeal for support of the regime against the developing insurrection and promised a return to the moderate governmental policies for which he had stood. Nagy appeared, however, to be endorsing the use of Russian troops, and, although later evidence indicated that he himself was at this point acting under the physical duress of the AVH, listeners in Hungary and the outside world assumed that he shared full responsibility for the Soviet intervention. American-sponsored broadcasters, especially the unofficial Radio Free Europe's Hungarian announcers at its station in Munich, Germany, adopted a policy of denouncing Nagy repeatedly and bitterly until November 2. The effect was to contribute to undermining his authority, tending rapidly toward the complete collapse of the power of the whole Communist party in Hungary.

The momentum of the revolution surged toward more and more far-reaching demands in those final days of October, and it met less and less resistance from anyone in Hungary except the Russian Army. Communication between the government and the insurgents took the form largely of a dialogue between the government radio in Budapest and the provincial radio stations, which were mostly in the hands of revolutionaries. Outside Hungary, Radio Free Europe and other Western senders picked up the varied demands coming from different radio voices within the country and repeated them, tending to reduce them to some programmatic order but allowing—indeed, in effect, encouraging—them to push far beyond "national Communism." Demands produced concessions—yet the whetted appetites brought further demands—and obtained further concessions. During October 24, 25, and 26, Gërö, not Nagy, participated in the discussions with Mikoyan and other Russians where the government's policy was actually determined; Nagy's premiership was nominal. But on October 25, Gërö was replaced as Party Secretary by János Kádár, a former Interior Minister who had suffered severe torture as a "nationalist deviationist" in the heyday of Salinism. On October 26, Gërö actually left for Russia, and the following day Nagy, physically free, formed a coalition cabinet that included noncommunists.

Outside, the Western powers appealed to the United Nations Security Council for consideration of the Hungarian situation, and Secretary of State Dulles offered new economic aid to nations like Poland and Hungary, promising that it would not be conditioned "upon the adoption by these countries of any particular form of society" and that "we do not look upon these nations as potential military allies." But Dulles still affirmed, rather mildly to be sure, a goal that went beyond "national Communism." "Our unadulterated wish is that these peoples . . . should have governments of their free choosing." * The United States was sending Red Cross medical supplies. If these American actions, belated and limited, aroused any Soviet anxiety that further Western intervention was in the offing, it was only such as to quicken Soviet withdrawal, not provoking an immediate heightening of the Red Army's repressive action; thus Dulles's ambiguities still appeared to be harmless, or perhaps even beneficial. On October 28, the Nagy government declared a cease-fire against the rebels and began negotiating for removal of Russian troops. The West allowed the Security Council debate to be adjourned without pressing for formal action; there was optimism about the course of events—and sudden distraction by the Israeli invasion of Egypt near Suez.

Russian troops actually began withdrawing from Budapest on October 30, while concentrating their forces warily in other parts of Hungary. The Kremlin announced its willingness to negotiate jointly with all the East European satellites ("the great commonwealth of socialist nations") whether Russian troops should continue to be stationed in Hungary at all. But this pronouncement was full of references to equality of "socialist countries," meaning only Communist countries, and the Soviet government formally expressed its "confidence that the peoples of the socialist countries will not permit foreign and internal reactionary forces to undermine the basis of the people's democratic regimes." Such warnings should have been given greater heed in America and in Hungary. Only a government in Hungary that would be Communist and able to maintain itself internally without the presence of Russian troops would be allowed to do so. In Hungary, however, the Communist regime under Imre Nagy was rapidly dismantling itself in continuing concessions. The AVH was disbanded on October 29. A multi-party coalition cabinet took office on October 30, with early free elections in view. On October 31, Nagy announced the beginning of negotiations for Hungary's withdrawal from the East European alliance system (the Warsaw Pact) and observed that the question of Hungary's becoming the nucleus of a neutral bloc would come up sooner or later.

The prospect that the whole apparatus of Russian domination in Eastern Europe would crumble brought Mikoyan and other Kremlin

* New York *Times*, October 28, 1956.

chiefs back hastily to Budapest. But on November 1, Nagy formally proclaimed Hungary's neutrality and announced in an appeal to the General Assembly that Hungary "turns to the United Nations and requests the help of the four Great Powers in defending the country's neutrality." Nagy may have hoped that the Kremlin would be able to content itself with neutralization—Russia being one of the guarantors, possessing a right to prevent Hungary from being used militarily by the West—and also that the Kremlin would be deterred by a show of Western support from intervening to keep Hungary fully in the Eastern bloc. But Moscow was more likely to regard the proposal as provocative, especially since it came simultaneously with the military intervention at Suez by the British and French, which the Kremlin must at first have presumed had American approval; with the mushrooming of noncommunist administrative structures in Hungary to replace the utterly disintegrated Communist apparatus; and with even the emergence of fragments of right-wing Hungarian parties as well as those of the center. The Russians knew that free elections in Hungary would produce an anticommunist government there, geographically situated like a knife through the very center of the buffer zone of satellites to the border of the Soviet Union itself. The coincidence of events in Hungary with those at Suez (against the anti-Western government in Egypt) may have seemed to the Kremlin to evince a devious Western strategy for the much talked-of "rollback," especially when the British government, conceivably fronting for the Americans in regard to Hungary, as supposedly at Suez, announced on November 3 that it was "supporting Hungary's request for the recognition of her neutrality." So it may have appeared to Moscow; actually Britain was probably hoping to distract a little attention from Suez, and the American government, with the Secretary of State suddenly undergoing an operation for cancer and the President immersed in the final days of an election campaign, was disabled by fear of war over Hungary and was concentrating in angry frustration on halting its own wayward allies at Suez. Thus the pressure of events was such as to provoke Russia hotly, without deterring her for more than a few days of watchful waiting, at the very most. Whether the Kremlin would have been finally deterred if the United States government had threatened to use force in support of the neutrality of a prospectively anticommunist regime in Hungary will never be certain, but it seems unlikely that any threat would have sufficed to induce the Soviet Union to take the chance of having an unfriendly regime in Hungary. Basically the political orientation of Hungary, geographically located where it is, was so much more important to Moscow than to Washington that the Russians were willing to run risks of total war most Americans would not run.

On November 4, the Soviet Army began a reconquest of Hungary, nominally at the invitation of Communist Party Secretary János Kádár, who had deserted Nagy and agreed to form a new Soviet-backed substitute government. Hungarian "Freedom Fighters" put up heroic resistance in fierce combat, but against such odds as to be suicidal. President Eisenhower dispatched urgent appeals to Premier Khrushchev. The General Assembly registered its nearly unanimous disapproval of the Soviet action and voted to send official UN observers, but they were never permitted to enter Hungary. No severe sanctions were imposed. The Freedom Fighters were soon crushed.

Again the various radio voices receiving American subsidy had failed to articulate a coherent line of advice for the Hungarians. Just as the broadcasters had, in the main, previously encouraged Hungarians to seek more freedom than the Russians were likely to allow, to the point of demolishing the internal structure of the Communist regime and Hungary's alliance with Russia, so after the renewed Soviet military onslaught they failed to make clear to the Freedom Fighters that their further resistance was martyrdom.

The glory of martyrdom, to be sure, might have been painted in glowing tones; Secretary of State Dulles later observed rather poignantly that the demonstration needs to be made from time to time that people are willing to die for freedom. The Hungarian resistance could not be crushed by the Russians without their having to resort to overt brutality so flagrant as to produce major defections from the ranks of Communist sympathizers in Western Europe, India, and elsewhere. And the demonstrated unreliability of the Hungarian Army for Soviet purposes made it possible for the West to recalculate in a favorable direction the entire balance of military manpower on the continent of Europe; the satellite armies could be treated as almost useless to Moscow in case of a war. These gains did accrue to the West from the bitter-end struggle of the Freedom Fighters, but there is no reason to suppose that the limited and almost always ambiguous encouragement the Hungarians received was deliberately designed to trick them into the martyrdom. A few American supervisors of the broadcasts may have been consciously, rather desperately, trying to outbluff the Russians; probably most were simply unable to shift gears, psychologically and institutionally, when the "hard-sell" American drive for "Liberation" by propaganda, in which they had been engaged for years, suddenly overachieved its goals in Hungary and contributed to provoking a horrible disaster. These Americans were unable to act effectively to call a halt to some of their other associates (particularly émigrés who were actually at the microphones), who hoped that if the heroic resistance could be kept going long enough after the American election, the Eisenhower Administration, secure in office, would finally resolve to risk war by intervening with force to prevent Hungarian annihilation.

The fact that such highly unrealistic attitudes became very widespread inside Hungary must, to be sure, be blamed mainly on the Hungarians themselves. It was they who interpreted the ambiguities in the external radio voices in the way most encouraging to themselves, failing to measure the balance of forces abroad with the prudent self-restraint the Poles exercised in the same period. To ascribe to American-sponsored propaganda broadcasts the principal blame for the Hungarian debacle would be merely to repeat, by inversion, the fallacy of the many Americans who imagined that the Iron Curtain could be rolled back by the right kind of blast from Joshua's new trumpets, the radio loudspeakers. And that fallacy was but a new manifestation of the historically familiar American tendency to overestimate in general the efficacy of righteous verbalization. Actually, of course, what happens in any overseas area must depend mostly upon what the people there have been doing, far more than upon what Americans have been saying—especially if the listeners feel there is much of a gap between what Americans are saying and how Americans are acting or are likely to act. However, where words cannot be tested by deeds they may carry unusual weight for a time in the minds of special audiences; this happened in Hungary until mid-November 1956. Even in such situations the audience responds selectively and interprets for itself what it hears. But surely it is the responsibility of top officials of propagandizer governments to keep at least themselves conscious of any gap that may exist between their spokesmen's words and their own probable deeds, especially when the listeners lack that consciousness. The oral crusaders have an obligation to avoid becoming intoxicated with the sound of their own voices, in the way that Americans tended to become bemused by their campaign for liberation. Authorities engaged in directing propaganda should soberly calculate their own prospective actions and not carelessly raise expectations about their country's conduct that they can anticipate are unlikely to be fulfilled. In situations where whatever marginal persuasiveness the words do exert would tend to spur listeners somewhat further than they would otherwise have gone along a blind alley that will probably lead to disaster (when deeds are not forthcoming), the scale of the disaster itself may potentially be so enormous for the listeners that its avoidance should be an end in itself for the propagandist, outweighing the limited gains that he might yet hope to salvage from the wreckage. Probably such ought to have been the American attitude toward the idea of liberating Hungary.

## Words, Deeds, and Symbolic Acts

Even much lesser hopes, rashly roused and then foully dashed, may so gravely impair the future persuasiveness of the propagandist that only rarely can he afford for long to yield to the familiar temptation to treat words as a substitute for deeds. This lesson is slowly being learned by American policymakers as their brief experience with official overseas propaganda becomes gradually assimilated to their older preferences for simply being observed face-to-face in everyday activities that require no elaborate ideological explanations, or for venting their righteous indignation in soul-stirring verbiage that takes little heed of the consequences it may have.

The essential principle of the lesson is very old, though it is still not fully applied: the most persuasive propaganda is usually that which highlights attractive deeds. The deeds are primary; the disseminated interpretations of them are secondary. The deeds of course must be relevant, and the march of the cold war changes the categories of relevance; but it is principally because the relevant deeds cannot be directly observed by a sufficient proportion of those foreigners whose opinions are politically significant that interpretations of the deeds must be formulated and disseminated systematically under official auspices. To communicate action, in a frame of meaning, has increasingly come to be the job-concept of the most experienced American professionals in the field of propaganda.

When they themselves avoid the term "propaganda" and label their arm of policy the "informational" (alongside the diplomatic, military, and economic arms), they are not merely eschewing a nasty word in favor of a euphemism. Their usage rather accurately reflects their relaxed image of what propaganda should be like to be effective when the propaganda comes openly from official sources. It should be informative—more than hortatory, denunciatory, or precariously misleading. This concept of their task is emphatically clear in the mandate from President Eisenhower that the United States Information Agency adopted as its official statement of purpose: "To submit evidence to the peoples of other nations by means of communications techniques that the objectives and policies of the United States are in harmony with and advance their legitimate aspirations for freedom, progress, and peace." Notice that this evidence was only to be "submitted" by the propagandists; they would not take the responsibility for creating it.

> This purpose is to be carried out primarily: (a) by explaining and interpreting to foreign peoples the objectives and policies of the United States Government; (b) by depicting imaginatively the correlation between United States policies and the legitimate aspirations of other peoples of the world; (c) by unmasking and countering hostile attempts to distort or to frustrate the objectives and policies of the United States, and (d) by delineating those important aspects of the life and culture of the people of the United States which facilitate understanding of the policies and objectives of the Government of the United States.

This mild statement of purpose—combative only in (c), and there defensive—can be recognized as envisaging a role for propaganda that is adapted to the requirements of the basic national policy of containment. The image is positive but not aggressive; within the limits of containment the concrete acts on which propaganda from avowed official sources ought to be grounded cannot be expected to support a much more forceful line than this. Fundamentally the posture is defensive, except in its long run hopes.

Related to this objective is the fact that the American government's official propaganda efforts are increasingly concentrated in the underdeveloped countries, more than in Communist and noncommunist Europe, even though the idea of containment itself, if frankly expounded as doctrine, can find much more sympathetic audiences in Europe. Containment must be implemented where the Communist challenge is strongest, even though doing so requires that American communications there emphasize themes other than containment. A majority of all the hours of direct American broadcasting in native languages is to Asia and Africa; about two-thirds of the United States expenditure for prepackaged radio shows to be presented on local stations is for Asia, Africa, and Latin America. More than half of all official American libraries and other cultural mission-stations are in Asia, Africa, and Latin America; and more than half of all persons coming to the United States on American government-subsidized exchange-of-persons and official visitors' programs are from those regions. The government's efforts at international persuasion through these techniques of propaganda are becoming oriented toward those nations outside the Iron Curtain whose commitment to the West seems shaky or nonexistent. Propaganda directed to them can be more purposeful—because less lamed by a nagging sense of futility and much better able to be interrelated with other concrete instruments of Western policy, notably economic aid programs—than is possible in the Soviet orbit. The use of the United Nations as a sounding board for these efforts at persuasion of underdeveloped countries has become particularly important to American policymakers, and with it comes an awareness of the greater need for relevant American deeds that are associated with the UN in such a way at least as to maintain the listening audience.

The growing understanding that the shouts of propagandists—however righteous and however much amplified—cannot be heard in a vacuum of concrete actions, and that information is usually most persuasive when it concerns a positive program like the Marshall Plan in Europe or the "Alliance for Progress" in Latin America, can lead further to an insistence that prospective policies be judged by the decision-makers partly in terms of their probable "propaganda effect." Projecting this attitude further through the notion of "propaganda of the deed," the publicity impact may even be urged as the principal purpose of an action. And at the extreme are the "symbolic acts" that are performed solely as gestures of persuasion. Along this spectrum, sophisticated practitioners are learning to find many varied combinations of words and deeds that are appropriate for particular purposes. The heightened sensitivity to the wide range of possibilities is generally beneficial. But toward the extreme, some danger may exist that "symbolic acts" may be performed that are as empty as mere verbalization, and subject to the same pitfalls. Deeds and words confirming each other—not words that purport to be deeds or deeds that amount to mere words—are the elements crucially required for the effective conduct of contemporary American foreign relations.

## SELECTED BIBLIOGRAPHY

AMERICAN ASSEMBLY. *The Representation of the United States Abroad.* New York: Columbia University, 1956.

BARGHOORN, FREDERICK C. *The Soviet Cultural Offensive: The Role of Cultural Diplomacy in Soviet Foreign Policy.* Princeton: Princeton University Press, 1960.

BEAL, JOHN ROBINSON. *John Foster Dulles: 1888–1959.* New York: Harper, 1959.

BRZEZINSKI, ZBIGNIEW K. *The Soviet Bloc: Unity and Conflict.* Cambridge: Harvard University Press, 1960.

DAUGHERTY, WILLIAM E., and MORRIS JANOWITZ. *A Psychological Warfare Casebook.* Baltimore: Johns Hopkins Press, 1958.

DIZARD, WILSON P. *The Strategy of Truth: The Story of the U.S. Information Service.* Washington, D.C.: Public Affairs Press, 1961.

DYER, MURRAY. *The Weapon on the Wall: Rethinking Psychological Warfare.* Baltimore: Johns Hopkins Press, 1959.

HOLT, ROBERT T. *Radio Free Europe.* Minneapolis: University of Minnesota Press, 1958.

———, and ROBERT W. VAN DE VELDE. *Strategic Psychological Operations and American Foreign Policy.* Chicago: University of Chicago Press, 1960.

STEPHENS, OREN. *Facts to a Candid World: America's Overseas Information Program.* Stanford: Stanford University Press, 1955.

UNITED NATIONS GENERAL ASSEMBLY. *Report of the Special Committee on the Problem of Hungary.* New York: United Nations, 1957.

URBAN, GEORGE. *The Nineteen Days: A Broadcaster's Account of the Hungarian Revolution.* London: Heinemann, 1957.

VALI, FERENC A. *Rift and Revolt in Hungary.* Cambridge: Harvard University Press, 1961.

# PART FOUR

# ECONOMIC RELATIONS AND FOREIGN POLICY

# CHAPTER 13

# AMERICA'S INVOLVEMENT IN THE ECONOMIC DISINTEGRATION OF THE EUROPEAN WORLD

Among the impressive deeds that affluent America is best fitted to perform are those that advance the well-being of the noncommunist world in economic terms. Conducting foreign relations, Americans typically feel more at home with the productivity of economic instruments than with the burdens of the military or with the subtleties of international communications.

## America's Predilection for Economic Foreign Relations

Traditionally, Americans have lacked the inhibitions about involvement in international commerce that they have felt about involvement in international power politics (see pp. 18–20). Economic isolationism, even on a continental scale, was never as prevalent in United States public opinion as was military isolationism. The interests of Americans in the world economy have always been important, both objectively and subjectively, to large and influential elements of the population. Typically, these people did not psychologically belong to the world of power politics but they did psychologically belong to the world of commerce. They could even occasionally be drawn into the former via the latter and become involved very deeply even if uncomprehendingly; in the early twentieth century the origins of the American entanglement with China and of United States entry into World War I were deeply rooted, respectively, in the aspirations and the realities of American export trade.

The reverse relationship between commerce and diplomacy—when commerce becomes an instrument for noneconomic foreign policy objectives—has been one to which Americans were traditionally much less accustomed. But the skills, institutions, contacts, and outward orientation many Americans formed while participating in the network of private and semi-public international economic relationships became potentially available for cooperation with compatible United States governmental purposes abroad, if such purposes ever took any definite shape. At least they were more fit for foreign policy purposes than were the highly informal and nonpolitical channels of American cultural and informational contacts abroad, or than was the military readiness of the American public, which was generally unprepared to provide muscle for any very coercive policies a President might have favored overseas.

Traditionally also, Americans became emboldened with immense self-confidence about their country's economic potential. Achievement in this sphere of the nation's life has historically been so stupendous and has mitigated so many other kinds of problems that Americans have naturally been inclined to seek for economic solutions to manifold tensions at home and abroad, often with grossly exaggerated hopes for the social and political efficacy of some particular set of economic arrangements.

Interlocked with this confidence has been the nation's affluence. Wealth has been amassed and continues to be amassed; it can be used for many purposes, with great confidence that there will later be more for other purposes. The affluence probably also contributes to a preference for relying on expending wealth rather than on enduring more arduous forms of personal sacrifice as means for furthering the nation's foreign policy. Military service in particular, even in peacetime, is typically felt to be much more painful than even quite high rates of taxation. If a draft must be endured at all, it must be accompanied by levels of in-service comfort, enlistee pay scales, and veterans' benefits that are unprecedented in the world's history, even though these amenities tend to deplete the military budget and limit the mobility of the armed forces. If there is need to keep some country within the Western sphere of influence, Americans tend to take for granted that it will be less burdensome for them to foot the bill for an aid program sufficient to keep its leaders in line than to maintain a military occupation sufficient to hold it by force. Of course, other considerations are involved in such a policy choice, but if modes of sacrifice were the only concern, a less wealthy power than the United States could not so readily afford to assume that dollars would nearly always be cheaper than its soldiers' lives (and probably easier to supply than seductive propaganda).

But the United States is indeed wealthy, and in economic activity is adept, self-confident, and experienced at home and abroad. In the economic dimension of coexistential competition between East and West these are major assets. How effectively they are and can be actually utilized is the subject of the next seven chapters of this book.

## The Nineteenth-Century World of Economic Liberalism

An understanding of the manner in which Western statesmen since World War I have grappled with international economic problems requires some acquaintance with the particular pattern of world commerce that came to be regarded as natural in the latter half of the nineteenth century, and from which twentieth-century developments have commonly been viewed as deviations, often undesirable.

The nineteenth-century world of economic liberalism took for granted and gave approval to a high degree of freedom from governmental interference for private enterprise, individual and corporate. It was a world that also felt a willingness to endure painful economic readjustments when they occurred in accordance with what were considered to be the automatic and inexorable "laws" of economics. Inflation and depression, urban overcrowding and technological unemployment, labor migration and ruinous competition were ills to be borne, not to be doctored by dubious governmental remedies. Besides, their severity was softened by the tremendous economic expansion typical of the era and by the extraordinary prevalence of peace, both externally between nation-states and internally between social groupings. World wars were avoided, and internecine civil wars were rare. The burden of healing the scars of one war and preparing for the next did not lie as a heavy extra weight upon the unsteady course of economic development.

The process of growth itself could be envisaged as proceeding on an intermeshed worldwide scale; it did not always have to be subject to jealous calculations of the relative advantage of individual nations. In a sense, what seemed to be coming into being was a "world economy," not just an "international economy"; national boundaries would make little more difference to commercial intercourse than do the sectional boundaries within a single nation.

Indeed, the theoretical case for free trade (the removal of tariffs and other barriers to persons, goods, and investment capital) was not made primarily on national grounds. It was cosmopolitan rather than national. The doctrine directed attention to long-run amelioration in an unfenced world economy rather than to short-run distempers in separate national economies. The creation of worldwide abundance was the theme, not a parsimonious allocation of scarcity. Local specialization would improve productive efficiency and bring increased output around the world. Each locality, without much regard for national boundaries, would specialize in the production of those goods and services in which it had the greatest "comparative advantage" (that is, those in which it was least inefficient relative to its other possible products), and would then exchange its surplus for the other products in which other localities were relatively more efficient. This concentration of effort and resources on particular products in each locality would further enhance productivity, and in the world as a whole, total production would grow. By competitive efficiency each locality, each industry, and indeed, each individual would earn a fair share of this increased production. Each country would also earn its share, the country being viewed as a thoroughly entangled aggregation of individual participants in an intermeshed world economy aiming at potential abundance, not at jealous national self-sufficiency. Romanticized, the vision became one of peoples allowing themselves to become so interdependent in their cooperative pursuit of the advantages of world trade that they would no longer find it practicable, much less desirable, to engage in power-political conflicts and wars.

To be sure, such a concept presumed that some countries and regions would have to specialize in agricultural production and raw materials while others did the manufacturing, and it would be unlikely that the raw materials suppliers could defend themselves militarily if the manufacturing powers did choose to use their advanced weapons for conquest. In the long run, the "backward" peoples might be unwilling to rely on the self-restraint of the "advanced" industrial powers, but in the nineteenth century such concerns seemed remote. The advanced white nations did in fact have the global hegemony that could give a firm framework to the worldwide division of labor. In particular, the hegemony was most widely held by Great Britain, which found self-interested advantages in utilizing it rather benignly for this purpose. Britain had taken the lead in industrialization in the early eighteen hundreds, and by the middle of the century had deliberately adopted national policies which would put an end to her self-sufficiency in food as well as other raw materials. In pursuit of the prosperity which came with industrialization, she allowed herself to become absolutely dependent upon world trade for survival. With enlightened self-interest, therefore, her bankers and traders presided solicitously over the central transactions which set the tone for the worldwide promotion of the free-trade cause upon which England's own well-being depended. They largely guided the semiautomatic equilibration whereby international commerce was kept in balance.

For example, consider the following simplified sequence of relationships: (1) Because prices of Swedish exports become less attractive than previously, Sweden has trouble earning enough to pay for imports; technically there is said to be a temporary disequilibrium in the "balance of payments" between Sweden and other countries. (2) British bankers indicate their willingness to make a short-term loan of their pounds sterling—interchangeable with gold and acceptable at full value almost anywhere in the world—if the Swedish interest rate is raised a little; the loan helps the Swedes to pay for imports. (3) But the increased interest rate in Sweden causes a contraction of credit internally; Swedes at home borrow less money with which to make purchases from other Swedes. (4) They purchase fewer goods; sales decline; some unemployment may occur; prices fall in Sweden. (5) Therefore, foreigners find the prices of Swedish exports attractive once more and resume purchases; also they have less incentive than recently to try to sell their own goods in Sweden in competition with the reduced Swedish prices. Equilibrium returns in the Swedish balance of payments.

Suppose no British loan is sought—or made. Sweden, under nineteenth-century conditions, would have to ship out some gold (or the equivalent) to pay for the "extra" imports. When the supply of gold in Sweden is thus reduced, Swedish banks have smaller reserves on which to make loans to Swedish borrowers. Consequently, in the absence of other influences, the interest rate in Sweden has to rise. Steps 3, 4, and 5 follow just as in the previous case. The equilibration has taken place apparently automatically.

In practice, of course, the process would not be so neatly isolated as in the above illustrations. Actually, the interpenetration of numerous national economies was extremely subtle and complex, coming in the pre-World War I period to resemble more and more the manifold interacting credit adjustments within a single country. The "foreign" factors were increasingly indistinguishable from the domestic. Trade and investment across national frontiers became comparable to intra-national business in its freedom from governmental interference, its guaranteed safety and predictability, and its spontaneous flexibility in adjusting to change. Reliance throughout was essentially upon private institutions to make the decisions.

## The Barricading of the World Economy

For good and ill this trading system is finished. It has been killed—subject only to partial resurrection—by the interaction of four major elements in the life of the twentieth century: democratization, national self-determination, war, and depression. Each of these forces has powerfully promoted the other three in important instances since 1900. Together they produced a prevailing attitude after World War I that the workings of market forces, as free of government control as they had been in the late nineteenth century, were simply intolerable. Neither the business cycle internally nor the pressure of gold shipments externally could be allowed to perform its painful purgative function at the expense of large segments of the population in particular nations. The new path was clear, and between the great wars nearly every country embarked upon it. (1) Its government increasingly intervened in its internal economy. (2) Interference with imports and exports was necessary, if only to shelter the internal experimentation from the effects of foreign competition. (3) This interference drove other countries to impose counterrestrictions in "economic self-defense"; and in turn counter-counterrestrictions emerged in each country. (4) The quest for autarky, economic self-sufficiency, spread contagiously as each nation sought to immunize itself against other shocks. The essence of the process was a revolt against interdependence. One might contend that the countries were all in the same boat and hence should not rock it, but if the boat appeared to be sinking, as it did after 1929, the ready answer was that the time had come for each country to get out of it fast and struggle independently in a sink-or-swim ordeal. The scars of World War I and the dread of World War II further stimulated the drive of governments to control and plan the allocation of their own people's resources, with a minimum of suspect interference from outside forces of the world economy.

The imaginations of economists have been fertile in devising numerous new techniques for manipulating international commerce. The simplest are restrictions on movements of persons across international boundaries. The requirement of passports and the imposition of limits on immigration became nearly universal after World War I. Migration barriers were refined with various degrees of discrimination on grounds of national origin, health, literacy, occupation, political associations, purpose of travel, and so on.

A second category of governmental intervention takes the form of subsidies to domestic producers who face competition from imports and to exporters who are seeking markets abroad.

The tariff itself is a third technique, an old one of course, but subject to elaboration. The "most favored nation" principle, for which the United States has long taken a stand, prescribes that any tariff concession granted to one country (the "most favored nation") must be granted on a nondiscriminatory basis to every other country that is party to a treaty to that effect. Some virtual discrimination can be arranged through crafty, narrow definition of the product on which a tariff is being lowered. But in the less inhibited spirit of twentieth-century trade restrictionism, there has been a tendency to resort to the forthright imposition of preferential tariffs to discriminate among outsiders and adjustable tariffs to facilitate manipulation of the preferences, abruptly and flexibly.

The tariff constitutes a kind of price control device, but one which is not sufficiently direct and precise in its quantitative restrictiveness to satisfy all advanced protectionists. They have often preferred a more detailed technique: specific quotas on the amounts of particular goods which will be permitted to be imported or exported. These import and export quotas, which are to be enforced by government licensing of individual shipments, are particularly useful in enabling planners to determine exactly how much of any particular commodity will be allowed to come in or go out, without upsetting price relationships as much as tariff fluctuations would do. If discrimination between countries as sources of the imports or markets for the exports is desired in order also to secure advantages in the international balance of payments, the government can also undertake to manipulate the rates at which its own currency may be exchanged for those of other countries. The foreign exchange rates may be "fixed" at particular levels and held there until a new decision is made to revalue one's own currency upward or, more commonly, downward ("devaluation"); or the rates may be allowed to fluctuate up and down ("float") between set limits; or a system of "multiple exchange rates" may be established such that the permissible rates will differ depending on the purpose for which the currency is being sought by a particular trader. These various patterns of exchange manipulation can be maintained partly by the government's simply buying and selling the currencies of other countries. But the multiple rates—and often the fixed or floating rates—cannot be maintained without direct exchange controls.

Exchange controls allow only a sharply limited convertibility of currencies one for another. One typical form of control is the following: The government requires that all exports be licensed; foreign currencies earned by exporters are not then allowed to be used without approval of the government. The government issues licenses permitting the use of this foreign exchange for the purchase of particular approved imports or for other authorized payments abroad. Obviously this system facilitates highly flexible, detailed control of imports, travel, and so on. Another use of exchange controls may be to tell a foreigner who has deposits in your banks that he will not be free to make withdrawals at his discretion or to change the assets into the currency of some third country and spend it there, but will instead have to accept payment in a form that is satisfactory to your officials. (This is a system of "blocked balances.")

These systems of exchange control require a high degree of administrative discretion in the hands of the particular officials who issue the licenses. It is they who decide in marginal cases what shall be the precise purposes for which currency may be converted and who shall be the favored persons allowed a share of the foreign exchange that is to be made available for a particular purpose. Obviously the opportunities for arbitrariness, graft, bribery, and evasion are enormous in such efforts to regulate the huge range of separate transactions in the international commerce of a modern state. The precision of the administrative tools may only worsen the trade distortion unless they are in the hands of officials who show prudence and probity. Therefore, in countries where the bureaucracy is unreliable, even staunch protectionists may see advantages in contenting themselves with the more impersonal nonquantitative kinds of barriers, notably tariffs.

A further extension of twentieth-century trade restrictions appears in the various manifestations of "bilateralism." One, known as "bilateral clearing," is an arrangement whereby total payments back and forth between two particular countries must be made to balance directly, or be within a fixed limit of imbalance, every so many months (as agreed), irrespective of the trade of either of them with other countries. The result is a narrow, security-minded channeling of commerce and a limit on wider potentialities of growth, if they exist. But a government which sees advantages in keeping very tight control in its own hands can find ways to go even further in dictating the terms of bargains between its own businessmen and the traders in any particular foreign nation. The government may itself take over the purchase and sale of commodities, thereby by-passing the private businessmen through "state trading." It may use this power for "bulk purchasing" (or bulk sales), thus exerting concentrated bargaining pressure to win advantageous deals in international markets. At the ultimate extreme, governments may resort to the barter of specific quantities of specific commodities by negotiated agreements, which can circumvent the complications of international finance but cannot suffice to arrange any very diverse and large volume of international trade. The tendency of such detailed regulation is to constrict the amount of trade that is administratively feasible to conduct at all.

Clearly the arsenal of modern weapons for pursuing national economic independence is richly varied: immigration restrictions, subsidies, tariffs, quotas, exchange controls, bilateral state trading—all in manifold combinations, and all available for use offensively, defensively, or in retaliation. Many of the procedures were improvised for economic warfare during World War I, but during the Great Depression most countries learned to adapt them for the peacetime demands of economic recovery, military revival, and international political pressure. Nazi Germany in the thirties may have shown more ingenuity and ruthlessness in manipulating the new techniques than did other capitalist powers; but the needs of self-defense induced imitation, and thereafter the experience of another world war made government control of trade in a spirit of national rivalry seem more natural to many participants than the by-gone system of nineteenth-century multilateralism with its many-sided free exchanges of currencies, goods, and services, revolving among numerous nations.

## Planning during World War II for Postwar Commerce among Nations

But jealous control did not seem more natural than multilateralism to the fathers of the first major set of institutions that emerged during World War II in the planning of the Western Allies for the postwar world economy. These institutions were those developed in the prolonged negotiations (mainly Anglo-American) climaxed by the international conference at Bretton Woods, New Hampshire, in mid-1944. The spirit of Bretton Woods was directly parallel to the contemporaneous spirit of exaggerated optimism that surrounded the formation of the United Nations Organization. In sharp reaction to the autarkic economic belligerency which had embittered and probably prolonged the depression of the 1930's, the framers of the charter for the new "International Monetary Fund" looked forward at the Bretton Woods meeting with considerable confidence to a postwar swing away from trade restrictionism toward a revived multilateralism. The feeling, however, was that the new multilateralism would have to be somewhat more consciously organized and internationally planned than that of the generation before World War I, even if due note were given to the quiet supervision British bankers had deliberately exercised over the "automatic" equilibration of that era. An International Monetary Fund was therefore to be established to operate a central pool of the currencies of participating nations, each of which would contribute its fixed share; these foreign currencies would be made available to any one of these nations that needed them to help tide it over any relatively minor, temporary adverse swings in its balance of payments with the other countries. When the equilibrium was restored, the funds would be returned. Meanwhile (after a postwar transitional period), the country suffering from the shortage of foreign exchange was not to be allowed to use exchange controls to "correct" the imbalance, unless the Fund organization declared the existence of a general shortage of a particular currency, dollars, for example, an international scarcity too widespread to be met by the resources of the Fund. Nor would a participating country be allowed to revalue its currency except after consulting the Fund, which would grant approval only in specified circumstances designed to prevent a competitive depreciation of currencies, but to permit the correction of fundamental disequilibria by orderly revaluations.

Unless the USSR should choose to join this basically capitalist trading system (it did not), the United States was assured by the voting arrangements of the preponderant voice in Fund decisions. But the United States could not make it work in the manner intended by the framers unless the economic authorities in the other major trading nations did in fact prove willing to put the creation and preservation of freedom of international payments ahead of local cares like full employment and special group interests. The Fund agreements did contain substantial concessions to twentieth-century political-economic realism in such matters. There was latitude for the postwar "transition period"; the "scarce-currency clause" was another escape hatch from the general ban on exchange controls; devaluation was not absolutely prohibited; indeed, the very existence of the currency pool itself showed awareness of the inadequacy of traditional mechanisms of international finance. This system was not simply an effort to recreate the gold-based multilateralism of 1906. If it were, it would surely have been stillborn in the world of 1946. The exceptions actually proved to be broad enough to keep the organization alive in the postwar decade, albeit almost inactive, until, after 1955, the revived economies of the Western world managed at last to make substantial use of the Fund. Nevertheless, the fact that the institution did survive and has finally found an international environment in which it can function usefully should not disarm criticism, for essentially its mechanisms had very little relevance to the world in which they came into being in 1946. And the effort which had gone into its creation during the war tended (as in the case of the UN itself) to raise hopes and distract attention from the stupendous problems that were actually emerging from the turmoil of the greatest war in history.

To be sure, some of the early British and American versions of the Fund plan had envisaged the use of part of its resources for the really pressing requirements of postwar reconstruction. But the negotiators soon whittled these "middle-range" purposes away from the "long-range" functions for which they were designing the Fund, evidently without corresponding awareness on their part of how prolonged the transitional period would then be before their institution could perform any important function at all.

The Bretton Woods planners did make a somewhat more substantial gesture in the direction of postwar reconstruction problems with their other new creation, the International Bank for Reconstruction and Development (commonly known as the World Bank). Here again, however, the trend in negotiations ran toward the very long range, though not quite as remote as the plans for the Fund. The "reconstruction" functions were not wholly eliminated. But the loanable funds which member nations were to subscribe were much too small, and the terms on which loans could be made were much too conservative, to satisfy the politically effective demands of postwar Europe. The Bank did manage to find a useful function for itself sooner than did the Fund, but it has been a modest function. By guarding its reputation for safe and profitable lending, it has been able to attract private capital for the expansion of its operations and to elicit further participation by private investors in the loans it arranges, and its international economic advisory work in the underdeveloped countries has been valuable. Together, these techniques help to generate larger flows of private investment in and to these countries. But it has become obvious that the underdeveloped peoples, like those of Western Europe in the immediate postwar years, are too impatient for economic improvement to await the remote fruition of these modest arrangements alone.

Just as the Bretton Woods institutions (Bank and Fund) were too long range to fulfill middle-range needs, so also the wartime economic planning allowed the middle range to be whittled away from the work of the main organization that was being established to deal with the immediate short-range hardships of the war-torn world. This was the United Nations Relief and Rehabilitation Administration. UNRRA was developed in the years 1942–44 with a decided emphasis on mere "relief," especially food supplies; clothing and fuel also received attention, and so did some forms of shelter. But no extensive heavy reconstruction came within the purview of the agency. Nor did even relief for all the Allied countries that had suffered war damage, for a policy was set in 1943 not to grant UNRRA aid to countries possessing substantial foreign assets of their own; they would be expected to pay. As a consequence, most UNRRA aid went to Eastern Europe, including Russia. Why didn't the Western European countries offer strong resistance to this policy? Partly because their governments-in-exile at the time of decision (notably the French) failed to anticipate fully the severity of their postwar needs. (Most of the ultimate damage of the war in the West had not yet been inflicted.) Also they felt they could rely very largely on surplus supplies that would be brought along by the armies in the course of liberation operations; in effect, these goods would presumably be free to the recipients. Furthermore, there seemed good reason to hope that Lend-Lease aid would be continued in the immediate postwar period—or at least in the interim period, which was expected to be of many months' duration, between Hitler's defeat and the conquest of Japan. In this interval, many civilian supplies and military occupation expenses of the Allies could be tagged as Lend-Lease if the American Administration cooperated in this accounting device. The British government in particular relied on this possibility for an important contribution to England's postwar recovery. Finally, most of the countries concerned preferred to deal with the United States directly, since the United States would ultimately put up most of the relief money anyway. Their chances of making an advantageous deal seemed somewhat better if their requests did not have to be screened and scrutinized by an organization of their neighbors who were rival claimants.

Reconstruction assistance to Western Europe was thus to be essentially bilateral. The United States Administration proceeded to equip itself accordingly. Any notion of using the device of "tapered" Lend-Lease for this purpose was soon doomed by the rapid end of the war and by the reluctance of key American officials to attempt to circumvent congressional objections. But authority was secured from Congress in mid-1945 to increase the loans of the Export-Import Bank by nearly $3 billion, and a special loan of $3.75 billion to Britain was voted the following year. Consequently, the American government loans to Western and Southern Europe in the immediate postwar years reached the neighborhood of $6.8 billion, but still they did not suffice to meet the political-economic needs of the situation. A great gap remained in the multilateral international economic planning between the very short-range UNRRA and the very long-range Bretton Woods institutions; and the American government, making bilateral loans to help fill that gap, failed in those years to take full advantage of the leverage it possessed to insist that the reconstruction efforts of the various Western European countries be jointly coordinated and directed in a gradual and orderly fashion toward the benefits of extensive international economic exchange, rather than toward parsimonious autarky. There was a failure to appreciate the true depth of the problems of the postwar European economy—a failure compounded of wishful thinking and inadequate information.

## A Palliative: The Initial American Postwar Loans

Probably this blindness was most strikingly evidenced in the treatment of Great Britain. Her well-being had come to mean more to American leaders than that of any other power outside the Western Hemisphere. The United States contributed nearly $27 billion in Lend-Lease supplies to Britain during the course of the war (British "reverse Lend-Lease" to America was valued at $6 billion). Nevertheless, the cost of the war, particularly in the years before Pearl Harbor, had laid an enormous domestic and foreign burden upon Britain's postwar economy. Overseas investments totaling about $4.5 billion (at the then current rate of exchange) had been liquidated to pay for war needs, and another $600 million in gold and dollar currency reserves had been sacrificed for the same purpose. Net shipping losses left the United Kingdom with only two-thirds of her prewar tonnage, thus decreasing her international earning capacity, as did much of the severe bomb damage in the home islands. In addition, other poorer countries—conspicuously India, Egypt, Eire, and some in Latin America—had been induced to make substantial contributions in goods and services to the British war effort in the expectation of being repaid when peace came; London still recognized these obligations, which brought the total of England's external sterling indebtedness in mid-1945 to £3.4 billion. This would have amounted to nearly $14 billion if it had had to be exchanged at the rate then officially prevailing; but the British government was striving to discourage the overseas creditors from insisting on rapid repayment in sterling goods and services (English exports) and from converting the accumulated debt balances into dollars. However, the United Kingdom was extremely reluctant to repudiate or "scale down" the war debts as American officials tended to urge, having allowed a similar arrangement themselves in regard to Lend-Lease. Britain's prestige for international financial reliability (profitable in the long run) would suffer disastrously from such action, and so also would her political ties with India, Egypt, and other countries.

In this situation, England's leading economist, Lord Keynes, came to Washington in the fall of 1945 as head of a mission seeking the $5 billion his government estimated was needed to cover the country's international balance of payments during the next few years, while exports were gradually being expanded by 50 to 75 per cent above the prewar level. If such a dollar total were not forthcoming, the standard of living of the British people would have to be slashed even below the worst wartime levels, and a system of penurious controls would very probably be enforced, which would doom progress toward freer international commerce.

In Washington, the negotiators settled upon a smaller figure, $3.75 billion, all that the Truman Administration believed could be obtained from Congress; but Canada was persuaded to make up the difference. Britain's relative success in acquiring $5 billion was sharply qualified, however, by the American refusal to present it as a grant, or even as an interest-free loan. The money would have to be repaid with interest. But the length of time for repayment (fifty years, not to start until 1951) was liberal, and so also was the interest rate of two per cent, subject even to being waived entirely in any year in which Britain suffered specified kinds of balance-of-payments difficulties. The resulting debt charges (repayment and interest) after 1951 would normally be $140 million a year—manageable, though burdensome, for a country in Britain's precarious postwar position.

More serious was the lack of foresight shown by the American negotiators and British implementers of provisions in the loan agreements that together required the United Kingdom after July 1947 to allow convertibility for all sterling that persons who did not reside there earned from current transactions, and also to allow convertibility for whatever part of the war-swollen sterling balances Britain might subsequently release for her creditors to use. Both the American and the British officials, in different ways, overestimated the United Kingdom's capacity to return rapidly to nondiscriminatory multilateral commerce. The Americans were encouraged in their overoptimism by an illusion derived from vague English promises that Britain would be willing and able to extract enforceable guarantees from her creditors that the accumulated balances from the past would not be converted into dollars, indirectly and surreptitiously as "current account" (newly earned) sterling—and, furthermore, that Britain would also induce the creditors to accept complete cancellation of a large part of the war debts and long-term postponement of the release of most of the remainder. The English, however, having yielded skeptically to American urging to embark at all upon the path of quick convertibility, went further in 1946 and early 1947 to take extreme chances in moving toward an even looser system of controls than the United States had envisaged. Washington then failed to insist on sufficient continuing consultation with the British authorities as they proceeded to promote extravagant improvements in the domestic standard of living and negotiated very liberal agreements with creditor governments, relying heavily on good will and London's prestige to avoid harmful "runs on the bank." Even if the Labor government in Britain, uncurbed, had felt willing and able to make the preparations the Americans expected for the restoration of sterling convertibility, the July 1947 deadline would probably have been too early; the Americans, too, were underestimating the international difficulties. But the United States willingness to rely on the formal legal commitment rather than on continuous oversight and consultation contributed to Britain's failure even to attempt seriously the frugal policies without which the convertibility deadline was sure to be only a signal for an economic crisis.

The result was that several hundred million dollars of the American loan drained swiftly away from Britain to other countries when sterling became convertible (supposedly only for current transactions) in the summer of 1947. Indeed, had convertibility been maintained in the form then prevailing, which in effect permitted the war debts to be converted into dollars, the American loan, scheduled to last until 1951, would have disappeared altogether before the fall of 1947.

A halt was called, sharply and drastically. The American government was driven to acquiesce in an indefinite general suspension of sterling convertibility, not merely a rigid restriction of the convertibility to current transactions. British opinion was shifting severely against any effort to achieve nondiscriminatory multilateralism in the near future. In defiance of the Anglo-American Loan Agreement, the English made intensified efforts to induce countries with close political and economic ties to Britain (notably those in the "Sterling Area" that made a regular practice of banking their foreign currency earnings in London) to limit their purchases from dollar countries so that they would not have to make irreplaceable withdrawals from London's fluctuating pool of dollars. This discriminatory retrenchment was largely successful. At home in Britain, overdue steps were taken to check the artificial domestic boom and force the public to live more nearly within the bounds of the country's war-straitened external capacity. The lax overoptimism that had largely characterized both the American and the British handling of the loan was supplanted finally by a new realism concerning the fundamental weakness of England's position.

The loan agreement, by demanding multilateralism without insuring adequate preparations for it, had contributed to a debacle which tended to cast new doubt even upon the value of the objective. In the end, the most that could be said for the arrangement was that it had postponed a crisis, giving the British people temporary relief from wartime hardships, and that the final breakdown was focused dramatically for belated harsh enlightenment of responsible men on both sides of the Atlantic.

The loans that were made to other European countries in the same period were also unsuccessful in accomplishing more than a postponment of the dollar-payments crisis until 1947. About $2 billion in Export-Import Bank credits went to Western Europe, almost all of it to France and the Low Countries, and nearly $3 billion was contributed by the United States to UNRRA, mainly to supply Italy and Eastern Europe. Heavy relief expenditures were made in Germany. But in all there was a continuing failure to grapple with essentials and to do so through cooperation among the Western European nations. Particularly on the Continent, conditions in the Western countries in 1946–47 were a dismal compound of low production, swift inflation, and severe balance-of-payments difficulties.

Production was low, most obviously, because equipment had been destroyed or grown worn and obsolete during seven years of war; investment funds were unavailable for any rapid reconstruction. Workers, also, had been massively dislocated from normal paths of peacetime employment; their skills were rusty or unformed. They were undernourished because crops were poor and because farmers had little incentive to bring food to urban markets when there were so few manufactured goods to buy there. Other basic raw materials were very scarce, notably coal (a result in turn of the extreme paralysis that gripped the German Ruhr as the Big Four struggled vainly to agree on occupation policy).

Physical shortages contributed to severe inflation which in turn discouraged potential investors. France particularly failed to master inflation; her price level rose about 10 per cent a month, almost tripling in two years, 1946–47. There and in Italy the crippling Communist-led general strikes of late 1947 further aggravated the calamities.

Inflation and low domestic production put extreme strain upon Europe's balance of payments. Many more goods than before the war could be obtained in adequate quantities only from Western Hemisphere sources, which required payment in dollars, whereas Europe's capacity to produce dollar-earning exports was low, at least temporarily. The immediate problem was not only a "dollar shortage" but a generalized pattern of "bilateral clearing." Each country's government felt it had to be sure that its trading accounts balanced fully at short intervals with each other country; one could not afford, it seemed, to waste precious export goods in sales for credit, if that were at all avoidable; conversely, a potential debtor country could not allow itself to get in a position where it would be obliged to pay out its precious gold or dollars to settle trading accounts with another European power. Within Europe itself, therefore, a squeezing for scarce currencies through bilateral trade deals narrowed the stifling constrictiveness of the impoverished trade.

The United States aid arrangements—UNRRA, loans, and occupation expenditures—provided a palliative for the dollar shortage, but failed to require that reconstruction efforts in Europe be collaborative, permitting the establishment of an orderly, gradually widening freedom of international exchanges. Instead, the American reconstruction-assistance program at the end of World War II amounted to setting optimistic deadlines for the attainment of full nondiscriminatory multilateralism (for current transactions) and to feeding dollar aid (probably in insufficient amounts anyway) to individual needy governments, relying on them spontaneously to use the dollars to promote the multilateralism. In Europe's impoverished condition, the contrary immediate political pressures and short-run economic incentives actually made it highly unlikely that governments would adhere to nondiscriminatory trade policies unless new and continuing incentives for a more practicable trading system were forthcoming, probably from the United States.

### SELECTED BIBLIOGRAPHY

GARDNER, RICHARD N. *Sterling-Dollar Diplomacy: Anglo-American Collaboration in the Reconstruction of Multilateral Trade.* New York: Oxford University Press, 1956.

MYRDAL, GUNNAR. *An International Economy: Problems and Prospects.* New York: Harper, 1956.

PENROSE, E. F. *Economic Planning for Peace.* Princeton: Princeton University Press, 1953.

SCHELLING, THOMAS C. *International Economics.* Boston: Allyn and Bacon, 1958.

WOODBRIDGE, GEORGE. *UNRRA: The History of the United Nations Relief and Rehabilitation Administration,* 3 vols. New York: Columbia University Press, 1950.

CHAPTER 14

# THE ECONOMIC REVITALIZATION OF NONCOMMUNIST EUROPE

As GREAT BRITAIN began gradually in the fall and winter of 1946–47 to confront the fundamental seriousness of her situation, her first resulting move of direct and major importance to the United States was her decision to cut the costs of shoring up Greece and Turkey. This choice, with the consequent highly dramatized transfer of the burden from London to Washington, had an important effect: it virtually insured that any re-examination of Europe's broader economic problems by America would have a sharp anticommunist focus. Most of the problems would actually have existed anyway in the aftermath of major war, even if no Communist pressures had developed to exacerbate them. So also would America's economic and humanitarian interest in finding and contributing to real solutions, not mere continual palliatives, but such considerations would have been hard to dramatize to an American public which was growing tired of "hand-outs." Thanks largely to Britain's policy on Greece and Turkey, the Truman Doctrine in the United States preceded the Marshall Plan. Americans came to view their country's political-security interest as the primary justification for any further large-scale assistance that might be provided for Europe's economic recovery. On the whole, this emphasis was a valuable aid to realistic thinking about the whole problem in the United States, though there was later some tendency to go to extremes and forget the economic and humanitarian purposes also served by the massive American program that emerged to meet Europe's critical situation in 1947.

### The Marshall Plan

The plan associated with Secretary of State Marshall's name was, when he announced it on June 5, 1947, little more than an idea, a plan for a plan. The idea was that the United States would listen favorably to any joint collaborative initiative which might come from the European countries for "a program designed to place Europe on its feet economically," "a cure rather than a mere palliative." The emphasis thus was to be put upon planning, completeness, and collaboration. A calculated risk was taken in inviting Russian participation; the onus for dividing Europe would not be on the United States. Moreover, there was little chance anyway that the proud and secretive Kremlin would join in a scheme requiring extensive intercommunication of economic policy information and mutual criticism among the member states. The Western European countries, in contrast, leapt to embrace the opportunity Marshall offered. Within three weeks Russia had looked in and walked out; within six weeks sixteen nations had formed a Committee for European Economic Cooperation (CEEC); within two and a half months after Marshall's first bid the committee agreed upon a formal report to Washington, including a tentative estimate of $19.6 billion as the net aid requirement of all the member states during the next four years. (CEEC had managed to scale down an initial $29 billion compilation of aid requests from the various countries.) The American government was meanwhile conducting several intensive surveys of its own regarding aspects of the massive program; the studies included congressional and quasi-private surveys as well as bureaucratic analyses. Ultimately the Truman Administration presented Congress with an estimate that $17 billion over a four-year period would suffice to free Western Europe as a whole from the need for any further extraordinary economic assistance. This concept of "getting it over with," at a fixed price, was highly appealing to the American public, especially when the French strikes, the Czech coup, and a tense Italian election campaign, all occurring in the fall and winter of 1947–48, dramatized how great the Communist danger would be if Europe's economy were allowed to continue to totter.

There was some considerable risk of "overselling" the recovery program to the American public. The four-year time span might well prove too optimistic, and disillusioned congressional economizing might set in before a self-sustaining recovery was actually achieved in Europe. That outcome was avoided in the end because very substantial revival did come (although only $13.6 billion in economic aid was finally provided during the four-year period), but also because the Korean war aroused defense worries in America that permitted a continuation in Europe of other United States programs with important economic benefits, though ordinarily the economic objective was officially secondary to the military in these arrangements. The stationing of well-paid American troops in Europe and the partial provisioning of local defense forces there have meant that most of the countries have never since World War II had to get along entirely on ordinary commercial earnings to pay for their imports and rebuild their gold reserves. Yet conditions did improve so much that at least by 1960 nearly all of Western Europe would have been capable of doing so, without politically dangerous cuts in the standard of living, in the unlikely event that military strategic considerations eventually induced the United States to abandon the expensive effort to support the lagging preparations for local defense on the Continent.*

## The European Recovery Program in Operation

The Marshall Plan succeeded so well largely because it was not only adequately financed, but was administered with exceptional competence and enthusiasm. The objectives seemed clear-cut and important, even exciting, and the time limit practicable, indeed convenient. The enterprise was one into which private Americans could throw themselves with a promising sense of accomplishment. The Economic Cooperation Administration (ECA) attracted the temporary services of a galaxy of able businessmen and economists from outside the United States government, who contributed special skills and remarkable drive; the agency's morale was high. Professional diplomats were still able to insure the necessary degree of foreign policy guidance; the teams usually worked together effectively, with a friction that was more often stimulating than frustrating.

As a technique for eliciting productive effort from the Europeans, the idea of the Marshall Plan was also beneficial. The request for them

* On the military aspects of this problem, see Chapter 10.

to exercise joint initiative had an advantageous appeal to the self-respect and dignity of the impoverished nations. Implicit too in the arrangement was an American determination to foster prosperity through freer trade and wider markets within Europe, even if such expansion could only be achieved gradually by permitting prolonged extension of some forms of commercial restriction, including discrimination against purchases of dollar goods. This far-sighted and unselfish American policy became conspicuous and persistent, though often understandably somewhat grudging, as the years passed. At first the major novelty of the Marshall Plan was its insistence, as a minimum, that there be a joint confrontation of Western Europe's economic problems. The CEEC of 1947 became semi-permanently established as OEEC (the Organization for European Economic Cooperation) in 1948, and it was given by the American government the responsibility of recommending how the total Marshall aid should be allocated among the member countries during the first two years. Since OEEC operated under a unanimity rule, such a vital decision was very difficult to reach. But the pressure to do so helped to induce each nation to share long-secret information about its economy and even about particular industries. In an annual joint review each country's programs and projected aid requirements were formally scrutinized by all the other participating governments; informally, continuous consultation ensued, international collaboration among specialists was fostered, and most of the governments learned to improve their statistical analyses for the understanding of their own national economies in the process of preparing to justify themselves to the others.

United States officials, on the scene in each country, at OEEC headquarters in Paris, and in Washington, took an active part in these discussions. They retained the final say as to allocation of funds, but they worked in a spirit of collaboration, seeking to elicit useful recommendations from OEEC. Generally they followed the recommendations on broad allocations, but they retained direct control of special "incentive" funds to promote some particular programs. In each separate country, they also shared with its government joint control over "counterpart" funds. "Counterpart" was local currency equal in value, at official exchange rates, to the amount of dollar aid which the country received; the European government was required to deposit this sum of local currency in a special "counterpart fund" when the aid was received; the government could recoup the local currency by selling the aid goods to its own citizens, thereby incidentally withdrawing local currency from circulation and thus checking inflation. If the fund were later expended for special purposes (for example, road-building, at no cost to taxpayers, some of whom would thus be able to keep their money for new investment in industry), the approval of ECA representatives would have to be obtained. The government could probably still find (or manufacture) other money for disapproved purposes; but at least the very existence of the counterpart occasioned regular consultations with its American supervisors, and the arrangement furnished some inducement for European governments to pursue policies favoring heavy productive investment, public and private, at somewhat less cost in inflation than would otherwise have been probable. The ECA representatives in each country also engaged in direct "end-use checking" at factories and farms to evaluate physical progress and offer specific advice.

The economic supervision, of course, was bound to involve a degree of political influence also, but exactly how much varied from country to country. Apart from the special case of occupied Germany, a declining scale of American interference could be traced, for instance, ranging downward from Greece, to Italy, to France, to Britain.

The economic objectives the ECA was seeking to achieve with these various instruments were basically threefold: to check inflation, to expand production, and to increase intra-European trade. The three were interrelated.

Inflation needed to be checked largely in order to promote production by putting an end to hoarding of goods and by reducing the felt need of workers to engage in strikes or slow-downs for wage increases with which to keep abreast of the cost of living. A stabler price level was also needed to permit predictability in economic activity and thus to encourage investment in long-run productive enterprise and in complex international trading operations. ECA's particular anti-inflation instruments were (1) American goods (to relieve shortages), (2) counterpart, and (3) admonition. By the end of the Marshall Plan period the extreme inflationary pressures had been brought successfully under control in nearly all the OEEC countries.

Direct efforts to expand production involved a twofold emphasis upon heavy investment programs and upon improved efficiency of human work. The investment was to be planned by government officials and private businessmen acting together, and ECA again had goods, counterpart, and admonition to contribute. On productive efficiency, the Americans were confident that they had something special to bestow from their own national experience. They saw the problem as involving modernization of technology and technical training in Western Europe, but also as requiring improvements in the competence and training of the business managers—changes which would be resisted—and above all an alteration in the general attitudes and incentives of both management and labor toward a new emphasis upon productivity and competitiveness. ECA grappled boldly with these fundamental problems. True recovery would require reform, and for this purpose the agency established illustrative local pilot projects. It organized seminars and training programs for European managers and noncommunist labor leaders, supplying them with American lecturers who possessed roughly appropriate experience. It arranged for "productivity teams" of managers and labor leaders from Western Europe to tour comparable factories in the United States, and for American teams from these fields of endeavor to go to Europe to spread the word. Much of this educational effort by the United States seemed frustrating and even presumptuous, but it left its mark. It deserves some of the credit for the renovation of Western European industry and commerce, which has surged ahead in the years since the Marshall Plan ended.

In the drive for improved productivity in Europe, the American government also placed emphasis upon the desirability of expanding the international markets for goods so that the economies of mass production could be utilized more advantageously in export trade. Up to a point, also, the stimulus of competition from imports would be valuable in each country in helping to break down rigidities in local business practices. ECA was willing to concede that the American economy—already continental in size and undamaged by war—might for some time have such great competitive advantages that if Europeans generally removed their dollar-import restrictions and relied on trying to undersell America at home by price cuts and deflation, the result would be an unacceptably severe depression in many parts of Europe with an underutilization of local productive capacity, which would stall recovery. But ECA was much less ready to accept the argument that the European countries could not afford to compete with each other in their own local markets. On the contrary, a gradual approach to unrestricted trade throughout Western Europe was seen by the Americans as an essential stimulant to long-term economic and political vitality in the area. Consequently, especially after 1949, United States policy became emphatically directed toward fostering an orderly expansion of intra-European trade. The American government demonstrated a willingness to concentrate on promoting this form of regionalism while postponing indefinitely the old Bretton Woods-British loan pursuit of universal multilateralism.

Through OEEC there was in the early years of the Marshall Plan a progressive development of agreed arrangements for freer trade within Western Europe, but the decisive step came in 1950 when the American government reinforced its prodding by insisting that $350 million of the new annual aid installment be utilized as a special central fund for a European Payments Union (EPU), to be linked with an accelerated OEEC trade-liberalization program for reduction of quantitative restrictions on intra-European commerce. EPU worked this way: Once each month, each member nation's central bank reported to EPU agents how much it owed to each other member-nation's central bank, as the outcome of all the individual trade transactions of the month. The EPU agents then set the surpluses off against the deficits and announced how much, on balance, each country owed to all the others put together (or how much it was owed by them). Each country had a quota, which specified how much credit it would have to extend through EPU to debtor members if its own net position for the month were one of surplus, and how much credit it would be entitled to receive through EPU if it were running a current trade deficit. A debtor nation, therefore, would ordinarily receive some credit, but if its deficit were large it would have to pay hard cash—gold or dollars—to EPU to cover at least part of its debt; correspondingly, a creditor nation would ordinarily have to extend some credit, but if its credit outlays were large it would be able to collect gold and dollars from EPU for part of what it was owed. In order to bring pressure on members to correct persistent imbalances in their trade, the payment system was graduated, steadily increasing the required hard-currency component of the monthly settlements. And also over the course of years, the EPU agreement was renegotiated further to "harden" the terms of settlement, until by mid-1955 at least 75 per cent of a country's deficit would normally have to be met in gold and dollars, only 25 per cent by credits received. Escape clauses were always provided, however; exceptions were sometimes made; and the ultimate sanction for good behavior was the desire of each country to belong, in order to secure the advantages of relatively nondiscriminatory trade, and the desire of the other countries to make it possible for even the weaker members to continue to belong without temporarily suffering excessive hardship. The sense of common standards and traditions and extensive common purposes thus constituted a basic foundation for the system; the institutional arrangement (including continuous consultations in OEEC) provided special deterrents and incentives for proceeding to reconstruct upon that foundation a booming pattern of commercial intercourse.

The result by 1955 was that under EPU and OEEC trade liberalization a very large percentage of all trade between OEEC countries was entirely freed from quota restrictions. Furthermore, a country that retained quotas—or after consulting with OEEC temporarily reimposed them—was unable to be very discriminatory in applying them to other particular countries within OEEC. And free "transferability" of currencies among OEEC members was becoming common; "transferable" currencies could be converted into each other. A citizen of one country earning money in a second country could spend the equivalent in a third. In some cases virtual convertibility into dollars was also possible. The resulting intensification of multilateral trade competition gave marked stimulus to productive efficiency.

The culmination came late in 1958 when the noncommunist world agreed to increase substantially the pool of currencies available in the International Monetary Fund for balance-of-payments emergencies, and then to arrange that any person who, while not residing in a particular OEEC country, earned money in that country from current transactions would be permitted to change his earnings freely into any other currency, including dollars. Thus the "nonresident current-account convertibility" that Americans had unsuccessfully pressed upon Britain in 1945–47 finally became a reality. Most of the Western European countries now went further, in varying degrees, to allow convertibility to their own residents as well as to foreigners, and to permit it to be exercised in disposing of past accumulations and in making new investments abroad, not just in handling current transactions. These moves implied willingness to place heavy reliance upon the newly enlarged pool of currencies in the International Monetary Fund and upon a new "European Fund," which inherited the resources of the European Payments Union while eliminating EPU's lenient automatic-credits procedure.

The willingness to rely on these institutions to tide over any balance-of-payments emergency meant a corresponding willingness on the part of the Western European countries to implement internal economic policies which would prevent severe, persistent imbalances, even if the result of those policies were hardship, at least temporarily, for substantial segments of their own populations.

The general level of economic well-being in Europe had been raised to a point where one could anticipate that any hardships would not have to be so great as to be politically unbearable. Several years earlier at the formal termination of the Marshall Plan in 1952, industrial production in Western Europe had already risen 35 per cent above the prewar level; agricultural production in 1952 was already 10 per cent above the prewar level, matching an equivalent rise in the population to be fed. Each year of the Marshall Plan had seen an increase in the value of goods produced that was several times as great as the aid the United States was putting in to stimulate the flow. Trade among the OEEC countries had increased several-fold during the period, and extreme inflation had been generally halted. The momentum of growth had continued in the years after 1952.

Such are the economic statistics that give a partial measure of the total success of the four-year European recovery program. Yet the economic statistics cannot tell the whole story. The political consequences of the enterprise are more difficult to establish, but undeniably at the very least a major defensive victory was won. The economic recovery forestalled a spread of social and political demoralization in which extreme authoritarian movements of the Right or Left could thrive. Communism was blocked in Europe without the necessity of any new anti-communist dictatorship there. In France particularly, Marshall aid was probably indispensable for the preservation of constitutional government against the assaults of both De Gaulle and the Communists, at a time when there was much more danger than later that any showdown between the two extremes would have been ruinously violent. The Fourth Republic was enabled to steady itself and achieve remarkable internal progress, until it was finally shattered by colonial problems in 1958. Even then the re-emerging Gaullism was much more moderate than it probably would have had to be to have dominated a France as rebellious and impoverished as that of 1948.

Yet clearly, it is true that despite the successes of the European recovery program authoritarianism of the Right and of the Left have remained powerful in both France and Italy. Communists and fellow-travelers continue to hold the support of at least a fifth of the French electorate and a third of the Italian. The question may be asked whether the rate of economic growth will be great enough for another decade or two to preserve the necessary margin of adherence to free institutions in all the Atlantic democracies, against continuing challenges from Right as well as Left.

### The Economic Integration of Europe

From this perspective the most creditable feature of the European recovery program was that the momentum of growth to which it gave impetus has continued throughout the region, even though the American aid declined sharply in the 1950's and the Europeans' burden of defense expenditures mounted heavily. In particular, note should be taken of the progressive institutionalization of economic unification for Western Europe. Organized patterns of behavior, conducive to increasing trade and productivity, have been acquiring a fixity that defies a ready return to autarky in response to temporary crises. One of the most far-reaching of the new organizations is the European Coal and Steel Community (ECSC), which was proposed by the French Foreign Minister, Robert Schuman, in May 1950 (and therefore tagged the "Schuman Plan"). After appropriate negotiations, ECSC came into operation in 1952, embracing the coal and steel industries of France, West Germany, Italy, and the Low Countries. An international authority, located in Luxembourg, relatively independent of control by the member governments was established to direct a process of integration designed to eliminate all tariffs, quantitative restrictions, subsidies, discriminatory taxes or freight rates, and other barriers to an open market among the six countries, for coal and steel. During a transition period of five years the countries and industries that would enjoy a competitive advantage were to compensate the weaker elements (especially Belgian coal and Italian steel) and assist them to adjust to the new conditions. Gains in productivity were expected and were to be promoted. From a more directly political viewpoint, also, the French would achieve some share in control of the Ruhr, and they were glad to find that under Konrad Adenauer the Germans, chastened by war, were genuinely desirous of fostering the rapprochement between France and Germany.

This urge toward enhanced harmony between the age-old enemies was also an important consideration in the French scheme, approved too by Adenauer's government, to accomplish the rearmament of West Germany, which America was demanding after 1950, through the form of intermeshed units—French and German—locked together in a single army. This plan for a "European Defense Community" (EDC) ultimately failed, very largely because other Frenchmen feared that the bonds might restrict France more than they would check Germany. (See pp. 190–91.)

The failure to achieve military integration was a severe setback in the movement toward unity of "the Six"—France, West Germany, Italy, and the Low Countries. This grouping was often called "Little Europe," especially in a tone of deprecation by those who believed that unification would do more harm than good unless it also included Britain, which, however, stoutly resisted participating in any tight federal arrangement. The British were scarcely more willing to bind themselves to continental Europe than the United States has been, and for many of the same reasons, including a sense of geographic and cultural separateness and unique worldwide responsibilities—basically a sense of superiority to the Continent, such as that felt by many Americans. However, the groups favoring much closer ties among the countries of Western Europe—for the sake of economic progress and reconciliation with Germany—were unwilling to let Britain by nonparticipation put a veto upon the unity the Six could achieve for themselves without her. After the shock of EDC's failure wore off and Franco-German rapprochement managed to resume its remarkable advance, negotiations began for the establishment of a "European common market," which would embrace not just coal and steel, like ECSC, but all products (except perhaps those of agriculture, which in Europe, as in the United States, is surrounded by all sorts of special governmental protections demanded by powerful pressure groups). All Western Europe was invited to join, but at first only the Six proved willing to accept the closely coordinated arrangement that emerged in 1957.

By this treaty the Common Market acquired the more dignified title of "European Economic Community" (EEC), and as in ECSC the supervisory authority was rendered somewhat independent of continuous control by the member governments. Its range of supervision was to be very far-reaching. Over a long transitional period (twelve to fifteen years) all tariffs and quantitative restrictions were to be removed from trade among the EEC countries; the Six would become a customs union, maintaining free trade with each other and a uniform set of tariffs against the rest of the world. Each of the uniform external tariffs would rest about halfway between the highest level maintained at the outset by any one of the Six and the lowest level maintained by any other; but these figures would be subject to years of negotiation, including deals with nonmember countries, which would determine how discriminatory their eventual effects would be against the outside world. Inside the Common Market, restrictions on the movement of investment capital would be reduced while tariffs and quotas were being eliminated, and even labor migration would become free. Various business practices impeding commerce among the Six would be prohibited, and legislation setting standards for working conditions and social security would be harmonized wherever disparities affected business competition in the EEC. As in the European Coal and Steel Community, the competitive adjustments would be eased by a compensation fund established by the stronger elements, and special arrangements would be made for them to contribute investments for underdeveloped areas in the territories of members, at home (for example, southern Italy) and overseas (French possessions).

All these arrangements, of course, could be upset by drastic changes in world conditions during the dozen years of transition. And agriculture would in any case be subject to special consideration to keep it immunized from most of the trials of a true competitive Common Market. But at least the formal institutionalization of the plan and its systematically graduated procedures under international authority did give substantial promise that it could weather minor crises and would eventually produce most of the prosperous and harmonious integration among the Six that its builders hoped for.

The British became worried about being left out. Partly this fear was a survival of centuries of English opposition to any unification of the Continent, which might then be inclined to challenge her. But the Continent could not really be unified, of course, except by the USSR as long as Soviet power remained; so the more up-to-date and serious British worry was that individual members of the Six would deflect their trade from Britain to other EEC partners, when tariffs were eliminated among them while remaining fixed against England and the rest of the world. The United States also had some reason for concern at such a tendency, but Americans had become accustomed to enduring various forms of commercial discrimination and could comfort themselves with the argument, justified by the OEEC-EPU experience, that increased trade within a region improves the economic strength of the countries concerned and also makes them more willing and able eventually to relax discriminations against American goods. This kind of regional multilateralism, postponing universal multilateralism, had received America's general support (though somewhat begrudged) ever since the initial failure of sterling convertibility in 1947. EEC seemed largely a culmination of the trend, and the American government endorsed it.

Great Britain, however, being much more dependent on trade than was the United States, showed quiet disapproval of the ambitious plans for a Common Market that was to be so highly integrated that she herself would not wish to join. When, nevertheless, the Six proceeded to form it without her, Britain pressed the idea of a "Free Trade Area" that would supplement the Common Market and include also those OEEC-EPU countries that refused to join the latter.* The notion was that all these countries would remove tariffs and quotas from trade with each other—an extreme extension of the OEEC-EPU idea—but they would not all agree to establish the same tariffs against the rest of the world (the Six could proceed to set their uniform tariffs if they insisted, but Britain

* Notably Scandinavia, which was closely bound to Britain by economic and cultural ties; Switzerland and Austria, which feared that their neutrality would be compromised by participation in the federalistic EEC; perhaps also Iceland, Ireland, Portugal, Greece, and Turkey, which seemed too remote and underdeveloped for EEC; and of course Great Britain herself.

and the others would not do so). Britain would be able to continue to maintain her preferential tariff arrangements with Commonwealth countries, thus favoring some of her exports while also continuing to compete on equal terms, for example, with France in the German market (the equality there changing to one of no tariffs from the existing equality of nondiscriminatory high tariffs). The British government also wanted to exclude food, drink, and tobacco entirely from the freedom of the Free Trade Area so that Commonwealth farmers (notably Australians and New Zealanders) could continue to be favored in the United Kingdom market, and incidentally, so that Britain's own agricultural programs could be immunized from the effects of the EEC's agricultural arrangements. Furthermore, the British proposal evidently did not intend to seek a very wide-ranging coordination of general economic policies such as what the Six were undertaking among themselves; the British clearly wanted to keep their hands freer.

All of these aspects of the Free Trade Area plan involved advantages for Britain that elicited little enthusiasm among the Six; and the sheer technical difficulties would be formidable in attempting to reconcile and synchronize the gradual tariff reductions for a dozen Free Trade Area countries with those of the included half-dozen Common Market countries—all within a larger framework of worldwide tariff bargaining in which the United States would play a major role. Nevertheless, there were powerful political and economic arguments for enabling Britain (with Scandinavia) to participate just as closely in European economic integration as she would be willing to do. Consequently, intensive negotiations proceeded in 1958 and 1959, with the general approval of the American government, attempting to compromise the EEC and Free Trade Area plans.

Unable to put over their full Free Trade Area, the British in 1960 settled, at least temporarily, for a much smaller model, the European Free Trade Association (EFTA), comprising the United Kingdom, Switzerland, Austria, Portugal, and the three Scandinavian countries. Hope was still held out that these "Outer Seven" would be able to reach a compromise with the "Inner Six" (the EEC) and that Europe would not remain "at sixes and sevens." But no person in the West who was at all familiar with twentieth-century history could take lightly the rising prospect that two trade blocs would crystallize, one centered on Britain, the other on France and Germany and potentially just on Germany if her relative strength increases.

The United States government, which had endured discrimination against dollar goods as long as European unification seemed thereby to be promoted, was much less disposed to accept such discrimination in the course of an incipient bifurcation, especially since Europe's economic recovery seemed finally to have made her well able to meet most American competition anyway. Also the United States was eager to get the now prosperous Europeans to contribute more to the newly developing noncommunist countries overseas. Accordingly, the old OEEC was reorganized in 1960 to provide a permanent framework for harmonization among the Six and the Seven and also the United States, which agreed to become a full member of this reconstituted Organization for Economic Cooperation and Development (OECD). Yet the future of OECD's intra-European function remained somewhat obscure, partly because America's price for joining was the elimination of many of the formerly obligatory OEEC codes; reliance was to be more on voluntary cooperation.

The influence of the United States in and out of OEEC-OECD tended to favor the Six over the Seven as the more unifying group. Also, valuable American private investment capital was flooding into the Six, expecting to find there a bigger market than Britain and her Seven would be able to provide. Suddenly in 1961 London decided to make a serious attempt to find acceptable terms on which to become a member of the Common Market group itself. Thus the kingpin of the Seven offered to become part of the Six. The ominous incipient rivalry of the two blocs would be dissipated by Britain's cutting her Gordian knot and "joining Europe." Hopes rose high in Washington that workable arrangements could now be found for bringing this merger to increasingly prosperous fruition in a progressive consolidation of Western Europe, the British Commonwealth, and perhaps the former colonies of the Continental powers in Africa.

The leverage remaining at the disposal of the American government to promote these tendencies was, however, not unlimited. The United States' own trade policies are confined by domestic interests; she wishes to redirect her economic aid to meet the pressing requirements of newly emerging countries in the Southern Hemisphere; and since the late 1950's she has found her prospective overseas expenditures of all kinds circumscribed by an unfamiliar anxiety about her own balance of payments. These limitations upon America's economic power as a means of influencing the further evolution of European economic growth and integration have become conspicuous in recent years as Europe has become weaned. Fortunately the spontaneous local economic forces there, vitalized by past American contributions, seem to be moving forward on their own momentum in a manner conducive to prosperity and unity within the Western alliance, despite numerous potential pitfalls.

## SELECTED BIBLIOGRAPHY

ASHER, ROBERT E. *Grants, Loans, and Local Currencies: Their Role in Foreign Aid.* Washington, D.C.: Brookings Institution, 1961.

BENOIT, EMILE. *Europe at Sixes and Sevens: The Common Market, the Free Trade Association, and the United States.* New York: Columbia University Press, 1961.

DEWHURST, J. FREDERICK, *et al. Europe's Needs and Resources: Trends and Prospects in Eighteen Countries.* New York: Twentieth Century Fund, 1961.

DIEBOLD, WILLIAM, JR. *The Schuman Plan: A Study in Economic Cooperation, 1950–1959.* New York: Praeger, 1959.

LISTER, LOUIS. *Europe's Coal and Steel Community: An Experiment in Economic Union.* New York: Twentieth Century Fund, 1960.

MC NEILL, WILLIAM HARDY. *Greece: American Aid in Action, 1947–1956.* New York: Twentieth Century Fund, 1957.

MOORE, BEN T. *NATO and the Future of Europe.* New York: Harper, 1958.

MUNKMAN, C. A. *American Aid to Greece: A Report on the First Ten Years.* New York: Praeger, 1958.

MYRDAL, GUNNAR. *An International Economy: Problems and Prospects.* New York: Harper, 1956.

PRICE, HARRY BAYARD. *The Marshall Plan and Its Meaning.* Ithaca: Cornell University Press, 1955.

SCHELLING, THOMAS C. *International Economics.* Boston: Allyn and Bacon, 1958.

TRIFFIN, ROBERT. *Europe and the Money Muddle: From Bilateralism to Near-Convertibility, 1947–1956.* New Haven: Yale University Press, 1957.

WALLICH, HENRY C. *Mainsprings of the German Revival.* New Haven: Yale University Press, 1955.

ZURCHER, ARNOLD J. *The Struggle to Unite Europe: 1940–1958.* New York: New York University Press, 1958.

## CHAPTER 15

# TRADE, AID, AND GOLD: THE POLITICAL ECONOMY OF THE ATLANTIC COMMUNITY

THE SIGNIFICANT LIMITATIONS that came to the fore in the later 1950's upon America's readiness to confer further economic benefits upon her European allies may be classified under three policy headings: trade, aid, and gold. Europe's well-being and the closeness of her ties with the United States make the problems that arise in these political-economic relationships much less critical than the military dimensions of America's foreign policy or the political-economic dimensions of her relationships with the newly developing countries of Asia and Africa. But the same conscious interdependence of Europe and America that facilitates solutions of their mutual economic problems makes it important to seek the solutions diligently—not always to be distracted by some new crisis of a different kind. Economic strengthening of the NATO countries helps make the other crises also more manageable.

With regard to American trade policy, the aspect most affecting these advanced industrial countries is America's reluctance to accept huge imports from them. But there is another aspect, which will be noted here first because it seems to United States policymakers to be much the simpler of the two, namely the problem of sales from the West to the Soviet bloc.

## Controls upon Trade with the Soviet Bloc

East-West trade affects Americans much less than it does Western Europeans. The degree of commercial interdependence of Eastern and Western Europe before World War II was never remotely equalled by any trading ties between the United States and the area that later became the Soviet bloc. In the early stages of the Marshall Plan, Americans, like Western Europeans, looked forward to substantial re-establishment of prewar East-West patterns of exchange, particularly in the hope of enabling the western Continent to secure coal, timber, grain, and some other raw materials without having to pay scarce dollars for them. But exportable commodities were in short supply in the Soviet sphere also in the aftermath of World War II, and soon the onset of the cold war reduced any desire on the part of either side to become in any way dependent upon trade with the other.

Presumably the autarkic nature of Communist totalitarianism anyway would never allow the Soviet economy to develop any avoidable dependence on trade with any country outside the Kremlin's sphere of preponderant political influence (better to avoid trade altogether than become subject to threats that it might be suddenly disrupted or subtly distorted). Similar anxieties apply also in free market economies, with the added factor that there are numerous private groups with interests in imports and exports who could separately be lured by the bait of special trading opportunities that might be dangled before them by totalitarian powers. If ever allowed by democratic governments to become accustomed to any very wide commercial intercourse of this kind, these private interests could be expected to develop some special stake in promoting policies at home that would be agreeable to the totalitarian bloc; in effect their livelihood may become hostage for their government's "good behavior." And in a democratic system their influence could not easily be suppressed in the event of renewed cold-war requirements.

In 1950 the outbreak of military action in Korea sharpened these points and added to them a powerful American public resentment toward the symbolism of any "trading with the enemy." A series of restrictive moves in Congress culminated in the Battle Act of 1951, named for its sponsor, Representative Battle; the apparent symbolism of the title was mainly coincidental, but it did reflect the spirit of the legislators. The measure required the President to establish lists of "strategic goods" and prohibit them from being exported to the Soviet bloc by any country which was receiving American aid, on pain of having the aid cut off. Except for shipments of actual munitions or atomic-energy materials, there was some leeway for the President to accommodate the export requirements of friendly nations by adjusting the lists or by making formal exceptions in particular cases on grounds of a lesser relative cost to American national security.

A large measure of collaboration in the enforcement of export controls was in fact secured by the American government from countries it was aiding, with no need to go beyond threats and really impose the Battle Act sanctions against them. To be sure, the lists of goods embargoed by the aided nations tended to be less restrictive than the bans the United States established for its own exporters, especially in prohibiting trade with Communist China and North Korea, and in the years since the end of the Korean war, most embargoes have been considerably relaxed. Also the adequacy of enforcement procedures for the export controls in several countries has left much to be desired from the viewpoint of the United States; unauthorized trans-shipments to the Soviet bloc via third countries are particularly difficult to forestall. Nevertheless, despite all these qualifications, American pressure upon noncommunist countries must be credited with holding East-West trade substantially below the levels it would otherwise have reached.

Clearly this policy has limited the degree to which vested interests in such trade might develop in Western countries and erode the determination to resist the cold-war adversary, much sooner than would be likely to happen in the politicized commerce of the Soviet sphere.*

Whether, in addition, the limitation of East-West trade has slowed to any important degree the economic or military growth of the Communist powers is much less certain. It can be argued that any possibilities of growth-retardation—slight at best in a system as richly endowed by nature and as well equipped to squeeze consumers as is the Soviet—can adequately be achieved merely by restricting export of genuinely secret products and by avoiding any long-term loans, so that the immediate gains from trade, East and West, would never be very far out of balance in case it were shortly disrupted for political reasons.

Near the other extreme of the argument, one might concede the in-

* In the special case of the Eastern European satellite states, there is some slight chance that local regimes might be willing to permit an expansion of trade to the level of genuine interdependence with the West, reducing their dependence on Moscow. The political gains for the West would be important, but Russia seems most unlikely to allow any such culmination. In the case of Communist China, even if she were willing and able to expose herself to the temptations of massive East-West trade, the West has special reason to fear and avoid any acceleration at all in the development of this colossus of Asia. (See pp. 110, 128–29, 385–86, 500–501.)

ordinate difficulty for the West to determine, from year to year, what products really are the most "strategic" for the Soviet economy, as the Kremlin's own next plans may envisage the problem—but conclude from this that the safest thing to do is simply to cut off all exports that can possibly be cut off. The United States itself could clearly afford to maintain such an attitude. Countries like Britain, West Germany, and Japan, dependent on extensive trade for the very survival of their populations, cannot so easily avoid attempting a calculus of the gains that they, as well as the Soviet bloc, might be making from expanded exchanges. In effect, therefore, a Western export-controls policy designed to thwart dependence upon the USSR has a much more general acceptability in the noncommunist world than one designed to retard Soviet economic growth. The latter objective raises in particularly difficult form the complex questions of "net advantage," which involve the politically articulate interests of individual firms and localities as well as entire national economies.

Furthermore, the estimate reached, say, by the British government of the "net advantage" England derives over the benefits Russia derives from a trade agreement involving possibly "strategic" exports from Britain is likely to be affected by the availability of alternative sources for needed imports and alternative markets for English goods in other parts of the world. To what extent is it reasonable to anticipate that the alternative export markets may be found in the United States or be opened up elsewhere by the power of example and the influence of an America strongly committed to expansionist trade policies throughout the noncommunist world?

## American Tariffs

The main problems in this connection center upon American tariffs. There are other protectionist devices, including the rule that 50 per cent of all foreign aid tonnage must be shipped in American-flag vessels, and the provisions of the "Buy American" Act which have been interpreted to require government contractors to purchase domestic goods unless foreigners can offer delivery at prices 6 to 10 per cent lower.* But the impact of tariffs is vastly wider both at home and abroad. Therefore,

* This interpretation represents a liberalization by the Eisenhower Administration of previous 25 per cent differentials. But if the lowest domestic bid comes from an area with high unemployment it may still be accepted although the price differential is well above 10 per cent.

some estimate of its real dimensions is badly needed as the foundation for any sober judgments concerning it; but unfortunately, the existing detailed studies are few and incomplete. They do suggest that a permanent removal of all American tariffs and import quota restrictions would bring more than $4 billion a year in additional imports after a few years.* This would amount to an increase of about one-fourth above recent levels of merchandise imports. Inasmuch as American tariffs tend to restrict imports of manufactures much more than of raw materials, it is reasonable to suppose that the increase in imports from Western Europe would be proportionately higher than the worldwide figure, perhaps totaling 50 per cent above the $3 billion of goods that were coming from there in 1956. Calculations at that time indicated that such an increase in shipments to the United States would in itself raise Western Europe's exports to the non-European world as a whole by about one-tenth; if the American tariff-cutting example were followed widely in other countries the increase would be even greater, substantially so. Clearly the chance of so large a rise continues to be of importance to Europe, particularly to Britain whose exports going outside Western Europe are equivalent to almost one-eighth of her gross national product.

The total impact on the United States economy would be much less. In 1957, a year of high domestic production and high imports, America's farms, mines, and factories produced more than twenty-five times as much as the merchandise imported, and not much more than one-twentieth of the American product found its market abroad or was given away in nonmilitary foreign aid. Projections of America's economic development in the next two decades indicate that imports would at least keep pace with the rise of the American national product, doubling over the levels of the early 1950's, even if tariffs and quotas were not changed from those prevailing in 1955.† In absolute terms these import increases would be impressive—but Europe's share might well be disproportionately low unless the tariffs on manufactures are lowered, and of course the import and export totals would still be only a small percentage of the American national income.

Superficially, it is easy to contend that the very fact that the import sector is so small means that an increase would not cause much disruption of the domestic economy if tariff cuts were extended. The trou-

* See Howard S. Piquet, "Tariff Reductions and United States Imports," *Foreign Trade Policy: A Compendium of Papers for the Committee on Ways and Means* (Washington, D.C.: Government Printing Office, 1958), especially p. 254, and the same author's *Aid, Trade and the Tariff* (New York: Thomas Y. Crowell, 1953).

† Henry G. Aubrey, *United States Imports and World Trade* (New York: Oxford University Press, 1957).

ble, of course, is that the hardships of new foreign competition would be concentrated. The total number of Americans who would be displaced from present employment by a permanent removal of all United States tariffs and import quota restrictions would be counted in the hundreds of thousands, probably not in the millions, but with their families these do constitute a sizable group, even in a national working force of over 60 million. Job turnover in the United States from other causes is much greater than these tariff cuts would cause, but unemployment and bankruptcy that are caused by "impersonal" market forces within an already established regulatory framework seem much more tolerable than similar hardships resulting from the government's changing the rules. Moreover, in mining, agriculture, and handicrafts, major production cuts caused by newly admitted imports would require the workers to seek employment in wholly new social environments. A cultural transformation would be demanded which might be unbearable, except perhaps for the young. And the impact on one-industry communities can be pervasive and staggering, reducing them to ghost towns. Agriculture poses enormous special difficulties because of the maze of special protective devices that have been built to shelter it even against the winds of the domestic economy; these props virtually require bracing by quota fences against the products of foreign farms. Other industries are sometimes in a position to demand protection on grounds that they are essential to national defense.

## Elements of Instability in American Restrictions upon Imports

These various resistances to increased imports have produced a tentativeness and apparent unreliability in American commercial policy, in general and with regard to particular commodities, that is a persistent barrier to foreign sales efforts even when tariffs themselves have been lowered.

Instability can operate in several forms. One is the extreme complexity of United States customs regulations, enforced at ports of entry by a customs service inculcated for many decades with a protectionist spirit of resolving doubts against the importer. Recent congressional legislation has finally simplified the procedures and reduced the costly delays and uncertainties, but they can still be discouraging, especially for smaller firms. Another source of unpredictability has been Congress's practice of never renewing the basic legislation (the Reciprocal Trade

Agreements Act) for more than a very few years at a time. The President's authority to bargain with other countries for a reciprocal lowering of tariffs is thus kept subject to frequent reappraisal and possible alteration by Congress if influential domestic interests suffer much injury.

Increasingly since World War II, Congress has further insisted on a trade agreements procedure that would not require domestic hardship cases even to wait for new legislation. The independent Tariff Commission has been authorized to hear complaints alleging that competitive imports are causing "or threaten serious injury to the domestic industry producing like or directly competitive products." The Commission is to make recommendations to the President if it finds that the conditions warrant an imposition of quotas or the rescinding of previous tariff reductions. The President retains generally free discretion to reject the proposed restrictions or modify them, but if he does so he could legally be overridden by two-thirds of both houses of Congress, under the latest (1958) Reciprocal Trade Agreements Act. (Insofar as any new restrictions mean that the United States is "escaping" from international agreements that had been reached in earlier tariff bargaining, American negotiators are committed to try to compensate the country whose exports are being excluded: barriers might be lowered on other commodities that country would like to sell in the United States, or retaliatory increases in barriers to American exports might be tolerated.)

This whole "escape clause" procedure, though subject to congressional alteration, clearly makes the attitude of the President the principal determinant from month to month as to whether enterprising foreigners will actually be allowed to succeed in developing markets in the United States at the expense of domestic producers. Even the initial tariff reductions negotiated with foreign countries have been allowed to go below what the Tariff Commission determines to be the "peril point" of prospective injury to a domestic industry, if the President personally takes responsibility in a public justification. In actual practice, both President Truman and President Eisenhower emphatically avoided a protectionist slant in the use of their wide discretion. For example, less than a dozen escape-clause applications in the decade 1947–57 ultimately resulted in increased restrictions; ten times as many were rejected, either by the Tariff Commission or by the President. This record indicates that foreigners seeking markets in the United States are likely to find the instability of American tariffs under the escape clause to be more apparent than real. But if heavy investments would be required in new equipment to produce goods adapted to American tastes and to develop a sales and advertising organization designed for wide market penetration, the risk of depending on antiprotectionist sentiments in the White House may well seem too great in cases where there is much danger that an American competitor will cry "Injury!"

What is "injury" anyway? And what is an "industry" producing "like or directly competitive products"? The degree of protectionism in escape-clause actions depends very largely on the criteria developed to define these terms. A few sample questions may indicate some of the complexities: When a variety of products is made by the same firms, and some of those products are losing out to foreign competition while sales of the others are expanding, is an "industry" being harmed? Suppose this ambiguous pattern prevails in the larger firms, but some smaller ones have specialized more in the lines where foreign competition is severe; is it then fair to define the "industry" so broadly that their hardships are disregarded—or, if pressed, could they too actually manage to shift over to the other lines of production? Any attempt to measure "serious injury" is affected by the choice of a base year from which to draw comparisons. For example, is an American industry that became established during the war, when traditional European suppliers were physically unable to compete, now entitled to claim injury when they successfully re-enter the market? Also, what if the domestic producers are managing to expand their sales profitably despite the competition of foreign imports, but the imports are expanding even more rapidly, so that the domestic producers' *share* of the total market is declining? Are such relative declines "serious injury," or are only absolute declines? If declines are the result of a combination of factors of which import competition is only one, is there sufficient reason to cut such competition off? How can the various factors be weighed? Is the "seriousness" of the injury such that it could scarcely be remedied without restricting imports, or are there practical and preferable alternative means of relief? Could any new restrictions be made only temporary, perhaps scaled down on a fixed schedule to give the domestic industry a limited time in which to readjust as well as strong incentives to do so?

Claims for protection become most compelling under contemporary conditions if they can be convincingly pressed in the name of national defense. A "national security amendment" to the Reciprocal Trade Agreements Act, embodying the concept of "defense essentiality," has consequently been added to the existing complexities of American customs regulations and the escape clause, as an additional discouragement to foreigners seeking access to the American market.

Cut off from foreign sources of supply in time of war, the United States might have need for special employee skills, ready production lines, and actively functioning raw materials industries. If these enterprises are not kept in operation in peacetime, it may not be possible to develop them fast enough in war. But what kind of war? It may well be argued that in a total thermonuclear holocaust no industrial production would be possible anyway, and that in a limited war, transportation routes would in fact be kept open for essential imports. Conservative hedging against all eventualities may require one to admit the possibility of some in-between forms of future war for which self-sufficiency in particular lines of import-vulnerable production would really be a national necessity. But such calculations must be made and revised soberly and expertly in the light of the latest war plans if "defense essentiality" is to be more than an excuse for an alternative appeals channel to the President via the Office of Emergency Planning (OEP, formerly the Office of Civil and Defense Mobilization), adopted by businesses that have failed to win escape-clause action via the Tariff Commission route.

The domestic watch industry, for example, facing pressure from Swiss imports, won a tariff increase under the escape clause in 1954, only to find that its share of the American market (but not its total sales) continued to decline. It appealed for import quotas under the national security amendment. The appeal failed in 1958, but new attempts are likely to be made as the OEP rules and business conditions change. The argument centers on the "essentiality" of the precision skills of the watchmakers and to a lesser extent on the essentiality of their usual products. Even if timepieces could be collected from other sources in wartime, perhaps the watchmakers would be needed to construct miniature fuzes and gyroscopes. The Office of Emergency Planning, responsible for overall war production plans, has thought not. Indeed, the argument has been made that the pressure of import competition has stimulated increased initiative on the part of top management in the clock and watch industry, with results that are indirectly beneficial to national defense—although the question might remain open whether enough new workers are being attracted into the high-skilled segment of the industry directly competitive with Swiss imports.

In some other cases where defense essentiality is claimed, a probability arises that increased protection for one industry would damage another equally essential industry. Quotas on raw material imports, for example, would raise costs for the manufacturers that use them. Carry the argument a little further: Import restrictions may reduce the earning power of foreign nations and hence weaken some American export industries that are "defense essential." Besides, American national defense under a strategy of alliances requires vigorous and cooperative allies; is not their productivity and access to wide markets a matter of defense essentiality to the United States?

Clearly, defense essentiality is a highly elastic concept which could be made to justify an approach to autarky at one extreme, or free trade (at least within the noncommunist world) at the other. The congressional elaboration of criteria in 1958 was protectionist in spirit, but retained enough ambiguities so the President could still find excuses for using his ultimate discretion to avoid trade restrictions if he wished.*

## Some Domestic Subsidy Devices

It may be possible in some instances to provide alternative forms of government assistance to industries that are judged to be of critical importance. Purchases of commodities to be accumulated in official stockpiles can ameliorate short-term hardships, but except for perishables, the stored accumulation inevitably becomes larger within a few years than any conceivable emergency requirements for its use. At the beginning of 1959 the existing American defense stockpiles were valued at about $8.5 billion, more than twice the amount that the Office of Civil and Defense Mobilization still believed to be essential, having revised its preparations downward from plans for a five-year war to one lasting three years.

The domestic lead and zinc mining industry, for example, had been kept operating profitably despite large imports, by means of government

* The Director of the Office of Emergency Planning "and the President shall, in the light of the requirements of national security and without excluding other relevant factors, give consideration to domestic production needed for projected national defense requirements, the capacity of domestic industries to meet such requirements, existing and anticipated availabilities of the human resources, products, raw materials and other supplies and services essential to the national defense, the requirements of growth of such industries and such supplies and services including the investment, exploration and development necessary to assure such growth, and the importation of goods in terms of their quantities, availabilities, character and use as those affect such industries and the capacity of the United States to meet national security requirements. In the administration of this section, the Director and the President shall further recognize the close relation of the economic welfare of the nation to our national security, and shall take into consideration the impact of foreign competition on the economic welfare of individual domestic industries; and any substantial unemployment, decrease in revenues of government, loss of skills or investment, or other serious effects resulting from the displacement of any domestic products by excessive imports shall be considered, without excluding other factors, in determining whether such weakening of our internal economy may impair the national security." Trade Agreements Extension Act of 1958, section 8c.

purchases for stockpiles during and after the Korean war. When the stockpiles had been filled beyond any reasonable anticipation of emergency defense requirements—and when, correspondingly, the aboveground accumulations made it very difficult to argue that the mines themselves would be "defense essential" and hence deserve quota protection under the national security amendment—the industry turned to seek escape-clause protection (tariffs or quotas); and the Administration attempted to establish an alternative in the form of direct subsidies to the mines.

Subsidies have the advantage that they are in principle more precisely adjustable than tariffs or quotas to the real needs of the nation, the industry, and particular firms, and that the costs are kept in public view and spread out over all taxpayers instead of being inflicted mainly on the consumers of particular imports. The disadvantages of subsidies correspond to the advantages: Precise adjustability means political and bureaucratic interference, perhaps on bad grounds as well as good, whereas tariffs—but not quotas—are relatively impersonal; and subsidies require appropriations, which seem of doubtful political reliability to the producer because the taxpayer finds them exasperatingly conspicuous and would more readily endure the hidden costs of tariffs.

The proposed subsidy program for lead, zinc, and a few other minerals was estimated to cost about a half-billion dollars over a five-year period, depending on how far the market prices fell below "fair" prices. The plan passed the Senate in 1958 but was killed by a narrow margin in the House of Representatives. Other than subsidies there was no means whereby importers and domestic producers could both be furnished profitable access to the American market for lead and zinc. President Eisenhower, finally feeling obliged to act under the escape clause, imposed quotas. They cut imports by about one-third from the 1957 level, a drastic reduction, but the Administration hoped the quotas could be liberalized again more readily than could tariffs, if international agreements could be reached limiting supplies on the world market, by which prices would be raised to a level where Americans could compete.*

In the case of wool, unlike lead and zinc, Congress was willing to follow the Eisenhower Administration's requests for authority to use subsidies instead of tariff increases as a means of preserving domestic production. Customs receipts under the pre-established wool tariffs were designated in 1954 as available for direct compensatory payments to American producers of the same commodity. Congressional amenability

* On international commodity agreements, see pp. 340–43.

to such a project for wool but not for lead and zinc reflects in part the greater public inurement to multifarious aid for agriculture. By 1960 Congress was willing to add subsidies for small lead-zinc mining outfits, but Eisenhower concluded that the new import barriers gave them adequate protection; the small-producer subsidies were finally enacted under President Kennedy. These various precedents remain inconclusive as to the future prospects of subsidy devices as acceptable substitutes for import quotas or tariff increases where foreign competition is unbearable.

Considerable discussion but very little action, at least until 1962, has surrounded the notion that the government should sponsor relocation and retraining programs and provide tapering subsidies for workers, firms, and communities hit by import competition. Such a program would be designed to facilitate economic and social readjustment through a temporary combination of advice, incentives, and compensation, so that those without adaptability can decline gently while the more energetic are steered into new fields. Under such a concept, any subsidies would not be a substitute for trade restrictions to keep old industries alive indefinitely, but a means to enable them to transform themselves (labor, capital, and management) into new industries, perhaps differently located.

This "adjustment assistance" program would be difficult to administer, particularly in determining legitimate levels of government aid for the conversion of plants and the liquidation of capital for reinvestment elsewhere. Some whole communities could be helped only to disperse themselves, if no new economic base could be established locally. Even if such planned dislocations are acceptable, it may well be doubted that the special aid should go to hardship cases caused by imports, when other more severe hardship cases, resulting from other causes, exist in the country. The implication of special assistance is that past tariff protection implies some right to future protection, or to compensation for the loss of it, that is greater than the reliability attached to other kinds of privileged positions in the dynamic American economy. Perhaps so. But under Eisenhower, Congress was not willing to enact the appropriate legislation, and the President was unwilling to see such transitional compensations generalized to cover many other types of exceptionally depressed communities and industries. Kennedy is willing; so "adjustment assistance" may now become linked to tariff reductions, as a partial or complete substitute for renewed protectionist trade barriers, on behalf of some industries that cannot find a way on their own to cope with problems arising from import competition.

### "Voluntary" Quotas on Exports to the United States

In cases where imposition of quantitative restrictions on imports appears, however, to be the only remedy that could be politically tolerable, there may be some attractions for the foreigners as well as for the American government if the restraints take the form of export quotas imposed in the country of origin rather than import quotas imposed in the United States. "Voluntary" export quotas enforced abroad permit the United States somewhat hypocritically to avoid open violation of its international trade agreements; they also permit the foreign government or producers' association ("cartel") to determine, in the light of its own policy preferences, which particular firms are to get the restricted American business through the allocation of the quota; and, if the particular licenses are to be *sold* or favors and bribes be provided, the proceeds will accrue to those foreigners who control the licenses, not to Americans. Also, being "voluntary," the quota regulations can be somewhat more readily altered than if the American government had established full authority over them. A penetration of the American market may be made gradually, at a rate designed to avoid shocking the domestic producers into frantic political resistance, yet with the hope of slow but steady gains. Thus the Japanese and some other Asian producers have succumbed to American pressure for "orderly marketing" by agreeing to limit their own exports of textiles to the United States as the only realistic alternative to confronting escape-clause barriers against the goods if they ever arrived as imports at American ports.

The durability of such self-restraint on the part of the exporters and its sufficiency for appeasing the still suffering American producers remain to be proved. In the case of oil imports, where "voluntary" quotas were established in 1957, they had to be replaced by mandatory quotas in 1959. This was true even though petroleum exports to the United States are mainly controlled by great American companies which also have important domestic oil interests of their own to protect, and although the "voluntary" quotas were backed by official pressure in the form of government refusal to make purchases from companies that violated the restrictions. Even so, the remaining leakage was unendurable to the competing coal interests and to those whose stake in oil production was principally domestic. They contended successfully under the national security amendment that domestic oil-producing capacity, to be readily available for an emergency, would require constant new exploration and investment in the industry, not a mere conservation of supplies already discovered. The necessary incentives, the President finally agreed, would be lacking if oil from outside the United States and Canada for which the sea-lanes might be vulnerable in war were allowed to fill a growing part of America's peacetime consumption. If "voluntary" quota restrictions were too porous, even with some government hardening, they would have to be made adamant and compulsory.

A similar outcome seems likely in the case of other experiments with voluntarism as a means of implementing American protectionism. Nevertheless, the test may often be worth making if, for a time at least, it appears to be the lesser of evils in the eyes of the foreigners the United States must try to conciliate.

### The Indecisiveness of American Commercial Policy

The uncertainties associated with American commercial policy—the complexities of customs regulations, the escape clause, the national security amendment, the experiments with stockpiling, subsidies, voluntary quotas—all create a hazardous legislative framework for import trade, one in which the foreigner's opportunities are heavily dependent on sympathetic administration of the statutes. This in general he has fortunately had since before World War II. What Congress gives with one hand—tariff reduction authority—it often seems to be prodding the President to nullify with the other—the escape clause, and the like. But the range of discretion accorded the Chief Executive remains large, and venturesome foreigners have been justified in taking the chance of pushing exports to the United States steadily upward.

The argument has often been made that a firm stabilization of existing tariffs and import quotas, closing the various escape hatches so that liberal commercial policy would not depend so continually on presidential decisions, would be even more valuable to friendly nations than additional tariff cuts would be, since the cuts would always be reversible later. Congress, however, in its 1958 temporary extension and revision of the basic legislation, the Reciprocal Trade Agreements Act, repeated the previous pattern. The statute gave the President new authority, in bargaining with other powers for reciprocal tariff reductions, to lower the American rate on any commodity by two percentage points or by one-fifth of the existing percentage, by stages over a maximum period of eight years. The process would still be subject, however, to prior scrutiny for "peril points" and to subsequent reversal through the escape clause and the national security amendment.

Here, if anywhere in the early 1960's, lay the Administration's bargaining power by means of trade policy to modify the evolution of European economic growth and integration in a fashion conducive to prosperity and unity in the Western alliance. Yet the margin of manipulable trade concessions available to the Administration, even if buttressed by a convincing degree of consistency on its part in avoidance of protectionist "escapes," did not really seem large enough to have major influence on European economic policies. A high degree of parallelism between (on one side) American purposes and (on the other) the spontaneous thrust of prevailing political and economic forces in Western Europe was the essential premise for the achievement of any important response to the merely marginal commercial benefits that Eisenhower or Kennedy could confer, unless Kennedy obtained new, far-reaching trade legislation.

Fortunately, the parallelism between American purposes and European tendencies has existed anyway, and perhaps a closer convergence can be attained by more vivid leadership and articulate persuasiveness on the part of American policymakers. To a relatively increasing degree, they have been finding themselves dependent on these qualities of mind and spirit, and less able to utilize virtual economic bribery in dealing with their allies concerning the political economy of Western Europe. Not only has this been true with regard to trade; it is also true with regard to economic aid. And the limitations that emerge are not only in domestic American politics but in the nature of international finance under the modified gold standard that prevails. America becomes increasingly aware of a need to secure economic cooperation from Europe, not just within Europe, and simultaneously aware of tighter restrictions on her own ability to purchase that cooperation.

### Persisting Forms of American Economic Assistance to Europe

Economic aid from the United States has lost much of the important role it once played. Though by no means ended, it has declined sharply.

The formal assistance, which has in fact continued since the end of the Marshall Plan in 1952, has taken the form principally of gifts of military equipment produced in the United States, some of which might have been purchased from America by the more self-reliant European countries who would have resolved to pay for it themselves rather than risk doing without the matériel. In such cases the military equipment grants are economically beneficial to Europe. And other American military programs confer even more obvious economic advantages upon the NATO allies.

For example, the United States has been spending about $75 million a year in Europe as her share of the continuing cost of NATO "infrastructure" (bases, pipelines, and communications networks). Also, particularly before 1955, large contracts were placed for "offshore procurement" whereby the American government bought matériel in Europe (usually paying out dollars), instead of producing it in the United States, and then gave it to European armed forces; thus the NATO country got the arms and also the dollars to improve its balance of payments. Offshore procurement in the mid-fifties was running at a rate of around a half-billion dollars a year, although it was scheduled to decline sharply by the end of the decade. But the decline was offset by the high level of expenditures of American armed forces in Europe obtaining goods and services there for their own use. The latter amount has been more than $1 billion a year—a kind of guaranteed "tourist" expenditure. United States troops, dependents, and hired civilians spending their pay; local procurement of "soft goods" such as food; various contractual services including construction and some maintenance—all of these items are inherent in the very fact of the stationing of a large American military component to participate in the defense of Western Europe. Actually the rise in these costs was more rapid than the decline in offshore procurements in the NATO countries for military assistance, with the result that the American military presence in Europe continued to benefit the balance-of-payments position of the OEEC nations by more than $1.5 billion a year. How important an item this is may be judged from the fact that Europe's other earnings from all the goods she exports to the United States are only about twice as great as the dollar receipts that result from playing military host.

In the early fifties, extensive use was made in Europe of "defense support," an American aid program devised by ECA in the Korean war period, with its characteristic emphasis on rearmament, specifically in order to justify economic aid programs as contributing to the capacity of the recipient countries to assume arms burdens by easing the impact of the diversion of resources in national economies and government budgets to military purposes. This kind of assistance to Europe, however, declined rapidly; only in Greece, Turkey, and Spain was it still continuing in the late fifties (totaling about $150 million a year), and these three countries all have special reasons for economic weakness —Spain, under Franco, was previously refused Marshall Plan assistance; Greece and Turkey have had to maintain exceptionally burdensome military forces.

Also, in the late fifties more than $150 million of surplus agricultural commodities were going to southern Europe each year under special programs which clearly constituted economic assistance (partly emergency relief but mostly deals for local currency to be used like "counterpart" for economic development and multilateral trade promotion).*

Together, therefore, the American aid programs to noncommunist Europe that have an explicitly economic, not only military, usefulness may be considered to have run in the general neighborhood of $300 million a year in the late 1950's. In the late forties the sum had been ten times as great, but gradually the outright economic grants went out of fashion except for the Mediterranean countries, which are less developed than the northwest. Small-scale "defense support" and surplus crops can still be useful tools of American policy in these weak nations of the south, but there is slight impact on Europe as a whole, where the decline in aid has been drastic and continuing. Yet its impact on the balance of payments has been cushioned by the immense expenditures of the American armed forces in Europe (more than $1.5 billion a year, as noted), not to mention the huge gifts of military supplies, some of which might otherwise have been purchased in the United States by European governments.

### The Tether of Gold

It can be calculated that in the half-decade after the Marshall Plan ended, the dollar receipts of Western Europe as a whole from her role of military host were roughly equivalent to a huge increase that simultaneously occurred in her holdings of gold and readily convertible ("liquid") dollar assets. No longer suffering from acute shortages of goods, the Continental countries were banking their earnings, replenishing and expanding their "hard" reserves to almost the proportion of the world's total that they had held in 1928. Correspondingly, the gold holdings of the United States declined, and more and more of what remained became subject to being drained away if the European financial author-

* Since this program was mainly of concern to underdeveloped countries outside Europe, it will be analyzed in a later chapter. See pp. 365–71.

ities decided to exercise their right to convert the liquid dollar assets to gold in the United States. For the first time since the rise of Hitler, the American government found itself obliged in the late fifties to give some anxious attention to a deficit in its own balance of payments. Vis-à-vis continental Europe, the "dollar shortage" of the early postwar years was turning into a "dollar glut"—an alarming increase in the liabilities of the United States to pay out gold on potentially short notice, particularly to West Germany.

The favorable position of Germany underscored again the importance of American military expenditures as a major cause of this problem, because the strategic location of United States troops there was enabling the Germans to obtain almost half of all the funds that the American forces were spending in Europe. Yet there were additional factors besides military expenditures that explained Germany's financial success and the American losses of gold. And an alleviation of the United States payments problem sufficient to insure that it would not put a grave tether upon activist American policies and programs overseas could be seen to require a wider range of measures than merely pressing Germany to disgorge her gains.

The foremost reason for the relative improvement in the gold-reserve position of the Continental countries (particularly Germany) vis-à-vis the United States was the corresponding superiority of their rates of economic growth and enhanced productivity during the 1950's. They were becoming able to outcompete the United States in many export markets around the world, while in their home markets various residual forms of discrimination against dollar goods still hindered American access. Germany in particular, with a docile labor force and without expensive international commitments, was able to sustain a prolonged and booming industrial expansion despite her high interest rates and other orthodox capitalistic financial policies that were designed to block inflation. Investments were allowed to earn a rich return in Germany. Capital flowed in from other countries, including the United States, and goods flowed out, their prices not inflated by extreme wage demands. The German policy examples were imitated elsewhere in Western Europe.

Meanwhile, in the United States the general withering of American isolationism was dissipating also the special trauma of the Great Depression and its autarkic phobias about overseas investments and foreign imports. Americans were spending more time abroad in service and travel, and their consumer fads began to run in favor of imported goods. Also, United States investors found their attention attracted by the profitability and improved security of foreign enterprise; and the continuing inflation and wage pressures in America, coupled with the residual barriers in Europe against goods exported from America, suggested to American manufacturers that they would be able to sell more if they produced their goods in European factories, close to the markets; the American owner would still reap the profits and could, if he wished, repatriate them to the United States. He might even produce in Europe for sale in America, joining the European manufacturers in surmounting the slowly sinking American tariff wall.

The confidence of international financiers in the prospects of business in Europe rose till occasionally it outclassed even their confidence in the prospects of business in the United States herself, for the short term at least.

Superimposed upon Europe's receipts of investment capital from America and upon Europe's earnings of gold and dollars by successful competition in worldwide export markets was the $1.5 billion or so of United States military expenditures, about one-fifth of it in Britain and the remainder on the Continent. (In 1960, for example, West Germany obtained $650 million from this source.) American exports were not being absorbed in the European markets in any such amounts. Continental financiers, official and private, together converted a few billion dollars of their new holdings to gold, thus reducing the gold reserves of the United States, and invested another several billion dollars in short-term American notes, readily convertible into gold. The resulting obligations upon the United States to pay out gold on short notice gradually approached $20 billion and thus in 1960 came to exceed the supply of gold that would be available in the United States to meet those obligations if all payments were demanded simultaneously. The United States position in this respect was a common one for any country to be in, but it was one to which the United States had been unaccustomed for many years. Gone now was the familiar certainty on the part of foreign holders of dollars that they would all be able to get gold for their dollars any time they might want it—and hence that there was little reason for them to cause inconvenience by ever converting any abnormally large sums into gold. Confidence in the dollar was now impaired.

This state of affairs had already been nearly reached by the year 1958, when almost all the Western European countries made their own currencies convertible into dollars (for most purposes) and hence, directly or indirectly, into gold. Thereafter, there was a rapid revival of the private international market for short-term capital. Freed from governmental restraints, private investors began switching their holdings back and forth among advanced capitalist countries to take advantage of temporarily higher interest rates in one nation or another. Speculators expanded their operations in attempts to profit from manipulating this flow. The United States found, for the first time in many years, that its own domestic policy of combating periodic recessions at home by increasing the supply of funds through the lowering of interest rates and the running of federal budget deficits would impel foreigners to make a severe "run on the dollar." Their action would be partly in simple quest of higher interest rates again in Europe, and partly in fear that the American means of combating recession would turn out to be too inflationary, pricing United States goods more than ever out of world markets, and soon causing other foreigners to accumulate, unspent, more dollars than ever out of earnings from America, and consequently to aggravate (unintentionally) the risk of a panic in which there would not be enough gold to convert all the foreign-held dollars.

Such fears of a future panic could even stimulate a present panic, as the United States discovered in 1960, when concern over a weakening of American exports in the previous two years became combined with a temporary widening of the interest-rate gap between the United States and Europe as the American government strove to combat a recession, while the level of American military and tourist expenditures in Europe and long-term American private investment there remained very high. All these elements, plus perhaps some anxiety about the American election campaign, combined to produce talk in Europe of a possible devaluation which would make the dollar convertible for only a smaller amount of gold than previously. Spurred by speculators, in October 1960 a minor panic of dollar-holders, taking precautions by converting hastily into gold, West German marks, and British pounds, occurred. This run on the dollar was soon halted by the vigorous collaborative action of financial authorities in the major Western countries, but the possibility of its recurrence remained real and suggested the need for considerable revision in international financial mechanisms, if United States freedom of action in foreign and domestic policy were not to become seriously restrained by anxiety about her gold.

These worries would not merely be selfish. An American economy left stagnant because the government dared not stimulate it would be a calamitous drag upon the entire Western world. A drastic curtailment of United States overseas activities because the government found that foreigners were in effect not content with American goods and American dollars, but were insisting on getting American gold, would be a similar emasculation of United States potency in the larger struggle for the containment of Communism. Fear on the part of the major trading nations to hold their currency reserves for international commerce in any form except gold would lead to a paralyzing constriction of that commerce, much like the constriction of trade inside Europe in the Marshall Plan era when the countries there feared to hold reserves in any form except gold and dollars. Thus the noncommunist world as a whole has a stake in cooperating with Washington's efforts to preserve confidence in the dollar—and to do so by means that impose a minimum of restraint upon the constructive employment of America's general economic power at home and abroad.*

In important measure, the new confined position of the United States in international finance is akin to the one that Britain has endured since World War II. Unlike the Continental countries, although England's receipts from the United States have been very large she has not greatly increased her gold (or convertible dollar) reserves. Her earnings have had to be paid out promptly to meet her overseas obligations and responsibilities, and her economy at home, since the disastrous splurge of 1946–47, has been kept under considerable restraint for fear that any inflationary expansion would produce another great "run on sterling." Like New York, London is a "reserve center." Many countries treat their holdings of British currency as their reserves (in the way that the dollar is treated), and as the pound sterling became increasingly convertible again in the 1950's Britain was accepting the liability of further runs—and was doing so with relatively much less gold available to meet them than the United States still has for meeting possible runs on the dollar.

Britain has been exercising financial responsibilities for the entire "Sterling Area." Most of Britain's present and former dependencies and dominions have continued since World War II to change the greater part of their foreign currency earnings into sterling to be deposited in London. Their willingness to do so enhances Britain's world prestige and facilitates profitable trading relationships and cooperative political ties. But the price for England is that she can only try to persuade, not command, the independent countries in this Sterling Area to pursue policies of a kind that will prevent their citizens from changing excessive amounts of the sterling back into the foreign currencies, consequently reducing the reserves of the entire area, including Britain herself. In general order of magnitude, the overseas Sterling Area has been keeping deposits in London equivalent to $7.5 billion; roughly the equivalent of another $2.5 billion is banked in London by non-Sterling Area depositors. If these worldwide holders of sterling panic and demand to change it suddenly,

* Some ears may sense faint echoes from William Jennings Bryan's restless ghost: "Thou shalt not crucify mankind upon a cross of Gold."

as has recently happened in 1956, 1957, and 1961, the drain on London's hard-currency reserves can reach the staggering dimensions of $300 million a month (as during the Suez crisis, 1956). Only by massive use of the resources of the International Monetary Fund was it possible to restore the confidence of the holders of sterling on those occasions and to reverse the outflow; indeed, the drain even upon IMF was so great as to necessitate the great enlargement of the Fund in 1959. Yet the fact that these huge IMF operations did succeed and that the Fund is now bigger should help to deter any future "run on sterling." Nevertheless, the convertibility of sterling remains somewhat precarious as long as Britain, for good reasons, must allow the investment and foreign currency requirements of the less-developed parts of her Commonwealth to compete with her own needs in sharing the limited financial resources of the whole Sterling Area.

As the United States by 1960 came to find herself in similar though much less-straitened circumstances, increasing doubts were felt as to the adequacy of the International Monetary Fund for underpinning the reserves of the major commercial powers. The pool of currencies in the IMF had been greatly expanded in 1959, and another large increase was suggested. But recently much consideration has also been given to a wide variety of other devices for mitigating the swings in the reserve positions of the United States and Britain particularly, as the principal reserve-centers for other countries, for clearly the economy of the Western world is such that both the dollar and the pound sterling need to be kept strong together. A strengthening of one at the expense of the other (a run from the dollar to the pound or from the pound to the dollar) adds little if any basic strength to the West as a whole. The "sloshing around" of short-term capital among the major financial centers of Europe and America, stimulated by temporary interest-rate differentials and agitated by speculative hopes and fears, can be destabilizing and even disruptive.

To be sure, some of the discipline that capital mobility imposes upon the internal economic policies of individual countries is salutary, as the United States used to insist when she pressed to bring about free exchanges among the Western European countries in the dozen years after World War II. An anti-inflationary discipline may now be useful also for the United States herself. The need to hold down prices in order to keep American goods competitive in foreign markets and also to meet import competition in home markets should spur the United States to greater productivity and should help distribute more of its fruits to all consumers in the form of price cuts, instead of permitting the workers in each particular industry to capture the productivity gains in the form of wage increases for themselves. Moreover, a disinflationary approach to interest rates and budget deficits can be compatible with reliable and impressive economic growth, as the Continental countries have been demonstrating.

### Economic Integration of the Atlantic Community

However, the relative success of the Continental countries also indicates that a great deal depends upon not relying on automaticity in the disciplinary impact of the balance of payments, country by country, but rather upon supplementing that discipline by a manifold, deliberate, internationally-organized coordination of numerous aspects of economic policy among and within the countries concerned—in short, upon increasing general economic integration. Thus, to cite one example, speculative excesses of the "sloshing around" may be countered by some kinds of cooperative official action among nations; and, to cite another, responses to the business cycle by interest-rate adjustments in different countries may be synchronized somewhat in order to minimize extreme differentials that would suck short-term capital from one country to another. The 1960 reorganization of the Marshall Plan's OEEC into the Organization for Economic Cooperation and Development (OECD) was partly designed to furnish a framework for this kind of coordination in which the United States would be willing to participate fully. Also, the International Monetary Fund is increasingly active and far-reaching in its range of concern and influence in Western economies. Additional organizations—or additional authority for the existing organizations—may yet be needed.

The Common Market countries may decide to adopt a common currency; Britain may join the Common Market. Increasingly, the United States finds herself in a position where even to maintain the existing level of her international activities and responsibilities, she too is impelled to move further toward systematic forms of integration of the direction of her economy with that of the other advanced Western countries and Japan. She needs their sympathetic cooperation herself so as to be able to go on bearing burdens for the defense of the noncommunist world and for its economic development outside Europe, without becoming excessively inhibited by anxiety about America's gold supply. And also, the United States wants their cooperation in making increasing contributions of their own to that defense and to that development assistance. Of course too, she wants them to unite among themselves for the same reasons that she has wanted them to do so since the days of the Marshall Plan—to become more prosperous and stronger bulwarks against totalitarian penetration.

What is new is the greater sense that the promotion of integration among Western countries is not something that benefits the United States only indirectly, albeit very importantly. Increasingly, it is seen as something of which the United States stands in such direct need, in order to accomplish her other vital international purposes, notably containment, that she should consent to become more tightly involved in the European integrative arrangements herself in order to influence and utilize them constructively. Unable any longer to confer upon Europe the benefits of much economic aid, or even, at least in the short run, the benefits of much more access for trade in American markets, and unsure of her international financial ability even to continue indefinitely her military expenditures on the order of $1.5 billion annually in Europe, the United States seeks to elicit a generous response there through a widening sense of commitment to true partnership in shared goals of prosperity and containment and in shared burdens of defense costs and Afro-Asian aid.

The newly heightened awareness in America that her interdependence with Western Europe is truly mutual, not utterly one-sided, if her economic instruments of policy are to be effective, bears a resemblance to the heightened awareness in America after the advent of "sputnik" that her interdependence with Western Europe is likely to have to be truly mutual, not utterly one-sided, if her military instruments of policy are to be effective (cf. pp. 182–84).

As in that instance, the United States public has been slow to absorb the lessons but willing to allow wide latitude to policymakers to respond as they see fit. And as in that instance, the early reaction of the policymakers was near panic—a frantic 1957 grab for European launching sites for primitive American IRBM's and a frantic 1960 Treasury Department grab for foreign cash where the hoard seemed most tempting, West Germany. In neither instance was the initial European response to American importunities cordial; yet in neither instance were the United States officials turned away empty-handed. America eventually got IRBM launching sites in Britain, Italy, and Turkey, though not until after they had ceased to seem so urgent. America in 1960 got large temporary German contributions to bolster her balance of payments, though not a satisfactory annual arrangement for regular contributions—at least not at the time when it seemed most urgent. Yet the negotiations have continued, and there has been good reason to believe that the United States—even if she finds little new to offer—can gradually manage to evoke sufficiently generous cooperation from the OECD in general and the Germans in particular in mitigating her gold problems, just as there has been reason to believe that she can gradually manage to evoke sufficient military cooperation in NATO. In neither case is the American ascendancy great enough any longer for the United States to put much reliance on pressure or virtual bribes. But an increasingly institutionalized sense of mutual commitment to shared purposes, cooperatively determined, gives reasonable hope that adequate solutions can be reached—even more in the economic sphere than in the military, because in European-American relations the economic problems are much the less critical, thanks largely to what was accomplished jointly in the Marshall Plan era. The economic integration of Europe may be broadening out toward integration of all the Atlantic Community, and perhaps even to include Japan.

No such confidence, however, can be expressed for the future of the less-developed, non-European countries of the noncommunist world. Yet Europe's continued access to an expanding trade with them is as vital to her economic viability and democratic political health as is the intermeshing of the advanced economies themselves. America's own reasons for concern about the emerging nations include interests that derive from her partnership with Europe as well as some that may be peculiarly her own. All the way from such narrow anxieties as arise from the first waves of a rising flood of simple Asian manufactures, with which neither American nor European light industry can hope to compete unprotected for long, right up to such paramount concerns as the future allegiance of a "middle billion" uncommitted people, the principal interests that America shares with Europe are linked to interests that the developed countries each have in their relations with the less-developed. Under the strains of competitive coexistence in a revolutionary world, what has been gained since 1946 in Europe could still be lost in Asia and Africa and even in Latin America.

The United States has been entitled to take much satisfaction from the abundant evidence of her success in Western Europe during the postwar decade in establishing a momentum of advance that was destined to go forward toward new heights of achievement, not merely to stagnate on a new plateau with standards like those of the interwar era. The contribution made by American aid and advice was on the whole a signal demonstration of what can be accomplished by intensive economic assistance when dealing with countries already possessing most of the habits, skills, and material structures of industrial civilization. Where these habits, skills, and structures are wholly lacking and must be created rather than renewed, the obstacles are much greater, and only a few of the same techniques may be applicable. Herein lie the most pressing current problems in the use of economic instruments for American foreign policy.

## SELECTED BIBLIOGRAPHY

ALLEN, ROBERT L. *Soviet Economic Warfare.* Washington, D.C.: Public Affairs Press, 1960.

AUBREY, HENRY G. *United States Imports and World Trade.* New York: Oxford University Press, 1957.

BALL, M. MARGARET. *NATO and the European Union Movement.* New York: Praeger, 1959.

BIDWELL, PERCY W. *Raw Materials: A Study of American Policy.* New York: Harper, 1958.

———. *What the Tariff Means to American Industries.* New York: Harper, 1956.

CAMPBELL, ROBERT W. *Soviet Economic Power: Its Organization, Growth, and Challenge.* Boston: Houghton Mifflin, 1960.

GORTER, WYTZE. *United States Shipping Policy.* New York: Harper, 1956.

HUMPHREY, DON D. *American Imports.* New York: Twentieth Century Fund. 1955.

MIKESELL, RAYMOND F., and JACK N. BEHRMAN. *Financing Free World Trade with the Soviet Bloc.* Princeton: Princeton University International Finance Section, 1958.

PIQUET, HOWARD S. *Aid, Trade, and the Tariff.* New York: Thomas Y. Crowell, 1953.

SALANT, WALTER S., and BEATRICE N. VACCARA. *Import Liberalization and Employment: The Effects of Unilateral Reductions in United States Import Barriers.* Washington, D.C.: Brookings Institution, 1961.

SCHELLING, THOMAS C. *International Economics.* Boston: Allyn and Bacon, 1958.

TRIFFIN, ROBERT. *Gold and the Dollar Crisis: The Future of Convertibility.* New Haven: Yale University Press, 1960.

UNITED STATES CONGRESS HOUSE COMMITTEE ON WAYS AND MEANS. *Foreign Trade Policy: Compendium of Papers on United States Foreign Trade Policy.* Washington, D.C.: Government Printing Office, 1958.

UNITED STATES DEPARTMENT OF COMMERCE. *Balance of Payments Statistical Supplement.* Washington, D.C.: Government Printing Office, 1958.

———. *Survey of Current Business,* monthly. Washington, D.C.: Government Printing Office, 1956– .

CHAPTER 16

# THE UNDERDEVELOPED WORLD BETWEEN

AMERICA'S traditional interests in the non-European world centered on cultural and commercial relations, but in a more limited sector, East Asia, Latin America, and particularly the Caribbean, she had long been sensitive also to national security considerations. Cultural relations were mainly conducted through various forms of missionary activity: religion, health, education, and social uplift in Asia and Africa. Commercial relations included some overseas investment—mainly in the extraction of raw materials and most of that in Latin America—and the exchange of goods, with the maintenance of trading opportunities for Americans equivalent, if possible, to those of the "most favored nation." Such privileges were difficult to preserve in colonial areas when the non-Western world was being partitioned in the nineteenth and early twentieth centuries. Thus American policy was remarkably active with regard to Morocco in the decade before World War I and in respect to Persian Gulf oil in the decade after that war, and, of course, the Open Door to China held entangling lures. But after the mid-nineteenth century there was no "backward area" toward which American policy was colored mainly by national defense considerations, except the Caribbean and the general area of the Philippines.

This picture has changed almost beyond recognition in the past quarter-century. Interests in profitable trade persist, and the quest for raw materials grows keener than ever as American economic expansion outruns domestic sources of supply. Private philanthropic activity continues to be vigorous, though less specifically religious in impetus and objective than in previous generations. The expanded overseas work of the great foundations like Rockefeller and Ford is important in this respect, and even the more traditional missionary enterprises reflect a heightened awareness of the social and political consequences of their efforts in backward lands.* Yet the main concerns now shaping American policy toward underdeveloped areas are not private philanthropic or commercial concerns at all, but are national security concerns centered on the containment of Communism. Lenin's directive that the "road to London and Paris runs through Peking and Calcutta" (see pp. 167–69) has become a guidepost for Americans moving to block that path, not just at Calcutta, but everywhere else in the less-developed regions.

## Dynamics of Multiple Revolution

To do so, moreover, in a world of revolution, for the underdeveloped areas are in rising turmoil, churned by the varying impact of ideas and techniques that have already kept Europe unstable for 300 years.

There are the technological revolutions: in agriculture, in communications, in commerce, in industry. The perspective of the tumultuous history that has attended these transformations in Europe may teach some patience to contemporary observers in countries now advanced when they survey the less developed—but surely there is little comfort to be derived from such comparisons for men of this century or the next.

Linked to the technological revolutions is the "population revolution," in which the death-rate falls unaccompanied (at least until much later) by any corresponding fall in the birth-rate. The result in Europe has been a fivefold increase in the population in 300 years; for the white race

* One social and political consequence of which Americans have become aware is a widespread resentment against the term "backward." For several years the standard euphemism was "underdeveloped." But native inferiority complexes cannot be so easily assuaged, so now the official American usage is "less developed." Nearly everyone is less developed than someone else, but probably the hypersensitive will impute some aura of contempt to this term also, and it will have to be changed, perhaps to "newly emerging" or simply "emerging." Meanwhile, the author of this book, after some hesitation, has decided for stylistic reasons to use "backward," "underdeveloped," "less developed," "newly developing," "newly emerging," and "emerging"—all interchangeably. Absolutely no invidiousness is intended—only the desire to avoid wearisome repetition. The terms refer to all the world except for these areas: the United States, Canada, Australia, New Zealand, South Africa, Japan, Russia, and the rest of Europe. (Obviously the "advanced" countries also differ importantly among themselves in degree of backwardness or development.)

as a whole, scattered worldwide by emigration, a sevenfold increase in that period. Japan's population has tripled in about one century; the Indian subcontinent (India and Pakistan) has multiplied almost as fast and now gets 50 million additional mouths to feed every decade. The one-decade increase in India is almost as great as the total intercontinental migration of persons in all directions during the entire period from 1800 to 1924, the great period of migration before the United States shut her doors. Obviously, even if America's gates were reopened to Asians on a scale vastly wider than the present token numbers (a few hundred a year), scarcely any alleviation would be noticed in the population pressure of South Asia.*

The technological revolutions and the pacification and public health measures which brought the population revolution were in most underdeveloped areas the products initially of Western colonial domination. Typically, the first response of native groups to contact with the white man was a kind of amused curiosity; but this was followed shortly by a recognition of the white man's contempt, which commonly led to a zealous nativistic reaction moving in the direction of ultra-orthodox traditionalism. However, if the Westerner for any reason were determined to entrench himself, such a primitivist attitude could not long prevail. Increasing numbers of native leaders would come to respect the West for certain kinds of superiority and acquire a determination to borrow for themselves whatever skills were needed to win "freedom from contempt," to be able to insist on being treated as equals. In this new imitativeness was embedded perhaps the most far-reaching change of all: a borrowed concept of the possibility of progressive change itself —the West's "idea of Progress," new to much of the rest of the world and metamorphic in its kindling impact. The spreading consequences have been tagged a "revolution of rising expectations." Thus employed, the word "expectations" is deliberately ambiguous in meaning, suggesting both anticipation and demand. "Rising aspirations" may be a more universally applicable expression, if the aspirations are understood to be mainly concerned with prestige and material welfare. But soon, in any case, a recognition of the possibility of lasting progress does turn aspirations into demands, a surging pressure for change, which must be granted, repressed, or deflected by the men in power whose authority is no longer simply taken for granted.

The demand is for progress to be speeded up. The three or four centuries that modernization took in Europe is utterly unacceptable as

* The quota for each of the Asian countries except China and Japan is 100 persons a year. The Chinese and Japanese quotas are each about twice as large as that.

a goal to awakened elements in the backward regions. They are looking for short-cuts, but where they look and what they find alluring varies greatly. Even in the same society at the same time different groups pick different models, and the significant groups and their models change as the society develops through the passage of time. The consequences are of course unsettling. They are also highly unpredictable, because, among other reasons, even similar societies hit by similar outside influences confront the impact at different stages of world history. A newer pattern of technology is "the latest thing" to copy, and its social and political effects may be novel. The prevailing ideological and power-political climate of the world at large changes: rebellious Arabs who found nazism attractive in the thirties may for almost the same reasons look to Moscow today. The later "maturing" groups and societies can always find in the experience of others notions of what, from their own perspective at least, should now be avoided.

A recurrent factor among these elements of unrest and unpredictability is the long span between current realities at any time in any underdeveloped country and the blueprints for social transformations that are available for copying from existing achievements in advanced nations. The gap between what is actual and what has come to be believed to be practicable by important segments of the population is much greater in backward lands in the twentieth century than it was in previous centuries in Western Europe. The development of the countries of the North Atlantic was of necessity rather tentative and gradual in spirit. No concrete vision of an industrial utopia guided the men who began the wave of enclosures of commonly-shared landholdings in eighteenth-century England; their recognized goal of a more productive agriculture was not so extremely remote from the status quo. Although the transformations they wrought were vast and painful, the time historically "allowed" for the changes nationwide was several decades, not just several years. Yet once the ultimate outcome of such a process could be clearly discerned to be some desirable pattern of industrialization, the temptation would become great elsewhere in less-developed countries to push the public through the preliminary stages much more rapidly. Even if intended to be merciful, hasty disruptions may thus ensue that are so bewildering as to engender unforeseen nonrational responses in the emerging lands. Similarly, although the process of laying a foundation for modern industrialism required an initial inventiveness and experimentalism in the West that accorded well with the comparatively permissive and mobile societies there, the structure once established can later be copied by authoritarian societies. Indeed, authoritarians can accomplish the imitation with swifter efficiency, at least in the short run, than can be achieved in other underdeveloped societies that try to be permissive.

Thus the fact that change is to be imitative yet accelerated means that it must largely be planned, not spontaneous. But change according to plan tends to widen the gulf between dominant elites and submissive masses that has always been a characteristic of most of the backward societies. Portions of the older elites are converted to modernization; others are overthrown and replaced. For a time, at least, there is a greater difference than ever between the outlook of the actual or potential leaders, oriented toward one or another advanced external model, and the mass of the population, which remains tradition-bound and illiterate, often poverty-stricken, and usually heterogeneous and particularistic in loyalties. The most convenient tendency is for the progressive elites, lacking a traditionally accepted legitimacy for their objectives and perhaps even for themselves, to resort to more modern forms and instruments of authoritarian rule. From this, it is often an easy step for them to consider most congenial among the various creeds that emanate from advanced countries those which justify ruthless tutelage of backward masses by elite vanguards.

Even if authoritarianism is not accentuated, the personal position of the native elite gives it a very powerful impetus for seizing upon the gospel of nationalism which Europeans have disseminated. Within his own society the member of the elite has a status from which he feels entitled to derive high self-esteem, but he recognizes that no such respect is accorded him in international society. To be treated as an inferior is personally galling. Yet since it commonly results from that fact that he is identified with his backward compatriots in the eyes of the foreigners, he can win his own "freedom from contempt" only if his whole nation can be rallied to achieve its emancipation as far as possible from every form of dependency. Thus the drive for equality for the nation among nations almost always has top priority in a newly-developing area. The status of the native elite is thereby enhanced, even if this sometimes be at the cost of postponing social and economic improvements for the masses. Yet the masses must somehow be aroused, at least with the national revolutionary sentiment, if the drive is to succeed, and the inculcation of nationalism does have some important constructive consequences. Particularistic barriers of class, caste, localism, and ethnicity are combated and largely surmounted by a wider sense of community and of civic obligation. The idea of progress is spread more broadly, and the demands it soon engenders are not limited to anticolonialism.

In particular, the more far-reaching demands will be pressed forward by the growing element of the elite that is commonly called the native intelligentsia: the lawyers, journalists, teachers, students, and the like, who have acquired at least the rudiments of a modern education—enough to make them misfits in traditional society. Increasing numbers of them come from origins that are closer to the masses than to the traditional elites, and may lack connections to obtain suitable white-collar positions in the underdeveloped community. Bitter resentment, aroused by the feeling that their own talents are unutilized and unappreciated in the existing society, is also kindled by their direct familiarity with the condition of the masses, to which they are sensitized by an awakening social conscience that is also largely European in derivation. All this bitter brew is compounded with the sense of national humiliation they share with the other more successful native elites. The passions roused for the anticolonial national revolution are turned toward the extreme of full-fledged social revolution.

## Affinities for the East

Thus, finally, in the twentieth-century contact of advanced and backward peoples, the various technological revolutions tend to interlock with a revolution of rising expectations, with national revolutions, and with social revolutions. Obviously, in all the ferment there are enormous dangers of nationalist excesses of all kinds, including aggression, and of intensified authoritarianism to the extreme of totalitarianism. And there is the related danger that among the great industrial nations the one with which these embittered elites in underdeveloped countries will find it most congenial to associate will be the one most unashamedly authoritarian; the one most rapid in building its power among nations; the one most revolutionary, both in its own professions and in the truer reality of its opposition to the international status quo and particularly to the Western colonial powers; and the one able to create the conviction that it is the most likely to succeed in making what it wants of the world.

To be on the winning side for a change, to ride on the "wave of the future"—this is a profoundly exhilarating lure for millions of persons in the modern world who know that their passion for complete independence can only be formally, not actually, achieved. If they must be dependent upon greater powers, they would prefer at least to avoid the penalties of futile resistance and perhaps even to share some psychic and material benefits from collaborating with the winners.

A generation ago, Hitler's Germany possessed for much of the Middle East and some of Latin America the magnetic attractions of revolutionary success attained through surging authoritarian power. The image of

Japan had a similar appeal in much of Southeast Asia. But the Axis success turned to failure, and in the case of Japan her own behavior as conqueror contributed further to disillusionment in the underdeveloped areas. They wanted new models and found them ready at hand: the Soviet Union and the Communist regime in China, both able to offer also a more plausible and pertinent ideology than the Axis powers had been able to extend to less-developed peoples, and superficially a more relevant image of development from backwardness than Germany or any major Western democracy could present.*

### Affinities for the West

A refusal on the part of Americans to recognize these attractions its adversaries have for the less-developed peoples in this revolutionary age would be a perilous delusion. A realistic view must be one of grim foreboding. But there are other important factors at work that remain favorable for the West.

1. The sense that Communism is the "wave of the future," that its ultimate triumph is assured, and that the resister has nothing to hope for but martyrdom, is not, at least not yet, the prevailing current of thought outside the Soviet sphere. To forestall a fatal spread of infectious defeatism is correspondingly the foremost object of an uncompromising and almost indiscriminate application of the policy of containment.

2. Although the drives of nationalism in underdeveloped areas take as their first main targets the locally predominant foreign powers, which because of past history must nearly always be Western European or American, this urge for independence can later find satisfaction in a neutralism that regards all great powers with suspicion and jealously avoids subordination by trying actively to keep one power balanced against another or by withdrawing cautiously into a quasi-isolationist position. In such situations the West can usually manage to content itself with knowing that the less-developed nation is at least not firmly committed to preferential collaboration with the enemy Soviet bloc. The most menacing challenge of Communist power arises from its massively integrated and readily maneuverable worldwide structure. In most circumstances even the countries that become deeply alienated from the West need cause only limited anxiety if they can be expected to avoid becoming dependably submissive to Soviet dictates in the cold war.

3. Indigenous cultures in most underdeveloped countries possess

* On the ideology, see pp. 167–70.

important humanistic and religious elements that persistently moderate the tendencies toward ruthless extremes of revolution and dictatorship.

4. In those colonial and semi-colonial areas where Western tutelage, mainly British and American, has been directed toward giving preparation for an early grant of independence, some of the heritage of principles and practices of constitutional democracy is implanted. The shoot is tender and likely to wither, but some traces can be expected to survive, tending to promote a Western orientation.

5. If the West conceives its immediate purpose as one of forestalling a spread of totalitarian regimes rather than everywhere seeking quixotically to establish capitalistic, constitutional, democratic welfare states, many different types of indigenous regimes may be seen to warrant some support. The United States is not contravening its ultimate preferences when it puts its interim emphasis simply upon keeping open the possibility of political change, in the hope that such change, if it comes, may be advantageous for the local population and for the West. Only the true totalitarian regime, by its ruthless and omnipresent controls, wholly forecloses until some remote period the possibility of its own transformation. Under other regimes, even the milder forms of dictatorship, there is opportunity from outside and inside to exert influence for improvements, reinforced through the possibility that the government itself could be changed, whether by election, *coup d'état,* or revolution. Thus as long as the irreversible choice of totalitarianism is averted, the West can usually afford to take a patient and rather tolerant view of the painfully lurching processes of political evolution in backward lands. Dictatorship, if nontotalitarian, may even be welcomed as a useful temporary phase for certain countries. But wherever and whatever a particular phase of development may be, the West does feel an interest, under the containment policy, to exert its active influence against totalitarianization—by means of advice, assistance, and sometimes pressure. Beyond the minimum, antitotalitarianism, other values of course also deserve promotion by the West. But overall success need not be measured mainly by the limited progress that may be achieved in immediately effectuating the high ideals of democratic liberalism.

6. The leading capitalist powers possess a great capacity for aiding the economic development of the backward countries. To be sure, if durable constitutional democracy depends on the prevalence of living standards anywhere near as high as those in the Western countries where it has in fact evolved and thrives, no conceivable amount of assistance would make transplanted democracy viable in this century. The standards in backward lands cannot be pushed so high so soon. But if mere evidence of a trend toward economic advancement and the prospect of

gradual improvement are sufficient to check the impatience of the politically active elements of a society, the West may be able to supply from outside the margin of hope that does make nontotalitarian institutions workable, even borrowed institutions of true constitutional democracy.

7. Economic assistance alone would rarely be enough; other kinds of aid and guidance, direct and indirect, would normally be needed. The process of modernization is highly complex, involving many closely interrelated changes in the psychological, social, and political dimensions of life in the emerging nation. Economic aid cannot be expected by itself to produce economic development, nor can economic development normally be expected by itself to produce political development of a kind the West could find congenial or beneficial. Thus the outside influences likely to be needed for an advantageous course of development are not only economic but may have to intrude more directly into indigenous politics. Yet the advanced Western countries can probably learn to provide these ticklish influences also, while showing whatever degree of tact is required locally.

## SELECTED BIBLIOGRAPHY

ALMOND, GABRIEL A., and JAMES S. COLEMAN, eds. *The Politics of the Developing Areas.* Princeton: Princeton University Press, 1960.

BAUER, PETER T., and BASIL S. YAMEY. *The Economics of Under-Developed Countries.* Chicago: University of Chicago Press, 1957.

KINDLEBERGER, CHARLES P. *Economic Development.* New York: McGraw-Hill, 1958.

LERNER, DANIEL. *The Passing of Traditional Society: Modernizing the Middle East.* Glencoe, Ill.: Free Press, 1958.

MILLIKAN, MAX F., and DONALD L. M. BLACKMER, eds. *The Emerging Nations: Their Growth and United States Policy.* Boston: Little, Brown, 1961.

MYRDAL, GUNNAR. *An International Economy: Problems and Prospects.* New York: Harper, 1956.

———. *Rich Lands and Poor: The Road to World Prosperity.* New York: Harper, 1958.

ROSTOW, W. W. *The Stages of Economic Growth: A Non-Communist Manifesto.* New York: Cambridge University Press, 1960.

STALEY, EUGENE. *The Future of Underdeveloped Countries: Political Implications of Economic Development,* rev. ed. New York: Harper, 1961.

THEOBALD, ROBERT. *The Rich and the Poor: A Study of the Economics of Rising Expectations.* New York: Potter, 1960.

# CHAPTER 17

# PRIVATE FOREIGN INVESTMENT FOR ECONOMIC DEVELOPMENT

Any American program designed to contribute to the economic development of the emerging countries, whatever its motives, must reckon with the fact that the United States herself is a capitalist nation in which the main reservoir of industrial and commercial talent is not in government but in private corporations—a country also where experienced welfare agencies function largely under independent auspices. Of course, specialist personnel can commonly be obtained by the government for temporary service, for consultation, and for missions at home and abroad. And in the fields of health, education, and agriculture—and to some extent even in construction, transportation, and communications—the regular civil service is traditionally equipped to play a much larger role than it is in mining, manufacturing, or distribution. Of course, too, the government can use its tax powers to secure vast financial resources for its foreign development programs. Yet despite these reservations the importance of individual and corporate initiative is so pervasive in America that one can hardly ponder the situation for long without recognizing that private or semi-private enterprise is bound to play the major role in most of the large-scale economic development operations the United States promotes overseas—assuming that Americans try to retain close supervision of them at all. Whether an operation is directed by a private outfit acting mainly on its own initiative, by a private outfit under United States government contract, or by personnel temporarily hired by a government overseas agency, the government is heavily dependent upon the private sector of the American economy and upon voluntary agencies for the success of any extensive overseas development program.

In some of the underdeveloped countries, at least temporarily, a fear of political entanglements with the United States government may make wholly private voluntary agencies the most acceptable sources of assistance in various fields of relief, education, and even administrative guidance. The overseas work of the Ford and Rockefeller foundations and their various adjuncts and of CARE and the church missions makes a highly valued contribution in many places.

In some kinds of activity, the need to use private channels arises from prejudices not in the receiving countries but in the United States. One of the most important of all the problems connected with economic development, for example, has thus far received utterly inadequate attention by any major American overseas unit, public or private: the need to check population growth in most of the underdeveloped countries, if there is to be much prospect that increased production can produce a significant rise in the standard of living and thus help to relieve social and political tensions internally and internationally. Religious influences in American politics must be expected to continue to make this crucial problem taboo as an object of United States government activity. Even the major private organizations have thus far generally shied away from sponsoring the research and dissemination of birth control techniques for the impoverished peasant peoples. Yet there is at least a possibility that the nongovernmental institutions could undertake to play a major role in this field of activity, which, despite its importance, is presumably foreclosed to officialdom. Without their participation, economic development, even industrialization, may provide the future elites in the partly industrialized countries with more home-produced cannon and more home-grown cannon fodder—but the latter, the masses, will be still so depressed as to require stimulation by chauvinists, who may even raise alluring hopes of spoils from recurrent wars.

These dangers, however, would arise even sooner if the West were not already providing other material assistance for economic development, without waiting for such improved prospects of ultimate success as would flow from an effective limitation of population growth. Most of the assistance is a mixture of public and private activities, but clearly, if there is to be any remotely adequate transfusion of resources from advanced capitalist America to the underdeveloped nations, wide opportunities must be provided for the profit-seeking as well as for the philanthropic impulses of United States citizens to operate, with various degrees of linkage to government programs for development assistance.

## Some Financial Comparisons between American Governmental and Private Contributions in the Less-Developed Countries

Epitomizing the actual relationship of the two sources of assistance in crude financial terms, in the late 1950's the principally private American contributions (investment expenditures and gifts) totaled about as much to the underdeveloped noncommunist countries annually as did all the various United States government nonmilitary grants and loans to those countries. The general order of magnitude of each source was about $2 billion per year. In the 1960's, governmental contributions have pulled ahead, but the private remain very important.

The nongovernmental total for the less-developed countries includes the outflow of capital for investment in branches and subsidiaries of American firms, minus any such capital that was brought back to the United States; the part of the earnings of those overseas units not distributed to the stockholders, but reinvested in the overseas units for expansion; the maintenance investment in the units made from their own depreciation and depletion reserves; private loans and purchases of securities, minus repayments of past loans and sales of securities; * and finally, any personal and institutional remittances, such as gifts to relatives and donations through church missions.

All of these contributions to underdeveloped areas from private American sources, approaching $2 billion a year in the late fifties, were matched by a roughly equivalent amount (now much more) in United States government nonmilitary grants and loans, which were also very largely channeled through private American firms. The totals, however, somewhat conceal important differences in emphasis among the various less-developed regions. The private investment contribution by Americans in southern Asia from West Pakistan eastward to the Pacific is much smaller than the government aid programs. In Israel and the oil countries of the Middle East the opposite situation prevails, as it did, until recently, in Latin America, though not so much as in the Middle East.

* Some of the loans are guaranteed by the government's Export-Import Bank.

More specifically: In 1956 about 35 per cent of the total value of United States investments abroad in branches and subsidiaries of American firms (so-called "direct investments") was in Latin America; about 10 per cent was in all the other underdeveloped areas of the world combined.* The same 10 per cent figure appeared for long-term loans outstanding from American banks to underdeveloped areas outside the Western Hemisphere. And in regard to such American bank loans abroad, fully 57 per cent went to Latin America, an even greater proportion of the total than the 35 per cent figure for direct investments. This heavy preference of American investors and bankers for Western Hemisphere outlets was also vividly evident in the 1956 statistics which showed that of the direct United States private investments in underdeveloped areas outside Latin America (small anyway), 90 per cent were in oil, mining, or smelting. Consolidated, these figures mean that only about 1 per cent † of all United States direct investments abroad went to backward lands outside the Western Hemisphere for anything but extractive industries. Apparently there is a large potential there for expansion of American private activity if satisfactory incentives can be provided.

Meanwhile, especially in the noncommunist borderlands of South and Southeast Asia, the need for development at a pace fast enough to meet the immediacy of the Communist challenge makes extensive United States government economic assistance programs also essential. The coexistence of government and private American programs obviously calls for cooperation between them so that they will complement not compete with each other, and promote genuine economic development. It would be naïve to suppose that anywhere near all the government grants and loans are truly "developmental," or that general economic development is now made to flow from every private gift and investment expenditure. A great deal is wasted. Nor is development the only legitimate purpose for United States public or private spending in the backward countries. Sheer charity and numerous special advantages in politics and profits are also worth some cost. But with the demand for economic development now clearly irreversible in the modern world and capable still of being steered in a direction favorable to the survival of the West, there is need for precise understanding of what kinds of contributions toward nontotalitarian economic advance are made by American companies operating abroad on their own initiative, and what kinds

* Values here are "book values"; "market values" would be much higher in totals, but the regional ratio would probably not be very different.

† Book value about $230,000,000.

of complementary developmental activity by United States government agencies can vitally assist and guide both the companies and the foreign country, with particular concern for its political tendencies.

## The Contribution of the Private Foreign Investor

The portion of American private investment abroad in short-term speculative ventures, anticipating quick profit-taking and early repatriation of the capital to the United States, makes a much less significant contribution to economic development than do the ventures that are intended to continue indefinitely, and that reinvest a considerable portion of their earnings in expanding existing enterprises or establishing new ones. The long-term investments can provide important continuity in the transfer of American capital and technology. A government agency that depends upon Congress for annual appropriations or periodic extension of its borrowing and lending authority often finds great difficulty in adopting and preserving a perspective with so fixed an emphasis upon long-term economic advantages and in sustaining the related flow of necessary interim assistance. The businessmen, who have a concrete, repeated measurement of success in the form of profit figures, may also be more persistent in seeing to it that each particular enterprise becomes deeply rooted in the overall economy, that it is not just one of numerous scattered projects which have been completed, but which must await the completion of many others according to an official master plan before any of them can really become integral to the developing economy. In this connection, the far-sighted American corporation can deliberately encourage a diversification of the local economy to supply its needs. Many kinds of goods and services the company feels obliged to provide for its employees and for its operations must at first be imported from outside by the company itself, but can gradually be obtained from local sources that have been consciously fostered for this purpose. A related and valuable result is that the businessmen of the district, finding a reliable market in the company and its personnel, are stimulated directly, and also indirectly by the force of example, to expand their own operations; in addition, wealthy persons are induced to risk investing their own capital in shares of local enterprises. The traditional reluctance of such persons to invest in anything except real estate, safe currencies, or foreign securities may be partly overcome. If it could ever be fully overcome, very important resources of capital would thereby become available for economic development in most of the backward areas, and correspondingly, the need for both private and governmental capital from advanced countries would be much less pressing than it is.

In addition to inculcating a new spirit of enterprise and providing stimulating opportunities for it in a widening range of complementary businesses, the Western firm is often in a position to give examples of the profitability of low-mark-up, high-turnover merchandising, thereby broadening the size of local markets, fostering mass production, and making the benefits available to larger segments of the population.

The process of stimulating more local enterprise helps to increase the number of distinct centers of initiative in the society as a whole. The traditional notion in backward countries that everything that is new and big must be done by top officials in the nation's capital is of course a fundamental brake upon the pluralization (or "democratization") of decision-making that is a necessary social concomitant of constitutional political practices. Even where no meaningful political democracy is yet in view (but totalitarian methods are being avoided), an underdeveloped society is not so rich in top-level planning and managerial skills and communications techniques that it can wisely afford to try to advance its economy without taking advantage of widely dispersed local initiative, spontaneous effort, and fresh imagination.

Of course this multiplication of spheres of enterprise tends not only in general to foster economic development and improve the social foundations for democracy, but also in particular it helps to establish local vested interests in private capitalism which can be a helpful buffer for an American company in the local political setting, at least as long as they regard themselves mainly as complementary to, rather than competitive with, the Americans. But the usefulness of this identification for the United States firm depends very largely upon the degree of respect that local capitalists themselves enjoy in their society. Unfortunately, this is likely to be low, with the dual result that contempt or resentment toward local businessmen is superimposed upon other xenophobic kinds of hostility toward foreign businessmen, and the native capitalists themselves are tempted to join or even lead the angry chorus in hope of deflecting the opposition away from themselves. A growing awareness of these tendencies gives special incentive for the American companies to show an increasingly far-sighted and humane social conscience in their operations, seeking by publicity, guidance, and example to create a favorable image not only of themselves but also of wider possibilities for "business statesmanship" on the part of local private enterprise.

The image, however, is not, at least not yet, anywhere near so rosy in most underdeveloped areas as the foregoing paragraphs by themselves might imply. Private investments from advanced countries have historically not been guided usually by such far-sighted social and political considerations. Very deep and widespread resentments have in fact been generated. Partly, these reflect and are exacerbated by native inferiority complexes and xenophobic reactions, but they also tend to center upon certain concrete complaints that demand attention. Foremost is the extreme emphasis upon extractive industries and plantation agriculture in foreign investment in the less-developed areas. Outside the Western Hemisphere, 90 per cent of direct American investments in backward countries in 1956 were in oil, mining, or smelting; even in Latin America 46 per cent * of United States direct investments were also in those extractive industries, and only about one-fifth was in manufacturing. Yet manufacturing, particularly heavy industry, is the principal criterion of economic advance for progressive native elites in most underdeveloped countries. Above all, to have their own national steel mill has become a paramount symbol of their equality among nations. This will to industrialize seethes with all the complexes of adolescent nationalism. It rails with particular venom at the "plunder of our resources" that is involved in mining and drilling operations (the minerals and petroleum being irreplaceable). "Better to leave everything in the ground; eventually our own future generations will learn how to exploit it for themselves!" Plantation agriculture does not raise the same specter of the nation denuded—only a specter of the nation dependent upon world markets. But that in itself is enough to arouse local anxiety and resentment.

Americans can point to the willingness of their large raw-materials enterprises to construct for their own operations essential facilities that are also of use to other broad segments of the developing local economy —including, for example, seaports, railroads, highways, power stations, and other utilities, provided by a fruit company; and the American companies' payments of local taxes and royalties can enable the backward country's government to finance a much wider and more diversified development than what the Americans are willing to promote on their own. They can also point to the availability of native personnel trained in the American enterprises, but free to leave to contribute the newly acquired skills to other phases of local economic development. The extraction of raw materials for export also provides the emerging country with foreign currency, for example, dollars, that can be used to buy

* Book value about $3.57 billion. Cf. p. 334, including footnotes.

imports for a balanced development program that will in time obviate the temporary dependence on world markets that is considered excessive.

Pursuing this case for gradual economic development, Americans are further inclined to argue that the diversification of the local economy that will then proceed in the field of manufacturing should first put emphasis upon the lighter forms of industry, for example, textiles, for which the requirements of native managerial and labor skills and of capital investment are relatively low, and which produce mainly consumer goods quickly available to improve local living standards.

But this whole line of argument is vehemently condemned as a Western capitalist rationalization for indefinitely keeping the less-developed countries in a dependent status economically and hence, in everything but legal form, politically subordinate also. The fact that American businesses do intentionally derive special advantages for themselves from this pattern of economic development can scarcely be denied, nor can the superior existing power and prosperity of the advanced industrial countries, which cannot soon be matched by any underdeveloped country relying mainly on the gradual pattern of progress through private profit incentive. Americans point to the other values for the developing country that are inherent in this arrangement and insist that their fruition would be worth waiting for, but bitterly jealous impatience remains an inescapable factor in the situation in most of the backward lands. Modern "chastened" American capitalism does have its wealth and its virtues to contribute in these countries, but usually it must also be supplemented and somewhat guided by the United States government if there is to be much durable receptivity among progressive native elites toward its performance.

## Commodity Price Fluctuation and Stabilization

There is a particular hardship felt with embittering keenness in most of the less-developed countries that find their economies becoming heavily specialized as the result of a concentration of foreign investments upon the production of a limited number of raw materials: prices of basic commodities in the world market tend to fluctuate severely.

Probably the prices of such materials, considered as a group, do not actually fluctuate much more over a period of decades than do the prices of the manufactures that the advanced countries have for sale, but such comforting averages would be misleading, because particular countries tend in early stages of development to be extremely dependent on particular commodities. If the price of rubber breaks, Malaya cannot offset the loss by taking advantage of a simultaneous rise in the price of coffee, any more than she can benefit from an increase in machine-tool prices. What counts for Malaya is rubber and tin. If those prices fluctuate sharply, the whole economy suffers. A similar situation obtains in many other underdeveloped countries, and the efforts to cope with it by international action, private or governmental, are still fragmentary.

The fluctuations severely limit the usefulness the raw materials exports can otherwise have as a means of financing more diversified economic development in a particular backward nation. When prices of its key commodities fall, the decreased overseas earnings leave a smaller margin available to be spent to import development goods (for example, factory equipment), above and beyond the indispensable continuing imports of consumer goods. Also the reduction of total imports cuts down the customs receipts, and the slowed economic activity lowers general tax revenues—with the result that the local government has less money itself to spend on development projects. A decision then has to be made whether to cut back the development program correspondingly, or to persevere by relying more heavily on inflationary deficit financing and on locally improvised sources of equipment. Both alternatives are frustrating.

Theoretically, these difficulties should be overcome when the cycle turns and the key commodity prices rise again. Reserves of foreign currency earnings should accumulate once more, and the government should be able to tax away the new prosperity incomes and establish long-term funding of the development projects, thereby hedging against the next down-turn. However, the political and administrative structure in many underdeveloped countries is too weak to assure that prosperity earnings are managed so prudently. Temptations are great to spend the foreign currencies quickly in order to obtain imports of all kinds, and to use the higher internal incomes in order to increase consumption expenditures—with the result that demand for goods again pushes up the domestic price level. Thus inflation continues endlessly, no matter whether the overspecialized, underdeveloped economy is temporarily in a state of boom or one of bust.

Perpetual inflation further reduces for private investors the attractiveness of the kind of long-term, productive, nonspeculative investment that is most needed if the economy is ever to become diversified. Persons with available funds, natives as well as foreigners, take refuge in short-term ventures of limited usefulness. An active role for government in promoting important development projects thus comes to seem more essential than ever—but the government also finds its plans disrupted by the unpredictable lunges of the inflationary spiral. Tendencies toward inflation would, of course, be present in a developing economy even if the problem of commodity price fluctuation did not exist. But the fluctuations do seriously aggravate the inflationary situation, giving much concrete justification for the diffuse resentments of native elites toward raw materials exploitation.

The quest for cures or palliatives for the commodity instabilities can emphasize action by the producer countries alone or by producers and consumers jointly. The producer country's government, as suggested above, may try to improve its own system of administrative controls so as, in effect, to tax away additional export earnings when they are on the rise and pay the accumulation out, at home and abroad, when the earnings are falling. Or the government may cooperate with individual producers in trying to hold supplies off the world market, thus exerting upward pressure on the commodity price when it is tending downward; in theory the accumulations would later be released for sale in time to prevent the price from rising to heights that could not be sustained. However, this kind of stabilization by an accumulation of "buffer stocks" through the unilateral action of producers is very vulnerable to the advent of new producers in other countries, who will offer the same basic commodity for sale in world markets at lower prices, and also to active resistance in the consumer countries that have not been conciliated. Once, for example, when the Brazilian government tried to support coffee prices by buying and storing the country's vast "surpluses," and then attempted to close loopholes by combining other major Latin-American countries in an international export quota agreement, the effort was largely undermined by a rapidly stimulated rise in coffee production in Africa, outside the restricted area, and by a "consumers' revolt" in the United States, which culminated in a congressional investigation.

Coffee agreements have been revised and expanded and continue to operate, helping somewhat to soften the shocks of price drops. However, except perhaps for a few commodities that come from such a small number of producers that nearly all the sales outside the Soviet bloc can be monopolized by tight cartels (as is the case, for example, with diamonds and nickel), some systematic cooperation on the part of major consumers, not just producers, would be required for any durable stabilization of prices.

Consumer countries may be willing to cooperate for a variety of reasons. They may wish to promote stability and predictability in the related sectors of their own economies. They may have foreign-policy motives for deliberately enabling a friendly country to increase its earnings even at the expense of their own populations. They may simultaneously be major producers and major consumers of a particular commodity, for example, wheat and oil in the United States. (In such cases the American government programs, state and national, that are designed principally to protect selling prices for the domestic producers, by arranging for limitation of production and of imports, have the incidental but obviously important consequence of tending to stabilize world market prices for the same commodities produced abroad.)

Formal stabilization by agreement between producer and consumer nations may take several forms. Simple bilateral bulk-purchasing contracts were common in British commercial relations with Commonwealth countries and with Argentina after World War II; the Soviet Union handles much of its growing trade with underdeveloped countries on a somewhat similar basis today. Agreements of this kind tend to be discriminatory against outsiders, who are likely to resist the privileged trading relationship that has been established between the signatories. And even the signatories, unless extraordinary mutual confidence prevails, must anticipate that the temporary stability they have achieved will suffer shocks when the time comes to renegotiate the contract; cumulative gradual changes elsewhere on the world market will then probably have to be reflected with disturbing suddenness in any extension of the bilateral arrangements.

A more embracing form of stabilization agreement, the multilateral contract, seeks to establish among many nations, producer and consumer, a limited range of permissible fluctuation in the price of a particular commodity. Thus the International Wheat Agreement (until relaxed in 1959) provided that a specific quantity of wheat, amounting to about one-third of world exports, would be made available by producing countries at a fixed maximum price even if market prices should happen to go still higher, and correspondingly, that the same amount would be bought by consumer countries at a fixed minimum price even if the world market price had dropped so far that they could obtain wheat elsewhere at even lower prices. For a few years these arrangements proved to be practicable (though not very important) even in the absence of membership by the world's largest importer, Great Britain, because the United States government was engaged anyway in massive price support operations for its own wheat growers, which had worldwide impact.

For tin, an international agreement has operated that is much less dependent on American self-interest, though the United States as a leading consumer does informally cooperate. An International Tin Council tries to hold price fluctuation within fixed limits by setting export quotas for the six main producing countries, and also by using a joint fund to intervene in the market, buying or selling, and by maintaining "buffer stocks" of tin.*

Tin, of course, is storable, but international buffer stocks for perishable commodities are harder to maintain. The allocation of depreciation is one of the difficulties. It compounds the other problems of financing, which in any case are very complex. Who is to underwrite the joint operations if they are not in fact self-sustaining? The research, planning, and executive decision-making for operating the joint fund and buffer stocks and adjusting the quotas call also for a very high order of managerial talent. Even the periodic efforts to agree on a proper permissible range of price fluctuation, to be preserved by the experts, is likely to be complicated by suspicion on the part of underdeveloped producer countries that the "normal" prices of their exports have been kept unfairly low by the superior power of the advanced consumer countries —hence any stabilization should be around some higher norm—and suspicion on the part of the advanced consumer countries that this kind of upward manipulation of norms is being attempted even if it actually is not. A high degree of mutual confidence among participating powers is needed, but any grouping large enough to dominate the world market for a particular commodity is not likely to share such trust.

The importance of cooperation by consumer countries is so great, in order to deter outside producers, that situations may arise, at least temporarily, where the United States as major consumer could undertake to stabilize the prices of some of its raw materials imports by its own purchasing decisions, with a minimum of formal international agreement. Thus the government's strategic stockpile (valued at about $8.5 billion in 1959) could to some extent be operated as a buffer, with purchases and sales adjusted to the state of the market.

Concerning items that Americans produce themselves as well as import, there would be so much continuing pressure on stockpile managers to make purchases and avoid sales that few if any flexible stabilizing operations would be possible. But for nonperishable items of which the United States is only a consumer, a government import-price stabiliza-

* The Soviet Union put severe strain on the International Tin Agreement in 1958 by sharply increasing its exports to the noncommunist world. The resulting hardship for Southeast Asian tin producers handicapped Soviet relations with those countries, and in 1959, while still refusing to join the Council formally, the USSR offered to send observers to its meetings and agreed to restrict its tin exports to a figure amounting to about 75 per cent of its 1958 level.

tion program relying on adjustable stockpiles might sometimes be helpful.

There is danger, however, that the relative convenience of unilateral (or near unilateral) stabilization by consumer-government action, even if it worked, would have such a monopolistic flavor of dictation by the advanced country that underdeveloped countries would resent it as much as they do the existing instabilities. Therefore, despite all the extreme difficulties, perseverance in seeking formal commodity agreements embracing producer as well as consumer countries is probably of central importance for many underdeveloped countries, in making the conditions of the world market tolerable to progressive elites and making them more receptive to a pattern of development that utilizes private investments from advanced countries in the fields to which the venture capital is obviously most ready to flow—extractive industries and plantation agriculture—with the hope that it will gradually move on into factory enterprises.

## Overcoming the Obstacles to Private Foreign Investment

Whatever the reasons, emotional or calculated, for which the less-developed countries are often reluctant even to try to attract private capital from abroad, the specific barriers the investor usually finds most discouraging are those that are as much inherent in the general social and political situation in the backward area as they are the product of any deliberately organized semi-official discrimination against the outsiders. There is typically an atmosphere of political instability and a socialistic climate of opinion that prefers governmental to private large-scale economic enterprise. These stormy winds may for a time be quieted, but the American investor, from bitter experience in many places, has some reason to fear in almost every underdeveloped country that the gales will rise again to aggravate his other difficulties. Usually deep in his mind is repressed anxiety that everything may suddenly be lost. His investment may be confiscated outright. He may be prohibited from engaging in his particular line of business, for example, oil production, finding it reserved for local firms or the local government. His holdings may be expropriated or nationalized with compensation that amounts to merely a token either because it grossly undervalues his property or because he cannot effectively transfer the proceeds out of the country or both.

A generalized fear of the unknown pervades these common American attitudes toward investment in backward lands. Yet even a sober appraisal of the situation in quiet times must cause the enterpriser to reckon with specific obstacles that may or may not be intentionally discriminatory; often he cannot be sure. Frequently a high standard of social security, labor, and welfare arrangements, appropriate to a much more advanced country, beyond the economic capacity of local business, is nonetheless enacted into law and then enforced with more or less deliberate discrimination against the American companies, on the theory that they can afford it; they must conform—and very likely try to be even more obliging, in hope of creating a tolerably favorable impression. Local taxation also, by policy or at least in implementation, is probably weighted against the outsiders. Currency exchange controls, repeatedly imposed in efforts to cope with the effects of price instability, may suddenly prevent the Americans from taking any proceeds back to the United States if they sell out their investments, or from even taking their profits out of the country. Exchange controls can also unpredictably disrupt the schedules of American firms intending to import into the underdeveloped country the equipment, spare parts, and raw materials essential to their operations there, but not available locally.

This latter risk is less for the firm that can afford to commit itself to a long-term investment, undertaking to foster the development of local sources of supply. Indeed, most of the other specific risks are also less frustrating to large enterprises that can hire expert researchers, investigators, and political contact men to find their way with comparative safety through the maze of local politics and bureaucracy, learning to anticipate future developments, to mollify disgruntled influentials, and to appease the appetites of troublesome officials. But overseas political activity of this kind is beyond the capacity of most small and middle-sized companies in the United States; indeed, it is beyond their very ken as businessmen. They are reluctant to rely on the United States Foreign Service for this kind of assistance, and in truth they need much that the diplomats cannot with propriety furnish them. For these and other reasons the smaller firms very often tend to overestimate the risks they would actually confront if they were to expand their operations into the less-developed nations.

The American investors must always weigh the uncertainties against the attractive profit opportunities existing within the comparative safety of the United States herself and of some other well developed countries. The rate of return on investments at home is usually so high as to leave little incentive to take great risks abroad.

Taking frank note of all the qualms of the Americans and the skepticism of the native elites—but also remembering that an extensive United States contribution to foreign economic development has to be very largely private—what if anything can the American government do to make investment in underdeveloped areas more attractive for its own citizens?

The main approaches involve efforts to reduce the obstacles or to hurdle them. Reducing the obstacles is a continuous process of diplomatic suasion directed at the foreign government, seeking the elimination of particular nuisances or negotiating for broad improvements in the general framework of taxation and regulation by means of treaties and other agreements. The use of economic assistance grants in this bargaining to get a more favorable investment climate is difficult. Some American aid can be steered toward projects likely to facilitate an additional inflow of private capital from abroad. But if the local officials are inclined to regard foreign private investment as tantamount to ruthless exploitation, the American aid administrator can scarcely convince them that they would derive dual benefits from making special concessions to the investors in return for larger government grants (gaining from the investment flow as well as from the aid). To insist on linking the two items may only destroy whatever good will the United States might otherwise be able to derive from the aid program alone. Still, from a longer perspective, the effort may be worth making; and an international agency like the World Bank or the International Monetary Fund, preserving an aura of jointness although under heavy American influence, can sometimes succeed in making such conditions acceptable in circumstances where the United States government by itself would arouse adamant resentment.

Further efforts can also be made to induce underdeveloped countries collectively to agree among themselves on codes of fair treatment for foreign investors; advanced countries would then be somewhat less often in the awkward position of pressing their own standards unilaterally on less fortunate peoples. But, of course, the agreed code to be meaningful would still in substance have to confirm the standards which foreign investors consider essential. Not until native elites in key underdeveloped countries have come to look with greater favor than hitherto upon private investments from advanced lands is there much prospect of manifesting that attitude in a code precise enough to be very useful.

If the strategy of United States government action to increase the flow of American investment abroad be directed at hurdling the obstacles rather than waiting to get them reduced, the most obvious available tool is adjustment of the federal tax rates on corporate profits and dividend income so as to favor companies operating abroad, everywhere or in a particular class of countries. The American investor, in effect, could be enabled to overleap higher obstacles overseas because he would be carrying a smaller burden at home. The range and variety of possible adjustments is tremendously complicated, but all of them can be challenged as evincing favoritism toward a particular category of investors, and most of them would cause some reduction of government revenue and would not in fact increase the flow of investment unless the foreign governments could effectively be dissuaded from multiplying their obstacles proportionately. If they can be dissuaded, the case for preferential treatment of these investments is markedly improved. Under existing United States tax law, American investments in Western Hemisphere countries have already been given some advantages; these geographically restricted incentives are related, both as cause and effect, to the existing concentration of United States foreign investment in Canada and Latin America.

In addition to promoting foreign investment by providing tax benefits, the United States government can make or guarantee loans to American firms on favorable terms for their operations overseas. In particular, there are large accumulations of local currency that come under the control of United States officials as a consequence of foreign aid and farm surplus disposal operations (see pp. 366–68, 371–72); productive uses need to be found for this money locally. Such uses include supplementing the limited loanable funds of the locality by lending out the United States holdings. Both native and American enterprises are among the appropriate recipients. Local financial institutions can be protected and fostered by a United States government practice of making the loans to the banks, thus increasing their reserves for further lending, or of guaranteeing the repayment of loans they make to American and native businessmen.

A program of direct guarantee of the value of the investments themselves through low-premium government insurance is the most prominent existing device for encouraging United States citizens to hurdle the obstacles to overseas investment. The usual annual charge is one-half of one per cent of the face value of each insurance contract during the years that it is fully in force. Most of the contracts are drawn to cover losses resulting specifically from expropriation, inconvertibility, or war, or (since 1961) from revolution, insurrection, and related civil disturbances. The country in which the guaranteed investment is to be made must have signed a general agreement with the United States government authorizing the latter to take over the local claims of any American investor who has suffered damages there that have been met through the United States government insurance—on any particular project for which both governments have given specific advance approval before the insurance contract was ever issued. The foreign government thus accords to the United States government in these special instances the formal right to use diplomatic pressure to obtain compensation for the loss its citizens have suffered (and for which they would be reimbursed through the insurance, leaving the State Department to pursue the claim). The whole arrangement is complex—a far cry from the pre-World War I days when it was taken for granted that a government had the right to seek redress for injury imposed upon the legitimate business undertakings of its citizens abroad, without being required to have any formal, limited advance authorization to do so. In the changed international climate one may even question whether a regime that comes to power at some future time determined to expropriate would in fact feel bound by the pledge of its predecessor to allow the United States government to intervene and obtain compensation. Nevertheless, the formal agreements are probably better than nothing. They may even give a certain confirmed legitimacy to private investments in general, as well as to the particular projects for which insurance contracts are ultimately approved.

Agreements had been reached with thirty-seven underdeveloped countries as of May 1961, providing an acceptable general framework for convertibility insurance, and with thirty-three underdeveloped countries for expropriation insurance; only fourteen for war risk.* Actually, however, the program was decidedly slow in catching on during its first ten years and not until 1958 was there a major spurt of specific applications and approved projects for insurance. Even then, most guarantees went for investments in Western Europe, until Congress in 1959 confined the program to the underdeveloped countries, where need was much greater. The Kennedy Administration in 1961 introduced a new option, experimentally, under which the United States government would insure only a part (up to 75 per cent) of the value of an investment but do so against almost all kinds of risks, as an alternative to the regular practice of insuring all of an investment against only part of the possible risks (those specified are expropriation, inconvertibility, war, and, since

* As "underdeveloped" is consistently applied in this book (see p. 323n.). The United States insurance administrators themselves prefer to use for this program a broader categorization labeling some European countries underdeveloped. American guarantees are also available in the overseas underdeveloped dependencies of six European countries; and prior to 1959 they could be obtained in most of Western Europe.

1961, revolution and insurrection). On projects the Administration considered of high priority, the new arrangement was designed to expedite contract-writing by minimizing legalistic quibbling and hedging. Together with the new inclusion of revolution and insurrection on the regular list of specific insurable risks, the extension of coverage gave prospect that the investment guarantees would finally be used widely in the less-developed countries—but prospect also that the costs to the United States government, heretofore negligible, would eventually become substantial.

## The Additional Need for Contributions from Europe and the American Government

There are several highly developed countries in Western Europe with traditions of foreign investment that are vastly stronger than those that prevail among Americans. The pattern of incentives in which the European tradition of overseas investment took root was closely linked with the special protections provided by direct colonial administration or at least by the disciplinary reprisals of "gunboat diplomacy." Those forceful business safeguards are, of course, rapidly disappearing, but the Western European powers and Japan, having recovered from World War II, now possess at least the resources for a substantial expansion of their investments, direct and indirect, in underdeveloped countries. Even during most of the years of postwar economic stringency, this flow remained considerable, with the encouragement of the governments concerned. It was, however, matched by much disinvestment from the most disturbed regions overseas. Increased incentives in Europe to step up the outward flow could facilitate important gains for the noncommunist world as a whole. Most of what has been said in the previous pages about United States private investment in the less-developed countries applies equally to venture capital from northwest Europe and Japan. The emphasis here on Americans in the description of the problems reflects only the general perspective of this book, not any actual preeminence of the United States in this aspect of contemporary international relations. Americans in fact have excellent reason to seek confidently for valuable cooperation with private and semi-private European and Japanese sources of investment capital in filling and expanding the role that corporate enterprise can play worldwide in developing the emerging lands.

That role, however, cannot by any available means be brought into full play fast enough to meet the politically effective demands for accelerated advance in Asia and Africa and probably even in Latin America. Incentives for sufficient foreign investments will in all probability not be created within the limited span of years the cold war will allow. Ultimate results that will not be decisively disadvantageous to the West cannot be achieved in this period unless corporate enterprise, itself government-fostered, is supplemented by massive direct foreign aid from the United States government, and probably also from the governments of other major allied powers. Principally, this requirement is produced by political pressures in many backward countries for a standard of living so high that it could not otherwise be attained, or a rate of development so bold that it could not otherwise be achieved, or both, within a framework of local institutions and policies that American officials consider, on balance, to be somehow much more advantageous to the West than the locally available alternatives. Sometimes also, the need to give aid is more crudely political, providing a channel for American influence (usually not short-sightedly selfish) or even amounting to indirect bribery of particular political factions that will cooperate with the United States government.

On the other hand, a less political consideration occasionally militates against reliance on private loans and investments even for a country to which they might conceivably be induced to flow in temporarily adequate amounts. In the long run the burden of repayments and of American profit-taking may seem likely to put a dangerous strain on the country's balance of dollar payments and hence upon its economic and political relations with the United States.

And in the case of agricultural commodities and obsolescent military equipment, the American government's desire to get rid of surpluses for almost any useful purpose constitutes a special incentive for foreign aid programs.

But, in general, the problem is one of impatience in the emerging countries, coupled with the availability of totalitarian alternatives to the irksome Western private enterprise methods of economic development. In this situation, the direct participation of Western governments in economic assistance is essential in order to maintain in the backward lands a sober hope and perseverance, with reliance on noncommunist internal practices and principally noncommunist international connections.

## SELECTED BIBLIOGRAPHY

BAUER, PETER T., and BASIL S. YAMEY. *The Economics of Under-Developed Countries.* Chicago: University of Chicago Press, 1957.

COALE, ANSLEY J., and EDGAR M. HOOVER. *Population Growth and Economic Development in Low-Income Countries: A Case Study of India's Prospects.* Princeton: Princeton University Press, 1958.

HAUSER, PHILIP M., ed. *Population and World Politics.* Glencoe, Ill.: Free Press, 1958.

MYRDAL, GUNNAR. *An International Economy: Problems and Prospects.* New York: Harper, 1956.

PIZER, SAMUEL, and FREDERICK CUTLER. *U.S. Business Investments in Foreign Countries.* Washington, D.C.: U.S. Department of Commerce, Office of Business Economics, 1960.

SCHELLING, THOMAS C. *International Economics.* Boston: Allyn and Bacon, 1958.

STALEY, EUGENE. *The Future of Underdeveloped Countries: Political Implications of Economic Development,* rev. ed. New York: Harper, 1961.

UNITED STATES CONGRESS SENATE SPECIAL COMMITTEE TO STUDY THE FOREIGN AID PROGRAM. *Foreign Aid Program: Compilation of Studies and Surveys.* Washington, D.C.: Government Printing Office, 1957.

UNITED STATES DEPARTMENT OF COMMERCE. *Balance of Payments Statistical Supplement.* Washington, D.C.: Government Printing Office, 1958.

———. *Survey of Current Business,* monthly. Washington, D.C.: Government Printing Office, 1956– .

CHAPTER 18

# ECONOMIC AID PROGRAMS IN THE LESS-DEVELOPED COUNTRIES

In most of the advanced countries of the noncommunist world where the rate of savings would permit extensive investments in underdeveloped areas, there is no sharp, clear line between private investment and governmental assistance going abroad. As should be expected in capitalist economies, the intermediate forms of public and private cooperation in external ventures is immensely variegated.

## Hard Loans from the Export-Import Bank, the World Bank, and the International Monetary Fund

The United States, since the early days of the New Deal, has had an official agency, the Export-Import Bank, designed principally to provide supplementary financing to facilitate foreign purchases of American exports. The United States trade thus financed is mainly with Latin America. Loans are made directly to foreign governments or to business firms or may involve partial guaranteeing of the loans of private banks for international transactions. Either American or foreign firms may receive the loans, but they are "tied" and thus cannot be used to import into a foreign country goods that were not produced in United States territory, even if the products could be purchased more cheaply in a third country. This provision, of course, reflects the original promotional objectives of the entire institution. Any resulting burden on underdeveloped countries has been somewhat reduced with the re-emergence on a large scale of similarly attractive credit arrangements by European firms, notably West German; the resulting competition among industrialized seller nations improves the range of choice open to persons in backward countries arranging the import of development goods.

The Export-Import Bank loans can be long term or short term, but they must be repaid in full in dollars with interest at a rate set high enough to avoid severely undercutting the private lending institutions. Thus these are "hard" loans. Ordinarily "Eximbank" authorities prefer to assure the "hardness" by loaning only for specific projects that have good prospects of profitability, but in emergencies general loans may be made to governments to cover their balance-of-payments crises.

Congress has treated the Bank with special favor, mainly because of its financial "soundness" and usefulness to American exporters. The latest increase, $2 billion, in its lending authority was granted by Congress in 1958, bringing the total the Bank is permitted to have outstanding at one time to $7 billion in loans, guarantees, and commercial insurance. The increased authority brought a new spurt of lending, largely to underdeveloped areas, with the result that in the late fifties actual Eximbank disbursements to such countries averaged about $300 million a year greater than the repayments that those countries were currently having to make on the Eximbank loans they had previously received. In the 1960's, however, unless the Administration and Congress agree to a further increase in lending authority, the Bank will have to revert to the policy that prevailed in the mid-fifties of roughly balancing its annual disbursements with its repayment receipts.

Comparable to the Export-Import Bank in the hard-loan field is the International Bank for Reconstruction and Development (IBRD, the so-called World Bank). Its lending policies are highly conservative, related to its practice of raising a very large part of its capital for loans by borrowing from private financial institutions. The interest rates and prospects of hard-currency repayment on IBRD loans must therefore be high enough to enable it to meet the interest and repayment demands of the private financiers from whom it borrows. World Bank loans, as a result, are for specific projects that have been very carefully analyzed by the Bank's own experts, at home and on the scene abroad, and that have indeed often been initially formulated upon their advice, after one of the wide-ranging survey missions they are invited to make by underdeveloped countries. The Bank takes pride in this technical assistance side of its work, covering all kinds of economic problems. The actual IBRD financial contribution to development has, however, been less impressive, rising slowly to a net rate (disbursements over repayments) in the late fifties that approached $250 million a year to the less-developed countries. Yet by working to induce the recipients to improve their general climate for private investment and by sharing loans for particular projects with private lenders and guaranteeing the loans of private banks, the World Bank has also established the incentives for a considerable share of the increasing flow of private capital going to underdeveloped lands.

In hope of contributing specifically to an expansion of the risk capital (as distinguished from fixed-interest loans), the Bank's staff won international approval in 1956 for the establishment of an "International Finance Corporation," which would, in effect, buy stock in corporations going into new lines of enterprise in the less-developed countries. The IFC would not purchase a controlling interest in any company, but its influence would be important, and the evidence of its confidence would be valuable for attracting other investors. When the profitability of a novel venture had been successfully demonstrated, the IFC would sell its holdings and use the proceeds to back other new enterprises. The actual degree of usefulness of this institution remains to be seen, since it was barely getting started in the late fifties.

The World Bank's operations are essentially determined by its own staff with remarkably little detailed supervision by any government or other international agency. But its headquarters are in Washington, and the voting arrangements insure that the United States can ordinarily exercise predominant influence when it chooses; close working relationships prevail among the World Bank, the International Monetary Fund, the Export-Import Bank and other American agencies as well as frequently the major Wall Street banks.

In the late fifties the United States government took increasing advantage of these ties to prevail upon several Latin-American countries to adopt severe anti-inflation policies as the price for emergency balance-of-payments loans in which the Fund, the Eximbank, and private United States financiers would jointly participate. The onus was put upon the International Monetary Fund, with an aura less objectionable than that of Wall Street or Pennsylvania Avenue to Latin nationalists, to insist upon a reduction of local deficit financing and a liberalization of foreign trade even though the result of this drastic medicine was temporarily increased hardship in some underdeveloped country.

The relatively orthodox capitalist economic prescription for overcoming a disequilibrium in the balance of payments would have been extremely difficult for the United States government to impose alone. Even the Fund's susceptibility to American influence was soon aired, and there was agitation against "the dictatorship of the Fund." Brazil in 1959 refused to follow the policies stipulated by the Fund's experts; the United States government found itself confronted with the awkward alternatives of letting Brazil default on previous international loans or else lending her money to use for the regular repayments even though the Brazilian government was defying the Fund, whose authority the United States was trying to foster and utilize to guide economic policies in the underdeveloped countries. The American government found a temporary way out of this dilemma by offering to postpone the repayments due, without canceling them or allowing Brazil to do so. The debt moratorium would still leave Brazil with the need to pay cash for current imports, unable to obtain much new credit; this penalty would help sustain the Fund's mandates elsewhere, at least for a while. By coincidence an improvement in coffee exports postponed a showdown, and a more amenable Administration was elected in Brazil a year later. But the whole episode demonstrated the limitations as well as some of the potentialities international agencies possess as supplements and instruments for economic development programs that fit the inherent conditions of a world economy remaining under capitalistic leadership.

The chief advantage of the international agency is that it usually arouses less suspicion than a single government like the United States in the country being aided and advised. Equality among the nations rather than subordination to a particular power is central in the usual symbolic connotations of an international organization. The multinational character of the personnel at various levels gives reassurance that devious political manipulation is not likely to be accomplished in the guise of economic advice; any unwelcome implications of the proposals will probably be brought to light somewhere within the organization by elements sympathetic with the recipient nation. The embarrassments of being on the receiving end are also soothed with a balm of supposed mutuality that seems to be more convincing than similar labels Americans have devised for their bilateral aid arrangements, like the "mutual security program." The emphasis on international sharing of effort does elicit greater cooperation on the part of some potential donor countries that would not otherwise engage in foreign aid, including those that are themselves mainly recipients. The greater diversity of donors means that recipients are less tied to particular sources of loans, goods, and advice; greater selectivity is possible, enhancing precise suitability to the economic needs.

But much of the greater attractiveness of the international program over bilateral arrangements from the viewpoint of the underdeveloped country derives, not only from the lack of economically irrelevant political strings, but from a hope that administrative procedures and economic development policies will be less confining. Conversely, from the viewpoint of the United States there is danger that the substance of a development program formulated so as to utilize private initiative and avoid coercion and waste will be lost through the multilateral operational need of conciliating a wider variety of viewpoints held by nationals of many countries. If their national backgrounds become submerged in an international civil servant's type of "agency viewpoint," this danger of a loosening of standards is much less, and the solidarity, not obviously serving the purposes of any one country, is particularly persuasive to those who receive the prescriptions; the World Bank and International Monetary Fund have achieved much of this character. But even then, the general usefulness of the international agencies for United States development assistance is limited by the extreme proliferation of such agencies, each concentrating on different economic and social problems without much close coordination among them, either at United Nations headquarters or in the individual countries being aided.

The international organizations that implement economic assistance are specialized agencies useful for specialized purposes. They could be transformed for effectiveness in overall economic development operations only if the United States, as the major donor, resolved to place her principal reliance on an amalgamation of them and to route most of her own foreign economic assistance through them. To do so would run considerable risks for the efficiency and success of the development programs, as has been suggested. But more important, the United States does not share the enthusiasm of many underdeveloped countries for being deprived of her freedom to pull some political strings, at least subtly, in connection with her aid abroad. From historical precedents, Americans lack confidence that even a successful economic development program will by itself eventually lead an emerging country to align itself internationally in a fashion that is not disadvantageous to the United States. Efforts may have to be made, as tactfully as possible, to influence the developing country's political evolution more directly. As long as Americans view the economic advancement of emerging mankind not primarily as an end in itself for the United States to pursue at enormous expense to herself, but rather more as an impelling worldwide urge that commands assistance in order that its outcome may not be ruinously unfavorable to the Americans, the United States government will prefer to take its own chances of making its own blunders in employing economic aid in conjunction with other instruments of its own policy, utilizing international agencies in the special instances where they fit these requirements.* Such an outlook does not preclude a much wider use of multilateral arrangements than characterized the 1950's; but it does mean that the major United States contribution to economic development will probably continue to be made through public and private instrumentalities that are under the control or paramount influence of Americans and perhaps Western Europeans.

## Point Four, the Peace Corps, and United Nations Technical Assistance

In addition to hard loans from the Export-Import Bank, the most prominent American government program for economic development assistance in the first decade after World War II was the technical assistance program—Point Four; just as the Eximbank was paralleled by the World Bank, so also Point Four had an international cognate in the United Nations Expanded Program of Technical Assistance (EPTA) and in other missions of the specialized agencies of the UN.

President Truman had announced in his 1949 inaugural address that there were four main points in American foreign policy: support for the UN, the European recovery program, military defense assistance, and "a bold new program for making the benefits of our scientific advances and industrial progress available for the improvement and growth of underdeveloped areas." This fourth point, however, remained to be formulated, and by the time it obtained congressional approval the fol-

* Formal congressional enactments during the Korean war stipulated the conditions for assistance in such a way as to imply that every kind of aid bound its recipient with strings to an alignment, at least symbolic, with United States foreign policies. Since the war, the legislative language applying to different types of aid programs has been more carefully differentiated so that alignment-strings are explicit only for military aid. The change to a less blatant formulation has relieved much embarrassment, but, of course, American aid administrators in most of the underdeveloped countries have continued to be keenly concerned with the internal politics and foreign policies in each locale.

lowing year it had dwindled to the Eximbank loans, investment insurance, and a worldwide expansion of American technical advisory missions financed through congressional appropriations. The missions were considered the distinctive element of what continued to be called "Point Four."

The United States had already been operating technical assistance systematically in Latin America since the beginning of World War II. The Institute of Inter-American Affairs (IIAA), headed by Nelson Rockefeller, had taken root in the fears that had then been engendered by the Nazi penetration of South America through attractive German offers of assistance and economic deals. The United States government wished to strengthen an uncoerced joint alignment of Western Hemisphere countries by means that would be appropriate to the label of "Good Neighbor Policy." There is a striking parallel here to the combination of anticommunism with humanitarianism and internationalist idealism in the implementation of Point Four a decade later. IIAA also had a distinctive mode of operation, which influenced the later direction of Point Four, known as a cooperative service, a *servicio,* an agency to which both the local government and the United States government contributed money and personnel to carry out a special program, usually in the fields of public health, education, or agriculture. As the program took root, the United States contribution was to be reduced and the newly trained native specialists were to take over responsibility. Emphasis was on demonstration projects, "pilot projects," that would not be so ambitious as to require a long-term dependency upon the United States in order to reach fruition. This propensity toward smallness, at least when measured against the immensity of the problems of economic development, can in turn be traced to some extent to the influence through Nelson Rockefeller of the pioneering work of the Rockefeller Foundation in Latin America. By the standards of private charities, this work was very extensive and ambitious; a government program largely modeled upon it was bound to become a favorite of those who shared those standards—the values, essentially, of American missionaries, except for their religious evangelism. But whether such a program could be considered a sufficient contribution from the United States government for the promotion of general economic development was more doubtful. The question, however, was not asked at all when IIAA began; no such objective was yet envisaged. And later, even when the objective was accepted, the limited nature of Point Four continued to have an appealing similarity to traditional American modes of private activity; it seemed merely an extension and supplement to them. Its charitable role looked familiar side by side with the financier's role of the hard-loan Export-Import Bank that was supplementing private banking and with the investment guarantees that were insuring private investments.

Point Four also won favor in the United States because of its relative cheapness. Until the Kennedy Administration, it never rose much above $150 million a year to underdeveloped countries—nor could it, without undergoing a transformation, because in its established fields of emphasis the program faces severe shortages of available American personnel. Before Kennedy, two-thirds of its activity was still in the fields where IIAA began—health, education, and agriculture—although geographically this "technical cooperation program" was expanded to cover nearly every underdeveloped noncommunist country and many colonial areas. These fields of emphasis, too, helped to win approval for the program in Congress. All the reassuring familiarity of the United States Public Health Service, the agricultural extension services, and the American public education system underpinned its acceptability. And from the standpoint of effective administration of a governmental technical aid program, it was convenient to draw upon specialists who are accustomed to low government salaries; such men are most available (for typical two-year assignments abroad) in the fields of health, education, and agriculture, and also in public administration, housing and city planning, and labor relations. But in industry, mining, and transportation, on which the remaining one-fifth of technical assistance money was being spent, the government finds greater difficulty in hiring technically competent personnel for overseas missions, not only because of the higher salary scales of private business, but also because specialists are more reluctant to jeopardize their competitive positions by interrupting their careers to go overseas.

The administrators of technical assistance have tried to get around this bottleneck by "contracting out" many of their missions to private firms for operation. Particular emphasis has been placed upon long-term contracts with American universities to carry on technical programs abroad; this concept is, of course, related to the familiar agricultural extension services long operated by the state universities at home. Advice given through a university program, rather than directly by United States government employees, may also get a better reception in suspicious, nationalistic underdeveloped countries. The university tie, providing a nongovernmental framework within which there can be a regular continuing rotation of specialists from their classrooms at home to posts abroad and back again upon replacement, contributes important stability to the program in an underdeveloped country. And, of course, the university acquires a vested interest of its own and may lend support to the agency's appropriations before Congress, reconfirming the "health service, missionary, extension service" image there.

Thus the technical assistance program of the United States government, as it developed out of the overseas work of the Rockefeller Foundation, was shaped by the perspectives of American society that emphasize private activity supplemented in limited fields by official cooperation. Within these limits, Point Four became well entrenched at home and overseas, less controversial than any other United States economic aid program, concentrating abroad on specialized face-to-face demonstration work and avoiding most major development projects that would require heavy investment of capital.

Even at this level of activity, however, the personnel shortage was a severe handicap. A fully effective Point Four field worker ought to possess technical competence plus local language skills, familiarity with the customs and problems of the district, an adaptable and sympathetic disposition, and a contagious enthusiasm for his mission (often he must work among people who do not want advice at all). In a sense, he should simultaneously be a fine technician, linguist, teacher, cultural anthropologist, diplomat, and missionary. Familiarity with the locale and its language can probably be acquired by nearly any technical specialist if he stays long enough; the usual two years, however, may be too short a time. But even time brings little improvement in the basic qualities of his temperament, and these are in most underdeveloped countries as important as his technical competence for the effective transmission of skills to the natives. Indeed, in many situations it may be wiser for Americans not to involve themselves in a technical assistance project at all than to undertake it using specialists who irritate and offend the people they are supposed to be helping. Yet the paragons who are needed are extremely difficult to find.

Here again, the American cultural setting fixes limits. The prevailing educational pattern fails to foster facility in foreign languages (or even technology as much as in Communist countries). Even more important is the lack of encouragement to young Americans to make their careers abroad. The capacity of the United States to play a role of leadership in the noncommunist world is badly crippled by the nearly universal presumption of her people that an overseas job, if taken at all, must be only a temporary "tour of duty" and that the natural limits of American mobility, great as it is, are those of the continental United States. This traditional kind of isolationistic escapism can be expected to change gradually as Americans find themselves forced into foreign contacts by foreign challenges and as a new prestige attaches to today's "frontiersmen." But the elements of this cultural change that increase the pool of potential qualified recruits for overseas service evolve much more slowly than even the corresponding changes in attitudes about taxation and overseas spending. And recruitment of technical specialists for Point Four is probably even harder to expand than the other services, public and private, in which the United States is now engaged abroad, because Point Four has even less overseas tradition than business or the military on which to build—except for a groundwork laid by American missionaries.

A secular missionary spirit can be an enormous asset for a technical assistance program as long as it does not produce indifference to wider problems of economic development in a revolutionary age—or worse yet, a fanatic preoccupation with some pet technique (for example, well-digger devices) as the panacea for a whole area. In the early days of the Korean war, there was a marked tendency in the so-called Technical Cooperation Administration (TCA) to try to stand apart from power politics and from concern with the containment of Communism—apart even from the thrust of backward countries toward industrialization. This "Point Four purism" was overcome in organizational terms when the Eisenhower Administration insisted on merging TCA with most other United States economic aid units in a new International Cooperation Administration (ICA).* The move was made at some cost in enthusiasm of TCA personnel, at least initially, and the amalgamation certainly blemished Point Four's reputation abroad for idealistic political virginity. But in the light of overall American foreign policy, concerned with general economic development and also with political evolution, the effort to integrate aid programs was thoroughly appropriate, even if it led to some reduction in the flow of "missionary types" for recruitment into Point Four.

The Kennedy Administration, however, decided to reclaim and rechannel some of the earlier enthusiasms by supplementing the established technical assistance program with a new "Peace Corps," which would deal with the fundamental problem, the personnel scarcity, by accepting a lower level of professional competence and hoping to offset it by a higher level of youthful dedication and self-sacrifice. Peace Corps volunteers might not yet have acquired the experience that would

* Despite its name, ICA was wholly within the American government. It was the successor, via several transformations, of the United States agency for the Marshall Plan European recovery program—the Economic Cooperation Administration (ECA). And in turn ICA became, under President Kennedy, the Agency for International Development (AID).

qualify them to function as technical advisers, but they would be willing to fill jobs at "middle manpower" levels themselves, doing the work and in the process teaching local people to replace them. Emphasis would be placed on the volunteers' accepting Spartan conditions of life for themselves and mixing on equal terms with the native population in accordance with the characteristic American supposition that face-to-face contact, "people to people," is the strongest selling point of the United States. There has been hope also that Peace Corps veterans will become a widening pool of potential recruits for longer tours of duty in the regular overseas agencies of the American government, and that the remainder who return to private occupations in the United States will at least help to build there a favorable public image of foreign aid.

The ultimate size of the Corps was left undetermined, but early planning envisaged a few thousand men in service at any one time, on an annual budget of $40 million. The number might be roughly equivalent to the total American overseas personnel of the main economic aid agency and its private contractors in 1956 (the International Cooperation Administration, which included the Point Four technical assistance workers). Such a size for the Peace Corps appeared rather ambitious, but from another perspective, one which is perhaps more relevant to its own spirit, the total would still be less than 20 per cent of the number of American religious missionaries abroad.

The Peace Corps, in making its recruitment appeals to similar kinds of idealism, humanitarianism, and adventurousness, was aware that it might be incorporating excessive political naïveté (and insufficient specialist competence), and the managers claimed to be on their guard against this occurrence in the selection process. Yet there appeared to be some tendencies toward a "Peace Corps purism" that would be akin to the earlier "Point Four purism," but lacking even the technical expertness of that previous incarnation. If so, the value of the Corps as an instrument of American foreign policy might still be great, but it would rank principally in the class of cultural exchange programs, albeit exceptionally large, rather than in the class of economic development programs. Indeed, the director of the Peace Corps affirmed that his enterprise was to be conceived more as a cultural and welfare operation than as an economic aid program. In light of previous experience, it should not be surprising that the principal fields in which the Peace Corps in its first year was able to pair available recruits with available overseas jobs were the old Point Four stand-bys: health, education, and agriculture. Classroom teaching assignments were particularly prominent. Not yet, however, could the Corps's ultimate role be confidently predicted.

To the extent that some complete divorce may be desired between technical assistance and international political rivalries, it probably finds a more appropriate setting in the United Nations Expanded Program of Technical Assistance (EPTA) and in regular UN agency programs of a similar character (the World Health Organization, Food and Agriculture Organization, United Nations Educational, Scientific, and Cultural Organization, and so on). Together these projects now spend over $30 million a year in underdeveloped countries. The United States contributes up to 40 per cent of the funds, and has agreed to do likewise for a further expanded United Nations Special Fund, which is seeking contributions totaling $100 million a year (with EPTA included) from member governments. About a sixth of all the American aid that is formally designated as grants for economic development would thus then be handled through UN channels. The general arguments for and against such an internationalization of aid have already been discussed.* But it should further be noted that in the case of technical assistance both the personnel shortage and its inherently limited scope tend to weight the balance more heavily in favor of multilateral aid than would be the case with some large-scale development programs. Also, any fears that American dollar contributions would be used to subsidize Communist agents are somewhat tempered by the fact that each recipient country can use the criterion of nationality to select the experts it will obtain under UN auspices; no Soviet technician need be accepted merely because he is a proficient specialist, and few have been.

## American Grant Aid

Technical assistance, through whatever channel it is administered, is very likely to raise hopes it cannot satisfy, unless it is linked to programs of capital investment, public or private. To show people how to use tools is useless (or worse than useless) unless somehow they can be provided with the tools to use, not just in quantities adequate for demon-

* Pp. 354–56. The other major UN agencies operating grant programs in underdeveloped countries are the United Nations Children's Fund (UNICEF) and the United Nations Relief and Works Agency for Palestine Refugees (UNRWA). UNICEF spends $15–20 million a year; UNRWA $30–35 million. The total United States contribution to these and to UN technical assistance, prior to the latest expansion (the United Nations Special Fund), was about $50 million a year; if the Fund gets fully into operation this latter figure might be nearly doubled. Also, if the Congo is effectively stabilized, the United Nations expenditures there would become more economic and less overwhelmingly military in character than heretofore; the United States has been contributing about half of the UN's Congo budget of over $100 million per year.

stration purposes, but in quantities sufficient to produce some discernible benefits, at least locally. Otherwise, most people will slip back into their old ways, while the more progressive native elites grow bitter at the tantalization. Villagers can be taught to provide certain kinds of new tools for themselves; other kinds can come from larger domestic investment projects within the underdeveloped country itself. But unless the population is to be squeezed by totalitarian methods or the course of development is to be extremely slow, much of the capital must come from other countries. If it comes in the form of investments or hard loans, governmental or private, one result is to burden the future balance of payments. The debtor nation, until its export capacity (including available markets) becomes much greater, will have to keep on borrowing abroad in order to use the proceeds of new loans to pay back the old ones, and the creditor nation will have to keep lending in order to get repaid; similarly, new foreign investment will be necessary in order to permit profit-taking and repatriation of old foreign investments. The exasperation of such a treadmill for all parties concerned led the United States government to prefer to use grants (gifts) rather than loans for most of the European recovery program (as had been the case with Lend-Lease during World War II). Most military assistance since the war has also continued to be on a grant basis.

In the late fifties about $1 billion of American economic aid each year was likewise going as grants to underdeveloped noncommunist countries. In some, with which the United States also had military assistance agreements, the economic aid grants were labeled "defense support"; in the others, "development assistance" and "special assistance." The Truman, Eisenhower, and Kennedy administrations changed the labels frequently as the supposed preferences of Congress and the American public shifted from aid with an economic look to aid with a military look and back again to the economic; the rubrics also reflected public-relations problems abroad. The Kennedy Administration chose to attempt a distinction between "development grants" and "supporting assistance," based on the degree to which economic progress, contrasted with mere survival, was the central purpose of a program in a particular country. Thus most of Eisenhower's "defense support" and "special assistance" became "supporting assistance" under Kennedy, and most "development assistance" (plus the former "technical assistance") became "development grants."

Most commonly in the 1950's, the Executive Branch presented an image of the economic grant aid as enabling the United States to obtain sites for overseas bases and as enabling impoverished allies to keep large standing armies in the field. But the real significance of aid depends on what the individual countries would have done without it. Some of their governments would have done their best to maintain the armies anyway, at whatever cost to the local economy and social structure; for those nations the "defense support" really meant principally economic relief, and perhaps it even permitted some development. In others the local impoverishment, if aid were cut off, would have literally meant chaos and Communist triumph from within; such countries, nominally independent but incapable of economic self-support at politically tolerable levels, may depend indefinitely upon doles from an outside power. Using the money to keep men in uniform can at least be a way for them to help relieve local unemployment. In still other countries, "defense support" could justly be criticized as forming pernicious habits of excessive militarization and perennial dependency on the United States where real potentiality for economic development otherwise existed, if energies were redirected. It was in those countries particularly that the Kennedy Administration, with its emphatic distinction between "development" and "support," hoped to focus attention on the possibilities for a shift in the emphasis of American aid programs.

On the theory that particular projects can be supervised more effectively than overall fiscal policy in the less-developed countries, the suggestion has often been made that a fair estimate of the "developmental" component of American grant aid can be obtained by computing the subtotals for *project*-aid in comparison with general "budgetary assistance." By this calculation, considerably less than half of the total $1 billion a year was developmental. But the standard is unreliable. For general economic development, the ultimate value of specific local projects (a road, a factory, an irrigation system, and the like) depends on their interrelationships with other aspects of the evolving economy that can be determined only centrally.

To be sure, in the most backward countries, where central programing of economic development is clearly beyond the administrative capacity of the government—with or without American guidance—the safer course may indeed be to concentrate on specific local projects that have at least a presumptive usefulness, which may turn out to be genuine. In helping to plan the separate aid projects and then in checking up on them, United States advisers can incidentally work to improve local procedures in administration, finance, and engineering. Since the Americans are unable to coordinate the entire alien economy themselves or to rely on the central government to do so in this early stage of development, they make a contribution through official aid to separate projects, in a spirit of providing diversified opportunities and incentives to individual enterprises, public as well as private, and of relying on these independent groups, thus stimulated, to exert initiative and accomplish more than central programing could do.

Much of the manifold project-aid thus provided, however, must be expected to be written off eventually as irrelevant and wasteful. When enough of the foundations of native administrative competence have been established—together with basic mechanical facilities of transport, communications, and power—to permit a meaningful overview of the whole economy and a more purposeful patterning of the incentives for accelerated change, then United States assistance can be most constructive (the Kennedy Administration came to insist) if the aid is allocated to support a well-formulated, wide-ranging development program for the country as a whole, trusting the local government, with which the program has been devised, to transfer the aid resources flexibly among various projects. Even in this case, American influence would presumably continue to be used to promote a large measure of decentralization and private initiative, and much specific project assistance would still seem appropriate. But "program assistance," or even "general budgetary assistance," would be respected as developmental also, and would cease to be regarded askance as perhaps a mere "pay-off," more or less voluntary, from the United States to a cooperative government in a backward country, or at best a mere relief operation to sustain a stagnant status quo.

## The Usefulness of American Agricultural Surpluses

In addition to the $1 billion a year that was going as grants to underdeveloped noncommunist countries, with diverse consequences for economic progress, the American government in the mid-fifties began shipping abroad its stocks of surplus agricultural products at a rate of $1.5 billion a year under special disposal programs. In the last year of the Eisenhower Administration the rate rose steeply, and the Kennedy team projected further increases, perhaps to more than 150 per cent of the earlier level. That level had involved about $200 million worth of commodities going annually to underdeveloped countries as part of the above mentioned $1 billion of grants ("defense support," "special assistance," and so forth), plus more than a half-billion going there under provisions of Public Law 480, first enacted in 1954 and regularly renewed ever since. Efforts have been made to attach the more attractive label "Food for Peace" to the P.L. 480 program, but somehow, like "Point Four," the initial tag sticks. Under Titles II and III of the act, about $125 million of surplus crops per year were going as special gifts to the less-developed lands, either in emergency grants through official channels or in operations of private charities, to which the United States government gives the food for them to give away abroad, and for which the government pays most of the transportation costs. Under Title I of the act, yet an additional vast quantity of farm surpluses was "sold" to the underdeveloped countries for their own respective currencies, not for dollars; each currency, piling up in deposits to the United States account in local banks, is, for the time until it is used, available to the local financial authorities like a short-term loan of their own country's currency. This item in the late fifties was, in general order of magnitude, the equivalent of $500 million a year in the less-developed countries. The short-term nature of these deposits would, however, sharply reduce their significance as aid were it not for the further fact that the United States government in the late fifties was disbursing these currencies, which it had accumulated from the crop "sales," in this generous ratio: two parts grants in the foreign country, two parts soft loans in that country, and only one part American expenses there (for example, embassy maintenance). Thus, about $200 million worth a year of the P.L. 480, Title I "sales" to the underdeveloped countries should really be considered grants, and another $200 million a year amounted to soft loans, repayable in the local currency (not dollars) on a long-term, low-interest-rate schedule, with some prospect that local inflation would be allowed to whittle even these repayments away. Or, refiguring these latter items for their cost to the American taxpayer in light of the higher prices that were originally paid for these same crops under the farm support programs in the United States, about $300 million a year of the crop "sales" to the underdeveloped countries amounted to grants, and another $300 million amounted to soft loans.*

As has been noted, the Kennedy Administration's plans envisaged increasing the assistance under Titles I, II, and III, of P.L. 480 by perhaps 50 per cent or more above these levels. There is reason to doubt, however, that the economic and political benefits abroad that can be derived

* Our concern here is with P.L. 480 as a tool for economic development assistance. Consequently, the figures given in this paragraph (except for the overall annual disposal total, approximately $1.5 billion) relate only to the countries this book identifies as underdeveloped; that is, the noncommunist world except for Europe, Japan, the United States, Canada, Australia, New Zealand, and South Africa. More than one-third of all P.L. 480 shipments in fact went to countries already relatively developed—mainly southern Europe and Japan. Cf. pp. 312, 370–71.

from overseas disposal programs for farm surpluses are at any real expense to the American taxpayer. As long as the domestic economics and politics of agriculture in the United States insure that surpluses will continue to accumulate in government warehouses, the real choice very probably is whether to give them away or let them rot. Even if Congress could be persuaded to make major reductions in the existing pattern of price incentives to farmers, the improvements and investments already under way in American agriculture will create surpluses, at least until the late sixties, that will be as large or larger than those that steadily increased the federal stockpiles in the fifties (even faster than P.L. 480 and other programs virtually gave the produce away). About one-third of all United States agricultural exports in recent years has been financed through foreign aid and P.L. 480. No realist can anticipate an early end to the production of government-subsidized farm surpluses in the United States, virtually requiring government subsidization of exports. The question becomes one of making the most effective use of the surpluses abroad. Here are resources for foreign policy which in a sense cost the American taxpayer nothing (nothing extra), but which can be of real benefit in foreign countries. The Kennedy Administration, unlike Eisenhower's, was not inhibited about accepting this conclusion and making the best of it.

The "sale" for local currency should be recognized as an adaptation, rather dressed up for American public consumption, of the Marshall Plan's grant-"counterpart" device. (Cf. pp. 284–85.) The accumulation of local currency to the United States government account is akin to a counterpart fund. There is a superficial difference in that the currency is initially formally subject to full American control, whereas United States authorities had only a veto over the use of counterpart funds. But Americans cannot diplomatically afford anyway to disregard the preferences of the local government when huge accumulations of currency are to be spent in a country (and the money cannot be converted for expenditure anywhere else). The currency might theoretically be allowed to pile up in local banks; in that case the banks would have relatively greater deposits on which to make loans, but the overall effect on the economy would depend on other aspects of the fiscal policy the local government was pursuing, over which the currency accumulation does not itself give Americans any control. Even so, the mere existence of huge United States government accounts in local banks around the world would become a serious political irritant in most countries. Anti-Americanism would thrive on it. Therefore, in practice United States officials have little choice but to get rid of most of the currency in a fashion that is agreeable to the local government as well as to the Americans; this relationship is essentially like the counterpart-fund arrangements, but with a somewhat greater power of initiative remaining in the hands of the United States officials.

In the long run, another distinctive feature will be the return to the United States government account of the repayments made on the portion of the initial currency accumulation that is disbursed in soft loans, rather than outright grants. Presumably whatever is repaid will simply have to be loaned out again, since it will still probably be inconvertible and embarrassing for the United States to reamass; thus the American involvement in details of the underdeveloped economies through the loaned proceeds of P.L. 480 "sales" will potentially last much longer than the involvements in Europe through the counterpart that was released there with American approval to local agencies for lending. But the longer involvement outside Europe is a characteristic feature of most aspects of foreign aid anyway, and in this instance it could at any time be mitigated by disbursing the accumulated currencies as grants rather than loans. There has been a marked tendency in that direction.

The portion of foreign currency proceeds from P.L. 480 "sales" used by the United States government for its own official expenses in an underdeveloped country should be noted as reducing the flexibility the local authorities would otherwise have had in using dollars that would have come from American officials who were, say, traveling on expense accounts. The local dollar earnings could have been utilized anywhere abroad for development goods that might be of greater benefit than the American food, which now comes instead of the "untied" dollars. On the other hand, if the local government would not have used the flexibility wisely, and if—a big if—the other United States crop-surplus disposal plans are in most respects carefully directed toward contributing to economic development, the overall advantages can far outweigh the disadvantages.

P.L. 480 "sales" are governed by a principle of "additionality"; that is, they should be in addition to, not in place of, normal commercial imports into the country that is allowed to use its local currency to make the purchases. Insofar as this principle is firmly applied, it means that the population of the receiving country is actually enabled to consume more food (and cotton) than it would otherwise have had. In countries where the masses live at a mere subsistence level, their physical strength and vitality may be somewhat increased, and thereby also their capacity for productive labor in economic development—unless population growth merely produces more mouths to eat up the extra food, and unless there is a decline in local farm production resulting from the imports. The latter problem has in fact been a reality only in the sense that some recipient governments are inclined to use the availability of American surpluses, however uncertain, as encouragement to concentrate their own limited resources on industrial development, as they would prefer to do anyway, rather than emphasize additional irrigation, fertilizers, and the like; however, local farm prices are not reduced by competition from United States surpluses to a point where the farmers themselves would have smaller incentives to use whatever productive facilities they already possess. The other problem—population growth—is central to all economic development difficulties in most of the backward lands. If the numbers of people naturally continue to increase to the limits of subsistence, an increase in those limits can provide only short-term amelioration, at some risk of long-term dependency on the United States to prevent starvation. This risk is substantial enough to warrant greatly increased attention to the feasibility of birth control measures in conjunction with economic development programs.

In the meantime, any heightened vigor of laborers in surplus-recipient countries should be put to use in productive work projects. The P.L. 480 "sales," when the proceeds are used for development grants and loans, have the potential merit of concentrating in the hands of progressive authorities local currency with which they can channel local resources, without having to create the initial accumulation themselves by inflationary deficit financing or having to obtain it by taxation, which for political or administrative reasons in the underdeveloped country might have to be imposed, depressingly, on enterprises that are themselves important for economic advance. Furthermore, the food itself can help to meet the direct consumption demands of newly employed workers without requiring a sudden widespread disruption of traditionally patterned economic relationships; in an underdeveloped country almost any such change, whatever its implications might be for the behavior of an advanced economy, would result in aggravated inflation.

To be sure, the newly mobilized labor force will seek other consumption goods besides food, and will bid up the prices in local markets if the other goods are unavailable; hence an integrated American contribution to economic development requires that more be contributed than mere food surpluses. Also, neither the food consumed nor the local currency accumulations spent to channel local labor and local materials can supply industrial equipment that must be obtained abroad, from the United States or elsewhere. Capital goods which must be imported for economic development require other forms of financing—investments, loans, and grants, public and private. This consideration must not be forgotten. But in conjunction with such financing, a carefully planned American farm surplus disposal program can make a very valuable contribution to economic development.

The main trouble has been that optimum use of American farm surpluses to promote economic advance in backward areas has at least until recently been a secondary objective for the United States government—secondary in the minds of most key congressmen to spurring the greatest possible haste in emptying federal warehouses, and secondary in the minds of key officials of the Eisenhower Administration to avoiding any admission that the surpluses would be permanent and would permanently need some special overseas disposal, like P.L. 480. Thus Congress demanded the largest possible "sales" in the shortest possible period of time, and the Administration refused to allow the operation to be authorized at all for more than a year or two at a time. So the whole undertaking proceeded in a series of spurts. Precipitancy and discontinuity: hardly a formula for a very effective program of economic development. Major usefulness requires a perspective of years, not months.* That perspective was just beginning to shape the program in the last years of the Eisenhower Administration, and not until Kennedy did the whole enterprise begin to be conducted with enthusiasm.

P.L. 480 was also encountering resistance from foreign countries producing crops for export. Many of these complaints, too, were clearly the result of the excessive haste and discontinuity with which the program was being implemented. There has been an acute need for the administrators of the program to establish long-term perspectives and make them known, for the sake of the exporting countries as well as the importing countries, if the P.L. 480 programs are not to be disruptively spasmodic. The perspectives should, of course, take account of the needs of both groups of nations, and especially of the underdeveloped countries in each group (they are mainly among the importers).

Some friction is unavoidable. For America to press her P.L. 480 "sales" at all, even under the rule of additionality, means that any actual increase in demand for these products abroad is less likely to find expression in higher prices and greater earnings for the producer countries; the

* Special time-consuming complexities emerge when efforts are made to arrange useful "three-cornered" deals; for example, P.L. 480 "sales" of cotton to Japan for yen, which would in turn be loaned by American officials to Japanese agencies to be reloaned to Thailand so that Thailand could use the yen to buy Japanese machinery for her economic development.

United States herself squeezes her way into the world market by filling much of any increased demand abroad with her own cut-rate "sales." At least these P.L. 480 "sales" do not, however, usually depress the world market price; relative to other factors they merely tend to keep it from rising. And since the largest "sales" are of wheat (40 per cent of all P.L. 480 "sales" in the first four years), any adverse effects for producers have mainly been born by well developed countries like Canada that are probably able to endure greater sacrifices than they are otherwise making to assist the development of backward nations. But some American surpluses are of commodities that are important among the exports of certain underdeveloped countries as well as of the United States, for example, cotton, rice, and tobacco. Here special caution would be warranted.*

Actually the principal complaint of the exporting countries has not been about "unfair" American competition through the "sales," but rather through another section of P.L. 480 that provided for roundabout "barter" deals. "Barter" of United States surpluses for foreign commodities, mainly to expand the strategic materials stockpile, was not governed by the additionality rule. Consequently, for a time in the mid-fifties, normal markets for commercial exports, particularly Canadian and Australian markets in Europe, were largely pre-empted by American surpluses. Protests from producers at home as well as abroad caused the Agriculture Department to reduce this program drastically in 1957, and, although Congress insisted on its revival the following year, bartering since then has been rather circumspect.

### Soft Loans from the American Government and International Agencies

P.L. 480 "sales" for local currency that is in effect inconvertible, and likely to remain so, reflect, as outright grants do, the reluctance of the recipients and the American government to burden the balance of payments by demanding dollars or other convertible currencies as commercial exporters normally must do. The same considerations affect the decision whether to use grants or loans in disbursing the local currency accumulations that result from the "sales." If the currency is disbursed in loans (or not disbursed at all) a remote possibility exists that the United States might be able to get some of its money back in dollars by converting it at some time in the future when the local economy has grown

* Cf. p. 393n. on Burma and rice.

relatively stronger and international trading relationships have changed.

A similar possibility is involved when the American government, instead of "selling" commodities under P.L. 480 or making direct grants of any kind, makes a "development loan" of dollars, to be spent for capital goods and to be repaid eventually in local currency (not in dollars, unless convertibility were to become feasible by the time repayment is required). The possibility, however remote, that the United States will get its money back is important mainly for its political-psychological consequences. American taxpayers may find it a comforting illusion, and, more important, it seems to gratify the pride of progressive elites in the underdeveloped countries, who prefer the pretense that the aid they are taking is a mutually profitable business arrangement between equals—hence, that it is not a true gift, embarrassingly beyond their power to reciprocate, or a humiliating payment for supposed political services, past, present, or future, on their part.

From the viewpoint of economic efficiency, even if the soft currency payments can never actually be converted for the financial advantage of the original donor, the United States, they may at least tend to stimulate a sense of fiscal responsibility and reduce wastefulness on the part of the recipients. In regard to the other principally economic effects, however, grants would be preferable to soft loans. Grants would not, even as a theoretical potentiality, burden the balance of payments of the less-developed country. They would not result in awkward, indefinitely protracted "revolving fund" lending and relending of American-owned foreign currency accumulations. And they would not undercut the hard loans that are available from the Export-Import Bank, the World Bank, and private sources. (The hard loans would also be rendered rather unattractive by any increased availability of what may be termed "semi-soft" loans, repayable in dollars but at very low interest rates and over very long periods of time.) To preserve the integrity of capitalist international lending, and thereby foster private American participation in economic development, the preferable course normally would be to combine hard loans, up to the limit of the backward country's capacity to repay in convertible currency, with outright grants (dollars, food, or other goods) for any development requirements beyond those limits. However, other considerations at home and abroad that are principally political make it necessary to experiment with a middle ground of soft and semi-soft loans—in effect a variable mixture of grants and loans that utilizes the prestige of the loan concept while attempting to avoid attenuating it.

This experimentation has been the function principally of a Development Loan Fund (DLF) established by Congress in 1957. The DLF was directed not to compete with private investment capital, the Export-Import Bank, or the World Bank. If adequate financing is not available from any of them, the DLF may lend on its own softer terms, but must require some form of repayment. ("Semi-soft" is the Kennedy Administration's preference.) Problems of coordination are inherent in this arrangement, and they have inevitably caused delays. Congress has also kept much closer rein on DLF purse-strings than the chief sponsors intended. Until 1960 the Fund's actual disbursements were extremely small—less than Point Four expenditures in the same years, despite the DLF's much more ambitious role.* But appropriations nearly doubled under Kennedy to $1.1 billion for the fiscal year 1962. If this rate continues—or is even further increased as Kennedy desires—the DLF will become the major element in American economic assistance.

One of the arguments for preferring loans to grants, even if the loans have to be rather soft, is that other advanced countries have shown a much greater reluctance than has the United States to contribute to economic development through grants (outside their own colonial dependencies). The chance of getting significant contributions from and to a widening range of countries seems to be somewhat greater if loans, however soft, are emphasized. Within the Western Hemisphere, the United States government finally agreed in 1959, after a decade of hesitation, to establish a special Inter-American Development Bank (IADB), for hard loans, like the World Bank, and also a related Fund for Special Operations, a kind of multilateral Development Loan Fund for soft loans.

The United States agreed to put up $165 million out of the total $400 million capital the American Republics were to make available to the Bank for hard loans; and another $185 million of the $450 million guarantees that the governments were to provide for the Bank's own securities when sold to private investors. The arrangement closely resembled the World Bank. As in the latter, the voting rules preserved for the United States the paramount influence when she chose to act; Americans might be a small minority of the personnel, but the IADB was located in Washington and the United States member on its Board of Directors would cast 40 per cent of the votes. Presumably, the Inter-American Development Bank's operations would be carefully coordinated

* In those early years of the DLF, considerable stretching of the value of the loans was possible by taking advantage of their lack of a requirement that the proceeds be spent in the United States; the loans were only loosely "tied." But the increased anxiety about gold outflow in the 1960's has led to tighter tying.

with the World Bank and the Export-Import Bank, which were already working closely together in Washington without much of a national-international institutional barrier.

The Inter-American Fund for Special Operations was a more novel organization. Its initial resources were only $150 million, of which the United States contributed $100 million. Since repayments of its loans could be in nonconvertible currencies, the Fund would not be able to sustain greater flexible operations unless it were periodically provided with more governmental dollars. The United States Treasury would cast only 40 per cent of the votes, as in the hard-loan Inter-American Bank, but it was taking precautions against too rapid a drain on the Fund (with resulting demands for quick dollar replenishment) by insisting that Special Operations loans be approved by a two-thirds majority of the votes cast; thus the United States representative would have a veto. Yet the careful safeguards should not minimize the fact that the United States was giving an international organization the initiative in determining how American dollars should be spent on a soft-loan basis for general economic development, not just on a basis of safe hard loans as in the case of the World Bank or for relatively small-scale technical assistance programs with an aura of missionary work and welfare services, like those of the UN. Furthermore, the intent of the Eisenhower Administration was to expect the Latin Americans to try to obtain soft-loan financing from the Special Operations Fund of the Inter-American Bank before they approached the Development Loan Fund of the United States government. Coordination would presumably be close here also, but the move reflected a significant tendency to weigh the advantages heavier than the disadvantages in deciding to channel an increasing—though still a minor—amount of United States economic development assistance through international agencies, especially through ones in which the American government normally could wield preponderant voting power.

There has been some resulting reduction in the suspicions that have been felt by the underdeveloped countries. Mutual criticism and firm expert advice through the international institution may be more acceptable than direct guidance by the United States. And the Inter-American Special Operations Fund itself has confirmed a remarkable spirit of mutuality: even underdeveloped countries thereby agreed to contribute initially up to $50 million to each other's economic advance. To be sure, the contributions were stimulated by the $100 million that is coming from the United States, but the precedent could be expected at least to be somewhat conducive to forming good habits of economic cooperation. In recent years the Eisenhower and Kennedy administrations have both sought earnestly to strengthen this cooperation, building toward an "Alliance for Progress" in the Western Hemisphere.

Another major step in the direction of internationalization of development assistance came in 1959 when the International Development Association (IDA) was established as a semi-soft-loan affiliate of the hard-loan World Bank, comparable on a much larger scale ($150 million a year) to the Fund for Special Operations adjunct to the new Inter-American Development Bank. Once again the United States was reconciled to internationalization largely because the special-agency voting arrangements insured that American influence would normally be preponderant if there seemed to be need for the government to exercise it. The link to the International Bank for Reconstruction and Development gave further assurance that the Soviet Union, which is not a member of this capitalistic World Bank, would play no role in distributing the aid. The institutional focus would still be on Washington, but some of the onus of giving and of nagging would be taken off the shoulders of American administrators and shared with other cooperative governments. The resulting nest of agency relationships may be illustrated this way, with a hard-loan core and a soft-loan periphery:

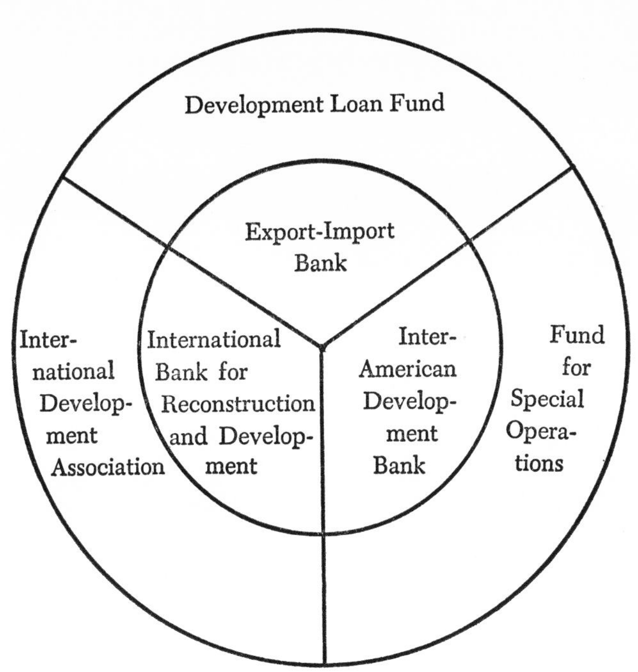

### Sharing the Aid Burden

Other Western governments that are contributing to the World Bank and Fund, IDA, and the UN assistance programs also operate their own bilateral loan programs, with varying degrees of hardness. But the United States government, partly as a result of its new anxiety about gold since 1959, has been ill-satisfied with the total amount that some of the Western Allies have been contributing to economic development in the newly emerging countries. Only $300 to $400 million of the annual increase in American obligations to pay out gold on short notice is directly attributable to economic aid activity of the United States government; most of the dollar aid is "tied" to American goods or spent voluntarily for them by recipient countries. But unless the gold problem is now somehow corrected, no very large increase in untied United States aid is likely to be feasible. Much the same holds true for Britain. The Algerian strife has restricted France's capacity. But Canada, Germany, and Scandinavia could do much more by almost any standard, as could Italy and Japan by some standards. A major difficulty has been to determine which standards can be made mutually acceptable among the industrialized countries of the noncommunist world. What constitutes "aid" and what constitutes the capacity to bear aid burdens?

The revamped Organization for Economic Cooperation and Development (OECD)—formerly the Marshall Plan unit, OEEC—has a Development Assistance Committee (DAC) designed to recommend answers to these questions in reviewing the programs of the member nations. DAC does not conduct aid operations itself but seeks to persuade laggard governments to make more useful contributions than they have previously in the emerging countries. The OECD is also the organization responsible, along with the IMF, for harmonizing economic relationships among the advanced Western nations in ways that will minimize payments-imbalances and disruptive gold flows. This linkage should improve the prospects of getting some reassuring cooperation among the donor countries, so that they will not feel obliged to protect their respective gold holdings with such extreme caution that they tie so much of their aid—each country to its own exports—that the trade patterns of the recipients are distorted and their development programs burdened with extra expenses. Tying can often be mutually convenient to both donor and recipient; probably too much opprobrium has been heaped on the notion of "dumping surpluses abroad"; newly developing countries now frequently shy away from chances to obtain real bargains. But excessive, generalized tying could greatly reduce the potential value of transfers of resources from the advanced to the developing nations. OECD should be of help in establishing criteria.

OECD's purposes also involve other, much larger questions. How immense really is the aid burden the United States seeks to share? What is the total of aid going to the emerging countries from the United States and others, and how is its adequacy to be evaluated? Is the need in the backward areas so pressing that any further contributions that can be elicited from Europe, Japan, and the advanced Commonwealth countries must be additional to, not substitutes for, a continuation and even some increases in the levels of United States aid?

The types of American contribution to the economic development of backward countries are so diverse that some totaling is necessary before any overall appraisal of their significance can be reasonably attempted. In the late fifties, United States government contributions of nonmilitary grants, loans, and technical assistance to the underdeveloped countries of the noncommunist world were running at a rate somewhat above $2 billion a year. Major increases in congressional appropriations during Eisenhower's last years as President and in the first year of the Kennedy Administration now make it possible for these expenditures in the early 1960's to run at a rate $1.5 billion higher, perhaps even more—a total figure of roughly $3.5 billion per year (unless future appropriations are reduced or net lending by the Export-Import Bank declines).*

* Cf. p. 352. In the late fifties the approximate order of magnitude of separate items annually was as follows: Export-Import Bank, $300 million net outflow; United States technical assistance, $150 million; American contribution to UN technical assistance and relief programs, $50 million; "mutual security" grant programs ("defense support," "special assistance," "development assistance," and some of the "contingency fund"), $1 billion; P.L. 480, Title II and Title III emergency and charitable commodity-grants, $125 million; P.L. 480, Title I "sales" proceeds disbursed as grants and loans, $400 million (the crops themselves having cost the American taxpayer $600 million). In addition, the International Bank for Reconstruction and Development had an annual net outflow of nearly $250 million; its operations were partly guaranteed by the United States government's contributions.

By 1962 the United States technical assistance funds and the "mutual security" grant programs had been redivided among a "contingency fund," "supporting assistance," "development grants," "food for peace," and the Development Loan Fund, the latter emerging with about $1 billion a year of net new money beyond its share of the split. Increases in annual P.L. 480 programs, perhaps on the order of 50 per cent, were likely to add another $200 million or more (net of the "food for

The OEEC-OECD calculated that the advanced European countries in 1956–59 were providing an annual total of $1.8 billion through governmental channels, compared with an American rate of $2.3 billion of official contributions in those years. (The $2.3 billion figure is essentially the same as the $2 billion computed in the note on p. 377 because OEEC-OECD included some European countries—Spain, Yugoslavia, Greece, and Turkey—as underdeveloped recipients, and they were jointly receiving about $300 million annually.) OEEC-OECD's calculations must be interpreted as giving all the member nations the benefit of the doubt as to their generosity. Actually, only the United States, Britain, and France (in Algeria) were making large grants; and the other main item in the European account was short-term German loans. There was presumably a need to put the non-American contributions on a more long-term and/or grant basis, but just how generous should they be? And how much more might be expected from the United States?

A comparison with the European recovery program may lend perspective. American economic aid to the underdeveloped countries in the early sixties was finally reaching about the same rate as the nonmilitary grants and loans to Western Europe during the Marshall Plan years 1948–52. And the economic development assistance was certain to continue much longer than did the European recovery program.

Another perspective for evaluating the foreign economic aid total is to measure it against an estimate of total need in the underdeveloped countries and of their capacity to absorb assistance with a minimum of waste. Of course, such comparisons would be highly desirable if they could be made reliably, but the data are extremely elusive. In the 1950's, research teams at the Massachusetts Institute of Technology attempted to work out the figures with great effort but with limited success. The MIT group concentrated on "absorptive capacity," supposing an international program were devised whereby the advanced industrial powers would agree to contribute all the additional development assistance that could genuinely be utilized anywhere in the noncommunist world on condition that strict nonpolitical standards of economic efficiency be applied. The resulting 1956 estimate of absorptive capacity was no more than $3.5 billion a year above the 1953 level of actual contributions. If that much more were promised, probably only 60 per cent of the increase would in fact meet the strict standards and be spent in any year,

peace" share of the split.) Also an additional sum of about $350 million per year was available for the Peace Corps, the International Development Association, the Inter-American Development Bank, and for an expansion of the contingency fund and UN assistance programs.

and of the new spending only 60 per cent would have to come from the United States government: about $1.2 billion a year more than in 1953.* In actual fact, the economic aid from America to the underdeveloped countries did increase by several hundred millions under Eisenhower; but so also—by new MIT calculations—did the absorptive capacity of many countries. Thus in 1961 the researchers still felt that an additional $1 billion a year beyond the Eisenhower peak was needed from the United States government, plus a similar amount from other OECD governments. Total official American nonmilitary loans and grants to underdeveloped countries would then run between $3.5 and $4 billion annually, of which the MIT group called $3 billion "developmental."

It should be remembered that these figures concern an economic development program for the entire noncommunist world, conditioned only upon strict nonpolitical standards of economic efficiency. One might conclude that more selective concentration on key countries like India would achieve successes in limited areas even with the present aid expenditures, which after all are only a few hundred million smaller in total than the MIT ideal (though somewhat differently distributed). To be sure, the MIT group's sharp upward revision of its own calculations after a half-decade is a reminder that estimates on such matters are inevitably speculative. And the MIT figures rest on an hypothesis that a 2 per cent rate of annual growth is an adequate goal. The question of whether it is or not is actually a problem in local political levels of endurance, which vary greatly over time as well as space. But at least the MIT study shows that the dollar dimensions even of highly ambitious development assistance programs would not be astronomical when measured on an annual basis, though the expense would continue for decades. These estimates remain impressionistic, but they furnish some reassurance that American economic assistance is not absurdly small in relation to the current developmental potentialities of a third of mankind—especially if substantially enlarged contributions are forthcoming from Europe and Japan. Increased amounts of American aid would be useful—perhaps a few hundred million more dollars annually—together with several hundred million more from other OECD governments. But existing amounts are already great enough to lend encouragement to any reorganization efforts that are designed to get a maximum benefit from

* Details of the MIT studies are conveniently available in Max F. Millikan and W. W. Rostow, *A Proposal: Key to an Effective Foreign Policy* (New York: Harper, 1957), *passim*, and in Max F. Millikan and Donald L. M. Blackmer, eds., *The Emerging Nations: Their Growth and United States Policy* (Boston: Little, Brown, 1961), pp. 121–23, 149–59.

the current expenditures alone. Reformist hopes need not be frustrated by any utter incommensurateness between the available tools and the looming problems.

## SELECTED BIBLIOGRAPHY

ASHER, ROBERT E. *Grants, Loans, and Local Currencies: Their Role in Foreign Aid.* Washington, D.C.: Brookings Institution, 1961.

BINGHAM, JONATHAN B. *Shirt-Sleeve Diplomacy: Point 4 in Action.* New York: Day, 1953.

BLACK, EUGENE R. *The Diplomacy of Economic Development.* Cambridge: Harvard University Press, 1960.

CURTI, MERLE, and KENDALL BIRR. *Prelude to Point Four: American Technical Missions Overseas, 1838–1938.* Madison: University of Wisconsin Press, 1954.

GLICK, PHILIP M. *The Administration of Technical Assistance: Growth in the Americas.* Chicago: University of Chicago Press, 1957.

LISKA, GEORGE. *The New Statecraft: Foreign Aid in American Foreign Policy.* Chicago: University of Chicago Press, 1960.

MILLIKAN, MAX F., and DONALD L. M. BLACKMER, eds. *The Emerging Nations: Their Growth and United States Policy.* Boston: Little, Brown, 1961.

———, and W. W. ROSTOW. *A Proposal: Key to an Effective Foreign Policy.* New York: Harper, 1957.

MYRDAL, GUNNAR. *An International Economy: Problems and Prospects.* New York: Harper, 1956.

ORGANIZATION FOR EUROPEAN ECONOMIC COOPERATION. *The Flow of Financial Resources to Countries in Course of Economic Development, 1956–1959.* Paris: Organization for European Economic Cooperation, 1961.

PRESIDENT'S COMMITTEE TO STUDY THE UNITED STATES MILITARY ASSISTANCE PROGRAM. *Composite Report and Annexes,* 2 vols. Washington, D.C.: Government Printing Office, 1959.

SCHELLING, THOMAS C. *International Economics.* Boston: Allyn and Bacon, 1958.

SHARP, WALTER R. *Field Administration in the United Nations System.* New York: Praeger, 1961.

SHONFIELD, ANDREW. *The Attack on World Poverty.* New York: Random House, 1960.

STALEY, EUGENE. *The Future of Underdeveloped Countries: Political Implications of Economic Development,* rev. ed. New York: Harper, 1961.

TEAF, HOWARD M., JR., and PETER G. FRANCK. *Hands Across Frontiers: Case Studies in Technical Cooperation.* Ithaca: Cornell University Press, 1956.

UNITED NATIONS ECONOMIC AND SOCIAL COUNCIL. *Economic Development of Underdeveloped Countries: International Economic Assistance to the Underdeveloped Countries, 1956–1957.* New York: United Nations, 1959.

UNITED STATES CONGRESS SENATE SPECIAL COMMITTEE TO STUDY THE FOREIGN AID PROGRAM. *Foreign Aid Program: Compilation of Studies and Surveys.* Washington, D.C.: Government Printing Office, 1957.

UNITED STATES DEPARTMENT OF COMMERCE. *Balance of Payments Statistical Supplement.* Washington, D.C.: Government Printing Office, 1958.

———. *Survey of Current Business,* monthly. Washington, D.C.: Government Printing Office, 1956– .

WOLF, CHARLES. *Foreign Aid: Theory and Practice in Southern Asia.* Princeton: Princeton University Press, 1960.

CHAPTER 19

# ECONOMIC DEVELOPMENT ASSISTANCE AND THE IMPACT OF EAST-WEST RIVALRY

THE MANIFOLD Western instruments, public and private, for promoting and guiding economic development have obviously become so numerous as to pose increasing problems of overall coordination; and in the individual programs there is need for constant experimentation with the specific improvements that are suggested through wider and wider experience.

## Some Comparisons with the Marshall Plan

Naturally the image of the Marshall Plan is often recalled in Europe and America as a pioneer model for success in such endeavors, and indeed many of its features have obviously been imitated in programs for the less-developed countries. The "counterpart" device, for example, has been adopted through local currency accumulations to become a major instrument for influencing the use of indigenous economic resources. Also of growing importance has been the process of exchanging technical specialists, with emphasis upon fostering an expansionist, competitive economic psychology, as well as upon teaching specialist skills. The

Marshall Plan's pressure for checking inflation and establishing predictability in economic relationships has also been a conspicuous element in American programs for the underdeveloped countries—with a similar objective of eliciting more productive, less speculative contributions from the acts of private economic enterprise. The very notion that American officials should encourage major foreign governments to engage at all in national economic planning was novel with the European recovery program; and with it came the basic precedents that the Americans themselves should be allowed to participate in a foreign government's planning; and correspondingly, that they must truly become competent to make a valuable contribution as advisers, not just as donors. Even the general presumption that the United States government may legitimately impose taxes upon its citizens in order to attempt a massive economic intervention is a heritage of the European recovery program.

On the other hand, the emphasis of the Marshall Plan on regional collaboration, a central feature, has had a cool reception when proposed likewise for the less-developed countries. In none of the major underdeveloped regions of the world is there a sense of traditional common standards and extensive common purposes so strong as those that survived in Western Europe even after two world wars; moreover, the network of trading relationships, which helped to knit Europe together in the nineteenth and early twentieth centuries, tended in contrast to link each of the backward countries and colonies across wide oceans to one or more of the advanced Western nations—not to its own neighbors. Except to some degree in Latin America and in the Arab world, the sense of belonging to a regional community and the willingness to make important mutual concessions to preserve that partnership simply does not exist as it did in Europe. Consequently, that foundation is lacking for an expansion of complementary trade within the area and for the location of new industry largely according to criteria of competitive economic efficiency in the region as a whole.

Rushing the industrialization and diversification of their own national economies, with the special object of escaping a cyclical buffeting in the world market by overcoming their countries' excessive dependence on single-commodity specialization, the sensitive elites of the less-developed nations are understandably hesitant to take shelter at any intermediate regional level, which would, for example, offer special incentives and protections to commerce within the area, but would avoid the extremes of mere national autarky on the one hand and of nondiscriminatory worldwide import competition for their weak infant industries on the other. The pattern of the Organization for European Economic Cooperation and of the European Payments Union (and now the Common Market)—to seek prosperity through competitive mass production in regional markets as populous as the United States itself, but not to expose oneself to the full blast of American industrial competition—is a pattern that has not yet been congenial to most of the less-developed countries.

The United States government offered, for example, to subsidize joint regional development programs in southern Asia (1955) and in the Middle East (1958), but both attempts foundered upon the rocks of economic and political friction among the countries of the area. Most of the nations preferred to seek aid through bilateral agreements with the United States or multilaterally through worldwide agencies. Only in Latin America (1959) has the regional approach been finally institutionalized on a major scale.

Probably the stage of development in Asia and Africa is still too early to permit great productive gains from intraregional specialization. The joint regional confrontation of economic problems is important to the United States usually more as a framework for the growth of noncommunist political solidarity than for immediate development goals. For this reason, the kind of decided American pressure for regional economic integration that usefully characterized the Marshall Plan in Europe would probably defeat its purpose in Asia and Africa. But when the impetus genuinely arises from an underdeveloped area itself without great prodding from the United States, as in the case of the Inter-American Development Bank and its Fund for Special Operations, encouragement from Washington seems desirable. As with the European economic organizations of the Marshall Plan days, such regional groupings can provide a framework for mutual criticism and counsel that relieves the United States of some of the onus of "dictation." They also tend to foster political solidarity and may help to provide markets in emerging countries for each other's products, as development proceeds. Otherwise, a rising flood of simple Asian-African manufactures threatens to cause a reversal of the notable tendency toward liberalization in trade relations among the Western powers that has hitherto prevailed since the onset of the European recovery program.

However hesitant, there has at least been some approach toward regionalism in American aid programs toward the underdeveloped countries. But one other central feature of the Marshall Plan has conspicuously been almost entirely lacking in these later ventures—at least until the Kennedy Administration: the emphasis upon laying complete foundations for self-sustaining growth within a limited period of years, and the willingness to provide extra-large grants in order to achieve this status rapidly. This kind of expense was regarded as too great for the United States to bear when dealing with countries that do not already possess the habits, skills, and material facilities of industrial civilization, and hence must create them, not merely renew them as in Europe. The American refusal to make the effort, however, has deprived the United States economic development programs of a chance to elicit the extraordinary enthusiasm the exciting short-term achievement potentialities of the Marshall Plan aroused, attracting remarkably talented, forceful administrators from private life for temporary service, and forestalling most of the routinization tendencies of bureaucracy. It is even possible that the "endlessness" of economic development assistance at the limited levels of the late fifties might become less bearable politically in the United States than a larger and more decisive program would be. At least a concentration of effort upon certain developing countries would offer more appealing prospects for success, as did the previous focusing upon Western Europe. The Kennedy Administration took office with hope of acting in accordance with this logic.

## The Case of India

In particular, there seemed good reason to believe that concentration of development assistance could achieve important gains in the most important of all the underdeveloped countries: India. Strategically located, possessing nearly half the population of noncommunist Asia, with substantial armed forces, an experienced bureaucracy, relatively stable political institutions, a large educated elite, and much of the basic substructure (railroads, ports, communications, and the like) for economic development, India constitutes a hopeful model for nontotalitarian advance. A crucial contrast in methods is the forced industrialization of Communist China. With a per capita output roughly the same as India's in the mid-fifties, China is estimated to have been leaving not more than 75 per cent of it for household consumption, whereas in India households consume 85 per cent.* The Indian government's feeling of greater immediate concern for the standard of living was also reflected in a faster

* These estimates and those that follow were derived by the staff of the India Project at MIT's Center for International Studies. They are based on unreliable Chinese and Indian data that can only be taken as indicative of relative orders of magnitude. Wilfred Malenbaum, "India and China: Contrasts in Development," *American Economic Review*, June 1959, pp. 284–309.

growth of industrial production of consumer goods in India, even though total industrial production rose much faster in China. Yet overall, the fact remains that the capacity and willingness of the Communist regime to drive and squeeze its population scored the more impressive material results. MIT researchers have estimated that the annual rate of increase of gross national product (at constant prices) in India, 1950–58, was almost 3.5 per cent; the growth rate in mainland China was estimated to have been at least three times as great, though slackening since then.

Proposals to use foreign assistance to match fully, under a nontotalitarian regime in India, the forced pace of Chinese Communist industrialization cannot realistically be regarded as practicable. But the United States did undertake to provide India with grants and loans of more than $1.5 billion during the 1950's, and toward the end of the decade the rate was over $300 million each year. Heavy contributions have also come from Western Europe and, presumably with competitive motivations, from the Soviet bloc (see pp. 387–97). A substantial increase in the rate of assistance is projected and will be necessary if Indian production is to continue to grow faster than India's population by a margin sufficient at least to make the preservation of political stability possible there. Anything less would destroy the valuable image of Indian nontotalitarian development in the eyes of impatient elites in other underdeveloped countries; the image has already suffered in cold-blooded comparisons with Communist China. Worse yet, if political turmoil in India were to degenerate to the point of an internal Communist takeover, a wave of defeatism would surely sweep over the noncommunist elites in much wider areas of Asia and Africa, where foundations for progress are much weaker than India's. These dangers are great, but correspondingly great would be the worldwide impact of a continued economic advance in India under democratic auspices, even though it were slower than China's. The chances for success are probably fair enough and the stakes certainly high enough to warrant the increasing Western assistance now centering there.

### The Aid Component of Soviet Loans to the Underdeveloped Noncommunist Countries

India's importance and the potential usefulness of foreign aid in shaping her development have been recognized also by the Soviet bloc by its concentration there of a large proportion of the loan programs the Communist countries have begun offering in competition with the West.

These programs suggest a new standard by which to evaluate the adequacy of American economic aid activity in underdeveloped areas.

Between the start of the Soviet assistance in 1955 and the end of 1960, according to tabulations made by the American State Department, the Sino-Soviet bloc had signed agreements with twenty-three noncommunist countries outside Europe to provide about $4.7 billion in grants and loans for military and economic aid. Of this, $3.5 billion was economic, almost $600 million a year on the average—less than one-third the contemporaneous American rate for underdeveloped countries in the noncommunist world. Furthermore, these figures measure the nonmilitary contribution of the United States government alone against the entire nonmilitary grant-and-loan activity of all the thoroughly socialized economies of the entire Communist bloc. Public and private investments, loans, and grants from all the OECD powers should be added to make the comparison appropriate. Thus, among noncommunist nations in the late fifties, public and private contributions from the more- to the less-developed countries were running at a rate five times as great as the Soviet bloc's commitments for nonmilitary contributions in the same part of the world, and more than twenty times the rate of actual deliveries the new Soviet programs had yet achieved.

However, in some of the individual countries the Soviet contribution has overshadowed that of the West: Egypt, Syria, Yemen, Afghanistan, Indonesia, Guinea, Cuba, and probably Iraq. Moreover, the physical capacity of the Soviet Union to expand its aid programs is unquestionably large. A careful study under the auspices of the Council on Foreign Relations has estimated that the USSR could raise its economic assistance outside the Soviet bloc to a level of $1.6 billion a year (roughly comparable to the Eisenhower program by some calculations) if it would devote just 12 per cent of the annual *increase* of its own total production to this purpose. No general reduction of internal consumption would be required. However, for a time there might well be a strain on the branches of industry that produce machinery; the program might take 5 per cent of total production in this area, and more assistance might have to go to China and some of the Eastern European satellites to mollify their protests.* The symbolism of rapid industrialization being as important as it is to Soviet leaders, any such drain on the USSR's own rate of growth would be painful to the Kremlin. Significant perhaps is the fact that remarkably little detailed publicity is given in the Soviet Union to the aid that is going to the less-developed peoples; at home

* Joseph S. Berliner, *Soviet Aid Policy: The New Aid and Trade Policy in Underdeveloped Countries* (New York: Praeger, 1958), pp. 104–18.

any pride shown in these arrangements for "mutual benefit" or even in this form of "peaceful competition" with the West is somewhat dimmed by generalities. Nevertheless, the capacity for expansion of aid clearly exists, and aid is in fact expanding substantially.

But is it aid? That question is more serious than it sounds, for the Soviet Union persistently claims to be merely selling or lending, in a manner more friendly than the West, but with financial repayment expected. The Communist bloc countries are usually unwilling to engage in grant aid (perhaps feeling unable yet to do so) and have spread the doctrine that grants are a demeaning form of charity which must have some political "strings" attached to them by implication. In any case the West does impose "strings" relating to economically efficient use of its grants or loans. The Soviet Union, in contrast, usually offers to try to provide whatever it can of anything that is requested, whether the request is developmentally wise or not, but to provide it in loans, not in grants. The apparent attitude of "letting the backward country hang itself" can backfire in disillusionment against unproductive Soviet aid, but there is some safeguard against this in that economic planners in many underdeveloped countries have previously learned valuable lessons from the World Bank and from other Western advisers concerning what is worth asking for; now they may be able to get it sooner from the USSR and then use it with reasonable success. The Soviet Union professes to trust the judgment of the government of its "trading partner," flattering its ego, while dramatizing the contention that a well-planned Communist regime by its very nature does not produce surpluses that it must then get rid of by making gifts abroad, and that it does not find itself unable to make room in its home market for imports taken in proper payment for what it loans and sells abroad. Capitalist economies, of course, are alleged to tend inevitably to overproduction. Thus any grants from Western countries must be just devices for turning an inescapable "dumping of surpluses" into a form of political bribery that is not only demeaning but even cheap. The P.L. 480 program, of course, fits this caricature with an exceptional veneer of plausibility.

Communist leaders may well take their own doctrine seriously enough to be inhibited by it from outright grant programs themselves, but there is some grant component in what is officially a program of loans between equals. Exactly how large cannot reliably be estimated.

One technique for making grants in the guise of loans would be to set artificially low prices on the Communist products for which the less-developed countries spend the loans they receive. Actually the Soviet bloc usually—not always—quotes prices that are competitive with those of Western suppliers for comparable goods and services. But how these prices relate to the structure of production costs in the Communist countries themselves is very difficult to judge. There may in some instances be considerable subsidization of exports, in others perhaps high profit-taking. Even if a detailed commodity study could permit generalization on this point, however, there would also be other room for quasi-grant assistance under the Soviet pattern of loans: the eventual repayment of the loans may be in commodities that are then priced higher than their relative economic value for the Communist bloc, or the rate of repayment and interest may be too low to cover the costs to the Communist countries of the immobilization of that part of their capital resources in the meantime, or the loan may actually be forgiven in whole or in part before it is ever repaid. Under the latter possibility, familiar in past Soviet practice, the USSR may gain political credit twice—first for "businesslike equalitarianism" in making a loan instead of a grant, and then later for extraordinary generosity in canceling the debt; at least until familiarity breeds contempt, a tacit understanding that that is the intention may be highly attractive to proud leaders of needy countries. One effect is the further corruption of the international debtor-creditor relationship upon which private international capitalism has to be able to rely.

The Communist bloc aid program has largely pioneered the soft-loan devices with which the Western world also is increasingly driven to experiment as a middle ground between hard loans and grants—but one on which private participation is exceptionally difficult. Soviet loans are usually repayable largely in local currencies that are likely to be nonconvertible, or even in specific local commodities (for example, Egyptian cotton), with interest rates of 2 to 2.5 per cent over 10-to-30-year periods. The interest rate approximates a very short-term rate inside the USSR, although these loans are long term; it is about half the hard-loan World Bank and Export-Import Bank rates. Yet again, the apparent subsidization involved in such cheap loans may possibly be offset in the Soviet economic view by privileged access over a long period of time to certain commodity exports of the less-developed country; a centrally planned economy may find such predictability especially valuable. On the other hand, the degree of privilege and the length of time are subject to renegotiation, which may result in expensive acts of generosity for political purposes.

It is important to recognize that the Communist countries, particularly the relatively industrialized Eastern European satellites, do have important uses for raw materials and agricultural imports of the kind that the underdeveloped countries can supply. Large quantities are required for industry. Consumers can also be provided with increased amounts of "soft goods" while internal production remains concentrated on heavy industrial growth; in any time of emergency, the civilian consumption could be cut back without severe disruption of the allocation of domestic resources. Thus the Communist protestations that their "aid" is actually trade, for "mutual benefits" of an economic character, should not be dismissed hastily. In large part the contention is probably true, but how large will become clear only after some years when large-scale repayments will have been made under this program, which is still very new. Only then will it be possible to judge accurately how big and how legitimate is the economic advantage that the Soviet bloc derives in recompense for its lending program.

### Soviet Loans as a Means of Political Influence

Meanwhile, there are political gains to be won. In the long run, the Soviet objectives must run counter to the American purpose in providing similar aid, which intends through development assistance to foster higher living standards and a sense of achievement under noncommunist governments, and thus to check the spread of totalitarianism. The Kremlin may reject this Western opinion concerning the probable political consequences of an economic development program—at least if the program is one to which Communist powers have made an important contribution themselves. The contrary Soviet view may be that development shrewdly aided in the direction of accelerated industrialization will produce an urban proletariat that will be organizable for Communist ends, and that, if economic advance is actually achieved, the prestige of Communism will be enhanced by the contribution the Soviet bloc made; even a failure to develop the economy despite generous Soviet assistance could later be cited as proof that only full-fledged Communist totalitarian methods can ever succeed.

On the other hand, if the Kremlin privately concedes that the long-run consequences of economic development assistance are more likely to be what the Americans have been hoping, the Communists can nevertheless find short-term political considerations that make the long-run risk worth taking. The advantages the United States derived from successful aid programs in Western Europe and in East and Southeast Asia demonstrated the usefulness of this policy instrument. A mighty and prosperous nation like the Soviet Union would not inveterately continue to deny herself the advantages of using it, once she had achieved her own postwar recovery and had become more than ever convinced that peaceful competition is more promising than a test of arms with the major capitalist powers. In particular, the United States had been gaining through its economic aid programs a reputation for constructiveness that could not always be offset by hostile allegations of "dumping" and "militarism." To some extent the Kremlin could continue trying to give itself credit for anything constructive the Western powers furnished to the less-developed nations: "You might as well say we Communists are providing it, since the West's gestures are only motivated by the cold war." But such arguments would become more convincing if the Soviet bloc made somewhat more constructive gestures itself, perhaps even beginning to provide a preferable alternative source for large-scale assistance.

In a sense then, some "propaganda of the deed" was needed to reinforce the "propaganda of the word." Important also among the noneconomic advantages to be derived by the Soviet bloc from its trade-aid program are the direct and indirect channels of access it provides, for influencing political tendencies in the less-developed countries. A program of massive loans is indispensable in virtually every underdeveloped noncommunist country that will accept them, if there is to be much chance of weaning such a country away from its established trading ties, which are almost exclusively with noncommunist countries. All of the vested interests in trade at all are in noncommunist trade. To break into this circle and redirect the economic interests and the related political tendencies toward greater cooperation with the Soviet Union, rather large bait must be held out; it is not usually accepted very readily. For example, the State Department has estimated that during 1957, of all the underdeveloped noncommunist countries in the world (excluding Europe as usual), only two, Afghanistan and Egypt, conducted more than 20 per cent of their total foreign trade with the Soviet bloc; only four more, Syria, Iran, Burma, and Ceylon, exceeded 10 per cent. The relationship to areas of concentrated Soviet lending is clear; so also is the failure of the Communist bloc thus far to establish politically reliable relationships of commercial dependency outside its own zone of total control.

The Soviet countries have also used their loans to pave the way for a large influx of technical advisers from the Communist bloc to underdeveloped countries outside. In the second half of 1957, for example, about 1,500 industrial, agricultural, and other professional specialists from the Soviet bloc worked for a month or more in sixteen of the less-developed countries (not counting the personnel who were solely in military or trade-promotion activities). In only eight of those countries were there more than fifty such specialists; the list was almost the same as the list of countries where Communist financial assistance was important. A year later the number of Soviet experts had nearly doubled, but they were still concentrated almost exclusively in countries also receiving loans and grants.

Partly, this situation is a result of the Communist preference for assistance that is directed toward industrialization, a preference that is of course shared by most of the progressive elites in the backward countries themselves. The development of factories, transport, communications, electric power, and mining requires importation of heavy equipment; technical advisers are not self-sufficient. But correspondingly, when the imports are financed by "tied" loans that must be spent for Soviet bloc products, Soviet bloc technicians must normally go along also, at least on short-term contracts, to instruct on installation and use. The arrangement in many respects resembles the overseas investments and the exports on credit that are made to the underdeveloped countries by private firms in the advanced Western nations. Governmental technical assistance from the United States has been heavily concentrated in the fields of health, education, and agriculture—fields which Soviet advisers scarcely touch. The American preference for private activity—with varying degrees of government cooperation—in the development of industry, mining, and transportation is condemned as swaddling and exploitation by the Communists and by large segments of the socialistically-inclined native elites. An emphasis on rapid industrialization under chiefly governmental auspices is often welcome, but the Soviet personnel who become involved in such operations have tended to function rather like their counterparts in the employ of private American firms. They serve abroad for short terms in self-contained teams for specific projects—coordinated, presumably, at a home office in Moscow, Prague, or Peiping, not through an annoyingly conspicuous large mission of personnel on two-year contracts overseas like American officials. The ominous shadow of a foreign bureaucracy has also been minimized by the avoidance of detailed supervision over native authorities, of the kind that the United States government is inclined to exercise for the sake of efficiency and economy in the use of its aid. Soviet bloc officials for their part have tried to establish a reputation for expeditious "businesslike" fulfillment of aid contracts with a minimum of red tape, in hope of dramatizing a contrast with the more cautious methods of American officialdom.

In these respects Soviet aid administration has both the advantages and the disadvantages of centralized authority. When outstanding executives can give aid projects their full attention, remarkable speed can be achieved; otherwise, there is danger of Soviet bureaucratic entanglements becoming at least as cumbersome as the American ones, with the added weakness that Soviet officials lack the rich experience, contacts, and trading facilities the West has at its disposal after generations of intercourse in the world market.* Since almost all Communist loans are "tied," the bloc also incurs displeasure comparable to that caused by Export-Import Bank and other "tied" loans from the West—multiplied by the fact that Communist governments have to accept blame and endure some adverse political reaction from anything that goes wrong in the availability of goods, their quality, terms of purchase, schedule of delivery, and so on, criticisms that in the West would more readily be deflected against particular private firms. All Soviet commerce with the outside world has to be carried on under the partial cloud of suspicions (or hopes) concerning its possible political motives. This sensitivity can seriously handicap a program that is barely getting started under administrators who have had a decade less experience than their American counterparts in learning from inevitable early mistakes.†

* This is probably one reason why the Eastern European satellites are encouraged to play a major role in the bloc's commerce with the outside world.

† The classic Soviet bungle, which must not be regarded as typical, was the Burma cement case of 1956. Burma had suffered from declining prices on the world market for her major export, rice, after the end of the Korean war. The Eisenhower Administration felt unable to help Burma sell her crop in view of the surplus disposal problems confronting the American government itself; and Burma rejected grant aid because of a dispute with the United States over responsibility for the depredations caused by armed Chinese Nationalist refugees in Burma.

The Soviet bloc came to the rescue with a series of barter agreements that were to absorb more than a third of Burma's total rice exports by 1956. The apparent generosity backfired, however, when Burma encountered delays in obtaining delivery of goods, some of which arrived in damaged condition or contrary to specifications, and were often extremely overpriced. Cement, on the other hand, arrived in excessive quantities, largely because of the mistakes of the inexperienced Burmese purchasers themselves. As it began to pile up on the docks and clog the harbor with nondebarkable cargoes, the Burmese tried to divert some of the additional shipments to India, offering to take a financial loss on the resale. But the Soviet sea captains had orders that specified Rangoon as their destination, and therefore refused to take the risk of diverting their cargoes to Calcutta. The cement was landed, though it could not be used and though there was no place to store it except a big shed in Rangoon—where it lay in porous paper sacks while the monsoon season came and went! Mountains of cement, hardened, caked, lumpy, spoiled—such an outcome could hardly have been the Kremlin's intention; but the aftertaste in Rangoon was bitter.

The Soviet Union did its best to make amends for the errors by freeing the

The resemblances existing between the Soviet trade-aid programs and typical Western commercial relationships, in fields that private enterprise finds profitable in capitalist countries—industry, mining, and transportation—should be recognized as offering partial confirmation, perhaps designedly, for the Soviet claim that its aid programs are essentially oriented toward mutual economic benefits for the Communist bloc and the underdeveloped countries. But also under these programs collateral opportunities are increasing for penetrating various kinds of local factions and institutions and infiltrating into them the Kremlin's policy preferences. Years of relatively good behavior on the part of technical personnel, producing few complaints in the recipient countries, may pay off politically better than overt proselytizing. Thus far, at least, the channels of access opened up to Soviet agents through the aid program have been used discreetly, with principal emphasis upon creating a relatively favorable image of the Soviet Union and Communist China in contrast to the West, by establishing Communist loan programs as a genuinely attractive alternative to the familiar trade and aid arrangements of the United States and its major allies. The very existence of such an alternative undermines the traditional relationships of economic interdependence that link the underdeveloped countries mostly with the West. The alternative also provides pleasing bargaining counters to uncommitted leaders in the underdeveloped countries, and tends to force the Western powers to some extent to raise the financial bids for their cooperation—an exasperating and somewhat expensive source of friction that may provoke lasting resentments in many quarters.

Thus far, in short, from the viewpoint of competitive coexistence, the Sino-Soviet "economic offensive" is more a series of probing operations and localized guerrilla raids against "targets of opportunity" than an all-out organized worldwide drive like the international Communist movement's political-action and propaganda activities, or an intensely sustained massing of diversified means of coercion like the Soviet bloc's military-industrial growth. The Soviet trade-aid program is basically imitative of things that the West had been doing for a decade or more; and even the

---

Burmese in turn from their full commitment of rice shipments, even cancelling most of the advance sales agreements altogether when the world market price rose and made Burma anxious to get a better deal elsewhere. But the pattern of commercial contact as a whole won the Communist bloc little if any lasting good will in Burma.

Presumably the Soviet administrators have learned not to repeat the administrative errors in a future aid program. Has the West for its part learned the importance of commodity price stabilization in order to prevent similar openings for Soviet initiative from arising again, with greater prospects that they would be exploited with ruthless efficiency the next time?

modifications the Communist powers have made have not revealed brilliant originality as their weapons development and their techniques of political action and subversion have. Only the emphasis on soft loans has been so attractive to the underdeveloped countries as to prompt extensive counterimitation by the Western powers.

## Soviet-American Competition in Economic Development Assistance

Indeed, the argument is sometimes made that the United States has so many advantages in productive capacity, worldwide economic ties, and foreign-aid experience that this sphere of competition is one in which the Soviet bloc could be utterly blanketed; a "token" aid program, expecting to get real economic compensations and only isolated political gains of importance, would, from this perspective, be all that the Kremlin would be likely to feel it could afford to attempt in the next decade. Such an argument is probably overoptimistic. Certainly the material means will be available in the USSR for a great expansion of existing trade and aid. With these known "reserves," the Kremlin's negotiators can afford to bluff, offering assistance even when they do not want to have to deliver on their promises, and thereby encouraging the leaders of the underdeveloped countries to make greater demands upon the United States, pressing her to outbid the Soviet offers.

The American government cannot always do so. Domestic political resistance would surely block any policy of topping each Communist bid, and, indeed, even the theoretically available wealth of the United States would truly be strained if both the Soviet bloc and the underdeveloped countries knew that they had only to connive at any barely plausible bargaining proposal to arouse the American government to surpass it in generosity. No such assurance exists, but bluffing does still have some chance to succeed, and even if it fails, its effects upon international relationships are corrosive in a way that injures the West. Even if no Soviet offer is made, or if the Soviet offer that is made (or is rumored to have been made) is not considered reliable by the underdeveloped country or by the American negotiators, an atmosphere of competitiveness and uncertainty is established that is likely to improve somewhat the bargaining position of the backward nation vis-à-vis the United States (though not so much as the USSR will claim it does)—at much cost in a heightening of suspicion and friction between the noncommunist countries concerned.

Superficially, the appropriate American answer to this kind of maneuver would be to cut off all aid—or some large penalizing proportion of the aid—from any underdeveloped country indicating willingness to accept assistance from the Soviet bloc. This sort of ruthless bluff-calling, however, is self-defeating when one cannot be certain that the Kremlin is really bluffing. If in fact the Communist countries did proceed to "pick up the check," the United States would be left with a much reduced local counterweight to the expanding Soviet economic penetration of the backward nation. If, on the other hand, the Kremlin should back away, the underdeveloped country would be left floundering without needed assistance from anywhere. It might be said to have been taught a lesson (though the Kremlin would presumably contrive as long as possible to keep false hopes up), but meanwhile, at no expense, the Communists would at least have managed to provoke a severance or suspension of important political-economic ties in which, before the Soviet intervention, both the less-developed nation and the United States had seen prospects of mutual advantage. By the pin-pointed concentration of fairly realistic offers of assistance, the Soviet bloc could thus conduct shrewder raids upon targets-of-opportunity—for example, countries in special balance-of-payments difficulties from commodity price fluctuations—secure in the knowledge that if it could only manage to score a first local breakthrough, the United States would largely withdraw for a while from that field of competition and leave the Soviets a relatively free hand.

No such self-defeating policy of systematic penalization can be risked with reasonable assurance by the American government. On the other hand, in the context of the cold war it would not be safe either to give the underdeveloped countries the impression that the United States is indifferent, or even usually sympathetic, to their acceptance of substantial links of "mutual benefit" with the Soviet bloc.

Any penalization should presumably be based on an overall appraisal of each individual situation, with an eye to the broad long-run political tendencies of the particular country and its capacity to weather close relationships with Soviet advisers—with and without Westerners also competing on the scene. Deep, long-term hostility can rarely be countenanced, but occasional separations and even some flirtations with the Communist powers should be appraised coolly. Western offers of assistance can provide an available alternative, preventing excessive dependency-relationships from forming toward the Soviet bloc, just as Soviet offers of assistance somewhat increase the room for maneuver of the underdeveloped countries vis-à-vis the Western powers. Indeed, in some situations the American officials might try to do some bluffing themselves, but they must recognize that in a free society their bluff is likely to be called by other Americans; in the use of such tactics dictatorships have decided advantages.

All these complicated bargaining strategies resulting from the Kremlin's intervention in the trade and aid relationships of the noncommunist world inevitably produce increasing distortion and politicization of economic development assistance as a whole. Yet it seems naïve to assert that America's stake in the economic advance of the backward nations has ever been nonpolitically "pure," or to imagine that the East-West competition could now be repressed and sublimated by making all development assistance a United Nations enterprise. Competitive coexistence spreads pervasive suspicions that would permeate any UN enterprise on the massive scale of the present unilateral or bloc economic assistance arrangements—a scale that can scarcely be reduced if there is to be much prospect of achieving the widespread development that effective political pressures in the backward countries now demand as they look eastward and westward for development models and aid. In this field the United Nations may gradually be assigned operations of limited but increasing dimensions by agreement among the industrial powers. But the bulk of economic development assistance will continue to be motivated and shaped to an important degree by the pressures of the cold war. The immediate political-psychological impact of particular aid projects can often have far-reaching consequences in setting a course for the recipient country's evolution and in determining its receptivity to other outside influences, so much so that conventional standards of economic efficiency or sometimes even of international comity would be frustrating encumbrances upon the competing donor powers.

In this inevitable adjustment, however, it is important not to lose sight entirely of the development process itself, which does require that injections of outside assistance be allocated at a sustained tempo in such a way as to have the maximum catalytic effect in stimulating greater and more efficient local efforts, until adequate foundations have been laid upon which the growth in local income, savings, and investment can become correlatively self-sustaining. The process is likely to require many years. If outside aid programs lack even an appearance of dependability, they can scarcely impress the recipients as offering much prospect of success. Development assistance would then be likely to acquire an ethos of cynicism that would corrode even the short-term political deals for which the programs may be manipulated. And when élan declined, Soviet disruption would become still easier. Clearly, whatever the tactical requirements may be for adjusting American economic assistance to meet Soviet competition, the long-term direction of strategy must steadily continue to be toward accomplishing genuine development under the nontotalitarian regimes. Without the development, no such regime—no matter how carefully propped, bribed, and insulated by the West—would be likely to survive for many years.

### SELECTED BIBLIOGRAPHY

ALLEN, ROBERT L. *Soviet Economic Warfare.* Washington, D.C.: Public Affairs Press, 1960.

AUBREY, HENRY G. *Coexistence: Economic Challenge and Response.* Washington, D.C.: National Planning Association, 1961.

BERLINER, JOSEPH S. *Soviet Economic Aid: The New Aid and Trade Policy in Underdeveloped Countries.* New York: Praeger, 1958.

CAMPBELL, ROBERT W. *Soviet Economic Power: Its Organization, Growth, and Challenge.* Boston: Houghton Mifflin, 1960.

HARRISON, SELIG S. *India: The Most Dangerous Decades.* Princeton: Princeton University Press, 1960.

KOVNER, MILTON. *The Challenge of Coexistence: A Study of Soviet Economic Diplomacy.* Washington, D.C.: Public Affairs Press, 1961.

LISKA, GEORGE. *The New Statecraft: Foreign Aid in American Foreign Policy.* Chicago: University of Chicago Press, 1960.

MIKESELL, RAYMOND F., and JACK N. BEHRMAN. *Financing Free World Trade with the Soviet Bloc.* Princeton: Princeton University International Finance Section, 1958.

MORAES, FRANK. *India Today.* New York: Macmillan, 1960.

PRESIDENT'S COMMITTEE TO STUDY THE UNITED STATES MILITARY ASSISTANCE PROGRAM. *Composite Report and Annexes,* 2 vols. Washington, D.C.: Government Printing Office, 1959.

SCHLESINGER, JAMES R. *The Political Economy of National Security: A Study of the Economic Aspects of the Contemporary Power Struggle.* New York: Praeger, 1960.

UNITED STATES CONGRESS SENATE SPECIAL COMMITTEE TO STUDY THE FOREIGN AID PROGRAM. *Foreign Aid Program: Compilation of Studies and Surveys.* Washington, D.C.: Government Printing Office, 1957.

# PART FIVE

## OVERT AND COVERT INTERVENTION IN THE INTERNAL POLITICS OF FOREIGN LANDS

CHAPTER 20

# DIPLOMATIC INTERVENTION AND SECRET POLITICAL ACTION: AMERICA AND THE PHILIPPINES AFTER WORLD WAR II

DEEP CURRENTS in the American tradition of foreign relations repeatedly press to sweep aside efforts of the American government to prop, bribe, and insulate particular foreign regimes, not only because such efforts may be futile, but because they are adjudged politically immoral.

## American Inhibitions about Meddling in Foreign Politics

It will be recalled that the antimilitarist reaction in the United States following World War I virtually destroyed American unilateral interventionism in the interwar period, at the same time that multilateral interventionist (collective security) sentiment was being gravely weakened (see pp. 33–35). In the heyday of self-abnegation under the "Good Neighbor" policy that was begun, in effect, by the Hoover Administration and sanctified by Franklin Roosevelt, a noninterventionist policy even for the Caribbean took very firm roots, supported, not only by the multilateral noninterventionists in whose benevolent spirit it was chiefly conceived, but also by the unilateral noninterventionists (the strict isolationists) whose strength in America was at its twentieth-century peak (see pp. 31–44). By 1937 the United States Senate, without even bothering for a debate or recorded vote, was willing to ratify a treaty among the American republics which declared "inadmissible the intervention of any one of them, directly or indirectly, and for whatever reason, in the internal or external affairs of any other" of the republics. "Every question concerning the interpretation" of this prohibition would have to be settled through diplomatic channels, by conciliation, arbitration, or by international adjudication, not by unilateral sanctions. The uncoerced cooperation Washington received from nearly all the Western Hemisphere during World War II seemed to attest the enlightened self-interest of this policy of nonintervention on the part of the United States. In 1948, through the Charter of the Organization of American States, she accepted an even more extreme self-denying ordinance (see p. 431).

With regard to the Eastern Hemisphere, the United States did not tie her hands so tightly by treaties, even in the United Nations Charter (see p. 473 and n.). But by the mid-1940's nonintervention (except in a context of ongoing general war and military occupation) had become so rooted in the value system of most Americans active in foreign affairs that they were very hesitant to interfere, openly or covertly, in the politics of foreign countries, even where the other instruments of American policy —economic, informational, and military—were failing to prevent local Communist gains. Interference would bring censure at home and personal pangs of conscience from a feeling that one was adopting indecent Communist methods, albeit for the high purpose of resisting Communism. So dead was the earlier interventionist tradition in the United States that only rarely did Americans refer to that experience as a relevant historical precedent for new acts of political meddling—much less suggest that the reaction toward extremes of noninterventionist self-denial of the 1930's was an isolationist excess bound to be left behind again, together with other aspects of isolationism, when the pendulum swung away to new international involvements much vaster than ever before. Americans were still ashamed of ever brandishing a "big stick" at backward nations.

Yet the slim residues of previous active United States involvement in the politics particularly of the Caribbean and the China-Philippines area certainly contributed to making those regions prominent among the places where the onset of the cold war produced before long a hesitant recrudescence of American willingness to interfere in local politics, without relying on the excuse of a general war or of military occupation. And the Latin Americans at least, with longer memories than "the Colossus of the North," were ready at once to draw bitter parallels to the earlier era of "Dollar Diplomacy."

Actually, of course, the pattern of American motivation and action continued in the 1950's and early sixties to be much more inhibited than were the Caribbean and Far Eastern interventions of the United States in the first third of the twentieth century. And national security considerations were more clearly transcendent over commercial considerations than in the earlier era. Overt and covert American intervention in the internal politics of foreign lands re-emerged only when the conclusion began to be drawn, reluctantly, that worldwide totalitarian penetration could not be effectively resisted by economic, informational, and military means alone; that even if military security and economic development could somehow be achieved for a country, its political evolution might well be in a direction fundamentally disadvantageous to the United States unless its politics were influenced directly (with as much tact and circumspection as the particular situation might require); and that anyway, military security and economic development themselves could scarcely be achieved under noncommunist auspices unless some kinds of political structures and tendencies were favored over others; indeed, that the very operation of economic and military aid programs was bound to favor certain local elements. The American political involvement would be logically inescapable and hence had better be managed consciously, deliberately, and more astutely.

Some of the methods and immediate objectives of the new American interventionism were sure to appear to be "neocolonialist" and some, like the old, "economic imperialist." But actually, in the spirit of containment the central purpose of this complex of policy instruments was simply to keep open the possibility of political change in countries outside the Iron Curtain by forestalling their totalitarianization. Only a true totalitarian regime by its omnipresent controls would wholly foreclose until some remote period the possibility of its own eventual internal transformation in a direction advantageous to its own population and to the West. Any other regime, however unattractive, might be improved gradually or even be overthrown from within. Hence, in the meantime, such leadership could be tolerated by the United States (at least if it were not becoming a breeding ground for Communists who, if they were once allowed to gain control, would be too ruthlessly totalitarian to be overthrown for generations).

Contemporary American meddling in foreign politics, therefore, has been basically defensive and widely permissive in character. If it can be said to be "neocolonialist" at all in any meaningful sense, then it is surely quite benign. Yet the defensiveness can be condemned as self-centered, and the permissiveness as cynical. Ambitious local elements are as likely to condemn the United States for not exerting influence on their own behalf as for exerting influence at all, and the two kinds of complaints are likely to fuse in a logic that is twisted but understandable. The path of interference in foreign politics is so full of pitfalls and so uncongenial still to the contemporary American conscience that there is need to explore more fully how one can get led into such a troublesome course of action in the first place. Then it may be possible to consider realistically how far the United States can afford to go and what some useful safeguards might be.

An appropriate means for first acquiring a feel for the deepening of American involvement is to trace the route of American policy in the Philippines in the decade after World War II.

## Emancipation and Decay: The Philippines, 1945–50

Independence was formally proclaimed in the Philippine Islands on July 4, 1946, the date promised by Congress a dozen years before; none of the toil and bloodshed of the intervening period caused Americans to hesitate to redeem the pledge. Indeed, the war against Japan had demonstrated the willingness of many Filipinos to make a fighting contribution, which was generally felt to strengthen their claim to independence. Among Americans, the war fostered a spirit of comradeship toward their distant wards in the Far Pacific; in this frame of mind the idea of sovereign equality was congenial, and with it went an attitude of protectiveness that was also directly heightened by the war experience.

The Filipinos, however, would not simply be cast off to make their own way in a perilous world. After the proclamation of independence, agreements were soon negotiated that established an American military advisory mission to assist the Philippine armed forces and also provided for United States retention of bases in various parts of the islands. Any external armed attack on the Philippines would almost automatically involve an attack upon the United States forces, bringing America into the war. Besides these arrangements for military defense, the United States provided advisory missions in economic fields, including agriculture and finance, and softened the painful prospect that the Philippines, when independent, would lose their former free-trade access to the American market. The shift toward normal trade barriers would be only very gradual. The island republic would not confront the full United States tariff wall until 1974; in the meantime the leading Philippine agricultural exports would be given fixed quotas in the American market.

In return for this privileged position, the islanders were obliged to amend their constitution to assure that during the transition period Americans would have absolute equality with Filipinos in the development of the natural resources of the new republic; any discrimination that might be inflicted upon other foreigners could not be applied against Americans. This provision was widely resented in the islands as an impediment to economic independence; more than half of the registered voters stayed away from the polls in the referendum, when the necessary constitutional amendment was finally adopted by a three-to-one margin among those who did vote.

Yet, all in all, it was fair for Americans to continue to believe that as a whole their postwar arrangements for the Philippines constituted a model example of equitable and cooperative colonial emancipation. In particular, there was not very much residue of virulent antipathy to the colonial masters, on which Communists would be able to capitalize for building an anti-imperialist coalition as a bridge to power for themselves.

Nevertheless, conditions in the Philippine Republic deteriorated steadily and drastically in the first four years of independence. By 1950 there was very real danger of early takeover of the islands by domestic Communists and their supporters. The "Hukbalahaps," an outgrowth of one of the main wartime guerrilla forces (its name was a condensation of the native words for "People's Army against the Japanese") and under Communist leadership, effectively controlled a large area in the center of the main island, Luzon. "Huk" units had also spread to other islands, and in every important city a Huk shadow-government waited for the opportunity to take power. There was open terrorist violence even in the capital city, Manila. The Huks were avoiding clashes with Americans, but United States forces considered the situation so insecure that they felt it necessary to use armed convoys to supply their main air base, Clark Field, not far north of Manila itself.

This critical situation had developed despite extraordinary special advantages the Philippines presumably possessed in comparison with other countries in Southeast Asia. Not only had independence been achieved harmoniously, but the young republic had at its disposal enormous American economic assistance—between one and two billion dollars' worth, depending on how one valued the World War II surplus stores left in the islands. Most of the postwar assistance was intended to be directed toward reconstruction of war-ruined cities and industries in the Philippines, a great opportunity in fact wasted, the aid being frittered away largely on luxury imports. The war had bred a spirit of laxity and lawlessness that also tolerated the scandalous dissipation of many tens of millions of dollars' worth of the war-surplus goods in crude corruption and outright thievery; the Huks themselves were able to lay their hands on large quantities of the stores. Urban unemployment remained a grave problem in the absence of efficient reconstruction efforts.

Similarly, the advantage the Philippines should have derived from possessing a large American-trained army officer corps that had had substantial experience in guerrilla warfare against the Japanese was dissipated by the republic's failure to utilize these forces against the Huks until very late. The Philippine Army numbered 35,000—probably two or three times the size of the Huk forces. Yet for many months this regular army was kept out of action lest the Huk rebellion be dignified as a civil war. Instead, the government tried to repress the insurgents by means of the national police, the Philippine Constabulary, which proved to be incompetent for the task. Often while pursuing the elusive Huks it was so brutal in its treatment of the peasantry that it created more enemies for the government than it overcame; and even worse reaction was caused by "special police" forces that local landlords were permitted to maintain, virtually as private armies. Meanwhile, the regular army, out of action, became debilitated through the general inertia and corrupted by political favoritism. Thus when finally the mounting intensity of the rebellion required that the army be thrown into combat, it commonly lacked the will to do more than conduct show battles and phony campaigns against the Huks. Contributing to the demoralization was the army's growing realization that the expanding Huk units were often better armed than the regulars sent against them, and that Huk agents had so thoroughly infiltrated the government forces that army drives were likely to find the enemy gone or, worse, ready with traps.

In the ensuing struggle against the Communist-led guerrillas the army was not much more effective than the constabulary in winning the confidence of the peasantry among whom the Huks found refuge. Here again, an exceptional potential advantage of the Philippines over many other Asian countries was not being put to use: the islands as a whole include a wealth of undeveloped land. But in some rural areas there is severe local overpopulation, the result largely of immobility of the peasantry and primitiveness of agricultural technology. The difficulty of moving between islands and establishing oneself on virgin land was further complicated after World War II by delays in determining land titles where records had been destroyed. There was also a justifiably widespread suspicion that powerful landlords in control of local administration were manipulating the process against the peasants' interests. Likewise, a progressive land-reform law the Philippine government enacted to regulate share-cropping was effectively sabotaged at the local level.

The population of the islands as a whole was 80 per cent Catholic—another feature unique in Asia that might be expected to strengthen opposition to Communism. But again, as in many Mediterranean and Latin-American countries, the advantage was seriously impaired by a traditional identification of the Catholic Church with vested landlord interests. The Church in the Philippines was only just beginning to demonstrate major concern for the economic betterment of the masses.

Next, the public educational system was noteworthy by Asian standards. More than half the adult population was considered literate, after about forty years of exposure of the islands to American-organized schooling; only Japan, Formosa, and Soviet Siberia possessed higher rates of literacy in the Far East. However, education was still grossly inadequate, amounting in the rural areas to an average of only four years of half-time classes in an unfamiliar language, English. The teaching of this common language was some aid to national unity, and with it went some inculcation of democratic political ideas, but the educational experience of peasant children was too brief to leave much residue. In the cities, on the other hand, where education went further, it tended, as elsewhere in underdeveloped countries, to produce a dangerous pool of unemployable "intelligentsia."

Probably the greatest advantage the Philippines possessed to buttress the viability of democratic political institutions was the sheer length of experience the islands had acquired in operating them by permission of the American government, particularly in the Commonwealth period of the 1930's. Among southern Asians, Filipinos were uniquely familiar with the workings of constitutional politics, including nationwide elections. Yet again, as in so many other respects, the advantages were being wasted. Corruption and general incompetence were eating the heart out of the system. Most taxes were not collected. Individual legislators auctioned immigration permits to desperate Chinese Nationalist refugees. The 1949 presidential election, which swept the Liberal party's candidate,

Elpidio Quirino, back into office over his Nationalist party opponent, José Laurel, became quickly notorious in the islands as "the Dirty Election." Violent intimidation and fraudulent vote-counting characterized this Liberal victory; without them Quirino might have lost the presidency. The inevitable result was rising public disillusionment with the potentialities of democratic elections as channels for governmental change.

Thus it was that constitutional democracy in an underdeveloped country was falling prey to Communist insurrection despite the fact that the country was Catholic and more than half literate, with a generation of experience in constitutional politics, an army experienced in guerrilla warfare, a wealth of underdeveloped land and some land-reform legislation, massive foreign economic aid, and few anticolonial grievances. The value of each of these assets was proving to be limited or declining. Yet both the American and the Philippine governments were slow to confront the handwriting on the wall. Not until 1950, when the Communist victory in China spurred the confidence of the Huks to new peaks of activity, did Washington and Manila face frankly the real gravity of the situation. And then it took the onset of the Korean war to create finally a true sense of urgency about conditions in the Philippines, as about so many other long neglected aspects of the containment policy.

## The Road from Economic and Military Assistance toward Political Action

Once aroused, however, the American and Philippine governments moved with impressive speed and harmony. United States Treasury Under Secretary Daniel Bell was sent by President Truman, with the approval of President Quirino, as head of a special new Economic Survey Mission to the Philippines to appraise the needs. The Bell Mission reported in October 1950 with a recommendation of heavy additional American economic assistance—but also with exceptionally forceful criticism of the existing administration of economic affairs in the islands and a demand for close supervision by United States officials of the new expenditures and of reforms—social and political as well as economic—that were prescribed as essential if a development program were to be successful. Drastic changes were proposed for a very wide range of aspects of Philippine public life. The United States government would not simply wait for these necessary reforms to be accomplished before granting massive assistance; experience with the Nationalist Chinese government in its latter years on the mainland had showed that prolonged bait-dangling could be self-defeating in an emergency situation. But neither would the aid be granted with no strings attached; if it were, it would simply be dissipated as before. Instead, on the Bell Mission's recommendation, the American government decided to accept the risk of having an "imperialist" label pinned upon it by insisting on close, continuous supervision of expenditures and reforms during the course of their implementation, in direct conjunction with the application of the new economic assistance.

Fortunately, in the Philippines more than in most of the underdeveloped countries trust in America existed, at least to the degree that, facing the urgent need for foreign aid, the heads of the recipient government dared to undertake publicly by formal agreement with the United States to achieve specific major legislative and administrative reforms as the price of an "equal partnership" with America in a joint development effort for their country, to which the United States would contribute $250 million. American officials themselves in this relatively friendly political climate dared to withhold the assistance until some of the main reforms requiring action by the Philippine Congress had been enacted into law (including sharply increased taxes and a minimum-wage statute). Thereafter, American advisers were active in devising new administrative arrangements and drafting further legislation, opportunities that enabled them to influence many sectors of the Philippine economy and bureaucracy. At its peak in 1952, American financial aid accounted for almost 11 per cent of all the government's revenue—substantial leverage for persuasion, in light of the moderation of any overt Philippine resistance to the guidance.

Similarly, in 1950 American military assistance to the Philippines increased sharply, including more equipment and a more active role for the Joint United States Military Advisory Group (JUSMAG), moves clearly directed to promoting greater Philippine Army efforts against the increasingly active Huks. The defeat of Communism in the islands was clearly seen to be a military problem as well as an economic and social problem. Thus the programs that took shape jointly between Filipino and American officials in 1950–51 were intended to deal energetically with both interrelated sets of problems.

Gradually, moreover, the increasing involvement of Americans aroused by a sense of urgency in the frustrating quest for solutions to these problems led them to the consciousness of a need for, and finally to a sense of responsibility for, a creative solution of a third set of problems: those of political leadership in the Philippine Republic. Americans found themselves involved in influencing Philippine elections as a means of securing a leadership that would make their other proposed reforms effective. This remarkable development had not been premeditated far in advance, but was an outgrowth of the other more conventional cold-war programs in the islands.

The central figure whose attractions made political action seem desirable was a young Filipino politician named Ramon Magsaysay. Magsaysay in 1950 was the 43-year-old chairman of the National Defense Committee of the Philippine House of Representatives. He was a self-made man whose father had been a farmer, storekeeper, blacksmith, and carpentry teacher—a little above the level of poverty, but far below the Spanish-descended aristocracy from which most of the top Philippine political leadership has been drawn. Ramon himself worked his way through one of the lesser colleges in the Philippines to a commercial science degree, joined the staff of a bus company, and after successes and setbacks, was one of its branch managers at the time of the Japanese invasion.

The war gave Magsaysay the opportunity to become a recognized leader. In his native province in west central Luzon he emerged during the years of Japanese occupation as the commander of 10,000 guerrillas. His effectiveness there so impressed the American liaison officers that he was appointed military governor of the province when the Japanese were driven out in 1945. The following year Magsaysay was elected to the House of Representatives on the Liberal party ticket and in 1948 was chosen to go to America to lobby for Philippine veterans' benefits. Presumably, this appointment reflected his colleagues' belief that his manner and his wartime record would be persuasive to Americans, a confidence well founded. And the benefits did not accrue only to Filipino veterans. Magsaysay himself had opportunities to establish many contacts in Washington during the several months he spent there, particularly in the Pentagon and on Capitol Hill. Americans found him congenial, fitting more the familiar image of a self-made small-businessman in the United States than that of an emissary from the Orient. He was affable, direct, even blunt in speech, sincere in manner. His 1948 mission was such a success that in April 1950 he was sent back to the United States as chairman of the House Committee on National Defense to seek increased military aid when the Philippine and American governments both began to recognize the seriousness of the Huk rebellion. Later that summer, when the expanded military effort gathered momentum, the JUSMAG commander and the American ambassador gave support to Filipinos who were promoting Magsaysay for the post of

Secretary of National Defense in Quirino's cabinet. The appointment was made in late August.

Magsaysay was fortunate at the onset of his secretaryship to be instrumental, through personal courage, in securing the defection of a Huk who was able to entrap most of the central Manila leaders of the movement. Thereafter, Magsaysay proceeded to take full advantage of his own guerrilla experience to wage the struggle through more efficient military operations in the field and also by techniques of psychological warfare. The army command was progressively purged and revitalized; the Defense Secretary saw to it that effective junior and noncommissioned officers received quick promotions in the field. He spent much of his own time out with them, actually directing campaigns in the hinterland. The strategy was to concentrate the government's forces to clean out the guerrillas region by region by aggressive action, rather than to continue to allow strength and spirit to be dissipated in scattered defensive skirmishes. Communist infiltration was combated by counterinfiltration which was facilitated by liberal payment of bribes and rewards (for Huk bodies as well as for information). The unorthodox expenditures were partly financed through the collection of a "Peace Fund" from private persons in the Philippines in 1951. The fund was also available for anti-Huk propaganda, much of it "black" (the true source concealed).

Constructive alternatives for those Huks who might be willing to surrender were also emphasized. The army was put under firm discipline to avoid looting and brutality that would alienate the peasantry, and was allowed to undertake the popular task of resettling thousands of ex-Huks on new land of their own in the distant, large, and less-developed island of Mindanao. This Economic Development Corps ("Edcor") was given lavish publicity in the effort to form a creative image to offset the inevitably ugly aspects of antiguerrilla warfare. Similarly, the defections of individual groups of Huks were dramatized to set in motion a wave of defeatism among those who still held out.

Much of the inspiration and guidance for the efforts to glamorize the depressingly sordid conflict came from an American, Colonel Edward Landsdale, one-time journalist and advertising man who had been chief public relations officer at Clark Field and later for JUSMAG, and who was detailed to advise the Philippine Army on effective psychological warfare. Landsdale appreciated the possibilities of further popularizing the Philippine Army's mission by personalizing its leadership and publicizing the appealing image of Ramon Magsaysay as top commander. Thus, quietly but effectively, a public relations build-up of Magsaysay became a conspicuous part of the activity of large numbers of United

States officials and unofficial American observers in the Philippines. At first they were publicizing Magsaysay as a means of publicizing effective anticommunism; most of them were slow to understand and then to welcome the impact the build-up was also having upon Magsaysay's standing relative to other democratic politicians in the Philippines.

### Getting Magsaysay Nominated for President

A turning point came in the congressional election campaign of 1951. Magsaysay, with the encouragement of his American advisers, was determined to make every effort to recreate public confidence in the electoral process, removing thereby another of the demoralizing conditions in the Philippines on which Communism was feeding. The "Dirty Election" of 1949 must not be allowed to repeat itself. A National Movement for Free Elections ("Namfrel") was founded with unofficial American backing. Philippine army personnel were carefully instructed by Magsaysay that their mission was to guard the citizens' access to the polls, free from force and intimidation. When at the last minute the Liberal party leaders sought to use army planes to carry stuffed ballot boxes to politically strategic points, Magsaysay himself managed to ground the planes. The army's communications system was used to publicize the vote count as soon as possible after the polls closed, in order to discourage fraudulent tabulations. The outcome was an orderly "Clean Election." And it was no coincidence that Quirino's (and Magsaysay's) party lost every contested Senate seat; even the President's own brother was defeated.

Such an outcome could not fail to offend Magsaysay's fellow Liberals, especially since he personally had received more benefit than the rest of the Quirino Administration from a vast publicity campaign that gave the army credit for assuring clean elections. Almost inevitably there developed a process of alienation of Quirino from Magsaysay, and with it went a process of American alienation from Quirino.

One factor was that a jealous belittling of Magsaysay's achievements and an effort to reduce his influence became increasingly evident on the part of many Liberal leaders, while the Americans, who remained preoccupied with the war against the Huks and were idolizing Magsaysay's contribution to it, were offended by this resistance to the build-up of their hero. More far-reaching doubts about the Quirino Administration were engendered in the Americans' minds by the sweeping nature of the Liberal defeat under clean electoral conditions in 1951. Quirino had certainly been cooperative toward the United States and was personally honest—but perhaps he was simply a mediocre leader who would never really be able to capture the popular imagination or overcome the selfish, corrupt, and reactionary influences that were still effectively impeding the implementation of most of the nonmilitary reforms on which the American aid program was founded. Without strong leadership from the presidential palace for such reforms, they might never be lastingly effectuated, and without them the war against the Huks might never be fully won. The 1953 presidential election might be the last opportunity for a change before it would be too late. Thus Americans found reasons for raising the level of the aspirations they held for their favorite, Magsaysay, and he in turn was gratifyingly appreciative of the assistance they had been giving him. He did not adopt a distant mien even when his enemies began to hurl such abusive epithets as "JUSMAGsaysay." His friendship with Colonel Landsdale and other Americans was very close indeed.

Landsdale's admiration for Magsaysay had reached the point by early 1952 that he wanted to publicize the Defense Secretary around the world as an inspiration for anticommunists in every land, and also to make him an "available" candidate for the presidency itself in the Philippine Republic. Landsdale's chief associate in this soaring enterprise was Manuel Gonzales, a Philippine advertising executive whose brother was the husband of Quirino's only surviving daughter and who was himself a top publicity man for the Quirino Administration. Gonzales, who happened also to be an organizer of the local chapter of Lions International, collaborated with Landsdale on the idea of seizing upon a visit to Manila by the worldwide president of that organization to wangle an invitation for Magsaysay to be keynote speaker at its international convention in Mexico City, June 1952. Gonzales then induced his sister-in-law to persuade her father, President Quirino, very reluctantly, to allow Magsaysay to take leave from the war against the Huks in order to make the trip, despite the President's forebodings concerning the political consequences that would flow from a well-publicized elevation of his Defense Secretary to this role of "international statesman."

Quirino's worst fears must have been confirmed. To be sure, as the American ambassador had urged, the Magsaysay tour won much favorable publicity in the United States for the Philippine government's new successes against Communism, but most of the credit went to Magsaysay personally. Through Jesuit contacts in the Philippines an honorary degree was arranged for the Defense Secretary at Fordham University. He was welcomed grandly to New York by a former JUSMAG commander, with a nineteen-gun salute, a troop review, and an elaborate reception. Traveling with Magsaysay to smooth the way were Landsdale and Gonzales. The former escorted him to Washington—using the pretext of a physical check-up for Magsaysay at Walter Reed Hospital, in deference to the desire of Quirino and his ambassador that the prestigious official aspect of Magsaysay's journey be minimized. In conferences at the Pentagon, Magsaysay was offered a half-million dollars in secret funds for which he would not have to account, for use in his undercover operations against the Huks. Gonzales, who acted as advance agent in Mexico, had Landsdale's aid in securing a public donation of a couple of million dollars' worth of agricultural equipment by individual Lions in California and Nevada, to be announced dramatically at the end of Magsaysay's stirring keynote address in Mexico City. Traveling Filipinos traditionally return home with gifts—and their statesmen with American aid; Ramon Magsaysay would not go back empty-handed. And of course the arrangements for publicity were carefully made.

During the summer, fall, and winter of 1952–53 a further projection of Magsaysay into the Philippine presidential arena took the form of specific negotiations among the leaders of both major parties. Magsaysay himself was hesitant, particularly when it became clear that Quirino would refuse to allow the Defense Secretary to be his successor on the Liberal party ticket. Magsaysay would have to be willing to run as the opposition party's candidate, and the Nationalists, for their part, would have to be willing to give him the nomination. Nationalist leaders could discern in Magsaysay a potential winner, even in a presidential campaign against the entrenched power of the Liberal party machine; furthermore, he would obviously be attractive to Americans and hence to the Philippine business interests linked with the American market, and to other Filipinos concerned with the flow of United States aid. But perhaps he would be too pro-American and "too" honest to be trusted with the powers of the presidency; and even the least selfish Nationalist leader was bound to be somewhat offended by the idea of giving the party's top nomination to so recent and dubious a convert as Magsaysay would be. The Defense Secretary, for his part, had natural qualms about behaving disloyally toward Quirino and probably had some doubts about his own capacity to perform the demanding and varied tasks of the presidency.

But by February 1953 the die was cast. The mutual resentment between Magsaysay and Quirino, largely provoked by their respective confidants, had reached the point in November 1952 where the Defense Secretary secretly pledged in writing that he would resign his office in return for a virtual guarantee of his nomination for the presidency by the Nationalist party. The resignation came on February 28 after rumors of the secret pact had produced further friction with Quirino. The Nationalists kept the bargain at their nominating convention in April. With Quirino's renomination on the Liberal ticket apparently assured, the November election took shape as a contest between him and the new Nationalist, Ramon Magsaysay.

## Getting Magsaysay Elected President

Then, at the end of April, Carlos P. Romulo, the perennial and distinguished head of the Philippine diplomatic corps in the United States, returned home and found himself the center of a rising "dump-Quirino" movement in the Liberal party. Romulo had not been tarred with the brush of corruption, was at least as respected and popular among Americans as Magsaysay, and outshone Magsaysay in intellectual distinction and international experience. He seemed to have a better chance than did Quirino of beating Magsaysay in November. Meanwhile, his candidacy could function as a convenient tool in some political infighting that was in progress between two powerful sugar-producer factions in the Liberal party. Romulo, somewhat naïve about the politics of his own country, where he had not resided in the previous decade, accepted the leading role in the struggle against Quirino's control of the Liberal convention in May, but the insurgents were quickly beaten down. Romulo, however, refused to accept defeat. His faction withdrew and founded a new party, with him as its candidate for the presidency. Promptly Romulo flew to the United States to sound out support for his candidacy, but found very little. Despite the high regard in which they held him personally, most Americans who were actively concerned with Philippine affairs had by this time gravitated so far toward support for Magsaysay over Quirino that they feared any third-party candidacy that would split the anti-Quirino votes and thus probably enable the President to be re-elected. Evidence of this American attitude soon shook the loyalty of Romulo's Philippine backers also, particularly the sugar interests. They too were alarmed that Quirino would be likely to emerge victorious; he might punish them for deserting the Liberal party. Consequently, during the summer they proceeded to negotiate a coalition with Magsaysay's Nationalists and then persuaded Romulo to abandon his own candidacy in favor of a joint campaign against Quirino. Romulo was later to describe Magsaysay's campaign as a "crusade" and to write two books in praise of the former Defense Secretary. The world-famous diplomat's enthusiasm had thus been captured almost as much as was that of the Philippine masses.

Presidential candidate Magsaysay waged a campaign unprecedented in the extent of his travels through the villages of the Philippine hinterland and in the cordiality of his face-to-face approach to peasant voters. His personal appeal was huge. In the end he received more than 70 per cent of the votes that were cast for president. Such a result was undeniably the expression mainly of internal Philippine political forces, perhaps overwhelmingly so. But Americans did continue to be involved, and the nature of that involvement must be given the emphasis here.

First, as indicated, Americans played some part in helping to discourage Romulo's candidacy, thereby assisting Magsaysay to monopolize the support of the principal anti-Quirino forces.

Second, most of the Americans resident in the Philippines, a colony of 10,000, made it perfectly obvious that their preferred candidate for the presidency was Ramon Magsaysay. Many of them also contributed financially to his campaign, but he does not appear to have been dependent on such contributions. At least after the formation of the coalition with Romulo, when Magsaysay's chances of victory seemed much improved, he found himself amply supplied with funds from Philippine sources, including many hedgers who were also financing Quirino. A more interesting question is whether at any stage American *government* money was routed indirectly into the Magsaysay coffers to bolster his election campaign, but conclusive evidence on this point is not available. Magsaysay made much of the fact that he had never used at all the half-million dollars that had been put at his disposal when he was in Washington in 1952, but the possibility may remain that he or his supporters, perhaps unknowingly, had access to other United States government funds. (The point is made here not to cast any aspersions on Magsaysay, but simply to illustrate in an appropriate setting one available technique of feasible secret political action by United States personnel.)

Third, Americans played a prominent role in efforts to assure that the election would be free from fraud and intimidation; indirectly, of course, the candidate most likely to benefit from such efforts was the immensely popular candidate of the party that was out of power: Ramon Magsaysay. Thus Americans were able to benefit their favorite candidate by simply protecting the democratic electoral process. In this convenient situation the motives of different Americans were mixed and often unclear even to themselves; one cannot reach a reliable overall judgment as to how much less American activity there would have been on behalf of clean balloting if Magsaysay's own chances had stood to suffer from it.

The State Department defined its policy as "absolute impartiality. . . . Yet, as one of our major objectives is political stability, we cannot deny that we are concerned that the democratic processes function so that the people may freely express their will. The eyes of the world will follow the elections in the Philippines. We are confident, on the basis of statements from Philippine leaders, that the elections will be conducted in such a way as to prove a blow to the aspirations of international communism and will advance the cause of the free world in the Far East." * A few days before the balloting, the prestige of President Eisenhower himself was put behind this line of expressed American policy. The White House made public an exchange of letters between Eisenhower and a former American ambassador to the Philippines in which the latter expressed "alarm" as to "the possibility of fraud and violence in the conduct of the elections." Eisenhower couched his reply diplomatically to appeal to "justifiable" Philippine pride:

> I too regard the forthcoming Philippine elections as a vital test of democracy for the Philippine Republic. However, I am confident that the people of the Philippines will meet this test, as they have met others, in a manner that will fully justify the esteem in which they are held throughout the free world. I know I speak for the American people when I say that all of us are indeed interested in what is happening in the Philippines. . . . I know that [Filipinos] will wish to stand before the world on their election day as having made full use of their political freedom and enlightened laws to elect a representative government of their own free choosing and dedicated to their service.†

Other Americans were also taking definite action, official and unofficial, to promote the cause of free elections, usually with the assumption that Magsaysay would be the immediate beneficiary. "Namfrel," for example, was again in action on a wide scale, crusading for honest elections, and incidentally facilitating Quirino's defeat; unofficial assistance in some of its activities was forthcoming from Americans. Also, an exceptionally large number of American journalists, many of whom were stranded in Korea in search of new assignments after the war ended there in July, were encouraged to go to the Philippines to cover what Magsaysay's supporters were anxiously predicting would be an excitingly violent and corrupt election. In Manila, the reporters were further im-

* *Department of State Bulletin*, October 19, 1953, p. 524.

† *Department of State Bulletin*, November 16, 1953, pp. 676–77.

pelled by their most convenient information sources, the resident Americans, to concentrate their campaign coverage on exposing the alleged plots of the Liberal party to rig the election. The most likely locations for electoral fraud and intimidation in various parts of the country were specifically identified. Thus aroused, the journalists dispersed themselves widely and observed vigilantly. The result, of course, was that their very presence tended to deter the evils that they intended to publicize. Magsaysay benefited.

On election day the chief of JUSMAG deployed twenty-six of his officers in a dozen teams to likely political trouble spots to observe Filipino army personnel as the latter watched the polls. Thus, in effect, the United States Army provided some small reinforcement in the function that the American journalists were performing. Much less tactful was the sending of an American naval flotilla to pay a courtesy visit at Manila shortly before election day. These actions lent some color to the charges made with increasing vigor by the Liberal party that the United States was preparing to promote a military coup on Magsaysay's behalf.

Carlos Romulo has since disclosed that the Nationalists did in fact have elaborate preparations ready for a *coup d'état.* Relying on Magsaysay's popularity with the Philippine Army, the blow was to be struck if a Quirino election victory were won by dishonest methods. Evidence that the United States government had decided to give active encouragement to any such violent enforcement of the Philippine people's will is lacking, but presumably some hint of the preparations did reach the Americans who were close to Magsaysay. Their apprehension concerning the possibility of military dictatorship or even civil war between anti-communist factions must have increased the determination of the informed Americans to take the steps that were taken to deter any rigging of the election at the very outset.

Some important Nationalist leaders were even urging Magsaysay to launch his military *coup d'état* without waiting for the election to be held at all, in order to "forestall inevitable fraud." Clearly, such a scheme would have met emphatic objection in Washington, and Magsaysay himself firmly discouraged the idea. But he and other Nationalist leaders showed very little hesitation about encouraging Americans to show open interest in the cleanness of the election. Even after the Liberals, chafing at the prospect of defeat, began to attempt to arouse the general public to some resentment over Yankee interventionist activities, the Nationalist strategists continued to allow Magsaysay's campaign to be identified with the United States. Indeed, the Nationalists seemed less worried, on the whole, than many Americans were about the Philippine public's reaction to the conspicuous affinity between Magsaysay and the United States. In the end, the electoral results did seem to vindicate the view that at that particular time and place anti-Americanism was indeed so weak a force that United States personnel could play an active role in a foreign election without necessarily having to conceal most of their moves in order to forestall a xenophobic reaction. As Manila wags observed, "JUSMAGsaysay's" prominent public-relations adviser had in this instance turned out to be a "Colonel Landslide."

No doubt such quips exaggerated the role of Landsdale and his fellow Americans. So also to some extent does this study in focusing attention upon them. In all, the American interventionist activity was essentially limited to giving extensive publicity and some financial backing to Magsaysay, while exerting considerable pressure on behalf of a clean electoral decision to be rendered by an absolute popular majority that would have a simple two-way choice between the incumbent Administration and a united (noncommunist) opposition. Magsaysay's own personal magnetism and creed of social justice clearly were the means by which he raised for himself a flood of mass support that no amount of American publicity could have created for him. Similarly, even if the United States had shown complete indifference, it is entirely possible that by 1953 his own hold on the affection of the Philippine Army would have enabled him by his own prestige to compel a fair count of the ballots.

## A Reappraisal in Light of Magsaysay's Brief Tenure as President

Probably the role of United States officials in promoting Magsaysay's career was actually of greater importance in the earlier years, before and shortly after he became Defense Secretary. But at that stage the process was one of individual Americans' becoming captivated by the hero-image they found themselves helping to form, while their government as a whole saw no reason yet for scorning the cooperativeness of the existing Quirino regime.

Indeed, one may still question whether the break with Quirino was necessary, or whether it was needlessly provoked by an overeager perfectionism on the part of the individual Americans who had attached themselves to Magsaysay. Were the goals in view really important enough to warrant the United States' adopting a favorite and thus allowing a wedge to be driven between two staunchly anticommunist constitutional democratic leaders in an underdeveloped, newly independent country (granted that the time and place were uniquely such as to permit Americans to escape most of the opprobrium usually attached to intervention)? Those who regard political interference in any form, even for the protection of free elections, as a meddlesome and patronizing derogation of the weaker nation's sovereignty, would naturally be found on the negative side of this question—with reinforcement from others who might be willing to support some interventionist activities if directed exclusively to insuring free elections without regard to the probable victor, and from still others who would support some intervention against enemies of the United States but not between foreign democratic leaders well-disposed to this country.

Another group of Americans would wish to know how much better Magsaysay turned out to be than Quirino in the presidency before rendering judgment on the favoritism the United States showed. Actually, Magsaysay's performance, while it lasted, may fairly be described as approaching the extravagant expectations of his pre-election American supporters. He combined a gratifyingly pro-Western orientation on the international scene with a vast surge of energy at home in idealistic enthusiasm for governmental probity and social justice, which produced a swelling tide of popularity for the Philippine government and a decisive weakening of the Huk rebellion. The Communist leader, Luis Taruc, even surrendered and went to jail. The various programs of pacification and reform on which the Quirino Administration had reluctantly embarked with American prodding during its last years in office were now accelerated by Magsaysay's leadership. But progress remained slow; too much of the new impetus depended on the President's own personal supervision; he was somewhat lacking in administrative competence, particularly in ability to devolve responsibility. He tended to try to popularize the government by personalizing it, hearing and handling the grievances of individual citizens himself. This practice, under the circumstances, had some obvious advantages as well as disadvantages. The latter might have been overcome. But on March 17, 1957, after only three years in office, President Ramon Magsaysay was killed in an airplane crash. Suspicions of sabotage have generally been discounted.

The United States found itself with a successor, Vice President Carlos Garcia, who soon proved to have many of the characteristics of Elpidio Quirino, both good and bad. The situation in the Philippines proceeded to deteriorate again.

Does this discouraging experience give added reason for contending that Americans ought to have continued to try to work with and through, not against, Quirino—that the difference between him and Magsaysay was not great enough to warrant even the limited amount of political interference that was undertaken on the latter's behalf? Or does the denouement in the Magsaysay story only suggest that the struggle for self-sustaining economic and political development that will be reliable enough to produce a permanent immunity to the virus of totalitarianism is a long-enduring struggle, entailing setbacks as well as gains, in which the loss of one progressive political leader only creates a need to find another one, and to promote his cause with whatever degree of tact and circumspection the special circumstances of the case require?

In this context the question may seem more debatable than if the electoral situation were not one in which both of the principal immediate alternatives were fully anticommunist, differing little except in their relative effectiveness. In a situation where the choice was between procommunists and nontotalitarian anticommunists, some forms of American political action might well be undertaken with little hesitation.

But what if the exertion of American influence between pro- and anticommunists might involve not just meddling in an election but subverting a legitimately constituted government, to prevent its being wholly subverted by Communists? This situation was the one that arose in Guatemala in the early 1950's. The way it was handled and the ensuing state of affairs there can shed further light upon the difficulties the United States confronts in becoming identified with particular foreign leaders and factions, and also upon other problems of overt and covert intervention in the internal politics of foreign lands.

## SELECTED BIBLIOGRAPHY

COQUIA, JORGE R. *The Philippine Presidential Election of 1953*. Manila: University Publishing Company, 1955.

QUIRINO, CARLOS. *Magsaysay of the Philippines*. Manila: Alemar's, 1958.

ROMULO, CARLOS P. *Crusade in Asia: Philippine Victory*. New York: Day, 1955.

———, and MARVIN M. GRAY. *The Magsaysay Story*. New York: Day, 1956.

SCAFF, ALVIN H. *The Philippine Answer to Communism*. Stanford: Stanford University Press, 1955.

SMITH, ROBERT AURA. *Philippine Freedom: 1946–1958*. New York: Columbia University Press, 1958.

CHAPTER 21

# SUBLIMITED WAR TO OVERTHROW THE GOVERNMENT OF GUATEMALA

FOR OBVIOUS REASONS much less is publicly known of the Guatemala episode than of the Philippine story. But the parts of the record that can be pieced together do call attention usefully to several facets of American undercover operations, including even those violent forms of subversion and countersubversion that have come to be labeled "sublimited war," that is, combat just a little beneath the scale of outright limited war. The United States involvement in Guatemala was very deep, but still not uninhibited.

## American Concern over Communist Penetration of Guatemala's Government

The population of the Guatemala Republic in the early 1950's passed the three million mark. More than three-fourths of these people derived their livelihood from agriculture. Most farming was for local subsistence; the population was poor but not in extreme misery. Land ownership, however, was extremely concentrated; 60 per cent of the cultivated area of the country was in the hands of about 2 per cent of the landholders, while at the other extreme, two-thirds of the owners held a total of only 10 per cent of the farmland. The great plantations were devoted mainly to two crops, coffee and bananas, which together produced almost 90 per cent of the country's overseas earnings, coffee being much more important than bananas in this "banana republic."

Hugely conspicuous in these operations was an American corporation, the United Fruit Company (UFCO), which, in Guatemala, employed almost 10,000 natives and owned more than a half-million acres, most of which was kept as fallow or undeveloped land. The company also had full control of Guatemala's only railroad and its only major seaport. These holdings, which loomed so large in the little country, actually constituted only about 10 per cent of the $600 million worth of assets the company held worldwide, mainly elsewhere in the Caribbean area. The size of this economic empire and its willingness to engage in political manipulation in order to secure its interests in the turbulent backward countries had made it the arch-symbol of iniquitous imperialist capitalism to generations of radical nationalists throughout Latin America. Even though the company's actual behavior was becoming much more far-sightedly responsive to the needs and aspirations of the countries where it operated, United Fruit had the heavy onus of its past reputation to live down. In Guatemala in the 1940's it was slow to yield to a moderate leftist government the privileges it had acquired from previous amenable regimes. In the early 1950's it found itself confronted by a radical challenge to its fundamental position in the country. An increasingly leftist regime embarked on a drastic land-reform program and expropriated more than 70 per cent of United Fruit's holdings, offering as compensation a mere $600,000 in twenty-five year bonds, which were almost certain to depreciate much further still. Meanwhile, the UFCO-controlled railroad was saddled with an unanticipated tax claim of $10.5 million.

The radical reformers aimed, not only to overcome the evils of excessive landlordism and foreign domination of the nation's economy, but also to integrate into the mainstream of Guatemalan life the Indians in backward villages who constituted a substantial majority of the total population but were passive nonparticipants in a society dominated by persons of European or mixed descent. Other objectives of the social revolution included the organization of a labor movement, the establishment of a social security system, and, of course, the acceleration of economic development. The opportunity to attempt these changes had come in 1944 with the overthrow of the traditionalist military dictatorship of General Jorge Ubico and of the army junta that briefly succeeded him.

The insurrection succeeded because of the defection of young officers, notably Captain Jacobo Arbenz and Major Francisco Arana, who accepted as the reform movement's presidential candidate and supported in office a politically inexperienced forty-year-old Guatemalan educator of mildly socialistic outlook named Juan José Arévalo. But when the question of succession arose toward the end of Arévalo's term, the traditionally recognized Latin necessity that politicians maintain the good will of the army contributed to focusing upon Arbenz and Arana the civilians' disagreements about the direction of the social revolution. Arbenz became leader of the more leftist faction, Arana of the more moderate. In 1949 friends of Arbenz assassinated Arana, partly in fear that otherwise Arana would stage a *coup d'état*. The way was clear for Arbenz to be elected president in late 1950, in a partly rigged election against disorganized opposition. Although he probably possessed genuine support on the part of more voters than any of the other candidates, he would be rash to forget the degree to which his power ultimately depended on the continuing benevolence of most of the key personnel of his army.

Under Arbenz Guatemala took a sharp turn to the far left. The Arévalo Administration, partly checked until 1949 by the moderating effects of Arana's influence with the army, had lacked a clear doctrinal impetus and had failed to develop a specific practical reform program of a kind for which competent leadership would be available in Guatemala. The United States government and the United Fruit Company wasted whatever opportunity might have existed to play a constructively cooperative role in the regime's reform efforts. Arévalo had succeeded in balancing the extremely disparate elements of his political coalition, which had little in common except their antipathy to the old elite of landowners, army officers, Catholic dignitaries, and foreign businessmen; he had survived numerous overthrow plots and had managed to complete his full six-year term, while preserving a degree of freedom of expression that was unprecedented in Guatemala. But little progress had been made toward a solution of the social and economic problems of the country. To a disturbing degree, he had begun to find in the tiny Guatemalan Communist minority the competence and dedication his regime badly needed. Consequently, his official tolerance gave the party an opportunity to strengthen its organization and expand its influence, particularly in the nascent labor movement. But under Arévalo, Communists had never achieved a principal role in the Administration. They did under Arbenz.

During 1951 they completed their gradual acquisition of control of

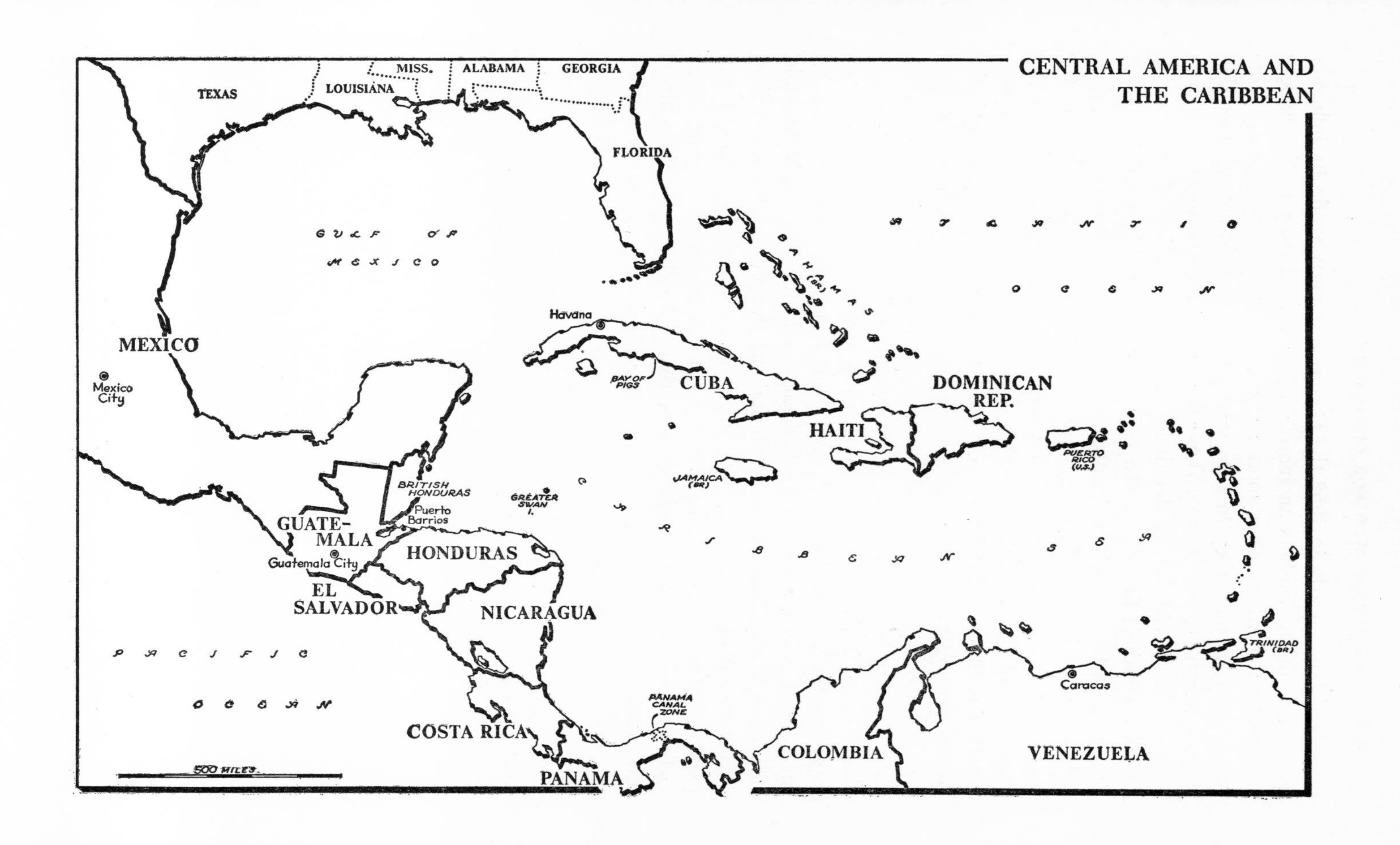
CENTRAL AMERICA AND
THE CARIBBEAN
TEXAS
LOUISIANA
MISS.
ALABAMA
GEORGIA
FLORIDA
GULF OF
MEXICO
ATLANTIC
OCEAN
BAHAMAS
(BR.)
MEXICO
Mexico City
Havana
BAY OF PIGS
CUBA
DOMINICAN
REP.
HAITI
PUERTO
RICO
(U.S.)
JAMAICA
(BR)
BRITISH
HONDURAS
Puerto
Barrios
GREATER
SWAN
I.
GUATE-
MALA
Guatemala City
HONDURAS
CARIBBEAN SEA
EL
SALVADOR
NICARAGUA
PACIFIC
OCEAN
PANAMA
CANAL
ZONE
COSTA RICA
PANAMA
COLOMBIA
VENEZUELA
Caracas
TRINIDAD
(BR)
500 MILES

organized labor. Early in 1952 the Communist party made its influence among the urban workers an increasingly essential prop to the Arbenz Administration, while other, more moderate elements drifted away from him largely because they in turn were alienated by the very fact that he was becoming increasingly dependent on the Communists. When in the summer of 1952 Arbenz decided to tackle the problem of agrarian reform in a radical fashion as the central feature of his Administration's program, he found the Communists to be the most enthusiastic workers for the cause, both in Congress and especially as administrators in the hinterland. They were rewarded with full legal status for their party at the end of 1952 and the inclusion of its leaders on the official Administration coalition ticket for the congressional elections of January 1953. In the ensuing year they eagerly exploited their leadership of the land reform program in order to build organizational support among the peasantry akin to what they had already achieved among urban workers. They saw to it that Arbenz's agrarian law was implemented in a much more drastic fashion than its formal provisions prescribed, exacerbating the conflict with the United Fruit Company in the process. Meanwhile, they brought the National Radio Station and the press of the pro-Arbenz parties effectively under their control, though anti-Administration newspapers continued to publish. The pro-Administration media were regularly utilized for propaganda themes that coincided with the current international campaigns of world Communism, such as "germ warfare in Korea"; and prominent officials in the Administration played leading roles in Communist front organizations having international ties. The Guatemalan police chiefs showed sympathy toward the rising Communist influence in the country, while treating the emergent resistance groups with brutality.

The army, however, was never effectively penetrated with reliable supporters of the far left. Many kinds of honors, perquisites, and favors were showered on key officers; some were given free trips behind the Iron Curtain. Arbenz's personal ties with these men were close. But the Communists themselves, in the few years that were available to them, seem to have put greater hope and effort into creating a peasant militia as an armed force on which they might eventually base a bid for power, rather than into slowly infiltrating and subverting the army officer corps. Probably they chose to rely on Arbenz to keep the army in line until their militia and mass organizations were in a position to challenge its ultimate authority.

The trend of events in Guatemala aroused little interest in the United States, preoccupied with the Korean war and the rearmament of Western Europe. American liberal observers were inclined to regard most of the reforms in progress in Guatemala as long overdue. Not until outright Communists assumed effective command of the vast new agrarian program in late 1952 was it possible to make a very convincing case in the United States for the proposition that a Latin republic "on her very doorstep" was soon likely to become a Soviet satellite. But finally during 1953 the prospect became real enough to stir the American government to grave concern and action.

One should emphasize that the action taken was preventive. Guatemala was not yet, when the plans took shape late in 1953, a fully Communist-dominated country; even in 1954, when police-state regimentation did develop, it was probably mainly a defensive reaction of the regime to the organization of military forces aiming to overthrow it. Very likely the Communists would have preferred not to be driven to show their hand repressively until they had established a much more secure base of popular support, particularly among the politically inert peasants who were beneficiaries of the land redistribution for which the Communists were taking credit. But the Eisenhower Administration was not disposed to wait until Communists were so firmly in control of Guatemala that the fact was undeniable; they might by that time have also become irremovable. Yet if the United States were to take prior action to prevent the Sovietization of Guatemala, Latin Americans would keenly suspect that the real concern of the American government was to protect the economic privileges of the United Fruit Company, the foremost American interest already suffering from the action of the Arbenz regime well before the Communists had achieved any general mastery in Guatemala. The Eisenhower Administration was not unmoved by the expropriation of UFCO, particularly because of the precedent that would be set if the seizure went unchallenged by the United States government and further damage were to result in other backward countries to legitimately profitable American enterprises and to the flow of private capital for foreign economic development. But the protection of American investments was decidedly secondary to other concerns when the Administration framed its policies against the Arbenz regime in Guatemala.

1. If Communists were allowed to achieve domination there, so close to the United States, American prestige would be seriously impaired. In international relations the impairment of prestige does not merely cause hurt pride but also reduced influence; what one group has shown it can get away with another group elsewhere may feel encouraged to attempt. When a sense of the futility of opposing a great power erodes,

the smooth applicability of its power erodes also, and can be reimposed only at considerable cost, if at all. Thus no precedent should be allowed to develop, even in tiny Guatemala, the Administration concluded, that would indicate that a Western Hemisphere country would be allowed to be subverted by Communists.

2. A Communist-controlled government in any Latin-American country would be continually able to irritate United States relations with other Western Hemisphere countries, exacerbating by propaganda and diplomacy all the manifold controversies that regularly arise between the great North American power and the sensitive weaker nations of the area. And similar trouble-making would disturb the intercourse among other Latin republics even where the United States was not directly involved. Indeed, Communists might decide to provoke their enemies, particularly the United States, to the extreme of outright military intervention, in the opinion that the martyrdom of a Communist regime in tiny Guatemala would redound to the worldwide advantage of Communism at least as much as would the regime's survival in power.

3. The further spread of Communist subversion in Latin America could be organized conveniently in hospitable Guatemala and conducted by various techniques from its territory and through its missions, official and unofficial, in other countries. Guatemala would become a base for Soviet political action in the Western Hemisphere.

4. Conceivably, Guatemala might even become a military base, more or less secret, for Russian use at the outset of a general war. Presumably a base so located could not hope to survive more than a very few days after the onset of fighting, but even in that short time a few planes or submarines from its territory might inflict great damage on the continental United States or the Panama Canal.

## Preparing for an Invasion of Guatemala by Insurgents

In October 1953, the American government took the first overt steps indicating that these prospects were causing alarm and that resistance to the Arbenz regime was being organized under United States auspices. A new American ambassador was appointed to Guatemala, a State Department administrator named John Peurifoy, whose only prior diplomatic experience abroad had been the period 1950–53 when he was ambassador to Greece. There he had been prominently active and fairly successful in efforts to influence internal Greek politics in such a way as to advance effective new and progressive leaders and help consolidate the deeply divided anticommunist elements. Peurifoy appeared to have been collaborating in these efforts with personnel of the Central Intelligence Agency who functioned more or less undercover in and out of his embassy. Consequently, his transfer to Guatemala in October 1953 implied interesting possibilities, especially when it was combined with a formal pronouncement by the Assistant Secretary of State for Latin-American Affairs that Guatemala was "openly playing the Communist game" and could expect no "positive cooperation" from the United States.

Indeed, the American government was submitting a new agenda item for action at a highly important meeting scheduled for March 1954 in Caracas, Venezuela, the Tenth International Conference of American States (five to ten years elapse between these imposing gatherings). The item was entitled "Intervention of International Communism in the American Republics." An implication was that the United States would seek some cooperative action by Western Hemisphere countries to counter such intervention—and where else in the near future but Guatemala? Already, since 1951, the United States had been exerting influence wherever possible to prevent Guatemala from purchasing arms, on the ground that the regime there had connived in revolutionary incursions into other Caribbean republics. The arms embargo might now help to deprive the Guatemalan government of the means of defending itself if the other countries were to connive in incursions against it. The United States was explicitly declaring its unwillingness to give "positive cooperation" to Arbenz. Guatemalan exiles and refugees in the Central American republics plotted against him with increasing openness, coalesced with growing capability, and acquired more men, more money, and more arms. Suspicions were bound to mount in the Arbenz Administration that the "positive cooperation" of the American government was in fact being given, undercover, to the new revolutionaries, or, from the Communist perspective, to the "counterrevolutionaries."

The emerging leader of these outside forces was a former Guatemalan army officer named Carlos Castillo Armas, who had served the Arévalo Administration as commander of Guatemala's "West Point" until Arana's assassination. The following year he had led an unsuccessful coup and in 1951 had escaped prison and fled abroad. His association with the moderate phase of Guatemala's social revolution and his own vaguely articulated liberal views gave promise that under his leadership a movement to overthrow Arbenz would not simply turn back the clock in pure reaction. Among the divergent, often unsavory exile factions he was the kind of person whose progressivism had appeal for

Americans as well as for moderate Guatemalans. Whether in fact United States money or advice played any part in his achievement of coordinating authority among the exiles in late 1953 cannot positively be ascertained, but clearly, by early 1954, when he did become recognized as the principal alternative to Arbenz, Americans found it easy to accord Castillo Armas their sympathy.

So did Hondurans, much more openly. The dominant elements in that "banana republic" on Guatemala's southern border were fearful of the contaminating influence of their neighbor's radical course. In the short run, their immediate apprehension was that Honduran workers would for the first time be effectively organized through the inspiration and guidance of Guatemala's Communist-dominated labor movement. In a longer perspective, they feared that the Communist fire would spread in Honduran tinder as readily as it had in Guatemala. The similarities were ominous. When in fact labor unrest did develop increasingly in Honduras after January 1954 and reached the level of a general strike by May, with clear evidence of encouragement and financial assistance given to the strikers by Guatemalan labor, the Honduran government felt that it had some justification for allowing Castillo Armas to build and arm his insurgent force on Honduran territory, not far from the Guatemalan border. Arbenz, in turn, may have considered that his continued toleration of Guatemalan labor's support of a Honduran strike was a justified reprisal for the Honduran government's toleration of the Guatemalan would-be insurgents. Arbenz even increased the number of his official consulates in some Honduran towns in the troubled area. Presumably the consular staffs would be able to serve the dual function of watching Castillo Armas and aiding the Honduran strikers. Whether such activity was properly to be regarded as aggressive or defensive was a question for which the Honduran and, further south, the Nicaraguan governments were not disposed to await final proof before retaliating.

Similarly, in January 1954, there came a brutally severe police crackdown in Guatemala. Anticommunist elements suffered heavily, and the effect was a sharp increase in the relative power of Communists in the country. But whether this result was the principal objective, or only the consequence of a need to strengthen the regime's ability to ward off an incipient invasion by the rebel forces, as Arbenz's supporters claimed, was a question that received an increasingly unsympathetic response from other Caribbean governments toward Arbenz's predicament. Most of these governments were right-wing dictatorships anyway. But the United States too was becoming angry and impatient.

However, the Good Neighbor policy and nonintervention treaties of the thirties and forties, soothing the continuing suspicions and resentments of Latin Americans, had set limits restricting the means available to non-Guatemalans to resist the course of events in that country. For example, Article 15 of the Charter of the Organization of American States (the "Bogota Charter," 1948) set down in binding treaty form: "No state or group of states has the right to intervene, directly or indirectly, for any reason whatever, in the internal or external affairs of any other state. The foregoing principle prohibits not only armed force but also any other form of interference or attempted threat against the personality of the state or against its political, economic, and cultural elements." The United States government had no desire to defy that prohibition itself or to allow others to set new precedents for defying it in the turbulent Caribbean, particularly if Communist subversion of Guatemala could be prevented by other, legitimate methods.

An essential step would be the establishment of a new foundation in the international law of the Western Hemisphere that would permit some forms of intervention (preferably collective) at least in extreme circumstances. Therefore, in March 1954 Secretary of State Dulles exerted the maximum influence of the United States government, including implied promises of economic aid programs, to obtain from the Tenth International Conference of American States what came to be called the Caracas Declaration:

> The domination or control of the political institutions of any American State by the international communist movement, extending to this Hemisphere the political system of an extracontinental power, would constitute a threat to the sovereignty and political independence of the American States, endangering the peace of [North and South] America, and would call for a Meeting of Consultation to consider the adoption of appropriate action in accordance with existing treaties.

In effect, applying Article 19 of the Bogota Charter, together with other relevant treaty provisions,* if two-thirds of the American republics, acting through their foreign ministers, were later persuaded to endorse some specified form of sanctions, the enforcement would "not constitute a violation of the principles set forth in" Article 15, cited above, on nonintervention. The Guatemalan representative was the only foreign minister who actually voted against the adoption of this declaration, but it was not at all popular among Latin Americans; Mexico and Argentina

* The Inter-American Treaty of Reciprocal Assistance (the "Rio Treaty," 1947), Articles 6, 8, and 17; and the Bogota Charter, Article 25.

abstained from the vote at Caracas. The net result was that legal forms were successfully instituted for a future collective intervention; they would be available for possible utilization—but the resistance encountered in creating even a mere framework of law demonstrated to the American government how unlikely the arrangement was to be workable in time to forestall Communist domination of Guatemala.

Arbenz's regime, however, could not take much comfort from these developments. At the very least they gave additional ominous evidence of official United States hostility and portended some kind of outside sponsorship for the already mounting opposition to his Administration. In light of the events of the winter, including the emergence of Castillo Armas, the mounting friction with Honduras and Nicaragua, the Caracas Conference, and the envenoming police crackdown in January inside Guatemala, it is hardly surprising that Arbenz felt he could no longer endure the arms embargo the United States had organized against his government among noncommunist countries. He finally turned to the Soviet bloc. In the months between January and mid-May a secret shipment of weapons was negotiated and transported circuitously from Czechoslovakia to Guatemala.

Arbenz's principal motivation in this action remains uncertain, but a range of mixed objectives can be discerned. The army's loyalty might be reinforced by the prestige of new military hardware; as a full or partial alternative, the workers' and peasants' militia might be armed to defend the regime in case the army were politically unreliable; some weapons might be allowed to filter across the border into the hands of Honduran strikers. On the other hand, clearly, the mere fact of the source of the arms was bound to infuriate Washington; the arming of Guatemala at all was sure to alarm Central America; and any transfer of the arms to a militia corps would probably offend the Guatemalan army chiefs who would discern a relative weakening of their own power within the state. These risks were obviously very great—so much so that it seems highly unlikely that Arbenz would have accepted them if he had not felt that the forces taking shape against him, in and out of Guatemala, were so menacing that counteraction was required, however provocative it might appear. And among the Communist advisers on whom he was becoming increasingly dependent, there may even have been some who were already resigned to thrusting his Administration into a stance of martyrdom.

During April, the rapid growth of Communist influence, in an atmosphere of growing crisis, produced vigorous resistance from the highest authorities of the Catholic Church in Guatemala. The Archbishop's

Easter pastoral letter, read from all the pulpits, was a bitter denunciation of Communist infiltration and influence and a severe rebuke to the government for condoning it. Guatemala's traditional anticlericalism set limits to the political impact of such a pronouncement, but it was unquestionably of considerable importance in rallying opposition to Arbenz.

The United States government took another step toward strengthening the external opposition by announcing a new military aid agreement with Nicaragua on April 23. Clearly, the possibility would then arise that the Nicaraguan government might feel that it could spare some old weapons to be smuggled to Castillo Armas, if it were about to get new ones from the United States. At least Nicaragua would now be in a stronger position to deter any aggressive ventures on Arbenz's part.

Unfortunately, the State Department allowed its anticommunist intent to appear colored by its concern for the Guatemalan interests of the United Fruit Company, which were certainly unpopular in Latin America, however legitimate they may have been. Only three days before publicizing United States military aid for Nicaragua, the American government formally pressed upon the Arbenz regime the claim of UFCO for nearly $16 million for part of its expropriated land (Guatemala was offering $600,000 in depreciating bonds). This action appears to have reinforced the myopic view of Caribbean leftists that the ultimate issue between Guatemala and the United States was the profitability of United Fruit; in forming for themselves an image of the State Department's motivation, they tended to project into the mind of American officialdom their own preoccupation with UFCO and their own incredulity about any worldwide danger from Communist imperialism. Any government action in behalf of the company was taken as confirmation of overall narrow selfishness in American attitudes toward Arbenz. Obviously this presumption produced a negative response toward all aspects of United States Guatemalan policy. A more subtle consequence may have been to induce some elements in the Arbenz Administration to suppose that if worse came to worst the hostility of the United States could be bought off by a favorable settlement of the United Fruit claim, thus avoiding an ultimate showdown over Communist infiltration. Such an attitude may have positively promoted the blatant Red penetration—at least among those members of the Administration who were not actually courting political martyrdom.

A climax was rapidly approaching. By the end of April, a clandestine "Radio Liberation" was broadcasting from the Honduran border region, calling upon the Guatemalan opposition to "arm and be ready to fight at a given moment." The Swedish freighter *Alfhem,* which had been detected by CIA agents being loaded with Czech arms in Stettin, Poland, was finally approaching Guatemala after thrice changing her course (probably to elude detection but possibly because of policy shifts in Moscow and Guatemala City). On May 17 the United States government was able to break the news that 2,000 tons of Soviet bloc arms had just been secretly unloaded from the *Alfhem* onto Guatemalan soil. Secretary of State Dulles himself emphasized that armaments of such quantity in a Central American country would enable it to dominate the area militarily. Danger suddenly came to seem very real, not only in Honduras and Nicaragua, but in more liberal republics like Costa Rica and Mexico. United States diplomats and propagandists did their best to heighten the alarm. The Nicaraguan government responded swiftly by breaking diplomatic relations with Guatemala, but found that the State Department was not ready to risk rebuff elsewhere in Latin America by attempting yet to call a formal consultative conference of Western Hemisphere foreign ministers to plan action against Arbenz within the framework of the Caracas Declaration. Instead the American government moved to bolster Nicaragua and Honduras directly. On May 20 a military aid agreement was signed with Honduras; under its authority and that of the April 23 agreement with Nicaragua, the United States Air Force dramatically airlifted matériel to Honduras and Nicaragua on May 24. Shipments by sea to the two countries began on June 2, and the chief of the United States military mission in Honduras announced that he had readied a training program for a new 800-man battalion there. Three B-36 intercontinental bombers paid a courtesy visit to Nicaragua.

It has never been proved that the arms rushed from America to Honduras and Nicaragua found their way into the hands of Castillo Armas, or that his troops received American guidance. Indeed, he never did manage to accumulate any very large force, either in manpower or in matériel. Some jeeps, a couple of DC-3 cargo planes, and a few P-47 piston-driven World War II fighters were the only big items, and they could have been purchased more or less clandestinely in many parts of the Caribbean or further afield, as could the smaller weapons. The overall cost has been very loosely estimated at a figure on the order of $1 million—small enough to have come from any number of official, semiofficial, or even truly private sources in North and South America. The actual sources were effectively concealed, as were any transfers of arms that may have been made to Castillo Armas from the Honduran or Nicaraguan governments. In the end, his total force remained little more than symbolic. Arbenz could not be overthrown without international sanctions unless an internal revolution could be generated or unless the Guatemalan Army defected.

International sanctions were encountering resistance; even the calling of a consultative conference was further delayed for lack of adequate support. The Guatemalan Foreign Minister, Guillermo Toriello, attempted to quiet the United States government by indicating a willingness to come to terms with the United Fruit Company, but Ambassador John Peurifoy re-emphasized the issue of Communist penetration. The Guatemalan government then proposed a personal conference at the presidential level between Arbenz and Eisenhower, but the State Department was not receptive, professing a preference for multilateral action though still not satisfied with the prospects for action under the Caracas Declaration. Other possibilities existed, as both the American and the Guatemalan governments were keenly aware. Inside Guatemala, the police cracked down more severely than ever before to destroy internal opposition; among the refugees who fled was the former chief of the Guatemalan Air Force. Flying out of the country, he took with him to safety the ex-deputy chief of the United States Air Force mission to Guatemala. The implied liaison was interesting, and the Guatemalan airman soon joined Castillo Armas.

## Helping Deprive the Government of Support for Its Survival

Finally, on June 18, the band of a few hundred insurgents crossed the frontier from Honduras into Guatemala, and Castillo Armas raised his standard of "Liberation" inside his country; yet his success would depend mainly upon whether the Guatemalan Army would decide to resist his motley force. The army chiefs seemed to be in no hurry to act, but they did establish a front line that contained the insurgents' advance not much more than twenty miles inside Guatemalan territory. Desultory skirmishing ensued. Token air raids were made over Guatemala City by Castillo Armas's few P-47's, and a few jets from Nicaragua made nonviolent overflights for psychological effect; fear was aroused. However, the crucial moves were not military, but diplomatic and political.

In a word, although the invasion from Honduran territory was a patent violation of international law, under which the Honduran government should have done all that it could to prevent the movement of any organized armed bands across its frontier, the United States government obviously wanted this invasion to succeed; and success came to depend on providing time for the Guatemalan Army to be persuaded to desert Arbenz. American efforts were therefore directed as inconspicuously as possible to the task of preventing Arbenz from getting any early outside aid against the Honduran-based insurrection. In particular, the United Nations and the Organization of American States must be prevented from acting against the invaders until the Guatemalan army chiefs had had ample time to reach a decision against Arbenz.

On June 19, the day after Castillo Armas moved into Guatemala, the Arbenz Administration made separate appeals to the United Nations Security Council and to the Inter-American Peace Committee of the Organization of American States. The Peace Committee, consisting of delegates of the United States and four Latin republics, granted an immediate hearing. But the Guatemalan government soon decided to rest all its hopes upon the Security Council, where it could have the vigorous support of the Soviet Union, and, therefore, on June 20 withdrew its request for OAS action. Conversely, however, that same day, when the Security Council met—conveniently chaired by the United States representative, Henry Cabot Lodge, who was entitled to preside that month, according to the standard rotation procedure of the council—the Americans were happy to find that the delegates of Brazil and Colombia would help by offering a resolution that would refer Guatemala's complaint away from the council, over to the OAS. All of the members of the Security Council except Soviet Russia were willing to accept the reasoning that the OAS should be given a chance to try to settle the conflict even if the government under attack, Guatemala, trusted the UN more than the OAS. The Russian delegate was still able to veto a formal referral of the dispute to the OAS, but no positive action could be obtained from the Security Council on Arbenz's behalf. The Guatemalan government, despite this stalemate, continued to prefer to rely on the UN. Appeals came to Lodge from Guatemalan Foreign Minister Toriello and the Soviet representative to call another meeting of the Security Council. Lodge managed to delay this second meeting until June 25.

A full week thus passed from the time that Castillo Armas began his invasion. Meanwhile, if Arbenz did not want OAS action, Honduras and Nicaragua decided that they did want it. At their insistence the Inter-American Peace Committee decided on June 24 to send a five-member observation committee to the troubled area. When the Security Council finally met again on June 25, the fact that this investigation had now actually been instituted provided Lodge with a convenient new basis for opposing any United Nations action until the Inter-American Peace Committee had had a chance to report. This time the Security Council vote was extremely close, five to four. Supporting the United States effort to keep Guatemala off the agenda were only Nationalist China, Turkey, and the two Latin-American members, Brazil and Colombia.

Voting with the Soviet Union on behalf of Arbenz's right to a full UN investigation were some of America's friends and allies, Denmark, New Zealand, and Lebanon; even Britain and France abstained.

The American government found itself in an awkward posture. The following day, June 26, it finally gave public support to the convening of a full consultative conference of the foreign ministers of all the American republics, as provided in the Rio, Bogota, and Caracas compacts. Simultaneously, the Arbenz Administration, probably despairing of the UN, again reversed its position in regard to the Inter-American Peace Committee and announced that it would welcome that group's investigation.

By this time, however, the Guatemalan army leaders had had eight full days to make up their minds about Arbenz—long enough to reach a decision against him. They refused to allow him to arm the workers' and peasants' militia the Communists had organized. The Soviet bloc's weapons remained firmly in the Guatemalan Army's hands, and they would no longer be used to defend Arbenz. He bowed to military pressure and resigned the presidency on June 27.

There followed several hectic days of political maneuver, into the midst of which United States Ambassador John Peurifoy emerged openly, playing the central mediating role through four successive changes in the personnel of the governing junta, moving with a dramatic flair between Guatemala City and El Salvador, wearing shoulder holster and green tyrolean hat. On July 2 a place was secured for Castillo Armas as a member of the ruling junta, and within a week he was head of the government.

Simultaneously, on July 2, the Eisenhower Administration went to court in the United States with a drastic antitrust prosecution of the United Fruit Company, thus dissociating the American government more emphatically than ever from UFCO's interests, which were very likely to benefit from the turn of events in Guatemala.*

In Washington, the permanent council of the OAS voted on July 2, on a motion of the Honduran delegate seconded by the United States, to call off the consultative conference of foreign ministers that had been arranged only a week before. And the Inter-American Peace Committee never got further than Mexico City, accepting the argument that its presence would now be superfluous to the political negotiations between Guatemalan army officers and Castillo Armas that were being mediated by Puerifoy and El Salvadoran officials. The success of the anticommunist

* The case was settled in 1958 when UFCO consented to sell the Guatemala railroad and to assist in creating new competition throughout the banana trade.

insurrection was accepted as a *fait accompli*, and the complicity of other countries was not investigated further, either in the OAS or in the UN.

## The Aftermath: Gains, Resentments, and Reverses

Ambassador Peurifoy's task in Guatemala was considered accomplished as leftists were driven into exile and Castillo Armas's regime gradually established its authority. Peurifoy was very soon replaced by one of America's most distinguished and experienced diplomats, Norman Armour, as ambassador to guide the formidable task of reconstruction in Guatemala. Peurifoy himself was assigned to Thailand, where he succeeded Ambassador William Donovan, the World War II head of the predecessor agency to CIA, the Office of Strategic Services (OSS). Donovan's continuing consultative connection with CIA and his lack of diplomatic experience were among the factors that had led many observers to consider the United States embassy in Bangkok to be a CIA outpost. Peurifoy evidently found the new post congenial. His conduct continued to be unorthodox. He enjoyed speeding around the countryside in a robin's-egg-blue Ford Thunderbird. Unfortunately, on August 12, 1955, on a narrow rural road outside Bangkok a heavy truck ran into his jaunty equipage and killed him. Assassination charges could not be substantiated.

In Guatemala, the reconstruction of the country's polity and economy made slow but steady progress under the leadership of Castillo Armas, with intensive American advice and economic assistance. As had been hoped, the new regime did not prove to be reactionary; many of the reforms of the preceding administrations were preserved. Some progress began to be made in forging around the President a new moderate political movement that could hope to survive the extremism of Guatemalan politics.

These successes tended somewhat to limit the impact of a prolonged campaign of leftist propaganda throughout Latin America and elsewhere, designed to denigrate "the Guatemalan affair" as a cardinal example of United States imperialism and to glorify the Arbenz Administration as political martyrs. After the Soviet suppression of the Hungarian revolt in 1956, Communists had new reason not to let die the memory of what they asserted to have been the cognate, though, of course, in their view less justifiable, action the United States had supported against Arbenz. They minimized (while Americans emphasized) the fact that the United States had at least not felt a need to use its own troops to re-establish a cooperative regime in a nearby foreign country. Yet whatever degree of responsibility a particular observer might accord to the United States for Castillo Armas's initial rise to power, the fact that he was now dependent on continued American economic assistance was obvious. This situation, inescapable in the short run, was an irritant to undercover leftists in Guatemala, to traditionally anti-Yankee nationalists, and to a large section of the pre-1944 ruling elite that felt that it was being cheated of the fruits of counterrevolution by the American pressure for liberal moderation.

Many other currents of intrigue and disaffection also agitated Guatemalan public life while Castillo Armas and his aides strove with some success for stabilization. In the end no one was able to establish reliably the full range of disgruntled elements that collaborated in a plot that produced the assassination of Castillo Armas by one of his own bodyguards on July 27, 1957.

The Magsaysay tragedy seemed to be repeating itself. Guatemala after Castillo Armas, like the Philippines after Magsaysay, had a president who was anticommunist, moderate, cooperative with America, but lacking in magnetism and in driving progressivism. An atmosphere of stagnation and an ominous polarization of politics tended to develop, despite continuing aid from the United States. Thus additional reasons emerged for questioning the benefit to the Western Hemisphere of having overthrown Arbenz by interventionist tactics. As years passed, a hostile interpretation of the events of 1954 became sufficiently well popularized in most underdeveloped countries to make any flagrant repetition of those tactics extremely unpromising, for example, when the United States found herself confronted with a very similar situation in Cuba after 1959.

## Problems of American Favoritism in Politics Abroad

The failure of Magsaysay and Castillo Armas to survive more than three years in office and the subsequent relapse in their countries does, of course, highlight a danger that confronts American purposes when they become intimately wedded in deep commitment to a particular foreign leader or small faction. The attachment may offend the beneficiaries, their fellow countrymen, and the outside world; it may make American policy toward the favorites too uncritical; the men in whom American hopes are placed may die, naturally or by assassination, or may fall from power in other circumstances. The prestige and influence of the United States is thereby impaired, inevitably and perhaps gravely. A conclusion can be drawn that the American government ought to avoid such embarrassments by avoiding the internal political entanglements altogether, playing no favorites and keeping its distance in dignified impartiality toward all the nontotalitarian political leaders. There are indeed certain countries where local sensitivity and/or adequate political stability makes such a posture eminently appropriate for United States representatives. And everywhere there are limits set by tact and discretion about the degree to which a display of American preferences among local leaders can be made frankly and openly, and the means by which the United States can inoffensively promote her favorites. Much experience is needed for determining these limits in particular settings. But to reject utterly the exercise of even a discreet partiality is to deny the American government a range of policy instruments that competitive coexistence requires. Some forms of political intrigue are needed to supplement other programs of assistance and influence.

Yet the United States government has no free hand simply to create foreign leaders who will conform exactly to its own ideal image of what would be good for their respective countries and who will at the same time be effective locally and also responsive to American counsel. Usually there are only a very limited number of men who are both receptive and in a position to become useful as national leaders at all. The temptation for Americans, in their own minds, to identify themselves with their local favorites is often strong and hazardous. So far as possible, it behooves American policymakers to keep themselves prepared with alternative promising local collaborators in case of emergencies. On the other hand, a sense of American commitment is sometimes gratifying to the recipients, builds confidence, and is even essential to their success. They may resent and even resist American connections with other leaders. In such situations American dedication to a very few leaders, not hedging widely against their failure, may offer the only substantial hope of grappling with the other manifold risks of the cold war situation.

In the specific case of Guatemala, assuming that the overthrow of Arbenz was essential to American interests because he would not rid himself of the Communist penetration of his Administration, a significant question would be whether a *coup d'état* could have been engineered without requiring Americans to give assistance to a flagrant invasion from foreign soil. The Guatemalan army chiefs held the keys to power in the country and were not procommunist. It is possible that more patient and skillful political intrigue and quiet diplomatic intervention, perhaps on a multilateral basis, could have sufficed to accomplish

Arbenz's removal. His successor in such circumstances would almost certainly have had to be one of the army chiefs themselves or someone very close to them. If mere political intrigue and/or diplomatic intervention had in fact achieved this overturn, however, the United States might not afterwards have had available to it in the presidency as promising a leader as Castillo Armas seemed to be, through whom to bring about fundamental long-term improvements in the Guatemalan economic, social, and political order. A *coup d'état* wrought by intrigue alone, whether abetted by diplomats or by secret agents, might not have sufficed to engender the impetus for drastic reconstruction that was needed if a lasting stabilization of the country's politics were to be achieved on a nontotalitarian basis. The changes Guatemala seemed to need were genuinely revolutionary. The limited potential for such creative leadership available in the upper echelon of the army could well have made reliance on Castillo Armas seem the wisest policy, even though the kind of interventionist activities that would be required to bring him to power would have to be more flagrant than those necessary to put over an army coup.

If these considerations, together with the possibility of the premature eclipse of Castillo Armas himself, were thoroughly explored by American policymakers, in secret of course, their approach to the problem can be commended as appropriate. The distinctly limited success they finally achieved would then caution policymakers in a future similar situation only to probe for fuller information and perhaps to weight the factors somewhat differently.*

* Thus with regard to Cuba, 1960–61, the United States government had much less reason to attempt to rely upon any internal military coup. Castro had learned the lesson of Guatemala from one of Arbenz's own aides, Ernesto "Che" Guevara, who became Castro's chief assistant. The pre-existing armed forces in Cuba were virtually disbanded when Castro took power, and replaced with picked units of supporters of the new regime; furthermore, they were soon supplemented with armed militia. Any possible overthrow of Castro, therefore, would probably have to be accomplished from outside the country, and the fact that Cuba is an island magnified the scale of operations that would be required for any reasonable prospects of success—whether by air-sea supply and liaison for guerrillas, or by an invasion to seize a beachhead and establish a revolutionary government there (like Castillo Armas's initial crossing into Guatemala), or preferably by both of these strategies simultaneously.

This situation was largely understood by the American government in 1960–61, but the related hazards of political favoritism were not adequately calculated, particularly in light of the heightened determination of the Kennedy Administration, as preparations proceeded in the early months of 1961, not to use American armed forces in combat against Cubans under any conditions (short of military aggression by Castro)—yet at the same time to hasten to establish the beachhead before Castro should acquire a large air force from behind the Iron Curtain with which to resist a

We are not concerned here to render a final judgment on the Guatemalan affair in particular; secrecy precludes a definitive history of the whole episode. Sufficient for our purposes is the availability in this case of enough evidence to exemplify the potential uses and the limitations of certain instruments of policy that are available to the American government and characteristic problems that are associated with their employment. It remains for us now to categorize the forms of intervention more precisely.

## SELECTED BIBLIOGRAPHY

ALEXANDER, ROBERT J. *Communism in Latin America.* New Brunswick: Rutgers University Press, 1957.

JAMES, DANIEL. *Red Design for the Americas: Guatemalan Prelude.* New York: Day, 1954.

MARTZ, JOHN D. *Communist Infiltration in Guatemala.* New York: Vantage Press, 1956.

SCHNEIDER, RONALD M. *Communism in Guatemala: 1944–1954.* New York: Praeger, 1959.

TAYLOR, PHILIP B., JR. "The Guatemalan Affair: A Critique of United States Foreign Policy," *American Political Science Review,* September 1956.

---

Cuban refugee invasion. These developments in the planning were making the projected beachhead operation more vulnerable, hence less self-sufficient; if the United States would not fly air cover, much less send in American troops, the beachhead would be correspondingly more dependent upon some simultaneous internal Cuban uprisings, sabotage, guerrilla action, and defections (at least from the militia if not from Castro's army). But the precipitancy was making an internal upheaval difficult to achieve, in light of the particular pattern of favoritism in which the American Central Intelligence Agency had meanwhile become involved.

CIA had devoted very elaborate efforts to training and equipping the volunteers that had made themselves available at an early stage from among the advance waves of refugees from Cuba, who for the most part were relatively conservative in their politics. In partial consequence, CIA did not manage to adapt its projected operations to incorporate—except as a formal facade—the other Cuban refugee groups who, being more radical, had been late in defecting from Castro. Yet it was these latter elements that had the stronger underground connections inside Cuba.

CIA, in short, became so preoccupied with the military-technical requirements of the invasion phase of the "sublimited war" (even yielding in part to the temptation to use some experienced soldiers from Batista's former army) that it neglected to construct more than the facade of a popular successor regime for Cuba—and discovered too late that there had been a need to make that facade an advance reality, if only to obtain information and participation from enough Cubans to be able to effectuate even the beachhead operation in the eventuality that most key Americans became characteristically inhibited about executing any deep military intervention themselves.

CHAPTER 22

# METHODS AND LESSONS OF NONMILITARY INTERVENTION

"POLITICAL WARFARE" is probably the term most commonly employed to denote various forms of overt and covert interference in the internal politics of foreign lands. But no consensus of experts has yet established conceptual limits to this highly elastic term. Consequently, the catch-all expression "political warfare" is being avoided here in favor of a fivefold categorization of particular forms of interference: *information* (including "white" propaganda, that is, from an avowed source), *diplomatic intervention, secret political action, raiding,* and *shows of force* (see pp. 479–80, 512). Raiding and shows of force, together are sometimes called "sublimited war," then shade on into gradations of limited war toward total war. Illustrations may help to clarify the categories.

In the story of Guatemala, diplomatic intervention was observed going beyond what had appeared in the Philippines, as Peurifoy mediated junta shifts to bring Castillo Armas to power, while Lodge and the State Department maneuvered to keep the UN and the OAS stalled.

## The Forms of Secret Political Action and Raiding

Secret political action in the Guatemalan affair took the forms of *intrigue, "black" propaganda,* and probably *bribery,* and Americans may have played a part in all of these.

There were, for example, hints of what may strictly be called "intrigue" in the defection of the Guatemalan air chief in the company of an American officer two weeks before Castillo Armas's invasion (the insurgents later encountered no Guatemalan military planes). The presence of other American liaison agents exercising influence upon Castillo Armas before his invasion can only be conjectured; one interesting aspect was the scheduling of the attack the month when Lodge would be chairman of the Security Council. At the very least, American espionage must have been keeping the United States government informed of the insurgents' preparations, whether or not American political-action operations personnel were guiding them, directly or indirectly.

Black propaganda was also in evidence in Guatemala. Propaganda is said to be "white" when its true source is avowed, as in the Voice of America, "black" when the declared source is fictitious. In such activities there are many shades of gray. The veracity of the contents usually declines also as the source darkens, because there is less desire to protect the credibility of the source for the future; indeed, the very purpose may be to destroy the credibility of an enemy source by simulating its form while distorting its content. In Guatemala, black propaganda appeared most obviously in the form of *clandestine radio.* A station calling itself "Radio Liberation" purported to be broadcasting from within Guatemala but probably was actually nearby in Honduras.

For Guatemala, clandestine propaganda in print was much less needed than on radio because of the considerable freedom of the press that was maintained even under Arbenz. Yet there were leaflet campaigns against the Administration, and presumably some of them were "black." In the Cuban affair several years later, air drops of leaflets were needed more, Castro having seized all the mass media in his country, but planes were scarce to make the deliveries. Principal reliance had to be put on radio. "Gray" relay transmitters were placed aboard boats moving in and out of Cuban waters, and the American Central Intelligence Agency established a 50,000-watt, medium-wave station for the use of favored Cuban refugees, on sparsely inhabited, American-controlled Greater Swan Island, 100 miles off the northern coast of Honduras. Honduras objected, having long claimed sovereignty over the island herself. And "Radio Swan's" performance on the day of the invasion of Cuba became particularly controversial. The broadcasters were instructed by CIA to exaggerate the scale of the insurrection, in hope of intimidating Castro's supporters and encouraging the opposition inside the country. But in the absence of any effective liaison with the underground in Cuba (see p. 441n.), the words alone did not avail to precipitate an upheaval. Their chief effect was to deepen the ensuing disillusionment. The need to unite words and deeds is almost as true of black propaganda as of white.

Bribery as a technique of secret political action by the anti-Arbenz forces was effectively concealed in Guatemala. Certainly in many instances elsewhere, for example, the Philippines, if not in Guatemala, American agents, like those of other countries, have resorted to "silver bullets" ("payoffs") to wage political struggles. In many places such procedures are simply taken for granted; the "bribes" can more accurately be considered a form of blackmail extorted by local men of influence in return for little or no service to the donor. But precise evidence is not available in the Guatemalan case, nor are details of specific instances that presumably arose in the massive American financial assistance to the Cuban insurgents.

The devices of secret political action—intrigue, black propaganda, and bribery—shade on through gradations of violence into *assassination, terrorism,* and *guerrilla operations* (all of which together may be termed "raiding"), thence to shows of force and into limited war. In the Guatemalan episode Americans clearly condoned and may in fact have helped to operate all these instruments of policy with the single exception of political assassination.

Assassination is evidently one of the few feasible means of waging the contemporary international struggle that still remain taboo for the United States. Only in cases involving the punishment of secret agents whom it would otherwise be extremely difficult to discipline (for example, foreigners in American employ who have turned "double agent" for the enemy) is there much reason to suppose that Americans have deliberately tried to arrange murder (to "terminate the agent").

But other less personalized sorts of raiding have certainly had American encouragement, both in the form of matériel and also of personnel. In Guatemala, the inconclusive evidence of matériel assistance to Castillo Armas's rebels has been noted, as have the indications of the presence of some liaison men for the United States. Nothing in this episode, however, was as blatant as the offer made in 1958 by a San Francisco firm, the American Sales Company, to supply up to 50,000 automatic rifles and 15 million rounds of ammunition to the anticommunist, local-autonomist forces that revolted against the government of Indonesia. In order to fill such an arms order, the managing partners, named Hirsch, would have had to obtain the weapons from other sources and rely on the American government not to prosecute them under the regulations that govern private export of arms. Actually, the deal was not allowed to go through after the Indonesian government discovered and publicized it, but other arms were airlifted from Formosa and the Philippines to the rebels in the outer islands of Indonesia. The Civil Air Transport Company, a Nationalist Chinese outfit largely staffed by Americans, conducted many of the supply operations; some foreign surface craft also appear to have been used, but evidently no submarines in this instance. Unofficial American aid to the rebels included, not only conniving at this smuggling of arms from lands where United States influence is strong, but also participation by some former American air force personnel flying actual bombing missions against Indonesian government strongholds. As long as the insurrection seemed to have any chance of success, it had the obviously keen sympathy of most Americans who were at all attentive to Southeast Asian affairs. It was seen as a possible means to prevent President Sukarno's regime in Indonesia from falling under Communist domination, which was beginning to appear alarmingly imminent.

### "Soldiers of Fortune" or Secret Agents?

In this context, the American government formulated an official rationale that can serve as a veneer for many kinds of undercover activities, including even joining foreign armies and engaging in raiding.

On May 5, 1958, shortly after the tide of the civil war had begun to turn decisively against the anticommunist rebels, President Eisenhower was asked at a news conference for his comments on an Indonesian government appeal that Washington take action to prevent United States citizens from aiding the insurrection. Eisenhower replied:

> When it comes to an intrastate difficulty anywhere, our policy is one of careful neutrality and proper deportment all the way through so as not to be taking sides where it is none of our business.
>
> Now, on the other hand, every rebellion that I have ever heard of has its soldiers of fortune. You can even start back to reading your Richard Harding Davis, and people were going out looking for a good fight and getting into it, sometimes in the hope of pay and sometimes just for the heck of the thing, and . . . that is probably going to happen every time you have a rebellion, and I do not believe there could possibly be anybody involved, anything, . . . more than that.
>
> Now, as I say, we will unquestionably assure the [Indonesian] Foreign Minister, through the State Department, that our deportment will continue to be correct.

President Eisenhower's notion of the uncontrollability of American "soldiers of fortune" is only partly justified by the present United States statutes. Relevant provisions prescribe fines or jail sentences for "whoever, within the United States, knowingly . . . prepares a means for, or furnishes the money for, or takes part in, any military or naval expedition or enterprise to be carried on from thence against the territory . . . of any . . . people with whom the United States is at peace," and also for "whoever, within the United States, enlists or enters himself . . . in the service of any foreign . . . people as a soldier or as a marine or seaman aboard any vessel of war." These prohibitions do leave apparent loopholes for the American soldier of fortune who avoids initiating his venture from within the United States. But other statutes provide that an American citizen may be deprived of his citizenship for simply "entering or serving in the armed forces of a foreign state unless, prior to such entry or service, such entry or service is specifically authorized in writing by the Secretary of State and the Secretary of Defense." After losing his citizenship, he would enjoy within the United States only the limited rights and privileges of aliens, and like them might even be deported. The denaturalization penalty has not yet been directly tested in Supreme Court litigation, but it has been effectively imposed.* Thus an American soldier of fortune would be rash to serve "in the armed forces of a foreign state" (it might be safer to serve insurgents) unless he has cleared with the State and Defense departments. To some extent this legal barrier implicates the American government in the activities of United States soldiers of fortune. Presumably CIA agents have no difficulty getting the necessary clearances; other persons have to take the chance that no Administration, present or future, will choose to prosecute. But if foreign governments seize on this situation to claim that an avowed "soldier of fortune" must secretly be an authorized agent, the United States government can of course point to the notorious recklessness of all true heroes in that old romantic world of Richard Harding Davis. As a "cover" story for relatively small numbers of agents, the soldier of fortune concept is probably workable, though familiarity may breed contempt, as happened to the corresponding image that the Soviet bloc likes to use, the "volunteer," which was applied to whole armies of Chinese serving North Korea.

One of the American pilots who were avowed "soldiers of fortune" against the Indonesian government was shot down in action on a bombing mission. His confession when he was eventually brought to trial pro-

* For example, against a one-time American aide to Fidel Castro, Major William Morgan, who was later put to death by Castro.

vides an excellent illustration of the ambiguities of such a role. Allan Lawrence Pope, of Miami, had received the Distinguished Flying Cross and Air Medal with two oak-leaf clusters for service in the United States Air Force in Korea. At the end of that war he was one of the American aviation specialists who became civilians, at least nominally, in order to help the French in the war against Communist-led rebels in Indochina; he aided in the vain effort to defend the fortress of Dienbienphu. When that war was settled by partition of the country, Pope worked with the international rescue teams that shepherded refugees out of North Vietnam. He served for a time with the Civil Air Transport Company of the Nationalist Chinese regime. In Saigon, South Vietnam, he was enlisted during December 1957 to join "an anticommunist movement," and then agreed to fly for Indonesian insurgents at a minimum of $200 per flight. On April 28, 1958, Pope piloted a B-26 bomber of World War II vintage from the United States Air Force base at Clark Field in the Philippines. Thereafter, he flew about a dozen flights in three weeks (training, reconnaissance, bombing, strafing) before he was shot down, and with a broken leg, was captured, carrying large sums in the distinctive currencies of the Philippines, Indonesia, and American military post exchanges. "Your Honor," he told his Indonesian judges, "I have been fighting Communists since I was 22 years old [Korea, 1950]. . . . I have done all this for the freedom of the individuals concerned and also for the states which had been threatened by Communist domination."

Allan Lawrence Pope—an unsung hero of the cold war? Surely in some sense a soldier of fortune, but ultimately under whose command?

A similar pattern of ambiguity surrounded the process whereby Americans, some of them affiliated with the Formosan regime's Civil Air Transport Company, participated in air supply and liaison with some Nationalist Chinese army units that had retreated from southwest China into Burma when the Communists conquered China. The objective during the Korean war was to enable these forces to raid back across the frontier into China. Burma, however, was an unwilling host and became so seriously alienated from the United States by the whole operation that the American government finally agreed in 1953 to press the Nationalist Chinese to call it off and arrange an evacuation of the troops, before they had accomplished much, if anything, that was useful.

This affair, like the Indonesian and the Guatemalan, illustrates further how the ambiguities inherent in any participation by Americans overseas in political action and raiding are likely to be compounded by the involvement of third countries that appear to have more immediate responsibility: Honduras and Nicaragua vis-à-vis Guatemala;

Nationalist China and the Philippines in re Indonesia; Nationalist China in re Burma; Guatemala and Nicaragua in re Cuba.*

If the American government is trying to cover its tracks, small allies who have their own reasons for being eager to act and are not timid about being detected can be immensely useful channels through which the various sorts of interventionist activities can be conducted. On the other hand, the claim that is usually made may be true in a particular instance, that the "third country" is honestly the "second country" and that any individual Americans who participate with it are doing so without the secret authorization of Washington. In that case the problem as always is to make the disclaimer of official American responsibility convincing—best of all to make it so convincing that it will carry over to provide cover for another operation where United States responsibility may actually be greater. The gravest difficulties occur in cases in which American officials not only have not authorized but positively disapprove, for whatever reason, the participation of United States citizens. In such instances, the damage that can be caused to the conduct of American foreign relations by wholly private meddling may be so great as to call for urgent counteraction from Washington.

## Risks of Overreliance on Nonmilitary Intervention

The risk in American support of raiding, whether official, unofficial, or truly private, is not only that its objectives may be unsound, that they may not be achieved, and that discovery is likely to be embarrassing, but also that a competitive shipment of "volunteers" from other powers may ensue, tending toward a limited war between countries of the East and West blocs. The costs of such a war can obviously be great, even if they are not enormously compounded by a later inability to keep the war itself limited. Consequently, unless the key Americans calculate that the establishment of a friendly government in a particular foreign country is worth their running considerable risk of a limited war's arising and "escalating" without much further premeditation on either side, they should probably not make any initial attempt to install that regime by supplying matériel undercover to insurrectionists, in any place where the local balance of forces is so one-sided that the aid required for likely success would be greater than could be given under even minimally decent cover. Usually, in other words, undercover operations are useful instruments of competitive coexistence only if there is some reasonable prospect

* See pp. 451–53 for the case of Cuba.

that they can be kept essentially undercover and still succeed. If not, official calculations must turn attention to the possible merits of deliberate, outright military intervention, just as a need for secret political action and raiding themselves develops from the inadequacy of economic, diplomatic, and informational influences.

But outright military intervention is vastly more repugnant to articulate opinion in most parts of the world than are the other techniques for manipulating politics abroad. This is one reason why it is often possible to maneuver in the United Nations to keep the spotlight of international debate and investigation turned away from disconcerting exposures of mere undercover operations, even extensive ones as in Guatemala, but very difficult to do so if the scale of fighting becomes conspicuously large. When it does, the existence of the United Nations Organization as an international forum, according heavy representation to the attitudes of the uncommitted nations, the very objectives of the whole struggle of competitive coexistence, ensures that their dread of war and hatred of military occupation (the most blatant form of foreign domination) will make military intervention by any power very costly to its general international attractiveness, whatever immediate local success may be achieved.

Thus the fear of escalation up to limited war and beyond and the fear of alienating the uncommitted peoples by any outright military intervention constitute substantial reasons for not engaging at all heavily in even *sub*limited war except in situations where, along with less violent instruments of policy, it is likely to be self-sufficient for accomplishing American purposes. This principle seems to have guided policy adequately—though only barely so—in the Indonesian affair. In contrast, the debacle of the American-sponsored invasion of Cuba points to the international political costs of overreliance on mere sublimited war, costs that may only grow greater when policymakers see a need to increase the scale of these raiding operations dramatically, while also feeling a need to continue studiously to avoid outright American participation in a limited war.

In Indonesia, the United States carefully refrained from putting all its reliance on the insurrection. Probably no decision was ever made to provide even matériel support on any massive scale to the rebels if they were unable to accomplish a successful revolt using limited, indirect, and surreptitious aid. Massive direct aid might well have involved the United States in an escalable limited war. The aid that was given was only on a scale such that if it failed it could be disavowed ("soldiers of fortune") with a fair degree of persuasiveness at least outside Indonesia; it was not too conspicuous to the rest of the world. Nonetheless, one may question whether a revolt aided on this restricted scale was not so likely to fail and the Indonesian government so likely to be alienated even further from the United States that no American aid at all should have been given to the insurgents. Fortunately, the actual response of Sukarno's regime, particularly the army leaders, to the revolt and to the ambiguous United States policy associated with it, was to accept the crisis as a signal that they themselves should move to obviate further troubles of this sort by establishing some barriers against the internal Communist penetration that had provoked the conflict. Thus, although the anticommunist rebel leaders were unable to rally their forces to fight pitched battles, the war itself dwindling quickly to mere guerrilla skirmishing as a result, a large part of the United States policy objective—to get Indonesia's Communists somehow restricted—was indirectly achieved; in effect, the Sukarno government was prodded finally into imposing its own restraints.

In Cuba, however, where the United States chose to wage a sublimited war so massive as to be virtually undisguisable (as the only possible means of overthrowing the government short of American military intervention), the venture failed despite its hugeness; its blatant immensity only magnified the damage to American prestige. Moreover, there was little in Castro's policies in the first months thereafter to entitle the American government to take any comfort, as in the Indonesian affair, from a notion that the surviving regime had at least been taught a lesson not to let itself become so procommunist as to provoke future insurrections and interventions against it. Communism simply tightened its grip in Cuba, if not in Indonesia, after the local government's victory over the United States in sublimited war.

The Cuban affair, 1960–61, of course constitutes the postwar extreme of American involvement in raiding. Indeed, the operation expanded into a genuine limited war, but one in which United States citizens were restrained from actually fighting. They did, however, provide most of the financing, the recruiting, the bases, the training, the supplies, the major operations plan (strategy and tactics), the air and sea transport (including a United States naval escort for the 1,600-man invasion force, from Nicaragua up to Cuban territorial waters), and the bombers, which were under United States command but were not jets and did not have American crews. In addition, Americans were responsible for much of the political and military structure that developed among the Cuban refugees, besides much of the gunrunning by sea and air to guerrillas back on the island and much of the black propaganda disseminated by radio and by leaflets.

To be sure, there were some American restrictions upon the Cubans' activity that were designed to mitigate a little the international embarrassments the United States was bound to endure from this involvement, even if it did succeed in overthrowing Castro. The most important limitation was one placed at a late hour upon the number of pre-invasion bombing missions that Cuban refugees would be allowed to fly from Guatemala in the old American planes. Some critics have blamed the failure of the whole invasion on this confining change of plan.

But much more likely, the general situation in Cuba was simply not ripe yet for any successful insurrection, no matter how heavily backed by the United States—unless American-manned planes were sent to shoot down Castro's aircraft, and probably also, unless an American air or naval bombardment of Cuba were unleashed, and perhaps, unless American troops were landed too. President Kennedy, however, had decided and (somewhat dishearteningly to anti-Castro Cubans) had told the world through his news conference a few days before the invasion that there would not "under any conditions, be intervention in Cuba by United States armed forces." And he stuck to this position, in effect, although sober reports contend that he finally resolved to try to salvage the beachhead by relaxing slightly the restraints upon United States military participation; when the landing had clearly begun to founder, he is said to have authorized American jets from the nearby aircraft carrier *USS Boxer* to fly air cover for one brief hour over the battle zone while the Cuban insurgents would make a final effort to bomb Castro's forces and get more supply ships ashore. This American military intervention never actually took place, but supposedly only because of a last-minute failure in synchronizing the operations correctly.* Assuming that much of this report is true,† American policy did very briefly extend beyond all-out raiding and beyond a show of (naval) force—in short, beyond sub-limited war—to participation in a limited war. But the American military participation was to be subject to extreme restraint on the part of the United States forces—one hour of overflights, probably hoping to

* C. J. V. Murphy, "Cuba: The Record Set Straight," *Fortune*, September 1961, pp. 235–36. See also p. 441n.

† An authoritative White House source has confirmed that United States planes were ordered into action at a late stage in the battle but recollects that the President's intention was merely to cover a withdrawal from the beachhead. However, subordinate officials might have had a different understanding of the objective.

intimidate Castro's planes into keeping away, without actually having to shoot them down. And ultimately, the Americans never fought even this limited war at all. The invasion was allowed to fail. It is hardly likely that it would have succeeded even if the Cubans had been allowed to make their one more pre-invasion bomb-run and the *USS Boxer*'s jets had spent one hour in the air over the beachhead.

Kennedy was somewhat concerned about possible Soviet intervention or (more likely) retaliatory action elsewhere. He was more concerned about the American image abroad. He publicly warned Khrushchev that "in the event of any military intervention by outside force we will immediately honor our obligations under the inter-American system to protect this hemisphere against external aggression"—a hands-off notice to the Russians. But in the same paragraph he coupled this warning with a reassurance that "the United States intends no military intervention in Cuba." In the absence of that intervention the whole enterprise collapsed so rapidly that there was neither time nor need for retaliation by the Soviet Union or by an aroused United Nations.

Essentially, the Kennedy Administration was accepting the view that even this debacle would be less damaging to the long-run prestige and influence of the United States around the world than the kind of victory over a Soviet-supported government in Cuba that could be won only by American military force. Significantly, that judgment was in line with the one that had guided the policy of the Eisenhower Administration in the tortuous politics of the Middle East in the mid-1950's. But there were differences in the degree to which the two administrations took this abjuration of ultimate uninvited invasions by Americans as grounds for avoiding extreme preliminary involvements in secret political action and sublimited war. The Kennedy Administration's measures in support of the Cuban refugees were vastly greater than any Eisenhower-Dulles nonmilitary intervention in the Middle East, though, to be sure, much of the Cuban preparations, but not their implementation, did take place under Eisenhower too.

On the other hand, the Eisenhower Administration was somewhat more willing than Kennedy proved himself to be in 1961 to edge toward the brink through the use of American shows of force as part of sublimited war. A closer analysis of the Middle Eastern experience should thus help to clarify further the American position on the availability of military techniques as overt supplements to other forms of United States influence upon the internal politics of foreign lands.

## SELECTED BIBLIOGRAPHY

CROZIER, BRIAN. *The Rebels: A Study of Post-War Insurrections.* Boston: Beacon Press, 1960.

DAUGHERTY, WILLIAM E., and MORRIS JANOWITZ. *A Psychological Warfare Casebook.* Baltimore: Johns Hopkins Press, 1958.

DYER, MURRAY. *The Weapon on the Wall: Rethinking Psychological Warfare.* Baltimore: Johns Hopkins Press, 1959.

HOLT, ROBERT T., and ROBERT W. VAN DE VELDE. *Strategic Psychological Operations and American Foreign Policy.* Chicago: University of Chicago Press, 1960.

RANSOM, HARRY H. *Central Intelligence and National Security.* Cambridge: Harvard University Press, 1958.

SCOTT, JOHN. *Political Warfare: A Guide to Competitive Coexistence.* New York: Day, 1955.

## CHAPTER 23

# TOWARD MILITARY INTERVENTION: AMERICA AND THE MIDDLE EAST IN THE 1950's

In 1956 the United States government virtually imposed economic sanctions against its most important ally, Great Britain, in order to prevent her from waging a limited war as the means of preventing defeat in a struggle that was being lost when confined to techniques of diplomacy, propaganda, secret political action, and raiding.

In 1958 the United States government took the initiative in sending its own troops to intimidate the opposition to a pro-Western regime and thus to prevent defeat in a struggle that was being lost when confined to techniques of diplomacy, propaganda, secret political action, and raiding.

The first episode was Suez, the second Lebanon.

They differed at the outset in that the British officials responsible for the Suez intervention were hoping their action would contribute to the overthrow of the established local regime; the Americans who undertook the Lebanese intervention were providing troops at the urgent request of the locally established regime, though it was, to be sure, an unpopular regime of dubious constitutionality.

The Suez and Lebanon cases also differed in their sequels, in that the British, together with the French, encountered armed resistance and were obliged to kill and be killed; the casualties of the Suez war were heavy considering how short a time it lasted before outside pressure, mainly American, brought it to an end. In contrast, the Americans in Lebanon succeeded in stabilizing the local situation by their mere presence in strength; any casualties were accidental. Thus the American military intervention constituted in the end only an extreme form of the policy instrument that is called a "show of force"; it did not constitute limited war. But that outcome was mainly because of the lack of resistance.

Thus the surface lesson of the Suez and Lebanon episodes is that at least in the Eisenhower Administration's view outright military intervention that may lead to fighting is a legitimate instrument in the cold war if—but only if—it comes at the request of the government of the country in which the intervention is to take place and that government enjoys general recognition among nations. The latter criterion would preserve the purity of Western doctrine to the effect that at least the Soviet military intervention in Hungary was of a kind too wicked to be imitated. However, there are some indications that even that criterion might have been set aside in July 1958 at the time of the Lebanon affair if a pro-Western nucleus had survived in control of any part of Iraq after the coup that overthrew the monarchy there; American and British troops might then have gone on from Lebanon and Jordan to rescue that nucleus as "the government of Iraq" and to re-establish its rule over the oil-rich country.

Thus there is reason to scrutinize the international relations of the Middle East in the mid-fifties for fuller evidence of the standards the American government has been applying to decide about the deliberate use of Western force on Western initiative as a means of forestalling the erosion of Western strongholds under subversive encroachment.

## Polarization in the Middle East, 1949 to Mid-1956

The Suez crisis took shape in the estrangement between the governments of Britain and Egypt and France and Egypt that gathered momentum after 1954. Combustible tinder was being piled increasingly close to the smoldering fires of Arab-Israeli hostility that had never been quenched since 1948, the year in which Israelis had established their independence by means of the military defeat of the Arab states and had conquered a third more territory than had been recommended for

THE EASTERN MEDITERRANEAN AND THE MIDDLE EAST

them by the United Nations General Assembly a year earlier; thereafter, Israel had refused to readmit the near-million indigenous Arabs who fled from the fighting in 1948. The Arab resentments were natural, and were kept burning by fanatical and cynical political leaders. Israel's attitude was also understandable: her total territory, even after the wartime conquests, remained a minute portion of the Arab world; her *raison d'être* as a state was to provide a homeland in this tiny area for all the world's Jews who might decide to immigrate; hundreds of thousands of Arab returnees would surely constitute an unassimilable Fifth Column, amounting to perhaps a third of her total population, imperiling her national existence.

The major Western powers could see no solution that would be worth using their own forces to impose. Peace in the Middle East was only a truce that rested nominally on a tripartite agreement of Britain, France, and the United States to maintain a balance in the flow of arms to Israel and the Arab countries and to "immediately take action, both within and outside the United Nations," if they should find "that any of these states was preparing to violate frontiers or armistice lines," in order "to prevent such violation." However, the Western powers lacked mobile forces close enough to be able to stop immediately at the frontier a heavy attack by either side; and both sides doubted the ultimate willingness of the West to send troops at all or even to impose economic sanctions that would hurt Jews or oil companies (at least sanctions would probably work too slowly to stop a military drive). For these reasons, the truce actually rested on a preponderance of Israeli armed strength over that of any practicable combination of Arab countries. In this situation, the Arabs were restricted to economic pressures and raiding as the instruments of protracted conflict.

Thus, when Israel struck back with a particularly heavy and bloody raid directly against an Egyptian army post in Gaza in February 1955, she failed to intimidate the military dictatorship in Cairo, but instead provoked it to put high priority on redressing the balance of armaments, even though that goal could be achieved only by circumventing the Western Tripartite Declaration limits and making a deal with the Russians. The bargain was made in the summer of 1955—Egyptian cotton for Soviet bloc weapons and training staffs. Israel would soon be outclassed as a military power by her avowed enemies and would have to become truly dependent on the Western powers for her defense through the Tripartite Declaration; or she would have to engage in an arms race against Soviet-backed Egypt, which would be burdensome even if the Western powers could be persuaded to permit it and subsidize her in it; or else she would soon have to wage a preventive war against Egypt. This last course too would require some degree of Western support.

The Soviet Union was helping Egypt to upset the Middle East balance, not only because she desired generally to stir up trouble in a Western sphere of influence, but also because the Western powers were engaged in reinforcing their own military defenses in that area with formal alliances that carried the prospect of a continuation, perhaps a proliferation, of the air bases closely encircling southern Russia. The Soviet government claimed provocation.

The military dictatorship in Egypt, for its part, was not Communist, but it shared with Communists an intense hatred of the pattern of "unequal alliances" that had developed in the aftermath of British domination of the area. An alliance between mighty Britain and a feeble Arab country typically gave the forms of sovereign equality to the Arabs while preserving for Britain the legal right and usually the physical facilities for military intervention in an emergency. Such an arrangement seemed unendurably demeaning to President Gamal Abdel Nasser, the youthful officer who had just become dictator of Egypt. Egyptian fear of Soviet Russia was not great enough to persuade his regime that defense requirements justified such treaties for Arab countries. In order to induce the British to evacuate their base at Suez, he did agree late in 1954 to allow British army technicians in civilian clothes to remain to keep the base in order, and he granted British troops the right for seven years to reoccupy it in a war for the defense of Turkey or any Arab country against any non-Middle Eastern power. But he would not go further than these concessions toward treaty links with the West.

The Eisenhower Administration was disposed to be tolerant and patient, while concentrating on lining up an alliance of the "Northern Tier" of Middle Eastern countries: Turkey, Iran, and Pakistan. But the oil-rich kingdom of Iraq also, under the autocratic pro-British premiership of Nuri as-Said, was in favor of joining this alliance. Nuri genuinely feared Russia and Communism and was positively attracted by the notion of keeping British bases in his country by "internationalizing" them as the facilities of a multilateral "Baghdad Pact," while appeasing the Anglophobes in the Arab world by the gesture of doing away with the "unequal" bilateral Iraqi treaty with Britain. However, since that treaty would soon have expired anyway and Britain would have lost all right to the bases, the Anglophobes were not appeased.

Gamal Nasser bitterly opposed Nuri's plan as a betrayal of the cause of true independence for all Arab peoples. As an Egyptian he also particularly resented the special favors that Iraq could now expect to receive from Britain; the rivalry between Mesopotamia and Egypt has been perennial since biblical times. And he was angry that Britain was thus managing to entrench herself militarily in the Persian Gulf countries for many more years. Embittering frustration would confront Egyptian hopes for a union of Arab countries of a kind that would make it possible to relieve Egypt's appalling poverty by means of the oil wealth of the Persian Gulf (and perhaps also by emigration). No other solution, even the Aswan Dam, could have seemed so promising (see p. 462).

Iraq and Britain formed bonds with the Turkish-Pakistani entente, through alliance with Turkey in the winter of 1955. Nasser's Egypt, declining membership for herself, bitterly opposed the adherence of Iraq or any other Arab state to this "unequal," "imperialist" Baghdad Pact system. The United States government, somewhat disconcerted by the haste with which Iraq and Britain had acted and the extreme alienation of Egypt, and reluctant to offend Israel by fully joining any Middle Eastern alliance from which Israel was excluded, adopted a stance of somewhat uneasy benevolence toward the new anti-Soviet coalition. The trouble was that the coalition was being considered as provocative by both Egypt and Russia and, simultaneous with the Gaza raid, was contributing to bringing those two powers together. Also it intensified the rivalry between Egypt and Iraq for leadership of the Arab world, as well as another old rivalry dating from World War I between the ruling family of Saudi Arabia (the Saudis) and the ruling family of Iraq and Jordan (the Hashemites). And finally, the coalition exacerbated more recent friction between the Saudis and Britain over territorial boundaries in the eastern part of the Arabian peninsula.

United States influence in Saudi Arabia was insufficient to prevent the oil revenues of the regime there from being used in bribery to help Egypt block the Baghdad Pact powers from expanding their membership. The Cairo Radio's "Voice of the Arabs" inflamed riotous resistance by anti-Hashemites, Anglophobes, "anti-imperialists," Communists, and those pan-Arab nationalists whose first loyalty is not to their own state (which is in most cases an artificial creation of the British and French) but to the "nation" of Arabic-speaking peoples that has yet to achieve the unified political statehood these "Arab nationalists" demand.

Turkey, concerned with the defense of her southern frontier against Communism, was particularly eager to achieve the adherence of Syria. The British government, hopeful of refurbishing its unequal treaty with Jordan as had been done with Iraq, was eager to achieve the adherence of Jordan. Likewise, Nuri himself had affirmative reasons for wanting Syria and Jordan in, but he was further spurred by the negative aim that

Iraq not become isolated from the rest of the Arab world. Thus an alliance that had been conceived as a means of containing Russia actually contributed to making an enemy of Egypt, and then became directed toward containing Egypt.

The alliance did not succeed in either objective. Russian influence overleaped the Northern Tier and penetrated the Middle East through Egypt. Likewise, no Arab government except the Iraqi was eventually able to take its country into the alliance. Nasser's influence spread mightily as he acquired symbolic leadership of the pan-Arab nationalists throughout the area. Britain not only achieved no advantages in the Arab world by means of the Baghdad Pact (even in Iraq, ultimately), but soon lost even her pre-existing dominant position in Jordan in consequence of anti-Pact agitation. In March 1956 the Jordanian government felt obliged to appease the opposition by dismissing the British commanders of the country's army, the Arab Legion; still the further erosion of British influence proceeded at a headlong pace in Jordan for another year. The British government was fiercely embittered. Nasser appeared as a new Hitler, collaborating with Soviet Russia, dedicated to mastering the whole Arab world and destroying British interests there.

To the British government it was no comfort, as it was to some Americans, to conclude that the Nasser-Khrushchev entente could not survive, that ultimately an Arab world unified under Nasser would be a bulwark against Soviet expansionism. Britain's own interests were so much more immediate that they would be likely to suffer almost as much from a pan-Arab dictatorship as from a Communist dictatorship. The immense profitability of British oil holdings in the Persian Gulf countries benefits the British balance of payments in every part of the world where Middle Eastern oil is sold; even a large part of the payments made to the local sheiks is redeposited by them in London. British officials have estimated that the country's overall balance of payments would be impaired by as much as $1 billion a year if British-owned companies ceased to be allowed to produce Middle Eastern oil and Britain herself were obliged to supply her own oil requirements by purchase elsewhere in the world. Sterling convertibility would presumably then have to be abandoned; progress toward freedom of commerce in the whole Western world would be severely set back. Britain herself might be driven to a dependence on new forms of American aid that would be financially costly to the United States and would cause resentment and friction for the alliance in both countries. Of course, a more gradual erosion of British control and profit-taking in the Persian Gulf oil lands might be much less catastrophic. But the British balance of payments since World War II has usually been so delicate that no loss can be regarded lightly in London. And in 1956, Nasser's success in upsetting the Baghdad Pact and destroying British influence was so sudden and shocking as to assume whirlwind proportions in British eyes. All seemed doomed if he were not stopped.

The mounting fury in London was communicated to Americans in the winter and spring of 1956. But it was slow to make headway against the hope that had developed in Washington in the fall of 1955, with British assent, that Nasser could be deflected from his obsession with Israel and the Baghdad Pact, from making Egypt dependent on the Soviet Union, and from casting covetous eyes upon the oil lands, if he could be given some realistic hope for a constructive and peaceful solution to Egypt's abysmal economic condition.

That hope was to be given him through Western financing that would make possible the construction of a high dam on the Nile at Aswan in southern Egypt. The river would be controlled for vast irrigation and hydroelectric projects. Over a generation these enterprises would suffice to preserve the miserable standard of living of the Egyptian people, which otherwise would decline still further because of increasing overpopulation. If anything could be done to limit population growth, the standard would rise. In any case the elite would be enabled to bask in the prestige of industrialization.

The dam project, as carefully worked out by World Bank officials, would not be very costly to the United States Treasury—about $15 million a year for ten years—if Egypt adhered to the development plan. On December 16, 1955, the American government announced a decision to proceed with the preliminary arrangements in conjunction with the British government and the World Bank, which were also to contribute to the financing. But Egypt would be left with enormous costs to bear for the internal expenses of constructing the dam facilities. Nasser was inclined to shop around for a better offer, even in the Communist camp —or at least to pretend to do so in hope of inducing the Western powers to improve their bid. The Soviet Union made this game feasible by not publicly denying the possibility of an attractive Communist offer.* Secretary of State Dulles very much resented the Egyptian maneuvers. The Eisenhower Administration was also becoming increasingly concerned about whether Egypt would be able at all to make her expected contribution to the dam project, in light of the increased commitments she was assuming to send exports to the Soviet bloc in order to pay for the new armaments. Instead of deflecting Nasser from external aggressive-

* For a general analysis of such situations, see pp. 395–98.

ness, American financing of part of the dam seemed to be only whetting his various appetites in the winter of 1955–56: insurrections in Jordan, raids against Israel, affinity with Russia, recognition of Communist China. Would it, after all, really be worth the Administration's effort in an election year to wage a legislative struggle to get an Aswan aid appropriation through Congress?

Dulles decided it would not, particularly from concern about Nasser's solicitation of competitive aid bids. The British government also wanted to withdraw the Western offer, particularly from concern about his Anglophobic agitation of pan-Arab nationalism. The manner, however, in which the project was cancelled in mid-July 1956 was wholly Dulles's. He administered an open rebuff to the Cairo government, instead of engaging in mere stalling. Nasser felt highly insulted, and to recoup his prestige, hastened action on a scheme that probably would have come eventually anyway: nationalization of the Suez Canal Company, more than 40 per cent of whose shares were owned by the British government. Nasser offered to reimburse the shareholders at the price that had been quoted on the Paris Stock Exchange on the day before his sudden announcement of nationalization, but he would be able to cover nearly all of this compensation cost by selling the assets that the Company held outside Egypt. These he insisted must first be turned over to him, since he was physically able to seize only those assets that were on Egyptian soil. The current earnings from Canal tolls were to go to finance the Aswan Dam, now that the West had withdrawn its aid.

## American Efforts to Head Off an Allied Invasion of Egypt

Nasser's ability to seize the Canal facilities in Egypt had resulted from the departure a few weeks previously of the last uniformed British troops from Suez. Their treaty-right of re-entry did not cover this contingency—indeed the English seem to have shown surprisingly little foresight concerning it; so had the United States, though less directly involved.

Under international law, the Egyptian government would not have committed a clear violation unless free navigation of ships through the Canal were impeded on its responsibility. Actually, Britain during both world wars and Egypt in her continuing "war" with Israel had barred enemy shipping from the Canal without suffering effective reprisals, but there was no reason to believe that Egypt would now be immune from very forceful reprisals if Nasser were in fact to impede Western shipping. The British government, moreover, had been bitterly stung by the succession of blows in the Middle East, and the French government was aroused by Nasser's support of the Algerian rebels and was pained by the injury inflicted on the Company, which had been very largely French in management and in private ownership since the days of the Canal promoter De Lesseps. Therefore, neither regime was in any mood to take the chance of waiting until Nasser actually chose to utilize his new control over the operations of the Canal as a means to obstruct Western Europe's oil "lifeline," whether he were doing so in pursuit of his foreign policies or simply in order to make foreigners pay higher tolls to finance Egypt's economic development. The British and French cabinets wanted to take prompt, forceful, preventive action to impose upon Nasser an international regime for the Canal, thus depriving him of any opportunity to exert pressure on their shipping, and, perhaps as important, deflating his newly acquired prestige in the Arab world and, hopefully, even producing his complete overthrow.

The United States government, for its part, evinced surprise not only at Nasser's action but at the violence of the Anglo-French reaction. Americans were aware that grave consequences must follow from any actual blockage of the Canal, but the symbolism of nationalization, without obstruction, was much less shocking to Washington than to London and Paris. The relevant treaties were so doubtfully adverse to Nasser that, as the United States noticed, even the Europeans were not venturing to take their case to the World Court. But probably there was some failure in the State Department to appreciate the full danger to the position in the Middle East, and hence throughout the world, of America's foremost ally, Great Britain, if Arab nationalism under Nasser were allowed to gallop forward unchecked—whether or not he was behaving with technical legality, and even whether or not he was pro-Soviet. Yet there is considerable reason to believe, in the light of America's diplomacy toward Iran (1951–53), Guatemala (1953–54), and Jordan in 1957, that given sufficient time—well beyond the November 1956 election at least—the State Department could have been persuaded of a need to attempt to organize an *internal* coup against Nasser.

Judging from relations with Iran under the anti-Western fanatic Mossadecq, 1951–53, the overthrow of Nasser would have seemed to Washington to be worth attempting only if a specific alternative local regime were known to be ready to take over and that regime were clearly likely to be less objectionable than the one in power, and only if the overthrow could be accomplished without military intervention from outside the country. Exertion of pressure by prolonged economic attrition might well have become acceptable to Washington as a preliminary step; Iran was prevented for years from selling her oil in the world market, and the American government's economic aid to her was kept very small (without being wholly cut off) until the Iranian economy declined to the point where the Shah and some army officers could be emboldened to overthrow Premier Mossadecq. Yet this precedent indicated that even economic sanctions would not be adopted against Nasser by Washington except after long consideration—and in the summer of 1956 London and Paris were frantically impatient.

As for the military intervention those governments wanted, the Americans were certain, and were likely to remain so, that neither Britain's oil objectives nor France's Algerian aims (with the latter they had basically less sympathy) could be advanced by extending the zone of armed conflict from Northwest Africa to the Middle East. Prolonged Western military rule in Egypt would be as costly as it had previously been for the British in Palestine and Suez before they abandoned those positions in the late forties and early fifties, perhaps even as costly as Algeria still was for France. Nasser would be likely to achieve some kind of martyrdom in the cause of pan-Arab nationalism; Anglophobic raiding would spread throughout the Middle East; the oil wells, pipelines, and refineries, for which ultimately any military venture against Nasser was mainly intended as a safeguard, would actually prove to be exceptionally vulnerable to sabotage by even small bands of fanatically aroused Arab raiders.

The strength with which these views were held in Washington was probably heightened by a somewhat exaggerated tendency of Americans to regard outright military occupation as essentially the only form of true "imperialism," rather than as just its most obnoxious form. The line between military occupation and domination by economic pressure and political intrigue seems sharper to Americans—more ethically significant—than it does to imperial powers that are poorer than the United States, powers, for example, for whom it seems less painful to use their drafted troops to exert necessary influence abroad than to do so through taxes levied for foreign aid. Comfortable Americans, who tend to take the opposite preference for granted where there is any choice, are sometimes inclined toward hypocrisy in reacting to forceful imperialism. But they are entitled to observe from the experience of the modern world that not only they but also most other peoples regard military occupation by foreigners as much the most intolerable form of domination from abroad. However questionable may be the logic behind the distinction between military and nonmilitary means of control, the distinction is emotionally compelling to most peoples who have had recent experience of prolonged foreign occupation. And in the large degree that competitive coexistence is a struggle for the adherence, or at least the neutrality, of these peoples, the United States herself has tactical grounds for generally upholding the distinction, even against her European allies who point the finger of hypocrisy.

Unfortunately, Secretary of State Dulles chose indirect tactics in his efforts to head off the use of military force by Britain and France in the summer and early fall of 1956. The heads of government in both countries were so eager to strike that they grasped eagerly at the slightest straw of encouragement from Washington and failed to give heed until too late to the fact that, as Prime Minister Anthony Eden later wrote, they were really being "strung along over many months of negotiation from pretext to pretext, from device to device, and from contrivance to contrivance." * Actually, such tactics may at first have seemed to Dulles a more proper manner of dealing with America's foremost allies than simply issuing a peremptory command to them not to go to war. But in that case he ought to have seen it as essential that the existence of a prohibition be understood by the allies as implicit in the American attitude, even if it were diplomatically left unspoken. Actually, Dulles's equivocations left open the possibility that the British and French cabinets could choose to interpret them as signs that the prohibitory command would never be pronounced categorically, instead of as signs simply that they would have further opportunities to persuade the American government gradually of a need to apply some other kinds of pressure against Nasser—as long as military intervention remained merely a matter of tough talk and threatening gestures.

The American government did take steps to arrange with the major oil companies in the summer of 1956 a system whereby Western Hemisphere oil would be supplied in an orderly fashion to Europe if Middle Eastern shipments were impeded in the Canal at Nasser's initiative. Dollar loans were also to be made available then, if needed to enable the Europeans to pay for the American and Venezuelan oil. Even under these circumstances, Dulles explained, the United States herself would not shoot her way through the Canal, but he did concede the right of Britain and France then to decide for themselves about using force. His disagreement with the allied governments concerned the preventive measures they wanted to take against Nasser. Very slowly during the summer and

* Anthony Eden, *The Memoirs of Anthony Eden: Full Circle* (Boston: Houghton Mifflin, 1960), p. 564.

early fall, the American government did move step by step toward the imposition of economic sanctions against Egypt; yet the possibilities in this direction were far from exhausted at the time the British and French governments finally lost patience.

## The Aborting of the Invasion under American Pressure

The French government moved well ahead of the British in organizing the final resort to arms, but whether the French were also responsible for nudging the claustrophobic Israelis into staging their violent eruption is much less clear. What is known is that in late September and early October, French and Israeli military planning moved beyond the supplying of matériel from France, especially fighter planes, to the joint consideration of an Israeli preventive invasion of Egypt that would be reinforced by air cover provided by the French Air Force itself.

The French leaders, proceeding to explore various possibilities with the British in mid-October, were encouraged to find that Prime Minister Eden, despite the Tripartite Declaration, was not inclined to resist an Israeli offensive, provided it were not against Jordan but against Egypt and did not require protracted hostilities. Both provisos were related to British concern for Persian Gulf oil and specifically for Nuri's Iraqi government; any British failure to honor her defense pact with Jordan against Israel would be intolerable to the other Hashemite kingdom, and a protracted Israeli-Egyptian war would probably embroil Jordan anyway in the Arab cause, with the same disastrous results for the London-Baghdad alliance, unless Britain aided the Arabs, including Nasser's Egypt, against Israel. The convenient solution, which Eden reports was discussed among British and French leaders during the third and fourth weeks of October and was approved in principle by the British cabinet on October 25,* was to prevent Israel from attacking Jordan, allow her to invade Egypt, but stop her from going beyond the Suez Canal by interposing British and French troops there, thus separating the Israeli and Egyptian belligerents, supposedly putting an end to hostilities while leaving the Western Allies in occupation of the Canal and, hopefully, deflating Nasser's prestige even to the point of his overthrow. The policymakers imagined that the United States government, in the final weeks of a presidential election campaign, would not rouse itself or stir the UN into action against Jews and Englishmen until at least two

* *Ibid.*, pp. 572–73, 584–85.

weeks had passed—time enough for Israel, Britain, and France to become so strongly entrenched in Sinai and the Canal Zone that they could thereafter bargain with the rest of the world from a strong position concerning the conditions surrounding any withdrawal.

The scheme aimed at confronting the American government with a *fait accompli.* The failure to consult Washington in advance was deliberate. Both the British and French heads of government have since admitted that their reason was anticipation that, as Eden put it, "there would be attempts to modify our proposals." * The secrecy was even reinforced by calculated deception, such as Anglo-French participation in special meetings that were called in Washington to plan action under the Tripartite Declaration.

Such maneuvers on the part of America's close allies later profoundly offended key officials in Washington, including perhaps particularly President Eisenhower with his military sense of honor. He and Secretary Dulles, as well as important Republican politicians, were also offended by Eden's participation in this international plot that produced war during the final week of an election campaign which the Republican party was waging on a platform of "peace and prosperity"; Eden's action was unfavorably compared with the boost the Americans had given to Eden's own use of the "peace" issue in his 1955 election, when they had acceded to the British desire for a summit conference with Khrushchev, at Eden's urging, despite the poor prospects for its success. Eden's re-election chances had been improved by the effort to which Eisenhower went in journeying to Geneva in 1955; now, Eisenhower's re-election chances might be impaired by Eden's action in 1956. The Prime Minister, from many years of personal and diplomatic relationships with American leaders, ought to have been sensitive to the degree to which such slights as these would tend to offset the domestic political difficulties that an American Administration would obviously confront in deciding to take any drastic action against Israel and Britain just before an election.

Eden's more fundamental blind spot, however, was indicated by his refusal to agree that the British government must in such circumstances as a military intervention in Suez "secure agreement from our American ally before we can act ourselves in what we know to be our own vital interests." † Actually, as soon became entirely evident, Britain lacked the economic strength to accomplish her objective without positive American financial assistance and probably lacked the military strength to

* *Ibid.,* p. 588. For Premier Guy Mollet's statement see the New York *Times,* December 10, 1956.

† *Ibid.,* p. 596, quoting his own House of Commons speech, October 31, 1956.

deter the Russians from coming to Egypt's aid if the United States were simply to stand aside and give Russia a free hand. Even America's strongest ally, Great Britain, in alliance with France, is virtually incapable of going to war without depending to some degree on America to rescue them both in case of catastrophe. The United States government, in turn, aware of this situation, has not felt in recent years that it could afford to tolerate ever being dragged into the imminent risk of having to fight a rescue war as a result of action taken in secret by its allies. The reins that have been held against aggressive action by such regimes as Chiang Kai-shek's on Formosa and Syngman Rhee's in Korea would become even harder to hold if a precedent were set of allowing Britain and France to have their head in any war where Soviet intervention was a distinct possibility. The converse may be debatable—whether the United States, somewhat less dependent on her major allies than they ultimately are on her, should have to secure their consent before dragging them to the brink of war. But in the 1950's not even a prime minister of Great Britain should have been so unrealistic as to look askance at "an apparent disinclination by the United States Government to take second place even in an area where primary responsibility was not theirs." *

The American attitudes that have been mentioned up to this point (pp. 468–69) were all so well known that they surely entered to some extent—though not enough—into the policymaking process in London and Paris, even though Dulles and Eisenhower failed to close every loophole of possible misunderstanding. In the final event, two additional considerations emerged, however, which could not have been foreseen and which further reduced whatever remote chance may have remained that the United States government would adopt a posture of neutrality toward military aggression undertaken in secret by its allies. (1) The heroic popular revolution against Soviet control in Hungary suddenly gathered swift momentum after October 21, and the United States government became deeply involved in trying to invoke the publicity pressures of the UN against Russian military operations there. How could these efforts be pursued with any hope of achieving the desired degree of unanimity in the General Assembly if hesitation were shown about arraigning Israel, Britain, and France for their military intervention also? (2) Since the United States government was determined to take some strong action to stop its allies, it seemed desirable, particularly in view of the Hungarian affair, that the action be taken with maximum dramatic effect in the UN spotlight, in order to differentiate the United

* *Ibid.*, p. 284.

States from Russia, Britain, France, and Israel in a way that would be most persuasive to the uncommitted countries; thus the Eisenhower Administration hoped to salvage some Asian-African regard for the West. Clearly, the impact of events in Hungary could not have been foreseen by the Israeli, French, and British planners. But there is little reason to believe that the ultimate response of the American government to the Suez invasion would have differed in substance even if the Hungarians had not simultaneously revolted, or that possible differences in the form of response to Suez, without Hungary, would have been so great as to enable Britain and France to accomplish their main objectives.

In essence the American pressure on Britain and France was economic, and it was more than Britain could stand. As soon as Britain decided to stop, France and Israel did likewise.

Eisenhower personally issued a series of last-minute appeals to heads of government not to invade Egypt (the United States had finally penetrated their veil of secrecy). These appeals were defied and hostilities began, with a resulting destruction of shipping in the Suez Canal, which blocked it. Thereupon, blaming the allies more than Egypt, the American government refused to begin official implementation of the arrangements that had been made the previous summer for allocating Western Hemisphere oil on a systematic basis to meet Europe's needs. The private oil companies could still take the chance of going ahead on their own to put into effect the plans for joint emergency operations; but in that case, not having received official authorization, they would be running the risk of later prosecution under the antitrust laws. Furthermore, under these circumstances would the United States government loan any dollars when needed, with which to pay for the oil? The international financial markets were shaken by sudden awareness of these problems, resoundingly amplified by the propaganda campaign that the United States government had immediately begun to lead against Israel, France, and Britain in the UN. Speculators and Anglophobes proceeded to stage a run on sterling, converting their holdings into harder currency at a rate that totaled tens of millions of dollars a day. The British balance of payments was much too delicate to endure any such strain for more than a few days without a resulting devaluation of the pound—unless the United States government could now be persuaded to act to restore confidence in sterling by seeing to it that many hundreds of millions of dollars were soon put at Britain's disposal, at least on a temporary basis, in order to convince the speculators that it would not be worth their while to continue selling pounds as a hedge against devaluation. On November 6, 1956, when the war had been in progress only one week, this financial appeal had to be made to Washington with desperate urgency. The answer was categorical: loans only if a cease-fire were accepted that very day. It was.

Other considerations, of course, also entered into the British decision to call a halt immediately even though the allies had thus far captured only part of the Canal. Two were the very angry reaction of Commonwealth countries, particularly Canada and Moslem Pakistan, and serious unrest in the oil lands of the Middle East. Another was the nearly unanimous resentment expressed in the UN. Another was all-out opposition on the part of the Labour party organizations, which reflected the intense objections of so large a minority at home that the conduct of any protracted war would be difficult for democratic Britain. A fourth special consideration was the appeal from Moscow for a joint military intervention by the USA and the USSR to block Britain, France, and Israel in Suez, combined with threats of missile attacks on London and Paris and with publicity about would-be Soviet volunteers for Egypt. The appeals for joint Soviet-American intervention and the missile threats, which anyway were somewhat cautiously delayed by the Russians for several days (perhaps because of their preoccupation with Hungary), were instantly countered by stern American warnings to the Communist powers that they must stay out; the Soviet threat to send volunteers was not even made official in Moscow until after the cease-fire had already been reached. The United States, in short, was still relieving her allies of the main burden of deterring Russia. It seems evident that the Soviet warnings therefore played a very small part in ending the Suez invasion. What happened essentially was that the wayward allied strays were roped in by America herself while she held off any potential Russian rustlers.

The Soviet presence in the Middle East that had been established through the 1955 arms deal with Egypt did, however, have a major indirect impact on the conduct of the invasion operations in 1956. British army leaders and Prime Minister Eden himself were convinced that the Russian airmen who were already in Egypt to train Nasser's forces in the use of the new Soviet jets would be likely to fly those jets themselves in combat, whether as "volunteers" or as "Egyptians." The conclusion was that the allied forces, avoiding complete collusion with the Israelis, might not have such absolute command of the air that they could afford to depend solely upon an airborne and air-supplied invasion force to fight successfully at Suez for the six days that would elapse before seaborne reinforcements could arrive from the nearest adequate base, Malta. The shipping, it was thought, could not leave Malta for Suez much before the Israelis actually attacked Egypt lest the collusion be too obvious. Thus a vain hope that world opinion, if confronted with a *fait accompli*, would be somewhat conciliated by the fiction that Britain and France were merely "separating the belligerents," coupled with a fear of exposing airborne forces alone to possible Russian-Egyptian strafing, together produced a military plan that kept the British from going ashore at all until the fleet finally arrived on November 5. Then, in a sense, their "money ran out" unexpectedly on November 6, and the war ended with them in possession of only one end of the Canal.

Prime Minister Eden contends that even this outcome would have been tolerable if the United States government had not thereafter continued to exert pressure, financially and through the UN, to force the contending parties to return to the status quo ante, that is, to withdraw, leaving Nasser as much in control of the Canal as he had been in mid-October. Israel alone reaped some final benefits, notably the ending of the Arab blockade at her outlet to the Indian Ocean from her southern port Eilat, as well as the destruction of huge (but replaceable) Soviet-Egyptian army stores. Britain's only consolation came, not about Suez, but from American attitudes toward any further expansion of Nasserism in the Middle East. American leaders had finally been shocked into a full awareness of how strongly their most important ally felt about the need to prevent the unification of the Middle East under an Anglophobic Egyptian dictatorship, particularly one that was willing to forge close ties with Moscow. The dramatic demonstration of Britain's determination probably contributed something to the early hardening of an American policy line for the containment of Nasser in the winter and spring of 1957. But this limited development in Washington could almost certainly have been produced in a similar length of time merely by patient British persuasion and the obstreperous behavior of Nasserites, without requiring the Anglo-French military demonstration at Suez.

## The "Eisenhower Doctrine" for the Middle East

The evolving American policy took the form of an active and successful effort to woo the Saudi Arabian government away from its policy of giving financial support to radical pan-Arab nationalists in neighboring countries, partly from antipathy to Britain and to the Hashemite kings of Iraq and Jordan and partly as a purchase, in effect, of "protection" against Nasserite agitation in Saudi Arabia itself. King Saud was persuaded that this policy was short-sighted, and moves finally began toward ending the long feud between his royal family and the Hashemites along his northern borders.

The other main prong of the more active American policy was the promulgation and legislative enactment of an "Eisenhower Doctrine" for the Middle East. The key section of the resolution as finally passed by Congress on March 9, 1957, was as follows:

> The United States regards as vital to the national interest and world peace the preservation of the independence and integrity of the nations of the Middle East. To this end, if the President determines the necessity thereof, the United States is prepared to use armed forces to assist any such nation or group of nations requesting assistance against armed aggression from any country controlled by international communism: *Provided,* that such employment shall be consonant with the treaty obligations of the United States and with the Constitution of the United States.

This pledge is elaborately but ambiguously hedged about. (1) The victims of aggression must be "requesting assistance"; the Eisenhower Doctrine does not provide for American intervention in case a regime would prefer to be defeated. (2) The aggression must come from a "country controlled by international communism"; the President might conceivably choose to identify radical pan-Arab nationalism with international Communism in this context, but not without indulging in some propagandistic distortion. (3) The aggression must be "armed"; and (4) it must be launched "*from* [the] country controlled by international communism," but not necessarily *by* that country; thus movements of "volunteers" could be a basis for American military intervention. (5) The relevant "treaty obligations" of the United States are principally those under the United Nations Charter, which provide that "the threat or use of force" is lawful as an exercise of "the inherent right of individual or collective self-defense if an armed attack occurs," but otherwise can be lawful only if somehow the force would not be "inconsistent with the Purposes of the United Nations," which are, however, themselves defined in such a way as to constitute only very ambiguous limitations.*

* United Nations Charter: Article 51; Article 2, section 4; and Article 1, section 1. One of the "Purposes" is "to take effective collective measures for the prevention and removal of threats to the peace, and for the suppression of acts of aggression or other breaches of the peace."

Thus, while a "threat or use of force" in "self-defense" against "armed attack" may be either "individual or collective" (that is, undertaken by one state or a group of states), the "threat or use of force" must apparently at least be "collective" if used "for the prevention and removal of threats to the peace" or "for the suppression" of "breaches of the peace" when the action is taken without waiting for any "armed

The ambiguities are such that the British government, for example, would not admit to international illegality in its Suez action. The Eisenhower Administration, having promoted in such cases the establishment of much more restrictive interpretive precedents by the General Assembly, would ordinarily feel bound by them in the future. Yet it is hard to see that Congress, when it emphasized the UN in drafting the final form of the Eisenhower Doctrine resolution, did much thereby to clarify the legislation.

As reassurance to any Middle Eastern state that felt threatened by the Soviet Union, the doctrine was admirable and would have probably been even more useful three years earlier, as a substitute for the incipient Baghdad Pact; now, at least, no Middle Eastern state was being pressed to join a formal "unequal alliance." The doctrine was specifically directed against international Communism and not designed apparently to keep the Arab world divided. But the scars of three years could not suddenly be removed. Actually, the doctrine was bound to be interpreted suspiciously by Arab nationalists as a new instrument in the struggle that had meanwhile developed between Western-backed Iraq and Soviet-backed Egypt. So also it was hopefully interpreted by the British. And so it even came to be interpreted, increasingly, by American policy-makers, who tended to dovetail it with their new Saudi Arabian policy as a means of containing Nasser. A special presidential envoy was even sent around the Middle East to "explain" the doctrine and to seek endorsements of it from the governments there, but such endorsements appeared to Arab nationalists as a new form of "unequal alliance." The Nasserites were further alienated. Then their hostility in turn reconfirmed the new tendency of the Americans to think of this Eisenhower Doctrine as a means of containing Nasser as well as the Russians—and then even to complain at the realization that the language of the doctrine, however well suited it might be to containing Khrushchev, especially if it had been announced in 1953, was actually not at all convenient for the new immediate purpose of stopping the spread of Nasser-inspired pan-Arab nationalism in 1957.

---

attack" by the enemy to occur. But the framers of the Charter did not make clear that the "collective" forceful action must be under the specific auspices of the United Nations Organization to be lawful in the absence of a prior armed attack. Thus a loophole remains that can be claimed to legitimize such "collective measures" as the Anglo-French military intervention at Suez.

## Nonmilitary Intervention plus
## Shows of Force: Success in Jordan, 1957

The first test came in the spring of 1957, and not surprisingly it came in Jordan. That kingdom, the very existence of which had always been dependent on subsidies from Britain (ever since she created it after World War I as the means of providing a throne for one of her wartime Hashemite allies against the Turks), was rapidily slipping away into Nasser's sphere of influence. Youthful King Hussein's dismissal of the British commanders of his army in March 1956 had not satisfied the leftists and pan-Arab nationalists abounding among the Palestinians Jordan had absorbed during the 1948 war with Israel, who now constituted an embittered majority of the total population. Passionate for the launching of war of revenge against Israel, they blamed Britain and the British-educated king for frustrating it and exploded with enthusiasm for Nasser as a leader for their cause. Most of them had acquired under the former British Palestine mandate a somewhat better education than the Bedouin tribesmen of pre-1948 Jordan, on whose continued loyalty the king's power depended, but who had now become a minority of the whole population. The Palestinians felt that ability as well as numbers entitled them to shape Jordan's policy, and they wanted to do so in a direction that would be anti-Israeli, anti-British, and pro-pan-Arab, and therefore, if necessary, anti-Hashemite, anti-American, pro-Egyptian, and pro-Soviet.

Among these Jordanians there certainly were some whose personal order of priorities would have placed foremost a specifically pro-Soviet attitude. The American policymakers, who still were much more reluctant than the British to appear, even in their own minds, to be opposed to pan-Arab nationalism as such, probably tended somewhat to exaggerate the relative importance of this dedicated pro-Soviet minority. Its existence was ultimately used as the main justification for buttressing King Hussein against his entire democratically elected opposition.

This opposition had been voted into power in October 1956 in elections that had contributed mightily to the Israeli decision to break loose from encirclement and attack somewhere, but not against Jordan herself, largely because of Britain's continuing determination to defend Hashemite kingdoms. However, Britain's loyalty to her Jordanian ally was insufficient to preserve her own influence in Amman: after all, she did side with Israel in a war against another Arab state, Egypt. She found the new Jordanian cabinet insisting upon ending the "unequal alliance" with Britain (even rejecting the subsidies) and substituting for it a close joint defense arrangement with Egypt, Syria, and Saudi Arabia, including the stationing of some Saudi and Syrian troops on Jordanian soil. Jordan would get her needed subsidy from her three Arab allies, and the arrangement was announced in Cairo in January 1957.

One effect was to increase the sense of urgency in Washington and London about trying to weaken King Saud's old ties with this Nasserite coalition—thus also, hopefully, to prevent Jordan's new ties with it from becoming too close. Eventually, these efforts bore fruit, but not until the situation in Jordan had deteriorated much further.

The leftist cabinet of Jordan, under Premier Suleiman Nabulsi, was taking steps to infiltrate its strong supporters into the upper ranks of the army and civil service. Quite rightly, Nabulsi considered the holdovers from the old conservative regime to be unreliable lieutenants for his new order. In particular, the Arab Legion, though no longer British-officered, still evinced a degree of loyalty to the young king that might prove troublesome. Efforts were made by Nabulsi's associates to manipulate army promotions in such a way as to establish closer bonds between the army and the cabinet. If this objective were finally secured, the throne and the very existence of Jordan would be in jeopardy, for there was little reason to suppose that the Nabulsi regime or the Palestinian Arab majority in Jordan for which it spoke cared very much about preserving the artificial country in which they found themselves, preoccupied as they were with the wider quest for Arab unity against Israel.

American officials, conscious of this developing situation and reluctant to condone it as simply "democratic," were increasingly roused to resistance by focusing their attention, as if with magnifying glasses, upon the Communist element among Nabulsi's supporters, and also by noting that if Jordan collapsed, not only would Nasser's dominion spread to Britain's injury, but Israel would probably go to war again against the Arab countries over division of Jordan's territory. Therefore, in the winter of 1957, American Middle Eastern policy acquired its third feature: interrelated with the Eisenhower Doctrine and the wooing of King Saud came diplomatic intervention and secret political action to encourage King Hussein to take a stand against the Nabulsi forces in his own country before it became too late.

The showdown began in early April when Premier Nabulsi announced his intention to establish diplomatic relations with Russia at the ambassadorial level and to accept any foreign aid that might be unconditionally offered, by which he specifically meant Soviet, not Western, assistance. The Jordanian cabinet, already oriented toward Nasser in other respects, would now be accepting links with Moscow like those that Egypt maintained. American alarm was heightened and was communicated to King Hussein. He himself sensed an immediate threat to his own position in the proposal that came from the Nabulsi cabinet on April 10 that about twenty senior officials be removed from office, including several of his close personal supporters. Hussein thereupon took the risk of exercising his constitutional authority to dismiss Nabulsi from the premiership. Immediately, the Cairo-Moscow entente threw a massive weight of propaganda and political action of all types into a fierce struggle to make it impossible for the king to form any more cooperative cabinet. Violent rioting began abetted by foreign raiders. Only the army would be strong enough to quell it. But was the army still loyal to the king?

On April 13 fighting broke out at an army base near Amman between a royalist Bedouin regiment and other troops under leftist command. Hussein himself, in a dramatic show of personal courage, went out and faced down the insurgent units. Their incipient mutiny, thus overcome, provided the king with ample justification for holding the army chief of staff responsible and removing him from command. The general's personal loyalty had been in doubt anyway. Indeed, the sequence of events was so neat that those who refused to believe the official version (that the local mutineers had risen prematurely, thus by chance alerting the king to nip in the bud a general conspiracy of the High Command) could circulate a different version: that royalist *agents provocateurs* had manufactured the situation to permit the young king to exhibit courage and to give him an excuse to purge the army of the officers to whom Nabulsi had shown favoritism, thus restoring its reliability as an instrument of royal rule. Whichever version was true, the king had shown deft leadership and continued to demonstrate it by seizing upon news of a precautionary Israeli troop movement toward Jordan to convince the divided politicians of the need to form a compromise cabinet of national unity. Nabulsi came back into office, but only as foreign minister, not as premier. The king had won remarkable political victories among the military and civilian leaders, but had not yet mastered them. The rioting soon resumed.

Whatever degree of responsibility Westerners may secretly have had, if any, for the advice on which the king had acted and for the rather convenient Israeli troop movements is still uncertain. The American ambassador and the military attachés of the United States and Britain seemed to be playing a prominent role. Certainly the next stage of the struggle evinced very obvious American diplomatic intervention.

The situation had reached a point where there was reason for optimism about the king's chances of overcoming the rebellious Palestinian Arabs among his population through the use of his freshly purged army —if neighboring countries did not intervene militarily on behalf of an insurrection against him. In particular, the Syrian forces already stationed in northern Jordan under the Egyptian-Syrian-Saudi-Jordanian defense pact would have to be deterred from aiding any revolution in the country. The Saudi forces similarly stationed in southern Jordan had constituted a similar kind of danger to the Hashemite monarchy, but now King Saud had been successfully wooed away from the Egyptian camp, and his forces inside Jordan suddenly became for Hussein a partial offset to the Syrians in the north. In any military showdown Hussein would, if he chose, also be able to secure some aid from army units that were concentrating near his eastern borders in the other Hashemite kingdom, Iraq. However, Israeli troops had been threatening to enter Jordan in the west if Iraqis entered in the east for any purpose whatever. Therefore, if Hussein had to depend on units from Iraq and Saudi Arabia, plus his own Jordanian army, to overcome units from Syria and possibly from Egypt, plus any Jordanian revolutionaries, he stood to lose much of his kingdom in the end anyway, to Israel. The morale and loyalty of his own army might crack in defeatism, despite the recent purges. In this situation, still more outside deterrents to prevent any military intervention by the Syrians seemed desirable if the king was to be free to use his army to crush his domestic opposition. A concentration of Turkish forces along Syria's northern frontier might be helpful; so also would some demonstration of American support—under the Eisenhower Doctrine if the Administration in Washington were obstinate in preferring this formula to any other, though Hussein recognized that the label had become a red flag to Arab nationalists and therefore hoped to avoid using it explicitly.

As the rioting in Jordan swelled on April 23, Secretary of State Dulles was asked at a press conference whether "it would help or hinder the situation" if Jordan embraced the Eisenhower Doctrine, now that she seemed to be in "imminent danger of falling under the more direct influence of Cairo or Moscow or both." Dulles replied:

> We have great confidence in and regard for King Hussein, because we really believe that he is striving to maintain the independence of

> his country in the face of very great difficulties. . . . It is our desire to hold up the hands of King Hussein in these matters to the extent that we can be helpful. He is the judge of that.*

A more pointed invitation could scarcely be imagined. King Hussein's public response took the form of a press conference in Amman the following day, in which he declared that he was prepared to take all necessary measures, including the imposition of martial law, to save Jordan, blaming the crisis on "international Communism and its followers inside and outside the country." The suggestion of possible applicability of the Eisenhower Doctrine implied in this carefully phrased half-truth was promptly adopted in America, where the White House announced that both President Eisenhower and Secretary of State Dulles regarded the "independence and integrity of Jordan as vital." A few hours thereafter in Jordan, King Hussein felt confident enough to throw his army suddenly into action to impose martial law throughout the kingdom, dissolving all political parties, arresting hundreds of oppositionists including Nabulsi himself, establishing tight censorship—making himself, in effect, the military dictator of his kingdom. In Washington, a State Department spokesman coolly called attention to the obvious connection between the Hussein press conference, the Eisenhower-Dulles statement, and the Eisenhower Doctrine.

American warnings went out to all the neighboring countries not to use force in Jordan. Clearly, if foreign "volunteers" or the Syrian troops already inside Jordan were to go into action at all, "armed aggression" could be said to have thereby occurred; and it would be "from" a "country controlled by international communism" if the President should choose to stretch that label to cover Syria or Egypt (or if the volunteers were Soviet). American military intervention on behalf of Hussein might then be expected to follow if the latter chose to request it specifically.

In the Mediterranean the United States Sixth Fleet, with fifty ships, was ordered to speed eastward toward Lebanon, where a Western-oriented government had once endorsed the Eisenhower Doctrine. By April 29, 1,800 Marines were anchored in the Beirut harbor, close to Syria and Jordan.

The United States was indulging in a "show of force" to supplement its successful political action by deterring any foreign military intervention that might reverse Hussein's coup. In one sense, the demonstration of American interest was akin to the efforts that had been taken to prevent illegitimate interference with Magsaysay's prospective victory in

* *Department of State Bulletin,* May 13, 1957, p. 768.

the Philippines in 1953. But in that case the demonstration of United States interest was mainly by diplomacy and propaganda and the immediate beneficiary was the genuine favorite of the local majority. Here the beneficiary probably lacked majority popular support for the action he was taking, and the form of American help was an outright show of force. To be sure, the show of force was directed mainly against the force that Jordan's neighbors were themselves showing, not against the Jordanian people; Hussein's own army could probably be relied upon to handle the latter, especially since the show of force would also probably help to steel the soldiers' nerves.

If not, would the United States government have accepted an invitation from Hussein to use its forces to put down a Jordanian domestic revolution that received only nonmilitary forms of support from outside? Probably not, in fact, but the show of force was being allowed to serve the function of intimidation in the domestic political struggle as well as in the interrelated contest among Arab states. A political-military technique that had fallen into disuse as an instrument of American foreign policy was suddenly being brandished with dramatic shock-effect—explicitly to deter foreign military aggression, implicitly to reinforce the diplomatic intervention and secret political action in which the United States had become involved, with at least some temporary success, inside the kingdom of Jordan.

If the Eisenhower Administration was to some degree bluffing, its bluff was not called. As in the Guatemalan affair, a speedy consolidation of the political victory in Jordan seemed essential before the United Nations spotlight could become too embarrassing. The Security Council met on April 26, the day after Hussein imposed martial law and after the American Mediterranean fleet started toward Lebanon. The Soviet delegate had denunciations and threats, but Russia sent no "volunteers." A round-robin of hasty summit talks among Arab leaders disclosed continued firmness on King Saud's part in his new policy of supporting Hussein. Egypt and Syria had been unable to accomplish the overthrow of Jordan's Western-oriented leaders by propaganda, political action, and limited raiding. They now chose not to attempt to do it alone by military force even in support of the rebellious Jordanian population.

The lid stayed tightly on in the Hashemite kingdom. The American government announced a grant of $10 million to Jordan on April 29, "recognizing the brave steps taken by His Majesty King Hussein and by the Government and people of Jordan to maintain the independence and integrity of their nation." The country had a new subsidizer.

## Nonmilitary Intervention plus Shows of Force: Failure in Syria, 1957

Syria, however, remained as a thorn in the flesh of the Baghdad Pact powers. The containment of Nasserite and Soviet influence, which had now been accomplished at the governmental level, at least temporarily, everywhere else in the Middle East, was showing no success in Syria. There, indeed, a faction that was even more pro-Soviet than pan-Arab nationalist was edging toward the principal role among the kaleidoscopic radical cliques that controlled the government. Syria was acquiring huge stockpiles of Soviet arms. United States efforts to reverse these tendencies failed. Having failed, they were never disclosed from Western sources with evidence anywhere near as reliable as the details available on the Jordanian operation.

Radical Syrian, Egyptian, and Soviet sources presented a particularly dramatic picture of American intrigue when three officers of the United States Embassy were expelled from Syria on August 13, 1957. The allegations of secret political action were mostly plausible in light of the Jordanian precedent, the Syrian situation, and the previous posts of at least one of the officers, but acceptable proof was lacking, and the American government of course issued denials.

Early September found United States officials flexing muscles in public with new shows of force. The Sixth Fleet was maneuvering in the area. American planes were airlifting arms into Jordan with a fanfare of publicity. Turkish troops were concentrating again on Syria's northern frontier. The official American version of events was that the increasingly radical Syrian regime might be planning to try again to overthrow Hussein in Jordan, this time by force. A Syrian-Egyptian-Soviet version, also plausible, was that United States officials were hoping that some Syrians would make the attempt, so that the Eisenhower Doctrine could then be implemented in retaliation by Turkish, Iraqi, and American forces to change the government in Syria, or, better still, so that other, more conservative Syrians would take heart and forestall these events by overthrowing the radical regime themselves. Secretary of State Dulles himself lent substantial color to the latter interpretation by announcing that the President had "affirmed his intention to carry out the national policy expressed in the" Eisenhower Doctrine referring to international Communism and the need to help Middle Eastern nations to "defend their independence," and "expressed the hope that the international Communists would not push Syria into any acts of aggression against her neighbors and that the people of Syria would act to allay the anxiety caused by recent events." * Such an appeal to a people over the head of its government in that context was bound to be considered a call for revolution. But there was no traditional focus of conservative leadership in republican Syria equivalent to the Bedouin-based monarchy in Jordan. Turkish and American intervention would have had to be too blatant to be tolerable to other Arab countries; the Saudi government this time sided with the Syrian regime. Soviet propagandists swung into action. Dulles's trial balloon was shot down, or, more likely, he was bluffing and his bluff was called. Tension relaxed, while Syria continued to drift leftward.

Then in early October a frontier shooting incident between Syrians and Turks was seized upon by Syrians and Russians as the cue for a propaganda offensive in and out of the UN far exceeding anything that had gone before. Khrushchev may have wanted to dramatize to American policymakers how difficult it would be to conduct any subversive operations in the future now that he had learned to manipulate the UN spotlight himself and had just orbited the first sputnik. He may also have simply wanted by repetition to keep alive among the Arabs the image of the Soviet Union saving them from Western imperialists. He may have had some hope of influencing the upcoming Turkish election on October 27. He may also have wanted to create a war scare while simultaneously completing final arrangements at home for the political downfall of the leader of the Soviet armed forces, Marshal Zhukov; what patriotic officer would be able to put personal loyalty to the old commander first and divide the government by resisting Khrushchev's coup at a time when Mother Russia seemed to be on the brink of war?

Any of these reasons would suffice to explain the sudden Communist propaganda barrage, nor did it contain any new, incontrovertible evidence about American conduct of Syrian affairs. Yet it is interesting to note what the main Soviet themes were, if only as evidence of Moscow's awareness of certain possible strategies that the United States had not previously attempted in the Middle East. One, the Communists alleged, was an adaptation of the Guatemalan strategy, using Turkey as an Honduran-like base from which a corps of Syrian refugees would march back into Syria, link up in the north with dissident elements (who seemed relatively more numerous in that part of the country than elsewhere), and form a new government there, relying on Turkish and even American military assistance if needed. (The Eisenhower Doctrine, if it was neces-

* *Department of State Bulletin,* September 23, 1957, p. 487.

sary, could perhaps be stretched by the President's recognizing the new government and heeding its request for military aid against "armed aggression" from Egypt, a "country controlled by international communism," the "aggressors" being Egyptian troops invited in by the regular Syrian government. The other strategy of which the United States and Turkey were accused was a plan to multiply the frontier incidents between Turks and Syrians to the point where cumulatively they would become so serious that Turkey could allege that she was suffering "armed attack" and must exercise her "inherent right of individual or collective [with the United States] self-defense" under the UN Charter—using force long enough to put a friendly government into power in Syria. It is possible the Kremlin really believed that such a plan was under way. It is even conceivable that it was—but highly unlikely in view of the Eisenhower Administration's previous response to the Suez intervention on the part of its other allies. Yet, suspicions were scarcely allayed by strenuous American efforts to keep the matter out of the UN and to block the Syrian-Soviet efforts to have the UN send a commission of inquiry to the border area; the United States government wanted Saudi Arabia to mediate, objecting that a formal UN investigation would unduly dignify ridiculous Soviet allegations.

The whole crisis passed quietly at the end of October, the Turkish election and Marshal Zhukov being both out of the way and Syria continuing to drift leftward beneath the apparently paralyzed gaze of Western observers. Suddenly in mid-winter the drift was halted, but by Nasser and pan-Arab nationalists, not by conservatives or pro-Westerners. Noncommunists in Syria turned their country over to Egyptian domination. The "United Arab Republic" of Egypt and Syria was formed, and Syrian Communists, like those in Nasser's Egypt, were firmly suppressed in the new federation. From the American viewpoint, the lesser of two evils had triumphed; it was even possible to find some quiet satisfaction in this ironical turn of events. The greatest humiliation was King Saud's. Still trying to find cooperative Syrian politicians who would keep their country independent of both Cairo and Moscow, as Washington had previously been urging, he had allowed his agents to pay about $5 million to the Syrian army intelligence chief. The latter accepted the money and then turned it over to Nasser, alleging that he was being bribed to balk the proposed union with Egypt and to have Nasser assassinated. This embarrassing tale was so plausible, at least to Arabs, that Saud was obliged to turn over the reins of active government in his kingdom to his brother, who could attempt a reconciliation with Nasser.

Tiny Yemen, at the foot of the Arabian peninsula, went even further than the Saudis in swimming with the new tide of pan-Arab nationalism. Yemen and the new United Arab Republic (UAR) formed a loose federation called the "United Arab States." Striving not to be outdone as Arab nationalists, the Hashemite monarchies of Jordan and Iraq established a loose federation of their own and named it the "Arab Union." Jordan thereby became indirectly associated with the Baghdad Pact, and there was some hope that the apparently greater stability of Iraq would help to strengthen Hussein's position in Jordan. But the Arab Union was unable to elicit enthusiasm from pan-Arabists; to them its posture was negative and defensive, doomed to "be scattered like dry leaves before the wind," as Nasser confidently proclaimed.

The Middle East line-up was thus once more close to what it had been in late 1955, except that the bonds between Egypt and Syria on the one side and between Iraq and Jordan on the other had become tighter. The Saudi Arabian government tended again to side with Egypt and Syria, the Lebanese government with the pro-Western regimes of Iraq and Jordan. But Lebanon was vulnerable, even more so than Jordan now that Hussein had clamped down martial law, whereas little Lebanon was maintaining more general internal freedom than any other Arab country.

### The American Military Intervention in Lebanon, 1958

Lebanon, with a total population like that of Jordan (about one and a half million), depended for political stability upon a delicate adjustment of power among religious and tribal groups, particularly between the Christian and the Moslem Arabs, who were almost equally numerous. President Camille Chamoun and Foreign Minister Charles Malik, both Christians, feared that their group would be submerged in a unified Arab republic of the radical Nasserite stripe, and they were strongly inclined anyway toward the West in the world conflict. They had welcomed the Eisenhower Doctrine, thereby embittering their relations with Egypt and incurring the intensified propaganda assaults of Cairo-led, pan-Arab nationalism.

The Moslem segment of the Lebanese population was particularly susceptible to these denunciatory themes, especially after the formation of the UAR in February 1958, which left Lebanon surrounded by Nasser's domain and by Israel. Chamoun responded by indicating his determination to perpetuate himself in the presidential office for a second term of six years by methods that were of dubious constitutionality. The effect was a further exacerbation of the controversies that were wracking Lebanon. The mounting friction exploded into rebellion and civil war in mid-May 1958.

The Lebanese Army would probably have been strong enough to put down the motley rebel forces if its commander, General Fuad Chehab, had been willing to have it used for that purpose. But he lacked full sympathy with President Chamoun's maneuvers, and feared that the army itself would tend to split along religious lines if required to suppress a revolt that was principally Moslem and largely pan-Arab nationalist in inspiration. Chamoun, therefore, wanted the United States to suppress the revolt, inasmuch as he had endorsed the Eisenhower Doctrine and there was some gun-running and movement of volunteers across the frontier from the UAR (Syria). Washington felt uncomfortable about Chamoun's behavior, but was willing to experiment with another show of force by the Sixth Fleet and another speed-up of arms deliveries to the threatened government. The situation did not, however, seem urgent enough to warrant stretching the Eisenhower Doctrine to send in United States troops (to combat the "armed aggression" of the volunteers "from" the UAR, a "country controlled by international communism"), or to rely upon an American president's controversial powers to use force as constitutional Commander in Chief without congressional authorization. Yet, shows of force now no longer sufficed. Like cries of "Wolf!" they had lost their impact; the United States was seen to be bluffing.

Still hopeful that the sporadic fighting in Lebanon would be brought under control domestically if infiltration from the UAR were deterred, Washington next encouraged Chamoun to appeal to the UN for assistance before pressing for more drastic American action. The UN agreed to provide a few hundred observers to check on the flow of arms and men across the frontier.

Perhaps the presence of the investigating teams did constitute some deterrent; perhaps the flow was proceeding in places where the rebels prevented them from inspecting; probably the relative importance of the foreign contribution to the raiding in Lebanon had been exaggerated. At any rate, the UN observers sent back reports that minimized it and emphasized the domestic character of Chamoun's opposition. The device of starting up the UN machinery with its spotlight had largely backfired. Now it would be even harder, not easier, for the United States to justify any intervention of its own. Chamoun was particularly unhappy.

The impasse showed no sign of clearing in the early summer of 1958—until suddenly, on July 14, events in Iraq shook the very foundations of the position that Britain and lately the United States had been struggling to build to contain the spread of anti-Westernism in the Middle East. The old reliable, Nuri as-Said himself, was overthrown and assassinated, together with the Hashemite king of Iraq and the other notables of Nuri's regime. The new rulers, army officers headed by Iraqi Brigadier General Abdel Karim Kassim, seemed likely to affiliate their country with the UAR, perhaps even with the Soviet Union. Also, if Nuri had proved to be so vulnerable, how many leaders in Lebanon or Jordan or Saudi Arabia would venture any longer to sustain their current rulers against the flood of radicalism and Arab unification? In the circles around Chamoun and Hussein the reaction was near-panic. Frantic appeals went out to the United States and Britain to send Western troops to Lebanon and Jordan.

Chamoun's appeal was wholly defensive; he would be overthrown, he said, unless the United States intervened within forty-eight hours. Not only would Lebanon thus be lost to Nasser, American policymakers noted, but the precedent of a refusal of United States assistance in such an emergency would dishearten the other surviving Middle Easterners who had adhered to the Eisenhower Doctrine, the most important of whom was the Shah of Iran, who had no other alliance with the United States to rely on and who must be at least as vulnerable as the Hashemites in Iraq had been.

Hussein's appeal, in partial contrast to Chamoun's, had overtones of counteroffensive, not mere defense. On July 14, Hussein had immediately assumed titular authority as the head of the Arab Union, "due to the absence of His Majesty the King of Iraq," the juridical formula in the constitution of the five-month-old confederation. A case could then be made in international law for his undertaking to "recover" the dissident Iraqi region of the Arab Union and for his receiving the assistance of Western troops in this venture, either directly in the front lines or indirectly by their policing Jordan for him while his own Jordanian army marched into Iraq. The practicability of such an operation, without its requiring such massive Western intervention as to provoke Soviet counterintervention, would, however, depend largely upon the amount of support that still survived in Iraq for the Hashemites. At the time of Hussein's initial appeals this factor was still unknown. What was clear was that Iraq was vastly more important than Jordan or Lebanon. The manifold Western efforts made in Jordan and Lebanon in 1955–58 to contain Nasserism had mainly been designed to keep it away from the oil fields of Iraq and the Persian Gulf. If the barrier of containment had now been overleaped and Iraq was irrevocably lost, efforts to shore up

Hussein and Chamoun by themselves could have little more than symbolic significance; yet, even that might have some value in the oil-rich lands of Iran and of the tiny Persian Gulf principalities, and possibly even help to discourage the new Iraqi regime from uniting completely with Nasser.

An alternative course, to abandon the effort to prevent the unification of the Arab world under the Egyptian president, seems scarcely to have been considered by the Eisenhower Administration. Confidence was lacking that Nasser would become cooperative once he was in full control, and such a reversal of recent American policy, even if desired, could not be accomplished so suddenly in light of attitudes and expectations that had theretofore been nourished in Britain, Israel, and the United States. At the opposite extreme, a policy aimed at seeking an excuse for war with Nasser, to destroy him, seems also to have been ruled out almost automatically. United States action might be provocative as seen from Cairo, but it would not be deliberately so.

What then would be the function of American troops if they were landed in Lebanon at Chamoun's urgent invitation and did not move on to invade Iraq or Egypt? They could scarcely be expected to police Lebanon indefinitely on Chamoun's behalf, confronting Arab nationalism under Soviet instigation in the full light of UN publicity; and yet there was no reason to suppose that the Lebanese Army under Chehab would be any more likely to take on a task of repression after a period of American occupation than before it. Consequently, the best obtainable solution would be a compromise, which would necessarily involve reorientation of the Lebanese government to some degree away from the West. American troops would be able to do little more than stabilize the situation to permit some peaceful change of government in which Chamoun himself would probably lose out. The Eisenhower Administration showed that it was resigned to this outcome by including in its initial explanation of the intervention decision to the American people a statement that "Chamoun has made clear that he does not seek re-election," * and by declaring soon afterward that "forces of the United States now in Lebanon at the specific request of the lawfully constituted Government of Lebanon would not remain if their withdrawal is requested by that Government." † Presumably a lawfully *re*constituted government would have the same eviction authority.

Even for the limited objective of stabilizing the situation, however, the United States could scarcely hope to get prior approval from the

* *Department of State Bulletin,* August 4, 1958, p. 184.
† *Ibid.,* p. 196.

Security Council for an American landing in Lebanon; the Russians would veto. The General Assembly too, hypersensitive about colonialism, would probably be unwilling to vote a positive authorization—certainly not in time to meet Chamoun's forty-eight-hour deadline. The Eisenhower Administration decided that it would be best to send troops first and then confront the UN with this *fait accompli*. No effort would be made, however, to avoid the UN spotlight altogether; at least for a while the United States would be able to rely heavily on Chamoun's invitation as legitimation for the intervention.

The Eisenhower Doctrine still would not quite apply, unless President Eisenhower were willing to brand the UAR as "controlled by international communism." Administration leaders now preferred not to do so, in view of Nasser's recent repression of Communists in Syria, and perhaps the issue could be avoided unless new large-scale clashes occurred with rebels in Lebanon, who might be said to be volunteers from the UAR, engaged in "armed aggression." The hope was that such clashes would be deterred, not provoked, by heavy concentrations of American troops. Congressional authorization through the Eisenhower Doctrine resolution was not, therefore, the President's immediate reliance for the landing of troops, though broad hints were given that it was considered to be available, somewhat stretched if necessary, if a question of constitutionality were to develop from the involvement of American troops in much actual fighting. Otherwise, in the first instance, the President was simply acting through his constitutional powers as Commander in Chief, relying on Chamoun's request, on an asserted need to protect the lives of American civilians in Lebanon, and on the absence of strenuous protests from the congressional leaders who were consulted.

July 15, 1958, twenty-six hours after word of Nuri's overthrow in Iraq reached Washington, American Marines began to go ashore in Lebanon. Ultimately, 14,300 United States troops arrived by sea and air. The speed of the operation left much to be desired if it were envisaged as a training exercise for an actual limited war. But as a show of force for the mixed purposes of internal political intimidation in Lebanon and external military deterrence of the UAR, including deterrence of raiding from the UAR, the size of the American build-up was large enough and even fast enough to be impressive in the Middle East. Essentially a show of force was what it was, but this time on such a scale and so very close to the brink of war that there could be no contemptuous talk about "crying wolf" or "just bluffing." The troops were planting themselves in the midst of a civil war with orders to fire if necessary to defend themselves. To be sure, *only* to defend themselves, but surely some hostilities were very probable. The American ambassador in Cairo was instructed to warn Nasser personally:

> We hope to complete our military assistance to Lebanon in a way which will not adversely affect our relations with other states, including the United Arab Republic. At the same time, it must be recognized that any attack on United States forces by military units of the United Arab Republic or under United Arab Republic control could involve grave consequences seriously impairing our relations.*

Almost any rebel military units in Lebanon could be alleged to be "under United Arab Republic control"; thus Washington was virtually threatening in diplomatic language to go to war against the UAR if anyone at all fired on Americans. The intent of this warning was as a deterrent, and it may even have been largely a bluff. Nasser, on the other hand, may well have supposed it to be a deliberate provocation; in any case, his actual conduct was prudent, taking the course that one would expect him to have followed if in fact he was being deterred and was doing his best to restrain his Lebanese and Syrian admirers from shooting at Americans. Whether or not as the result of his influence, only one American serviceman was deliberately killed in the fighting in Lebanon.

By mutual agreement between Washington and London, it was Britain that responded to the Jordanian appeal, as the United States had responded to the Lebanese; American planes flew sorties as gestures of cooperation lest the onus of intervention in Jordan fall wholly on England. The purpose of sending troops, as in Lebanon, was to prevent the overthrow of the government. Any additional hopes that existed in the minds of some Jordanian and Western policymakers that without great effort it might be possible to reverse by force the revolution in Iraq—on the pretext that Hussein was now head of the Arab Union and entitled to recover all his territories—dimmed very rapidly as reports arrived of the almost total lack of resistance inside Iraq to Kassim's new regime. The logistical difficulties of moving troops across the desert from Amman to Baghdad would also be immense, and Hussein's own army was unlikely to have much stomach for a war of conquest against Iraqi Arabs. As early as July 17, when Prime Minister Harold Macmillan first announced to Parliament the airlifting of a couple of thousand British troops to Jordan, he stated as one basis of his action that Hussein had denied to London having any intention that British or Jordanian troops enter Iraq. Macmillan did not, however, wholly foreclose a future move in case of a counterrevolution there. In a press conference on July

* *Ibid.*, p. 197.

20, Hussein himself gave strong indications that he was still thinking in terms of a counterthrust. No military operation could succeed, however, without Western participation, and the London government was not disposed to attempt such a venture on its own or even, apparently, to urge Washington to collaborate in it. British tempers were cooling quickly with regard to Kassim, as it very soon became clear that he did not intend to join Nasser's United Arab Republic or to put extortionate pressure on British oil interests in Iraq. London and Washington therefore decided to try to get along with him; in this decision they were ironically joined by the Communists, who now feared that they would be repressed like the Syrian Communists if Nasser's UAR took over Iraq. The British and the Communists each developed greater hope for their own survival under a weak Kassim than under a strong Nasser. The game was a hazardous one, but Washington was willing to let London play the hand. Hussein was firmly persuaded to give up his claims to headship of the Arab Union and declare it dissolved on August 2.

In Lebanon also, quick progress was made toward a compromise. State Department officials turned to General Chehab himself as the logical compromise candidate to succeed Chamoun as president of Lebanon, and he was elected by the parliament on July 31. Three more months of negotiation and sporadic violence ensued before a similar compromise acceptable to both sides could be worked out for the composition of the cabinet.

By the end of October 1958, the British forces were withdrawn again from Jordan and the American forces from Lebanon. Nasserite and Communist pressure on the two countries had been somewhat reduced, partly no doubt so that the troops would leave. (As long as they remained there was constant danger of an incident that would lead to war.) The compromise solutions the Western powers were being required to swallow in Lebanon and Iraq constituted such severe weakening of the West's position in the Middle East that a little patience must have seemed tolerable to Cairo and Moscow. The erosion, they had good reason to anticipate, would ultimately go further—and faster, they may well have supposed, if Western troops were out of the way.

The opposite view prevailed in Washington, but clearly as a choice of evils. United States policy in the Middle East had in four short years experimented with almost the full range of policy instruments from the least coercive to the most coercive—short only of preventive limited war and military occupation. Those techniques of overt violence, when exercised by America's allies in the Suez affair, had been blocked by Washington. If that taboo still operated, there was nothing to be done but to retreat somewhat from an overextended position that could not indefinitely be maintained in the Arab world without some actual overt use—not just show—of regular military force. Fortunately the partial retreat could be made with some dignity as part of a temporary compromise settlement—partly explicit, partly implicit—for Lebanon, Jordan, and Iraq. How long the new Middle East balance would survive would then depend on the skill with which the other, less violent instruments of policy were employed in the inevitable renewal of coexistential competition.

## SELECTED BIBLIOGRAPHY

ADAMS, SHERMAN. *First-Hand Report: The Story of the Eisenhower Administration.* New York: Harper, 1961.

BEAL, JOHN ROBINSON. *John Foster Dulles: 1888–1959.* New York: Harper, 1959.

BROMBERGER, MERRY and SERGE. *Secrets of Suez.* London: Pan Brooks, 1957.

CAMPBELL, JOHN C. *Defense of the Middle East: Problems of American Policy,* rev. ed. New York: Praeger, 1960.

CHURCHILL, RANDOLPH S. *The Rise and Fall of Sir Anthony Eden.* New York: Putnam, 1959.

EDEN, ANTHONY. *Full Circle: The Memoirs of Anthony Eden.* Boston: Houghton Mifflin, 1960.

GLUBB, JOHN BAGOT. *A Soldier with the Arabs.* New York: Harper, 1958.

IONIDES, MICHAEL. *Divide and Lose: The Arab Revolt of 1955–1958.* London: Bles, 1960.

KNEBEL, FLETCHER. "Day of Decision [the Lebanon intervention]," *Look,* September 16, 1958.

LENCZOWSKI, GEORGE. *The Middle East in World Affairs,* rev. ed. Ithaca: Cornell University Press, 1956.

———. *Oil and State in the Middle East.* Ithaca: Cornell University Press, 1960.

MARLOWE, JOHN. *Arab Nationalism and British Imperialism: A Study in Power Politics.* New York: Praeger, 1961.

QUBAIN, FAHIM I. *Crisis in Lebanon.* Washington: Middle East Institute, 1961.

ROYAL INSTITUTE OF INTERNATIONAL AFFAIRS. *British Interests in the Mediterranean and Middle East.* New York: Oxford University Press, 1958.

SHWADRAN, BENJAMIN. *Jordan: A State of Tension.* New York: Council for Middle Eastern Affairs Press, 1959.

———. *The Power Struggle in Iraq.* New York: Council for Middle Eastern Affairs Press, 1960.

STEWART, DESMOND S. *Turmoil in Beirut: A Personal Account.* London: Wingate, 1958.

WINT, GUY, and PETER CALVOCORESSI. *Middle East Crisis.* Harmondsworth (Eng.): Penguin, 1957.

WRIGHT, QUINCY. "United States Intervention in the Lebanon," *American Journal of International Law,* January 1959.

# PART SIX

## SUMMATION

CHAPTER 24

# THE WAYS OF CONTAINMENT

IN THE ENSUING three years Gamal Nasser's power and influence did in fact subside in the Middle East. The region began to experience an unaccustomed era of relative peacefulness and unaggressiveness. Syria even broke away from the United Arab Republic. Oil continued to flow freely from the Persian Gulf fields, and Israeli cargoes moved unmolested to the Red Sea through the Gulf of Aqaba. These remarkable developments largely stilled the voices of the minority in the West who had favored a more drastic military intervention in 1958 against Kassim and/or Nasser, or against Nasser in 1956. Most of whatever sustained criticism there was of the Lebanon operation tended to be in the opposite direction: that it had been more forceful than the situation really warranted.

In 1961 the United States again held back at the very brink of violent military intervention in two situations that were badly deteriorating politically from the American viewpoint—Cuba and Laos. This time, however, there was nothing in the immediately ensuing events to raise much hope that there would be few regrets later about the Administration's restraint. And soon a guerrilla war crisis in South Vietnam was posing the same issues once more.

They seemed bound to recur again and again, in country after country.

Outside the small remaining nucleus of diehard advocates of preventive total war, few would deny in principle that it is desirable to go on trying to avoid as long as possible initiating a violent military intervention that would amount on the part of the United States to "preventive limited war" (hopefully limited). The risks in such a deliberate employment of American force in order to prevent local defeat by Communist subversion are so great that there has been general agreement that the possible ultimate availability of this violent instrument of policy must at least not be taken by Americans as an excuse for their failing to show the utmost effort and imagination in applying the other, more "peaceful" techniques of "coexistential competition." But whether the United States should absolutely never wage a preventive limited war as a means of containment—and with it the question whether the government should actually be willing to deprive itself of the physical capacity ever to make that choice by agreeing to total disarmament in advance—are questions which do not evoke any deep consensus in contemporary America, and which can help, when raised, to give a focus to some of the very diverse problems of the United States participation in the ongoing worldwide struggle for the preservation of nontotalitarian societies.

## Heritage and Doctrine

The United States finds herself obliged by her relative strength to provide a great part of the sinews and most of the overall leadership in that global effort to defend both the part of the free world that already has democratic institutions and also the other part that may eventually acquire them if in the meantime it can be prevented from succumbing completely to totalitarianization.

Yet the historical experience of the United States until quite recently gave her very little preparation for any exacting role of international leadership: so long had isolation been both the ideal and the reality of her national security position, with the related military and naval and balance-of-power factors remaining almost invisible to her people; so long had she been free from any profound conflict of classes and ideologies, which might give her understanding of similar antipathies abroad; so profound and so rewarding had been her faith in the self-sufficiency of principally private enterprise at home and abroad; so long and so widely had she been able to extend her commerce and even her empire while becoming involved only spasmodically in military conflicts; so painlessly had she reached her own position of territorial satiation and thus come to imagine that others could content themselves with only peaceful change of the status quo; so ignorant and apathetic (and, if ever somehow aroused, so addicted to panaceas) she had been able to afford to be concerning the affairs of the outside world; most profoundly of all, so blessed with good fortune as to acquire very little sense of the life of men or nations as compounded of strife and tragedy. All these aspects of her heritage were cramping brakes upon her acceptance and effective exercise of the responsibilities of international leadership in the mid-twentieth century.

But the responsibilities have, for the most part, in fact been shouldered. And now, after nearly a generation of deep involvement, the United States has been at it long enough to have become acquainted (sometimes reacquainted) with almost the entire range of instruments of an activist foreign policy and to have taken them to hand around the world.

Effectiveness is still encumbered by many residues of the old heritage, not least by the only slowly eroding reluctance of Americans to make their careers abroad and by the tendency of top policymakers to become absorbed, successively, in each particular crisis in particular parts of the world, thus failing to maintain the global overview that would alert them to incipient trouble in some other area.

But at least there does exist an overall perspective for American foreign policy—one that could in general be applied and usually is, though often not deliberately or thoroughly. The perspective is that of *containment until mellowing,* for the Soviet Union and (much less optimistically) for Red China.

The duality of aspects—containment and mellowing—furnishes some indication of certain emphases in the means that can hopefully be applied toward the end. If at all possible the means of containment must, from this policy outlook, not be such as to preclude the ultimate relaxation of Communist tensions and pressures both within the Soviet orbit and abroad, which is the ultimate hope for truly peaceful, noncompetitive coexistence of the great powers. On the other hand, the overall policy perspective suggests that the West cannot safely seek relaxation (mellowing) by a series of concessions and by self-incapacitation that might naïvely be designed to produce satiation and remove all anxiety in the Kremlin and/or Peiping.

## No Precise Prescriptions

These propositions about the containment policy are very broad and general. The experience the United States has had since the end of her invulnerability in 1940 permits one to refine them considerably further.

Much of the body of this book has had that intention, and the ensuing summation will undertake to draw together some of the conclusions that can at least tentatively be drawn. But the author does not believe that the record of two-score years of continuous American great-power responsibilities, only about a dozen of them clearly in the cold war context, can yet enable him to find a large enough number of sufficiently comparable, adequately documented episodes as examples from which to generalize useful "mixes" (recipes) for the precise combination of particular kinds of policy instruments in particular kinds of cold war situations. Much more contemporary history needs to be appraised first, especially with regard to the military dimension of competition among the industrial powers and with regard to all the dimensions of competition in the less-developed countries.* There is much to be digested too from the experience of other powers situated similarly to the United States. The aim would be to learn more reliably how to mix the various instruments of American policy abroad in response to different forms of psychological status-seeking (classes and their leaders within nations, nations and their leaders in international society); also, simultaneously and interrelatedly, how to do the mixing in response to different types and stages of economic development; likewise to different kinds of non-totalitarian political organization; and to different forms of military organization, training, and equipment—all in relation to different contours of Communist challenge. The product might perhaps be a handbook of types of recommended responses to types of likely situations. But one must recognize that even if eventually it were compilable, its usefulness would probably be ephemeral, if only because the Communist adversary would soon take each prescribed mix into account in planning his counterstrategy too.

Yet in the absence of some such detailed, cross-indexed manual there is need for caution in appraising any particular policy instrument per se. Its value can only be determined in a situational context, particularly if

* In rare instances, an "insider" in the top policymaking process manages to approach with insightful detached perspective his own long and varied personal experience which has acquainted him in direct detail with a great number of such episodes. The perspicacious product of such a background comes probably as close as is yet attainable to constituting a convincing book of rules and recipes of a high order. A masterly example of such a study was the one by John H. Ohly, longtime Defense Department authority on aid programs, written for the Draper Committee in 1959: "A Study of Certain Aspects of Foreign Aid: Some Observations on the Assumptions Which Should Be Made and the Analytical Techniques Which Should Be Employed in the Field of Foreign Aid," Annex G, *Supplement to the Composite Report of the President's Committee to Study the United States Military Assistance Program*, vol. 2 (Washington, D.C.: Government Printing Office), pp. 181–335.

it is a kind of instrument that can be closely associated with any of a wide variety of other types of instruments, when there is a desire to emphasize one or another special aspect of overall policy.

For example, there has been much controversy in America over the value of military pacts with countries outside the NATO area. Secretary of State Dulles was frequently accused of "pactomania." But these links cannot validly be approved or condemned in any blanket fashion. It would be necessary at least to analyze types of situations in which pacts might be utilized and types of pacts which might be appropriate to them. Since 1947 the United States has woven a net of outright treaties of military alliance and a gossamer web of military assistance agreements (to furnish training and matériel) linking most of the countries outside the Soviet sphere. From the American viewpoint these arrangements serve a variety of purposes. They blend the prestige of numbers of governments and supposedly of masses of peoples to foster policies the United States approves. They clarify, for all the world to see, the determination of the American government to prevent specific countries from succumbing to aggression. In some cases, notably NATO, they provide for the development of local forces really able to make some substantial contribution together with the United States against the heaviest possible enemy attacks. The worldwide influence of massed political prestige and massed military strength—either or both—especially when directed toward the defense of specified positions, can contribute heavily to deterring any invasion and thus help to keep Communist power contained. In some countries where the local military potential is unpromising, the alliances at least permit the establishment of base installations that would expedite the arrival of American forces in time to defend the country in case it be invaded; without such bases a local defense would be impracticable. In other countries where bases are rejected, military staff conversations can at least help to coordinate advance preparations for the joint defense effort that might later, on short notice, be required. Where guerrilla warfare constitutes an internal menace, American military supplies and advice may be an important supplement to local effort, even if geographical or international political considerations put outright invasion outside the realm of necessary apprehension. In these and other countries, the frequent contacts between Americans and officers of the local government may be useful channels for political influence from the United States; and the supply of heavy matériel often makes the local regime conveniently dependent on the United States for spare parts to keep it running. In some countries, such supplies together with appropriate flattery may amount to diplomatic bribes to the chiefs of the local armed forces to induce them to allow particular regimes compatible with American purposes to remain in office. Or, if the men of arms take power themselves, they may operate their military dictatorships in a fashion less objectionable than otherwise to the United States and to their own peoples, by reason of some lessons in responsible administration they have learned from American military advisors, who are themselves becoming increasingly alert to their own opportunities and responsibilities in these formative contacts. Also, just beginning to be explored fully is a related sphere of activity that has a wide range of potential advantages for many underdeveloped countries and for the United States: creating inducements for the relatively modern military organizations in those countries to associate themselves closely with useful and popular projects of economic development for which they are well suited to contribute talent and facilities without slighting their conventional military tasks (for example, Magsaysay's EDCOR in the Philippines).

All of these purposes are under certain circumstances important reasons to establish and maintain military agreements; they need not be disapproved except in a case where a particular kind of pact is not in fact conducive to its purposes in a particular kind of situation, or where the purposes themselves are outweighed by others that are more important. But detailed generalization on this subject would require the scrutiny of a wider range of experience than has yet been sifted.

The same gap in information obtains concerning many other specific instruments of policy. On a broader level of analysis, however, the cold war record of the West can already lend itself to considerable clarification. Relationships between means and ends can frequently be adduced in a fashion that makes a useful contribution to an understanding of the competitive processes of "peaceful coexistence."

## To Contain and Also to Let Mellow: How to Do Both?

A central problem is the dilemma involved in the West's setting its policy sights on the day when the major Communist countries will act like satiated powers, yet not daring to provide them with more of the means of satiation than they acquired before 1950—partly lest their appetites, hopefully becoming jaded, might be whetted once again. An integral element of this dilemma is the difficulty of even imagining a satiable China, with a population that is increasing by a dozen millions every year. How can the United States avoid for very long a military clash with mainland China over the riches of Southeast Asia, except conceivably by surrendering them to conquest by the Communists through sublimited wars? (Could the clash at least be postponed until Russia would feel complacent enough to stand aside from such a Chinese-American military conflict? Or conceivably even until Russia would be willing to give America some assistance in it?) The prospects for a nonmilitary containment of China are so bleak that by themselves they would constitute substantial reasons for the perpetuation of a military dimension in "peaceful competitive coexistence."

But this stark dimension is also stiffened by Khrushchev's continued endorsement of "popular uprisings" and "wars of liberation" and by his failure to dissociate the North Korean invasion of South Korea from this category of supposed legitimacy. In calling attention to revolutions and civil wars Khrushchev can of course correctly affirm that they are much more nearly inevitable in the contemporary world than are wars between independent countries. Certainly not all the turmoil in the emerging nations is Communist-inspired. Probably the greater part of it would exist anyway even if there were no global cold war. But in insisting, overtly as well as covertly, on embracing this violent ferment within the framework of "peaceful competitive coexistence," the Kremlin makes any Western policy much more dubious than it otherwise might be of taking a chance on waiting patiently and tolerantly for the outcome of decades of spontaneous eruptions in the newly developing lands whatever the results turn out to be. If the Kremlin will not even pretend to keep its own hands off completely, how can the West risk not interfering also —perhaps even intervening militarily, especially if the Kremlin continues to consider it legitimate even for one half of a divided country to launch a massive invasion of the other half. There are many divided countries today, and there may have to be more tomorrow as the most practicable means of compromising civil conflicts for a while. Here again is a reason for the United States not to risk putting her own military preparations at any very low level.

Hence some degree of lasting militarization of the cold war would have to be anticipated, if only because of China's doubtful satiability and because the other interventionist activities of competitive coexistence will for an indefinite period continue to heat up the already combustible tinder of political and economic development among a third of mankind. But the gravest aspect of this situation is that these factors, which by their interrelationship would make it very difficult anyway to confine containment to nonmilitary means, are furthermore operating in an environment of extremely rapid and secret technological change, the military applicability of which has acquired an absorbing fascination compounded of avidity and fear. Set at work initially because of the West's resentment of Communist interventionist activities and because of the Kremlin's resentment of America's atomic superiority, and still kept at work in the respective blocs for these reasons among others, the military innovators have found such vast intellectual and material resources available to them and have been able to present an image of the future that is so promising if they succeed and so forbidding if they fail—with so little predictability either way—that their arms race has acquired an additional momentum of its own, tending cumulatively to absorb the attention of decision-makers on both sides of the Iron Curtain and to subsume nations' policies to the dictates of a groping technology.

Unpredictability, incomprehensibility, instability, tension—these characteristics of the war of the laboratories produce so much exacerbation of the other anxieties and frustrations of the cold war that the manifold provocations seem very likely to produce irrepressible temptations for leaders on one side or the other to seize upon an apparent temporary superiority as the opportunity for an all-out military showdown. Correspondingly, this fact makes it seem essential that no such tempting margin of superiority ever be allowed to emerge, either in fact or in supposition. Thus the existence of an arms race shrouded in secrecy produces a kind of objective requirement for a continuance of that same race still shrouded in secrecy. But subjectively how long can the tense uncertainty be endurable? Can it ever be compatible with the hoped-for growth of a mood of complacency in the Communist powers? If it can be—and in the case of Russia (not China) it has been for the past several years—can the insecurity be prevented from getting so much worse in the future as to become incompatible with still further mellowing? And if actually even the present levels of tension associated with the technological competition cannot indefinitely be sustained without producing an overall hardening of Soviet policy, could they possibly be reduced enough to forestall such a reaction toward "Stalinism"?

The notion of "arms control" asserts an affirmative answer to this final question and sets out to devise and negotiate an adequate pattern —but without eliciting any encouraging response from the Kremlin (at least as recently as 1961). Moscow has claimed to be interested only in rapid total disarmament as an answer to the overall East-West confrontation. (Less ambitious arrangements have been offered for particular regions of the globe.) But rapid total disarmament would produce new uncertainties, perhaps even more upsetting than the old, concerning the military-technological balance during the period of arms reduction. And, if ever the total disarmament were achieved, it would leave wide scope for subversion and sublimited war (including perhaps mass interventions on the part of Chinese "volunteers" armed with police weapons), unless an international force strong enough to prevent such takeovers were created. But in that case the dimensions of the risk would simply shift to encompass the potential subvertibility of the international force itself.

## The Economic Arm of Policy for Europe

The analysis that has been made in this book has suggested that even these risks would come somewhat closer to being bearable—and hence that the arms race and its particular dangers would become more escapable (so far as America bears responsibility for it)—if the United States were to manage to improve her capacity for sustained, effective implementation of containment by means of the nonmilitary instruments of policy. Most of this book has therefore been devoted to the achievements and potentialities of available techniques that are relatively peaceful. For overall clarity it may now be helpful to take a few steps backward and make a more methodical approach to the policy problems of "containing and yet letting mellow," by retracing the path that has led many Americans from dissatisfaction with economic and propaganda techniques toward the temptation to experiment with degrees of military intervention by the United States and correspondingly to hesitate to rely solely on the nonmilitary techniques that alone would remain available to either side in a totally disarmed world.

To begin, few would deny that the nonmilitary instruments of policy would at least be more likely to be self-sufficient in Western Europe to forestall totalitarianization there (under conditions of complete disarmament) than they would be in the underdeveloped parts of the noncommunist world. Deprived of the threat of a Soviet invasion of Europe, it is hard indeed to imagine any devices that would remain at the disposal of the Communists that would be very likely to enable them to take over any European country presently outside the Iron Curtain, unless first they were able to win most of the rest of the world and thus indirectly create economic pressures and a mood of defeatism in Europe itself. The struggle for Europe would thus become, even more simply than it is now, the indirect culmination of the struggle for the newly developing nations.

The conventional problems in United States economic relations with Western Europe—notably American tariffs and import controls—do not any longer seem to be of a critical nature, although they are not approaching a full solution either. European exports are at least managing increasingly to overcome the persistent barriers in the American market. (The great difficulty in this instance, as in most others, is with the newly developing countries.) * Europe, for her part, is lowering most barriers against American exports, though the uniform external tariff of the Common Market may pose problems unless United States delegates are given authority to reduce American tariffs somewhat further in reciprocal trade negotiations.

There are, to be sure, some serious nonmilitary problems remaining in America's relations with Europe, which stem principally from the burdens of the cold war; for instance, disagreements over trade with the Communist bloc and over the sharing of foreign aid and other expenditures abroad, in light of their impact upon gold flows and also in light of the related need to coordinate financial policies among the Western industrial powers. And even these problems have been lending themselves increasingly to genuinely cooperative efforts at alleviation and solution—efforts that more and more become regularized, systematized, and institutionalized, so that not only Western Europe but also the Atlantic Community as a whole moves in the direction of full economic and political integration. If France and Germany can so completely bury the primordial hatchet, who can be so pessimistically "realistic" as to be sure that Britain will not be able to join with them? And, if so, perhaps within a decade may not the United States do likewise? These at least are the trends, and they meet real needs that are increasingly recognized by all the partners in the Western community of industrialized nations: to stabilize and advance their mutual prosperity and to share the burdens of the coexistential competition against the totalitarians. These needs, as a whole, are unlikely to decline, even if possibly the military aspect were to become less demanding. That the response to them will culminate in unity is of course not foreordained, but prospects are good enough to warrant sustaining the efforts in an optimistic frame of mind.

* More and more they will be able to overcome the barriers "too easily" if allowed to do so; but if not permitted access, how can they ever manage to earn their way in world markets?

## The Economic Arm of Policy for the Emerging Nations

The great doubts concerning the self-sufficiency of nonmilitary means for preventing the totalitarianization of a disarmed world—as well as many doubts about the efficacy even of a "mix" of instruments that includes the military—concern the "middle billion" of the emerging nations. Their impatience with gradualism; their antipathy to the status quo both at home and in international relationships (a status quo that included the Western hegemony); their traditions of authoritarianism, which may be amenable to totalitarianism; their urge to be on the winning side for a change, to leap on the "wave of the future" if ever they see one rising—all these tendencies are conducive to affinities for the Communist powers, much more widespread among men of actual or potential influence than is the case in the industrialized West. Yet there are also contrary, hopeful aspects in this baleful situation. Infectious defeatism is not yet epidemic; adolescent nationalism has shown that it can produce a neutralism which is as sensitive to dictation from the East as from the West; totalitarianism runs athwart some humanistic and religious elements in the indigenous traditions and often athwart the residues of Western institutions and values inculcated in the colonial period; and the West has the capacity to continue to furnish discreet guidance and to supply massive material assistance that may provide evidence of a trend toward improvement at a pace sufficient to check the impatience of the politically active and give them the margin of hope that makes nontotalitarian institutions viable.

In this situation, as earlier in the Marshall Plan days in Europe, America's traditions fit her better for ambitious economic efforts than for featuring the other instruments of coexistential competition. In the economic sphere she has vast self-confidence, manifold resources, and a large measure of relevant experience and proficiency. Most of this accumulated capacity, to be sure, has been in the nongovernmental sector of American life. Any very large American contribution of aid and advice to newly developing countries has to draw heavily on private enterprise—at least indirectly via government contracts. And to the extent that private American participation is acceptable directly in an emerging nation, in the form of investments and philanthropic and educational activities, fuller use can be made of what the United States potentially has to offer. But the sensitivity to exploitative abuses is so acute that barriers develop, which are so large that they cannot be overcome sufficiently—even through efforts of the United States government—to raise the flow of private foreign investment from the advanced to the newly developing countries of the noncommunist world to a level high enough to meet their needs within the limited time-span that cold-war pressures permit. Supplementary assistance is required from governmental sources in the United States and the other Western industrial powers—much of it in grants in order to minimize the long-run burden on the international balance of payments. America has been acquiring extensive experience in this realm for more than a decade; her economic aid involvement has been steadily increasing.

In some spheres, which have long been very largely within the province of governmental authority in the United States, notably health, education, and agriculture, the provision of technical assistance in the form of American specialists and demonstration facilities was expanded abroad under official auspices at an early stage in the cold war and has since been maintained near the limit of the number of competent personnel available for such service. But the American governmental commitment to furnish the wherewithal for newly developing countries to purchase massive amounts of capital equipment they cannot or will not obtain through private investment or through commercial-rate loans has only taken shape since the mid-1950's, and is only now beginning to reach levels that offer real promise of making impressive economic development possible, at least in a few key countries—most important of all, India.

In these countries the competence of the local authorities and the basic facilities of transport, communications, and power are sufficiently well established so that the American government can furnish its aid increasingly for overall development programs instead of concentrating on the individual projects appropriate to the lesser capacity of nations that have not yet managed to build this "infrastructure" for their general advancement. But success now in "program assistance," even more than in "project aid," depends on sustained, reliable, large-scale support over a period of years, in order that meaningful economic plans can be drawn and executed in the emerging nations. Actually American aid has been more durable in many of the newly developing countries than it has appeared to be; appearances, however, are very important in this regard. The Kennedy Administration has been particularly anxious to put the funding of its economic aid programs on a multi-year basis.

Such efforts tend, of course, to reduce the freedom of United States officials to manipulate economic assistance month by month or even year by year as a tool to influence the internal politics and foreign policies of the recipient nations. Emphasis is put instead upon achieving economic progress as uninterruptedly as possible under nontotalitarian auspices, in the broad faith that the political outcome of such an accomplishment would not be disadvantageous for the United States, even though Americans have to endure temporary serious irritations along the way.

However, this faith has not been so absolute that any American president has yet been willing to take the bulk of United States economic assistance and route it through international agencies that would irretrievably neutralize the political influence the American government might come to regard as essential in connection with an aid program in a particular country. The reasons are clear. Even if it were possible to assume (which, of course, it is not) that UN authorities would always allocate aid by nonpolitical criteria of efficiency that would achieve maximum economic development, and even if it were possible to assume that maximum economic development would nearly always tend to produce regimes that were not prototalitarian (which cannot simply be assumed—note Nazi Germany), still there would presumably be short-run circumstances in which the disruptive maneuverability of the Communist adversary in the ongoing coexistential competition would have to be countered by flexible alterations in the aid arrangements, imposed by the United States unilaterally. Economic development assistance cannot be expected to be simply divorced from the exercise of deliberate influence upon short-term political evolution in the recipient countries and internationally—unless in a cold war the donor were somehow indifferent to long-run political outcomes, or unless there were sufficient substantiation for a faith that the long-run outcome of successful modernization would be favorable even if the other side did continue to meddle. Neither prerequisite obtains. Hence the use of international agencies as channels for development aid seems bound to continue to depend on an expediential calculus of relative impressiveness, convenience, and efficacy, as compared with bilateral aid in particular kinds of situations.

## The United Nations

Even as such, multilateral assistance will probably grow, including aid through UN agencies that the Western powers do not themselves control as they do the World Bank and the International Monetary Fund. The symbolic connotations of equality among nations and the apparent neutralization of political influence are so attractive to many emerging nations that the purposes even of the containment policy can be advanced by muting it to the extent of giving much assistance to and through the "impartial" UN agencies.

This process of associating American purposes with UN institutions is one which has only spasmodically received the earnest attention of United States policymakers in the cold war years. Useful opportunities have probably been neglected, not only in the implementation of economic aid, but also in America's employment of her other peaceful instruments of international persuasion.

The UN has never been able to function as the strong bridge that its originators intended between East and West, but neither has it ever utterly collapsed. It has found a new role of great significance in serving as a bridge for communication and reciprocal influence between the underdeveloped countries and the West, and, much less reciprocally, between those newly emerging nations and the East. In a sense the balance of voting strength in the General Assembly, with its heavy representation of underdeveloped countries, has come to reflect the realities of a likely future balance of power among nations more than the present realities of power. But in doing so, it has tended to induce the American government to accord respectful attention to that future in forming its own present policies. On the whole this incentive is a useful prod to farsightedness, and one to which the American government's responsiveness (at the very least its show of responsiveness) might well be made more consistent in the formation and implementation of a wide range of foreign policy.

But it must be remembered that the General Assembly is like a model of the future structure of world power only if totalitarianism actually is contained in the meantime. Otherwise, a Kremlin conference of the heads of the world's Communist parties would be a closer model of the future. Therefore it seems unlikely that any idealization of the UN and its present voting majorities should be allowed, by itself, to stand in the way of measures that the United States otherwise deems virtually essential for the prosecution of the cold war. The UN is in part an end of United States foreign policy, not just an instrument with which other means can and should be associated. But even as an end it is not paramount in the sense that antitotalitarianism and perhaps some other goals of American policy are paramount. Both as a symbol and in operation, the UN is worthy of American succor and use, but there are purposes that can not be sacrificed for its sake.

## Image-Building

In seeking to identify a large part of the conduct of American foreign relations with the ethos of the UN, as in seeking through aid programs to identify the image of the United States with the achievement of the aspirations of the emerging peoples for economic development, the American government has gradually been learning ways of associating its deeds and its words in a more persuasive and appealing composite presentation of the national posture. The primacy of deeds, but the related need to interpret them and to disseminate an awareness of them by words and symbolic acts, and to adjust both deeds and words enough to keep them relevant to the changing patterns of the cold war—these are among the precepts that are slowly becoming accepted in the operation of American informational (white propaganda) activities abroad.

Long-term persuasiveness cannot ordinarily be sustained by sloganeering, exhortation, invective, and deception. Official sources of United States propaganda must take a long view, not only because of the protracted nature of the cold war in which short-term gains may bring long-term losses, but also because of the very nature of an avowed official information source, which needs to preserve its credibility for years and cannot be replaced like black-propaganda outlets, one after another, whenever exposure destroys usefulness. To remain persuasive the white propagandists need to strive to communicate action in a frame of meaning with an eye on the outcome years beyond, rarely indulging in current distortions except in cases where immediate circumstances would otherwise tend to foreclose the ultimate attainment.

And since so much of the action of Americans, with their form of civilization, is pursued under nongovernmental auspices through the initiative and enterprise of private individuals, voluntary associations, and autonomous corporate entities, the function of government in communicating the activity of the United States is mostly one of furnishing subsidization and supplementation, with some measure of guidance, to the private contacts and the private communications. The pluralistic reality of American life is bound to be reflected in the image that gets abroad; this condition is frustrating to those who believe that a more uniform presentation would be more convincing to foreigners, both because it would be intrinsically less complicated and more intelligible and also because it would not encounter continuous contradiction by other American voices abroad. But pluralism and moderate dissonance within a familiar but ill-defined consensus are so close to the essence of the American way that efforts to get agreement in the United States on any precisely articulated program for changing the country itself or the rest of the world seem futile. Diversity (without pugnacity) is itself considered by so many Americans to be an end as well as a means in the development of civilization that those who conduct American foreign policy must probably accept this condition and make the best of it. They must try to give a favorable connotation abroad to the very facts that Americans have no exact blueprint for their own society or for any other society; indeed, that they resist totalitarianism largely because it does presumptuously and arrogantly want to impose blueprints; and that accordingly, if only totalitarianism be avoided, the United States will continue to be tolerant of other forms of social and political organization, will welcome a pluralistic pattern of states in the international arena, while also giving a little encouragement of its own, undogmatically, to pluralistic patterns of society within states.

Such an image of the containment policy may help somewhat to offset its native air of self-centered and complacent status-quo-mindedness. In most details American purposes abroad are flexible and are likely to remain so. The permissiveness may, to be sure, largely derive from indifference and indecisiveness; nevertheless, there is reason to believe that advocacy of the permissiveness can be made persuasive, especially if any American interventionist actions that may have to be undertaken reflect a discreet caution that accords with the verbal permissiveness.

### Intervention

Admittedly, the United States is not and cannot be wholly permissive. In seeking at least to forestall totalitarianization, the American government has been finding that the indirect influences upon political evolution associated with economic assistance and information programs are often insufficient, or can even produce gravely adverse consequences, unless the aid and propaganda are tactfully but deliberately supplemented with direct techniques of political influence. Such techniques must, of course, operate within the indigenous political setting, restricted to the locally available forces and leadership potential; the United States has no free hand simply to create the kind of men of authority whom it would like. The choice of whom to back is circumscribed, and there are dangers in becoming too closely identified with particular leaders or factions that may prove to be ephemeral. But sometimes the involvement is inescapable. Discreet circumspection and genuine respect for a diversity of political systems are not to be equated with paralyzing inhibitions and self-doubt on the part of Americans in the midst of the cold war.

The available methods of nonmilitary interference have here been categorized as (besides white propaganda) diplomatic intervention, secret political action (intrigue, black propaganda, and bribery), raiding (assassination, terrorism, and guerrilla operations), and shows of force. Near the end of this band of the spectrum military intervention appears and shades next into official violence, which ranges through gradations of limited war toward total war. The United States has had some experience in the last two decades with operating all these instruments of policy with the exception, of course, of total nuclear war and probably of assassination. But the experience with secret political action and raiding is still scant, particularly in what has come to light for analysis by outsiders.

Few generalizations on this subject can scrupulously be hazarded as yet. They would include emphasizing the sensitivity of any close relationship between Americans and their foreign political favorites, and would extend to recognizing the tenuousness of the "soldier of fortune" cover story and to giving attention to the uses and possible abuses of "third country" relationships when the United States government wants to cover its own "second country" traces in a possible intervention in some "first country."

But most important of all would be the implication that can be drawn from the Cuban disaster and other episodes that undercover operations should rarely be attempted at all except in situations where there is reasonable prospect that they can be kept essentially undercover and still succeed—unless, also, the United States government is ready to accept the risks of violent military intervention as a necessary follow-up in case the "sublimited war" operations ultimately fail.

The arguments against such a "preventive limited war" center, of course, on the risk of its expansion and escalation, the further militarization of the cold war, and the alienation of the uncommitted and the partly committed. The first two fears reflect not only anxiety about major war itself but also apprehensiveness about producing a further postponement of the hoped-for mellowing of the Communist powers. And the fear of causing alienation of neutrals is linked directly to the requirements of containment itself: as the result of a violent military intervention, more may be lost in other parts of the world than can be won in the local fighting. Any power that generally stands opposed to rapid drastic changes in the status quo is bound to hesitate to be the one to set any precedents for the coercive use of outside military force as a means of imposing political arrangements that cannot be obtained by other techniques. The ethos of the UN also contravenes such action. To be sure, the Charter itself does have loopholes, but the pattern of representation at the UN guarantees that its halls will resound with condemnation of any blatant military intervention, except perhaps one that happens to have been made at the request of the generally recognized local government.

Thus far, the United States government has in fact avoided any uninvited military intervention. Indeed, it has avoided obvious opportunities to be offered such invitations, even in deteriorating situations (for example, Laos in 1961). And it has also pressured its major allies into avoiding or halting uninvited military interventions of their own (Suez, for example) that were designed to recover positions being lost politically. The United States government has shown itself, not only to be determined not to be dragged to the brink of war by a secret venture of this kind on the part of its allies, but to be unwilling to authorize their intervention if they were to make their proposal openly. Elaborate shows of force have been the limit for the United States thus far in noninvitational cold-war competitions. But shows of force are like the more conventional kinds of symbolic acts in that often they cease to impress unless they are given real substance—and the substance of a show of force is the use of force to overcome resistance. A lucky borderline situation like Lebanon is not likely to repeat itself often. Eventually, no matter how much improvement can reasonably be expected in the West's use of the nonmilitary instruments of policy in the underdeveloped countries, the United States government will almost certainly have to resolve somewhere to wage a preventive limited war (hopefully limited) or accept a breach in the dike of containment.

It's possible that some breaches could be accepted, in light of the manifold danger of such a war. They might be deemed to be not irreversible, or at least to be recontainable by new dikes further back. The concept of "containment till mellowing" suggests that one ought to consider intently both aspects (the effects upon containment and upon mellowing) in making any major policy choice, and particularly, in deciding whether to unleash an outright war, even a small one. Often it is difficult to discern other than symbolic reasons for struggling to hold some particular piece of territory. But one must also recognize that the course of the cold war itself brings changes in the criteria of usefulness, and the objective of forestalling a spread of infectious defeatism endows with portentous significance even the "merely" symbolic concern. The United

States cannot afford to allow an image of Communism as the Wave of the Future to acquire general currency.

Moreover, a military intervention may actually succeed, not only in preventing a local Communist sublimited-war victory, but also in forestalling a future outright invasion that would then have to be resisted by at least a limited-war defense under circumstances that would have become even less conducive than before to furthering both containment and mellowing. Thus, conceivably, an American military intervention on behalf of Chiang Kai-shek soon after World War II might not only have saved most of China from Communism, but might have done so with less real danger of an expanded conflict or even of an extraordinarily intensified arms race than was the case in the ensuing defense of South Korea.

## Local Defense against Invasion in the Emerging Nations

To be sure, in some instances it might conceivably be reasonable even to allow an eventual outright invasion to succeed, without sending United States armed forces to block the victory and without mobilizing heavily. However, the symbolic connotations of America's permitting Communism to expand by military conquest from outside, rather than intervening with force to prevent it, would not be so comfortably equivocal and ambivalent as the symbolic connotations of her simply permitting the expansion to take place by sublimited-war techniques, rather than sending American troops from outside to prevent it, might be. So the presumption must be very strong indeed in favor of using United States military force to shore up the dike of containment, at least against any outright totalitarian invasion in any place where indigenous and allied defense forces cannot hold. And evaluation of a proposed American preventive local military intervention must take into account how probable such an eventual invasion showdown would be in the particular situation, weighing its estimated likelihood along with other considerations that together might offset the manifold risks of taking a United States military initiative.

Of course, a sufficient possibility may exist of preparing the indigenous forces with Western training and equipment so that they would be able to hold anyway in case of an eventual Communist invasion—or at least would be able to make a major contribution to a joint defense along with the Americans, if the latter were ultimately found to be indispensable. Much can be usefully accomplished in this direction through military assistance programs. But permanent "overmobilization" of an underdeveloped country poses peacetime dangers of its own that may often actually make it safer for the United States to take upon herself the major responsibilities for external defense. Insistence on participation by a considerable number of allies may also be unwise, in making it more difficult to keep the war localized, or in stymieing vigorous action through the hesitations of lukewarm members of the coalition, or in exhausting them prematurely where their exertions may not be essential (for example, Nationalist China in Burma during World War II).

The Kennedy Administration has been showing greater interest than Eisenhower's in readying the kinds of American military forces (with equipment and rapid transport) that would be required if the United States were resolved to participate fully in local defense of the underdeveloped borderlands between East and West. But the domestic trauma of the Korean war in United States politics still reinforces the inhibitions of leaders of both parties about ever waging a limited war again, even against an outright invasion. Their misgivings are not only about being voted out of office, as the Democrats were in 1952 partly as a result of the unpopularity of the Korean war, but also about a resurgence of right-wing radicalism ("McCarthyism"), which had its roots largely in the bloody frustrations of Korea. Both of these misgivings are surely legitimate in a democracy, and the latter one (apprehensiveness about the radical right) is legitimate also within the particular policy framework of containment: limited-war means of effectuating the containment policy might be so exasperating at home as to make the policy itself unacceptable to so many Americans that no Administration would be able to maintain it in the long run. Yet only in the long run can it hope to succeed. There is no evident way out of this dilemma except education and courageous political leadership and a willingness to experiment with limited war again (if ever, apart from the domestic political reaction, it comes to seem desirable) to see whether by then the American people will have finally become reconciled to the costly responsibilities of world power in this regard as they have in others. The Kennedy Administration seemed to be moving toward such an experiment in South Vietnam in early 1962.

### The Military Defense of Europe

With regard to Western Europe, the military defense problems are, of course, even more complicated. And they give a much more direct impetus to the surging momentum of technological innovation, which in turn has been exacerbating all the other tensions of competitive coexistence. Measures to ease this intercontinental arms race may perhaps be attainable without requiring such complete disarmament as would incapacitate the United States for police actions in backward countries.

The Western practice in the first decade of the cold war of relying on ever more massive and more automatic strategic nuclear retaliation capabilities as the principal means of deterring a Soviet invasion—the need that Americans felt to go on developing, one after another, more and more devastating versions of a first-strike capability or at least of a virtually pre-emptive second-strike capability—was a major stimulus to the accelerated efforts of the Russians in turn to acquire comparable strategic striking power for themselves.* Unwilling to draft enough manpower to form what seemed, after the Korean experience, to be the minimum conventional force adequate to deter a conventional invasion, the NATO powers initiated the reliance on nuclear deterrents. Since then the USSR has gradually come to approach NATO's level of nuclear power also, and recognition has therefore spread that the struggle to keep abreast of Russia enough for nuclear deterrence will be for NATO more dangerous and no cheaper (except perhaps in manpower) than an effort at last to pull abreast of her in conventional forces sufficient for deterrence. But by now the previous commitment to nuclears cannot be undone. A new effort to avoid the use of nuclear weapons by building a conventional force adequate to deter conventional invasion would not any longer suffice. There would also have to be an intercontinental nuclear strike force adequate to deter intercontinental attack and a tactical nuclear force adequate to deter a tactical nuclear attack. Of course, some feasible overlaps could function to reduce the total burden, but it would still be immense. And since extensive nuclear capabilities would continue to be indispensable anyway, there will remain a strong incentive not to undertake the general conventional capability but to continue to rely principally on the nuclears—possibly attempting a modified strategy like graduated limited retaliation, but more likely trying once more to renew the deterrent persuasiveness of intercontinental massive retaliation by reinforcing its apparent automaticity, and to do so by becoming

* From the Kremlin's viewpoint, Western leaders would be likely to use SAC's strike force not just to deter a Soviet invasion but to launch a preventive total war, unless Russia managed to develop a similar force of her own that could function at least as a deterrent against the desperate final lunge that Communists expect "doomed" capitalists will want to make as the world erodes away from "imperialist" control in the competition of peaceful coexistence.

fully committed to a tactical nuclear defense and then deliberately emphasizing its probable propensity to escalate (rather than its possible limitability).

But how much automaticity of response is conceivably practicable over an indefinite period—automatic enough to make a threat of mutual annihilation convincing as a deterrent to Russian encroachments upon Europe, yet not so tightly strung as to produce unacceptable risks of accidental unleashing?

The growing dispersion of nuclear capabilities among individuals and groups as well as among nations has already been heightening the dangers of "accidental nuclear war," "anonymous nuclear war," "catalytic nuclear war," and "nuclear war by proxy." Correspondingly, a sense of urgency is spreading about a need for greater efforts, both unilateral and multilateral, to forestall the progression of "N + 1" nuclear countries and to improve the reliability of central command and control over the individual units of nuclear force. But such concerns contravene in considerable measure any efforts to etch more deeply an image of the automaticity of nuclear response and thus (but, of course, unavowedly) to deter by intensified bluffing—feigning a kind of "mad" willingness to go over the brink of thermonuclear war if provoked, in order to deter the adversary by using the "rationality of irrationality." *

Such desperate expedients, when imitated by both sides, would be bound to end in bluff-calling, if the stakes were not so appallingly great that neither side may dare to take even a small chance that the enemy might not just be bluffing. Even a small possibility of a thermonuclear holocaust may suffice to deter—but for how long?

Can the tension be sustained indefinitely? Could other, less ghastly modes of defense still be found for the Western industrial powers, even in an age of nuclear plenty?

### Arms Control

The honest answer seems to be that no one can yet know for sure, and that what hope there is lies chiefly in the realm of further experimentation with schemes of arms control, directed (at a minimum) toward making the tension subjectively a little more bearable and, insofar

* Imagine two drivers racing toward each other on a one-lane road, neither of them willing to pull off and let the other go by. If A can manage, by horn-blowing or otherwise, to convince B that A has let his own car get completely out of control, then B may feel forced to yield the right of way, lest both B and A perish.

as may yet become practicable, toward abating at least a few of its most immediate exacerbating objective causes.

Moscow has in fact shown some fairly persistent interest in obtaining regional arrangements for arms control—redeployment and disengagement—but not for control systems that would operate to any very large extent within the Soviet homeland. (One could scarcely regard the variable Russian offers about nuclear test inspection as an exception to this rule.) On the other hand, the West too has not yet really devoted much effort to exploring the possibilities that may actually exist for mutually acceptable arms control, either regionally or worldwide. Nor has the West developed the diversified forces—particularly the conventional forces—that would be prerequisites to most forms of nuclear arms control (essential because without these forces the defense would not possess any reasonably adequate substitutes for the weapons systems that were to be restricted).

To be sure, no well-informed person can deny the extraordinary difficulty of ever estimating the net consequences of any proposed scheme of arms control. The exploration is likely to be at least as labyrinthine as evaluating a whole new weapons system, and much the same sort of intellectual process is involved. But a commitment of brains and effort in this direction, even if, inescapably, other brains and effort must simultaneously continue to be committed in the other, cataclysmic direction of the arms race, may generate some ideas that can be communicated to Moscow persuasively enough to stimulate a comparable diversion of some attention in Russia toward controlling the competition. The result might be some reduction of mutual suspicions, at least with regard to the supposed eagerness of each side to seize upon any opportunity that might apparently arise to unleash preventive total war. Therewith might come some de-emphasis of any further acceleration of the arms race, some check upon the cumulative momentum of the war of the laboratories that has been rushing toward ever more nearly suicidal weapons systems.

Further benefits might ensue: perhaps some concrete inspectable agreements, reached because the increasingly shared focus of attention upon these problems might at last make East-West communication of the possible answers prompt enough so that opportunities would not continue to pass by in the rush of the arms race and be gone before both sides could manage to comprehend that a chance to agree did exist, as has probably happened in the past.

With inspection, if it eventually worked to the satisfaction of both sides, might gradually come opportunities to expand the range of contacts between Russians and Westerners and to shrink the sphere of Soviet secretiveness, perhaps even to do so cumulatively, and thus to reduce the urgency of the remainder of the arms race and contribute directly to a mellowing of Communist society. Such vistas are, of course, very appealing, and in the perspective of containment no path that appears to lead in that direction should be dismissed as a mirage, except after very conscientious exploration.

Much will have been gained, however, even if all that can be accomplished is to prevent the arms race from becoming yet more compulsive and absorbing than it already is—during whatever span of years may yet be required for Russian society, without an increasing sense of insecurity, to continue to grow prosperous, stable, and (hopefully) ultimately complacent. In that same period of prolonged containment, increasing evidence about Peiping should emerge to substantiate comparisons between mellowing and outright break-up as practicable goals of American policy toward the Red Chinese regime, and the emphasis of United States activity in the Far East, heretofore ambiguous, could eventually be appropriately adjusted.

In the meantime, short-term gains and losses in the coexistential competition must be taken seriously; they may foreclose the hoped-for future. The struggle, probably including some form of weapons race, must be expected to endure for many years, during which any extreme unilateral weakening might well produce irresistible temptations to the adversary, as in Korea in the spring of 1950. Limited wars may have to be fought again. They too are a kind of arms control.

## Prospects and Counsel

Competitive coexistence will thus continue to make enormous demands on America's resources, brains, nerves, and will power. Actually, it is a remarkable thing that a nation as ill-prepared historically for world leadership as is the United States of America has managed to learn so much and to do so much as she has in less than one generation of deep international involvement. But the question remains whether she still has time enough remaining to learn what more needs to be learned and to do what more needs to be done.

Nikita Khrushchev is sublimely confident that America does not, that Western institutions will slowly die and be buried and be resurrected for the grandchildren, if at all, only in a Communist mold, and that the East will retain its expansionist élan and its widespread magnetism long after the West has declined and succumbed, probably not with a bang but a whimper—"better Red than dead."

It is hard for any objective Westerner to match Khrushchev's degree of confidence when affirming an opposite, noncommunist view of the destiny of civilization. Often the odds do seem to be precariously balanced. Yet there is much resourcefulness remaining to give new substance to the nontotalitarian image of man's future prospects, enough promise at least to sustain the striving certainly required if that image is to endure as the reflection of reality.

Probably the most cogent precept for the West in the protracted conflict of competitive coexistence was given by Winston Churchill in 1955. For the first time in public he was unfolding the picture of the balance of terror, that canopy of fear under which the two worlds coexist and compete, until, if the West holds out, long contained totalitarian ambitions may finally mellow: "The day may dawn when . . . tormented generations [will] march forth serene and triumphant from the hideous epoch in which we have to dwell. Meanwhile, never flinch, never weary, never despair." *

## SELECTED BIBLIOGRAPHY

BLOOMFIELD, LINCOLN P. *The United Nations and U.S. Foreign Policy: A New Look at the National Interest.* Boston: Little, Brown, 1960.

HERZ, JOHN H. *International Politics in the Atomic Age.* New York: Columbia University Press, 1959.

KAPLAN, MORTON A., and NICHOLAS DE B. KATZENBACH. *The Political Foundations of International Law.* New York: Wiley, 1961.

ROSTOW, W. W. *The United States in the World Arena: An Essay in Recent History.* New York: Harper, 1960.

SETON-WATSON, HUGH. *Neither War nor Peace: The Struggle for Power in the Post-War World.* New York: Praeger, 1960.

STILLMAN, EDMUND, and WILLIAM PFAFF. *The New Politics: America and the End of the Postwar World.* New York: Coward McCann, 1961.

STRAUSZ-HUPÉ, ROBERT, *et al. Protracted Conflict.* New York: Harper, 1959.

———, WILLIAM R. KINTNER, and STEFAN T. POSSONY. *A Forward Strategy for America.* New York: Harper, 1961.

WOLFERS, ARNOLD, ed. *Alliance Policy in the Cold War.* Baltimore: Johns Hopkins Press, 1959.

* *Parliamentary Debates,* March 1, 1955, p. 1905.

# INDEX

# INDEX